PRINCIPLES OF BIOCHEMISTRY:
General Aspects

Previous Editions of *Principles of Biochemistry*

First Edition, 1954: Abraham White, Philip Handler, Emil L. Smith, and DeWitt Stetten, Jr.

Second Edition, 1959: Abraham White, Philip Handler, Emil L. Smith, and DeWitt Stetten, Jr. Japanese edition, 1961; Spanish edition, 1964.

Third Edition, 1964: Abraham White, Philip Handler, and Emil L. Smith. Japanese edition, 1966.

Fourth Edition, 1968: Abraham White, Philip Handler, and Emil L. Smith. Japanese edition, 1970; Spanish edition, 1970.

Fifth Edition, 1973: Abraham White, Philip Handler, and Emil L. Smith. Japanese edition, 1974; Portuguese edition, 1976.

Sixth Edition, 1978: Abraham White, Philip Handler, Emil L. Smith, Robert L. Hill, and I. Robert Lehman. Japanese edition, 1979; Russian edition, 1981.

Seventh Edition

PRINCIPLES OF BIOCHEMISTRY: General Aspects

Emil L. Smith, Ph.D.

Emeritus Professor of Biological Chemistry
School of Medicine, University of California
Los Angeles

Robert L. Hill, Ph.D.

James B. Duke Professor and Chairman
Department of Biochemistry
Duke University School of Medicine

I. Robert Lehman, Ph.D.

William M. Hume Professor
Department of Biochemistry
Stanford University School of Medicine

Robert J. Lefkowitz, M.D.

Investigator, Howard Hughes Medical Institute
James B. Duke Professor of Medicine
Duke University School of Medicine

Philip Handler, Ph.D. (deceased)

James B. Duke Professor of Biochemistry
Duke University School of Medicine

Abraham White, Ph.D. (deceased)

Distinguished Scientist, Syntex Research
Consulting Professor of Biochemistry
Stanford University School of Medicine

McGraw-Hill Book Company

New York   St. Louis   San Francisco   Auckland   Bogotá   Guatemala   Hamburg   Johannesburg
Lisbon   London   Madrid   Mexico   Montreal   New Delhi   Panama   Paris
San Juan   São Paulo   Singapore   Sydney   Tokyo   Toronto

PRINCIPLES OF BIOCHEMISTRY: General Aspects

Copyright © 1983, 1978, 1973, 1968, 1964, 1959 by McGraw-Hill, Inc. All
rights reserved. Printed in the United States of America. Except as permitted
under the United States Copyright Act of 1976, no part of this publication may
be reproduced or distributed in any form or by any means, or stored in a data
base or retrieval system, without the prior written permission of the publisher.

34567890   DOCDOC 8987654

ISBN 0-07-069762-0

This book was set in Baskerville by York Graphic Services, Inc.; the editors
were Richard S. Laufer, Ellen Warren, and Donna McIvor; the production
supervisor was Thomas J. LoPinto; the designer was Jules Perlmutter, Off-
Broadway Graphics. New drawings were done by Vantage Art.
R. R. Donnelley & Sons Company was printer and binder.

Library of Congress Cataloging in Publication Data
Main entry under title:

Principles of biochemistry, general aspects.

   Companion vol. of: Principles of biochemistry,
mammalian biochemistry.
   Includes index.
   1. Biological chemistry.   I. Smith, Emil L.,
date.
QP514.2.P754   1983   574.19′2   82-13109
ISBN 0-07-069762-0

In memory of

ABRAHAM WHITE
(March 8, 1908–February 14, 1980)

and

PHILIP HANDLER
(August 13, 1917–December 29, 1981)

Contents

Contents of *Mammalian Biochemistry* (companion text)

Preface

The publication of the seven editions of this book over the last thirty years has paralleled an explosive growth of biochemistry. During this time we have witnessed not only an increase in knowledge of the traditional subjects of biochemistry, but also an understanding of the molecular basis of various biological phenomena previously inaccessible to study. Moreover, biochemistry has become the language of much of biology, as evident from the diversity of subjects considered in Chapter 1, "The Scope of Biochemistry." The principles and methods of biochemistry now provide the underpinnings of all of the basic biological sciences and are the rational language for discourse in such diverse areas as ecology, clinical medicine, and agriculture. Indeed, the boundaries between biochemistry and much of the rest of biology have become blurred. Consequently, the preparation of this edition has been a formidable task since it has become necessary more than ever before to limit or even omit certain subjects considered in past editions in favor of the inclusion of new developments.

Our goal continues to be not only an exposition of the well-established principles that are generally acknowledged as the subject matter of biochemistry, but also an introduction to the new knowledge of the numerous biological structures and processes that has emerged in recent years. As in the past, the primary emphasis is on mammalian biochemistry, particularly of human beings. We have, however, been unwilling to omit large and important areas of fundamental biochemical interest and have decided for the first time to present the subject in two distinct but related books, each of which can stand by itself. We have retained the name, *Principles of Biochemistry* for both books, with one subtitled *General Aspects,* and its companion subtitled *Mammalian Biochemistry.* Clearly, *General Aspects* contains material essential for the beginning student and encompasses the subjects that are usually considered in an introductory course in biochemistry. The material in *Mammalian Biochemistry* requires some previous exposure to basic biochemical principles, but also illustrates how these principles have been applied to gain great insight into the molecular functions of diverse mammalian organ systems. Although each book is separately indexed for the convenience of the reader, each also contains numerous cross-references denoted by G for general and by M for mammalian aspects, so that the reader can correlate the relevant subject matter in different parts of the two texts.

The text of *Principles of Biochemistry: General Aspects* is divided into four parts: (I) "The Major Constituents of Cells," (II) "Catalysis," (III) "Metabolism," and (IV) "Molecular Genetics." The chapters in each part were largely rewritten, and in many cases shortened considerably by omission of material that has been judged to be less essential for the beginning student interested in mammalian biochemistry. Such important processes, however, as photosynthesis (Chapter 19) and nitrogen fixation by microorganisms (Chapter 22) have been retained since they are of such significance to all of biology.

Part I, "The Major Constituents of Cells," considers the chemistry of carbohydrates, lipids, proteins, and nucleic acids, as in past editions. Two new chapters, however, are included. Chapter 2 introduces the reader to the general nature of cellular constituents, including water, the solvent for all biological processes. Chapter 9, "Isolation of Compounds of Biological Interest," presents the principles which underlie the highly selective methods for analysis and preparation of various cellular constituents. Some of the subject material of these chapters was included in several other chapters in earlier editions; however, the current unified presentation permits better focus on the essential concepts. Among the most important revisions in this Part are those in Chapter 8, "The Nucleic Acids," which includes a description of the rapid micromethods that have been developed recently for sequence analysis of DNA, and which have provided information beyond the imagination of the authors of the first edition.

Part II, "Catalysis," considers the nature and properties of enzymes, that large class of proteins that catalyzes biochemical reactions. The mechanisms of action of a large number of enzymes have now been elucidated, but limitations of space permit only a few to be considered in detail. Nevertheless, from the examples given, the beginning student will be able to glean the important principles necessary for understanding the behavior of the many other enzymes to be encountered in subsequent parts of the text.

Part III, "Metabolism," has been extensively revised and reorganized, especially the introductory chapters. Thus, Chapter 13, "Membranes and Subcellular Organelles," is a new chapter written in response to the recent advances in our knowledge of the molecular basis of many aspects of cellular action, including membrane structure and function. In addition, Chapter 14, "Receptors and Transport," treats systematically the biochemical principles required to understand these important phenomena. Indeed, if we were to hazard a choice of one area of biochemistry that has had a greater impact than any other since the sixth edition was published, it is the biochemistry of receptors. This is reflected in the fact that new knowledge of membrane receptor function pervades the remaining chapters in Part III on carbohydrate and lipid metabolism to an extent not possible in earlier editions, and it is at the heart of many chapters in *Mammalian Biochemistry,* especially in Part III, "Biochemistry of the Endocrine Systems."

Part IV, "Molecular Genetics," has been almost entirely rewritten to reflect the rapid pace of discovery in this area in recent years. The differences in the structure and the expression of the genomes of prokaryotes and eukaryotes have necessitated treating these two subjects separately (Chapters 28 and 29). Indeed, the rapid advances in sequence analysis of genes and the ability to examine eukaryotic genes in the laboratory have brought not only unexpected insights into eukaryotic molecular genetics, but also enormous opportunities for gene manipulation with potentially great benefit to humanity.

Although we expect that the subject matter in *General Aspects* will suffice for many students whose interests may lie in other scientific disciplines, we recognize that many of the topics covered cannot be included in all university courses. It is our hope, however, that students will subsequently seek to broaden and deepen their knowledge of other areas of biochemistry, including the substance of *Mammalian Biochemistry*.

The companion text, *Principles of Biochemistry: Mammalian Biochemistry* is also divided into four parts: (I) "Body Fluids and Their Constituents," (II) "Specialized Tissues," (III) "Biochemistry of the Endocrine Systems," and (IV) "Nutrition." This book will be of special interest to students of human biology and medicine, although the subjects considered illuminate the behavior of molecules and cells of great general importance to all students of biology. Most chapters in Parts I and II have been extensively revised as required by the continuing important progress in many of the subjects treated, for example, blood coagulation, the complement system, erythrocytes and leukocytes, and the control of body fluids. Chapter 2, "The Immune System," has been extensively revised to incorporate the rapidly accumulating advances in molecular and cellular immunology. Perhaps no other chapter better illustrates how the approaches of molecular and cellular biology have impinged to give great insight into the functions of a specialized organ system. Chapter 10, "The Gastrointestinal Tract," is a new chapter that systematically treats the many advances in our knowledge of this organ system, especially its hormonal control by peptides, several of which may also function as neurotransmitters in the brain (Chapter 8).

Part III, "Biochemistry of the Endocrine System," has been completely reorganized and rewritten, and several new chapters introduced. These focus on the general principles of endocrine biochemistry (Chapter 11) and the mechanisms of hormone action (Chapter 12). Since its beginnings as a physiological science, endocrinology has advanced to a remarkable extent into the realm of biochemistry. Although diverse in their actions, all hormones are now recognized as acting by receptor-mediated mechanisms with similar principles underlying their functions.

Part IV, "Nutrition," remains central to all of biochemistry. Few areas of science have been applied more triumphantly to the benefit of humankind than the principles of nutrition. Failure to nourish adequately the human population results from inadequate production and distribution of food, rather than an incomplete knowledge of the nutritional requirements of human beings.

Although the separation into two distinct books has permitted introduction of new information, there have had to be many painful omissions. Regretfully, we have had to omit presentation of the historical developments of biochemistry and the names of the untold cadre of scientists who have contributed to our knowledge. Few descriptions of actual experiments could be included. The serious student will have to go beyond these texts to gather an appreciation of the arduous process by which knowledge has been gained, bit by bit, in countless laboratories utilizing a remarkable panoply of biological materials, techniques, instruments, and experimental approaches. References have been limited to more recent monographs and reviews that can help the student find additional information, as well as the relevant details of the experimental findings and the historical background.

In the presentation of this edition, we acknowledge our great loss in the deaths of two of our coauthors. Abraham White was the initiator of this book and its senior author through six editions. While this edition was in its first phases of preparation, Philip Handler was struck by the difficult illness which terminated his life. In dedicating this edition to them and retaining their names as coauthors, we acknowledge the large residue of their scholarship, wisdom, and perspective that remains in this seventh edition. Their contributions as scientists and teachers will not be forgotten. It is painful to accept the termination of a happy collaboration and friendship with them during which we shared the joy of interpreting and presenting the ever-accelerating growth of our science; for the present senior author (E.L.S.), this began more than thirty years ago.

One of us (R. J. L.) is a new coauthor for this edition, but as in past editions, we continue to take collective responsibility for the content of the entire work. Although each chapter was the primary responsibility of one author, all chapters were circulated among, criticized by, and even rewritten by one or more of the other authors.

We are grateful to the colleagues and friends too numerous to mention, who have generously contributed information and suggestions for the preparation of this edition. We acknowledge with thanks those teachers and students who have made valuable comments and have noted our occasional past errors. We wish also to recognize and thank Richard S. Laufer, Ellen J. Warren, Editors; Donna McIvor, Editing Supervisor; and other members of the staff of the McGraw-Hill Book Company for their pleasant cooperation and help in the task of preparing and publishing these books.

Emil L. Smith
Robert L. Hill
I. Robert Lehman
Robert J. Lefkowitz

PRINCIPLES OF BIOCHEMISTRY:
General Aspects

Chapter 1

The Scope of Biochemistry

The word *life* is employed to characterize objects as diverse as grass, trees, insects, worms, birds, fish, and humans. Each proceeds through a life cycle, reproduces its own kind, and responds in a variety of ways to external stimuli. Over the course of a few millennia, "living" forms were classified, first in terms of their gross comparative anatomy and later with the aid of the light microscope. In 1838, Schleiden and Schwann recognized that all these forms were constructed of unit cells of rather similar dimensions and general appearance. This relatively primitive body of information, together with increased understanding of the fossil record, sufficed to permit formulation by Darwin of the most sweeping and compelling biological generalization of all, the concept of historic and continuing biological evolution.

Meanwhile, progress in the physical sciences led to increasingly sophisticated questioning by students of biology. Identification of the major atmospheric gases was soon followed by demonstrating the use of oxygen and production of carbon dioxide by animals and the photosynthetic reversal of this relationship in green plants. The laws of conservation of energy and matter applicable to the physical world were shown by Lavoisier and Laplace in 1785 to be equally valid in a biological system that they could examine experimentally. Isolation of increasing numbers of purified materials from living forms and recognition that all contained carbon gave birth to *organic chemistry*. This actually fortified vitalistic thinking until Wöhler synthesized urea in 1828, thereby demonstrating that carbon compounds need not necessarily be formed in living organisms. Formulation of the principles of catalysis by Berzelius led to recognition that the ptyalin of saliva, pepsin of gastric juice, and amylase of sprouted malt were biological catalysts. Yeast was believed to be an inert cata-

---

Any cross-references coded M refer to the companion text, *Principles of Biochemistry: Mammalian Biochemistry.*

lyst at that time, and hence early studies of the chemistry of fermentation failed to contribute to the decline of vitalistic thinking. Indeed, it is ironic that chemical synthesis of ethanol, by Hennell, had preceded the synthesis of urea but did not serve as an equivalent philosophical milestone because the living nature of yeast failed to gain acceptance until the work of Pasteur.

Until the major laws of physics and chemistry governing inanimate matter had been elucidated, it was not possible to formulate the more penetrating questions concerning the nature of life. These questions, which we shall consider shortly, were not expressed until the first quarter of the twentieth century. Meanwhile, chemistry flourished, the laws of thermodynamics were enunciated, and it became possible to examine, in detail, whether living systems obey the laws of physics and chemistry. The doctrine of evolution gained acceptance, the principles of inheritance were formulated by Gregor Mendel, and the list of compounds obtained from living organisms grew ever larger. The conducting role of the nervous system was described and the role of glycogen as a storage form of glucose in liver and muscle was demonstrated by Claude Bernard, who also recognized the constancy of the *milieu interieur*. The germ theory of disease was established and systematic microbiology introduced.

At the turn of the century, Emil Fischer established the structures of many carbohydrates, amino acids, and other compounds and initiated much of contemporary biochemical thought by recognizing the optical configurations of carbohydrates and amino acids and by demonstrating the specificity of enzymic action. In postulating the "lock-and-key" concept of enzymic action (Chap. 10), Fischer began the study of the relation of the topography of macromolecules to the phenomena of life. With these studies and the exploitation by Harden and Young of the accidental observation by the Buchner brothers that a cell-free extract of yeast could ferment glucose with the production of alcohol, modern biochemistry began. The term *biochemistry* was introduced by Carl Neuberg in 1903.

Ever since, information and understanding have been increasing exponentially. Intellectual curiosity and philosophical questions have given general direction to the course of these biochemical investigations. In large measure, however, the quickening tempo of this effort reflects not only a fundamental human drive for self-understanding but the belief that the knowledge gained could improve agricultural practice and with it animal and human nutrition, as well as assist in the alleviation of human disease. In significant measure, these goals have been realized.

Research in biochemistry has been addressed to a series of major questions, each of which continues to command attention. These are briefly considered below.

Of what chemical compounds are living things composed? A catalog of such compounds is a *sine qua non* for the understanding of life in chemical terms. New compounds are, however, continually being recognized during investigations directed toward studying the fate of a well-recognized chemical entity or by isolation of a substance responsible for some physiological event. The ubiquitous distribution of many of these compounds results in a high degree of similarity in the *qualitative* composition of most organisms, with differences among these forms, as well as among their own tissues and organs, being primarily *quantitative*. In addition, specialized cells make certain compounds uniquely related to their special functions, e.g., hemoglobin in erythrocytes, antibodies in lymphocytes, and so on.

What are the structures of the macromolecules characteristic of living organisms? The pioneers of biochemistry recognized substances they named proteins, nucleic acids, polysaccharides, and lipids. The rate of progress of biochemical understanding has, in no small part, been paced by development of procedures to purify such materials. New physicochemical methods revealed molecular weights of from 10,000 to more than 100 million for individual substances. For years, the seemingly herculean task of establishing the structures of such molecules appeared to be experimentally unapproachable. However, development of new methods and instruments, including the ultracentrifuge, electrophoretic apparatus, recording spectrophotometers, spectropolarimeters, and amino acid analyzers, helped reveal the general structures of these molecules. Improved techniques and, in particular, diverse chromatographic methods permitted the separation and quantitative determination of the micro amounts of complex mixtures required to deduce the covalent structures of diverse macromolecules. With the development of x-ray crystallographic methods, detailed three-dimensional models of many smaller proteins and nucleic acids are now available, and there is a rapidly growing understanding of the forces by which these molecules fold themselves into highly specific three-dimensional structures which are essential for their functions.

Understanding of the structures of these large molecules is rapidly expanding, thereby providing the basis for a more penetrating insight into the mode of operation of enzymes, the structural basis for genetic phenomena, and the fine structure of living cells. Indeed, this is a major theme of this book.

How do enzymes accomplish their catalytic tasks? In the nineteenth century, degradation of proteins, starch, and fats to smaller constituents in the digestive tract was recognized as being due to enzymic activity. That fermentation is also the result of such catalysis was later shown by the Buchners. Earlier, Kühne had coined the name *enzyme* (Gk., "in yeast") to designate the unorganized "ferments," as distinguished from bacteria, which were also called ferments. The studies of Fischer on the specificity of enzymes were followed by the formulation by Michaelis and others of the elementary rules describing enzymic catalysis and by Sumner's isolation in 1926 of the enzyme urease as a crystalline protein. Since then, hundreds of enzymes have been isolated and many crystallized; each has proved to be a distinct, unique protein.

How these proteins function as catalysts is one of the central problems in biochemistry. In the last few decades, the phenomena operative in enzyme catalysis and their bases in protein structure have been revealed in considerable detail. This fascinating aspect of science, in many respects the heart of biochemistry, is the subject of Part 2 of this book.

What substances are required to satisfy the nutritional requirements of humans and other organisms? The small catalog of these substances, now perhaps complete for human beings, is presented in Part M4. This knowledge is adequate to manage the nutritional affairs of humankind; the inadequate nutrition of much of humanity reflects failures of production and inequities of distribution, not lack of nutritional understanding.

In the course of studies of the nutrition of bacteria, powerful experimental tools were forged that have influenced all aspects of biochemical research. The ability to estimate bacterial growth quantitatively has been utilized as the basis of sensitive analytical procedures. The fact that a given compound is an essential nutrient because the organism cannot make it and yet requires that compound for further metabolic transformations has been exploited in the elucidation both of metabolic pathways and of genetic mechanisms.

By what chemical processes are the dietary materials transformed into the compounds characteristic of the cells of a given species? Rather large quantities of a small group of organic compounds are ingested daily. The growing child retains some of these as a collection of compounds that is rather different from the composition of the ingested mixture. Similarly, plants "ingest" only water, minerals, and CO_2 yet accumulate a remarkable ensemble of diverse compounds. Most of the carbon of the ingested food of the child is released with the expired air while the nitrogen appears in the urine as urea; the remarkably sensitive regulation of these processes is even more evident in adults, who maintain constant weight and composition while they process about a pound of mixed solids per day.

Since, in attempting to understand these processes, one cannot readily sample the reaction mixture, how can one ascertain the reactions to which ingested foodstuffs are subjected? This experimental impasse was broken by the availability of radioactive isotopes, particularly ^{14}C, and apparatus with which to measure their abundance. With the use of these tools, increased skill in the handling of tissue preparations in vitro, and the powerful separatory capabilities of chromatographic procedures an elaborate, interwoven network of metabolic pathways was quickly exposed. Although the outlines of most major processes appear to have been revealed, specialized organs and tissues reveal remarkable differences both in composition and metabolism.

How is the potential energy available from the oxidation of foodstuffs utilized to drive the manifold energy-requiring processes of the living cell? Among such processes, we need note only the synthesis of hundreds of new molecular species, intracellular accumulation of inorganic ions and organic compounds against concentration gradients, and the performance of mechanical work. The impossibility of utilizing thermal energy to accomplish useful work at constant temperature makes untenable a simple analogy between food-burning animals and fuel-burning heat engines. Understanding of the biological solutions to this problem is cardinal to the understanding of living cells (Chap. 12).

The corollary problem, elucidation of the mechanism by which light energy is harnessed to achieve fixation of CO_2 into carbohydrate, has been a challenging major question in its own right (Chap. 19). Understanding of the primary photochemical events and the subsequent reactions that lead to carbohydrate accumulation has expanded greatly in recent years.

What is the structure of a living cell, and how is it organized to perform its characteristic chemical functions? The general topography of cells—an outer membrane, an inner nucleus, and numerous smaller bodies—was early revealed in the compound light microscope. Electron microscopy has provided much more detailed stereoscopic visualization of the finer structure. Isolation of each of the cellular components, free of others, has revealed the partition of functions within the cell: the nucleus as the site of genetic control and cellular duplication; ribosomes as loci for protein synthesis; mitochondria as membranous structures in which oxidative metabolism generates adenosine triphosphate; the membranous endoplasmic reticulum as the site of metabolism of many complex carbohydrates and certain lipids and proteins; the cell membrane as the site of vectorially organized mechanisms for controlling the general electrolyte composition of the cytoplasm, bringing required nutrients into the cell proper (Chaps. 13 and 14) and possessing numerous specialized receptors that receive chemical messages from other cells or from the general environment; intracellular contractile fibers; a variety of highly specialized structures, each unique to a specific cell type; and the cytoplasm, a solution of hundreds of individual enzymes that direct the multitudinous reactions by which nutrients are converted into cell constituents. It is the sum of all these chemical activities that constitutes the "life" of the cell.

How do cells divide to yield identical daughter cells? What is the chemistry of inheritance? What is a gene and how does it function? No chapter in the history of science has unfolded with such great rapidity or engendered such widespread interest as the answers to these questions. Few hold deeper or more significant implications for our future. This is the subject of Part 4.

It is the presence and activities of its complement of proteins that determine the form, organization, and functions of a cell, viz., its life. It follows that the genetic "instructions" to a cell must provide the information required to achieve the precise synthesis of the ensemble of proteins characteristic of that cell. This information is encoded in the structure of the very large molecules of deoxyribonucleic acid. Cell duplication requires perfect reproduction of these molecules with subsequent equal distribution of the information between the cells. Utilization of this information requires its transmittal from nucleus to the ribosomal protein factories. Changes in the structure of deoxyribonucleic acid (DNA) become evident as mutations in subsequent generations. How these processes operate was first disclosed by utilizing a nonpathogenic enterobacterium, *Escherichia coli,* and by studying its viruses, each of which is a limited bit of genetic information wrapped in a specific protein coat and capable of self-duplication only by utilizing the synthetic apparatus of a living cell. Almost daily, new information is becoming available concerning the structure of human chromosomes, the constituent genes, and the controls on the expression of these genes. This information has made intelligible the laws of genetics, the nature of hereditary diseases, and the biochemical operation of the process of evolution.

Progress in understanding these problems has been spectacular, thanks in part to the development of methods that make possible the rapid sequencing of DNA. Use of these techniques has resulted in knowledge of the complete sequences of some viral and large parts of bacterial and human genomes. The work continues at an explosive pace (Part 4).

Had evolution not been deduced earlier on other grounds, it would surely have become obvious to the biochemist. Whereas the unaided eye reveals the diversity of life, the qualitative answers to each of the foregoing questions are essentially identical for *all* living forms. The impressiveness of this oneness of the cardinal aspects of all forms of life is matched only by the remarkable manner in which subtle variations on these themes have given rise to the rich variety and abundance of living forms.

Since the life of a cell is the totality of thousands of different chemical reactions, how are these synchronized into a harmonious whole? Clearly it is advantageous to the cell to match the pace of energy-yielding reactions to the requirement for that energy and to provide the requisite monomeric units (amino acids, nucleotides, sugars) at a rate commensurate with the demands for polymer synthesis (proteins, nucleic acids, polysaccharides). Examples of such rate regulation can be found throughout this book. Included are arrangements analogous to both the negative and positive feedback systems of electronic engineering; these are intrinsic in the structures of some enzymes that participate in synthetic processes and help to ensure a steady flow, but not a surplus, of necessary synthetic intermediates. In other instances, regulation involves repression or derepression of the synthesis of enzymes that participate in synthetic processes.

In the vertebrate, not only must the diverse aspects of the metabolism of each cell be synchronized, but the various organs, muscle, liver, brain, etc., must also operate in harmony. Information concerning the metabolic state of muscle, for example, must be transmitted to the liver as required. In large part, this is the role of the endocrine system. Endocrine cells, responding to changes in the composition of the blood, which in turn reflect changes in some tissue or organ, synthesize and release hormones, which are carried in the circulation indiscriminately to all cells. Only target cells possess the specific receptors which bind the hormone. In this way only certain cells respond to a particular hormone which can then modulate specific cellular metabolic activities.

A host of compounds transmits information in other ways: the neurotransmitter substances, which signal the arrival of a nervous impulse from nerve cell to nerve cell or from nerve cell to muscle cell; the prostaglandins made in many organs and tissues, which modulate the activities of numerous other cell types; and cyclic adenylic acid, ubiquitous in living cells, which is formed upon receipt of some message from the cell's environment, e.g., a peptide hormone or a change in the nutritional properties of that environment, and then elicits a cellular response that varies with both the cell type and the message. Such processes will be encountered frequently in this book and form the major theme of Part M3.

How do the specialized cells of animal tissues make their unique contributions to the total animal economy? Osteoblasts make bone, muscle cells contract, nerve cells conduct, kidney cells make urine, and endocrine cells make hormones, all by mechanisms specific for these cell types. Because the systems involved are more readily accessible, biochemists were initially more successful in learning the generalities of how all cells live than in ascertaining the details of this group of specialized activities. More recently, this field of inquiry has proceeded apace. Rather detailed understanding of the biochemistry of specialized mammalian cells is presented in Part M2 and elsewhere. Much of present biochemical effort is directed to expansion of such understanding.

How does an animal regulate the volume and composition of the fluids that constitute the environment of its cells and of the blood that interconnects them? The information that has been gathered in this area of inquiry has contributed significantly to the management of diverse human disorders and to the achievements of modern surgery. The mechanisms involved are extraordinarily sensitive and, as in critically engineered systems, frequently redundant. The high degree of perfection of these mechanisms in human beings permits people to range the earth from the equator to the poles, from ocean depths to mountain peaks, and to survive despite enormous variations in the composition and quantity of the food and drink they ingest. These regulatory mechanisms are also discussed in Part M1.

Central to much of this aspect of vertebrate life are the physiology of the erythrocyte (Chap. M3) and the chemistry of hemoglobin, perhaps the most thoroughly studied of all proteins and our most detailed source of insight into the correlation of protein structure with physiological function (Chap. M4).

How is the genetic information in a totipotent, fertilized egg utilized to direct development of a differentiated organism? Thoughtful statement of this problem has been possible for decades; bits of information concerning chemical aspects of the developing embryos of various species have been accumulated for many years. Only now, with deeper understanding of molecular genetic mechanisms, does it appear possible to attack this problem successfully. A growing literature describes various aspects of the mechanisms involved in the developmental changes observable in simple organisms. Nevertheless, this wondrous process cannot yet be described in a satisfying way.

Accidental entry of microorganisms or foreign macromolecules into an animal evokes a series of complex responses. How is the foreign nature of such materials "recognized" by the body and how are they rendered innocuous? These processes have proven to be remarkably complex, requiring participation of several types of cells and are rapidly being disclosed. A summary is presented in Chap. M2.

By what mechanisms do cells "recognize" other cells? In the course of development there are countless instances in which cells of one type must "recognize" and closely unite with other, similar cells. In some instances this is necessary to form a parenchymatous organ, e.g., liver; in others, a specific nerve cell must synapse with only one other specific nerve cell to form a tract leading from the periphery of the body to an area in the central nervous system. The basis for such cellular discrimination is now under intensive investigation.

Can behavior be described in chemical terms? With the evolutionary appearance of nerve cells, the subsequent development of primitive nervous systems, and their ultimate elaboration in the brain of *Homo sapiens*, there emerged increasingly complex behavior patterns. Albeit neurally mediated, responses to stimuli in most species are invariant and species-specific, e.g., the patterns of spider webs, courting behavior in birds. Patently, cognitive processes are limited to relatively few species. For all forms, however, the fundamental questions concerning individual nerve cells are much the same: What chemical events underlie conduction along the nerve axon, stimulatory or inhibitory transmission between nerve cells, and between nerve and muscle cells? When a single nerve cell makes multiple contacts with other both stimulatory and inhibitory nerves, how is the "information" that is received integrated into a single command decision? These are the simplest elements of activity, the sum of which constitutes "behavior"; the current understanding of such processes is summarized in Chap. M7.

Can disease be adequately described in molecular terms? The great triumphs in this area were the isolation of the substances needed for adequate human nutrition (Part M4) and the hormones produced by endocrine glands (Part M3). Such understanding has prevented many millions of deaths due to malnutrition and made possible successful therapy of hundreds of thousands of patients with endocrine disorders. Detailed understanding of metabolism and genetics has permitted the identification of approximately 1500 distinct abnormalities of humanity that are transmitted by the genetic apparatus. Of these, more than a third reflect mutations of some specific single gene. This enormous chapter in human self-understanding is only presently being written. Examples of these metabolic disorders are described throughout this work.

Literally dozens of new genetically transmitted disorders of metabolism are recognized annually (Chap. 31). However, only in a few instances has this led to successful therapy. Nonetheless, by a combination of genetic counseling and examination of fetal cells obtained from amniotic fluid, it is increasingly possible to prevent such disorders or to abort fetuses that would otherwise be doomed to limited physical or mental capability. It is hoped that, as fundamental knowledge expands, it will become possible equally successfully to address such questions as the genesis of atherosclerosis, the nature of the cancerous transformation of previously normal cells, the origins of a host of neurological diseases, and perhaps even some of the major psychoses.

The partial answers presently available to the above questions have largely been obtained in the last few decades. Progress has, in large measure, been determined by the rate at which new tools have become available. A single example will suffice to illustrate this fact. The most important biological concept established in recent years, that the genetic apparatus encodes instructions for the precise ordering of the amino acids in the primary strand of a protein, was most clearly demonstrated by the single amino acid substitution in the β chain of hemoglobin (Chap. M4) from individuals with sickle cell anemia, a genetic disorder. However, this study could not have been undertaken until electrophoretic analysis had shown a difference in the net charge of hemoglobin from normal human beings and those with this disease. Thereafter, the precise amino acid substitution could not have been sought until there were techniques for deciphering the amino acid sequences of proteins. To appreciate why a single amino acid replacement exerts so profound an effect on the physical properties of hemoglobin required understanding of its three-dimensional structure. X-ray diffraction analysis could be attempted only after development of a technology for preparation of large protein crystals to which heavy metals are specifically bonded and which are also crystallographically isomorphous (Chap. 5). Even then, analysis of the massive body of x-ray data was impossible before the development of high-speed digital computers. Only then could one discover and appreciate the consequences of the point mutation in DNA structure that results in sickle cell anemia. Regardless of the specific problem under study, current research in biochemistry almost invariably requires a complement of sophisticated methods. Accordingly, it may be anticipated that the next burst of research progress will follow the next significant addition to the armamentarium of techniques and instruments.

In addition to application of physical instrumentation, biochemistry has developed its own complement of sensitive and specific techniques that depend on use of such tools in combination with biological methods. To mention only a few: the use of microorganisms for assay of unknown nutrients; immunochemical and radioimmunochemical methods for assay of enzymes, hormones, receptors, and other molecules; modified strains of microorganisms for producing large quantities of a specific enzyme; specific ligands for affinity chromatography in order to isolate cellular constituents present in only minute amounts; the use of specific enzymes for fragmenting large molecules to smaller ones more amenable to structural analysis. These and other methods were developed because of the need to separate, identify, and measure the miniscule quantities of materials in cells. The study of cultures of pure mammalian cell types has given great insight into the metabolic events in specialized cells and of abnormal metabolism in various disorders.

The current explosive developments in molecular genetics rest not only on the use of sophisticated instruments but on biochemical methods: enzymes to split nucleic acids; enzymes to insert other bits of nucleic acid including whole genes; enzymes to copy sections of nucleic acids to obtain large amounts of materials. This has permitted transfer of genes of bacterial origin into mammalian cells and vice versa. These procedures are yielding great insight into the operation of toxic substances, mutagens, carcinogens, and the ways in which viruses and bacteria modify host cells.

Biochemistry is not an isolated discipline; it has become the language of biology and is basic to the understanding of phenomena in the biological and medical sciences. Since the time of Aristotle, students of biology have sought to correlate structure and function. This endeavor continues; the correlation of biological function and molecular structure is the main theme of biochemistry.

Part 1

The Major Constituents of Cells

Chapter 2

The Major Chemical Constituents of Cells

General chemical composition. Water and its properties. Acids and bases.

GENERAL CHEMICAL COMPOSITION

The chemical composition of a rapidly dividing, bacterial cell, *Escherichia coli*, given in Table 2.1, illustrates several aspects of living things. The composition is typical of that of many prokaryotic cells and it is not remarkably different from that of a eukaryotic cell, except for some highly differentiated cells that may contain somewhat higher or lower amounts of some of the substances listed. The bulk of all cells is water, in which relatively small amounts of inorganic ions and several organic compounds are dissolved. The number of distinct organic chemical species in a cell is large, but most may be classified as carbohydrates, lipids, proteins, nucleic acids, or derivatives thereof. Those few compounds not so classified account for only a small fraction of the mass of the cell and are metabolically derived from substances in one of the four major classes. Proteins account for about one-half, and nucleic acids about one-quarter, of the dry weight of the cell.

The four major classes of organic compounds have markedly different structures, properties, and biological functions, but each class contains two types of compounds: *polymeric macromolecules,* with molecular weights ranging from 10^3 to 10^{10}, and *monomeric species,* with molecular weights of about 10^2 to 10^3. The monomeric species are the repeating units, or the building blocks, of the polymeric macromolecules in each class. Thus, high-molecular-weight carbohydrates, called *polysaccharides,* are macromolecules formed by covalent combinations of monomers (monosaccharides), such as glucose, the principal repeating unit of the macromolecules starch (from plants) and glycogen (from animals). Proteins are macromolecular, covalent combinations of α-amino acids, and the

Any cross-references coded M refer to the companion text, *Principles of Biochemistry: Mammalian Biochemistry.*

TABLE 2.1 The Approximate Chemical Content of a Rapidly Dividing Cell of _E. coli_

Substance	% of total cell weight	Average MW	Approximate number of molecules per cell	Approximate number of different kinds
Water	70	18	5×10^{10}	1
Inorganic ions (Na^+,K^+,Ca^{2+}, Fe^{2+},Cu^{2+},Cl^-,SO_4^{2-},HCO_3^-, P_i, etc.)	1	40	2×10^8	18
Proteins	15	50,000	1×10^6	2,500
Amino acids	0.5	110	3×10^7	20
Carbohydrates:				
Monosaccharides	0.5	180	2×10^7	20
Polysaccharides	1.5	10^6	12×10^3	5
Lipids	2	700	3×10^7	50
Nucleic acids:				
DNA	1	2.5×10^9	1	1
RNA	6	$0.02\text{–}1 \times 10^6$	4.5×10^5	1,100
Nucleotides	0.5	300	1×10^7	50

Adapted from J. D. Watson, _Molecular Biology of the Gene_, 3d ed., W. A. Benjamin, Inc., Menlo Park, Calif., 1976.

nucleic acids are macromolecules formed from monomeric repeating units, the nucleotides. Lipids are not covalently combined to form macromolecules. However, lipids associate through noncovalent interactions to form high-molecular-weight aggregates, such as the biological membranes, which, in a sense, behave as macromolecular structures.

The unique chemistry of living things results in large part from the remarkable and diverse properties of biological macromolecules. Macromolecules from each of the four major classes may act individually in a specific cellular process, whereas others associate with one another to form _supramolecular structures_, such as ribosomes, microtubules, and microfilaments, all of which are involved in important cellular processes (Chap. 13). Such structures are very large. For example, ribosomes may have effective molecular weights ranging from 3.0 to 4.5 $\times$ 10^6, the exact value depending on the organism in which they occur. A typical bacterial ribosome contains one or more molecules of each of about 55 proteins with molecular weights from about 9000 to 65,000 and three molecules of ribonucleic acid, which together contain a total of about 4600 nucleotides. More complex mixtures of other macromolecules in eukaryotic cells assemble to give _subcellular organelles_, such as mitochondria, the Golgi apparatus, plasma membranes, and the endoplasmic reticulum (Chap. 13), each of which contains dozens of different macromolecules.

WATER

Dilute aqueous solutions of inorganic ions are the intra- and extracellular liquid media of all cells, and the properties of biological compounds in all living things are uniquely suited to water as a solvent. Many compounds are completely water-soluble and the reactions they undergo proceed in an aqueous medium. But water serves as much more than the solvent in which biochemical reactions proceed. Because of its unique properties, water aids in maintaining the structures of macromolecules as well as the assemblies of macromolecules in supramolecular structures and subcellular organelles. Moreover, some compounds, in particular those in certain supramolecular structures or subcellular organelles, are largely water-insoluble and react at a liquid-solid interface formed by water and insoluble macromolecular complexes. Accordingly, before considering the various kinds of biological molecules to be described in subsequent chapters, it is essential to understand some of the properties of water.

Water Structure—The Hydrogen Bond

The molecules of a pure liquid interact with one another in definite ways through intermolecular interactions with each molecule constantly changing molecular partners. Thus, the structure of any liquid is difficult to describe exactly; it is best considered to be the average of many possible intermolecular arrangements. Water is no exception. Although its structure as a liquid is incompletely understood, the general features of the structure of different forms of ice have offered clues about liquid water. As illustrated in Fig. 2.1a, each water molecule in one form of ice is thought to have a tetrahedral arrangement in space, with each hydrogen atom directed to one corner of the tetrahedron and the two unshared electron pairs on the oxygen atom directed to the other corners. Each water molecule may interact with four neighboring molecules by noncovalent *hydrogen bonds* to form a continuing tetrahedral array of indefinite size. Liquid water is thought to exist as much smaller clusters of molecules similar to those in ice, each containing as many as several dozen molecules arranged as depicted in Fig. 2.1b.

Hydrogen bonds are weak compared with covalent bonds. In water, each such bond contributes only about 5 kcal/mol to the stability of the system as compared with 110 kcal/mol for the covalent H—O bond in water. The hydrogen bonds formed by any water molecule with its partners are continuously breaking and re-forming with other partners. The tendency for water molecules to form hydrogen bonds is a function of the orientation of its hydrogen atoms and its unshared electron pairs. The oxygen atom is a center of negative charge since it attracts electrons from the two hydrogen nuclei, which have a partial positive charge (Fig. 2.1). The partially positively charged hydrogen atoms in one water molecule thus form hydrogen bonds by bridging with the partially negatively charged oxygen atoms of another molecule. The hydrogen atoms in a few molecules occasionally are lost to neighboring water molecules as follows:

Hydronium ion

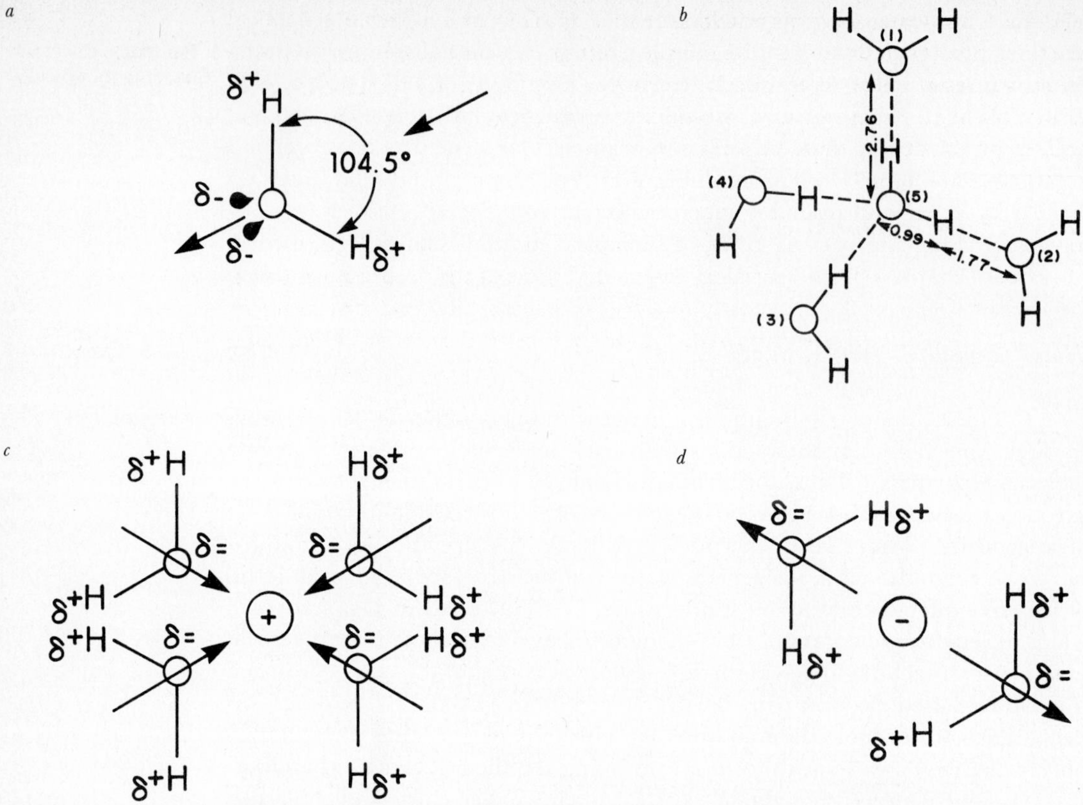

Figure 2.1 Features of water structure. (*a*) The structure of a water molecule, indicating the bond angle between the two hydrogen atoms and the polar character of the molecule, which has two partially negative charges (δ^-) on the oxygen atom and a partial positive charge (δ^+) on each hydrogen atom. This electronic orientation provides water with a dipole moment in the direction of the arrows ($+ \rightarrow -$). (*b*) The tetrahedral arrangement of water molecules in one form of ice. Molecules 1, 2, and 5 are in the plane of the paper with molecule 3 above and molecule 4 below the plane. Molecules 1, 2, 3, and 4 are positioned at the corners of a regular tetrahedron. The bond distances for the hydrogen bond are in angstroms. A schematic depiction of water dipole-ion interactions formed by water with either cations (*c*) or anions (*d*).

giving rise to a hydrated proton, called a *hydronium ion*, H_3O^+, and a hydroxide ion, OH^-. All protons in water are hydrated to some extent, but it is unnecessary to write them as such; the symbol H^+ refers to any proton in water irrespective of its degree of hydration. At 25°C, the concentration of H^+ and OH^- in pure water is each about 10^{-7} mol/liter, in comparison with water itself, which is 55.6 mol/liter.

From the foregoing discussion, any body of water may be likened to a single giant molecule, in which intermolecular hydrogen bonds are being constantly broken and re-formed. No single hydrogen bond contributes a great deal to the stability of water, but, taken together, such bonds cooperate to give great stability to the system.

Many substances, in addition to water, are capable of forming hydrogen bonds. The —OH, —NH, and —SH groups in many biological compounds may serve as donors of the hydrogen atom in a hydrogen bond, and different kinds of substituted oxygen and nitrogen atoms may serve as acceptors of a hydrogen atom. Hydrogen bonds are particularly important in stabilizing the structures of macromolecules and many examples of the specific groups involved will be given later (Chaps. 3, 5, and 8).

Some Unique Properties of Water

Because of hydrogen bonding, water differs remarkably in its properties from other compounds with closely related electronic structures. For example, the melting point of water is $0°C$, notably higher than that of methane $(-184°C)$, ammonia $(-78°C)$, or hydrofluoric acid $(-92°C)$. Similarly, the boiling point $(100°C)$ of water is much higher than that of each of the same compounds: CH_4 $(-161°C)$; NH_3 $(-33°C)$; and HF $(+19°C)$. The high values of these constants for water reflect its strong, three-dimensional hydrogen-bonded network, which requires much more energy to disrupt on change of state from solid to liquid, or liquid to gas, than those in electronically similar liquids.

Other physical constants of water also differ markedly from those of other liquids, as listed in Table 2.2. The *specific heat of water*—the amount of heat required to raise 1 g of water from 15 to $16°C$—is 1 cal, whereas the specific heats for such liquids as methanol, acetone, chloroform, and benzene are considerably lower. These values reflect the heat capacity of each liquid, with water having the largest heat capacity by virtue of its extensive network of hydrogen bonds. Water also has the highest *heat of vaporization* (Table 2.2). If 1 kg of water absorbs 1 kcal of heat, its temperature rises by $1°C$, but vaporization of only 2 g of water will lower the temperature of the remaining 998 g of water by $1°$. This unique "cooling effect" of water not only minimizes water loss from a system in the face of large temperature changes, but also provides biological systems with a unique thermostat, ensuring that the rates of chemical reactions are only slightly changed by external temperature variations.

The latent *heat of fusion* of water is also larger than that of other substances. Thus, the heat released by water on freezing minimizes temperature changes in winter and the heat absorbed on melting ice diminishes temperature changes in spring. Were liquids other than water to form the oceans of the earth, the seasonal temperature extremes would be far greater than those observed. The *surface tension* of water at a water-air interface is also considerably larger than the surface tensions of other liquids against air. This provides not only the high capillary rise of water, but also the tendency of dissolved substances to concen-

TABLE 2.2

Substance	Specific heat, cal	Heat of vaporization, cal/g	Heat of fusion, cal/g	Surface tension	Dielectric constant
Water	1	585 (20°C)	79.7	72.8	80
Methanol	0.6	289 (0°C)	16	. . .	33
Acetone	. . .	125 (56°C)	23	23.7	21.4
Chloroform	0.24	59 (61°C)	. . .	27.1	5.1
Benzene	0.5	94 (80°C)	30	28.9	2.3

Some Physical
Properties of
Water and
Other Solvents

trate at an interface and thereby lower the surface tension—two properties of importance in considering biological membranes (Chap. 13).

The *dielectric constant* of water far exceeds that of most other liquids. The dielectric constant D for any medium is related to the charges e_1 and e_2 of two particles in that medium and the distance r between them by the relationship

$$F = \frac{e_1 e_2}{Dr^2} \qquad (1)$$

where F is the force of interaction between the charges. For water $D = 80$, whereas for benzene $D = 2.3$. This indicates that the force of interaction between two ions, such as Na^+ and Cl^-, is about 40 times greater in benzene than in water. The relatively weak interaction of ions of opposite charge in water is related to the ability of water to orient around the ions. The negatively charged oxygen atoms tend to cluster around cations and the positively charged hydrogen atoms around anions through ion-dipole interactions (Fig. 2.1). In some instances a shell of water molecules is immobilized around the ion, and the entire network is stabilized by hydrogen bonding with the molecules adjacent to the shell. Such effects account for the high solubility of electrolytes in water. Polar, uncharged molecules, such as ethanol or acetone, also readily dissolve in water, primarily by formation of hydrogen bonds to water molecules. Nonpolar substances, such as hydrocarbons, are poorly soluble in water, since they can neither form hydrogen bonds nor interact ionically with water molecules. Indeed, dissolution of a hydrocarbon requires considerable energy to break the hydrogen bonded network of liquid water. For this reason, the hydrocarbon groups in many biological compounds, especially in macromolecules, tend to cluster together by formation of so-called hydrophobic bonds, noncovalent bonds of great importance in maintaining structures of macromolecules. A more thorough description of such bonds is given later (Chaps. 3 and 5).

ACIDS AND BASES

The acidic character of a compound is often critical to its biological function. In turn, the acidic character of a compound is markedly influenced by solvent, which in living things is water. Thus, it is important to consider some aspects of the behavior of acids in aqueous solution.

Acids are substances that are capable of dissociating to liberate protons (H^+) in solution according to the following reaction for the general acid HA:

$$HA \rightleftharpoons H^+ + A^- \qquad (2)$$

Bases are those substances capable of accepting a proton as follows:

$$B + H^+ \rightleftharpoons BH^+ \qquad (3)$$

Since reaction (3) can proceed to the left and BH^+ is by definition an acid, consideration need only be given to the acidic character of any substance, since it is reciprocally related to its basicity. Hence, the following discussion will consider only properties of acids.

The strength of an acid depends upon the equilibrium position of reaction

(2). A strong acid is essentially completely dissociated and exists almost entirely as H^+ and A^-, whereas a weak acid is dissociated to only a small extent, with as few as 1 in 10^3 to 10^{12} of all HA molecules dissociated into H^+ and A^-. But water influences acid strength considerably. For example, HCl is a strong acid in water; it dissociates fully because both H^+ and Cl^- are readily accommodated into the structure of water. HCl in benzene is not a strong acid, since it cannot participate in polar interactions or in hydrogen bonding with benzene as it can with water. Stated differently, because of the high dielectric constant of water, the force of interaction (Eq. 1) between H^+ and Cl^- is much weaker in water than in benzene. The acidity of many biological compounds may differ considerably from that observed in aqueous media, especially when a compound is placed in a hydrophobic, nonaqueous environment. Examples of such effects will be discussed later (Chaps. 3 and 5).

Most acids encountered in living things are weak acids, and the strength of a weak acid is indicated by its dissociation, or ionization, constant K_a, which is obtained by applying the mass-action equation to reaction (2) above. Thus,

$$K_a = \frac{[H^+][A^-]}{[HA]} \tag{4}$$

where brackets indicate the concentration of each species. The higher the value of K_a, the greater the number of hydrogen ions liberated per mole of acid in solution and hence the stronger the acid. Consequently, different acids may be compared in terms of their K_a values. It should be noted that the discussion here is of concentrations of H^+, A^-, and HA. More correctly, Eq. 4 applies rigorously only to activity values. Activity is related to concentration by the equation $a = \gamma c$, where a is the activity, c the concentration, and γ the activity coefficient, which reflects the deviations from ideal behavior due to interionic and other intermolecular forces. At low concentrations where such forces are negligible, differences between concentration and activity values are minimal. Most measurements of $[H^+]$ are measurements of activity rather than concentration.

As noted above, water contains equal amounts of OH^- and H^+ ($\sim 10^{-7}$ mol/liter each). Its dissociation according to the reaction

$$H_2O + H_2O \rightleftharpoons H_3O^+ + OH^- \tag{5}$$

is defined by the equation

$$K_{eq} = \frac{[H_3O^+][OH^-]}{[H_2O]^2} \tag{6}$$

The ion product of water K_w is defined as follows:

$$K_w = (55.4)^2 K_{eq} = [H_3O^+][OH^-] \tag{7}$$

At 25°C, its value is about 10^{-14}. The ion product of water is of fundamental importance in considering the behavior of acids and bases in solution, since Eq. 7 must always be satisfied in aqueous solutions. Alteration of $[H^+]$ or $[OH^-]$ must always result in an immediate compensatory change in the concentration of the other ion.

The values of $[H^+]$ and K_a of weak acids vary from about 10^{-2} to 10^{-12}. Such small numbers were difficult to manipulate prior to the advent of rapid, inexpensive hand calculators, and a logarithmic formulation of $[H^+]$ and K_a was introduced years ago as an aid in handling these small numbers. It is still used today and is the basis of the so-called pH scale. By definition, the $-\log [H^+]$ is designated the pH and the $-\log K_a$ is designated pK_a. Thus,

$$pH = -\log [H^+] \tag{8}$$

and

$$pK_a = -\log K_a \tag{9}$$

Similarly, the ion product of water (Eq. 7) is expressed as

$$pK_w = pH + pOH \tag{10}$$

For neutral solutions, or pure water, pH = 7, since $pK_w = 14$ and pOH = 7. To obtain the pH of a solution from the hydrogen ion concentration, it is merely necessary to take the logarithm of the $[H^+]$ and change the sign. For example, to calculate the pH of a solution when $[H^+] = 2 \times 10^{-5} M$, take the logarithm of 2, which is 0.30, and add −5.00, which gives −4.70. The pH is 4.70. It must be kept in mind that a change of one pH unit is a tenfold change in hydrogen ion concentration since *decimal* logarithms are used for the pH scale.

Buffers

An extremely important property of any weak acid is its behavior as a buffer. A buffered solution resists the marked changes in $[H^+]$ that would otherwise result from the addition of H^+ or OH^-. Buffer action is readily understood from the titration curves of weak acids, as illustrated in Fig. 2.2, which shows the change in pH as a function of the addition of OH^- to aqueous solutions of two weak acids with pK_a values of 4.7 and 9.3, respectively. At point a in the figure, the weak acid HA will be only slightly dissociated as defined by the relationship (Eq. 4)

$$K_a = \frac{[H^+][A^-]}{[HA]} = 10^{-4.7} \tag{11}$$

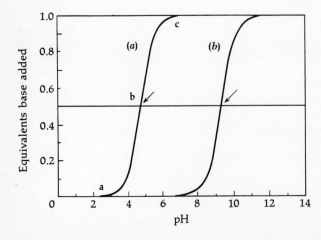

Figure 2.2 The calculated titration curves for 1 equiv of an aqueous solution of two weak acids. (*a*) is acetic acid (HAc), $pK_a = 4.7$, and (*b*) is ammonium ion (NH_4^+), $pK_a = 9.3$. At a, only HAc is present, whereas at b $[HAc] = [Ac^-]$ and at c only Ac^- is present. At b, where $[HAc] = [Ac^-]$, pH = pK_a. The shape of the curves is the same for any univalent (monobasic) acid, but the position of the curve is determined by the pK of the acid.

As OH^- is added to the solution of HA, the following reaction occurs:

$$HA + OH^- \rightleftharpoons A^- + H_2O \qquad (12)$$

with [HA] decreasing and $[A^-]$ increasing. At point b, one-half of the acid has been titrated so that $[HA] = [A^-]$; at point c, there is essentially no HA, but only A^-, the salt of the weak acid. The titration curve shows that the system is behaving as a buffer around pH 4.7, since considerable quantities of OH^- are being added to the solution with only small changes in pH. Similarly, the acid in curve *b* is acting as a buffer about pH 9.3. Titration of a weak acid with base thus forms a buffered solution, and the maximal buffering action exists at the pK_a of the weak acid where $[HA] = [A^-]$. It should be noted that buffers can be prepared not only by titration of a weak acid with OH^- (or titration of a solution of A^- with H^+), but also by addition of A^-, the salt of HA, to a solution of HA.

It is possible to relate the change in pH to changes in [HA] and $[A^-]$ as shown in Fig. 2.2 as follows. Rearranging Eq. 4 for any weak acid gives

$$[H^+] = \frac{K_a[HA]}{[A^-]} \qquad (13)$$

and taking the negative logarithm of Eq. 13 gives

$$pH = pK_a + \log\frac{[A^-]}{[HA]} \qquad (14)$$

Equation 14 can be rewritten to give the Henderson-Hasselbalch equation as follows:

$$pH = pK_a + \log\frac{[salt]}{[acid]} \qquad (15)$$

where [acid] and [salt] refer to the concentrations of HA and A^- (the salt of the weak acid HA), respectively. Thus, if [acid] = [salt] in a buffered system, $pH = pK_a$ since log [salt]/[acid] = 0.

There are many weak acids that can be used as buffers, several of which are

TABLE 2.3
pK_a Values of Some Useful and Biologically Important Compounds

Compound	pK_a*	Compound	pK_a*
Phosphoric acid (pK_1)	2.0	Citric acid (pK_3)	6.4
Citric acid (pK_1)	3.1	Phosphoric acid (pK_2)	6.7
Formic acid	3.8	Imidazole	7.0
Lactic acid	3.9	Diethylbarbituric acid	8.0
Benzoic acid	4.2	Tris(hydroxylmethyl)aminomethane	8.1
Acetic acid	4.7	Boric acid	9.2
Citric acid (pK_2)	4.7	Ammonium ion	9.3
Pyridinium ion	5.3	Ethylammonium ion	9.8
Cacodylic acid	6.2	Triethylammonium ion	10.8
Maleic acid	6.2	Carbonic acid (pK_2)	10.4
Carbonic acid (pK_1)	6.3	Phosphoric acid (pK_3)	12.4

*Many of these values are dependent on temperature and ionic strength.

listed in Table 2.3. Polybasic acids can form buffers at more than one pH; thus phosphoric acid forms buffers near each of its three pK_a values: 2, 6.7, and 12.4.

REFERENCES

Books

Edsall, J. T., and J. Wyman: *Biophysical Chemistry,* vol. 1, Academic, New York, 1958.

Eisenberg, D., and W. Kauzmann: *The Structure and Properties of Water,* Oxford, 1969.

Tanford, C.: *The Hydrophobic Effect,* 2d ed., Wiley, New York, 1980.

Vinogradov, N., and R. H. Linnel: *Hydrogen Bonding,* Van Nostrand Reinhold, New York, 1971.

Review Articles

Frank, H. S.: The Structure of Water, *Fed. Proc.* **24**(suppl. 1):1–11 (1965).

Klotz, I. M.: Role of Water Structure in Macromolecules, *Fed. Proc.* **24**(suppl. 1):24–33 (1965).

Sinanoglu, O., and S. Abdulnur: Effect of Water and Other Solvents on the Structure of Biopolymers, *Fed. Proc.* **24**(suppl. 1):12–23 (1965).

Stillinger, F. H.: Water Revisited, *Science* **209**:451–457 (1980).

Chapter 3

Proteins I

General structural design.   Amino acids and peptides.

FUNCTIONAL DIVERSITY OF PROTEINS

Although all proteins are built of the same 20 amino acids, proteins are the most functionally diverse of all biological compounds.   Their functional versatility is perhaps best exemplified by the fact that each of the thousands of *enzymes* catalyzing a specific biological reaction is a specific protein.   Thus, different enzymes participate in such unrelated reactions as formation of a carbon-carbon bond, hydrolysis of an ester, oxidation of an alcohol, or hydroxylation of an aromatic ring, to mention only a few.   But the versatility of proteins extends beyond their roles as catalysts.   Proteins may also serve as *carriers,* each transporting a specific substance such as oxygen, a metal ion, or a metabolite throughout the body.   Other proteins anchored in biological membranes are *receptors* for specific compounds that function in cellular regulation, including the transport of molecules in and out of the cell.   Some *hormones* are proteins and aid in regulating a variety of cellular processes.   The proteins of the *immune system* defend an organism against infections whereas those of *blood coagulation* defend against blood loss.   Other specialized proteins are involved in providing *motility* to an organism.   Some proteins serve primarily *structural* roles in cytoskeletal structures and the extracellular matrix of tissues.

Notwithstanding their diversity of functions, all proteins have structural similarities that are important to understand before considering the many specific proteins to be encountered later in the text.   In this chapter we shall consider these structural similarities as well as the chemistry of the amino acids contained in proteins.

Any cross-references coded M refer to the companion text, *Principles of Biochemistry: Mammalian Biochemistry*.

**The Major
Constituents of Cells**
Covalent Structure

Proteins are macromolecules with molecular weights ranging from approximately 5000 to many millions. Each is a polymer with α-*amino acids* as the repeating units. The α-amino acids contain an α-carbon, to which an amino group and a carboxyl group are attached.

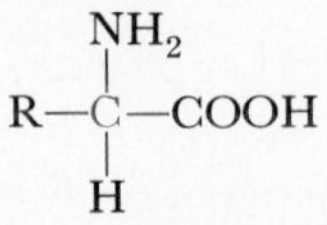

$$R-\underset{\underset{H}{|}}{\overset{\overset{NH_2}{|}}{C}}-COOH$$

Twenty α-amino acids are commonly found in proteins; these differ therefore in the structure of the R group, which may be hydrophilic or hydrophobic, acidic, basic, or neutral.

The amino acids in proteins are linked together by *peptide bonds,* which are amide bonds formed by the α-carboxyl and α-amino groups of adjacent residues, as shown below (peptide bonds enclosed).

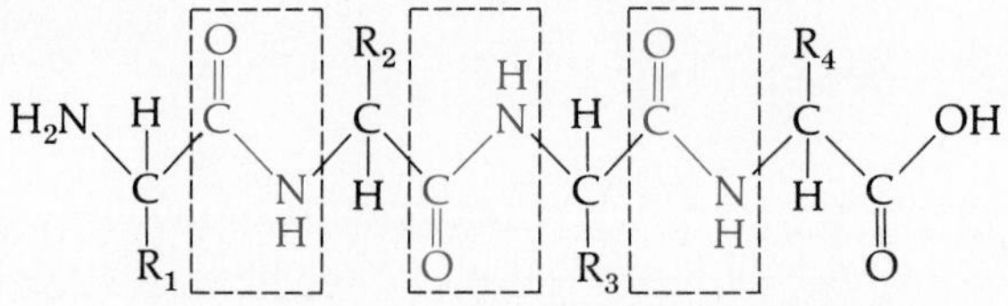

The resulting polymers are called *peptides;* a prefix, such as *di-, tri-, tetra-,* etc., indicates the number of residues; thus, a dipeptide contains two residues, a tripeptide three residues, etc. In contrast to these *oligopeptides,* the term *polypeptide* refers to peptides containing about 8 or more residues; proteins have polypeptide chains containing from about 50 to as many as 2500 residues, the exact number depending on the specific protein. An enormous number of differing orders of the amino acids in the polypeptide chains of proteins is possible since for a 20-residue peptide containing one of each of the 20 different amino acids there is the possibility of 20! or about 2×10^{18} different sequences! Obviously, living things utilize but a minute fraction of the possible isomers of proteins; nevertheless, the functional diversity of proteins is possible because of the remarkable potential for structural variation.

Three-Dimensional Structure

If peptide bonds were the only structural linkages in proteins, all proteins would exhibit random spatial structures. This is clearly not the case; the polypeptide chain of a protein is folded into a specific three-dimensional structure, which is referred to as the *conformation* of the protein. The exact conformation of a protein is determined by its amino acid sequence and is stabilized by the noncovalent interactions between the peptide bonds and the R groups in the polypeptide chain. These noncovalent interactions are primarily *hydrogen bonds, hydrophobic bonds,* and to a lesser extent *electrostatic interactions.* Typical are hydrogen bonds formed by the carbonyl oxygen and the amide group in peptide bonds.

$$\diagdown \text{C}\!\!=\!\!\text{O}\cdots\text{HN}\diagup$$

Many other types of hydrogen bonds are formed between these same atoms and certain of the polar R groups.

Hydrophobic bonds form between nonpolar R groups of the aliphatic or aromatic side chains of certain amino acid residues, as discussed below. Such groups are poorly soluble in water, since they cannot orient themselves into the hydrogen bonded matrix of water molecules without great expenditure of energy (Chap. 2). Thus, they tend to cluster together, usually in the interior of a protein molecule, and form hydrophobic bonds among themselves. Hydrogen and hydrophobic bonds are individually much weaker than a covalent bond, but because of the large number of these bonds in a protein molecule, sufficient energy is available to stabilize the conformation. Electrostatic bonds are found less frequently in proteins than hydrogen or hydrophobic bonds, and although they would be more stable in the hydrophobic interior of a protein, they are usually located in surface regions or between different protein subunits, as discussed below.

There are many types of molecular models used to depict the three-dimensional structure of a protein, and each reveals different aspects of the molecule. Space-filling models, such as that for the enzyme lysozyme (Chap. 11), shown in Fig. 3.1, show the general globular shape of a protein and the details of some surface structures, e.g., the site where substrate binds. Ball-and-stick models, such as that for cytochrome c (Chap. 15) in Fig. 3.2, open up the structure and give an overall view of the relationship in space of one residue to another and in a general way show how the polypeptide chain is folded. In this type of model, a ball represents the α-carbon atom of a residue; a stick joins each pair of balls and its length represents the distance between two α carbons. The buried *heme prosthetic group* of cytochrome c, in which an electron is added to or abstracted from this carrier protein (Chap. 15), is also revealed. Figure 3.3 shows a third type, the ribbon structural model of cytochrome c. This model also indicates the folding of the polypeptide chain in space and gives a diagrammatic view of regularly folded structures, such as the α-helix. Another model shows the bond distances and bond angles between each of the atoms in the molecule except hydrogen atoms. This type is too complex to reveal much structural detail when photographed at a single angle and reproduced in a textbook. Portions of this kind of model, however, such as those shown in Figs. 5.11, 5.12, 5.14, and 5.15 (Chap. 5), reveal considerable structural detail. Moreover, they are essential in practice for analysis of the three-dimensional structure and the structure-function relationships of a protein.

The different models illustrate several other general features of protein structure. There is a large array of hydrogen bonds within the molecule as well as internally placed aliphatic and aromatic R groups that form hydrophobic bonds. The hydrophobic interior of the molecule is generally devoid of water.

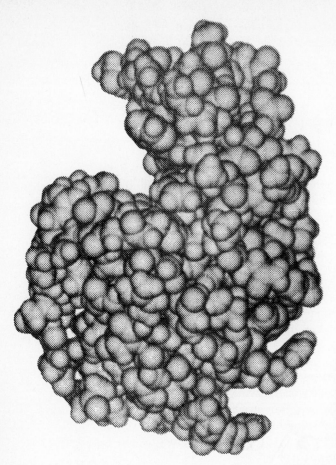

Figure 3.1 A space-filling model of hen egg-white lysozyme (Chap. 11). Hydrogen atoms are not shown. This type of model reveals only surface structures, but at the angle shown here, the active site cleft (upper left), where substrate binds, is clearly evident. [*Courtesy of Dr. Richard J. Feldman, National Institutes of Health.*]

In contrast, the surface of the molecule is mainly hydrophilic, and many polar groups, including positively or negatively charged R groups of the amino acids, are hydrated and in contact with water molecules surrounding the protein. Thus, the surface of the protein has a net electric charge, which may be neutral, positive, or negative depending on the pH of the solvent and the number and kinds of ionic groups in the protein. It is presently impossible to predict the conformation of a protein solely from knowledge of its amino acid sequence, but the different conformations that are found among proteins with diverse sequences represent the most energetically favorable and stable structures for each given protein with its unique sequence. It should be emphasized that proteins are not static but are dynamic structures that can undergo limited changes in conformation during biological function. In addition, R groups of the amino acids at the surface may have considerable freedom of movement within the solvent surrounding the molecule.

It is convenient to consider four aspects of protein structure. The *primary structure* is the amino acid sequence, often referred to as the *covalent structure*. The *secondary structure* refers to the regular structural arrays found in proteins, principally *helices,* which are formed by folding a single segment of polypeptide chain,

Figure 3.2 A ball-and-stick model of the protein cytochrome c. The R groups are not shown except for those for the two half-cystine residues and a methionine and histidine residue, which are combined with the heme prosthetic group (Chap. 24). Individual amino acids are numbered beginning with the NH$_2$-terminal residue (1, at top) to the COOH-terminal (104, lower right). The iron atom in the molecule is contained in the heme prosthetic group, which is attached covalently to the polypeptide chain through the R groups of cysteine. Ligands to the heme iron are formed by the sulfur atom of Met-80 and an imidazole nitrogen of His-18. [*From R. E. Dickerson and R. Timkovich, in P. D. Boyer (ed.): The Enzymes, vol. XI, Academic, New York, 1975, p. 407.*]

and β *structures,* or pleated sheets, formed by two or more chains. The *tertiary structure* is the total conformation of the molecule, its disposition in three-dimensions. *Quaternary structure* refers to the spatial arrangement of the subunits, when such exist. The secondary, tertiary, and quaternary structures are often collectively referred to as the *noncovalent structure* and will be discussed more fully in Chap. 5.

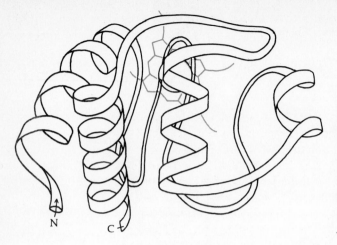

Figure 3.3 A ribbon model of cytochrome c. This model shows only the folding of the polypeptide chain, which is depicted as a ribbon. Four stretches of helical regions are evident, one near one end of the molecule and another at the other end. The view shown differs from that in Fig. 3.1 since the molecule has been rotated about 90° along two axes. [*Courtesy of Jane Shelby Richardson.*]

Subunit Structure

Some proteins consist of only one polypeptide chain, but others contain two or more polypeptide chains combined either through covalent or noncovalent bonds. An example is human hemoglobin, which contains four chains, two α chains (141 residues), and two β chains (146 residues) that differ in sequence. Its structure is shown schematically in Fig. 3.4. The chains are combined by the same types of noncovalent bonds that maintain the conformation of a single polypeptide chain, primarily hydrogen and hydrophobic bonds, although electrostatic bonds of critical importance to the function of hemoglobins are also located between the chains (Chap. M4). Proteins with more than one polypeptide chain are said to contain *subunits,* and the subunit structure of a protein indicates the number and kinds of chains in the molecule. Thus, the subunit structure of hemoglobin is designated $\alpha_2\beta_2$. Other proteins contain subunits combined covalently by the disulfide bond, —S—S—, formed by R groups of one amino acid, cystine. The monomeric fibrin in blood clots (Chap. M1) is one example, and its subunit structure is designated $\alpha_2\beta_2\gamma_2$ since it contains two each of three chains that differ in sequence. The Greek letters α, β, γ, etc., are

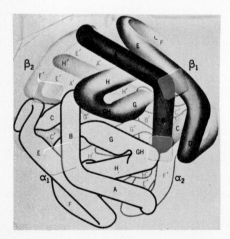

Figure 3.4 A schematic representation of hemoglobin showing the folding of the two α and β chains and the spatial relationships of the four chains. The capital letters on the chains refer to segments of sequence that have conformational similarities. Both α chains and both β chains are identical, but the subscripts 1 and 2 are used to distinguish between the two chains. The diameter of the molecule is about 64 Å. The rectangular shaded structures depict the heme group (Chap. M3, G) in each chain. [*From R. E. Dickerson and I. Geis, The Structure and Action of Proteins, Harper & Row, New York, 1969, p. 56.*]

to the protein considered, i.e., the α chains of hemoglobin are entirely different from the α chains in fibrinogen.

Biological Activity and Protein Structure

The specific biological function of a protein, e.g., enzymes and hormones, depends upon its conformation; perturbations in the conformation may lead to loss of biological activity.   Thus, a protein in its normal conformation is said to be in the *native state,* and a *native protein* is one which necessarily has the conformation that endows the molecule with its specific biological function.   Rather mild changes in physical conditions can disrupt that conformation, including alteration of pH, changes in temperature, or exposure to aqueous solutions of some organic compounds, such as detergents, ethanol, urea, and guanidine · HCl (Chap. 4).   Proteins exposed to such conditions undergo *denaturation;* the conformation of a *denatured protein* is drastically altered from the native protein and is devoid of biological activity.

Functional Sites in Proteins

Each protein contains one or more specific regions or sites on its surface that directly participates in its biological activity.   These sites may represent only a small fraction of the total surface of the molecule.   Thus, every enzyme has an *active site* that interacts directly with the substances that take part in the reaction it catalyzes (Chap. 10).   Carrier proteins contain binding sites that specifically and reversibly combine with only one or a few ligands.   For example, hemoglobin (Fig. 3.4) contains four heme groups, each of which can reversibly combine with oxygen.   Proteins that serve as cell receptors have binding sites for specific substances.   Moreover, protein hormones contain specific surface structures that interact directly with hormone receptors on a cell surface (Chap. 14).   Thus, the functional versatility of proteins results from the vast number of sequences that the 20 amino acids can generate, and from the unique way in which a polypeptide chain with a given sequence folds three dimensionally so as to generate a specific functional site.   Each site, irrespective of its function, has a structure that permits interaction with one or at most a very few specific structures in other molecules.   The geometry of a functional site is clearly only one feature of protein conformation; other features serve to stabilize the functional site, and provide sites of interaction of the protein with other molecules, or may serve other functions.

Classification of Proteins

Because of the diversity in the structures and functions of proteins, it is not possible to classify the many kinds of proteins satisfactorily.   Two broad classes of proteins are generally acknowledged, the *globular proteins* and the *fibrous proteins*.   The globular proteins are spherical or ellipsoid and have a general structural design as described above for cytochrome c (Fig. 3.2) and hemoglobin (Fig.

3.4). That shape may result from the conformation of one or more subunit polypeptide chains. It may be solely protein or conjugated with a characteristic non-amino acid substance called the *prosthetic group*. Hemoglobin is an example of a conjugated protein; each chain in the protein moiety, globin, is combined with a heme group (Chap. M4) to form the biologically functional molecule.

Proteins are often described in terms of their prosthetic group; thus, hemoglobin is a *hemeprotein,* and proteins containing lipid, carbohydrate, or metals are called *lipoproteins, glycoproteins,* and *metalloproteins,* respectively. The great majority of known proteins are globular proteins. When only a few pure proteins were known, attempts were made to classify globular proteins on the basis of their solubility in aqueous media, but most of this nomenclature is not meaningful today and only of historical interest. For example, *albumin* is the name used to describe proteins readily soluble in water and coagulable by heating; *globulins* are proteins sparingly soluble in water but soluble in dilute solutions of salts such as NaCl. These names persist in the literature for some proteins, e.g., serum albumin and γ-globulin, and refer to specific proteins.

Fibrous proteins may also contain one or more polypeptide chains; they are elongated, very asymmetric molecules whose length may be many times their diameter. Fibrous proteins are found primarily as structural proteins of connective tissue, elastic tissue, contractile tissue, or as the insoluble substance of hair and skin. Proteins of this type will be considered in detail in Part M2.

Closely related proteins serving the same function are widely distributed among plants and animals. For example, the enzyme lactate dehydrogenase is found in most prokaryotic and eukaryotic organisms; however, each is species-specific and differs somewhat in amino acid sequence.

Some General Principles of Protein Chemistry

Three major principles pertaining to protein structure, function, and assembly may be recognized. (1) As will be described later (Chap. 5), the sequence of amino acids in a polypeptide chain is determined by the DNA template that directs its synthesis (Chap. 28). (2) The three-dimensional conformation that a polypeptide chain assumes is determined entirely by the specific amino acid sequence of the chain. Changes in sequence resulting from genetic mutations yield conformationally altered, and often less stable, less active, or inactive, proteins. (3) The specific activity of a protein as an enzyme, hormone, oxygen carrier, etc., depends on its specific conformation. Although proteins are dynamic structures, even a limited alteration of conformation can lead to loss of biological activity.

AMINO ACIDS

There are many amino acids in nature, but those found in proteins are the most abundant. The α-amino acids of proteins participate in a variety of essential metabolic reactions that are considered in Chaps. 22 to 25, but their structures are most conveniently considered in view of their individual structural roles in proteins.

The structures of the 20 α-amino acids generally present in proteins are listed in Table 3.1. They may conveniently be classified into three groups, reflecting the character of their side chains or R groups: (1) those with ionic side chains, (2) those with polar, non-ionic side chains, and (3) those with nonpolar, aliphatic, or aromatic, side chains.

Seven amino acids contain R groups that can have a negative or a positive charge. Two, *aspartic acid* and *glutamic acid,* the so-called acidic amino acids, may carry a negative charge on their β- and γ- carboxylate groups, respectively.

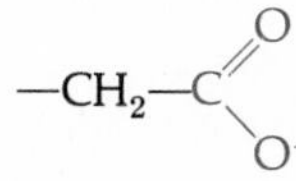

β-Carboxylate 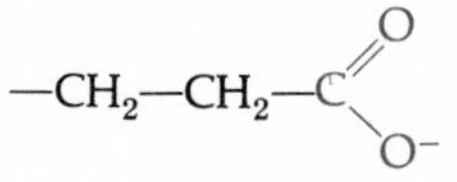   γ-Carboxylate

Three amino acids, *lysine, arginine,* and *histidine,* the so-called basic amino acids, may carry a positive charge since their R groups accept a proton. The ε-amino group of lysine, the guanidine group of arginine, and the imidazole ring of histidine are protonated to form the following structures:

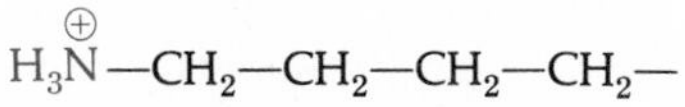

Ammonium group 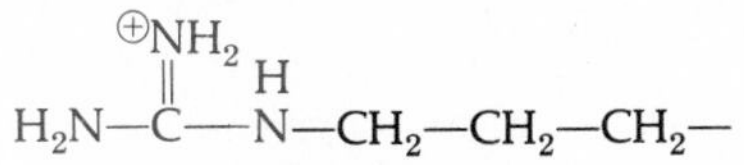   Guanidinium group 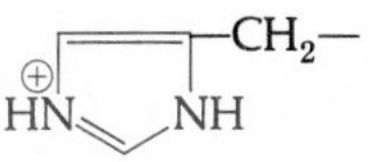   Imidazolium group

Two additional amino acids, *tyrosine* and *cysteine,* have R groups that may also possess a negative charge:

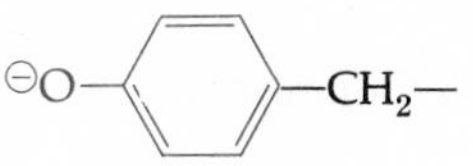

Phenolate group 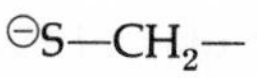   Thiolate group

These ionic groups are found mainly on the surface of globular proteins and are solvated in their immediate vicinity through hydrogen bonds or ion-dipole interactions with water molecules. These groups provide the positive and negative charges on protein surfaces and together determine the electrostatic properties of proteins in solution. Occasionally, ionic groups form electrostatic bonds (salt bridges); e.g., deoxyhemoglobin contains a few electrostatic bonds, such as one between an aspartic acid residue and an arginine residue in two different α chains.

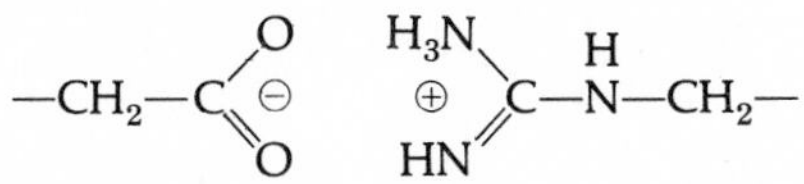

Both tyrosine and cysteine residues may be buried in the hydrophobic interior of proteins since as un-ionized species their R groups may participate in hydrophobic or hydrogen bonding, e.g.,

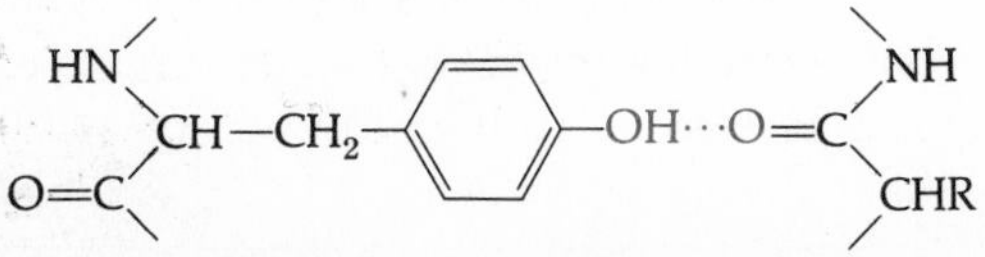

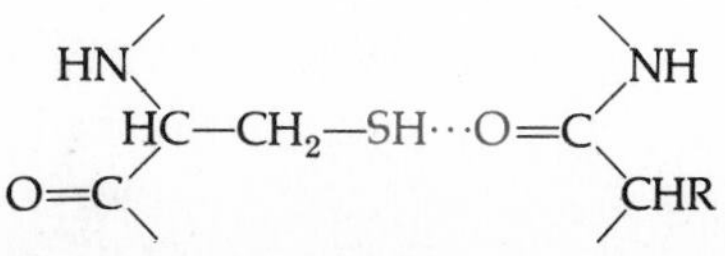

A.   Amino Acids with Ionic Side Chains

Name and 3- and 1-letter abbreviations*	Structure	pK_1 α-COOH	pK_2	pK_3	pI†
Aspartic acid (Asp, D)	$HOOC-CH_2-\overset{\overset{\displaystyle NH_2}{\vert}}{\underset{\underset{\displaystyle H}{\vert}}{C}}-COOH$	1.88	3.65 β-COOH	9.60 α-NH_3^+	2.77
Glutamic acid (Glu, E)	$HOOC-CH_2-CH_2-\overset{\overset{\displaystyle NH_2}{\vert}}{\underset{\underset{\displaystyle H}{\vert}}{C}}-COOH$	2.19	4.25 γ-COOH	9.67 α-NH_3^+	3.22
Lysine (Lys, K)	$H_2N-CH_2-CH_2-CH_2-CH_2-\overset{\overset{\displaystyle NH_2}{\vert}}{\underset{\underset{\displaystyle H}{\vert}}{C}}-COOH$	2.18	8.95 α-NH_3^+	10.53 ε-NH_3^+	9.74
Arginine (Arg, R)	$H_2N-\overset{\overset{\displaystyle NH}{\Vert}}{C}-\overset{\overset{\displaystyle H}{\vert}}{N}-CH_2-CH_2-CH_2-\overset{\overset{\displaystyle NH_2}{\vert}}{\underset{\underset{\displaystyle H}{\vert}}{C}}-COOH$	2.17	9.04 α-NH_3^+	12.48 guani-dinium	10.76
Histidine (His, H)	$CH_2-\overset{\overset{\displaystyle NH_2}{\vert}}{\underset{\underset{\displaystyle H}{\vert}}{C}}-COOH$ (imidazole ring)	1.82	6.00 imida-zolium	9.17 α-NH_3^+	7.59
Tyrosine (Tyr, Y)	$HO-\langle ring \rangle-CH_2-\overset{\overset{\displaystyle NH_2}{\vert}}{\underset{\underset{\displaystyle H}{\vert}}{C}}-COOH$	2.20	9.11 α-NH_3^+	10.07 phenolic —OH	5.66
Cysteine (Cys, C)	$HS-CH_2-\overset{\overset{\displaystyle NH_2}{\vert}}{\underset{\underset{\displaystyle H}{\vert}}{C}}-COOH$	1.96	8.18 thiol	10.28 α-NH_3^+	5.07

B. Amino Acids with Non-ionic Polar Side Chains

Name and 3- and 1-letter abbreviations*	Structure	pK_1 α-COOH	pK_2 α-NH_3^+	pI†
Asparagine (Asn, N)	$H_2N-\overset{\overset{\displaystyle O}{\Vert}}{C}-CH_2-\overset{\overset{\displaystyle NH_2}{\vert}}{\underset{\underset{\displaystyle H}{\vert}}{C}}-COOH$	2.02	8.80	5.41
Glutamine (Gln, Q)	$H_2N-\overset{\overset{\displaystyle O}{\Vert}}{C}-CH_2-CH_2-\overset{\overset{\displaystyle NH_2}{\vert}}{\underset{\underset{\displaystyle H}{\vert}}{C}}-COOH$	2.17	9.13	5.65
Serine (Ser, S)	$HO-CH_2-\overset{\overset{\displaystyle NH_2}{\vert}}{\underset{\underset{\displaystyle H}{\vert}}{C}}-COOH$	2.21	9.15	5.68

B. Amino Acids with Non-ionic Polar Side Chains

Name and 3- and 1-letter abbreviations*	Structure	pK_1 α-COOH	pK_2 α-NH_3^+	pI†
Threonine (Thr, T)	CH_3—$\overset{\overset{NH_2}{\mid}}{CH}$—$\overset{\underset{H}{\mid}}{C}$—COOH, with OH below	2.09	9.10	5.60

C. Amino Acids with Nonpolar Aliphatic or Aromatic Side Chains

Name and 3- and 1-letter abbreviations*	Structure	pK_1 α-COOH	pK_2 α-NH_3^+	pI†
Glycine (Gly, G)	H—C—COOH with NH_2 above and H below	2.34	9.60	5.97
Alanine (Ala, A)	H_3C—C—COOH with NH_2 above and H below	2.34	9.69	6.00
Valine (Val, V)	$(CH_3)_2$HC—C—COOH with NH_2 above and H below	2.32	9.62	5.96
Leucine (Leu, L)	$(CH_3)_2$HC—CH_2—C—COOH with NH_2 above and H below	2.36	9.60	5.98
Isoleucine (Ile, I)	CH_3—CH_2—HC(CH_3)—C—COOH with NH_2 above and H below	2.36	9.60	6.02
Methionine (Met, M)	CH_3—S—CH_2—CH_2—C—COOH with NH_2 above and H below	2.28	9.21	5.74
Proline (Pro, P)	ring structure CH_2—CH_2 / C—COOH / CH_2—NH	1.99	10.60	6.30
Phenylalanine (Phe, F)	C_6H_5—CH_2—C—COOH with NH_2 above and H below	1.83	9.13	5.48
Tryptophan (Trp, W)	indole—CH_2—C—COOH with NH_2 above and H below	2.83	9.39	5.89

*The abbreviations are used only in presenting the sequence of a polypeptide or protein.
†pI = pH at isoelectric point (see page 36).

Ribonuclease is one example of a protein with buried tyrosine residues. When a protein with buried groups of this type is denatured, the groups are exposed as the polypeptide chain unfolds and readily lose a proton. The four methylene groups of lysine and the three of arginine may also be located below the surface of the molecule with hydrophobic contacts, the charged ϵ-ammonium and guanidinium groups being on the surface.

The four amino acids with polar, non-ionic R groups include *asparagine*, the β-amide of aspartic acid; *glutamine*, the γ-amide of glutamic acid; and *serine* and *threonine*, aliphatic hydroxyamino acids, which contain a β-hydroxyl group. Acid or alkaline hydrolysis of proteins liberates ammonia, which is derived from amides of asparagine and glutamine. These amino acids are sufficiently polar for their R groups to be on the surface, solvated by water, or located internally, with the polar amide or hydroxyl groups forming hydrogen bonds, e.g.,

$$
\begin{array}{cc}
\text{HN} & \text{R—CH} \\
\quad \text{CH—CH}_2\text{—OH}\cdots\text{O=C} & \text{NH} \\
\text{O=C} & \text{CHR}
\end{array}
\qquad
\begin{array}{cc}
\text{HN} & \text{CHR} \\
\text{HC—CH}_2\text{—C} & \text{O}\cdots\text{HN} \\
\text{O=C} & \text{C=O} \\
 & \text{NH}_2 \quad \text{HCR}
\end{array}
$$

Glycine is unique in not possessing an R group. Thus, glycine residues can be accommodated both in the hydrophobic interior and on the exterior surface of a protein. Further, the absence of an R group permits flexibility of folding of peptide chains at glycine residues since there is no bulky group to interfere with such bends or folds. Many reactive residues of enzymes are frequently surrounded by glycine residues, thus permitting easy access of substrates to the enzyme surface.

Eight amino acids have nonpolar aliphatic or aromatic side chains. *Alanine, valine, leucine, isoleucine,* and *methionine* have aliphatic side chains of sufficient hydrophobicity for them to participate in hydrophobic interactions between themselves and other R groups in this class. The *phenyl* side chain of *phenylalanine* and the *indole* ring in *tryptophan* also provide structures for hydrophobic interactions. The hydrogen on the nitrogen of the indole ring can also participate in hydrogen bonding with internally located atoms. The remaining member of this group is *proline,* an α-imino acid since it contains a secondary amine in a pyrrolidine ring. This structure, like glycine, permits the polypeptide chain to bend, allowing the chain to fold back on itself.

Although the nonpolar aliphatic and aromatic amino acids contribute to the total structure of proteins by forming internal hydrophobic interactions, some of the R groups of these residues may be located near the surface and can be detected by reaction with group-specific reagents. For example, some tyrosine residues in proteins are on the surface of the protein and react normally with various reagents for phenolic groups, whereas other tyrosine residues, present in the hydrophobic interior of the molecule, are unreactive. Hydrophobic residues at the surface of the protein are unlikely to be hydrated and exist in surface regions by virtue of the stabilization of the region by the polar and ionic groups that are solvated in their immediate vicinity.

The aromatic amino acids are responsible for the ultraviolet absorption of

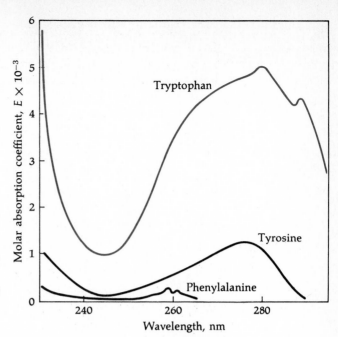

Figure 3.5 The absorption spectra of tryptophan, tyrosine, and phenylalanine at pH 8. The spectrum for tyrosine undergoes a shift in position at alkaline pH values; this is completed above pH 12. The contribution of phenylalanine to absorption by proteins at 275 to 280 nm is negligible. [*After J. S. Fruton and S. Simmonds, General Biochemistry, 2d ed., Wiley, New York, 1958, p. 73.*]

most proteins with maxima between 275 and 285 nm. Tryptophan has a much greater absorption than the other aromatic amino acids (Fig. 3.5). The molar absorption coefficients of proteins are useful in providing a means for spectrophotometric determination of the concentration of a protein in solution.

One additional amino acid of low polarity is *cystine*, the disulfide formed from two cysteine residues. However, by virtue of its structure, cystine can form four peptide bonds through its two carboxyl and two amino groups.

$$HOOC-\underset{\underset{NH_2}{|}}{\overset{\overset{H}{|}}{C}}-CH_2-S-S-CH_2-\underset{\underset{H}{|}}{\overset{\overset{NH_2}{|}}{C}}-COOH$$

Cystine

$$O=C\diagup \quad HC-CH_2-S-S-CH_2-CH \quad \diagdown NH$$
$$HN\diagdown \qquad\qquad\qquad\qquad\qquad\qquad \diagdown C=O$$

**Disulfide bond linking
two polypeptide chains**

This allows two regions of a single polypeptide chain remote from each other in the sequence to be linked covalently through a disulfide bond (intrachain disulfide bond). The disulfide bond is also the most common covalent bond that joins two polypeptide chains (interchain disulfide bond). These bonds are broken by reduction. Thus, by treating a protein with an excess of a reducing agent, e.g., mercaptoethanol or dithiothreitol at alkaline pH, cysteine residues are formed.

Stereochemistry of Amino Acids

With the exception of glycine, all α-amino acids are optically active since the α-carbon atom is a center of asymmetry. The amino acids of proteins all possess the same absolute configuration as L-alanine, which has been shown to be related to L-glyceraldehyde. The symbols D or L here, as in the sugar series (Chap. 6), do not refer to the sign of rotation but indicate configurational relationships of similar compounds.

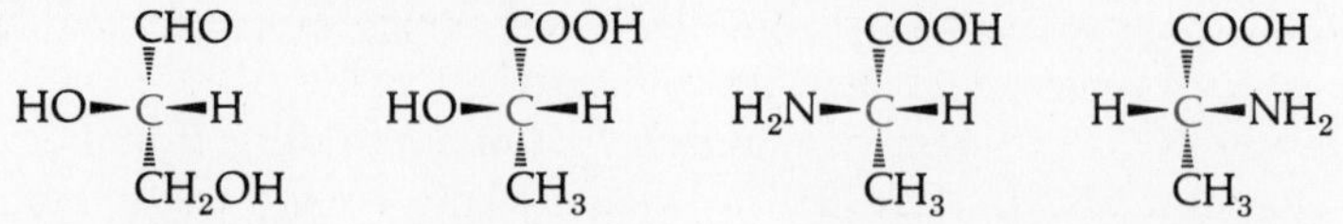

$$
\begin{array}{cccc}
\text{CHO} & \text{COOH} & \text{COOH} & \text{COOH} \\
\text{HO}\!-\!\text{C}\!-\!\text{H} & \text{HO}\!-\!\text{C}\!-\!\text{H} & \text{H}_2\text{N}\!-\!\text{C}\!-\!\text{H} & \text{H}\!-\!\text{C}\!-\!\text{NH}_2 \\
\text{CH}_2\text{OH} & \text{CH}_3 & \text{CH}_3 & \text{CH}_3 \\
\text{L-Glyceraldehyde} & \text{L-Lactic acid} & \text{L-Alanine} & \text{D-Alanine}
\end{array}
$$

Threonine, cystine, and isoleucine possess two asymmetric centers. Hence the synthetic compounds are mixtures of four diastereoisomers. Two are designated as L and D, respectively. The two additional diastereoisomeric forms are described as L-*allo*- and D-*allo*-; thus the terms *allo*-threonine and *allo*-isoleucine.

Amino Acids as Electrolytes

α-Amino acids in aqueous solution exist largely as *dipolar ions*. The ratio of dipolar ions to uncharged molecules in the equilibrium

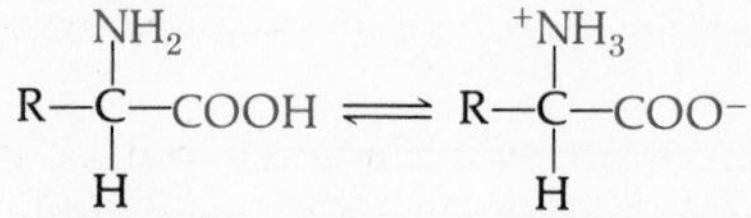

$$
\underset{\text{H}}{\overset{\text{NH}_2}{R-C-\text{COOH}}} \;\rightleftharpoons\; \underset{\text{H}}{\overset{\text{+NH}_3}{R-C-\text{COO}^-}}
$$

has been estimated to be 10^5 to 10^6 for several α-monoamino monocarboxylic acids. Although, for convenience, amino acids are often denoted as uncharged molecules, all α-amino acids exist in solution either as negatively charged anions, positively charged cations, electrically neutral dipolar ions, or some mixture of these forms, the predominant species depending on the pH of the solution. The pH at which a dipolar ion does not migrate in an electric field is called the *isoelectric point* (pI) and is that pH at which the number of positive and negative charges is equal. When a pure amino acid is dissolved in water, the pH of the resulting solution is the *isoionic point*. The isoelectric point and isoionic point of an amino acid are approximately the same in dilute salt solution.

α-Amino acids have at least two potentially acidic groups, the undissociated carboxyl group and the protonated ammonium group. An amino acid hydrochloride is a strong electrolyte and dissociates completely in water as follows:

$$
\text{Cl}\cdot\text{NH}_3\text{—CHR—COOH} \;\rightleftharpoons\; \text{Cl}^- + \text{+NH}_3\text{—CHR—COOH}
$$

to give the fully protonated, positively charged cation shown on the right. This cation possesses two weakly acidic groups, each with its characteristic pK value; by convention they are designated pK_1, pK_2, etc., the lowest number referring to the strongest acidic group. Thus, the dissociation constant pK_1 for the above amino acid reflects the dissociation of the carboxyl group

$$
\text{+NH}_3\text{—CHR—COOH} \;\rightleftharpoons\; \text{H}^+ + \text{+NH}_3\text{—CHR—COO}^-
$$

and

$$
K_1 = \frac{[\text{H}^+][\text{R}^\pm]}{[\text{R}^+]}
$$

where R^+ is the cationic species (on the left) and $R^\pm$, the dipolar ion (on the right). If a solution of an amino acid hydrochloride is titrated with 2 equiv of sodium hydroxide, the change in pH as a function of the number of equivalents of OH^- added is as shown in Fig. 3.6. On addition of 0.5 equiv of base, the pH of the solution is equal to pK_1, the pK of the carboxyl group, and equal amounts of both R^+ and $R^\pm$ are present. On addition of 1 equiv of base the predominant species is $R^\pm$, the dipolar ion, and the pH of the solution is at the pI. The dipolar species is also a weak acid by virtue of the weakly acidic ammonium group and dissociates as follows:

$$^+H_3N-CHR-COO^- \rightleftharpoons H^+ + H_2N-CHR-COO^-$$

and

$$K_2 = \frac{[H^+][R^-]}{[R^\pm]}$$

where $R^\pm$ is the dipolar ion and R^- is the anionic species (on the right). Addition of 1.5 equiv of base to a solution of the amino acid at its isoelectric point results in titration of the ammonium group, and the pH of the resulting solution is equal to pK_2, the pK of the ammonium group. Addition of 2 equiv of alkali fully titrates the amino acid, and the solution contains only the anionic species, R^-.

Three additional important points emerge from consideration of the titration of the monoamino monocarboxylic acid shown in Fig. 3.6. (1) All α-amino acids behave like strong electrolytes irrespective of the pH range shown. The species in solution (whether the cationic, positively charged amino acid hydrochloride, the electrically neutral dipolar ion, the negatively charged salt, or any combination of these species) are ionic salts. Many properties of the amino acids are more characteristic of salts than non-ionic organic compounds, including their high melting points, generally good solubility in water, and low solubility in nonpolar solvents such as ether and chloroform. (2) The isoelectric point of the amino acid is determined by the magnitude of the two dissociation constants. It is evident by inspection of the curve in Fig. 3.6 that for a monoamino monocarboxylic acid, the isoelectric point is equal to the numerical average of pK_1 and pK_2. (3) All amino acids are buffers and exert their maximal buffering efficiency at a pH equal to the pK value of either of the two acidic groups. Thus, the amino acid in Fig. 3.6 is an excellent buffer at pH 2.3 and at pH 9.6, the values of pK_1 and pK_2, respectively. Amino acids are not buffers at their pI.

The titration curves of amino acids containing three acidic groups have three inflection points reflecting the pK of each group, and complete titration of a fully protonated species requires 3 equiv of base, as with aspartic acid, for example.

The isoelectric form of aspartic acid is structure II, since it is the only electrically neutral species. Similarly the ionic forms of lysine are

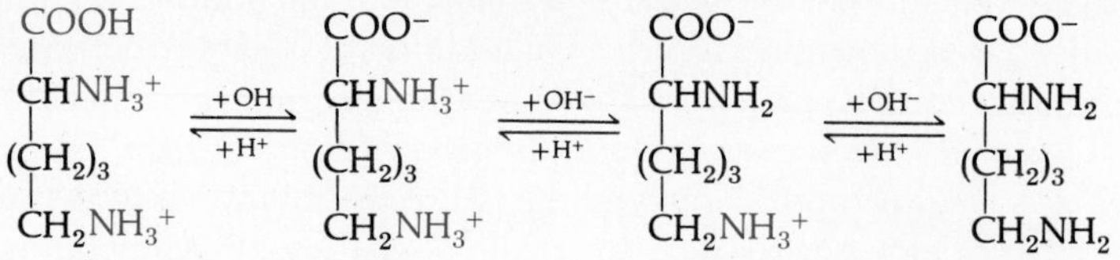

$$
\begin{array}{cccc}
\text{COOH} & \text{COO}^- & \text{COO}^- & \text{COO}^- \\
| & | & | & | \\
\text{CHNH}_3^+ & \text{CHNH}_3^+ & \text{CHNH}_2 & \text{CHNH}_2 \\
| & | & | & | \\
(\text{CH}_2)_3 & (\text{CH}_2)_3 & (\text{CH}_2)_3 & (\text{CH}_2)_3 \\
| & | & | & | \\
\text{CH}_2\text{NH}_3^+ & \text{CH}_2\text{NH}_3^+ & \text{CH}_2\text{NH}_3^+ & \text{CH}_2\text{NH}_2 \\
\text{I} & \text{II} & \text{III} & \text{IV}
\end{array}
$$

with $\xrightleftharpoons[+\text{H}^+]{+\text{OH}}$ between I and II, $\xrightleftharpoons[+\text{H}^+]{+\text{OH}^-}$ between II and III, and $\xrightleftharpoons[+\text{H}^+]{+\text{OH}^-}$ between III and IV.

and III is the isoelectric form. A useful equation for calculating the pI of amino acids is

$$ pI = \frac{pK_n + pK_{n+1}}{2} $$

where n is equal to the maximum number of positive charges on the fully protonated amino acid. For aspartic acid $n = 1$, and for lysine $n = 2$. For a monoamino monocarboxylic α-amino acid, $n = 1$ also; from the pK values for the acidic groups in amino acids listed in Table 3.1, the pI for any amino acid can easily be calculated.

PEPTIDES

Nomenclature

Peptide nomenclature lists the amino acid residues as they occur in the chain starting from the free amino group, which is conventionally shown at the left-hand part of the structure. The ending -yl is used for each of the residues except the terminal one, which bears the unsubstituted carboxyl group. The tetrapeptide alanylglycyltyrosylglutamic acid has the structure

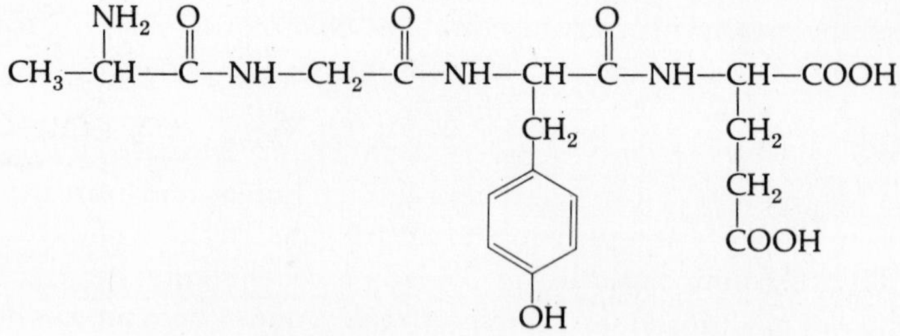

Alanylglycyltyrosylglutamic acid

The residues are named, as indicated above, by using the ending -yl. Thus, glycyl, alanyl, seryl, etc., are used for residues of glycine, alanine, serine, respectively. The distinction between glutamyl and glutaminyl, as well as aspartyl and asparaginyl, should be noted. In writing the sequences of longer peptides, however, it is useful to use either the three- or one-letter abbreviations instead of the complete structure. Table 3.1 gives the accepted abbreviations; thus the above tetrapeptide could also be written either as $\text{H}_2\text{N-Ala-Gly-Tyr-Glu-COOH}$ or AGYE. These abbreviations are used only for the residues in a sequence and not for the free amino acids.

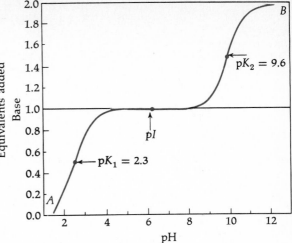

Figure 3.6 Diagrammatic representation of the titration curve of a monoamino monocarboxylic acid such as glycine with $pK_1 = 2.3$ and $pK_2 = 9.6$. In the acid solution A the substance present is glycine hydrochloride; B, sodium glycinate. The isoelectric point pI (page 36) is 6.

Peptide Synthesis

The chemical reactions of α-amino acids as either amines or carboxylic acids are not given here except for those illustrating the chemical synthesis of peptides. Peptides have been synthesized in solution by a variety of procedures. In the *solid-phase* method, which permits the synthesis of very large peptides and even proteins, an α-amino acid whose amino group is blocked is attached through its carboxyl group in ester linkage to an insoluble resin, as shown in Fig. 3.7.

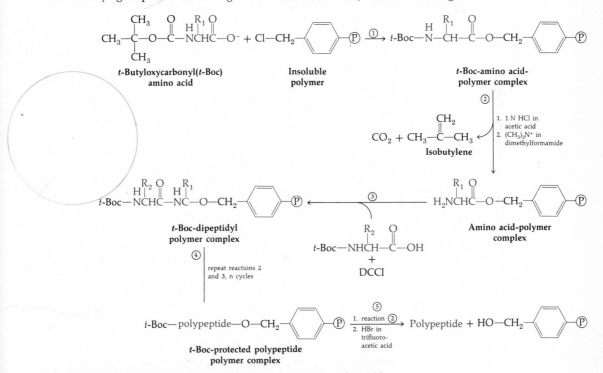

Figure 3.7 Reactions involved in the solid-phase synthesis of a polypeptide.

There are many ways to block the α-amino group; the convenient *tert*-butyloxy-carbonyl (*t*-Boc) blocking group is attached as follows,

$$CH_3-\underset{\underset{CH_3}{|}}{\overset{\overset{CH_3}{|}}{C}}-O-\overset{\overset{O}{\|}}{C}-Cl + H_2N-\underset{\underset{R_1}{|}}{CH}-\overset{\overset{O}{\|}}{C}-O^- \longrightarrow CH_3-\underset{\underset{CH_3}{|}}{\overset{\overset{CH_3}{|}}{C}}-O-\overset{\overset{O}{\|}}{C}-\overset{\overset{H}{|}}{N}-\underset{\underset{R_1}{|}}{CH}-\overset{\overset{O}{\|}}{C}-O^- + H^+ + Cl^-$$

t-Butyloxycarbonyl   *t*-Boc amino acid
chloride (*t*-Boc)

After then attaching the *t*-Boc-amino acid to the resin, the *t*-Boc group is removed as indicated to free the amino group for reaction with the carboxyl group of a second amino acid. A convenient procedure to render carboxyl groups reactive with amino groups, involves use of N,N^1-dicyclohexylcarbodiimide (DCCI), which reacts with a *t*-Boc-amino acid as follows,

$$\text{*t*-Boc}-\overset{\overset{H}{|}}{N}-\underset{\underset{R_2}{|}}{CH}-\overset{\overset{O}{\|}}{C}-OH + \text{\Large$\bigcirc$}-N=C=N-\text{\Large$\bigcirc$} \longrightarrow \text{*t*-Boc}-\overset{\overset{H}{|}}{N}-\underset{\underset{R_2}{|}}{CH}-\overset{\overset{O}{\|}}{C}-O-\overset{\overset{HN-\bigcirc}{}}{\underset{\underset{N-\bigcirc}{}}{C}}$$

                              DCCI   DCCI-activated
                                                               t-Boc-amino acid

to give an "activated" derivative, i.e., one in which the carboxyl group is rendered reactive with amines. Such activated derivatives react with amino groups of the amino acid on the resin as follows,

$$\text{*t*-Boc}-\overset{\overset{H}{|}}{N}-\underset{\underset{R_2}{|}}{CH}-\overset{\overset{O}{\|}}{C}-O-\overset{\overset{HN-\bigcirc}{}}{\underset{\underset{N-\bigcirc}{}}{C}} + H_2N-\underset{\underset{R_1}{|}}{CH}-\overset{\overset{O}{\|}}{C}-O-\text{\large$\textcircled{P}$} \longrightarrow$$

$$\text{*t*-Boc}-\overset{\overset{H}{|}}{N}-\underset{\underset{R_2}{|}}{CH}-\overset{\overset{O}{\|}}{C}-\overset{\overset{H}{|}}{N}-\underset{\underset{R_1}{|}}{CH}-\overset{\overset{O}{\|}}{C}-O-\text{\large$\textcircled{P}$} + \text{\Large$\bigcirc$}-\overset{\overset{H}{|}}{N}-\overset{\overset{O}{\|}}{C}-\overset{\overset{H}{|}}{N}-\text{\Large$\bigcirc$}$$

                                                                           Dicyclohexylurea

where $\textcircled{P}$ indicates the resin with benzyl groups attached to the amino acid. This effects the synthesis of a peptide bond, yielding a protected dipeptide still coupled to the resin. The protecting group can be removed and the procedure repeated many times until the desired product is achieved. The peptide is then liberated from the resin. An advantage is that, at each chemical step, impurities and by-products are readily removed by simply washing the resin with appropriate solvents. In addition, by reaction with an excess of the amino acid to be coupled at each step, the yields at each step approach 100%. An automated machine has been developed to carry out the entire procedure for as many steps as necessary.

The reactions shown in Fig. 3.7 may also be performed entirely in liquid phase, i.e., without use of a solid phase matrix. Such reactions permit the

40

isolation and characterization of each stable intermediate produced during the entire synthesis. In practice, a combination of solid and liquid phase syntheses is often used, the exact choice of method depending on the product desired and the choice of blocking and activating groups. With present methods of peptide synthesis, many naturally occurring polypeptides and several protein hormones, including insulin (Chap. M16), gastrin (Chap. M10), and parathormone (Chap. M15), and the enzyme ribonuclease, have been synthesized. In so doing, peptide synthesis has served to confirm the structures that were deduced by degradative methods. Synthesis has also been used to prepare analogs of the natural compounds to investigate the role of each side chain and residue in the biological activity and to obtain compounds with more desirable properties for therapeutic purposes.

REFERENCES

Books

Cohn, E. J., and J. T. Edsall: *Proteins, Amino Acids and Peptides as Ions and Dipolar Ions,* Reinhold, New York, 1942.
Greenstein, J. P., and M. Winitz: *Chemistry of the Amino Acids,* vols. I–III, Wiley, New York, 1961.

Review Articles

Erickson, B. W., and R. B. Merrifield: Solid-Phase Peptide Synthesis, in H. Neurath and R. L. Hill (eds.): *The Proteins,* 3d ed., vol. II, Academic, New York, 1976, pp. 255–527.
Finn, F. M., and K. Hofmann: The Synthesis of Peptides by Solution Methods with Emphasis on Peptide Hormones, in H. Neurath and R. L. Hill (eds.): *The Proteins,* 3d ed. vol. II, Academic, New York, 1976, pp. 105–253.
Vickery, H. B.: The History of the Discovery of the Amino Acids, II, *Adv. Protein Chem.* **26:**81–171 (1972).

Chapter 4

Proteins II

Properties and structure in solution. Molecular weight and shape. Amphoteric properties. Spectral analysis in solution. Denaturation.

GENERAL CONSIDERATIONS OF PROTEIN STRUCTURE

Several different experimental methods are required to determine the structure of a protein, since not all desired information can be obtained by any one method. When used in concert, presently available methods reveal considerable structural detail. In principle, the complete amino acid sequence of any protein can be determined (Chap. 5); indeed, hundreds of complete sequences of different proteins have been established. The size and shape of a protein and the number and kinds of subunits it may contain can be determined with great accuracy. Analysis of single protein crystals by x-ray diffraction (Chap. 5) affords great insight into three-dimensional structure, especially when the complete sequence of the protein is known. Crystallographic methods can identify the exact location in space of each of the atoms in a molecule, but for proteins, only carbon, nitrogen, oxygen, and sulfur atoms can be identified, and some of these only poorly in certain regions of a molecule. X-ray methods do not reveal hydrogen atoms, and thus the presence of hydrogen bonds can be deduced only from the relative positions of potential donor and acceptor atoms that form such bonds. Unfortunately, many proteins cannot be crystallized in a form suitable for x-ray analysis. Other methods used to determine structural features of a protein may reveal no more than the environment of certain residues. However, such information is useful in correlating protein structure with function, especially when x-ray analysis is impossible.

In this chapter some methods are described for examining the structures of proteins in solution, each of which also illustrates important properties of proteins. These methods, when used in conjunction with those presented in Chap. 5, have provided considerable insight into protein structure.

Any cross-references coded M refer to the companion text, *Principles of Biochemistry: Mammalian Biochemistry*.

Schilieren optical system
瞄线照相光学系统

The molecular weights of some typical proteins, listed in Table 4.1, illustrate the wide range in sizes of proteins. Only the more important methods that can be used to obtain an accurate molecular weight are considered here.

boundary 界之

Sedimentation Velocity-Diffusion

The rate at which a particle in solution sediments in a centrifuge tube under the action of centrifugal force depends on (1) the force applied, (2) the size, shape, and density of the particle, and (3) the density and viscosity of the solvent. Large molecules can be sedimented at high centrifugal forces, whereas small molecules cannot be sedimented at forces presently attainable in the laboratory. Modern ultracentrifuges attain speeds as high as 75,000 r/min and forces in excess of 400,000 times gravity by using high-speed electric motors. The rotors are operated in high vacuum to minimize air resistance and consequent heating due to friction.

In understanding how proteins sediment in ultracentrifugal fields, it is useful to consider the nature of an ultracentrifuge rotor (Fig. 4.1) and the changes in concentration of the protein in the ultracentrifuge cell during sedimentation, shown diagrammatically in Fig. 4.2. As a result of the changes in concentration of protein with time, a sharp boundary is formed between the solvent and the protein in solution. By measuring the position of the boundary at different times, a sedimentation constant, which is a function of molecular weight, can be measured. The rate of sedimentation can be monitored by several means. In one method, ultraviolet light (230 to 290 nm) is directed through the cell and the absorbance of the protein, which is proportional to its concentration, is recorded from the top to the bottom of the cell. The boundary formed between protein and solvent can also be observed by schlieren optics. In this optical

Protein	MW	p*I**	
			TABLE 4.1
			Molecular Weights
			and Isoelectric
Cytochrome c	13,000	10.6	Points of Some
Ribonuclease	14,000	7.8	Proteins
Myoglobin, horse	17,000	7.0	
Growth hormone (somatotropin), human	21,500	6.9	
Carboxypeptidase	34,000	6.0	
Pepsin	35,500	<1.0	
Ovalbumin, hen	40,000	4.6	
Hemoglobin, horse	65,000	6.9	
Serum albumin, human	66,500	4.8	
Serum γ-globulins	160,000	6.4–7.2	
Catalase	250,000	5.6	
Fibrinogen	330,000	5.5	
Urease	480,000	5.1	
Thyroglobulin	660,000	4.6	
Hemocyanin, octopus	2,800,000		

*pH at the isoelectric point (see page 50).

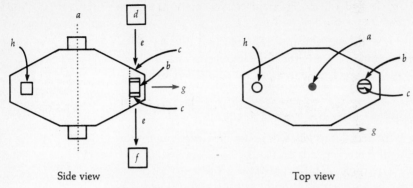

Side view   Top view

Figure 4.1 An ultracentrifuge rotor, a = axis of rotation, b = sample cell, c = window, d = light source, e = direction of light, f = optical monitor, g = direction of sedimentation, and h = counterbalance.

system, light is passed through the cell, and the refractive-index changes, which are proportional to the differences in protein concentration, are measured from the top to the bottom of the cell. The optical system is so constructed that the alteration in the refractive index at the boundary is the concentration gradient dc/dx, which gives peaks as shown in Fig. 4.2. A number of boundaries are observed when the solution contains a mixture of substances of different particle size. It is thus possible to determine homogeneity with respect to particle size and at the same time to estimate molecular weight.

Particles moving at a uniform speed in a circular path are accelerated toward the periphery. The acceleration is $\omega^2 x$, where ω is the velocity of rotation in radians per second and x is the radius from the axis of rotation in centimeters. The product of the mass and the acceleration gives the force F, where

$$F = V(\rho_2 - \rho_1)\omega^2 x = V\rho_2\omega^2 x - V\rho_1\omega^2 x \tag{1}$$

Figure 4.2 Schematic diagram of (a) the protein-concentration c profile and (b) the protein-concentration gradient dc/dx from the top to the bottom of the sedimentation velocity cell after various centrifugation times. The direction of sedimentation is from left to right. A = top of cell; B = bottom of cell; x = distance from axis of rotation; t = time. The protein concentration is the same throughout the cell before centrifugation, and dc/dx is 0. After a given time t_1 the protein has partially sedimented, accumulating in increasing concentrations at the bottom of the cell, and its concentration is near zero at the upper portion of the cell. At subsequent times, t_2, t_3, and t_4, under a constant centrifugal field, c and dc/dx behave as shown.

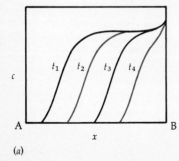

(a)

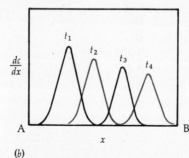

(b)

V is the volume of the particle and ρ_2 and ρ_1 are the densities in grams per milliliter of the particles and the solvent, respectively.

For 1 mol of solute, where M is the molecular weight of the solute and $\bar{v}$ is partial specific volume in milliliters per gram,

$$F = M\omega^2 x - M\bar{v}\rho_1\omega^2 x = M\omega^2 x(1 - \bar{v}\rho_1) \tag{2}$$

The force resisting sedimentation F is $f(dx/dt)$, where dx/dt is the rate of sedimentation and f is the frictional coefficient per mole, and

$$f\frac{dx}{dt} = M\omega^2 x(1 - \bar{v}\rho_1) \tag{3}$$

Rearranging gives

$$\frac{dx/dt}{\omega^2 x} = \frac{M(1 - \bar{v}\rho_1)}{f} = s \tag{4}$$

where $(dx/dt)/\omega^2 x$ is the rate of sedimentation per unit field of force, termed s, the sedimentation constant. Solving for M gives

$$M = \frac{fs}{1 - \bar{v}\rho_1} \tag{5}$$

The molecular weight cannot be calculated from this equation since f is unknown and not easily measured. However, f is related to the diffusion constant D (in square centimeters per second) by the relationship

$$D = \frac{RT}{f} \tag{6}$$

where R is the gas constant (8.31×10^7 erg/mol/deg) and T the absolute temperature in kelvins; thus, the equation for molecular weight becomes

$$M = \frac{sRT}{D(1 - \bar{v}\rho)} \tag{7}$$

In Eq. 7 all parameters can be determined experimentally and M calculated.

The sedimentation constant s has the dimensions of centimeters per second per unit of force and usually lies between 1×10^{-13} and 200×10^{-13}. The factor 1×10^{-13} is called the *Svedberg unit* (S).

The diffusion constant can be measured by observing the spread of an initially sharp boundary between a protein solution and the solvent in which it is dissolved as the protein diffuses into the solvent layer. The boundary can be analyzed by schlieren optics just as in the ultracentrifuge.

Rate of sedimentation depends on the difference in density of solute particles and the solvent. If the particles are of lower density than the solvent, they will rise to the top (as in a cream separator) and flotation will occur. For most proteins the partial specific volume $\bar{v}$ has the value of 0.70 to 0.75. The value depends on the amino acid composition and can be calculated when this is known, or it can be determined experimentally.

Highly purified proteins show homogeneous boundaries in the ultracentrifuge; representative sedimentation curves obtained with schlieren optics are shown in Fig. 4.3. The ultracentrifuge is a useful tool for studying the stability of proteins since under many conditions, e.g., extremes of pH or temperature,

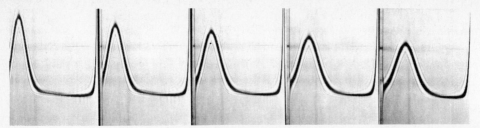

Figure 4.3 Sedimentation of crystalline ribonuclease at 59,780 r/min in the analytical ultracentrifuge. The direction of sedimentation is from left to right. The first picture was taken 1 h after attaining speed; subsequent exposures were made at 16-min intervals. It will be noted that as the boundary moves down the cell, the height of the boundary decreases and its width increases. This is due to normal diffusion. The boundary at the bottom of the cell represents sedimented protein. When a preparation contains a mixture of proteins or other solutes of different sizes, multiple peaks are found.

aggregation or dissociation may occur, and this can be detected by changes in sedimentation constants. Molecular weights of some proteins obtained by the sedimentation method are given in Table 4.1. Present methods estimate the molecular weight to within 5 percent of the true value.

Sedimentation Equilibrium

Sedimentation constants can be measured accurately, but diffusion constants are more tedious to determine. Hence, the technique of sedimentation equilibrium, which does not require knowledge of the diffusion constant, is more widely used for determining molecular weights of proteins. In this procedure, the ultracentrifuge is operated until an equilibrium distribution of the protein is achieved throughout the length of the cell. At equilibrium, there is no net migration of protein since its movement to the bottom of the cell due to the centrifugal force is exactly balanced by its movement to the top of the cell by diffusion. The equilibrium distribution depends upon the molecular weight of the protein. The concentration of the protein from the top to the bottom of the cell is measured when equilibrium has been achieved by the optical methods discussed above for sedimentation velocity.

The equation for calculating the molecular weight from sedimentation equilibrium is

$$M = \frac{2RT \ln c}{\omega^2 x^2 (1 - \overline{v}\rho)} \tag{8}$$

where R, T, ω^2, $\overline{v}$, and ρ are defined as described above and c is the concentration of protein at any distance x in the ultracentrifuge cell. Rearranging the equation gives

$$\ln c = \frac{M(1 - \overline{v}\rho)\omega^2 x^2}{2RT} \tag{9}$$

Inspection of Eq. (9) indicates that the slope of the line in plots of $\ln c$ vs. x^2 is equal to $M(1 - \overline{v}\rho)\omega^2/2RT$. Because c, x, $\overline{v}$, ρ, ω^2, R, and T are easily measured

swinging bucket rotor
水平式离心转子

rotor 转子 (离心机中旋转的部件.)

or calculated, M can be determined from the slope. Alternatively, the molecular weight at any single concentration at any distance x can be measured directly and provides an indication of the homogeneity of the protein if it is the same throughout the ln c vs. x^2 plot. If the protein solution analyzed is heterogeneous, ln c vs. x^2 plots will not be linear. This method also has the advantage that very small concentrations of proteins may be employed.

Sedimentation in Sucrose Gradients

沉降系数

meniscus
界面

It is possible to determine sedimentation constants and hence to estimate *approximate molecular weights* with impure solutions if a suitable technique is available for determining relative concentrations, e.g., enzymic activities. The sample to be studied is layered on a linear sucrose gradient and centrifuged at a high speed in a swinging-bucket rotor. Generally, a protein of known s value is added as a standard or marker. Substances of different sedimentation properties will separate from one another as bands in the gradient. At the end of the run a small hole is punched in the bottom of the centrifuge tube and emerging fractions are collected and analyzed. If samples are collected at different times of centrifugation, a plot of distance of peak activity from the meniscus vs. time should be linear. For a specific time of centrifugation in comparison with the protein of known s as standard, the following approximations obtain for substances that are nearly spherical and for which $\bar{v}$ is similar:

$$\frac{\text{Distance from meniscus of unknown}}{\text{Distance from meniscus of standard}} = \frac{s_{20,w} \text{ of unknown}}{s_{20,w} \text{ of standard}} = \left(\frac{M_{\text{unknown}}}{M_{\text{standard}}}\right)^{\frac{2}{3}} \quad (10)$$

Filtration on Molecular Sieves

calibrate 校准.

Cross-linked polysaccharide or polyacrylamide gels can be used to estimate molecular weights by the technique of *gel filtration* (Chap. 9). The gels are used in columns similar to those employed for chromatography. This technique permits evaluation of purity by molecular size and allows separation of substances of different molecular weights. When a column has been calibrated with substances of known molecular weights, the position of emergence can be used to estimate the molecular weight.

Figure 4.4 shows a plot of elution volume V_e against molecular weight on a log scale for a number of proteins. A single smooth curve fits all the data except for some proteins whose shape is quite asymmetric, e.g., fibrinogen and ovomucoid.

A rapid way of measuring the molecular weight of protein subunits is provided by gel filtration by the same methods as in Fig. 4.4 but with aqueous $6\,M$ guanidine $\cdot$ HCl as the chromatographic solvent. This solvent completely disrupts the conformation of proteins (Chap. 5); thus the shape of the molecules does not limit the accuracy of the measurement as severely as in gel filtration of native proteins. Disulfide bonds must be reduced before analysis of proteins by this method.

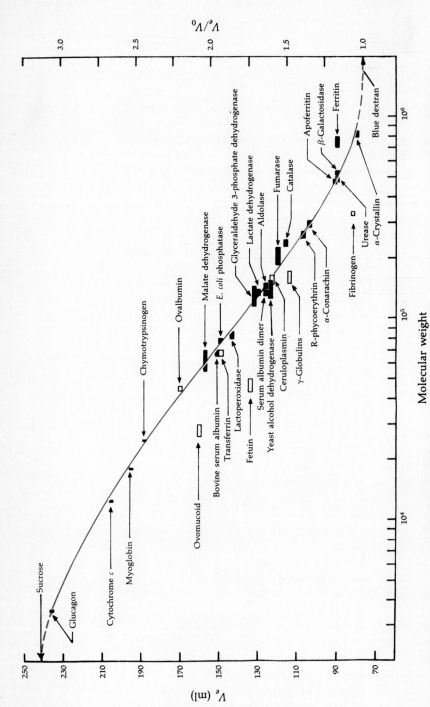

Figure 4.4 Relationship of elution volume V_e to molecular weight for proteins on a column of cross-linked dextran (Sephadex G-200) (2.5 by 50 cm) at pH 7.5. Open bars represent glycoproteins. V_0 represents the void volume. [*After P. Andrews, Biochem. J.* **96**:595 *(1965).*]

intrinsic
内在的固有的,

Proteins are denatured as well as dissociated into their subunits by the detergent sodium dodecyl sulfate (SDS). On exposure to solutions of SDS for a few minutes at 100°C, protein molecules unfold and bind about one molecule of SDS for every two amino acid residues. The resulting complexes contain approximately 1.4 g SDS per gram of protein. Aliphatic dodecyl groups are in the interior of the complex, whereas sulfate groups are on the surface, and the complexes have a net charge that far exceeds the intrinsic charge of the amino acid side chains. The SDS-protein complexes are rod-shaped, with a diameter of about 18 Å and a length proportional to the molecular weight of the polypeptide chain. The SDS-protein complexes have about the same negative charge per unit mass and thus migrate on zone electrophoresis (Chap. 9) in gels of uniform porosity, the rate of migration increasing as the molecular weight decreases. Figure 4.5 shows a typical gel pattern for a protein mixture and the

Figure 4.5 Behavior of a mixture of proteins on SDS-gel electrophoresis in polyacrylamide gels. (*a*) The protein mixture was subjected to electrophoresis in a phosphate buffer, pH 7.2, containing 0.2% SDS. The proteins migrate from the top to bottom. The gel was removed and stained with a dye, Coomassie Blue, to detect the proteins, shown as sharp zones or bands. The protein and its molecular weight are indicated next to each band. The mobilities of a protein are calculated relative to that of the dye, which migrates with a rate of 1 (arrow). (*b*) Comparison of the mobilities of a mixture of 24 proteins on SDS-gel electrophoresis as a function of their molecular weights. The molecular-weight range examined was 11,700 to 68,000, with cytochrome c migrating the farthest and serum albumin the least. [*From K. Weber and M. Osborn, p. 180, in H. Neurath and R. L. Hill (eds.): The Proteins, vol. 2, Academic, New York, 1975.*]

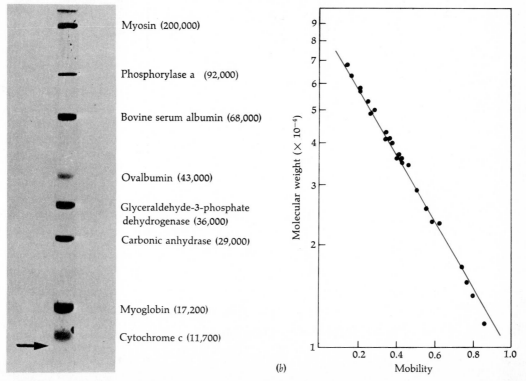

Myosin (200,000)

Phosphorylase a (92,000)

Bovine serum albumin (68,000)

Ovalbumin (43,000)

Glyceraldehyde-3-phosphate dehydrogenase (36,000)

Carbonic anhydrase (29,000)

Myoglobin (17,200)

Cytochrome c (11,700)

(*a*)

(*b*)

logarithmic relationship between molecular weight and mobility. If disulfide bonds are present, the proteins must be reduced before analysis. However, some proteins behave aberrantly, e.g., those with a high carbohydrate content. This method has also had wide application for assessing the complexity and size of membrane-bound proteins, e.g., those of erythrocytes (Chap. M3), which are largely water-insoluble.

Alternatively, the molecular weight of protein subunits can be estimated by gel filtration with solutions of SDS as the chromatographic solvent by methods similar to those in Fig. 4.4.

SHAPE OF PROTEIN MOLECULES

From the sedimentation constant s and the diffusion constant D a frictional ratio f/f_0 can be calculated from the formula

$$\frac{f}{f_0} = 10^{-8}\left(\frac{1 - \bar{v}\rho}{D^2 s \bar{v}}\right)^{\frac{1}{3}} \tag{11}$$

where the various terms have the same meaning as above. The molar frictional coefficient for a compact spherical and unhydrated particle is f_0. When f/f_0 is equal to 1.0, the molecule is essentially an unhydrated sphere. When f/f_0 is greater than 1.0, the molecules may be asymmetrical or hydrated or both. Many globular proteins appear to be spherical or nearly spherical molecules, including ribonuclease, insulin, chymotrypsinogen, and pepsin. However, other proteins are highly asymmetrical and exist in solution as pronounced ellipsoids or rods. Fibrinogen, the plasma protein that is the precursor of fibrin of blood clots, and myosin, the main protein of muscle fibers, are examples of long ellipsoids. It is estimated that human fibrinogen has a cross-sectional diameter of 38 Å and a length of 700 Å. Solutions of such molecules show pronounced light scattering, and the shape of molecules, as well as molecular weight, can be estimated from light-scattering measurements.

The *viscosity* of a solution depends on the molecular weight and shape of the solute molecules at a given solute concentration. Highly asymmetrical molecules show a high intrinsic viscosity compared with spherical molecules of the same molecular weight. When the latter is known, the shape can be estimated from the variation of viscosity with solute concentration.

AMPHOTERIC PROPERTIES

Proteins are polyelectrolytes, containing positively and negatively charged groups on their surfaces, the exact number depending upon the amino acid composition and the pH of the medium. At one pH, called the *isoelectric point* (p*I*), the numbers of positive and negative charges are equal. Isoelectric points vary from one protein to another, as illustrated in Table 4.1 for some typical proteins. Irrespective of its exact p*I*, a protein has a net positive charge at pH values below its p*I*, and a net negative charge above the p*I*.

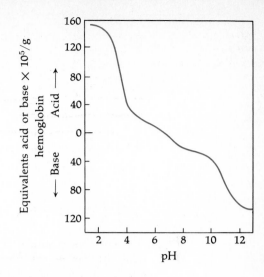

Fig. 4.6 The titration curve of crystalline horse hemoglobin. [*After E. J. Cohn, A. A. Green, and M. H. Blanchard, J. Am. Chem. Soc.* **59**:590 (1937).]

The number of ionic groups in a protein can be estimated by titration from its pI with acid and base; the titration curve for horse hemoglobin, shown in Fig. 4.6, is typical of globular proteins. Table 4.2 gives the characteristic pK values for the ionic groups found in proteins. The exact pK value for any individual ionic group in a specific protein may differ somewhat from the value shown for such groups in Table 4.2, depending on the nature of its local environment; e.g., each γ-carboxyl group in a protein may have a slightly different pK value, but together they titrate to give an apparent pK of about 4.4. Indeed, local environmental effects also result in extreme pK values for some ionic groups, also listed in Table 4.2.

lattice 晶格

SOLUBILITY

Each protein has a characteristic solubility in a given solvent. Solubility is markedly influenced by pH and is usually at a minimum at the pI, as shown in Fig. 4.7 for β-lactoglobulin. This probably results from the fact that at the pI, electrostatic repulsion forces are at a minimum and crystal-lattice forces in the

Group	pK_a at 25°C	Extreme pK values	
α-Carboxyl (terminal)	3.0–3.2	1–6.8	TABLE 4.2
β-Carboxyl (aspartic)	3.0–4.7	1–6.8	Characteristic pK Values for Acidic and Basic Groups in Proteins
γ-Carboxyl (glutamic)	$\approx$4.4	1–6.8	
Imidazolium (histidine)	5.6–7.0	5.6–7.5	
α-Amino (terminal)	7.6–8.4	7.3–>12	
ϵ-Amino (lysine)	9.4–10.6	7.3–>12	
Guanidinium (arginine)	11.6–12.6	11.5–12.6	
Phenolic hydroxyl (tyrosine)	9.8–10.4	9.4–>12	
Sulfhydryl (cysteine)	$\approx$8–9	8–9.5	

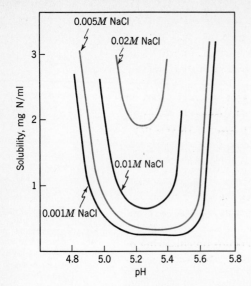

Figure 4.7 The solubility of β-lactoglobulin as a function of pH at four different values of ionic strength.

solid state at a maximum. When proteins are either predominantly positively or negatively charged, at either side of the pI, their solubility increases.

Salts may either decrease or increase the solubility of a protein. The effect of salts in increasing the solubility of proteins is called the *salting-in effect,* as shown in Fig. 4.8 for horse hemoglobin. The solubility is a function of the ionic strength, which is calculated from the molar concentrations of the ions and their charge using the expression

$$\mu = \tfrac{1}{2}\Sigma m Z^2 \tag{12}$$

where μ is the ionic strength, m the molality, and Z the charge of the ion.

It has not been possible to predict the solubility of a protein, even when crystal analysis has revealed the complex array of surface structures. In general, however, the solubility of any substance depends on the relative affinity of solute

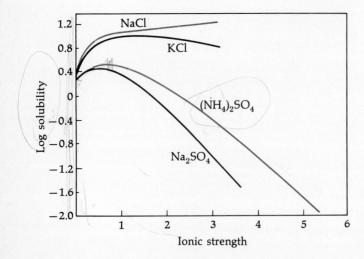

Figure 4.8 The solubility of horse carbon monoxide hemoglobin in salt solutions of different ionic strength. Both salting in and salting out are observed with this protein. [*From E. J. Cohn and J. T. Edsall, Proteins, Amino Acids, and Peptides as Ions and Dipolar Ions, Reinhold, New York, 1942.*]

molecules for each other (crystal-lattice formation) and for the solvent molecules. Any factor that decreases interaction of solute molecules will tend to increase solubility. In salting in, the small ions of neutral salts will interact with the ionic groups of the protein molecules, diminishing protein-protein interactions and therefore increasing solubility.

Proteins are precipitated from aqueous solution by high concentrations of neutral salts. This is the *salting-out phenomenon*. Di- and trivalent ions are more effective than univalent ions (Fig. 4.8). Commonly used salts are $(NH_4)_2SO_4$, Na_2SO_4, magnesium salts, and phosphates. The most effective region of salting out is at the isoelectric point of the protein. Salting out occurs not only with proteins but with gases, uncharged molecules, and electrolytes. The mechanism of salting out is complex. Salt ions attract the polarizable water molecules (Chap. 2), making less water available for the proteins since at high salt concentrations the number of charged groups contributed by the salts is enormous compared with those of the proteins. Since solubility depends on clustering of water molecules around the hydrophilic ionic groups, removal of water molecules to other ions will decrease protein solubility.

Proteins are precipitated from solution by nonpolar solvents that are miscible with water, e.g., methanol, ethanol, and acetone. Protein solubilities in these solvents are markedly affected by neutral salts. Most satisfactory results are obtained by working at low temperatures, where proteins are most stable.

Prior to the development of cascade methods for protein purification (Chap. 9), the major methods for protein purification were differential precipitation with salts and organic solvents. Such procedures are not nearly as selective as cascade methods but are useful since they permit separation from the bulk of other proteins in biological material prior to use of more selective purification methods. The rather pure single crystals required for x-ray crystallography (Chap. 5) are obtained by carefully adjusting the temperature, the pH, and the salt or organic solvent concentrations of a protein solution so as to control the rates of crystal nucleation and growth.

SPECTRAL ANALYSIS OF PROTEINS

Most of the spectroscopic methods that have been used to obtain detailed structural information about small molecules have also been applied to proteins. Rapidly oscillating electromagnetic fields ranging in wavelength from 10^{-13} to 10 cm have been used to irradiate proteins, and the ways in which the radiation is altered or the proteins are perturbed have given useful structural information. The methods of x-ray diffraction of crystals (Chap. 5) have given the most information, but those used to analyze proteins in solution have also been useful, particularly in following perturbations of structure that occur when proteins bind with other molecules or that occur when they are denatured.

Even a qualitative discussion of each of the various spectroscopic methods presently used is beyond the scope of this text, but a few methods will be considered to illustrate the kinds of information that can be obtained.

Absorption Spectra of Proteins

chromophore

For any range of wavelengths, certain groups, called *chromophores,* dominate the observed spectrum. Proteins generally absorb light at wavelengths between 180 and 300 nm because of the presence of certain chromophores; thus between 250 and 300 nm with a peak near 280 nm, absorption is due almost entirely to the R groups of aromatic amino acids, as shown in Fig. 3.3. The absorption spectrum of a protein solution in this region often changes between pH 8 and 12. This is the result of the ionization of the phenolic side chains of tyrosine. Often the ionization of only a fraction of the tyrosine residues may be detected, although after denaturation complete ionization of all phenolic groups is observed. Those residues that ionize only after denaturation are usually buried in the protein molecule and may participate in hydrogen bonding or hydrophobic interactions that aid in maintaining the three-dimensional structure. Measurement of the absorption of a protein solution at 280 nm is also used to determine protein concentration, since the molar absorption coefficient can be measured accurately at this wavelength. Approximate estimates of protein concentration are also made by assuming that the absorbance of a 1-cm path length of a protein solution at a concentration of 1 mg/ml is 1.1 ± 0.5, an average value for several proteins picked at random.

Between 190 and 200 nm, the spectrum of a protein reflects absorption by peptide bonds and is influenced by the conformation of the native state of the protein. Thus, as shown in Fig. 4.9, the polyamino acid poly-L-lysine has a different absorption spectrum when it is dissolved in solvents that permit it to fold in different conformations. Native proteins have more complex spectra reflecting the amount of helix and pleated sheet structures which they contain.

Optical Rotation of Proteins

Protein solutions rotate polarized light, and the observed optical rotation results in part from the optical activity of the constituent amino acids and partly from the conformation of the polypeptide chain. The specific rotations $[\alpha]_D$ of proteins are usually in the range of -30 to $-60°$; on denaturation, the specific rotations become more negative. Thus, the change in the optical rotation usu-

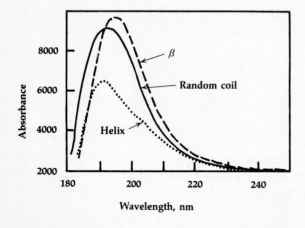

Figure 4.9 The ultraviolet absorption spectra of poly-L-lysine in different solvents that permit the polymer to fold in different conformations. Random coil, pH 6.0, 25°C; helix, pH 10.8, 25°C; β structure, pH 10.8, 52°C. [*From K. Rosenheck and P. Doty, Proc. Natl. Acad. Sci. U.S.A.* **47**:1775 *(1961).*]

ally reflects a change in conformation. Some insight into the conformational changes can be gained by measuring the variation in optical rotation with the wavelength of the monochromatic light, a method called *optical rotatory dispersion* (ORD). The ORD curves for poly-L-lysine shown in Fig. 4.10 illustrate how conformation influences rotation. Native and denatured proteins have different ORD curves as well, and the exact curves reflect their content of α-helix and β structures. Empirical methods to analyze the complex ORD curves of native and denatured proteins have been developed and permit estimates of the helical contents of proteins. Such methods often give helical contents in excellent agreement with those determined by x-ray crystallography, but for some proteins the agreement is rather poor. Nevertheless, ORD analysis of proteins provides a quick and sensitive measure of conformation changes in a protein in solution under a variety of experimental conditions.

Circular dichroism (CD) measures the *difference in absorption* of left and right circularly polarized light, as indicated by the relationships indicated in Fig. 4.11. Thus, the CD curves of proteins between 190 and 300 nm are also very sensitive to changes in conformation of a protein and the environment of the chromophores. Such curves, like ORD curves, can also be used to estimate the helical content of native proteins.

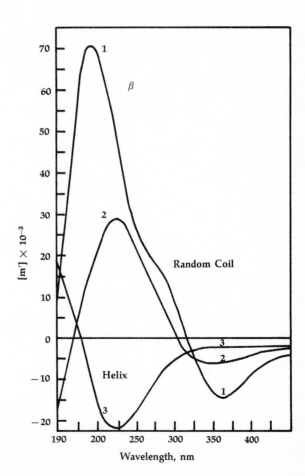

Figure 4.10 The optical rotatory dispersion curves of poly-L-lysine in the different conformations obtained as indicated in Fig. 4.9. [*From N. Greenfield, B. Davidson, and G. B. Fasman, Biochemistry 6:1630 (1967).*]

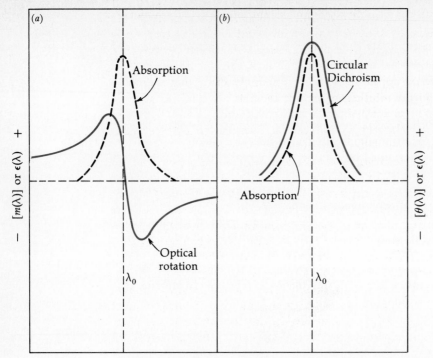

Figure 4.11 Theoretical relationships for (*a*) an optically active band for line absorption or optical rotation and (*b*) circular dichroism. The ordinates are as follows: $m(\lambda)$, optical rotation per mole at wavelength λ; $\epsilon(\lambda)$, extinction coefficient of absorption band; $\theta(\lambda)$, ellipticity or difference in absorption for left and right circularly polarized light. [*From S. Beychok, Science 154:1288 (1966).*]

DENATURATION OF PROTEINS

The term *denaturation* applies to a wide variety of structural changes and involves differing effects on individual proteins. Denaturation is a chemically complex process, and the interactions between denaturing agents and proteins that lead to conformational changes are not easily described. All denaturing agents disrupt the noncovalent structure of the native protein to some degree. For some proteins modest changes in conformation may not destroy the biological activity of the molecule, whereas in others biological activity is lost with only small conformational changes that are not easily detected by available methods.

Guanidine · HCl and urea are two generally used denaturing agents. Most

$$
\begin{array}{cc}
\overset{\displaystyle O}{\underset{\|}{}} & \overset{\displaystyle NH}{\underset{\|}{}} \\
H_2N{-}C{-}NH_2 & H_2N{-}C{-}NH_3{}^{+}Cl^{-} \\
\textbf{Urea} & \textbf{Guanidine hydrochloride}
\end{array}
$$

proteins are denatured in aqueous solvents of 6 to 8 M guanidine · HCl or in 8 to 10 M urea. Under these conditions the native structure is usually completely disrupted, the polypeptide chain is a random structure in which all noncovalent bonds are broken, and there is continuous fluctuation between different local and overall conformations. The hydrophobic R groups tend to become ex-

56

posed in concentrated solutions of urea and guanidine · HCl, suggesting that denaturation of proteins by these solvents involves disruption of hydrophobic interactions. This is in accord with the increased solubility of hydrophobic substances in concentrated aqueous solutions of urea and guanidine · HCl.

Nonaqueous solvents appear to denature proteins by disrupting hydrophobic bonding, as do aqueous solutions of organic alcohols. In some nonaqueous solvents, proteins appear to adopt a conformation containing a high helix content. Detergents bind directly to polypeptides, causing major structural changes, as noted earlier for SDS (page 49).

Proteins usually are denatured when the temperature of a protein solution is raised to about 50 to 60°C for a given period, but the exact temperature that will induce conformational changes varies widely among different proteins. Indeed, some enzymes are inactivated on cooling from 30 to 0°C (*cold-labile enzymes*). Often heat denaturation is reversible, but in many cases aggregation of protein occurs with precipitation. Proteins are also denatured by pH-induced conformational changes. As polymers, all proteins may participate in electrostatic interactions that will differ as the pH varies, but it is difficult to predict why a given protein is denatured at one pH and not another. Pepsin, with a p*I* less than 1, denatures rapidly at neutral pH, but ribonuclease and lysozyme, with p*I*s of 7.8 and 10.6, repectively, undergo no change in conformation at pH 2. Most proteins, however, are stable only within a narrow pH range.

REFERENCES

Books

Cantor, C. R., and P. R. Schimmel: *Biophysical Chemistry,* parts I, II, and III, Freeman, San Francisco, 1980.
Haschemeyer, R. H., and A. E. V. Haschemeyer: *Proteins. A Guide to Study by Physical and Chemical Methods,* Wiley, New York, 1973.
Schachman, H. K.: *Ultracentrifugation in Biochemistry,* Academic, New York, 1959.
Svedberg, T., and K. O. Pedersen: *The Ultracentrifuge,* Oxford, Fair Lawn, N.J., 1940.

Review Articles

Tanford, C.: Protein Denaturation, *Adv. Protein Chem.* **23:**122–282 (1968); **24:**1–95 (1970).
Van Holde, K. E.: Sedimentation Analysis of Proteins, in H. Neurath and R. L. Hill (eds.): *The Proteins,* 3d ed., Academic, New York, 1976, pp. 225–241.

Chapter 5

Proteins III

Amino acid sequences and three-dimensional structure.

This chapter will consider further aspects of the primary, secondary, tertiary, and quaternary structures of proteins. Special attention is given to sequence analysis and to the determination of three-dimensional structure by x-ray diffraction of protein crystals, the two methods that have given the greatest insight into protein structure. Since the amino acid sequence of a protein determines its conformation and biological function depends upon conformation (Chap. 3), the entire molecular architecture of a protein must be understood to describe how it functions in molecular terms.

PRIMARY STRUCTURE

The amino acid sequences of hundreds of proteins have been established since the first complete sequence, that of insulin, was reported in 1955. Although in principle x-ray crystallography could be used for sequence analysis, it is presently insufficiently accurate with proteins to be used for this purpose. Thus, the methods considered in this section remain the principal means to establish primary structure. However, amino acid sequences may also be deduced from the nucleotide sequences of DNA and RNA, which can be established by the very rapid, ultramicro methods developed recently (Chap. 8). Most protein sequences have been established by direct analysis, but in the future nucleotide sequencing may become an excellent alternative approach if the appropriate polynucleotide can be obtained and certain aspects of primary structure are obtained by other means.

Any cross-references to chapters coded M refer to the companion text, *Principles of Biochemistry: Mammalian Biochemistry*.

The amino acid composition of a protein is required prior to sequence analysis. Hydrolysis of a protein in 6 N HCl at 110°C for 20 to 96 h suffices to cleave most peptide bonds. The amino acids in the resulting hydrolysate can be identified and quantitated by ion-exchange chromatography, as shown in Fig. 9.8. Since some amino acids (serine, threonine, tyrosine) may be partially destroyed on hydrolysis whereas others (isoleucine, valine) form peptide bonds that are slowly hydrolyzed, the time of hydrolysis of separate samples is varied. Tryptophan is destroyed by hot acid while asparagine and glutamine are converted quantitatively to aspartic acid and glutamic acid, respectively. Nevertheless, special methods are readily available to determine the amino acid composition of a protein within an accuracy of ±5 percent.

Establishing the composition of a protein reveals little concerning specific structural information but will detect the presence in certain proteins of unusual amino acids, e.g., thyroxine in thyroglobulins (Chap. M14), and 5-hydroxylysine and two isomers of hydroxyproline in collagens (Chap. M6).

Subunit Structure

The number of distinct polypeptide chains in a given protein must be established prior to sequence analysis; if more than one kind of chain is present, they must be separated and sequenced individually. Determination of the molecular weight of a protein by sedimentation equilibrium (Chap. 4) is especially useful for identifying the number of subunits. As indicated in Table 5.1, the subunit structure of a protein can be predicted by measuring its molecular weight under three different conditions. Guanidine·HCl (6 M) usually disrupts all noncovalent interactions between chains whereas β-mercaptoethanol breaks all disulfide bonds between chains by disulfide interchange as follows:

$$P_1\text{—S—S—}P_2 + RSH \rightleftharpoons P_1\text{—S—S—R} + P_2\text{—SH}$$
$$P_1\text{—S—S—R} + RSH \rightleftharpoons P_1\text{—SH} + R\text{—S—S—R}$$

where P_1—S—S—P_2 represents a protein with two chains, P_1 and P_2, combined by a disulfide bond, and RSH is mercaptoethanol, HS—CH_2—CH_2—OH. Thus, if a protein contains only one polypeptide chain, its molecular weight will

TABLE 5.1 The Molecular Weight of Hypothetical Proteins with Molecular Weights of 100,000 and Different Subunit Structures under Three Different Conditions of Ultracentrifugation

| Protein | Molecular weight | | | Subunit structure |
	Native	6 M guanidine·HCl	6 M guanidine·HCl mercaptoethanol	
A	100,000	100,000	100,000	1 chain
B	100,000	50,000	50,000	2 chains no —S—S— bonds
C	100,000	100,000	50,000	2 chains combined by —S—S— bonds
D	100,000	50,000	25,000	4 chains; pairs combined by —S—S— bonds

be the same in $6\,M$ guanidine · HCl (G-HCl) or $6\,M$ G · HCl plus RSH (protein A). If it contains two subunits of identical size that are combined solely by noncovalent bonds, then its molecular weight in G · HCl is one-half that of the native protein and it is unaltered by RSH (protein B). A protein that contains two chains of the same size that are combined by disulfide bonds will have the same molecular weight as the native protein in G · HCl, but one-half the molecular weight in G · HCl plus RSH (protein C). Similarly, a protein with four chains, in which pairs of chains are combined by disulfide bonds, will give the molecular weights shown for protein D.

Another simple method for determining subunit number employs crosslinking of the subunits with a bifunctional reagent such as dimethyl suberimidate. A protein containing two subunits (P_1 and P_2) of equal size that are bound noncovalently reacts with suberimidate through ϵ-amino groups in P_1 and P_2 as follows:

$$\begin{array}{c} \underset{\textstyle \overset{|}{P_1}}{NH_2} \quad \underset{\textstyle \overset{|}{P_2}}{NH_2} \; + \; HN{=}\overset{\textstyle OCH_3}{\overset{|}{C}}{-}[CH_2]_4{-}\overset{\textstyle OCH_3}{\overset{|}{C}}{=}NH \; \longrightarrow \; \underset{\textstyle \overset{|}{P_1}}{HN}{-}\overset{\textstyle \overset{NH}{\|}}{C}{-}[CH_2]_4{-}\overset{\textstyle \overset{NH}{\|}}{C}{-}\underset{\textstyle \overset{|}{P_2}}{NH} \; + \; 2CH_3OH \end{array}$$

Dimethyl suberimidate

Analysis of un-cross-linked protein by SDS-gel electrophoresis reveals only monomeric subunits, but after partial cross-linking, dimeric cross-linked species also appear with about twice the molecular weight of P_1 and P_2. A protein with three subunits gives monomers, dimers, and trimers, and one with four subunits gives tetramers as well, each of which can be detected electrophoretically.

Subunits of different sizes can be identified easily by SDS-gel electrophoresis (Chap. 4). But size determinations are insufficient to establish whether the subunits in a protein differ in sequence from one another. Two approaches can be used for this purpose. (1) NH_2- or COOH-terminal amino acids in a protein can be determined by end-group methods (see below). This may be ambiguous, however, since different subunits may, nevertheless, have the same terminal residues. Such is the case for human hemoglobin, in which valine is NH_2-terminal in both the α and β chains (Table 5.2). (2) *Peptide mapping* usually gives unambiguous results by determination of the number of peptides in an enzymic hydrolysate of a protein. The enzyme trypsin, which hydrolyses only those peptide bonds formed by the COOH groups of lysine and arginine, is most often used to degrade the protein. The resulting peptides are then separated two-dimensionally on paper (Chap. 9) and the number of peptides counted. Figure 5.1 shows the tryptic peptides from human hemoglobin after separation by such means. The pattern for each protein is unique; hence, it is likened to a *fingerprint* as a means for identification. If the number of tryptic peptides is equal to one plus the lysine and arginine contents of the protein, then the protein contains one chain; if the number is one-half that value, there are two chains with identical sequences; and if one-quarter that value, there are four identical chains. Several variations in the number of peptides are possible depending on the number and sequences of the subunits; thus hemoglobin (Fig. 5.1), with two pairs of nonidentical chains, gives one-half the number of peptides expected for a protein with four nonidentical chains.

TABLE 5.2 The NH$_2$-Terminal End Groups and Number of Subunits in Some Proteins

Protein (source)	MW	NH$_2$-terminal groups	Number of chains	Sequences of subunit chains
Insulin (Bovine)*	5,700	1 Phe, 1 Gly	2	2 different
Ribonuclease T1 (*Aspergillus*)*	11,100	1 Ala	1	
Cytochrome c (Vertebrate heart)*	12,400	1 Acetyl-Gly	1	
Ribonuclease (Bovine pancreas)*	13,700	1 Lys	1	
Trypsinogen (Bovine pancreas)*	24,400	1 Val	1	
α-Chymotrypsin (Bovine pancreas)*	24,300	1 Ile, 1 Ala, 1 Cys	3	3 different
Hemoglobin (Human erythrocytes)*	65,000	4 Val	4	2 different
Albumin (Human serum)*	66,500	1 Asp	1	
Alcohol dehydrogenase (Equine liver)*	80,000	2 Acetyl-Ser	2	Same
Hemerythrin (*Sipunculid* worm)*	108,000	1 Gly	8	Same
Glyceraldehyde-3-phosphate dehydrogenase (Rabbit muscle)*	140,000	4 Val	4	Same
Immunoglobulin γG (Human)	150,000	(see Chap. M2)	4	2 different
Fibrinogen (Human plasma)*	333,000	2 Ala, 2 Glu, 2 Tyr	6	3 different
Phosphorylase a (Rabbit muscle)*	370,000	4 Met	4	Same
Glutamate dehydrogenase (Bovine liver)*	336,000	6 Ala	6	Same
Fatty acid synthetase (Rat)	500,000	Unknown	2	†
Turnip yellow mosaic virus*	5,000,000	180 N-acetyl-Met	180	Same

*The amino acid sequences have been established.
† Not determined.

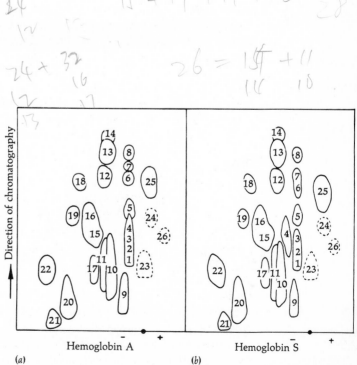

Figure 5.1 Peptide maps of human hemoglobin. Tryptic digests of (*a*) human hemoglobin A and (*b*) hemoglobin S were placed at the origin (●) on a piece of filter paper, which was subjected to electrophoresis in one dimension and then chromatography in the other, as indicated. The peptides were detected with ninhydrin to give the peptide map of (*a*) hemoglobin A and (*b*) hemoglobin S. Native hemoglobin contains 56 lysine and arginine residues, and since only 26 peptides were observed, it must contain two peptides with different sequences. If all four chains had identical sequences, 15 peptides would be found, whereas if all chains differed in sequence, about 60 peptides could be expected. The single amino acid that differs in hemoglobin A and S is contained in peptide 4. [*From W. A. Schroeder, The Primary Structure of Proteins, Harper & Row, New York, 1968.*]

Direction of chromatography

Hemoglobin A   (a)

Hemoglobin S   (b)

Table 5.2 lists the subunit structures of several different proteins. Many different combinations of subunits are possible.

General Approach to Sequence Analysis

When the composition of a polypeptide chain is known, its sequence can be determined as follows: (1) partial hydrolysis of the chain by enzymes or chemical reagents, (2) separation and isolation of the resulting peptides, (3) determination of the amino acid sequences of the peptides, and (4) deduction of the complete sequence from peptides with overlapping sequences.

To determine the complete sequence, at least two different forms of partial hydrolysis must be used. A hypothetical case is illustrated below, in which the points of hydrolysis by one enzyme are shown by the arrows, yielding peptides I through VI.

$$H_2N-A \cdot B \cdot C \cdot D \cdot E \cdot F \cdot G \cdots\cdots\cdots\cdots\cdots\cdots T \cdot U \cdot V \cdot W \cdot X \cdot Y \cdot Z-COOH$$

I   II   III   IV   V   VI

Even when the sequence of each peptide is known, this will not permit positioning the peptides. However, when hydrolysis of the protein is accomplished by another enzyme which hydrolyzes a different set of peptide bonds, a different series of peptides will be obtained (Table 5.3). If it yields a peptide having the sequence $C \cdot D \cdot E \cdot F \cdot G \cdot$, etc., this will overlap with the sequences of peptide I and peptide II, thus establishing their relationship. By this means the entire sequence can be deduced.

Before sequence analysis, it is essential to modify the cystine and cysteine residues in the protein since they may be destroyed during sequencing. This is accomplished by several means including reduction of the protein with β-mercaptoethanol followed by carboxymethylation of the resulting thiol groups with iodoacetate, as follows:

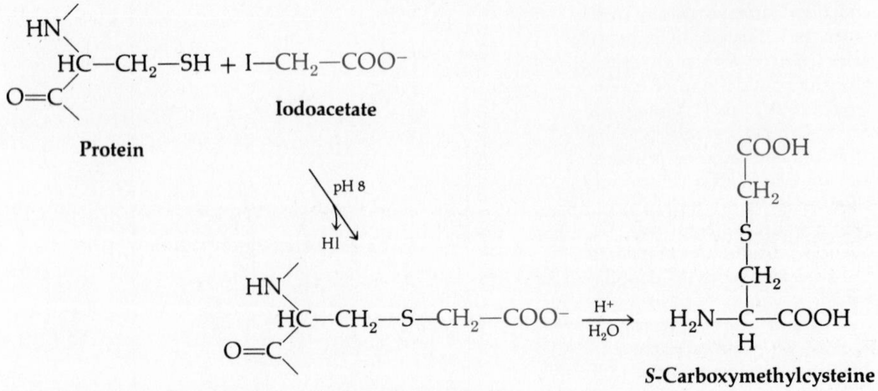

Alternatively, the protein may be oxidized with performic acid to convert all cysteine and half-cysteine residues to cysteic acid.

TABLE 5.3 Specificity of Some Proteolytic Enzymes

Proteolytic enzyme	Source	Major sites of action*	Other sites of action
Trypsin	Pancreas	Arg, Lys	None at X—Pro†
Chymotrypsin	Pancreas	Trp, Phe, Tyr	Leu, Met, Asn, His
Elastase	Pancreas	Neutral aliphatic residues	
Pepsin	Gastric mucosa	Trp, Phe, Tyr, Met, Leu	Various, acidic, etc.
Carboxypeptidase A	Pancreas	C-terminal bond of Tyr, Trp, Phe, etc.	Does not act at Arg, Lys, Pro
Carboxypeptidase B	Pancreas	C-terminal Arg, Lys	None
Carboxypeptidase Y	Yeast	C-terminal bond	All residues, but Gly is slow
Leucine aminopeptidase	Kidney, intestinal mucosa, etc.	NH$_2$-terminal bond of various residues	No action at X—Pro bond†
Papain	Papaya	Arg, Lys, Gly, etc.	Wide specificity; does not act at acidic residues
Subtilisin	*Bacillus subtilis*	Aromatic and aliphatic residues	Various
Thermolysin	*B. thermoproteolyticus*	Amino-linked bonds of aliphatic residues	Ala, Phe
Staphylococcal protease	Staphylococcus	Glu	Some Asp bonds

*Except for the carboxypeptidases and thermolysin, the sites of action refer to the residues bearing the carbonyl group of the peptide bond; e.g., trypsin catalyzes hydrolysis of arginyl and lysyl bonds.
†X = any other amino acid residue.

Performic acid oxidation has the disadvantage that it destroys tryptophan and converts methionine to its sulfone. Cysteic acid, as well as *S*-carboxymethyl-cysteine, are stable throughout sequence analysis.

Partial Hydrolysis

Proteolytic enzymes catalyze limited cleavage of a polypeptide chain, producing peptides of various sizes, the exact size depending on their substrate specificities, as listed in Table 5.3, and the sequence of the polypeptide. Thus, hydrolysis by trypsin is limited to bonds formed by either lysine or arginine; by chymotrypsin to those formed primarily by aromatic residues, etc. The choice of enzymes is somewhat arbitrary, but trypsin is the enzyme most frequently used to obtain one of the two sets of overlapping peptides, since it cleaves the potentially susceptible bonds almost quantitatively and usually provides a set of peptides of sizes amenable to further sequence analysis. The second set of peptides is usually obtained with chymotrypsin or pepsin, which often give more peptides than trypsin, but usually not as complex a mixture as those obtained with papain, thermolysin, or elastase.

Various methods have been devised to extend the action of enzymes. For example, transformation of cysteine residues to *S*-aminoethylcysteine residues by treatment with ethyleneimine permits tryptic action at the carbonyl bonds of such residues.

| Cysteine residue | Ethyleneimine | *S*-Aminoethylcysteine residue |

Modification of lysine or arginine residues can limit the action of trypsin. Acylation of the ε-amino groups of lysine residues at neutral pH by treatment with maleic anhydride or citraconyl anhydride renders the maleyl or citraconyl lysine residues resistant to trypsin, limiting the proteolysis to arginine residues.

| Maleic anhydride | Lysine residue | Maleylated lysine residue |

Maleyl or citraconyl groups are readily removed at room temperature by treatment with weak acid (pH 2.5 to 3.0), and the lysine bond can then be hydrolyzed with trypsin.

Similarly, arginine residues react at pH 8 with cyclohexanedione, butanedione, or other dicarbonyl reagents, forming an adduct with the guanidine group. The action of trypsin is then restricted to hydrolysis at lysine residues.

A valuable chemical method of specific cleavage at methionine residues utilizes *cyanogen bromide* (CNBr) in acid solution.

The carboxyl-terminal residue of the peptide is now homoserine lactone. Since most proteins contain relatively few methionine residues, the method offers an excellent approach to obtaining large peptides.

Isolation of Peptides

Peptides derived by partial hydrolysis are separated by a variety of methods, but the most widely used are gel filtration, ion-exchange chromatography, and paper electrophoresis (Chap. 9). Often only a few peptides can be obtained pure by one method, and mixtures must be separated by another method. Large hydrophobic peptides are often difficult to purify, but gel filtration in aqueous organic acids, e.g., 50% acetic or formic acids, or in 2 to 8 M urea or guanidine · HCl have been used successfully.

Sequence Analysis of Peptides: End-Group Methods

Amino terminal end-group methods are used to determine the sequence of most peptides. In principle, these methods depend upon derivatizing the α-amino group of the NH_2-terminal residue with a group that resists removal under the conditions for complete hydrolysis of proteins or peptides. After derivatization and hydrolysis of the peptide, the single amino acid derivative is identified by cascade methods, such as thin layer chromatography or gas-liquid chromatography (Chap. 9).

By far the most versatile and widely used end-group method employs *phenylisothiocyanate* (Edman's reagent) for derivatization of the α-amino group as follows:

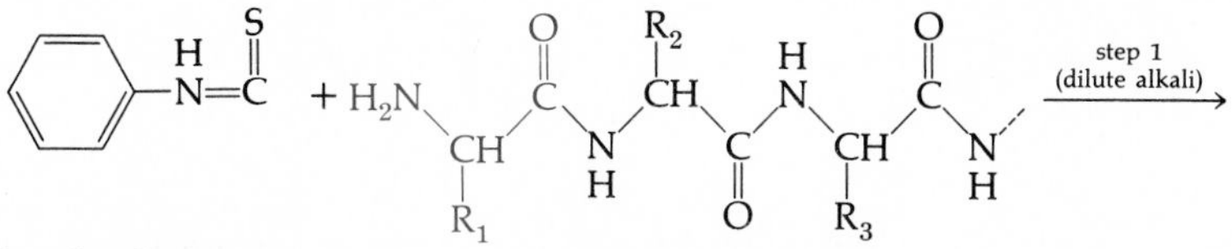

Phenylisothiocyanate   **Polypeptide (n residues)**

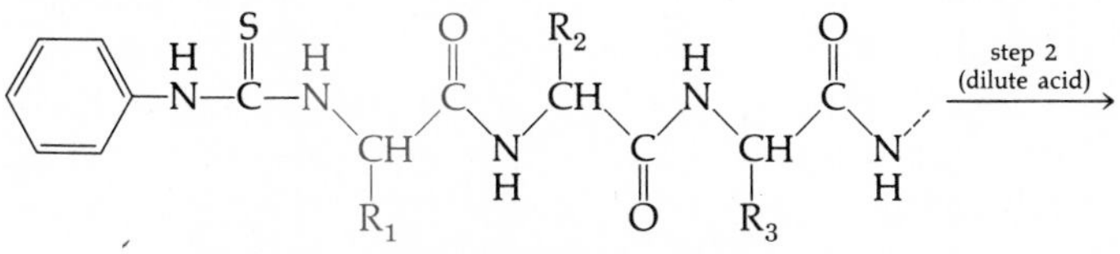

step 1
(dilute alkali)

Phenylthiocarbamoyl (PTC) peptide

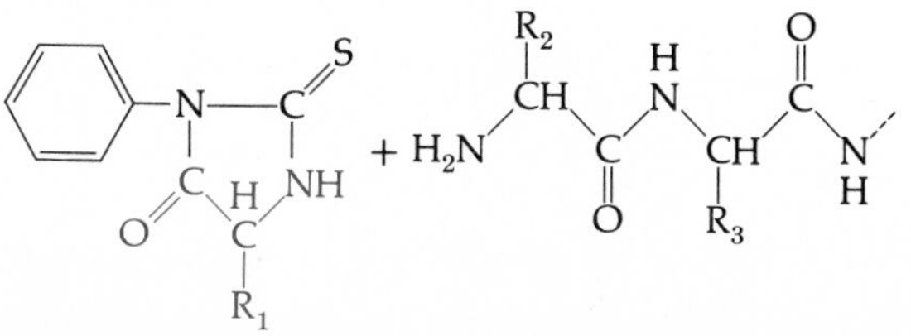

step 2
(dilute acid)

Phenylthiohydantoin   **Polypeptide (n-1 residues)**

In step 1, the peptide (or protein) reacts to form the PTC derivative, which in step 2 is cleaved to release a phenylthiohydantoin derivative of the NH_2-terminal residue. Equally significant, the intact peptide minus its original NH_2-terminal residue is also formed in step 2. Each of the phenylthiohydantoins of the 20 amino acids is readily identified. The power of the method, however, is the fact that after one cycle of steps 1 and 2, the peptide can be reacted in a second cycle, and a second residue identified. In fact, many residues may be placed in sequence by this method. Automated equipment that permits as many as 50 to 60 cycles of reaction with Edman's reagent has been developed and is used with proteins and peptides of varying lengths. Under favorable circumstances the complete sequence of peptides containing up to 50 residues can be determined.

Another sensitive method utilizes the reagent *dansyl chloride* (1-dimethyl-aminonaphthalene-5-sulfonyl chloride). The dansyl group is strongly fluorescent, and dansyl amino acids can be detected and estimated in ultramicro amounts.

Dansyl chloride   Peptide   Dansyl peptide   Dansyl amino acid

This method allows only a single residue to be identified, since hydrolysis to free the dansyl amino acid also hydrolyzes the polypeptide. However, the dansyl method is frequently used in combination with Edman degradation. As each successive residue is removed, an aliquot can be treated by the dansyl method to establish the nature of the next residue.

In addition to chemical methods for determining NH_2-terminal residues, an enzymic method can be used. The enzyme *leucine aminopeptidase* requires a terminal free α-amino group for its action. Therefore, in a peptide whose sequence is $H_2N—A \cdot B \cdot C \cdot D \cdots$, the action of the aminopeptidase will liberate the free amino acids sequentially. At any given time the amount of A liberated will be greater than that of B, and the amount of B greater than that of C, etc. Therefore, the sequence ABC can be deduced from rate measurements of amino acid release. The aminopeptidase acts rapidly on terminal leucine residues but also liberates all others found in proteins; it does not act at X—Pro bonds (Table 5.3).

Some large peptides cannot be sequenced completely by end-group methods and it may be necessary to hydrolyze them into smaller fragments more amenable to sequence analysis.

Carboxyl-terminal end-group methods are also used but are generally less satisfactory than NH_2-terminal methods. One method depends on treatment of protein or peptide with hydrazine under anhydrous conditions at $100°C$.

Cleavage of the peptide bonds by *hydrazinolysis* converts all amino acid residues to amino acid hydrazides except the carboxyl-terminal residue, which remains as the free amino acid and can be isolated and identified chromatographically.

$$\text{Protein} + n\text{H}_2\text{N---NH}_2 \longrightarrow n\text{H}_2\text{NCHRCONHNH}_2 + \text{amino acid}$$

An enzymic method involves the use of pancreatic *carboxypeptidases*. *Carboxypeptidase A* liberates only the residue from a protein or peptide that bears a free α-COOH group. As with the analogous aminopeptidase, information concerning the COOH-terminal sequence can be obtained by the rate of liberation of successive residues. Carboxypeptidase A has little or no action on COOH-terminal proline, arginine, or lysine residues. *Carboxypeptidase B*, a different enzyme, liberates COOH-terminal arginine or lysine residues (Table 5.3).

Location of Disulfide Bonds

The half-cystine residues that form disulfide bonds are identified after the complete sequence of the protein is deduced. For proteins containing only disulfide bonds, it is usually possible to hydrolyze the protein with one or more proteolytic enzymes and purify only the cystine-containing peptides. Enzymes particularly suitable for this purpose are pepsin and thermolysin, which act at acidic pH where disulfide bonds remain intact. Once a pure peptide is obtained, it is then reduced with mercaptoethanol and alkylated with iodoacetic acid; the two peptides are separated, and the composition of each peptide is determined. Because the sequence around each half-cystine residue is known, the two residues forming the disulfide bond are identified.

The Amino Acid Sequence of Porcine Pancreatic Trypsin Inhibitor I

The methods described above have been used to establish the sequences of many proteins. To illustrate their use, the proof of sequence of a protein inhibitor of pancreatic trypsin is summarized in Fig. 5.2. The protein was first reduced with mercaptoethanol and aminoethylated. The *S*-aminoethyl-protein was maleylated with maleic anhydride to block the amino groups of lysine and *S*-aminoethylcysteine and digested with trypsin, and the peptides were separated by gel filtration, as indicated. The remainder of the sequence analysis is shown, including methods of partial hydrolysis and separation procedures for peptides. The final sequence with the disulfide bonds is shown in Fig. 5.3.

Further Comments on Sequence Analysis

Additional methods are often required to determine the complete sequence of a protein. Unusual amino acids may be acid labile, e.g., γ-carboxyglutamate in the blood coagulation enzymes (Chap. M1). This amino acid decarboxylates in aqueous acid to give glutamate, but it is quite stable in alkaline hydrolysates. Some proteins contain blocked α-amino groups, most frequently an acetyl group, whereas others contain an α-keto acid, such as pyruvate, as the terminal residue. Several proteins contain NH_2-terminal pyrrolidone carbox-

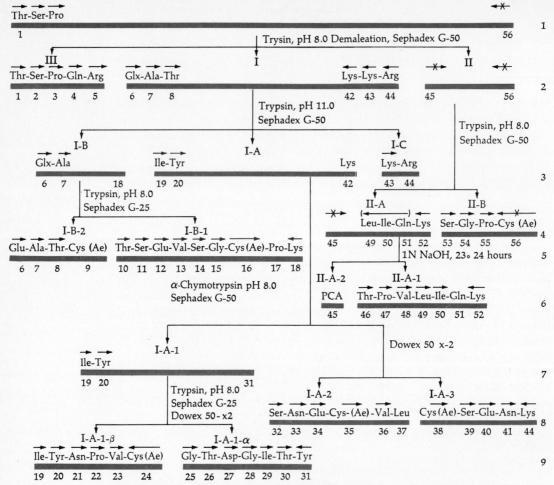

Figure 5.2 Flow diagram for the selective enzymic cleavage and amino acid sequence determination of maleyl-*S*-2-aminoethylcysteinyl derivative of porcine pancreatic secretory trypsin inhibitor I. Arrows above the amino acid residues indicate the results of Edman degradation (→) and digestion with carboxypeptidase A and B (←). Arrows with crosses (X→, ←X) show unsuccessful attempts to identify the residue by the procedure indicated by the direction of the arrow. Glx indicates either Gln or Glu; Cy(Ae) = *S*-aminoethylcysteine; PCA = pyrrolidone carboxylic acid. [*From D. C. Bartelt and L. J. Greene, J. Biol. Chem. 246:2218 (1971).*]

ylic acid, which arises from cyclization of NH_2-terminal glutamine with elimination of ammonia (as in Fig. 5.2, residue 45).

$$H_2N-\underset{O}{\overset{CH_2-CH_2}{C}}\quad \underset{NH_2}{\overset{}{CH}}-\overset{O}{C}-\overset{H}{N}- \longrightarrow O=\underset{NH}{\overset{CH_2-CH_2}{C}}\quad CH-\overset{O}{C}-\overset{H}{N}- + NH_3$$

NH_2**-terminal glutaminyl**   **Pyrrolidone carboxylyl**

Carboxyl-terminal groups are also blocked in some cases; e.g., certain polypeptide hormones contain a terminal α-carboxamide group. All proteins with

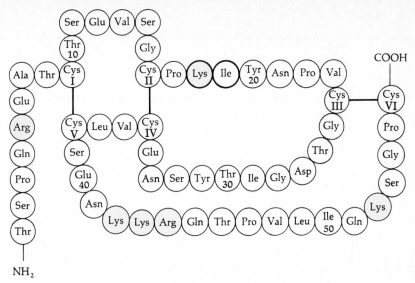

Figure 5.3 The complete amino acid sequence and location of the disulfide bonds in porcine pancreatic secretory trypsin inhibitor I. [*From O. Guy, R. Shapanka, and L. J. Greene, J. Biol. Chem.* **246:***7740 (1971).*]

blocked termini resist end-group analysis, and enzymic hydrolysates must be examined to identify the blocked residue.

Rapid, ultramicro methods have been developed recently for determining the sequences of nucleotides in DNA and RNA (Chap. 8). Since nucleotide sequences specify amino acid sequences, an amino acid sequence can be deduced from a nucleotide sequence. These methods are extremely valuable for determining protein sequences providing the appropriate polynucleotide can be obtained. However, nucleotide sequences code only for the 20 amino acids commonly encountered in proteins. Thus modified residues, disulfide bonds or other types of cross-linked residues, non-amino acid constituents, and the number and kinds of subunits in a protein can only be identified by analysis of the protein itself.

THREE-DIMENSIONAL STRUCTURE

X-ray Crystallography

X-rays have wavelengths of the order of interatomic distances; e.g., the x-ray beam of 1.542 Å, produced by electron bombardment of copper, is used for much of the work on protein structure. When x-rays strike an atom, they are diffracted (reflected) in proportion to the number of electrons in the atom. Hence atoms of higher atomic number produce more diffraction than lighter ones. A crystal may be regarded as a three-dimensional pattern of electron density which has high values near the centers of atoms and low or zero values in between.

Figure 5.4 shows a typical x-ray diffraction pattern of a single protein crystal. The spots form a regular two-dimensional lattice, and, clearly, there is a

69

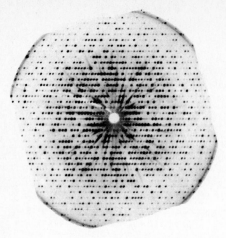

Figure 5.4 An x-ray diffraction pattern of crystalline whale myoglobin. The spots form a regular two-dimensional lattice in symmetry. Only a portion of the complete pattern is shown. [*Courtesy of Dr. John C. Kendrew.*]

marked symmetry in the pattern of spots in the four quarters of the photograph. Such a photograph has been obtained by mounting a small crystal in a known orientation in the path of a fine beam of monochromatic x-rays. The x-rays, scattered by the crystal, impinge on a photographic plate mounted behind the crystal. The picture in Fig. 5.4 is a two-dimensional lattice, since the photograph has been taken in a single plane, whereas the crystal is three-dimensional. From a series of electron-density photographs in different planes, the three dimensions of molecules can be constructed.

By measurement of a series of plates, such as shown in Fig. 5.4, with respect to both distance and intensity of the spots, it is possible to calculate electron-density maps which, at a relatively gross level, indicate the general conformation of a molecule and, at a highly refined level, may permit the assignment in space of each of the heavier atoms of a protein molecule, providing its amino acid sequence is known.

The remaining sections of this chapter consider the kinds of structural information that have been obtained from studies of protein crystals. The three-dimensional structures of more than 100 proteins have been determined by x-ray diffraction, and although the structure of a given protein is necessarily unique, most proteins contain similar kinds of regular, three-dimensional structural features. Just as the amino acids are the building blocks of a polypeptide chain, secondary and tertiary structures such as α-helices, pleated sheets, β-bends, and hydrophobic regions are the three-dimensional building blocks for folding polypeptides into biologically functional molecules. The three-dimensional structures of only a few proteins can be considered here, although others will be presented in subsequent chapters.

Geometry of the Peptide Bond

The bond distances and bond angles in amino acids and small peptides have been established with great accuracy by crystallographic analysis. Figure 5.5 shows that the length of the C—N bond in peptide linkage is shorter than that of C—N bonds in many other compounds, indicating that the peptide bond has a partial double-bond character and is not as free to rotate as the other

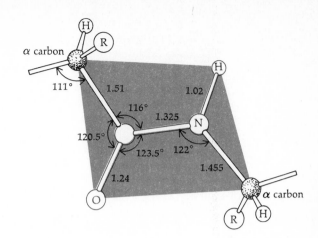

Figure 5.5 The bond distances in angstroms and bond angles of the peptide bond and its adjacent C_α atoms. The atoms within the shaded area are in the same plane. [*After R. E. Dickerson and I. Geis, The Structure and Action of Proteins, W. A. Benjamin, Menlo Park, Calif. Copyright 1969 by Dickerson and Geis.*]

Figure 5.6 The geometry of two adjacent peptide bonds and the allowed rotation of the two planes formed by the peptide bonds. ψ and ϕ angles are zero when the two (shaded rectangles) peptide bonds are coplanar. The structures on the right show the maximum forbidden overlap imposed by the allowable contact radius of unbonded atoms when $\psi = 0°$, $\phi = 180°$, $\psi = 180°$, and $\phi = 0$. [*From R. E. Dickerson and I. Geis, The Structure and Action of Proteins, W. A. Benjamin, Menlo Park, Calif. Copyright 1969 by Dickerson and Geis.*]

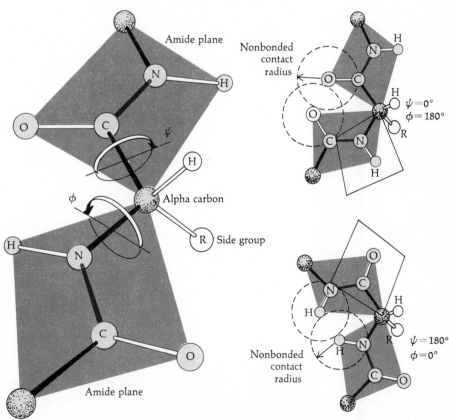

71

two types of bonds ($N-C_\alpha$ and $C-C_\alpha$) that form the polypeptide backbone. The peptide bond is in a planar, rigid arrangement of either the cis or trans configuration with respect to the α-carbons on either side of the peptide bond. The trans configuration is preferred, however, because it is less sterically hindered by R groups than the cis form. The trans configuration of a peptide bond and the bond distances and bond angles of a peptide are shown in Fig. 5.5. Because of the structural properties of the peptide bond, adjacent peptide bonds may be considered as forming a series of planes joined by the $N-C_\alpha$ and $C-C_\alpha$ bonds, as shown in Fig. 5.6. Free rotation about the $N-C_\alpha$ and $C-C_\alpha$ bonds is also constrained by the unbonded atoms. In addition, the rotation of the $N-C_\alpha$ bond limits the degree of rotation allowed around the $C-C_\alpha$ bond, and vice versa. The angle of rotation about the $C-C_\alpha$ bond is designated the ψ *angle* and that about $N-C_\alpha$ bond, the ϕ *angle* (Fig. 5.6). The actual number of ϕ and ψ angles that may exist in proteins is limited, and the secondary structure accommodates only a few types of folding. In all proteins in which ϕ and ψ angles are known, they are within the expected limited range.

Solely from the geometry of the peptide bond, helical and pleated sheet structures were predicted to occur as regular secondary structures in proteins well before they were first found by analysis of protein crystals.

Secondary Structure

The α-Helix The *right-handed α-helix* is one of the most frequently encountered secondary structures in proteins. For example, calcium-binding protein of carp muscle, whose ribbon structure model is given in Fig. 5.7, has a substantial

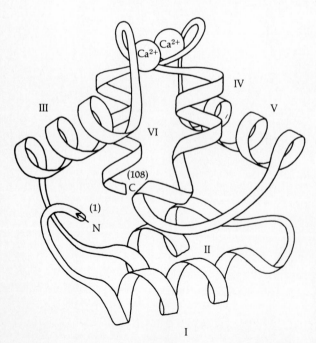

Figure 5.7 Schematic diagram of the folding of the polypeptide chain of carp muscle calcium-binding protein. The NH_2- and COOH-termini are indicated by the numerals 1 and 108, respectively. The R groups are not shown, including those that bind the two Ca^{2+} atoms. Six helical segments are present in the I–VI molecule. The details of helix III are shown in Fig. 5.8. [*Courtesy of Jane Shelby Richardson.*]

α-helical content.   There are six helical segments in the molecule, and the detailed structure of one, shown in Fig. 5.8, illustrates the basic features of a typical α-helix.   There are 3.6 residues per turn and all side chains point away from the helix.   The length per turn is 5.4 Å, or 1.5 Å per residue, and neglecting side chains, the α-helix has a diameter of about 6 Å. Except at the ends of a helix, every peptide carbonyl oxygen atom serves as a hydrogen bond acceptor for a hydrogen atom on an N—H group 4 residues away.   Such hydrogen bonds stabilize the helix.   The lengths of α-helices vary in proteins, but on the average there are about 10 residues per helical segment, the exact number depending on the amino acid sequence.   As noted below, some amino acids are not accommodated into helices, whereas others tend to lead to helix formation.

Other types of helices are found in proteins.   The most common is a variant of the α-helix called the *3_{10} helix*, which contains 3.0 residues per turn and is usually formed by shorter stretches of sequence than the α-helix.   Another important type is the *left-handed polyproline helix*, which is not found in globular proteins but forms the major secondary structure in collagens (Chap. M6).

Figure 5.8 A model of one of the helical segments (residues 40 to 51) in carp muscle calcium-binding protein.   The ribbon with the arrow follows the polypeptide backbone for one turn of the α-helix.   The bond angles and the bond distances are indicated by the sticks joining C, N, or O atoms, which are represented by the balls. Hydrogen bonds are indicated by the dotted lines and are formed between the O and N atoms indicated.   The hydrogen atoms are not shown.   The side of the helix nearest the reader is hydrophilic and exposed to solvent; the opposite side is hydrophobic and contains side chains contributing to the hydrophobic region in Fig. 5.15. [*Courtesy of Jane Shelby Richardson.*]

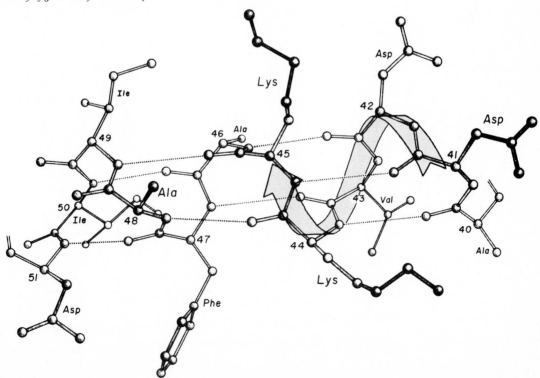

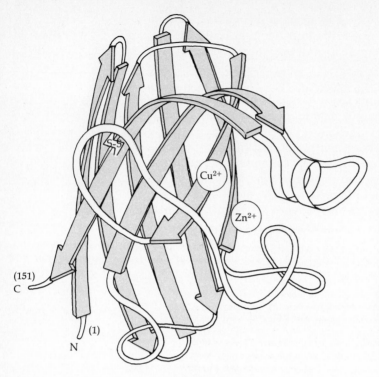

Figure 5.9 Schematic diagram of the folding of the polypeptide chain of superoxide dismutase (bovine erythrocyte), which contains 151 residues, a single disulfide bond (intrachain), and a single atom each of copper and zinc. The arrows indicate the NH_2- to COOH-terminal direction of those regions forming pleated-sheet strands. The native enzyme has two identical chains, or subunits, each with the same conformation as the one shown. Pleated sheets in globular proteins are not flat; antiparallel sheets, as in this molecule, are often arranged to form a cylinder or barrel. [*Courtesy of Jane Shelby Richardson.*]

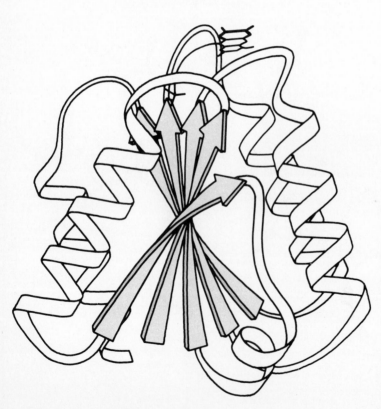

Figure 5.10 Schematic diagram of the folding of the polypeptide chain of flavodoxin (*Clostridium MP*), which contains 138 residues and a molecule of noncovalently bound flavin mononucleotide (FMN) (Chap. 16). The arrows indicate the direction of the parallel pleated-sheet strands (NH_2 to COOH terminus). Parallel pleated sheets usually do not form cylinders. [*Courtesy of Jane Shelby Richardson.*]

Pleated Sheets Two types of *pleated sheet,* or *β-structure,* occur in proteins. The ribbon models of *superoxide dismutase* (bovine erythrocyte) and *flavodoxin* in Figs. 5.9 and 5.10, respectively, illustrate the structural features of each type. Dismutase contains *antiparallel pleated sheets,* formed by extended, adjacent segments of polypeptide chain, whose sequences with respect to the direction from NH_2- to COOH-terminus run in the opposite direction. The other *β*-structure, the *parallel pleated sheet,* is found in flavodoxin and is formed by chains running in the same direction. Antiparallel sheets may be formed by a single chain folding back on itself or by two or more segments of chain from remote parts of the molecule. The sheets in each molecule are not flat, but twisted, as predicted by the ϕ and ψ angles for the peptide bond. The sheet in dismutase has a barrel-like shape, called a *β-barrel,* whereas that in flavodoxin resembles somewhat a propeller with a left-handed twist. Figures 5.11 and 5.12 show molecular models of each type of *β*-structure and illustrate how the sheets are stabilized by an extensive array of hydrogen bonds among the adjacent chains.

There are several *β*-sheet topologies shared by proteins. They differ from one another in the number and the direction (parallel or antiparallel) of the segments that form the sheet. Thus two, three, or four segments of chain can

Figure 5.11 Molecular model of an antiparallel pleated sheet in superoxide dismutase (Fig. 5.9). Bond distances and bond angles are indicated by the sticks joining C, N, and O atoms. The arrows indicate the direction (from NH_2 to COOH terminus) of the three segments of the chains that form the antiparallel sheet. The dotted lines joining the atoms indicate hydrogen bonds, which help stabilize the structure. Hydrogen atoms are not shown. One side of the sheet (nearest the reader) is hydrophilic and on the surface of the molecule, whereas the opposite side is hydrophobic and buried. [*Courtesy of Jane Shelby Richardson.*]

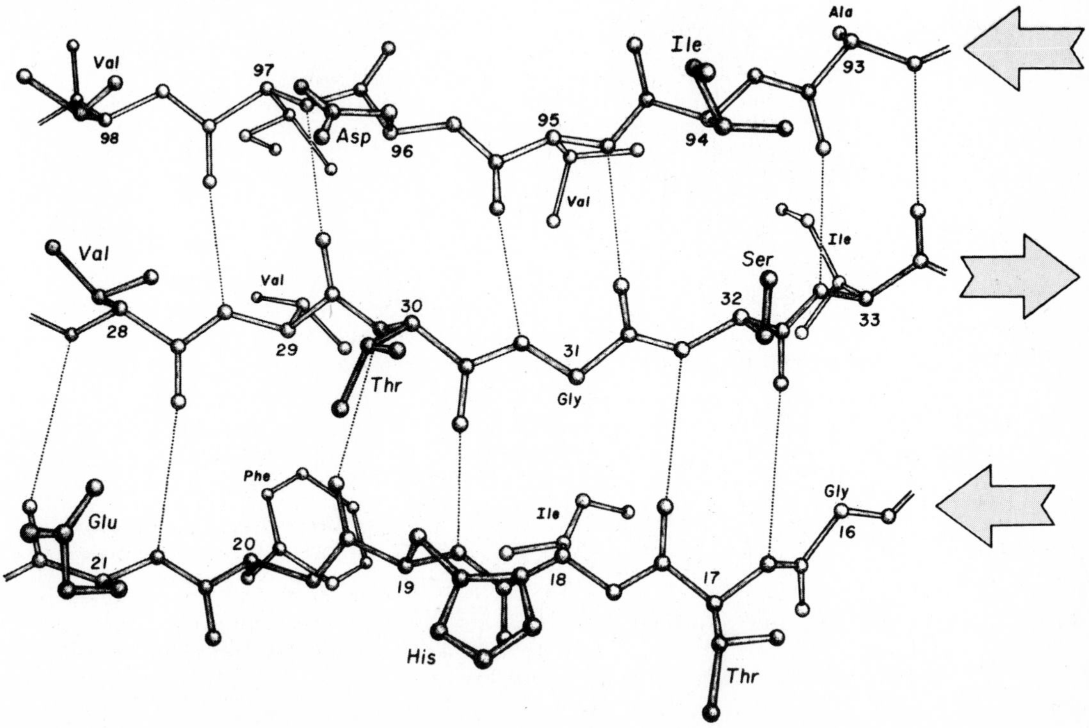

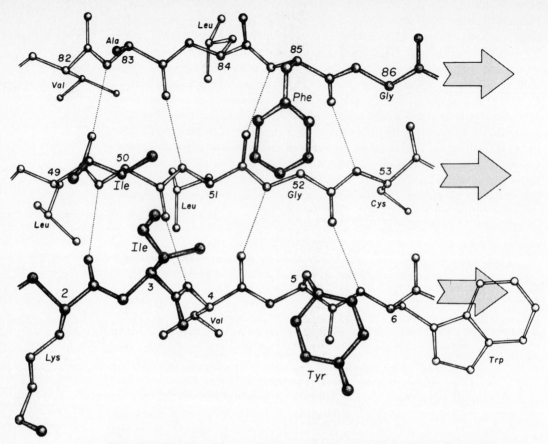

Figure 5.12 Molecular model of a parallel pleated sheet in flavodoxin (Chap. 16). The arrows indicate the direction from the NH_2- to the COOH-terminus of the three segments that form the sheet. The symbols used are the same as in Fig. 5.11. Parallel pleated sheets are typically buried in the center of the molecule and are formed by residues with hydrophobic side chains. [*Coordinates courtesy of Martha L. Ludwig and drawing courtesy of Jane Shelby Richardson.*]

form an antiparallel sheet, but five or more segments are required to form parallel sheets. Pure parallel or antiparallel sheets are favored over a mixture of the two, although some proteins, such as *carbonic anhydrase,* contain both types. In addition, a segment of chain in the NH_2-terminal region is more likely to be in the center of a sheet than one in the COOH-terminal region, consistent with the view that chains can fold before their biosynthesis is complete (Chap. 28).

β-Bends The polypeptide chains in many proteins often fold sharply back upon themselves, giving rise to secondary structures called *β-bends.* Two types of β-bends are possible, as illustrated in Fig. 5.13. Each β-bend contains four residues and is stabilized by a hydrogen bond. Many amino acids are found in β-bends, but glycine, proline, aspartic acid, asparagine, and tryptophan occur most often.

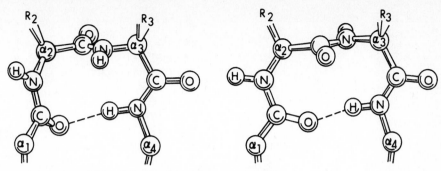

Figure 5.13 The structures of two types of β-bends in proteins. The two types differ primarily in the orientation of the peptide bond joining α_2 and α_3 and the orientation of the atoms forming the hydrogen bond (dotted line). In type II, R_3 is generally glycine. [*Courtesy of Jane Shelby Richardson.*]

Secondary Structure in Some Proteins Table 5.4 lists the secondary structural content of typical globular proteins as determined by crystallographic methods. Note that some proteins contain none or only small amounts of one of the structures whereas other proteins contain each of the three. At least 50 percent of the residues in a polypeptide chain of a globular protein are usually folded into secondary structures, although some have considerably more, e.g., myoglobin. Those segments of a chain that are not contained in a regular secondary structure may make up only a small portion of a molecule, but such segments may contribute a great deal to the tertiary structure through noncovalent interactions. Indeed, the small protein *ferredoxin* from *Pseudomonas aerugenosa* has no recognizable secondary structure.

Prediction of Secondary Structure It is presently impossible to predict the conformation of a protein from its amino acid sequence; however, the amounts of α-helix and β-structure can often be predicted from amino acid sequences. The rules for predicting such structures were derived empirically by determining

	Helix		β-Structure		β-Bends	TABLE 5.4
Protein	%	No. of helices	%	No. of segments	No. of bends	Secondary Structures in Some Globular Proteins
Carbonic anhydrase	20	7	37	13	4	
Carboxypeptidase A	38	9	17	8	13	
Chymotrypsin	14	3	45	12	17	
Cytochrome c	39	5	0	0	6	
Insulin	52	3	6	1	0	
Lysozyme	40	6	12	6	6	
Myoglobin	79	8	0	0	6	
Ribonuclease	26	3	35	6	2	
Thermolysin	38	7	22	10	0	

From A. Liljas and M. G. Rossmann, *Annu. Rev. Biochem.* **43**:480 (1974).

TABLE 5.5

Classification of
Amino Acids in
Formation of
α-Helical or
β-Sheet Regions

	α-Helix*		β-Structure*
Hα	Glu, Ala, Leu, Met	Hβ	Tyr, Val, Ile
hα	Ile, Lys, Gln, Trp, Val, Phe	hβ	Cys, Met, Phe, Gln, Leu, Thr, Trp
Iα	His, Asp		
iα	Thr, Ser, Arg, Cys	iβ	Arg, His, Asn, Ala
bα	Asn, Tyr	bβ	Ser, Gly, Lys
Bα	Pro, Gly	Bβ	Glu, Pro, Asp

*H = strong former; h = former; I = weak former; i = indifferent; b = weak breaker; B = strong breaker. From P. Y. Chou and G. D. Fasman, *Adv. Enzymol.* **47**:45 (1978).

the location of amino acids in the crystal structures of many different proteins and estimating the frequency of occurrence of each amino acid in an α-helix or a β-structure. On this basis each amino acid can be classified as a *former*, a *breaker*, or *indifferent* for α-helical and β-sheet regions, as listed in Table 5.5. β-Bends may also be predicted in a similar fashion. Thus the following sequence in thermolysin from residues 137 to 152 was predicted to be α-helical and not part of a β-structure.

```
136          140                145                150                155
Gly-Ile -Asp-Val-Val-Ala-His-Glu-Leu-Thr-His-Glu-Leu-Thr-His-Ala-Val-Thr-Asp-Tyr
Bα  hα  Iα  hα  hα  Hα  Iα  Hα  Hα  iα  Iα  Hα  Hα  iα  Iα  Hα  hα  iα  Iα  bα
bβ  Hβ  Bβ  Hβ  Hβ  iβ  iβ  Bβ  hβ  hβ  iβ  Bβ  hβ  hβ  iβ  iβ  Hβ  hβ  Bβ  Hβ
```

X-ray crystallography revealed that residues 137 to 155 are indeed in an α-helix. Six other helical structures were predicted correctly in thermolysin as well as six of the seven segments of chain that form β-structures. Application of these rules may provide useful information when used in concert with methods that give less exact structural information than crystallography (Chap. 4).

Tertiary Structure

The noncovalent interactions between helixes and β-structures in proteins, together with the side-chain and backbone interactions unique to a given protein, provide the totality of the tertiary structure. Essential for the tertiary structures are the hydrophobic regions formed by nonpolar R groups. One such region of hydrophobic bonding created by several R groups in the calcium-binding protein (Fig. 5.7) is shown in Fig. 5.14. The R groups forming such a region are often contributed by residues remote from one another in sequence.

The driving force generating the tertiary structure of a protein is partly hydrophobic and is similar to micelle or bilayer formation by amphiphiles (Chap. 7); 25 to 30 percent of the residues in *globular proteins* have significantly hydrophobic side chains, and 45 to 50 percent have ionic or polar side chains. The hydrophobic R groups tend to disrupt water structure far less when the protein is in its native state than in the completely unfolded state; thus the native state is thermodynamically the more favored structure. However, the native tertiary structure is in dynamic reversible equilibrium with other possible

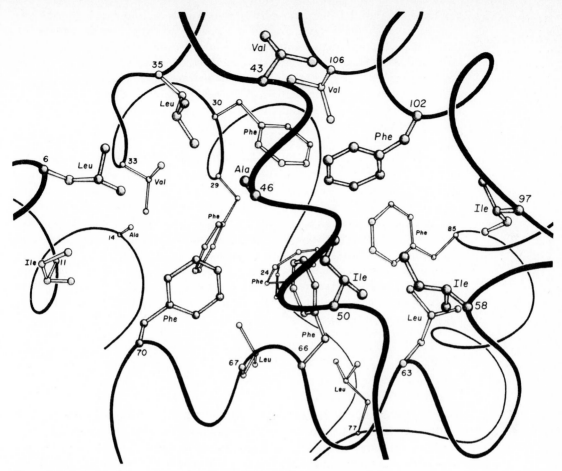

Figure 5.14 A model of the amino acid residues and the folding of segments of the polypeptide chain that form a hydrophobic region of the carp muscle calcium-binding protein (Fig. 5.8). The polypeptide backbone is schematic, and only R groups participating in hydrophobic interactions are shown. The R groups in hydrophobic regions of globular proteins are in close contact and have a strict orientation with respect to one another [*Courtesy of Jane Shelby Richardson.*]

conformations, depending on the pH, the composition, and the temperature of the aqueous medium. Because the hydrophobic residues are interspersed with ionic or polar residues in the sequence of a protein, formation of hydrophobic interactions results in the incorporation of some polar peptide bonds into the interior of the native structure. Even so, the hydrophobic interactions, together with hydrogen bonds between R groups and the secondary structures, must aid in nullifying the affinity for water of peptide bonds that are also in the interior.

An equally strong driving force in the attainment of the native conformation is provided by the ionic R groups, especially those from residues of aspartic acid, glutamic acid, arginine, and lysine, whose R groups require considerable energy to be buried. As in micelle formation, these ionic groups must remain largely exposed to the aqueous medium. Since ionic and nonpolar side chains are not separated in sequence from each other in globular, water-soluble proteins, it is likely that there may be a limitation to the number of possible sequences that can form stable conformations when the various amino acids are in their normal proportions. Indeed, an arbitrary sequence of amino acids may not form globular structures, and the conformations observed in native proteins may be the result of an evolutionary selection process that has preserved those sequences which allow formation of internally placed hydrophobic domains stabilized by hydrogen bonding and the hydrophilicity of ionic groups on the surface of the molecule. Thus, formation of the specific native globular structure characteristic of a given protein is a cooperative process involving different types of noncovalent interactions. Disulfide bonds are not a major force leading to correct folding but undoubtedly stabilize the conformation after folding occurs. These bonds form spontaneously when the appropriate thiol groups are brought into juxtaposition as the result of the cooperative interactions of the R groups that lead to correct folding. This has been demonstrated for several proteins, including the enzymes ribonuclease and lysozyme, which contain four disulfide bonds each. When these enzymes are fully reduced and dissolved in 8 M urea, they lose all enzymic activity and assume a random form. There are $7 \times 5 \times 3 = 105$ theoretically possible arrangements of the disulfide bonds in each of these enzymes, but when the urea is removed by dialysis, only the four correct bonds form as the molecule refolds in the presence of oxygen and the original enzymic activity is regained.

Although globular proteins generally possess a hydrophobic interior and a hydrophilic surface, some hydrophobic regions are at or near the surface of proteins. For example, many proteins interact with hydrophobic molecules, e.g., enzymes whose substrates are lipids. Indeed, hydrophobic bonds may contribute to the binding of a substrate with an enzyme. Undoubtedly the tertiary structure of such proteins stabilizes the inherently unstable patches of hydrophobic surface groups.

There are often regular tertiary structures in proteins. For example, superoxide dismutase (Fig. 5.9) contains a so-called *β-barrel,* a barrel-like structure formed by the pleated sheets in the molecule. A remarkably similar structure is found in the immunoglobulins, indicating that such structures can occur in molecules with very different functions. Many proteins are divided into domains; i.e., a single polypeptide chain may fold in such a way that two or more globular domains with similar tertiary structures are formed. The immunoglobulins (Chap. M2) are the most striking examples of proteins with such domains. Other proteins, such as NAD-dependent dehydrogenases, also have domains (Chap. 16).

Quaternary Structure

When subunits are present in a protein, the amino acid sequences of the subunits must determine the quaternary structure. This is evident from crystal structural analysis as well as from the fact that the dissociated subunits can recombine to give a functionally competent native protein. Hemoglobin is one such molecule. At acid pH values in acetone, the heme is separated from each of the four unfolded subunits of globin. Nevertheless, at neutral pH values, native hemoglobin will be reformed from heme and globin. Thus, the complex

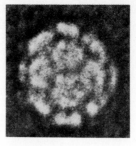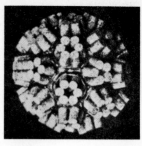

Figure 5.15 Electron micrograph of turnip yellow mosaic virus (a) and model of the virus deduced from electron microscopy (b). [From J. T. Finch, in H. Neurath and R. L. Hill (eds.): The Proteins, vol. I, Academic, New York, 1975, p. 479.]

(a)   (b)

native conformation is reattained with a heme group in the proper position in each of the four chains. Thus, not only the secondary and tertiary structures but also the quaternary structure are determined by the specific amino acid sequences of the polypeptide chains.

Most large proteins, including many enzymes, can be dissociated into inactive subunits by various agents. Thus, although the active protein molecule may be very large, on the order of several hundred thousand to millions in molecular weight, the primary peptide chains are usually much smaller (see Table 5.2). The forces involved in forming the native proteins by an aggregation of smaller units are the same as those involved in folding the individual peptide chain into its proper conformation.

Electron microscopy provides a useful means for examining the quaternary structure of proteins, particularly large molecules with many subunits. The main limitations of the method are the relative transparency of proteins to electrons and the disruption of structure when proteins are dried in a vacuum in the microscope and bombarded by an electron beam. Despite these problems, structural detail can usually be obtained to 20-Å resolution and, if care is taken, to about 10 Å.

Because of their size, viruses are particularly amenable to study by electron microscopy, and considerable structural detail is revealed. Figure 5.15 shows an electron micrograph of turnip yellow mosaic virus and a model of the virus deduced from detailed electron microscopy. The model depicts 32 protein structural units on the surface of the virus, 12 of which are rings containing five protein subunits and 20 rings with six subunits. There are a total of 180 subunits, each with a molecular weight of 21,000 (Table 5.2).

REFERENCES

Books

Anfinsen, C. B., Jr., M. L. Anson, J. T. Edsall, and F. M. Richardson (eds.): *Advances in Protein Chemistry,* Academic, New York. (An annual series beginning with vol. I in 1944.)

Dayhoff, M. (ed.): *Atlas of Protein Sequence and Structure,* vol. 5, suppls. 1 and 2, National Biochemical Research Foundation, 1972. (A compilation of protein and nucleic acid sequences published through early 1976.)

Dickerson, R. E., and I. Geis: *The Structure and Action of Proteins,* W. A. Benjamin, Menlo Park, Calif., 1969.

Fasman, G. D. (ed.): *Handbook of Biochemistry and Molecular Biology: Proteins,* 3d ed., 3 vols., The Chemical Rubber Co., Cleveland, Ohio, 1976.

Hirs, C. H. W. (ed.): Enzyme Structure, in S. P. Colowick and N. O. Kaplan (eds.): *Methods in Enzymology,* vol. XI, Academic, New York, 1967.

Hirs, C. H. W., and S. N. Timasheff (eds.): *Methods in Enzymology,* vol. XXV, pt. B, Academic, New York, 1972.

Means, G. E., and R. E. Feeney: *Chemical Modification of Proteins,* Holden-Day, San Francisco, 1971.

Neurath, H., and R. L. Hill (eds.): *The Proteins,* 3d ed., vols. I–V, Academic, New York, 1975–1982.

Schroeder, W. E.: *The Primary Structure of Proteins,* Harper & Row, New York, 1968.

Review Articles

Chou, P. Y., and G. D. Fasman: Prediction of the Secondary Structure of Proteins from Their Amino Acid Sequence, *Adv. Enzymol.* **47:**45–148 (1978).

Kendrew, J. C.: The Three-Dimensional Structure of a Protein Molecule, *Sci. Am.* **205:**96–110 (1961).

Phillips, D. C.: The Three-Dimensional Structure of an Enzyme Molecule, *Sci. Am.* **215:**78–90 (1966).

Richardson, J. J.: The Anatomy and Taxonomy of Protein Structure, *Adv. Protein Chem.* **34:**168–339 (1981).

Chapter 6

The Carbohydrates

GENERAL BIOLOGICAL ROLES

Carbohydrates, the simplest of which are called *sugars,* participate in virtually all aspects of cellular life. By degradative processes they yield the energy that maintains life. As the primary products of plant photosynthesis, carbohydrates are the metabolic precursors of all other organic compounds. Carbohydrates may also be covalently bound to protein, and they can account for up to 90 percent of the weight of such substances, called *glycoproteins.* Carbohydrates combined with lipid, designated *glycolipids,* are constituents of biological membranes. Glycolipids and glycoproteins with their component oligosaccharides are collectively referred to as *glycoconjugates.* Many glycoconjugates have structural roles, including the proteoglycans of the extracellular matrix in animals, the celluloses of plants, and the peptidoglycans of bacterial cell walls. Glycoconjugates coat cell surfaces, and their carbohydrate groups may participate in a variety of cellular phenomena, including growth, adhesion, transformation, fertilization, and endocytosis.

CLASSIFICATION

Carbohydrates are polyhydroxyaldehydes or polyhydroxyketones that are classified into four groups: *monosaccharides, derived monosaccharides, oligosaccharides,* and *polysaccharides.* Monosaccharides have low molecular weights and, as the name *carbohydrate* implies, may be considered as hydrates of carbon with the general formula $(CH_2O)_n$; as found in nature, n is 3 to 9. Derived monosaccha-

Any cross-references coded M refer to the companion text, *Principles of Biochemistry: Mammalian Biochemistry.*

rides are derivatives of monosaccharides that contain functional groups other than the carbonyl and the hydroxyl groups. The monosaccharides and derived monosaccharides are the simplest forms of carbohydrate and are the repeating units (Chap. 2) for the oligo- (Gk. *oligo,* "few") and polysaccharides (Gk. *poly,* "many"). The simplest oligosaccharide, a *disaccharide,* is formed from two monosaccharides combined through a glycosidic linkage; those formed by three, four, or five monosaccharides are called *tri-, tetra-,* and *pentasaccharides,* respectively. Each individual monosaccharide in an oligosaccharide is called a *residue.* Polysaccharides are macromolecular polymers with molecular weights of up to 1 million; some contain monosaccharide residues of only a single type (*homopolysaccharides*) while others consist of two or more different types (*heteropolysaccharides*).

MONOSACCHARIDES

Glyceraldehyde and dihydroxyacetone, the only possible monosaccharides containing three carbon atoms, are called *trioses.* Noteworthy is the single asymmetric center in glyceraldehyde, giving rise to two different configurational forms, or *enantiomers,* D-glyceraldehyde and L-glyceraldehyde.

$$
\begin{array}{ccc}
\text{CHO} & \text{CHO} & \text{CH}_2\text{OH} \\
\text{HCOH} & \text{HOCH} & \text{C}=\text{O} \\
\text{CH}_2\text{OH} & \text{CH}_2\text{OH} & \text{CH}_2\text{OH}
\end{array}
$$

D-Glyceraldehyde   L-Glyceraldehyde   Dihydroxyacetone

Monosaccharides containing a free aldehyde function are called *aldoses;* those with a ketonic group are *ketoses.* Glyceraldehyde (D or L) is an *aldotriose,* and dihydroxyacetone is a *ketotriose,* names that not only distinguish important functional groups but also the number of carbon atoms. Generic names for aldoses designate the number of carbon atoms in the molecule; thus, tetroses, pentoses, hexoses, and heptoses contain four, five, six, and seven carbon atoms, respectively. Generic names for the ketoses are formed by insertion of *ul* in the names for the corresponding aldoses, e.g., pentulose, hexulose, and heptulose.

Designation of Configuration

Aldoses with three or more carbon atoms and ketoses with four or more carbon atoms contain asymmetric centers formed by carbon atoms with four different substituents. Thus, the nomenclature for monosaccharides is based upon the configuration about each center of asymmetry in the molecule. Figures 6.1 and 6.2 show the linear structural formulas for the D-aldoses and the D-2-ketoses and give the generally accepted trivial names for each compound. In each formula, carbon atom 1 is at the top, and the other atoms are numbered successively. These structures are called *Fischer projections* because they were first introduced by Emil Fischer many years ago. Note that this use of D and L is based upon the configurational differences between monosaccharides and not on optical rotational properties. The Roman small capitals D and L are used to designate configuration, but if the sign of rotation of a specific monosaccharide is to be included in naming the compound, it is designated by the italic letters

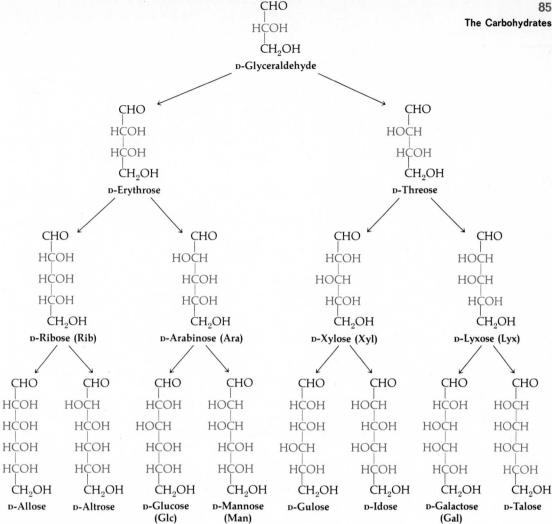

Figure 6.1 Relationships of the D-aldoses. The formulas of the L-aldoses are in each case the mirror images of the structures in the figure. The symbols in parentheses are the three-letter abbreviations for the sugars indicated and are used to write structures of oligosaccharides.

d and *l* or by (+) and (−); thus D-glucose, which is dextrorotatory, and D-fructose, which is levorotatory, may be written

CHO	1	CH₂OH
HCOH	2	C=O
HOCH	3	HOCH
HCOH	4	HCOH
HCOH	5	HCOH
CH₂OH	6	CH₂OH
d(+)-Glucose		*l*(−)-Fructose

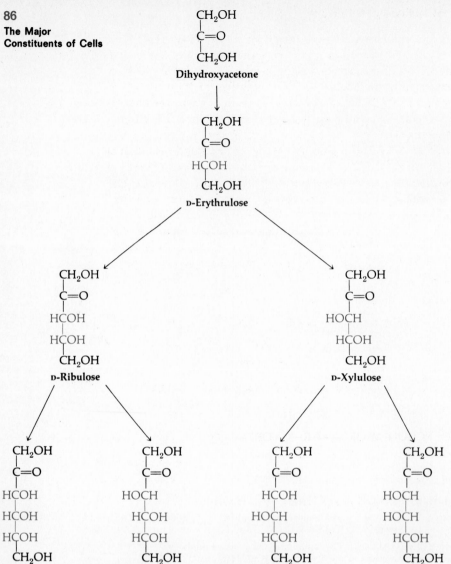

Figure 6.2 Relationships of the D-ketoses. The formulas of the L-ketoses are in each case the mirror images of the structures in the figure.

By convention, the configurational designations D and L for any monosaccharide refer to that center of asymmetry most remote from the carbonyl group of the molecule, which for the aldohexoses is carbon atom 5, and for the 2-keto-hexoses is carbon atom 4. The absolute configurations apply to the two glyceraldehydes, and all sugars terminating in these configurations are said to belong to either the D or L configurational series. Thus, the structures for the tetroses, pentoses, and hexoses shown in Fig. 6.1 can be constructed from D-glyceralde-hyde by adding one carbon atom at a time in either of two configurations, as indicated by the arrows.

The linear structures shown in Figs. 6.1 and 6.2 are a convenient means for distinguishing the different stereoisomers of the monosaccharides but do not show an additional center of asymmetry present in many sugars. The discovery that an additional center of asymmetry occurs in glucose came from the observation that the optical rotation of freshly prepared solutions of glucose changes with time. Two forms of D-glucose may be isolated with different specific optical rotations; one form, with $[\alpha]_D^{20} = +112.2°$, is designated α-D-glucose, and the other, with $[\alpha]_D^{20} = +18.7°$, β-D-glucose. However, the specific rotation of freshly prepared solutions of either the α or β form changes to a value of $+52.7°$. This phenomenon, termed *mutarotation,* is not unique to glucose and has been observed with a variety of hexoses and pentoses as well as certain disaccharides.

The explanation for mutarotation is that monosaccharides, like other aldehydes and ketones, readily form hemiacetals and acetals or hemiketals or ketals, respectively. When an aldehyde reacts with one equivalent of an alcohol, a hemiacetal is formed.

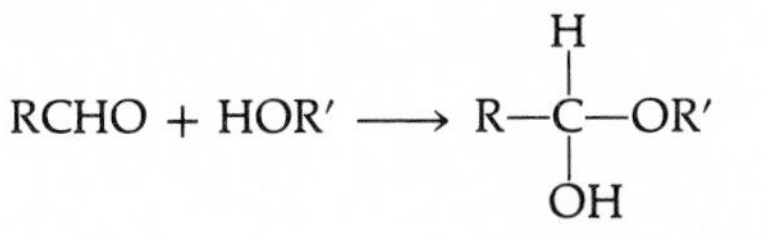

$$RCHO + HOR' \longrightarrow R-\overset{\overset{\displaystyle H}{|}}{\underset{\underset{\displaystyle OH}{|}}{C}}-OR'$$

Hemiacetal

On reaction with two equivalents of an alcohol, aldehydes form acetals.

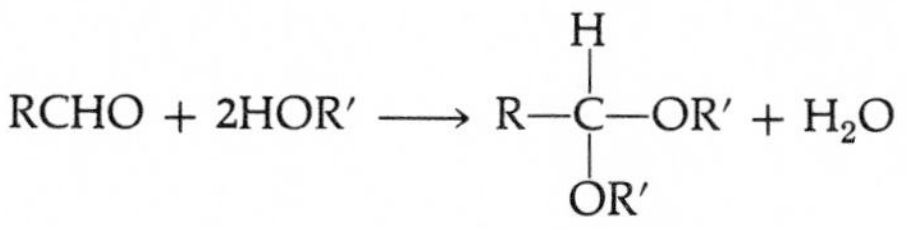

$$RCHO + 2HOR' \longrightarrow R-\overset{\overset{\displaystyle H}{|}}{\underset{\underset{\displaystyle OR'}{|}}{C}}-OR' + H_2O$$

Acetal

Similarly, ketones form hemiketals and ketals.

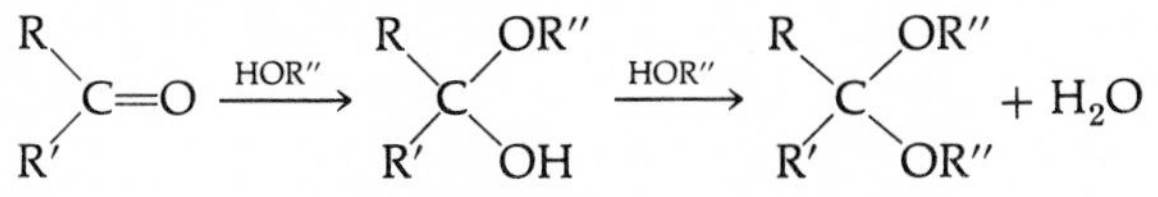

Hemiketal   **Ketal**

Hemiacetals have a center of asymmetry on what was formerly the aldehydic carbon atom, as do hemiketals formed from unsymmetrical ketones, e.g., methyl ethyl ketone. Acetals and ketals would also have an asymmetric center if two different alcohols had reacted with the carbonyl functions. The aldehydic group in glucose forms hemiacetals, not only through intermolecular reactions with alcohols, as shown above, but also through intramolecular reactions with hydroxyl groups in glucose itself. Thus, glucose forms stable internal hemiacetals with the hydroxyl group on carbon atom 5 to give the two different stereoisomers that undergo mutarotation, α-D-glucose and β-D-glucose. Like other hemiacetals, the α and β forms of glucose possess a center of asymmetry on the original aldehydic carbon atom.

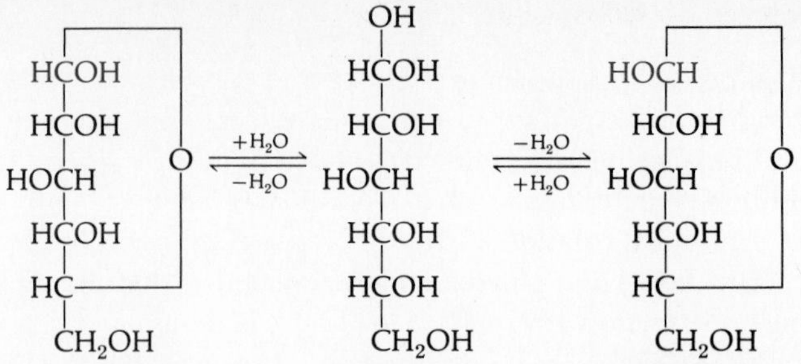

$$
\begin{array}{ccccc}
 & & \text{OH} & & \\
\text{HCOH} & & \text{HCOH} & & \text{HOCH} \\
\text{HCOH} & & \text{HCOH} & & \text{HCOH} \\
\text{HOCH} \quad \text{O} & \underset{-H_2O}{\overset{+H_2O}{\rightleftharpoons}} & \text{HOCH} & \underset{+H_2O}{\overset{-H_2O}{\rightleftharpoons}} & \text{HOCH} \quad \text{O} \\
\text{HCOH} & & \text{HCOH} & & \text{HCOH} \\
\text{HC}\!\!- & & \text{HCOH} & & \text{HC}\!\!- \\
\text{CH}_2\text{OH} & & \text{CH}_2\text{OH} & & \text{CH}_2\text{OH} \\
\alpha\text{-D-Glucose} & & \text{D-Glucose} & & \beta\text{-D-Glucose} \\
[\alpha]_D^{20} = +112.2^\circ & & \text{monohydrate} & & [\alpha]_D^{20} = +18.7^\circ
\end{array}
$$

Carbon 1, the new center of asymmetry, is the *anomeric carbon*, and aldoses that differ only in configuration at the anomeric carbon atom are called *anomers*. Thus α- and β-D-glucose are anomers.

It is supposed that mutarotation involves mutual interconversions of α and β forms through the intermediate formation of the open-chain aldehyde or its hydrate. In aqueous solution the open-chain aldehyde is a very minor constituent of the equilibrium mixture; for glucose, it is approximately 0.024 percent of the total.

The number of stereoisomers of any compound is 2^n, where n is the number of asymmetric centers in the molecule. Thus, in the Fischer projection formulas (Fig. 6.1) glucose has $2^4 = 16$ stereoisomers. In the above formulation there are 5 centers of asymmetry, and hence there are $2^5 = 32$ possible isomers, permitting an α and β modification of each aldohexose. The α designation is used to indicate that in plane projection, the hydroxyl group on C-1 is on the same side of the structure as the ring oxygen; the β modification refers to the form in which the C-1 hydroxyl group is on the side of the structure opposite the ring oxygen. Among sugars of the D configuration, the α isomer always has a more positive specific optical rotation than that of the β isomer. In the L series, the mirror images of the D structures shown above, the anomer with the more negative rotation is the α anomer.

Glucose also forms acetals when treated with mineral acid in the presence of methanol, and the resulting derivatives are designated as methyl glucosides, with either the α or β configuration retained on C-1.

It is apparent that α- and β-D-glucose as well as their corresponding methyl glucosides possess a six-membered ring formed on joining the first and fifth carbon atoms by an oxygen bridge. Sugars containing such a ring are termed *pyranoses* because of their resemblance to the pyran ring. Aldoses containing this type of ring are termed *aldopyranoses*. α-D-Glucose and β-D-glucose may thus be designated as α-D-*glucopyranose* and β-D-*glucopyranose*, respectively, and their methyl glucosides are designated *methyl-α-D-glucopyranoside* and *methyl-β-D-gluco-pyranoside*.

The methyl glucosides are members of a group of compounds known as *glycosides*. Glycosides are named in accordance with the monosaccharides from which they are derived, e.g., α- and β-glucosides, α- and β-fructosides, and α- and β-ribosides. Biologically important glycosides are formed by reaction of the hemiacetal or hemiketal hydroxyl group of monosaccharides with a hydroxyl group of another monosaccharide, and sugars combined in this manner

HCOCH₃ structure → Methyl-α-D-glucoside

$$\begin{array}{c} \text{HCOCH}_3 \\ \text{HCOH} \\ \text{HOCH} \\ \text{HCOH} \\ \text{HC} \\ \text{CH}_2\text{OH} \end{array}\Bigg] O$$

$$\begin{array}{c} \text{H}_3\text{COCH} \\ \text{HCOH} \\ \text{HOCH} \\ \text{HCOH} \\ \text{HC} \\ \text{CH}_2\text{OH} \end{array}\Bigg] O$$

$$\begin{array}{c} \text{CH} \\ \text{CH} \\ \text{CH}_2 \\ \text{CH} \\ \text{HC} \end{array}\Bigg] O$$

Methyl-α-D-glucoside $[\alpha]_D^{20} = +158.9°$

Methyl-β-D-glucoside $[\alpha]_D^{20} = -34.2°$

Pyran

form many types of oligo- and polysaccharides. It should be noted that the reactivity of the hemiacetal hydroxyl group is quite different from that of an alcohol, and the glycosidic oxygen bridge, though bearing a superficial resemblance to an ether bridge, has none of the chemical stability found in aliphatic ethers.

Aldoses may also exist largely as a stable five-membered ring similar to that of furan and are commonly referred to as *aldofuranoses*. Five-membered rings are common among aldopentoses as they occur in oligosaccharides. Thus, β-D-arabinose may be named β-D-*arabinofuranose* and β-D-ribose, β-D-*ribofuranose*. Ketohexoses also form stable five-membered rings; thus β-D-fructose is β-D-*fructofuranose*.

$$\begin{array}{c} \text{HOC}-\text{CH}_2\text{OH} \\ \text{HOCH} \\ \text{HCOH} \\ \text{HC} \\ \text{CH}_2\text{OH} \end{array}\Bigg] O$$

$$\begin{array}{c} \text{HOCH} \\ \text{HOCH} \\ \text{HCOH} \\ \text{HC} \\ \text{CH}_2\text{OH} \end{array}\Bigg] O$$

$$\begin{array}{c} \text{CH} \\ \text{CH} \\ \text{CH} \\ \text{HC} \end{array}\Bigg] O$$

β-D-Fructofuranose

β-D-Arabinofuranose

Furan

These hemiketals also form ketals on reaction of the hemiketal hydroxyl with an alcohol and form the corresponding *furanosides*, whose structures are analogous to the pyranosides.

Inasmuch as pentoses as well as longer aldoses and ketoses may form furanose or pyranose rings, the most stable ring form depends not only on the configuration about each asymmetric center but also on the nature of the substituents on the carbonyl and hydroxyl groups. Structural analysis is required to establish the exact ring structures formed by any given aldose or ketose.

To depict the cyclic structures of the pyranoses and the furanoses, they are often written as hexagons or pentagons, a convention called *Haworth projections*. The plane of the ring is shown as perpendicular to the plane of the paper; this is emphasized by shading the bonds that are nearer to the reader.

or more simply

In this convention the two forms of D-glucopyranose are

α-D-Glucopyranose   β-D-Glucopyranose

Similarly, the furanoses are represented as

β-D-Fructofuranose   β-D-Arabinofuranose

The transition from one convention to the other relates the left and right sides of the carbon chain structure to the upper and lower aspects, respectively, of the plane of the ring formulation. Exceptions do occur, as will be seen by comparing the arrangement about C-5 of glucose in the two conventions. In the linear formulation the hydrogen at this position is to the left of the carbon, whereas in the hexagonal formulation it is below the plane of the ring. The reader is referred to texts on carbohydrate chemistry for a detailed discussion of this transition.

Conformation

The Fischer and Haworth projections fail to depict accurately the bond distances and bond angles of the ring atoms and the substituents on each carbon; i.e., they do not indicate the conformation, or three-dimensional structure, of monosaccharides. Structural analysis of hexoses reveals that they may have two different types of conformation, called the *chair* and *boat* forms, similar to those of cyclohexane, in which the valence-bond angles are completely unstrained.

Chair   *Boat*

Cyclohexane

In both forms of cyclohexane, at each carbon atom, one hydrogen is directed upward, and the other is directed downward. The six hydrogens that are up are represented by solid lines; those which are down are represented by dotted lines. The 12 hydrogen atoms of cyclohexane also fall into two classes: those on C—H bonds parallel to the axis of symmetry of the ring are *axial,* and those parallel to the nonadjacent side of the ring are *equatorial.* It is apparent that the boat and chair forms will have different properties because of the differences in possible interactions between the various groups on the molecule. In the boat form, the axial and equatorial substituents on the ring are as close together as possible (*eclipsed*), and there is maximal interaction between them. In contrast, the ring substituents in the chair form are arranged in space (*gauche*) so that minimal interactions occur between them. Thus, on this basis the chair form is usually much more stable than the boat form.

The conformational method of representation is far more than an exercise in projectional geometry. It reveals properties of the molecule that are not at once apparent from other representations of structure. A generalization of importance is that a substituent, particularly a large substituent, is at a lower energy state in the equatorial than in the axial location. Therefore, for example, in the equilibrium between the conformational isomers of methylcyclohexane,

Methylcyclohexane

the equatorial form preponderates.

The presence of an oxygen atom in the pyranoses does not markedly strain the ring, and monosaccharides also exist in chair and boat forms. α- and β-D-Glucopyranose exist in two chair forms. One chair form has the substituents on carbon atoms 2 to 4 in equatorial orientation and is designated the *normal conformer* (C1).

(*normal conformer*, C1)   (*alternate conformer*, 1C)

α-D-**Glucose**

The other chair form is an *alternate conformer* (1C) in which the hydroxyl groups on carbon atoms 2 to 4 are in the axial orientation. The alternate conformer for glucose is much less stable than the normal form because the accumulation of axially oriented polar substituents on the same side of the ring gives it much less stability than the normal conformer.

The furanose ring is also not planar and can exist in either an *envelope* (E) or *twist* (T) conformation. In the envelope conformation, four atoms, including the ring oxygen, are coplanar with the fifth atom out of the plane. In the twist conformation three adjacent atoms are coplanar, and the other two lie above or below the plane of the ring.

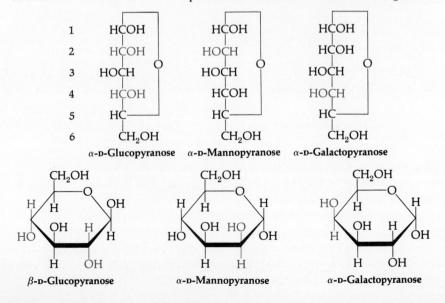

**Envelope conformation
of β-D-fructofuranose**

**Twist conformation
of methyl-β-D-lyxofuranoside**

BIOLOGICALLY IMPORTANT MONOSACCHARIDES

Only 10 monosaccharides are found in the oligosaccharide groups of glycoconjugates in higher eukaryotic organisms and include three neutral hexoses (glucose, galactose, and mannose) (Fig. 6.1); three amino sugars (N-acetylglucosamine, N-acetylgalactosamine, and sialic acid); two derivatives of glucose containing a carboxyl group (glucuronic acid and iduronic acid); one 6-deoxy sugar (fucose), and one neutral pentose (xylose). In addition, the nucleic acids contain one of two neutral pentoses, either ribose or deoxyribose. All are of the D configuration except fucose and iduronic acid, which have the L configuration. In addition, various other monosaccharides are intermediates in the metabolic conversion of hexoses to carbon dioxide and water and in the photosynthetic process, as will be evident subsequently (Chaps. 17 and 19).

Neutral Hexoses

Comparison of the α forms of glucose, mannose, and galactose in different structural conventions shows important similarities between these sugars.

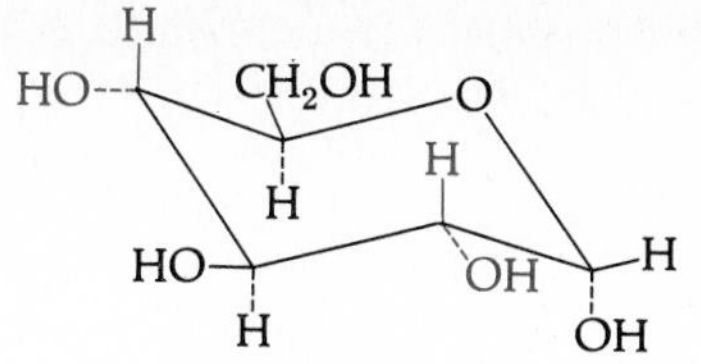

α-D-Glucopyranose 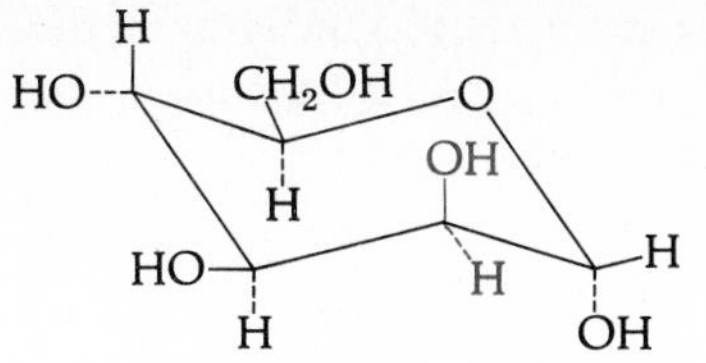   α-D-Mannopyranose 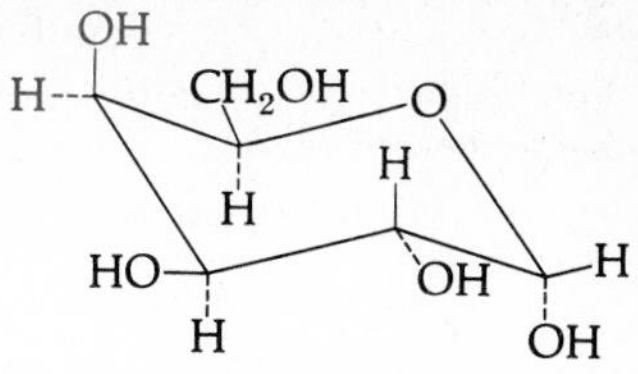   α-D-Galactopyranose

Glucose and mannose are *epimers* (differ only in the configuration of a single carbon atom) with respect to carbon atom 2. Glucose and galactose represent an epimeric pair with respect to carbon atom 4.

Sugar Acids

The most common compounds of this group are formed by oxidation of aldoses to carboxylic acids at either the C-1 aldehydic carbon, the C-6 hydroxymethyl carbon, or both. These acids have the generic names aldonic acid, uronic acid, and aldaric acid, respectively, and the general structures

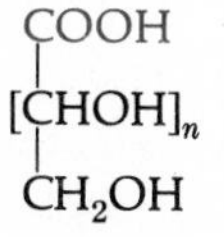

$$\begin{array}{ccc}
\text{COOH} & \text{CHO} & \text{COOH} \\
\text{[CHOH]}_n & \text{[CHOH]}_n & \text{[CHOH]}_n \\
\text{CH}_2\text{OH} & \text{COOH} & \text{COOH}
\end{array}$$

Aldonic acids  | Uronic acids  | Aldaric acids

Oxidation of glucose gives rise to the following acids:

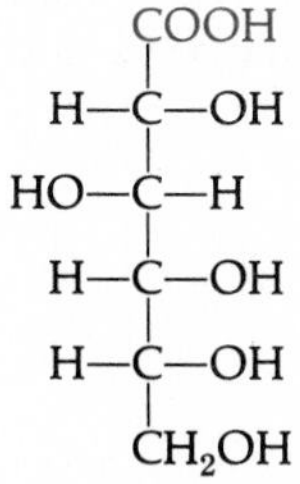

D-Gluconic acid 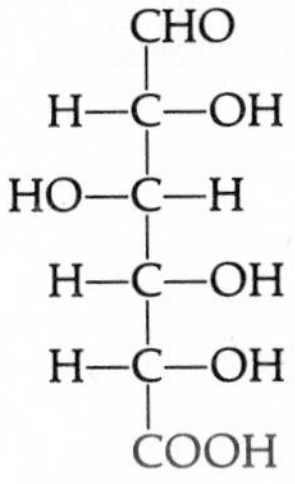   D-Glucuronic acid 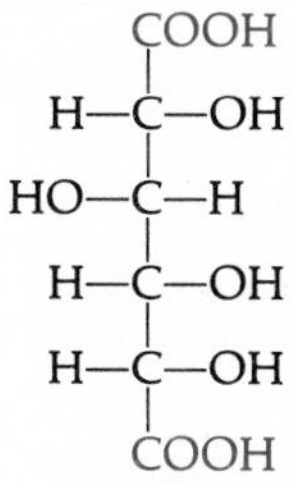   D-Glucaric acid (saccharic acid)

Like other γ- and δ-hydroxy acids, these acids tend to form inner esters, or lactones, generally resulting in a five- or, more rarely, a six-membered ring.

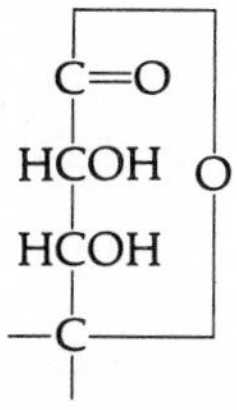

γ-Lactone 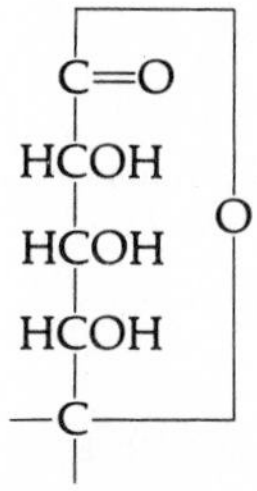   δ-Lactone

D-Glucuronic acid (GlcUA) and L-iduronic acid (IdUA) occur in the proteoglycans of connective tissue (Chap. M6). They differ from one another only in

93

the configuration at carbon atom 5.   Glycosides of glucuronic acid with phenols, steroids, or bile pigments are normal constituents of urine.

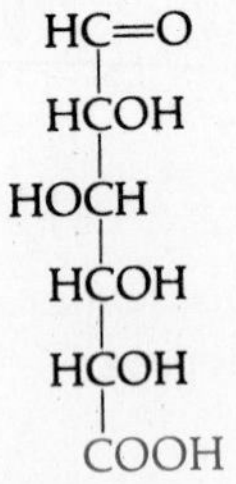

HC=O
HCOH
HOCH
HCOH
HCOH
COOH

D-Glucuronic acid

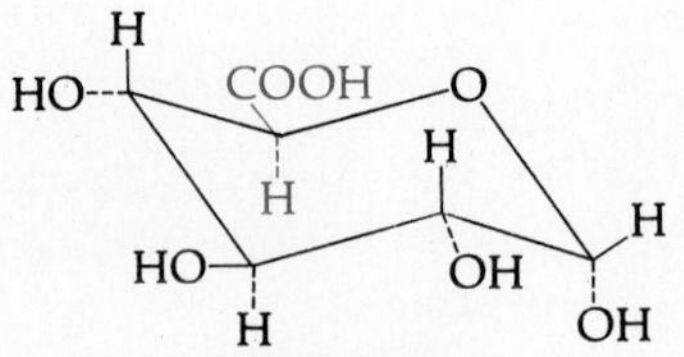

**D-Glucuronic
acid (GlcUA)**

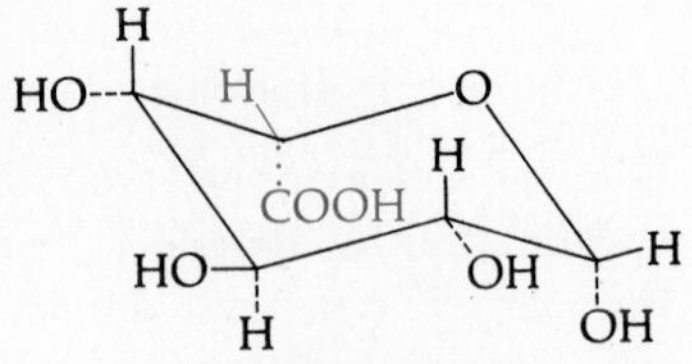

**L-Iduronic acid
(IdUA)**

A sugar acid of great biological importance, widely distributed in animals and plants, is vitamin C, ascorbic acid (L-xyloascorbic acid), considered in detail in Chap. M22.

Amino Sugars

In these compounds a hydroxyl group on one of the pyranose-ring carbon atoms is replaced by an amino group.   Widely distributed in a variety of glycoconjugates (Chap. M6) are the 2-amino aldohexoses D-glucosamine and D-galactosamine, which occur as the N-acetyl derivatives.

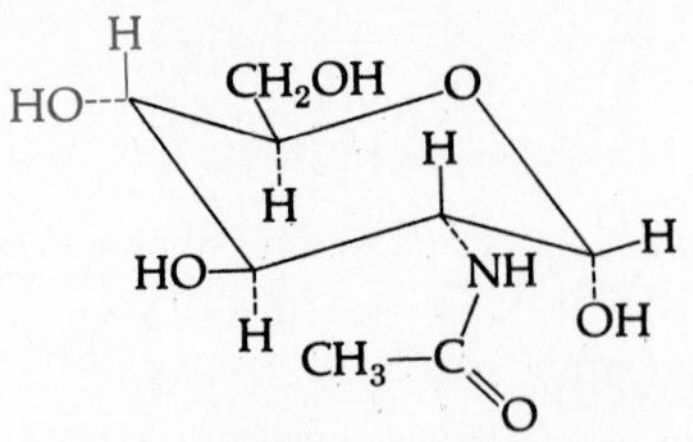

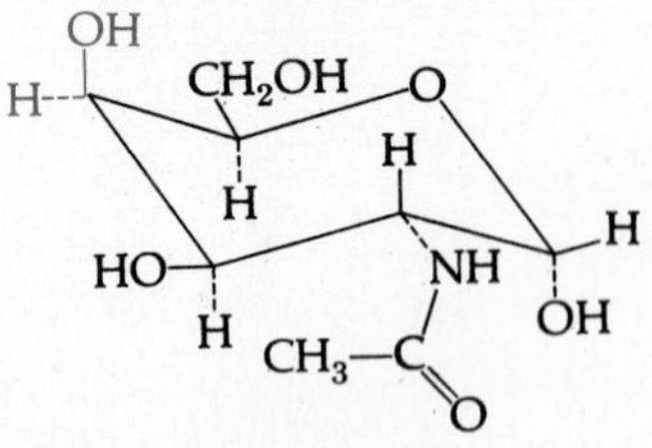

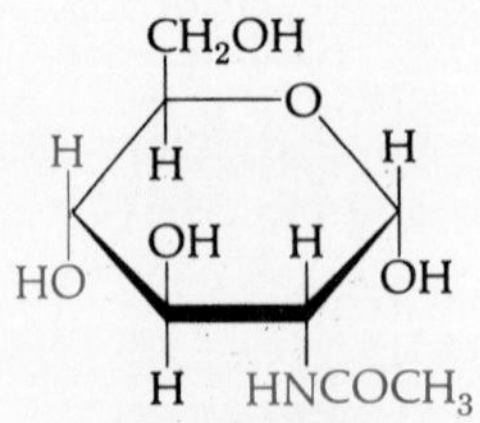

**N-Acetyl-α-D-glucosamine
(2-acetamido-2-deoxy-
α-D-glucosamine)
(GlcNAc)**

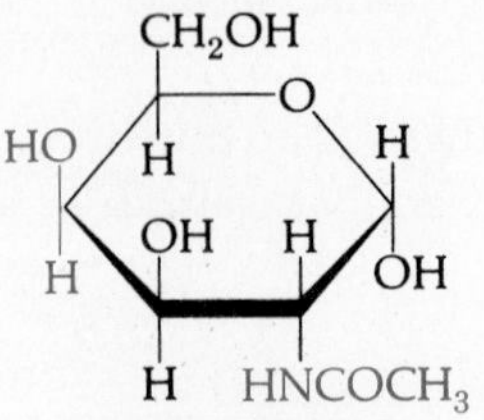

**N-Acetyl-α-D-galactosamine
(2-acetamido-2-deoxy-
α-D-galactosamine)
(GalNAc)**

Sialic acids (Sia) form a class of important ketoses that contain nine carbon atoms (ketononoses) and are acylated derivatives of 3,5-dideoxy-5-amino-nonulosonic acid, called *neuraminic acid*.   N-Acetylneuraminic acid, abbreviated NeuNAc, has the following structure shown in the three different conventions:

I   II   III

N-Acetyl-D-neuraminic acid (NeuNAc)

Although the pyranose ring corresponds to that of an L sugar, neuraminic acids are strictly D sugars because C-8, the configurational carbon atom, is of the D-glycero configuration.

Sialic acids are widely distributed in bacteria and animal tissues and in most mammals may be either *N*-acetyl or *N*-glycolyl ($-\overset{\text{O}}{\underset{\|}{\text{C}}}-CH_2OH$) deriva-

tives, the ratio of the two differing in various tissues with different species. For example, bovine, ovine, and porcine mucins of saliva contain mostly the *N*-acetyl derivative, whereas the *N*-glycolyl group predominates in erythrocyte stroma of these species. Several diacetyl derivatives have also been isolated in which the additional acetyl is present on either the C-4, C-7, or C-8 hydroxyl group. An 8-*O*-methyl derivative of *N*-acetylneuraminic acid is also known.

Pentoses

Among the naturally occurring pentoses are the aldoses L-arabinose, D-ribose, and D-xylose and the ketopentose L-xylulose. α-D-Xylose forms the pyranose ring and in this form is structurally very similar to glucose except that it lacks the hydroxymethyl group at carbon atom 5.

L-Xylulose   α-D-Xylopyranose (Xyl)

α-D-Arabinofuranose (Ara)   α-D-Ribofuranose (Rib)

Deoxy Sugars

These sugars include compounds with one or more hydroxyl groups on the pyranose or furanose rings replaced by hydrogen. 2-Deoxyribose is a component of the repeating unit in the polymeric deoxyribonucleic acids (Chap. 8).

2-Deoxy-α-D-ribose   α-L-Rhamnose (Rha)   α-L-Fucose (Fuc)

L-Rhamnose (6-deoxy-L-mannose) and L-fucose (6-deoxy-L-galactose) are among the few monosaccharides of the L configuration found in plants and animals.

OLIGO- AND POLYSACCHARIDES—GLYCOPROTEINS

Structural Diversity

In theory, the 10 different monosaccharides found in mammalian glycoconjugates can be combined by glycosidic bonds to give about 700 different disaccharides, each of which differs from another in the configuration of its glycosidic bond and the groups forming these bonds. Thus, the potential for variability of oligo- and polysaccharide structures is almost unlimited. In fact, however, only a few of the possible combinations of monosaccharides do occur in nature.

Disaccharides

Typical disaccharides frequently isolated from natural sources are those shown in Fig. 6.3, whose structures illustrate several points about oligosaccharides. *Maltose, isomaltose,* and *cellobiose* are homooligosaccharides containing only glucose. The anomeric carbon atom (C-1) of one residue is in glycosidic linkage with a hydroxyl group in the other residue. That residue with an unsubstituted anomeric carbon atom is at the *reducing end* of the molecule, since sugars with a free anomeric carbon react as reducing agents for such substances as Cu^{2+}, Ag^+, or $Fe(CN)_6^{3-}$. That residue with a substituted anomeric carbon atom is at the *nonreducing end*. In maltose and cellobiose, the glycosidic linkage is formed by the C-1 hydroxyl group of one residue and the C-4 hydroxyl group of the other residue, but in maltose it is in the α configuration, whereas in cellobiose it is the β configuration. In isomaltose, the glycosidic bond is formed by the C-1 hydroxyl of one residue and the C-6 hydroxyl of the other residue. To write shorthand structures of these disaccharides, the abbreviations (Fig. 6.1) for each sugar are used along with the symbols α or β to designate the configuration of the glycosidic bond and the numbers of the atoms that form the bond. In this

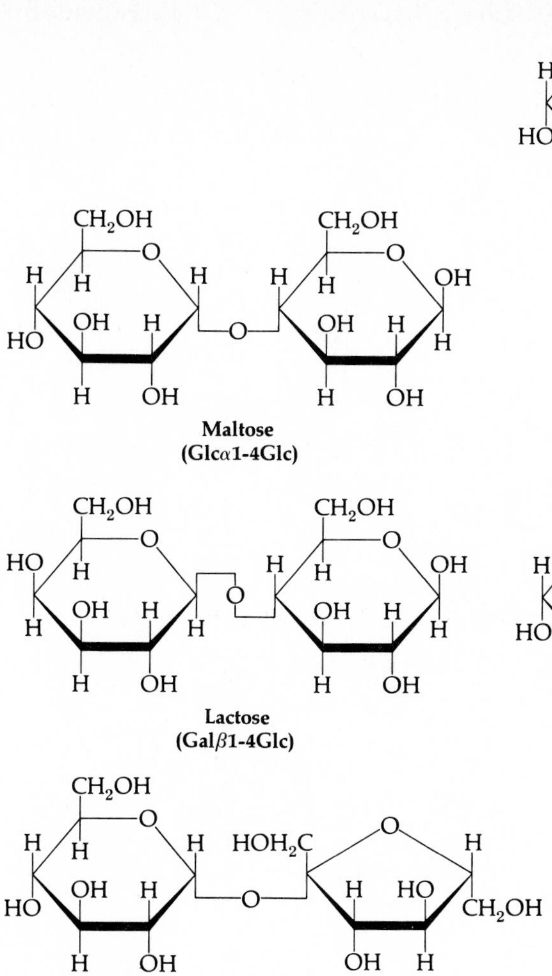

Maltose (Glcα1-4Glc)

Isomaltose (Glcα1-6Glc)

Lactose (Galβ1-4Glc)

Cellobiose (Glcβ1-4Glc)

Sucrose (Glcα1-2Fru)

Figure 6.3 The structures of some typical disaccharides.

system, maltose is Glcα1-4Glc; isomaltose is Glcα1-6Glc; and cellobiose is Glcβ1-4Glc. The glycosidic linkage may be written alternatively with an arrow, and thus maltose is designated as Glcα1 → 4Glc.

Lactose and *sucrose* are heterooligosaccharides; i.e., they contain different monosaccharide residues. Lactose is a reducing sugar whose shorthand structure is Galβ1-4Glc. In contrast, sucrose, designated as Glcα1-2Fru, is a nonreducing sugar since the anomeric hydroxyls of both residues form the glycosidic bond.

Lactose and sucrose are examples of free disaccharides in nature that are not degradation products of higher molecular weight oligo- or polysaccharides. Lactose is synthesized by the mammary gland only during lactation and is present in milk to the extent of 2 to 6 percent by weight, the concentration varying with the species. Sucrose, the common sugar of commerce and the kitchen, is obtained from either cane or beet; it is also present in all other plants, but in lesser amounts.

Maltose and isomaltose are disaccharides obtained on partial hydrolysis of both *starch* (in plants) or *glycogen* (in animals), the widely occurring polysaccharides that serve as storage forms of glucose (Chap. 18). The glucose repeating units in starch and glycogen are linked primarily by α1,4-glycosidic bonds, and thus maltose is the principal disaccharide found in hydrolysates of either polysaccharide. Isomaltose is found in much lesser amounts because glycogen (MW = 3 to 100 $\times$ 10^6) and a form of starch called *amylopectin* are branched structures, the branches being formed by α1,6-glycosidic bonds as follows:

Branch point of amylopectin and glycogen
$A = \alpha$ 1,4-glucosidic bond
$B = \alpha$ 1,6-glucosidic bond

Cellobiose is obtained on hydrolysis of *cellulose* and is the disaccharide repeating unit of this widely distributed plant polysaccharide.

Repeating cellobiose unit of cellulose

Cellulose is the most abundant organic compound in the world, constituting 50 percent or more of all the carbon in vegetation. Soluble cellulose preparations containing from 150 to 1250 cellobiose units per molecule with molecular weights ranging from 50,000 to 400,000 have been isolated, but these materials are much smaller than the water-insoluble cellulose from which they are derived.

Glycoproteins

The vast majority of types of oligo- and polysaccharides in living things are present in glycoconjugates in covalent combination with either proteins or lipids. Thus it is important to consider some of the structural features of these substances, particularly glycoproteins, which show considerable diversity and are widely distributed in living things. Glycolipid structures will be considered later (Chap. 21).

Carbohydrate–Protein Linkages Some typical oligosaccharides found in glyco-proteins are shown in Fig. 6.4. In mammalian glycoproteins the oligosaccha-

Figure 6.4 Typical oligosaccharide structures found in various glycoproteins. Structures (a), (b), and (c) are, respectively, asparagine-linked bi-, tri-, and tetraantennary oligosaccharides. The exact linkages of sialic acid in structure (c) are unknown but are either 2-3 or 2-6. Structures (a) and (c) are the complex type and structure (b) is a high mannose type. Examples of the proteins in which each structure is found are as follows: (a) human transferrin; (b) bovine thyroglobulin; (c) human α_1-acid glyco-protein; (d) porcine submaxillary mucin; and (e) human glycophorin, an erythrocyte membrane protein. (g) Shows only the linkage region of a proteoglycan oligosaccha-ride; the actual chain is far longer (Chap. M6). Hyl = hydroxylysine.

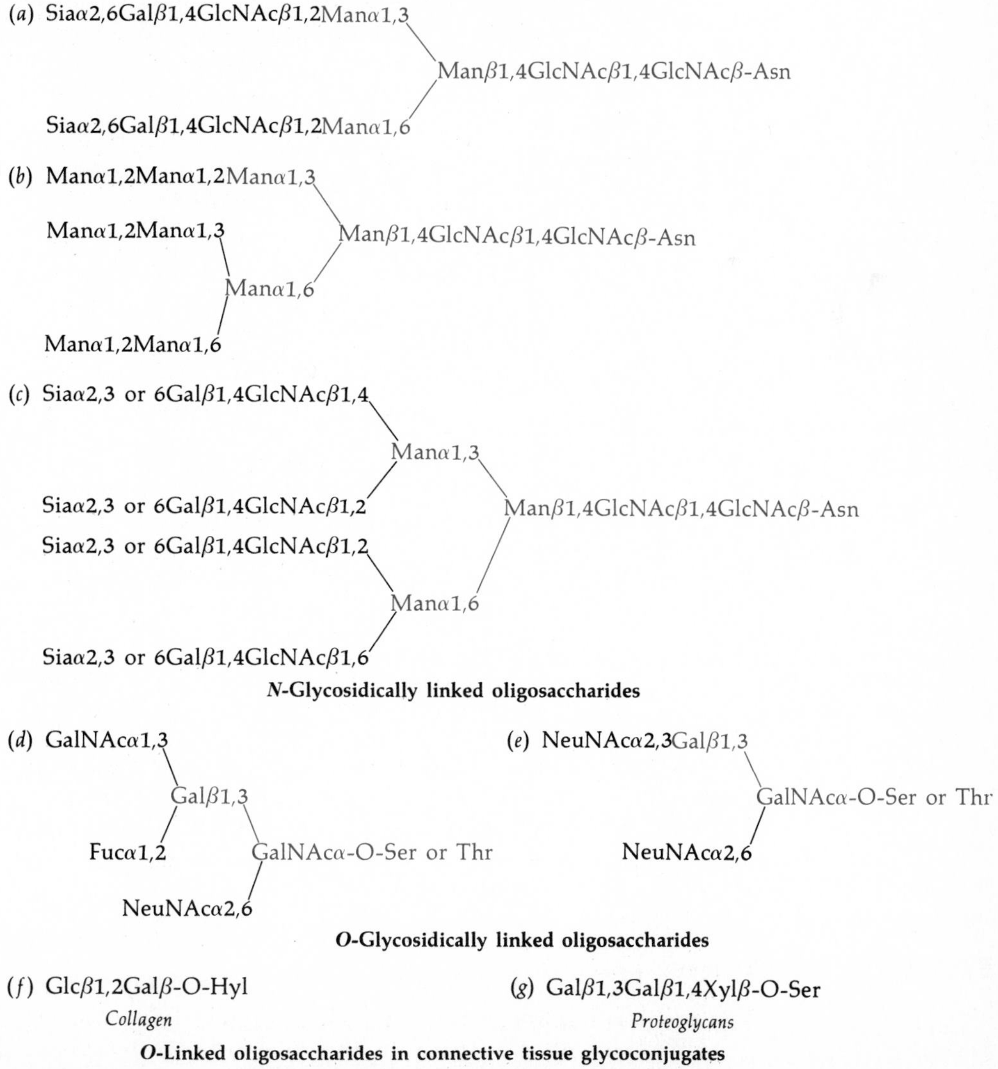

(a) Siaα2,6Galβ1,4GlcNAcβ1,2Manα1,3

Manβ1,4GlcNAcβ1,4GlcNAcβ-Asn

Siaα2,6Galβ1,4GlcNAcβ1,2Manα1,6

(b) Manα1,2Manα1,2Manα1,3

Manα1,2Manα1,3

Manβ1,4GlcNAcβ1,4GlcNAcβ-Asn

Manα1,6

Manα1,2Manα1,6

(c) Siaα2,3 or 6Galβ1,4GlcNAcβ1,4

Manα1,3

Siaα2,3 or 6Galβ1,4GlcNAcβ1,2

Siaα2,3 or 6Galβ1,4GlcNAcβ1,2

Manβ1,4GlcNAcβ1,4GlcNAcβ-Asn

Manα1,6

Siaα2,3 or 6Galβ1,4GlcNAcβ1,6

N-Glycosidically linked oligosaccharides

(d) GalNAcα1,3

Galβ1,3

Fucα1,2   GalNAcα-O-Ser or Thr

NeuNAcα2,6

(e) NeuNAcα2,3Galβ1,3

GalNAcα-O-Ser or Thr

NeuNAcα2,6

O-Glycosidically linked oligosaccharides

(f) Glcβ1,2Galβ-O-Hyl
Collagen

(g) Galβ1,3Galβ1,4Xylβ-O-Ser
Proteoglycans

O-Linked oligosaccharides in connective tissue glycoconjugates

rides are covalently bound to polypeptide chains by one of four different types of linkages. In the asparagine-linked type, which occurs in a wide variety of glycoproteins, an *N*-acetylglucosamine residue is in *N*-glycosidic linkage to the *β*-amide nitrogen of an asparagine residue in the peptide chain.

The *N*-acetylglucosamine-asparagine linkage

In the other three types, the carbohydrate is in *O*-glycosidic linkage to a hydroxyamino acid in the peptide. The most widely distributed *O*-glycosidic linkage in glycoproteins is formed between an *N*-acetylgalactosamine residue and the *β*-hydroxyl group of a serine or threonine residue.

The *N*-acetylgalactosamine-serine linkage

The remaining two *O*-glycosidic types of linkage are found primarily in connective tissue glycoconjugates (Chap. M6). In collagens of different types, the galactose residue in the disaccharide Glcβ1-2Gal is linked to the hydroxyl group of 5-hydroxylysine in the peptide chain, whereas in proteoglycans, such as heparin and chondroitin sulfate, the polysaccharide chains are linked through a xylose residue to the hydroxyl group of serine in the polypeptide.

The galactose-hydroxylysine
linkage

The xylose-serine linkage

An oligosaccharide in thyroglobulin has also been found to be linked through xylose to serine, but it is presently the sole example to date of such a linkage in a non-connective-tissue glycoprotein.   The glycopeptide Glcβ1-3Fuc-O-Thr has been isolated from human urine and rat tissues and may represent a fifth linkage type in yet other glycoproteins.   The oligosaccharides of glycolipids are also linked O-glycosidically to a hydroxyl group, primarily that in the complex lipid called *ceramide,* as described later (Chap. 26).   The cell wall polysaccharides of yeast are bound to peptide through Man-Ser/Thr linkages and the cuticle collagen of the earthworm contains the Gal-Ser/Thr linkage.   The plant cell wall structural protein extensin contains oligosaccharides O-glycosidically linked to peptide through the pentose arabinose.

High Mannose and Complex Oligosaccharides   Most N-linked oligosaccharides have the same sequence of three sugars, called the *core region,* linked to asparagine, viz., Manβ1-4Glc$N$Acβ1-4Glc$N$Acβ-Asn.   However, many different N-linked oligosaccharides varying in structural complexity have been identified and can be classified into two different groups.   One, the *high mannose* group, is found in glycoproteins such as ovalbumin, thyroglobulin, and certain cell surface membrane-bound glycoproteins and contains up to six mannose residues in addition to those in the core region.   The second group, called the *complex type*, contains several different monosaccharides in addition to mannose, including sialic acid, fucose, galactose, and N-acetylglucosamine.   Examples of each type of structure are shown in Fig. 6.4.

The high mannose and complex types of oligosaccharides may have two, three, or four branches and accordingly have been designated as *bi-*, *tri-*, or *tetraantennary* oligosaccharides.   Examples of these types of structures are also shown in Fig. 6.4.

Structural Microheterogeneity   In contrast to other macromolecules, oligosaccharides and polysaccharides in glycoconjugates may have considerable structural *microheterogeneity;* i.e., many of the N- and O-linked oligosaccharides as isolated from glycoproteins are devoid of one or more residues at the nonreducing end.   Thus, one or more sialic acid residues may be missing from the N-linked oligosaccharide chains shown in Fig. 6.4, or one of the two branches in an N-linked oligosaccharide may be devoid of both sialic acid and galactose, although the other chain contains both sugars in the sequences shown.   Similar microheterogeneity is observed in O-linked oligosaccharides; e.g., porcine submaxillary mucin, which contains structure D in Fig. 6.4, also contains almost all of the mono-, di-, tri-, and tetrasaccharide chains that can be derived from the parent pentasaccharide structure.   One of the most thoroughly studied glycoproteins, α_1-acid glycoprotein of blood plasma, contains at least 16 different oligosaccharide structures in four different positions in its single polypeptide chain.   The basis for the occurrence of microheterogeneity in glycoconjugates will be discussed in Chap. 18.

Structural Analysis

Determination of the sequence of residues in an oligosaccharide differs little in principle from sequence analysis of proteins (Chap. 5), except that branching and the type of glycosidic bond between each residue must be identified.   After determining the number and kinds of monosaccharides, the oligosaccharide is degraded by chemical or enzymic means into smaller fragments whose sequences can be established and then used to reconstruct the complete sequence of the parent molecule.

Because most oligosaccharides are present in glycoconjugates, sequence analysis usually requires degradation of the glycoconjugate and separation of the noncarbohydrate portion of the molecule from the oligosaccharide. Glycoproteins can be degraded by a number of proteolytic enzymes to yield the oligosaccharide attached to a small peptide, called a *glycopeptide*. Glycopeptides are readily separated from nonglycosylated peptides by the methods employed for peptide separation (Chap. 5). However, they are often selectively removed from peptides by affinity chromatography (Chap. 9) on insoluble adsorbents containing covalently bound plant lectins (Chap. 18), proteins that bind specific types of sugars in oligosaccharides. Thus, the lectin *concanavalin A* from jack beans will adsorb some mannose-containing glycopeptides, which are then eluted with solutions of α-methylmannoside. The composition of a pure glycopeptide can be obtained by a variety of chromatographic methods; e.g., monosaccharides and borate ion form ionic complexes, which can be separated and quantitated by ion-exchange chromatogaphy.

The oligosaccharide groups of *O*-linked glycopeptides can be obtained free of peptide by treatment with dilute aqueous hydroxide followed by reduction with sodium borohydride as follows:

$$
\text{Gal}\beta\text{1-3GalNAc}\alpha\text{—O—CH}_2\text{—CH} \overset{\text{(1) OH}^-}{\underset{\text{(2) NaBH}_4}{\longrightarrow}} \text{Gal}\beta\text{1—O—CH} + \text{H}_2\text{C=C} + \text{acetate}
$$

$$
\begin{array}{c}
\text{CH}_2\text{OH} \\
\text{H—C—NH}_2 \\
\text{Gal}\beta\text{1—O—CH} \\
\text{HOCH} \\
\text{HC—OH} \\
\text{CH}_2\text{OH}
\end{array}
$$

Dehydroalanyl residue

The hydroxide catalyzes β elimination of the disaccharide, which is then reduced by borohydride to give a sugar alcohol at the residue formerly linked to serine. Serine is converted to dehydroalanine on β elimination and then reduced to DL-alanine. In the example shown, the alcohol is 2-deoxy-2-aminogalactitol. The *N*-linked oligosaccharides do not undergo β elimination, but they can be freed of peptide enzymically with *endo-β-N-acetylglucosaminidases*, which hydrolyze between the two *N*-acetylglucosaminyl residues in the core region, as shown in Fig. 6.5, to give a free oligosaccharide.

Peptide-free oligosaccharide groups or glycopeptides can be sequenced by a variety of means. Very useful are the *exoglycosidases*, enzymes widely distributed in microorganisms, plants, and animals that remove specific sugars from the nonreducing termini of oligosaccharides. Many different glycosidases with different substrate specificities are known; the actions of a few are illustrated in Fig. 6.5. Thus, for the oligosaccharide shown, *sialidases*, often called *neuraminidases*,

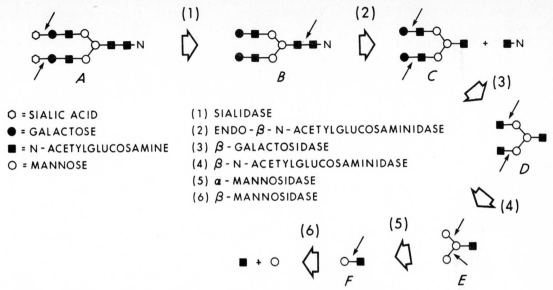

Figure 6.5 Sequential degradation of an asparagine-linked oligosaccharide by glycosidases. Structure (*a*) is identical to structure (*a*) in Fig. 6.4. Each of the exoglycosidases in steps 1, 3, 4, 5, and 6 hydrolyze only glycosidic bonds formed by a sugar at the nonreducing end of the molecule. The endoglycosidase in step 2 hydrolyzes an internally located glycosidic bond between the two *N*-acetylglucosamine residues. The exoglycosidases in steps 3, 4, 5, and 6 could also be used sequentially to degrade structure (*b*). The endoglycosidase in step 2 from a basidiomycete hydrolyzes structure (*a*) poorly. Endo-β-*N*-acetylglucosaminidases from other organisms have somewhat different substrate specificities, although each hydrolyzes the same glycosidic bond in the core region.

specifically hydrolyze glycosidic bonds formed by nonreducing terminal sialic acid. Treatment of the resulting desialyzed oligosaccharide with *β-galactosidase* releases galactose, and subsequent sequential hydrolysis with *β-N-acetylhexosaminidase* and *α-* and *β-mannosidases* releases *N*-acetylglucosamine and α- and β-linked mannose residues, respectively. The glycosidases have strict substrate specificities for the kind of monosaccharide as well as for the configuration of the glycosidic bond; e.g., *β*-galactosidases hydrolyze *β*- but not α-galactosides. The many available α- and β-specific glycosidases serve to establish the configuration of glycosidic bonds.

Enzymic degradation is usually inadequate to define the nature of branch points in an oligosaccharide, necessitating use of chemical methods. In *methylation analysis,* all unsubstituted hydroxyl groups are methylated by reaction of an oligosaccharide with methyl iodide and methyl sulfinyl carbanion in dimethylsulfoxide. Thus, methylation of the tetrasaccharide

$$
\begin{array}{c}
\text{Man}\alpha 1\text{-}3 \\
\qquad\qquad \searrow \\
\qquad\qquad\qquad \text{Man}\beta 1\text{-}4\text{GlcNAc} \\
\qquad\qquad \nearrow \\
\text{Man}\alpha 1\text{-}6
\end{array}
$$

gives the following methylated product:

103

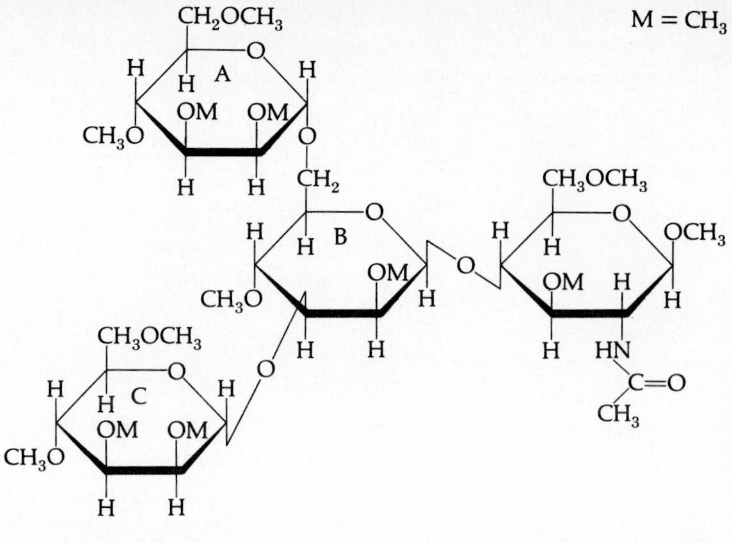

M = CH$_3$

The fully methylated product is then hydrolyzed in aqueous acid, and after acetylation of the free hydroxyl groups in the resulting methylated mannose residues, the different kinds of methylated mannose residues are separated by gas-liquid chromatography and then identified by mass spectrometry. Unmethylated hydroxyl groups reflect those groups in glycosidic linkage, and thus it is evident that one mannose residue (ring B) in the above structure has unsubstituted 3 and 6 hydroxyl groups, indicating that they are at a branch point with the other two mannose residues (rings A and C), each of which is fully methylated except for its anomeric hydroxyl.

A recent approach to structural analysis employs high resolution ^{1}H nuclear magnetic resonance (NMR) spectroscopy, which in conjunction with methylation analysis can provide detailed structural information quite rapidly.

Conformation

Many oligo- and polysaccharides in solution appear to behave as random coils, in which there is continuous fluctuation between different local and overall conformations of a molecule. Individual monosaccharides in a chain may not only assume transiently different conformations between boat and chair forms, but also rotate somewhat around their glycosidic bonds, as permitted by ϕ and ψ angles analogous to those of the peptide bond in proteins (Chap. 5). However, carbohydrate polymers in general are not as flexible as other natural biopolymers and form somewhat more rigid structures.

Many homopolysaccharides in the cell walls of plants reside in a largely water-free environment and assume characteristic conformations. Thus, the polysaccharide chains of cellulose form helix-like structures known as *ribbons* and when packed together lengthwise, give rigid aggregates with great structural strength. Ribbon-like structures are also found in the polysaccharide chains of *chitin*, a linear polymer of β1,4-linked N-acetylglucosamine that is the basic exoskeletal material of crustaceans, insects, and spiders and is one of the most abundant substances in nature—it is estimated that 10^{10} to 10^{13} tons of new chitin and cellulose are synthesized annually on the earth's surface.

Starch and glycogen have so-called hollow helix structures, which permit individual molecules to be packed together closely and accumulated as concentrated deposits. Patently, individual glucose residues in such deposits are sufficiently accessible to the enzymes required for synthesis and degradation.

Vertebrate oligo- and polysaccharides do not seem to possess conformations analogous to those found in plants and invertebrates. Those of connective tissue (Chap. M6) are random coils whose structural functions are associated with the binding of water in the extracellular matrix. The *N*- and *O*-linked oligosaccharides in mammalian glycoconjugates (Fig. 6.4) are very hydrophilic and in close contact with water at the surface of a molecule, contributing little structural stability to such molecules. Oligosaccharide groups are partially buried in crystal structures of immunoglobulins (Chap. M2) and appear to interact with polypeptide groups, but since glycosidases can hydrolyze the oligosaccharides in immunoglobulins in solution, it is likely that the groups do not lend much stability to immunoglobulin structure.

Notwithstanding their lack of a regular conformation, specific groups in oligosaccharides may interact highly specifically with other structures, such as cell surface receptors, and in this respect contain considerable "biological information." Such aspects of carbohydrate function are considered in Chap. 18.

REFERENCES

Books

Aspinall, G. O.: *Polysaccharides,* Pergamon, New York, 1970.
Barker, R.: *Organic Chemistry of Biological Compounds,* Prentice-Hall, Englewood Cliffs, N.J., 1971.
Jeanloz, R. W. (ed.): *Chemistry of Amino Sugars,* vol. IA, Academic, New York, 1969.
Jeanloz, R. W., and E. A. Balazs (eds.): *The Amino Sugars,* vol. IIA, Academic, New York, 1965.
Pigman, W. W., and D. Horton (eds.): *The Carbohydrates: Chemistry/Biochemistry,* vols. IA, IIA, IIIB, Academic, New York, 1972.
Rees, D. A.: *Polysaccharide Shapes,* Chapman & Hall, London, 1977.
Rosenberg, A., and C-L. Schengrund, (eds.): *Biological Roles of Sialic Acid,* Plenum, New York, 1976.
Sharon, N.: *Complex Carbohydrates: Their Chemistry, Biosynthesis, and Functions,* Addison-Wesley, Reading, Mass., 1975.
Stanek, J., M. Cerny, J. Kocourek, and J. Pacek,: *The Monosaccharides,* Academic, New York, 1963.
Stoddart, J. F.: *Stereochemistry of the Carbohydrates,* Wiley, New York, 1971.
Whistler, R. L., and M. L. Wolfrom, (eds.): *Methods in Carbohydrate Chemistry,* vol. I: *Analysis and Preparation of Sugars,* 1962; vol. IV: *Starch,* 1964, vol. V: *General Polysaccharides,* 1965, Academic, New York.

Review Articles

Bentley, R.: Configurational and Conformational Aspects of Carbohydrate Biochemistry, *Annu. Rev. Biochem.* **41**:953–996 (1972).
Kirkwood, S.: Unusual Polysaccharides, *Annu. Rev. Biochem.* **43**:401–417 (1974).

Kornfeld, R., and Kornfeld, S.: Comparative Aspects of Glycoprotein Structure, *Annu. Rev. Biochem.* **45**:217–237 (1975).

———: Structure of Glycoproteins and Their Oligosaccharide Units, in W. J. Lennarz, (ed.): *The Biochemistry of Glycoproteins and Proteoglycans*, Plenum, New York, 1980.

Many special phases of carbohydrate chemistry have been reviewed in *Advances in Carbohydrate Chemistry and Biochemistry*, Academic, New York, 1945–current.

Chapter 7

The Lipids: Biological Amphiphiles

GENERAL PROPERTIES

The lipids include a wide variety of compounds that are characteristically poorly soluble in water. However, most lipids have structural features in common that endow them with important biological properties as well as characteristic solubilities. They are mainly ionic or polar derivatives of hydrocarbons and belong to the class of compounds called *amphiphiles*. Amphiphiles (Gk. *amphi*, "both"; *phile*, "affinity") contain ionic or polar groups that are *hydrophilic* and have an affinity for water as well as nonpolar hydrocarbon groups that are *hydrophobic* and lack an affinity for water. The properties of an amphiphile are influenced greatly by the nature of these groups. Thus, some lipids are very weak amphiphiles with little affinity for water. Such lipids are stored in tissues largely in a water-free state but are mobilized and transported to other tissues where their oxidation provides the major source of energy for vital metabolic processes (Chap. 20). Other, more polar lipids are stronger amphiphiles and form the major structural components of the various biological membranes that serve to compartmentalize cells and tissues (Chap. 13).

The purpose of this chapter is to describe the structures and some properties of the various lipids and consider how the amphiphilic character of some lipids endows them with the ability to self-associate and form biological membranes.

CLASSIFICATION

The lipids, often referred to as *fats,* are the major compounds extracted from animal and plant tissues by the so-called fat solvents, viz., methanol, etha-

Any cross-references coded M refer to the companion text, *Principles of Biochemistry: Mammalian Biochemistry.*

nol, ether, chloroform, benzene, or mixtures thereof. They are conveniently classified into five major groups: (1) *fatty acids,* which are long-chain aliphatic carboxylic acids; (2) neutral esters of glycerol and fatty acids, the *neutral fats,* or the *acylglycerols;* (3) ionic esters of glycerol, fatty acids, and phosphate, the *phosphoglycerides;* (4) lipids not containing glycerol, including *sphingolipids, aliphatic alcohols, waxes, terpenes,* and *steroids;* and (5) lipids combined with other classes of compounds, such as protein and carbohydrate. Lipids in classes 1 to 4 will be considered in this chapter whereas those in class 5 are considered elsewhere (Chaps. 3, 6, and 21).

THE FATTY ACIDS

Only small quantities of free fatty acids are present in nature. They are found in abundance, however, in either *ester* or *amide* linkage in several of the classes of lipids listed above as well as in many compounds that are formed transiently during lipid metabolism. The fatty acids of biochemical importance have, with occasional exceptions, the following properties: (1) they are mainly monocarboxylic acids that contain an ionizable carboxyl group and a nonpolar, acyclic, unbranched hydrocarbon chain; (2) they usually contain an even number of carbon atoms, although fatty acids with an odd number of carbon atoms are found in nature; (3) they may be saturated or may contain one or more double bonds.

Saturated Fatty Acids

The common names and formulas of some of the saturated fatty acids are given in Table 7.1. In numbering the carbon atoms, the carboxyl group is always C-1 and the other atoms are numbered successively so that the highest-numbered carbon atom is in the terminal methyl group.

Boiling points and melting points of fatty acids rise with increasing chain length. Even-numbered-carbon-atom saturated fatty acids of less than 10 carbon atoms are liquids at room temperature; longer-chained members are solids.

The fatty acids are weak acids and dissociate in aqueous solution, $RCOOH \rightleftharpoons RCOO^- + H^+$. Except for the first member of the series, formic acid ($pK = 3.75$), all the saturated fatty acids resemble acetic acid ($pK = 4.76$) in their dissociation constants ($pK \approx 4.85$).

The mixture of fatty acids obtained upon hydrolysis of lipids derived from various sources generally contains both saturated and unsaturated fatty acids. In typical animal lipids, the most abundant saturated fatty acid is usually palmitic (C_{16}), with stearic (C_{18}) second in amount. Shorter-chain fatty acids (C_{14} and C_{12}) occur in small quantities, as do longer-chain members (up to C_{28}). Fatty acids of 10 carbon atoms or fewer are rarely present in animal lipids.

Unsaturated Fatty Acids

The names and structures of certain of the more common unsaturated fatty acids are given in Table 7.2. The systematic names contain the same prefix as that of saturated fatty acids, which indicates the number of carbon atoms in the

TABLE 7.1

Molecular formula	Common name	Systematic name	Structural formula	Some Naturally Occurring Saturated Fatty Acids
CH_2O_2	Formic	. . . .	HCOOH	
$C_2H_4O_2$	Acetic	Ethanoic	CH_3COOH	
$C_3H_6O_2$	Propionic	n-Propanoic	CH_3CH_2COOH	
$C_4H_8O_2$	n-Butyric	n-Butanoic	$CH_3[CH_2]_2COOH$	
$C_6H_{12}O_2$	Caproic	n-Hexanoic	$CH_3[CH_2]_4COOH$	
$C_8H_{16}O_2$	Caprylic	n-Octanoic	$CH_3[CH_2]_6COOH$	
$C_9H_{18}O_2$	Pelargonic	n-Nonanoic	$CH_3[CH_2]_7COOH$	
$C_{10}H_{20}O_2$	Capric	n-Decanoic	$CH_3[CH_2]_8COOH$	
$C_{12}H_{24}O_2$	Lauric	n-Dodecanoic	$CH_3[CH_2]_{10}COOH$	
$C_{14}H_{28}O_2$	Myristic	n-Tetradecanoic	$CH_3[CH_2]_{12}COOH$	
$C_{16}H_{32}O_2$	Palmitic*	n-Hexadecanoic	$CH_3[CH_2]_{14}COOH$	
$C_{18}H_{36}O_2$	Stearic*	n-Octadecanoic	$CH_3[CH_2]_{16}COOH$	
$C_{20}H_{40}O_2$	Arachidic	n-Eicosanoic	$CH_3[CH_2]_{18}COOH$	
$C_{22}H_{44}O_2$	Behenic	n-Docosanoic	$CH_3[CH_2]_{20}COOH$	
$C_{24}H_{48}O_2$	Lignoceric	n-Tetracosanoic	$CH_3[CH_2]_{22}COOH$	
$C_{26}H_{52}O_2$	Cerotic	n-Hexacosanoic	$CH_3[CH_2]_{24}COOH$	
$C_{28}H_{56}O_2$	Montanic	n-Octacosanoic	$CH_3[CH_2]_{26}COOH$	

*These are the most abundant saturated fatty acids encountered in animal lipids.

molecule, but the suffix is *-enoic* for compounds with one double bond, *-dienoic* for those with two double bonds, *-trienoic* for those with three double bonds, etc. The position of the double bond is indicated by a number that signifies the lowest-numbered carbon atom forming the bond; thus, the double bond in *cis*-9-hexadecenoic acid is between carbons 9 and 10, and the two double bonds in *cis,cis*-9,12-octadecadienoic acid are between carbons 9 and 10 and 12 and 13.

Unsaturation in the cis series of fatty acids markedly alters certain properties of a fatty acid. In general, the melting point is greatly lowered and solubility in nonpolar solvents is enhanced. All the common unsaturated fatty acids of nature are liquids at room temperature.

The double bond in the singly unsaturated fatty acids of animal lipids is generally in the 9,10 position. Thus, the two most abundant singly unsaturated fatty acids of animal lipids are oleic acid and palmitoleic acid (Table 7.2).

TABLE 7.2 Some Naturally Occurring Unsaturated Fatty Acids

Molecular formula	Common name	Systematic name*	Structural formula
$C_{16}H_{30}O_2$	Palmitoleic†	9-Hexadecenoic	$CH_3[CH_2]_5CH=CH[CH_2]_7COOH$
$C_{18}H_{34}O_2$	Oleic†	*cis*-9-Octadecenoic	$CH_3[CH_2]_7CH=CH[CH_2]_7COOH$
$C_{18}H_{34}O_2$	Vaccenic	*trans*-11-Octadecenoic	$CH_3[CH_2]_5CH=CH[CH_2]_9COOH$
$C_{18}H_{32}O_2$	Linoleic†	*cis,cis*-9,12-Octadecadienoic	$CH_3[CH_2]_4CH=CHCH_2CH=CH[CH_2]_7COOH$
$C_{18}H_{30}O_2$	α-Linolenic	9,12,15-Octadecatrienoic	$CH_3CH_2CH=CHCH_2CH=CHCH_2CH=CH[CH_2]_7COOH$
$C_{18}H_{30}O_2$	γ-Linolenic	6,9,12-Octadecatrienoic	$CH_3[CH_2]_4CH=CHCH_2CH=CHCH_2CH=CH[CH_2]_4COOH$
$C_{18}H_{30}O_2$	Eleostearic	9,11,13-Octadecatrienoic	$CH_3[CH_2]_3CH=CH-CH=CH-CH=CH[CH_2]_7COOH$
$C_{20}H_{32}O_2$	Arachidonic†	5,8,11,14-Eicosatetraenoic	$CH_3[CH_2]_4-[CH=CH-CH_2]_4-[CH_2]_2-COOH$
$C_{24}H_{46}O_2$	Nervonic	*cis*-15-Tetracosenoic	$CH_3[CH_2]_7CH=CH[CH_2]_{13}COOH$

*Except where indicated, the double bonds are in the *cis* geometrical configuration.
† These are the most abundant unsaturated fatty acids in animal lipids.

The presence of a double bond offers the possibility of cis-trans isomerism. The isomeric forms of 9-octadecenoic may be written

$$H-\underset{\|}{C}-[CH_2]_7-CH_3 \qquad CH_3-[CH_2]_7-\underset{\|}{C}-H$$
$$H-C-[CH_2]_7-COOH \qquad H-C-[CH_2]_7-COOH$$

Oleic acid (cis)   **Elaidic acid (trans)**

In this and in certain other known cases in which ethylenic double bonds occur in the fatty acid series, it is the cis configuration that is predominantly found in nature.

Fatty acids containing more than one double bond are also common. When two double bonds occur in a carbon skeleton in the relationship

$$-CH=CH-CH=CH-$$

they are said to be *in conjugation*. Although some fatty acids with such conjugated double bonds are found in the fatty acid series of the plant world (eleostearic acid, Table 7.2), multiple unsaturation as it occurs in the fatty acids of animal lipids is in general not conjugated but of the divinylmethane type:

$$-CH=CH-CH_2-CH=CH-$$

The polyunsaturated fatty acids most frequent in mammalian tissues are linoleic acid, containing two double bonds, linolenic acid, with three double bonds, and arachidonic acid, with four double bonds (Table 7.2).

LIPIDS CONTAINING GLYCEROL

Neutral Fats

Mono-, Di, and Triacylglycerols The neutral fats constitute by far the most abundant group of lipids in animals. These are esters of fatty acids with the trihydroxyl alcohol glycerol, $CH_2OH-CHOH-CH_2OH$. One, two, or three of the hydroxyl groups of glycerol may be esterified with a fatty acid; hence the designations mono-, di-, and triacylglycerols. The terms *mono-, di-* and *triglycerides* are often used to designate these glycerol esters but are not the preferred names and will not be used in this text. The triacylglycerols represent the most abundant group of neutral fats, although mono- and diacylglycerols are present in nature and are of importance in lipid metabolism.

The generic formula for a triacylglycerol is

$$CH_3-[CH_2]_n-\overset{O}{\underset{\|}{C}}-O-\underset{|}{C}H \quad \begin{matrix} O \\ \| \\ CH_2-O-C-[CH_2]_n-CH_3 \\ O \\ \| \\ CH_2-O-C-[CH_2]_n-CH_3 \end{matrix}$$

in which the fatty acids may be the same or different depending on the value of n. The system for naming the neutral fats is based upon the names of the constituent fatty acids. Thus tristearin contains three stearic acid residues per molecule, and oleodistearin contains one residue of oleic acid and two of stearic acid.

Physical and Chemical Properties The neutral fats are weak amphiphiles because the ester linkages as well as the free hydroxyl groups in mono- and diacylglycerols are nonionic and only weakly polar. Thus, the properties of these substances are dominated largely by the hydrophobic alkyl groups in the molecule (see "Lipids as Amphiphiles" below).

Table 7.3 contains examples of the fatty acid composition of neutral fats from various sources. Most samples of animal fat contain predominantly esters of palmitic, stearic, palmitoleic, oleic, and linoleic acids in various proportions. Fat from diverse portions of the same organism may differ widely in composition. Thus, human subcutaneous fat contains more saturated fatty acids than the fat from the liver, which is richer in unsaturated acids.

Hydrolysis and Saponification Hydrolysis of the neutral fats yields three molecules of fatty acid and one of glycerol.

$$
\begin{array}{c}
\text{CH}_2\text{—OOC—R} \\
| \\
\text{R'—COO—CH} + 3\text{H}_2\text{O} \xrightarrow{\text{(H}^+\text{)}} \\
| \\
\text{CH}_2\text{—OOC—R''} \\
\textbf{Neutral fat}
\end{array}
\quad
\begin{array}{c}
\text{R—COOH} \\
+ \\
\text{R'—COOH} + \\
+ \\
\text{R''—COOH} \\
\textbf{Fatty acids}
\end{array}
\quad
\begin{array}{c}
\text{HOCH}_2 \\
| \\
\text{HOCH} \\
| \\
\text{HOCH}_2 \\
\textbf{Glycerol}
\end{array}
$$

The rate of hydrolysis is greatly accelerated by catalytic concentrations of H^+ or OH^- or enzymes called *esterases,* or, more specifically, *lipases.*

Hydrolysis of neutral fats by OH^- is termed *saponification.* The carboxylate ions formed, in the presence of a cation, become soaps.

TABLE 7.3

	Depot fat, mol %				Liver fat, cow, mol %	Milk fat, cow, mol %	Approximate Composition of the Fatty Acid Mixtures
Acid	Human	Cow	Pig	Sheep			Obtained from Triacylglycerols from Various Sources
Butyric	. .	. .	. .	. . .	. . .	9	
Caproic	. .	. .	. .	. .	. . .	3	
Caprylic	. .	. .	. .	. .	. . .	2	
Capric	. .	. .	. .	. .	. . .	4	
Lauric	. .	. .	. .	. .	. . .	3	
Myristic	3	7	1	2	3	11	
Palmitic	23	29	28	25	35	23	
Stearic	6	21	10	26	5	9	
Palmitoleic	5	. .	. .	. .	10	4	
Oleic	50	41	58	42	36	26	
Linoleic	10	2	3	5	8	3	
C_{20-22} unsaturated	. .	. .	. .	. .	3*		

*Average number of double bonds per molecule = 3.8.
Source: Data adapted from T. P. Hilditch, *The Chemical Constitution of Natural Fats,* Wiley, New York, 1940, and from the literature.

Glycerol and Glycerol Phosphate Glycerol forms esters with inorganic as well as organic acids, and one of its phosphate esters, L-glycerol-3-phosphate, is a component of the phosphoglycerides as well as an important intermediate in carbohydrate (Chap. 17) and lipid (Chap. 20) metabolism.

The nomenclature for the glycerol phosphates often poses difficulties because of alternative ways of designating the position of the phosphate with respect to the asymmetric carbon atom. The ambiguity and confusion that may result from the use of different configurational prefixes (D or L) for the same compounds have led to a different system of nomenclature in which the prefix *sn* (for *s*tereospecifically *n*umbered) is used. Thus, L-glycerol-3-phosphate (D-glycerol-1-phosphate) is designated *sn*-glycerol-3-phosphate,

$$
\begin{array}{cc}
\text{① CH}_2\text{OH} & \text{① CH}_2\text{—O—P—O}^- \\
\text{HO—② C—H} \quad \text{O} & \text{HO—② C—H} \\
\text{③ CH}_2\text{—O—P—O}^- & \text{③ CH}_2\text{—OH} \\
\text{O}^- &
\end{array}
$$

sn-**Glycerol-3-phosphate**   *sn*-**Glycerol-1-phosphate**

and its enantiomer, D-glycerol-3-phosphate (L-glycerol-1-phosphate), is called *sn*-glycerol-1-phosphate. This system clearly shows the stereospecificity between the two enantiomers of glycerol phosphate, but the name L-glycerol-3-phosphate rather than *sn*-glycerol-3-phosphate will be used throughout the text unless otherwise clearly stated, since it is the major natural phosphate ester of glycerol and its D enantiomer is seldom encountered.

Glyceryl Ethers and Glycosyl Glycerols Neutral lipids called *alkyl glyceryl ethers* that have been isolated from tissues contain an ether linkage between a long-chain (saturated or unsaturated) alcohol and glycerol.

$$
\begin{array}{cc}
\text{CH}_2\text{—O—[CH}_2]_n\text{—CH}_3 & \text{CH}_2\text{—O—[CH}_2]_n\text{—CH}_3 \\
\text{HCOH} & \text{R—C—O—CH} \\
\text{CH}_2\text{OH} & \text{CH}_2\text{—O—C—R}
\end{array}
$$

Alkyl glyceryl ether   **Alkyl diacylglyceryl ether**

Present evidence suggests that these ethers do not occur naturally in free form but as esterified derivatives of long-chain fatty acids, called *alkyl diacylglycerols*.

Glycosyl diacylglycerols have been isolated from plants and bacteria. Representative of such lipids is that isolated from chloroplasts and chromatophores of photosynthetic bacteria, which contains 6-sulfo-6-deoxyglucose (*quinovose*) in glycosidic linkage with a diacylglycerol (Chap. 19).

Phosphoglycerides

This group of glycerol-containing lipids comprises compounds that are derivatives of L-glycerol-3-phosphate. They are classified on the basis of the type of linkage between the hydrocarbon moiety and glycerol and the nature of the polar groups in the molecule other than phosphate. The phosphoglycerides are

113

The Lipids:
Biological Amphiphiles

important biological amphiphiles since they usually represent up to 50 percent of the lipids in biological membranes.

Phosphatidates This group of phosphoglycerides is widely distributed among living things. They may be viewed as derivatives of L-phosphatidic acid.

$$
\begin{array}{ll}
\text{H}_2\text{COOCR} & \alpha' = 1 \\
\text{R}'\text{COOCH} \quad \text{O} & \beta = 2 \\
\text{H}_2\text{C}-\text{O}-\overset{\parallel}{\text{P}}-\text{OH} & \alpha = 3 \\
\quad\quad\quad\quad \underset{\text{O}^-}{|}
\end{array}
$$

L-**Phosphatidic acid**
(**1,2-O-diacyl-L-glycerol-3-phosphate**)

$$
\begin{array}{l}
\text{H}_2\text{COOCR} \\
\text{R}'\text{COOCH} \quad \text{O} \quad\quad\quad\quad \text{CH}_3 \\
\text{H}_2\text{C}-\text{O}-\overset{\parallel}{\text{P}}-\text{OCH}_2\text{CH}_2\overset{+}{\text{N}}-\text{CH}_3 \\
\quad\quad\quad\quad \underset{\text{O}^-}{|} \quad\quad\quad \underset{\text{CH}_3}{|}
\end{array}
$$

L-**Phosphatidylcholine**

$$
\begin{array}{l}
\text{H}_2\text{COOCR} \\
\text{R}'\text{COOCH} \quad \text{O} \\
\text{H}_2\text{C}-\text{O}-\overset{\parallel}{\text{P}}-\text{OCH}_2\text{CH}_2\overset{+}{\text{NH}}_3 \\
\quad\quad\quad\quad \underset{\text{O}^-}{|}
\end{array}
$$

L-**Phosphatidylethanolamine**

$$
\begin{array}{l}
\text{H}_2\text{COOCR} \\
\text{R}'\text{COC}\quad\text{O} \\
{}_2\text{C}-\text{O}-\overset{\parallel}{\text{P}}-\text{OCH}_2\text{CHNH}_3^+ \\
\quad\quad\quad\quad \underset{\text{O}^-}{|} \quad\quad \underset{\text{COO}^-}{|}
\end{array}
$$

L-**Phosphatidylserine**

They most frequently contain a nitrogenous base, choline, ethanolamine, or serine, giving phosphatidylcholine, phosphatidylethanolamine, and phosphatidylserine, respectively. The phosphatidylcholines are also called *lecithins*. Saturated fatty acids (C_{16} to C_{18}) predominate at the C-1 position and unsaturated fatty acids (C_{16} to C_{20}), with one to four double bonds, at the C-2 position in most naturally occurring phosphatidates, although exceptions are known.

The phosphatidates are dipolar ions at physiological pH and exist largely as the ionic species shown above. Thus, the ionic groups provide a highly polar, hydrophilic region in the molecule and the two acyl groups, strongly hydrophobic regions.

Plasmalogens These phosphoglycerols contain an α,β-unsaturated alcohol in ether linkage (alkenyl) rather than a fatty acid in ester linkage (acyl) on C-1 of L-glycerol-3-phosphate. Thus, the plasmalogens containing ethanolamine have the general structure

$$
\begin{array}{l}
\quad\quad\quad\quad \overset{\text{H}\ \ \text{H}}{\text{H}_2\text{COC}=\text{CR}} \\
\text{R}'\text{COOCH} \quad\quad \text{O} \\
\text{H}_2\text{C}-\text{O}-\overset{\parallel}{\text{P}}-\text{OCH}_2\text{CH}_2\overset{+}{\text{NH}}_3 \\
\quad\quad\quad\quad \underset{\text{O}^-}{|}
\end{array}
$$

L-**Phosphatidalethanolamine**

The three principal classes of plasmalogens are phosphatidalcholines, phosphatidalethanolamines, and phosphatidalserines. The α,β-unsaturated ether linkage in these compounds is stable in dilute alkali, but in dilute acid the linkage is hydrolyzed to yield an aldehyde of the corresponding α,β-unsaturated alcohol. The α,β-unsaturated alcohols have the cis configuration and vary in chain length from C_{12} to C_{18}. In pure phosphatidalcholine from beef heart the fatty acyl chain on C-2 is predominantly unsaturated whereas the alcohol moiety on C-1 is predominantly saturated, except for the α,β double bond.

1-Alkyl-2-acylphosphatidylcholine derivatives are widely distributed in animal tissues, although generally constituting less than a few percent of the total phosphoglycerides. The generic formula for these ether phosphatidates is

$$
\begin{array}{l}
\overset{\displaystyle H_2C-O-[CH_2]_n-CH_3}{\underset{\displaystyle |}{}} \\[2pt]
\overset{\displaystyle O}{\underset{\displaystyle \|}{}} \\
R-C-O-\overset{\displaystyle CH}{\underset{\displaystyle |}{}} \\
\overset{\displaystyle O}{\underset{\displaystyle \|}{}} \\
H_2C-O-P-O-CH_2-CH_2-\overset{+}{N}\equiv(CH_3)_3 \\
\underset{\displaystyle O^-}{}
\end{array}
$$

Diphosphatidylglycerols As their designation implies, the diphosphatidylglycerols present in animal and, particularly, plant tissues are composed of 1 mol of glycerol and 2 mol of L-phosphatidic acid. These compounds have the generic formula

$$
\begin{array}{l}
RCOOCH_2 \qquad\qquad\qquad\qquad\qquad H_2COOCR \\
R'COOCH \quad O \qquad\qquad\qquad O \quad HCOOCR' \\
H_2CO-P-OCH_2-CHOH-CH_2O-P-O-CH_2 \\
O^- \qquad\qquad\qquad\qquad O^-
\end{array}
$$

A diphosphatidylglycerol

An important example is *cardiolipin*, initially isolated from heart tissue.

Phosphoinositides A second group of nitrogen-free derivatives of α-phosphatidic acid contains inositol. On hydrolysis, phosphoinositides usually yield 1 mol of glycerol, 1 mol of the hexahydroxy alcohol *myo*inositol, 2 mol of fatty acid, and 1 to 3 mol of phosphate. A triphosphoinositide, 3-phosphatidyl-5,6-diphospho-myoinositol, has the generic structure

A phosphoinositide   Myoinositol

Myoinositol is one of nine stereoisomeric forms of the carbocyclic hexitols and belongs to that class of compounds called *cyclitols*, which are cycloalkanes containing one hydroxyl group on each of three or more ring atoms. It is the most important cyclitol and is widely distributed in microorganisms, higher plants, and animals. In plants it is found as its hexaphosphate, *phytic acid,* or as the mixed magnesium-calcium salt of phytic acid.

Phosphatidylglycerol and Its Derivatives Certain bacteria contain substantial amounts of phosphatidylglycerol and its aminoacyl derivative, L-lysylphosphatidylglycerol, or the corresponding L-alanine derivative.

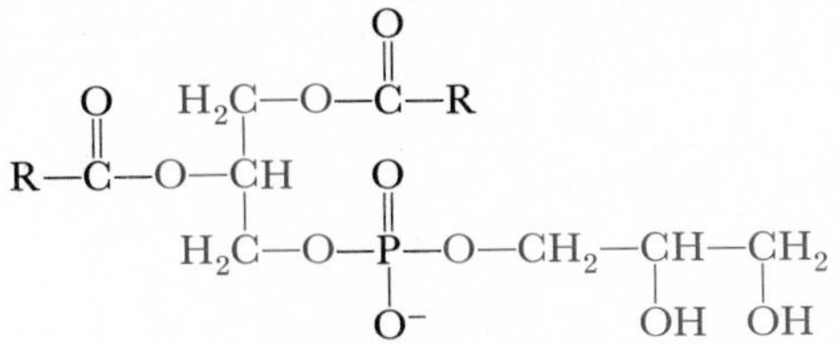

$$
\begin{array}{l}
\quad\quad\quad\quad\quad O \\
\quad\quad\quad\quad\quad \| \\
O\quad\quad H_2C{-}O{-}C{-}R \\
\| \quad\quad\quad\quad | \\
R{-}C{-}O{-}CH \quad\quad O \\
\quad\quad\quad\quad | \quad\quad \| \\
\quad\quad\quad\quad H_2C{-}O{-}P{-}O{-}CH_2{-}CH{-}CH_2 \\
\quad\quad\quad\quad\quad\quad\quad | \quad\quad\quad | \quad\quad | \\
\quad\quad\quad\quad\quad\quad\quad O^- \quad\quad OH \quad OH
\end{array}
$$

Phosphatidylglycerol

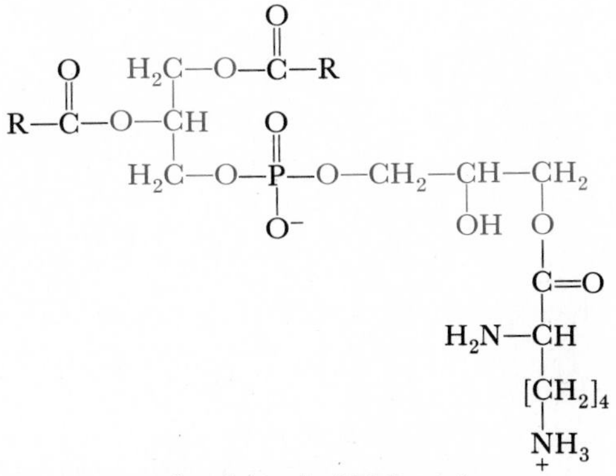

L-Lysylphosphatidylglycerol

Glycosaminylphosphatidylglycerol, which contains D-glucosamine in β-glycosidic linkage with either the 2' or 3' hydroxyl groups of the glycerol moiety of phosphatidylglycerol, has also been isolated.

LIPIDS NOT CONTAINING GLYCEROL

Sphingolipids

A large number of lipids contain the aliphatic base, *sphingosine* or *dihydrosphingosine.*

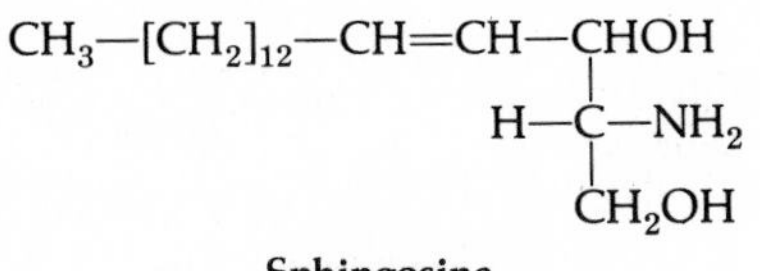

$$CH_3{-}[CH_2]_{12}{-}CH{=}CH{-}CHOH$$
$$H{-}C{-}NH_2$$
$$CH_2OH$$

Sphingosine
(D-4-sphingenine)

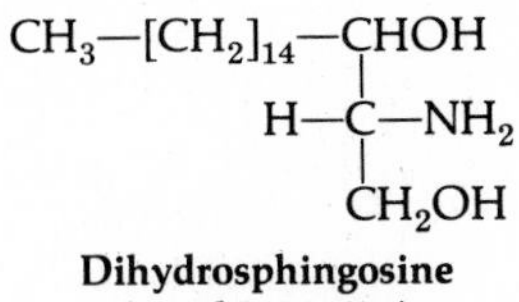

$$CH_3{-}[CH_2]_{14}{-}CHOH$$
$$H{-}C{-}NH_2$$
$$CH_2OH$$

Dihydrosphingosine
(D-sphinganine)

Lipids containing these bases are termed *sphingolipids.* Although the C_{18} sphingosines are most abundant in sphingolipids, other homologous C_{16}, C_{17}, C_{19}, and C_{20} sphingosines also are found among naturally occurring sphingolipids.

The double bond in sphingosine has the trans configuration, and the asymmetric centers C-2 and C-3 are of the D-configuration.

Ceramides and Sphingomyelins Ceramides are *N*-acyl fatty acid derivatives of sphingosine. They are widely distributed in plant and animal tissues but only in small amounts. The fatty acid residue, present in amide linkage, is of the C_{16}, C_{18}, C_{22}, or C_{24} series, and the various ceramides differ in their constituent fatty acids. The sphingomyelins are phosphocholine derivatives of ceramides and are a major group of phospholipids.

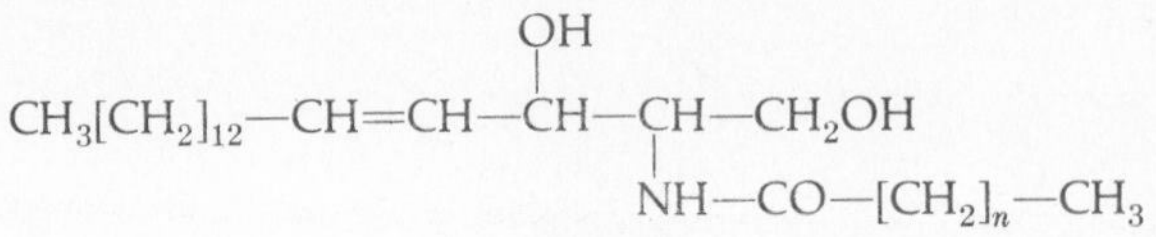

$$CH_3[CH_2]_{12}-CH=CH-\overset{\overset{\displaystyle OH}{|}}{CH}-CH-CH_2OH$$
$$\underset{\displaystyle NH-CO-[CH_2]_n-CH_3}{|}$$

A ceramide

$$CH_3[CH_2]_{12}-CH=CH-\overset{\overset{\displaystyle OH}{|}}{CH}-CH-CH_2-O-\overset{\overset{\displaystyle O}{||}}{\underset{\underset{\displaystyle O^-}{|}}{P}}-OCH_2CH_2\overset{+}{N}\equiv(CH_3)_3$$
$$\underset{\displaystyle HN-CO-[CH_2]_n-CH_3}{|}$$

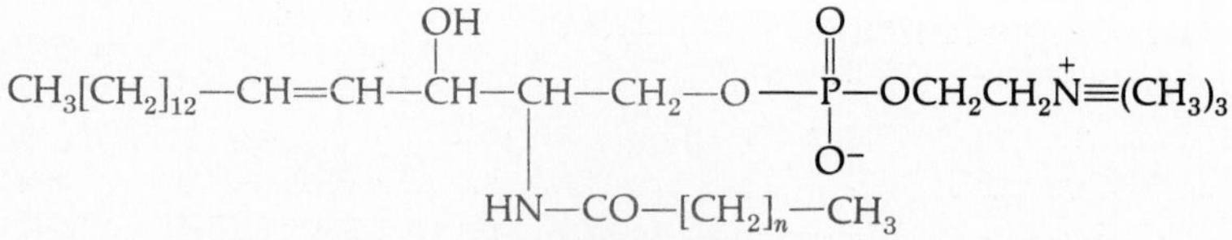

A sphingomyelin

Glycosphingolipids Several classes of lipids are derivatives of a ceramide, as are the sphingomyelins, but unlike the latter are lacking in phosphorus and the additional nitrogenous base; they do, however, contain one residue or more of carbohydrate. The *cerebrosides* are ceramide monosaccharides occurring most abundantly in the myelin sheath of nerves. On hydrolysis, cerebrosides yield one molecule each of a sphingosine, a fatty acid, and a hexose (most frequently D-galactose, less commonly, D-glucose). The generic formula for glucocerebroside and the structural formula for the specific galactocerebroside, *phrenosine*, are

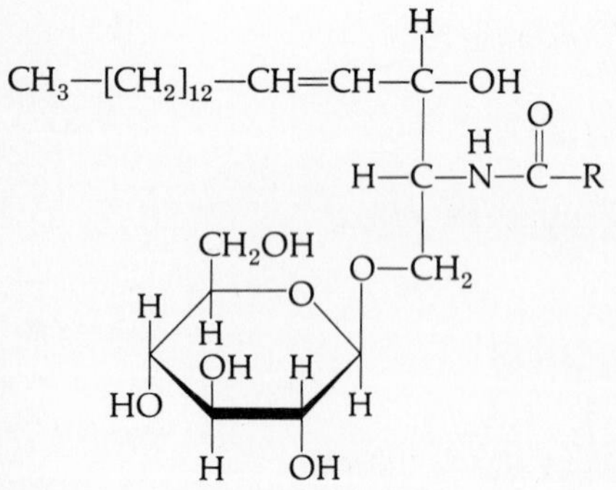

A glucocerebroside

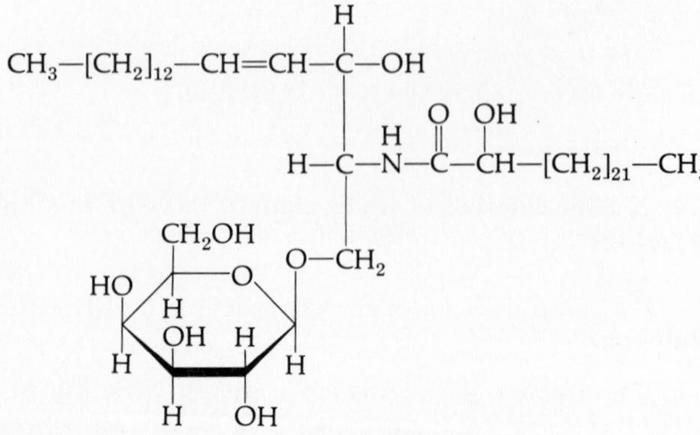

Phrenosine

Cerebroside sulfatides are sulfate derivatives of galactocerebrosides; the sulfate is present in ester linkage on the C-3 hydroxyl group of the galactosyl residue. Other sulfate esters of cerebrosides have been reported and are designated as sulfatides. The presence of the negatively charged sulfate enhances the polarity of these amphiphilic lipids.

Ceramide oligosaccharides are sphingolipids that contain a hetero-oligosaccharide attached in glycosidic linkage to ceramide. Hence the terms *ceramide disaccharide, ceramide trisaccharide,* etc., are used to designate these compounds. These lipids are considered further in Chap. 21.

Gangliosides are ceramide oligosaccharides containing at least one residue of sialic acid in addition to the other sugars. The nomenclature and biosynthesis of these compounds are presented in Chap. 21.

Diol Lipids

Although most lipids are derivatives of glycerol, it is now apparent that many organisms—animal and plant—also contain small amounts of lipids that are derivatives of diols, including ethylene glycol, the 1,2- and 1,3-propanediols, 1,3-, 1,4-, and 2,3-butanediols, and 1,5- pentanediol. The derivatives are mono- and diacyl esters of various fatty acids, diethers, mixed alkyl and alk-1-enyl ether esters, the diol analogs of phosphatidylcholine, phosphatidylethanolamine, diol choline plasmalogen, acylated diol galactosides, and diol lipoamino acids. Usually, these constituents are present at a concentration of 0.5 to 1.5% that of glycerol derivatives.

Aliphatic Alcohols and Waxes

Significant quantities of aliphatic alcohols have been recovered from certain lipid sources. Thus in the feces there are detectable amounts of cetyl alcohol, the primary alcohol corresponding to palmitic acid, $CH_3[CH_2]_{14}CH_2OH$. Cetyl alcohol occurs in various highly specialized lipids, and such an alcohol occurs as an ester of a fatty acid. The *head oil* of the sperm whale yields spermaceti, which is largely cetyl palmitate, and beeswax is rich in myricyl palmitate.

$$CH_3[CH_2]_{14}-\overset{\text{O}}{\underset{\|}{C}}-OCH_2[CH_2]_{14}CH_3 \qquad CH_3[CH_2]_{14}-\overset{\text{O}}{\underset{\|}{C}}-OCH_2[CH_2]_{28}CH_3$$

Cetyl palmitate   **Myricyl palmitate**

The generic name *wax* is given to a naturally occurring fatty acid ester of any high-molecular-weight alcohol other than glycerol.

Terpenes

Throughout the biological world there are numerous compounds whose carbon skeletons suggest a structural relationship to isoprene, 2-methylbutadiene.

$$CH_2=\overset{\overset{\textstyle CH_3}{|}}{C}-CH=CH_2$$

Isoprene

Many of these compounds contain multiples of five carbon atoms so related to each other as to make dissection of their structures into isoprene-like fragments possible. This class of compounds, called *terpenes* (from turpentine), includes essential oils from plants and a variety of other compounds. Examples of open-chain terpenes are phytol, the alcoholic fragment obtained on hydrolysis of chlorophyll,

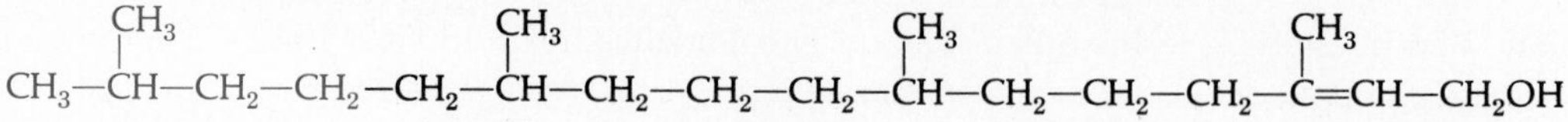

$$CH_3-\overset{\overset{\displaystyle CH_3}{|}}{CH}-CH_2-CH_2-CH_2-\overset{\overset{\displaystyle CH_3}{|}}{CH}-CH_2-CH_2-CH_2-\overset{\overset{\displaystyle CH_3}{|}}{CH}-CH_2-CH_2-CH_2-\overset{\overset{\displaystyle CH_3}{|}}{C}=CH-CH_2OH$$

Phytol

and squalene, a hydrocarbon intermediate in the biosynthesis of cholesterol.

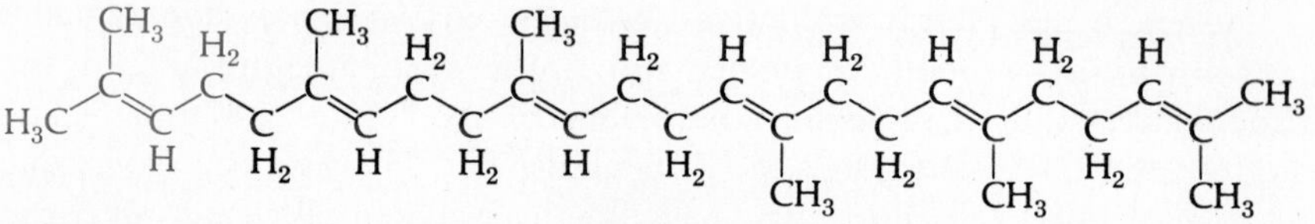

Squalene

The shorthand of open-chain terpene chemistry transforms this formula into

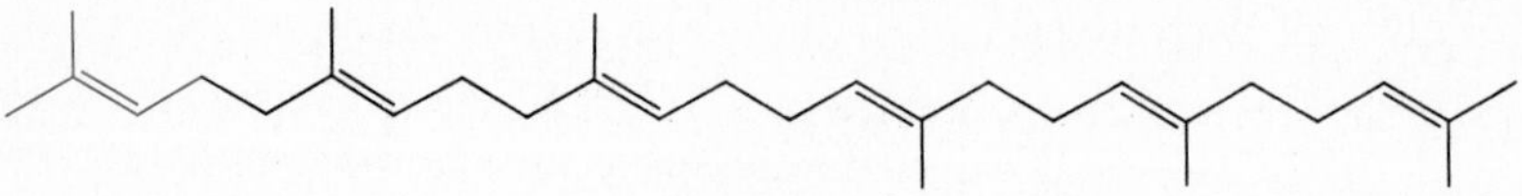

Squalene

The carbon and hydrogen atoms are not shown, and the methyl groups are indicated as branch points.

Other compounds containing isoprene-like units will be encountered in subsequent chapters, where they are best introduced in conjunction with their biological roles, e.g., the plant pigment β-carotene, vitamin A (Chap. M23), vitamin E (Chap. M23), vitamin K (Chap. M1), and coenzyme Q (Chap. 15).

Steroids

The members of this group may be considered as derivatives of a fused, reduced ring system, cyclopenta[α]-phenanthrene, comprising three fused cyclohexane rings (A, B, and C) in the nonlinear, or phenanthrene, arrangement,

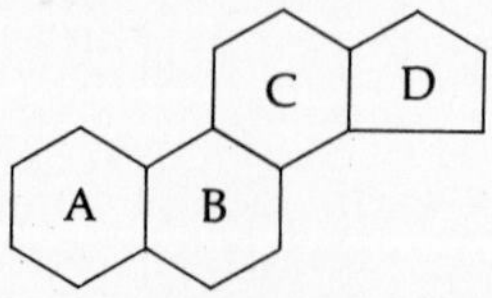

Perhydrocyclopentanophenanthrene 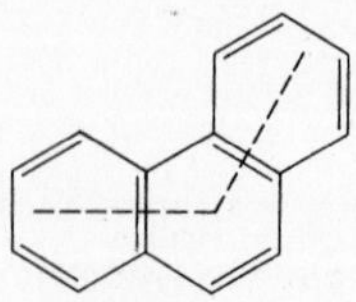   **Phenanthrene**

and a terminal cyclopentane ring (D). Certain general characteristics of the group may be considered with reference to a typical member, cholestanol; the conventional numbering of the carbon atoms is included in the formula.

Cholestanol

Note that each ring is completely saturated. In this shorthand, whenever double bonds occur, they are specifically indicated.

Certain facts of general applicability to steroids are illustrated by the structure of cholestanol. There is an oxygenated substituent on carbon atom 3, a characteristic shared by almost all naturally occurring steroids. There are "angular" methyl groups, numbered 19 and 18, on carbon atoms 10 and 13. This also is a general characteristic. However, in some steroids ring A is aromatic, and under these circumstances carbon atom 10 cannot bear a methyl group. There may be an aliphatic substituent on carbon atom 17. This substituent serves as a convenient basis for classification, and steroids are grouped according to the number of carbon atoms in this side chain. It contains 8, 9, or 10 carbon atoms in the sterols (total carbon atoms 27, 28, or 29, respectively), 5 carbon atoms in the bile acids (total of 24 carbon atoms), 2 in the adrenal cortical steroids and in progesterone (total of 21 carbon atoms), and none in the naturally occurring estrogens or androgens (total of either 18 or 19 carbon atoms).

The conformational formula of cholestanol shows that the molecule is rigid. The small letters in the formula, a and e, refer to axial and equatorial substitutions, respectively. Not all hydrogen atoms have been represented; some are indicated by solid or dashed lines. The chair conformation of cyclohexane itself is more stable than the boat conformation. In cholestanol, rings B and C are locked rigidly in the chair conformation by the trans fusions to rings A and D. Ring A is free to assume the boat form, but the instability associated with the boat form of cyclohexane itself would be enhanced by a strong interaction between methyl and hydroxyl groups at the bow (C-10) and stern (C-3) positions.

Conformational structure of cholestanol

In cholestanol there are nine centers of asymmetry in the molecule; these are at carbon atoms 3, 5, 8, 9, 10, 13, 14, 17, and 20. In the structural representation of the steroid (see formulas above), substituents that are β-oriented (to the front) are indicated with a solid bond; those that are α-oriented (to the back) are represented with a dashed line. Thus, the angular methyl groups at carbon atoms 10 and 13 are β-oriented, as are the 3-hydroxyl, the 8-hydrogen, and the side chain at carbon atom 17.

Two positions of asymmetry warrant further comment. The hydrogen atom on carbon 5 may be either on the same side of the plane of the molecule as the methyl group on carbon 10 or on the opposite side. In the former case, rings A and B will be cis to each other, and the molecule is said to belong to the *normal* configuration. If the hydrogen and methyl are on opposite sides, rings A and B are trans to each other, and the molecule is of the allo configuration.

A/B cis normal   A/B trans allo

The hydroxyl group on carbon 3 may similarly be α- or β-oriented.

α, allo   β, allo

Several steroids will be considered in subsequent chapters in the text, where their structures are more appropriately introduced along with their functions. This includes cholesterol (Chap. 21), the most abundant steroid in animal tissues, the bile acids (Chap. M10), and a number of steroid hormones, such as progesterone (Chap. M17), the androgens (Chap. M17), the estrogens (Chap. M21), and the adrenal cortical steroids (Chap. M18).

LIPIDS AS AMPHIPHILES: MICELLES AND BILAYERS

Solubility of Amphiphiles

Lipids do not form macromolecules analogous to the proteins, the polysaccharides, or the nucleic acids, but because of their amphiphilic properties, lipids form highly organized structures of macromolecular dimensions. When combined with proteins, these structures form the biological membranes that serve to contain and compartmentalize cells (Chap. 13).

Lipids and all amphiphiles are extended compounds with long hydrocarbon chains extending away from the polar groups, as shown in Fig. 7.1.

The polar, or ionic, group of a lipid is called the *head,* and the nonpolar hydrocarbon chain or chains the *tail.* The differing nature of the interaction of the head and tail groups with water largely determines the amphiphilic proper-

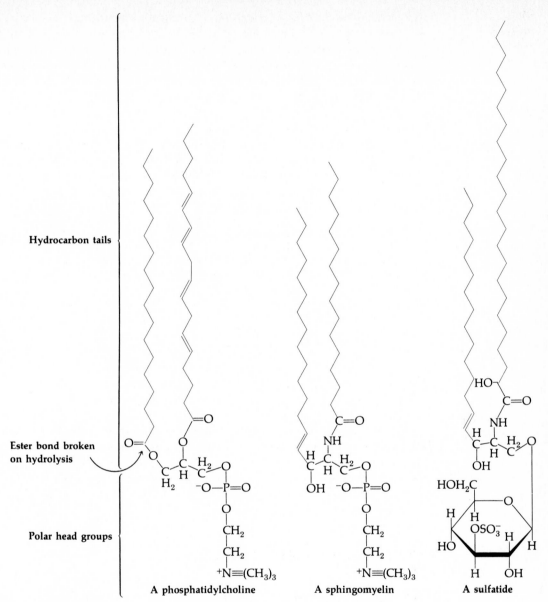

Hydrocarbon tails

Ester bond broken on hydrolysis

Polar head groups

A phosphatidylcholine   A sphingomyelin   A sulfatide

Figure 7.1 The structures of three phospholipids that illustrate important features of typical amphiphilic lipids.

ties of a lipid. The head groups of lipids are hydrophilic and quite soluble in water. By itself, phosphorylcholine is soluble in water but almost insoluble in alcohol or ether.

In contrast, simple hydrocarbons of chain length similar to those found in the tails of lipids are hydrophobic and they are virtually insoluble in water. Thus, in aqueous solution, hydrocarbon groups in amphiphiles tend to gather together and cluster around themselves or with apolar groups of other molecules to form *hydrophobic bonds* (Chap. 2). The solubility behavior of lipids in water may therefore be considered as the sum of the independent contributions of the head and tail groups. The addition of one or two methyl groups to a head group does not markedly lower its solubility in water, but as the hydrocarbon chains in tail groups are lengthened, their effects become more influential and the resulting lipid becomes less water-soluble.

Lipid Micelles

The solubilities of salts of long-chain fatty acids (soap) and of *surfactants* (detergents), such as sodium dodecyl sulfate (Chap. 4), demonstrate important general features of the behavior of amphiphiles in aqueous media, including the more complex lipids. Sodium and potassium salts of fatty acids and sodium dodecyl sulfate are somewhat soluble in water. The salts of divalent or trivalent metal ions, however, for example, Ca^{2+} and Fe^{3+}, are quite insoluble. Potassium salts are more soluble than sodium salts. As shown diagrammatically in Fig. 7.2, when small amounts of a soluble surfactant (or soap) are added to water, a portion is dissolved as monomers and some forms a monolayer at the air-water interface, the polar carboxylate groups projecting into the water and the hydrocarbon groups protruding from it. The molecules in solution are in equilibrium with those forming the monolayer at the air-water interface. The surface tension of the water at the interface decreases because the strong adhesive forces of water at the interface are weakened by the presence of the hydrocarbon chains. As additional surfactant is added, the concentration of monomer cannot be exceeded, and at this critical concentration, the surfactant molecules begin to associate to form *micelles*. Micelles are stable colloidal aggregates which are formed by amphiphiles above a very narrow concentration range, called the *critical micelle concentration* (cmc), which differs for various amphiphiles. Micelle formation is a cooperative process, and the micelles formed above the cmc contain about the same number of molecules; no significant amounts of micelles containing only a few molecules or markedly different numbers of molecules are formed.

The driving force for micelle formation is hydrophobic since the tail groups cannot break the hydrogen bonds between water molecules and as a result cluster together in close proximity. The interior of the micelle therefore consists of the hydrophobic tail groups that, in effect, provide a small volume of liquid hydrocarbon. The hydrophobic interior is sequestered from the aqueous medium by the polar groups, which cover the surface of the micelle. The

TABLE 7.4

Some Properties of Common Surfactants	Surfactant	Number of molecules per micelle	Micelle MW	cmc, mM	Conditions
	Sodium dodecyl sulfate	62	18,000	8.2	Water
		126	36,000	0.52	0.5 M NaCl
	Cetyltrimethylammonium bromide	169	62,000	. . .	0.013 M KBr
	Triton-X-100	140	90,000	0.24	Water

detergency of surfactants depends in large part on their ability to dissolve lipid-soluble substances in the hydrophobic interior of the micelles.

Table 7.4 lists some of these molecular parameters for micelles formed by the surfactants *sodium dodecyl sulfate,* an anionic surfactant, *cetyltrimethylammonium bromide,* a cationic surfactant, and *Triton-X-100,* a neutral surfactant. Clearly the size and shape of the micelles depend upon the structure of the surfactant and the composition of the aqueous medium.

$$
\begin{array}{c}
\text{O} \\
\| \\
\text{O}-\text{S}-\text{O}^- \quad \text{Na}^+ \\
\| \\
\text{O}
\end{array}
$$

Sodium dodecyl sulfate

$$
\begin{array}{c}
\text{CH}_3 \\
+| \\
\text{N}-\text{CH}_3 \quad \text{Br}^- \\
| \\
\text{CH}_3
\end{array}
$$

Cetyltrimethylammonium bromide

$$-\text{O}-[\text{CH}_2-\text{CH}_2-\text{O}]_n\text{H}$$

Triton-X series

Lipid Bilayers

The amphiphilic properties of biological lipids differ in important respects from the surfactants. When extremely small amounts of a phosphatidate are dissolved in water, very little exists in solution as monomers and micelle formation may occur at concentrations as low as $10^{-10}\,M$. For example, the cmc for dipalmitoyl phosphatidylcholine is about $5 \times 10^{-10}\,M$, and the concentration of monomers, air-surface monolayers, and micelles, analogous to those depicted in Fig. 7.2, is extremely small. The micelles, in addition, appear to have large molecular weights of about 2×10^6. If dry phosphoglyceride is placed in water and becomes swollen, the predominant structures are not micelles but multilay-

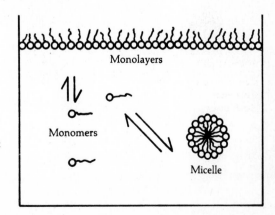

Figure 7.2 Diagrammatic representation of the structures of a surfactant in aqueous media. Molecules of surfactant exist in equilibrium between monomer, monolayers, and micelles. Micelles are globular structures (which, of course, exist in three dimensions), and the micelle is depicted in a cross-sectional view to show the relative location of the head groups (O) and the tails (⌒).

ered structures. Under certain conditions, the multilayer structures can be disrupted to give vesicles of different shapes and sizes that are bounded by lipid and contain an internal cavity filled by water. An electron micrograph of one such vesicle in cross section is shown in Fig. 7.3a. Structural analysis of the lipid in these vesicles reveals that it is a *lipid bilayer,* a planar structure containing polar head groups on the internal and external surfaces of the vesicle with a double layer of hydrocarbon in between, as shown diagrammatically in Fig. 7.3b. Thus, in contrast to surfactants, which form micelles, biological lipids tend to form bilayers, which are self-sealing, and thus form vesicles. The dimensions of the bilayer in such vesicles is almost identical to that seen in the lipid bilayer of biological membranes (Chap. 13). Thus, biological lipids self-assemble under appropriate conditions to form the bilayer matrix of the biological membranes, ultrastructures that compartmentalize living cells and their subcellular organelles.

The driving force for bilayer assembly is again hydrophobic, just as in micelle formation by surfactants, discussed above. Biological lipids form

Figure 7.3 (*a*) Electron micrograph of a thin section of a phosphoglyceride vesicle isolated from an ultrasonicated lipid vesicle from a phosphoglyceride mixture derived from soybeans. The vesicles were negatively stained after fixation in osmium tetroxide. The thickness of the lipid wall of the vesicle is 45 ± 5 Å. (*b*) Diagrammatic representation of a phosphoglyceride bilayer found in vesicles of the type shown in part *a*. The dimensions of the individual lipid are not on the same scale as actually observed in part *a*. The head groups and tails are designated as (○) and (⌒), respectively. It should be emphasized that vesicles of this type are spherical or ellipsoidal in three dimensions and contain solvent inside as well as outside. [*Part a from V. K. Miyamoto and W. Stoeckenius, J. Membr. Biol. 4:257 (1971).*]

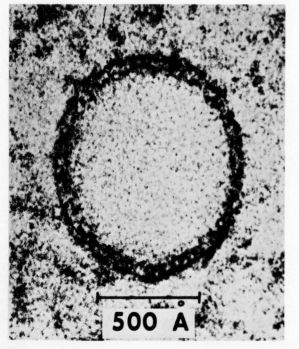

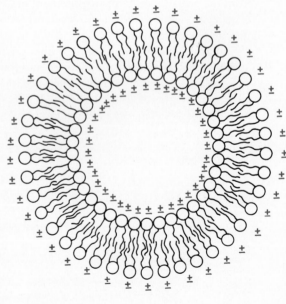

(*a*)   (*b*)

bilayers because these molecules contain two hydrophobic tails rather than
one. Surfactants form micelles of different sizes and shapes, but the maximum

125

**The Lipids:
Biological Amphiphiles**

size a globular micelle can attain is limited by the structure of the surfactant;
when the surface area per head group is about 60 Å^2, micelles approach their
maximal size in aqueous media.   In contrast, the surface area per head group in
two-chain amphiphiles in a bilayer arrangement is about the same as that of a
single-chain amphiphile in a globular micelle.   Thus, the bilayer arrangement is
optimal for two-chain amphiphiles since globular micelles of these compounds
of a size similar to the surfactants shown in Table 7.4 would have too large a
surface area per head group.

Although a more detailed discussion of biological membranes is given in
Chap. 13, it should be emphasized here that lipid forms the matrix of biological
membranes and that proteins are anchored to the membrane lipid.   Many
membrane-bound proteins are virtually impossible to solubilize into aqueous
solution without first disrupting the membrane.   Detergents (surfactants) are
widely used for this purpose, especially the neutral detergents such as the
Triton-X series.   It is thought that detergents of this type act by first being
incorporated into the membrane bilayer and that when the membrane is saturated with detergent, mixed micelles of detergent, membrane lipid, and protein
become soluble.   Another widely used detergent is the bile salt *deoxycholate*,
which is thought to act by breaking up the bilayer into small disklike pieces in
which the bile salt molecules cover the exposed hydrophobic ends.

REFERENCES

Books

Ansell, G. B., J. N. Hawthorne, and R. M. C. Dawson (eds): *Form and Function of Phospho-
lipids,* 2d ed., Elsevier, Amsterdam, 1973.
Gundstone, F. D.: *An Introduction to the Chemistry and Biochemistry of Fatty Acids and Their
Glycerides,* Chapman & Hall, London, 1967.
Gurr, M. I., and A. T. James: *Lipid Biochemistry, An Introduction,* 2d ed., Cornell, Ithaca,
N.Y., 1975.
Heftmann, E.: *Steroid Biochemistry,* Academic, New York, 1970.
Hilditch, T. P., and P. N. Williams: *The Chemical Constitution of Natural Fats,* 4th ed.,
Chapman & Hall, London, 1964.
Snyder, F. (ed.): *Ether Lipids: Chemistry and Biology,* Academic, New York, 1972.
Tanford, C.: *The Hydrophobic Effect: Formation of Micelles and Biological Membranes,* 2d ed.,
Wiley, New York, 1980.

Review Articles

Bangham, A. D.: Lipid Bilayers and Biomembranes, *Annu. Rev. Biochem.* **41**:753–776
(1972).
Helenius, A., and K. Simons: Solubilization of Membranes by Detergents, *Biochim.
Biophys. Acta* **415**:29–79 (1975).
Kates, M., and M. K. Wassef: Lipid Chemistry, *Annu. Rev. Biochem.* **39**:323–351 (1970).
Ledeen, R.: The Chemistry of Gangliosides: A Review, *J. Am. Oil Chem. Soc.* **43**:57–
66 (1966).
Stoffel, W.: Sphingolipids, *Annu. Rev. Biochem.* **40**:57–82 (1971).

Chapter 8

The Nucleic Acids

Structure and properties.

Nucleic acids are the macromolecules that carry genetic information and are responsible for transmission of the information for the synthesis of proteins. These aspects of molecular genetics are described in Chaps. 27 through 29. In this chapter the structure and physical properties of nucleic acids are considered.

The nucleic acids are of two kinds: those containing ribose (ribonucleic acids, RNA) and those containing deoxyribose (deoxyribonucleic acids, DNA). Both are linear polymers of *nucleotides* (see below) that are formed by phosphodiester linkages between the 5'-phosphate of one nucleotide and the 3'-hydroxyl group of the sugar of the adjacent one.

COMPONENTS OF NUCLEIC ACIDS

Nucleotides consist of three components: a pyrimidine or purine base linked to a sugar, either ribose or deoxyribose, and phosphate esterified to the sugar at carbon 5. Esterification may also occur at carbons 2 or 3. In addition to the nucleic acids, derivatives of the nucleotides include nucleoside polyphosphates; several coenzymes, and various activated intermediates in polysaccharide, lipid, and protein biosynthesis.

Pyrimidines

These are all derivatives of the heterocyclic compound *pyrimidine*. Its structure and the convention of numbering the positions in the ring are

Any cross-references coded M refer to the companion text, *Principles of Biochemistry: Mammalian Biochemistry.*

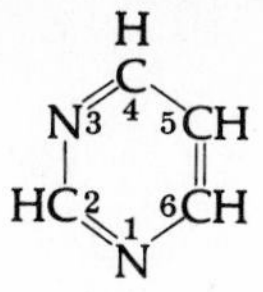

$$
\begin{array}{c}
\overset{H}{\underset{}{C}} \\
\underset{3}{N}\ 4\ 5CH \\
HC\underset{2}{\overset{}{}}\ 1\ 6CH \\
N
\end{array}
$$

Pyrimidine

The principal pyrimidines found in RNA are uracil and cytosine; in DNA they are thymine and cytosine. Methylated and other pyrimidine derivatives are found in some nucleic acids (page 152).

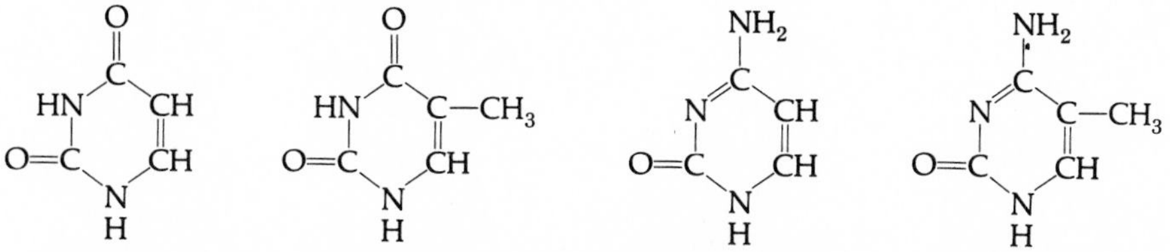

$$
\begin{array}{cccc}
O & O & NH_2 & NH_2 \\
\parallel & \parallel & & \\
C & C & C & C \\
HN\quad CH & HN\quad C-CH_3 & N\quad CH & N\quad C-CH_3 \\
O=C\quad CH & O=C\quad CH & O=C\quad CH & O=C\quad CH \\
N & N & N & N \\
H & H & H & H \\
\text{Uracil} & \text{Thymine} & \text{Cytosine} & \text{5-Methylcytosine}
\end{array}
$$

The pyrimidines all show *lactam-lactim tautomerism* and may be written in either form, as illustrated for uracil. At neutral and acid pH values, the lactam forms predominate.

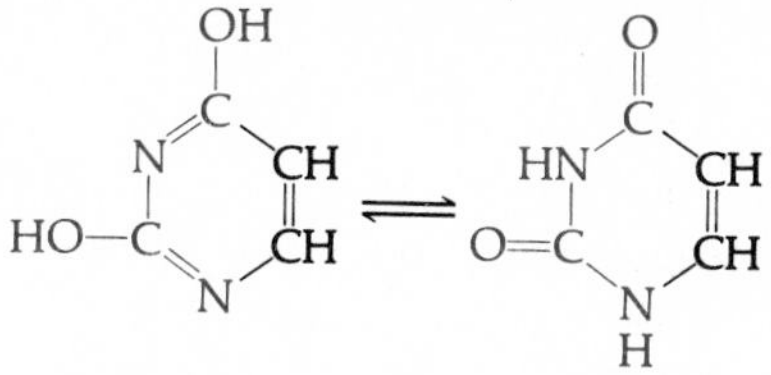

$$
\begin{array}{ccc}
OH & & O \\
| & & \parallel \\
C & & C \\
N\quad CH & \rightleftharpoons & HN\quad CH \\
HO-C\quad CH & & O=C\quad CH \\
N & & N \\
& & H
\end{array}
$$

Lactim form   *Lactam form*

Uracil

Purines

The parent substance, purine, contains the six-membered pyrimidine ring fused to a second ring formed by two additional nitrogen atoms and one carbon atom. Adenine and guanine are the major purines of both types of nucleic acids.

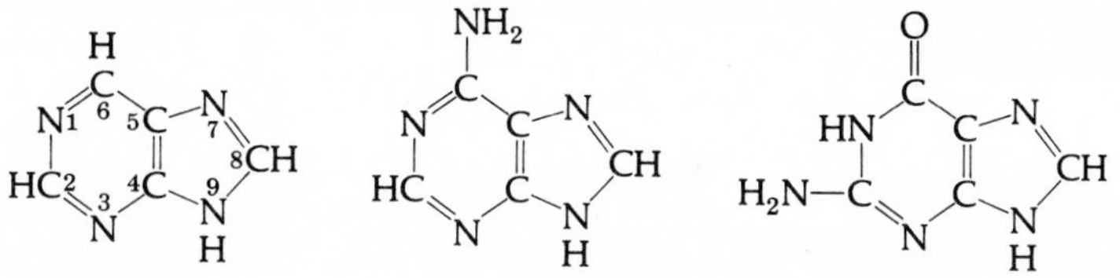

$$
\begin{array}{ccc}
H & NH_2 & O \\
C & C & \parallel \\
N1\ 6\ 5C-N & N\quad C-N & C \\
\quad\quad 7 & \quad\quad & HN\quad C-N \\
HC2\ 3\ 4C\ 9\ 8CH & HC\quad C\quad CH & H_2N-C\quad C\quad CH \\
N\quad N & N\quad N & N\quad N \\
H & H & H \\
\text{Purine} & \text{Adenine} & \text{Guanine}
\end{array}
$$

Of more limited distribution are hypoxanthine and various methylated and other derivatives of adenine and guanine (page 152). Other important purines are xanthine and uric acid.

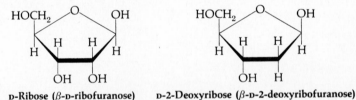

Hypoxanthine   **Xanthine**

Uric acid (lactim form)   **Uric acid (lactam form)**

Like the pyrimidines, purines show lactam-lactim tautomerism, illustrated above for uric acid. The lactim form of uric acid is weakly acidic ($pK = 4.5$, 10.3) and forms salts such as mono- and disodium or potassium urates. Uric acid and its salts are sparingly soluble in water. Other purines include caffeine (1,3,7-trimethylxanthine) and theobromine (3,7-dimethylxanthine), which have important pharmacological actions.

Sugars

The sugar in RNA is D-ribose, which occurs in the furanose form (Chap. 6); the sugar in DNA is D-2-deoxyribose.

D-Ribose (β-D-ribofuranose)   D-2-Deoxyribose (β-D-2-deoxyribofuranose)

Nucleosides

The combination of a purine or pyrimidine with a pentose is referred to as a *nucleoside*. Adenine linked to ribose is called adenosine, the guanine ribonucleoside is guanosine, and, correspondingly, the pyrimidine ribonucleosides are cytidine and uridine. The analogous nucleosides formed with deoxyribose are deoxyribonucleosides; thus, adenine deoxyribonucleoside or deoxyadenosine, deoxycytidine, etc. Note that the deoxyribonucleoside of thymine, which is found primarily in DNA, is called thymidine, not deoxythymidine. Thymine also occurs in one type of RNA, termed transfer RNA (page 152). In this instance, the names thymine ribonucleoside or ribosylthymine are used.

The purine ribonucleosides have a β-glycosidic linkage from carbon 1 of the sugar to the nitrogen in position 9, as shown for adenosine. The pyrimidine ribonucleosides are N-1 glycosides, as shown for cytidine.

Adenosine
(9-β-D-ribofuranosyladenine)

Cytidine
(1-β-D-ribofuranosylcytosine)

Like *O*-glycosides, these *N*-glycosides are stable in alkali. Purine nucleosides are readily hydrolyzed by acid whereas pyrimidine nucleosides are hydrolyzed only after relatively prolonged treatment with concentrated acid.

Nucleosides containing deoxyribose possess the same types of glycosidic linkages and are the 9-β-D-2'-deoxyribofuranosides of guanine and adenine and the 1-β-D-2'-deoxyribofuranosides of cytosine and thymine. In the terminology of nucleosides and their derivatives, the primed numbers refer to positions in the sugar moiety.

Nucleotides

As phosphoric acid esters of nucleosides, the nucleotides are strong acids with pK' values near 1 for the primary phosphate ionization. Phosphorylation of the ribo- and deoxyribonucleotides occurs at position C-5' to yield the corresponding nucleoside 5'-phosphates; ribonucleoside 2'- and 3'-phosphates and the deoxyribonucleoside 3'-phosphates are also known.

All cells contain adenylic acid, adenosine-5'-phosphate, as well as the polyphosphate compounds adenosine-5'-diphosphate and adenosine-5'-triphosphate. Analogous nucleoside 5'-mono-, di-, and triphosphates of other purines and pyrimidines, as both the ribose and deoxyribose derivatives, are also widely distributed. The structures of deoxyadenosine-5'-phosphate and inosine-5'-phosphate (inosinic acid) are shown.

Deoxyadenosine-5'-phosphate

Inosine-5'-phosphate

TABLE 8.1

Nucleotide	Abbreviation
Adenosine monophosphate (adenylic acid)	AMP
Adenosine diphosphate	ADP
Adenosine triphosphate	ATP
Guanosine monophosphate (guanylic acid)	GMP
Guanosine diphosphate	GDP
Guanosine triphosphate	GTP
Cytidine monophosphate (cytidylic acid)	CMP
Cytidine diphosphate	CDP
Cytidine triphosphate	CTP
Uridine monophosphate (uridylic acid)	UMP
Uridine diphosphate	UDP
Uridine triphosphate	UTP
Thymine ribonucleoside monophosphate (ribothymidylic acid)	TMP
Deoxyadenosine monophosphate (deoxyadenylic acid)	dAMP
Deoxyguanosine monophosphate (deoxyguanylic acid)	dGMP
Thymidine monophosphate (thymidylic acid)	dTMP
Deoxycytidine monophosphate (deoxycytidylic acid)	dCMP

*This nomenclature refers only to 5' substituents. Other monophosphates are 3'-AMP, 2'-AMP, 3'-GMP, etc. For additional abbreviations and symbols, see inside back cover, Table 8.6, and *J. Biol. Chem.* **245:**5171, 1970.

Some of these mono-, di-, and triphosphates are listed in Table 8.1, with the commonly used abbreviations for their designation.

STRUCTURE OF DEOXYRIBONUCLEIC ACIDS

Internucleotide Linkages

The deoxyribonucleic acids are polynucleotides in which the phosphate residues of each nucleotide form diester linkages between the deoxyribose moieties. Proof that the internucleotide bonds are between C-3' and C-5' is derived from examination of enzymic hydrolysates of DNA (see Table 8.3 below). Hydrolysis of DNA by pancreatic *deoxyribonuclease* (DNase) followed by *snake venom phosphodiesterase* yields a mixture of deoxyribonucleoside-5'-monophosphates; the combined action of DNase II and *spleen phosphodiesterase* yields deoxyribonucleoside-3'-monophosphates. Thus, DNA is a polymer with internucleotide linkages between C-3' and C-5', as illustrated in Fig. 8.1 for a portion of a DNA chain.

To show the DNA structure schematically, the type of diagram given in Fig. 8.1 is used, in which the horizontal line represents the carbon chain of the sugar with the base attached at C-1'. The diagonal line indicates the C-3' phosphate linkage near the middle of the horizontal line; that at the end of the horizontal line denotes the C-5' phosphate linkage. This diagrammatic notation is used for both DNA and RNA.

For long sequences of polynucleotides, another shorthand system is also used. The letters A, G, C, U, and T represent the nucleosides, as in Fig. 8.1. The phosphoric group is shown by the letter p; when p is placed to the right of

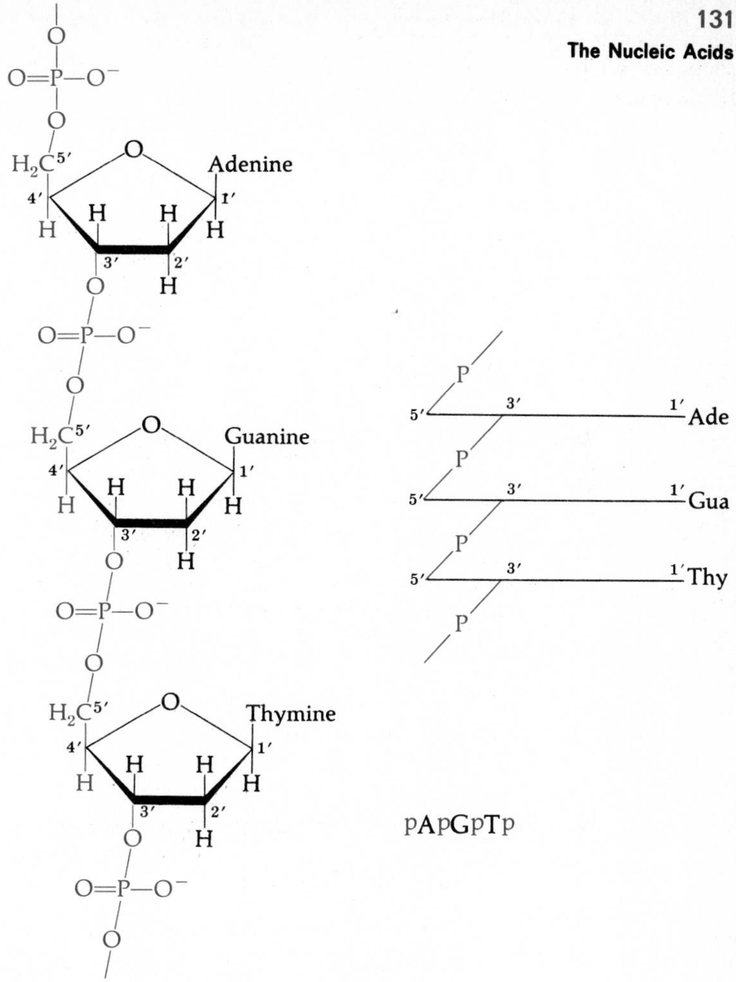

Adenine

Guanine

Thymine

pApGpTp

Figure 8.1 Representation of a portion of a DNA chain showing the position of the internucleotide linkage between C-3′ and C-5′. Schematic representations are given at the right.

the nucleoside symbol, esterification is at C-3′; when p is placed to the left, esterification is at C-5′. Thus ApUp is a dinucleotide with a monoester at C-3′ of a uridine and a phosphodiester bond between C-5′ of U and C-3′ of A. Unless it is evident that deoxynucleotides are under discussion, it is useful to specify this, as in d-ApTpGpTp, etc., or as in d-ATGT, etc., indicating that all these nucleotides contain deoxyribose.

Composition of DNA

Table 8.2 gives the base compositions of DNA from a number of species. In all cases, the amount of purines (Pur) is equal to the amount of pyrimidines (Pyr); that is, Pur/Pyr is equal to 1, as is also the ratio of adenine (A) to thymine (T). Similarly, the ratio of guanine (G) to cytosine (C) (plus methylcytosine, where it occurs) is equal to 1.

The DNA of each species shows a characteristic composition that is unaffected by age, conditions of growth, various environmental factors, etc. Indeed,

TABLE 8.2 Composition of DNA of Various Species

Species	Base proportions,* mol %				$\dfrac{A + T}{G + C}$	$\dfrac{A}{T}$	$\dfrac{G}{C}$	$\dfrac{Pur}{Pyr}$
	G	A	C	T				
Sarcina lutea	37.1	13.4	37.1	12.4	0.35	1.08	1.00	1.02
Alcaligenes faecalis	33.9	16.5	32.8	16.8	0.50	0.98	1.03	1.02
Brucella abortus	20.0	21.0	28.9	21.1	0.73	1.00	1.00	1.00
E. coli K12	24.9	26.0	25.2	23.9	1.00	1.09	0.99	1.08
Wheat germ	22.7	27.3	22.8†	27.1	1.19	1.01	1.00	1.00
Bovine thymus	21.5	28.2	22.5†	27.8	1.27	1.01	0.96	0.99
Human liver	19.5	30.3	19.9	30.3	1.54	1.00	0.98	0.99
Saccharomyces cerevisiae	18.3	31.7	17.4	32.6	1.80	0.97	1.05	1.00
Clostridium perfringens	14.0	36.9	12.8	36.3	2.70	1.02	1.09	1.04

*G = guanine; A = adenine; C = cytosine; T = thymine; Pur/Pyr = purine/pyrimidine.
† Cytosine + methylcytosine.
Source: Compiled from the work of several investigators.

in higher organisms, samples of DNA from different organs or tissues of the same species are identical in composition. The characteristic composition of DNA from a given source can be indicated by the ratio $(A + T)/(G + C)$. In bacteria a wide range of compositions is encountered, some being high in $A + T$, others in $G + C$. In higher organisms the range is more limited; in most animals the ratio $(A + T)/(G + C)$ is found to be from 1.3 to 1.5, and in higher plants from 1.1 to 1.7.

In addition to the two purines adenine and guanine and the two pyrimidines cytosine and thymine, low levels of 6-methyladenine and 5-methylcytosine have been identified in the DNA of a wide variety of bacteria. Some of these methylated bases protect the DNAs in which they are present from attack by restriction deoxyribonucleases (Chap. 27). Methylcytosine is also a characteristic component of the DNA of certain higher plants and animals, the DNA of plants being richer in this pyrimidine than the DNA of animals. Wheat germ, the richest source yet found, contains 6 mol of methylcytosine per 100 mol of bases. In all cases, however, it replaces an equivalent amount of cytosine. The function of methylcytosine in these instances is unknown.

Action of Nucleases

Nucleases hydrolyze specifically DNA or RNA, or they may attack both kinds of polynucleotide. These enzymes are generally of two types: (1) *exonucleases,* which require a terminus at which to initiate hydrolysis, and (2) *endonucleases,* which do not require a terminus and which may attack at one or at many sites within a polynucleotide.

Exonucleases may act from either the 3′ or the 5′ terminus to yield mononucleotides or a mixture of mono- and oligonucleotides. They show no base or sequence specificity; however, they may be highly specific in their ability to attack single- or double-stranded polynucleotides (see below).

Endonucleases may also discriminate between single-stranded and duplex structures. They may be specific for a certain base sequence, e.g., the restriction endonucleases (Chap. 27), or they may be base-specific. Thus, the ribonuclease (RNase) of bovine pancreas hydrolyzes internucleotide linkages at a point distal to the phosphate of a pyrimidine nucleotide that is esterified at the 3′ position. The points of cleavage by this enzyme are shown by the vertical dotted lines marked a.

In this schematic structure, Pyr represents a pyrimidine. The final products of *pancreatic RNase* action are pyrimidine nucleoside 3′-phosphates and oligonucleotides terminating in a pyrimidine 3′-P, that is, . . . GpApPyrp. *Ribonuclease T1* from *Aspergillus oryzae* cleaves RNA to yield guanosine 3′-phosphate and oligonucleotides terminating in guanosine 3′-P. The sites of cleavage (above) are marked b. Both are very useful reagents in RNA sequence analysis. The use of restriction endonucleases in DNA sequence analysis is described below. The specificities of some of the nucleases are given in Table 8.3.

Determination of DNA Sequence

The determination of nucleotide sequences has been greatly simplified by the discovery of site-specific restriction endonucleases that can cleave large DNA molecules into small fragments (Table 8.3) and the development of rapid and accurate methods, both chemical and enzymic, for the sequence analysis of these fragments. A DNA chain is first cleaved by one or more restriction nucleases into segments that can be separated by electrophoresis in agarose gels. If required, these segments can be purified and their amount amplified by molecular cloning (Chap. 27). The nucleotide sequence can then be determined by several methods. In one of these, the DNA segment is labeled at its 5′ terminus with *polynucleotide kinase,* an enzyme that catalyzes transfer of ^{32}P from $\gamma\text{-}^{32}P$-labeled ATP to the hydroxyl group of the 5′ terminus. Partial cleavage at each of the four bases yields a set of labeled fragments extending from the labeled terminus to each of the positions of that base in the sequence. This is accomplished by chemical modification of a specific purine or pyrimidine base, elimination of the base from its linkage with deoxyribose, and scission of the DNA strand at the sugar without its base. Cleavage at guanine is achieved by methylation with dimethylsulfate (step A), followed by piperidine-catalyzed elimination of the methylated base (step B), and alkaline scission of the chain by β elimination of phosphates from the modified sugar (step C).

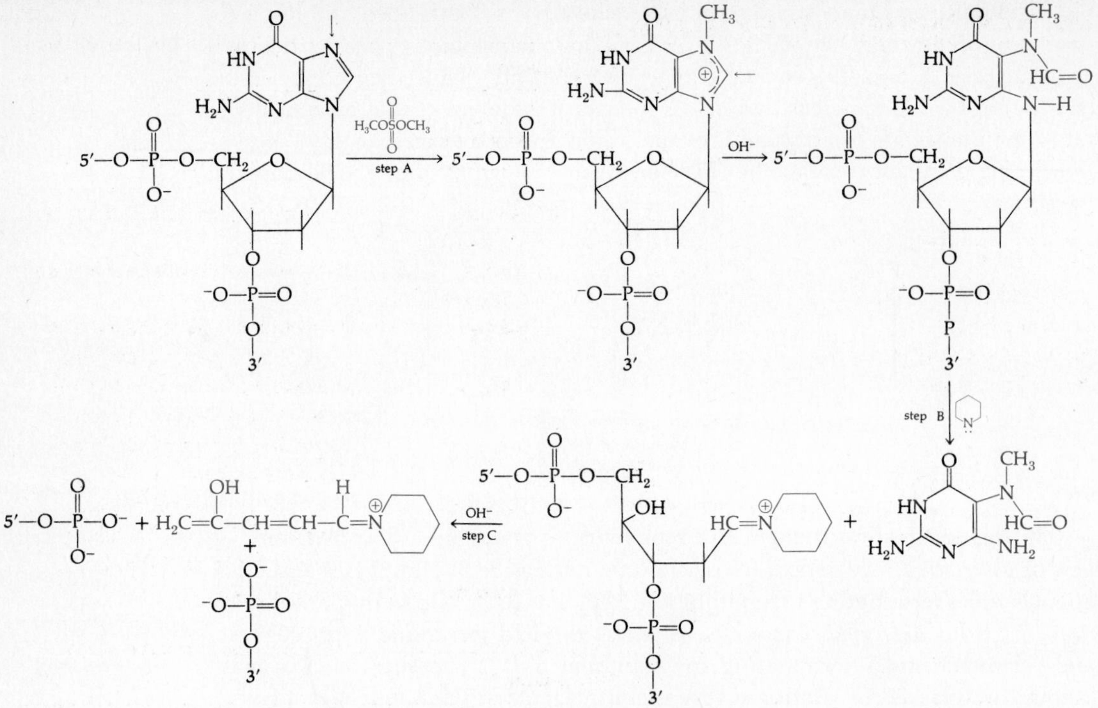

Hydrazinolysis modifies cytosine and thymine at low ionic strength but only cytosine at high ionic strength; specific cleavage is achieved as indicated in the following reactions for cleavage at thymine:

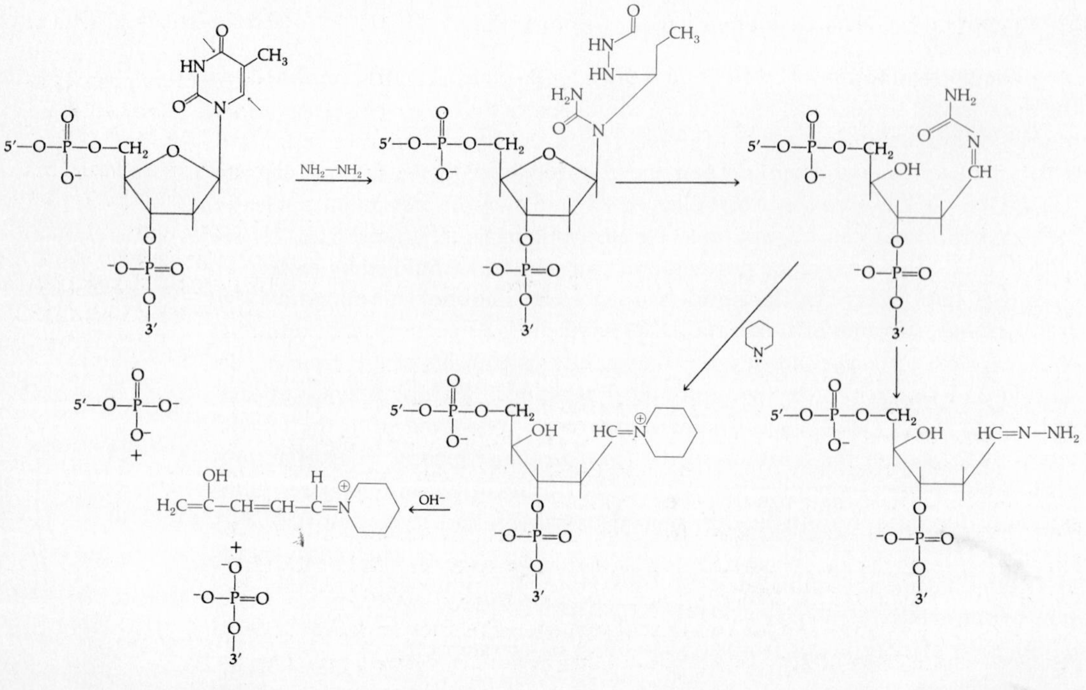

TABLE 8.3 Specificity of Some Nucleases

Endonucleases			
Enzyme	Substrate	Poly-nucleotide structure	Products
DNase I (bovine pancreas)	DNA	Single, duplex	5′-P-terminated oligonucleotides.
DNase II (calf thymus, spleen)	DNA	Single, duplex	3′-P-terminated oligonucleotides
S$_1$ endonuclease (*Aspergillus oryzae*)	DNA, RNA	Single	Nucleoside 5′-phosphates; 5′-P-terminated oligonucleotides
Pancreatic RNase	RNA	Single	Pyrimidine nucleoside 3′-phosphates and oligonucleotides terminated with pyrimidine nucleoside 3′-phosphate
RNase T1 (*Aspergillus oryzae*)	RNA	Single	Guanosine-3′-phosphate and oligonucleotides terminated with guanosine-3′-phosphate
Restriction endonucleases	"Unmodified" DNA	Duplex	See Chap. 27

Exonucleases				
Enzyme	Substrate	Poly-nucleotide structure	Terminus attacked	Products
Snake venom phospho-diesterase	RNA, DNA	Single	3′	Nucleoside 5′-phosphates
Spleen phospho-diesterase	RNA, DNA	Single	5′	Nucleoside 3′-phosphates
E. coli exonuclease I	DNA	Single	3′	Nucleoside 5′-phosphates and dinucleotide from 5′ terminus
E. coli exonuclease III	DNA	Duplex	3′	Nucleoside 5′-phosphates and single-stranded DNA
E. coli exonuclease VII	DNA	Single	3′ or 5′	5′-P-terminated oligonucleotides
Phage λ-induced exonuclease	DNA	Duplex	5′	Nucleoside 5′-phosphates and single-stranded DNA

Other chemical reactions provide additional kinds of specific cleavage, for example, removal of purines with acid followed by piperidine treatment cleaves at G + A, whereas alkaline treatment cleaves at A and, to a lesser extent, at C.

If the ^{32}P-labeled products for each of the four types of cleavage are separated on the basis of their size by gel electrophoresis (page 136), then it is possible to read the nucleotide sequence from the gel simply by noting which of the four base-specific agents gave cleavage fragments of a given size, as illustrated in Fig. 8.2 for a hypothetical DNA segment containing 15 nucleotides.

As in the determination of amino acid sequence, the nucleotide sequence of the DNA chain from which the restriction fragments are derived can then be

(a)

```
 * 1    2    3    4    5    6    7    8    9    10   11   12   13   14   15
 p Cp  Tp   Gp   Cp   Tp   Ap   Cp   Gp   Ap   Cp   Gp   Tp   Ap   Gp   C
```

G
```
 *
 p Cp  Tp  (1-2)
 *
 p Cp  Tp   Gp   Cp   Tp   Ap   Cp  (1-7)
 *
 p Cp  Tp   Gp   Cp   Tp   Ap   Cp   Gp   Ap   Cp  (1-10)
 *
 p Cp  Tp   Gp   Cp   Tp   Ap   Cp   Gp   Ap   Cp   Gp   Tp   Ap  (1-13)
```

G + A
```
 *
 p Cp  Tp  (1-2)
 *
 p Cp  Tp   Gp   Cp   Tp  (1-5)
 *
 p Cp  Tp   Gp   Cp   Tp   Ap   Cp  (1-7)
 *
 p Cp  Tp   Gp   CP   TP   Ap   Cp   Gp  (1-8)
 *
 p Cp  Tp   Gp   Cp   Tp   Ap   Cp   Gp   Ap   Cp  (1-10)
 *
 p Cp  Tp   Gp   Cp   Tp   Ap   Cp   Gp   Ap   Cp   Gp   Tp  (1-12)
 *
 p Cp  Tp   Gp   Cp   Tp   Ap   Cp   Gp   Ap   Cp   Gp   Tp   Ap  (1-13)
```

C
```
 *
 p
 *
 p Cp  Tp   Gp  (1-3)
 *
 p Cp  Tp   Gp   Cp   Tp   Ap  (1-6)
 *
 p Cp  Tp   Gp   Cp   Tp   Ap   Cp   Gp   Ap  (1-9)
 *
 p Cp  Tp   Gp   Cp   Tp   Ap   Cp   Gp   Ap   Cp   Gp   Tp   Ap   Gp  (1-14)
```

C + T
```
 *
 p
 *
 p Cp  (1)
 *
 p Cp  Tp   Gp  (1-3)
 *
 p Cp  Tp   Gp   Cp  (1-4)
 *
 p Cp  Tp   Gp   Cp   Tp   Ap  (1-6)
 *
 p Cp  Tp   Gp   Cp   Tp   Ap   Cp   Ap   Ap  (1-9)
 *
 p Cp  Tp   Gp   Cp   Tp   Ap   Cp   Ap   Ap   Cp   Gp  (1-11)
 *
 p Cp  Tp   Gp   Cp   Tp   Ap   Cp   Ap   Ap   Cp   Gp   Tp   Ap   Gp  (1-14)
```

(b)

	G	G + A	C	C + T
1-15	▬	▬	▬	▬
1-14			▬	▬
1-13	▬	▬		
1-12		▬		
1-11				▬
1-10	▬	▬		
1-9			▬	▬
1-8		▬		
1-7	▬	▬		
1-6			▬	▬
1-5			▬	
1-4				▬
1-3			▬	▬
1-2	▬	▬		
1	▬			
*P			▬	▬

Figure 8.2 (*a*) The sequences of nucleotides that would be obtained from a DNA fragment (top) containing 15 nucleotides after chemical cleavage by four different procedures that cleave at G, G + A, C, or C + T. (*b*) A "ladder" gel that resolves the mixtures of nucleotides (shown in A) on the basis of size. The numbers at the left indicate the number of nucleotides in each band on the gel. The sequence of the parent nucleotide can be read from the 5′ end to the 3′ end by noting which cleavage procedure gives a band on the gel from bottom to top. Thus the lanes for C + T and C contain [^{32}P]PO$_4$ (*P) indicating the sequence is pC. Since the lane from C + T but no other lane contains a band migrating just above [^{32}P]PO$_4$, the sequence is pCpT. The third nucleotide is pG, and the sequence is pCpTpG, since lanes G and G + A contain bands. The remainder of the sequence is deduced in a similar fashion, reading from bottom to top of the gel.

determined from overlapping sequences which establish the arrangement of the fragments within the molecule.

DNA segments containing up to 250 nucleotides can be sequenced by this means, as shown in Fig. 8.3.

DOUBLE-HELICAL STRUCTURE OF DNA

DNA is a right-handed double helix in which two polynucleotide strands are wound around each other so that there are 10 base pairs for each turn. The double helix which has a diameter of 20 Å has both major and minor grooves (Fig. 8.4).

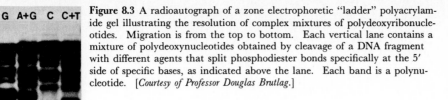

| G | A+G | C | C+T |

Figure 8.3 A radioautograph of a zone electrophoretic "ladder" polyacrylamide gel illustrating the resolution of complex mixtures of polydeoxyribonucleotides. Migration is from the top to bottom. Each vertical lane contains a mixture of polydeoxyribonucleotides obtained by cleavage of a DNA fragment with different agents that split phosphodiester bonds specifically at the 5′ side of specific bases, as indicated above the lane. Each band is a polynucleotide. [*Courtesy of Professor Douglas Brutlag.*]

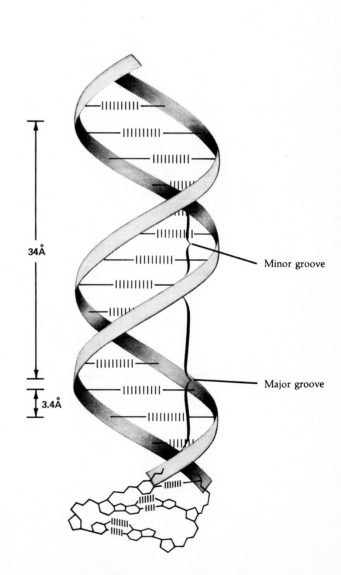

34Å

3.4Å

Minor groove

Major groove

Figure 8.4 Diagrammatic model of the DNA double-helix (B form). [*From A. Kornberg, DNA Replication, p. 11, Freeman, San Francisco, 1980.*]

The two chains are held together in part by hydrogen bonding, each amino group being joined to a keto group, i.e., adenine to thymine, guanine to cytosine, etc. The glycosidic bonds joining each base of a pair to the sugar-phosphate backbone are the same distance apart for each pair (10.85 Å) and are symmetrically related about a dyad (twofold) axis located in the plane of the pair. Hence, the connection between two C-1′ atoms (at the same level in each helix) can be made by any of four base pairs: A-T, T-A, G-C, C-G. The types of hydrogen bonding are shown in Fig. 8.5. *The bonding is always of a purine to a pyrimidine.*

The two chains are not identical but are *complementary* in terms of the appropriate base pairing, A to T and C to G, and they are *antiparallel.* If, for example, one chain is linked 5′ to 3′ with respect to AG, GT, TC, etc., then the complementary chain is linked 3′ to 5′ with respect to TC, CA, AG, etc. (Fig. 8.6).

The planes of the adjacent base pairs are 3.4 Å apart and there are 10 base pairs for each turn of the helix; thus, each base pair is rotated 36° relative to its neighbors and each full turn has a length of 34 Å (pitch of the helix). These

Figure 8.5 Dimensions and hydrogen binding of (*a*) thymine to adenine and (*b*) cytosine to guanine in double helix of DNA. Note the two hydrogen bonds between A and T and three hydrogen bonds between G and C. [*From M. H. F. Wilkins and S. Arnott, J. Mol. Biol.* **11**:291 (1965).]

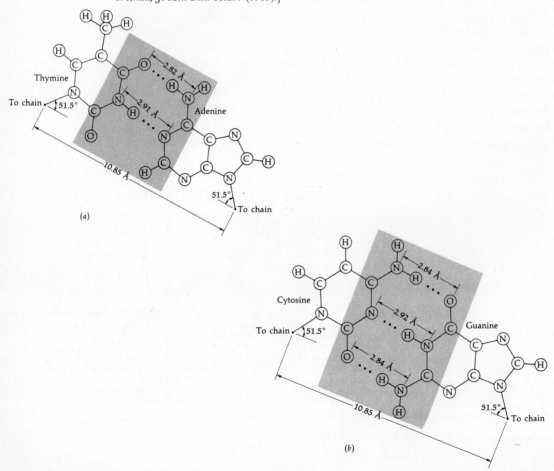

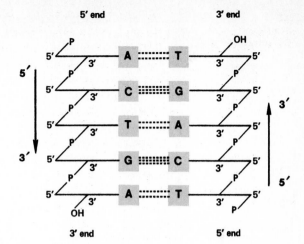

Figure 8.6 Segment of duplex DNA showing the antiparallel orientation of the complementary chains. [*From A. Kornberg, DNA Replication, p. 12, Freeman, San Francisco, 1980.*]

parameters apply to the B structure of the sodium salt of DNA found in fibers at 92% humidity. Under physiological conditions, the number of base pairs per turn is increased from 10.0 to 10.4.

Three similar forms of double-helical DNA have been observed: A, B, and C. The B form is that described above and appears to be the one commonly found in solution and in vivo. The B form of DNA is converted to the A form when the humidity of the fiber environment is reduced below 75%. The A form differs from the B in that the base pairs are not perpendicular to the helical axis but are tilted about 20°. As a result, the pitch is reduced to 28 Å (from 34 Å) and there are 11 pairs per turn, resulting in a shortening of about 25 percent (Fig. 8.7). The C form, which is very much like the B structure, has a pitch of 33 Å with 9 base pairs per turn.

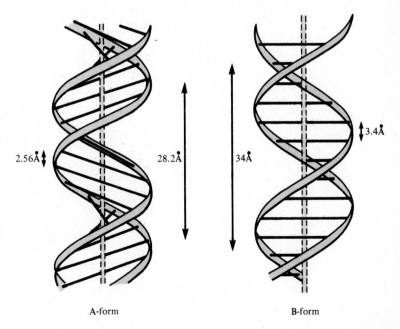

Figure 8.7 A schematic representation of the difference in structure between the A and B forms of DNA. The two ribbons represent the phosphate-sugar chains, and the horizontal rods represent the bonding between the pairs of bases. The vertical line indicates the fiber axis.

A-form

B-form

Yet another form of DNA has been observed in crystals of the hexanucleo-tide d(CpG)$_3$ and in fibers of the alternating d(GC)$_n$ polymers (see below). This structure, which is in the form of a left-handed rather than a right-handed helix, has been termed *Z DNA*. In the Z helix the conformations of the guanine and cytosine deoxyribonucleotides differ considerably from one another, so that the repeating unit becomes a dinucleotide, unlike the B DNA helix in which the repeating unit is a single nucleotide. The result is a staggered zigzag course for the deoxyribose-phosphate backbone, and hence the term Z DNA (Fig. 8.8). Although thus far observed only in synthetic polymers, it is possible that the Z conformation may exist in DNA in vivo, in particular in stretches of alternating guanosine-cytosine or possibly other kinds of alternating purine-pyrimidine se-quences.

Synthetic Polynucleotides

Polynucleotides containing only one or two bases have been synthesized enzymically with the use of polynucleotide phosphorylase and RNA polymerase

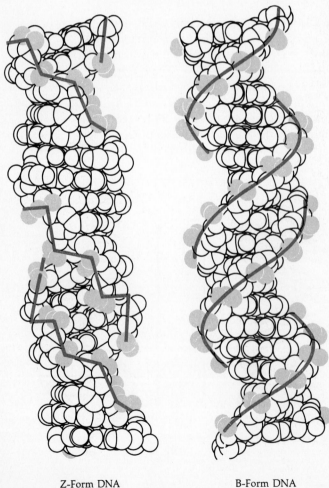

Z-Form DNA   B-Form DNA

Figure 8.8 Space filling model of the Z and B forms of DNA. The irregularity of the Z DNA backbone is illustrated by the heavy lines that go from phosphate to phosphate along the chain. [*From A. H. J. Wang, G. J. Quigley, F. J. Kolpak, J. L. Crawford, J. H. Van Boom, G. van der Marel, and A. Rich, Nature* **282**:*680* (*1979*).]

(Chap. 28) for the polyribonucleotides, and DNA polymerase I (Chap. 27) for
the polydeoxyribonucleotides.

141
The Nucleic Acids

When polyadenylic acid, poly(A), is mixed with polyuridylic acid, poly(U), a double-stranded helical structure, poly(A)(U), similar to natural DNA is formed in which the hydrogen bonding is between A and U pairs. Under certain conditions, triple-stranded helices consisting of one poly(A) and two poly(U) chains can be generated. Two- and three-stranded structures consisting of polyinosinic or polyguanylic and polycytidylic acids have also been synthesized.

The synthetic polydeoxyribonucleotide polyd(AT) has a double-helical structure in which each chain possesses a strictly alternating base sequence . . . dA $\cdot$ dT $\cdot$ dA $\cdot$ dT . . . . X-ray diffraction analysis has demonstrated that the lithium salt exists in the B form with helix dimensions identical to those of lithium DNA. As expected for a DNA composed of only adenylic and thymidylic acid residues, its melting temperature T_m (page 142) is very low.

Simple repetitive polymers, similar to those described above, have been found in the centromeric regions of eukaryotic chromosomes (Chap. 29). During equilibrium centrifugation of animal DNA in cesium chloride density gradients, they appear as a *satellite* band of low density. One of these satellites present in the DNA of several crab species is very similar in composition and sequence to the alternating polyd(AT) copolymer. It is composed of approximately 97% deoxyadenylic and thymidylic acids, almost all of which are arranged in an alternating sequence; the 3% of deoxyguanylic and deoxycytidylic acid residues are distributed throughout the satellite DNA. Similar DNAs have been isolated from other animals.

Density of DNA

When a concentrated solution of cesium chloride is centrifuged in the analytical ultracentrifuge at high speeds until equilibrium is attained, the opposing processes of sedimentation and diffusion (Chap. 4) produce a stable concentration gradient of the CsCl; that is, there is a continuous increase in density along the direction of centrifugal force. The density gradient formed is proportional to the centrifugal force according to the equation

$$\frac{d\rho}{d\tau} = \alpha\omega^2\tau$$

where ρ is the density as a function of the distance τ from the center of rotation, ω is the angular velocity, and α is a constant depending upon the nature of the salt. When the CsCl solution contains a small amount of DNA, at equilibrium the molecules of DNA will band at those zones of the centrifuge cell where their density and the density of the medium are exactly equal (*isopycnic*). The position of the DNA in the cell can be established by ultraviolet absorption photography. Since the gradient of solution density can be precisely estimated throughout the cell, the density of the DNA sample can be established. The technique is termed *isopycnic density-gradient centrifugation*. The buoyant density of DNA can be correlated with its G + C content by the empirical equation

$$\rho = 1.660 + 0.100(\text{G + C frequency}) \qquad \text{g/cm}^3$$

This property permits the fractionation of DNA molecules according to their G + C content.

Denaturation of DNA

In addition to hydrogen bonds, hydrophobic forces between the stacked purines and pyrimidines contribute to maintenance of the rigid two-stranded structure of DNA. Thus, reagents like formamide and urea, which increase the solubility of the aromatic groups in the surrounding aqueous medium, also tend to denature DNA. Similarly, the effect of a variety of agents and conditions, e.g., acid, alkali, heat, and low ionic strength, on DNA structure can be explained on the basis that the DNA, initially in a stiff helical two-stranded native structure, can be converted to a denatured state, which is a single-stranded flexible structure.

Several methods are used to assess the transition from the native to a denatured state, as well as to determine other properties of DNA.

Ultraviolet absorption All nucleic acids show a strong absorption in the ultraviolet with a maximum near 260 nm. When native DNA is altered, there is a marked *hyperchromic effect,* or increase in absorption. This change reflects a decrease in hydrogen bonding and is observed not only with DNA but with RNA and with synthetic polynucleotides that have a hydrogen bonded structure.

Optical rotation Native DNA shows a strong positive rotation, which is markedly decreased by denaturation.

Viscosity Solutions of native DNA possess a high viscosity because of the relatively rigid double-helical structure and long, rodlike character of the DNA. Disruption of the hydrogen bonds produces a marked decrease in viscosity.

Effect of Temperature Heating a sample of DNA in a given ionic environment produces an increase in ultraviolet absorption and a decrease in optical rotation and viscosity, reflecting the disruption of the hydrogen bonds between the two strands in the double helix.

Since the interactions between the bases in the two strands are cooperative in a manner similar to the interactions between molecules in a crystal, the disruption of the ordered helical structure occurs over a small temperature interval, much like the melting of a crystal. The more uniform the G + C content of the DNA, the narrower the width of the thermal transition. For this reason, the heat denaturation of double-stranded DNA is often referred to as the "melting" of the DNA and the temperature at which 50 percent of the DNA is denatured is termed the *melting temperature* T_m.

DNA preparations from diverse sources possess different T_m values, which depend on the absolute amounts of G + C and A + T (Fig. 8.9). The higher the content of G + C, the higher the transition temperature between the native two-stranded helix and the single-stranded form (Fig. 8.10). This is expected since the G-C pair can form a triply hydrogen bonded structure whereas the A-T pair can form only a doubly hydrogen bonded structure (Fig. 8.5). Indeed, T_m determinations, using careful calibration with DNA preparations of known

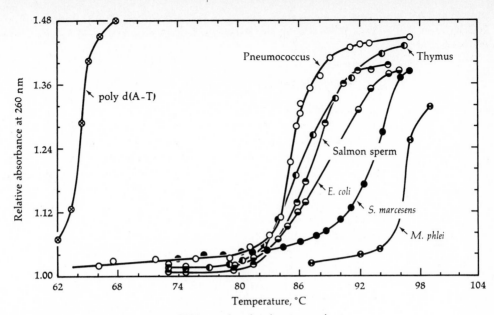

Figure 8.9 The absorbance-temperature curves for DNA samples of various composition. [*After P. Doty, in D. J. Bell and J. K. Grant (eds.), The Structure and Biosynthesis of Macromolecules, p. 8, Biochemical Society Symposia, no. 21, Cambridge, New York, 1962.*]

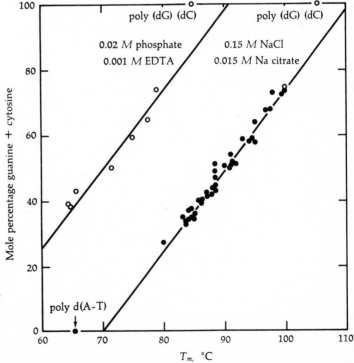

Figure 8.10 Dependence of the temperature midpoint T_m on the content of G + C. The points represent samples of DNA from various sources. [*After P. Doty, in D. J. Bell and J. K. Grant (eds.), The Structure and Biosynthesis of Macromolecules, p. 8, Biochemical Society Symposia, no. 21, Cambridge, New York, 1962.*]

143

composition, permit estimations of the relative G + C content of an unknown DNA by means of the empirical relationship

$$\% \text{ G} + \text{C} = 2.44(T_m - 69.3) \qquad \text{in } 0.2 \, M \, \text{NaCl}$$

Such studies must be performed at fixed ionic strength and pH, since these have a marked effect on the stability of DNA. The increased stability of DNA with increasing ionic strength is given by

$$\% \text{ G} + \text{C} = 2.44(T_m - 80.2 - 15.4 \log_{10} [\text{Na}^+])$$

The T_m is lowered by the addition of urea, an agent which disrupts hydrogen and hydrophobic bonds. For example, in $8 \, M$ urea the T_m is decreased by nearly 20°C. In 95% formamide, DNA is completely separated into single strands at room temperature.

Effect of pH Disruption of the double-stranded structure of DNA also occurs when ionic groups are present on the purine and pyrimidine bases. Near pH 12, ionization of enolic hydroxyl groups prevents the keto-amino group hydrogen bonding. Similarly, in acid solutions near pH 2 to 3, amino groups are protonated and the helix is disrupted.

Renaturation of DNA

The denaturation of DNA is a reversible process. If DNA is only partially denatured, e.g., by heat treatment, each molecule will renature in a rapid first-order reaction when the temperature is lowered. If DNA is completely denatured, the two complementary strands will reassociate (anneal) by a slower process consisting of two component reactions. The first is of the second order and involves nucleation of complementary sequences on two strands followed by a rapid first-order "zippering" reaction. If the initial concentration of denatured DNA is C_0 (moles per liter of DNA phosphate), the change in concentration dC of single-stranded DNA at any time obeys second-order kinetics according to the equation

$$\frac{dC}{dt} = K_2 C^2$$

the integrated form of which is

$$\frac{C}{C_0} = \frac{1}{1 + K_2 C_0 t}$$

The rate of renaturation is usually displayed as a plot of C/C_0 vs., log $C_0 t$ (Fig. 8.11). From such a plot $C_0 t_{\frac{1}{2}}$, the time at which $C/C_0 = 0.5$, and the second-order constant K_2, which is equal to $1/C_0 t_{\frac{1}{2}}$, can be determined. K_2 is characteristic of the DNA and is inversely related to N, the number of base pairs in the DNA when the DNA has no repeated sequences (Chap. 29). The $C_0 t_{\frac{1}{2}}$ is therefore directly proportional to N, which is called the *complexity* of the DNA.

The complexity N calculated from the slowest renaturing species [$C_0 t_{\frac{1}{2}} = 10^2$ to 10^3 (mol · s)/liter for a typical eukaryote] is consistent with the total amount of DNA in the genome. This slowly renaturing species which consists of DNA sequences represented only once in the genome can constitute

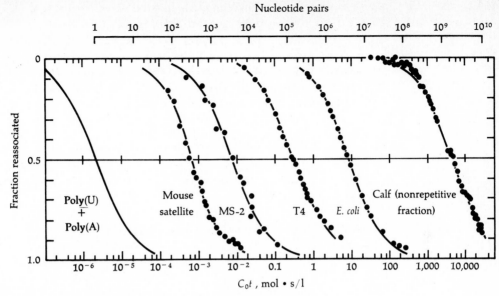

Figure 8.11 Renaturation of double-stranded nucleic acids from various sources. *[From R. J. Britten and D. E. Kohne, Science* **161***:529 (1968). Copyright 1968 American Association for the Advancement of Science.]*

up to 90 percent of the genome. These sequences represent structural genes, i.e., genes that code for proteins (Chap. 28). The material that renatures at an intermediate rate ($C_0 t_{\frac{1}{2}} = 10^{-2}$ to 10) consists of sequences repeated 100 to 1000 times in the genome. The function of these sequences is not known, but they are believed to be involved in the regulation of gene expression. The most rapidly renaturing sequences ($C_0 t_{\frac{1}{2}} < 10^{-2}$) are short nucleotide sequences present 100,000 to 1,000,000 times in the genome. These sequences are usually repeated in tandem and form satellites (Chap. 29).

Size of DNA

DNA molecules are long and unbranched. They may be linear, i.e., contain two ends, or they may be circular (Table 8.4). Their size can be stated in

TABLE 8.4

Size of Various DNA Molecules	Organism	Average number of base pairs per chromosome, thousands	Length, cm	Number of chromosomes (haploid)	Shape
	Simian virus 40 (SV40)	5.1	0.00017	1	Circular
	Bacteriophage φX174	5.4	0.00018	1	Circular single-strand
	Bacteriophage λ	46	0.0015	1	Linear
	E. coli	4,000	0.13	1	Circular
	Yeast	1,000	0.033	17	
	Drosophila melanogaster	41,000	1.4	4	
	Homo sapiens	125,000	4.1	23	

145

three types of units (length, number of base pairs, and mass). Thus 1 μm (10^{-4} cm) of DNA contains 3000 base pairs and has a mass of 2×10^6 daltons (660 daltons per base pair, Na^+ salt). DNA was originally thought to be no longer than 15,000 base pairs (5×10^{-4} cm) until it was discovered that it is very sensitive to hydrodynamic shear. When care is taken to avoid shear, considerably longer DNA molecules can be isolated (see Table 8.4).

The length of DNA molecules containing 3×10^2 to 3×10^5 base pairs can be measured directly by electron microscopy (Fig. 8.12). DNA molecules in the size range of 2×10^2 to 2×10^4 base pairs can be resolved by making use of the molecular-sieving effect of porous agarose gels. The mobility of DNA in these gels is directly proportional to the logarithm of the molecular weight over a limited size range, which can be altered by changing the density of the agarose gel (Fig. 8.13).

The classical method of determining the size of DNA is by sedimentation in a centrifugal field. The sedimentation coefficient s (page 43) is related to molecular weight (MW) of the DNA by the empirical equation

$$s = 0.0882 \times 10^{-13} MW^{0.346}$$

Determination of size by sedimentation is limited to DNA in the range 3×10^2 to 1.5×10^6 base pairs because of the effect of shear forces on larger molecules.

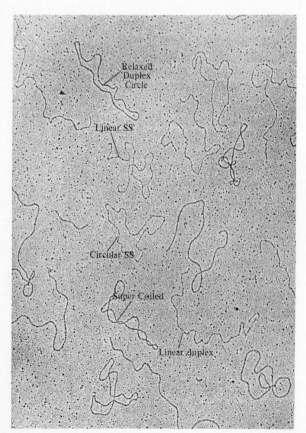

Figure 8.12 Electron micrograph of single-stranded circular and linear, duplex linear, relaxed closed circular, and supercoiled DNA. The DNA shown is from papilloma virus, type 1, and contains 7800 bases (or base pairs). SS = single-stranded. [*Courtesy of Dr. Louise Chow.*]

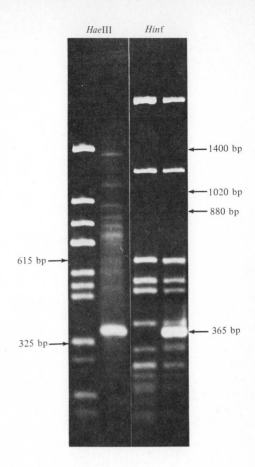

Figure 8.13 Separation by agarose gel electrophoresis of *E. coli* plasmid DNA (pSC101) cleaved with restriction endonucleases *Hae* III and *Hinf*. [*Courtesy of Professor Douglas Brutlag.*]

The technique of viscoelastic retardation can be used to measure the size of DNA molecules in excess of 10^8 base pairs. In this procedure, a rotor is floated in the solution of DNA to be measured and is rotated very slowly by an externally applied magnetic field. When the magnetic field is released, the rotor returns to its initial position because of the energy stored in the elastically extended DNA molecules in solution. The rate at which the rotor returns to its initial position is a function of the MW of the largest DNA molecule in the solution according to the equation

$$MW = 2.2 \times 10^8 \tau^{0.6}$$

where τ is the viscoelastic retardation time in seconds. τ is calculated by means of the equation

$$\theta(t) - \theta(\infty) = \theta(0)e^{-t/\tau}$$

where θ is the angle of the rotor measured at times t, ∞, and 0.

The length of the DNA in the largest chromosome of *Drosophila melanogaster*, as measured by this method, is in good agreement with the mass of DNA determined cytologically. These measurements have further shown that animal chromosomes, like bacterial and viral chromosomes, contain a single, uninterrupted DNA helix. However, DNA does not normally exist in an extended

147

form but is highly condensed within the chromosome of a cell or in the head of a virus (Chaps. 29 and 30). Thus, while the length of DNA in the largest human chromosome may be 8 cm, it is condensed in a mitotic chromosome whose length is only 5 nm.

Topology of DNA

An important physical property of DNA is its superhelical density. This parameter is a property of both linear and circular molecules; however, it is most readily observed and understood in circular DNA. A circular DNA molecule that possesses precisely one helical turn per 10 base pairs is said to be "relaxed," with no superhelical turns. Such a structure can lie flat on a plane without tension. If, however, the two strands of the duplex are overwound so that the duplex contains more than one helical turn per 10 base pairs, the molecule twists about itself in space and becomes positively supertwisted. Conversely, if the DNA is underwound and possesses less than one turn per 10 base pairs, the molecule becomes negatively supertwisted. Thus, if there are 1.1 turns per 10 base pairs, the superhelix density is -0.1. Because circular DNA molecules become more compact with the introduction of each additional supertwist, molecules with different superhelix density can be separated by electrophoresis in agarose gels (Fig. 8.14).

Supercoiling can be relieved by cleavage of one of the two strands by the action of a deoxyribonuclease. In addition, *nicking-closing enzymes* (Chap. 27) can relax superhelical DNA by transiently cleaving one of the strands and then rapidly resealing the break. Another way in which supertwists can be relieved is by intercalation of a planar molecule such as *ethidium bromide* between the stacked base pairs.

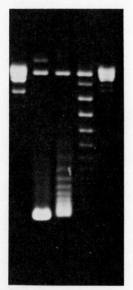

Figure 8.14 Separation by agarose gel electrophoresis of closed circular *E. coli* plasmid DNA (pBR322) with varying degrees of supercoiling. The relaxed DNA is at the top of the gel and the most highly supercoiled is at the bottom. Supercoiling was introduced by treatment of relaxed DNA with a type II topoisomerase (Chap. 27). [*Courtesy of Professor Douglas Brutlag.*]

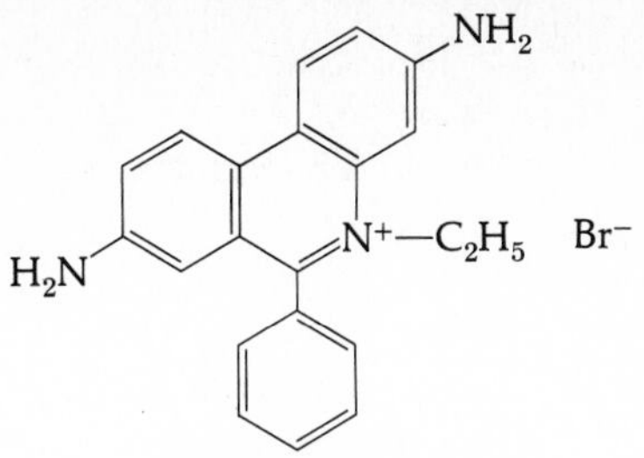

Ethidium bromide

Intercalation of one ethidium molecule untwists the supercoil by 12°, thus requiring 30 ethidium molecules per superhelical turn. When the negative superhelical turns are completely relieved, further addition of ethidium causes the introduction of new superhelical turns but now in the positive direction (Fig. 8.15). Most circular DNAs possess negative superhelical turns. In eukaryotes, negative supertwists arise from the association of DNA with histones (see below). In prokaryotes negative supertwists result from the action of an enzyme *DNA gyrase* (Chap. 27).

PROTEINS ASSOCIATED WITH DNA

In vivo, DNA does not exist as the free acid, but rather it is associated with a variety of proteins. In *eukaryotes,* DNA is complexed with a group of basic proteins, the *histones,* which because of their high concentration of the basic amino acids lysine and arginine, can effectively neutralize the phosphate groups in DNA. Because of their specific association with DNA, histones play a significant role in the condensation of DNA within eukaryotic chromosomes. Smaller amounts of acidic proteins are also present. A DNA-histone complex termed *chromatin* can be isolated from interphase nuclei (see below).

The DNA of *prokaryotes* also exists within a highly condensed structure. Although lacking histones found in eukaryotes, DNA of the bacterium *Escherichia coli* complexes with substantial quantities of small histone-like proteins.

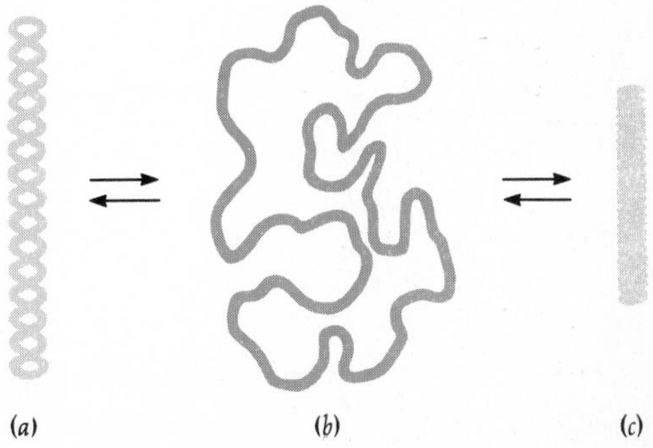

Figure 8.15 Representation of the effects of the binding of ethidium bromide to SV40 DNA. The diagram presents three stages in the reversible binding of dye to SV40 DNA. (*a*) The dye-free molecule with 14 negative superhelical turns. (*b*) The addition of 420 molecules of ethidium bromide completely unwinds the superhelical turns to form the relaxed molecule. (*c*) The addition of a further 720 dye molecules leads to the formation of a positive superhelical molecule with 24 superhelical turns. [*Redrawn from W. Bauer and J. Vinograd, J. Mol. Biol.* **33**:*141* (*1968*).]

(*a*)　　　　(*b*)　　　　(*c*)

Histones

These proteins are usually obtained by extracting nuclei or chromatin with dilute acids (0.2 M HCl) and can be precipitated from solution with alkali at about pH 10 or by salting-out techniques. Separation of individual types of histones has been accomplished by gel filtration, ion-exchange chromatography, and electrophoresis. There are five major classes of histones, with a number of subtypes in certain of these classes. The same classes have been found in all animal and plant cells studied. All histones are relatively small proteins (MW = 12,000 to 20,000), contain approximately 25 mol percent arginine plus lysine, lack tryptophan, and, with one exception, lack cysteine and cystine. Classification of histones has been based mainly on the relative amounts of lysine and arginine. *Histone H1* is very rich in lysine. *Histone H2* is moderately rich in lysine; there are two types, *H2A* and *H2B*. *Histone H3* is moderately rich in arginine and contains cysteine. *Histone H4* is rich in arginine and in glycine (Table 8.5).

The same type of histone obtained from various animals and plants is very similar in sequence. This conservatism in evolution presumably reflects a need to conserve a sequence intimately related to essential and specific functions. This is best exemplified by the fact that the amino acid sequences of histone H4 from pea seedlings and bovine thymus differ in only two residues of the 102 present in the molecule. In histone H4, the NH_2-terminal region contains most of the basic residues whereas the COOH-terminal half of the molecule contains most of the hydrophobic residues and is only slightly basic. This unusual distribution is also reflected in the known sequences of histones H2A, H2B, and H3. In contrast, for histone H1 the converse is the case, the COOH-terminal region being the more basic and rich in proline and the NH_2-terminal region containing most of the acidic and hydrophobic residues.

Histones contain many modified amino acid side chains, e.g., *O*-phosphorylserine, mono-, di-, and tri-ϵ-*N*-methyllysine, ϵ-*N*-acetyllysine, and various methylated arginine derivatives. The possible roles of the histones and the significance of the modified residues are discussed later (Chap. 29).

A tetramer consisting of two molecules of histone H3 and two of histone H4 can be isolated from chromatin by extraction with 2 M NaCl rather than with acid. Under these conditions, histones H2A and H2B can be isolated together as dimers. A current model of chromatin structure suggests that one tetramer

TABLE 8.5

Properties of
Histones*

Histone	Basic residues		Acidic residues	Basic† acidic	Number of residues	Molecular weight
	Lys	Arg				
H1	63	3	10	6.6	213	≈23,000
H2A	14	12	9	3.3	129	13,960
H2B	20	8	10	3.1	125	13,775
H3	13	18	11	3.0	135	15,324
H4	11	14	7	3.9	102	11,282

*Compiled from the work of various investigators. H1 is for the rabbit, the others are for calf thymus.
†Includes histidine.

and two dimers form an octamer which interacts with 200 base pairs of DNA to form a repeating spherical structure termed a *nucleosome* in which histones form a central core around which the duplex DNA is wound. As a consequence, the 200 base pairs of DNA, approximately 700 Å in length, are condensed into a spherical structure approximately 100 Å in diameter (Chap. 29).

Protamines

These basic proteins are found in the mature sperm of certain families of fish after completion of all cell divisions, when the histones are replaced. They are relatively small proteins, lacking many amino acids but being extremely rich in arginine. Fowl sperm contain a protamine-like basic protein, *galline,* which resembles fish protamines with its high arginine content and lack of acidic residues. Mammaliam sperm also contain small basic proteins associated with DNA, but these proteins are unrelated to the protamines. It appears that all mature sperm lack histones but contain basic proteins bound to DNA. They are probably involved in the tight packing of DNA within the sperm head.

STRUCTURE OF RIBONUCLEIC ACIDS

There are three types of RNA, distinguishable by characteristic composition, size, functional properties, and their location within the cell. Some general aspects of the structure of RNA will be given first, followed by a brief discussion of the properties of the presently recognized types of RNA. Their special functions will be considered later (Chap. 28).

Internucleotide Linkage

In RNA, the hydroxyl groups at C-2', C-3', and C-5' are available for esterification. When RNA is treated with the phosphodiesterase from snake venom, the main products are nucleoside 5'-phosphates. When treated with the phosphodiesterase from spleen, nucleoside 3'-phosphates are formed. Hence, as in DNA, the internucleotide linkages are between C-3' and C-5'.

Unlike DNA, RNA is hydrolyzed by weak alkali (pH 9 at 100°C or pH 12 at 37°C). This treatment leads to translocation of one bond to form 2',3' cyclic phosphates, of the type shown below, for each of the purine and pyrimidine nucleotides present in RNA.

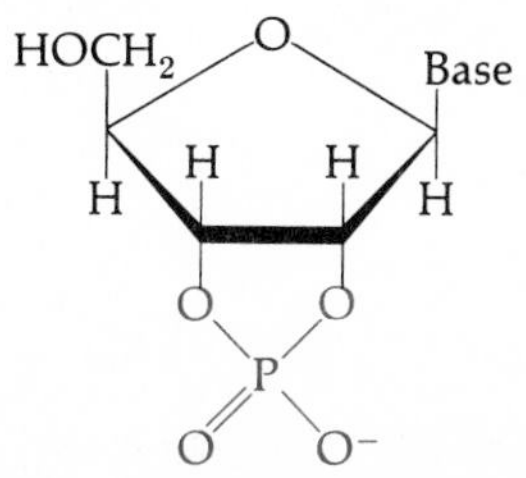

A 2',3' cyclic nucleoside monophosphate

The further action of alkali on the cyclic compounds produces a random hydro-lytic cleavage to give a mixture of the isomeric 2'- and 3'-mononucleotides.

The possibility of formation of a cyclic intermediate exists only with RNA, since in DNA there is no hydroxyl group at C-2'; this explains the relative stability of DNA in weak alkali. Pancreatic and T1 ribonucleases (Table 8.3) also form and hydrolyze cyclic 2',3' phosphates, which then yield only the 3' phosphates. The structure and mechanism of action of pancreatic ribonuclease are presented later (Chap. 11). A different type of cyclic nucleotide, cyclic 3',5'-adenosine monophosphate, cyclic AMP, is of considerable importance in metabolism (Chap. 12).

Transfer RNA

Transfer RNA (tRNA) comprises approximately 10 to 20 percent of the cellular RNA and is composed of relatively small molecules of chain lengths of 75 to 90 nucleotides. tRNA functions as an "adapter" in peptide-bond synthe-sis, at least one specific tRNA serving for each amino acid; this role of tRNA is presented in Chap. 28.

Although tRNA is composed largely of the four main ribonucleotides (ade-nylic, guanylic, cytidylic, and uridylic acids), there are, in addition, smaller quantities of other nucleotides. These include the nucleotides of pseudouridine (see below), various methylated adenines and guanines, methylated pyrimi-dines, such as thymine and 5-methylcytosine, and others. Not all these are

Figure 8.16 Structures of some of the minor bases and nucle-osides from tRNA. R = ribose.

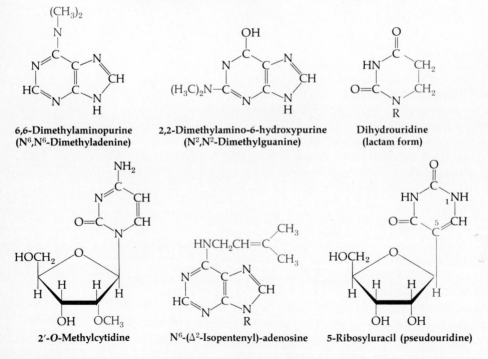

6,6-Dimethylaminopurine
(N⁶,N⁶-Dimethyladenine)

2,2-Dimethylamino-6-hydroxypurine
(N²,N²-Dimethylguanine)

Dihydrouridine
(lactam form)

2'-*O*-Methylcytidine

N⁶-(Δ²-Isopentenyl)-adenosine

5-Ribosyluracil (pseudouridine)

TABLE 8.6

Name	Abbreviation	
Adenosine	A	Some Nucleosides Found in Transfer RNAs from Various Sources
1-Methyl-	m^1A	
2-Methyl-	m^2A	
N^6-Methyl-	m^6A	
N^6,N^6-Dimethyl-	m_6^6A	
N^6-Isopentenyl-	i^6A	
N^6-Isopentenyl-2-methylthio-	ms^2i^6A	
N^6-(N-Threonylcarbonyl)-	t^6A	
2'-O-Methyl-	Am	
Cytidine	C	
5-Methyl-	m^5C	
N^4-Methyl-	m^4C	
N^4-Acetyl-	ac^4C	
2'-O-Methyl-	Cm	
2-Thio-	s^2C	
Guanosine	G	
1-Methyl-	m^1G	
N^2-Methyl-	m^2G	
N^2,N^2-Dimethyl-	m_2^2G	
7-Methyl-	m^7G	
2'-O-Methyl-	Gm	
Inosine	I	
1-Methyl-	m^1I	
Pseudouridine	ψ	
2'-O-Methyl-	ψm	
Ribosylthymine	T	
Uridine	U	
5,6-Dihydro-	hU or D	
3-Methyl-	m^3U	
2'-O-Methyl-	Um	
2-Thio-†	s^2U or ^2S	
4-Thio-	s^4U or ^4S	
5-Carboxymethyl-5-oxyacetic acid	cm^5U, aco^5U	

† 2-Thiouridines substituted at the 5 position with methyl, methylamino-methyl, and 5-acetic acid methyl ester groups have been found.

present in any one tRNA molecule, but pseudouridine and dihydrouridine are the most abundant and are universally distributed. A list of some known minor components of tRNA and their abbreviations is given in Table 8.6; the structures of some of them are shown in Fig. 8.16.

Pseudouridine is of special interest in that the usual N-glycosidic bond is absent; the ribose is directly linked at C-1' to the 5 position of uracil by a carbon-carbon bond. The 2'-O-methyl nucleotides provide nucleotide linkages that are stable both to alkaline and enzymic hydrolysis. The possible functional significance of the presence of pseudouridine and the other unusual bases is discussed later (Chap. 28).

The linear base sequences of more than 100 tRNAs have been determined.

$$\text{m}^2 \qquad\qquad\quad \overset{\text{h}}{|} \qquad \overset{\text{h}}{|} \qquad\qquad \text{m}^2_2 \qquad\qquad\qquad \text{m}^1$$

pG-G-G-C-G-U-G-U-G-G-C-G-C-G-U-A-G-U-C-G-G-U-A-G-C-G-C-G-C-U-C-C-C-U-U-I-G-C-I-
1   10   20   30

ψ-G-G-G-A-G-A-G-U*-C-U-C-C-G-G-T-ψ-C-G-A-U-U-C-C-G-G-A-C-U-C-G-U-C-C-A-C-C-AOH
40   50   60   70   77

Figure 8.17 Nucleotide sequence of a yeast adenyl tRNA. Abbreviations are given in Table 8.6. U* = a mixture of U and UV.

The three-dimensional structure of the phenylalanine-specific tRNA of yeast is described later (Chap. 28).

The method in most common use for the sequence analysis of tRNAs is the following. The tRNA molecules uniformly labeled in vivo with ^{32}P are hydrolyzed to small fragments by the action of pancreatic RNase or RNase T1 or a combination of the two (Table 8.3). The radioactive oligonucleotide components of the digest are then fractionated by electrophoresis in two dimensions (Chap. 9). Separation in the first dimension is by high-voltage ionophoresis on cellulose acetate at pH 3.5. The products are then transferred to diethylaminoethyl (DEAE) cellulose paper, and ionophoresis is performed in the second dimension at acid pH. All the oligonucleotides in an RNase T1 digest of most tRNAs can be separated by this procedure. The nucleotide sequence of each of the isolated oligonucleotides can then be established by degradation with the appropriate endo- and exonucleases followed by fractionation and identification of the products of degradation.

The linear sequence of a yeast alanyl tRNA is shown in Fig. 8.17. The 77 residues represent a molecular weight of 26,600 as the Na$^+$ salt. The usual conventions are followed in that the 5′ phosphate is shown at the left and the 3′ hydroxyl on the right.

Ribosomal RNA

Ribosomes (Chap. 28) contain a large portion of the total RNA of a cell, representing as much as 80 percent in some bacteria. This type of RNA (rRNA) is strongly associated with the ribosomal proteins. The RNA of ribosomes constitutes about 40 to 50 percent of the dry weight of these particles and is generally present as three components with S values of 5, 16 to 18, and 23 to 28, respectively. The properties of these particles will be considered in connection with their biological role (Chaps. 28 and 29).

Ribosomal RNA preparations from various sources, e.g., rat liver or *E. coli,* yield similar values for nucleotide content. Guanylic acid is invariably most abundant; uridylic and cytidylic acids are present in approximately equal amounts and are least abundant. Pseudouridine is present only in trace amounts, and methylated bases are present in small quantities.

Messenger RNA

This type of RNA (mRNA) consists of a single strand of variable length; its role is presented later (Chaps. 28 and 29).

154

Books

Adams, R. P. L., R. H. Adams, A. M. Campbell, and R. M. S. Smellie (eds.): Davidson's *The Biochemistry of the Nucleic Acids,* 8th ed., Wiley, New York, 1977.

Bloomfield, V. A., D. M. Crothers, and I. Tinoco, Jr.: *Physical Chemistry of Nucleic Acids,* Harper & Row, New York, 1974.

Cantor, C. R., and P. R. Schimmel: *Biophysical Chemistry,* parts I, II, and III, Freeman, San Francisco, 1980.

Fasman, G. D. (ed.): *Handbook of Biochemistry and Molecular Biology Nucleic Acids,* 3d ed., vols. I and II, CRC Press, Cleveland, 1975.

Hall, R. H.: *The Modified Nucleosides in Nucleic Acids,* Columbia, New York, 1971.

Kornberg, A.: *DNA Replication,* Freeman, San Francisco, 1980.

Review Articles

Air, G. M.: Rapid DNA Sequence Analysis, *CRC Crit. Rev. Biochem* **6:**1–33 (1979).

DeLange, R. J., and E. L. Smith: Chromosomal Proteins, in H. Neurath and R. L. Hill (eds.): The Proteins, 3d ed., vol. 4, Academic, New York, 1979, pp. 229–243.

Kornberg, R. D.: Structure of Chromatin, *Annu. Rev. Biochem.* **46:**931–954 (1977).

Roberts, T. M., G. D. Lauer, and L. C. Klotz: Physical Studies on DNA, from "Primitive" Eukaryotes, *Crit. Rev. Biochem.* **3:**349–449 (1976).

Sanger, F: Determination of Nucleotide Sequences in DNA, *Biosciences Reports* **1:**3–18 (1981).

Wells, R. D., T. C. Goodman, W. Hillen, G. T. Horn, R. D. Klein, J. E. Larson, U. R. Müller, S. K. Neuendorf, N. Panayotatos, and S. M. Stirdivant: DNA Structure and Gene Regulation, in W. Cohn (ed.): Progress in Nucleic Acid Research and Molecular Biology, vol. 24, Academic, New York, 1980, pp. 167–267.

Wu, R.: DNA Sequence Analysis, *Annu. Rev. Biochem.* **47:**607–634 (1978).

Boyer, P. D. (ed.): *The Enzymes,* 3d ed., vol. IV, Academic, New York, 1971, pp. 205–336; vol. XIV, 1981.

Nomenclature

IUPAC-IUB Commission on Biochemical Nomenclature, Abbreviations and Symbols for Nucleic Acids, Polynucleotides and Their Components, *J. Biol. Chem.* **245:**5171 (1970).

Chapter 9

Methods of Separation and Purification of Compounds from Biological Sources

THE PROBLEMS OF PURIFICATION

The methods employed for the purification of substances from biological sources were developed because the conventional methods of organic chemistry are often not applicable to the isolation of most of the constituents of living things. The problems encountered in isolating pure biochemical compounds will become evident from the following considerations. Each substance must be separated from the other components with which it occurs, viz., from material containing several thousands of different molecular species, varying in molecular weight from 10^2 to 10^8, and present in concentrations varying from 10^{-15} to $10^{-2}\,M$. The desired substance is often very labile and cannot be exposed to extremes of pH, temperature, or pressure; except for some lipids and a few other substances, most are poorly soluble in organic solvents and must be purified in aqueous solutions. Further in any one class of biological compounds there are usually many others that differ from it only slightly. For example, glycine is but 1 of the 20 amino acids found in protein hydrolysates, and its general properties resemble those of other α-amino acids. These requirements have led to the development of very selective methods with a high degree of resolution that can be used with micro amounts of compounds. Many of these methods are also used in extremely sensitive and quantitative analytical procedures and for establishing the purity of a compound.

Principles and Classification of Purification Methods

The most powerful methods that have been developed for purification of biological compounds may be viewed as *cascade* operations, whose essential fea-

Any cross-references coded M refer to the companion text, *Principles of Biochemistry: Mammalian Biochemistry*.

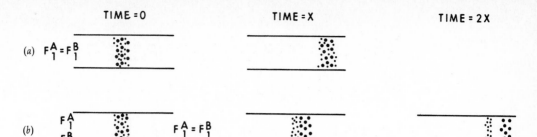

Figure 9.1 Principles of a cascade method for the separation of a mixture of solutes. (a) A mixture of two different solutes, A ($\bullet$) and B ($\cdot$), is placed in an appropriately arranged channel with a cross-sectional area small compared with its length. If the mixture is impelled through the channel by a constant force F_1, which acts equally on both A and B, then A and B will move at the same rate and will remain unseparated. If, however, there are retarding forces F_2 which act differently on solutes A and B, as shown in (b), then as the molecules are driven through the channel, A will tend to separate from B. The rate of movement of A is given by the expression $\Delta F^A = F_1{}^A - F_2{}^A$ and that of B is $\Delta F^B = F_1{}^B - F_2{}^B$; therefore, the separation achieved depends on the ratio $\Delta F^A / \Delta F^B$. As A and B are kept under the influence of constant impelling and retarding forces, they tend to separate more completely as they pass from one stage to another along the length of the channel, until after time 2X, complete separation is achieved. An arbitrarily short length of the channel may be considered a single stage and the entire length of the channel a series of stages, with molecules A and B passing from one stage to another in a cascading fashion.

ture is the passage of a mixture of substances through a sequential arrangement of purification stages. As a mixture passes through a single stage, some degree of separation of the constituents in the mixture is obtained, but as the mixture continues to pass through hundreds or thousands of stages, it is separated into its components and a high degree of purification is achieved. The series of events that occurs during the passage from one stage to another in a cascade method is illustrated diagrammatically in Fig. 9.1 for a mixture of two solutes. The molecules in the mixture can be considered to be propelled through a channel of solvent by a force F_1, which acts equally on each molecule. Other forces F_2, however, retard the molecules and act in a direction opposite to that of the driving force. If the retarding forces are different for different molecules in the mixture, then as the mixture passes through a significant length of the channel, separation of the components starts and can be complete if the process continues. An arbitrary small length of the channel can be considered as a single stage in the cascade of purification stages.

Table 9.1 lists some common cascade methods for purification of mixtures of biochemical compounds and classifies each method on the basis of the impelling and retarding forces acting during purification. The ultracentrifugal methods discussed earlier (Chap. 4) are also cascade methods.

Countercurrent Distribution

Countercurrent distribution illustrates how a typical cascade method of purification operates. This method involves the repetitive distribution of a mixture of solutes between two immiscible solvents in a series of tubes in which the two immiscible phases are in contact. To illustrate its use, consider the

TABLE 9.1

Classification of Methods of Purification	Method	Impelling force F_1	Retarding force F_2	Separation depends on
	Countercurrent distribution	Mechanical	Partition	Solubility of solutes in immiscible phases
	Chromatography:			
	Adsorption	Hydrodynamic	Surface energy adsorption	Structural properties of solute and adsorbent
	Partition (liquid-liquid)	Hydrodynamic	Partition	Solubility of solutes in immiscible phases
	Partition (gas-liquid)	Mechanical	Partition	Solubility of gaseous substances in liquid phase
	Ion exchange	Hydrodynamic	Electrostatic	Ionic nature of exchanger and solutes
	Gel filtration	Hydrodynamic	Partition	Partition determined by size and shape of solutes
	Affinity	Hydrodynamic	Noncovalent interactions	Specific interaction of solute with adsorbent
	Electrophoresis:			
	In free solution	Electrostatic	Molecular friction	Ionic properties
	In porous aqueous media (zone)	Electrostatic	Molecular friction	Ionic properties
	In porous aqueous detergents	Electrostatic	Molecular friction	Molecular weight

Source: Adapted from C. J. O. R. Morris and P. Morris, *Separation Methods in Biochemistry,* 2d ed., Pitman, New York, 1976.

behavior of a single compound when transferred sequentially between two immiscible solvents in a series of separatory funnels. The amount of the compound that will be distributed between the two immiscible solvents is determined by its partition coefficient K_p, defined as

$$K_p = \frac{\text{concentration in upper phase}}{\text{concentration in lower phase}}$$

If $K_p = 1$, equal amounts of the substance will be present in the upper and lower phases when the substance is dissolved in either phase and the phases thoroughly mixed and then allowed to separate, provided that the upper and lower phases are equal in volume. If the substance is then transferred in a countercurrent manner, as shown in Fig. 9.2, it will be distributed between the upper and lower phases. The lower phase remains stationary and is considered the *stationary phase,* whereas the upper phase is transferred and is referred to as the *mobile phase.* At the end of a given number of transfers, the substance will be distributed among the series of tubes as determined by its partition coefficient and the number of transfers. Thus, if a compound with a partition coefficient of 1 is transferred in a countercurrent manner 100 times, the maximum amount

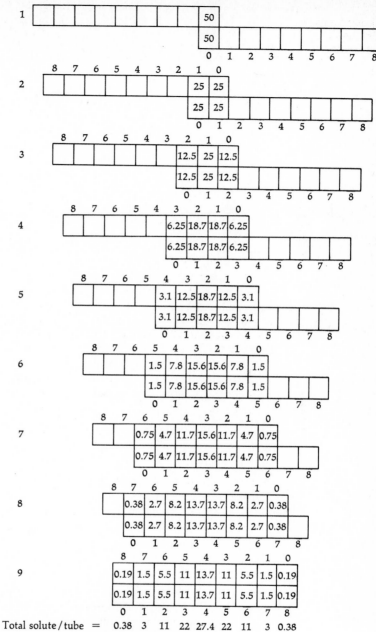

Figure 9.2 The distribution of a solute with a partition coefficient K_p of 1.0 when distributed in a countercurrent manner between equal volumes of upper and lower phases of an immiscible pair of solvents. Step 1: 100 mg of solute is placed in tube 0; the upper and lower phases, which are in contact, are mixed and allowed to separate. Because $K_p = 1$, 50 mg of solute will be in each of the upper and the lower phases. Step 2: the first countercurrent transfer: the upper phase of the tube 0, containing 50 mg of solute, is brought into contact with lower phase in tube 1, and the lower phase of tube 0 is brought in contact with fresh upper phase. The tubes are mixed and the phases separated. Each upper and lower phase contains 25 mg of solute at this stage. Steps 3 to 9: the distribution of solute after seven additional transfers as described for step 2.

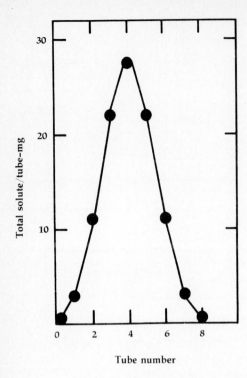

Figure 9.3 The total amount of solute in each tube (upper and lower phases) after nine transfers as shown in Fig. 9.2 plotted against the tube number.

of solute will be in tube 50 and progressively lesser amounts in tubes adjacent to tube 50, so that the actual distribution of the solute approaches a Poisson distribution, as shown in Fig. 9.3. Compounds with a partition coefficient <1 will distribute in a series of tubes between 1 and 50, and compounds with a partition coefficient >1, in tubes between 50 and 100. The separation of a mixture of organic acids is shown in Fig. 9.4.

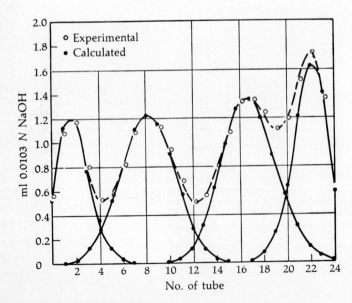

Figure 9.4 Separation by countercurrent distribution of a mixture of four organic acids. The maxima, from left to right, are acetic, propionic, butyric, and valeric acids, respectively. The acid content of each tube was determined by titration with sodium hydroxide. [*From L. C. Craig and D. Craig, p. 171, in Technique of Organic Chemistry, vol. 3, Interscience, New York, 1950.*]

161

**Methods of
Separation and
Purification of
Compounds from
Biological Sources**

Each tube in the countercurrent series is a single stage in the purification cascade, and solutes are passed from one stage to another mechanically (F_1, impelling force) and are retarded to an extent (F_2, retarding forces) depending on the partition coefficient of the solute. Two compounds with different partition coefficients will separate from each other after a sufficient number of countercurrent transfers, the exact number of transfers required depending upon the absolute difference in partition coefficients. Substances with partition coefficients differing only by a small fraction, for example, 0.05 to 0.1, will require many more transfers for complete separation than those differing by large values.

For some time, countercurrent distribution found wide application in biochemistry, but it is not widely used today. Nevertheless, it is a powerful separation method since, compared with other cascade processes, gram quantities of a mixture can be separated by this means. In addition, apparatus containing from 50 to 2000 tubes is available that automatically carries out each transfer.

CHROMATOGRAPHY

Chromatography is the term now generally employed to describe any separation method that involves percolation of a mixture of dissolved substances through a porous solid support irrespective of the forces that lead to separation of the mixture, as shown in Fig. 9.5. Because of the variety of chromatographic procedures that are now utilized, the basic principles of each technique are worthy of consideration.

Adsorption Chromatography

The separation of the pigments extracted from plants on a variety of solid adsorbents was first described in 1903. Extracts of leaves contain two green pigments (chlorophylls a and b) and several yellow pigments (carotenoids). The method was named chromatography (Gk. *chromo*, "color"; *graphein*, "to

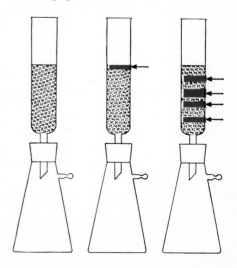

Figure 9.5 A diagrammatic representation of the simplest type of chromatogram. A column of a suitable adsorbent is prepared (left). A solution of a mixture of four substances is poured on the column and is allowed to move slowly by gravity. The mixture is adsorbed as a narrow band at the top of the column (middle). When an eluting solvent is poured on the column, the substances are separated as four discrete zones (right).

write"). Adsorption chromatography on charcoal, alumina, diatomaceous earth, and silica gel has wide application and is used for separation of many types of compounds. The separation depends upon the differential adsorption of compounds to the surface of specific adsorbents by means of such forces as dipole interactions, hydrogen bonding, and hydrophobic interactions.

Partition Chromatography

immiscible 不混合的

This method was developed as an alternative to countercurrent distribution. It was reasoned that if the stationary phase of two immiscible solvents could be immobilized on an inert solid support in a cylindrical column and the mobile phase, in equilibrium with the stationary phase, were passed through the column, a mixture of substances dissolved in the solvents would separate as the mobile phase passed through the column. In the initial studies, silica gel was used as the solid support and was saturated with water containing methyl orange. Chloroform saturated with water was used as the mobile phase. When *N*-acetylamino acids were dissolved in a small amount of mobile phase, applied to the column, and then developed with the mobile phase, they were found to move at different rates. The *N*-acetylamino acids are sufficiently strong acids to change the color of methyl orange, so that they were visualized as red bands traveling against a yellow background, as shown diagrammatically in Fig. 9.5. The bands were collected as they emerged from the column. This method did not separate each *N*-acetylamino acid, but by the use of several solvent systems and different columns, some *N*-acetylamino acids were obtained in pure form. The mechanism for the separation was partition of the acids between the stationary (aqueous) and mobile (chloroform) phases as in countercurrent distribution, and a short length of the column was equivalent to one stage of a countercurrent distribution train. Although some adsorption to the silica gel influenced the mobility of the acids, the method was named *liquid-liquid chromatography*.

Sheets of cellulose filter paper instead of columns filled with solid supports were soon found to be excellent matrices in liquid-liquid chromatography. *Paper chromatography* proved to be one of the most powerful of all analytical techniques and was applicable for the separation of unsubstituted amino acids as well as many other complex mixtures of small-molecular-weight biological compounds. For separation of amino acids, for example, a strip of filter paper bearing a small amount of mixture near the top is saturated with the aqueous stationary phase and hung from a trough containing the mobile phase; the entire apparatus is then enclosed in a chamber saturated with vapors of the stationary and mobile phases. The mobile phase is siphoned out of the trough by capillarity and flows down the strip. The individual compounds move down the paper at rates dependent on their partition coefficients in the two phases, just as in countercurrent distribution. The order of migration of compounds down the paper differs for different solvents. This method is *descending paper chromatography*. Alternatively, *ascending paper chromatography* works equally well in either one or two dimensions. Paper chromatography in one dimension is often inadequate for the complete separation of a complex mixture, but by chromatogra-

mobile phase 流动相

stationary phase 静止相

trough 槽

siphon 虹吸

phy in two dimensions first with one solvent and then with a second solvent run at a right angle to the first, complete separation may be achieved.

163

**Methods of
Separation and
Purification of
Compounds from
Biological Sources**

Many variations in the technique have been introduced in which cellulose is replaced by synthetic polymers, silica, alumina, and cellulose derivatives. An almost limitless range of solvents may be used for development. In *thin-layer chromatography,* a thin film of solid support of uniform thickness is prepared on a glass or plastic surface, and mixtures of substances are separated with appropriate solvents by the same methods of ascending chromatography.

High-performance liquid chromatography (HPLC) is another variation of partition chromatography that employs very high pressures to propel the solvent through a thin column. It has a very high resolving power and finds application both for the analysis and the separation of complex mixtures.

Another partition method, called *gas-liquid chromatography* (GLC), is particularly suitable for separation of volatile substances.

In GLC, a glass or metal column, 1 to 2 m long and 0.2 to 2 cm in diameter, is filled with a finely divided inert solid and impregnated with a nonvolatile liquid. A mixture of volatile compounds is flash-evaporated at one end of the column, which is maintained at an elevated temperature (170 to 225°C). The volatilized compounds are swept through the column by a stream of inert gas, e.g., nitrogen, flowing at a constant rate. Each component of the mixture moves on the column at a different rate determined by its partition coefficient between the gas (mobile) phase and the nonvolatile liquid (stationary) phase. Individual compounds in the gas emerging from the column are detected by physical or chemical means. The resolution of a mixture of the methyl esters of fatty acids by this procedure is illustrated in Fig. 9.6. This technique permits quantitative resolution and analysis of nanogram quantities of volatile substances.

Figure 9.6 Gas-liquid chromatographic analysis of the methyl esters of the fatty acids present in normal rat plasma as cholesterol esters. The numbers at the top or at one side of the peaks, to the left and right of the colon, indicate, respectively, the number of carbon atoms and double bonds present in each fatty acid. The areas under each peak reflect the relative quantities of each fatty acid present. [*Courtesy of Dr. L. I. Gidez.*]

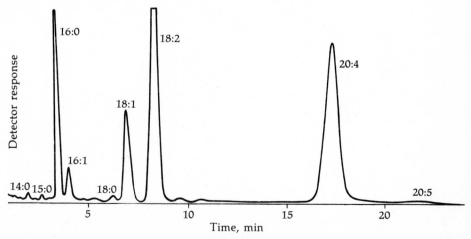

Ion-Exchange Chromatography

In this method a mixture of ionic solutes is placed on a suitable charged insoluble matrix in equilibrium with an aqueous buffer, and the buffer is percolated down the column by hydrostatic pressure (impelling force). Separation is achieved by ionic interactions (retarding forces) between the solutes and the charged matrix. The migration of an ionic solute down the column depends upon the nature of the ionic matrix, the ionic properties of the solute, and the solvent.

Several types of ion-exchange materials (ion exchangers) are employed, including synthetic resins, cellulose, dextran, and agaroses in which either negatively or positively charged groups are introduced. Negatively charged exchangers that bind cations are called *cation exchangers,* and positively charged exchangers are called *anion exchangers.* Table 9.2 lists some typical commercially available cation and anion exchangers.

The principles of ion exchange are illustrated in Fig. 9.7, which considers the interactions of two hypothetical substances, PH^+ and QH^+, with a sulfonic acid cation exchanger. The exchanger is washed with a large volume of a buffer containing equal concentrations of the weak acid HB ($pK = 4$) and its sodium salt (Na^+B^-), so that the buffer is pH 4 and the Na^+ concentration is $0.1\ M$. The resin is then in the "sodium cycle" and suspended in a column in the same buffer at pH 4. Substances PH^+ and QH^+ are weak acids, each with a single acidic group, $pK_P = 6$ and $pK_Q = 7$. If very small amounts of PH^+ and QH^+ are dissolved in a small volume of the buffer (pH 4) and applied to a column of the resin in the same buffer, PH^+ and QH^+ exchange with Na^+ on the cationic exchanger. When this occurs, PH^+ and QH^+ are in the stationary phase, i.e., on the insoluble cation exchanger. The binding of PH^+ and QH^+ is very tight at pH 4, in the buffer containing $0.1\ M\ Na^+$. The binding will be

TABLE 9.2 Some Common Cation and Anion Exchangers

Cation exchangers		Anion exchangers	
Matrix	Functional group	Matrix	Functional group
Sulfonated polystyrene	$-SO_3H$	Triethylamino polystyrene	$-\overset{+}{N}(CH_3)_3$
Polymethacrylic acid	$-COOH$	Dimethyl-(hydroxy-methyl)-amino polystyrene	$-\overset{+}{N}-(CH_3)_2$ $\quad\vert$ CH_2-OH
Cellulose phosphate	$-O-PO_3H$	O-(Diethylaminoethyl)-cellulose (DEAE-cellulose)	$-(CH_2)_2-\overset{+}{N}\overset{\diagup CH_2-CH_3}{\diagdown CH_2-CH_3}$
Carboxymethyl cellulose (CM-cellulose)	$-CH_2-COOH$	Diethyl-(2-hydroxypropyl)-amino ethyl dextran	$-(CH_2)_2-\overset{+}{N}\overset{\diagup (CH_2-CH_3)_2}{\diagdown CH_2-\underset{\vert}{C}-CH_3}$ $\qquad\qquad\qquad OH$
Sulfopropyl dextran	$-(CH_2)_3-SO_3H$		

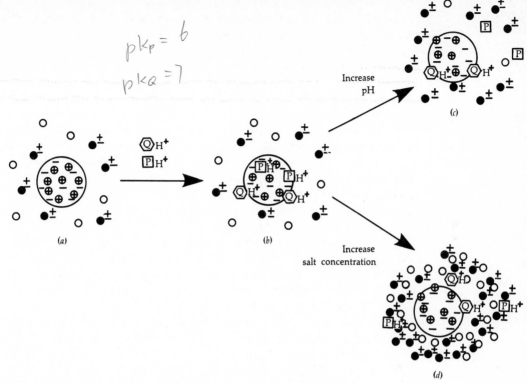

$pk_P = 6$

$pk_Q = 7$

Figure 9.7 General properties of ion exchange. In (*a*) a particle of cation ion-exchange resin is depicted by the large circle; this resin has fixed negative charges and is in the Na⁺ cycle (Na⁺ = ⊕). The Na⁺ ions on the resin are rapidly exchanging with Na⁺ of the sodium salt (●±) of the weak acid (O) HB. (*b*) If substances PH⁺ (⊡⁺) and QH⁺ (⬡⁺) are added, they will exchange with sodium ions and bind to the resin. (*c*) The affinity of PH⁺ for the resin can be weakened by raising the pH to near 6, the pK of substance PH⁺. (*d*) The affinity of both PH⁺ and QH⁺ is weakened by increasing the Na⁺ concentration.

weakened, however, if the composition of the buffer is altered either in pH or Na⁺ concentration. If the buffer is made more alkaline, approaching pH 6, PH⁺ will become partially dissociated to give P, which binds poorly to the exchanger. At still more alkaline pH, Q will be formed as well, bind more weakly, and pass down the column. Alternatively, if the Na⁺ concentration is increased considerably while maintaining the pH at 4, e.g., from 0.1 to 0.5 M, both PH⁺ and QH⁺ will bind less tightly to the resin, since more Na⁺ is available to compete with PH⁺ and QH⁺ for the sulfonic acid groups. Thus, to separate P and Q in this system, a buffer with an exact pH and Na⁺ concentration is constructed so that PH⁺ and QH⁺ have different affinities for the exchanger, and when a mixture of PH⁺ and QH⁺ is passed down a column, separation will occur. Other factors, such as temperature and organic solvents, also influence the chromatographic behavior of mixtures of ionic compounds. Although the principles are illustrated with a cation exchanger, the same mechanisms of exchange apply to anion exchangers except that the fixed cation of the ion-exchange matrix binds anions rather than cations.

165

Ion-exchange chromatography is widely used for separation of biological compounds. An example of its high degree of resolution is illustrated in Fig. 9.8, which shows the separation of amino acids. Two columns are required, one (150 cm) for the acidic, neutral, and aromatic amino acids and another (15 cm) for the basic amino acids. Clearly, the degree of retardation of an amino acid depends on its basicity. Thus, aspartic and glutamic acids emerge before most of the neutral amino acids and well ahead of the basic amino acids. Even the neutral amino acids separate readily from one another since, despite their similar pK values, their side chains possess different nonpolar affinities for the resin. In the system used in Fig. 9.8, the effluent from the column was mixed automatically with a ninhydrin reagent, the color developed by passage through a hot-water bath, and the absorbancy recorded as the effluent passed through a light beam directed on a photocell. The change in photocurrent with volume of effluent was plotted automatically by a recorder. Integration of the area under each peak permits quantitative estimations. Identification of each amino acid is readily made since, under controlled conditions, the position of emergence is constant for each amino acid. These principles have been utilized for construction of amino acid analyzers.

The separation of amino acids by ion-exchange chromatography as illustrated in Fig. 9.8 is an example of resolution of a mixture at a constant Na^+ concentration and three fixed pH values. Often it is impossible to find such exact conditions for suitable separation of an unknown mixture of ionic compounds. Accordingly, many ion-exchange separations are performed by gradi-

Figure 9.8 Automatically recorded chromatographic analysis of a synthetic mixture of amino acids on a sulfonated polystyrene resin. [*From D. H. Spackman, W. H. Stein, and S. Moore, Anal. Chem.* **30:**1190 (1958).]

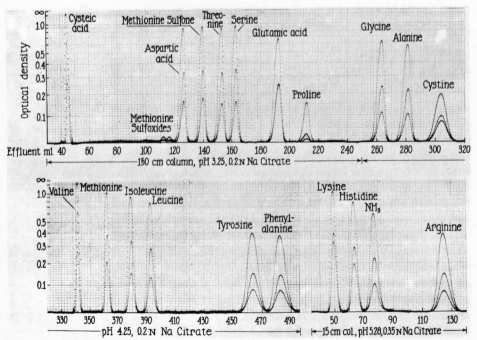

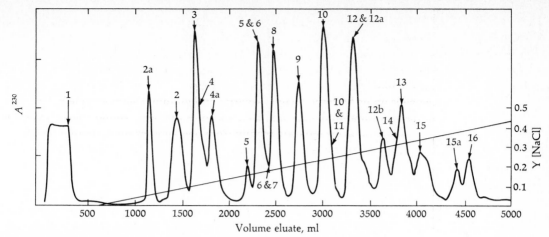

Figure 9.9 Chromatographic pattern of a mixture of proteins from 30S ribosomes of *E. coli* (Chap. 28). The mixture was applied to a column (2.8 × 60 cm) of phospho-cellulose equilibrated with phosphate buffer at pH 6.5, containing 6 *M* urea. A linear gradient from 0 to 0.6 *M* NaCl was then applied as shown at a flow rate of 45 ml/h. The protein concentration in fractions collected automatically was estimated from absorbance at 230 nm (A^{230}) and is plotted against volume of eluant. Some of the peaks contain single proteins; peaks indicated by more than one number at their apex or to one side of the peak are mixtures of proteins. [*From S. J. S. Hardy, C. G. Kurland, P. Voynow, and G. Mova, Biochem.* **8:**2897 (1967).]

ent elution, in which the buffer entering the column is slowly changed in composition with respect to either the inorganic ion exchanging with the functional groups on the resin or the pH, which alters the affinity for the exchanging solutes for the resin. Figure 9.9 shows an example of gradient elution of a mixture of proteins that has been purified by this means.

Gel-Filtration Chromatography

Gel-filtration chromatography separates compounds on the basis of molecular size and shape. If granular preparations of diverse insoluble materials such as synthetic resins, porous glass, or the polysaccharides dextran or agarose are suspended in an appropriate solvent, they form two solvent phases, one within the internal space of the granules and the other outside the granules, as shown diagrammatically in Fig. 9.10. A column of the granular matrix acts as a molecular sieve, because molecules of small molecular weight may be distributed between both phases, whereas molecules of high molecular weight are excluded from the internal phase. The granular matrix can be chemically modified to a predetermined porosity and thus can establish the approximate molecular weight of substances that are able to penetrate the matrix and enter the internal phase.

As indicated in Fig. 9.10, if a solution of a mixture of solutes with different molecular weights is applied to a column of a matrix that excludes substances above a given molecular weight, then as the mixture passes down the column, solutes of low molecular weight are included in the internal phase and separated from those of higher molecular weight that are confined to the external phase.

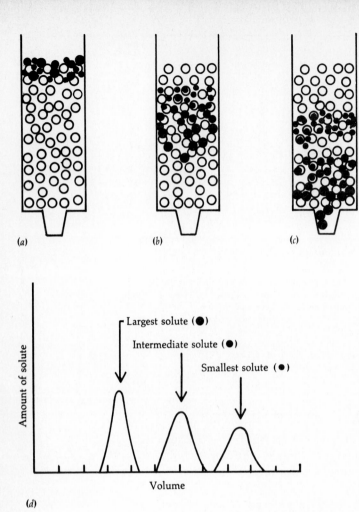

(a) (b) (c)

(d)

Amount of solute

Largest solute (●)

Intermediate solute (●)

Smallest solute (●)

Volume

Figure 9.10 Diagrammatic illustration of the separation of three solutes with different molecular weights by gel filtration. The smallest solute (·) can freely enter the internal space within the gel particles; the solute with a higher molecular weight (•) can enter the inner space but, because of its size, cannot penetrate as freely as the smallest solute. The solute of highest molecular weight cannot enter the internal space of the gel (●). The open circles represent the gel matrix. Different stages of development are represented by (a) when sample is applied, (b) about halfway through development, and (c) when the largest solute molecules begin to emerge from the column. After complete development, the elution profile, obtained by measuring the amount of solute in each fraction of eluate collected, is shown (d). If V_o is the volume of solvent between the gel matrix and V_s the volume of available solvent within the matrix, the relationship $V_e = V_o + KV_s$ relates the volume V_e in which any solute will emerge from the column with V_o, V_s, and K, the partition coefficient of the solute between the mobile phase V_o and stationary phase V_s in the system. $V_e = V_o$ for solutes that cannot enter the gel since $K = 0$. V_e is larger than V_o for solutes that enter the gel and is equal to $V_o + V_s$ for solutes entering the gel freely, where $K = 1$.

Some substances that differ only slightly in molecular weight can be separated from one another; e.g., mono-, di-, and trisaccharides are readily separated from higher-molecular-weight oligosaccharides, as illustrated in Fig. 9.11. Clearly, the shape of a molecule as well as its molecular weight determines whether it can enter the internal phase, and, for macromolecules, the exact elution position of a substance is often markedly influenced by shape as well as size.

Affinity Chromatography

This method can be used for purification of any protein that binds with considerable specificity to another substance. The principle is as follows: a ligand L, which binds specifically to the protein to be purified, is firmly attached to an insoluble matrix M, to give the adsorbent M—L. The adsorbent is suspended in an appropriate solvent and placed in a column, and a mixture containing the desired protein percolated through the column. The protein P is

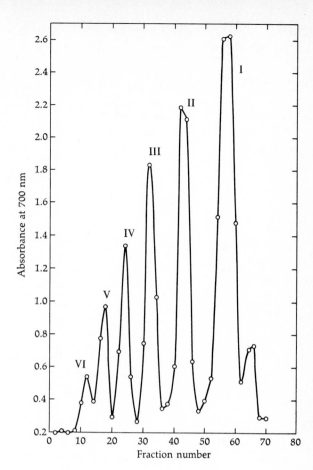

Figure 9.11 Gel filtration of a mixture of oligosaccharides of *N*-acetylglucosamine. The oligosaccharide mixture was applied to the column in 0.1 *M* NaCl and developed at a flow rate of 600 ml/h. Each fraction contained 125 ml. The numerals refer to the number of residues of *N*-acetylglucosamine in the carbohydrate in each peak from I, containing only the monosaccharide *N*-acetylglucosamine through VI, which contains only chitohexaose. [*From M. A. Raftery, T. Rand-Meir, F. W. Dahlquist, S. M. Parsons, C. L. Borders, R. G. Wolcott, W. Beranek, and L. Jao, Anal. Biochem.* **30**:*429* (*1969*).]

retarded by interacting with the specific adsorbent by noncovalent interactions as follows:

$$M—L + P \rightleftharpoons M—L \cdots P$$

If the applied mixture contains no other protein that will bind to the immobilized ligand, all proteins will pass through the column but P. After washing the column thoroughly, P is eluted with a solvent that promotes dissociation from the specific adsorbent, i.e., shifts the equilibrium of the above reaction to the left.

The chemical nature of the ligand depends on the protein to be purified as well as on the means for coupling it to the insoluble matrix. For an enzyme, the ligand is usually structurally related to its substrates, products, or competitive inhibitors since the specificity of binding is provided by interaction of the ligand with the active site of the enzyme. For a specific antibody, it should be structurally related to its antigen: for lectins to a specific sugar and for receptor proteins to the specific ligand that binds the receptor. The major problem in designing the ligand, however, is the chemical means for coupling it to the matrix, and although affinity chromatography was thought to be possible for many years, it became feasible when a general method for coupling ligands to dextrans and agaroses was discovered. At alkaline pH, cyanogen bromide reacts with the dextrans and agaroses commonly used for gel filtration and the resulting derivative can react with primary and secondary amines and bind them covalently. It is believed that the reaction of cyanogen bromide with free hydroxyl groups on agaroses is as follows (M refers to the polymeric granular agarose):

$$\text{(M)} \underset{\text{OH}}{\overset{\text{OH}}{\bigg\langle}} \quad \overset{\text{OH}^-}{} \quad \underset{\text{N}}{\overset{\text{C}-\text{Br}}{\|}} \longrightarrow \text{(M)} \underset{\text{OH}}{\overset{\text{O}-\text{C}\equiv\text{N}}{\bigg\langle}} + H_2O + Br^- \tag{1}$$

Agarose

$$\text{(M)} \underset{\text{OH}^-}{\overset{\text{O}-\text{C}\equiv\text{N}}{\bigg\langle}} \longrightarrow \text{(M)} \underset{\text{O}}{\overset{\text{O}}{\bigg\langle}} \text{C}=\text{NH} \tag{2}$$

Activated agarose

The "activated agarose" then reacts with amines by the following kinds of reactions:

$$\text{(M)} \underset{\text{O}}{\overset{\text{O}}{\bigg\langle}} \text{C}=\text{NH} + H_2N-R \longrightarrow \text{(M)} \underset{\text{O}}{\overset{\text{O}}{\bigg\langle}} \text{C}=\text{N}-R + NH_3 \tag{3}$$

$$\text{(M)} \underset{\text{O}}{\overset{\text{O}}{\bigg\langle}} \text{C}=\text{NH} + H_2N-R \longrightarrow \text{(M)} \underset{\text{OH}}{\overset{\text{O}-\overset{\overset{\text{NH}}{\|}}{\text{C}}-\text{NH}-R}{\bigg\langle}} \tag{4}$$

Other methods of coupling to different kinds of matrices have been developed, but cyanogen bromide activation of agarose followed by coupling with amines remains the most widely used and versatile method for preparing specific adsorbents for affinity chromatography. Of course, this method requires synthesis of specific ligands containing an amino group if one is not present on the ligand itself.

Since proteins contain free amino groups, they are readily coupled to activated agarose. The protein soybean trypsin inhibitor, when coupled to agarose, provides an excellent specific adsorbent for trypsin and chymotrypsin, as illustrated in Fig. 9.12. Both enzymes in pancreatic juice bind the inhibitor under conditions in which they are catalytically active and are most likely to interact specifically with the inhibitor. In affinity chromatography, adsorption is usually performed under conditions that promote maximal binding. Thus, after most of the other proteins of pancreatic juice emerge unretarded from the column, chymotrypsin is specifically eluted with buffer containing tryptamine, which is an excellent inhibitor for this enzyme. Trypsin is then eluted with buffer containing benzamidine, which strongly inhibits trypsin but not chymotrypsin. Thus, these reagents are specific desorbants for these enzymes because

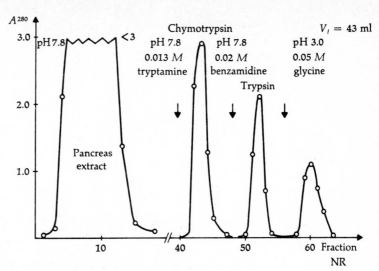

Figure 9.12 Isolation of chymotrypsin and trypsin from crude pancreatic juice by one-step specific displacement with a specific inhibitor of the two enzymes. Fraction NR emerged from elution at pH 3, contained neither trypsin nor chymotrypsin, and represents nonspecifically bound protein that probably was retained by ionic interactions with the soybean inhibitor. This is often found in affinity chromatography, since many ligands are ionic and the affinity adsorbents behave like ion exchangers. Thus, specific elution of an enzyme with inhibitors gives more highly purified enzymes than nonspecific elution achieved by changing the ionic strength or pH during desorption. [*From J. Porath and T. Kristiansen, pp. 95–175, in H. Neurath and R. L. Hill (eds.), Proteins, vol. I, Academic, New York, 1975.*]

they also bind at the active sites of the enzymes and therefore can dislodge the soybean trypsin inhibitor.

ELECTROPHORESIS

Electrophoresis refers to the migration of ionic solutes in an electric field and is particularly useful for the analysis of proteins, nucleic acids, peptides, and nucleotides. The velocity v of a macromolecule with a net charge q in an electric field E is given by the relationship $v = qE/f$, where f is the frictional coefficient relating to size and shape of the molecule (Chap. 3). Since migration depends upon charge, it will also depend upon pH, as illustrated in Fig. 9.13, which relates the mobility $[(\text{cm}\cdot\text{s})/(\text{V}\cdot\text{cm})]$ of egg albumin to its titration curve. Thus, there is no migration at the isoelectric point. Below the isoelectric point, where all molecules have a net positive charge, migration will proceed to the cathode; above the isoelectric point, where all molecules have a net negative charge, migration will proceed to the anode.

Moving Boundary or Free Electrophoresis

This type of electrophoresis is performed by analyzing the electrophoretic migration of a boundary formed between a buffer solution and a protein solution dissolved in the buffer. Such boundaries are formed by placing the buf-

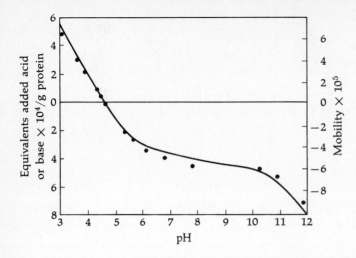

Figure 9.13 The electrophoretic mobility [(cm·s)/(V/cm)] (shown on dots) and the titration curve of crystalline egg albumin. [*Mobility data from L. C. Longsworth, Ann. N.Y. Acad. Sci. 41:275 (1941); titration data from R. K. Cannan, A. Kibrick, and A. H. Palmer, Ann. N.Y. Acad. Sci. 41:243 (1941).*]

fered protein solution in a U tube and carefully layering buffer over the protein in each arm of the tube. The tube is then submitted to an electric field and the migration of the protein-buffer boundary is observed with time by the optical methods employed to analyze a moving boundary in the ultracentrifuge (Chap. 4).

Figure 9.14 shows the electrophoretic pattern given by a typical protein mixture, human plasma. Each protein in the mixture migrates as a peak; the area under a peak is proportional to the amount of protein in plasma. This method is not a good preparative technique since separation takes place only at the two boundaries in the U tube. Moreover, experimental difficulties such as convection and heating often disturb the boundary and limit accurate analysis. Thus, zone electrophoretic methods, in which such difficulties are minimized, are preferred.

Zone Electrophoresis

In this method, a narrow band of buffered solution containing the substances to be separated is applied to a porous inert medium containing buffer and the entire system is placed in an electric field. A piece of filter paper or a thin layer of such materials as starch or polyacrylamide gel placed on a glass or plastic sheet will usually suffice for the porous medium. After sufficient time to permit electrophoretic resolution, the resolved mixture on the support medium

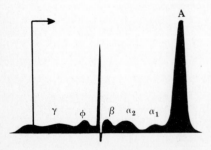

Figure 9.14 The electrophoresis pattern of the complex mixture of proteins found in human plasma. The vertical line indicates the starting point of the boundary. The peak at the initial position is due to the presence of the buffer salt. The largest peak is albumin (A), which has the highest mobility; the peaks labeled α, β, and γ are globulins; ϕ is fibrinogen (Chap. M1). A pure substance gives only one peak.

is analyzed by a staining or spectral method that detects substances such as proteins or nucleic acids. Each compound with a unique charge in the mixture is observed as a discrete zone; hence the name *zone electrophoresis*. This method requires very small amounts of sample and has great resolving power.

There are several kinds of zonal electrophoretic methods. One, called *disc* or *discontinuous electrophoresis,* is widely used for analysis of proteins. In this method a protein mixture is applied to a column or slab of polyacrylamide gel divided into two sections, with each section containing buffer at a different pH. When voltage is applied, the protein mixture migrates first through a short porous section (stacking gel), and as it enters a less porous, longer section (running gel) at a different pH, each protein becomes concentrated into a sharp, thin band. The use of discontinuous buffers permits much greater resolution of the protein bands as they migrate through the running gel than that obtained in other zonal methods where the bands are more diffuse. An example of disc gel electrophoresis is given in Fig. 4.5, which shows the separation of a mixture of proteins in sodium dodecyl sulfate (SDS).

Other zonal electrophoretic methods employ a gel matrix that acts as a molecular sieve, enabling separation of substances on the basis of differences in size as well as charge. One such method has been applied to the separation of the complex mixtures of deoxyribonucleotides encountered on sequence analysis of DNA (Chap. 8).

Zone electrophoresis on SDS gels (Chap. 4) also gives separation on the basis of size and permits accurate estimations of the molecular weight of polypeptides. This method is especially useful for analyzing proteins from biological membranes, which are usually insoluble in aqueous salt solutions but soluble in dilute aqueous solutions of detergents such as SDS.

Many of the cascade methods are combined to give two-dimensional separation of mixtures. Thus, zone electrophoresis in one dimension followed by chromatography or electrophoresis in a second direction provides a powerful means for resolving compounds with similar properties. This is illustrated by the two-dimensional resolution of the tRNAs from *Escherichia coli* shown in Fig. 9.15.

Figure 9.15 Two-dimensional separation of a mixture of about 80 tRNAs by slab gel electrophoresis. The mixture was applied to a 9.6% polyacrylamide gel containing a 7 M solution of urea, having a pH of 8.3, and electrophoresed in one direction. The tRNA-containing strip was removed and polymerized at right angles to a second gel of 20% polyacrylamide containing a 4 M solution of urea, having a pH of 8.3, and electrophoresed in a second direction. The urea-denatured tRNAs separate on the basis of different net charges as well as on size, with the gel matrix acting as a molecular sieve. [*Courtesy of Deborah A. Steege.*]

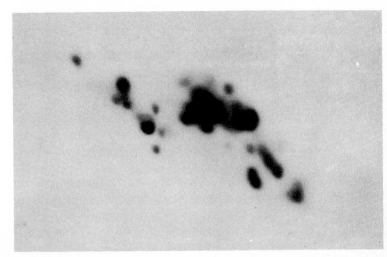

The zonal method called *isoelectric focusing* depends upon formation of a pH gradient throughout the length of a column. The gradient results from the electrophoretic distribution of mixtures of ampholytes, polymers containing numerous carboxyl and amino groups with a wide range of isoelectric points. On electrophoresis of a mixture of proteins and ampholytes, each protein migrates to that position in the column where the mixture of ampholytes provides a pH equal to the pI of the protein. Once at its isoelectric point, the protein ceases to migrate.

CRITERIA OF PURITY

The different cascade methods are useful in assessing the purity of a compound. The ordinary criteria of purity for small organic molecules of biological origin include retention of physical constants and elementary composition after repeated purification. Analysis by one or more of the cascade methods is also useful; e.g., the purity of small ionic compounds can be assessed by zone electrophoresis, by ion-exchange chromatography, or by paper chromatography. Similarly, the purity of non-ionic substances can often be evaluated by paper chromatography, by adsorption chromatography, or by gas-liquid chromatography.

The usual criteria of purity of small molecules are not readily applicable to proteins, nucleic acids, and polysaccharides. These macromolecules are usually difficult to crystallize and may form mixed crystals with other substances, and none gives a sharp melting point. Elementary compositions are not meaningful because of the large number of atoms in macromolecules.

There are really no tests for purity of macromolecules, only methods for the detection of impurities or inhomogeneity. It is assumed that a substance is pure only after all possible tests have failed to reveal inhomogeneity. Thus, the cascade methods are relied upon primarily to detect impurities when used to assess purity.

Homogeneity in molecular weight can be determined by ultracentrifugal methods or by gel filtration, but neither method is as sensitive as those methods in which separation is dependent on the net charge of a macromolecule. Zone-electrophoretic methods are particularly useful, especially by determining the electrophoretic mobility in different buffers at several pH values. Ion-exchange chromatography may also be useful in detecting impurities. Partition chromatography and countercurrent distribution have been employed successfully with only a few proteins because of the limited solubility and stability of most proteins in nonpolar solvents. However, the purity of peptides containing up to 40 to 50 residues can often be determined by paper chromatography. Often macromolecules that appear to be pure by one method are found to be impure by another. For example, a protein may be homogeneous in size on ultracentrifugation but unhomogeneous on electrophoresis. In evaluating the purity of any substance, different methods should be employed in which separation depends upon different properties of the molecule, such as size, charge, solubility, etc.

Tests for functional purity are extremely valuable when used in conjunction with cascade methods, since only minute quantities of a substance are

175

**Methods of
Separation and
Purification of
Compounds from
Biological Sources**

usually needed to assay its biological activity. Thus, an excellent criterion of purity is whether the *specific activity* of a substance, i.e., its biological activity per unit weight of the substance, remains constant after repeated purification. For example, if an enzyme acts on compound A 10 times faster than on B but after purification by one means or another acts on A 50 times faster than on B, it is likely that the enzyme is contaminated with another enzyme that acts on B. Frequently impurities present in as little as one part per 10^3 to 10^6 in a preparation can be detected by tests for functional properties, a sensitivity far greater than that given by most physical or chemical methods.

Many substances are antigens (Chap. M2) that stimulate the production of specific antibodies when injected into an experimental animal. If a single substance, or antigen, is allowed to react with its antibody, as in immunoelectrophoresis (Chap. M2), only one precipitin line will be observed. In contrast, if the substance used as antigen is impure, multiple precipitin lines will be seen corresponding to the interactions of each antigen with the corresponding antibodies it elicited. Such methods provide an extremely sensitive means for detecting impurities that are antigenic.

REFERENCES

Books

Cantor, C. R., and P. R. Schimmel: *Biophysical Chemistry, Part II, Techniques for the Study of Biological Structure and Function,* Freeman, San Francisco, 1980.

Heftmann, E. (ed.): *Chromatography,* 3d ed., Rheinhold, New York, 1975.

Morris, C. J. O. R., and P. Morris: *Separation Methods in Biochemistry,* 2d ed., Pitman, New York, 1976.

Williams, C. A., and M. W. Chase (eds.): *Methods in Immunology and Immunochemistry,* 3 vols., Academic, New York, 1967–71.

Work, T. S., and E. Work: *Laboratory Techniques in Biochemistry and Molecular Biology,* North-Holland, Amsterdam, from 1969 to present, a continuing series.

Review Articles

Ackers, G. K.: Molecular Sieve Methods of Analysis, in H. Neurath and R. L. Hill (eds.): *The Proteins,* 3d ed., vol. I, pp. 1–94, Academic, New York, 1975.

Porath, J., and T. Kristiansen: Biospecific Chromatography and Related Methods, in H. Neurath and R. L. Hill (eds.): *The Proteins,* 3d ed., vol. I, pp. 95–178, Academic, New York, 1975.

Weber, K., and M. Osborne: Proteins and Sodium Dodecyl Sulfate: Molecular Weight Determination on Polyacylamide Gels and Related Procedures, in H. Neurath and R. L. Hill (eds.): *The Proteins,* 3d ed., vol. I. pp. 179–223, Academic, New York, 1975.

Part

2

Catalysis

Chapter 10

Enzymes I

Nature. Classification. Kinetics. Metabolic inhibitors. Regulation of enzymic activity.

Enzymes are catalysts that enhance the rates of biochemical reactions from 10^6 to 10^{12} times those of uncatalyzed reactions. All enzymes share certain structural and functional features irrespective of the reaction catalyzed. They are all proteins and contain a functional site, called the *active site*, where reactants are converted to products. Each enzyme is highly specific, catalyzing one or at most a few reactions. It is the high degree of specificity of enzymes that permits the coordinated network of chemical reactions that occurs in living cells, the sum of which constitutes metabolism. The actual velocity of various critical enzymic reactions is regulated by special mechanisms that modify their catalytic activities, a phenomenon of great importance to metabolism.

The purpose of this chapter is to consider the general features of enzymes. Particular attention will be given to enzyme kinetics, the study of enzyme-catalyzed reaction rates. Kinetic analysis under a variety of conditions may not only reveal many properties of an enzyme but also give important information about the reaction catalyzed.

GENERAL NATURE OF ENZYMES

All Enzymes Are Proteins

Nonprotein catalysts enhance the rate of chemical reactions, a few to about the same extent as enzymes, but they are not found in living things. Enzymes vary in size. While a few have molecular weights as low as 10^4, usually they are larger and range in molecular weight from about 1.5×10^4 to 10^6. Like all proteins, they are labile and rendered inactive if denatured.

Any cross-references coded M refer to the companion text, *Principles of Biochemistry: Mammalian Biochemistry*.

Enzymes Increase the Rate of a Chemical Reaction But Do Not Influence the Equilibrium

Enzymes are highly efficient: an enzyme molecule may transform 10^2 to 10^6 molecules of substrate per minute. For a chemical reaction $A + B \rightleftharpoons C + D$, the velocity in the forward direction is proportional to the concentrations of A and B, or

$$v_1 = k_1[A][B]$$

For the reverse reaction

$$v_2 = k_2[C][D]$$

where k_1 and k_2 are the individual velocity constants. At equilibrium $v_1 = v_2$; thus

$$k_1[A][B] = k_2[C][D] \qquad \text{or} \qquad \frac{k_1}{k_2} = \frac{[C][D]}{[A][B]} = K_{eq}$$

The equilibrium constant gives the relationship between the velocity constants, and at a given temperature, it is the same for a reaction in the presence or absence of an enzyme. Enzymes do, however, markedly influence the rate of conversion of A and B to C and D, or the reverse reaction, so that an equilibrium mixture is obtained very rapidly.

Enzymes Exhibit a High Degree of Specificity for Their Substrates

Some enzymes catalyze a reaction with only a single set of reactants. Thus, *fumarase* catalyzes the interconversion of fumarate and malate.

L-Malate   Fumarate

Neither maleate, which is the cis stereoisomer of fumarate, nor D-malate is a substrate. Other enzymes have a somewhat broader substrate specificity. For example, each proteolytic enzyme listed in Table 5.3 hydrolyzes peptide bonds but is specific for bonds formed by different amino acids, and also shows a strict stereospecificity, catalyzing hydrolysis of peptide bonds formed by L- but not D-amino acids. Enzymes are also specific for the type of reaction catalyzed; thus fumarase catalyzes hydration or dehydration of its substrates, the proteolytic enzymes catalyze a hydrolytic reaction, and none catalyzes other possible reactions of their substrates, e.g., oxidation-reduction or decarboxylation reactions. Some enzymes, however, have a somewhat broader substrate specificity; e.g., proteolytic enzymes can also catalyze hydrolysis of esters and thioesters.

In contrast to laboratory experience with organic chemical reactions there are few side reactions in enzyme-catalyzed reactions. This aspect of the efficiency and specificity of enzymes is even more remarkable because enzymes act in aqueous media, at atmospheric pressure, usually near 37°C, and between pH 2 and 10, conditions in the laboratory that seldom permit organic chemical reactions of similar complexity.

The formation of enzyme-substrate complexes as intermediates in enzyme-catalyzed reactions has been demonstrated by several means, including kinetic analysis, chemical modification by R-group-specific reagents, inhibition of enzymes by specific compounds that interact with active sites, detection of characteristic spectral absorption bands when enzymes act upon their substrates, and x-ray crystallography of enzymes combined with compounds similar in structure to their substrates.

The Region of an Enzyme That Specifically Interacts with the Substrate Is Called the Active Site

The conformation of an enzyme brings certain R groups in the polypeptide backbone into juxtaposition in a highly specific manner to form the active site. The spatial arrangement of the structures in the active site determines not only which compounds can fit stereospecifically into the site but also the nature of the subsequent events that lead to conversion of the substrate to products. Substrate binding to the active site may occur through formation of specific noncovalent bonds and, in some instances, covalent bonds. On binding to this site, the substrate is brought into close proximity to specific groups on the enzyme that cooperatively destabilize certain bonds in the substrate, making them more chemically reactive. In many instances, a covalent bond is formed transiently between the substrate and enzyme.

Enzymes Lower the Activation Energy Required for a Chemical Reaction

At constant temperature a population of molecules has a kinetic energy that is distributed among the molecules, as shown diagrammatically in Fig. 10.1a. At a temperature T_1, the population of molecules has insufficient energy to undergo some specific chemical reaction, but if the temperature is raised to T_2, the energy distribution profile changes as shown. At T_2 there is now sufficient energy to increase the number of collisions between the molecules so that a chemical reaction can proceed. Thus, when the temperature is raised from T_1 to T_2, the increase in reaction rate is largely the result of an increase in the number of *activated molecules*, i.e., that fraction possessing the required energy of activation.

Figure 10.1c shows a very simplified view of the energy level of a population of molecules during the course of the reaction A $\rightleftharpoons$ B. As the reaction proceeds, enough molecules have sufficient energy to become activated and enter a transition state, at which they decompose to products. The energy required to achieve the transition state, or activated state, is E_a, *the energy of activation*. For any reaction to proceed, the energy content of the reactants must be greater than that of the products. During the course of reaction the energy of activation E_a reemerges, and the total energy change resulting from the reaction is the difference in the energy levels of A and B (Fig. 10.1c).

Enzymes, like all catalysts, accelerate the rates of chemical reactions by

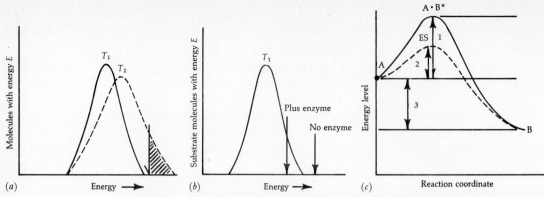

Figure 10.1 (*a*) The kinetic-energy distribution of a population of molecules at temperature T_1 and at a higher temperature T_2. The arrow indicates the minimum energy content required for the molecules to react; thus at T_1 no reaction occurs, but at T_2 reaction proceeds. (*b*) The kinetic energy of a population of substrate molecules at a temperature T_1. The arrows indicate the energy content required for reaction in the absence and presence of enzyme. Note that in the absence of enzyme no reaction could occur, but in the presence of enzyme reaction may proceed without altering temperature. (*c*) The energy profile of an uncatalyzed reaction and a catalyzed reaction, A ⇌ B. In the uncatalyzed reaction the energy level of A must be raised sufficiently to activate molecules of A and bring them to the transition state A·B*, where they can react to form B. The energy required to bring molecules to the transition state is the energy of activation E_a, the difference in the energy level of A and A·B*, indicated by 1. In the catalyzed reaction the E_a required to form an activated ES complex, indicated by 2, is much less than 1 for the uncatalyzed process. The difference in energy levels between A and B, 3, is the same in either the catalyzed or uncatalyzed reaction.

lowering the activation energy for a specific reaction. Figures 10.1*b* and 10.1*c* show this effect in terms of the energy content of a population of molecules as well as the energy levels of reactants and products throughout the course of the reaction. The means for estimating the energy of activation and the effects of temperature on enzyme-catalyzed reactions are considered later in this chapter.

Some Enzymes Aid in Regulation of Rates of Reaction

Most organisms do not alter rates of metabolic reactions by changing temperature; accordingly, catalyzed reactions are necessary to make a process go fast enough at the temperature of an organism. Moreover, if biological reactions proceeded without catalysts, no control could be exercised over their rates.

Numerous regulatory mechanisms are employed to regulate metabolism (Chap. 12); some operate at the level of the enzyme itself. A substance that either increases or decreases the rate of a catalyzed reaction by acting directly on the responsible enzyme is called an *effector*. Effectors exert their action by altering the structure of the enzyme in a way that affects only reaction rates. Mechanisms of enzymic regulation are considered later in this chapter.

Some Enzymes Are in Multienzyme or Multifunctional Complexes

High-molecular-weight *multienzyme complexes* are known that contain three or more different enzymes which are tightly combined by noncovalent interactions which dissociate only under conditions that disrupt noncovalent bonds. Each of the enzymes catalyzes a separate reaction, but together the enzymes in a complex catalyze a single overall reaction. *Pyruvate dehydrogenase* (Chap. 15) is one example of a multienzyme complex to be encountered subsequently. Other enzymes are in *multifunctional complexes,* in which two or more separate enzymes are contained in separate domains of a single polypeptide chain. Each of the separate domains catalyzes one step in a single overall reaction catalyzed by the multifunctional complex. *Fatty acid synthase* (Chap. 20) is an example of this type of complex.

NOMENCLATURE AND CLASSIFICATION OF ENZYMES

Enzymes are usually named in terms of the reactions they catalyze. It is customary to add the suffix *-ase* to a major part of the name of the substrate acted upon, e.g., *urease* acts on urea, and *tyrosinase* on tyrosine. Trivial names persist, such as trypsin and pepsin. An international commission has devised a rather complex nomenclature and classification of all enzymes. Table 10.1 lists the major classes of enzymes in this system, each of which has subclasses and sub-subclasses. Thus, the oxidoreductases include all enzymes catalyzing an oxidation-reduction reaction, the transferases, a transfer reaction, and so forth.

There are ambiguities in the classification; however, use of the classification scheme in the research literature ensures precision in understanding communications, oral or written. Some of the recommended names have not yet gained general usage and are not uniformly used in this book. Thus, it is best to learn the name of an enzyme and the reaction it catalyzes as it first appears in the text. It will become apparent that the names of enzymes are descriptive of a general type of reaction, e.g., dehydrogenases catalyze dehydrogenation reactions, hydrases catalyze hydration reactions, decarboxylases catalyze decarboxylation of a compound, etc.

TABLE 10.1
Classification of Enzymes

Class	Action
1. Oxidoreductases	Enzymes catalyzing oxidation-reduction reactions. The hydrogen donor is regarded as the substrate.
2. Transferases	Enzymes catalyzing reactions of the general form $$X-Y + Z \rightleftharpoons X + Z-Y$$
3. Hydrolases	Enzymes catalyzing a hydrolytic cleavage of C—O, C—N, C—C, and other bonds.
4. Lyases	Enzymes catalyzing cleavage of C—C, C—O, C—N, and other bonds by elimination, leaving double bonds, or alternatively by the addition of groups to double bonds.
5. Isomerases	Enzymes catalyzing a change in the geometric or spatial configuration of a molecule.
6. Ligases	Enzymes catalyzing the joining together of two molecules with the accompanying hydrolysis of a high-energy bond.

COFACTORS AND COENZYMES

Many enzymes contain prosthetic groups that are non-amino acid in nature. The conjugated protein is known as the *holoenzyme* and can be dissociated into a protein component, the *apoenzyme,* and its nonprotein prosthetic group, the *cofactor*. For example, reddish-brown *catalase* (Chap. 16) dissociates in acid to give a colorless apoprotein and a ferriprotoporphyrin. Other enzymes contain metals, which are often tightly bound and not readily removed unless the enzyme is denatured. Superoxide dismutase (Chap. 16) contains two metals, one atom each of Cu^{2+} and Zn^{2+}, in each of its two subunits. The metals are difficult to remove without disrupting the conformation of the enzyme; the apoenzyme, however, can be recombined with the two metals to regenerate the holoenzyme. Some metals participate directly in the catalytic process, for example, Cu^{2+} and Zn^{2+} in superoxide dismutase. Metals may aid in maintaining the native structural conformation of the enzyme and are often called *activators*.

Some organic cofactors act as acceptors or donors of atoms or of functional groups that are removed from or added to the substrate of the enzyme. These cofactors are often easily dissociated from the enzyme and are referred to as *coenzymes*. These should be regarded more properly as *cosubstrates,* since they participate stoichiometrically in the reaction catalyzed and are consumed with the substrates. For example, some enzymes (E) contain a coenzyme (C) and react with a substrate (AH_2) as follows:

$$E \cdot C + AH_2 \rightleftharpoons A + E \cdot CH_2$$

$E \cdot CH_2$ must be converted to $E \cdot C$ before further reaction with AH_2. This can be accomplished by another enzyme catalyzing the reaction

$$E \cdot CH_2 + B \rightleftharpoons E \cdot C + BH_2$$

or by dissociation of CH_2 from the enzyme and conversion to C by another enzyme.

There are hundreds of enzymes that require a metal or an organic cofactor, but the number of metals and cofactors in organisms is limited. Various organic cofactors will be encountered throughout the text; several are derived from components that cannot be synthesized by mammals and are essential nutritive factors, or *vitamins*. The action of these and many other cofactors is best understood as they are considered in the enzymic reactions of metabolism.

KINETICS OF ENZYMIC REACTIONS

Study of the rates of enzyme-catalyzed reactions reveals considerable information about the mechanism of action of an enzyme as well as features of the reaction catalyzed that are helpful in understanding the metabolism of the reactants. Accordingly, it is essential to understand some fundamental concepts concerning the rates of enzyme-catalyzed reactions and the factors that may influence the rate.

It is useful before considering enzyme-catalyzed reactions to recall the effects of concentration on noncatalyzed reactions. The rates of chemical reactions are generally estimated as the change in concentration c of substrate or product per unit of time t. Customary units are moles per liter and seconds. Thus, if c is the initial concentration, which decreases with time, the reaction velocity is $-dc/dt$. This rate may depend upon the instantaneous value of c in various ways. It may be independent of c,

$$-\frac{dc}{dt} = k_0 = k_0 c^0 \qquad \text{zero-order reaction}$$

proportional to c,

$$-\frac{dc}{dt} = k_1 c = k_1 c^1 \qquad \text{first-order reaction}$$

or proportional to the second or, rarely, a higher power of c,

$$-\frac{dc}{dt} = k_2(c \times c) = k_2 c^2 \qquad \text{second-order reaction}$$

In each case, k is known as the *reaction velocity constant* or *rate constant*. The dimensions of k depend on the order of the reaction and are given by $k = c^{1-n}t^{-1}$, where n is the order of the reaction. Thus k_0 has dimensions of ct^{-1}; k_1 of t^{-1}; and k_2 of $c^{-1}t^{-1}$. From this it follows that velocity constants for reactions of different orders cannot be added or subtracted. Also, only in first-order reactions is the velocity constant devoid of the dimensions of concentration. Thus only for first-order reactions is the half-time, i.e., the time required to halve the initial concentration, a constant at all concentrations.

For enzymic reactions in which the molecular weight of the enzyme is known, rate constants are given in terms of moles of enzyme. When the molecular weight of the enzyme is unknown or the enzyme preparation is impure, the rate is frequently expressed per milligram of protein per milliliter. By international convention, one unit of any enzyme is that amount which will catalyze the transformation of substrate at the rate of 1 mol/s under specified conditions. One unit is a very large quantity of enzyme; nano- and picounits are used generally to express unit activity.

Rate Equations

For all enzymic processes, the rate of the reaction depends upon the concentrations of both the enzyme and its substrate, other conditions being constant. Figure 10.2 shows the relationship between the initial velocity of a reaction and the substrate and enzyme concentrations. Although the velocity increases linearly with enzyme concentration, at constant enzyme concentration it increases hyperbolically as the substrate concentration increases toward a limiting maximal velocity. This indicates that the enzyme has a finite number of sites to combine with substrates; when all sites are occupied, no further rate enhancement occurs and the enzyme is saturated with substrate. A general rate equa-

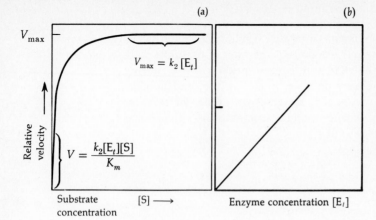

$$V_{\max} = k_2\,[E_t]$$

$$V = \frac{k_2[E_t][S]}{K_m}$$

(a)

(b)

Relative velocity

Substrate concentration

[S] →

Enzyme concentration [E$_t$]

Figure 10.2 The relative initial velocity (*a*) as a function of substrate concentration and (*b*) as a function of enzyme concentration. $V_{\max}$ is 1.0, and the substrate concentration required to achieve $\tfrac{1}{2}V_{\max}$ is equal to K_m. In contrast to the hyperbolic curve relating V to [S], V as a function of [E$_t$] is linear. When $V = V_{\max}$, the rate is independent of substrate concentration (zero-order reaction). When the substrate concentration is small compared with K_m, the rate is proportional to the substrate concentration and a first-order expression is obtained.

tion, known as the *Michaelis-Menten equation*, describes the reaction in which only a single substrate and single product are reversibly interconverted. It is assumed that an enzyme E reacts with substrate S to give a substrate-enzyme complex ES, which then gives enzyme and product P as follows:

$$\text{E} + \text{S} \underset{k_{-1}}{\overset{k_1}{\rightleftharpoons}} \text{ES} \underset{k_{-2}}{\overset{k_2}{\rightleftharpoons}} \text{E} + \text{P} \tag{1}$$

where k_1, k_{-1}, k_2, and k_{-2} are the respective velocity constants of the assumed reaction steps. Because of the importance of this equation, it will be derived according to the assumption of Michaelis and Menten that an ES complex is formed and with additional assumptions, some of which are introduced subsequently.

It is assumed that the concentration of S is much greater than that of E and that only *initial velocities* are measured, i.e., velocities estimated under conditions where only a very small fraction of S has been converted to P. Under these conditions, conversion of P to ES is almost nil, and the step P → ES can be ignored. The total enzyme in the system is E$_t$, and

$$[E_t] = [E] + [ES] \tag{2}$$

where [E] is the uncombined enzyme concentration, [ES] the ES concentration, and [E$_t$] the total enzyme concentration. Further, the velocity V of the reaction will be equal to

$$V = k_2[ES] \tag{3}$$

This is the actual rate equation for reaction (1), but it is not useful since neither k_2 nor [ES] can be measured directly. The *maximal velocity* $V_{\max}$ is equal to

$$V_{\max} = k_2[E_t] \tag{4}$$

This must obtain since $V_{\max}$ cannot exceed that rate at which all enzyme is saturated with substrate. Finally, it is assumed that the reaction proceeds at a *steady state*, i.e., where the rate of change of [ES] is zero ($d[ES]/dt = 0$), and thus [ES] does not change during the period the velocity is measured.

The rate of formation of ES, v_f, is proportional to [E] and [S], as in any second-order reaction:

$$v_f = k_1[E][S] = k_1([E_t] - [ES])[S] \tag{5}$$

The rate of disappearance of ES, v_d, is

$$v_d = k_{-1}[ES] + k_2[ES] = (k_{-1} + k_2)[ES] \tag{6}$$

In the steady state since $d[ES]/dt = 0$, $v_f = v_d$; thus

$$k_1([E_t] - [ES])[S] = (k_{-1} + k_2)[ES] \tag{7}$$

Rearranging Eq. 7 gives

$$\frac{[S]([E_t] - [ES])}{[ES]} = \frac{k_{-1} + k_2}{k_1} = K_m \tag{8}$$

where the term containing the three velocity constants, K_m, is called the *Michaelis constant*, a useful parameter characteristic of each enzyme and a substrate.

Rearranging Eq. 8 by solving for [ES] gives

$$[ES] = \frac{[E_t][S]}{K_m + [S]} \tag{9}$$

Then by substituting V/k_2 for [ES] (from Eq. 3) and V_{max}/k_2 for [E_t] (from Eq. 4) in Eq. 9, the final Michaelis-Menten rate equation is

$$V = \frac{V_{max}[S]}{K_m + [S]} \tag{10}$$

Let us now consider methods for determining V_{max} and K_m from measurements of V at different values of [S]. K_m could be estimated from the hyperbolic curve in Fig. 10.2, but V_{max} cannot be obtained accurately since substrate concentrations often cannot be achieved experimentally to saturate the enzyme. However, by taking the reciprocal of Eq. 10, the following linear equation, termed the *Lineweaver-Burk equation*, is obtained:

$$\frac{1}{V} = \frac{K_m}{V_{max}} \frac{1}{[S]} + \frac{1}{V_{max}} \tag{11}$$

A plot of $1/V$ vs. $1/[S]$ as shown in Fig. 10.3 (a double reciprocal plot) yields a straight line with slope K_m/V_{max} and ordinate intercept $1/V_{max}$. Since the slope and intercept are readily measured from the graph, V_{max} and K_m can be accurately determined. Alternatively, multiplying Eq. 11 by [S] gives

$$\frac{[S]}{V} = \frac{K_m}{V_{max}} + \frac{[S]}{V_{max}} \tag{12}$$

Multiplying Eq. 11 by V_{max} and rearranging gives

$$V = -K_m \frac{V}{[S]} + V_{max} \qquad V_{max} = K_m \frac{V}{[S]} \tag{13}$$

Equations 12 and 13 can also be used graphically for determining K_m and V_{max}, as shown in Fig. 10.3.

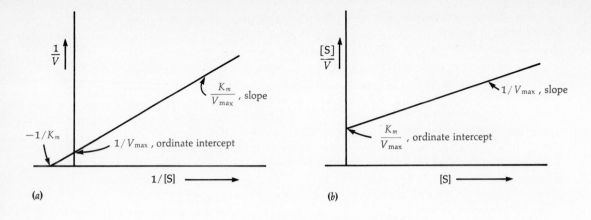

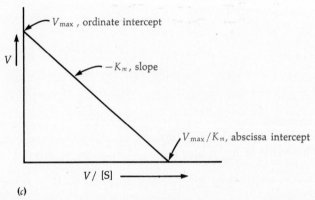

Figure 10.3 Reciprocal plots used for determining K_m and V_{max} of an enzymic reaction. The velocities at several different concentrations of substrate are measured and plotted as shown. The slopes and intercepts of the resulting straight lines give values for K_m and V_{max} as indicated. (*a*) Double reciprocal plot according to Eq. 11. (*b*) Single reciprocal plot according to Eq. 12. (*c*) Single reciprocal plot according to Eq. 13.

Nature of K_m

Of particular interest is the following equation obtained by solving Eq. 10 for K_m:

$$K_m = [S] \left(\frac{V_{max}}{V} - 1 \right) \tag{14}$$

At the concentration of S at which $V = \frac{1}{2}V_{max}$, numerically solving Eq. 11 for K_m gives

$$K_m = [S]$$

Thus, K_m is equal to the substrate concentration (moles per liter) that results in one-half the numerical maximum velocity V_{max}.

K_m is a complex constant as $K_m = (k_{-1} + k_2)/k_1$. Three conditions are possible: (1) If k_{-1} is very much greater than k_2, we may neglect k_2 and $K_m = k_{-1}/k_1$. K_m is then the thermodynamic equilibrium constant for the reversible formation of ES. (2) If k_2 is very much greater than k_{-1}, then $K_m = k_2/k_1$ and K_m is a steady-state constant containing two independent velocity constants. (3) When k_{-1} and k_2 are of the same order of magnitude, all three reaction constants determine the value of K_m. All three conditions have been observed to obtain, depending on the particular enzyme-substrate system being studied.

The Michaelis-Menten rate equation (Eq. 10) was derived assuming that the rate of the transformation of $P + E \rightarrow EP$ is essentially zero because of the initial velocity assumption. However, this equation also applies to the enzymic catalysis of reactions with an overall equilibrium constant near 1, in which the interconversion of S and P is catalyzed reversibly. This is shown by the Haldane equation, which relates K_{eq} to rate constants for the forward and reverse reactions and to the K_m and V_{max} values as follows:

$$K_{eq} = \frac{[P]_{eq}}{[S]_{eq}} = \frac{k_1 k_2}{k_{-1} k_{-2}} = \frac{V_{max}^S K_m^P}{V_{max}^P K_m^S} \tag{15}$$

where the superscript S pertains to the substrate S and P to the product P. K_m^P and V_{max}^P are measured for the reverse reaction in the same manner as already described above for the substrate S. Measurements for the reversible system of Eq. 11 offer a useful check on the kinetic constants, particularly as K_{eq} can be determined from concentrations alone. For the enzyme *fumarase*, which catalyzes reversibly the interconversion: fumarate + $H_2O \rightleftharpoons$ malate, $K_{eq} = 4.4$, as determined by measuring [malate] and [fumarate] in equilibrium mixtures. This value of K_{eq} is also obtained from the following experimentally determined parameters for K_m and V_m: $K_m^{mal} = 2.5 \times 10^{-5} M$, $K_m^{fum} = 0.5 \times 10^{-5} M$, $V_m^{mal} = 54,000 \ min^{-1}$, and $V_m^{fum} = 48,000 \ min^{-1}$, where $K_{eq} = V_m^{fum} K_m^{mal}/V_m^{mal} K_m^{fum}$.

Equation 1 is also an oversimplification of the mechanism of an enzyme-catalyzed reaction, since there is generally more than one enzyme complex formed during catalysis. For example, an enzyme may first bind substrate to form ES, and then enter the activated transition state ESP, to give EP and finally $E + P$ as follows:

$$E + S \rightleftharpoons ES \rightleftharpoons ESP \rightleftharpoons EP \rightleftharpoons E + P \tag{16}$$

The inclusion of various enzyme complexes as in Eq. 16 does not change the form of the Michaelis-Menten equation (Eq. 10) derived by using the simplification shown in Eq. 1.

Order of Enzymic Reactions

The course of the reaction with time will be determined by the region of the curve in Fig. 10.2a represented by the initial substrate concentration. It is therefore of interest to examine the relationships indicated in Eq. 10 for circumstances of high and low initial substrate concentration, as these are the conditions that obtain at the beginning and at the end, respectively, of an enzyme-catalyzed reaction. When [S] is much greater than K_m, that is, at constant enzyme and very high substrate concentrations, on the plateau portion of the curve of Fig. 10.2a,

$$-\frac{d[S]}{dt} = V = V_{max} = k_2[E_t] \tag{17}$$

The rate of the reaction is constant for a given enzyme concentration. Under these conditions, the enzyme is saturated with substrate, and the reaction is proceeding at maximal velocity. In other words, a further increase in [S] will not alter the velocity.

With a fall of the substrate concentration to values much less than K_m, at the bottom portion of the curve in Fig. 10.2a,

$$-\frac{d[S]}{dt} = V = \frac{V_{max}[S]}{K_m} = \frac{k_2[E_t][S]}{K_m} \tag{18}$$

That is, the velocity of the reaction is directly proportional to the substrate concentration for a given enzyme concentration $[E_t]$.

Kinetics of Two-Substrate Enzyme-Catalyzed Reactions

Most enzymes have more than one substrate and catalyze reactions of the form $A + B \rightleftharpoons C + D$. These reactions involve formation of enzyme-substrate complexes, just like one-substrate reactions, and their kinetics can be examined to give the K_m for each substrate and the V_{max} for the reaction. The K_m values for each substrate are measured by graphical analysis (Eq. 11 to 13) of initial velocities obtained by varying the concentration of one substrate at a fixed, usually saturating concentration of the other substrate.

Equations for the kinetics of bisubstrate reactions analogous to the Michaelis-Menten rate equation have been derived and provide a great deal of insight into the general mechanism of the reaction as well as values for kinetic constants such as K_m and V_{max}. These equations are too complex to be developed here, and more advanced texts should be consulted for details. It is useful, however, to consider the basic mechanisms of bisubstrate reactions, which are of two general types, *double-displacement and sequential mechanisms*.

In a *double-displacement mechanism* for the reaction $A + B \rightleftharpoons C + D$, one substrate, A, binds to the enzyme to give an EA complex. E and A then react to form a new complex, FC, and then one product, C, is released to give a stable enzyme-substrate intermediate, F, that differs from E. The intermediate F then reacts with the second substrate, B, to give the enzyme-substrate complex FB, which yields the second product D and the enzyme E. Reactions of this type can be written in several ways, but shorthand notation illustrates well a double-displacement mechanism:

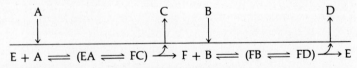

where A, B, C, and D are the reactants and products, E the enzyme, F the stable enzyme intermediate, and the letters in parentheses the putative intermediate complexes. The horizontal line shows the enzyme forms during the reaction and the arrows indicate the combination or dissociation of substrates and products as indicated. These mechanisms have also been described as *ping-pong mechanisms*. Transamination by the enzyme glutamic-aspartic aminotransferase (Chap. 22) is an example of a ping-pong mechanism.

Sequential mechanisms are of two types: *ordered* and *random*. In contrast to the ping-pong mechanism, in sequential mechanisms all substrates must combine to form a ternary complex before product is formed. Ordered reactions can be written in shorthand notation as

Substrate A binds to E to form EA, which in turn binds B. The ternary complex EAB becomes complex ECD, which then gives products in the order shown. The reactions catalyzed by phosphofructokinase (Chap. 17) and glyceraldehyde-3-phosphate dehydrogenase (Chap. 17) are examples of ordered reactions.

In other instances, E carries binding sites for both A and B and reaction velocity is unaffected by whether A or B binds first. The mechanism is stated to be random. If there is also no "preferred" order of release of the products C and D after the ternary complex EAB has become ECD, the random mechanism can be depicted as

Examples of random-ordered reaction mechanisms are UDP-galactose:*N*-acetylglucosamine galactosyl transferase (Chap. 18) and creatine kinase (Chap. 15).

More complex kinetic mechanisms involving three and even four reactants are known. They may have sequential or ping-pong mechanisms or components of both types.

Effect of Temperature

The equilibrium constant for any chemical reaction as well as the rate of the reaction depends strongly on temperature, and enzyme-catalyzed reactions are no exception. The effect of temperature on an equilibrium constant for a chemical reaction is given by the van't Hoff equation

$$2.3 \log K = C - \frac{\Delta H}{RT} \tag{19}$$

where ΔH is the heat of the reaction in calories per mole, R is the gas constant, equal to 1.98 cal/mol/°C, and T is the absolute temperature. C is an integra-

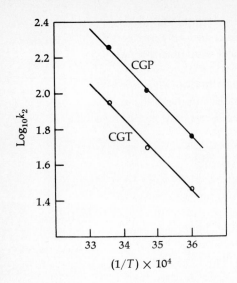

Figure 10.4 Effect of temperature on the rate of hydrolysis k_2 of benzyloxycarbonylglycylphenylalanine (CGP) and benzyloxycarbonylglycyltryptophan (CGT) by crystalline carboxypeptidase. The data are plotted as log k_2 vs. the reciprocal of the absolute temperature. The apparent activation energies E_a are 9900 cal/mol for CGT and 9600 cal/mol for CGP. [*Data from R. Lumry, E. L. Smith, and R. R. Glantz, J. Am. Chem. Soc.* **73**:*4330 (1951).*]

tion constant. It follows from this equation that graphs of log K plotted against the reciprocal of the absolute temperature $(1/T)$ should give a straight line. The slope of the line is $\Delta H/2.3R$. The effect of temperature on the velocity of a reaction is given by the Arrhenius equation

$$2.3 \log k = B - \frac{E_a}{RT} \tag{20}$$

This equation is of the same form as Eq. 19 but relates k, the rate constant for the reaction, with T, R, E_a, the energy of activation in calories per mole, and a constant B that qualitatively expresses the frequency of collisions and the requirement for specific orientation between colliding molecules.

Figure 10.4 shows the effect of temperature on the velocity constant k_2 for the hydrolysis of two substrates by carboxypeptidase. This is a typical *Arrhenius plot* for an enzyme-catalyzed reaction showing that log k_2 varies linearly with $1/T$ between 5 and 25°C and that E_a can be calculated from the slope of the line. For enzyme-catalyzed reactions, the rate increases with temperature until a maximal rate is achieved, but at temperatures above the maximum, decreased rates are observed because of thermal denaturation of the enzyme.

Energy of Activation

The energy of activation is a measure of the energy needed for the conversion of molecules to the reactive state. A catalyst lowers the activation energy necessary for a reaction that can proceed spontaneously without a catalyst. Table 10.2 shows the activation energies for a number of processes. The decomposition of hydrogen peroxide requires 18,000 cal/mol; this is lowered to 11,700 when colloidal platinum is the catalyst and is much lower for the enzymic reaction. It is obvious that catalase is far more efficient than the inorganic catalyst of this reaction. In fact, catalase is so efficient that only a small activation energy is required in the process. This is consistent with the fact that the

192

decomposition of hydrogen peroxide by catalase proceeds at one of the highest rates known for enzymic reactions (Chap. 16).

Inspection of the other data in Table 10.2 shows that the same relationship holds, viz., that a catalyst lowers the required activation energy and that an enzyme decreases E_a more than an inorganic catalyst does. The effectiveness of enzymes as catalysts is indicated by the high reaction velocities at physiological temperatures. In other words, the decreased E_a permits extremely rapid reaction rates at temperatures much lower than for the uncatalyzed reaction. This may be readily shown by comparing the relative values of the velocity constants for the same reaction at a given temperature ($37°C$) when the activation energies are different. The data given in Table 10.2 for the hydrolysis of sucrose by yeast invertase and by hydrogen ion may be taken as an example. Equation 20 may be written for the enzyme (e) as

$$\log k_e = \frac{B_e}{2.3} - \frac{8000}{2.3RT} \tag{21}$$

and for the hydrogen ion h as

$$\log k_h = \frac{B_h}{2.3} - \frac{25,600}{2.3RT} \tag{22}$$

Let us assume that the values for B_e and B_h are approximately the same. In order to calculate the relative rates of the enzyme-catalyzed and hydrogen ion-catalyzed reactions, i.e., the ratio of the two rate constants, Eq. 22 is subtracted from Eq. 21, with the values for R (1.98) and T ($37 + 273 = 310$) substituted. This gives

$$\log \frac{k_e}{k_h} = \frac{25,600 - 8000}{2.3 \times 1.98 \times 310} = \frac{17,600}{1415} = 12.4 \tag{23}$$

and

$$\frac{k_e}{k_h} = 2.5 \times 10^{12} \tag{24}$$

In other words, the rate constant for the enzymic reaction may be expected to be approximately a trillion times greater than for the hydrogen ion–catalyzed reaction at the same temperature.

TABLE 10.2

Energy of Activation for Enzymic and Nonenzymic Catalyses

Process	Catalyst	E_a, cal/mol
Decomposition of hydrogen peroxide	None	18,000
	Colloidal platinum	11,700
	Catalase	<2,000
Hydrolysis of ethyl butyrate	Hydrogen ion	16,800
	Hydroxyl ion	10,200
	Pancreatic lipase	4,500
Hydrolysis of casein	Hydrogen ion	20,600
	Trypsin	12,000
Hydrolysis of sucrose	Hydrogen ion	25,600
	Yeast invertase	8,000–10,000
Hydrolysis of β-methylglucoside	Hydrogen ion	32,600
	β-Glucosidase	12,200

TABLE 10.3

Optimal pH Values for Some Hydrolytic Enzymes	Enzyme	Substrate	Optimal pH
	Pepsin	Egg albumin	1.5
		Casein	1.8
		Hemoglobin	2.2
		Benzyloxycarbonylglutamyltyrosine	4.0
	α-Glucosidase	α-Methylglucoside	5.4
		Maltose	7.0
	Urease	Urea	6.4–6.9
	Trypsin	Proteins	7.8
	Pancreatic α-amylase	Starch	6.7–7.2
	Malt β-amylase	Starch	4.5
	Carboxypeptidase	Various substrates	7.5
	Plasma alkaline phosphatase	2-Glycerophosphate	9–10
	Plasma acid phosphatase	2-Glycerophosphate	4.5–5.0
	Arginase	Arginine	9.5–9.9

Effect of pH

The pH markedly influences reaction rates. Many enzymes have optimal rates at one pH, called the *pH optimum,* whereas others act equally well over a limited range of pH values. Table 10.3 lists the pH optima for several enzymes, illustrating the spread of such values over the pH scale. Several factors influence the pH optimum, including acidic groups in the active site. If an enzyme requires a protonated acidic group for its catalytic action, the enzyme may show maximal activity at pH values below the pK of the group; conversely if a dissociated form of an acid is required, maximal activity may occur above the pK of the group. Often two or more dissociable acidic groups participate in the active site, and the curves relating pH to activity will reflect the pH dependence of each group. Indeed, study of the effects of pH on reaction rate may aid identification of acidic groups in an active site, although other information is usually required (Chap. 11).

Some substrates are weak acids or contain ionic substituents, and the pH optimum may reflect the fact that the enzyme requires a particular ionic or nonionic form of the substrate. In addition, the rapid decline in reaction rates often found at high or low pH values may reflect enzyme denaturation or dissociation of an essential cofactor.

INHIBITION OF ENZYMES

The rates of enzyme-catalyzed reactions are decreased by specific inhibitors, compounds that combine with the enzyme and prevent enzyme and substrate from combining normally. The toxicity of many compounds, such as HCN and H_2S, results from their action as enzyme inhibitors. Many drugs also act to inhibit specific enzymes. Thus, knowledge of enzyme inhibitors is vital to understanding drug action and toxic agents. Moreover, information about enzymes themselves is revealed by studying enzyme inhibition.

Three types of reversible inhibition of enzymes are recognized, *competitive inhibition, noncompetitive inhibition,* and *uncompetitive inhibition;* each can be distinguished from the others by kinetic analysis.

Competitive Inhibition

Competitive inhibitors can combine reversibly with the active site of the enzyme and compete with the substrate for the active site. While the site is thus occupied, it is unavailable for the binding of substrate. The combination of a competitive inhibitor I with enzyme E can be written in the same manner as combination with substrate, although the inhibitor is not chemically transformed to products.

$$E + I \rightleftharpoons EI \tag{25}$$

K_i, the *dissociation constant* for the enzyme-inhibitor complex EI, is

$$K_i = \frac{[E][I]}{[EI]} \tag{26}$$

Because formation of EI depends upon [I] just as formation of ES is dependent on [S], the actual rate of a competitively inhibited reaction is strictly dependent upon the relative concentrations of S and I at a fixed concentration of E.

One of the first known and now classic examples of competitive inhibition is illustrated by the effect of malonic acid on *succinate dehydrogenase,* which catalyzes the following reaction in the presence of a suitable hydrogen acceptor A:

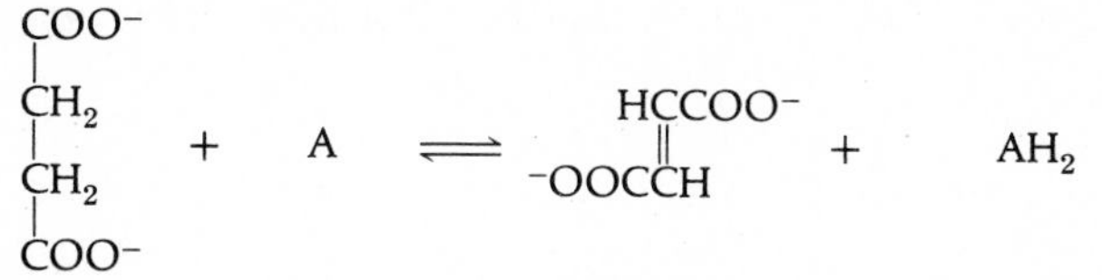

| Succinate | Acceptor | Fumarate | Reduced acceptor |

Many compounds which are structurally similar to succinic acid are competitive inhibitors of the dehydrogenase, including the following:

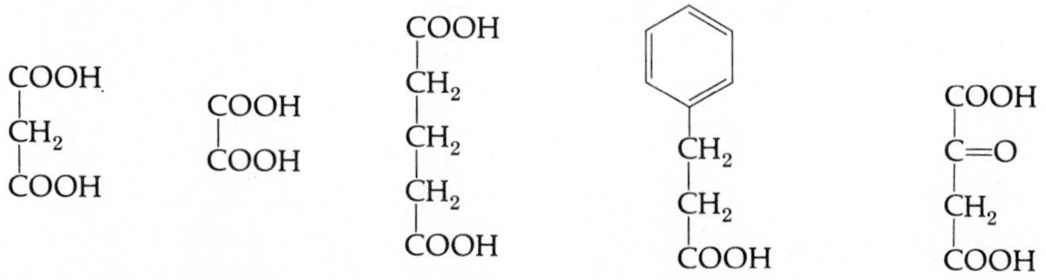

| Malonic acid | Oxalic acid | Glutaric acid | Phenylpropionic acid | Oxaloacetic acid |

The most potent of these inhibitors is malonic acid. When the ratio of [I] to [S] is 1:50, the enzyme is inhibited 50%. Increasing the concentration of substrate at constant [I] decreases the amount of inhibition, and, conversely, decreasing the substrate concentration increases the inhibition. If succinic and malonic acids were bound at different sites on the enzyme, it would be difficult to explain why they should compete with each other. Since they do compete, it is concluded that they combine with the enzyme at the same locus, the active site. The structure of each competitive inhibitor is similar in some respect to that of the substrates.

Competitive inhibitors can be recognized kinetically by the effect of inhibitor concentration on the relationship between V and [S], as shown graphically in Fig. 10.5 in the form of the Lineweaver-Burk equation (Eq. 11). The action of a competitive inhibitor obeys the following equation, which incorporates K_i, the dissociation constant of EI.

$$\frac{1}{V} = \frac{K_m}{V_{\max}}\left(1 + \frac{[I]}{K_i}\right)\frac{1}{[S]} + \frac{1}{V_{\max}} \tag{27}$$

Figure 10.5 Double reciprocal plots illustrating different types of inhibition of an enzymic reaction: (*a*) competitive inhibition (Eq. 27); (*b*) noncompetitive inhibition (Eq. 32); (*c*) uncompetitive inhibition (Eq. 34). K_m and $V_{\max}$ are estimated from the slopes and intercepts of the uninhibited reactions and K_i from the slopes and/or intercepts of the inhibited reactions.

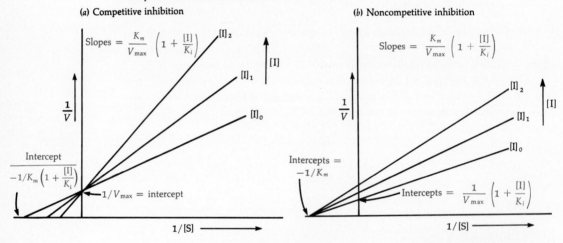

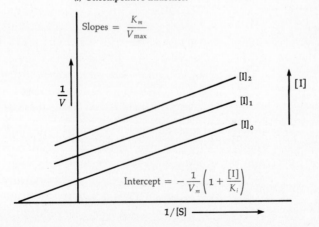

All other terms in Eq. 27 are identical to those in Eqs. 11 and 26. Competitive inhibition characteristically shows the same ordinate intercept, $1/V_{max}$, as the uninhibited reaction but a slope that differs from that of the uninhibited reaction by the term $1 + [I]/K_i$. This is to be expected since at saturating concentrations of S, obtained by extrapolation of $1/[S]$ to the ordinate intercept, the concentration of S is so high that no inhibitor binds the active site. At lower concentrations of S, however, inhibition occurs, the extent depending on the ratio of $[I]$ to $[S]$ and the value of K_i.

Noncompetitive Inhibition

In this instance, there is no relationship between the degree of inhibition and the concentration of substrate. Inhibition depends only on the concentration of the inhibitor. In contrast to the competitive type, it is assumed that the formation of EI occurs at a locus on the enzyme other than where the substrate binds:

$$E + I \rightleftharpoons EI \qquad (28)$$

and

$$ES + I \rightleftharpoons ESI \qquad (29)$$

where EI and ESI are inactive. There are two dissociation constants:

$$K_i^{ES} = \frac{[E][I]}{[EI]} \qquad (30)$$

and

$$K_i^{ESI} = \frac{[ES][I]}{[ESI]} \qquad (31)$$

The double reciprocal equation for noncompetitive inhibition is

$$\frac{1}{V} = \frac{K_m}{V_{max}}\left(1 + \frac{[I]}{K_i}\right)\frac{1}{[S]} + \frac{1}{V_{max}}\left(1 + \frac{[I]}{K_i}\right) \qquad (32)$$

where all terms are the same as in Eq. 27 and K_i reflects the combined effects of K_i^{EI} and K_i^{ESI}, which may be different or equal. Graphical analysis of Eq. 32 is shown in Fig. 10.5, where both the slope and intercept differ from those of the uninhibited reaction by the term $1 + [I]/K_i$. Noncompetitive inhibitors do not react at the active site but elsewhere on the enzyme, leading to a sufficient alteration of the conformation of the enzyme to prevent the active site from combining normally with substrate. Examples of noncompetitive inhibitors are heavy-metal ions such as Ag^+, Hg^{2+}, and Pb^{2+}, which react reversibly with thiol groups of an enzyme, and chelating agents, which combine with essential metal ions.

Many compounds combine irreversibly with enzymes and form covalent derivatives either at the active site or at another part of the molecule not directly involved in the enzyme-substrate interaction. These are not noncompetitive inhibitors in the strict sense because they irreversibly inactivate an enzyme. For example, papain has a single thiol group in its active site that reacts very rapidly with iodoacetate to form an S-carboxymethylcysteine residue. The extent of inactivation of papain by this inhibitor is directly proportional to the extent of S-carboxymethylation. Iodoacetate also inactivates some enzymes which contain thiol groups that are not in the active site and impairs catalytic activity by disrupting conformation. Additional reagents that form covalent derivatives with specific groups in enzymes and have found wide use as probes for detecting structures essential for enzyme activity will be considered in Chap. 11.

Uncompetitive Inhibition

This type of inhibition occurs when an inhibitor combines reversibly only with ES to form ESI, which cannot yield products. Thus,

$$K_i = \frac{[ESI]}{[ES][I]} \tag{33}$$

and the double reciprocal equation is

$$\frac{1}{V} = \frac{K_m}{V_{max}}\frac{1}{[S]} + \frac{1}{V_{max}}\left(1 + \frac{[I]}{K_i}\right) \tag{34}$$

where all terms have the same meaning as in Eq. 27 or 32. Graphical analysis (Fig. 10.5) shows that this type of inhibition characteristically results in a change in the ordinate intercept but not in the slope when compared with the uninhibited reaction. Just as in noncompetitive inhibition, uncompetitive inhibition is not reversed by increasing substrate concentration. This type of inhibition is often found in enzymic reactions with two or more substrates.

METABOLIC INHIBITORS: ANTIMETABOLITES

Metabolites are compounds that are normally metabolized by an organism, whether or not they are provided in the diet or in the nutrient medium. *Antimetabolites* are substances whose structures are related to a particular metabolite and, when administered to an organism, inhibit the utilization of that metabolite. Antimetabolites inhibit the growth and may eventually kill an organism, although inhibition of growth may be overcome by supplying the necessary metabolite. Antimetabolites have long been used experimentally to identify essential metabolites, or growth factors, especially in microorganisms. Current interest in antimetabolites, however, often focuses on their use as growth inhibitors of pathogenic microorganisms and tumors. Thus, if an antimetabolite can be found that inhibits the growth of a pathogenic microorganism without markedly altering the metabolism of the host, then it has the potential for use as an *antibiotic*. Similarly, if it inhibits tumor growth to a greater extent than that of host tissues, it may become an effective *antitumor agent*.

Many antimetabolites are inhibitors of specific enzymes; thus knowledge of such enzymes aids in designing antimetabolites, although many antibiotics and antitumor agents were discovered by empirical research. Several antibiotics will be encountered in subsequent chapters, but to illustrate how they exert their effects by specifically acting on enzymes, the mechanism of one group of antibiotics, the "sulfa drugs," is considered here.

The rational investigation of antimetabolites began with the observation that the inhibition of bacterial growth by sulfanilamide is competitively overcome by *p*-aminobenzoic acid, whose role as a growth factor was first recognized in this way. The structural similarity of the two substances is obvious, and the phenomenon is akin to that of competitive inhibition of an isolated enzyme.

p-Aminobenzoic acid   Sulfanilamide

Indeed, *p*-aminobenzoic acid will competitively overcome the inhibition of all sulfonamides of the structure $NH_2—C_6H_4—SO_2NHR$, for example, sulfaguanidine, sulfathiazole, sulfapyridine, and sulfadiazine.

Organisms that require *p*-aminobenzoic acid (PABA) for growth utilize it for synthesis of folic acid. Growth of these organisms is inhibited by sulfonamides, and this inhibition can be reversed by PABA. Organisms that need folic acid for growth and cannot utilize PABA are not inhibited by sulfonamides. Thus the sulfonamides inhibit the enzymic step or steps involved in the synthesis of folic acid from *p*-aminobenzoic acid and other metabolic precursors. The effective utilization of the sulfonamides in combating bacterial infections in human beings probably depends on the fact that they require folic acid and cannot synthesize it from PABA. Thus, the sulfonamides block a metabolic reaction essential for certain bacteria without influencing the metabolism of the host who does not make folic acid from PABA.

Folic acid (pteroylglutamic acid)

Certain of the folic acid antagonists have been used experimentally and, to some extent, clinically in the treatment of leukemia and other neoplastic diseases. Thus 4-aminopteroylglutamic acid (aminopterin) inhibits growth of certain types of tumors.

REGULATION OF ENZYMIC ACTIVITY

Allosteric Enzymes

Physiologically, the rates of some enzyme-catalyzed reactions are regulated by reversibly binding compounds, termed *effectors*, at specific sites other than their substrate binding sites, which, accordingly, are called *allosteric sites*. At constant enzyme and substrate concentrations, binding of a *negative effector* reduces the reaction rate (*allosteric inhibition*); binding of a *positive effector* increases the rate (*allosteric activation*). Allosteric inhibition may be achieved either by reducing the binding affinity of the enzyme for its substrate, evident as an increase in K_m, or by increasing the time required for each catalytic turnover, evident as a decrease in V_{max}, or both. Conversely, allosteric activation may occur either by reduction in K_m or by an increase in V_{max}, or both.

If the allosteric effector is the substrate itself, the effect is said to be *homotropic;* if the effector is other than the substrate, it is said to be *heterotropic.* Some enzymes have multiple allosteric sites, some specific for positive effectors, others for negative effectors. Allosteric sites exhibit a range of binding specificities analogous to those of substrate binding at the catalytic site, i.e., either absolute specificity for a single compound or variable affinity for a class of related compounds. Most importantly, in vivo the effectors are all compounds that occur normally in the life of the cell and, by their presence, can modulate the rate of a given reaction. The magnitude of the effect is a function of their own concentration in the cell at any instant.

All known allosteric enzymes have two or more subunits, which may have the same or different amino acid sequence, and contain more than one substrate binding site per molecule. The interaction of an effector or substrate with an allosteric enzyme produces a change in conformation of the subunits that alters the catalytic properties of the enzyme. A schematic representation of conformational changes that may occur on binding of effectors to a single subunit is given in Fig. 10.6. When the conformation of one subunit changes on binding the effector, it may also influence the conformation of a second subunit that has no bound effector and alter the catalytic activity of the second subunit, as illustrated in Fig. 10.7.

The fact that an enzyme molecule is built of several subunits does not mean that it necessarily is subject to allosteric modification. Many multisubunit enzymes with more than one active site obey Michaelis-Menten kinetics according to Eq. 10, have no effectors, and have active sites unaffected by partial saturation of active sites on neighboring subunits. Allosteric enzymes, however, may have more complex kinetics, as illustrated in Fig. 10.8. They often given sigmoidal curves in V vs. [S] plots, and double reciprocal plots are nonlinear.

Figure 10.6 Schematic illustration of ligand-induced conformational changes in a monomer. (*a*) Protein conformation before ligand binding. (*b*) Protein conformation after substrate has induced conformational change leading to proper alignment of catalytic groups A and B. (*c*) Activator J induces conformational change, aligning catalytic groups properly in the absence of substrate. (*d*) Inhibitor I induces conformational change maintaining peptide chain with B group in such a position that the substrate has too low an affinity to bind effectively. (*e*) Substrate and activator bind to protein in active conformation. [*From D. E. Koshland, p. 345, in P. Boyer (ed.), The Enzymes, vol. I, 3d ed., Academic, New York, 1970.*]

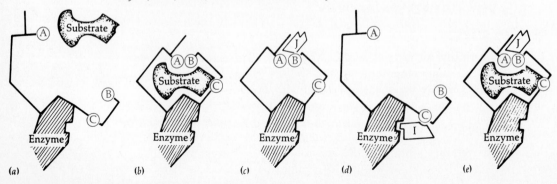

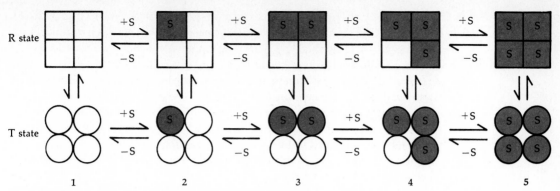

Figure 10.7 Schematic representation of the conformational states of an allosteric enzyme and the changes in conformation induced by an effector S, which may or may not be a substrate. Each square or circle represents one of four subunits. Each state is in dynamic equilibrium with the other, as indicated, but the amounts of each species will depend on the specific enzyme considered. Assume that the enzyme is more active in the R ("relaxed") state than in the T ("tight") state, that S is a positive effector, and that the equilibrium between the R and T states, when no S is bound, is greatly in favor of the T state. Then as the concentration of S increases, T states 1, 2, and 3 will be the species in major amount, but as T state 3 is formed, the equilibrium between T state 3 and R state 4 is such that R state 4 is favored, resulting in an effector-induced conformational change that enhances binding of substrate to R states 4 and 5. Thus, the most favored forms are T1, T2, T3, R3, R4, and R5.

Their kinetic behavior is described by an equation similar to Eq. 14, which can be rearranged to give

$$K = [S]^{n_H} \frac{V_{max} - V}{V} \qquad (35)$$

or in logarithmic form

$$-\log \frac{V_{max} - V}{V} = n_H \log [S] - \log K \qquad (36)$$

Figure 10.8 Schematic illustration of types of kinetics shown by allosteric enzymes. (a) V vs. [S] plot; (b) double reciprocal plot; (c) plot according to Eq. 36. Each graph shows the relationships expected when $n_H = 1$, no cooperativity; $n_H = 2$, positive cooperativity; and $n_H = 0.5$, negative cooperativity. Note that when $n_H > 1$, a sigmoid curve is seen in plots of V vs. [S]. [*From D. Koshland, p. 352, in P. D. Boyer (ed.), The Enzymes, vol. I, 3d ed., Academic, New York, 1970.*]

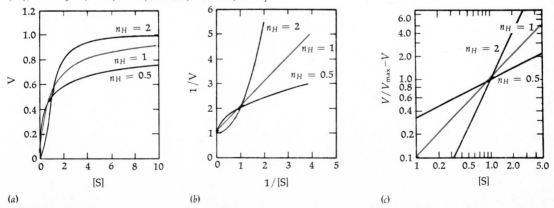

(a)   (b)   (c)

where n_H, the only parameter that differs in Eqs. 35 and 36 from Eq. 14, is the *Hill coefficient*, first introduced to describe the nonlinear oxygen saturation curves for hemoglobin (Chap. M4). The numerical value of n_H cannot exceed n_m, which is the maximum number of ligands (effectors or substrates) that can be bound to the protein, a value usually equal to the number of subunits in the allosteric enzyme. When $\log[V/(V_{\max} - V)]$ is plotted against $\log[S]$, the result is a straight line with slope equal to n_H (Fig. 10.7). The numerical values of n_H reflect the nature of the allosteric properties of an enzyme in the presence of effectors. As depicted in Fig. 10.7, when an effector or substrate binds to one allosteric site, the conformation of a neighboring subunit changes to activate substrate sites on the neighboring subunits, which is an indication of the *cooperativity* among the catalytic sites in the enzyme.

Positive cooperativity results when n_H is >1 and reflects the fact that the binding of the first substrate or effector molecule enhances the binding of a second substrate molecule and activation occurs. This is illustrated graphically for pyruvate kinase in Fig. 10.9, with its positive effectors phosphoenolpyruvate (PEP) and fructose-1,6-bisphosphate (FDP). In the plot of V vs. [PEP], a hyperbolic curve is obtained, whereas the log plot with the same data with Eq. 36 gives a straight line at concentrations above about 0.25 mM, with a slope of 2, that is, $n_H = 2$. Thus, occupancy of one allosteric site by PEP enhances the catalytic activity of other sites for PEP. Moreover, at a constant [PEP], increasing concentrations of the effector FDP, which binds at other allosteric sites, results in activation of the enzyme with PEP as substrate. The logarithmic plot shows that the value of n_H is 1.98, a measure of the degree of positive cooperativity for FDP.

Negative cooperativity is also known in which effectors or substrates decrease the binding of subsequent substrate molecules to the substrate sites or subunits of an allosteric enzyme. The value of n_H in negative cooperativity is always <1 (Fig. 10.8).

Much is known about aspartate transcarbamoylase (ATCase) of *Escherichia coli*, and its structure and function illustrate some properties of allosteric enzymes. ATCase catalyzes the first of six steps in the biosynthesis of CTP, cytidine triphosphate.

Carbamoyl phosphate   Aspartate   Carbamoyl aspartate

CTP is a negative modifier of ATCase; thus when the supply of CTP is high, the formation of carbamoyl aspartate is inhibited. Conversely, when the concentration of CTP is low, the rate of formation of carbamoyl phosphate may proceed more rapidly, enhancing CTP synthesis. ATP is a positive effector of ATCase, and in high concentrations it stimulates the catalytic activity of the enzyme. The action of CTP on ATCase is one of the many examples of the way the final product of a metabolic pathway that involves several consecutive enzymic reactions may temporarily slow down its own synthesis; conversely

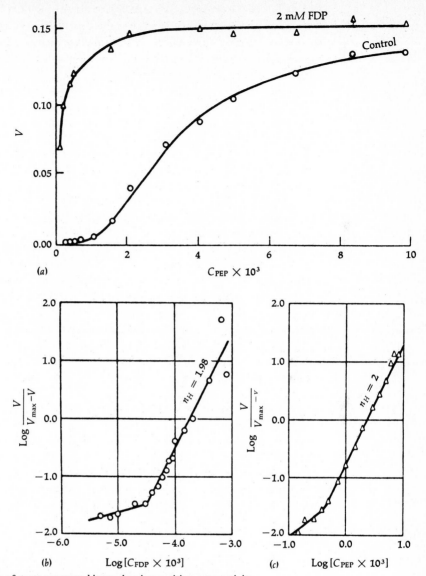

Figure 10.9 Kinetic analysis of yeast pyruvate kinase showing positive cooperativity in binding of fructose-1,6-bisphosphate (FDP), a positive effector, and phosphoenolpyruvate (PEP), one substrate. (*a*) *V* vs. [PEP] plot in presence and absence of FDP. (*b*) and (*c*) Plots according to Eq. 36 that give n_H values for FDP and PEP. [*After H. Gutfreund, Enzymes: Physical Principles, Wiley, New York, 1972, p. 85.*]

ATCase also illustrates that a compound, ATP, which acts together with CTP in some metabolic processes, can also influence CTP synthesis. This type of metabolic control at the enzymic level is discussed further in Chap. 12.

ATCase (MW = 300,000) dissociates on reaction of its thiol groups with mercurials into two types of subunits, as shown schematically in Fig. 10.10. One type of subunit, C_3, a trimer (MW = 99,000), possesses catalytic activity and is insensitive to CTP; there are two C_3 catalytic subunits per molecule of native enzyme. Each trimer binds succinate, an unreactive analog of the sub-

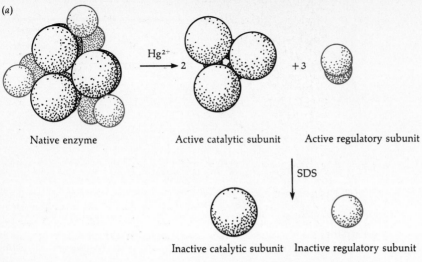

(a)

$\xrightarrow{\text{Hg}^{2+}}$ 2   + 3

Native enzyme   Active catalytic subunit   Active regulatory subunit

$\downarrow$ SDS

Inactive catalytic subunit   Inactive regulatory subunit

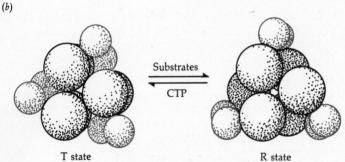

(b)

Substrates $\rightleftharpoons$ CTP

T state   R state

Figure 10.10 A schematic model (*a*) of the subunit structure of aspartate transcar-bamoylase, showing the dissociation by mercurials to give active trimeric catalytic subunits (MW = 99,000) and dimeric regulatory subunits (MW = 34,000), which on treatment with denaturing agents, for example, SDS, give inactive monomeric C and R subunits. (*b*) Models for possible conformational differences in the R and T states of the enzyme (Fig. 10.7). Substrates induce the R state and CTP the T state.

strate aspartate, in accord with the finding that each molecule of native enzyme can bind six molecules of succinate. There are three active dimeric *regulatory subunits* (MW = 33,000) that bind CTP but not substrates, in accord with the binding of six CTP per native molecule. Re-formation of the native enzyme requires removal of the mercurial and the presence of both types of subunits. Six atoms of Zn^{2+} are present in the native enzyme and are essential for reconstitution from the isolated subunits. Strong denaturing agents dissociate active C_3 and R_2 into monomeric inactive subunits, C (MW = 33,000) and R (MW = 17,000).

When the enzyme is dissociated by treatment with mercurials, the catalytic activity is enhanced approximately fourfold. Under these conditions, V as a function of [S] exhibits a hyperbolic function (Fig. 10.11; curve *A*), whereas in the intact native enzyme a sigmoidal relationship is found and the entire curve is shifted to the right (curve *B*). In the presence of CTP, the curve is shifted

204

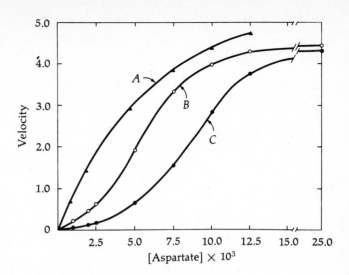

Figure 10.11 Velocity of aspartate transcarbamoylase as a function of substrate concentration. Curve A illustrates the behavior of the enzyme in the presence of a mercurial. In this condition, the enzyme exhibits normal Michaelis kinetics and $n = 1$. Curve B shows the behavior of the native enzyme; it is evident that $n > 1$. Curve C shows the effect of $2.0 \times 10^{-1} M$ CTP. The curve is shifted on the abscissa; K is increased, reflecting a decreased affinity of the enzyme for substrate; i.e., a higher concentration of aspartate is required to attain the same velocity.

further on the abscissa (curve C). In the presence of the positive modifier, ATP, CTP is displaced, and the curve is shifted to the left. Thus ATP and CTP compete for the same regulatory site. The sigmoidal curve for aspartate transcarbamoylase in the absence of CTP shows that n_H is greater than 1; the negative modifier increases the value of K (higher concentrations of S being required for saturation of the enzyme).

Some evidence suggests that ATCase has different conformational forms; in the presence of substrates it exists in the *relaxed*, or R, state, whereas in the presence of CTP it is in the *tight*, or T, state. The 32 thiol groups in the enzyme react more rapidly with mercurials in the presence of carbamoyl phosphate and succinate than in their absence or in the presence of just one of the substrates. CTP diminishes the rate of reaction of mercury with the thiols. Thus, in the R state, the enzyme has a more relaxed conformation that renders thiol groups more accessible for reaction. In addition, the sedimentation coefficient of the enzyme is about 4 percent lower when measured in the presence of carbamoyl phosphate and succinate than in the presence of CTP. Since the enzyme does not dissociate under these conditions, the difference in the values of the sedimentation coefficient suggests that the R state is more asymmetrical with an increased frictional coefficient.

Only three molecules of carbamoyl phosphate (CP) bind native ATCase in the absence of succinate, but six bind in its presence, which suggests that ATCase also has homotropic positive cooperativity as follows:

$$C_6R_6 \xrightarrow{3CP} C_6R_6(CP)_3 \xrightarrow{S} C_6R_6(CP)_3S \rightleftharpoons C_6R_6^*(CP)_3S$$

$$C_6R_6^*(CP)_3S \xrightarrow[3CP]{6S} C_6R_6^*(CP)_6(S)_6$$

where C_6R_6 indicates the 12 subunits in native enzyme, CP is carbamoyl phosphate, S is succinate, and $C_6R_6^*$ the relaxed state. The value of n_H for CP is 1 in the presence or absence of succinate, but once succinate binds, it allows full saturation with CP and additional succinate binds cooperatively.

Cascade Systems—Covalent Modification

Regulation of enzyme activity is also achieved by the cyclic interconversion of enzymes between covalently modified and unmodified forms. Modification and demodification are achieved by *converter enzymes,* which together with the modified and unmodified enzymes and their effectors form a *cascade system.* A covalently modified enzyme may be either more active or less active than the unmodified form, or the two forms may respond differently to effectors.

The simplest cascade system is said to be *monocyclic* and operates as shown in Fig. 10.12 for a hypothetical enzyme that is covalently modified by phosphorylation. To begin the cascade, an inactive converter enzyme E_i is activated by a specific allosteric effector e_1, forming the active converter enzyme E_a. The unmodified enzyme I_o is then phosphorylated by E_a in an ATP-dependent reaction giving rise to modified enzyme I_m and ADP. The process is reversible, since a second activated converter enzyme R_a, which is formed by the activation of inactive converter R_i and a second allosteric effector e_2, dephosphorylates I_m to give I_o and P_i. Different monocyclic cascades may be coupled to give *bi-, tri-* and *multicyclic cascades,* which, with great sensitivity to the concentrations of effectors, can finely regulate the flux of substrates and products through interlocking metabolic pathways. Some cascades such as those in blood coagulation (Chap. M1) and complement fixation (Chap. M2) operate to turn on, or totally turn off, a series of reactions, but cascades functioning in major metabolic pathways lead to a steady-state concentration of the interconvertible I_o and I_m forms of enzyme, thereby ensuring the required flux of metabolites through a pathway.

Cascade systems have the capacity to regulate the rates of reactions in several ways (Fig. 10.12): (1) They provide *signal amplification;* i.e., only a small amount of converter enzyme (E_a or R_a) produces large amounts of modified (I_m) or unmodified (I_o) enzyme. (2) They can modulate the *amplitude* of the maximum response that I_m can achieve with saturating amounts of e_1. (3) They can modulate the *sensitivity* of modification to changes in effector concentrations. (4) They serve as *integration systems,* sensing minute changes in intracellular concentrations of metabolites and adjusting the concentrations of I_o and I_m accord-

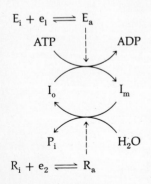

Figure 10.12 Schematic representation of a monocyclic cascade system involving phosphorylation of unmodified enzyme I_o to give phosphorylated enzyme I_m under the influence of converter enzymes E_a and R_a, which are formed by interaction of effectors e_1 and e_2, respectively, with inactive converter enzymes E_i and R_i. E_a and R_a catalyze phosphorylation and dephosphorylation of I_o and I_m, respectively. [*From P. B. Chock, S. G. Rhee, and E. A. Stadtman, Annu. Rev. Biochem. 49:813 (1980).*]

ingly. (5) They are *flexible systems*, capable of various responses to allosteric stimuli. (6) They are *rate amplifiers*, responding within milliseconds to changes in intracellular concentrations of metabolites.

Phosphorylation of the hydroxyl groups of serine, threonine, or tyrosine in enzymes occurs in many cascade systems; more than 20 different enzymes are known to undergo reversible phosphorylation by mechanisms similar to that shown schematically in Fig. 10.12. Each of these systems uses ATP, which provides the energy to maintain the proper amounts of interconvertible enzymes needed to regulate rates of reaction. Such diverse processes as glycogen synthesis and degradation (Chap. 18), cholesterol biosynthesis (Chap. 21), and amino acid transformations (Chap. 23) are regulated in this manner.

Some cascade systems are regulated by covalent modifications other than phosphorylation. Thus, glutamine synthetase of *Escherichia coli* operates in a cascade involving nucleotidylation of enzymes or their regulatory proteins by reactions with ATP or UTP (Chap. 22). ADP ribosylation from NAD may be involved in regulating gene expression (Chaps. 28 and 29). Discussion of different cascade systems is presented in subsequent chapters.

Zymogen Activation

Many enzymes that function extracellularly are synthesized as inactive *zymogens*, and are activated only after secretion from their site of synthesis and storage. Activation is achieved by proteolysis of one or a few peptide bonds in the zymogen. The active sites in zymogens are incomplete but become fully formed on activation. Figure 10.13 shows the events leading to activation of pancreatic chymotrypsinogen by trypsin. Cleavage of the bond in chymotrypsinogen produces NH_2-terminal isoleucine, which completes the formation of the active site (Chap. 11).

Zymogen activation is involved not only in intestinal digestion but also in the cascade systems that regulate blood coagulation (Chap. M1) and complement fixation (Chap. M2).

Protein-Protein Interactions

Allosteric enzymes exercise control of enzymic reactions by cooperative interactions between subunits. Other enzymes, such as *protein kinase*, exist in an inactive state through protein-protein interactions of their subunits. Protein kinase of skeletal muscle exists as an inactive holoenzyme with the subunit structure R_2C_2, where R is the regulatory subunit and C is the catalytic subunit. The kinase is a converter enzyme and is activated by $3',5'$ cyclic AMP (cAMP) as follows:

$$R_2C_2 + 2cAMP \rightleftharpoons R_2(cAMP)_2 + 2C$$

The binding of cAMP to R subunits releases tightly bound C subunits, which catalyze the phosphorylation of a number of enzymes in cascade systems (Fig. 10.12).

Another example of protein-protein interactions in enzymic control is pro-

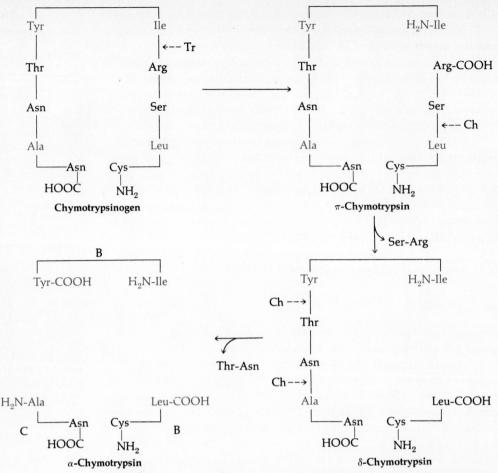

Figure 10.13 Scheme for activation of chymotrypsinogen A. Hydrolysis of the Arg-Ile bond produces π-chymotrypsin. This is followed by liberation of the dipeptide, Ser-Arg, producing δ-chymotrypsin, which is then converted to α-chymotrypsin by liberation of Thr-Asn. Chymotrypsinogen possesses a single peptide chain, whereas α-chymotrypsin contains three chains called A, B, and C, which are combined by five disulfide bridges, but not shown in the diagrams. Tr = trypsin; Ch = chymotrypsin.

vided by *lactose synthase,* the enzyme responsible for synthesis of lactose in lactating mammary glands. Lactose synthase catalyzes the reaction

$$\text{UDP-galactose} + \text{glucose} \longrightarrow \text{galactosyl-}\beta1{\rightarrow}4\text{-glucose (lactose)} + \text{UDP}$$

The synthase requires the enzyme *UDP-galactose:N-acetylglucosamine galactosyl transferase,* which is present in many tissues other than mammary glands and participates in synthesis of the oligosaccharide prosthetic groups of some glycoproteins as follows:

$$\text{UDP-Gal} + \text{Glc}N\text{Ac} \cdots \text{protein} \longrightarrow \text{Gal-}\beta1{\rightarrow}4\text{-Glc}N\text{Ac} \cdots \text{protein} + \text{UDP}$$

In lactating mammary secretory cells, the membrane-bound galactosyl transferase interacts with a soluble protein, *α-lactalbumin,* uniquely found in milk, to form lactose synthase, which catalyzes lactose synthesis. On interaction of α-lactalbumin with the galactosyl transferase the substrate specificity of the transferase is so altered that glucose becomes an acceptor substrate. Glucose is a very poor acceptor ($K_m = 1$ to $2\,M$) for the transferase itself, but in the presence of α-lactalbumin it becomes a very good acceptor ($K_m = 10^{-3}\,M$).

REFERENCES

See the list following Chap. 11.

Chapter 11

Enzymes II

Substrate specificity. Rate enhancement. Active sites and mechanism of action.

Enzymes differ most strikingly from ordinary chemical catalysts in their substrate specificity and catalytic efficiency. Most enzymes have only a few natural substrates, which are converted to single products in remarkably high yields. The unique structures of the active sites of enzymes provide this specificity and not only allow favorable binding of specific substrates but also exclude the unfavorable binding of many substances that are not substrates. There are strong attractive noncovalent forces between the active site and a substrate, and enzymes may be thought to act by "attracting" the substrate into the site, where the structural transformations of the substrate occur. This high degree of specificity is maintained, with the reaction proceeding 10^6 to 10^{12} times faster than the spontaneous, uncatalyzed reaction in aqueous solution.

This chapter will consider the general features of substrate specificity, the possible mechanisms that effect rate enhancements, the nature of active sites, and the reaction mechanisms for a few enzymes whose actions illustrate important principles of enzymic catalysis. These subjects are too extensive to be treated fully; only a few examples have been selected to illustrate general concepts. However, the remarkable chemistry that is involved in enzymic action will be encountered repeatedly in succeeding chapters.

SUBSTRATE SPECIFICITY OF ENZYMES

Experimentally, enzymic specificity is determined by testing as substrates compounds whose structures are systematically varied from that of a natural substrate. Many enzymes have *absolute specificity* for a single substrate and prod-

Any cross-references coded M refer to the companion text, *Principles of Biochemistry: Mammalian Biochemistry.*

uct. Such is the case for *succinate dehydrogenase* and *fumarase* (Chap. 10), which catalyze the reversible conversion of succinate to fumarate and of fumarate to malate, respectively. Other enzymes have a wider specificity, e.g., *trypsin*, which hydrolyzes peptide, amide, or ester bonds formed by lysine or arginine (Table 5.3). Although different types of bonds can be hydrolyzed, trypsin has strict specificity for the R groups of lysine and arginine. Thus, the α-*N*-benzoylamide derivatives of homoarginine and ornithine are not substrates whereas those of arginine and lysine are rapidly hydrolyzed to the α-*N*-benzoylamino acid and NH_3. Homoarginine has one more $-CH_2-$ group in its R group than arginine, and ornithine one less $-CH_2-$ group than lysine.

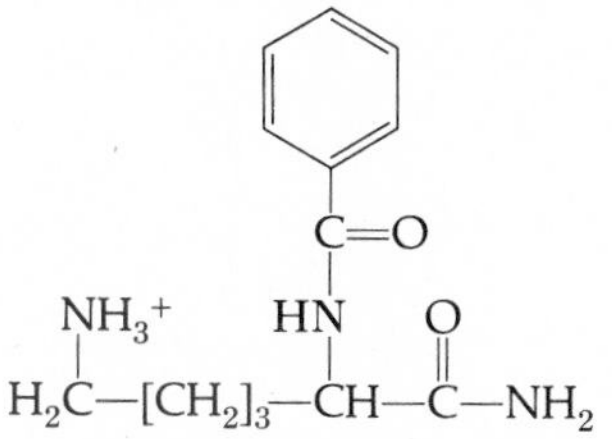

α-**Benzoyl-L-lysinamide**

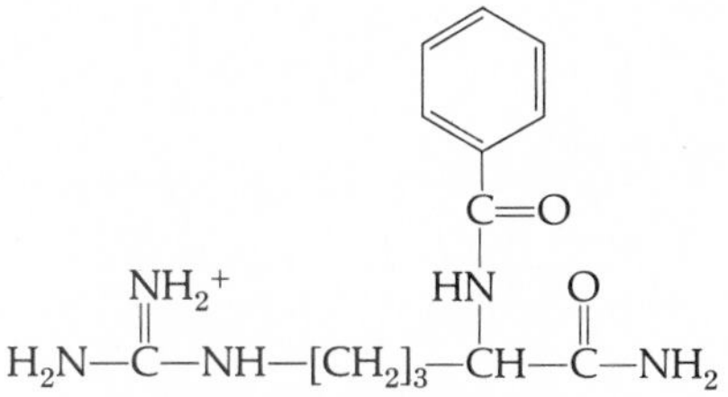

α-**Benzoyl-L-argininamide**

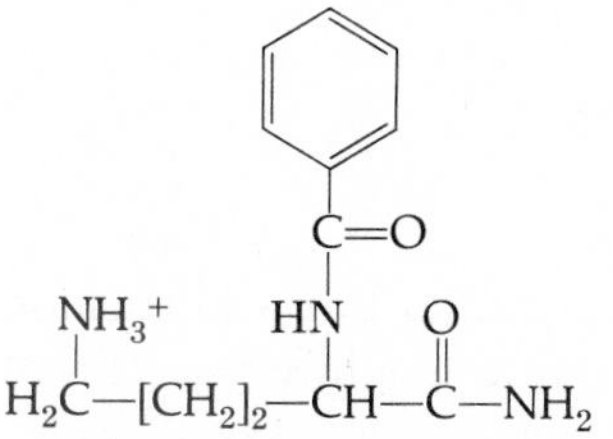

α-**Benzoyl-L-ornithinamide**

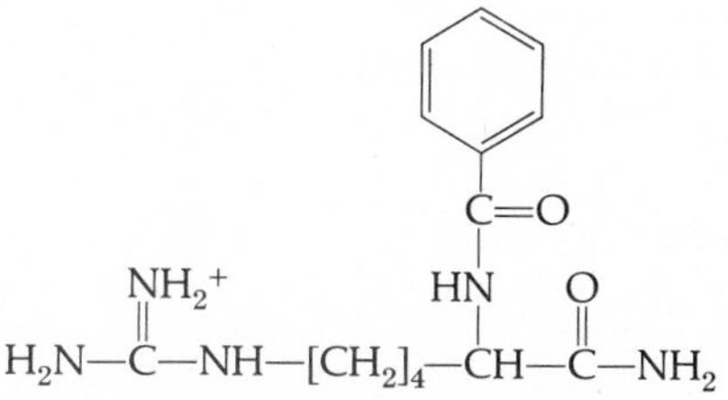

α-**Benzoyl-L-homoargininamide**

Crystallographic studies showed that the specificity resides in the reaction of the cationic group of the substrate with an aspartyl COO^- of the active site. Displacement of the cationic group by one $-CH_2$ group from the sensitive amide bond does not permit proper substrate binding and thus does not allow enzyme-catalyzed deamidation.

Other enzymes react with a variety of compounds and have a broad *relative substrate specificity*. An example is *leucine aminopeptidase*, which hydrolyzes many α-L-amino acid amides and dipeptides at different rates (Table 11.1), as follows:

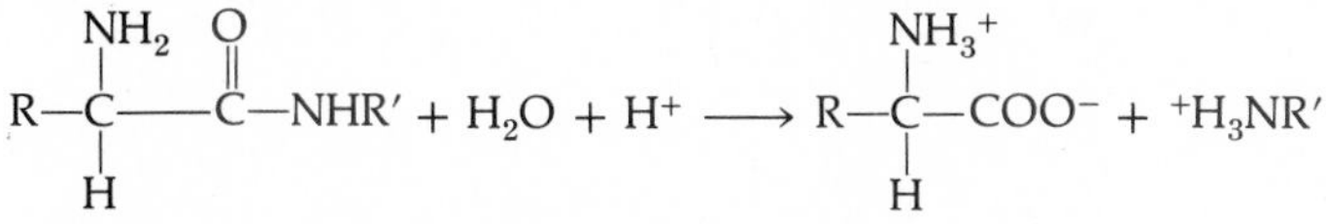

where R is the amino acid side chain and R′ is either an H atom (in amides) or another amino acid (in dipeptides). Each substrate must have an unsubstituted amino group and a hydrogen atom on the α-carbon adjacent to the susceptible amide or peptide bond. The NH_2- terminal residue must be of the L configuration, except for glycine, which forms very poor substrates (Table 11.1).

The specificities of the foregoing enzymes illustrate that the size, shape, and chemical character of the groups on a substrate determine the rate at which a

TABLE 11.1

The Rate of Hydrolysis of Various Substrates by Leucine Aminopeptidase	Substrate	Relative rate*	Substrate	Relative rate*
	L-Leucinamide	100	L-Leucyl-L-leucine	100
	L-Phenylalaninamide	26	L-Leucyl-L-isoleucine	64
	L-Isoleucinamide	20	L-Leucyl-L-alanine	64
	L-Histidinamide	19	L-Leucyl-L-phenylalanine	26
	L-Lysinamide	7.1	L-Alanyl-L-leucine	93
	L-Alaninamide	3.4	L-Alanylglycine	9.4
	L-Aspartic acid diamide	2.9	Glycyl-L-leucine	10
	L-Serinamide	0.8	Glycylglycine	1.1
	L-Prolinamide	0.7	D-Leucylglycine	0
	Glycinamide	0.1	Acetyl-L-tyrosinamide	0
	D-Leucinamide	0	α-Benzoyl-L-argininamide	0
	β-Alaninamide	0		

*The value for leucinamide is given as 100 and the other substrates as the relative rate. The enzyme is from porcine kidney, and all rates were measured with Mn^{2+}-activated enzyme at pH 8.0 to 8.5.
Source: From R. J. DeLange and E. L. Smith, in P. Boyer (ed.), *The Enzymes,* vol. III, Academic, New York, 1971, p. 81.

substrate is acted upon. Present evidence suggests that the complementary binding of substrates in the active site of an enzyme may involve hydrophobic and electrostatic interactions as well as hydrogen bonding. In some cases, covalent intermediates may be transiently formed in enzyme-substrate complexes. Thus, all groups of a substrate fit snugly into the active site in such a manner that each is exactly juxtaposed to complementary groups in the site. A major goal of biochemistry is to understand how such specific binding ultimately leads to the chemical transformation concomitant with binding of the substrates in the active sites of enzymes. Before considering plausible mechanisms, however, it is desirable to first consider mechanisms of rate enhancement by enzymes.

MECHANISMS OF RATE ENHANCEMENT OF CHEMICAL REACTIONS BY ENZYMES

Enzymes lower the energy of activation of a reaction (Chap. 10) and, in so doing, increase the reaction rate. Such rate enhancements must be explained in terms of the chemical events that occur on interaction of a substrate and an enzyme. This section considers mechanisms that can contribute to the enhancement of the rates of reactions by enzymes. In fact, it appears that such mechanisms are closely related to those that determine substrate specificity.

Rate Acceleration and Substrate Specificity

The types of chemical reactions that proceed during enzyme catalysis have analogs in organic chemical reactions. The unique aspect of enzymic action, however, is substrate specificity, and the behavior of many enzymes suggests that the specific binding energy of the multiple interactions which occur between complementary groups in a substrate and an active site is utilized to provide the driving force for catalysis and contributes substantially to the reduction in the activation energy of the reaction and thus to the enormous rate of all enzyme-catalyzed reactions. This is manifested by many enzymes and impressively by *phosphoglucomutase* (Chap. 17), which catalyzes the reversible reaction

$$\alpha\text{-D-Glucose-1-phosphate} \rightleftharpoons \alpha\text{-D-glucose-6-phosphate}$$

This reaction proceeds only in the presence of Mg^{2+} and glucose-1,6-bisphosphate; a phosphoryl-enzyme intermediate is formed in which a phosphate group on a single serine residue of the enzyme is exchanged with substrates and glucose-1,6-bisphosphate,

$$\text{Enz—OH} + \text{glucose-1,6-bisphosphate} \rightleftharpoons$$
$$\text{Enz—O—PO}_3^{2-} + \text{glucose-1-(or -6-) phosphate}$$

By repetition of the reaction, either of the glucose phosphate esters can be converted to the other. The phosphoryl enzyme is stable in aqueous solution but reacts with a variety of substances of the general structure ROH (listed in Table 11.2) according to the reaction

$$\text{Enz—O—PO}_3^{2-} + \text{ROH} \longrightarrow \text{Enz—OH} + \text{ROPO}_3^{2-}$$

TABLE 11.2   Relative Rates of Phosphorylation of Some Substances by Phosphoglucomutase

Substrate (ROH)	Relative rate*	Substrate (ROH)	Relative rate*
HOH	1		2×10^5
Phosphite	580		4.4×10^5
			1.4×10^6
α-D-Xylose-1-phosphate	1.7×10^5		4.4×10^7
			9×10^5
D-Xylose	7×10^4	α-D-Glucose-1-phosphate	3×10^{10}
D-Xylose	2×10^9		

* The rate of phosphorylation of water is given as 1 and for the other compounds the relative rate compared with water as substrate. The first-order rate constant for the phosphorylation of water by the phosphoryl enzyme is 3.2×10^{-8} s^{-1} at 30°C and pH 7.5.
Source: W. P. Jencks, *Adv. Enzymol.* **43**:219 (1975).

Comparison of the rates of phosphorylation of the different substrates reveals the interesting relationship of substrate binding and catalytic effectiveness. Transfer of the phosphoryl group on the enzyme to water (hydrolysis of the enzyme phosphate) is very slow and proceeds at a rate only about 60 times faster than the nonenzymic hydrolysis of serine phosphate. In contrast, phosphite and xylose-1-phosphate, neither of which is a substrate, accelerate the transfer of phosphate to water by factors of 580 and 1.7×10^5, respectively. This indicates that the reactivity of the enzyme phosphoryl group is markedly enhanced by substances structurally similar to substrates that can bind at the active site; phosphite binds at a phosphate site and xylose at a sugar site. Xylose is phosphorylated by the phosphoryl enzyme much more rapidly than water is, but in the presence of phosphite it is phosphorylated to give xylose-1-phosphate 2×10^9 times faster than phosphorylation of water, which is within a factor of 15 of the rate of phosphorylation of the natural substrate, glucose-1-phosphate. Finally, several acyclic monophosphodiols, in which the hydroxyl group is four atoms removed from the phosphoryl group, can be readily phosphorylated by the enzyme at rates of the order of about 10^5 to 10^7 times greater than the phosphorylation of water. Thus, binding at a phosphate site and a sugar site, by compounds with appropriate structures, leads to large decreases in the activation energy of the phosphorylation reactions by phosphoglucomutase. The data in Table 11.2 clearly support the view that the strong forces associated with binding of a substrate to an active site are also involved in the chemical events that lead to the extraordinary rate enhancement of transformation of substrate.

With the foregoing concepts in mind, we can now consider the kinds of mechanisms which contribute to the enhancement of enzyme-catalyzed reactions and which depend upon the binding energy of the substrate to an enzyme.

Induced Fit and Enzyme Catalysis

It has been proposed that many enzymes in the absence of substrates exist in an inactive state and that not all groups in the active site are correctly oriented spatially to interact with complementary groups on the substrate. Binding of specific substrate, however, results in a conformational change in the enzyme and thus of the active site and shifts those R groups in the site into the correct position for proper binding so that catalysis can proceed. Such substrate-induced conformational changes, termed the *induced fit* of enzymic action, are illustrated diagrammatically in Fig. 10.6. Considerable evidence indicates that conformational changes occur on substrate binding, derived notably from comparison of the structures of enzymes obtained by x-ray crystallography in the presence and absence of inhibitors, e.g., in *carboxypeptidase A* and *lysozyme*. In addition, the properties of enzymes in solution suggest conformational differences in the presence and absence of substrates. For example, some enzymes fail to react with their specific antibodies in the presence of substrates; many enzymes are stabilized to heat denaturation, undergo changes in optical rotation, or fail to dissociate into subunits in the presence of specific substrates, and others show differences in sedimentation properties. It is generally acknowledged that an induced fit may enhance the rate of some enzymic reactions, but the overall rate acceleration is probably small compared with other mechanisms.

The most obvious way an enzyme could enhance the rate of a bimolecular reaction is to bring together in the active site the two reactants, correctly oriented, and in such close proximity to one another that the effective concentrations are much greater than in dilute solution. This is inherently a highly improbable process, but because of the strong and multiple binding forces between substrate and active-site structures, enzymes appear to increase the probability that two substrates can indeed come together for reaction and effectively convert a bimolecular reaction to a monomolecular intramolecular reaction. This effect is known under a variety of names (orientation, propinquity, proximity, orbital steering, the commonsense phenomenon, etc.) but will be called here *approximation.*

Approximation is most effectively illustrated by model intramolecular reactions and the effect of structure on reaction rates. Table 11.3 gives the structures of several *p*-bromophenyl esters of succinates and glutarates and their relative rates of hydrolysis compared with the rate of the bimolecular acetate-catalyzed hydrolysis of *p*-bromophenylacetate. Each is hydrolyzed by an intramolecular nucleophilic attack of a neighboring carboxylate ion according to the general reaction

Esters that are more rigid increase the probability that the attacking carboxylate group is properly oriented and is in close proximity to the ester bond, and they are cleaved more rapidly than those with greater freedom of rotation and less rigidity. Many of these compounds have strained bond angles, and they may be considered akin to springs whose tension is partly released as they enter

TABLE 11.3 **The Structures and Relative Rates of Hydrolysis of Monophenyl Esters of Dicarboxylic Acid Anions**

Structure*	Relative rate of hydrolysis	Structure	Relative rate of hydrolysis
$CH_3COO^- + CH_3COOR$	1.0	COOR / COO⁻ (cis)	1×10^7
—COOR / —COO⁻	$\sim 1 \times 10^3$		$\sim 5 \times 10^7$
R' —COOR / R' —COO⁻	$3 \times 10^{3\dagger}$ 1.3×10^6	COOR / COO⁻	
—COOR / —COO⁻	$\sim 2.2 \times 10^5$		

*R = *p*-Br—C$_6$H$_4$—.

†Rate depends upon nature of R'.

Source: T. C. Bruice, *Annu. Rev. Biohem.* **45:**331 (1976).

the transition state. From these rate data, it has been estimated that the *effective concentrations* of the carboxylate groups around the ester groups may be as high as 10^5 to 10^8 M. These are physically impossible concentrations, but serve to illustrate the advantage of an intramolecular reaction over an intermolecular reaction and how bringing reactants together at an active site allows enormous rate increases. Indeed, it is estimated that rate enhancement by approximation may be as high as 10^8 and along with the enhancement provided by other mechanisms may well account for the rates observed for enzymic reactions.

Destabilization

A relatively old hypothesis to account for enzymic catalysis suggested that binding of the active site induces strain, deformation, or destabilization of the bonds that are to be broken, implying the intermediate formation of a compound or complex in which the susceptible bond is inherently less stable than in the original reactants. An example of strain in this sense is afforded by the fact that the base-catalyzed hydrolysis of ethylene phosphate occurs 10^7 times more rapidly than that of dimethyl phosphate.

Ethylene phosphate   Dimethyl phosphate

The strain hypothesis has seemed an attractive explanation for enzymic catalysis. An actual example appears to be reflected in the behavior of horse *liver esterase*. For the hydrolysis of a series of esters of *m*-hydroxybenzoic acid,

K_m is almost independent of the chain length of R, whereas V_{max} increases by several orders of magnitude as the chain length is increased. Since the bond to be hydrolyzed is the same in each instance, it can be inferred that the increased binding energy of the longer-chain esters reduces the activation energy for the reaction; i.e., the energy of the tighter bonding of the hydrocarbon moiety is offset by the strain energy induced in the acyl portion of the molecule. Other examples of destabilization in enzymes by strain are known, and one of the most interesting is provided by lysozyme, discussed later in this chapter.

Destabilization of a substrate can also occur when charged substrates are desolvated; i.e., on combination with an active site, the substrate may be transferred from an aqueous medium into a relatively hydrophobic environment in the active site, thereby permitting great rate accelerations. A model reaction illustrating this effect is the decarboxylation of an adduct of pyruvate and an analog of thiamine pyrophosphate (Chap. 15), which proceeds as follows:

The structures I, II, III are shown (adduct equilibrium with CO$_2$ evolution).

$$\text{I} \rightleftharpoons \text{II}^{2+} \rightarrow \overset{O}{\underset{O}{C}} + \text{III}$$

I   **II**   **III**

The adduct (I) is relatively stable in water, and decarboxylation is slow, but in ethanol it decarboxylates 10^4 to 10^5 times faster, to give CO_2 and III. In dimethyl sulfoxide, decarboxylation is even faster, and solutions froth due to CO_2 evolution when I is added to this solvent. The increase in rate of decarboxylation is thought to result from the decrease in localized charge in II, the presumed transition state, relative to the adduct I. There is evidence that this type of rate enhancement may occur in *pyruvate decarboxylase* (Chap. 15), which requires enzyme-bound thiamine pyrophosphate as a cofactor, since the cofactor appears to be in a relatively hydrophobic region of the enzyme.

Concerted General Acid-Base Catalysis

There are many acid- or base-catalyzed reactions in organic chemistry; e.g., the formation of hemiacetals is catalyzed by either acids or bases.

$$\overset{CH_3}{\underset{H}{C}}=O + CH_3OH \rightleftharpoons \overset{CH_3}{\underset{H}{C}}\overset{OH}{\underset{OCH_3}{}}$$

Acetaldehyde   Methanol   Hemiacetal

The base OH^- accelerates hemiacetal formation as follows:

$$CH_3-OH + OH^- \rightleftharpoons CH_3O^- + H_2O$$

$$\overset{CH_3}{\underset{H}{C}}=O \quad \bar{O}-CH_3 \rightleftharpoons \overset{CH_3}{\underset{H}{C}}\overset{OCH_3}{\underset{O^-}{}} \underset{-H_2O}{\overset{+H_2O}{\rightleftharpoons}} \overset{CH_3}{\underset{H}{C}}\overset{OCH_3}{\underset{OH}{}} + OH^-$$

Acid catalysis involves formation of the oxonium salt, followed by reaction with the alcohol.

$$\overset{CH_3}{\underset{H}{C}}=O + H^+ \rightleftharpoons \overset{CH_3}{\underset{H}{C}}=\overset{+}{O}H$$

$$\overset{CH_3}{\underset{H}{C}}=\overset{+}{O}H + \overset{H}{\underset{}{O}}-CH_3 \rightleftharpoons \overset{CH_3}{\underset{H}{C}}\overset{\overset{CH_3}{+O}}{\underset{OH}{\underset{H}{}}} \rightleftharpoons \overset{CH_3}{\underset{H}{C}}\overset{OCH_3}{\underset{OH}{}} + H^+$$

Present evidence suggests that many groups in active sites of enzymes may serve as general acid or base catalysts acting upon different groups in a substrate and thereby contributing to rate acceleration. Concerted general acid-base catalysis is particularly effective, and an illustrative model reaction is the muta-rotation of tetramethylglucose, which proceeds as follows:

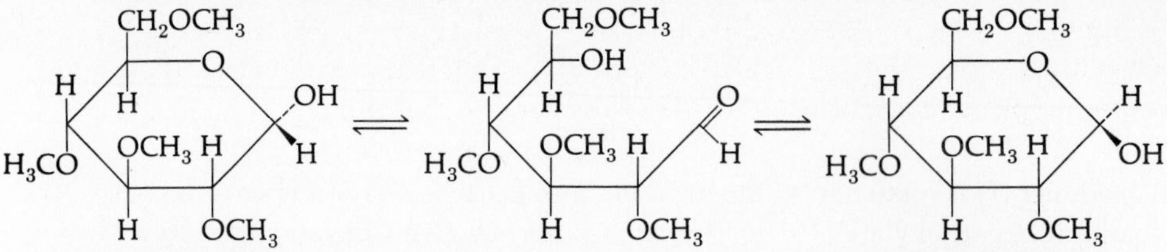

2,3,4,6-*O*-Tetramethyl-
β-D-glucopyranose

2,3,4,6-*O*-Tetramethyl-
α-D-glucopyranose

Acids and bases catalyze the rate of mutarotation; thus, when either anomer is dissolved in benzene and a mixture of phenol and pyridine added, mutarotation proceeds very rapidly. The kinetics of the reaction show that the rate is depend-ent on the concentrations of phenol and pyridine as well as that of tetramethyl-glucose, suggesting that phenol and pyridine are acting as acid and base cata-lysts at the same time. Further, if the functional groups of phenol and pyridine are incorporated into a single molecule, as in α-pyridone (α-hydroxypyridine), a much more effective catalyst is obtained, even though the catalyzing groups in α-pyridone are much weaker acids and bases than phenol and pyridine. α-Pyridone is thought to exert its catalytic effect as follows:

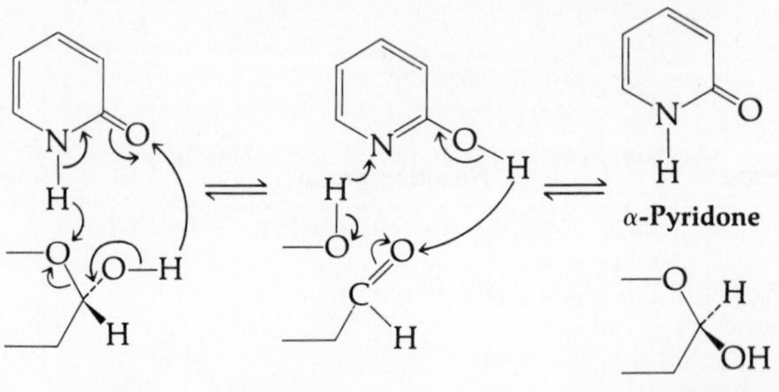

α-Pyridone

β anomer

α anomer

Thus, the hydrogen on the pyridine nitrogen (an acid) donates a proton, and the carbonyl oxygen (a base) accepts a proton when the pyranose ring of the β anomer is opened. The pyridine nitrogen accepts a proton from the open-ring intermediate, whereas the phenolic hydroxyl donates a proton when the ring closes to form the α anomer.

Such concerted general acid-base catalysis occurs in the catalytic mecha-nisms of several enzymes including ribonuclease. It is unlikely that this type of catalysis causes rate enhancements greater than a factor of 10 to 100 but to-gether with other mechanisms contributes to increases in the rate of enzymic reactions.

Many amino acid R groups may act as general acid-base catalysts in en-zymes, including those of glutamic acid, aspartic acid, histidine, lysine, tyrosine, and cysteine. In their protonated forms they are acidic catalysts, and in their unprotonated forms they are basic catalysts. Obviously, the effectiveness of these R groups in catalysis will depend on the pK of each functional group and the pH at which the enzymic reaction is observed.

Table 11.4 lists some enzymes that form covalently linked enzyme-substrate complexes transiently during catalysis. Often such intermediates can be readily identified; for example, the acetyl-chymotrypsin complex that forms during chymotryptic hydrolysis of *p*-nitrophenylacetate is stable at acid pH and can be isolated.

$$O_2N\!-\!\langle\ \rangle\!-\!O\!-\!\overset{\overset{\displaystyle O}{\|}}{C}\!-\!CH_3 + \text{(C)}\!-\!OH \longrightarrow \text{(C)}\!-\!O\!-\!\overset{\overset{\displaystyle O}{\|}}{C}\!-\!CH_3 + O_2N\!-\!\langle\ \rangle\!-\!OH$$

      p-Nitrophenylacetate   Chymotrypsin   Acetyl-enzyme   *p*-Nitrophenol

This type of intermediate is discussed further below.

Thiol groups of cysteine in proteolytic enzymes *papain* and *ficin* participate in forming covalent intermediates with substrates. Enzyme-bound thiol esters are formed with the carboxyl group of the peptide bond to be cleaved, as follows:

$$\text{Enz}\!-\!SH + R\!-\!\overset{\overset{\displaystyle O}{\|}}{C}\!-\!NH\!-\!R' \rightleftharpoons \text{Enz}\!-\!S\!-\!\overset{\overset{\displaystyle O}{\|}}{C}\!-\!R + H_2N\!-\!R'$$

$$\text{Enz}\!-\!S\!-\!\overset{\overset{\displaystyle O}{\|}}{C}\!-\!R + H_2O \rightleftharpoons \text{Enz}\!-\!SH + R\!-\!\overset{\overset{\displaystyle O}{\|}}{C}\!-\!OH$$

TABLE 11.4

Some Enzymes That Form Covalent Enzyme-Substrate Intermediates

Enzyme class	Reacting group	Type of covalent intermediate
1. Chymotrypsin, trypsin, subtilisin, elastase, thrombin, acetylcholine esterase, plasmin	$HO\!-\!CH_2\!-\!CH$ Serine	$R\!-\!\overset{\overset{O}{\|}}{C}\!-\!O\!-\!CH_2\!-\!CH$ Acyl ester
2. Phosphoglucomutase, alkaline phosphatase	$HO\!-\!CH_2\!-\!CH$ Serine	$^-O\!-\!\overset{\overset{O}{\|}}{\underset{\underset{O^-}{\|}}{P}}\!-\!O\!-\!CH_2\!-\!CH$ Phosphoryl ester
3. Papain, ficin, glyceraldehyde-3-phosphate dehydrogenase	$HS\!-\!CH_2\!-\!CH$ Cysteine	$R\!-\!\overset{\overset{O}{\|}}{C}\!-\!S\!-\!CH_2\!-\!CH$ Acyl thioester
4. Succinic thiokinase, glucose-6-phosphatase	Histidine	$O^-\!-\!\overset{\overset{O}{\|}}{\underset{\underset{O^-}{\|}}{P}}\!-\!N$ Phosphorylimidazole
5. Aldolase, transaldolase, pyridoxal-phosphate enzymes	$H_2N\!-\![CH_2]_4\!-\!CH$ Lysine	$R\!-\!\overset{\overset{R}{\|}}{C}\!=\!N\!-\![CH_2]_4\!-\!CH$ Schiff base

Another form of covalent bonding of substrates is in the formation of Schiff bases (aldimines) between carbonyl compounds and the ϵ-amino group of lysine in some enzymes. Thus when dihydroxyacetone phosphate is incubated with *aldolase* (Chap. 17) in the absence of glyceraldehyde-3-phosphate (its normal partner in the reaction catalyzed by this enzyme), the following reaction occurs:

$$
Enz-[CH_2]_4-NH_2 + O{=}C\overset{\displaystyle CH_2OH}{\underset{\displaystyle CH_2OPO_3{}^{2-}}{|}} \rightleftharpoons Enz-[CH_2]_4-N{=}C\overset{\displaystyle CH_2OH}{\underset{\displaystyle CH_2OPO_3{}^{2-}}{|}} + H_2O
$$

The major advantage in the formation of covalent substrate-enzyme intermediates is that they increase the probability that a given reaction will proceed. An intermediate covalently attached to the enzyme has great constraints upon its motion within the active site and can be brought into a more favorable position for further reaction with the appropriate groups in the site required to complete the reaction. The rate enhancement provided by formation of covalent intermediates may be large, as with *aldolase,* but in other cases it may be no greater than a factor of 10^2 to 10^3.

NATURE OF ACTIVE SITES AND ENZYMIC MECHANISMS

Multiple approaches are needed to identify the functional groups in an active site, the types of interactions that occur between substrate and enzyme, and a plausible mechanism of action. Comparison of the complete three-dimensional structure of an enzyme, in the absence and presence of inhibitors, gives much information. Such structures can be obtained only by x-ray crystallography coupled with complete sequence analysis, but not all enzymes can be analyzed easily by these means. Thus, other experimental approaches are required and indeed are necessary for interpretation of the crystallographic structures. Much information about an enzyme and its mechanism can be obtained by determination of its substrate specificity, the nature of inhibitors, and by kinetic analysis of the catalytic activity as a function of pH, temperature, and solvent composition. Additional insight is gleaned by assessing how enzymic activity is affected by chemical modification, either with reagents that covalently modify specific R groups in the enzyme or with proteolytic enzymes that specifically cleave one or a few peptide bonds.

Four different hydrolytic enzymes will be considered here, *chymotrypsin, ribonuclease, lysozyme,* and *carboxypeptidase A,* which serve as excellent examples of the multiple experimental approaches that are used to probe the structure-function relationships of enzymes. The complete structure of each has been established, and it has been possible to observe the structure of the active site and the mode of substrate binding. The catalytic properties of each have also been studied in detail, and with this additional information, plausible mechanisms of action have been suggested.

The serine proteases (Table 5.3) include chymotrypsin, trypsin, elastase, and several enzymes of blood coagulation which share many structural and functional properties. α-Chymotrypsin is the most thoroughly studied member of the group and will be considered here to illustrate the mechanism of action of these proteases.

The three-dimensional structure of α-chymotrypsin is shown diagrammatically in Fig. 11.1. It contains three polypeptide chains, designated A, B, and C (Fig. 10.13). This structure is very similar to those of elastase and trypsin, although overall it is very different from that of subtilisin, another type of serine protease. All serine proteases, however, act by essentially the same mechanism, in which an acyl-enzyme covalent intermediate is transiently formed. The mechanism of action of one serine protease differs from another primarily in the interactions that determine substrate specificity. Before discussing the variety of structural, kinetic, and chemical modification studies from which this mechanism was deduced, the mechanism itself, shown in Fig. 11.2, should be considered.

Structure I in Fig. 11.2 is the substrate and structure VII is the hydrolysis product which correspond to the following reactants in a typical chymotrypsin-catalyzed reaction.

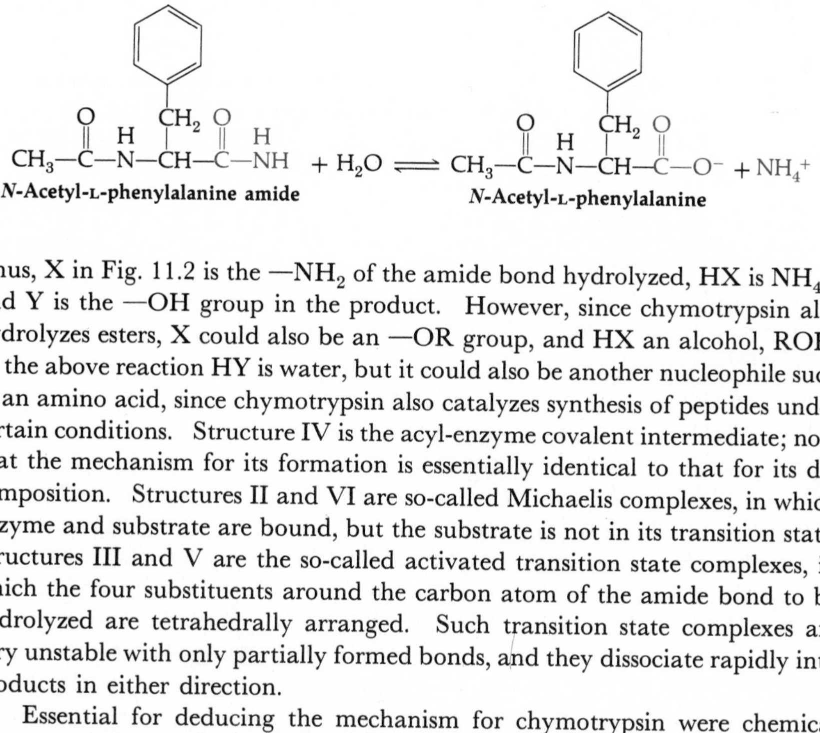

$$CH_3-\overset{\overset{O}{\|}}{C}-\overset{\overset{H}{|}}{N}-\overset{\overset{CH_2}{|}}{CH}-\overset{\overset{O}{\|}}{C}-\overset{\overset{H}{|}}{NH} + H_2O \rightleftharpoons CH_3-\overset{\overset{O}{\|}}{C}-\overset{\overset{H}{|}}{N}-\overset{\overset{CH_2}{|}}{CH}-\overset{\overset{O}{\|}}{C}-O^- + NH_4^+$$

N-Acetyl-L-phenylalanine amide   N-Acetyl-L-phenylalanine

Thus, X in Fig. 11.2 is the $-NH_2$ of the amide bond hydrolyzed, HX is NH_4^+, and Y is the $-OH$ group in the product. However, since chymotrypsin also hydrolyzes esters, X could also be an $-OR$ group, and HX an alcohol, ROH. In the above reaction HY is water, but it could also be another nucleophile such as an amino acid, since chymotrypsin also catalyzes synthesis of peptides under certain conditions. Structure IV is the acyl-enzyme covalent intermediate; note that the mechanism for its formation is essentially identical to that for its decomposition. Structures II and VI are so-called Michaelis complexes, in which enzyme and substrate are bound, but the substrate is not in its transition state. Structures III and V are the so-called activated transition state complexes, in which the four substituents around the carbon atom of the amide bond to be hydrolyzed are tetrahedrally arranged. Such transition state complexes are very unstable with only partially formed bonds, and they dissociate rapidly into products in either direction.

Essential for deducing the mechanism for chymotrypsin were chemical modification studies that identified the amino acid side chains in the active site. Serine was first implicated by examining the reaction of enzyme with diisopro-

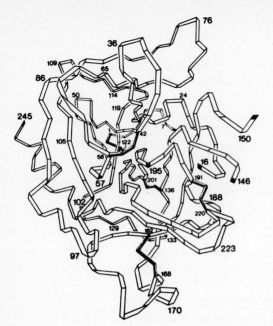

Figure 11.1 Drawings of the main-chain conformation of α-chymotrypsin. The creases in the tape indicate the position of α-carbon atoms; no R groups are shown. [*From B. S. Hartley and D. M. Shotten, p. 323, in P. Boyer (ed.), The Enzymes, vol. III, Academic, New York, 1971.*]

pylphosphofluoridate (DIPF) and similar compounds, which react stoichiometrically as follows:

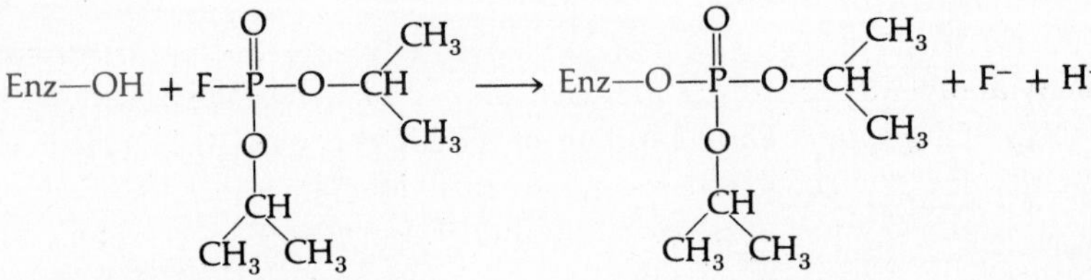

$$Enz-OH + F-\overset{\overset{O}{\|}}{\underset{\underset{\underset{CH_3}{|}}{\overset{|}{CH}}}{P}}-O-\overset{\overset{CH_3}{}}{\underset{CH_3}{CH}} \longrightarrow Enz-O-\overset{\overset{O}{\|}}{\underset{\underset{\underset{CH_3}{|}}{\overset{|}{CH}}}{P}}-O-\overset{\overset{CH_3}{}}{\underset{CH_3}{CH}} + F^- + H^+$$

Diisopropylphosphofluoridate   **DIP-chymotrypsin**

Figure 11.2 The general mechanism of action of serine proteases. E—OH is the protease in which —OH is the serine hydroxyl group that is acylated during catalysis as indicated by structure IV. K_s and K_p are the Michaelis constants for the reactions, and k_1 and k_2 etc. are the rate constants for the reactions indicated. The other structures and symbols are explained in the text. [*From J. Kraut, Annu. Rev. Biochem. 46:331–358 (1977).*]

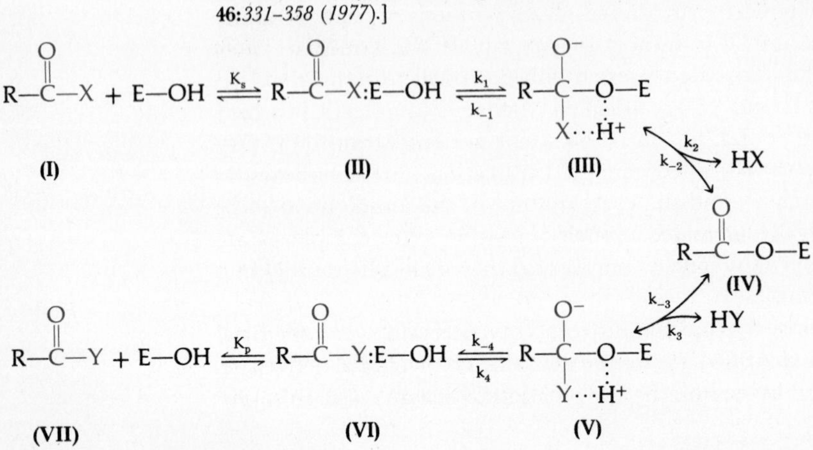

$$R-\overset{\overset{O}{\|}}{C}-X + E-OH \underset{}{\overset{K_s}{\rightleftharpoons}} R-\overset{\overset{O}{\|}}{C}-X{:}E-OH \underset{k_{-1}}{\overset{k_1}{\rightleftharpoons}} R-\overset{\overset{O^-}{|}}{\underset{X\cdots\overset{\cdot\cdot}{H}^+}{C}}-O-E$$

(I)   **(II)**   **(III)**

$$\underset{k_{-2}}{\overset{k_2}{\rightleftharpoons}} \quad HX$$

$$R-\overset{\overset{O}{\|}}{C}-O-E$$

(IV)

$$\underset{k_3}{\overset{k_{-3}}{\rightleftharpoons}} \quad HY$$

$$R-\overset{\overset{O}{\|}}{C}-Y + E-OH \underset{}{\overset{K_p}{\rightleftharpoons}} R-\overset{\overset{O}{\|}}{C}-Y{:}E-OH \underset{k_4}{\overset{k_{-4}}{\rightleftharpoons}} R-\overset{\overset{O^-}{|}}{\underset{Y\cdots\overset{\cdot\cdot}{H}^+}{C}}-O-E$$

(VII)   **(VI)**   **(V)**

222

DIP-chymotrypsin is completely inactive. Because only one of the 28 serine residues (Ser-195) in the enzyme so reacts and because enzymically inactive heat-denatured chymotrypsin does not react at all, a single serine residue is suggested to be in the active site. All serine proteinases that form covalent acyl intermediates also react similarly with DIPF, and sequence analysis of DIP-enzymes reveals the reactive serine in an identical or similar sequence. The same sequence, Gly-Asp-Ser·P-Gly-Gly-Pro, is present in chymotrypsin, trypsin, elastase, thrombin, and plasmin but differs in subtilisin. The high reactivity of these serine residues can only reflect their unique environment in the active site of each enzyme.

Reaction of chymotrypsin with L-1-(*p*-toluenesulfonyl)-amido-2-phenyl-ethyl-chloromethyl ketone (TPCK) also results in parallel loss of activity and covalent incorporation of the inhibitor.

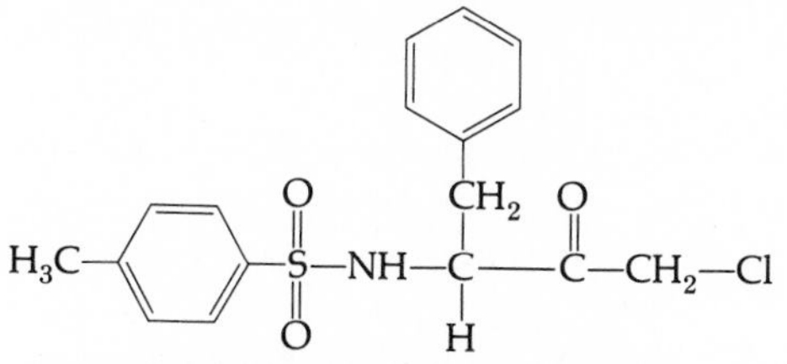

L-1-(*p*-Toluenesulfonyl)-amido-2-phenylethylchloromethyl ketone

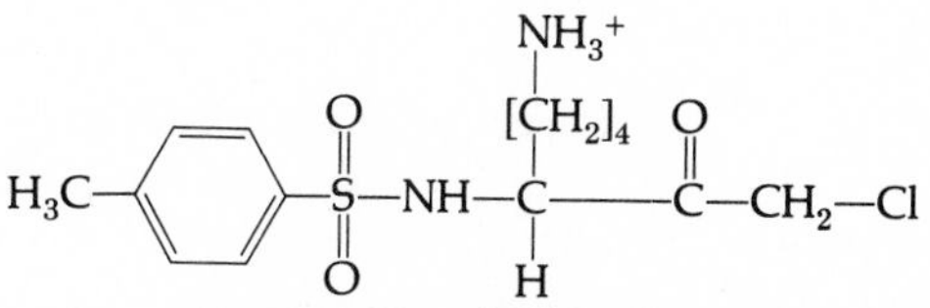

L-5-Amino-1-(*p*-toluenesulfonyl)-amidopentylchloromethyl ketone

TPCK has all the structural requirements of a good substrate and resembles an acylphenylalanyl amide or ester in which the —NH_2 or —OCH_3 is replaced by the —CH_2—Cl group. Thus, binding of TPCK occurs with as much energy and affinity for the active site as a substrate does, but because of the high reactivity of the O=C—CH_2—Cl grouping it reacts with nucleophiles in the active site. Structural studies reveal that TPCK reacts with 3-nitrogen of the imidazole ring of His-57 as follows:

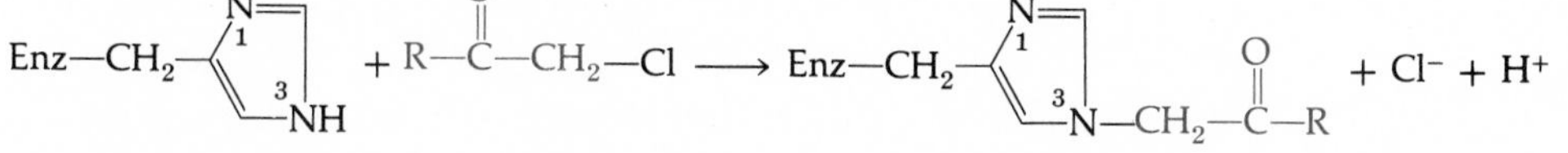

where R represents the remainder of the structure of TPCK. His-57 is in the sequence Ala-Ala-His-Cys in chymotrypsin; the same sequence is present in trypsin, thrombin, and elastase.

The specificity of TPCK for chymotrypsin is shown by the fact that it does not react with another serine proteinase, trypsin, but trypsin reacts with a similar inhibitor, L-5-amino-1-(*p*-toluenesulfonyl)-amidopentylchloromethyl ketone (TLCK), whose structure contains the —$[CH_2]_4$—NH_3^+ side chain required for productive binding to the active site of trypsin. TLCK modifies a specific histidine residue in trypsin but will not inactivate chymotrypsin. The reactions of TPCK and TLCK are examples of *affinity labeling* of an active site, and the inhibitors themselves are *affinity labels* or, more picturesquely, *Trojan Horse inhibitors*. Affinity labels are known for various other enzymes, and are particularly useful in identifying residues present in the active site.

Crystal structures of α-chymotrypsin confirmed that Ser-195 and His-57, as shown in Fig. 11.3, are in close proximity, just as expected for active site structures. But other important structures were also found. First, the α-ammonium group of Ile-16, which is the NH_2-terminus of the B chain and is formed on activation of chymotrypsinogen, makes an ion pair with the carboxylate group of Asp-194. The active site in chymotrypsinogen is very similar structurally to that in chymotrypsin, and indeed chymotrypsinogen is actually about 10^{-7} as active as chymotrypsin. Thus, formation of the ion pair, the critical step in chymotrypsinogen activation, leads to small but important rearrangements in the site. The side chains of Met-192 and Ser-195 change significantly. Met-192 moves to a position that permits binding of the aromatic ring of substrates, and the —OH group of Ser-195 is directed towards the nitrogen atom 3 of His-57, but not sufficiently close to hydrogen-bond the two atoms. Nevertheless, the distances between other atoms suggest that other hydrogen bonds are formed. One is between an oxygen atom of the carboxylate group of Asp-102 and the nitrogen atom 1 of His-57, and another between an oxygen atom of Asp-102 and a nitrogen atom of a peptide nitrogen. This arrangement of hydrogen bonded structures forms the so-called charge-relay system, which is suggested to aid transfer of a proton from Ser-195 in the conversion of structure II to structure III in Fig. 11.2.

Comparison of the crystal structures of chymotrypsin in the presence and absence of α-N-formyl-L-tryptophan (a so-called virtual substrate, since its carboxyl oxygen exchanges with water oxygen in the presence of chymotrypsin) reveals some additional details of the active site.

α-**N-Formyl-L-tryptophan**

As shown in Fig. 11.4, α-N-formyl-L-tryptophan is positioned in a crevice, one side of which is composed of planar peptide bonds formed by Ser-214, Trp-215,

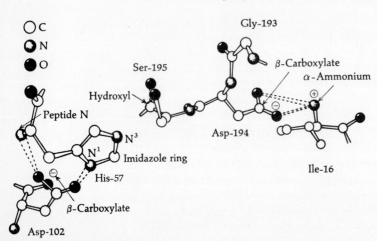

Figure 11.3 The charge-relay complex in the active site of chymotrypsin as deduced from x-ray crystallography. [*From D. M. Blow, p. 185, in P. Boyer (ed.), The Enzymes, vol. III, Academic, New York, 1971.*]

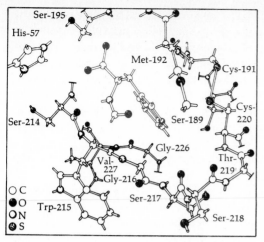
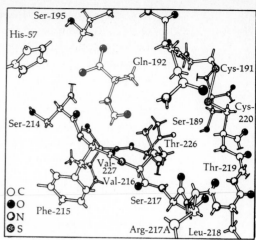

Figure 11.4 The active-site regions of models of α-chymotrypsin (left) and elastase (right) showing the binding of virtual substrates. Structures shown were established by x-ray crystal analysis: that for α-chymotrypsin in the presence of *N*-formyl-L-tryptophan and that for elastase in the presence of *N*-formyl-L-alanine. [*From B. S. Hartley and D. M. Shotten, p. 323, in P. Boyer (ed.), The Enzymes, vol. III, Academic, New York, 1971.*]

and Gly-216 and the other by peptide bonds of Ser-189, Cys-191, and Met-192. The indolyl ring is almost parallel to the polypeptide chain on either side and makes hydrophobic contacts, most notably with Met-192. The oxygen atom of α-*N*-formyl-L-tryptophan, which corresponds to the carbonyl oxygen of the susceptible peptide bond, is directed towards Ser-195 and His-57, into a site often called the *oxyanion hole*.

The structures in Fig. 11.4 give plausible explanations for the differences in the substrate specificities of the serine proteases. The active site in α-chymotrypsin easily accommodates a planar aromatic ring. This site in elastase, also shown in Fig. 11.4, is occluded by more bulky groups from Val-216 and Thr-226, which do not permit binding of aromatic rings. Smaller groups such as methyl groups in alanine can fit into the elastase site, consistent with its substrate specificity (Table 5.3). This site in trypsin contains Asp-189, whose carboxylate group permits tight binding of the cationic side chains of lysine or arginine.

From the foregoing structural information, the chemical events that transpire during chymotrypsin catalysis by the mechanism in Fig. 11.2 can now be considered to proceed in the following steps. (1) The binding of the substrate to the active site to form structure II includes the many interactions between specific groups of the enzyme and the substrate as discussed above and illustrated in Fig. 11.4. Most of these interactions occur on the acyl side of the substrate and include binding of the aromatic ring as well as hydrogen bonding of the reactive carbonyl oxygen to polypeptide —NH— groups of Ser-195 and Gly-193. This position of the reactive carbonyl group crowds the binding site, causing the reactive —OH of Ser-195 to take the position of the transition state. (2) The transition state, structure III, then forms with the OH of Ser-195, attacking the reactive carbonyl carbon. Simultaneously, a tetrahedral intermediate forms with a proton from Ser-195 migrating to the imidazole ring of His-57 while the other proton on the imidazole of His-57 moves to the carboxylate of Asp-102. The negative charge on the reactive carboxyl oxygen forms a shorter, stronger hydrogen bond with the peptide NH group of Gly-193 and aids stabilization of structure III. (3) The transition state structure III then dissociates to give the acyl-enzyme covalent intermediate with the leaving group free to diffuse from the site. (4) The acyl enzyme (structure IV) then deacylates when the charge-relay system activates water in a manner similar to reversal of steps (2) and (3), to give the transition state structure V, which then dissociates to give the enzyme-product complex (structure VI).

Many kinetic studies were required to deduce the foregoing mechanism, but they are too detailed to consider here. Nevertheless, it should be evident that catalysis is a dynamic process, and to deduce an enzyme mechanism from a static x-ray structure requires additional information. Thus, the detailed mechanism for chymotrypsin will undoubtedly be refined as additional information becomes available.

Implicit in the mechanism of action of chymotrypsin is the formation of the activated transition state complexes, structures III and V in Fig. 11.2. Such complexes are thought to form because the active site furnishes a template complementary to the transition-state configuration of the substrate. The proposed tetrahedral intermediate in the transition state is an unstable structure of high energy. The effectiveness of TPCK and DIPF as inhibitors of chymotrypsin is thought to result from the fact that they possess the tetrahedral geometry of the transition state and fit tightly to the complementary configuration of the active site. Thus, these inhibitors are also called *transition-state inhibitors*. Inhibitors of this type are known for several enzymes in addition to the serine proteases, suggesting that transition state complexes form with many enzymes.

The mechanism of subtilisin is very similar to that considered for chymotrypsin, notwithstanding the fact that its overall three-dimensional structure is very different from those of chymotrypsin, trypsin, and elastase. This results from the fact that subtilisin also has the same charge-relay system as the other serine proteases.

Ribonuclease-A

Bovine pancreatic *ribonuclease-A* was the first enzyme whose complete sequence was established (1963). Before its three-dimensional structure was determined, much had been learned about its active site and mechanism of action.

Ribonuclease (RNase) consists of a single polypeptide chain of 124 residues with four intrachain disulfide bonds. Subtilisin hydrolyzes only one peptide bond in RNase, that between Ala-20 and Ser-21. The cleaved enzyme (*RNase-S*) is enzymically active with residues 1 through 20 (*S-peptide*) bound noncovalently to the remainder of the molecule (*S-protein*). At acid pH the S-peptide and the S-protein dissociate, and neither is enzymically active, but when mixed at pH 7, they combine to generate active RNase-S. These events are illustrated schematically as follows, where the symbols (H) refer to histidine residues 12 and 119:

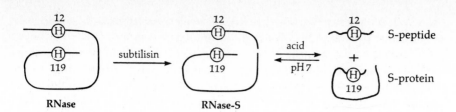

A synthetic peptide with the same sequence as residues 1 to 13 restores about 70 percent of the activity of RNase-S when combined with S-protein, but another peptide with residues 1 to 11 does not restore activity. This suggests that either His-12 or Met-13 or both play a key role in RNase activity, whereas residues 14 to 20 are not directly involved in the active site.

Additional modification of native RNase implicated other residues essential for activity. Pepsin cleaves the peptide bond between Phe-120 and Asp-121 and the enzyme, devoid of residues 121 to 124, is inactive. Carboxypeptidase A removes Val-124 from the S-protein with no loss of activity on recombination with S-peptide. Further hydrolysis removes Ser-123 with activity falling to 45 percent. More extensive digestion through Phe-120 results in complete inactivation.

The involvement of His-12 in supporting enzymic activity and forming a critical group in the active site was revealed by carboxymethylation of RNase with iodoacetate (IAA) at pH 5.5, conditions that allow carboxymethylation principally of histidine. The extent of carboxymethylation was proportional to the degree of inactivation of the RNase, and the completely inactive enzyme contained only one carboxymethyl group per molecule of RNase. Ion-exchange chromatography of carboxymethylated RNase revealed two chemically distinct forms of enzyme, one obtained in about 15 percent yield, called 1-Cm-His-119-RNase (Cm = carboxymethyl), and the other, called 3-Cm-His-12-RNase, found in about 85 percent yield. Both derivatives were inactive, and significantly, neither form contained two modified histidine residues. The two forms differ in structure, shown schematically as follows:

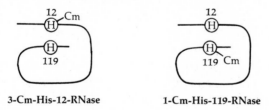

3-Cm-His-12-RNase   **1-Cm-His-119-RNase**

Analysis revealed 3-Cm-His-12-RNase to contain a single residue of 3-carboxymethylhistidine at residue 12 and 1-Cm-His-119-RNase a single residue of 1-carboxymethylhistidine at residue 119.

N^1-**Carboxymethylhistidine** N^3-**Carboxymethylhistidine**

Moreover, further reaction of either RNase derivative with IAA did not alkylate the unsubstituted histidine. These findings suggested that the histidines act as nucleophiles to displace the reactive iodine of IAA and form the carboxy-methylhistidine derivatives, but, more significantly, His-12 and His-119 were judged to be in close proximity in the active site, with His-119 facilitating the alkylation of His-12 and His-12 the alkylation of His-119 as follows:

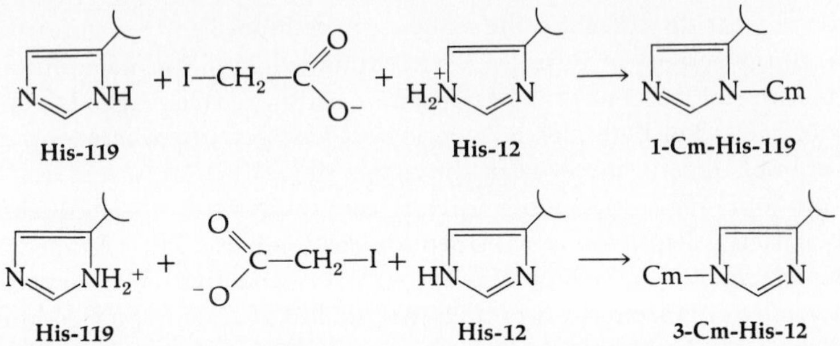

Thus, the orientation and distance between the two histidines and the electrostatic interaction between the charged imidazolium groups and the carboxylate group of IAA allow alkylation to proceed. The fact that His-119 is alkylated more extensively than His-12 suggests that the pK of the imidazolium group of His-12 is higher than that of His-119 and exists largely as a protonated species at pH 5.5; His-119 must therefore have a lower pK than His-12. The pK values for His-12 and His-119 were subsequently found to be 6.2 and 5.8, respectively; they are raised to 8.0 and 7.4 on binding an inhibitor.

Dimers of RNase are formed on lyophilization of the enzyme from 50 percent acetic acid and have full activity. Dimers are also formed when inactive 1-Cm-His-119-RNase and 3-Cm-His-12-RNase are mixed in equal amounts and lyophilized from acetic acid. Most remarkably, 25 percent of the alkylated dimers have about 50 percent of the specific activity of unmodified dimers. The activity is lost when the dimers are dissociated by heating at 67°C for 10 min. These results can be explained with the aid of the schematic reactions shown on page 229. When dissolved in 50 percent acetic acid, the noncovalent bonds that fix residues 1 to 20 to the remainder of RNase are broken, just as the S-peptide dissociates from S-protein in acid. On lyophilization of the dimers, however, the S-peptide regions on two different molecules bind tightly to the S-protein regions to form the dimers as illustrated; two of the three kinds of dimers are inactive since they contain either an alkylated His-119 or an alkylated His-12 residue at both active sites, but a third dimer forms one productive, noncarboxymethylated active site and one inactive site, the latter containing both 1-Cm-His-119 and 3-Cm-His-12.

The crystal structures of RNase and RNase-S have been established and differ only in the region where the peptide bond in RNase-S has been cleaved. As shown in Fig. 11.5, the molecule is kidney-shaped with a marked depression on one side, in which are located His-12 and His-119. Examination of crystal structures in the presence of inhibitors reveals that each binds in the same site near the two histidine residues, in accordance with expectations. The orientation of one inhibitor in the active site is shown in Fig. 11.6: It contains a methyl-

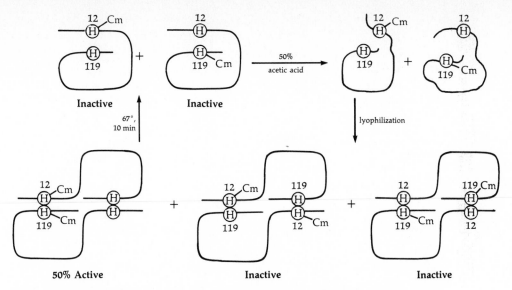

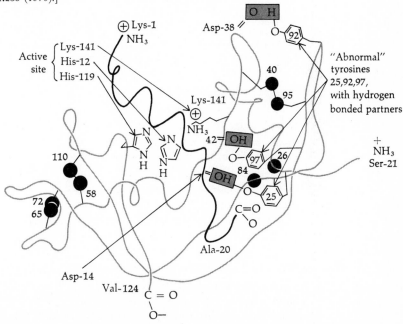

ene carbon atom replacing the oxygen between the phosphorus and the 5′ CH_2 of the ribose of adenine (A) and is designated as UpCH₂A. The orientation of the latter in the active site is probably somewhat different from that of an enzyme-substrate complex, but, as seen in Fig. 11.6, His-12 and His-119 are in close juxtaposition to the phosphate ester linkage that is hydrolyzed. The pyrimidine and purine rings fit into specific regions. Hydrogen bonds form be-

Figure 11.5 A schematic representation of the conformation of ribonuclease-S showing the relative locations of His-12, His-119, and Lys-41 in the active site. The four disulfide bonds are also shown. [*After T. F. Spande, B. Witkop, Y. Degani, and A. Patchornik, Adv. Protein Chem.* **24**:*235 (1970)*.]

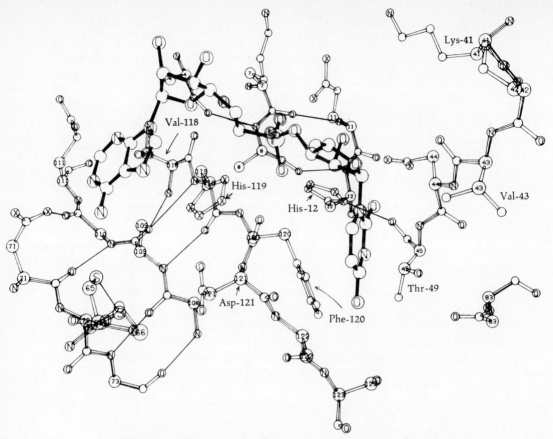

Figure 11.6 A schematic model of the binding of the dinucleotide phosphate UpCH₂A (solid bonds) to RNase-S. His-119 is in one of four possible locations found on crystallographic analysis. In the native protein the ε NH₃⁺ of Lys-41 is down and projects toward the phosphate group but is not in contact with it. The —CH₂— attached to the phosphate of the inhibitor has perturbed the position of His-119. [*From F. M. Richards and H. W. Wyckoff, p. 647, in P. Boyer (ed.), The Enzymes, vol. IV, Academic, New York, 1971.*]

tween the pyrimidine ring and the side chain —OH and peptide —NH— of Thr-45. Phe-120 is on one side of the pyrimidine ring and Val-43 on the other, and together they form a groove for pyrimidine binding. This may explain the inactivation of RNase when residues 120 to 124 are removed by pepsin.

Several mechanisms of catalysis have been proposed for RNase, including that shown in Fig. 11.7. Although it lacks unequivocal support, it is clear that rate enhancement for ribonuclease action must come from specific binding of different groups on the substrate in the active site and that His-12 and His-119 appear to act through concerted general acid-base catalysis.

Lysozyme

Good substrates for lysozyme include the polysaccharides of bacterial cell walls with the repeating disaccharide unit of *N*-acetylglucosamine in β-1,4 link-age with *N*-acetylmuramic acid (Chap. 18). A simpler substrate is a hexamer of

Figure 11.7 The substrate binds (*a*) at the active site with the purine and pyrimidine rings in tight binding subsites and the ε NH$_3^+$ of Lys-41 projecting toward the negatively charged phosphate along with His-12 and His-119. The concerted action of His-12, acting as a base to accept a proton from the ribose 2′ OH, and His-119, acting as an acid to form a hydrogen bond with a phosphate O atom, as in (*b*), leads to a transition-state complex with a pentacoordinate phosphorus, represented by a trigonal bipyramid. The formation of the cyclic 2′,3′-phosphoribose intermediate (*c*) is accompanied by loss of a proton from His-119 and the uptake of a proton by His-12. Water then enters the site (*d*), donating a proton to His-119 and an —OH to the phosphate to form the trigonal pyramid structure in (*e*), which rearranges with the aid of the acidic His-12 to form the pyrimidine ribose-3′-phosphate product (*f*). [*From G. C. K. Roberts, E. A. Dennis, D. H. Meadows, J. S. Cohen, and O. Jardetzky, Proc. Natl. Acad. Sci. USA* **62:***1151* (*1969*).]

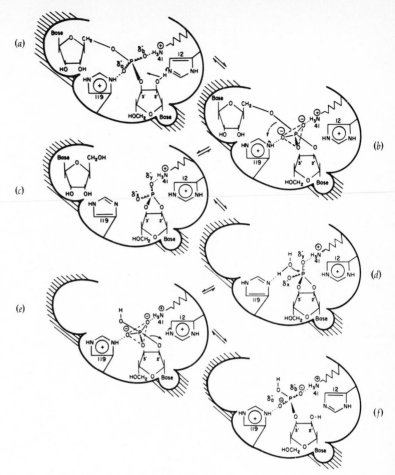

N-acetylglucosamine, hexa-*N*-acetylchitohexaose, in which one glycosidic bond is hydrolyzed to give a tetra- and a disaccharide:

Hexa-*N*-acetylchitohexaose

Chicken egg-white lysozyme, which contains 129 residues, has the structure illustrated schematically in Fig. 11.8. Studies of the enzyme in the presence of mono-, di, and trisaccharides showed the sugars bound into a cleft between two halves of the molecule. Tri-*N*-acetylchitotriose was bound with its nonreducing end (sugar A) at the top of the cleft and its reducing end (sugar C) pointing down the cleft. There were also small shifts of the order of 0.75 Å in the position of certain side chains in the cleft, suggesting some induced fit in the cleft on substrate binding. By model building, three additional sugars were added one at a time from the reducing end, and each additional sugar was maintained insofar as possible in the same spatial configuration as the first three. All but

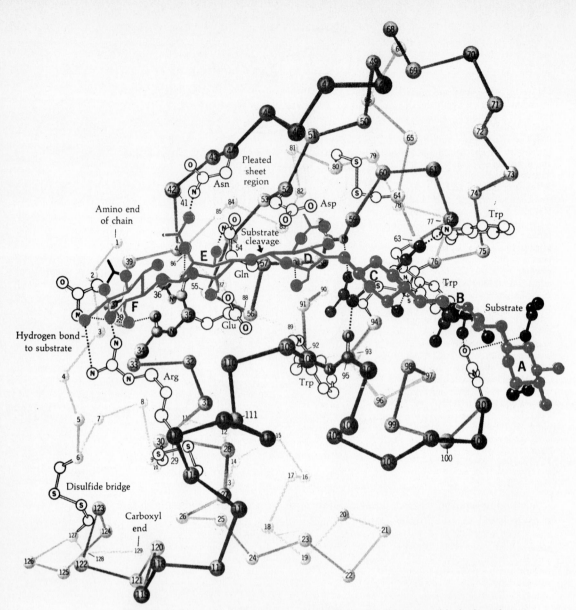

Figure 11.8 A representation of the structure of egg-white lysozyme with the bound substrate hexa-*N*-acetylchitohexaose. Only the distances between α-carbon atoms are shown, except for certain R groups lining the cleft, or active site, that provide the many noncovalent interactions with substrate. The sugar residues are designated A, the nonreducing end, through F, the reducing end. Hydrolysis occurs at the glycosidic bond between D and E. The presence of Asp-52 and Glu-35 on either side of the substrate cleavage site aids in catalysis, as described in the text. [*From R. E. Dickerson and I. Geis, The Structure and Function of Proteins, Harper & Row, New York, 1969, p. 71*]

one of the six sugars made excellent noncovalent interactions with R groups and peptide backbone structures, as illustrated in Fig. 11.8. A remarkable number of hydrogen bonds and nonpolar contacts was suggested. Since hydrolysis occurs between the fourth and fifth sugar residues, designated residues D and E, it was particularly important to examine the environment of the glycosidic bond to be cleaved. On one side of the bond (O atom), about 3 Å distant, was the carboxyl group of Glu-35, whereas on the other side was the carboxyl group of Asp-52. Glu-35 is in a particularly hydrophobic environment, and its carboxyl group is likely to be protonated, but Asp-52 is in a complex, hydrophilic network of hydrogen bonded groups, and it is thought to have an ionic carboxylate group. Since these were the only acidic and basic groups closest to the glycosidic bond, it was proposed that they serve in a concerted catalytic mechanism. Moreover, when sugar residue D was built into the cleft, it made too close contact with cleft structures if it was maintained in the normal chair structure. Residue D could be distorted from the chair into the half-chair conformation, however, and normal contact distances were found. Thus, residue D appears to be strained on binding, but the energy required to form this unfavorable half-chair ring could easily be overcome by binding the other sugar residues.

From the foregoing considerations, the following mechanism of hydrolysis of a hexasaccharide was proposed (Fig. 11.9). The enzyme binds the hexasaccharide as shown in Fig. 11.9, with distortion of sugar residue D. Glu-35 donates a proton to the glycosidic oxygen atom between sugars D and E, allowing the bond to break with formation of one product, a disaccharide (residues E and F, see Fig. 11.8), which dissociates from the enzyme and the tetrasaccharide formed by residues A to D. On cleavage of the glycosidic bond, however, sugar residue D is a carbonium ion (C^+) in a half-chair configuration. The negative charge on Asp-52 and the strain imposed on sugar residue D also promotes formation of the carbonium ion and helps stabilize it once formed. Carbonium ions of sugars in the half-chair configuration are known to be stabilized by sharing the positive charge with the pyranose-ring oxygen. An —OH group

Figure 11.9 A proposed mechanism for the catalytic hydrolysis of a lysozyme substrate facilitated by specific binding of sugars A to E in the active site. Only sugar residue D and part of E are shown. As the substrate binds (a), it comes into close proximity with Asp-52 and Glu-35; they aid in cleavage of the D—E glycosidic bond to form the transition-state complex (b), which reacts with OH$^-$ and H$^+$ from water to form product and the original enzyme (c). [*From R. E. Dickerson and I. Geis, The Structure and Action of Proteins, Benjamin, Menlo Park, Calif., p. 77. Copyright 1969 by Dickerson and Geis.*]

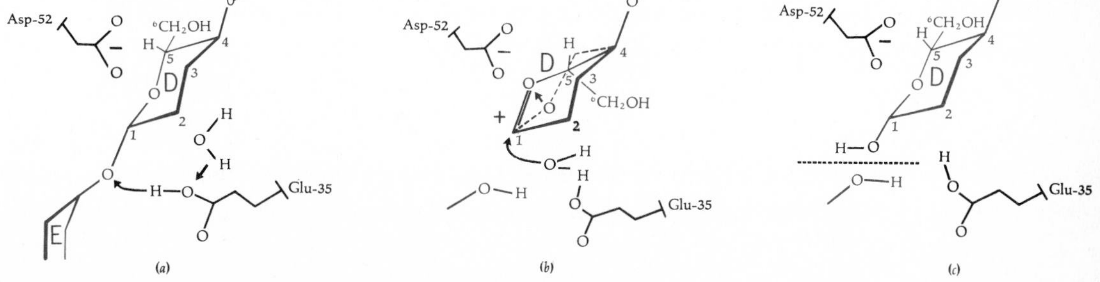

from water then adds to the C-1 atom of the carbonium ion of sugar residue D, and a proton from water adds back to Glu-35 to complete the reaction.

In this mechanism, rate enhancement is partly contributed by the specific binding energy of the substrate, the strain of one sugar ring, and the induced fit of the substrate, with Glu-35 serving as a general acid catalyst and Asp-52 aiding in formation of the carbonium ion in the transition state. This mechanism has received considerable support from subsequent studies of the reaction in solution. (1) The binding energy of each sugar has been measured, and only that of sugar D is unfavorable, in accord with its distortion into the half-chair. (2) In the presence of substrate, all carboxyl groups in lysozyme except those of Asp-52 and Glu-35 can be chemically modified without loss of activity. In the absence of substrate, Asp-52 but not Glu-35 is also modified with total loss of enzymic activity. (3) A *transition-state analog* of the following structure is a powerful inhibitor:

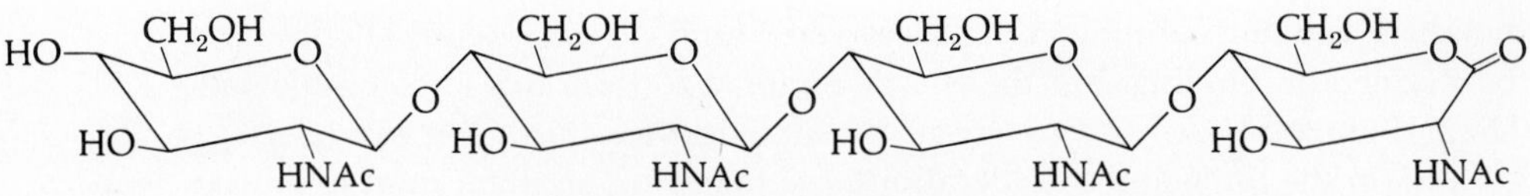

Lactone of tetra-*N*-acetylchitotetraose

Ring D is a lactone, which normally exists in the half-chair configuration; thus this tetrasaccharide has the same structure as residues A to D in substrates when bound to enzyme and fits snugly into the cleft with favorable binding energy. (4) Lysozyme acts optimally at pH 5, with activity decreasing on either side of the optimum, in accord with deprotonation of Glu-35 at high pH and protonation of Asp-52 at low pH.

Carboxypeptidase A

The specificity of this enzyme is shown with the following model substrate:

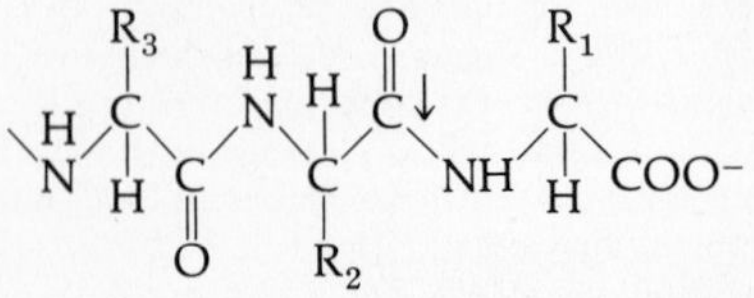

The bond hydrolyzed, indicated by the arrow, must be adjacent to a free α-carboxyl group, and the rate is enhanced if R_1 is a large aromatic or aliphatic group. Dipeptides are poor substrates, but α-*N*-acyldipeptides are good model substrates. The COOH-terminal residue must be in the L configuration, and D-amino acids in R_2 decrease the rate considerably. Esters similar in structure to the above peptides are also good substrates.

Carboxypeptidase A is a metalloenzyme containing one atom of zinc, which can be removed to give an inactive enzyme, but activity is restored on addition of zinc as well as other metals. A role for zinc in the active site was suggested many years ago, and mechanisms of hydrolysis were proposed that proved to be in good accord with subsequent crystallographic analysis of the enzyme and enzyme-inhibitor complexes. Chemical modification of the enzyme suggested that one of the tyrosine residues may be involved in catalysis.

The complete three-dimensional structure of carboxypeptidase A is shown diagrammatically in Fig. 11.10. Crystal structures in the presence of the inhibitor Gly-Tyr revealed the location of the active site, which is a shallow groove leading over the zinc atom into a hydrophobic pocket. The orientation of the inhibitor in the active site is shown in Fig. 11.11. Most strikingly, at least five molecules of water are expelled from the site on substrate binding, including one from the zinc atom. The essential carboxylate group is bound to the guanidinium of Arg-145, and the phenolic hydroxyl group of Gly-Tyr is in a hydrophobic pocket. The zinc atom is chelated to the enzyme through the side chains of His-69, Glu-72, and His-196. The fourth coordination position on zinc in the absence of inhibitor is occupied by water, and the three ligands and water are oriented approximately tetrahedrally around the zinc atom. In the presence of inhibitor, a remarkable number of structural changes occur in the site on substrate binding, suggesting a substrate-induced fit. Arg-145 moves ~2 Å to in-

Figure 11.10 A schematic model of carboxypeptidase A showing distances between α-carbon atoms and the location of the active sites, including the zinc atom (Z), Glu-270, Tyr-248, Glu-72, His-69, and Arg-145, which participate in binding and catalysis. [*From R. E. Dickerson and I. Geis, The Structure and Action of Proteins, Benjamin, Menlo Park, Calif., p. 90. Copyright 1969 by Dickerson and Geis.*]

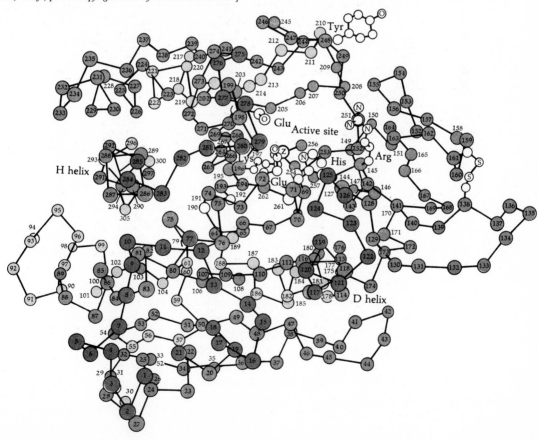

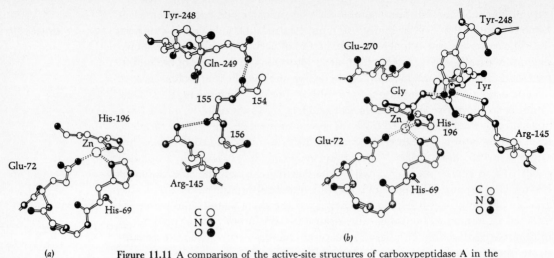

(a)

Figure 11.11 A comparison of the active-site structures of carboxypeptidase A in the absence *(a)* and presence *(b)* of Gly-L-Tyr. Binding of Gly-Tyr results in considerable reorientation of Glu-270, Tyr-248, and Arg-145, as shown. [*From D. M. Blow and T. A. Steitz, Annu. Rev. Biochem.* **39**:*63 (1970)*.]

Figure 11.12 Postulated events for the action of pancreatic carboxypeptidase A; the binding of the carboxyl terminal portion of the substrate is shown *(a)*. As many as six residues of a polypeptide substrate can bind to the cleft in the enzyme by a complex series of hydrogen bonds and hydrophobic interactions. The zinc atom of the enzyme is coordinated to two histidine and one glutamic acid residues. See text for description of events in *(a)* to *(d)*.

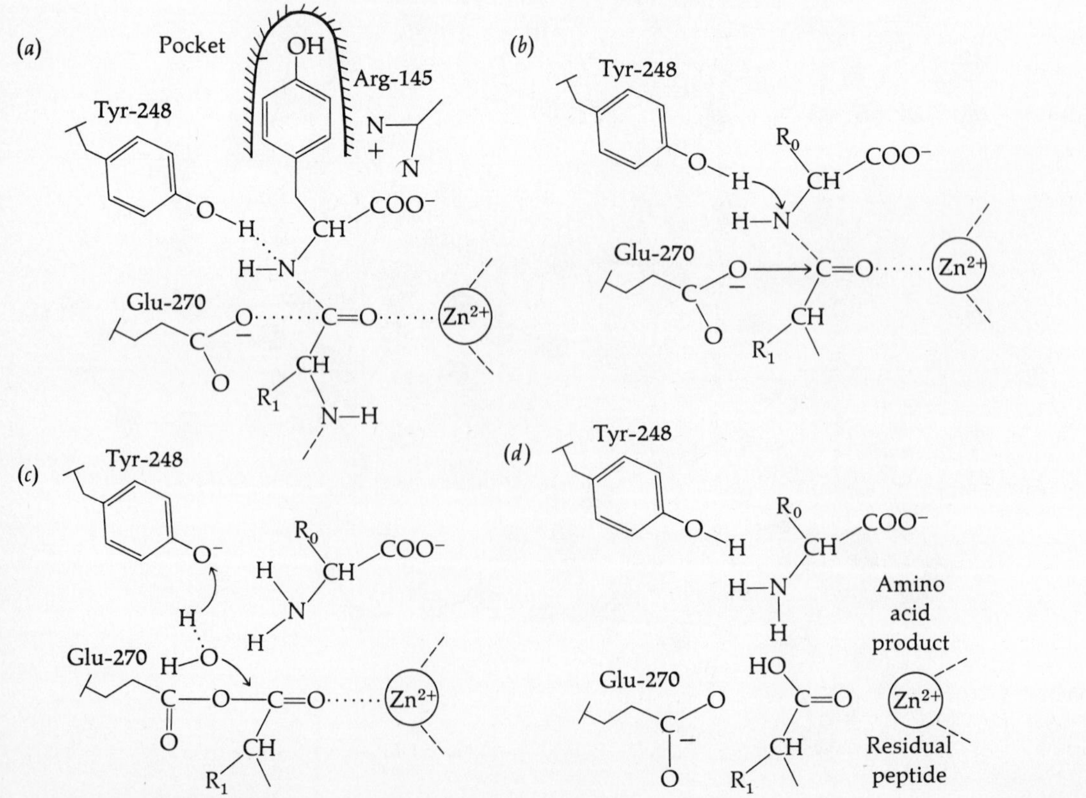

(a)   Pocket

(b)

(c)

(d)

teract with the α-carboxyl of the inhibitor, the carboxylate group of Glu-270 moves ~2 Å toward the substrate, and Tyr-248 moves about 12 Å, which brings its hydroxyl group near the —NH— group of the susceptible bond. The movement of tyrosine appears to close off the cavity so that it is largely inaccessible to solvent, suggesting that rate enhancement may partly result from desolvation. The carbonyl oxygen of the inhibitor is directed toward the zinc atom. These observations have led to a proposed mechanism for hydrolysis shown in Fig. 11.12. Glu-270 acts as a nucleophile to form an anhydride with the —C=O peptide bond to be cleaved, while Tyr-248 attacks the —NH— of the susceptible bond. The zinc atom acts as an electrophile to polarize the carbonyl oxygen. Intervening water then restores the proton of Tyr-248 and cleaves the anhydride, restoring the active site and yielding products.

REFERENCES

Books

Boyer, P. D. (ed.): *The Enzymes,* 3d ed., Academic, New York, 1970–.

Colowick, S. P., and N. O. Kaplan (eds.): *Methods in Enzymology,* Academic, New York, (a continuing series since 1958).

Dixon, M., E. C. Webb, C. J. R Thorne, and K. F. Tipton (eds.): *The Enzymes,* Academic, New York, 1980.

Fersht, A.: *Enzyme Structure and Mechanism,* Freeman, San Francisco, 1977.

Gutfreund, H.: *Enzymes: Physical Principles,* Interscience-Wiley, New York, 1972.

International Union of Biochemistry: *Enzyme Nomenclature,* Academic, New York, 1979.

Jencks, W. P.: *Catalysis in Chemistry and Enzymology,* McGraw-Hill, New York, 1969.

Meister, A. (ed.): *Advances in Enzymology,* Interscience, New York. (This is an annual series since 1941 and contains numerous review articles.)

Segal, I. H.: *Enzyme Kinetics,* Interscience-Wiley, New York, 1975.

Walsh, C.: *Enzymatic Reaction Mechanisms,* Freeman, San Francisco, 1979.

Review Articles

Bruice, T. C.: Some Pertinent Aspects of Mechanism as Determined with Small Molecules, *Annu. Rev. Biochem.* **45**:331–373 (1976).

Chock, P. B., S. G. Rhee, and E. A. Stadtman: Interconvertible Enzyme Cascades in Cellular Regulation, *Annu. Rev. Biochem.* **49**:813–844 (1980).

Jencks, W. P.: Binding Energy, Specificity and Enzyme Catalysis: The Circe Effect, *Adv. Enzymol.* **43**:219–410 (1975).

Kirsch, J. F.: Mechanism of Enzyme Action, *Ann. Rev. Biochem.* **42**:205–234 (1973).

Koshland, D. E., Jr., and K. E. Neet: The Catalytic and Regulatory Properties of Enzymes, *Annu. Rev. Biochem.* **37**:359–419 (1968).

Kraut, J.: Serine Proteases: Structure and Mechanism of Catalysis, *Annu. Rev. Biochem.* **46**:311–358 (1977).

Krebs, E. G., and J. A. Beavo: Phosphorylation-Dephosphorylation of Enzymes, *Annu. Rev. Biochem.* **48**:923–960 (1979).

Liljas, A., and M. G. Rossmann: X-ray Studies of Protein Interactions, *Ann. Rev. Biochem.* **43**:475–507 (1974).

Phillips, D. C.: The Three-Dimensional Structure of an Enzyme Molecule, *Sci. Am.* **215**:78–90 (1966).

Quiocho, F. A., and W. N. Lipscomb: Carboxypeptidase A: A Protein and an Enzyme, *Adv. Protein Chem.* **25**:1–78 (1971).

Sigman, D. S., and G. Mooser: Chemical Studies of Enzyme Active Sites, *Annu. Rev. Biochem.* **44**:889–931 (1975).

Metabolism

Chapter 12

Introduction to Metabolism: Principles of Bioenergetics

Thermodynamic principles. Oxidation and reduction. High-energy phosphates. Energy requirements. Mechanisms regulating metabolism. Experimental approaches to the study of metabolism.

Living organisms manifest continuous chemical change; this constitutes their *metabolism*. The magnitude of this metabolism and of the tasks involved may be appreciated from an overall consideration of metabolic phenomena in microorganisms and in human beings.

Bacterial cells in culture, e.g., *Escherichia coli*, can double in number every 40 min in a medium containing only glucose and inorganic salts, or in 20 min in a rich broth. The components of the medium are depleted, but little is added to the medium by cells. Each cell contains a complement of 5 to 10 to as many as a few thousand molecules of each of approximately 2500 different proteins, 10 to 300 million molecules of each of about 1000 kinds of organic compounds, and a variety of nucleic acids. It is evident that these cells participate in a remarkable biosynthetic performance: The net synthesis of each of the cell constituents proceeding at the rate required to ensure harmonious growth of the cells with no significant over- or underproduction of any component.

No less impressive are the metabolic activities of a human adult who maintains constant weight for perhaps 40 years, during which a total of about 6 tons of solid food and 10,000 gal of water are processed, yet both weight and body composition remain constant.

Mammalian cells are characterized by a dynamic steady state. Each cell is at all times engaged in fabricating from the materials of its environment a complex mixture of amino acids, fatty acids, and their many derivatives, steroids, mono- and polysaccharides, purines and pyrimidines, nucleotides, nucleic acids, and proteins while at the same time also degrading these polymers by hydrolysis and oxidizing carbohydrate and fatty acids to gain energy. Again it

Any cross-references coded M refer to the companion text, *Principles of Biochemistry: Mammalian Biochemistry.*

will be evident that fine control is exercised over the multitudinous metabolic processes and the energy-yielding reactions that make them possible.

These metabolic accomplishments of both simpler and more complex organisms are achieved even in the face of disturbing conditions in the environment. The fact that an organism can maintain a normal, constant internal state, termed *homeostasis,* while performing the numerous complex metabolic reactions and despite significant alterations in its environment is due to the sensitivity of specific regulatory mechanisms. Indeed, the failure of one or more of such mechanisms to function normally leads to the metabolic aberrations underlying various diseases.

An understanding of metabolism requires knowledge of the chemistry of the participating molecules (metabolites), the reactions in which they participate, the enzymes catalyzing these reactions, and the regulatory mechanisms that determine the rates of the sequential enzyme-catalyzed reactions by which any given metabolite A is converted to another B. Such a series of steps constitutes a *metabolic pathway,* and the operation of the manifold metabolic pathways and their integrative functioning constitute *metabolism.*

In order to discuss metabolism, it is necessary to review some of the basic principles of thermodynamics and to consider how the transformations of energy make possible the continuous synthesis of the major constituents of cells and their roles in the structure and life of the cell. The types of processes that regulate metabolism and the kinds of tools available for the study of metabolism will also be presented in this chapter.

The synthesis of the major cell constituents requires energy. In general, the energy derived by oxidation of foods is channeled into formation of high-energy phosphate compounds, mainly adenosine triphosphate, or ATP. These compounds are used in coupled reactions to effect the synthesis of all types of essential small molecules and macromolecules: polysaccharides, proteins, and nucleic acids. In addition, many compounds derived from the diet or fabricated by metabolic reactions have to be oxidized or reduced to be used by the cell. The types of substances involved in these coupled reactions of oxidation or reduction are described in Chaps. 15 and 16.

THERMODYNAMIC PRINCIPLES

Whereas chemistry generally is concerned with molecular transformations, the occurrence and extent of such phenomena are governed by the flow of energy. This is the subject matter of thermodynamics, the major concepts of which are stated in the first and second laws of thermodynamics. These laws permit one to understand the direction of chemical events, i.e., whether a reaction will proceed from left to right or from right to left as it is written, whether the progress of such reactions will permit the accomplishment of useful work, or whether, in order for the reaction to proceed, energy must be delivered from an external source. The principles of thermodynamics are stated in terms of parameters that permit a description of energy transformation: *enthalpy, entropy,* and *free energy.* Of these, the last has proved to be most useful for describing biochemical events.

The first law of thermodynamics is the law of conservation of energy. Even for a finite system of molecules within a container, it is generally impossible to ascertain the magnitude of U, the total energy of that system. But if energy is added to the system as heat Q, then sometime later

$$Q = \Delta U + W \qquad (1)$$

where ΔU is the change in total energy and W is the total work, if any, accomplished.

In some instances the addition of energy as heat Q to the system may result in a change of volume, the pressure remaining constant. This change, $P\,\Delta V$, is in effect a form of work and hence is a component of the term W in Eq. 1. However, since it is rarely a useful form of work, it is convenient to combine this component of W with the change in U, thereby defining the term H, *enthalpy,* or heat content. The change in enthalpy in any process at constant pressure is

$$\Delta H = \Delta U + P\,\Delta V \qquad (2)$$

The first law can then be restated as

$$Q = \Delta H + W' \qquad (3)$$

where W' is therefore *useful* work accomplished by input of the quantity of heat Q.

Equation 3 is an adequate description of an ideal, reversible system, i.e., one in which the energy utilized to alter the system can be released as an exactly equal amount of energy available to perform yet other work when the system reverts to its original state. In fact, however, perfect reversibility does not occur. Some fraction of the increase in enthalpy is not available to do useful work when the reverse process is allowed to proceed. Thus, most physical and chemical processes occur spontaneously in only one direction; e.g., water runs downhill and protons and hydroxyl ions combine, giving off heat. However, heating water does not drive it uphill, nor does it result in a net redissociation of water to a mixture of protons and hydroxyl ions. All such processes can be described in terms of the concept of equilibrium. *Spontaneous changes tend toward the equilibrium state, not away from it;* this is a manner of stating the second law of thermodynamics. The simplest description of the second law is in terms of another thermodynamic quantity, *entropy S,* that fraction of enthalpy which cannot be utilized for the performance of useful work; in most cases, this is because the random motions of the molecules in the system have increased. Hence, entropy is a measure of randomness, or disorder. The product TS, in which T is the absolute temperature, represents energy wasted in the form of random molecular motions. The second law of thermodynamics states that, given the opportunity, any system will undergo spontaneous change in that direction resulting in an increase in entropy. Equilibrium is attained when entropy has reached a maximum; no further change can occur spontaneously unless additional energy is supplied from outside the system.

Let us now consider the consequence of adding heat to a system. Since heat represents the kinetic energy of random molecular motion, the addition of heat increases the entropy. When the system is at equilibrium,

$$Q = T\,\Delta S \qquad (4)$$

If the system is not at equilibrium, however, change in the system may spontaneously increase the entropy even without the addition of heat. Thus, in general, for systems not at equilibrium,

$$T \Delta S > Q \tag{5}$$

If we combine Eqs. 3 and 4, that is, combine the first and second laws, for a system at equilibrium, we find that

$$\Delta H = T \Delta S - W' \tag{6}$$

However, in biochemistry we are rarely interested in the equilibrium state. Interest is in reactions proceeding, as they must, in the direction that approaches equilibrium and at constant temperature. For such systems, Eq. 5 modifies the statement of Eq. 6, so that

$$\Delta H < T \Delta S - W' \tag{7}$$

This permits introduction of the thermodynamic parameter of greatest general usefulness in biochemistry, the quantity called *free energy G*. It is defined as

$$G = H - TS \tag{8}$$

In general ΔG is the change in free energy that is available for the accomplishment of work, if there are appropriate means, as a system proceeds toward equilibrium. At constant temperature,

$$\Delta G = \Delta H - T \Delta S \tag{9}$$

This is the form in which the laws of thermodynamics are most useful for description of biochemical systems. For a system that is not at equilibrium,

$$\Delta G = -W' \tag{10}$$

Hence, systems not at equilibrium proceed spontaneously only in the direction of *negative* free-energy change. At equilibrium, no further change in free energy can occur spontaneously and the *available* free-energy content is zero. Conversely, a system already at equilibrium can be brought to a state remote from equilibrium only if, by some means, free energy can be made available to it. Utilization of free energy in this manner constitutes the performance of work.

Chemical Equilibria

The phenomenon of diffusion demonstrates that the free energy of a solution of a solute is a function of the concentration of the solute. Solutes in a concentrated solution placed in contact with a dilute solution diffuse into the latter until a uniform concentration is achieved. Since this occurs spontaneously, the *free-energy change* for dilution by such diffusion must be negative and, if means were available, this energy could be utilized to accomplish work. Indeed such mechanisms are important physiological devices. The variation of free energy with concentration is logarithmic,

$$G = G° + RT \ln [C] \tag{11}$$

where R is the gas constant, T absolute temperature, [C] the molar concentration of solute, and $G°$ the *standard free energy*, i.e., the free energy at a concentration of 1 mol/liter. Our interest, however, is not in such absolute values but in the changes associated with chemical reactions.

For any chemical reaction,

$$A + B \rightleftharpoons C + D$$

and

$$\Delta G = \Delta G^\circ + RT \ln \frac{[C][D]}{[A][B]} \tag{12}$$

where $\Delta G^\circ = G_C^\circ + G_D^\circ - G_A^\circ - G_B^\circ$. Equation 12 is applicable under all conditions. At equilibrium, ΔG is zero. Since $[C][D]/[A][B] = K$, where K is the equilibrium constant, then

$$\Delta G^\circ = -RT \ln K \tag{13}$$

or, at 37°C,

$$\Delta G^\circ = -1420 \log K \tag{14}$$

Hence a change in K by a factor of 10 results in a change in ΔG° of 1.42 kcal/mol. For example, if K were 10, 10^4, 10^7, or 10^{-5}, ΔG° would be -1.4, -5.7, -9.9, or $+7.1$ kcal/mol, respectively. Thus, the standard free energy of a chemical reaction, i.e., the free energy made available by reaction of a mole of each reactant to form a mole of each product under continuing standard conditions, can be calculated from measurement of the equilibrium constant. If ΔG° is negative, the process may proceed spontaneously; if it is positive, the reaction can be made to proceed only if, by some means, external free energy is made available. Values for ΔG° may be expressed in joules or calories per mole [1 kilocalorie (kcal) = 1000 cal].

Energy Transformation in Living Systems

In machines the energy released by oxidation of fuels is used to do work by taking advantage of a heat-induced temperature gradient, as in a steam engine. However, since cells do not tolerate significant temperature gradients, they must trap the energy released by oxidations in some chemical form before it is degraded into heat. In all species this is done by coupling to the oxidation process the formation of a single compound, adenosine triphosphate (ATP), from adenosine diphosphate (ADP) and inorganic phosphate (P_i).

$$ADP + P_i \rightleftharpoons ATP \qquad \Delta G^\circ = +8.4 \text{ kcal}$$

The standard free energy ΔG° of formation of ATP in this reaction is about $+8.4$ kcal mol. At 37°C, pH 7.4, and prevailing biological concentrations of ADP and P_i, the free energy ΔG actually required is about 12 kcal/mol. The energy of ATP, in turn, can be utilized by the cell to drive all known endergonic processes, such as chemical syntheses, ion transport, and mechanical exercise.

However, 12 kcal/mol is small compared with the energy released by the complete oxidation of glucose, 686 kcal/mol.

$$C_6H_{12}O_6 + 6O_2 \longrightarrow 6CO_2 + 6H_2O \qquad \Delta G^\circ = -686 \text{ kcal}$$

To take maximal advantage of this potential energy, the oxidation of glucose must be fractionated into the greatest possible number of lesser steps, each proceeding with a free-energy change of approximately the 12 kcal/mol necessary to drive synthesis of ATP. As we shall see, in living cells, 38 molecules of ATP are synthesized per molecule of glucose that is fully oxidized.

It is implicit in these concepts that, unless released as heat, energy *may* be transferred repeatedly under isothermal conditions. Thus, energy supplied to the organism as glucose next appears locked into a molecule of ATP, may then appear in a newly synthesized protein molecule, and may finally emerge as heat when the protein undergoes hydrolysis to component amino acids. The overall energy change is identical with the heat produced when glucose is oxidized in a bomb calorimeter. To the extent that each of the consecutive transfers of energy in a given series of reactions involves the production of heat and of unavailable energy as entropy, metabolic transformations may seem inefficient, insofar as the energy economy of the organism is concerned. However, it is this very "inefficiency" that gives *direction* to metabolic events. For example, in the reaction series

$$A \rightleftharpoons B \rightleftharpoons C \rightleftharpoons D \rightleftharpoons E$$

if ΔG for each step were zero, viz., if the equilibrium constant for each reaction were 1, A would be transformed into an equal mixture of A, B, C, D, and E. If, however, ΔG for each step were large and negative, particularly that for $D \rightleftharpoons E$, then A would be converted almost entirely to E. Wastage of some chemical energy is thus the price paid in order to drive net chemical transformation.

At this point one may ask why an animal requires a source of energy, apart from that required to do work on the environment. This question is particularly pointed in respect to the adult, who remains at constant weight and fixed body composition in an isothermal environment. In general, *a major use of the energy derived from biological oxidations is to maintain the body in a state remote from equilibrium.* Thus, cells contain large quantities of polysaccharides, proteins, lipids, and nucleic acids in the presence of relatively small concentrations of their constituents, i.e., glucose, amino acids, etc. Yet, at equilibrium, in the presence of appropriate enzymes, quite the opposite situation would prevail. Again, the ionic composition of the solution bathing body cells is remarkably different from that within the cell, despite permeability of the cell membrane to the ions on both sides. Free energy from the oxidation of glucose is employed to synthesize polysaccharides, proteins, etc., at a rate equal to that at which they are degraded as they tend toward equilibrium. Similarly, energy is employed to expel various ions from within the cell in opposition to the tendency to attain equilibrium across the cell membrane. As large molecules are hydrolyzed, or as ions return to the cell, energy is lost as a consequence of both entropy change and lack of means of utilizing the resultant free-energy release. Over a period of time, therefore, since the rates in both directions are equal, all the energy supplied appears ultimately as heat, but the disequilibrium has been maintained. Thus, the "order" of the foodstuffs is decreased, through oxidation, to maintain the high degree of "order" of the cell. The sum of such processes in the organism may be presumed to comprise a major fraction of the basal metabolic rate (page 257). If the supply of food or oxygen ceases, the tendency toward equilibrium is not counterbalanced and the expected equilibria are attained.

Chapters 15 and 16 will describe how the oxidation of foodstuffs by molecular oxygen occurs while the free energy of oxidation is made available for the endergonic processes of living cells.

OXIDATION-REDUCTION REACTIONS

Reactions involving the transfer of electrons are termed *oxidation-reduction reactions* or, more briefly, *redox* reactions. Electron loss is oxidation; electron gain is reduction. In some instances the electron transfer is explicit in the reaction, e.g.,

$$Fe^{3+} + Cu^+ \longleftrightarrow Fe^{2+} + Cu^{2+}$$

in which Cu^+ is oxidized and Fe^{3+} is reduced. The redox nature of reactions between covalent compounds is not quite so obvious and involves considerations of electronegativity and different degrees of sharing of electrons. Thus, consider the oxidation of methane:

$$CH_4 + \tfrac{1}{2}O_2 \longrightarrow CH_3OH$$

A C—H bond has been replaced by a C—OH bond. Because oxygen is more electronegative than hydrogen, it pulls electrons from the carbon. The dipole moment of methanol is a direct measure of this displacement of electrons from the carbon, which has been oxidized, toward the oxygen, which has been reduced, even though the two atoms "share" an electron pair.

Oxidation of organic compounds is often synonymous with *dehydrogenation*. Consider the oxidation of hydroquinone, which may be represented as

The reaction may proceed in steps, an acidic dissociation and then withdrawal of electrons.

In this example, protons and electrons depart independently from the molecule being oxidized. In other cases, the mechanism may involve transfer of a proton with an associated electron, i.e., a hydrogen atom, $H\cdot$, or of a proton with an associated pair of electrons, i.e., a hydride ion, $H\!:^-$.

The occurrence of oxidation-reduction reactions demonstrates that molecules and atoms vary in their affinity for electrons. This is found in the electro-

TABLE 12.1

Electrode Potentials of Some Reduction-Oxidation Systems

System	E_0', V*	pH
$OH\cdot + H^+/H_2O$	1.35	7.0
$\frac{1}{2}O_2/H_2O$	0.82	7.0
NO_3^-/NO_2^-	0.42	7.0
Ferricyanide/ferrocyanide	0.36	7.0
$\frac{1}{2}O_2 + H_2O/H_2O_2$	0.30	7.0
Fe^{3+} cytochrome a/Fe^{2+}	0.29	7.0
Fe^{3+} cytochrome c/Fe^{2+}	0.22	7.0
Methemoglobin/hemoglobin	0.17	7.0
Fe^{3+} cytochrome b_2/Fe^{2+}	0.12	7.4
Ubiquinone ox/red†	0.10	7.4
Fe^{3+} cytochrome b/Fe^{2+}	0.07	7.4
Fumaric acid/succinic acid	0.03	7.0
Methylene blue ox/red	0.01	7.0
α-Ketoglutaric acid + NH_4^+/glutamic acid	−0.14	7.0
Oxaloacetic acid/malic acid	−0.17	7.0
Pyruvic acid/lactic acid	−0.19	7.0
Glutathione ox/red	−0.23	7.0
Acetoacetic acid/β-hydroxybutyric acid	−0.27	7.0
1,3-Diphosphoglyceric acid/glyceraldehyde-3-phosphate + P_i‡	−0.29	7.0
Lipoic acid ox/red	−0.29	7.0
Flavodoxin ox/red (clostridial)	−0.31	7.0
$NAD^+/NADH + H^+$	−0.32	7.0
Pyruvic acid + CO_2/malic acid	−0.33	7.0
$H^+/\frac{1}{2}H_2$	−0.42	7.0
Ferredoxin ox/red (spinach)	−0.43	7.5
$O_2/O_2^-\cdot$	−0.45	7.0
α-Ketoglutaric acid/succinic acid + CO_2	−0.67	7.0
Pyruvic acid/acetic acid + CO_2	−0.70	7.0

*The values shown for E_0' (midpoint potential) are the potentials that would be exhibited by a potentiometer interposed between a standard hydrogen electrode and an inert electrode in a solution containing equimolar amounts, at the pH specified, of the oxidized and reduced member of each pair if the latter were electroactive.
†ox/red = oxidized form/reduced form.
‡P_i = inorganic orthophosphate.

motive series of the elements, which quantitatively expresses the tendencies of the elements to gain or lose electrons. The point of reference in electromotive-force (EMF) tables is the standard hydrogen electrode, which is arbitrarily assigned a standard redox potential E_0 of zero. A similar series, compiled for organic substances, expresses their relative affinities for electrons. In this case, the more effective a substance is as a reductant, the more negative its redox potential. A few systems of interest are presented in Table 12.1.

Oxidations Involving Molecular Oxygen

Reduction of oxygen to water requires four electrons. If this process were to occur by successive univalent steps, the intermediates would be superoxide anion ($O_2^-\cdot$), hydrogen peroxide (H_2O_2), and hydroxyl radical ($OH\cdot$). These are reactive species, particularly the radicals $O_2^-\cdot$ and $OH\cdot$, which are thought to be among the mutagenic radicals created by ionizing radiation (Chap. 27). It is fortunate, therefore, that the major pathways for the biological reduction of oxygen yield H_2O_2 or H_2O directly.

There are, however, reactions which effect the univalent reduction of oxygen. In some instances, the superoxide radical thus generated is utilized in a subsequent reaction, but in others the radical is simply liberated from the enzyme surface. Oxygen-utilizing organisms must therefore face the threat posed by the intracellular generation of both $O_2^- \cdot$ and H_2O_2. Enzymic defense mechanisms have evolved to deal with these reactive oxygen compounds. *Superoxide dismutase* (Chap. 16) converts the superoxide radical to hydrogen peroxide by the reaction

$$O_2^- \cdot + O_2^- \cdot + 2H^+ \longrightarrow H_2O_2 + O_2$$

Hydrogen peroxide is eliminated by *catalase* (Chap. 16).

$$H_2O_2 + H_2O_2 \longrightarrow O_2 + 2H_2O$$

Organisms that lack these defenses are restricted to anaerobic environments.

Energy Relations in Oxidative Reactions

Since interest in oxidative reactions derives in large measure from the fact that they yield energy, let us consider some quantitative aspects of oxidative changes in relation to energy changes.

If the reduced form of one system is mixed with the oxidized form of another, reaction proceeds according to the equation

$$AH_2 + B \rightleftharpoons A + BH_2$$

For this reaction, the standard free-energy change, in calories per mole, can be calculated from equilibrium data in the usual manner. K is the equilibrium constant,

$$K = \frac{[A][BH_2]}{[AH_2][B]} \tag{15}$$

Actual determination of K therefore depends on the availability of adequate analytical methods for the various components. When the difference in potential $\Delta E_0'$ between the two reacting systems is large, equilibrium may lie so far in one direction that accurate determination of the final concentration of AH_2 and B may be impossible. The free-energy change associated with the reaction can be calculated, however, from knowledge of the potentials of the two reacting systems by use of the *Nernst equation*

$$E = E_0' + \frac{RT}{nF} \ln \frac{[\text{oxidant}]}{[\text{reductant}]} \tag{16}$$

where $E_0' =$ the potential of the half-reduced system at a given pH and temperature; R is the gas constant 8.315 J/°K/mol; T is the absolute temperature; F is the faraday, 96,500 coulombs (C); and n is the number of electrons transferred per mole (J is the Joule). At 30°C, this simplifies to

$$E = E_0' + \frac{0.06}{n} \log \frac{[\text{oxidant}]}{[\text{reductant}]} \tag{17}$$

The E_0' values for some biologically important systems are listed in Table 12.1. By substituting the expression for K of Eq. 15 in Eq. 16, one obtains

$$\Delta E_0' = \frac{RT}{n\mathrm{F}} \ln K \tag{18}$$

or

$$n\mathrm{F}\,\Delta E_0' = RT \ln K$$

Since $-\Delta G° = RT \ln K$ (Eq. 13),

$$-\Delta G° = n\mathrm{F}\,\Delta E_0' \tag{19}$$

where $\Delta G°$ is the standard free energy of the reaction, and $\Delta E_0'$ is the difference between the E_0' values of two systems. The units of $\mathrm{F}\,\Delta E_0'$ are coulomb-volts, or joules, which are convertible to units of thermal energy, since $4.18\,\mathrm{J} = 1$ cal. The value obtained for $\Delta G°$ is that for the oxidation of 1 mol of reductant.

Consider as an example the oxidation of malate to oxaloacetate by cytochrome c under circumstances such that equimolar concentrations of each of the reactants of the two systems always exist. Since the E_0' values are -0.17 and 0.2 V, respectively (Table 12.1),

$$\Delta G' = -n\mathrm{F}\,\Delta E_0' = \frac{-2 \times 96{,}500 \times [0.2 - (-0.17)]}{4.18} = -17.1 \text{ kcal}$$

The oxidation of 1 mol of malic acid by cytochrome c, under these circumstances, results in the release of 17.1 kcal, which could then be available under physiological circumstances for doing useful work. If the malate were oxidized by O_2, 45.7 kcal would be released, since E_0' for the reduction of oxygen to water is $+0.82$ V (Table 12.1).

In the dynamic steady state of the living cell, the concentrations of each of the components of a given redox reaction system are held constant, e.g., malate, oxaloacetate, and the oxidized and reduced forms of cytochrome c in the example just considered, but rarely are the concentrations of the oxidized and reduced members of a redox pair identical. Hence the actual ΔE for a given reaction may differ significantly from $\Delta E_0'$.

HIGH-ENERGY PHOSPHATE COMPOUNDS

Types of Compounds

It was stated earlier that within cells mechanisms exist whereby the free energy available from oxidative reactions can be utilized to drive endergonic processes. This is accomplished largely by trapping this energy through the formation of a special class of compounds, most of which are anhydrides of phosphoric acid. The standard free-energy change associated with hydrolysis of the diverse phosphate derivatives found in biological systems ranges from about -2 to -13 kcal/mol.

$$\mathrm{R{-}O{-}PO_3^{2-}} + H_2O \rightleftharpoons \mathrm{R{-}OH} + P_i \qquad \text{and} \qquad K = \frac{[\mathrm{ROH}][P_i]}{[\mathrm{ROPO_3^{2-}}]}$$

P_i is the symbol used to designate the mixture of HPO_4^{2-} and $H_2PO_4^{-}$ that will exist at the pH of the system under consideration. pK_a for $H_2PO_4^{-}$ is 6.8. The

smaller values, −2 to −5 kcal/mol, are observed with simple esters, of which glycerol-3-phosphate and glucose-6-phosphate are typical.

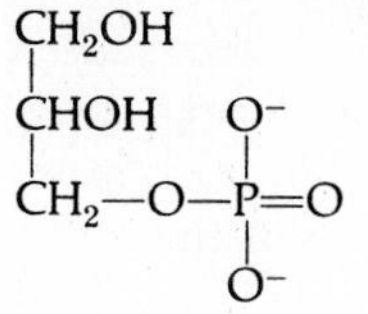
α-Glycerol-3-phosphate

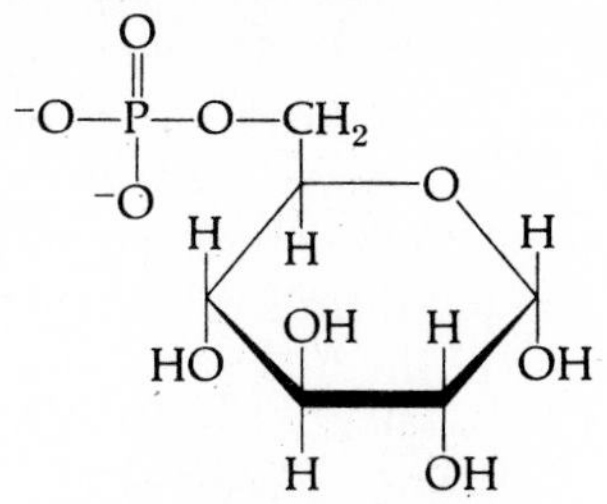
α-Glucose-6-phosphate

Numerous esters of this type occur in intermediary metabolism. A smaller group consists of those organic phosphates whose hydrolysis occurs with a $\Delta G°$ between −5 and −13 kcal/mol. Examples of such high-energy phosphate-containing compounds are the following:

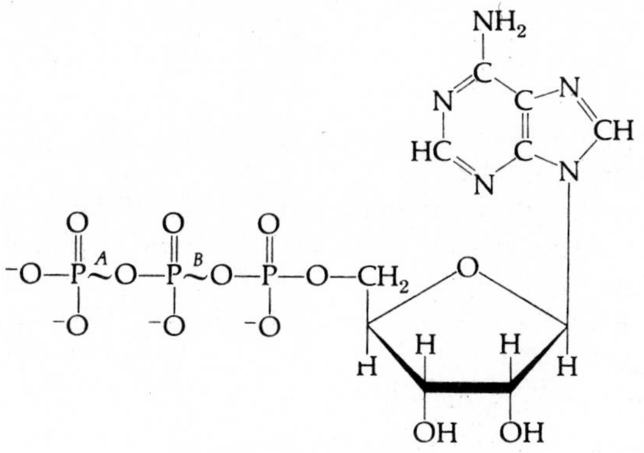
Adenosine triphosphate (ATP)

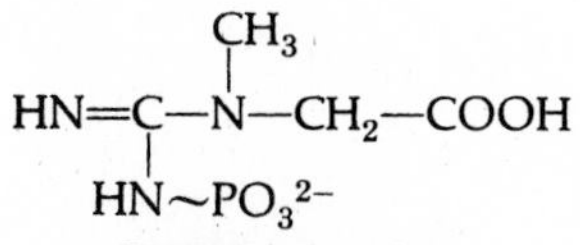
Creatine phosphate

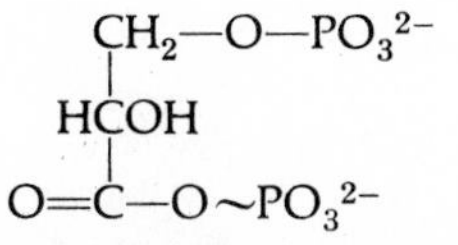
1,3-Diphosphoglycerate

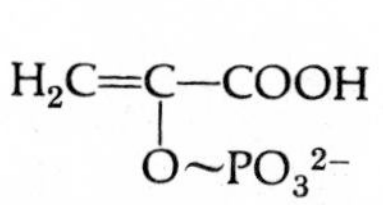
Phosphoenolpyruvate

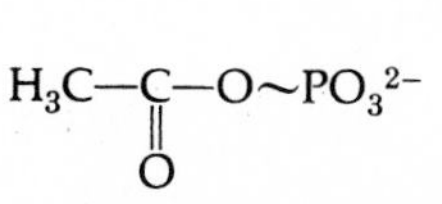
Acetyl phosphate

Several classes of high-energy phosphate compounds are known: acid anhydrides, phosphoric esters of enols, and derivatives of phosphamic acid, $R{-}NH{-}PO_3^{2-}$. The free energy of hydrolysis for phosphoenolpyruvate (−13 kcal/mol at 25°C) is among the highest of all known naturally occurring high-energy phosphate compounds. By convention, ~ denotes the bond whose hydrolysis is accompanied by release of a large amount of free energy. The relatively high potential energy that is made available on hydrolysis is a property of the structure of the phosphate compound as a whole and does not merely reside in the ~P bond ruptured by hydrolysis.

Several factors contribute to the large release of free energy associated with hydrolysis of this group of compounds. In some, such as the anhydrides, the electron-withdrawing properties of the phosphoric group make the electrophilic carbonyl carbon atom of the acyl group less stable. A second contribution arises from the fact that the resonance energy of the hydrolysis products may substantially exceed that of the high-energy compound. The number of resonant forms of creatine + phosphate considerably exceeds that of creatine phosphate; similarly the number of resonant forms of acetate ion + P_i exceeds the number possible in acetyl phosphate. In general, the greater the number of possible resonant forms, the greater the stability of the system. In addition, hydrolysis of many members of this group, at pH 7, results in an increase in charge. Compare the hydrolysis of glycerol and acetyl phosphates.

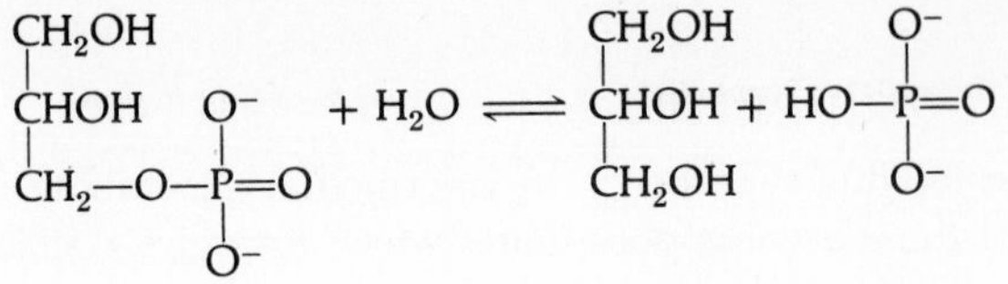

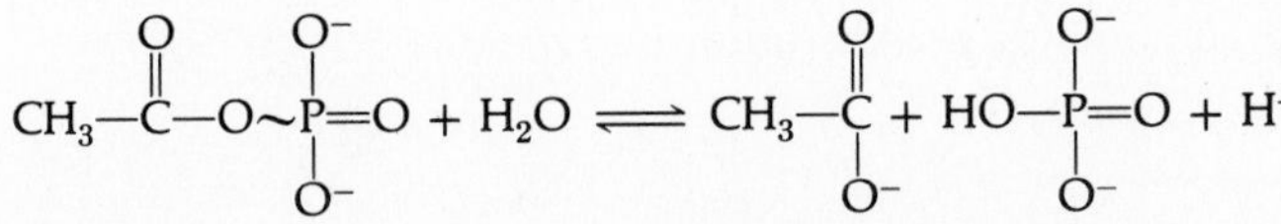

$$
\begin{array}{l}
CH_2OH \\
| \\
CHOH \quad O^- \\
| \qquad\quad | \\
CH_2-O-P=O \\
\qquad\qquad | \\
\qquad\qquad O^-
\end{array}
+ H_2O \rightleftharpoons
\begin{array}{l}
CH_2OH \qquad O^- \\
| \qquad\qquad\quad | \\
CHOH + HO-P=O \\
| \qquad\qquad\quad | \\
CH_2OH \qquad\quad O^-
\end{array}
$$

$$
\begin{array}{l}
\quad O \quad\; O^- \\
\quad \| \quad\;\; \| \\
CH_3-C-O{\sim}P=O + H_2O \rightleftharpoons
\end{array}
\begin{array}{l}
\quad O \qquad\;\; O^- \\
\quad \| \qquad\;\;\; \| \\
CH_3-C + HO-P=O + H^+ \\
\quad | \qquad\qquad | \\
\quad O^- \qquad\quad O^-
\end{array}
$$

Hydrolysis of 1 mol of acetyl phosphate liberates a proton, and its removal by the buffered medium makes a large contribution to the total change in free energy, driving the reaction, as shown above, to the right.

Two high-energy compounds found in virtually all cells are the cyclic nucleotides adenosine 3',5' cyclic monophosphate (cAMP) and guanosine 3',5' cyclic monophosphate (cGMP). Each is synthesized from the respective ribonucleoside triphosphates, ATP and GTP, in reactions catalyzed by plasma-membrane-bound enzymes, *adenylate cyclase* and *guanylate cyclase*.

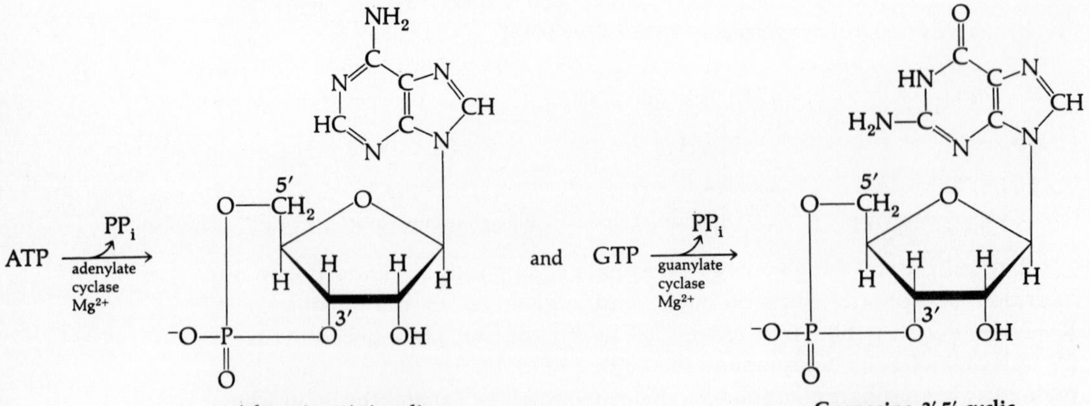

Adenosine 3',5' cyclic
monophosphate (cAMP)

Guanosine 3',5' cyclic
monophosphate (cGMP)

Presumably, the high free energy of hydrolysis of the cyclic nucleotides ($\Delta G° = -11.9$ kcal/mol) reflects strain in their anhydride structure. Reversal of the hydrolytic reaction is prevented and formation of cAMP and cGMP are ensured by the rapid removal of PP_i by the ubiquitous pyrophosphatase (see below). The significance of these compounds, however, lies not in their high-energy character but in the regulatory roles of cAMP and cGMP in metabolism, which will be encountered frequently in subsequent chapters of this book.

Other instances of high-energy compounds, in the sense that the free-energy change accompanying their hydrolysis is in the range -6 to -11 kcal/mol, are found in living systems. Particularly noteworthy are acyl thioesters, i.e., compounds of the general structure shown below, such as the fatty acyl esters of coenzyme A (Chap. 15). Other high-energy classes include aminoacyl esters of the ribose moiety of transfer RNA (Chap. 28), as well as sulfonium compounds of the general structure

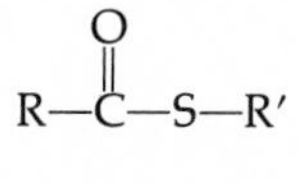

Acyl thioester 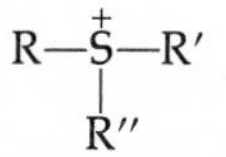   Sulfonium compound

S-Adenosylmethionine (Chap. 23) is an example of such a compound. Another group of high-energy compounds consists of the acyl esters of carnitine (Chap. 20), which are of importance in the transport of fatty acids across mitochondrial membranes. In this case the high free energy of hydrolysis of the acyl ester bond is probably due to the proximity of a positively charged quaternary nitrogen atom.

Types of Reactions

Table 12.2 presents the free-energy changes associated with hydrolysis of diverse compounds of biological interest.

The unique merit of phosphate anhydrides, other than requirements of enzymic specificity, resides in the fact that phosphate confers kinetic stability on thermodynamically labile molecules. Thus, although the magnitudes of $\Delta G°$ for hydrolysis of acetic anhydride, acetyl phosphate, and inorganic pyrophosphate are similar, these compounds are stable in water for a few seconds, several hours, and years, respectively. The fact that ATP is stable in water permits

TABLE 12.2

Free Energy of Hydrolysis of Some Compounds of Biological Interest

Compound	$-\Delta G°$ at pH 7, kcal/mol	Compound	$\Delta G°$ at pH 7, kcal/mol
Acetyladenylate (acetyl AMP)	13.3	Uridine diphosphate glucose	7.6
Phospho*enol*pyruvate	13.0	Sucrose	6.6
Cyclic AMP	11.9	Phosphodiesters	ca. 6.0
Acetyl phosphate	10.5	Glucose-1-phosphate	5.0
Acetoacetyl CoA	10.5	Alanylglycine	4.0
S-Adenosylmethionine	10.0	Glutamine	3.4
Phosphocreatine	9.0	Glycerol-3-phosphate	3.0
ATP, inner bond (*B*)	8.5	Lactose	3.0
ATP, terminal bond (*A*)	8.4	Peptide bond (internal in large polypeptide only)	0.5
Acetyl CoA	7.7		
Palmityl carnitine	7.7		

utilization of the energy of oxidation to drive endergonic reactions, since its spontaneous hydrolysis would be wasteful.

It should be noted that the various values for free-energy changes cited above are those for $\Delta G°$, the free-energy change when each of the reactants is in the standard equimolar state. The actual free-energy change is defined by

$$\Delta G = \Delta G° + RT \ln \frac{[C][D]}{[A][B]}$$

If the reactants in a biological system are maintained in a steady state in which [C][D] is not equal to [A][B], the true value for ΔG may be significantly greater or less than $\Delta G°$.

Let us now examine the mechanism by which an exergonic reaction can drive an endergonic process. Consider the synthesis of an ester and the hydrolysis of ATP to AMP + PP_i.

$$RCOOH + HO—R' \rightleftharpoons RCOOR' + H_2O \qquad \Delta G° = +4.0 \text{ kcal} \quad (20)$$

$$ATP + H_2O \rightleftharpoons AMP + PP_i \qquad \Delta G° = -8.4 \text{ kcal} \quad (21)$$

NET: $RCOOH + HO—R' + ATP \longrightarrow RCOOR' + AMP + PP_i \quad \Delta G° = -4.4 \text{ kcal}$

Under ordinary chemical circumstances, the occurrence of reaction (21) would be without influence on reaction (20), which would not proceed in the absence of the requisite 4 kcal/mol. The energy released in reaction (21) can be utilized to drive Eq. (20), however, if they are coupled by way of a common intermediate, as in the following hypothetical case:

$$RCOOH + ATP \rightleftharpoons RCOO—AMP + PP_i \qquad \Delta G° = +2.5 \text{ kcal} \quad (22)$$

$$RCOO—AMP + HO—R' \rightleftharpoons RCOOR' + AMP \qquad \Delta G° = -6.0 \text{ kcal} \quad (23)$$

NET: $RCOOH + HO—R' + ATP \longrightarrow RCOOR' + AMP + PP_i \quad \Delta G° = -3.5 \text{ kcal}$

Hydrolysis of the two anhydride bonds of ATP proceeds with essentially identical changes in free energy. In the example given, it is the bond between the pyrophosphate and adenylate moieties of ATP that is ruptured, so that pyrophosphate, rather than orthophosphate, appears as the final product. From the relatively small change in ΔG, it will be recognized that synthesis of the organic ester by this mechanism can proceed but is not strongly favored. However, synthesis is ensured by an additional factor. The major fate of pyrophosphate is hydrolysis to orthophosphate catalyzed by *pyrophosphatases* present in all cells. The $\Delta G°$ for pyrophosphate hydrolysis is -7 kcal; accordingly, hydrolysis of the pyrophosphate effectively renders ester synthesis irreversible. Processes in which pyrophosphate hydrolysis renders a system irreversible include the synthesis of nucleotides, polynucleotides, peptide bonds, and S-adenosylmethionine and fatty acid activation, all considered in later chapters.

Enzymes that catalyze transfer of phosphate from ATP to an acceptor are designated *kinases* and may be considered in two categories. Kinases catalyzing transfer among high-energy phosphate compounds can operate readily in both directions and actually do so in metabolism, e.g.,

$$ADP + \text{creatine phosphate} \rightleftharpoons ATP + \text{creatine} \qquad \Delta G° = -1.5 \text{ kcal}, K = 10$$

Phosphate transfer with formation of relatively low-energy compounds may be expected to proceed significantly in the forward direction only.

$$\text{ATP} + \text{glucose} \longrightarrow \text{ADP} + \text{glucose-6-phosphate} \qquad \Delta G^\circ = -4.5 \text{ kcal}, K = 4000$$

From these facts a basic concept emerges: To supply energy for endergonic biological processes, the respiratory process must somehow be coupled with the synthesis of ATP. The latter is the "unit of currency" in metabolic energy transformation; i.e., **ATP is the immediate source of energy for most endergonic biological systems.**

The equation describing the overall oxidation of glucose can now be stated as

$$C_6H_{12}O_6 + 6O_2 + x P_i + x \text{ADP} \longrightarrow 6CO_2 + 6H_2O + x \text{ATP} + \text{unavailable energy}$$

The fraction of the total ΔG for the oxidation of glucose or any other metabolic fuel employed for ATP synthesis represents the true efficiency of the cellular respiration process. ΔG° for ATP synthesis is about $+8.4$ kcal/mol. Under physiological conditions, the true ΔG exceeds $+12$ kcal/mol. Since ΔG° for complete oxidation of 1 mol of glucose is -686 kcal, this could *potentially* provide energy for the synthesis of about 50 mol of ATP from ADP and P_i. As we shall see in Chap. 15, the actual yield is 38 ATP per molecule of glucose. If the combustion of a molecule of glucose is to result in the formation of 38 ATP molecules, the total oxidation process must be fractionated into many lesser steps. How this is accomplished is described in Chap. 15.

ENERGY REQUIREMENTS

Transformation of energy necessarily accompanies the chemical reactions that make possible the characteristic properties, e.g., movement, reproduction, growth, and response to stimuli, which distinguish living cells from nonliving structures. The energy processed by the total metabolism, released by all chemical transformations in the animal and derived ultimately from the oxidation of foodstuffs, must appear either as heat or as external mechanical work. Even during muscular activity, the major portion of the energy appears as heat because of the relative inefficiency of the muscles as mechanical devices. During rest, practically all this energy appears as heat.

Subsequent chapters will present the details by which biological oxidations normally proceed only at the rate at which the free energy liberated (ΔG) is required for the performance of useful work. The latter takes many forms. Living cells transduce chemical potential energy into other forms of energy, viz., chemical, mechanical, electric, osmotic, and, in some organisms, even light. Thus the free energy derived from oxidation of glucose can be utilized for synthesis of proteins, nucleic acids, or steroids; for contraction of muscles, conduction of the nervous impulse, or generation of an electric charge; for secretion of hypertonic urine or maintenance of a large concentration gradient for Na^+ and K^+ inside and outside of cells; and, as in the firefly, for production of light.

If all such processes are summed in a 70-kg adult, about 2000 kcal is generated and released each 24 h. Since neither body weight, structure, nor composition changes significantly over this period, all this energy appears as heat, regardless of any intermediary transformations, except for the work that was done upon the environment, such as lifting weights, etc. Since the energy lost as heat is irretrievably dissipated to the environment, there is a continuing requirement for new external sources of energy, viz., foodstuffs that can be oxidized. The total process, which incidentally provides the heat for maintenance of body temperature in an environment cooler than 37°C, is not, as might appear, wasteful. It is the sum of these activities, made possible by transient use of the free energy of oxidation, that makes possible the highly ordered structures and vital activities of the living organism.

Caloric Values of Foodstuffs

The calorie used in metabolic studies of humans is the kilocalorie. The energy derived from oxidation of compounds, including those of food, can be measured in a bomb calorimeter. The overall energy released accompanying a chemical reaction ΔH is independent of the reaction mechanism. The ΔH for the reaction

$$\text{Glucose} + 6O_2 \longrightarrow 6CO_2 + 6H_2O$$

is identical whether it occurs in a bomb calorimeter or a human being.

For carbohydrate and lipid, the values are similar whether the foodstuff is burned inside or outside the body. However, the in vivo value for protein (4.1 kcal/g) is less than that obtained in the bomb calorimeter (5.3 kcal/g). This difference is due to the fact that the nitrogen of proteins is not oxidized, physiologically, but is excreted mainly as urea.

Since foodstuffs include mono- and polysaccharides, short- and long-chain fatty acids, saturated and unsaturated fatty acids, etc., the caloric values of individual members of each of these major classes are different. Thus, glucose yields 3.75 kcal/g, whereas glycogen gives 4.3 kcal/g. Again, animal proteins yield higher values than plant proteins, and most animal lipids liberate 9.5 kcal/g, although butter and lard give 9.2 kcal/g. Therefore, the caloric values for the three classes of foodstuff represent averages, viz., 4.1, 9.3, and 4.1 kcal/g for carbohydrate, lipid, and protein, respectively. If allowance is made for the possibility of incomplete digestion and/or absorption, the values can be rounded off to 4, 9, and 4 kcal/g.

Ingestion of food is followed by an increase in heat production above normal basal level (resting, postprandial state). This occurs immediately after eating and can be related to the digestion and absorption of foodstuffs. Additional heat production may then result as a consequence of subsequent metabolic transformations of absorbed products.

The ingestion of protein causes the greatest increase in heat production. This effect, termed the *specific dynamic action*, is the extra heat produced by the organism, over and above the basal heat production, as a result of food ingestion. In the case of protein, the specific dynamic action is approximately 30 percent, for carbohydrate 6 percent, and for lipid 4 percent, respectively, of the energy value of the food ingested. Thus, ingestion of 25 g of protein, equivalent to 100 kcal, leads to 30 kcal of extra heat production over the basal level. These calories are wasted as heat; only 70 kcal of potentially useful energy can be derived from the 25 g of protein. It is essential, in calculating the caloric value of a diet, to make provision for the specific dynamic effect.

The extra energy released incident to the metabolism of foodstuffs in the liver is apparently responsible in large part for their specific dynamic action.

Inasmuch as the first law of thermodynamics is obeyed by living organisms, the balance that obtains between caloric intake and energy expenditure is the prime factor, under normal circumstances, that determines whether weight gain or weight loss occurs over a period of time. Weight gain in relation to lipid metabolism is discussed in Chap. 20. Aspects of human daily caloric requirements are indicated in Chap. M21.

Basal Metabolism

It is not possible to assess, at a specified time in metabolism, the relative significance of the energy relationships between foodstuffs, heat production, metabolic energy, and stored energy. However, the significance of food and of heat produced as a result of work can be delimited. This is accomplished by measurement of energy exchange in a postabsorptive period and in the resting state, thus minimizing energy utilization due to work on the environment. Under these controlled conditions, heat production becomes the major means of energy loss from the body, and stored energy can be the only source of this heat. Since energy cannot be created or destroyed, the decrease in stored energy becomes equal to the heat loss, and measurements of the latter afford an estimate of the total metabolism. This measurement, made under resting, postabsorptive conditions, is the *basal metabolism;* i.e., it reflects the energy requirements of those cellular and tissue processes due to continuing activities of the organism, e.g., the metabolic activity of muscle, brain, kidney, liver, and other cells plus the heat released as a result of the mechanical work represented by contraction of the muscles involved in respiration, circulation, and peristalsis. The basal metabolism constitutes approximately 50 percent of the total energy expenditure required for the diverse activities of a normal 24-h day of relatively sedentary individuals. The *basal metabolic rate* (BMR) is not the minimal metabolism necessary for mere maintenance of life, since during sleep the metabolic rate may be lower than the basal rate.

From the volume of O_2 consumed by an individual, corrected to standard temperature and pressure, it is possible to calculate heat production by use of the value 4.825, the caloric equivalent of 1 liter of O_2 for a respiratory quotient (see below) of 0.82.

The basal metabolism is usually given in terms of *kilocalories per hour* (kcal/h). In practice, the basal metabolic rate is determined over a 10- to 15-min period and expressed per square meter of body surface.

Many factors affect basal metabolism, e.g., body size, age, sex, climatic conditions, diet, physical training, drugs, etc. The basal metabolism may deviate from normal values in a variety of pathological states, and in certain of these its measurement provides a useful diagnostic tool (Chap. M14).

There are four important factors in heat loss from the organism: (1) the temperature difference between the environment and the organism, (2) the nature of the surface that radiates the heat, (3) the area of that surface, and (4) the thermal conductance of the environment. Under the conditions of determination of the basal metabolism, surface area is the most important of these factors, and, under similar physiological conditions, the basal metabolism of various mammals is proportional to surface area. Although the *heat production per kilogram* may vary widely among various species and is *inversely* related to *body weight*, the *heat production per square meter of body surface is essentially constant.*

The Respiratory Quotient

Metabolic heat production is the consequence of the ultimate oxidation of foodstuffs by atmospheric O_2 with production of CO_2. The magnitude and nature of this gaseous exchange vary with the type of foodstuffs undergoing oxidation. The relationship between O_2 consumption and CO_2 production can be calculated from the stoichiometry of the equations for the oxidation of carbohydrate, lipid, and protein, respectively. For all carbohydrates, the molar ratio of CO_2 produced to O_2 utilized is 1. This ratio is the *respiratory quotient* (RQ). For lipids, the average RQ is 0.70. The fact that the value is less than 1 reflects the highly reduced nature of fatty acids compared with carbohydrates. Therefore, more O_2 must be consumed in the oxidation of lipid, per mole of CO_2 produced, than with carbohydrate. During fasting, when energy production is derived almost entirely from depot lipids, the RQ approaches 0.71. Conversely, when a marked degree of conversion of carbohydrate to lipid occurs, the RQ may even exceed 1.0. The average RQ for protein is 0.80.

MECHANISMS REGULATING METABOLISM

General Aspects of Metabolic Reaction Sequences

1. The synthesis of each of the compounds required for the life of the cell, such as amino acids, purine and pyrimidine nucleotides, and steroids, is accomplished by a series of specific, consecutive enzymic reactions, called a *sequence* or *pathway*. For each pathway the starting compound is one of the substances that arises in the metabolism of the ubiquitous carbohydrates or fatty acids.

2. In order to provide a substance for use by the cell, a metabolic pathway must be essentially irreversible; i.e., it must proceed with a substantial release of free energy. Although there may be several freely reversible reactions in a pathway, it is the essentially irreversible steps that render the process unidirectional.

3. There are numerous instances in which two metabolites are interconvertible. In almost all such instances that are physiologically meaningful, i.e., if in metabolism there is a requirement for conversion of A to B and at other times of B to A, this is made possible by reaction sequences that are totally or partially independent (Fig. 12.1) and each of which,

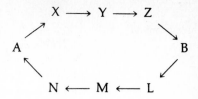

Figure 12.1 Independent reaction sequences for conversion of A to B and B to A.

overall, proceeds with a large negative ΔG. If these two sequences are also independently controlled, each can be utilized for differing metabolic functions.

4. In each metabolic pathway there is a *committed step*, the reaction producing the first metabolite which may have no role other than to serve as an intermediate in the biosynthesis of the end product of the sequence of reactions, e.g., reaction step c → 1 in the sequence shown in Fig. 12.2. In most instances the committed step proceeds with a large loss in free energy so that the reaction is essentially irreversible. It is apparent that a metabolic control intended to restrict the formation of end product would function most satisfactorily if it governed the committed step. Inhibition later in the sequence would result in accumulation of intermediates such as l, m, or n, which have no other roles in metabolism and might even be deleterious to the cell if they were to increase significantly in concentration. Synthesis of unneeded compounds wastes energy.

Control of the committed step is an almost invariant attribute of metabolism; control of intermediary steps is exercised less frequently. Although all five types of regulation to be cited below are known in mammalian systems, much information has been derived from studies of bacteria. They lend themselves readily to investigations of control mechanisms and participate in diverse reaction sequences that do not occur in mammals, such as synthesis of all the α-amino acids present in proteins.

Types of Regulation

Each individual reaction proceeds at a rate commensurate with the requirements of the cell, whether to synthesize a coenzyme, a complex lipid for the cell membrane, a nucleic acid, or a protein, or to provide the energy needed for these diverse endergonic activities. These phenomena occur at greatly differing rates,

Figure 12.2 Formulation of a synthetic pathway from glucose to an end product. a, b, and c may be intermediates in glucose metabolism common to many metabolic pathways. The reaction c → 1 is the committed step since 1, m, and n are intermediates with no metabolic role other than to serve as precursors of the final product of the pathway.

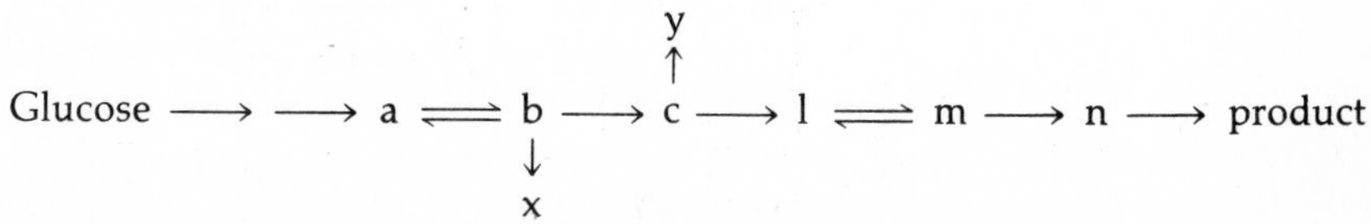

$$\text{Glucose} \longrightarrow \longrightarrow a \rightleftharpoons b \longrightarrow c \longrightarrow l \rightleftharpoons m \longrightarrow n \longrightarrow \text{product}$$

(with y above and x below intermediate b)

commensurate with their individual roles in the life of the cell. These specific rates are determined by *regulatory,* or *control* mechanisms, of which five categories are noteworthy.

Regulation of Entry of Metabolites into Cells Few substances other than water cross membranes by simple diffusion. Two types of processes effect transfer of substances across the membrane. The concentrations of many low-molecular-weight soluble metabolites are higher in cells than in the extracellular blood or lymph. Thus, entry of such metabolites into cells requires their transfer against a concentration gradient. *Active transport* therefore is a process for which ΔG is positive and energy, as ATP, is required for its functioning. In other instances, the translocated material moves inward with its concentration gradient, viz., ΔG is negative. However, such *passive transport* is usually made possible by specific mechanisms in the membrane. These transport systems not only ensure the constancy of internal cellular composition but also function in transport phenomena across the membranes of intracellular organelles. Transport processes are discussed in Chap. 14.

All other regulatory mechanisms operate by influencing either the amount of a given enzyme in the cell or the catalytic capacity, that is, V_{max} or K_m, of the enzyme molecules already present. In a general way, the intracellular concentrations of various enzymes are roughly proportional to the rate at which reactions proceed. Thus, the enzymes that make the oxidation of glucose possible are present in far greater concentration than those that catalyze synthesis of coenzymes.

Repression of Enzyme Synthesis by an End Product of a Metabolic Sequence (Chap. 22) This is a coarse mechanism for control of metabolism. Synthesis of the enzymes required for histidine formation is repressed by the presence of histidine in the medium of a bacterial culture. Newly formed cells will then contain decreasing quantities of these enzymes, but the enzymes initially present at the time of histidine addition can continue to function and are destroyed slowly. In mammalian systems, in the presence of a repressor, several days may be required for the enzymic level to diminish significantly as a result of its own continuing degradation.

An interesting problem arises in the case of branched metabolic pathways, which result in synthesis of more than one desired product, e.g.,

$$A \longrightarrow B \longrightarrow C \longrightarrow E \longrightarrow F \longrightarrow G$$
$$\searrow$$
$$D$$

where D, E, and G might all be useful end products. In bacteria, various solutions to this problem have been observed. Thus there may be several genetically independent enzymes for the step B → C that can be repressed by sufficient concentrations of D, E, and G, respectively. Alternatively, there may be one enzyme for this step repressible by one of the products and another that is not repressible by any.

Induction of Formation of One or More Enzymes by Substrates This is generally very rapid in bacteria and may occur within a few minutes after addition of the substrate to the medium in which the organism is cultured. Details of the mechanism of this phenomenon in microorganisms are discussed in Chap. 28. In mammals similar induction of a specific enzyme occurs over several hours or days in response to the inducer, representing an important but relatively coarse metabolic control.

Modulation of the Activity of an Enzyme Already Present This serves as the fine control mechanism for regulating metabolic events since it permits instantaneous response to changing intracellular environment. Feedback inhibition of the enzyme responsible for the committed step in a pathway by the product of the pathway is a common form of control of biosynthetic processes. The advantages to the cell of such self-regulation are evident. Allosteric modification of enzymic rates has been considered earlier (Chap. 10). Of particular significance is the typical sigmoid curve relating velocity and either substrate or modifier concentration (Fig. 10.11), since at low concentration little effect is evident, whereas in the intermediate concentration range, below saturation, relatively small changes in concentration result in large changes in velocity. Hence, allosteric feedback control can serve as an efficient regulatory mechanism. Activation or inactivation of an enzyme may also result from covalent modification, e.g., by phosphorylation and adenylylation reactions. Examples of these mechanisms for increasing and modulating the activity of an enzyme will be encountered in many chapters.

Stimulation by a Metabolite of Enzymes That Function in Pathways That Achieve Its Utilization This type of metabolic control is present in mammalian as well as bacterial cells. Many examples of these control devices in the regulation of metabolism will be given in the chapters that follow.

Regulation of Metabolism by Extracellular Agents

Each of the above regulatory mechanisms is effective at the cellular or subcellular level. However, in a multicellular animal, one group of cells, e.g., the liver, may produce a compound that is then utilized elsewhere, e.g., in skeletal muscle. If supply is to be regulated by demand, the producing cells require information concerning the metabolic state of the consumers. This information is provided by *chemical messengers,* e.g., hormones (Part M3), neural transmitters (Chap. M7), and other cell-produced mediators, e.g., prostaglandins (Chap. M13), that may transmit their "signals" to cells other than those in which they are made. On arrival at the target cell, the extracellular signals are transformed into intracellular actions. This transition occurs by binding the messenger to a specific receptor in the target cell (Chap. 14), causing activation of membrane-bound and/or intracellular enzymes, with resultant increased synthesis of compounds that serve as second messengers.

A well-defined second messenger is adenosine 3′,5′ cyclic monophosphate (cAMP). The metabolic role of guanosine 3′,5′ cyclic monophosphate (cGMP) is less clear. The synthesis of these nucleotides by specific membrane-bound *cyclases* has been described (page 252).

cAMP and cGMP often appear to undergo inverse alterations in their intracellular concentrations in response to the same stimulus. Thus, cellular proliferation is frequently preceded by an increased intracellular [cGMP], whereas augmented cAMP formation generally inhibits cell division and growth (Chap. M20). Although often the concentration of only one or the other cyclic nucleotide is measured, the *ratio* of their concentrations appears to be more significant to physiological function. This relationship probably reflects in part the fact that each of the cyclic nucleotides increases the activity of different types of protein kinases (see below). The influence of the cyclic nucleotides in the regulation of metabolic phenomena will be described in many of the succeeding chapters.

The intracellular [cAMP] and [cGMP] are regulated also by a second group of enzymes, the *cyclic nucleotide phosphodiesterases*, which catalyze inactivation of cAMP and cGMP by hydrolysis to 5′-AMP and 5′-GMP, respectively,

$$\text{cAMP (or cGMP)} + H_2O \xrightarrow{\text{Mg}^{2+}} \text{5′-AMP (or 5′-GMP)}$$

The concentration of cAMP in animal tissues is in the range of 0.1 to 1.0 μmol per kg of wet weight.

Free cAMP probably has a short half-life because of the efficiency and capacity of cAMP phosphodiesterase. The concentration of cAMP in resting rabbit skeletal muscle is 0.2 to 0.3 μM, approximately that required for half-maximal activation of cAMP-dependent protein kinase (see below). Thus, alterations in cAMP concentration regulate protein kinase activity.

Protein kinases link cAMP and cGMP to cell metabolism. These kinases catalyze the transfer of the γ-phosphate of ATP to serine or threonine hydroxyl groups of a wide variety of acceptor proteins, including glycogen synthase and phosphorylase b kinase (Chap. 18), ribosomal proteins (Chap. 29), as well as membrane proteins of subcellular organelles such as mitochondria (Chap. 15). The altered enzymic activity of the phosphorylated enzymes is the metabolic process "regulated" by the agent whose binding to the membrane receptor initiated this series of events. The nonenzymic substrates of the protein kinases, e.g., ribosomal proteins, may exert their influences by other mechanisms, viz., reactions affecting rates of synthesis of specific proteins (Chap. 29).

Many protein kinases are intracellular enzymes present in the cytosol. However, localization of both kinase activity and substrate on the outer plasma membrane of adipocytes, fibroblasts, and viral-transformed mammalian cells has been described. Membrane localization of both kinase and its protein substrate would allow a phosphorylation-dephosphorylation mechanism to operate in the interaction of the cell with its environment.

EXPERIMENTAL APPROACHES TO THE STUDY OF METABOLISM

Many techniques have been devised to study the chemical reactions that occur in living systems. In all experimental science, it is necessary to apply some disturbance to the system under observation and to measure the effect of that disturbance. Understanding of metabolism has been obtained from experiments in which the applied disturbance has been minor, e.g., the addition of a minute amount of an isotopically labeled nutrient to the diet of an animal, to experiments in which an isolated organ has been studied or, at the simplest level, the reaction catalyzed by a pure enzyme.

Levels of Organization of Experimental Systems

The Intact Organism The rates at which materials are delivered to, and removed from, a given tissue are affected or modified by the activities of other tissues. Many processes are regulated by products discharged into the bloodstream at remote points. Membranes serve to limit the rates at which substrates enter or leave certain compartments. Hence, the rate at which the normal liver in situ performs various reactions may differ from the rate at which the same organ, isolated and perfused, sliced or minced, conducts these processes. Although the intact animal offers many experimental difficulties, there are several techniques for exploring what is happening. These include administration of diets deficient in a normal nutrient and addition by dietary or parenteral route of some metabolite labeled in such a fashion that the subsequent distribution of the label can be studied. A variety of surgical procedures also gives the experimenter access to a specific tissue or body fluid.

Laboratory Animals The nature of the problem may determine which species is most useful. Nutritional data relevant to humans have been obtained with mice, rats, dogs, guinea pigs, chickens, and monkeys. Studies requiring large numbers of animals have generally employed mice and rats, which breed well under laboratory conditions, are omnivorous, and have a rapid rate of growth and development. Their relatively short life spans, viz., under 3 years, permit study of several generations of animals. Inbreeding is readily possible in these species; hence, pure strains can be established and the significance of genetic factors examined.

Naturally Occurring and Induced Metabolic Alterations Surgical extirpation of an organ is one of the oldest approaches to studies of metabolism. For example, the role of the pancreas in the etiology of *diabetes mellitus* was discovered in studies of the surgically depancreatized dog.

Several metabolic disorders were termed by Garrod *inborn errors of metabolism* because each is present throughout life and is hereditary. The inborn errors listed by Garrod in 1908 were cystinuria, alkaptonuria, pentosuria, and albinism. Hundreds of metabolic alterations of humans have since been described that have a genetic basis (Chap. 31). Individuals with such conditions provide an experimental approach to problems of metabolism, particularly the elucidation of metabolic pathways. These hereditary errors of metabolism arise as a consequence of mutations. In this sense, human mutants resemble microorganisms in which a mutation can be discerned as a consequence of the organism's inability to conduct a particular metabolic transformation. Such bacterial mutants have been singularly useful in elucidating metabolic pathways.

Organ Perfusion Perfusion of an isolated organ in vitro makes possible introduction of a substance and, by analysis of the fluid emerging from the organ, evaluation of its fate. In practice, the organ is placed in a closed system in which a suitable oxygenated fluid circulates under positive pressure. This approach has been used with liver, heart, kidney, and small intestine and has contributed to knowledge of the metabolic roles of these organs.

The usefulness of this technique has been greatly enhanced by development of procedures in which an *isotopically labeled metabolite* is perfused for a brief period and metabolism halted abruptly by instantaneous freezing devices ("freeze-clamp"), thus permitting determination of the isotope in a variety of metabolites within seconds after its introduction and at subsequent fixed intervals.

The Sliced Organ Liver, kidney, brain, and other tissues, cut into slices approximately 50 μm thick, can be maintained in an appropriate fluid for several hours. Most cellular constituents remain within the cells. Although reaction rates in the cell of a sliced liver may deviate from normal rates, this technique has proved extremely useful. There is complete experimental control of the composition of the bath fluid and of the gas phase with which it is equilibrated. Addition of possible precursors to the bath fluid may help delineate detailed pathways of metabolism.

Tissue Cultures Cells and tissues can be grown in accurately defined media. This technique permits examination of biochemical processes in *growing* animal cell populations and in successive generations of cells. Many different cell types have been grown in tissue culture. Tissue cultures have been used for studying viral replication (Chap. 30).

Studies with Microorganisms Microorganisms have relatively simple requirements for growth and development, they reproduce rapidly, and can be grown in large quantities. Thus, microorganisms provide sources for preparation of individual enzymes and for study of metabolic transformations. In addition, mutant strains of microorganisms can easily be obtained experimentally. The contributions of studies with microorganisms to our understanding of biochemical genetics are considered in Chaps. 27 and 28.

Cell-Free Systems Cell membranes can be disrupted by subjecting a suspension of cells in a suitable isotonic medium, e.g., 0.25 M sucrose, to ultrasonic vibration or by utilizing of mechanical devices that disrupt cells and yield broken cell preparations, loosely termed *homogenates*. These preparations have few unbroken cells and are suspensions of nuclei, mitochondria, microsomes, other cellular organelles, and fragments of plasma membranes, as well as the soluble phase of the cytoplasm, the *cytosol*. Differential centrifugation at low temperatures separates the individual fractions of the suspension, permitting assessment of the metabolic roles of diverse cellular organelles and subcellular particles.

The Enzyme in Solution Availability of a purified enzyme permits studies of its properties in detail. The thermodynamic and kinetic characteristics of the reaction can be established and the effects of inhibitors and activators can be ascertained.

Metabolism of Tissue Slices, Minces, and Broken-Cell Preparations Studies of the metabolism of tissue slices, using a manometric technique which measures quantitatively changes in gas volume or pressure, permits study of small segments of a particular tissue by determining the rate of O_2 utilization and CO_2 production (the respiratory quotient, page 258). In addition, the rate of utilization of a substrate added to the medium and the nature and amount of metabolites produced can be determined. This technique has been used with minces of tissue and cell-free preparations of tissues.

Application of Isotopic Tracers The availability of isotopes permits labeling of molecules in order to follow the metabolic pathways of specific portions of their structures. The labeling technique is particularly valuable for metabolic studies since the administered molecules are indistinguishable from the same molecules already present in the body, although differential rates of reaction between isotopic compounds are observed.

The isotopes most frequently used are those of the common elements of organic compounds, viz., H, N, C, S, P, and O. In addition, isotopes of I, Na, K, Fe, and Ca have been utilized. Except for the stable isotopes of H (^{2}H), N (^{15}N), and O (^{18}O), unstable isotopes that emit radiation as they decay (radioactive isotopes) are most generally employed. The quantitative measurement of radioactivity is technically simpler than is the estimation of a stable isotope. The availability of radioactive isotopes with relatively long half-lives of decay makes possible their use in long-term experiments. The sensitivity of methods for measuring radioactivity, when combined with high-resolution separation techniques (Chap. 9), has provided powerful tools for the identification and estimation of compounds.

In some experiments, the isotopic atom may be a constituent of a simple molecule: $^{14}CO_2$, ^2H_2O, ^3H_2O, $^{24}NaCl$, $NaH_2{}^{32}PO_4$, $K^{131}I$. In other experiments, labeled organic compounds must be prepared by organic synthesis or biosynthetically. Thus, isotopic serum albumin can be prepared by inclusion of a labeled amino acid in the diet of an animal and subsequent isolation of albumin from the animal's plasma. Similarly, ^{14}C can be incorporated into glucose by allowing photosynthesis (Chap. 19) to proceed in an atmosphere of $^{14}CO_2$, with subsequent hydrolysis of the starch that accumulates in the leaves. Not infrequently, it is desirable to include more than one isotopic label in the same substance. Thus, for study of the fate of both the carbon skeleton and the nitrogen atom of the amino acid glycine, one can synthesize ^{14}C glycine and ^{15}N glycine separately and by simply mixing the two products obtain what is, in effect, a doubly labeled material. Compounds with more than one isotopic label can also be synthesized by methods that result in incorporation of more than one isotope into a single molecule.

Isotopes are widely used in the study of the *precursor-product relationship,* i.e., to determine whether compound A is converted into B. Isotopic compound A is administered; compound B is purified and then analyzed for isotope. Application of this technique has been particularly useful in demonstrating reactions that had previously been susceptible to study only in a simpler system. By degradation of the product and determination of the distribution of isotope among its atoms, one can often procure information about the mechanism of the transformation.

The isotopic technique has been extremely important in the *analysis of rates of processes,* particularly in the intact animal. The quantity of any tissue constituent may be reasonably constant in the adult animal, but this constancy may result from a balance between rates of synthesis and degradation. Before the advent of the isotopic technique, no satisfactory method was available for measurement of these rates. Two approaches have been used. The body store of a given compound is labeled in a preliminary period by administration of the labeled material, and the subsequent disappearance of isotope is followed. Alternatively, a labeled precursor of the material is administered and a study made of the appearance of isotope in the product. From the rates of change in isotope concentration, the rates of synthesis and destruction can be calculated. From studies of this type, the concept of continuous *turnover* (synthesis, degradation, and replacement) of certain body constituents, reflecting a dynamic steady state even at constant composition, was evolved.

Another application of isotopes relates to the problem of *anatomical distribution.* The isotopic material is administered, and from subsequent analyses of various tissues, the distribution of the isotopic atom can be ascertained. With radioactive isotopes an additional tool is available, viz., autoradiography. In this procedure photographic film is applied to a cut section of a tissue. After adequate exposure and development, portions of the film that were close to radioactive areas in the tissue will have darkened. With isotopes of satisfactory radiation characteristics this technique can permit resolution at the intracellular level.

Isotopic labeling techniques have also found great use in quantitative measurements of substances present in tissues and fluids in very small quantities. By combining immunologically specific techniques with isotopic labeling, many blood constituents, e.g., protein and polypeptide hormones, as well as low-molecular-weight metabolites, can be estimated. This approach, termed *radioimmunoassay,* is based upon the production in an experimental animal, e.g., rabbit, of an antibody by injection of either an antigenic protein, e.g., a polypeptide hormone, or a low-molecular-weight substance, in itself nonantigenic but converted to a *hapten* by covalent linkage to a protein carrier (Chap. M2). The antibody formed, directed against either the administered antigen or the hapten, can then readily be labeled, generally with ^{125}I. The labeled antibody will combine specifically with its antigen in the biological fluid, and the amount of radioactivity added, compared to that bound, can then be estimated. Alternatively, the labeled antigen can be utilized to displace antigen or hapten in the biological fluid after treatment of the latter with excess unlabeled antibody (Chap. M11). The amount of isotope bound is readily estimated by reference to calibration curves obtained utilizing pure antigen in varying concentrations with its respective antiserum. These techniques permit quantitative determination of nanogram quantities of substances and, in addition, have provided solutions to problems such as the rate of disappearance (half-life) of an administered substance, the rate of degradation of substances within the vascular system, and the localization of administered substances in target tissues, including extent of binding to specific receptors.

REFERENCES

Books

Birnie, G. D. (ed.): *Subcellular Components: Preparation and Fractionation,* Butterworths, London, 1976.

Bray, H. G., and K. White: *Kinetics and Thermodynamics in Biochemistry,* 2d ed., Academic, New York, 1966.

Garrod, A. E.: *Inborn Errors of Metabolism,* reprinted with supplement by H. Harris, Oxford, Fair Lawn, N.J., 1963.

Klotz, I. M.: *Energy Changes in Biochemical Reactions,* Academic, New York, 1967.

Lehninger, A. L.: *Bioenergetics,* 2d ed., Benjamin, New York, 1972.

Umbreit, W., R. H. Burris, and J. F. Stauffer: *Manometric Techniques and Related Methods for the Study of Tissue Metabolism,* 4th ed., Burgess, Minneapolis, 1964.

Review Articles

Eagle, H.: Metabolic Studies with Normal and Malignant Cells in Culture, *Harvey Lect.* **54:**156–175 (1958–1959).

George, P., and R. J. Rutman: The High Energy Phosphate Bond Concept, in J. A. V. Butler and B. Katz (eds.): *Progress in Biophysics and Biophysical Chemistry,* vol. X, Pergamon, New York, 1960, pp. 1–53.

Huennekens, F. M., and H. R. Whiteley: Phosphoric Acid Anhydrides and Other Energy-rich Compounds, in M. Florkin and H. S. Mason (eds.): *Comparative Biochemistry,* vol. I, Academic, New York, 1960, pp. 107–180.

Ingraham, L. L., and A. B. Pardee: Free Energy and Entropy in Metabolism, in D. M. Greenberg (ed.): *Metabolic Pathways,* 3d ed., vol. I, Academic, New York, 1967, pp. 2–46.

Pullman, B., and A. Pullman: Electronic Structure of Energy-rich Phosphates, *Radiat. Res.,* suppl. 2, 1960, pp. 160–181.

Chapter 13

Membranes and Subcellular Organelles

FUNCTIONS OF MEMBRANES AND SUBCELLULAR ORGANELLES

All living cells are enveloped by a surface, or plasma, membrane 70 to 100 Å thick and composed primarily of lipid and protein. Because of its many and diverse functions, the plasma membrane plays a central role in cellular biology. An important feature of the plasma membrane is its selective permeability, which controls the transport of metabolites into and out of the cell and is responsible for the maintenance of its internal milieu. The plasma membrane also controls the flow of information from the cellular environment into the cell. This is accomplished by means of specific receptor molecules, many of which are imbedded in the plasma membrane and which respond to a variety of stimuli, both chemical and electric. In addition to the plasma membrane, eukaryotes possess an array of highly specialized intracellular organelles, each of which is enclosed by a membrane and which may differ from each other and from the plasma membrane in their composition and structure. The existence of such membranous organelles in eukaryotes permits the compartmentalization of cellular functions and promotes the higher level of cellular organization that is characteristic of eukaryotes. For example, fatty acids synthesized in the extramitochondrial *cytosol* are oxidized within the *mitochondrion*, permitting synthesis and degradation to be separated by the mitochondrial membrane and to be controlled in different ways (Chap. 20). As another example, transcription of the genetic message occurs in the nucleus, but translation of the resultant transcript takes place in the cytosol (Chap. 29). Separation of these two aspects of gene expression in this manner allows for more refined controls of these crucial processes.

Any cross-references coded M refer to the companion text, *Principles of Biochemistry: Mammalian Biochemistry.*

Although specialized cells may contain subcellular organelles and vesicles designed to support their specific functions, only a relatively small number of intracellular organelles is common to all eukaryotic cells. The *nucleus* contains the genetic apparatus and is the site where the chromosomes are replicated and the genes expressed. By electron microscopy the nucleus is seen to have two membranes. The inner one may be considered the true nuclear membrane; the intermembrane space appears frequently to be continuous with the channels of the endoplasmic reticulum. Also present is the *nucleolus*, where ribosomal RNA is synthesized (Chap. 29). Surrounding the nucleus is the *cytoplasm*, which consists of the *cytosol* in which are suspended the other *cytoplasmic organelles*. Many of the enzymes of intermediary metabolism are present in the cytosol; it is also the site of protein synthesis on *free ribosomes*. The *rough endoplasmic reticulum* with its associated ribosomes is the site of membrane protein and lipid synthesis. The *smooth endoplasmic reticulum* contains a variety of enzymes used to detoxify drugs and other potentially toxic compounds. The *Golgi apparatus* is involved in the glycosylation of proteins that are to be incorporated into the membranes of other subcellular organelles or are destined for secretion by the cell. The primary energy-conserving process, oxidative phosphorylation, is confined to the *mitochondria*, or more properly the mitochondrial membrane, which, because of its impermeability to protons, permits the generation of the electrochemical gradient needed to drive the endergonic synthesis of ATP from ADP and P_i (Chap. 15). *Chloroplasts* house the photosynthetic apparatus of plant cells (Chap. 19). *Lysosomes* contain a variety of degradative enzymes including *proteases, nucleases, glycosidases, lipases,* and *phosphatases* that are required for the intracellular degradation of macromolecules. The *peroxisomes* contain enzymes which relate to the metabolism of H_2O_2. *Urate oxidase, D-amino acid oxidase,* and *α-hydroxyacid oxidase* all generate H_2O_2, a potentially lethal oxidant; catalase, the major protein of peroxisomes, decomposes the H_2O_2 to yield H_2O and O_2. The *glyoxylate cycle* (Chap. 17) and the process of *photorespiration* in plants (Chap. 19) are also housed within the peroxisomes.

A rat hepatocyte, averaging 20 μm in diameter, has a cytoplasmic volume of approximately 5000 μm^3 and a nuclear volume approximately one-tenth to one-twentieth as large. Of the cytoplasmic volume, mitochondria normally occupy 15 to 20 percent, peroxisomes 1 to 2 percent, and lysosomes 0.2 to 0.5 percent. The volume of the endoplasmic reticulum cannot be estimated because of its extremely irregular shape; however, the area of its membranes may be as great as 25,000 μm^2 for the rough endoplasmic reticulum and approximately 15,000 to 20,000 μm^2 for the smooth endoplasmic reticulum. Each hepatocyte contains about 1000 to 2000 mitochondria, as well as about 500 peroxisomes and half as many lysosomes as there are peroxisomes. Several million ribosomes and a large but unknown number of microtubules and microfilaments (see below) are also present. Figure 13.1 is an electron micrograph of a rat hepatocyte showing the major subcellular organelles.

The basis for the diverse array of functions associated with the plasma membrane and the membranes of the cellular organelles must ultimately reside in their structure. We will therefore consider the composition of membranes and then go on to discuss the manner in which proteins and lipids are arranged to permit specific membrane functions.

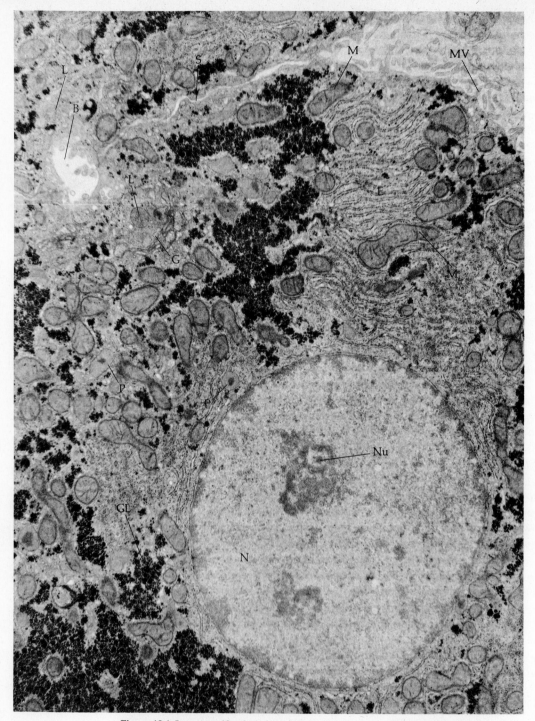

Figure 13.1 Low-magnification electron micrograph of a rat hepatocyte. Higher magnifications will show the detailed structures of organelles. (B) indicates a bile canaliculus and (MV) a microvillus at the sinusoid surface; some microvilli near (MV) were sectioned so as to appear unattached to the cell. The dense deposits at (GL) are masses of glycogen granules. (M) indicates mitochondria, (N) the nucleus, (Nu) a nucleolus, (P) peroxisomes, (L) lysosomes, (G) the Golgi apparatus, and (S) the intercellular space separating this cell from an adjacent hepatocyte. (E) indicates endoplasmic reticulum. ×25,000. [*Courtesy Dr. A. B. Novikoff.*]

As noted above, the principal components of membranes are lipids and protein. Membranes also contain carbohydrate covalently bound to either protein or lipid; however, the amount of carbohydrate is relatively small, usually less than 10 percent of the mass of the membrane. Most membranes contain approximately equivalent amounts of protein and lipid. For example, the erythrocyte membrane has a protein-to-lipid ratio of $1:1$. However, the protein-to-lipid ratio may vary widely; e.g., the ratio in the myelin membrane of nerve cells is $0:23$; in the inner mitochondrial membrane it is $3:2$.

Lipid Three types of lipids are found in membranes: phospholipids (which are the most abundant), neutral lipids, particularly cholesterol, and glycolipids (Table 13.1). The predominant phospholipids are phosphatidylcholine, phosphatidylethanolamine, phosphatidylserine, and sphingomyelin. Phosphatidylinositol is usually present but in lesser amounts than the others. Phosphatidylglycerol and diphosphatidylglycerol are found only in bacteria and in mitochondrial membranes. Cholesterol, found exclusively in eukaryotes, and particularly abundant in the plasma membrane, is absent from bacteria and the inner mitochondrial membrane (Table 13.1).

Protein In contrast to the relatively few kinds of lipids in membranes, each present in abundance, many different types of membrane proteins are present, each in relatively few copies. These proteins are involved in the enzymic, transport, receptor, and other functions associated with cellular membranes. Certain, highly specialized membranes contain only one or perhaps a few proteins. For example, the sarcoplasmic reticulum membrane of muscle, whose specialized function is the uptake of Ca^{2+}, contains a Ca^{2+}-dependent ATPase which promotes this process (Chap. M8). Similarly, the principal protein of the membrane of the retinal rod outer segment is rhodopsin, involved in visual reception (Chap. M9).

Carbohydrate The carbohydrate of membranes is present as oligosaccharide side chains linked covalently to proteins to form glycoproteins and, to a lesser

TABLE 13.1 Lipid Composition of Some Membranes (Percentage of Total)

Lipid	Myelin (bovine)	Rod, outer segment (bovine retina)	Erythrocyte (human)	Mitochondria (bovine heart)	E. coli
Phosphatidylcholine	10	32	17	37	
Phosphatidylethanolamine	15	16	16	35	72
Phosphatidylserine	9	6	8		
Phosphatidylinositol	1	1	0.8	3	
Phosphatidylglycerol		1		18	16
Sphingomyelin	8	1	16		12
Glycolipid	28	23	Trace	Trace	
Cholesterol	22	5	23	1	

Source: From M. N. Dewey and L. Barr, *Curr. Top. Mem. Transp.* 1:6 (1970).

extent, to sphingolipids to form glycolipids. The principal sugars found in glycoproteins and glycolipids are galactose, mannose, fucose, N-acetylgalactosamine, N-acetylglucosamine, glucose, and sialic acid (Chap. 18). The latter is generally the terminal residue and contributes to the net negative charge carried by all mammalian cells.

The carbohydrate in glycoproteins is usually linked to an asparagine residue (N-linked) or to the hydroxyl group of a serine or threonine residue (O-linked). The N-linked structures are of two general types, referred to as *complex* and *high mannose*. Their structures are given in Chap 18.

The glycolipids are glycosylated derivatives of N-acylsphingosine (ceramide) in which the carbohydrate ranges in size from a monosaccharide (glucose or galactose) to oligosaccharides of up to seven or more sugars (Chaps. 18 and 21).

STRUCTURE OF MEMBRANES

The Lipid Bilayer

A membrane can be considered to be a huge assembly of macromolecules which, unlike most other macromolecules, is held together by noncovalent interactions. The most important force stabilizing membranes is the hydrophobic interaction between lipids and between lipids and proteins. As discussed in Chap. 7, all membrane lipids are *amphipathic,* containing a hydrophobic portion and a hydrophilic polar head group. For a phospholipid, e.g., phosphatidylcholine, the hydrophobic portion consists of the long-chain fatty acids. The hydrophilic portion is the glycerol phosphorylcholine moiety. In the case of cholesterol, the sole hydrophilic portion of the molecule is the hydroxyl group at C_3 (Chap. 7).

When phospholipids or mixtures of phospholipids and cholesterol are placed in an aqueous environment, they spontaneously form a bimolecular bilayer in which the fatty acyl tails are buried and the polar head groups are exposed to water. The bilayer is stabilized by hydrophobic interactions (Fig. 13.2). Physical measurements, in particular, low-angle x-ray diffraction analysis (Chap. 5) of a variety of membranes, show that the bulk of the phospholipids within these membranes are arranged in the form of such a bilayer, which provides the structural framework for the membrane.

A variety of physical techniques, including nuclear magnetic and electron spin resonance, has been used to investigate the motion of individual lipid molecules within phospholipid bilayers. These studies have shown that lipid bilayers behave as two-dimensional fluids. The lipid molecules within the bilayer readily exchange sites with neighboring molecules in their own monolayer but rarely migrate from one monolayer to the next, a process that has been termed *flip-flop*. Thus, lateral diffusion of a lipid molecule is extremely rapid, each molecule exchanging places with its neighbor 10^6 times per second. In contrast, transmembrane diffusion, the exchange of neighboring phospholipid molecules across the bilayer by flip-flop, occurs with a half-time of months, in part because of the very high activation energy that would be required to bring the polar head group through the hydrocarbon core of the bilayer. A similar

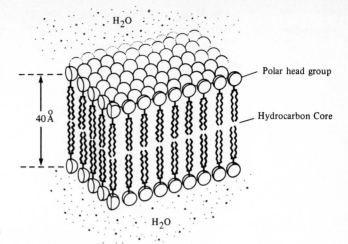

H$_2$O

Polar head group

40 Å

Hydrocarbon Core

H$_2$O

Figure 13.2 Orientation of phospholipids in a lipid bilayer.

situation has been found for biological membranes. The individual phospholipid molecules within a bilayer are not rigid. In fact, the rapid lateral diffusion of phospholipids in the lipid bilayer results in part from rapid thermal movements of the fatty acyl chains, which promote intermolecular collisions. These collisions are, in turn, caused by the rapid rotation around the carbon-carbon bonds in the fatty acyl chains. The degree of rotation is greatest near the end of the chain, e.g., the center of the bilayer, and least, adjacent to the polar head group.

The fluidity of a bilayer at a given temperature is determined by its composition. A bilayer composed of only one kind of phospholipid shows a sharp and characteristic phase transition at which it changes from its liquid state to a crystalline hexagonal lattice. This "gel" phase is much less fluid and displays little lateral diffusion. The phase transition temperature (T_m) depends upon the nature of the phospholipid head group and on the length and degree of unsaturation of the fatty acyl chains. As noted above, the speed of rotation around carbon-carbon bonds in a fatty acyl chain increases toward the terminal methyl group. Shorter chains therefore have an enhanced flexibility and thereby impart a high degree of fluidity to membranes. The presence of cis double bonds, which introduce kinks in the fatty acyl chains, interferes with the close packing of the chains within the crystalline lattice, also resulting in an increase in membrane fluidity. Thus, shorter chains and higher degrees of unsaturation result in a lower T_m. In lipid bilayers containing a mixture of phospholipids, there is a progressive phase change over a broad temperature range, indicating that a "phase separation" into clusters of gel and liquid occurs.

The importance of a fluid bilayer to membrane function is emphasized by the fact that cells possess specific mechanisms to maintain their lipid bilayers in the fluid state, i.e., above the T_m. In bacteria, the fatty acid composition of the membrane is altered in response to changes in growth temperature, so as to maintain the fluid state. Cholesterol serves this function in eukaryotes. When cholesterol is introduced into the bilayer, the planar steroid rings interact with and partially immobilize those regions of the fatty acyl chains closest to the polar head group, leaving the rest of the chain flexible and thereby preventing

273

the hexagonal close packing required for their crystallization, even at much lower temperatures. At the same time cholesterol reduces the mobility of the fatty acyl chains and increases the "microviscosity" in the interior of the bilayer. In contrast, lanosterol, the biosynthetic precursor of cholesterol (Chap. 21), has only a marginal effect on the fluidity. The reason for this difference is that the methyl groups of lanosterol render it sterically less favorable for interactions with fatty acyl chains than cholesterol. The α-methyl group of lanosterol at C_{14} bulges and protrudes from the planar underside of the steroidal ring system preventing effective contacts with the fatty acyl chain (Chap. 7). Indeed the methyl group at C_{14} is the first to be removed in the biosynthetic sequence leading to cholesterol (Chap. 21).

Membrane Proteins

Proteins imbedded in the lipid bilayer display the same dynamic properties as the bilayer itself. That is, lateral diffusion is rapid, a consequence of their being dissolved in a two-dimensional fluid. Similarly, proteins do not undergo flip-flop, a process that is even less favorable energetically than it is for phospholipids. The rapid lateral diffusion of proteins was first suggested by the demonstration that different antigens on the surfaces of nucleated mouse and human cells mixed rapidly after cell fusion by inactivated Sendai virus. It was subsequently observed that the rhodopsin molecules of the retinal rod disk membranes (Chap. M9) rotate about an axis perpendicular to the plane of the membrane. Moreover, the rate of diffusion of rhodopsin was consistent with the intrinsic viscosity of the lipid bilayer. Despite the fact that membrane proteins are free to diffuse laterally at great speeds, there are clearly instances where they are immobilized. These include gap junction proteins (page 284), junctional acetylcholine receptors (Chap M7), and the basolateral and apical plasma membranes of epithelial cells (Chap. M10). The nature of the restraints on these immobilized proteins is unknown; however, several mechanisms have been suggested. (1) Membrane proteins may be either immobilized or impeded in their mobility by membrane-associated components such as microfilaments and microtubules at the cytoplasmic surface (page 285). (2) They may be restrained by sequestration or exclusion of membrane components to (or from) specific lipid domains. (3) They may be restrained by peripheral membrane components at the cytoplasmic or extracellular membrane surfaces.

Membrane proteins fall into two distinct classes. The first, *integral proteins,* are intercalated directly into the lipid bilayer and interact hydrophobically with the lipids. That is, the hydrophobic surfaces are solvated by the hydrocarbon core of the bilayer and the polar surfaces are solvated by water (Fig. 13.3). The second class, *peripheral proteins,* are not imbedded in the bilayer but rather bind to membranes only through their interaction with integral membrane proteins. The integral proteins can only be extracted from the membrane by agents, e.g., detergents, that disrupt the lipid bilayer (Chap. 7); once isolated, they are generally insoluble in aqueous media. Peripheral proteins, in contrast, can be removed from the membrane by milder treatments which disrupt quaternary structure, e.g., high ionic strength urea. In contrast to integral proteins, the peripheral proteins are usually water-soluble.

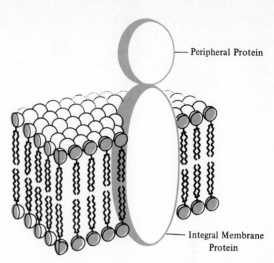

Figure 13.3 Diagram of lipid bilayer into which integral and peripheral membrane proteins have been inserted.

Peripheral Protein

Integral Membrane Protein

The hydrophobic surface of integral membrane proteins is, in general, a consequence of the clustering of hydrophobic amino acids to generate hydrophobic regions, or domains, which interact with the hydrocarbon core of the bilayer. In some instances, the hydrophobic surface results from a high content of hydrophobic amino acids. Such proteins may be extremely hydrophobic and are, in fact, soluble in nonpolar organic solvents. They have been termed *proteolipids*. One such proteolipid, termed *lipophilin*, is the major membrane protein of brain myelin. Of its amino acids, 66 percent are hydrophobic; in addition, it contains approximately 2 mol of fatty acids bound in ester linkage to either serine or threonine residues per mole of polypeptide. Although their precise functions are unknown, proteolipids are believed to participate in the ion fluxes that mediate ATP synthesis in mitochondria, possibly by controlling the entry of ions into the membrane transport site (Chap. 16).

The integral membrane proteins imbedded in the lipid bilayer can be visualized directly by a technique termed *freeze-fracture electron microscopy*. When frozen membranes are fractured at the temperature of liquid nitrogen, they are literally split in half through the center of the bilayer. The splitting, a consequence of the decay of hydrophobic interactions at very low temperatures, provides a view from the interior surface of the membrane at the interface between the two monolayers. When the fracture face is observed in the electron microscope, particles approximately 80 Å in diameter are observed on an otherwise smooth surface. The particles are integral proteins, and the smooth surface is the inside of the lipid bilayer (Fig. 13.4).

Many, and perhaps all, integral membrane proteins span the full width of the lipid bilayer, with amino acid sequences exposed on both sides of the membrane. This principle was first established for glycophorin and for "band 3," the major integral protein of erythrocyte membranes (Chap. M3), when it was demonstrated that different parts of the same polypeptide chain were covalently modified when the cytoplasmic and extracellular sides of the erythrocyte membranes were reacted selectively with a membrane-impermeable reagent. In the case of glycophorin, the NH_2-terminal portion is outside, the COOH-terminal

Figure 13.4 Electron micrograph of a freeze-fracture preparation of an onion root tip cell showing the protoplasmic face of the cell membrane (×36,000). [*Courtesy of Professor Daniel Branton.*]

portion is inside, and the hydrophobic part passes through the lipid bilayer (Fig. 13.5).

A detailed picture of an integral membrane protein has been obtained for the purple membrane protein *bacteriorhodopsin* of the halophilic *Halobacterium halobium,* which functions in vivo as a light-driven proton pump (Chap. 15). The purple membrane consists of a single protein species, bacteriorhodopsin (MW = 26,000), which makes up 75 percent of the mass, and phospholipid, which makes up the remaining 25 percent. Retinal (Chap. M9), covalently linked to each protein molecule in a 1:1 ratio, confers the characteristic purple color on the membrane. The bacteriorhodopsin molecules are disposed in a regular crystalline array. By image reconstruction from a large number of electron micrographs taken at extremely low beam exposures to prevent damage to the membrane, a low-resolution (7 Å), three-dimensional map of the purple membrane has been obtained. As shown in Fig. 13.6, the purple membrane

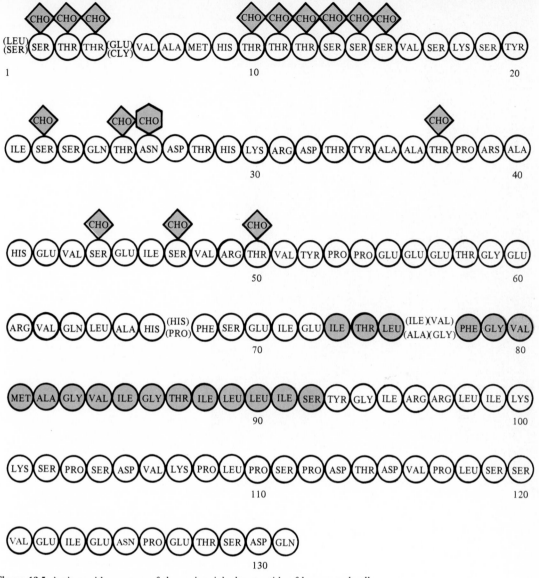

Figure 13.5 Amino acid sequence of the major sialoglycopeptide of human red cell membranes (glycophorin A). The hydrophobic part of the molecule, which is thought to extend through the lipid bilayer, is indicated by the colored circles. [*Adapted from M. Tomita and V. T. Marchesi, Proc. Natl. Acad. Sci. U.S.A. **72:**2964 (1976).*]

protein is globular. The polypeptide chain spans the full width of the lipid bilayer, asymmetrically winding back and forth seven times in α-helical segments, with a different portion of the polypeptide exposed at the cytoplasmic and extracellular surfaces of the membrane. Although a similar analysis has not yet been applied to other integral membrane proteins, it is likely that the arrangement found for bacteriorhodopsin is a general one.

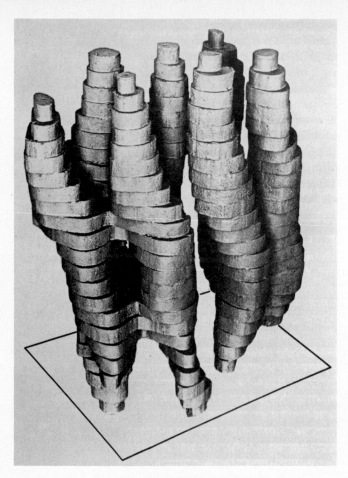

Figure 13.6 A model of a single bacterio-rhodopsin molecule in the purple membrane, viewed roughly parallel to the plane of the membrane. The top and bottom of the model correspond to the parts of the protein in contact with the solvent, the rest being in contact with lipid. The most strongly tilted α-helices are in the foreground. [*Adapted from R. Henderson and P. N. T. Unwin, Nature* **257**:28 (1975).]

ASYMMETRY OF MEMBRANES

Protein Asymmetry

Membranes are functionally asymmetric structures whose asymmetry is a consequence of the vectorial arrangement of their components, e.g., proteins, carbohydrates, and lipids. In the case of membrane proteins the asymmetry is complete; viz., each copy of the same polypeptide chain has the same orientation within the membrane. The NH_2-terminal portion is on the extracellular face of the membrane and the COOH-terminal portion is on the cytoplasmic side of the membrane.

The existence of integral membrane proteins that can span the lipid bilayer asymmetrically provides a means to receive various extracellular chemical signals (hormones, neurotransmitters, etc.) and convey them to the cytoplasm, for example, by inducing a change in conformation or quaternary structure. Transmembrane proteins often have oligomeric structures with their axes of symmetry perpendicular to the plane of the membrane. Such structures can provide a specific route or channel for small molecules to cross the membrane and in this manner function as transport proteins (Chap. 14). The asymmetry of membrane proteins poses a problem with respect to their synthesis. In particular, how are the membrane proteins inserted in the extracellular surface of the membrane when their synthesis occurs on the other side of the membrane, in the cytoplasm? This point will be considered in detail later (Chap. 29).

Carbohydrate Asymmetry

Carbohydrate is located exclusively on the extracellular side of the plasma membrane. This is true both for glycolipids, which are found only in the extracellular monolayer of the lipid bilayer, and for the glycoproteins. In the case of erythrocyte glycophorin all of the carbohydrate is linked to the NH_2-terminal portion of the polypeptide (Fig. 13.5) and is external to the cell. The carbohydrate portions of band 3, the major glycoprotein of the erythrocyte membrane, as well as the minor erythrocyte glycoproteins are also localized externally. All of the carbohydrate in the membranes of intracellular organelles is located on the lumenal side of the membrane; thus, as in the plasma membrane, no carbohydrate is found on the cytoplasmic surface.

Although there is yet no direct proof, it is likely that the external localization of the carbohydrates of plasma membranes serves a number of important functions. The terminal sialic acid residues impart a negative charge to the surface of the cell, thereby preventing nonspecific aggregation. Moreover, the surface carbohydrates can serve as hormone receptors and may mediate such highly specific processes as cell-cell recognition and adhesion (Chap. 18).

Lipid Asymmetry

Lipid asymmetry differs from protein and carbohydrate asymmetry in that it is not absolute. However, it is clear that different lipids are present in the two monolayers. In the erythrocyte membrane, which has been examined in most detail, the phospholipids of the external monolayer consist almost exclusively of phosphatidylcholine and sphingomyelin, the choline-containing phospholipids. In contrast, phosphatidylethanolamine and phosphatidylserine are confined to the cytoplasmic monolayer. The asymmetry of erythrocyte membrane phospholipids has been demonstrated in a number of ways, including susceptibility to amine-specific reagents and to phospholipases (Chap. 21). Cholesterol, however, is distributed equally between the two sides of the membrane, as might be anticipated from its high concentration (approximately one-half of the total lipid).

Lipid asymmetry also occurs in the membrane of influenza virus, which, like many other enveloped animal viruses, obtains its lipid bilayer by budding out from the plasma membrane of the host cell (Chap. 30). The viral membrane, which reflects that of its host cell, is readily purified and obtained in homogeneous form. As depicted in Fig. 13.7, two types of phospholipid asymmetry are evident in influenza virus grown in bovine kidney cells. Approximately twice as much total phospholipid is present inside as outside, and the phospholipid composition of the two sides differs markedly. The data shown in Fig. 13.7 demonstrate that the asymmetry of the influenza virus membrane (which reflects that of bovine kidney) is different from that of the erythrocyte

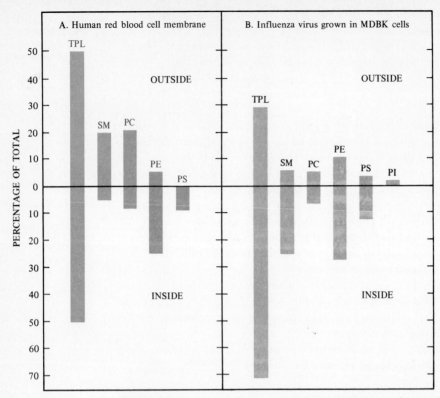

Figure 13.7 Asymmetric distribution of phospholipids in membranes of human red blood cells and of influenza virus grown in cultured bovine kidney (MDBK) cells, expressed as mole percent. The abbreviations are: *TPL*, total phospholipid; *PC*, phosphatidylcholine; *SM*, sphingomyelin; *PE*, phosphatidylethanolamine; *PS*, phosphatidylserine; and *PI*, phosphatidylinositol. [*From J. E. Rothman and J. Lenard, Science 195:743 (1977).*]

membrane. In particular, the amounts of phospholipid in the two monolayers differ. Sphingomyelin is present mainly in the internal monolayer, and phosphatidylethanolamine and phosphatidylserine constitute a considerably larger fraction of the external phospholipids. Thus, although lipid asymmetry is a general property of biological membranes, the lipids are not always unequally distributed and the specific details of lipid asymmetry vary in different membranes.

The maintenance of lipid asymmetry is clearly related to the extremely low frequency of reorientation of membrane components (flip-flop) discussed earlier. The existence of lipid asymmetry, in particular, the synthesis of phospholipids of the external monolayer, poses the same problem as has been indicated for the synthesis of the extracellular portions of integral membrane proteins; that is, the source of energy and biosynthetic precursors required for phospholipid synthesis are located on the opposite (cytoplasmic) side of the membrane. It would appear that after synthesis at the cytoplasmic side of the membrane, phospholipids can be translocated to the outer half of the bilayer by a special protein-facilitated mechanism. Features of this mechanism will be described in the following section.

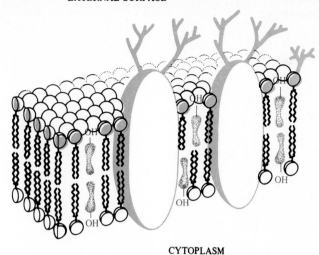

Figure 13.8 Generalized model of plasma membrane showing lipid bilayer into which cholesterol, glycosylated integral proteins, and glycolipids have been inserted. The colored head groups at the external surface refer to phosphatidylcholine and sphingomyelin; those without color (cytoplasmic surface) indicate phosphatidylethanolamine and phosphatidylserine.

CYTOPLASM

The structure of a typical membrane, indicating the various features of protein, carbohydrate, and lipid asymmetry, is depicted in Fig. 13.8.

ASSEMBLY OF MEMBRANES

General Principles

Membranes can be solubilized by treatment with detergents (Chap. 7) and their lipids and proteins separated. Upon mixing the separated lipids and proteins and removal of the detergent by dialysis, for example, the membranes will reform. However, the reconstituted membrane no longer shows the asymmetry that characterizes the original membrane. Thus, unlike certain viruses (Chap. 30), muscle fibers (Chap. M8), and ribosomes (Chap. 28), membranes do not undergo self-assembly. In the absence of a permeability barrier, insertion of lipids and proteins can occur equally well from both sides of the growing membrane, a process that is not possible in the cell. In vivo, *membrane assembly takes place only within the confines of a preexisting membrane.* Newly synthesized proteins and lipids are inserted into preexisting membranes which form closed vesicles from one side of the vesicle only. This process of membrane growth by expansion ensures the assembly of an asymmetric membrane.

Membrane components are, in general, synthesized or inserted into sites distinct from their ultimate destination. For example, the protein and lipids of the plasma membrane are synthesized in the endoplasmic reticulum (page 269) and are then transported to the plasma membrane *via* the Golgi apparatus (page 269), where they undergo modification. Indeed, the endoplasmic reticulum is the site of synthesis of essentially all of the macromolecules found in the membranes of the cell. It is the endoplasmic reticulum that is the membrane that grows by expansion. In a current model for the transport of membrane proteins and lipids, small vesicles bud off from the endoplasmic reticulum and then fuse with the Golgi apparatus. Membrane proteins and lipids destined for the plasma membrane then bud off from the Golgi and are transported to the plasma membrane, where they are inserted, again by the process of fusion. A vesicle transport process of this kind conserves transmembrane asymmetry since the lumenal surface of the membrane of the vesicle is the same side as the lumen of the organelle from which the vesicle budded and, following fusion, the same side as the lumen of the organelle with which the vesicle has undergone fusion.

When a vesicle fuses with the plasma membrane, its lumenal surface becomes the extracellular surface of the plasma membrane. Thus, the lumenal side of intracellular vesicles and organelles and the extracellular side of the plasma membrane are topologically equivalent. Recall that membrane carbohydrate is found only on the extracellular side of the plasma membrane and the lumenal side of intracellular membranes.

Assembly of the Lipid Bilayer

The enzymes catalyzing phospholipid synthesis (Chap. 21) in bacteria are integral membrane proteins with their active sites on the cytoplasmic side of the cytoplasmic membrane. Thus, the biosynthetic enzymes have access to their substrates, which are in the cytoplasm. The newly synthesized phospholipids are released into the cytoplasmic monolayer, where they mix with existing lipids by lateral diffusion. In this way, the cytoplasmic monolayer grows by continuous expansion as new phospholipids are synthesized. Transmembrane movement of the newly synthesized lipids is required for equivalent growth of the extracellular monolayer of the membrane. In fact this process has been found to occur at a rate 10^5-fold greater than in nongrowing cells, suggesting that rapid transmembrane movement requires the catalytic activity of a protein. As yet, no such protein has been found.

A similar process appears to be involved in bilayer assembly in eukaryotes. Thus, the synthesis of phosphatidylcholine and phosphatidylethanolamine occurs on the cytoplasmic surface of the endoplasmic reticulum, and there is rapid translocation of the newly synthesized phospholipids to the lumenal monolayer of the membrane bilayer.

Assembly of Proteins into Membranes

Proteins are synthesized on ribosomes, which translate the messenger RNA sequence into the amino acid sequence of a protein; synthesis proceeds from the NH_2- to the COOH-terminus (Chap. 28). All proteins, both soluble and those associated with membranes, are synthesized by this process, which occurs in the cytosol. What, then, is the mechanism by which integral proteins are inserted into the plasma and other cellular membranes? One hypothesis involves spontaneous insertion of the protein into the lipid bilayer because of a hydrophobic surface, or domain, and is exemplified by cytochrome b_5 and the NADH-cytochrome b_5 reductase of the endoplasmic reticulum (Chap. 16). A second hypothesis proposes that the protein is inserted into the membrane as it is being synthesized and before it has assumed its ultimate folded conformation. The details of this "signal hypothesis," for which there is considerable supporting evidence, will be given in Chap. 29. Briefly, however, the signal hypothesis holds that synthesis of integral membrane proteins is initiated on free unattached ribosomes. The nascent protein contains a special NH_2-terminal sequence of approximately 20 amino acid residues termed the "signal" sequence, which emerges first from the ribosome and directs the protein-synthesizing complex to the rough endoplasmic reticulum, or *rough ER*. The signal sequence is bound by a specific receptor found only in the rough ER, forming a membrane-bound ribosome. The polypeptide chain is then continuously inserted into the ER membrane in an extended form as it is being synthesized. Finally, a protease on the noncytoplasmic side of the ER removes the signal sequence.

Glycosylation of Membrane Proteins

As noted earlier (page 272), there are two types of asparagine-linked oligosaccharides in membrane glycoproteins, the high mannose and the complex. The synthesis of these oligosaccharides takes place by the combined action of enzymes localized in the rough ER and the Golgi apparatus. The mannose and core *N*-acetylglucosamine residues of the complex oligosaccharides are added in the rough ER, as are all of the mannose residues of the high-mannose type. The terminal trisaccharide sequence (*N*-acetylglucosamine-galactose-sialic acid) is then added to the core of the complex oligosaccharide upon its transfer to the Golgi apparatus.

Details of the glycosylation reactions in glycoprotein synthesis are given in Chap. 18; however, several features of this process should be noted in the context of membrane assembly. (1) Glycosylation of membrane proteins occurs only on the lumenal side of the endoplasmic reticulum. This side is the topological equivalent of the lumenal side of other organelles and of the extracellular side of the plasma membrane. The asymmetry of glycosylation thus accounts for the observation (page 279) that all oligosaccharides are found on the noncytoplasmic surface of membranes. (2) A lipid linked-activated oligosaccharide (Chap. 18) is transferred in a single step to an asparagine residue of the growing polypeptide chain. The completely folded protein cannot serve as an acceptor, presumably because the acceptor asparagine residue is accessible to the glycosyl transferase only when the polypeptide is in the unfolded state. (3) Attachment of the oligosaccharide to lipid in an activated form appears to serve two functions. First, it provides a means for the translocation of the oligosaccharide across the membrane in an activated form. The sugar nucleotide precursors are in the cytosol, but transfer of the oligosaccharide takes place in the lumen of the ER. Second, the lipid may anchor the activated oligosaccharide to the lumenal membrane surface of the ER, making it possible for the membrane-bound glycosyl transferase to catalyze transfer of the preformed oligosaccharide from its lipid carrier to the acceptor asparagine residue just as it emerges at the lumenal side of the membrane, but before it can fold up.

Assembly of Mitochondrial Membranes

The assembly of mitochondrial membranes differs from that of other cellular membranes. This point is exemplified by the multisubunit enzyme *cytochrome c oxidase,* an integral protein of the inner mitochondrial membrane (Chap.

16). Three of the seven subunits of cytochrome oxidase are made in the cytosol as water-soluble polypeptides. Moreover, insertion into the mitochondrial membrane of the three subunits occurs after synthesis on cytoplasmic ribosomes is complete. Inasmuch as mitochondria lack the lipid biosynthetic enzymes, a mechanism of this kind leaves unresolved the origin of the mitochondrial lipids. Recall that in the case of the plasma membrane both proteins and lipids are transported together in vesicles that have budded off from the endoplasmic reticulum. A class of cytoplasmic proteins known as *phospholipid-exchange proteins* appears to be involved in the insertion of phospholipids into mitochondrial membranes. These lipid transfer proteins, which display different activities depending upon the phospholipid to be transferred, bind 1 mol of lipid per mole of protein to form a soluble complex. Such a complex can participate repeatedly in the net transfer of phospholipids from the endoplasmic reticulum to the mitochondrial membrane. Lipid-exchange proteins may also be involved in the net transfer and exchange of phospholipids in other cellular membranes (Chap. 21).

Gap Junctions

Communication between cells in many eukaryotic tissues is effected through specialized regions of contact between opposing plasma membranes known as *gap junctions,* which are believed to mediate and regulate the passage of ions and small molecules from one cell to the other. The gap junctions are morphologically distinct structures composed of units, termed *connexons,* that are

Figure 13.9 Model of the connexon showing the transition from the open to the closed configuration at a gap junction. [*From P. N. T. Unwin and G. Zampighi, Nature* **283:**545 (*1980*).]

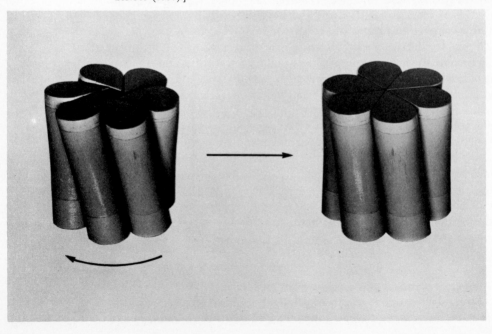

imbedded in the opposing membranes, in register and linked to each other. An analysis of the three-dimensional structure of isolated gap junctions by the high-resolution electron microscopic technique employed in the analysis of bacteriorhodopsin (page 276) has shown the connexon to be an annular oligomer composed of six protein subunits 25 Å in diameter and 75 Å long that span the membrane and protrude from either side. The channel between the subunits is approximately 20 Å wide at its exposed region but is narrower within the membrane. Such a 20 Å-diameter channel enables ions and a variety of small metabolites (e.g., sugars, amino acids, and nucleotides) to flow from the interior of one cell to the adjoining one. As demonstrated in Fig. 13.9, closure of the 20 Å-channel, a process regulated by $[Ca^{2+}]$ as well as pH, is achieved by the subunits sliding against each other, decreasing their inclination and hence rotating, in a clockwise sense, at the base.

INTRACELLULAR FIBRILLAR STRUCTURES

Microtubules

Microtubules are cytoplasmic elements that contribute to cell morphology, cytoplasmic streaming, intracellular transport, and cytoplasmic contractility. Thus, microtubules participate in maintenance or control of cell shape (a cytoskeletal function), participate in cellular movements, e.g., movement of chromosomes by the cell spindle, and act as channels of oriented flow. The function of microtubules in nerve axons is presented in Chap. M10.

Microtubules are hollow cylinders, approximately 20 to 30 nm in diameter with a wall thickness of 4.5 to 7.0 nm. They are composed of a globular glyco-protein, *tubulin,* containing two almost identical subunits (MW = 50,000 to 60,000), designated α and β. The stable form of isolated tubulin is an α-β dimer (MW = 110,000). Tubulin molecules are helically arranged to form the wall of the cylinder, whose diameter is determined by the number of tubulin molecules in one complete helical turn. The number of microtubular subunits (protofilaments) per turn is usually 13 but may vary from 10 to 14, depending on the organism. Figure 13.10 is a diagrammatic view of the arrangement in a microtubule of the tubulin subunits and the 13 protofilaments.

Microtubules are continuously being formed and dissociated; this is particularly striking in brain. The assembly of microtubules from tubulin is an example of self-assembly since it is initiated by activation of the protein by GTP that is present in tubulin as isolated and is bound tightly to each of the subunits. Assembly then proceeds by both lateral and longitudinal interaction and assembly of tubulin molecules. Attachment of each tubulin subunit is accompanied by hydrolysis, in the presence of Mg^{2+}, of GTP with formation of bound GDP and liberation of P_i. The drugs *colchicine, vinblastine,* and *vincristine* block microtubule formation by binding to tubulin with displacement of the nucleotide, thus preventing polymerization. This function of GTP in tubulin polymerization appears to mirror that of ATP in actin polymerization (Chap. M8). However, tubulin assembly without added GTP can occur in the presence of high concentrations of glycerol or sucrose. Ca^{2+} in millimolar concentrations inhibits polymerization of microtubules, and it has been suggested that changes

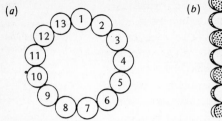

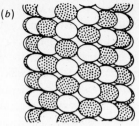

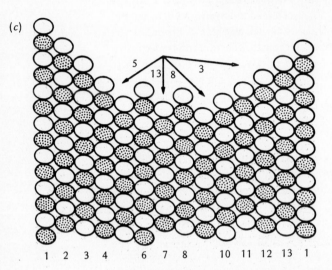

Figure 13.10 (a) Diagrammatic view of a cytoplasmic microtubule cross section showing arrangement of 13 protofilaments. (b) Diagrammatic view of cytoplasmic microtubule surface lattice showing the probable arrangement of the heterodimers. Stippled and clear ellipsoids represent α and β subunits. Monomers alternate along the three-start helix and within a protofilament. (c) Representation of the helix net as seen in an unfolded microtubule. Numbers at the bottom of the sheet correspond to the numbered protofilaments in (a). Each ellipsoid represents a monomer, colored and clear ellipsoids representing the α and β subunits, respectively. Subunits alternate along the three-start helix, and the α-β dimer axis is parallel to the axis of the protofilament. The arrow designated 3 shows the angle of the three-start helix and arrow 13, the angle of the 13-start helix. α-β dimers can be viewed as arranged in a five-start right-handed helix (arrow 5) or an eight-start left-handed helix (arrow 8). [From J. A. Snyder and J. R. McIntosh, Annu. Rev. Biochem. **45:**706 (1976).]

in intracellular $[Ca^{2+}]$ may play a regulatory role in the assembly-disassembly phenomena of microtubules. The role of microtubules in impeding lateral movement of plasma membrane components was described earlier (page 274).

Two groups of proteins, τ (MW = 55,000–62,000) and "HMW-MAP" (*h*igh *m*olecular *w*eight-*m*icrotubule-*a*ssociated *p*rotein) (MW = 300,000 and 350,000), copurify with tubulin and under certain conditions are required for tubulin assembly into microtubules. The τ factor appears to be required for both the initiation and growth of microtubules and is incorporated stoichiometrically into a microtubule throughout its length. The HMW-MAP family of proteins also appears to be incorporated stoichiometrically during microtubule assembly. Possibly these tubulin assembly-enhancing proteins act by shifting the monomer-polymer equilibrium toward the polymer by stabilization of the already assembled microtubules.

Microfilaments

Microfilaments occur in all eukaryotic cells as a fine filamentous structure with a diameter of approximately 50 Å. They are composed of subunits very similar, and perhaps identical, to skeletal muscle actin. Indeed they occur in association with cytoplasmic myosin and tropomyosin (Chap. M8). Microfilaments appear to exist in two forms: *lattice microfilaments,* which form a loose network of short, interconnected filaments which may be associated with the cytoplasmic side of plasma membranes and are sensitive to the drug *cytochalasin B,* which promotes their depolymerization; and *sheath microfilaments,* which exist as bundles of fibers running beneath the plasma membrane. The sheath microfilaments are less sensitive to the action of cytochalasin B than the lattice microfilaments. Both types of microfilaments are contractile and collectively have an important musclelike role in cell locomotion and in various forms of plasma membrane movement, in particular, *endocytosis* and *exocytosis.* The latter refers to the extrusion of materials from the cell by a process in which the membrane of a secretory vesicle fuses with the plasma membrane. Endocytosis is the mechanism whereby soluble (*pinocytosis*) and particulate (*phagocytosis*) substances are internalized as part of endocytic vesicles. Once in the interior of the cell, these substances may be subjected to intracellular digestion or they may be transported through the cytoplasm and undergo subsequent exocytosis. These processes are particularly well illustrated by the uptake of low-density lipoproteins at specific receptor sites situated on the plasma membrane of human fibroblasts (Chaps. 14 and 21).

REFERENCES

Books

Bronner, F., and A. Kleinzeller (eds.): *Current Topics in Membranes and Transport,* vol. I., Academic, New York, 1970.

Chapman, D. (ed.): *Biological Membranes: Physical Fact and Function,* Academic, London, 1968.

Fox, C. F., and A. D. Keith (eds.): *Membrane Molecular Biology,* Sinauer Associates, Stamford, Conn. 1972.

Novikoff, A. B., and E. Holzman: *Cells and Organelles,* 2d ed., Holt, New York, 1976.

Rothfield, L. (ed.): *Structure and Function of Biological Membranes,* Academic, New York, 1971.

Wallach, D. F. H., and H. Fisher (eds.): *The Dynamic Structure of Cell Membranes,* Springer-Verlag, New York, 1972.

Weismann, G., and R. Claiborne (eds.): *Cell Membranes: Biochemistry, Cell Biology and Pathology,* HP Publishing Co., New York, 1975.

Review Articles

Bergelson, L. D., and L. I. Barsukov: Topological Asymmetry of Phospholipids in Membranes, *Science* **197:**224–230 (1977).

Blobel, G., and B. Dobberstein: Transfer of Proteins across Membranes, *J. Cell Biol.* **67:**835–851, 852–862 (1975).

Bretscher, M. S.: Membrane Structure: Some General Principles, *Science* **181:**622–629 (1973).

Bretscher, M. S., and M. C. Raff: Mammalian Plasma Membranes, *Nature* **258:**43–49 (1975).

Edidin, M.: Rotational and Translational Diffusion in Membranes, *Annu. Rev. Biophys. Bioeng.* **3:**179–201 (1974).

Henderson, R., and P. N. T. Unwin: Three-Dimensional Model of Purple Membrane Obtained by Electron Microscopy, *Nature* **257:**28–32 (1975).

Melchior, D. L., and J. M. Stein: Thermotropic Transitions in Biomembranes, *Annu. Rev. Biophys. Bioeng.* **5:**205–238 (1976).

Op den Kamp, J. A. F.: Lipid Asymmetry in Membranes, *Annu. Rev. Biochem.* **48**:47–71 (1979).

Palade, G.: Intracellular Aspects of the Process of Protein Synthesis, *Science* **189**:347–358 (1975).

Rothman, J. E., and J. Lenard: Membrane Asymmetry, *Science* **195**:743–753 (1977).

Singer, S. J., and G. L. Nicholson: The Fluid Mosaic Model of the Structure of Cell Membranes, *Science* **175**:720–731 (1972).

Snyder, J. A., and J. R. McIntosh: Biochemistry and Physiology of Microtubules, *Annu. Rev. Biochem.* **45**:699–720 (1976).

Timasheff, S., and L. M. Gresham: *In vitro* Assembly of Cytoplasmic Microtubules, *Annu. Rev. Biochem.* **49**:565–591 (1980).

Wickner, W.: The Assembly of Proteins into Biological Membranes: The Membrane Trigger Hypothesis, *Annu. Rev. Biochem.* **48**:23–45 (1979).

Wirtz, K. W. A.: Transfer of Phospholipids Between Membranes, *Biochim. Biophys. Acta* **344**:95–117 (1974).

Chapter 14

Receptors and Transport

The ligand binding will produce
① transportation ② chennal open ③ activitate enzyme.

Many cellular processes require, as an initial event, the specific recognition and binding of biologically active molecules. The macromolecules (generally proteins) responsible for this recognition process are termed *receptors*. Some examples are those involved in cellular recognition of neurotransmitters and drugs (Chap. M7), hormones (Chap. M11), antigens (Chap. M2), viruses (Chap. 30), plasma lipoproteins (Chap. 21), and glycoproteins (Chap. 18), among other substances. Receptor binding is generally followed by a conformational change induced in the receptor protein which can lead to activation or inhibition of an enzymic activity with resultant changes in cellular metabolism. In some cases, binding of a molecule at the plasma membrane is followed by its translocation or *transport* to the other side of the membrane or by the opening of channels through which other molecules may move.

In addition to the types of substances mentioned above, many essential nutrients are taken up by cells of the intestinal tract for delivery to the blood plasma by similar transport processes (Chap. M10). These include Ca^{2+}, Mg^{2+}, Fe^{2+}, and other ions, as well as various essential organic compounds, including some vitamins and other nutrients. The uptake from the plasma of many of these substances by various organs or cell types is also in many instances controlled by specific receptor proteins. This chapter describes the general properties of such receptor and transport systems.

RECEPTORS

Receptors perform two essential functions. ① The first is that of recognition and binding of a particular chemical substance, presumably by complementarity

Any cross-references coded M refer to the companion text, *Principles of Biochemistry: Mammalian Biochemistry*.

in the structure of the molecule and the receptor itself; this is analogous to the formation of an enzyme-substrate complex. The second function is activation of specific biological processes.

Receptors may be studied in several ways. The conventional approach has been to define the characteristics of receptors by an examination of some consequence of the receptor interaction. For example, one can characterize hormone receptors by the nature and intensity of a specific physiological effect induced by a series of structurally related hormones or drugs (Chap. M11). Such studies describe the specificity of the hormone receptor, that is, the molecular structural features necessary for interaction with that particular receptor. Although such studies have provided much information about receptors, they are indirect. They describe the receptor (operationally defined as the first step in the sequence of hormone-stimulated events) in terms of what is perhaps the most remote event in the sequence that leads to a specific physiological effect. An unknown number of steps must intervene between the hormone-receptor interaction and the effect which is being measured. More recently, techniques for studying directly the binding of radioactively labeled hormones, drugs, and other substances to their specific receptors have been widely applied. In a few cases the receptor macromolecules have been isolated and characterized.

As noted above, many kinds of substances interact with specific receptors. For the purposes of this discussion, all such agents will be referred to as *ligands,* viz., an ion or molecule that binds with a high degree of specificity to a macromolecule.

The interaction of a ligand (L) with its receptor can be described in terms of its *affinity* and *activity*. The affinity of binding reflects the energy involved in the formation of the ligand receptor (LR) complex and can be directly assessed. It is also reflected in the concentration range over which the ligand elicits a graded physiological response. The remarkable affinities of many hormone-receptor interactions are exemplified by the fact that plasma concentrations of hormones are often in the range of 10^{-11} to 10^{-9} M (Chap. M11). Yet, even at these very low concentrations the agent binds specifically to its receptors. The second property of ligand interaction with receptors concerns the nature and intensity of the evoked physiological response, which is the biological activity of the ligand. This latter property is dealt with in Chap. M12.

Ligand-Receptor Interactions

Ligand-Receptor Binding The ligand-receptor interaction can be measured by radioligand binding studies or, somewhat less directly, by examination of dose-response curves for a biological effect (Chap. M12). The basic requirements for such studies are a radiolabeled form of the ligand and a source of receptors, such as a membrane fraction, intact cells, or solubilized preparations. The receptors and the ligand are incubated until equilibrium is reached, and the amount of labeled ligand bound to the receptors is determined. When particulate preparations are used, this is generally accomplished by millipore filtration, centrifugation, or equilibrium dialysis, which separates free from receptor-bound ligand. With soluble preparations receptor-bound and free ligand may be sep-

The main point of the interaction of a ligand with its receptor

arated by gel filtration, equilibrium dialysis, precipitation of the receptor-bound ligand by polyethylene glycol, differential adsorption of free labeled hormone to talc or charcoal, or other methods.

Criteria for Receptor Identification In general, several criteria must be met if true receptors are being labeled; the most important are the following:

1. Saturability There should be only a finite number of receptor sites; thus the receptor binding phenomenon should display saturability. Most cells contain only a few hundred to a few thousand receptors of a particular type. In contrast, "nonspecific" binding is that binding of a radioligand which occurs to various nonreceptor sites. It is generally nonsaturable and linearly related to free ligand concentration.

2. Affinity The concentration range over which the ligand occupies the receptors should be comparable to the range over which a biological response is elicited. For reasons discussed later (Chap. M12) the affinity of a ligand for its receptor assessed by direct binding studies may not agree exactly with the apparent affinity determined from a biological dose-response curve.

3. Specificity This is the most important criterion, since specificity is the hallmark by which receptors are defined. The details of specificity apparent from the ability of various ligands to stimulate (or inhibit) a biological response presumably reflect the specificity of binding of ligand and receptor. Thus the sites labeled by a radioligand should exhibit the specificity and stereospecificity of the biological response mediated by the receptor. For example, the order of potency of a series of hormones or hormone analogs in eliciting a biological response should be exactly reflected in their order of potency in competing for the binding sites. Any significant discrepancy in such potency series would raise questions as to the nature of the sites being labeled. It should be noted that in addition to physiological receptors, other biological macromolecules can potentially bind ligands with high affinity. These include transport proteins and degrading enzymes. If such proteins are labeled, however, their specificity would differ from that of the physiological receptors.

For some receptors, e.g., those for hormones or neurotransmitters, ligand binding studies *alone* cannot prove that a particular binding site is *equivalent* to a physiological receptor. This is because, in addition to their ability to bind ligands, receptors also transduce this binding into a biological signal. Thus the ultimate proof that an isolated binding site is a receptor requires the successful reconstitution of the hormonally responsive system from the isolated components. This means that when the isolated binding site is recombined with other biochemical effector components (e.g., enzymes, transport systems, etc.), it should convey upon the effector component sensitivity to the appropriate ligands. Such reconstitutions have been achieved in only a few systems.

Law of Mass Action and Graphical Analysis of Ligand Binding In receptor binding studies, the simplest assumption concerning the interaction of a radioactive ligand L with a receptor R is the formation of a complex RL, as shown in Eq. 1 for a bimolecular reaction:

$$L + R \rightleftharpoons RL \qquad (1)$$

At equilibrium,

$$\frac{[R][L]}{[RL]} = K_L \tag{2}$$

where K_L is the dissociation constant of the complex and [R], [L], and [RL] represent the concentrations of free receptor, free ligand, and receptor-bound ligand, respectively. Since the total concentration of receptors $[R_t]$ equals [R] + [RL], Eq. 2 may be rewritten as

$$\frac{[R_t - RL][L]}{[RL]} = K_L \tag{3}$$

which rearranges to

$$\frac{[RL]}{[R_t]} = \frac{[L]}{K_L + [L]} \tag{4}$$

The ratio $[RL]/[R_t]$ is the fraction of total receptors occupied by ligand. At one-half maximal occupancy of the receptors, $[RL]/[R_t] = 0.5$ and $K_L = [L]$. Thus, in the simplest case the concentration of ligand required for half maximal occupancy of the receptors is equal to K_L.

Equation 4 can be further rearranged to give

$$[RL] = \frac{[R_t][L]}{K_L + [L]} \tag{5}$$

This is the familiar hyperbolic function in which [RL] approaches $[R_t]$ as [L] becomes large. In radioligand binding studies Eq. 5 is useful for determining the number of receptor binding sites R_t and their affinity K_L for the ligand L. Experimentally, the radioligand is added over a range of concentrations to a fixed concentration of receptors, present in a membrane fraction, cell suspension, or in soluble or purified form. The level of binding which is approached asymptotically at high ligand concentrations is R_t, and the concentration of free ligand which fills half the receptors $[R_t]/2$ represents the K_L. The relationship is plotted in Fig. 14.1a. Alternatively, this relationship may be plotted as $[RL]/[R_t]$ vs. log [L] (Fig. 14.1b). This plot is a symmetrical sigmoidal curve that has a maximal slope at the midpoint of 0.58 per tenfold change in concentration.

tenfold ?

Plots such as that shown in Fig. 14.1a are often used in hormone receptor research to determine K_L and R_t; however, the horizontal asymptote is difficult to determine with accuracy. Accordingly, several linearized forms of such binding curves are generally used. An example is the Scatchard plot (Fig. 14.2a). Equation 3 can be rearranged as follows:

$$\frac{B}{F}\frac{[RL]}{[L]} = \frac{\overset{B_{max}}{([R_t]} - \overset{B}{[RL]})}{K_L} \tag{6}$$

Let us redefine [RL] as bound ligand or B, [L] as free ligand or F, and R_t as B_{max}, and then Eq. 6 becomes

$$\frac{R_2}{L}\frac{B}{F} = \frac{B_{max} - B}{K_L} = -\frac{1}{K_L}B + \frac{B_{max}}{K_L} \tag{7}$$

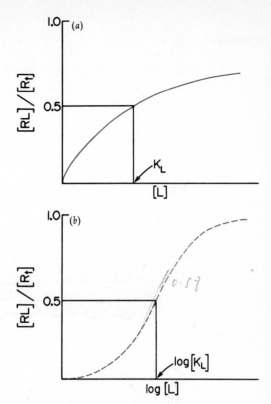

Figure 14.1 (a) Plot of the relationship of $[RL]/[R_t]$ to $[L]$ from Eq. 4. [RL], concentration of bound ligand; $[R_t]$, concentration of total receptors; [L], concentration of free ligand; K_L, the equilibrium dissociation constant defined by Eq. 2. (b) Plot of the same relationship as in (a) with the concentration of ligand plotted logarithmically. [*From L. T. Williams and R. J. Lefkowitz, Receptor Binding Studies in Adrenergic Pharmacology, Raven, New York 1978, p. 29.*]

Thus, a plot of the ratio of bound to free ligand vs. the concentration of bound ligand, B/F vs. B, yields a straight line with a slope of $-1/K_L$ and an abscissa intercept (i.e., when B/F = 0) of B_{max} (Fig. 14.2a). Several alternative linear forms of such data may be used, all of which derive from rearrangements of Eqs. 3 to 5. For example, from the reciprocal of Eq. 5 with the B and F notation introduced above we obtain

Figure 14.2 Graphical representation of ligand binding data in three different linear forms. B_{max} = total number of receptor binding sites; K_L = equilibrium dissociation constant of ligand L for receptor R; B = receptor-bound ligand concentration; F = free ligand concentration.

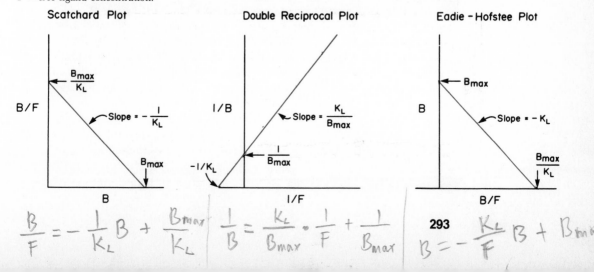

293

$$\frac{B}{F} = -\frac{1}{K_L}B + \frac{B_{max}}{K_L} \qquad \frac{1}{B} = \frac{K_L}{B_{max}} \cdot \frac{1}{F} + \frac{1}{B_{max}} \qquad B = -\frac{K_L}{F}B + B_{max}$$

$$\frac{1}{B} = \frac{K_L + F}{B_{max} F} = \frac{K_L}{B_{max} F} + \frac{1}{B_{max}} \qquad (8)$$

Thus a plot of $(1/B)$ vs. $(1/F)$ (Fig. 14.2b) yields a linear plot with slope of (K_L/B_{max}), an ordinate intercept of $(1/B_{max})$, and an abscissa intercept of $-1/K_L$. This equation is of the same form as the familiar Lineweaver-Burk equation (Chap. 10).

Alternatively, if Eq. 8 is multiplied by B_{max} and rearranged, we obtain

$$B = -K_L \frac{B}{F} + B_{max} \qquad (9)$$

A plot of B vs. B/F (Fig. 14.2c) yields a linear plot with slope of $-K_L$, an ordinate intercept of B_{max}, and an abscissa intercept of (B_{max}/K_L). Such graphs are referred to as Eadie-Hofstee plots. *All of these equations are algebraically interchangeable; they all follow from the definition of K_L.* The advantage of linear plots (as compared with Fig. 14.1) is that data can be collected at low or intermediate concentrations of ligand where nonspecific binding is low and the linear plot then extrapolated to estimate K_L and B_{max}.

The equations developed above for analyzing *equilibrium* ligand binding data are formally analogous to the equations developed in Chap. 10 for studying enzyme *kinetics*. There are certain pitfalls, however, in directly transferring the approaches of enzymology to the study of receptors. In enzyme studies, the [substrate]/[enzyme] ratio is usually very high (at least 100) so that one can assume "pseudo-first-order kinetics," i.e., that total and free substrate concentrations are equal. In the study of hormone-receptor interactions, [R] and [L] are often in the same concentration range. Hence, total and free ligand concentrations may be very different.

In many ligand-receptor systems, Scatchard plots are nonlinear. This may be due to a number of factors, other than certain technical artifacts. For example, there may be several types of receptor binding sites with different affinities for the ligand. Several hormones are capable of interacting with different kinds of receptors to produce different biological effects, with different affinities for the different receptors. This produces a "concave upward" Scatchard plot. Alternatively, the existence of "negative cooperativity" amongst receptors (see Chap. 10 and below) might also cause concave upward Scatchard plots. This occurs when the binding of one ligand molecule to a receptor decreases the affinity for binding subsequent ligand molecules. Figure 14.3 shows a concave upward Scatchard plot for binding of insulin to its receptors in human lymphocytes. Equilibrium binding studies, such as those depicted in Fig. 14.3, cannot distinguish between negative cooperativity and multiple classes of binding sites. Kinetic experiments can aid in distinguishing these possibilities.

An additional method of presenting equilibrium data for the binding of a ligand to a receptor is the Hill plot. This is based on the Hill equation originally developed to describe deviations from classic mass-action principles in the binding of oxygen to hemoglobin. An example of a Hill plot for allosteric enzymes is presented in Chap. 10. Its application to ligand binding is analo-

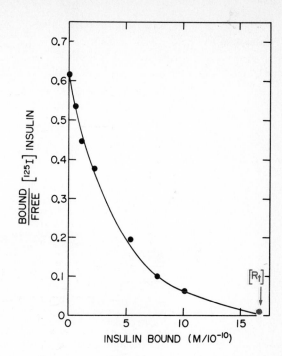

Figure 14.3 Curvilinear Scatchard plot for insulin binding to cultured human lymphocytes. [^{125}I]insulin $(7 \times 10^{-12}\,M)$ was incubated with 10^7 cells at 15°C for 90 min with a range of concentrations of unlabeled insulin. The bound/free ratio of [^{125}I]insulin is plotted as a function of the insulin bound. Nonspecific binding has been subtracted. [*From P. DeMeyts, Insulin and Growth Hormone Receptors, in M. Blecher (ed.), Methods in Receptor Research, vol. I, Dekker, 1976, p. 349.*]

gous to that for enzyme-substrate reactions. With the B, F, B_{max} terminology, Eq. 5 becomes

$$B = \frac{B_{max}[L]}{K_L + [L]} \tag{10}$$

In a sequential binding model when more than one ligand molecule can bind to each receptor molecule and the binding of each ligand molecule is influenced by the binding of the prior ligand molecules, an analogous equation can be written

$$B = \frac{B_{max}[L]^n}{K'_L + [L]^n} \tag{11}$$

where n is the "Hill coefficient," which cannot exceed the number of ligand binding sites per receptor molecule, and K'_L is a composite, or "average," dissociation constant composed of the intrinsic dissociation constants for each discrete binding step.

Rearrangement of Eq. 11 yields

$$\frac{B}{B_{max}} = \frac{[L]^n}{K'_L + [L]^n} \tag{12}$$

which is the Hill equation. A logarithmic transformation of this equation is often used for plotting ligand binding data. Equation 12 may be rearranged as follows:

$$B_{max}[L]^n = B\,K'_L + B[L]^n \tag{13}$$

Rearranging terms gives

$$\frac{B_{max}[L]^n - B[L]^n}{B} = K'_L \qquad \text{or} \qquad \frac{[L]^n(B_{max} - B)}{B} = K'_L \qquad (14)$$

In logarithmic form this becomes

$$n \log [L] + \log \frac{B_{max} - B}{B} = \log K'_L \qquad (15)$$

or

$$\log \frac{B}{B_{max} - B} = n \log [L] - \log K'_L \qquad (16)$$

When $\log [B/(B_{max} - B)]$ is plotted as a function of $\log [L]$, a "Hill plot" is obtained; the linear slope is n, the Hill coefficient. As in enzyme substrate interactions, $n > 1$ is consistent with positive and $n < 1$ with negative cooperativity (Chap. 10).

An example of a Hill plot for binding of human growth hormone (Chap.

Figure 14.4 Hill plot for binding of human growth hormone to cultured human lymphocytes. The Hill coefficient of 0.97 is not significantly different from 1.0. [*Modified from P. DeMeyts, Insulin and Growth Hormone Receptors, in M. Blecher (ed.), Methods in Receptor Research, vol. I, Dekker, 1976, p. 353.*]

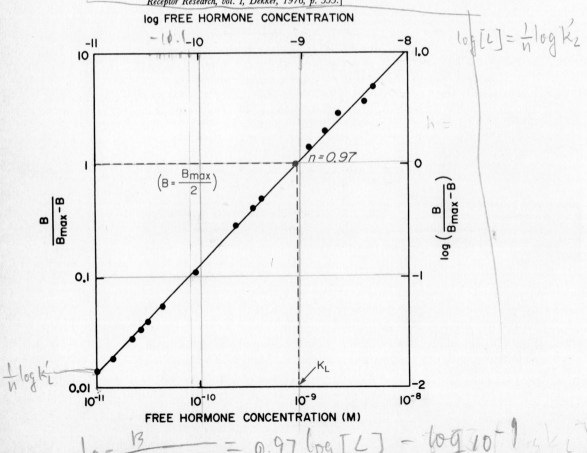

1. photosynthetic apparatus is composed of ___①___, ___②___, ___③___, (photosystem I, II, and a light-harvesting array)

2. In green plants, the array is composed of ___①___, ___②___, ___③___, in a form termed the ___④___ (① chl a ② chl b ③ carotenoid ④ light-harvesting Chl a/b - protein)

3. In deep water, the plant can use weak light by _____ (an additional light-gathering pigments, chlorophyll c and a carotenoid, peridinin).

4. ① Bacteriochlorophyll dimer (BChl)$_2$
   ② two molecules of Bacteriopheophytin
   ③ one atom of Fe^{2+}
   ④ two molecules of ubiquinone
   ⑤ a carotenoid

Ouabain ———— ⊕ Na^+, K^+ ATPase
Ionophores ———— Valinomycin
Ion channels ———— gramicidin.

Vit B$_{12}$. ① corrin core
② cobalt .
③ 5'-deoxyadenosyl group
④ 5,6-dimethylbenzimidazole

acetyl CoA Carboxylase

① biotin carboxyl carrier protein (BCP)
② biotin carboxylase
③ transcarboxylase

addition of insulin and adrenal steroid
accelerates protein synthesis to of ① ATP citrate l
② acetyl CoA carboxylase palmitate synthetas
④ malic enzyme .

linoleic ———— (w-6)
linolenic ———— (w-3)
Palmitoleic ———— (w-7)
oleic ————————— (w-9)

M20) to its receptors in lymphocytes is shown in Fig. 14.4. When the receptors are half filled with ligand,

$$B = \frac{B_{max}}{2} \quad \text{and} \quad \frac{B}{B_{max} - B} = 1 \quad \text{so} \quad \log \frac{B}{B_{max} - B} = 0 \quad (17)$$

$\frac{1}{n} \log k_L' = \log [L]$

At this point the corresponding value on the abscissa is log K_L'. When cooperative interactions are present, this intercept is $(1/n) \log K_L'$.

When $n \neq 1$, cooperative interactions may be present but multiple classes of binding sites will yield the same result. Also when $n \neq 1$, the plot is apparently linear over a limited range of log [L] values that are distributed symmetrically around log K_L'.

As noted, a Hill plot with $n < 1$ or a curvilinear Scatchard plot may indicate either negatively cooperative interactions or heterogeneity of receptor sites. Neither plot serves to distinguish these possibilities. A further test of whether negatively cooperative interactions are responsible for "shallow" Hill plots of receptor binding data can be accomplished with kinetic as opposed to equilibrium experiments. Thus, if negative cooperativity occurs, the rate of ligand dissociation should increase as B/B_{max} increases, i.e., as the fraction of receptors filled with ligand increases and the overall affinity of binding decreases. Such results have been described for several hormone receptor systems, most notably the insulin receptor (Chap. M16).

heterogeneity 多相性.

Competition Binding Studies The specificity of a ligand binding site is one of the major criteria for equating it with a physiological receptor. The specificity of a ligand binding site is generally determined by examining the ability of a series of structurally related analogs to inhibit competitively the binding of a radiolabeled ligand. From such data "competition curves" such as those shown in Fig. 14.5 can be constructed. In the simplest case, where there is only one

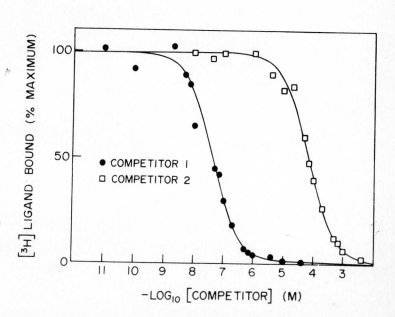

Figure 14.5 Competition curves of two competitors with a [³H]ligand for a uniform population of receptor binding sites. The competition curves shown conform closely to the law of mass action and the K_I's of the two competitors may be calculated from the concentrations producing 50 percent inhibition of radioligand binding according to Eq. 18.

● COMPETITOR 1
□ COMPETITOR 2

[³H] LIGAND BOUND (% MAXIMUM)

$-LOG_{10}$ [COMPETITOR] (M)

class of binding sites and no cooperative interactions, the concentration of a competitor, I, which inhibits 50 percent of the radioligand binding is directly related to its dissociation constant.

$$K_I = \frac{[I_{50}]}{1 + ([L]/K_L)} \tag{18}$$

Thus the concentration of competitor producing 50 percent inhibition of radioligand binding, $[I_{50}]$, can be used to determine the dissociation constant of the unlabeled competitor. It should be noted that these equations are similar to those for competitive inhibitors of enzymes and that [I] and [L] refer to the *free* concentrations of I and L at equilibrium.

The effects of a competitive inhibitor can also be described in a saturation binding experiment plotted in one of the ways shown in Fig. 14.2. A competitive inhibitor does not decrease B_{max} but will change only the apparent K_L of the radioligand. Thus, the effects of a *competitive* inhibitor on the various graphical presentations in Fig. 14.2 would be as follows: Scatchard plot—decrease in slope, no change in B_{max}; double reciprocal plot—increased slope, same ordinate intercept; Eadie-Hofstee plot—increased slope with same ordinate intercept. For a *noncompetitive antagonist* the plots will reveal a decreased B_{max} and no change in apparent K_L.

Rates of Ligand-Receptor Association and Dissociation From Eq. 19 it can be seen that for the simplest possible case the binding of a ligand to a receptor is a bimolecular reaction.

$$L + R \underset{k_2}{\overset{k_1}{\rightleftharpoons}} RL \tag{19}$$

Thus at any time t the rate of association is $k_1[L][R]$ and the rate of dissociation is $k_2[RL]$, where k_1 and k_2 are second- and first-order rate constants, respectively.

In some ligand-receptor binding studies the concentration of ligand is much greater than receptor, and hence as the binding reaction proceeds, [L] does not change appreciably while R (unoccupied receptors) decreases significantly. Thus in such cases the binding reaction can be considered a "pseudo-first-order" reaction. It is a relatively straightforward matter to determine these rate constants from experimental measurements of the rates of ligand association with and dissociation from the receptors. In a number of systems it has been possible to show that $k_2/k_1 = K_L$. In the presence of cooperative interactions, however, more complicated patterns may be observed. Thus the rate of dissociation may change, depending on the fractional receptor occupancy.

Receptor Regulation

The concentration of receptors in or on cells is subject to various forms of regulation (Chap. M12) and serves to regulate cellular responsiveness to the actions of various hormones and neurotransmitters.

A particularly dramatic example of the dynamic regulation of receptors by ligands is the phenomenon of ligand-receptor internalization. First demonstrated for the plasma membrane receptors for low-density lipoproteins (LDL) (see Chap. 21), this process has now been observed for a variety of ligands whose receptors are present at the cell surface, e.g., insulin (Chap. M16), epidermal growth factor (EGF) (Chap. M20), and human chorionic gonadotropin (HCG) (Chap. M20). This often occurs via a special form of endocytosis (Chap. 13), utilizing organelles termed *coated pits* (Fig. 14.6). In some cases the receptors, e.g., those for LDL, are located in the coated pits. In other cases, for example, the receptors for insulin, the binding sites are initially located diffusely over the cell surface and move to the coated pits after they become occupied with ligand. The "coated" pits are coated on their cytoplasmic surface with the protein *clathrin* (MW = 180,000). From this coated pit the receptor is transferred to a special intracellular vesicle which appears to avoid initial fusion with lysosomes while traveling to the Golgi region of the cell. Ultimately the vesicles fuse with lysosomes exposing the ligand and the receptor to the degradative actions of the lysosomal enzymes. Alternatively, some vesicles may escape lysosomal fusion and recycle their receptors back to the plasma membrane or release undegraded ligand at intracellular sites of action. Electron micrographs documenting the receptor-mediated endocytosis of LDL via coated pits are shown in Fig. 14.6. Although the full physiological significance of receptor internalization is not yet entirely clear, it probably serves a number of purposes, including: (1) "down regulation" of receptors, leading to decreased biological responsiveness to hormone or other ligand (Chap. M12); (2) ligand inactivation; (3) transport of ligand to intracellular sites of action, e.g., as with LDL (Chap. 21).

The Nature of Receptors and Ligand-Receptor Interactions

As yet, few receptors have been isolated and characterized. There are two main reasons for this. First, since the amount of receptors present in most tissues is minute, approximately a 10,000- to 100,000-fold purification is required. An exception is one type of receptor for the neurotransmitter acetylcholine (Chap. M7), which is found in the electric organs of certain fish, where they may make up as much as 20 to 40 percent of the total membrane protein.

A second problem is that the protein receptors must often be solubilized from membranes with detergents that may denature them. This is true for peptide, catecholamine, and neurotransmitter receptors, but not for steroid hormone receptors, which are soluble in the absence of detergents (Chap. M12). The membrane-bound receptors are transmembrane integral membrane proteins (Chap. 13), which are most effectively solubilized with non-ionic detergents. Several of these are glycoproteins and some may contain tightly bound lipid. Some receptors are composed of several subunits. Solubilized receptors generally behave as asymmetric proteins with relatively large hydrodynamic radii (60 to 70 nm) in relation to their sedimentation coefficients of 6.5 to 9.0 S. They bind large amounts of detergent, which suggests that they may possess a hydrophobic lipid-binding domain. The soluble cytoplasmic receptors are easier to purify since detergents are not required. They are in general asymmetric proteins several times longer than they are wide with molecular weights ranging from 50,000 to 300,000.

coated pit

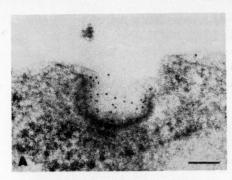

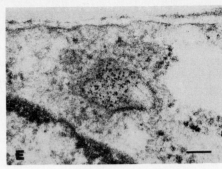

a coated pit being transformed into an endocytic vesicle

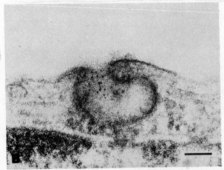

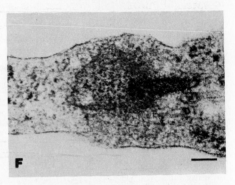

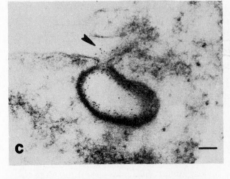

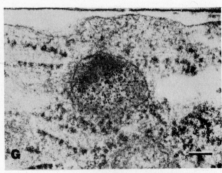

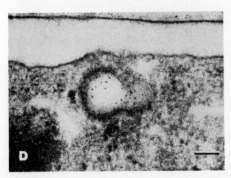

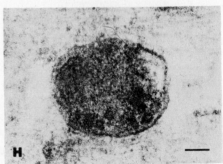

Because of the great specificity and high affinity of ligand-receptor interactions, affinity chromatography with immobilized ligands is useful for receptor purification (Chap. 9). For the glycoprotein membrane-bound receptors plant lectin-agarose chromatography may also afford a convenient purification procedure (Chap. 9). In some cases, naturally occurring antireceptor antibodies have been used for receptor purification, either for immunoprecipitation or as immunoaffinity adsorbants.

Since few receptors have been purified and characterized, little is known concerning ligand-receptor interaction. Its seems clear, however, that several types of interactions contribute to this noncovalent binding. For peptide interactions with membrane receptors as well as steroid interactions with cytoplasmic receptors, hydrophobic interactions may provide the major driving force. However, for some peptide hormone receptors (Chap. M12) hydrogen and ionic binding may also be important, especially for specificity of binding. For example, location of a charged group on a ligand in proximity to a hydrophobic portion of the receptor may serve to weaken the binding interaction. For some hormones charged groups actually appear to be important in the binding of the hormone to its receptor. In general the reversible interaction of hormones with their receptors does not alter the biological activity of the hormone molecule.

The structural features of a ligand which determine *affinity* for a receptor are distinct from those which determine *activity*, i.e., the ability to stimulate a biological response. Thus a ligand may bind very tightly to a receptor (high affinity) yet have little or no activity (potent antagonists). The affinity of a ligand is presumably the net result of the various interactions which occur between complementary portions of the ligand and its receptor. Activity is generally related to other structural features of the ligand molecule which tend to promote conformational changes in the receptor molecules that then initiate the cascade of interactions ultimately resulting in a biological response. The mechanisms whereby ligand occupancy of receptors is transduced into a biological response are discussed in Chap. M12.

Figure 14.6 Electron micrographs showing representative stages in the receptor-mediated endocytosis of low-density lipoprotein-ferritin and its subsequent delivery to the lysosome. Normal fibroblasts were incubated with LDL-ferritin for 2 h at 4°C, washed extensively, and then warmed to 37°C for various times. Scale bar = 1000 Å. (a) A typical coated pit (time at 37°C, 1 min). ×67,900. (b) A coated pit being transformed into an endocytic vesicle with LDL-ferritin included (time at 37°C, 1 min). ×56,700. (c) Formation of a coated vesicle. As the plasma membrane begins to fuse to form the vesicle, some of the LDL-ferritin is excluded from the interior and is left on the surface of the cell (arrow) (time at 37°C, 1 min). (d) A fully formed coated vesicle that appears to be losing its cytoplasmic coat on the right side (time at 37°C, 2 min). ×52,000. (e) An endocytic vesicle that has completely lost its cytoplasmic coat. Note the irregular shape of this vesicle (time at 37°C, 2 min). ×52,500. (f) An irregularly shaped endocytic vesicle that contains more LDL-ferritin than a typical coated vesicle and also has a region of increased electron density within the lumen (time at 37°C, 6 min). ×52,500. (g) An endocytic vesicle similar to (f) with more electron-dense material in the lumen (time at 37°C, 6 min). ×48,300. (h) A secondary lysosome that contains LDL-ferritin (time at 37°C, 8 min). ×52,500. [*From M. S. Brown and J. L. Goldstein, Harvey Lectures* **73**:*163–201* (*1979*).]

Living cells are surrounded by lipid bilayers, the plasma membranes in which various proteins are embedded (Chap. 13). These membranes are relatively impermeable to hydrophilic substances, such as ions and a variety of metabolites, and thus provide a selective permeability barrier. Nonetheless, hydrophilic substances present outside the cell must gain access to the cytoplasm and be maintained there within narrowly defined concentration ranges. The mechanisms by which such materials are moved across extracellular or intracellular membranes are generally termed *transport* processes.

Solutes may traverse membranes in the different ways that are listed in Table 14.1. These processes are either *mediated* or *nonmediated*. Nonmediated transport of a solute is simply physical diffusion of the solute down its concentration gradient. The diffusing molecule is neither chemically modified nor bound

branes mainly by simple diffusion are generally nonpolar molecules which are soluble in the lipid bilayer. This class of solutes includes many drugs.

In contrast most solutes traverse biological membranes by specific pathways characterized (1) by specificity for the substance transported and (2) by saturation kinetics, i.e., the transport system can become saturated with the solute transported. Such mediated transport is made possible by proteins capable of reversibly binding specific substrates; these transport molecules have been variously termed *transport systems, carriers, porters, translocases,* or *permeases.*

Several classes of mediated transport have been recognized (Table 14.1). *Facilitated diffusion* refers to processes whereby a compound moves across the membrane by mechanisms that *do not* require the explicit provision of energy. Entry of glucose into hepatocytes, erythrocytes, and muscle cells along its concentration gradient is such a process (Chap. 17).

Active transport is the term employed for processes by which a molecule moves across the membrane *against* its concentration or, for ions, its electrochemical gradient. The required energy may be supplied in either of two ways. (1) Energy is provided by simultaneous hydrolysis of ATP (ATPase activity) or of some other high-energy compound on the surface of the protein that is serving as the porter. Such processes are referred to as *primary active transport* and also as *pumps.* (2) Transport of a metabolite is coupled to the simultaneous movement of a second substance, which is moving down its concentration gradient, a process for which ΔG is large and negative. The second molecule may be moving in

---

TABLE 14.1

Classification of Transport Processes

I. Nonmediated transport—i.e., simple diffusion
II. Mediated transport
    A. Facilitated diffusion—i.e., carrier-mediated passive transport
    B. Active transport
        1. Primary (ATPase-linked pumps)
        2. Secondary (gradient-coupled)

the same direction as the first (*symport*) or in the opposite direction (*antiport*). Such processes are referred to as *secondary* or *gradient-coupled active transport*.

Transport processes perform a number of vital functions. They maintain cellular volume, pH, and tonicity. They maintain gradients of K^+ and Na^+ between the intra- and extracellular fluid compartments, which are necessary for the normal function of excitable cells (Chap. M7). They bring important metabolites into the cell while extruding products of cellular metabolism for excretion or for delivery to another type of cell.

Nonmediated Transport-Simple Diffusion

In simple diffusion, the transport rate or net flux is directly proportional to the concentration gradient of the substance across the membrane. Several types of forces may drive such nonmediated transport processes. If the membrane is permeable to the substance and its concentration is higher on one side of the membrane than the other, the solute will tend to be transferred across the membrane down its concentration gradient. Differing hydrostatic pressures of fluids on opposite sides of a membrane may provide the driving force for passive movement of fluid from higher to lower pressure. A third type of driving force for nonmediated diffusion is the electric potential developed by separation of charges across a membrane with different finite permeabilities for the ions involved. In each of the cases mentioned, transport against a gradient requires provision of energy (active transport). In other terms, regardless of the specific driving force involved, passive diffusion across permeable membranes occurs spontaneously only in the direction for which the change in free energy (ΔG) for the transport process is negative.

That the rate of solute transfer by free diffusion is proportional to the driving force is an important principle summarized in Fick's first law of diffusion which states

$$J = -DA\frac{dc}{dx} \tag{20}$$

where J is the rate of flux of solute, A is the area of membrane available for diffusion and dc/dx is the concentration gradient which provides the driving force for diffusion. The diffusion coefficient D depends on the physical properties of solute and solvent, such as the shape and size of the solute particle and the viscosity of the solvent. When a solute bears a net charge, then its rate of diffusion across a membrane is influenced not only by its concentration gradient but also by the electric potential gradient across the membrane.

Another factor regulating diffusion is the solubility of the substance in the membrane, as reflected by its partition coefficient between its aqueous environment and the membrane. Some solutes are weak acids or bases. For such substances the rate of diffusion into cells depends on the concentration of the nonionized, i.e., lipid-soluble, form. This is in turn critically dependent on the pH of the extracellular fluid, as can be discerned from the Henderson-Hasselbach equation (Chap. 3).

$$pH = pK + \log\frac{[A^-]}{[HA]} \tag{21}$$

If only the undissociated form, HA, can diffuse across the membrane, as is the case with most weak acids, then a 1 unit decrease in pH resulting in a tenfold increase in [HA] produces a tenfold increase in the rate of diffusion of the solute or drug into the cells.

Facilitated Diffusion

Transport by facilitated diffusion is analogous to other receptor-mediated phenomena discussed earlier in this chapter. It is mediated by specific interactions of solutes with transport proteins located in cellular membranes. Since the number of such units in the membrane is finite, facilitated diffusion processes are usually characterized by saturability (Fig. 14.7, upper curve). This is in contrast with the situation for nonmediated transport processes, in which the rate of transport shows a linear dependence on solute concentration (Fig. 14.7,

① have a character of saturability

② specificity

Another "receptor-like" characteristic of the transport by facilitated diffusion is their specificity. This is apparent in the competition for the transport sites between compounds of similar structure. In several nonelectrolyte transport systems of mammalian cells there is also *strict stereospecificity* with D sugars transported but not L sugars or L amino acids transported but not D amino acids.

An example of such a facilitated diffusion system is the hexose transport system of human erythrocytes. The specific transport protein (MW = 55,000) is a glycoprotein which has been purified and reconstituted into phospholipid vesicles. It binds sugars with stereospecificity for the D form. Each carrier molecule is able to translocate about 500 glucose molecules per second.

driven by the gradients of the transported solutes

Facilitated diffusion processes are driven only by the gradients of the transported solutes and do not require an external supply of energy. In most cases

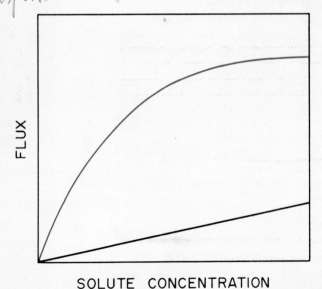

FLUX

SOLUTE CONCENTRATION

Figure 14.7 The dependence of transmembrane solute flux on solute concentration. The upper curve shows the dependence on the solute concentration of the flux of a solute that is mediated by a facilitated diffusion mechanism. The lower line shows the expected dependence of solute flux on concentration for a nonmediated transport process.

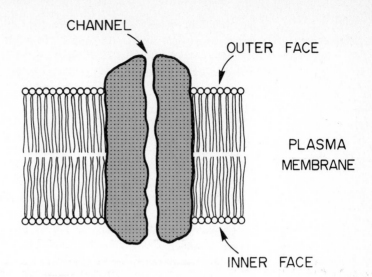

CHANNEL

OUTER FACE

PLASMA
MEMBRANE

INNER FACE

Figure 14.8 Model for an integral membrane protein involved in mediated transport. [*From S. J. Singer, J. Supramol. Struct.* **6***:313–323 (1977).*]

the molecular properties of the transport systems involved are unknown. In

such transport proteins may exist in at least two distinct conformational states. In one state the ligand binding site is accessible to the aqueous phase on one side of the membrane. In the other conformational state, the binding site is accessible at the other side of the membrane. The affinity of binding the transported ligand may be different in the two states. The energy necessary for conversion between the two states is derived from the negative free energy of the ligand's binding to its transport site and/or from the concentration gradient of the ligand (or from another energy source in the case of active transport). The term *gating* is used to describe the phenomenon whereby conformational changes triggered by substrate binding accomplish the transport of a solute molecule from one side of a membrane to the other by opening a channel (see below).

The integral membrane proteins involved in gating probably have structures similar to that shown in Fig. 14.8. Such proteins are oligomers with a small number (two to four) of similar or identical subunits spanning the membrane. Each polypeptide chain has the same orientation in the membrane. Such aggregates may generate a narrow water-filled *channel* down the center of the aggregate through which translocation of solute occurs. Hydrophilic groups on the protein presumably line this channel. If the aggregate is a dimer of two identical polypeptide chains, it would have a twofold axis of symmetry parallel to this channel and perpendicular to the plane of the membrane.

Active Transport

Active transport refers to the net movement of a solute against its concentration or (for ions) electrochemical gradient. The energy required for such "uphill" transport may be provided either by direct coupling to an energy-producing reaction (primary active transport) or by coupling to the net transport of another solute down its gradient (secondary active transport). In

305

mammalian tissues the *Na⁺ pump* and the *Ca²⁺ pump* are examples of primary active transport in which the "uphill" transport of ions is linked to simultaneous hydrolysis of ATP by specific enzymes (ATPases). Secondary active transport systems exist in certain tissues for glucose and other sugars and for neutral amino acids.

Secondary Active Transport In some mammalian cells the transport of sugars, neutral amino acids, and some other solutes is often accomplished by *sodium symport systems*. Examples are the transfer of glucose against its concentration gradient from the intestine into the blood (Chap. M10) and the reabsorption of glucose from the renal tubule (Chap. M5). Such mechanisms are closely related to facilitated diffusion but with the carrier representing a two-substrate rather than a single-substrate system. Thus the carrier has binding sites for both Na^+ and another solute. The free energy made available by the inward Na^+ movement *down* its electrochemical gradient is used for the symport (cotransport) of the other solute *against* its concentration gradient. As Na^+ and the other solute (e.g., glucose) are discharged into the cytoplasm, the Na^+ pump (see below) returns Na^+ to the extracellular environment, thus maintaining the Na^+ electrochemical gradient. Since this latter process requires the hydrolysis of ATP (one step removed), such processes are termed *secondary active transport*. Some secondary active transport mechanisms such as Na^+-Ca^{2+} exchange in cardiac muscle cells are *antiport* systems, in which a solute is transported against its concentration gradient by the same carrier which is transporting another solute in the opposite direction down its concentration gradient. In mammalian cells, antiports, like symports, are often driven by Na^+ gradients maintained by the Na^+ pump.

Primary Active Transport

The Na⁺ Pump and Na⁺,K⁺-ATPase In most animal cells the intracellular $[K^+]$ is relatively high and constant, between 120 and 160 mM, whereas $[Na^+]$ is much less, being 10 to 20 mM. In contrast, extracellular fluid contains a relatively high $[Na^+]$, about 150 mM, and $[K^+]$ is usually about 4 mM. Therefore, concentration gradients of these two ions exist across all cell membranes. The constancy of the high intracellular $[K^+]$ is maintained by an energy-requiring extrusion of Na^+ out of the cell with replacement by K^+.

The analogy to a mechanical pump is apt in that energy is utilized to move the ions against an opposing concentration gradient. In all cases studied, the source of energy for this work is ATP, which serves as substrate for a membrane-bound ATPase. Erythrocyte membranes have an ATPase that requires Mg^{2+} and both Na^+ and K^+ for activation. The enzyme is spatially asymmetric, i.e., it is stimulated by extracellular but not intracellular K^+ and by intracellular but not extracellular Na^+. Addition of either Na^+ or K^+ alone does not affect the rate of enzymic activity, but when both ions are present, ATP hydrolysis is markedly accelerated. Extensive evidence indicates that the Na^+,K^+-ATPase is the enzymic equivalent of the system responsible for active transmembrane Na^+ transport, the Na^+ pump.

The Na^+,K^+-ATPase (MW $\approx$ 250,000) is involved in the transport of Na^+ and K^+ across the plasma membrane of nearly all eukaryotic cells. The puri-

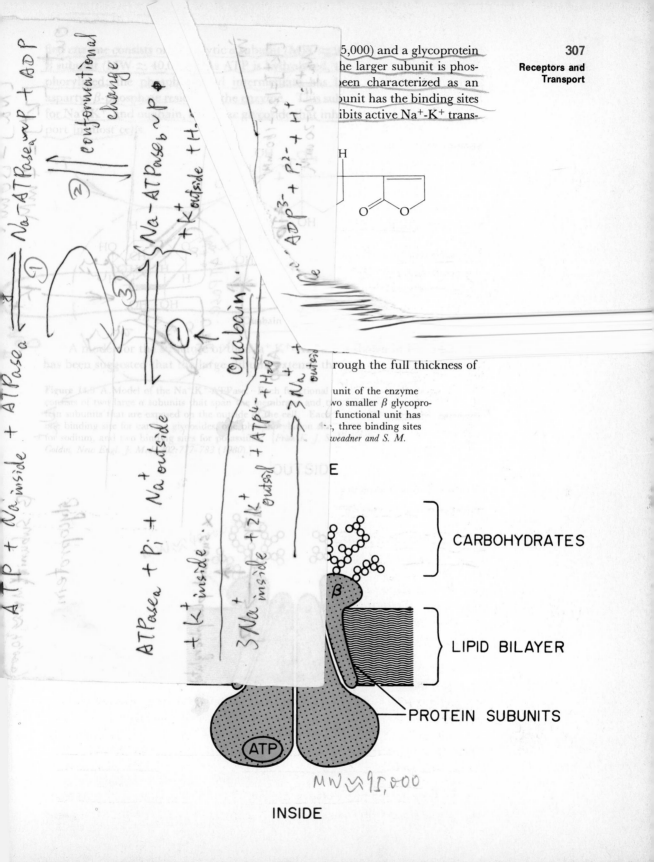

...5,000) and a glycoprotein ...he larger subunit is phos- ...been characterized as an ...unit has the binding sites ...ibits active Na^+-K^+ trans-

...rough the full thickness of

Figure 14-9 A Model of the Na^+K^+-ATPase. Each functional unit of the enzyme consists of two large α subunits that span the membrane and two smaller β glycoprotein subunits that are exposed on the outside. The e... Each ... functional unit has ... binding site for cardi... glycoside (ouabain) on ... side, three binding sites for sodium, and two binding sites for potassium. [From ...weadner and S. M. Goldin, New Engl. J. Med. ...]

OUTSIDE

CARBOHYDRATES

β

LIPID BILAYER

PROTEIN SUBUNITS

ATP

$MN \sim 95,000$

INSIDE

1. $ATP + Na^+_{inside} + ATPase_a \overset{Mg^{2+}}{\rightleftharpoons} Na\text{-}ATPase_a \smallsmile P + ADP$

2. $Na\text{-}ATPase_a \smallsmile P \rightleftharpoons Na\text{-}ATPase_b \smallsmile P$

3. $Na\text{-}ATPase_b \smallsmile P + K^+_{outside} + H_2O \rightleftharpoons ATPase_a + P_i + Na^+_{outside} + K^+_{inside}$

Figure 14.10 A schematic representation of the transport of Na^+ and K^+ by the Na^+,K^+-ATPase. The two forms of the enzyme, a and b, designate possible conformational alterations occurring during the exchange of ions. The action of ouabain inhibits the K^+-dependent reaction 3, the hydrolysis of phosphorylated Na^+-ATPase$_b$. Although depicted as an Na^+/K^+ of 1, experimental ratios of $3Na^+$ transported in to $2K^+$ transported out have been reported for the red cell membrane and the squid axon, as well as for experimentally prepared vesicles into which the purified enzyme has been incorporated. This is in accord with the stoichiometry indicated above.

the cell membrane with the K^+ and oubain sites on its outer surface and Na^+ and ATP sites on its inner surface. The glycoprotein may orient the enzyme when it takes its position in the membrane bilayer since oligosaccharides generally project from the outer surface of the plasma membrane, as described previously (Chap. 13). The isolated enzyme contains about one-third phosphoglycerides by weight; these are required for its activity.

The mechanism of action of the ATPase has been postulated as follows. Transfer of the terminal phosphate of ATP to the enzyme occurs in a reaction dependent upon Mg^{2+} and Na^+, in the absence of K^+. When K^+ is added to the intermediate, phosphate is released as P_i. The vectorial equation for the reaction is as follows, and the sequence suggested is shown in Fig. 14.10.

$$3Na^+_{inside} + 2K^+_{outside} + ATP^{4-} + H_2O \longrightarrow$$
$$3Na^+_{outside} + 2K^+_{inside} + ADP^{3-} + P_i^{2-} + H^+$$

Thus the energy made available by ATP hydrolysis drives the active transport process by leading to conformational changes in the enzyme. The mechanism is cyclical: binding of Na^+ causes phosphorylation of the enzyme leading to a conformational change which seems to move Na^+ to the other side of the membrane. Subsequent binding of K^+, associated with hydrolysis of the phosphorylated intermediate, leads to translocation of K^+.

The Na^+,K^+-pump of mammalian cells plays multiple roles. These include: maintenance of osmotic equilibrium across fragile cell membranes, thus preventing their rupture; maintenance of high intracellular $[K^+]$ required for various metabolic processes; maintenance of Na^+ and K^+ gradients required for propagation of nerve impulses; maintenance of the Na^+ and K^+ gradients which drive various secondary active transport processes, such as those involved in solute reabsorption by the kidney and intestinal absorption of nutrients.

Calcium, Calcium Transport, and the Calcium Pump The regulation of the free cytosolic $[Ca^{2+}]$ in mammalian cells is complex. A simplified, schematic portrayal of some of the features controlling free cytosolic $[Ca^{2+}]$ is presented in Fig. 14.11. Since $[Ca^{2+}]$ is much higher extracellularly ($\approx 1\ mM$) than intracellularly ($<1\ \mu M$), there tends to be a movement of Ca^{2+} down its electrochemical gradient into the cell. There are at least two distinct channels that move Ca^{2+} into cells, one being a relatively specific Ca^{2+} channel independent of membrane

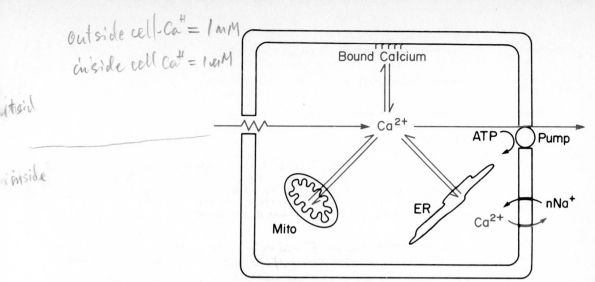

Figure 14.11 Factors regulating the cytosolic concentration of Ca^{2+} in cells. Pump refers to the energy-dependent Ca^{2+} extrusion mechanism. Mito and ER refer to mitochondria and endoplasmic reticulum, respectively; these contain high concentrations of calcium. "Bound calcium" refers to calcium bound to specific sites on plasma and other cell membranes. The precise stoichiometry of the Ca^{2+}-Na^{+} exchange mechanism is not known. [*From J. Exton, Amer. J. Physiol.* **238:***E3–E12* (*1980*).]

potential and the other being a potential-dependent calcium permeability channel. Calcium efflux against its electrochemical gradient occurs via at least two pathways. One is by a specific ATP-dependent Ca^{2+} pump which is linked to a Ca^{2+},Mg^{2+}-activated ATPase. The other is a Na^{+}-Ca^{2+} exchange mechanism which utilizes the energy of the inwardly directed Na^{+} gradient to accomplish "uphill" Ca^{2+} extrusion (gradient coupled active transport). This process depends on the maintenance of the Na^{+} gradient by the Na^{+},K^{+}-pump.

Within the cell there are reservoirs of Ca^{2+} maintained in the mitochondria and smooth endoplasmic reticulum by energy-dependent uptake processes. In secretory cells an additional pool of Ca^{2+} is found in secretory granules (Chap. M5). The mitochondrial pool appears to be the largest.

Certain hormones may alter cytosolic [Ca^{2+}]. They may promote an influx of extracellular Ca^{2+} or Ca^{2+} which is loosely bound to plasma membrane sites. Alternatively, they may lead to mobilization of Ca^{2+} from the intracellular stores discussed above. In striated (and probably smooth) muscle the pool mobilized is from the sarcoplasmic reticulum, while in liver cells it appears to be largely mitochondrial. Both mechanisms may be operative within the same cell.

An important example of the functioning of a Ca^{2+}-ATPase in the active transport of Ca^{2+} is the system in the membranes of the sarcoplasmic reticulum of striated muscle (Chap. M8). The sarcoplasmic reticulum is an intracellular network of tubules and vesicles which surrounds the myofibrils. Active transport of Ca^{2+} by the sarcoplasmic reticulum is driven by ATP hydrolysis in the presence of Mg^{2+} and serves to pump Ca^{2+} against its concentration gradient from the cytosol into the sarcoplasmic reticulum. The consequent lowering of

[Ca^{2+}] in the region of the contractile fibers of the muscle terminates the contraction cycle (Chap. M8). The major subunit of the Ca^{2+}-transporting ATPase (MW = 100,000) makes up ≈ 85 percent of all the proteins of the sarcoplasmic reticulum. As with other transport systems, the Ca^{2+}-ATPase of sarcoplasmic reticulum is an oligomer containing four such subunits. In addition to this subunit, which spans the sarcoplasmic reticulum membrane, the enzyme system also contains a lower molecular weight proteolipid subunit.

As with the Na^+,K^+-ATPase discussed above, the enzyme is transiently phosphorylated during the transport process. Each cycle of phosphorylation-dephosphorylation is associated with the transport of 2 Ca^{2+} for each ATP hydrolyzed; 1 Mg^{2+} appears to move in the opposite direction. The phosphorylation site is the β-carboxyl group of an aspartic residue of the 100,000 dalton subunit. Conformational changes in the protein caused by the cycle of phosphorylation-dephosphorylation presumably trigger the translocation of Ca^{2+}. The system can be solubilized with detergents and its Ca^{2+}-transporting and ATPase activities reconstituted in phospholipid vesicles.

Calmodulin

A variety of cellular metabolic processes is modulated by alteration of the activity of Ca^{2+}-sensitive enzymes. In many cases these effects are mediated by specific "receptor" proteins which first bind Ca^{2+} and then interact with enzymes to modify their activity. The most important "calcium receptor" is *calmodulin*, or Ca^{2+}-dependent regulatory protein (CDR). This protein is ubiquitous in eukaryotic cells (MW = 16,700). It is an acidic heat-stable protein with a structure that has been highly conserved throughout evolution. Its properties are very similar to those of troponin C (Chap. M8), and there is 45 percent direct homology in their amino acid sequences (78 percent for conservative replacements). The close sequence homologies between calmodulin, troponin C, and the parvalbumins (which are Ca^{2+}-binding proteins found in myoplasm) suggest that these molecules represent a family of Ca^{2+}-binding proteins which have evolved from a single ancestral calcium-binding protein. Each molecule of calmodulin contains four binding sites for Ca^{2+} with $K_D \approx 10^{-5}\,M$.

The sequence of calmodulin from bovine brain is shown in Fig. 14.12. The four proposed Ca^{2+}-binding domains are shown. From the crystal structure of the related parvalbumin, the protein is viewed as consisting of helical regions which are connected by loops of 12 residues which bind Ca^{2+}. These loops contain sequences favoring the formation of β turns. There is considerable homology between domains 1 and 3 and 2 and 4.

Calmodulin-Ca^{2+} complexes function as regulators of a wide variety of metabolic processes. Regulation of enzymic activities by Ca^{2+} and calmodulin occurs in a two-stage process as follows:

$$\text{Calmodulin}_{\text{inactive}} + Ca^{2+} \rightleftharpoons (\text{Calmodulin}^* \cdot Ca^{2+})_{\text{active}} \qquad (1)$$

$$E_{\text{less active}} + (\text{Calmodulin}^* \cdot Ca^{2+})_{\text{active}} \rightleftharpoons (E^* \cdot \text{Calmodulin}^* \cdot Ca^{2+})_{\text{active}} \qquad (2)$$

where E denotes a regulated enzyme and * denotes a new conformation. Thus, calmodulin by itself is inactive. It becomes active only after binding Ca^{2+},

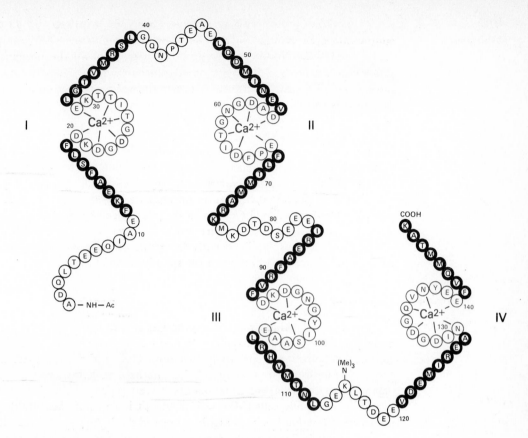

Figure 14.12 Sequence of calmodulin: The sequence of bovine brain calmodulin is shown in the one-letter codes for amino acid residues. A, Ala; D, Asp; E, Glu; F, Phe; G, Gly; H, His; I, Ile; K, Lys; L, Leu; M, Met; N, Asn; P, Pro; Q, Gln; R, Arg; S, Ser; T, Thr; V, Val; Y, Tyr. The four proposed Ca²⁺-binding domains with the stretches of α-helix (darker circles) are indicated. Domains III and IV contain tyrosyl residues; a single trimethyllysyl residue (115) lies between domains III and IV. Domain III also contains the single histidyl residue (107). Note the homology especially between domains I and III and between domains II and IV. [*From C. B. Klee, T. H. Crouch, and P. G. Richman, Ann. Rev. Biochem. 49:489–575 (1980).*]

which leads it to assume a more helical conformation, which is the active species capable of reversibly binding to and activating various enzymes. A list of some enzymes that are regulated by calmodulin is presented in Table 14.2. In addi-

TABLE 14.2

Phosphodiesterase—"high K_M for cAMP" (Chap. M12)	**Some Enzymes Regulated by Calmodulin**
Adenylate cyclase (brain) (Chap. M12)	
Ca²⁺-ATPase (erythrocyte, synaptosomal) (Chap. 14)	
Myosin light chain kinase (Chap. M8)	
NAD kinase (Chap. 26)	
Phosphorylase kinase (Chap. 18)	
Phospholipase A₂ (Chap. 21)	
? Other Ca²⁺-dependent protein kinases	
? Guanylate cyclase (Chap. M12)	

311

tion, calmodulin also appears to be involved in the processes of microtubule assembly and disassembly (Chap. 13), and neurotransmitter release (Chap. M7).

In the case of phosphorylase kinase (Chap. 18), not only does exogenous calmodulin $\cdot Ca^{2+}$ activate the enzyme, but one of the four different subunits of the enzyme is identical to calmodulin. This enzyme is completely dependent on Ca^{2+} for activity.

Ionophores and Ion Channels

Insights into the mechanisms of transport phenomena have been obtained from the study of certain substances termed *ionophores,* which can interact with metal ions and mediate their transport across both artificial and natural membranes. There are two major groups of such agents. The first are *ion carriers;* the second are *ion channels* (Fig. 14.13).

An example of an ion carrier is the macrocyclic peptide valinomycin (Fig. 14.14), which is an antibiotic synthesized by certain fungi. It contains six amide and six ester bonds with side chains consisting of hydrophobic alkyl radicals. Valinomycin binds numerous cations of the alkali, alkaline-earth, and transition metal series. Its complex with K^+ is 10^3 to 10^4 times more stable than that with Na^+. Valinomycin increases K^+ conductance in virtually all membrane systems studied and at concentrations as low as $10^{-8} M$. Valinomycin lowers the energy barrier for transport of K^+ across membranes and in a sense catalyzes its translocation in an enzyme-like fashion. Binding of the ion in the central cavity of the carrier is followed by a variety of conformational changes. Within the membrane the cation is screened from the hydrophobic lipid environment by the hydrocarbon skeleton and side chains of the carrier while held tightly by six ion-dipole bonds formed with the inwardly pointing ester carbonyls of valine. Thus, by forming a complex with valinomycin the cation is made lipid-soluble and its passage through the membrane is facilitated.

Examples of ion channels are the gramicidin antibiotics. The structure of gramicidin A, a linear peptide of 15 residues is shown in Fig. 14.14. It has the interesting feature of alternating residues of L and D configuration. Unlike ion carriers such as valinomycin, ion channels such as the gramicidins do not com-

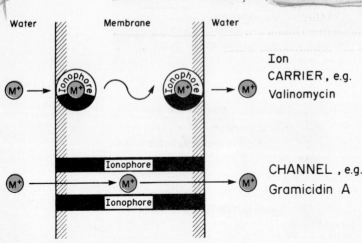

Figure 14.13 Schematic representation of the action of ionophores on membranes. $M^+ =$ metal ion. [*From Y. A. Ovchinnikov, Eur. J. Biochem.* **94:***321–336 (1979).*]

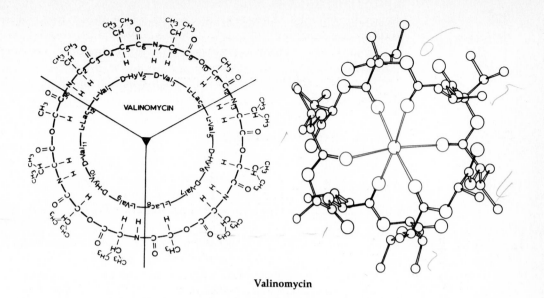

Valinomycin

$$L \quad\ \ L\ \ D\ \ L\ \ D\ \ L\ \ D\ \ L\ \ D\ \ L\ \ D\ \ L\ \ D\ \ L$$
HCO-Val-Gly-Ala-Leu-Ala-Val-Val-Val-Trp-Leu-Trp-Leu-Trp-Leu-Trp-NH(CH$_2$)$_2$OH

Gramicidin A

Figure 14.14 The structure of valinomycin (an ion carrier) is composed of three repetitions of the unit L-Val → D-hydroxyvalerate (HyV) → D-Val → L-lactate (L-Lac), so that ester and peptide bonds alternate. The ball-and-stick model indicates that K$^+$ is ligated to six carbonyl groups (in color). In gramicidin A (an ion channel) HCO = formyl and —NH(CH$_2$)$_2$OH = ethanolamine.

plex cations. Rather they induce ion permeability of membranes at concentrations as low as 10^{-10} M by forming dimers which in effect provide "tubes" which span the membrane and through which the cations pass. The exact structure of these dimers is unknown. There is a dynamic equilibrium in the membrane between ion-conducting dimers and nonconducting monomers. Such channels provide models for the naturally occurring channels which mediate, for example, the rapid changes in Na$^+$ and K$^+$ conductances which occur during action potentials in electrically excitable tissues (Chap. M7).

Transport in Bacteria

From an evolutionary point of view, transport systems developed very early and are present in bacteria. Many of the principles discussed above for transport in higher organisms are equally applicable to bacterial systems.

There are at least three distinct types of mechanisms by which bacteria accomplish active transport processes. One type, termed *membrane-bound transport systems,* are located entirely within the cell membrane. These systems are generally symports which, unlike the Na$^+$ symports of animal cells, are driven by a gradient of protons across the plasma membrane. An example is the *lactose*

313

permease of *Escherichia coli,* which transports lactose into the organism together with H$^+$.

A second type of bacterial transport is termed *binding-protein transport.* Such systems require, in addition to the integral membrane protein involved in transport, an additional hydrophilic protein which specifically binds the substrates of the transport system. The energy for such transport systems is generally derived from ATP. The individual binding proteins are distinct from the carrier proteins and are involved in the physiological function of these transport systems, presumably by concentrating substrates. However the mechanisms of their interactions with the transport proteins are largely unknown. An example of such a transport system is the high-affinity galactose transport system of *E. coli,* referred to as the *β-methylgalactoside transport system.*

A third type of bacterial transport system, referred to as *group translocation,* is characterized by the property that transport of a molecule across the cell membrane is coupled to its chemical modification. One of the best studied examples is the *phosphotransferase system* of bacteria (Chap. 17). This system moves monosaccharides across the cell membranes and phosphorylates them by using phospho*enol* pyruvate as the phosphate donor. Since sugar phosphates do not penetrate cell membranes, they are trapped within the bacterium.

REFERENCES

Books

Birnbaumer, L., and B. W. O'Malley (eds.): *Receptors and Hormone Action,* vol. 1, Academic, New York, 1976. (A series of volumes devoted to recent research in the area of receptors.)

Hoffman, J. F. (ed.): *Membrane Transport Processes,* vol. 1, Raven, New York, 1978.

O'Brian, R. D. (ed.): *The Receptors: A Comprehensive Treatise,* vol. 1: *General Principles and Procedures,* Plenum, New York, 1979.

Tosteson, D., Y. A. Ovchinnikov, and R. Latorre (eds.): *Membrane Transport Processes,* vol. 2, Raven, New York, 1978.

Williams, L. T., and R. J. Lefkowitz: *Receptor Binding Studies in Adrenergic Pharmacology,* Raven, New York, 1978.

Review Articles

Brown, M. S., and J. L. Goldstein: Receptor-Mediated Endocytosis: Insights From the Lipoprotein Receptor System. *Proc. Natl. Acad. Sci. U.S.A.* **76:**3330–3337 (1979).

Catt, K. J., and M. L. Dufau: Peptide Hormone Receptors, *Annu. Rev. Physiol.* **39:**529–557 (1977).

Hebert, S. C., J. A. Schafer, and T. E. Andreoli: Principles of Membrane Transport, in B. Brenner, and F. C. Rector, Jr.: *The Kidney,* Saunders, Philadelphia, 1981, pp. 116–143.

Kaplan, J.: Polypeptide-Binding Membrane Receptors: Analysis and Classification. *Science* **212:**14–20 (1981).

Klee, C. B., T. H. Crouch, and P. G. Richman: Calmodulin, *Annu. Rev. Biochem.* **49:**489–515 (1980).

Klingenberg, M.: Membrane Protein Oligomeric Structure and Transport Function. *Nature* **290:**449–454 (1981).

Ovchinnikov, Y. A.: Physical Chemical Basis of Ion Transport through Biological Membranes: Ionophores and Ion Channels. *Eur. J. Biochem.* **94:**321–336 (1979).

Roth, J.: Receptors for Peptide Hormones, in L. DeGroot (ed.): *Endocrinology,* vol. 3, Grune & Stratton, New York, 1979, pp. 2037–2054.

Sweadner, K. J., and S. M. Goldin: Active Transport of Sodium and Potassium Ions: Mechanism, Function and Regulation. *New Engl. J. Med.* **302:**777–784 (1980).

Wilson, D. B.: Cellular Transport Mechanisms, *Annu. Rev. Biochem.* **47:**933–965 (1978).

Chapter 15

Biological Oxidations I

Citric acid cycle. The mitochondrion. Electron transport. Oxidative phosphorylation.

intail [in'teil] vt 必需

aerobe 需氣生物

prokaryotic 散核C和的

The energy requirements of animal cells, of plant cells in the dark, and of aerobic prokaryotic microorganisms are met with the energy released by the oxidation of diverse organic compounds, using molecular oxygen. In eukaryotic cells these oxidative processes occur in the *mitochondria*, which receive from the cytoplasm a mixture of materials derived from the prior partial metabolism of carbohydrates, lipids, and amino acids. Within the mitochondria, a large fraction of the free-energy release associated with the oxidative process is conserved in the concomitant synthesis of ATP.

As shown earlier, the metabolism of glucose (page 245) entails reduction of six O_2 per molecule of glucose, viz., abstraction from each glucose of 12 pairs of electrons or hydrogen atoms. In the cytosol, each hexose is fragmented into two identical trioses, each of which is then converted to pyruvate by a reaction sequence in which a single pair of hydrogens is removed.

$$\text{Glucose} \longrightarrow 2 \text{ triose} \longrightarrow 2 \text{ pyruvate} + 2(2\text{H})$$

The latter, on a suitable shuttle, and the pyruvate then enter a mitochondrion, where the oxidation of each pyruvate yields five electron pairs on suitable carriers. Each of the 12 electron pairs derived from the original glucose then flows along an organized array of electron carriers from that of lowest to that of highest potential and thence to oxygen. Passage of each pair of electrons (as 2H) provides three opportunities to synthesize ATP, each by the same mechanism:

$$(2\text{H}) + \tfrac{1}{2}O_2 + 3\text{ADP} + 3P_i \longrightarrow H_2O + 3\text{ATP}$$

Any cross-references coded M refer to the companion text, *Principles of Biochemistry: Mammalian Biochemistry*.

Altogether, therefore, these oxidative events generate about 36 molecules of ATP per molecule of glucose.

In this chapter, we shall introduce the electron carriers and then discuss in order the conversion of pyruvate to acetyl CoA, which is the intermediate common to the metabolism of carbohydrate, fatty acids, and some amino acids; the citric acid cycle, wherein acetyl CoA is oxidized to CO_2; the structure of mitochondria; the electron-transport system, which conducts electrons from substrate to oxygen; the mechanism of ATP formation (oxidative phosphorylation); and the regulation of these processes.

OXIDATIVE ENZYMES AND COENZYMES

Biological oxidations are catalyzed by enzymes, each of which functions in conjunction with a coenzyme or electron carrier. These substances are discussed in some detail in the following chapter.

The protein moiety of an oxidative enzyme confers substrate specificity, activates both substrate and prosthetic group, and frequently alters the redox potential of the latter. Although there are many oxidative enzymes, there are only a few coenzymes; each enzyme is specific both for its substrate and its coenzyme. Reactions in which a metabolite (MH_2) reacts directly with O_2

$$MH_2 + \tfrac{1}{2}O_2 \longrightarrow M + H_2O$$

do occur, but they do not permit conservation of energy for ATP synthesis. Instead a *dehydrogenase* catalyzes electron transfer from substrate to enzyme.

$$MH_2 + Coenz \longrightarrow M + CoenzH_2$$

Oxidation of the reduced coenzyme is catalyzed by a different enzyme in an independent process.

Coenzymes of Dehydrogenases

The coenzyme most frequently employed as acceptor of electrons from the substrate is *nicotinamide adenine dinucleotide* (NAD^+) (page 318). On the dehydrogenase, a hydride ion is transferred from the substrate to the nicotinamide moiety (page 318) and a proton is liberated into the medium.

$$MH_2 + NAD^+ \longrightarrow M + NADH + H^+$$

Details of this process are described in Chap. 16. *Nicotinamide adenine dinucleotide phosphate* ($NADP^+$) (page 358) functions in the same manner. The reduced forms of these coenzymes are denoted NADH and NADPH, respectively. The reduced coenzyme is not further metabolized by a dehydrogenase. Both the oxidized substrate and the reduced coenzyme dissociate from the dehydrogenase; the reduced coenzyme is then reoxidized by transfer of electrons to an acceptor bound to a second enzyme.

Although NAD and NADP have almost identical midpoint potentials E_0' (Table 12.1), they are used differently in metabolism. *NAD is the coenzyme used in*

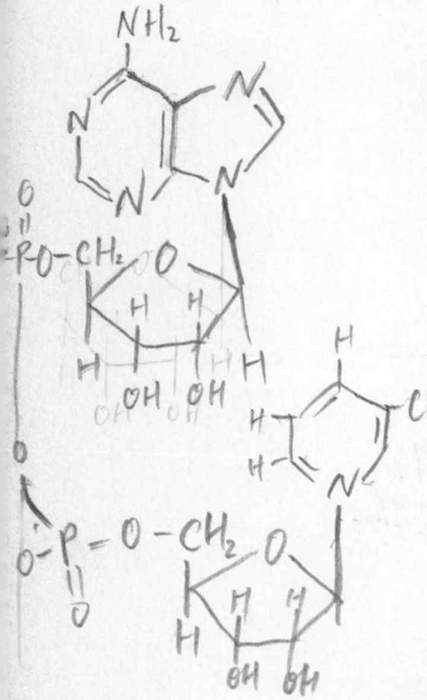

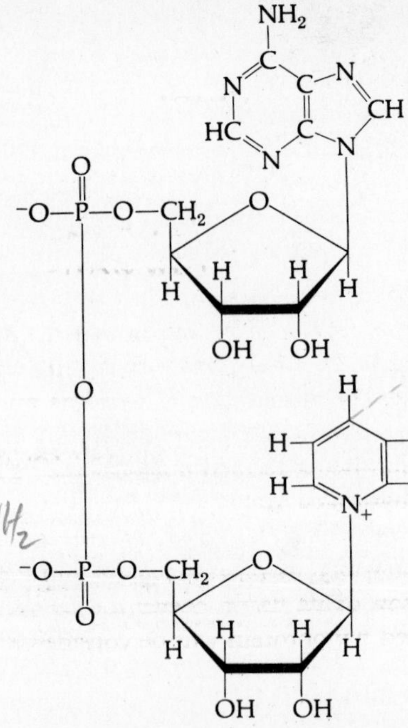

NH_2

$-O-P-O-CH_2$

OH OH

$-O-P-O-CH_2$

OH OH

H—CONH₂

CONH₂

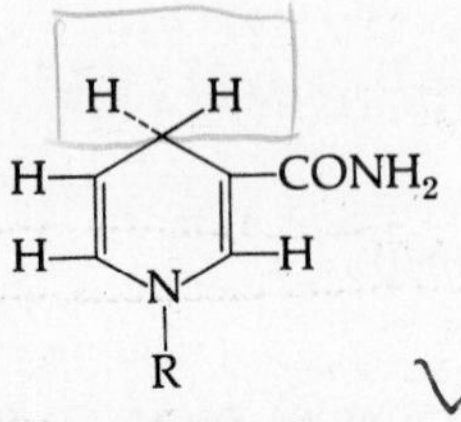

R = remainder of molecule as in NAD⁺

Nicotinamide adenine dinucleotide (NAD⁺)　　　　**NADH**

most reactions in which the free energy of substrate oxidation is conserved for subsequent ATP synthesis. Thus, NAD functions in five of the six dehydrogenations in the oxidation of glucose. NADP is employed in processes by which reducing power is utilized for synthesis of a cell metabolite or for fatty acid synthesis. In plants, NADP is used in the pathway for the reductive synthesis of glucose from CO_2 using the energy of light (Chap. 19). In both plants and animals NADP is utilized in reactions for synthesis of steroids and many amino acids. Both NAD and NADP are frequently called "pyridine" coenzymes.

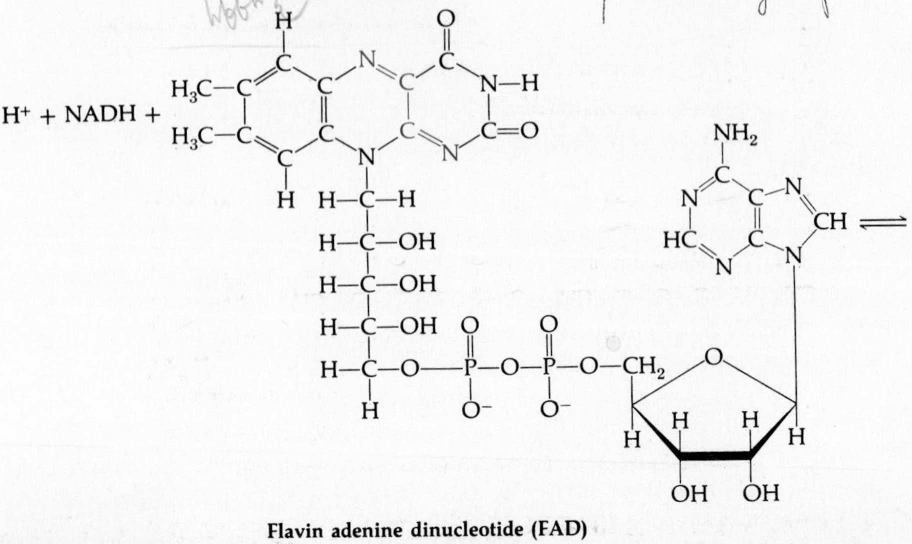

$H^+ + NADH +$

Flavin adenine dinucleotide (FAD)

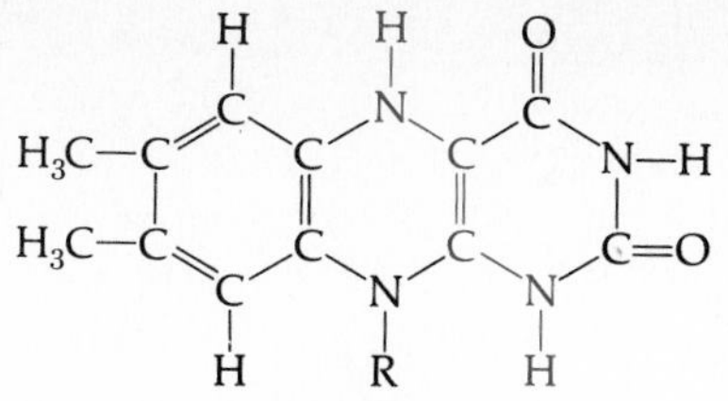

$+ NAD^+$

R = remainder of
molecule, above

FADH$_2$

A second group of oxidative enzymes employs as cofactor one of two derivatives of the vitamin riboflavin (Chap. M22), *flavin adenine dinucleotide* (FAD)
(page 318) and *flavin mononucleotide* (FMN) (Chap. 16), which lacks the adenylate portion of FAD. In contrast to the dissociation of pyridine nucleotides
from dehydrogenases, the two flavin nucleotides are invariably tightly bound, in
some cases covalently, the combination being termed a *flavoprotein*. Flavoproteins transfer electrons, effectively as hydrogen, from an organic substrate to the
riboflavin component of the coenzyme. Two flavoproteins are of particular
interest in the present context, the one which oxidizes succinate and the *NADH
dehydrogenase*, which catalyzes reduction of its flavin coenzyme by NADH, as
shown above. The fate of the reduced flavoprotein will be considered later in
this chapter.

Pyruvate Dehydrogenase and Formation of Acetyl CoA

As indicated above (page 316), the product of glucose metabolism in the
cytosol is pyruvate; the pathway of its formation from glucose is presented in
Chap. 17. Pyruvate is also formed from the carbon skeletons of a number of
amino acids, e.g., serine, alanine, cysteine, etc. (Chap. 25). Pyruvate from all of
these sources enters the mitochondria and is then converted to acetyl coenzyme
A and CO$_2$.

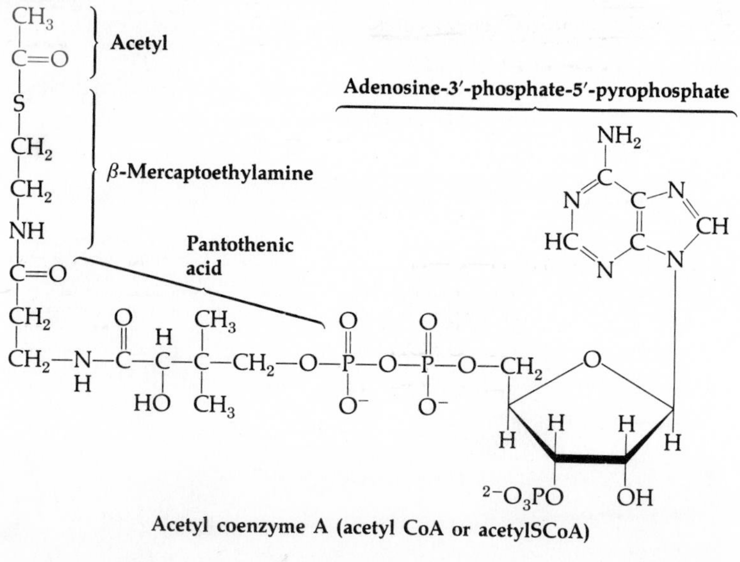

Acetyl coenzyme A (acetyl CoA or acetylSCoA)

Coenzyme A is composed of adenosine-3'-phosphate-5'-pyrophosphate bound in ester linkage to the vitamin pantothenic acid, which in turn is attached in amide linkage to β-mercaptoethylamine. The acetyl group is linked to the sulfur of coenzyme A as a thioester (see page 319). Coenzyme A is abbreviated CoA or CoA—SH. The —SH in the second abbreviation refers to the sulfhydryl group of coenzyme A.

Acetyl CoA is utilized in a variety of biological processes; it is a precursor for synthesis of fatty acids and sterols from carbohydrate and it is the acetylating agent for other synthetic processes. Acetyl CoA can be synthesized from acetate by heart muscle, kidney, and other tissues. The activation requires adenosine triphosphate (ATP) and yields adenylic acid (AMP) and inorganic pyrophosphate (PP$_i$). The reaction proceeds in two steps, catalyzed by a single enzyme, *acetate thiokinase*.

$$ATP + acetate \rightleftharpoons acetyl\text{-}AMP + PP_i \qquad (1)$$
$$Acetyl\text{-}AMP + CoA \rightleftharpoons AMP + acetyl\ CoA \qquad (2)$$
$$\text{SUM: } ATP + acetate + CoA \rightleftharpoons AMP + PP_i + acetyl\ CoA$$

Acetyl-AMP is the mixed anhydride of the carboxyl of acetate and the phosphate of adenylic acid; the bond is of higher energy than that of ATP (Table 12.2). No free acetyl-AMP appears during the reaction; it is bound tightly to the enzyme, and the subsequent reaction occurs with this enzyme-bound intermediate.

Hydrolysis of acetyl CoA proceeds with a free-energy change $\Delta G°$ of approximately -10 kcal/mol.

$$CH_3CO\text{—}SCoA + HOH \longrightarrow CH_3COO^- + HS\text{—}CoA + H^+$$

This is the energy that enables acetyl CoA to drive the formation of citrate. Acetyl CoA is a high-energy compound and the reaction catalyzed by acetate thiokinase is readily reversible. In vivo, acetyl CoA formation from acetate is ensured by the hydrolysis of pyrophosphate to two P$_i$ by *pyrophosphatase*, for which $\Delta G°$ is about -7.5 kcal/mol.

As indicated above, glucose is converted in cytoplasm by a series of reactions to two molecules of pyruvate (Chap. 17); this can be summarized as

$$C_6H_{12}O_6 + 2NAD^+ \longrightarrow 2CH_3COCOO^- + 2NADH + 2H^+$$

The link between pyruvate formation and the citric acid cycle is the formation of acetyl CoA. This occurs in the matrix space of mitochondria, catalyzed by the *pyruvate dehydrogenase complex*. The complex consists of a minimum of three proteins that utilize five coenzymes, viz., thiamine pyrophosphate, lipoic acid, CoA, FAD, and NAD$^+$. The overall reaction ($\Delta G° = -8$ kcal/mol) is effectively irreversible.

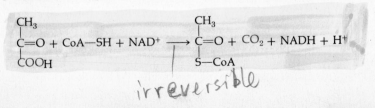

thiamine 硫胺素、 Vit.B₁.

The initial event is a reaction between pyruvate and enzyme-bound thiamine pyrophosphate (ThPP), the pyrophosphate ester of vitamin B_1 (Chap. M22), liberating CO_2 and forming α-hydroxyethylthiamine pyrophosphate.

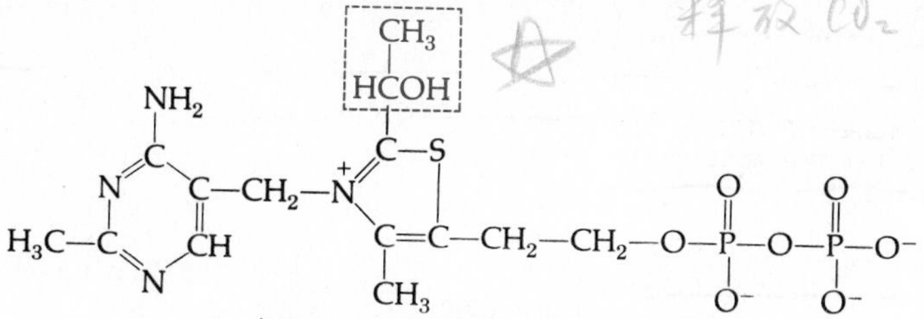

释放 CO₂ 形成 ThPP

α-Hydroxyethylthiamine pyrophosphate

The hydroxyethyl group, at the oxidation level of acetaldehyde and attached to ThPP on the same enzyme, then reacts with the disulfide form of lipoic acid, which is in amide linkage to the ε-amino group of a lysine residue (R_L) on a second protein of the complex. In this reaction, the disulfide is reduced to two sulfhydryl groups, one of which carries the acetyl resulting from the oxidation of pyruvate.

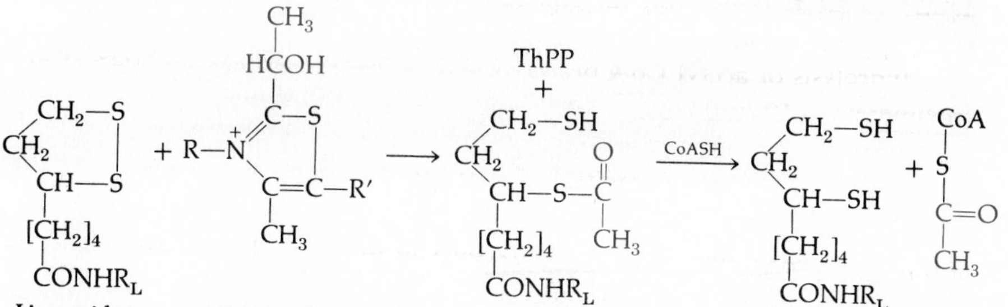

| Lipoamide | α-Hydroxyethyl thiamine pyrophosphate | Acetyl- lipoamide | Dihydrolipo- amide | Acetyl CoA |

The 2-carbon group is then transferred to the sulfhydryl group of CoA from the medium, forming acetyl CoA, which dissociates from the enzyme, leaving the lipoyl residue in its disulfhydryl form. The disulfhydryl is reoxidized to the disulfide form by the bound FAD on a third protein component of the complex and the resulting reduced FAD is oxidized by another enzyme, transferring the hydrogens to NAD^+ from the medium. This completes the process restoring the three enzymes to their original state.

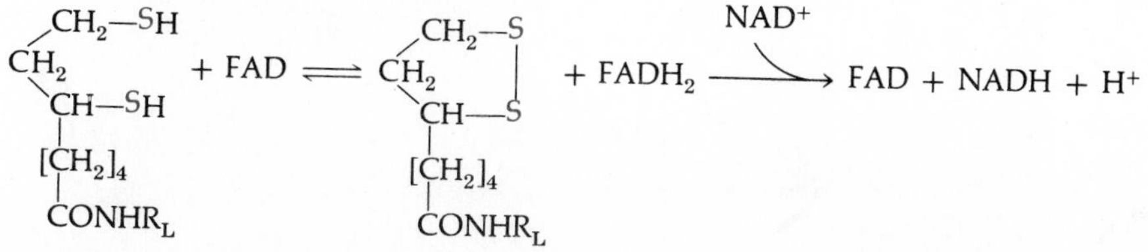

This reaction sequence, using three bound and two dissociable coenzymes, takes place on a large, organized multienzyme complex which, in mammalian cells, has MW $\simeq 9 \times 10^6$. It is composed of three distinct enzymes:

1. *Pyruvate decarboxylase* (MW = 152,000), which contains two β chains, each of which bears a thiamine pyrophosphate, and two α chains that catalyze transfer of the C$_2$ unit to a lipoyl residue.

2. *Lipoyl transacetylase* (MW = 52,000), possessing a covalently bound lipoyl group.

3. *Dihydrolipoyl dehydrogenase* (MW = 110,000), which oxidizes the reduced lipoyl group of the transacetylase to the disulfide form. The disulfhydryl moiety of this enzyme reduces an adjacent bound molecule of FAD, which in turn is reoxidized by NAD$^+$. The core of the complex comprises 60 molecules of the transacetylase and there are probably about 30 molecules each of the decarboxylase and the flavoprotein.

Figure 15.1 shows the pivotal role of the long (14 Å) hydrophobic arm of the lipoyllysyl of the transacetylase in delivering the C$_2$ unit from the decarboxylase to CoA and being reoxidized on the dehydrogenase, prior to repetition of the cycle. Control of the activity of the pyruvate dehydrogenase complex is presented later.

The pyruvate dehydrogenase complex is similar in structure and properties to two other mitochondrial α-keto acid decarboxylase complexes, one of which functions in the decarboxylation of α-ketoglutarate (page 325) and the other which acts on α-keto acids derived from branched-chain amino acids (Chap. 25).

Figure 15.1 A schematic representation of the possible rotation of a lipoyllysyl moiety between α-hydroxyethylthiamine pyrophosphate (TPP-Ald) bound to *pyruvate decarboxylase* (D), the site for acetyl transfer to CoA, and the reactive disulfide of the flavoprotein, *dihydrolipoyl dehydrogenase* (F). The lipoyllysyl moiety is an integral part of *lipoate acetyltransferase* (LTA). [*Courtesy of Dr. L. J. Reed.*]

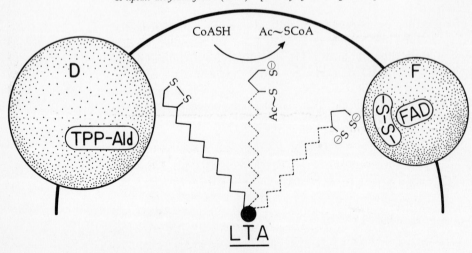

The major reaction sequence providing electrons to the transport system which reduces oxygen while generating ATP is known as the *citric acid cycle,* the *tricarboxylic acid cycle,* or the *Krebs cycle.*

Figure 15.2 shows the major intermediates of the citric acid cycle. The reaction sequence starts [reaction (1)] by condensation of acetyl CoA with oxaloacetate to yield citrate. For each revolution of the cycle, 1 mol of acetyl CoA is consumed and 2 mol of CO_2 are evolved. Oxaloacetate, which is utilized in the initial condensation, is regenerated, permitting the process to operate continuously, as long as acetyl CoA continues to enter the cycle and hydrogen atoms and CO_2 are removed. Four of the individual reactions are dehydrogenations, three of which result in the formation of NADH. The net accomplishment of the citric acid cycle, per revolution, may be represented as follows:

$$\text{Acetyl CoA} \longrightarrow 2CO_2 + \text{CoA} + 4(2e)$$

The electrons are delivered to O_2 via an electron-transport chain, considered later, and the free energy released in this process is conserved as ATP. It is therefore by oxidation of acetyl CoA via the citric acid cycle that cells obtain most of the energy potentially available from the oxidation of glucose, fatty acids, and amino acids. The individual reactions of the cycle may now be considered.

Reactions of the Citric Acid Cycle

Citrate Formation Formation of citrate is catalyzed by *citrate synthase.* Equilibrium for the reaction (Fig. 15.2) is far to the right; $K = 2.2 \times 10^6$ and $\Delta G° = -9$ kcal/mol. The mammalian enzyme (MW = 98,000) consists of two identical subunits; K_m for both substrates is in the micromolar (μmol) range. The reaction mechanism suggests that a carbanion is formed at the methyl group of acetyl CoA, so that it can engage in the aldol-type condensation. Citryl CoA is formed and hydrolyzed on the enzyme with only citrate and CoA departing.

Citrate, *cis*-Aconitate, and Isocitrate The reversible interconversions of these compounds, involving dehydration and hydration reactions, are catalyzed by *aconitase.* Pig heart aconitase (MW = 89,000) contains two identical subunits, each containing iron-sulfur clusters (Chap. 16). *cis*-Aconitate is an enzyme-bound intermediate.

The enzyme catalyzes trans addition of the elements of water to the double bond of *cis*-aconitate. As shown in Fig. 15.2, aconitase forms either isocitrate or citrate. At equilibrium, the relative abundances of the three compounds are 90 percent citrate, 4 percent *cis*-aconitate, and 6 percent isocitrate. Oxidation of isocitrate in the subsequent reaction drives the cycle forward.

Respiration is inhibited by fluoroacetate, with resultant accumulation of citrate. Fluoroacetate is converted to fluoroacetyl CoA, which condenses with oxaloacetate to give fluorocitrate. Fluorocitrate inhibits aconitase and thus prevents utilization of citrate. This is an example of many instances where a

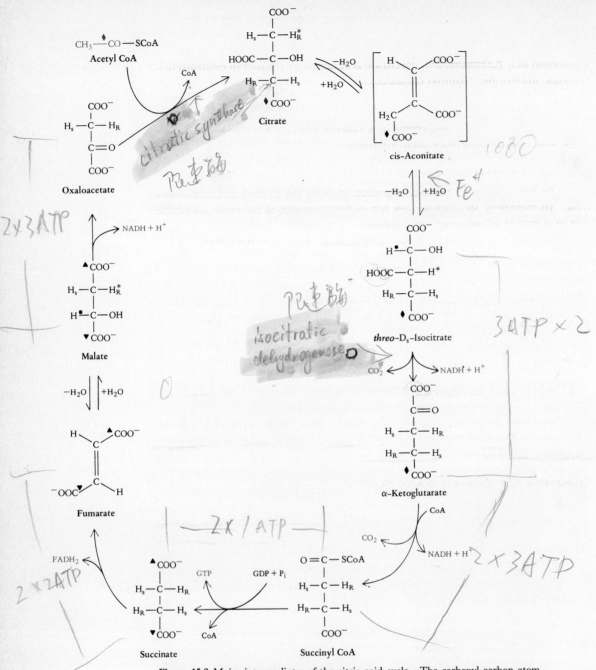

Figure 15.2 Major intermediates of the citric acid cycle. The carboxyl carbon atom of acetyl CoA (◆) is followed around the cycle. The two carboxyl groups of succinate are equivalent, so that if the carboxyl carbon of acetyl CoA is introduced as ^{14}C, the label becomes equally distributed in both carboxyls of the 4-carbon dicarboxylic acids (▲▼). Hydrogen atoms fixed into place in C—H bonds by addition of water to a double bond are noted by an asterisk (*), and hydrogen atoms transferred to NAD are noted as •. The subscripts indicate the chirality of the hydrogen atoms. The $2CO_2$ liberated (in color) are not derived from acetyl on the first cycle (see text). The four energy-yielding steps produce 3NADH, 1FADH₂, and 1GTP (shown in color).

nontoxic substance is metabolically converted to a deleterious compound, e.g., various potential carcinogens.

Formation of α-Ketoglutarate Isocitrate is oxidized by transfer of electrons to a pyridine nucleotide. Animal tissues contain two *isocitrate dehydrogenases*, specific, respectively, for transfer to NAD^+ and $NADP^+$. Both enzymes require either Mg^{2+} or Mn^{2+}. The mitochondrial NAD-specific enzyme (MW = 160,000) has three different subunits present in a ratio of $2:1:1$. Each molecule of enzyme has two binding sites each for metal ion, isocitrate, and NAD^+. The enzyme is markedly activated by ADP, which lowers K_m for isocitrate and is inhibited by NADH and NADPH.

The NADP-specific protein is present in both the cytosol and mitochondria. Its metabolic significance lies not in the operation of the citric acid cycle but as a source of reducing equivalents.

The action of the isocitrate dehydrogenases involves intermediary formation of enzyme-bound oxalosuccinate.

$$
\begin{array}{l}
H_2C-COO^- \\
HC-COO^- \\
HOC-COO^- \\
H
\end{array}
\; + \; NADP^+ \;\rightleftharpoons\;
\begin{array}{c}
NADPH \; + \\
+ \\
H^+
\end{array}
\left[
\begin{array}{l}
H_2C-COO^- \\
HC-COO^- \\
O=C-COO^-
\end{array}
\right]
\;\overset{Mn^{2+}}{\rightleftharpoons}\;
\begin{array}{l}
H_2C-COO^- \\
HCH \\
O=C-COO^-
\end{array}
\; + \; CO_2
$$

Isocitrate　　　　　　　　　　　**Oxalosuccinate**　　　**α-Ketoglutarate**

The isocitrate dehydrogenases are typical of a group of β-hydroxyl acid dehydrogenases. In no instance does the keto acid formed in the initial dehydrogenation dissociate from the enzyme; the second step, decarboxylation, follows immediately. Equilibrium favors α-ketoglutarate formation; under physiological conditions, $\Delta G° = -5$ kcal/mol.

Conversion of α-Ketoglutarate to Succinate The oxidation of α-ketoglutarate is analogous to that of pyruvate (pages 320 ff.). The process, for which $\Delta G° = -8$ kcal/mol, is catalyzed by a multienzyme complex (MW = 2.5×10^6). The complex consists of 24 *transsuccinylase* molecules (MW = 42,000) to which 12 *α-ketoglutarate decarboxylase* molecules (MW = 95,000) and 12 *dihydrolipoyl dehydrogenase* molecules (MW = 56,000) are attached. The dihydrolipoyl dehydrogenase is identical with that present in the pyruvate dehydrogenase complex.

$$
\begin{array}{l}
COO^- \\
CH_2 \\
CH_2 \\
C=O \\
COO^-
\end{array}
\; + \; HS-CoA + NAD^+ \;\longrightarrow\;
\begin{array}{l}
COO^- \\
CH_2 \\
CH_2 \\
C=O \\
S-CoA
\end{array}
\; + \; NADH + H^+ + CO_2
$$

α-Ketoglutarate　　　　　　　　　　　**Succinyl CoA**

α-Ketoglutarate reacts initially with thiamine pyrophosphate, to form α-hydroxy-γ-carboxypropylthiamine pyrophosphate and CO_2.

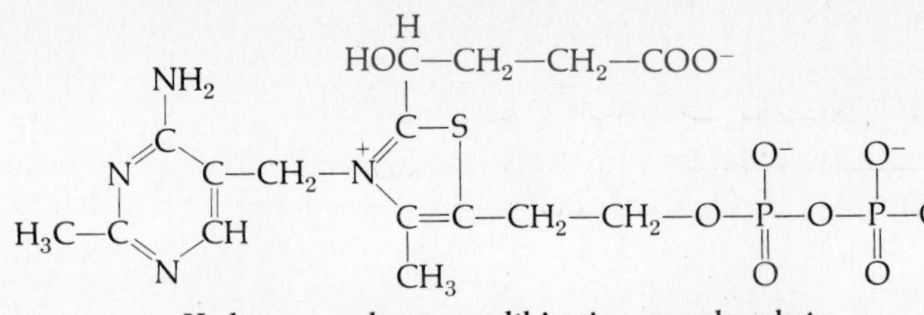

$$\alpha\text{-Hydroxy-}\gamma\text{-carboxypropylthiamine pyrophosphate}$$

The 4-carbon unit is then transferred to enzyme-bound lipoate, forming a succinyl lipoyl enzyme. The succinyl group, in turn, is transferred to CoA, and the dihydrolipoamide is reoxidized by FAD and NAD^+, as in the case of pyruvate oxidation. Succinyl CoA is itself an effective competitive inhibitor of the reaction.

In the next step CoA is removed, and the energy of the succinyl CoA bond is conserved by the *succinate thiokinase* reaction:

$$\text{Succinyl CoA} + \text{GDP} + P_i \longrightarrow \text{succinate} + \text{GTP} + \text{CoA}$$

GDP and GTP are guanosine di- and triphosphate, respectively. A single protein (MW = 140,000) catalyzes the following postulated steps:

$$\text{Enz} + \text{succinyl CoA} + P_i \rightleftharpoons \text{Enz—succinyl} \sim P + \text{CoA}$$
$$\text{Enz—succinyl} \sim P \rightleftharpoons \text{Enz} \sim P + \text{succinate}$$
$$\text{Enz} \sim P + \text{GDP} \rightleftharpoons \text{Enz} + \text{GTP}$$

A *nucleoside diphosphokinase*, in the intermembrane space (page 333), catalyzes phosphate transfer from GTP to ADP. In this process, and that catalyzed by succinate thiokinase, the N-3 of the imidazole group of a histidine residue of the enzyme accepts and transfers the phosphate group.

$$\text{GTP} + \text{ADP} \rightleftharpoons \text{GDP} + \text{ATP}$$

Dehydrogenation of Succinate The oxidation of succinate to fumarate is the only dehydrogenation in the citric acid cycle in which NAD^+ does not participate.

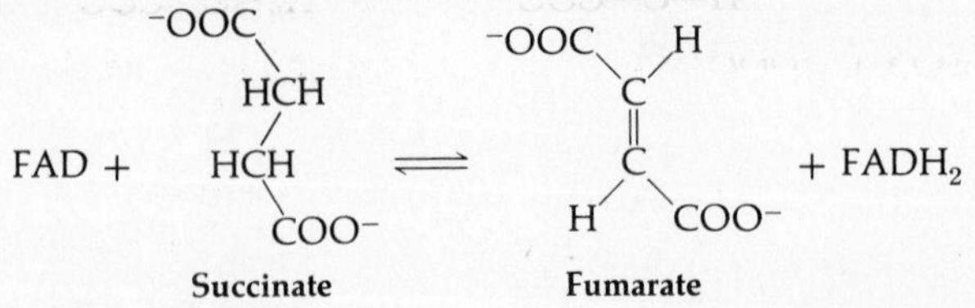

Succinate dehydrogenase from heart muscle (MW = 97,000) is an integral protein of the mitochondrial inner membrane. It consists of two subunits of MW 70,000 and 27,000, the larger member bearing the active site and FAD, which is covalently linked to a histidine residue, as shown below:

R_1 = remainder of enzyme protein
R_2 = remainder of FAD molecule

Both subunits contain *nonheme iron,* a molecular arrangement found in proteins of diverse electron-transfer systems (Chap. 16). In most instances, this structural grouping has a characteristic absorption in the visual region of the spectrum and, when reduced, a distinct electron-paramagnetic-resonance spectrum ($g = 1.94$). The most common arrangement is $(FeS)_2$, of which there are two in the 70,000 subunit of succinate dehydrogenase. The nonheme-iron chromophore consists of a pair of iron atoms close enough to share a single electron or otherwise engage in antiferromagnetic coupling. The structure that seems most reasonable for the nonheme-iron chromophore is shown later (Chap. 16).

The redox potential (Chap. 12) of nonheme iron is determined by the protein in which it is embedded. On the large subunit the two nonheme-iron groups differ in E_0' and are estimated to be $E_0' = 240$ and 30 mV, respectively. E_0' for the succinate/fumarate couple is about $+ 30$ mV. On the small subunit is another nonheme-iron arrangement, $(FeS)_4$; E_0' is about 120 mV. Electrons flow from substrate to flavin through the nonheme-iron groups on the large subunit to that on the small subunit. The two subunits are tightly associated within the mitochondrial membrane.

The reaction product is fumarate; the cis isomer, maleate, is not produced by this enzyme. Malonate is a specific competitive inhibitor which can interrupt the citric acid cycle, with resultant accumulation of succinate.

Malate Formation The hydration of fumarate to yield L-malate is catalyzed by *fumarase,* a tetramer with identical subunits (MW = 48,500).

$$H_2O + \ ^-OOC-\overset{\displaystyle H-C-COO^-}{\underset{\displaystyle H}{C}} \ \rightleftharpoons \ ^-OOC-\overset{\displaystyle H_2-C-COO^-}{\underset{\displaystyle H}{C}}-OH$$

Fumarate   **L-Malate**

For the reaction as written, $K = 4$; thus it is freely reversible. There is absolute specificity, for the trans-unsaturated acid and the L-hydroxy acid.

Regeneration of Oxaloacetate Malate is oxidized in the presence of *malate dehydrogenase* and NAD^+ to yield oxaloacetate.

$$\begin{array}{c} H_2-C-COO^- \\ | \\ {}^-OOC-C-OH \\ | \\ H \end{array} + NAD^+ \rightleftharpoons \begin{array}{c} H_2-C-COO^- \\ | \\ {}^-OOC-C=O \end{array} + NADH + H^+$$

Malate   **Oxaloacetate**

The enzyme (MW = 66,000) consists of two subunits. This reaction completes the cycle. The regenerated oxaloacetate can condense with acetyl CoA and allow repetition of the process. However, equilibrium greatly favors the reverse reaction, reduction of oxaloacetate by NADH.

Citric Acid Cycle: Steric Considerations

Biological Asymmetry of Citrate Citrate does not have an asymmetrically substituted carbon atom and is thus devoid of optical activity. It was therefore anticipated (Fig. 15.2) that introduction of [1-^{14}C]acetyl CoA into the pathway should give rise to α-ketoglutarate equally labeled in both carboxyl groups. However, the product proved to be labeled exclusively in the γ-carboxyl carbon.

$$\begin{array}{c} CH_3{}^{14}CO-CoA \\ + \\ O=C-COO^- \\ | \\ H_2C-COO^- \end{array} \xrightarrow[1-4]{\text{reactions}} \begin{array}{c} H_2C-{}^{14}COO^- \ {}^{\gamma} \\ | \\ H_2C \ \beta \\ | \\ O=C-COO^- \ {}_{\alpha} \end{array}$$

This led to the recognition that, as shown in Fig. 15.3, the two —CH_2COO^- groups of citrate are not truly geometrically equivalent and this nonequivalence becomes apparent upon three-point attachment to an enzymic surface.

Each step in the citric acid cycle is completely stereospecific, as indicated in Fig. 15.2. Aconitase discriminates between the two CH_2COO^- groups of citrate so that successive removal and addition of water results specifically in *threo*-$_{D_S}$-isocitrate. As shown in Fig. 15.2, it is possible to specify the fate of each hydrogen atom in the entire reaction sequence. Whereas 2 mol of CO_2 are produced as 1 mol of acetyl CoA is consumed, neither of the carbon atoms lost as CO_2 actually is derived from that acetyl CoA on the first turn of the cycle. At the level of free succinate, randomization does occur, and isotope, introduced as ^{14}C in the carboxyl carbon of the acetate moiety of acetyl CoA, will be symmetrically distributed about the plane of symmetry. On the subsequent revolutions of the cycle, with reappearance and reutilization of oxaloacetate, acetyl carbon will be eliminated as CO_2.

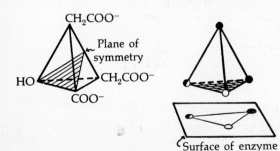

Figure 15.3 Asymmetry apparent upon three-point attachment of citrate to an enzymic surface. The central carbon atom of citrate is in the center of the equilateral tetrahedron. Because of the plane of symmetry, the molecule in free solution is optically inactive. If the surface of the enzyme is asymmetric and forces attachment by —OH, —COO^-, and —CH_2COO^- groups in the manner shown, only one face of the tetrahedron (cross-hatched) can be accommodated on the enzymic surface. Hence, the two —CH_2COO^- groups are not equivalent, and only one attaches to the enzymic surface.

The major processes that provide and conserve energy in cells can be summarized as

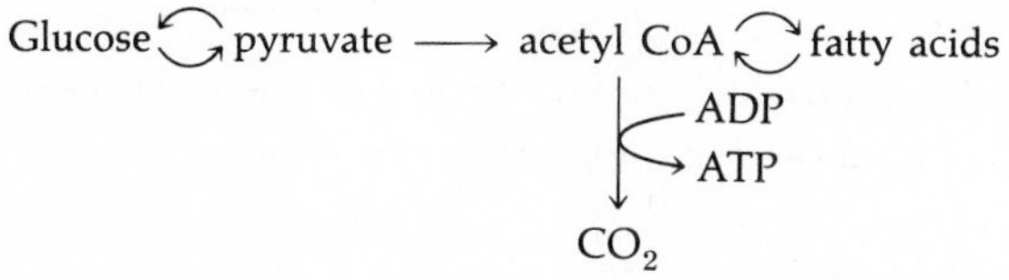

$$Glucose \rightleftarrows pyruvate \longrightarrow acetyl\ CoA \rightleftarrows fatty\ acids$$

$$\begin{array}{c} \\[-0.5em] \longmapsto ADP \\ \searrow ATP \\ CO_2 \end{array}$$

[handwritten margin note: obligatory 在尽站.7临村)加]

Regulation of this system must provide a continuous supply of ATP, as required for metabolism, permit carbohydrate, when in excess supply, to be converted to fatty acids via pyruvate and acetyl CoA, and, when carbohydrate is in short supply, permit oxidation of fatty acids to acetyl CoA to keep the citric acid cycle operating.

The citric acid cycle provides electrons to the electron-transport system, where flow is coupled to ATP synthesis, and, in lesser degree, provides reducing power for synthesis of intermediates that also flow into other metabolic pathways. In a general way, the cycle can proceed no more rapidly than is permitted by utilization of the ATP generated; in the extreme case, with electron flow and ATP synthesis obligatorily coupled, if all the ADP of a cell were converted to ATP, no further electrons could flow to O_2 from NADH, which would accumulate. Since no NAD^+ would be available to participate in the dehydrogenations of the cycle, it would cease functioning. In addition to this coarse relationship, there are finer controls that regulate the operation of the citric acid cycle itself.

[handwritten margin note: cycle depends on the utilization of ATP generated]

Succinate dehydrogenase is embedded in the inner mitochondrial membrane; all the other enzymes of the citric acid cycle are dissolved in the internal matrix space of the mitochondrion but may be associated in a loosely bound complex. Measurements of the relative amounts of these enzymes and the steady-state concentrations of their substrates in mitochondria supplied with ADP and P_i indicate that the individual reactions proceed at the rate of the rate-limiting reaction. Once pyruvate (or other potential source of acetyl CoA) enters the interior matrix of the mitochondrion, the entire cycle proceeds within that compartment. The inner membrane is an effective barrier to most organic compounds and to ions. Entry into the interior matrix of the mitochondrion, or departure from it, is accomplished either by *facilitated exchange* on a specific protein that serves as a *porter* or by active transport (Chap. 14). It is not yet possible to portray in detail how the battery of transporters in the inner membrane collectively ensures that the concentrations of the various substrates of the citric acid cycle will be maintained at appropriate levels in the mitochondrial matrix, but it is clear that these arrangements are effective and do not, for example, permit excessive intramitochondrial accumulation of pyruvate or citrate when they are not being oxidized or prevent entrance of external malate if it is required to sustain the cycle.

The major regulatory controls of the cycle are summarized in Fig. 15.4. At several loci, stimulation or inhibition is determined by the relative levels of $NADH/NAD^+$, ATP/ADP, acetyl CoA/CoA, or succinyl CoA/CoA. When

[handwritten numbers at bottom: ① ② ③ ④]

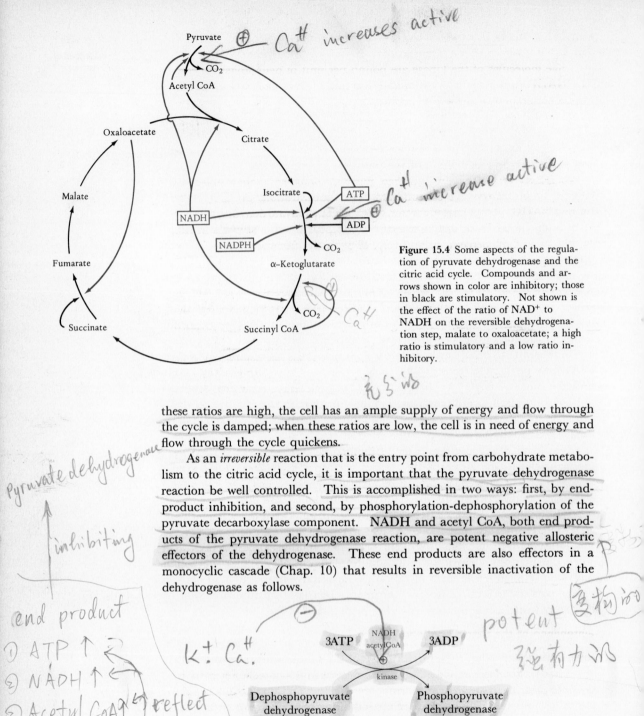

Figure 15.4 Some aspects of the regulation of pyruvate dehydrogenase and the citric acid cycle. Compounds and arrows shown in color are inhibitory; those in black are stimulatory. Not shown is the effect of the ratio of NAD^+ to NADH on the reversible dehydrogenation step, malate to oxaloacetate; a high ratio is stimulatory and a low ratio inhibitory.

these ratios are high, the cell has an ample supply of energy and flow through the cycle is damped; when these ratios are low, the cell is in need of energy and flow through the cycle quickens.

As an *irreversible* reaction that is the entry point from carbohydrate metabolism to the citric acid cycle, it is important that the pyruvate dehydrogenase reaction be well controlled. This is accomplished in two ways: first, by end-product inhibition, and second, by phosphorylation-dephosphorylation of the pyruvate decarboxylase component. NADH and acetyl CoA, both end products of the pyruvate dehydrogenase reaction, are potent negative allosteric effectors of the dehydrogenase. These end products are also effectors in a monocyclic cascade (Chap. 10) that results in reversible inactivation of the dehydrogenase as follows.

Pyruvate dehydrogenase kinase is a converter enzyme that phosphorylates three different serine residues in the decarboxylase component of the dehydrogenase. About five molecules of the kinase are bound per unit of dehydrogenase complex. NADH and acetyl CoA stimulate the rate of phosphorylation, thereby effecting conversion of the active dephospho-enzyme to the inactive phospho-enzyme. ·ATP also enhances the rate of phosphorylation. In contrast, pyruvate, Ca^{2+}, and K^+ inhibit phosphorylation. A few molecules of *pyruvate dehydrogenase phosphatase*, a Ca^{2+}-Mg^{2+}-stimulated enzyme attached to the dehydrogenase complex, stimulates reactivation of the phospho-dehydrogenase.

When fatty acids are being oxidized, pyruvate dehydrogenase is markedly inhibited. This reflects the concomitant high concentrations of ATP, acetyl CoA, and NADH. Most tissues contain an excess of pyruvate dehydrogenase so that in postprandial liver, and in muscle and adipose tissue in animals at rest, only 40, 15, and 10 percent, respectively, of pyruvate decarboxylase is in the active, nonphosphorylated form. When the demand for ATP increases, NAD^+, CoA, and ADP concentrations increase at the expense of NADH, acetyl CoA, and ATP and the kinase is deactivated. The phosphatase continues to function, however, reactivating the decarboxylase. An increased $[Ca^{2+}]$ can activate the mitochondrial phosphatase, decreasing the effective K_m of the phosphorylated enzyme as substrate for the phosphatase by as much as a factor of 20.

The NAD-isocitrate dehydrogenase reaction appears to be the rate-limiting step of the citric acid cycle (see Fig. 15.4). ADP is a strong positive effector, whereas NADPH is an allosteric inhibitor. NADH and ATP are competitive inhibitors for the NAD^+ site. NADH and NADPH bind to different sites on the enzyme. These comments pertain to the mammalian enzyme; in yeast and *Neurospora*, AMP rather than ADP is the major positive effector.

Some evidence suggests that $[Ca^{2+}]$ is an important regulator of the activity of the citric acid cycle. As noted above, the Ca^{2+}-Mg^{2+}-dependent phosphatase increases the activity of the pyruvate dehydrogenase. Ca^{2+} also stimulates the activity of the NAD^+-dependent isocitrate dehydrogenase and the α-ketoglutarate dehydrogenase. Whether $[Ca^{2+}]$ is regulated by the calmodulin system (page 310) is unclear.

The α-ketoglutarate dehydrogenase complex is inhibited by both succinyl CoA and NADH, the concentration of succinyl CoA probably being the principal rate-governing factor.

Maintenance of the Citric Acid Cycle

For the energy needs of the cells, the citric acid cycle must be kept operative at all times. This requires that oxaloacetate always be available for the first step of the cycle. Under some circumstances, one or more intermediates of the cycle may be drawn off for synthesis of other metabolites. The intermediates of the cycle can be replenished, however, by other reactions: Among the more important are (1) fixation of CO_2 into pyruvate to yield oxaloacetate (see *Anaplerosis,* Chap. 17), (2) formation of α-ketoglutarate by dehydrogenation of glutamate (Chap. 23), and (3) formation of fumarate in the operation of the urea cycle (Chap. 23). The relative magnitude and importance of these reactions varies with the type of cell and with the metabolic state of the tissue.

MITOCHONDRIAL STRUCTURE

When the citric acid cycle is complete, the status of glucose metabolism can be summarized as

$$\text{Glucose} + 10NAD^+ + 2FAD \longrightarrow 6CO_2 + 10NADH + 2FADH_2 + 10H^+$$

[The ATP available from glycolysis (Chap. 17) and GTP formed in the citric acid cycle (page 326) and here disregarded.] The reactions of the citric acid cycle are catalyzed by enzymes, which except for succinate dehydrogenase, are dissolved in the fluid of the mitochondrial matrix. In contrast, the reactions whereby NADH and $FADH_2$ are oxidized and much of the associated free energy release conserved by synthesis of ATP are catalyzed by enzymes that are bound to membranes of the mitochondria whose structure must be described to understand their functions in these processes.

Mitochondria, present in all animal and plant cells, are generally cigar-shaped bodies that appear to be of the order of 0.5 by 3 μm. In the living cell, individual mitochondria seem to alter in shape from filaments to rods, loops, or spheres. In some tissues, mitochondria are aligned in a manner that facilitates delivery of ATP to the energy-utilizing organelles, e.g., aligned with the contractile fibers of muscle cells, in the direction of secretion in acinar pancreatic cells, or coiled about the midpiece of a spermatozoon. The number of mitochondria varies with the size and energy requirements of the cell; for example, 250 per sperm cell, 500 to 2000 per liver cell, and 500,000 in the giant amoeba *Chaos chaos*. Mitochondria are major components of animal and plant cells that engage in oxidative metabolism; the mitochondrial protein accounts for approximately 20 percent of the total protein of liver cells.

In addition to the enzymes of the citric acid cycle mitochondria contain the electron-transport arrangements that deliver to oxygen the electrons abstracted from intermediates of the citric acid cycle and the means by which the energy of this process is conserved by formation of ATP. These organelles also contain the enzymes of fatty acid oxidation (Chap. 20), enzymes for the metabolism of phosphoglycerides and other structural lipids (Chap. 21), various enzymes of nitrogen metabolism such as *glutamate dehydrogenase* and several *aminotransferases,* and some of the enzymes required for urea synthesis (Chap. 23). Indeed a beef heart mitochondrion may have a complement of at least 100 different proteins (Table 15.1).

Figure 15.5 shows diagrammatically the structure of a mitochondrion. It contains two membranous structures; the innermost is separated by 50 to 100 Å from the outer, which constitutes the boundary between the cell sap and the mitochondrial contents. It is about half lipid and half protein. The inner membrane is a denser structure, about three-quarters protein and one-quarter mixed lipids by weight. The inner membrane generally runs parallel to the outer membrane along the circumference of the particle but is characterized by large infoldings into the interior, called *cristae*. In liver mitochondria, the cristae

TABLE 15.1 Location of Some Enzymes in Mitochondria

Outer membrane	Intermembrane space	Inner membrane	Matrix
Cytochrome b_5	Adenylate kinase	Cytochromes b, c_1, c, a, a_3	Pyruvate dehydrogenase
Cytochrome b_5 reductase	Nucleoside diphospho-	NADH dehydrogenase	Citrate synthase
Monoamine oxidase	kinase	Succinate dehydrogenase	Aconitase
Kynurenine hydroxylase	Sulfite oxidase	Ubiquinone	Isocitrate dehydrogenase
Fatty acyl CoA synthe-		Electron-transferring	(NAD)
tase		flavoprotein	Fumarase
Glycerophosphate acyl		ATPase	α-Ketoglutarate dehydro-
transferase		β-Hydroxybutyrate dehy-	genase
Choline phosphotrans-		drogenase	Malate dehydrogenase
ferase		Carnitine-palmityl	Fatty acid oxidation
Phospholipase A		transferase	system
			Glutamate dehydrogenase
			Aspartate-glutamate
			aminotransferase
			Ornithine transcarbamoy-
			lase

[handwritten: $2ADP \rightleftharpoons ATP + AMP$]

are continuous with the inner membrane. The interior surfaces of the cristae, facing the matrix, are studded with small, protruding structures called *elementary bodies,* or *inner membrane particles,* consisting of spherical headpieces connected to the membrane through narrow stalk pieces. Thus the major definable mitochondrial compartments are (1) the outer membrane, (2) the intermembrane space, and (3) the inner membrane, which lies between the intermembrane space and (4) the matrix (Fig. 15.5). Liver mitochondria contain many enzymes not related to the citric acid cycle or electron transport and have loosely packed cristae; heart muscle mitochondria, which are almost entirely concerned with the citric acid cycle, fatty acid oxidation, and electron transport, have extremely tightly packed cristae.

[handwritten margin note: protrude 突出]

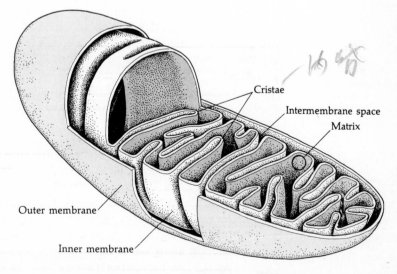

Figure 15.5 Somewhat diagrammatic representation of a mitochondrion. The inner wall of the inner membrane faces the matrix contents and is studded with protrusions, largely the ATP synthetase. In the matrix are dissolved the dehydrogenases of the citric acid cycle. Suspended in the matrix are ribosomes and the mitochondrial DNA, *inter alia.* [*From S. Wolfe, Biology of the Cell, Wadsworth, Belmont, Calif., 1974.*]

Cristae

Intermembrane space

Matrix

Outer membrane

Inner membrane

333

Several types of submitochondrial preparations have been useful in studies of mitochondrial function. The outer membrane can be ruptured by osmotic shock or by a *phospholipase*, thus releasing the intermembrane fluid for examination. Since it is denser than the outer membrane, the inner membrane and its contained matrix can then be separated by density-gradient centrifugation. The inner membrane can be ruptured by detergents such as deoxycholate to release the matrix contents, e.g., the enzymes of the citric acid cycle, much of the cytochrome c, and various components of the membrane. If the membrane is briefly exposed to ultrasonic energy, particles like those shown below are obtained. They are vesicles in an inside-out configuration relative to the disposition of the inner membrane of intact mitochondria. These vesicles are formed by rupture and resealing of the inner membrane, with the projecting stalks now on the external surface of the particles, enclosing a fluid representative of the mixed contents of the original matrix, intermembrane space, and experimental medium. By forming such vesicles in a medium of desired composition, it is possible to "load" the vesicle content with specific materials. These particles are still capable of electron transport from NADH and succinate to O_2, and they retain the ability to couple this transport to ATP formation. By these procedures, the location of various enzymes in the mitochondrial compartments (Table 15.1) has been ascertained.

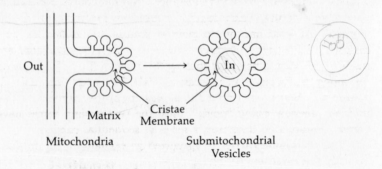

Mitochondria   Submitochondrial Vesicles

Composition of Mitochondria

The outer membrane of liver mitochondria serves as a protective barrier of the specialized functional apparatus, the cristae. It resembles the endoplasmic reticulum of hepatocytes in general chemical composition, viz., the distribution of phosphoglycerides, and contains some but not all of the enzymes associated with *microsomes* prepared from such cells. This membrane is easily penetrable by molecules as large as 10,000 daltons.

The inner membrane is a highly hydrophobic lipid bilayer consisting predominately of choline- and ethanolamine-containing phosphoglycerides and lesser amounts of those containing serine, inositol, and glycerol. Heart mitochondria (Table 13.1) contain an abundance of phosphatidylcholine and phosphatidylethanolamine, whereas these are virtually absent from liver mitochondria. About two-thirds of all the fatty acids are unsaturated in some degree, twice the relative concentration seen in the outer membrane. Almost all the cardiolipin (page 114) of the cell is in the inner mitochondrial membranes, making up about one-fifth of the total lipid. It is its intensely hydrophobic, lipid character that gives the membrane a coherent structure, determines its permeability, and accounts for the properties of its enzymes. Although the properties of the enzymes, when purified and examined in aqueous media, are of interest, it is their behavior in the anhydrous hydrophobic lipid milieu of the membrane that is significant. It is noteworthy that some of the enzymes, e.g., β-hydroxybutyrate dehydrogenase (Chap. 20), are inactive when freed of lipid.

Transport across the Inner Mitochondrial Membrane

Mitochondria suspended in a solution of impermeable solutes are excellent osmometers. The inner membrane is essentially impermeable to large molecules as well as to all charged ions. Water, small neutral molecules (O_2 and NH_3 but not H^+, OH^-, or NH_4^+), and a few "permeant" anions (Cl^- and acetate if accompanying a cation) may gain spontaneous entry into, or exit from, the matrix. Traffic into this matrix space is brisk but occurs largely either as active transport or as facilitated exchange on or through specific proteins highly specialized for this purpose. These proteins are termed *porters* or *translocases* (Chap. 14), most of which function in an *antiport* mode; viz., a substance is assisted in moving across the membrane only in exchange for some rather specific countermoving substance of similar charge, for example, ADP for ATP. No external energy supply is required for transport on a translocase (porter). At least one of the pair of transported molecules moving in the antiport (exchange) mode must be moving down a significant concentration gradient. It should be noted that by this means efflux of a major cell constituent along a concentration gradient—whether across the mitochondrial or plasma membrane—can drive the movement of the countersubstance against a gradient, thereby doing "work," until the two driving forces have been balanced. Figure 15.6 illustrates a series of sequential carriers.

The phosphate carrier is visualized as exchanging P_i for OH^- although formally this could not be distinguished from cotransport (*symport*) with H^+. P_i transport is saturable; K_m decreases with acidity. The carrier bears at least one

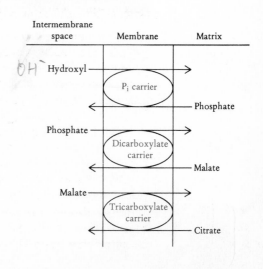

Figure 15.6 Scheme of sequential action of metabolite carriers across the mitochondrial inner membrane. [*From A. Fonyó, F. Palmieri, and E. Quagliariello, Horiz. Biochem. Biophys.* **2**:60–105 (1976).]

critical sulfhydryl group; reaction with such sulfhydryl reagents as *N*-ethylmale-imide abolishes P_i transport.

The metabolic role of the processes catalyzed by the porters of the di- and tricarboxylates are discussed below. Here it need only be noted that for operation of the citric acid cycle only the facilitated entry of pyruvate is required. Malate entry from the cytosol provides a source of additional reducing equivalents (page 327) and its exit from the mitochondria makes possible a mechanism for gluconeogenesis in the cytosol (Chap. 17). Citrate exit provides a means of exporting potential acetyl groups (Chap. 20). The dicarboxylate porter will exchange P_i for malate, malate for other dicarboxylates, and P_i for P_i but not OH^-.

The tricarboxylate carrier transports citrate, isocitrate, aconitate, and such dicarboxylates as malate; in each case, the two carboxyl groups are in the cis configuration. The distinct nature of this carrier is evident from its inability to transport P_i and its inhibition by various tricarboxylate compounds. A separate protein serves to exchange α-ketoglutarate for malate or succinate. The pyruvate carrier, in contrast, exchanges pyruvate only for the OH^- ion (or symports with H^+); this process is inhibited experimentally by α-cyano-4-hydroxycinnamate, which seems highly specific for this process. Yet another set of carriers appears to be specific for glutamate, exchanging this amino acid, respectively, for aspartate, glutamine, and most notably OH^-.

The *ADP-ATP translocase* imports into the matrix space the ADP that is to be phosphorylated by the mitochondrial device coupling that process to oxidation and then exports ATP back to the exterior of the mitochondrion. This carrier was first noted in attempts to understand the mode of action of *atractyloside* and *bongkrekic acid*, which interfere with mitochondrial oxidative phosphorylation. In nanomolar concentrations these toxic compounds bind tightly to the inner membrane, thereby halting oxidative phosphorylation. The ADP-ATP translocase is about 5 percent of the membrane protein; there are one or two atractyloside binding sites in the membrane per cytochrome a (page 339). As isolated, the carrier is a dimer of identical subunits (MW = 30,000).

The adenine nucleotide translocase participates in a system that maintains a higher ATP/ADP ratio in the cytosol than in the mitochondrial matrix; the necessary energy is provided by the arrangement whereby the ATP is synthesized. No special energy supply would seem required for the di- and tricarboxylate carriers, for example, but the P_i and pyruvate translocases, which either exchange for OH^- ions or cotransport H^+ ions, would continually generate a ΔpH across the membrane were it not for the counterpumping of protons to the intermembrane space by another means. It is that process, therefore, which supplies the energy needed for selective entry of P_i and pyruvate into the mitochondrial matrix.

Understanding of the movement of cations across the inner membrane is somewhat less satisfactory. It has been proposed that there are ionophores (translocases, perhaps) (Chap. 14) for Na^+/K^+ and for Ca^{2+}/Mg^{2+} exchanges, but little evidence supports this suggestion. The equivalent of a Na^+/K^+-ATPase (page 306) has not been found in mitochondria, but there is thought to be a porter for K^+/H^+ exchange. There is no Ca^{2+}-ATPase in the inner membrane comparable to that in sarcoplasmic reticulum (Chap. M8), for example, but there is a $Ca^{2+}/2H^+$ exchange mechanism and entry of Ca^{2+} into the matrix is also made possible by ATP hydrolysis. Once inside, Ca^{2+} is bound very tightly to a special glycoprotein in the matrix and can markedly influence metabolic events; e.g., in its presence the $NADH/NAD^+$ ratio in the matrix increases, and the pyruvate dehydrogenase phosphatase is activated (page 331).

Depsipeptides such as valinomycin (Chap. 14) have been extremely useful in exploring these relationships. K^+ spontaneously and reversibly enters the cage of this cyclic macrolide. Valinomycin penetrates across the lipid fraction of the inner membrane of the mitochondrion and of the plasma membranes of almost all cells. Hence, a small amount of valinomycin can "ferry" large quantities of K^+ across the membrane and equilibrate the $[K^+]$ on both sides. No natural equivalent of valinomycin is known in animal cells.

ELECTRON TRANSPORT

Electrons are transferred from substrates to NAD^+ by dehydrogenases in the mitochondrial matrix and to flavin nucleotides by enzymes that are either dissolved in the matrix (fatty acyl CoA dehydrogenase) or embedded in the inner membrane (succinate and glycerophosphate dehydrogenases). ATP is generated as electrons flow from these reduced nucleotides to O_2. For the reaction

$$H^+ + NADH + \tfrac{1}{2}O_2 \longrightarrow NAD^+ + H_2O$$

$\Delta E_0' = 1140$ mV and $\Delta G°$ is therefore about -51 kcal/mol. For the reduction of O_2 by $FADH_2$, $\Delta E_0' = 790$ mV, and $\Delta G°$ is about -35 kcal/mol. Under conditions prevailing in the mitochondrial matrix, viz., relatively low [ADP] and $[P_i]$ and relatively high [ATP], ΔG required for ATP synthesis is about 12 to 14 kcal/mol; hence, oxidation of the reduced nucleotides can support formation of about 3 and 2 mol of ATP, respectively. We can now consider how this is accomplished.

In addition to the pyridine and flavin nucleotide coenzymes, the inner membrane contains several other classes of compounds that undergo reversible reduction and oxidation. Nonheme-iron proteins [(FeS) proteins] have already been described in characterizing succinate dehydrogenase (page 327) and are considered in detail later. Perhaps a dozen such entities function in the mitochondrial electron-transport chain. *Ubiquinone* and its hydroquinone are a widely distributed redox couple but are not bound to specific proteins as coenzymes. Instead a small pool of this compound is in the lipid phase of the membrane, where it serves as electron acceptor for one group of enzymes and electron donor to the next enzymes in the chain. It is therefore a mobile lipid-soluble substrate, available to appropriate enzymes more rigidly locked in place in the membrane structure.

Ubiquinone (Coenzyme Q, CoQ)

In ubiquinone obtained from diverse biological sources, n in the above formula varies from 6 in some yeasts to 10 in mammalian liver. Ubiquinone forms a relatively stable semiquinone on partial reduction and hence could participate in 1-electron transfer as well as 2-electron reduction to the hydroquinone.

The cytochromes, discussed more fully in Chap. 16, are *hemoproteins*, viz., proteins to each of which is bound a *heme* or iron-porphyrin such as that shown in Fig. 15.7. (The synthesis of the porphyrins is discussed in Chap. 24 and their properties in Chap. M4.) Individual cytochromes differ from one another in their proteins, the nature of the side chains on their porphyrins, and the mode of attachment of the heme group to the proteins. Illustrative of the group is cytochrome c. Uniquely among the mitochondrial cytochromes, it is easily extracted by aqueous solvents, being loosely attached to the outer surface of the inner membrane. Mammalian cytochrome c is a single polypeptide chain of 104 residues (MW = 12,400), to which the heme moiety is covalently bound as shown in Fig. 15.7.

The iron of ferricytochrome c can be reduced and reoxidized. Ferrocytochrome c is diamagnetic (has no unpaired electrons); unlike ferrohemoglobin, it does not combine with O_2 or CO, nor does it spontaneously reoxidize in the presence of O_2. The evolution of the sequence of cytochrome c is discussed in Chap. 31. The three-dimensional structure of this protein is shown in Figs. 3.1 to 3.3.

The heme resides in a hydrophobic pocket, almost completely covered by protein. Since only the edge of the porphyrin plane is exposed to the medium

Figure 15.7 Protoporphyrin IX and cytochrome c. The imino groups of the pyrrole moieties are weak acids. Insertion of Fe^{2+} forms heme by displacing both imino hydrogen atoms; the metal ion is joined to the planar porphyrin ring by two covalent and two coordinate bonds. This iron-protoporphyrin IX compound is *heme;* the two charges of Fe^{2+} are electrically neutralized by the anionic form of two of the pyrroles. When oxidized to the Fe^{3+} state, ferriheme, the entire molecule bears a positive charge that must be balanced by an anion in the medium. In cytochrome c, the vinyl side chains at positions 2 and 4 have been reduced by the effective addition of H—S across the double bonds, the sulfhydryl groups being those of the cysteine residues at positions 14 and 17 of the protein chain, to form a covalent bond to the protein. The iron atom has available two additional liganding positions (represented by the dashed line), above and below the plane of the ring, respectively. These positions are filled by the sulfur atom of methionyl-80 and the imidazole nitrogen of histidyl-18 (Figs. 3.2 and 31.1). Oxidation of ferrocytochrome c to ferricytochrome c does not alter the distribution of ligands around the iron atom. Me = methyl; Pr = propionic; V = vinyl side chains.

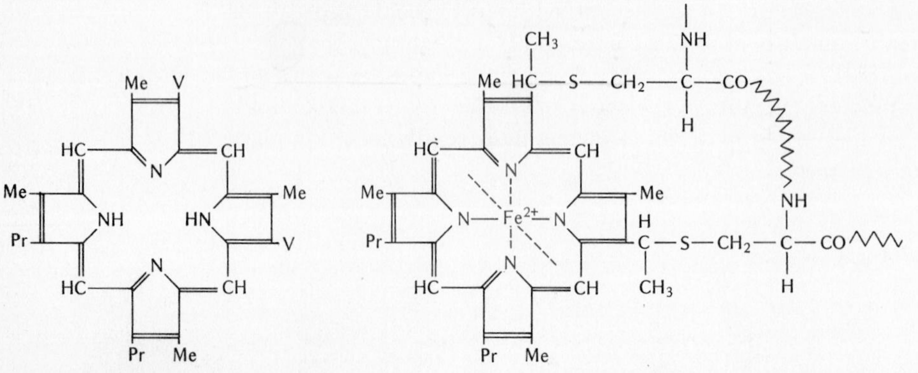

Protoporphyrin IX

Cytochrome c

and the iron is fully coordinated, electrons must enter and leave the structure by some other route, probably the protruding edge of the porphyrin.

The various cytochromes are identified by the absorption spectra of their reduced forms. They function by being alternately reduced by a membrane component of lower potential and reoxidized by a component of higher potential. Of the group, the cytochromes b, which also contain iron protoporphyrin IX as their prosthetic group, are those of lowest potential, with increases through c and c_1 to cytochromes a and a_3. The latter pair constitutes cytochrome oxidase, a copper-containing protein linked to a special form of heme (Chap. 16). Reduced cytochrome a_3 is the only component of mitochondria that can readily reduce molecular oxygen, the predominant mechanism for reduction of oxygen in mammalian tissues.

The Electron-Transport Chain

If mitochondria are incubated anaerobically with intermediates of the citric acid cycle, NAD^+, flavin, nonheme iron, ubiquinone, and cytochromes b, c_1, c, a, and a_3 undergo reduction. The absorption spectra of each of these compounds differs in the oxidized and reduced forms (Fig. 15.8). With the use of inhibitors that proved to be specific for a single step in the electron-transport

Figure 15.8 Difference absorption spectrum between anaerobic and aerobic guinea pig liver mitochondria. The curve traces the change in optical density of a mitochondrial suspension at each wavelength over the range shown, caused by complete lack of oxygen. Reduction of nonheme iron would result in varying decrease in optical density over the entire spectral range, with maximal drop at about 450 nm. Reduction of ubiquinone would not be visible in the spectral range shown. [*Courtesy of Dr. Britton Chance and Dr. Ronald W. Estabrook.*]

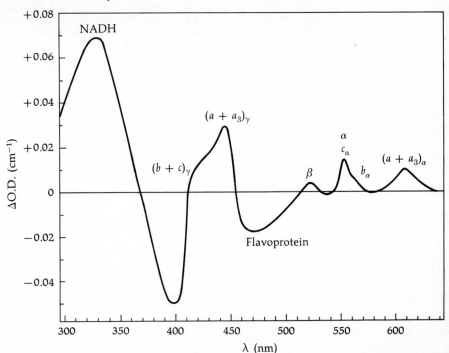

chain, it was noted which compounds were reduced and which remained oxidized. For example, rotenone, an insecticide, inhibits the NADH dehydrogenase; barbiturates, e.g., Amytal, prevent reduction of ubiquinone; piericidin A and nordihydroguaiaretic acid prevent reduction of the cytochromes by NADH; thenoyltrifluoroacetone prevents reduction of cytochrome b by succinate; antimycin A (an antibiotic from *Streptomyces griseus*) blocks reoxidation of cytochromes b; cyanide, H_2S, and azide combine with the oxidized heme iron of cytochromes a and a_3 and prevent their reduction without affecting the reduction of any other component.

From these and other experiments, it was determined that the principal flow of electrons in the mitochondrion occurs as follows:

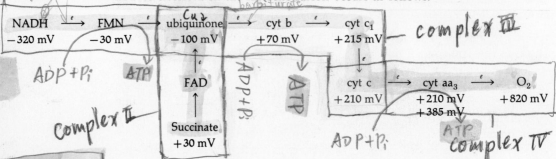

The values shown are those of E_0' for each redox component. All other reduced flavoproteins are assumed to be oxidized also by the pool of ubiquinone (CoQ) in the membrane and thus lead into the main chain.

As shown in Fig. 15.9, upon treatment with detergents, submitochondrial particles have yielded five principal fractions that catalyze the following reactions: complex I: NADH $\rightarrow$ ubiquinone; complex II: succinate $\rightarrow$ ubiquinone; complex III: dihydroubiquinone $\rightarrow$ ferricytochrome c; complex IV: ferrocytochrome c $\rightarrow O_2$; complex V (not shown, see page 349): ATP $\rightarrow$ ADP + P_i. Each fraction retains some of the membrane lipid; thus the isolated particle weights shown in Table 15.2 exceed those of the proteins which they contain. These fractions reflect the real structure of the membrane itself; each complex is composed of several different subunits that assemble themselves into a single functional entity, with the complexes separated by membrane lipids. If the functionally associated ATPase of complex V as well as the adenine nucleotide and P_i transporters are also included, the overall "molecular weight" of the

Figure 15.9 Complexes isolated from mitochondria and participating in mitochondrial electron transport. [*Modified from D. E. Green and R. F. Goldberger, Molecular Insights into the Living Process, Academic, New York, 1967.*]

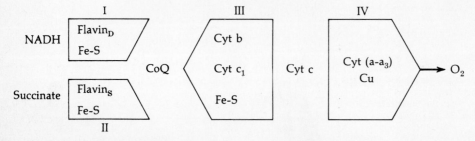

TABLE 15.2 Components of the Main Respiratory Chain

Complex	Components	Particle weight	Ratio in membrane	Reaction
I	1 FMN, 16 (FeS)	550,000	1	NADH $\xrightarrow{e}$ ubiquinone
	Ubiquinone (CoQ)		64	
III	4 cyt b, 2 cyt c_1, 8 or 12 (FeS)	250,000	4	Dihydroubiquinone $\xrightarrow{e}$ cyt c
	cyt c	12,000	8	
IV	cyt aa_3, 2 Cu^{2+}	200,000	8	Ferrocyt c $\xrightarrow{e}$ O_2

functional complex that includes one FMN-containing NADH dehydrogenase is about 5×10^6.

Electron-paramagnetic-resonance spectrometry has been used to detect the presence and behavior of specific nonheme-iron proteins. This led to the identification of a series of such reactive centers in complexes I, II, and III, including perhaps six distinct centers of varying potential in complex I alone. Since a given protein may have more than one center, it is not known how many proteins are in each complex. Indeed, when complex I is fractionated, it yields no fewer than seven different proteins ranging from 25,000 to 75,000 daltons. Additional groups of such nonheme-iron centers occur in complexes II and III.

Figure 15.10 portrays the relationships between the components of the elec-

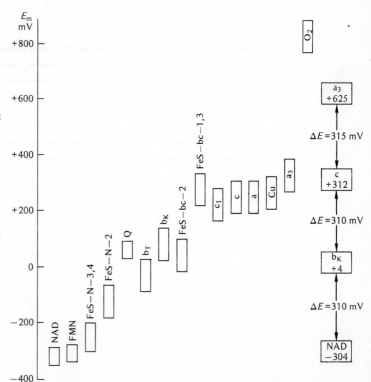

Figure 15.10 Electromotive potentials of components of the respiratory chain. Each box represents the redox potential of a given component over the range 9 to 91 percent of total reduction; the center of the box represents the midpoint potential E_m of that component on the scale at the left. On the right are shown the actual redox potentials E of several key members of the respiratory chain as observed in pigeon heart mitochondria in state 4 (see Table 15.5) respiring on β-hydroxybutyrate in the presence of a low [ADP], as well as the potential gaps ΔE between these component systems under these circumstances. [*After D. F. Wilson, P. L. Dutton, M. Erecinska, and J. G. Lindsay, in G. F. Azzone, L. Ernster, S. Papa, E. Quagliariello, and N. Siliprandi (eds.): Mechanisms in Bioenergetics, Academic, New York, 1973, p. 530, and D. F. Wilson, M. Erecinska, L. S. Owen, and L. Mela, in L. Ernster, R. W. Estabrook, and E. C. Slater (eds.): Dynamics of Energy-transducing Membranes, Elsevier, Amsterdam, 1974, p. 225; additional data courtesy of Professor Helmut Beinert.*]

341

tron-transport chain in terms of their redox potentials. Complexes I, II, and III make electrical contact with their respective oxidizing complexes via a non-heme-iron (FeS) center. This also is true of other flavoproteins that have access to the electron chain. The redox potentials of the centers in each complex are such that there is a voltage span of 250 to 300 mV with only a modest gap between the highest-potential member of one complex and the lowest-potential member of the next.

From the relation $\Delta G° = nF \Delta E$, at $\Delta E = 312$ mV, $\Delta G°$ for each major span is -14 kcal/mol. This value may be compared with $\Delta G°$ for ATP synthesis at the same conditions of [ATP], [ADP], and [P_i]. With [ADP] deliberately kept low, so that [ATP]/[ADP] is very large, $\Delta G°$ is $+14.6$ kcal/mol, calculated by accepting $+8.4$ kcal/mol as $\Delta G°$ for ATP $\rightleftharpoons$ ADP + P_i. It is remarkable that the free energy that could be made available by electron transfer across each complex is approximately equal to that required for ATP synthesis under these conditions. The picture that emerges is of a highly ordered physical arrangement in which, in the inner membrane, each participant is positioned between its reductant and oxidant, enabling electrons to flow from substrate through the successive electron carriers to O_2. The NADH dehydrogenase active site is on the matrix side, as is the site for oxygen on cytochrome a_3. Cytochrome c is attached on the outer side of the membrane, while ubiquinone (CoQ) is probably dissolved in the lipid phase of the membrane, as illustrated in Fig. 15.11.

Oxidation of Extramitochondrial NADH Mitochondrial NAD is largely restricted to the fluid in the matrix, where it links substrates to be oxidized with the electron-transport chain. Total mitochondrial NAD is present in about 40-fold excess with respect to cytochrome c. The fact that the cytoplasmic NADH is essentially denied passage across the inner membrane of hepatic mitochondria poses the problem of how NAD, reduced in cytoplasm, can be usefully reoxidized by the mitochondria. NADH is generated in some tissues, e.g., heart and skeletal muscle cytoplasm, in amounts that exceed any known use other than reoxidation. The problem is particularly evident in the metabolism of glucose; four of the six dehydrogenation reactions in glucose oxidation are those of the citric acid cycle and the fifth dehydrogenation is that of pyruvate, which

Figure 15.11 Schematic arrangement of the major components of the electron-transport chain, NADH → O_2. Only cytochrome c is readily removable by aqueous solvents; all other proteins are integral to the membrane. Cytochrome c is substantially exposed on the exterior surface of the membrane and NADH dehydrogenase and cytochrome a_3 on the inner surface, where they react with NADH and O_2, respectively. Only relative positions are indicated; components are not drawn to scale, nor are the lipid molecules shown.

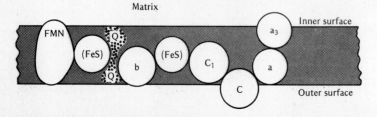

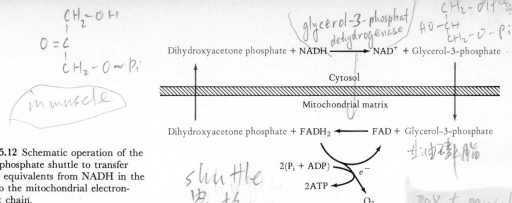

Figure 15.12 Schematic operation of the glycerol phosphate shuttle to transfer reducing equivalents from NADH in the cytosol to the mitochondrial electron-transport chain.

also occurs in the mitochondrial matrix. But the sixth, which actually precedes the others, is the cytoplasmic oxidation by NAD^+ of glyceraldehyde-3-phosphate (Chap. 17). Reoxidation of the NADH formed in that reaction is thought to be accomplished by operation of a shuttle; viz., some substance can be reduced by cytoplasmic NADH to yield a reduced form that can traverse the mitochondrial membranes, be reoxidized there by an enzyme linked to the electron-transport chain, and then return to the cytoplasm for repetition of the process. In muscle, this role is played by glycerol-3-phosphate. Dihydroxyacetone phosphate is reduced in cytoplasm by the *glycerol-3-phosphate dehydrogenase* and NADH. The resultant glycerol-3-phosphate is oxidized in the mitochondria, by a different glycerol-3-phosphate dehydrogenase, in this case a flavoprotein. Since the dihydroxyacetone phosphate can return to the cytoplasm, their coupled action accomplishes the mitochondrial oxidation of cytoplasmic NADH, as shown in Fig. 15.12. Further, the hormone thyroxine (Chap. M14), which increases the O_2 consumption of almost all tissues, induces the synthesis of mitochondrial glycerol-3-phosphate dehydrogenase in liver and probably other tissues.

There is also in liver and heart cells a more complex malate-aspartate shuttle that accomplishes the net transfer of reducing equivalents from NADH in the cytoplasm to NAD^+ in the mitochondrion. As shown in Fig. 15.13, this is made possible by the presence of a malate dehydrogenase and glutamate-aspartate aminotransferase (Chap. 23) in both cytosol and mitochondria, plus two

Figure 15.13 The malate-aspartate shuttle for transferring the reducing equivalents of cytosol NADH to mitochondrial matrix NAD^+. A and B are antiporters; C is malate dehydrogenase; D is glutamate-aspartate aminotransferase.

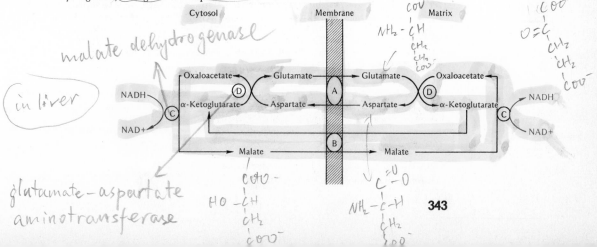

343

antiporters, one for glutamate and aspartate and one for malate and α-ketoglutarate. As shown, cytoplasmic NADH (assumed to arise as in Fig. 15.13) is used to reduce oxaloacetate to malate, which exchanges for α-ketoglutarate across the membrane and is then oxidized by the usual enzyme of the citric acid cycle. However, the oxaloacetate so formed cannot itself return across the membrane. This return can be achieved, however, by transamination from glutamate to form aspartate, which is exchanged for glutamate across the membrane; transamination in the cytosol regenerates oxaloacetate to initiate the next cycle. No energy is lost or required for this process, which involves readily reversible reactions; the driving force is the continuing reduction of NAD$^+$ in the cytosol by glyceraldehyde-3-phosphate derived from glucose (Chap. 17).

OXIDATIVE PHOSPHORYLATION

A principal function of the oxidation of carbohydrates and fatty acids is to supply the free energy released in the oxidation process, in a form utilizable for endergonic processes, viz., ATP. Mitochondrial oxidation of a molecule of NADH by an atom of oxygen results in net formation of three molecules of ATP, expressed as a P/O or P/2e of 3. Since the oxidation of substrates that transfer their electrons to NAD$^+$, such as β-hydroxybutyrate or malate, also proceeds with a P/O of 3, it is apparent that all the ATP is generated incident to the subsequent reoxidation by the organized electron-transport system of the NADH thus formed.

Many studies have been directed at establishing the "sites" of energy conservation along the path of electron transport from NADH to O$_2$. One molecule of ATP is formed as a pair of electrons moves from NADH to ubiquinone, a second as reduced ubiquinone reduces ferricytochrome c, and the third as ferrocytochrome c is oxidized by O$_2$. These correspond to the electron-transfer events accomplished by complexes I, III, and IV. Whereas E_0' for NADH/NAD$^+$ is -320 mV, that for succinate/fumarate is $+30$ mV. Hence, not only can succinate not reduce NAD$^+$, no ATP formation can be expected in the reduction of ubiquinone by succinate since, as already noted, ATP synthesis requires a ΔE of about 250 mV. Hence, the oxidation of succinate, which delivers electrons to the main chain at the level of ubiquinone, occasions formation of ATP only by electron transport at the two sites of higher potential. The elaborate structure of complex II, succinate dehydrogenase, seems required only to ensure reduction of ubiquinone while being irrelevant to the phosphorylating capability of the membrane.

Energy Yield of the Citric Acid Cycle The operation of the citric acid cycle includes three steps in which NADH arises. These are isocitrate → α-ketoglutarate; α-ketoglutarate → succinyl CoA; and malate → oxaloacetate. Each of these steps allows formation of 3 mol of ATP. An additional mole of ATP is derived from succinyl CoA (page 326). Oxidation of succinate to fumarate, by a flavoprotein in lieu of a pyridine nucleotide, yields 2 rather than 3 mol of ATP. The yields of ATP for the individual steps in the cycle are summarized in Table 15.3. Per mole of acetyl CoA consumed, 12 mol of P$_i$ are utilized and

TABLE 15.3

Energy Yield of the Citric Acid Cycle

Reaction	Coenzyme	ATP, yield per mole
Isocitrate $\rightarrow$ α-ketoglutarate + CO_2	NAD	3
α-Ketoglutarate $\rightarrow$ succinyl CoA + CO_2	NAD	3
Succinyl CoA + ADP + P_i $\rightarrow$ succinate + ATP	GDP	1
Succinate $\rightarrow$ fumarate	FAD	2
Malate $\rightarrow$ oxaloacetate	NAD	3
Total	. . .	12

12 mol of ATP are generated. Taking $\Delta G°$ for synthesis of ATP from ADP and P_i as approximately + 12 kcal/mol under physiological conditions, this represents a net energy yield, as ATP synthesized, of about 150 kcal/mol of acetate utilized. This energy yield can be compared with the total oxidation of acetic acid to CO_2 and H_2O,

$$CH_3COOH + 2O_2 \longrightarrow 2CO_2 + 2H_2O$$

for which $\Delta G°$ is 209 kcal/mol. The oxidation of acetyl CoA via the citric acid cycle, coupled to phosphorylation, can therefore be represented as follows:

$$CH_3CO—SCoA + 2O_2 + 12ADP + 12P_i \longrightarrow 2CO_2 + CoASH + 12ATP$$

We are now in a position to ascertain the total yield of ATP from the mitochondrial oxidative events associated with the reduction of three O_2 by the six pairs of electrons abstracted from one molecule of triose derived from one-half of glucose. The malate shuttle makes available one molecule of NADH per triose; pyruvate dehydrogenase results in a second. As each is oxidized, 3 ATP are formed. As noted earlier, one revolution of the citric acid cycle yields 12 ATP per acetyl CoA. The total, therefore, is 18 ATP per triose or 36 ATP per glucose.

Tissue Respiration The oxygen consumption (microliters of gas at standard pressure and temperature) per milligram of tissue is denoted as the Q_{O_2}. Table 15.4 reveals the considerable variation in animal tissues. Since the bulk of respiration in all tissues occurs over the cytochrome chain and is phosphate-linked, it is the demand for ATP that conditions the respiratory rate of each tissue. Oxidations catalyzed by aerobic dehydrogenases, copper-containing oxidases, oxygenases, and peroxidases (Chap. 16) are not phosphate-linked. Consequently, when a substance whose oxidation is catalyzed in this manner is administered to an animal or incubated with an isolated tissue, its oxidation will occur readily, thereby increasing the Q_{O_2}. In contrast, administration of a substance whose oxidation is coupled obligatorily with phosphorylation, e.g., feeding of excess carbohydrate, does not stimulate respiration. The superfluous calories are stored in tissues as glycogen or lipid until the demand for ATP initiates oxidation of these reserve sources of energy.

Respiratory Control Freshly prepared mitochondria with an ample supply of O_2 and substrate respire only if ADP and P_i are also present. O_2 consumption and ATP synthesis then proceed in parallel with constant P/O. Such mito-

bulk 是主体

obligatory

无条件的，强制的

TABLE 15.4

	Tissue	Q_{O_2}*	Tissue	Q_{O_2}*
Respiration of Various Tissues and Mitochondria	Flight muscle mitochondria	500	Duodenal mucosa	9
	Heart mitochondria	80	Lung	8
	Retina	31	Placenta	7
	Kidney	21	Myeloid bone marrow	6
	Liver, fasted animal	17	Thymus	6
	Fed animal	12	Pancreas	6
	Jejunal mucosa	15	Diaphragm	6
	Thyroid	13	Heart (resting)	5
	Testis	12	Ileal mucosa	5
	Cerebral cortex	12	Lymph node	4
	Hypophysis	12	Skeletal muscle (resting)	3
	Spleen	12	Cornea	2
	Adrenal gland	10	Skin	0.8
	Erythroid bone marrow	9	Lens	0.5

*Q_{O_2} = microliters O_2 consumed per milligram dry weight per hour. All values except the first two are for rat tissue slices in Ringer's phosphate + glucose.

chondria are described as tightly coupled and under respiratory control. This behavior of mitochondria is summarized in Table 15.5.

The tight coupling of oxidation to phosphorylation in normal mitochondria provides a means by which the rate of oxidation of foodstuffs is regulated by the requirements of the cell for useful energy. Utilization of ATP to drive the diverse energy-requiring processes of the cell automatically increases the available supply of ADP and P_i, which in turn become available to react in the coupling mechanism and permit respiration to proceed. It should be recalled that the activity of the enzymes of the citric acid cycle (Fig. 15.4) is mediated by the levels of ATP/ADP and NADH/NAD$^+$, particularly by the inhibitory effects of ATP and NADH and the activation by ADP of isocitrate dehydrogenase.

Adenylate kinase is a widely distributed enzyme that catalyzes the reaction

$$2ADP \rightleftharpoons ATP + AMP$$

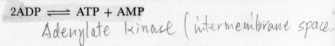
Adenylate kinase (intermembrane space

TABLE 15.5 Oxidation-Reduction Levels of Members of the Respiratory Chain in Various Metabolic Circumstances

State	Substrate level	O_2 level	ADP level	Respira-tion rate	Rate-limiting factor	NAD	Flavins	Cytochrome b	Cytochrome c	Cytochrome a
1	Low	+	Low	Slow	ADP	90	21	17	7	0
2	0	+	High	Slow	Substrate	0	0	0	0	0
3	High	+	High	Fast	Respiratory chain	53	20	16	6	4
4	High	+	Low	Slow	ADP	99	40	35	14	0
5	High	0	High	0	O_2	100	100	100	100	100

(Steady-state percentage reduction of:)

Source: Adapted from B. Chance and G. R. Williams, *Adv. Enzymol.* **17**:65 (1956).

This enzyme can serve as a means of maximizing utilization of the energy of ATP when the latter is being used more rapidly than it is synthesized. In the reverse direction, it is the first step in resynthesis of ATP from AMP. The latter is generated in a variety of synthetic processes, e.g., fatty acid and amino acid activations (Chaps. 20 and 28). It is AMP rather than ADP that serves as a positive or negative modifier controlling the rates of several enzymic processes, e.g., the phosphofructokinase, fructose bisphosphate phosphatase, and glycogen phosphorylase reactions (Chaps. 17 and 18), signaling a requirement for synthesis of ATP and accelerating the flow of substrate into and through the citric acid cycle.

Other Forms of Mitochondrial Energy Transformations Mitochondria engage in several types of energy transformations. Each represents another aspect of the fundamental coupling process; any hypothesis to explain oxidative phosphorylation must account for these other phenomena as well. Some of these processes are summarized below.

Transhydrogenation The cellular ratio NAD/NADP varies widely, being least in liver and greatest in skeletal muscle. Liver mitochondria contain about three times as much NADP as NAD. Most striking is the fact that although E_0' for the systems $NAD^+/NADH$ and $NADP^+/NADPH$ are essentially identical, in the living steady state most of the NAD is in the oxidized form whereas the bulk of the NADP is reduced. This reflects their metabolic roles. NADH is largely generated by mitochondrial dehydrogenases and reoxidized by the electron-transport system; NADPH is generated by cytoplasmic dehydrogenases, by mitochondrial transhydrogenation from NADH (see below), and by mitochondrial NADP-specific isocitrate dehydrogenase. NADPH is used for reductive biosynthesis in all cell compartments. NADPH cannot be directly reoxidized by a mitochondrial system analogous to that operative for NADH. The question is how does the reducing power of the citric acid cycle dehydrogenases contribute to the reduction of $NADP^+$.

Liver and heart mitochondria conduct a transhydrogenation that can be represented as

$$NADH + NADP^+ + energy \longrightarrow NAD^+ + NADPH$$

The process is unidirectional; i.e., equilibrium is far to the right, as written. The process can be demonstrated in respiring mitochondria in the virtual absence of ATP and can also be shown to operate anaerobically with the utilization of 1 equiv of ATP per mol of NADH. Accordingly, the energy is used to maintain the NAD/NADP systems remote from equilibrium. The energy-linked transhydrogenase is described below (page 353).

Translocation of ions The matrix of freshly isolated mitochondria is approximately isotonic with respect to $[K^+]$. If exposed to hypotonic media or, briefly, to *swelling agents*, e.g., soap, K^+ leaches from the interior but may be reaccumulated during respiration. This process is particularly impressive in the presence of valinomycin. The K^+ can be either exchanged for intramitochondrial H^+ in a ratio close to 1, or the requisite energy can be provided as external ATP. The ratio of K^+ translocated to $\sim$P utilized is about 3. Moreover, the reverse can also be demonstrated. If mitochondria are pretreated so that $[K^+]$

in the matrix exceeds that in the medium, the efflux of K^+ can drive formation of ATP from ADP and P_i. In the latter instance, the medium becomes alkaline as protons enter the mitochondrion; in the former the converse occurs.

Impressive also is the ability of mitochondria to concentrate Ca^{2+} or Sr^{2+} from the medium. In a medium rich in Ca^{2+}, accumulation can occur with a Ca^{2+}/O of about 4. If the anion of the medium is Cl^-, the Ca^{2+} is exchanged for H^+ and the matrix becomes alkaline. If the anion of the medium is P_i, the latter follows Ca^{2+} across the membrane. Concentration is so effective that granules of a hydroxyapatite (Chap. M15) may precipitate in the matrix. Like K^+ transport, this process can be driven either by respiration or extramitochondrial ATP.

MECHANISM OF OXIDATIVE PHOSPHORYLATION

Diverse reagents uncouple phosphorylation from mitochondrial electron transport, which then proceeds at a maximal rate without ATP formation. The unusual nature of such a process is made yet more apparent by the behavior of respiratory particles from the membranes of bacteria such as *Alcaligenes faecalis*. These particles exhibit no respiratory control; respiration proceeds entirely at the rate dictated by the availability of oxidizable substrate and at the same rate in the presence or absence of ADP + P_i. Yet if the latter are provided, ATP formation proceeds. Similar patterns are shown by many microorganisms, none of which exhibits P/O > 2. In contrast, maximal respiration of mitochondria occurs only if the coupling mechanism has been altered and ATP synthesis is impaired or abolished. Earlier theories concerning the mechanism of oxidative phosphorylation supposed that, as in glycolysis (Chap. 17), high-energy compounds are formed and utilized to drive ATP synthesis. No such intermediates have been found; it is likely that none exists.

However, an elegant alternative has found increasing experimental support and general acceptance, in principle. This is the *chemiosmotic*, or *proton-motive*, *hypothesis*. The principal thesis is that the physical arrangement of the components of the inner mitochondrial membrane permits movement of protons to establish an electrochemical gradient (downhill), which then accomplishes such work as the synthesis of ATP or movement of other solutes against their concentration gradients without forming intermediary high-energy compounds. To accomplish oxidative phosphorylation, the system is thought to function, in outline, as shown schematically in Fig. 15.14.

The proteins of the electron-transfer chain are thought to be so situated within the membrane that the system operates vectorially, i.e., as an integral part of the reduction and reoxidation of the successive electron carriers, and protons, without accompanying anions, are ejected on the outer side of the membrane, viz., into the intermembrane space. This would both lower the pH of the latter fluid and create an electric potential, $\Delta\Psi$, across the membrane with the inside negative. Return of protons back across the membrane occurs through the membrane protein to which is attached F_1, the *ATP synthetase* (a reversible ATPase) with its active site on the inner matrix surface. This proton flow is given direction and intensity by both the internal negative charge and

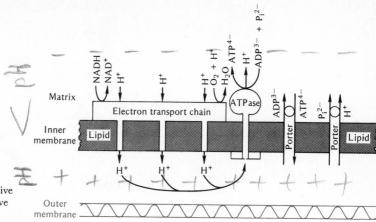

Figure 15.14 General plan of oxidative phosphorylation by the proton-motive mechanism.

the difference in proton concentrations, ΔpH, on the two sides of the membrane. It is this concentrated proton current that drives the reaction

$$ADP^{3-} + P_i^- + H^+ \longrightarrow ATP^{4-} + H_2O$$

The source of the protons is less certain. One hypothesis holds that the protons delivered to the outside of the membrane are specifically those generated as certain components of the electron transport chain are oxidized, viz., NADH, FMNH$_2$, and CoQH$_2$. This assumes a specific vectorial release of these protons due to the physical arrangement of the membrane components; each of the above components must then accept protons from the matrix solution as it accepts electrons from its specific reductant. An alternative possibility is that it is a vectorially arranged process. In their reduced states, various electron carriers would dissociate one or more protons, releasing them to the outside of the membrane, and recapture protons from the matrix when reoxidized, based on a change in conformation and dissociation constants of appropriate groups. A mechanism of this type is necessary for the terminal portion of the process, since no protons are involved in the electron transfers between cytochrome c_1, c, and cytochrome oxidase and since oxidation of the latter by oxygen actually consumes protons. Perhaps both mechanisms are operative.

Whereas the exact mechanism of proton secretion is uncertain, there is no doubt of its reality. The process is readily demonstrated with intact mitochondria, or submitochondrial particles. There is secretion of 4H$^+$ per electron pair at each phosphorylating "site," viz., from each of complexes I, III, and IV (Fig. 15.15). This suffices to synthesize ATP and to titrate P_i as it is transported into the matrix. When the protons have passed though the ATPase and return to the matrix, they have completed a proton circuit. This flow of protons, from high *protic* potential to low protic potential, is analogous to an electric circuit, and it is appropriate to refer to it as *proticity*.

Mitochondrial ATPase The major component of the elementary bodies protruding from the surface of the inner membrane is complex V, a $Ca^{2+} - Mg^{2+}$-dependent ATPase, originally identified as a *coupling factor* F_1, required for ATP synthesis. This complex (MW = 340,000) is built of two of each of five poly-

349

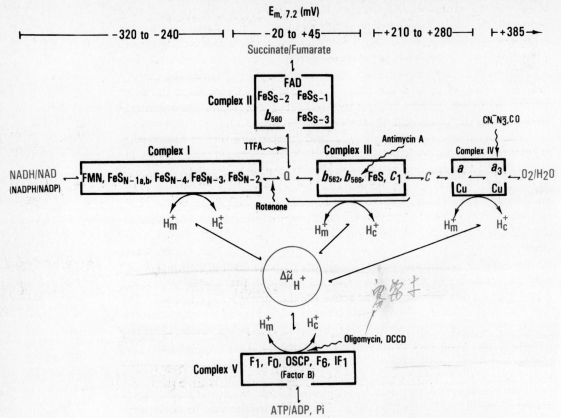

Figure 15.15 Profile of the mitochondrial electron transport and oxidative phosphorylation system, showing the components of complexes I to V, and the energy communication via $\Delta\mu_{H^+}$ and complexes I, III, IV, and V. The respective complexes are: I, *NADH: ubiquinone oxidoreductase;* II, *succinate: ubiquinone oxidoreductase;* III, *ubiquinol: ferrocytochrome c oxidoreductase;* IV, *ferrocytochrome c: oxygen oxidoreductase;* V, *ATP synthetase.* The various (FeS) centers which have been identified are designated by different subscripts. The abbreviations are: Q = ubiquinone or coenzyme Q; a, b, c are different cytochromes; F_1 = coupling factor I or F_1 — ATPase; F_0 = the membrane sector ATP synthetase which contains the dicyclohexylcarbodiimide (DCCD)-binding protein and possibly coupling factor B; OSCP = oligomycin sensitivity conferring protein; F_6 = coupling factor 6; IF_1 = ATPase inhibitor protein; H_m^+ and H_c^+ represent protons on the matrix (m) and cytosolic (c) sides of the inner mitochondrial membrane; TTFA = 2-thenoyltrifluoroacetone. Serpentine arrows show the sites of inhibition of various inhibitors. The E_m scale is not applicable to FeS_{N-1a} ($E_m = < -400$ mV), FeS_{S-2} ($E_m \sim -400$ mV), and cytochrome b_{560} ($E_m = < -100$ mV). [*Courtesy of Dr. Y. Hatefi.*]

peptide chains designated α, β, γ, δ, and ϵ with molecular weights of 56,000, 52,000, 32,000, 21,000, and 11,500, respectively. From the dissociation behavior of the enzyme, the subunits are arranged thus:

trypsin 胰蛋白酶

None of the separate subunits exhibits ATPase activity, but this activity can be observed with a combination of α and β subunits. The isolated enzyme contains two tightly bound ADP per molecule. They can be removed by several procedures, including brief exposure to trypsin, without effect on hydrolytic ATPase activity, but such preparations are devoid of ATP synthetase activity. Hence the form of F_1 to which ADP is bound is the active form in the membrane. By several criteria this protein undergoes a profound conformational change when the membrane is activated by electron flow.

The membrane also contains F_0, a hydrophobic substructure composed of at least four polypeptide chains. Its hydrophobic character locks F_0 into the lipid of the membrane. F_1 appears to be attached to F_0 specifically by the δ chains. Removal of F_1 from the membrane renders the latter "leaky" to external protons. The leak appears to occur through a proteolipid channel in F_0; it can be sealed either by readdition of F_1 or by dicyclohexylcarbodiimide (DCCD), which reacts covalently with F_0. Thus F_0, which extends across the entire membrane, offers the channel through which protons pass to the active site on the ATP synthetase. Conceivably the effective $[H^+]$ in this proton well may be much higher than in the solution on either side of the membrane, creating a microenvironment at the active site that drives ATP formation. The detailed mechanism by which this drives the formation of ATP is unknown. Observations suggest that the enzyme operates in a "flip-flop" mode, using two binding sites. With ATP bound to one, the second binds ADP and P_i. The change induced by at least two protons per molecule results in simultaneous formation of ATP at one site and discharge of ATP at the other, etc.

Confidence in the proton-motive hypothesis rests in large measure on the results of studies with model systems. Synthetic vesicles can be made with mitochondrial phosphoglycerides into which are incorporated F_1, or $F_1 + F_0$, and, for example, cytochrome oxidase, cytochrome c, and a reductant for the latter such as ascorbic acid. Preparations of this type secrete protons as the substrate is oxidized, and if ADP and P_i are available, they will synthesize ATP, whereas an ordinary solution of these proteins and substrates fails to make ATP.

Submitochondrial particles, being inside out (page 334), secrete protons into their interiors when oxidizing succinate or NADH. When synthetic vesicles are prepared so that their orientation is like that of inside-out submitochondrial particles, with the ATPase on the exterior surface, arrangements that similarly acidify the interior of the vesicle, e.g., the incorporation of functioning complex III, occasion ATP synthesis on the exterior. Particularly dramatic was the result of incorporation, into synthetic vesicles, of the purple membrane of the bacterium *Halobacter halobium*, a structure similar to the rhodopsin of the retina (Chap. M9). When illuminated, this structure pumps protons to the interior of the vesicle while accepting protons from the medium. When liver mitochondrial ATPase was included so that the headpiece was on the outside of such vesicles, synthesis of ATP was observed upon illumination.

In the inside-out vesicle systems described above and in photosynthesis (Chap. 19) the transmembrane ΔpH is the major force that drives ATP synthesis. But in mitochondria in situ and in respiring bacteria (see below), the observed ΔpH, somewhat greater than 1 pH unit, is too small to account for ATP

leak 渗

render 搜

concei

结合

视紫红质
retina 视网膜

situ

$$\Delta p = \Delta \Psi - \frac{RT}{nF} \Delta pH \qquad \Delta \Psi - \text{membrane potential}$$
$$\Delta p - \text{proton motive force}$$

synthesis. As noted earlier, a ΔE of about 250 mV is required for ATP synthesis under most physiological conditions. To translate ΔpH into electrical units,

$$\Delta E = \frac{-2.3RT}{F} \Delta pH = -60 \, \Delta pH \, \text{(mV)} \qquad \text{at } 30°C$$

Thus a ΔpH of 1 pH unit would make possible a potential of only 60 mV across the membrane, insufficient to drive ATP synthesis even if [ADP] and [P_i] were very high and [ATP] very low. However, the concomitant electric potential difference across the membrane, $\Delta \Psi$, is close to 175 mV in the functioning mitochondrion. The total effective electrochemical potential $\Delta \tilde{\mu}_{H^+}$ is expressed simply as

$$\Delta \tilde{\mu}_{H^+} = \Delta \Psi - 60 \, \Delta pH \, \text{(mV)}$$

The measured electrochemical potential, $\Delta \tilde{\mu}_{H^+}$, has been found to be 230 to 250 mV in mitochondrial and bacterial membranes, which suffices to drive ATP synthesis. Using special techniques for preparing planar membranes of soybean lipids or of phosphatidylethanolamine or phosphatidylserine (but not phosphatidylcholine) in which cytochrome oxidase is oriented, protons are secreted only on the side on which cytochrome c is placed and $\Delta \Psi$ can be measured directly with suitable electrodes. Such crude preparations have exhibited $\Delta \Psi$ of almost 200 mV. Similar observations have been made with illuminated bacteriorhodopsin in a planar membrane.

Uncouplers The mode of action of many of the classical *uncouplers* of oxidative phosphorylation is explained by the proton-motive hypothesis. In their presence, ATP synthesis does not occur, while substrate oxidation and oxygen consumption proceed maximally. Examples of these compounds are 2,4-dinitrophenol and carbonylcyanide trifluoromethoxyphenylhydrazone.

2,4-Dinitrophenol

Carbonylcyanide
trifluoromethoxyphenylhydrazone

These compounds are lipophilic weak acids that can migrate through the lipid phase of the membrane in both their ionized and un-ionized forms. Hence they short-circuit the proton current by ferrying protons directly through the lipid phase of the membrane.

There are also true inhibitors of oxidative phosphorylation, notably DCCD and the antibiotic oligomycin. As noted above, DCCD (page 351) reacts with F_0 to block proton passage across the membrane. Oligomycin in some similar way reacts with the stalk connecting F_1 to the membrane so that neither ATP synthesis nor proton flow is possible. Damming the flow of protons reverses the equilibria of the proton-secreting constituents of the electron-transfer chain so that neither ATP synthesis nor electron flow (respiration) is possible.

Earlier it was noted that anaerobically the presence of ATP enables the mitochondrion to utilize succinate to reduce NAD^+. This is made possible by the reversibility of the ATPase. Hydrolysis of matrix ATP drives protons through F_0 back into the intermembrane space, building up $[H^+]$ in that fluid. This has been observed with vesicle preparations, the H^+ flowing across the membrane from the side on which ATP hydrolysis occurs, depending on whether the vesicle is right side out or inside out. If one visualizes, simplistically, that the NADH dehydrogenase (complex I) catalyzes

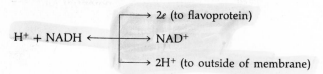

$$H^+ + NADH \longleftrightarrow \begin{cases} 2e \text{ (to flavoprotein)} \\ NAD^+ \\ 2H^+ \text{ (to outside of membrane)} \end{cases}$$

then the succinate dehydrogenase delivers $2e$ to the electron-transport chain, $E_0' = +30$ mV, while the ATPase applies proton pressure through the intermembrane space in a localized intense manner. Since, indeed, NAD^+ is reduced, this proton pressure deriving from ATP hydrolysis must have the effect of almost 300 mV in these experimental conditions.

This circumstance strikingly reveals the nature of the control that couples oxidation to phosphorylation. Respiration cannot proceed unless ADP and P_i are available to form ATP. If ATP is abundant in the matrix and ADP and P_i are unavailable, the reversible ATPase will ensure maximal proton pressure in the intermembrane space; the oxidative process is then unable to proceed since the oxidation reduction cycle of each of the working complexes obligatorily entails proton "secretion" as well. If the latter is impossible because of preexisting high $[H^+]$ created by ATPase activity, electron transfer is also impossible. The control therefore is the proton pressure outside the inner membrane.

One other component of the membrane, the NADP-NAD transhydrogenase, resembles the ATPase in consisting of a water-soluble protein (MW $\approx$ 10,000) bound to a protein complex that is integral to the membrane; this enzyme utilizes the energy of proton translocation in catalyzing the reaction

$$NADH + NADP^+ \longrightarrow NAD^+ + NADPH$$

In the functioning mitochondrion $\Delta\tilde{\mu}_{H^+}$ is generated by the mechanisms of electron transfer, NADH to O_2; experimentally it can be generated by addition of ATP to mitochondria in the absence of oxidizable substrate. The mitochondrion therefore is a transducer of various forms of chemical potential; in this system ΔE, ΔpH, $\Delta\Psi$, $\Delta\tilde{\mu}_{H^+}$, and $\sim P$ are equivalent and interchangeable.

The reinforcing aspect of $\Delta\Psi$ and ΔpH is illustrated by the behavior of mitochondria and submitochondrial particles with ionophores. Addition of valinomycin to a suspension of respiring mitochondria in a K^+-containing medium results in influx of K^+, carried by the valinomycin, thereby decreasing $\Delta\Psi$ but not affecting ΔpH. By further H^+ efflux $\Delta\tilde{\mu}_{H^+}$ is rebuilt in part, the matrix becoming more alkaline. The carboxylic polyether antibiotic *nigericin* acts as a K^+/H^+ antiport. Hence, it can decrease ΔpH across the membrane but not

$\Delta\Psi$. Accordingly, neither nigericin nor valinomycin alone uncouples phosphorylation or markedly affects $\Delta\tilde{\mu}_{H^+}$ of respiring inside-out submitochondrial particles that are secreting protons into their interiors. When the two are combined, however, nigericin exchanges inside H^+ for outside K^+, which then effluxes in the valinomycin so that both ΔpH and $\Delta\Psi$ collapse and ATP synthesis ceases.

Active Transport in Bacterial Cells

The roles of $\Delta\Psi$, ΔpH, and ΔE in ATP synthesis have also been observed with respiring plasma membranes of bacterial cells. The respiratory enzymes of *Escherichia coli* are present in the lipid membrane; when oxidizing a substrate such as D-lactate, protons are released to the outside, ΔpH and $\Delta\Psi$ are established, and intracellular [ATP] is maintained. In this state, the cell membrane can then engage in active transport of lactose (Chap. 14) from the medium. This inward transport occurs in the symport mode, in which galactoside entry accompanies that of either H^+ or K^+. With such membranes, ATP synthesis can be induced by inward flow of H^+ or by outward flow of K^+ and is dependent upon a $\Delta\tilde{\mu}_{H^+}$ greater than 200 mV but independent of whether ΔpH or $\Delta\Psi$ is the major contributor to $\Delta\tilde{\mu}_{H^+}$.

Conclusion

Although the chemiosmotic hypothesis rationally ties many observations together, there remain important questions. For example, when measured, $\Delta\tilde{\mu}_{H^+}$ is frequently no greater than 200 mV, yet more than 250 mV would be required to synthesize ATP under physiological circumstances. The mechanism and stoichiometry of proton secretion require explanation. The mechanism of action of the ATP synthetase is unknown, nor is it clear how events at its active site are given direction by $\Delta\tilde{\mu}_{H^+}$. Nevertheless, so many observations fit rationally into this elegant single construct that it has been generally accepted. The general principles of electron transport and oxidative phosphorylation are summarized in Fig. 15.15.

The Mitochondria of Brown Fat

All young vertebrates require a thermogenic device, in addition to the shivering mechanism, with which to generate heat. A device of this kind is particularly necessary for hibernating animals. Shivering muscles contract against no load, using the contractile proteins to hydrolyze ATP and releasing as heat the full energy potentially available from the hydrolyzed ATP. This arrangement is necessitated by the tightly coupled oxidative phosphorylation of normal mitochondria. If that process could be uncoupled, it would serve as a heat-producing device. This occurs in the mitochondria of brown fat. Although these mitochondria have the usual reversible ATPase, there appears also to be a transmembrane proton translocase by which protons can return to the matrix and short-circuit the ATPase. It seems probable that the porter really carries hydroxyl ions from the matrix to the intermembrane space; this, however, accomplishes the same result. If sufficient to hold $\Delta\tilde{\mu}_{H^+}$ well below 200 mV, ATP synthesis becomes impossible, the oxidative process runs free, and the resultant energy is released entirely as heat.

Evolutionary Origin of Mitochondria

An organelle equipped to use oxygen to oxidize diverse substrates and thereby synthesize ATP could not have evolved until oxygen had been placed in the primeval reducing atmosphere by photosynthetic organisms. Considerable evidence suggests that the present mitochondrion is derived from the successful symbiosis of a eukaryotic cell having a well-developed glycolytic (anaerobic) pathway of carbohydrate metabolism and a prokaryotic (nonnucleated) cell that had developed a membranous respiratory electron-transfer chain. Even now mitochondria contain their own genetic apparatus—DNA, ribosomes, and the ability to fabricate proteins based on the information in their own DNA (Chap. 29). However, only a fraction of all mitochondrial proteins, about a dozen polypeptide chains, is made by this apparatus, including, for example, three of the five polypeptide chains of cytochrome oxidase, one or two of cytochrome b, and two or three of the chains of ATPase, which are integral to the membrane. It appears that it is the unusually hydrophobic, water-insoluble proteins, which would otherwise self-aggregate, that are synthesized in the mitochondria, thereby avoiding the necessity of their moving any distance through the cytosol. Instructions for synthesis of the numerous other peptide chains of the mitochondrion are provided, as usual, in the chromosomes of the cell nucleus (Chap. 29).

An interesting evolutionary marker is *superoxide dismutase* (page 249), of which there are two types in mammalian cells. The cytoplasmic enzyme contains Cu^{2+} and Zn^{2+} and is distinct in its amino acid sequence from the mitochondrial Mn^{2+}-containing enzyme. The sequence of the latter is homologous to the Mn^{2+}- or Fe^{2+}-containing dismutases of prokaryotes like *E. coli* (Chap. 31).

A contemporary prokaryote that may resemble the ancestor of mitochondria is *Paracoccus denitrificans*. The respiratory chain of this organism (Fig. 15.16) is organized very much like that of mitochondria and functions with tight respiratory control and a P/O of 3. These cells have adaptive ability to use their respiratory chains with other substrates, but the basic constitutive chain and the

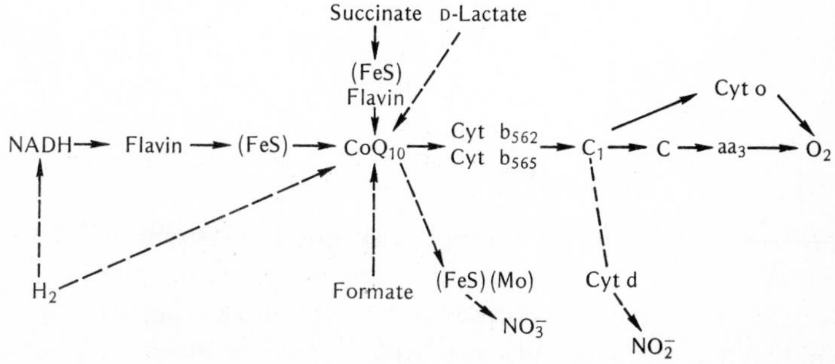

Figure 15.16 The electron-transfer chain of *P. denitrificans*. Solid arrows represent the constitutive enzymes. Broken arrows represent adaptive enzymes formed only under appropriate conditions.

associated phosphorylating capability could readily have served as the evolutionary forerunner of a mitochondrion (Chap. 31).

REFERENCES

Books

Azzone, G. F., M. E. Klingenberg, E. Quagliariello, and N. Siliprandi, (eds.): *Membrane Proteins in Transport and Phosphorylation,* North-Holland, Amsterdam, 1974.

Hatefi, Y., and L. Djavadi-Ohaniance, (eds.): *The Structural Basis of Membrane Function,* Academic, New York, 1976.

Lovenberg, W. (ed.): *Iron-Sulfur Proteins,* vol. 3, Academic, New York, 1977.

Munn, E. A.: *The Structure of Mitochondria,* Academic, New York, 1974.

Parsons, D. S. (ed.): *Biological Membranes,* Oxford, London, 1975.

Racker, E.: *A New Look at Mechanisms in Bioenergetics,* Academic, New York, 1976.

Sanadi, D. R. (ed.): *Current Topics in Bioenergetics,* Academic, New York (a continuing annual series).

Tzagoloff, A.: *Mitochondria,* Plenum, New York, 1982.

Review Articles

Boyer, P. D., B. Chance, L. Ernster, P. Mitchell, E. Racker, and E. C. Slater: Oxidative Phosphorylation and Photophosphorylation, *Annu. Rev. Biochem.* **46**:955–1026 (1977).

Chance, B.: The Nature of Electron Transfer and Energy-coupling Reactions, *FEBS Lett.* **23**:3–20 (1972).

Cross, R. L.: The Mechanism and Regulation of ATP Synthesis by F_1-ATPases, *Annu. Rev. Biochem.* **49**:1079–1114 (1980).

Fillingame, R.: The Proton-Translocating Pumps of Oxidative Phosphorylation, *Annu. Rev. Biochem.* **49**:1079–1114 (1980).

Flatmark, T., and J. I. Pedersen: Brown Adipose Tissue Mitochondria, *Biochem. Biophys. Acta* **416**:53–104 (1975).

Fonyo, A., F. Palmieri, and E. Quagliariello: Carrier-mediated Transport of Metabolites in Mitochondria, *Horizons Biochem.* **2**:60–105 (1975).

Mitchell, P.: Keilin's Respiratory Chain Concept and its Chemiosmotic Consequences, *Science* **206**:1148–1159 (1979).

Mitchell, P.: Vectorial Chemistry and the Molecular Mechanics of Chemiosmotic Coupling: Power Transmission by Proticity, *Trans. Biochem. Soc.* **4**:399–430 (1976).

Papa, S.: Proton Translocation Reactions in the Respiratory Chain, *Biochim. Biophys. Acta* **456**:39–84 (1976).

Pressman, B. C.: Biological Applications of Ionophores, *Annu. Rev. Biochem.* **45**:501–530 (1976).

Skulachev, V. P.: Energy Transformations in the Respiratory Chain, *Curr. Top. Bioenerg.* **4**:127–190 (1971).

Skulachev, V. P.: Transmembrane Electrochemical H^+-potential as a Convertible Energy Source for the Living Cell, *FEBS Lett.* **74**:1–9 (1977).

Smoly, J. M., B. Kuylenstierna, and L. Ernster: Topological and Functional Organization of the Mitochondrion, *Proc. Natl. Acad. Sci. USA* **66**:125–131 (1970).

Srere, P. A.: The Enzymology of the Formation and Breakdown of Citrate, *Adv. Enzymol.* **43**:57–102 (1975).

Vignais, P. V.: Molecular and Physiological Aspects of Adenine Nucleotide Transport in Mitochondria, *Biochim. Biophys. Acta* **456**:1–38 (1976).

Wikström, M., K. Krab, and M. Suraste: Proton-translocating Cytochrome Complexes, *Annu. Rev. Biochem.* **50**:623–655 (1981).

Chapter 16

Biological Oxidations II:
Oxidative Enzymes,
Coenzymes, and Carriers

Hundreds of enzymes catalyze oxidation-reduction reactions, but there is only a rather small number of functional groups that serve as acceptors or donors of the electrons and protons transferred between substrates and products. Thus, many *dehydrogenases* use nicotinamide nucleotides (NAD^+ or $NADP^+$) as coenzymes in reactions of the following form.

$$E \cdot NAD^+ + MH_2 \rightleftharpoons E \cdot NADH + H^+ + M$$

The reducing potential of the substrate (MH_2) is secured in NADH, which then participates in other oxidation-reduction processes. Other dehydrogenases employ flavin nucleotides and catalyze the general reaction (where F = flavin nucleotide)

$$E \cdot F + MH_2 \rightleftharpoons E \cdot FH_2 + M$$

The reduced enzyme, $E \cdot FH_2$, may transfer its electrons and protons to other substances, as in oxidative phosphorylation (Chap. 15), or to O_2 with the formation of H_2O_2,

$$E \cdot FH_2 + O_2 \rightleftharpoons E \cdot F + H_2O_2$$

Enzymes of the latter type are called *oxidases*. Some oxidases, however, employ only metals, such as Cu^{2+} or Fe^{2+}, as cofactors, whereas others use a more complex set of functional groups.

Like the oxidative enzymes, the carrier proteins of mitochondrial electron transport also employ a rather small number of functional groups. These include metals, such as Fe, Cu, or Mo, heme groups, and iron-sulfur clusters. The last are prosthetic groups formed by iron coordinated with inorganic sulfide and cysteinyl thiol groups of the carrier protein.

Any cross-references coded M refer to the companion text, *Principles of Biochemistry: Mammalian Biochemistry*.

Often, only one functional group of an oxidative enzyme or a carrier protein is combined with a single polypeptide chain to form a functional unit. Many enzymes, however, contain a combination of flavin nucleotides, iron-sulfur clusters, heme groups and metals to create a single, miniature electron-transport chain of varying length and complexity, uniquely suited to a particular metabolic need.

The nature of the functional groups in oxidative enzymes and electron-carrier proteins, many of which were introduced in Chap. 15, will be considered in further detail in this chapter, with emphasis on the principles that underlie their action.

NICOTINAMIDE ADENINE NUCLEOTIDES

Nicotinamide adenine dinucleotide (NAD$^+$) (page 318) and nicotinamide adenine dinucleotide phosphate (NADP$^+$), although termed *dinucleotides*, contain two nucleotides linked by a pyrophosphate bond rather than a diester bond, as in the nucleic acids (Chap. 8). NADP$^+$ contains a third phosphate group in ester linkage on the C-2' of the ribose of the adenosine moiety.

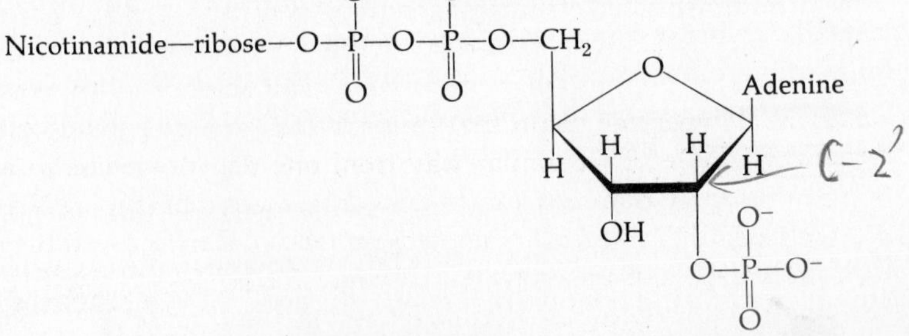

Nicotinamide—ribose—O—P—O—P—O—CH$_2$

Adenine

C-2'

OH

Nicotinamide adenine dinucleotide phosphate (NADP$^+$)

When NAD$^+$ or NADP$^+$ is reduced, the nicotinamide group accepts a H$^+$ and two electrons, the equivalent of a hydride ion (H : $^-$), from the substrate, while another H$^+$ from the substrate is released into the medium.

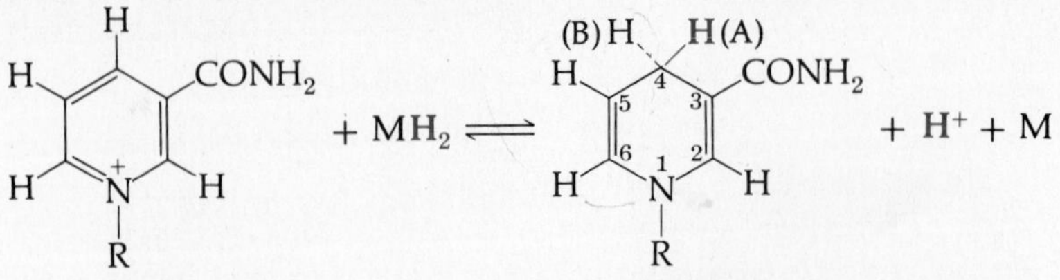

NAD$^+$ + MH$_2$ $\rightleftharpoons$ NADH   + H$^+$ + M

NAD$^+$ contains a positive charge on the ring; NADH has no charge and contains a H atom, derived from the substrate, on the carbon atom 4 of its quinoid ring. Hence, the abbreviation for reduced NAD$^+$ is NADH. Carbon-4 is a prochiral center; i.e., the two hydrogen atoms in NADH are not equivalent. The one in front of the plane of the ring is designated H$_A$ and the one behind the

TABLE 16.1

Substrate	Nucleotide	Source	Stereospecificity	
Isocitrate	NADP	Heart	A	**Stereospecificity**
Ethanol	NAD	Yeast	A	**of Some**
Lactate	NAD	Heart	A	**Nicotinamide**
Malate	NAD	Heart	A	**Nucleotide**
NADP⁺	NAD	Heart mitochondria	A	**Enzymes**
Glucose	NAD	Liver	B	
Glutamate	NADP, NAD	Liver	B	
Glutathione	NADP	Yeast	B	
3α-Hydroxysteroids	NAD	Liver	B	
6-Phosphogluconate	NADP	Liver	B	
Glyceraldehyde-3-phosphate	NAD	Yeast, muscle	B	

ring is H_B. Dehydrogenases that distinguish between H_A and H_B are of two types. Those of type A, such as yeast alcohol dehydrogenase, transfer H to the A position, whereas those of type B, e.g., liver glyceraldehyde-3-phosphate dehydrogenase, transfer H to the B position. Representative enzymes of each type are listed in Table 16.1.

The three-dimensional structures of several dehydrogenases reveal that the bound NAD⁺ or NADH is an extended structure, as shown in Fig. 16.1, rather than in a compact configuration observed in free solution. Moreover, that portion of the polypeptide chain that binds NAD⁺ in each subunit is usually, but not always, folded in a similar way from one dehydrogenase to another.

Some of the more than 250 known dehydrogenases that employ NAD⁺ and NADP⁺ are listed in Table 16.2. There is no relationship between the properties of the substrate and the enzyme preference for NAD⁺ or NADP⁺. Each dehydrogenase is usually highly specific for either NAD⁺ or NADP⁺. Different forms of the same dehydrogenase, however, may use one dinucleotide in one subcellular compartment and the other in another compartment; e.g., mitochondrial isocitrate dehydrogenase uses NAD⁺ whereas that in cytosol uses NADP⁺. Similar differences are found among the same enzymes from different species. By separation of these specificities the cell maintains NAD largely as NAD⁺, and NADP as NADPH, in keeping with the metabolic roles of each coenzyme. *In general, enzymes responsible for oxidations supplying energy to an organism*

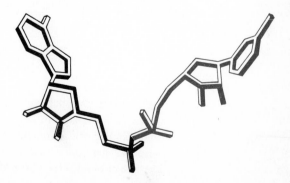

Figure 16.1 Extended conformation of NAD when bound to a dehydrogenase. The nicotinamide nucleotide moiety is shown in color. [*Adapted from L. J. Banaszak and R. A. Bradshaw, p. 384, in P. D. Boyer (ed.), The Enzymes, 3d ed., vol. XI, Academic, New York, 1975.*]

359

TABLE 16.2 Some Dehydrogenases Employing Pyridine Nucleotides

Enzyme	Substrate	Product	Source
1. NAD⁺-specific			
Aldehyde	Aldehydes	Carboxylic acids	Liver
Ethanol	Ethanol	Acetaldehyde	Liver, kidney, yeast
Glycerol-3-phosphate	Glycerol-3-phosphate	Dihydroxyacetone phosphate	Liver, muscle cytosol
Isocitrate	Isocitrate	α-Ketoglutarate + CO₂	Liver and muscle mitochondria
Lactate	Lactate	Pyruvate	Liver, muscle
Malate	Malate	Oxaloacetate	Muscle, liver
2. NADP⁺-specific			
Isocitrate	Isocitrate	α-Ketoglutarate + CO₂	Heart cytosol
Malate	Malate	Pyruvate + CO₂	Liver
Glucose-6-phosphate	Glucose-6-phosphate	6-Phosphogluconolactone	Liver, erythrocytes
Glutathione	Reduced glutathione	Oxidized glutathione	Erythrocytes, liver
3. NAD⁺ or NADP⁺			
Glycerol	Glycerol	Dihydroxyacetone	Liver, bacteria
Glutamate	Glutamate	α-Ketoglutarate	Muscle, liver
3-β-Hydroxysteroids	3-β-Hydroxysteroids	3-Ketosteroids	Liver

utilize NAD, whereas those providing the energy for reductive syntheses utilize NADP. The [NAD] in animal tissues is 0.5 to 3 mM, several times higher than the [NADP], which ranges from 0.01 to 0.1 mM. In liver, however, the [NADP] is about one-third the [NAD].

When the NAD-specific enzyme is in one cellular compartment and an NADP-specific enzyme in another, shuttling of the substrate back and forth across the membrane can serve as a mechanism for the process of *transhydrogenation*. Thus, mitochondrial *malate dehydrogenase* (Chap. 15) and cytosolic *malic enzyme* (Chap. 17) are coupled where malate produced in mitochondria passes the mitochondrial membrane and is reoxidized in the cytoplasm, in effect transporting "hydrogen" between compartments.

$$\text{Oxaloacetate}_{mit} + \text{NADH}_{mit} + \text{H}^+ \underset{\text{malate dehydrogenase}}{\rightleftharpoons} \text{malate}_{mit} + \text{NAD}^+_{mit}$$

$$\text{Malate}_{cyt} + \text{NADP}^+_{cyt} \underset{\text{malic enzyme}}{\rightleftharpoons} \text{pyruvate} + \text{NADPH}_{cyt} + \text{H}^+ + \text{CO}_2$$

SUM: $\text{Oxaloacetate} + \text{NADH}_{mit} + \text{NADP}^+_{cyt} \longrightarrow$
$$\text{pyruvate} + \text{CO}_2 + \text{NAD}^+_{mit} + \text{NADPH}_{cyt}$$

Other types of coupled reactions are given in Chaps. 17 and 20 and serve important roles in carbohydrate and fatty acid metabolism.

From the above considerations it is evident that the term *coenzyme* is a misnomer when applied to NAD and NADP. The K_m values for each are only slightly lower than those for substrates, and equivalent amounts of substrate and dinucleotide are consumed in the reaction. Thus, NAD and NADP are actually cosubstrates. Indeed, they may dissociate from an enzyme in one form and be converted to the other form by a second enzyme, in effect, permitting reduction of one metabolite by another, as follows:

$$AH_2 + NAD^+ \longrightarrow A + NADH + H^+$$
$$B + NADH + H^+ \longrightarrow BH_2 + NAD^+$$

SUM: $$AH_2 + B \longrightarrow A + BH_2$$

Examples of this type of coupled reaction are found in glycolysis (Chap. 17) and other metabolic pathways considered in later chapters.

Spectral analysis of the reaction catalyzed by NAD and NADP-requiring enzymes is particularly useful for deducing enzyme mechanisms, since, in contrast to NAD and NADP, NADH and NADPH strongly absorb light at 340 nm. Moreover, when activated by light at 340 nm, they fluoresce with maximal emission at 465 nm; fluorescence excitation and emission spectra also change on binding of the nucleotides to enzymes. Thus, spectroscopic analysis of the kinetics of dehydrogenases reveals that many of them effect catalysis by a compulsory ordered mechanism (Chap. 10) for substrate binding, as in lactate dehydrogenase.

Dehydrogenases often are composed of different ratios of subunit polypeptides with different sequences and form so-called isoenzymes. For example, five different lactate dehydrogenases are known in animal tissues, as described in Chap. 17, with subunit structures M_4, M_3H, M_2H_2, MH_3 and H_4, where M refers to one subunit type and H to another. Each form exhibits significantly different K_m and V_{max} values for substrates, in accord with its metabolic role.

NAD may also have other roles that do not involve oxidation-reduction reactions. Thus, *DNA ligase* of *Escherichia coli* catalyzes formation of phosphodiester bonds between two polydeoxynucleotides with the concomitant cleavage of NAD into AMP and nicotinamide mononucleotide (Chap. 27). In addition, diphtheria toxin catalyzes transfer of the adenosine pyrophosphate ribose group of NAD^+ to protein factors required for eukaryotic protein synthesis (Chap. 29). Cholera toxin and the enterotoxin of *E. coli* catalyze a similar type of ADP ribosylation of a regulatory protein of adenylate cyclase (Chap. M12).

FLAVOPROTEINS

Flavoproteins are ubiquitous, containing riboflavin-5'-phosphate (flavin mononucleotide, FMN) or flavin adenine dinucleotide (FAD), whose structures were given in Chap. 15. The mode of attachment of a flavin to the apoenzyme varies from enzyme to enzyme and may be covalent or noncovalent. FMN and FAD may be bound covalently through the 8-α-methyl group to the N-1 atom of histidine (Chap. 15) or to the thiol group of cysteine in the enzyme. Noncovalent binding of flavins is always considerably stronger than that of NAD or NADP to dehydrogenases. In addition, flavin enzymes may contain one or more metal ions, an iron-sulfur complex or a heme, to give a remarkable variety of catalytic capabilities.

Table 16.3 lists some flavoproteins, illustrating their extraordinarily diverse character. They vary widely in size (MW = 12,000 to $\approx$700,000), containing either a single flavin on one subunit polypeptide, as in flavodoxin (Chap. 5), or two or more subunits to which are bound not only several flavin groups, but also other functional prosthetic groups, which together form a minature electron-transport system.

$$H_3C\text{-}\underset{8}{\overset{7}{\diagdown}}\underset{9}{\overset{6}{\diagup}}\overset{5}{\underset{10}{N}}\overset{4}{\underset{1}{\overset{N}{\diagup}}}\overset{O}{\underset{2}{\overset{NH}{\diagdown}}}\underset{R}{\overset{3}{}} + MH_2 \rightleftharpoons H_3C\text{-}\diagup\diagdown\overset{H}{N}\overset{O}{\diagdown}NH + M$$

Like NAD$^+$- and NADP$^+$-dependent dehydrogenases, the flavoproteins exhibit characteristic spectral features, as exemplified by the absorption spectra for the oxidized, reduced, and partially reduced forms of glucose oxidase in Fig. 16.2. Such spectra reveal that on reduction, when the flavin moiety accepts two electrons and two protons from the substrate, as above, formation of the quinoid ring causes complete abolition of the absorption band at 450 nm and

TABLE 16.3 Some Flavoproteins

Substrate	Physiological electron acceptor	Flavin	Other functional components	Source
D-Amino acids	O_2	FAD	—	Liver, kidney
L-Amino acids	O_2	FAD or FMN	—	Kidney, snake venoms
Monoamines	O_2	FAD	—	Liver, kidney, brain
Diamines	O_2	FAD	—	Liver, kidney, brain
Acyl CoA(C_6–C_{12})	ETF*	FAD	—	Mitochondria
Dihydrolipoate	NAD$^+$	FAD	Internal disulfide	Mitochondria, *E. coli*
NADPH	Glutathione	FAD	Internal disulfide	Liver, yeast, *E. coli*
NADPH	Thioredoxin	FAD	Internal disulfide	Liver, *E. coli*
NADH	Cytochrome b_5	FAD	—	Liver microsomes
NADPH	Cytochrome P_{450}	FAD + FMN	—	Liver microsomes
Ferredoxin	NADP$^+$	FAD	—	Chloroplasts
NADPH	Adrenodoxin	FAD	—	Adrenal cortex mitochondria
NADH	Ubiquinone	FMN	$(FeS)_n$†	Mitochondria
Succinate	Ubiquinone	FAD	$(FeS)_n$	Mitochondria
α-Glycerophosphate	Ubiquinone	FAD	$(FeS)_n$	Mitochondria
Purines	NAD$^+$	FAD	Mo, $(FeS)_4$	Liver, milk
Aldehydes	O_2	FAD	Mo, $(FeS)_4$	Liver
NADPH	NO_2^-	FAD	Siroheme ‡	*Neurospora crassa*
NADPH	SO_3^{2-}	FAD + FMN	$(FeS)_4$, siroheme	*E. coli*

*ETF = electron-transferring flavoprotein (Chap. 20).
† $(FeS)_n$ = iron-sulfur center.
‡ See Chap. 22.

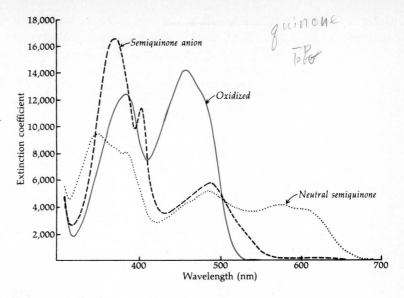

Figure 16.2 Absorption spectra of the two forms of flavin semiquinone, as seen with glucose oxidase. The red anionic form was obtained by irradiation with visible light anaerobically in the presence of ethylenediaminetetraacetate at pH 9.2. The blue form was produced by lowering the pH to 5.85. [*From V. Massey, R. G. Matthews, G. P. Foust, L. G. Howell, C. H. Williams, Jr., G. Zanetti, and S. Ronchi, p. 394, in H. Sund, (ed.), Pyridine Nucleotide Dehydrogenases, Springer-Verlag, New York, 1970.*]

partial diminution of those at 280 and 380 nm. By these and other spectral methods, two classes of flavoproteins are recognized: those that yield a red semiquinone on partial reduction and those that yield a blue semiquinone. As shown in Fig. 16.3, the former, which are usually oxidases, are reoxidized by O_2 with formation of H_2O_2, without the formation of the flavosemiquinone. The latter, which are usually dehydrogenases, react with O_2 in a 1-electron reaction

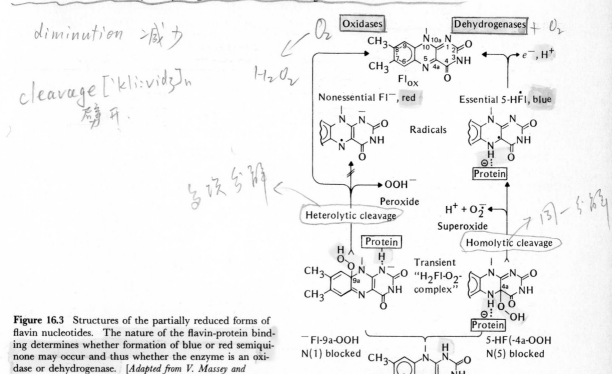

Figure 16.3 Structures of the partially reduced forms of flavin nucleotides. The nature of the flavin-protein binding determines whether formation of blue or red semiquinone may occur and thus whether the enzyme is an oxidase or dehydrogenase. [*Adapted from V. Massey and P. Hemmerick, p. 245, in P. D. Boyer, (ed.), The Enzymes, vol. XII, Academic, New York, 1976.*]

363

that forms the blue semiquinone and generates the superoxide anion O_2^-. Such spectral analyses permit considerable insight into the mechanisms of flavoprotein electron transfers and the flow of electrons from one carrier prosthetic group to another.

Oxidases are the simplest of the flavoenzymes and act by accepting a hydride ion, two electrons, and one proton from the substrate. The electrons are then transferred to O_2 to give H_2O_2. Other enzymes appear to abstract a pair of electrons in univalent steps, with the flavin shuttling between the oxidized and semiquinone forms, with the substrate radical remaining bound to the enzyme, as follows:

$$MH_2 + Enz\text{—}FAD \longrightarrow Enz\text{—}FADH \cdot \text{—} MH \cdot \xrightarrow{O_2} Enz\text{—}FAD + M + H_2O_2$$

Reactions of oxidases are the major source of peroxide, which is used as an oxidant by several peroxidases, considered later.

Other flavoproteins acting as dehydrogenases and oxidases, often in conjunction with other electron-carrier proteins or other functional groups, will be considered later in this chapter or in subsequent chapters of the text.

PROTEINS CONTAINING IRON–SULFUR CENTERS

These proteins contain iron and sulfur in a cluster that is capable of electron transfer; representative examples are listed in Table 16.4. Three types of clusters are known, as shown in Fig. 16.4. (1) Iron may be coordinated with the thiol groups of four cysteinyl residues in the simplest iron-sulfur containing proteins, such as the *rubredoxins* found in many microorganisms, but of unknown, specific functions. Rubredoxin of *Clostridium pasteurianum* contains 54 residues and one cluster with a single iron atom in tetrahedral coordination with four thiol groups. Other rubredoxins may contain two such clusters per polypeptide chain. (2) Iron-sulfur clusters containing two iron atoms coordinated to two sulfide sulfur atoms (Fe_2S_2), in addition to four thiol groups of cysteine in the

TABLE 16.4

Some Representative Iron-Sulfur Proteins	Center	MW	E_0', mV	Source
	Fe rubredoxin	6,000	−60	Clostridia (obligate anaerobes)
	(Fe_2S_2) ferredoxins	10,600	−420	Spinach chloroplast
		10,300	−400	*Microcystis* (blue-green alga)
		21,600	−350	*Azotobacter* (aerobic, N_2-fixing)
		12,600	−360	*E. coli* (enteric bacterium)
		12,500	−270	Adrenal cortex
	(Fe_4S_4) ferredoxins	6,000	−330	*Desulfovibrio* (anaerobic SO_4-reducing)
		8,800	−380	*Bacillus* (soil, N_2-fixing)
	($Fe_4S_4)_2$ ferredoxins	6,000	−395	Clostridia (obligate anaerobe)
		10,000	−490	*Chromatium* (red sulfur, photosynthetic)
		15,000	−420	*Azotobacter* (aerobic, N_2-fixing)
	(Fe_2S_2)-HiPiP	30,000	+280	Complex III, heart mitochondria
	(Fe_4S_4)-HiPiP	9,650	+350	*Chromatium* (red sulfur, photosynthetic)

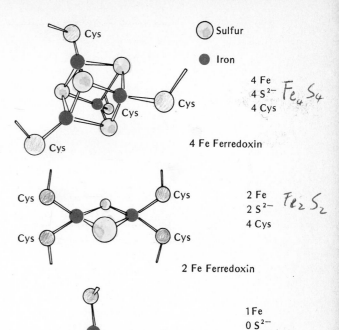

Figure 16.4 Structures of iron-sulfur centers. [*From D. O. Hall, K. K. Rao, and R. Cammack, The Iron-Sulfur Proteins: Structure, Function and Evolution of a Ubiquitous Group of Proteins, Sci. Prog. 62:286 (1975).*]

polypeptide chain, occur in proteins such as the *ferredoxins.* Acidification of proteins with this type of inorganic sulfur results in the release of H_2S in an amount stoichiometric with the iron content. Chloroplast ferredoxin (Chap. 19) and *adrenodoxin* of the adrenal cortex (Chap. M18) contain one such center each; the latter protein serves as an electron donor for cytochrome P_{450}, discussed below. Both iron atoms are Fe^{3+} in the oxidized state, and the center can accept one electron to become the Fe^{3+}-Fe^{2+} form. (3) Iron-sulfur clusters containing four iron atoms and four noncysteinyl sulfur atoms (Fe_4S_4), which are also coordinated to four thiol groups of cysteine (Fig. 16.4), are found in *ferredoxins* from some anaerobic bacteria (Table 16.4). More than one cluster of this type may occur in a single protein and can function independently of one another. The oxidized center, Fe_2^{3+}-Fe_2^{2+}, in such proteins accepts one electron on reduction to give the reduced center Fe^{3+}-Fe_3^{2+}. Similar Fe_4S_4 clusters are found in bacterial proteins designated *high-potential iron proteins* (HiPiP), which contain the Fe_3^{3+}-Fe^{2+} state in contrast to the clusters in the low-potential ferredoxins in the oxidized form. On reduction, these clusters accept one electron to give the reduced center Fe_2^{3+}-Fe_2^{2+}. The reduced cluster of a HiPiP is thus quite similar to the oxidized cluster of a ferredoxin. These iron clusters are buried deep in both the ferredoxins and HiPiP, but the mode of entry and exit of the electrons is unclear.

Iron-sulfur clusters of the Fe_2S_2 and Fe_4S_4 types are also found in enzymes that contain other functional prosthetic groups. Thus, they occur in flavoproteins with or without a heme group and in molybdenum-containing proteins with or without a flavin, as discussed below. Proteins containing a novel hexagonal Fe_3S_3 cluster are also known, but the role of the center is unknown.

Several flavo-iron-sulfur proteins are found in mitochondria and act in electron transfer to ubiquinone. Ubiquinone is the common entry point of electrons from NADH and succinate of the citric acid cycle (Chap. 15), from oxidation of fatty acyl CoAs (Chap. 20), and from other compounds such as dihydroorotate, α-glycerophosphate, choline, and sarcosine, as shown in Fig. 16.5. The NADH dehydrogenase complex contains 16 Fe and 16 S^{2-} per FMN and is thought to contain four distinct Fe_4S_4 centers (Fig. 16.5). The succinate dehydrogenase complex contains a protein with two covalently bound FAD and two Fe_2S_2 clusters and another protein with one Fe_4S_4 center, but no flavin. Examples of even more complex flavo-iron-sulfur proteins are discussed later.

It has been suggested that the sulfide for formation of an iron-sulfur center is made available by reaction with the ubiquitous iron-containing enzyme *rhodanese*, long known to catalyze a reaction between thiosulfate and cyanide.

$$SSO_3^{2-} + CN^- \longrightarrow SO_3^{2-} + SCN^-$$

Hepatic mitochondria contain rhodanese bound to protein rich in iron-sulfur centers. The tentative reaction scheme is as follows, where E_1 represents the functional site of rhodanese and E_2-Fe is the sulfide-free form of an iron-sulfur center.

$$SSO_3^{2-} \quad\quad E_1 \quad\quad E_2-FeS$$
$$SO_3^{2-} \quad\quad E_1-S_2^- \quad\quad E_2-Fe$$
$$CN^- \quad SCN^-$$

In the absence of the formed iron-sulfur center, cyanide attacks the rhodanese sulfide to form thiocyanate. The physiological role of this enzyme is not clear; its ability to detoxify cyanide is apparently only a secondary physiological benefit. Iron-sulfur centers are also found in a few enzymes that do not catalyze an oxidation-reduction reaction, e.g., mitochondrial aconitase (Chap. 15), gluta-

Figure 16.5 Iron-sulfur proteins in the electron-transport chain. Values shown are E_0' in millivolts; UQ = ubiquinone.

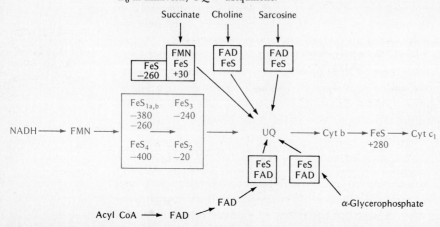

367

**Biological
Oxidations II:
Oxidative Enzymes,
Coenzymes, and
Carriers**

mate synthase (Chap. 22), and phosphoribosylpyrophosphate amidotransferase of *E. coli* (Chap. 26). The centers are required for activity but their role is unknown.

CYTOCHROMES

Cytochromes are found in all aerobic organisms, and the amounts in a mammalian tissue parallel its respiratory activity. Thus, heart muscle is rich in cytochromes, but liver, kidney, and brain have lesser amounts, and skin and lung the lowest amounts. The cytochromes are distinguished from one another by their absorption spectra and are designated as *a*, *b*, *c*, or a subscript thereof based on their unique absorption maxima, listed in Table 16.5. This designation does not imply that the structures and other properties of all cytochromes of one type are necessarily closely related, nor that they serve equivalent roles.

The cytochromes serve as electron-carrier proteins in mitochondria or the endoplasmic reticulum and perhaps other membranous structures also. Cytochrome P_{450} is an exception and acts as a monooxygenase, as described later. The electron-carrier cytochromes transfer electrons to and from the iron atoms of their heme prosthetic groups, which, except for cytochrome a and a_3, is the protoporphyrin IX-Fe complex also present in hemoglobin (Chap. M3). Thus, in mitochondria, electrons originating from NAD-dependent and flavoprotein enzymes are accepted one at a time by the heme iron atom of one cytochrome and transferred to the heme iron of others. The order of transfer in mitochondrial electron transport is b, c_1, c, a, to a_3 with the latter, a component of cytochrome oxidase (Chap. 15), reducing molecular oxygen.

Most cytochromes are firmly bound to membranes, and their solubilization requires extraction of membranes with aqueous solutions of detergents. Cytochrome c is the exception and is readily solubilized by dilute aqueous salt solutions. Many cytochromes are tightly associated with other cytochromes and electron carriers and are components of multisubunit complexes, which serve not only as electron carriers, but as proton pumps that form the proton gradients required for synthesis of ATP during oxidative phosphorylation. Thus, two major energy-transducing complexes, designated III and IV (Chap. 15), contain cytochromes b and c_1 and a and a_3, respectively. The two complexes

TABLE 16.5

Properties of
Mammalian
Cytochromes

| Cytochrome | Absorption maxima of reduced forms | | | E'_0, mV |
	α, nm	β, nm	γ, nm	
a_3	600	. . .	445	+200
a	605	517	414	+340
c	550	521	416	+260
c_1	554	523	418	+225
b	563	530	430	+30
b_t	565	535, 528	430	−30 (+245)
b_5	557	527	423	+0.03

are remarkably similar structurally, containing seven to eight subunits, some of which are encoded by mitochondrial genes, and others by nuclear genes. Both complexes traverse the inner mitochondrial membrane and are similar in shape.

Each of the major types of cytochromes will now be considered in turn, with emphasis on structure, mechanism of action, and biological function.

Cytochrome c

Cytochromes c are the most thoroughly studied electron-carrier proteins. They act as components of the electron-transfer chain of mitochondria, accepting electrons from the cytochrome bc_1 complex and donating electrons to the cytochrome oxidase (aa_3) complex.

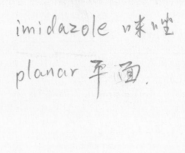

Complete amino acid sequences have been established for more than 70 animal cytochromes, as discussed in Chap. 31. The three-dimensional structure of horse heart cytochrome c has also been determined and analyzed in great detail. This small globular protein, whose structure is shown in Fig. 3.2, contains 104 residues in all mammals examined and contains a single heme (protoporphyrin IX-Fe) group attached covalently by two thioether linkages to cysteinyl residues in the polypeptide chain. The heme group is buried in the hydrophobic interior of the molecule, and its iron atom is fully coordinated; the four planar positions are filled by porphyrin nitrogens, one axial position by an imidazole N (His-18), and the other axial position by a sulfur atom (Met-80). Only one edge of the planar heme ring is accessible to the surface, and small channels on either side of the ring permit access of only small hydrophobic

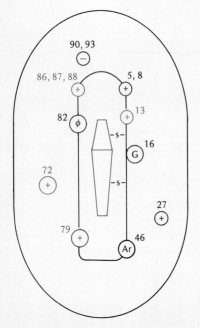

Figure 16.6 Schematic representation of the heme crevice of tuna cytochrome *c* (see Fig. 3.2 for complete structure). The numbers represent residue positions. + and − = positively and negatively charged residues; G = glutamine; φ = phenylalanine; Ar = aromatic. Only those residues are shown which are common to most cytochromes c. [*After R. E. Dickerson and R. Timkovich, p. 484, in P. D. Boyer (ed.), The Enzymes, 3d ed., vol. XI, Academic, New York, 1975.*]

molecules to the heme iron. Indeed, neither O_2 nor CO reacts with reduced cytochrome c at neutral pH, nor does oxidized cytochrome c react with CN^- or N_3^-. The mechanism of electron transfer by cytochrome c is unclear, since the structural basis for its interaction with either cytochrome oxidase or cytochrome c_1 is unknown, and the way in which electron transfer to and from the deeply buried iron atoms occurs is largely conjectural. Inspection of the three-dimensional structure of cytochrome c, however, suggests that features on the surface of the molecule near the heme may be involved in cytochrome-cytochrome interactions. As indicated diagrammatically in Fig. 16.6, the cationic side chains of several lysine and arginine residues are clustered at the surface on one face of the molecule and are thought to provide a binding site for anionic groups on cytochrome oxidase. On the opposite face of the molecule is a cluster of anionic residues (glutamic and aspartic acids) that may provide a binding site for a reductase or other components of the electron transport system. These structures are remarkably invariant among cytochrome c from several species, in accord with their suggested functional roles. Its exact interactions with other proteins, however, are probably more complicated since cytochrome c mixed with phospholipids forms a complex that serves as a substrate for cytochrome oxidase.

369

Biological
Oxidations II:
Oxidative Enzymes,
Coenzymes, and
Carriers

Cytochrome Oxidase

Cytochrome oxidase is one of the few enzymes capable of reducing molecular oxygen to water. It catalyzes the overall reaction

$$4 \text{ Cytochrome c } (Fe^{2+}) + O_2 + 4H^+ \longrightarrow 4 \text{ cytochrome c } (Fe^{3+}) + 2H_2O$$

Cytochrome oxidases from various sources contain seven subunits; the molecular weights and stoichiometry of the components of the beef heart enzyme are listed in Table 16.6. The functional oxidase in situ is thought to be dimeric

Subunits	Molecular weight	Relative amounts
Cytochrome aa$_3$ complex		
I—cytochrome a$_3$	35,000	1
II—cytochrome a, Cu	26,000	1
III—proton translocase	21,000	1
IV	17,000	1
V	12,000	1
VI	11,000	1
VII (heterogeneous)	5,000	>1
Cytochrome bc$_1$ complex		
I—core protein I	46,000	1
II—core protein II	43,000	1
III—cytochrome c$_1$	29,000	1
IV—cytochrome b	28,000	2
V—iron-sulfur center	24,000	1
VI	12,000	2
VII	8,000	2
VIII (heterogeneous)	≈6,000	≈2

TABLE 16.6

The Subunits of
the Beef Heart
Mitochondrial
Cytochrome aa$_3$
and bc$_1$ Complexes

(MW $\approx$ 400,000), with seven subunits per monomer. Heme A is one of the functional groups in the molecule and has the following structure, with a farnesyl ethyl side chain.

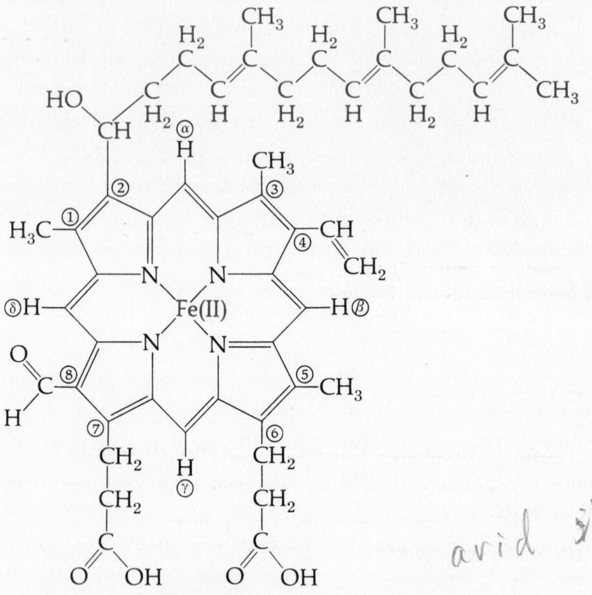

Heme A

The Fe^{2+} in heme A in the oxidase has an extremely high affinity for both CO and O_2. In the Fe^{3+} form it avidly binds CN^-, S^{2-}, and N_3^-; this explains the extremely high toxicity of these anions for all aerobic organisms. Subunits I and II each contain one molecule of noncovalently bound heme A, forming cytochrome a_3 and a, respectively. Two Cu atoms, the other functional groups of the oxidase, are also bound to subunit II. Subunit III is presumed to serve as a proton translocase. The complete amino acid sequence of subunit III from yeast cytochrome oxidase has been established and it is thought to form seven α-helical chains that traverse the membrane. Most of the sequence contains very hydrophobic residues, but there are 38 hydroxy amino acids evenly distributed across the thickness of the membrane. The hydroxyl groups may provide a network of hydrogen bonds that form channels through which protons move at rates similar to those observed for protons in ice.

A proposed mechanism for reduction of O_2 to water by cytochrome oxidase is given in Fig. 16.7. Cytochrome a of subunit II is the primary acceptor of electrons from cytochrome c and is in rapid redox equilibrium with one of the two Cu atoms, designated Cu_A. Subunit II is on the cytosolic face of the inner membrane, and its heme A group is in a hydrophobic crevice. Electrons from heme A-Cu_A are then transferred to cytochrome a_3 of subunit I, which is associated with the second Cu, Cu_B, in subunit II. Subunit I also contains a heme crevice, but on the matrix side of the membrane. Cytochrome a_3 and Cu_B of subunit I are directly involved in reduction of O_2 to water. The mechanism of the 4-electron reduction of O_2 by cytochrome oxidase is largely conjectural, inasmuch as no partially reduced intermediate forms of oxygen dissociate from

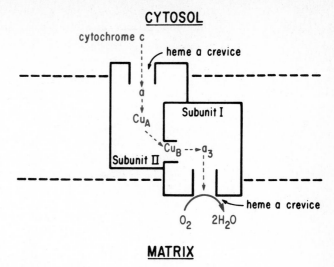

Figure 16.7 The suggested organization of subunits I and II of cytochrome oxidase in the mitochondrial inner membrane. There are two crevices in which heme A is bound, one where cytochrome c interacts on subunit II (cytochrome a), and another on subunit I (cytochrome a_3), which interacts directly with O_2. [*Adapted from D. B. Winter, W. J. Bruyninckx, F. G. Foulka, N. P. Grinich, and H. S. Mason, J. Biol. Chem. 255:11408–11414 (1981).*]

the enzyme. Since cytochrome oxidase is the site of reduction of most cellular O_2, this mechanism ensures that the toxic, partially reduced species of oxygen, such as O_2^- (discussed below), are not produced.

Cytochromes a and a_3 also function in conserving the released energy of electron transfer for formation of a proton gradient across the inner mitochondrial membrane. Such gradients are required for ATP synthesis during oxidative phosphorylation (Chap. 15). The mechanism for forming gradients is also conjectural, however; that shown in Fig. 16.8, or some variation of it, may be involved. Subunits I and II (a_3 and a cytochromes) are thought to be arranged in the membrane as shown. When an electron from cytochrome c is accepted by cytochrome a, the pK of an acidic group in cytochrome a changes and can take up a proton from the matrix of the mitochondrion. The arrangement of

Figure 16.8 A proposed mechanism for the H^+-translocating activity of cytochrome oxidase. The pair of a-a_3 cytochromes in the oxidase are arranged in the membrane (dotted lines) as shown, with a largely on the cytosolic side and a_3 on the matrix side. Cytochrome a contains an acidic group (black dot) whose pK differs in the oxidized and reduced states of the molecule. Thus, in 1, one electron from cytochrome c is transferred to cytochrome a, resulting in uptake of protons from the matrix by reduced cytochrome a. The subunits then rearrange in the membrane in a reciprocating fashion shown in 2 to give the orientation shown in 3. State 3 permits electron transfer from cytochrome a to a_3, with reversal of the change in pK of the acidic group of cytochrome a and loss of the proton into the space outside the membrane. Proton input and output is shown only for the aa_3 pair on the left, but they also occur alternatingly on the right-hand pair, to give, in effect, a reciprocating H^+ pumplike action of the two pairs. The arrangement of the Cu atoms and the steps in the reduction of O_2 by cytochrome a_3 are not shown.

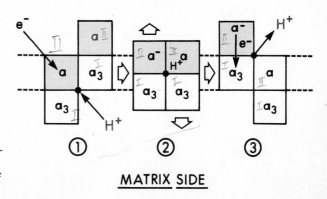

CYTOSOLIC SIDE

MATRIX SIDE

the subunits then switches as shown, and the reduced cytochrome a transfers its electron to cytochrome a_3, resulting in reversal of the change in pK of the acidic group and dissociation of the proton in cytochrome a into the cytosolic side of the membrane. This is an oversimplification of the process and does not indicate a role for other subunits, including that of subunit III, the proton translocase. Nevertheless, it is a useful model for the proton-translocating activity of cytochrome oxidase as well as other mitochondrial complexes.

Cytochrome oxidase is synthesized jointly on mitochondrial and cytoplasmic ribosomes. Thus, subunits I, II, and III are synthesized in the mitochondria and encoded by three different mitochondrial genes, whereas the other subunits are encoded by nuclear genes and, after synthesis on cytoplasmic ribosomes, are transported into the mitochondria for assembly of the oxidase (Chap. 29).

Cytochrome b Complexes

The b-type cytochromes are electron-carrier proteins with noncovalently bound heme (protoporphyrin IX-Fe) as a functional group. About 80 percent of cytochrome b is in complex III (Chap. 15), the mitochondrial *ubiquinol: cytochrome c reductase,* and about 20 percent is in mitochondrial *succinate: ubiquinone reductase* (complex II). The cytochrome b of complex III is intimately involved in the second site of phosphorylation of mitochondrial oxidative phosphorylation.

Complex III of beef heart contains eight subunits, whose molecular weights and amounts are listed in Table 16.6. The two largest subunits, the core proteins, span the mitochondrial inner membrane and contain no redox centers. Subunits III, IV, and V are electron-carrier proteins; III is cytochrome c_1, with covalently bound heme as in cytochrome c; IV is cytochrome b; and V contains one Fe_2S_2 center. A model for the subunit organization of complex III in situ is shown in Fig. 16.9.

The function of complex III is to accept electrons from reduced ubiquinone and transfer them to cytochrome c, with the simultaneous formation of a proton gradient across the mitochondrial membrane. A proposed mechanism for coupling electron transfer with gradient formation is shown in Fig. 16.10. In essence, it involves the cyclic transfer of electrons among components of the bc_1 complex and ubiquinone (Q) in reactions that consume H^+ from the mitochon-

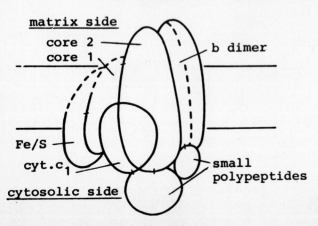

Figure 16.9 A model for the arrangement of the subunit polypeptide chains of complex III of mitochondrial membranes. [*From G. von Jagow and W. Sebald, Annu. Rev. Biochem.* **49**:*281–314 (1980).* © *1980 by Annual Reviews Inc.*]

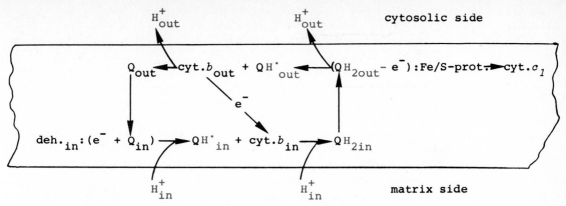

Figure 16.10 A proposed mechanism for a proton motive cycle involving ubiquinone and mitochondrial complex III. The arrows indicate the flow of electrons and the input or output of H^+ from the membrane. Q is ubiquinone; deh = dehydrogenase. [*From G. von Jagow and W. Sebald, Annu. Rev. Biochem.* **49**:281–314 (1980). © 1980 by Annual Reviews Inc.]

drial matrix and produce H^+ on the cytosolic side of the inner membrane. An electron from ubiquinone reductase (complex II) transfers one electron to Q, with uptake of a H^+ from the matrix to give $QH\cdot$, the semiquinone of ubiquinone. The $QH\cdot$ then reacts with and accepts an electron from cytochrome b and is converted to QH_2 as the result of uptake of another H^+ from the matrix. QH_2 subsequently releases one electron to the Fe_2S_2 center and thence to cytochrome c_1 with the loss of a H^+ into the cytosolic side. The resulting $QH\cdot$ is then transferred to another cytochrome b, which accepts an electron from $QH\cdot$ to produce another H^+ on the cytosolic side of the membrane. Oxidized ubiquinone (Q) is regenerated to restart the cycle. The key to the process is the transfer of electrons from one cytochrome b to another, as indicated (Fig. 16.10).

Complex II, the succinate:ubiquinone reductase, which transfers electrons from succinate to the mitochondrial ubiquinone pool, contains four polypeptide subunits and at least three redox centers. A 70,000-dalton subunit is the succinate dehydrogenase and contains one FAD and two Fe_2S_2 centers. A 27,000-dalton peptide contains a Fe_4S_4 center, whereas two peptides (MW = 13,000 and 15,000) are apocytochrome b. The exact mechanism of electron transfer among the functional groups in the complex is unclear.

The extremely hydrophobic apoprotein of cytochrome b in complex III is encoded by mitochondrial DNA, whereas all other polypeptides in the complex, including cytochrome c_1, are encoded by nuclear DNA. The single structural gene for yeast cytochrome b contains about 900 base pairs and is composed of three exons, or coding regions, interrupted by two introns (noncoding regions) containing from 2000 to 4000 base pairs (Chaps. 27 and 29).

Cytochrome b_5

Cytochrome b_5 (MW = 16,000) is tightly bound to the membranes of the endoplasmic reticulum but can be solubilized in aqueous detergent solutions. The protein contains a hydrophobic COOH-terminal domain of about 40 residues that anchors the molecule to the membrane, and a hydrophilic, NH_2-

terminal domain of about 80 residues that contains the noncovalently bound heme (protoporphyrin IX-Fe). The two domains of cytochrome b_5 can be separated after mild proteolysis. The three-dimensional structure of the hydrophilic domain reveals that the heme iron is coordinated axially to the N atoms of two histidine residues. It is presumed that this is the same coordination of the heme as in cytochrome b. Evidence suggests that the hydrophobic domain forms a loop into the membrane, thereby fixing the globular functional domain of cytochrome b_5 on the cytoplasmic surface of the endoplasmic reticulum.

Cytochrome b_5 is a component of the *NADH-dependent* Δ^9-*stearyl CoA desaturase* system (Chap. 20) and has been implicated in a number of NADPH-linked microsomal hydroxylation reactions. The natural electron donor for cytochrome b_5 is the *NADH-cytochrome b_5 reductase* (MW = 33,000), a flavoprotein that is tightly bound to endoplasmic reticulum membranes. The electron acceptor for cytochrome b_5 is the stearyl CoA desaturase (MW = 53,000), which is an oxidase tightly bound to the membrane; it contains one iron atom. The reductase, cytochrome b_5, and the desaturase can be reconstituted into the outer surface of liposomes and can catalyze the following conversion.

Cytochrome b_5 is also involved in hepatic microsomes in one type of elongation of fatty acids (Chap. 20), serving as the reductant in two steps of the pathway as follows:

375

**Biological
Oxidations II:
Oxidative Enzymes,
Coenzymes, and
Carriers**

Reduction of cytochrome b_5 in this process is achieved by *NADH-cytochrome b_5 reductase.*

Cytochrome b_5 is also found in erythrocytes and participates in reduction of methemoglobin to hemoglobin by NADH-cytochrome b_5 reductase (Chap. M3). Erythrocyte cytochrome b_5 is soluble and seems to be identical to the soluble, hydrophilic domain of membrane-bound cytochrome b_5. It appears to be formed by proteolysis of cytochrome b_5 during erythrocyte maturation.

Cytochrome P_{450}

This class of cytochromes includes a number of heme proteins that have a characteristic absorption maximum at 450 nm when combined with CO in the reduced state. They occur in most animal tissues, plants, and microorganisms and catalyze the monoxygenation of a vast variety of hydrophobic substances. Cytochrome P_{450} serves as the oxygenating catalyst in the presence of one or more electron-transfer proteins or redox enzymes in reactions of the following form, where RH is substrate.

$$RH + O_2 + NAD(P)H + H^+ \longrightarrow ROH + H_2O + NAD(P)^+$$

In some cases, such as the liver endoplasmic reticulum, a flavoprotein, *NADH-cytochrome P_{450} reductase,* directly catalyzes electron transfer from NADPH to P_{450}, whereas in other cases, such as in adrenal mitochondria, an iron-sulfur protein acts as an electron carrier between the FAD-containing reductase and P_{450}. The metabolite (RH) to be oxygenated is as varied as the ω-CH_3 group of an alkane or fatty acid, the ring carbons of steroids, or other polycyclic compounds such as benzpyrene and phenobarbital. The liver P_{450} system not only catalyzes hydroxylation of aliphatic and aromatic carbons, but also is involved in *N*-oxidation, sulfoxidation, epoxidation, *N*-, *S*-, and *O*-dealkylation, peroxidation, deamination, desulfuration, and dehalogenation reactions. Substrates may include normal metabolites; such as fatty acids and prostaglandins; a vast array of naturally occurring compounds present in foodstuffs; as well as *xenobiotics,* foreign synthetic compounds that are drugs, insecticides, anesthetics, or petroleum products, including many carcinogens. In addition, synthesis of many steroids also requires action of cytochrome P_{450} (Chaps. M17 and M18). The cytochrome P_{450} system usually leads to detoxification of xenobiotics, often rendering them water-soluble and more amenable to excretion, but in some cases compounds are formed with greater mutagenic, cytotoxic, or carcinogenic properties. The cytochrome P_{450} of liver is inducible, with the amount of enzyme increasing as much as 25-fold after administration of any one of a large number of xenobiotics, e.g., phenobarbital or the insecticide DDT.

Several different forms of cytochrome P_{450} may occur in a tissue, and often differ from those of other tissues. Each form appears to be encoded by a specific gene and, accordingly, they differ in their primary sequences, substrate specificities, immunological cross-reactivity, and inducibility by xenobiotics. The total number of forms that occur is unclear, but liver, for example, contains at least three forms inducible by phenobarbital that differ in size (MW = 48,000) from those induced by methylcholanthrene (MW = 53,000).

Irrespective of their structural and enzymic differences, the cytochromes P_{450} appear to have similar active sites. Each contains an iron-protophoryrin

IX (heme) group in a rather hydrophobic cleft, which is open to the substrate. The heme is bound to the apoprotein by a combination of noncovalent interactions as well as by a coordinate covalent bond to the iron atom. The iron has four bonds to the pyrrole N atoms, and a fifth to a thiolate group of cysteine in the polypeptide chain. The sixth coordination position is open in the ferrous form, but in the substrate-free, ferric state is occupied by water. Upon reduction, the resulting Fe^{2+} atom becomes the binding site for dioxygen.

The proposed mechanism for P_{450}-catalyzed hydroxylation reactions of the type given above (RH → ROH) is shown in Fig. 16.11 and is in accord with a variety of studies on the liver microsomal system as well as the soluble systems from certain bacteria. In this scheme, binding of a lipophilic substrate (step 1) to the oxidized Fe^{3+}-P_{450} on the endoplasmic reticulum is followed by reduction (step 2) to the reduced Fe^{2+}-P_{450}. Substrate binding is largely the result of hydrophobic interactions between substrate, the membrane-bound cytochrome, and phospholipids in the membrane. The 1-electron reduction (step 2) ultimately comes from NADPH with microsomal and mitochondrial P_{450}. In liver, the electron is provided by *NADPH-cytochrome P_{450} reductase,* a flavoprotein containing both FMN and FAD. In the mitochondrial system, electrons from NADPH are transferred to an iron-sulfur redoxin by a FAD-containing reductase, and the P_{450} is then reduced by the redoxin; e.g., adrenodoxin is the iron-sulfur protein acting in steroid hydroxylation by the adrenal gland (Chap. M18). After reduction, the heme of P_{450} binds O_2 (step 3), leading to formation of the transient Fe^{3+}-O_2^- complex. A second electron from the reductase is then accepted by the latter complex (step 4), giving rise to an Fe^{3+}-O_2^{2-} complex. Cytochrome b_5 may also supply the second electron in some systems. In the absence of substrate, this complex dissociates to give H_2O_2, accounting for the oxidase activity of the system. Even in the presence of substrate as much as half of the O_2 consumed by the system may be converted to peroxide. Splitting of the oxygen-oxygen bond occurs in step 5 with the release of one atom of oxygen in water; the other oxygen atom is "activated" oxygen, which accepts a H atom from RH (step 6), giving rise to ROH in step 7. Hydroxylation involves abstraction of a hydrogen from RH and binding of the carbon radical, R^-, to the

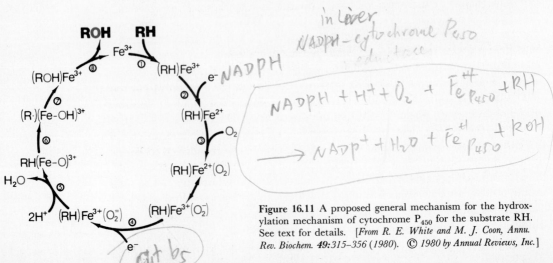

Figure 16.11 A proposed general mechanism for the hydroxylation mechanism of cytochrome P_{450} for the substrate RH. See text for details. [*From R. E. White and M. J. Coon, Annu. Rev. Biochem.* **49**:315–356 (1980). © 1980 by Annual Reviews, Inc.*]

$(Fe-OH)^{3+}$ intermediate.   Dissociation of ROH from the active site completes the cycle.

Oxidants such as peroxyacids or alkylhydroperoxides (R′OOH) can substitute for O_2 and the electron donor in cytochrome P_{450}-catalyzed hydroxylations by the overall reaction

$$RH + R'OOH \longrightarrow ROH + R'OH$$

The following series of reactions is one mechanism whereby this type of process occurs.

This mechanism suggests the transient formation of a "ferryl" ferric iron (Fe^{4+}), an unusual oxidation state of iron, whose electron distribution is unclear.

Bacterial Cytochromes and Oxidative Phosphorylation

Bacteria lack subcellular organelles resembling mitochondria; however, the bacterial cell membrane contains dehydrogenases, cytochromes, and iron-sulfur centers that conduct oxidative phosphorylation. On the basis of absorption spectra, cytochromes resembling a, b, and c are present in bacteria, but they differ markedly in their structural and functional properties from those of eukaryotes. A remarkable variety of electron-transport processes are found in bacteria with a wide range of substances ultimately being consumed or produced during oxidative phosphorylation. In photosynthetic bacteria and green plants, various cytochromes participate in electron transfer (Chap. 19).

OXIDASES, OXYGENASES, AND HYDROXYLASES

Most of the O_2 consumed by cells is in oxidative phosphorylation, but numerous enzymes also use O_2, although their contribution to the total O_2 utilization in most tissues is small. In liver, however, their contribution to the total O_2 consumption may be a third of the total. These enzymes act in a variety of metabolic pathways and will be encountered in subsequent chapters; therefore only a few examples of each class that illustrate principles of their actions are considered here.

Oxidases

As noted above, several flavoproteins are oxidases catalyzing oxidation of substrates with reduction of O_2 to H_2O_2. Other oxidases, widely distributed in plants, employ only metal cofactors. Thus, plant *ascorbate oxidase*, a Cu^{2+}-dependent enzyme, oxidizes ascorbic acid to dehydroascorbic acid (Chap. M22). Animal tissues contain *diamine* and *monoamine oxidases;* the former catalyzes oxidation of diamines to aminoaldehydes with production of NH_3 and H_2O_2,

$$H_2N-[CH_2]_5-NH_2 + O_2 \longrightarrow H_2N-[CH_2]_4-CHO + NH_3 + H_2O_2$$

and the latter converts monoamines to aldehydes.

Oxygenases

Monooxygenases catalyze reactions utilizing O_2 in which one atom of oxygen is transferred to product and the other is reduced to water. *Dioxygenases* also utilize O_2 but both oxygen atoms are transferred to products. Cytochrome P_{450}, discussed above, is one type of monooxygenase. Other mono- and dioxygenases also utilize a variety of functional prosthetic groups or cofactors. Monooxygenases require the input of 2 electrons from a reductant whereas dioxygenases have no such requirements.

Dopamine β-monooxygenase, a Cu^{2+}-enzyme of brain and adrenal medulla, catalyzes synthesis of norepinephrine from dopamine (3,4-dihydroxphenylethylamine, Chap. M18) as follows:

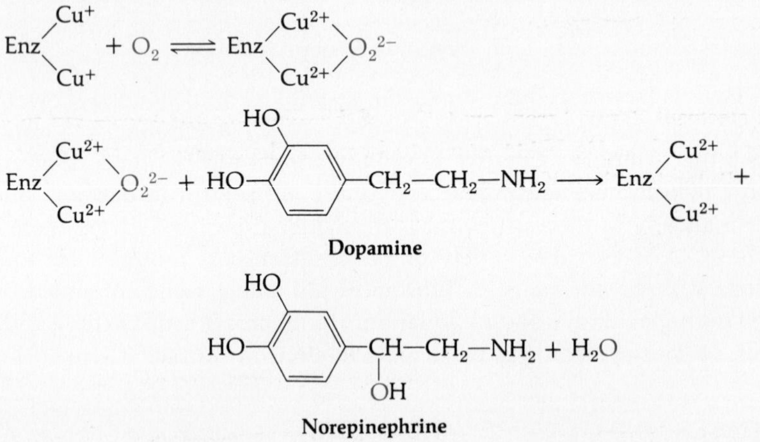

$$Enz \overset{Cu^+}{\underset{Cu^+}{\big<}} + O_2 \rightleftharpoons Enz \overset{Cu^{2+}}{\underset{Cu^{2+}}{\big<}} O_2^{2-}$$

$$Enz \overset{Cu^{2+}}{\underset{Cu^{2+}}{\big<}} O_2^{2-} + HO{-}\underset{HO}{\bigcirc}{-}CH_2-CH_2-NH_2 \longrightarrow Enz \overset{Cu^{2+}}{\underset{Cu^{2+}}{\big<}} +$$

Dopamine

$$HO{-}\underset{HO}{\bigcirc}{-}\underset{OH}{CH}-CH_2-NH_2 + H_2O$$

Norepinephrine

Oxygen is activated by the enzyme-Cu^+ complex, with one atom of oxygen transferred to dopamine to give product and the other reduced to water. Ascorbic acid is required for reduction of the inactive enzyme-Cu^{2+} complex to the active enzyme-Cu^+ complex.

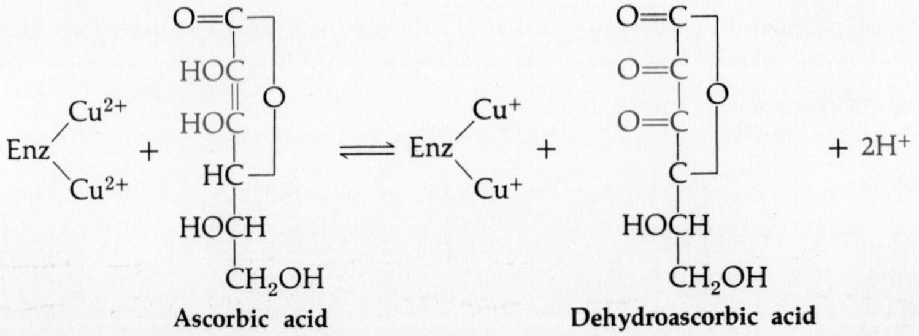

$$Enz \overset{Cu^{2+}}{\underset{Cu^{2+}}{\big<}} + \begin{matrix} O{=}C{-} \\ HOC \\ HOC \\ HC \\ HOCH \\ CH_2OH \end{matrix}\Big] O \rightleftharpoons Enz \overset{Cu^+}{\underset{Cu^+}{\big<}} + \begin{matrix} O{=}C{-} \\ O{=}C \\ O{=}C \\ C \\ HOCH \\ CH_2OH \end{matrix}\Big] O + 2H^+$$

Ascorbic acid   **Dehydroascorbic acid**

379

**Biological
Oxidations II:
Oxidative Enzymes,
Coenzymes, and
Carriers**

Other monooxygenases use different types of functional groups. Thus, bacteria contain many flavin enzymes that effect monooxygenations with either NADH or NADPH as external reductants. A small but important group of animal monooxygenases utilizes a pterin functional group, *tetrahydrobiopterin*, as described in Chap. 23 for the conversion of phenylalanine to tyrosine.

Another type of monooxygenase utilizes α-ketoglutarate as substrate and catalyzes reactions of the general form

$$M + O_2 + \alpha\text{-ketoglutarate} \longrightarrow M\text{—OH} + succinate + CO_2$$

This type of monooxygenase catalyzes synthesis of 5-hydroxylysine, 3- and 4-hydroxyproline (Chap. M6), and carnitine (Chap. 24). Enzymes of this class require iron as a cofactor and use ascorbate to reduce inactive Fe^{3+}-enzyme to active Fe^{2+}-enzyme in a reaction analogous to that for reduction of dopamine β-monooxygenase, described above. Dioxygenases also utilize several functional groups. Thus, *tryptophan 2,3-dioxygenase* (Chap. 25) utilizes a heme prosthetic group to effect cleavage of the indole ring of tryptophan.

Molybdenum-Containing Hydroxylases

Three animal enzymes in this class, *sulfite oxidase, xanthine oxidase,* and *aldehyde oxidase,* are unique because they employ a molybdenum-containing cofactor along with other functional groups. They are not strictly oxidases, as their names imply, but hydroxylases that catalyze reactions of the following form.

$$A + H_2O \longrightarrow A\text{—OH} + H^+ + 2e^-$$

The electrons derived from oxidation of the substrate are transferred to various acceptors such as O_2, NAD, and cytochrome c, depending on the specificity and the cellular location of the enzyme.

Sulfite is produced by the catabolism of sulfur-containing amino acids (Chap. 25) and is converted by mitochondrial *sulfite oxidase* to sulfate, the major form of urinary sulfur, as follows.

$$SO_3{}^{2-} + H_2O + 2 \text{ cytochrome c-}Fe^{3+} \longrightarrow SO_4{}^{2-} + 2H^+ + 2 \text{ cytochrome c-}Fe^{2+}$$

Rat liver sulfite oxidase (MW = 120,000) is composed of two identical subunits, each of which contains a noncovalently bound heme (protoporphyrin IX-iron), and a molybdenum-containing pterin, designated *molybdopterin*, which has the tentatively assigned structure

Molybdopterin

The proposed catalytic mechanism of sulfite oxidase is shown in Fig. 16.12. Initially sulfite is converted to sulfate by a 2-electron transfer at the molybdopterin site, with the extra oxygen atom being derived from water. During this process Mo^{6+} accepts 2 electrons giving Mo^{4+}, which then reduces the Fe^{3+} in

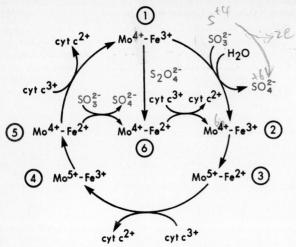

Figure 16.12 The proposed catalytic mechanism of sulfite oxidase showing the electron-transfer reactions among functional groups. [From K. V. Rajagopalan, Sulfite Oxidase, in M. P. Coughlan (ed.), Molybdenum and Molybdenum-Containing Enzymes, Pergamon, Elmsford, N.Y., 1980, pp. 241–272.]

heme to Fe^{2+} to give an intermediate with ferroheme and Mo^{5+}. The latter complex subsequently donates 2 electrons to 2 molecules of cytochrome c in different steps, as shown. The cytochrome c is thought to interact directly with the heme of the oxidase. Sulfite oxidase has two domains, one formed by the NH_2-terminal 100 residues that binds heme, and the other formed by the remainder of the molecule that binds molybdopterin. The absorption spectrum of the reduced heme domain, which is isolated from a limited tryptic hydrolysate of the oxidase, is much like that of reduced cytochrome b_5. Moreover, portions of its sequence are homologous to that of the beef liver cytochrome b_5.

The importance of sulfite oxidation is evident, since human beings with a deficiency of sulfite oxidase exhibit mental retardation, and other severe neurological problems. Multiple gene lesions can cause the enzyme deficiency, since mutations can occur in structural genes for the subunits as well as in genes controlling the assembly of molybdopterin.

Xanthine and aldehyde oxidases, from liver and several other tissues, also contain molybdopterin but, in addition, are flavoproteins with iron-sulfur centers. The native enzymes are similar in size (MW = 300,000, depending on the source) and are composed of two identical subunits. Each subunit contains one FAD and one molybdopterin molecule bound noncovalently and two Fe_2S_2 centers. Moreover, both catalyze very similar reactions of the following general form:

$$AH + H_2O + X \longrightarrow AOH + XH_2$$

AH is the reducing substrate which donates a pair of electrons to the enzyme. The electrons are subsequently accepted by a second substrate X, which may be either NAD^+ or O_2. Thus the enzyme exhibits a dehydrogenase as well as oxidase-like activities. AH may be the same substance, e.g., purines or pyrimidines, although the rates of utilization may differ for the two enzymes. The enzymes can be distinguished from one another since xanthine is a substrate for xanthine oxidase, but not for aldehyde oxidase, and 6-methylpurine is a substrate for aldehyde oxidase, but not xanthine oxidase. Thus, xanthine oxidase

380

oxidizes xanthine to uric acid (Chap. 26) and aldehyde oxidase oxidizes 6-methylpurine to 6-methylhypoxanthine.

PEROXIDE AND SUPEROXIDE

The reduction of O_2 by direct insertion of a pair of electrons is a rather slow process, since O_2 contains 2 unpaired electrons with parallel spin states in separate orbitals and reduction requires inversion of one electronic spin. Were it not for this slow reactivity, the coexistence of O_2 with organic matter would be highly unlikely. The spin restriction to reduction of O_2 can be obviated, however, by the successive additions of single electrons. Thus, reduction of one molecule of O_2 to H_2O requires the addition of 4 electrons.

$$O_2 + 4H^+ + 4e^- \longrightarrow 2H_2O$$

and results in the production, as intermediates, of highly reactive superoxide ions (O_2^-), hydrogen peroxide (H_2O_2), and hydroxyl radicals (OH^-). The continual presence of these substances would pose a serious threat to living systems because of their damaging effects on cellular components. Indeed, $OH\cdot$, the most active mutagen produced by ionizing radiation, is extraordinarily reactive and can attack all organic compounds.

Univalent reduction of O_2 initiates the following chain of reactions.

$$O_2 + e^- \longrightarrow O_2^- \tag{1}$$

$$O_2^- + H^+ \rightleftharpoons HO_2\cdot \tag{2}$$

$$O_2^- + HO_2\cdot + H^+ \longrightarrow H_2O_2 + O_2 \tag{3}$$

$$O_2^- + Fe^{3+} \longrightarrow O_2 + Fe^{2+} \tag{4}$$

$$H_2O_2 + Fe^{2+} \longrightarrow Fe^{3+} + OH^- + OH\cdot \tag{5}$$

The superoxide anion formed in reaction (1) can be protonated to the hydroperoxyl radical [reaction (2)] since $HO_2\cdot$ is an acid of pK_a close to 4.8. Reaction (3) is a spontaneous dismutation, yielding $H_2O_2 + O_2$. Taken together, these reactions ensure that any system producing O_2^- will also soon contain H_2O_2. Reactions (4) and (5) indicate that iron compounds can catalyze a reaction between O_2^- and H_2O_2, thereby forming $OH\cdot$.

As discussed above, numerous enzymes produce H_2O_2 and O_2^- as do the spontaneous oxidations of the iron atom in such substances as hemoglobin, cytochrome b_5, reduced ferredoxin, and other electron carriers or redox systems. The threat posed by the high reactivity of O_2^- and H_2O_2 is averted by enzymes that convert these substances to less reactive species.

Superoxide Dismutases

These scavengers of O_2^- are enzymes that catalyze the following dismutation reaction.

$$O_2^- + O_2^- + 2H^+ \longrightarrow H_2O_2 + O_2$$

Dismutases are found in all aerobic organisms and are absent from obligate anaerobic microorganisms. V_{max} is very high [$k_2 = 2 \times 10^9 \, (mol^{-1} \cdot s^{-1})$] and is limited only by the rate of diffusion of O_2^-. Three types of dismutases are recognized. (1) The cytosolic dismutase (MW = 31,200) of eukaryotic cells contains one atom each of Cu and Zn in its two identical subunits, whose three-dimensional structure is shown in Fig. 5.11. (2) The mitochondrial dismutase (MW = 75,000) in eukaryotes as well as the cytosolic form in bacteria (MW = 40,000) contains two Mn per molecule. (3) A Fe form (MW = 40,000), whose sequence is highly homologous to the Mn form, is found in bacteria, blue green algae, and certain plants. The sequences of the Cu-Zn forms are not homologous to the Mn and Fe forms (Chap. 31).

The catalytic activities of the dismutases are dependent on their Cu^{2+}, Mn^{2+}, or Fe^{2+} contents. Thus, on catalysis the metals undergo cyclic oxidation-reduction, with $n = 2$ for the Cu-Zn enzyme and 3 for the Mn and Fe enzymes.

$$Enz-Me^{n+} + O_2^- \longrightarrow Enz-Me^{(n-1)+} + O_2$$
$$Enz-Me^{(n-1)+} + O_2^- + 2H^+ \longrightarrow Enz-Me^{n+} + H_2O_2$$

Catalase and Peroxidases

Catalase is present in nearly all animal cells, plants, and bacteria and acts to prevent accumulation of noxious H_2O_2, which is converted to O_2 and H_2O.

$$H_2O_2 + H_2O_2 \rightleftharpoons 2H_2O + O_2$$

Peroxidases, which are less widely distributed, catalyze the following type of reaction.

Beef liver catalase (MW = 248,000) is a heme protein, containing one non-covalently bound protophorphyrin IX-iron (heme) group in each of its four identical subunits. Enzymes from other species are very similar. Its catalytic turnover rate is one of the highest known, with 44,000 molecules of H_2O_2 decomposed per second and is diffusion limited. Peroxidases are rather rare in animal tissues but plentiful in higher plants. Leukocytes contain a *verdoperoxidase* that utilizes a variety of organic substrates as well as a *myeloperoxidase* that oxidizes halides to halogens.

$$2I^- + 2H^+ + H_2O_2 \longrightarrow I_2 + 2H_2O$$

Halogens so formed are effective bacteriocidal agents. A thyroid peroxidase is involved in formation of thyroid hormone (Chap. M14). *Glutathione peroxidase* of erythrocytes is a selenium-containing enzyme that oxidizes glutathione (GSH) (Chaps. 25 and M3).

$$2GSH + H_2O_2 \longrightarrow GSSG + 2H_2O$$

383

**Biological
Oxidations II:
Oxidative Enzymes,
Coenzymes, and
Carriers**

All hemoproteins have a low peroxidase activity, catalyzing the oxidation of such compounds as benzidine to colored products, a reaction useful in forensic medicine for detecting traces of blood.

Perhaps the most spectacular utilization of peroxidases occurs in the bombardier beetle, which accumulates a concentrated solution of dihydroxyphenol in about 25% H_2O_2 in one sac and a suspension of crystalline peroxidase in another. When the beetle is attacked, the contents of these sacs are ejected through a turretlike mixing tube. The reaction is so intense that the fluid emerges at 100°C.

If the catalatic and peroxidatic reactions are written as

the analogy between the two reactions becomes apparent. In this sense, the catalatic splitting of H_2O_2 to H_2O and O_2 is a special case of a peroxidatic reaction in which hydrogen peroxide serves both as substrate and as acceptor. This analogy becomes more real on noting that at high concentrations of low-molecular-weight alcohols or formaldehyde and low peroxide concentration, catalase also exhibits peroxidatic activity. Both enzymes can utilize as substrates organic hydroperoxides with small aliphatic substituents, e.g., ethyl hydrogen peroxide and peracetic acid. Since, physiologically, there may exist high concentrations of other acceptors and low concentration of peroxide, it is conceivable that catalase serves almost exclusively as a peroxidase in animal tissues.

REFERENCES

Books

Boyd, G. S., and R. M. S. Smellie (eds.): *Biological Hydroxylation Mechanisms,* Academic, New York, 1972.

Boyer, P. D. (ed.): *The Enzymes,* 3d ed., vol. XI–XIII: *Oxidation-Reduction,* Academic, New York, 1975, 1976.

Coughlan, M. (ed.): *Molybdenum and Molybdenum-Containing Enzymes,* Pergamon, Elmsford, N.Y., 1980.

Hayaishi, O. (ed.): *Molecular Mechanisms of Oxygen Activation,* Academic, New York, 1974.

King, T. E., H. S. Mason, and M. Morrison (eds.): *Oxidases and Related Redox Systems,* University Park Press, Baltimore, 1973.

Lemberg, R., and J. Barret: *Cytochromes,* Academic, New York, 1973.

Singer, T. P. (ed.): *Flavins and Flavoproteins,* Elsevier, Amsterdam, 1976.

Smith, K. M.: *Porphyrins and Metalloporphyrins,* Elsevier, Amsterdam, 1975.

Sund, H. (ed.): *Pyridine Nucleotide Dependent Dehydrogenases,* Springer-Verlag, New York, 1970.

Thurman, R. G., T. Yonetani, R. Williamson, and B. Chance (eds.): *Alcohol and Aldehyde Metabolizing Systems,* Academic, New York, 1974.

Ambler, R. P.: Cytochromes c, *Syst. Zool.* **22**:554–565 (1974).

Bartsch, R. G.: Bacterial Cytochromes, *Annu. Rev. Microbiol.* **22**:181–210 (1968).

Beinert, H.: Iron-Sulfur Centers of the Mitochondrial Electron Transfer System, pp. 61–100, in W. Lovenberg (ed.): *The Iron-Sulfur Proteins,* vol. III, Academic, New York, 1977.

De Pierre, J. W., and G. Dallner: Structural Aspects of the Membrane of the Endoplasmic Reticulum, *Biochim. Biophys. Acta* **415**:411–472 (1975).

Diesseroth, A., and A. L. Dounce: Catalase, *Physiol. Rev.* **50**:319–375 (1970).

Fridovich, I.: Superoxide Dismutase, *Annu. Rev. Biochem.* **44**:147–159 (1975).

Hall, D. O., K. K. Rao, and R. Cammack: The Iron-Sulfur Proteins: Structure, Function and Evaluation of a Ubiquitous Group of Proteins, *Sci. Prog.* **62**:285–316 (1975).

Ross, E. M., M. E. Dockter, and G. Schatz: Synthesis and Degradation of Mitochondrial Cytochromes, in J. Cook (ed.): *Biogenesis and Turnover of Membrane Molecules,* Raven, New York, 1977.

Sweeney, W. V., and J. C. Rabinowitz: Proteins Containing 4Fe-4S Clusters: An Overview, *Annu. Rev. Biochem.* **49**:139–162 (1980).

Von Jagow, G., and W. Sebald: b-Type Cytochromes, *Annu. Rev. Biochem.* **49**:281–314 (1980).

White, R. E., and M. J. Coon: Oxygen Activation by Cytochrome P_{450}, *Annu. Rev. Biochem.* **49**:315–356 (1980).

Wikström, M., and K. Krab: Respiration-Linked H^+ Translocation in Mitochondria: Stoichiometry and Mechanism, *Current Topics in Bioenergetics* **10**:51–101 (1980).

Wikström, M., K. Krab, and M. Saraste: Proton-Translocating Cytochrome Complexes, *Annu. Rev. Biochem.* **50**:623–655 (1981).

Chapter 17

Carbohydrate Metabolism I

Cellular uptake and production of glucose.
Glycolysis. Anaplerosis.
The phosphogluconate pathway.
Some specialized pathways in plants
and microorganisms.

dietary 极量食物

Carbohydrates are the first cellular constituents formed by photosynthetic organisms and result from the fixation of carbon dioxide on absorption of light, the ultimate energy source of all biological processes (Chap. 19). The further metabolism of carbohydrates in these organisms yields a vast array of other organic compounds, many of which are subsequently used as dietary constituents by animals. Animals ingest large amounts of carbohydrates that can be stored, oxidized to obtain energy as ATP, converted to lipids for more efficient energy storage, or used for the synthesis of numerous cellular constituents.

Only a small fraction of the carbohydrate in plants is available for human nutrition, since cellulose (Chap. 6) and certain other plant polysaccharides cannot be degraded into monosaccharides because appropriate hydrolytic enzymes are not present in the intestinal tract. Most utilizable carbohydrate is ingested as plant starches; as animal glycogen; or as the disaccharides sucrose, maltose, or lactose (Chap. 6). Intestinal carbohydrate digestion and absorption are considered in Chap. M10; this chapter will describe how the absorbed sugars, primarily glucose, are transported into cells and then utilized for energy production and for the formation of numerous organic substances whose metabolism is the subject of subsequent chapters.

CARBOHYDRATE METABOLISM: AN OVERVIEW AND SOME GENERAL PRINCIPLES OF METABOLISM

Before considering the details, it is useful to present first a general overview of the metabolic transformations of glucose, including its synthesis and its

Any cross-references coded M refer to the companion text, *Principles of Biochemistry: Mammalian Biochemistry.*

didactic 教导的

degradation. Such an overview serves to illustrate several principles of metabolism in addition to those presented in earlier chapters. Moreover, it also emphasizes that the metabolism of an organism is the totality of innumerable reactions among diverse organic compounds. Although it is useful for didactic purposes to discuss the fate of an individual metabolite, such as glucose, the interrelationships among all cellular constituents must be considered in order to understand the metabolism of an organism.

Figure 17.1 shows a simplified guide to the metabolism of glucose and some of the interrelationships between glucose and other metabolites. Several general features of this metabolic "map" warrant special consideration before presenting the details of glucose metabolism.

1. **The consecutive reactions that lead from one key intermediate to some terminal compound are termed a metabolic pathway.** Thus, conversion of glucose-6-phosphate to lactate via reactions 8, 9, 11, 12, 13, and 14 is one pathway of glucose metabolism. Other pathways for glucose are evident.

consecutive 连续的

Figure 17.1 A guide to some major metabolic transformations. Not shown are the adenine nucleotides and pyridine nucleotides integral to these processes or the numerous intermediates lying along each sequence indicated by a broken arrow. A simple reversible arrow is intended to mean that the process, whether one reaction step or several, is catalyzed by the same enzymes in both directions. A unidirectional arrow represents a reaction or reaction sequence that, in practice, proceeds significantly only in the direction shown. A pair of oppositely oriented, curved arrows denotes that different enzymes are operative in the two directions and that usually some or all of the unspecified intermediates differ in the two processes.

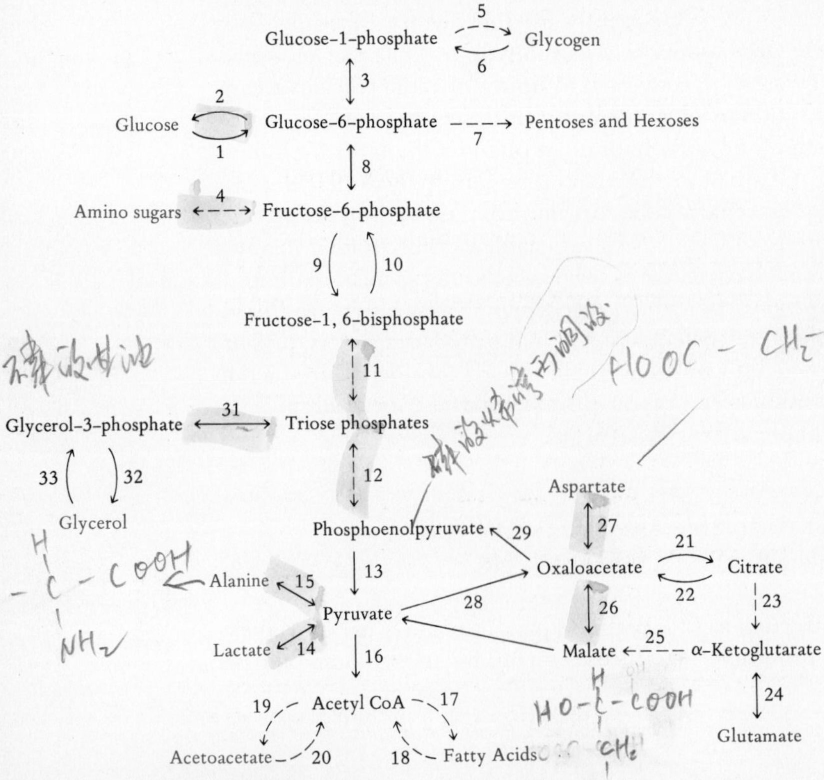

2. **All pathways proceed with a loss of free energy.** For example, the formation of 2 mol of lactate from glucose-6-phosphate occurs with the formation of 4 mol of ATP and an overall $\Delta G° = -22$ kcal/mol.

3. **Many pathways are reversible, but the actual paths in each direction may involve different enzymes and different intermediates.** The functional reverse pathway from lactate to glucose-6-phosphate involves reactions 14, 28, 29, 12, 11, 10, and 8 and entails the input of 6 mol of ATP per mole of glucose-6-phosphate formed. This situation stems in large part from the metabolic needs of a cell under different circumstances. Thus, lactate formation from glucose-6-phosphate is most likely when other processes are removing the ATP formed, whereas glucose-6-phosphate formation from lactate is unlikely unless there is an abudance of ATP produced by other pathways.

4. **The major form of utilizable energy in all cells is ATP.** In most cells, ATP is generated primarily by oxidative phosphorylation (Chap. 15). The reductants of NAD^+ for this process are the intermediates of the citric acid cycle, indicated by reactions 16, 21, 23, 25, and 26, and of fatty acid oxidation, reaction 18 (Chap. 20). The metabolism of other substances, however, results in production of much smaller amounts of ATP, and although it is quantitatively less important than that from oxidative phosphorylation, the ATP produced by conversion of glucose to lactate is especially important in some tissues, e.g., muscle (Chap. M8).

5. **Cells generally cannot store most substances in a pathway.** Compounds such as glucose, glucose-6-phosphate, and acetyl CoA are not stored in cells. The primary storage form of glucose is glycogen, mainly in liver and skeletal muscle, made from glucose-1-phosphate in one pathway (reaction 5) and returned by another (reaction 6). The major storage form of energy, however, is in the fatty acids of neutral triacylglycerols, synthesized from acetyl CoA (reaction 17).

6. **A price is paid for storage of energy as glycogen or triacylglycerols.** In animals, the reaction sequence 8, 9, 11, 12, 13, 16, and 17 permits net conversion of two-thirds of the carbon of glucose-6-phosphate to fatty acids. But the energy consumption in ATP by the processes of reaction 17 exceeds by about 20 percent the energy production as ATP of the processes in reaction 18. Thus, there is no mechanism in animals that permits the net conversion of fatty acids to glucose without energy consumption.

7. **Many of the interlocked metabolic conversions occur in different subcellular compartments.** Conversion of glucose to lactate and of acetyl CoA to fatty acids occurs in the cytosol, whereas the citric acid cycle and fatty acid oxidation to acetyl CoA occur in mitochondria. This compartmentalization of metabolism often necessitates the transport of key intermediates from one subcellular compartment to another and can provide a means of regulation.

8. **Many intermediates of a metabolic pathway are not free to leave the cell.** This is illustrated by glucose-6-phosphate, which is formed from glucose (reaction 1); it is essentially "locked" into the cell and cannot leave unless reconverted to glucose (reaction 2) or metabolized to a nonphosphorylated compound, such as lactate, pyruvate, or alanine.

commensurate
问书分沿

futile cycle
亿元元多场环

9. **Although the metabolic requirements differ from one organ to another, the metabolism of an organ is highly dependent on that of other organs.** This is exemplified by the massive ATP requirement of muscular contraction (Chap. 18), which is met by the conversion of glycogen to lactate. At rest, glycogen synthesis in muscle can be initiated from blood glucose, which is maintained at a fairly constant level by the liver, the kidney, and the intestine, the only organs that can convert glucose-6-phosphate to glucose. Lactate produced on muscle contraction, however, enters the blood and is reutilized in liver to form glucose-6-phosphate. In addition, glucose is the major energy source of the brain since most other potential energy sources in the blood cannot cross the blood-brain barrier.

10. **The stores of carbohydrate are rapidly depleted on brief fasting but may be restored in part by synthesis of glucose-6-phosphate from noncarbohydrate precursors, a process termed gluconeogenesis.** Amino acids by reactions 15, 24, 27, *et seq.*, and glycerol from neutral triacylglycerol (reactions 33, 31, *et seq.*) may be utilized by the liver to replenish its glucose-6-phosphate in times of dietary carbohydrate deprivation.

11. **The rate of each metabolic process is regulated commensurate with the needs of the cell.** Each of the pathways in Fig. 17.1 is proceeding to some degree at all times. The requirements of the cell, however, dictate the net flux of a metabolite through alternate pathways. Thus, the cell regulates its metabolism to prevent such an inherently wasteful situation as the simultaneous conversion of glucose-6-phosphate to lactate at rates equivalent to the conversion of lactate or amino acids to glucose-6-phosphate. Such a process would produce a *futile cycle,* and nothing would be accomplished but the waste of the differences between ATP production in one direction and the ATP utilization in the opposite direction. There are many possibilities for futile cycles in Fig. 17.1, the simplest example being represented by reactions 1 and 2,

$$\text{Glucose} + \text{ATP} \longrightarrow \text{glucose-6-phosphate} + \text{ADP} \qquad (1)$$

$$\text{Glucose-6-phosphate} \longrightarrow \text{glucose} + \text{P}_i \qquad\qquad (2)$$

$$\text{SUM:} \quad \text{ATP} \longrightarrow \text{ADP} + \text{P}_i$$

Both of the above reactions occur in the liver, but reaction (1) operates maximally only when glucose is plentiful and reaction (2) primarily when blood glucose must be replenished.

12. **Cells employ a variety of mechanisms to regulate their metabolism.** The many kinds of mechanisms employed to regulate metabolic processes include: (1) lack of necessary reactant, (2) inhibition of an enzyme by a product of the pathway, (3) allosteric control of an enzyme by some intermediate in the pathway (Chap. 10), (4) the presence of tissue-specific isoenzymes with differing properties, (5) hormonal signals from another organ, and (6) control of the rate of synthesis and degradation of an enzyme. These and other types of controls abound and will appear in the discussions that follow; they make possible the subtle and often very rapid modulation of the processes in question.

The first step in glucose metabolism is its conversion to glucose-6-phosphate after entry into the cell. Glucose-6-phosphate is at a major crossroad in the pathway of glucose metabolism, since it is involved in several subsequent pathways: (1) synthesis and degradation of glycogen, (2) production of lactate and ATP (glycolysis), (3) gluconeogenesis, and (4) synthesis of pentoses and other hexoses, each of which is considered in this chapter and in Chap. 18. Hydrolysis of glucose-6-phosphate to give glucose and P_i is also a primary metabolic fate of glucose-6-phosphate in liver, kidney, and intestine and is the major means of maintaining blood [glucose] within normal limits. Thus, it is essential to consider the mechanisms of glucose transport into cells, the enzymes leading to formation and hydrolysis of glucose-6-phosphate, and the regulatory mechanisms involved in these processes.

Glucose Entry into Cells

The intracellular free [glucose] of most animal cells is very low, whereas the blood [glucose] is close to 5 mM, or about 90 mg per dl of blood. Entry therefore occurs along a downhill gradient. It does not, however, occur by passive simple diffusion but as a facilitated process that requires specific transport systems (Chap. 14). Two major types of transport systems are recognized, one that is responsive to insulin and another that is not.

Insulin-independent transport systems are found in hepatocytes, erythrocytes and the brain. In hepatocytes, the rate of D-glucose uptake is inhibited by phlorizin, a glycoside of a toxic plant polyphenol, and cytochalasin B (Chap. M10). Galactose and fructose are taken up by the same system at rates slightly less than that of glucose; the intracellular [hexose] equilibrates with the external concentration within minutes. The rate of glucose uptake far exceeds its rate of metabolism, in contrast to fructose, which is metabolized rapidly as it enters the cell. The K_T (Chap. 13) for glucose is about 30 mM, and all hexoses compete with one another for uptake. The carrier-mediated uptake of hexoses by hepatocytes is very similar to that of erythrocytes, which exhibit $K_T = 8$ mM for glucose uptake, a value smaller than those for galactose and fructose. Glucose uptake is also inhibited by phlorizin but is not responsive to insulin. The erythrocyte glucose transport carrier appears to be a protein (MW $\approx$ 200,000) composed of four subunits of equal size.

Muscle and adipocytes contain *insulin-dependent glucose transport systems*. Studies with adipocytes indicate that insulin increases the V_{max} but not the K_T for glucose transport. Moreover, the binding of insulin to its plasma membrane receptors (Chap. M16) appears to affect the glucose transport system in two ways: (1) it recruits glucose transport proteins from intracellular microsomal membranes to the plasma membrane, thereby increasing the number of transport proteins on the cell surface, and (2) after recruitment, the transport proteins have about twice the transport activity of those from cells not treated with insulin.

Bacterial cells have several types of glucose uptake systems; some do not appear to require metabolic energy, whereas others do. One of the latter is found in *Escherichia coli, Salmonella typhimurium,* and *Staphylococcus aureus* and employs phosphoenolpyruvate as the energy source and three different protein components. In the first step of this process, which is catalyzed by enzyme I, phosphoenolpyruvate and a protein in the cytosol (MW = 9,400), called *HPr,* react to form phospho-HPr, with the phosphate bound to a histidine residue.

$$\text{Phosphoenolpyruvate} + \text{HPr} \xrightarrow[\text{Mg}^{2+}]{\text{enz I}} \text{pyruvate} + \text{P}\sim\text{HPr}$$

P~HPr reacts with the sugar to be transported, e.g., glucose, to give the 6-phosphosugar, in a reaction catalyzed by a membrane component, enzyme II.

$$\text{P}\sim\text{HPr} + \text{glucose} \xrightarrow{\text{enz II}} \text{HPr} + \text{glucose-6-phosphate}$$

In this way extracellular glucose becomes internal glucose-6-phosphate. Mutants deficient in HPr lack the ability to transport several sugars, whereas mutants with a defective subunit of enzyme II are unable to transport only a specific sugar.

Glucose Phosphorylation: Hexokinase and Glucokinase

All cells that metabolize glucose contain kinases that catalyze phosphorylation of glucose by ATP to produce glucose-6-phosphate.

Glucose α-D-Glucose-6-phosphate

Two kinds of kinases are involved, *hexokinase* and *glucokinase*; each catalyzes the same reaction but has different molecular and kinetic properties. A Mg^{2+}-ATP complex is employed as substrate, and the reaction is effectively irreversible, $\Delta G° = -5$ kcal/mol, because of the relatively low-energy character of glucose-6-phosphate and the lower stability of $\text{Mg}^{2+} \cdot \text{ADP}$ compared to $\text{Mg}^{2+} \cdot \text{ATP}$.

Hexokinase is found in all tissues and exists in three isoenzyme forms, types I, II, and III. Each form is composed of a single subunit (MW = 100,000). Brain contains primarily type I, and skeletal muscle type II, but all tissues except liver contain different amounts of each form. The kinetic properties of the three forms are very similar, exhibiting low K_m values for glucose in the range of 8 to 30 μM. Each form may utilize other monosaccharides as substrates to form the corresponding 6-phosphates, but they are much poorer substrates than glucose. Glucose-6-phosphate is a potent inhibitor of each form, and acts by binding at a site separate from the active site. Type II hexokinase is found primarily in the cytosol, whereas type I exhibits "ambiquitous" behavior; i.e., it may exist in the cytosol or bound to mitochondria, the exact amounts in each form depending on the metabolic status of the cell. Glucose-6-phosphate, the product of hexokinase action, is a potent inhibitor of both forms of hexokinase, but its action on each form differs. The half-time for inhibition of type II by glucose-6-phosphate is about 10 times smaller than that for type I. Thus, the inhibitory effects on the type II, soluble form occur after only a short time when a small amount of glucose-6-phosphate has been phosphorylated. In contrast, the membrane-bound mitochondrial type I form succumbs to inhibition well after potentially inhibiting concentrations of glucose-6-phosphate are produced. Moreover, the inhibitory effects of glucose-6-phosphate for the type I form, but not the type II form, are attenuated by P_i. The [glucose-6-phosphate] also determines the amount of type I hexokinase that is bound to mitochondria. High [glucose-6-phosphate] favors binding to mitochondria, presumably because the binding of a sugar phosphate to a site separate from the active site results in a conformation that favors binding to mitochondria. This complex type of regulation seems to be geared to the metabolic needs of the specific tissue. Thus, muscle, rich in type II hexokinase, requires adjustment of its metabolism to sudden changes in the intracellular [glucose] and [glucose-6-phosphate], and the rate of the hexokinase-catalyzed reaction responds rapidly to the ratio of the concentrations of its substrate and product. In contrast, brain, which contains almost exclusively the type I form, has an almost constant supply of glucose, and its phosphorylation of glucose is regulated primarily by alteration in the intracellular $[P_i]$.

Yeast hexokinase differs somewhat from the mammalian forms. It is a dimer of identical subunits (MW = 50,000), and its activity is unaffected by glucose-6-phosphate. Its complete three-dimensional structure has been established and shown to change significantly on binding glucose, thereby providing strong evidence that its mechanism of rate enhancement involves an induced fit (Chap. 11) by substrates.

Glucokinase, often designated hexokinase IV, is a monomeric protein (MW = 48,000) found almost exclusively in liver. Glucose and mannose are the only natural monosaccharide substrates and have K_m values of about 12 and 33 mM, respectively. In addition to exhibiting much higher K_m values for glucose, glucokinase differs from the hexokinases in that glucose-6-phosphate is not an inhibitor. These properties reflect the differences between the transport of glucose into liver compared to other tissues. Although normal metabolism of hepatic glucose is insulin-dependent, liver does not require insulin for glucose uptake. The liver cell is freely permeable to the inward and outward flow of glucose. Moreover, in contrast to such tissues as muscle and adipose tissue, in which glucose uptake is insulin-dependent, glucose is taken up by the liver only in the hyperglycemic condition. Thus, glucokinase serves to trap all available excess blood glucose in the liver cell regardless of the glucose-6-phosphate concentration, thereby permitting storage of glucose as glycogen or, after further metabolism, as fatty acids. Its enzymic properties are well adapted to this role.

Glucokinase levels in liver vary with age, with the diet, and with the hormonal status of the animal. Fetal liver contains only hexokinase, which declines at birth as glucokinase appears; adult levels of glucokinase are reached about 2 weeks after birth. This may be correlated with the fact that during fetal life glucose is delivered via the placenta. Glucokinase levels are also highly dependent on dietary carbohydrate. A 72-h fast results in a decrease in the glucokinase levels to about 25 percent that of well-fed animals. Refeeding glucose results in near normal levels within 4 h. Such dietary induction of glucokinase is a consequence of new enzyme synthesis, since recovery is blocked by several inhibitors of protein synthesis. Glucose *per se*, however, is not the only factor required for glucokinase synthesis, since diabetic animals, which are literally saturated with glucose, also have low glucokinase activities. This effect of insulin may be related to its general effects on protein synthesis (Chap. M16). Other hormones also regulate glucokinase levels; epinephrine and glucagon inhibit restoration of glucokinase in starved animals after glucose refeeding, whereas adrenal corticosteroids aid in induction of hepatic glucokinase synthesis after fasting.

Glucose Dephosphorylation: Glucose-6-Phosphatase

Glucose-6-phosphate, like other phosphorylated compounds, cannot leave the cell; however, it contributes to the maintenance of blood [glucose] when hydrolyzed to glucose and P_i. The responsible enzyme, *glucose-6-phosphatase*, which catalyzes the reaction

$$\text{Glucose-6-phosphate} + H_2O \xrightarrow{Mg^{2+}} \text{glucose} + P_i$$

is found only in liver, kidney, and intestine, the three tissues that aid in replenishing blood [glucose].

Glucose-6-phosphatase is a membrane-bound enzyme of the rough endoplasmic reticulum, and its solubilization requires extraction of hepatocytes or the renal cortex with aqueous detergent solutions. It acts, however, as a member of a multicomponent system, which, in isolated microsomal particles, is composed of four different integral membrane components, as shown in Fig. 17.2. (1) A *glucose-6-phosphate translocase,* called T_1, mediates transport of glucose-6-phosphate from the cytosol to the lumen of the endoplasmic reticulum. (2) The *phosphatase,* which is on the lumenal, noncytoplasmic side of the reticulum, hydrolyzes glucose-6-phosphate, although when examined in solubilized form, it has a rather broad substrate specificity, effecting hydrolysis or transphosphorylation of a number of phosphate compounds. (3) An *inorganic phosphate translocase,* designated T_2, mediates P_1 efflux from the reticulum to the cytosol. And (4) a *glucose translocase,* designated T_3, permits D-glucose to pass the microsomal membrane freely.

Several aspects of the glucose-6-phosphatase system are noteworthy. The translocase T_1 is highly specific for glucose-6-phosphate. Moreover, the activity of the translocase is rate limiting in the hydrolysis of glucose-6-phosphate. Penetration of glucose-6-phosphate is inhibited by phlorizin, and the rate of hydrolysis is proportional to the extent of phlorizin inhibition. Moreover, the rates of hydrolysis in the intact system are much less than those obtained with the soluble phosphatase in disrupted systems. The phosphatase is a *phosphohydrolase-transphosphorylase* that catalyzes such diverse reactions as

$$PP_i + \text{glucose} \longrightarrow \text{glucose-6-phosphate} + P_i$$
$$PP_i + H_2O \longrightarrow 2P_i$$

These reactions are probably of minimal importance in the intact system, and glucose-6-phosphate hydrolysis is the favored reaction. Although T_2, the phos-

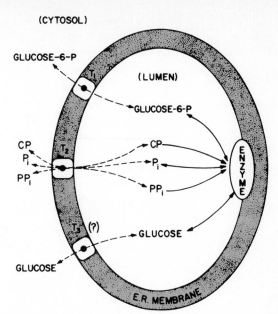

Figure 17.2 Diagram of the glucose-6-phosphatase system in a microsomal particle from the hepatic endoplasmic reticulum (E.R.). T_1 = glucose-6-phosphate translocase; enzyme = glucose-6-phosphatase; T_2 = inorganic phosphate translocase; and T_3 = glucose translocase. CP is carbamoyl phosphate. [*From W. J. Arion, A. J. Lange, H. E. Walls, and L. M. Ballas, J. Biol. Chem.* **255:** *10396–10406 (1980).*]

phate translocase, can permit reversible transfer of PP_i and carbamoyl phosphate, its major role in the system is to permit efflux of P_i. Finally, the properties of T_3, the *glucose translocase*, differ from the glucose transport system of hepatocyte plasma membranes, described above; e.g., it is not inhibited by phlorizin and many monosaccharides other than glucose are transported. Therefore, T_3 seems to act by simple diffusion.

In the liver, the potentially futile cycle provided by the glucokinase and the glucose-6-phosphatase reactions appears to be prevented by the metabolic status of the organ. When glucose-6-phosphate is required for other metabolic processes, its concentration is sufficiently low to prevent operation of the glucose-6-phosphatase system. In turn, the latter can operate only when glucose-6-phosphate is plentiful, as the result of decreased needs for further glucose metabolism. Therefore, there appears to be no need to control the activity of these enzymes by mechanisms other than regulation of their substrate concentrations, although the physical separation of the two enzymes, one in cytosol and the other in the endoplasmic reticulum with specific translocases involved in the hydrolytic path, also prevents recycling and waste of ATP.

GLYCOLYSIS

The ATP requirement for each tissue is fairly constant, and although it differs from one tissue to another, it is usually satisfied by mitochondrial oxidation of NADH, which is produced in conjunction with the metabolism of acetyl CoA in the citric acid cycle (Chap. 15). Cardiac and skeletal muscles (Chaps. 18 and M8) are the exceptions, since their requirements for ATP may increase enormously during exercise, a condition in which even maximal rates of mitochondrial oxidation are inadequate to the task. Thus, on exercise, skeletal muscle and to a lesser extent cardiac muscle obtain the necessary large incre-

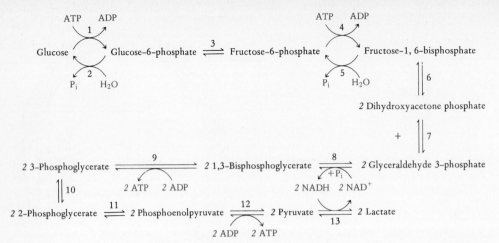

Figure 17.3 The reactions of glycolysis. Numbers adjacent to the arrows refer to the enzymes listed in Table 17.1.

ment in ATP from *glycolysis,* the sequence of reactions leading from glucose to lactate (Fig. 17.1). This pathway occurs in the cytosol and is unique in that it can proceed in the complete absence of O_2, which is at very low concentrations in exercising muscle and is essential for ATP production by mitochondrial oxidation. The individual reactions of glycolysis are shown in Fig. 17.3; however, the overall process is summarized by the net reaction,

$$\text{Glucose} + 2\text{ADP} + 2\text{P}_i \longrightarrow 2 \text{ lactate} + 2\text{ATP}$$

As discussed in Chap. 18, muscle can use its glycogen stores as the source of glucose-phosphate esters, early intermediates of glycolysis.

Glycolysis (Fig. 17.3) is identical with anaerobic, lactic acid fermentation by microorganisms and is closely related to other types of anaerobic, microbial fermentation that yield such products as ethanol or glycerol, as discussed later. Indeed, the mechanisms of bacterial fermentations and the various substrates, enzymes, and cofactors involved were elucidated in part well before the details of animal glycolysis were discovered.

The enzymes that participate in glycolysis and some of their properties are listed in Table 17.1. Glycolysis from glucose commences with glucose phosphorylation by hexokinase, as discussed above; each of the subsequent reactions leading from glucose-6-phosphate to lactate will now be considered in turn. In the following discussion, particular attention will be given to the reactions of anaerobic glycolysis in muscle, but since most of the same reactions are involved in several other tissues, important differences in glucose metabolism by the glycolytic pathway in various tissues, most notably the liver, will also be discussed.

Glucose-6-phosphate → Fructose-6-phosphate

This reaction is catalyzed by *phosphoglucose isomerase,* and at equilibrium the ratio of aldose to ketose is 7:3.

394

$$\text{Glucose-6-phosphate} \rightleftharpoons \text{Fructose-6-phosphate}$$

Glucose-6-phosphate   Fructose-6-phosphate

The human enzyme (MW $= 134{,}000$) is a dimer of identical subunits, and like many other enzymes in glycolysis, it requires Mg^{2+}. Fructose-6-phosphate has metabolic fates (Fig. 17.1) other than glycolysis, as discussed later (Chap. 18).

Fructose-6-phosphate + ATP → Fructose-1,6-bisphosphate + ADP

This reaction is catalyzed by *phosphofructokinase*, a highly regulated enzyme that controls the flux of glucose through the glycolytic path in all tissues.

$$\text{Fructose-6-phosphate} + \text{ATP} \xrightarrow{Mg^{2+}} \text{Fructose-1,6-bisphosphate} + \text{ADP}$$

Fructose-6-phosphate   Fructose-1,6-bisphosphate

tetramer 4聚

Since the reaction proceeds with $\Delta G^\circ = -4.5$ kcal/mol, it is essentially irreversible and is considered to be the *committed step* in glycolysis. The catalytically active muscle enzyme (MW $= 320{,}000$) is a tetramer of identical subunits but dissociates into inactive dimers in the presence of citrate; fructose-6-phosphate promotes reformation of tetramers. The muscle and liver enzymes are controlled by different genes and have distinct kinetic and structural properties. This may reflect, at least in part, the different roles of the two tissues in glucose metabolism. Anaerobic glycolysis in muscle is utilized for ATP production on exercise when $[O_2]$ is low and there is a large ATP requirement for muscle contraction; liver controls the rate of glucose flux through the glycolytic pathway as a function of its need to replenish blood [glucose] or to provide intermediates from glucose for the synthesis of other cellular constituents, e.g., fatty acids.

Phosphofructokinase aids in regulating the rate of glycolysis because of its properties as an allosteric enzyme (Chap. 10). In muscle, its effectors operate so that kinase activity is high when ATP is required and low when ATP is plentiful. As shown in Fig. 17.4, in the absence of other effectors, the enzyme exhibits virtually hyperbolic kinetics at low [ATP] and sigmoidal kinetics at high [ATP]. $Mg^{2+} \cdot$ ATP is the substrate, whereas uncomplexed ATP binds at an allosteric, inhibitory site, thereby increasing the K_m for fructose-6-phosphate. Rather high [citrate] concentrations (5 mM) also inactivate the enzyme by causing dimer formation, but citrate at lower concentrations accentuates the inhibitory effects of ATP, suggesting a possible role for this citric acid cycle intermediate in vivo. AMP, cAMP, ADP, and P_i reverse the inhibitory effects of ATP, but

*hyperbolic
双曲线.
sigmoidal
"S"形*

*accentuate
重读*

TABLE 17.1 Enzymes of Glycolysis

No.*	Enzyme	Coenzymes and cofactors	Activators	Inhibitors	K'_{eq} at pH 7.0	$\Delta G^{\circ\prime}$, kcal/mol
1	Hexokinase	Mg^{2+}	ATP^{4-}, P_i	Glucose-6-phosphate	$\dfrac{[\text{Glucose-6-phosphate}][Mg^{2+}\text{-}ADP^{3-}]}{[\text{Glucose}][Mg^{2+}\text{-}ATP^{4-}]} = 650$	-4.0
1A	Glucokinase	Mg^{2+}	—	—		
2	Glucose-6-phosphatase	Mg^{2+}	—	—	$\dfrac{[\text{Glucose}][P_i]}{[\text{Glucose-6-phosphate}]} = 210$	-3.3
3	Phosphoglucose isomerase	Mg^{2+}	—	—	$\dfrac{[\text{Fructose-6-phosphate}]}{[\text{Glucose-6-phosphate}]} = 0.5$	$+0.4$
4	Phosphofructo-kinase	Mg^{2+}	Fructose-2,6-bis-phosphate, P_i, AMP, ADP, cAMP, K^+, NH_4^+	ATP^{4-}, citrate	$\dfrac{[\text{Fructose-1,6-bisphosphate}]}{[\text{Fructose-6-phosphate}][Mg^{2+}\text{-}ATP^{4-}]} = 220$	-3.4
5	Fructose-1,6-bisphosphatase	—	—	Fructose-2,6-bis-phosphate	$\dfrac{[\text{Fructose-6-phosphate}][P_i]}{[\text{Fructose-1,6-bisphosphate}]} = 650$	-4.0
6	Aldolase	—	—	—	$\dfrac{[\text{Glyceraldehyde-3-phosphate}] \times [\text{dihydroxyacetone phosphate}]}{[\text{Fructose-1,6-bisphosphate}]} = 10^{-4}$	$+5.7$

	Enzyme	Cofactors/Activators	Inhibitors	Mass action ratio		ΔG
7	Phosphotriose isomerase	Mg^{2+}		$\dfrac{[\text{Glyceraldehyde-3-phosphate}]}{[\text{Dihydroxyacetone phosphate}]} = 0.075$	—	+1.8
8	Glyceraldehyde-3-phosphate dehydrogenase	NAD		$\dfrac{[\text{1,3-Bisphosphoglycerate}][\text{NADH}]}{[\text{Glyceraldehyde-3-phosphate}][\text{NAD}^+][\text{P}_i]} = 0.08$	—	+1.5
9	Phosphoglycerate kinase	Mg^{2+}		$\dfrac{[\text{3-Phosphoglycerate}][\text{Mg}^{2+}\text{-ATP}^{4-}]}{[\text{1,3-Bisphosphoglycerate}][\text{Mg}^{2+}\text{-ADP}^{3-}]} = 1500$	—	−4.5
10	Phosphoglycero-mutase	Mg^{2+}, 2,3-bisphospho-glycerate		$\dfrac{[\text{2-Phosphoglycerate}]}{[\text{3-Phosphoglycerate}]} = 0.02$	—	+1.0
11	Enolase	Mg^{2+}, Mn^{2+}		$\dfrac{[\text{Phosphoenolpyruvate}]}{[\text{2-Phosphoglycerate}]} = 0.5$	—	+0.4
12	Pyruvate kinase	K^+, fructose-1,6-bisphos-phate	ATP, alanine, acetyl CoA	$\dfrac{[\text{Pyruvate}][\text{Mg}^{2+}\text{-ATP}^{4-}]}{[\text{Phosphoenolpyruvate}][\text{Mg}^{2+}\text{-ADP}^{3-}]} = 2 \times 10^5$	—	−7.5
13	Lactate dehydrogenase	NAD		$\dfrac{[\text{Lactate}][\text{NAD}^+]}{[\text{Pyruvate}][\text{NADH}]} = 1.6 \times 10^4$	—	−6.0

*The numbers in this column correspond to the reactions of Fig. 17.3.

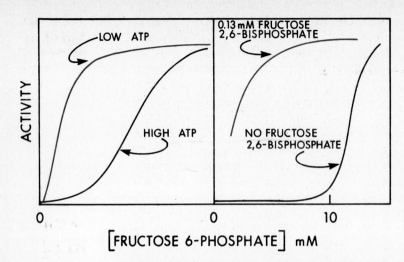

Figure 17.4 *Left*. Influence of the concentration of ATP on the kinetics of the reaction catalyzed by phosphofructokinase. *Right*. Effect of fructose-2,6-bisphosphate on the activity of rat liver phosphofructokinase.

at low [ATP] they are without effect on the activity of the kinase. Moreover, the muscle enzyme requires K^+ or NH_4^+, the latter lowering the K_m for both substrates. The sum of these effects can either enhance or depress the activity of the kinase as dictated by the concentration of various effectors or activators in the cell.

The most important means for regulation of liver phosphofructokinase, however, involves a specific activator, *fructose-2,6-bisphosphate*.

Fructose-2,6-bisphosphate

Figure 17.4 shows that the rat liver kinase is virtually inactive at low [fructose-6-phosphate], but that at these low concentrations it is highly active in the presence of fructose-2,6-bisphosphate. The 2,6-bisphosphate is also a potent inhibitor of fructose-1,6-bisphosphatase, which, as discussed below, effectively reverses the phosphofructokinase reaction. The origin of the 2,6-bisphosphate and its dual role as an activator of the kinase and an inhibitor of the phosphatase are also described below.

$$\text{Fructose-1,6-bisphosphate} + H_2O \xrightarrow{\ Mg^{2+}\ } \text{Fructose-6-phosphate} + P_i$$

As noted above, the phosphofructokinase reaction is essentially irreversible under physiological conditions, but there are circumstances when the reversal of glycolysis, viz., conversion of lactate to glucose, is required by the liver. The ability to reverse the glycolytic path depends upon *fructose-1,6-bisphosphatase*, which catalyzes hydrolysis of the 1-phosphate group of fructose-1,6-bisphosphate to give fructose-6-phosphate and P_i. The phosphatase is also an allosteric enzyme; it exhibits hyperbolic kinetics as a function of [fructose-1,6-bisphos-

398

phate] but sigmoidal kinetics with an increased K_m for substrate in the presence of mixtures of AMP and either ATP or ADP. Moreover, fructose-2,6-bisphosphate, the activator of phosphofructokinase, as discussed above, is a potent competitive inhibitor of the liver phosphatase ($K_I = 0.5\ \mu M$). In addition, the 2,6-bisphosphate enhances the inhibition of AMP. Thus, hepatic fructose-2,6-bisphosphate increases the flux of substrates through phosphofructokinase and inhibits the flux through the phosphatase, thereby preventing waste of ATP by the futile cycle of the simultaneous actions of the kinase and the phosphatase.

The dual action of fructose-2,6-bisphosphate in activating phosphofructokinase and inhibiting fructose-1,6-bisphosphatase is well adjusted to the metabolic needs of the liver. **Fructose-2,6-bisphosphate is synthesized only when glucose is in good supply in well-fed animals.** Under these conditions, it activates the kinase and inhibits the phosphatase, permitting glycolysis of glucose to pyruvate, which gives rise to the acetyl CoA that is used for synthesis of fatty acids and other substances. In fasting animals, however, whose glycogen stores are depleted, glucose must be replenished by hepatic gluconeogenesis from lactate, pyruvate, or amino acids. Under these conditions, fructose-2,6-bisphosphate levels are diminished, resulting in the simultaneous reduction in phosphofructokinase activity and the release of the inhibition of the fructose-1,6-bisphosphatase activity, conditions that favor gluconeogenesis. The exquisite regulatory actions of fructose-2,6-bisphosphate as an activator in one pathway and as an inhibitor of another, are reflected by the fact that the [fructose-2,6-bisphosphate] required for half-maximal activation of the kinase is about equal to that for half-maximal inhibition of the phosphatase ($K_I = 0.5\ \mu M$).

The molecular basis for control of the levels of fructose-2,6-bisphosphate resides in the action of glucagon (Chap. M16), and its control of the rate of formation of the 2,6-bisphosphate, as shown in Fig. 17.5. Glucagon, released from the pancreas in response to diminished blood [glucose], is a hormonal signal to the liver to replenish blood glucose. The liver, in effect, responds by

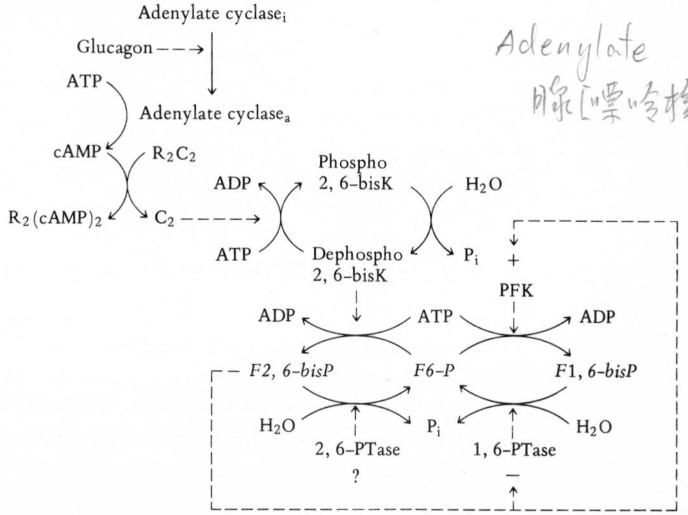

Figure 17.5 A proposed mechanism for the dual action of fructose-2,6-bisphosphate (F2,6-bisP) in regulating the catalytic activities of liver phosphofructokinase (PFK) and fructose-1,6-bisphosphatase (1,6-PTase). The cAMP-dependent protein kinase (C₂), which is activated as the result of glucagon activation of adenylate cyclase (Chap. 18), phosphorylates fructose-2,6-bisphosphate kinase (2,6-bisK), rendering it inactive and preventing formation of fructose-2,6-bisphosphate. The lack of the latter releases its inhibition of the 1,6-PTase and stops activation of PFK, thereby permitting fructose-1,6-bisphosphate to form fructose-6-phosphate, a critical step in reversal of glycolysis. The presence of a fructose-2,6-phosphatase (2,6-PTase) is speculative. + and − refer to activation and inhibition, respectively, of the reactions shown.

stopping glycolysis, enhancing gluconeogenesis, and replenishing blood [glucose]. This is accomplished by inhibition of the synthesis of fructose-2,6-bisphosphate by *fructose-2,6-bisphosphate kinase,* which catalyzes the following reaction with $Mg^{2+} \cdot ATP$ as substrate.

$$\text{Fructose-6-phosphate} + \text{ATP} \xrightarrow{Mg^{2+}} \text{fructose-2,6-bisphosphate} + \text{ADP}$$

Inhibition of this specific kinase is thought to involve the multicyclic cascade shown in Fig. 17.5. Thus, glucagon activates hepatic adenylate cyclase, which catalyzes formation of cAMP (Chaps. 13, 18, and M16); cAMP, in turn, activates cAMP-dependent protein kinase, which then phosphorylates fructose-2,6-bisphosphate kinase, thereby diminishing its activity. Vasopressin (Chap. M20) and phenylephrine (Chap. M18) stimulate the synthesis of fructose-2,6-bisphosphate in the liver of well-fed, but not fasted, animals, indicating that lack of glucose, or other precursors of fructose-6-phosphate, diminishes the synthesis of the 2,6-bisphosphate. Although fructose-2,6-bisphosphate is a potent activator of muscle phosphofructokinase in vitro, its role in muscle glucose metabolism is unclear.

Fructose-1,6-bisphosphate ⇌ Dihydroxyacetone Phosphate + Glyceraldehyde-3-phosphate

The next step in glycolysis is catalyzed by *aldolase,* which cleaves fructose-1,6-bisphosphate between C-3 and C-4 to give the phosphate esters of dihydroxyacetone and glyceraldehyde as follows.

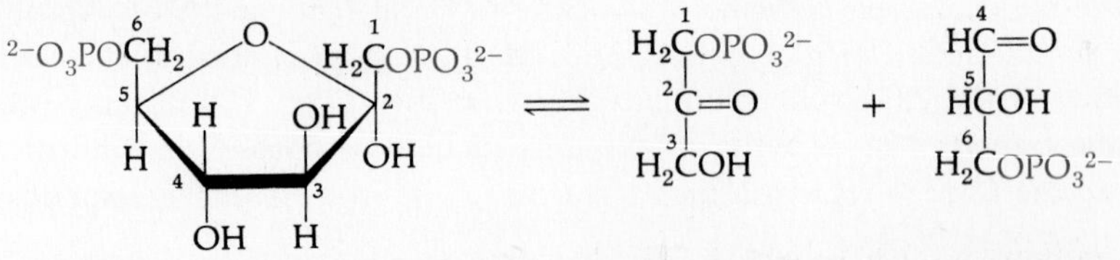

Fructose-1,6-bisphosphate   Dihydroxyacetone phosphate   D-Glyceraldehyde-3-phosphate

The reaction is far to the left at equilibrium (Table 17.1).

The aldolase from all animal tissues is a tetramer (MW = 160,000), but different tissues contain primarily one of three different forms characteristic of muscle, liver, and brain, respectively. All forms catalyze the above reaction, but at different rates; the liver, but not the muscle, enzyme catalyzes the cleavage of fructose-1-phosphate, which is derived from ingested fructose (Chap. 18), almost as rapidly as the bisphosphate.

Dihydroxyacetone Phosphate ⇌ D-Glyceraldehyde-3-phosphate

This reaction is catalyzed by *triose phosphate isomerase* (MW = 56,000), a dimer of two identical subunits, whose three-dimensional structure has been established.

[handwritten marginal notes:]
vasopressin
血吁加压素

phenylephrine
周极拟交感神经
加压剂
fast 替质

四种形式的
酶, 肝, 脑肌肉
各一种, 活性不同

$$
\begin{array}{ccc}
\mathrm{CH_2OH} & & \mathrm{HC=O} \\
| & & | \\
\mathrm{C=O} & \rightleftharpoons & \mathrm{HCOH} \\
| & & | \\
\mathrm{CH_2OPO_3^{2-}} & & \mathrm{CH_2OPO_3^{2-}}
\end{array}
$$

Dihydroxyacetone phosphate   D-Glyceraldehyde-3-phosphate

The two substrates are isomers, having the same relationship to one another as glucose-6-phosphate to fructose-6-phosphate. At equilibrium, dihydroxyacetone phosphate is the predominant species (Table 17.1).

The role of the isomerase is to direct the dihydroxyacetone phosphate into glyceraldehyde-3-phosphate, which is then converted to other intermediates of glycolysis. Another major metabolic fate is conversion to glycerol phosphate, a reaction essential for lipid metabolism (Chap. 20), by *glycerol-3-phosphate dehydrogenase.*

glycerol-3-phosphate dehydrogenase

$$
\begin{array}{ccc}
\mathrm{CH_2-O-PO_3^{2-}} & & \mathrm{CH_2-O-PO_3^{2-}} \\
| & & | \\
\mathrm{C=O} & + \mathrm{NADH} + \mathrm{H^+} \rightleftharpoons & \mathrm{HCOH} \qquad + \mathrm{NAD^+} \\
| & & | \\
\mathrm{CH_2OH} & & \mathrm{CH_2OH}
\end{array}
$$

Dihydroxyacetone phosphate   Glycerol-3-phosphate

The enzyme (MW = 62,000) in liver and muscle is composed of two identical subunits, but separate genes control the synthesis of the two tissue types. Each is an allosteric enzyme; the binding of NADH enhances binding of dihydroxyacetone phosphate while decreasing binding of glycerol-3-phosphate, whereas NAD⁺ has the opposite effect. The liver form has a much lower K_m for its substrates. In muscle, the major role of the dehydrogenase may be to act as a carrier by cytosolic NADH into mitochondria (Chap. 15), whereas in liver its major role is to generate glycerol for lipid synthesis (Chap. 20).

Glyceraldehyde-3-phosphate + NAD⁺ + Pᵢ →
1,3-Bisphosphoglycerate + NADH + H⁺

In this reaction, catalyzed by *glyceraldehyde-3-phosphate dehydrogenase,* the aldehyde group of glyceraldehyde-3-phosphate is oxidized to the level of a carboxyl group with the concomitant formation of a high-energy anhydride of the carboxyl group and phosphate.

$$
\begin{array}{ccc}
\mathrm{CHO} & & \mathrm{O=COPO_3^{2-}} \\
| & & | \\
\mathrm{HCOH} & + \mathrm{NAD^+} + \mathrm{P_i} \rightleftharpoons & \mathrm{HCOH} \qquad + \mathrm{NADH} + \mathrm{H^+} \\
| & & | \\
\mathrm{H_2COPO_3^{2-}} & & \mathrm{H_2COPO_3^{2-}}
\end{array}
$$

D-Glyceraldehyde-3-phosphate   1,3-Bisphosphoglycerate

The rabbit muscle and yeast enzymes (MW = 146,000) are tetramers of four identical subunits and represent 10 and 20 percent, respectively, of the total, soluble, cellular protein, providing these tissues with [enzyme] equal to or greater than intracellular concentrations of some glycolytic intermediates, e.g., fructose-6-phosphate and the triose phosphates. The three-dimensional structure of the lobster muscle dehydrogenase has been established, revealing that each of the four subunits contains two domains, one that binds glyceraldehyde-3-phosphate and the other that binds NAD^+. The latter domain is folded in much the same way as other NAD^+-dependent dehydrogenases, suggesting evolutionary relationships among various NAD^+-dependent enzymes.

The enzyme contains an essential thiol group in its active site on which an acyl-enzyme intermediate is formed. In the first step of catalysis, the enzyme thiol group and substrate react to form a thiohemiacetal (where R—CHO is glyceraldehyde-3-phosphate).

$$R-CHO + HS-E \rightleftharpoons R-\underset{H}{\overset{OH}{\underset{|}{\overset{|}{C}}}}-S-E$$

NAD^+ is then reduced with formation of

$$R-\underset{H}{\overset{OH}{\underset{|}{\overset{|}{C}}}}-S-E + NAD^+ \rightleftharpoons R-\overset{O}{\overset{||}{C}}-S-E + NADH + H^+$$

the acyl enzyme intermediate, which is a thiol ester, and NADH. Phosphorolysis of the acyl enzyme intermediate with P_i then gives 1,3-bisphosphoglycerate and free enzyme.

$$R-\overset{O}{\overset{||}{C}}-S-E + P_i \longrightarrow R-\overset{O}{\overset{||}{C}}-O-PO_3^{2-} + E-SH$$

Arsenate (As_i), which prevents ATP synthesis by glycolysis, acts by replacing P_i in this reaction.

$$R-\overset{O}{\overset{||}{C}}-S-E + As_i \longrightarrow R-\overset{O}{\overset{||}{C}}-O-AsO_3^{2-} + E-SH$$

The resulting 1-arsenate anhydride rapidly and spontaneously hydrolyzes to give 3-phosphoglycerate and arsenate, a process that bypasses the formation of the 1,3-bisphosphoglycerate that is essential for ATP synthesis in the next reaction of glycolysis.

1,3-Bisphosphoglycerate + ADP → 3-Phosphoglycerate + ATP

This essentially irreversible reaction is catalyzed by *phosphoglycerate kinase* (MW = 45,000) and results in the formation of ATP; since 2 mol of triose are produced per mole of glucose, 2 mol of ATP become available at this step per mole of glucose utilized.

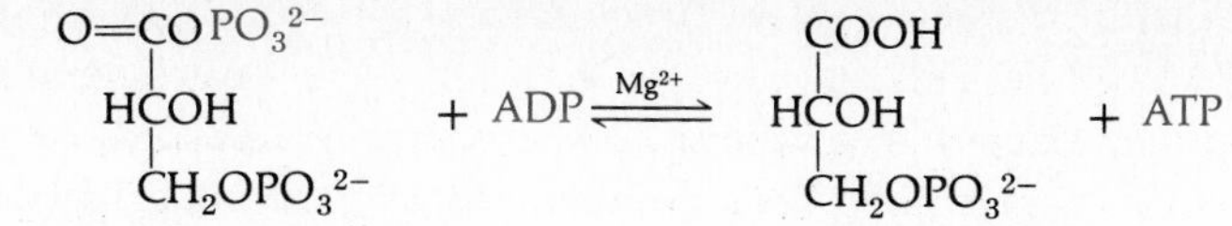

$$\underset{\text{1,3-Bisphosphoglycerate}}{\overset{\displaystyle O=COPO_3^{2-}}{\underset{\displaystyle CH_2OPO_3^{2-}}{|}}} \quad + \text{ ADP} \overset{Mg^{2+}}{\rightleftharpoons} \underset{\text{3-Phosphoglycerate}}{\overset{\displaystyle COOH}{\underset{\displaystyle CH_2OPO_3^{2-}}{|}}} \quad + \text{ ATP}$$

Because $\Delta G°$ for hydrolysis of 1,3-bisphosphoglycerate is -14.8 [+1.8] kcal/mol, more than sufficient energy is available for synthesis of ATP. The sum of the dehydrogenase and kinase reactions is as follows.

Glyceraldehyde-3-phosphate + P_i + NAD^+ + ADP $\longrightarrow$

$\qquad\qquad\qquad\qquad\qquad$ 3-phosphoglycerate + ATP + NADH + H^+

The equilibrium of the coupled reaction is far to the right, a condition that pulls the otherwise unfavorable aldolase and triose phosphate isomerase reactions. This reaction is the best-known example of the coupling of the exergonic oxidation of a metabolite with the generation of useful energy in the form of the pyrophosphate bond of ATP.

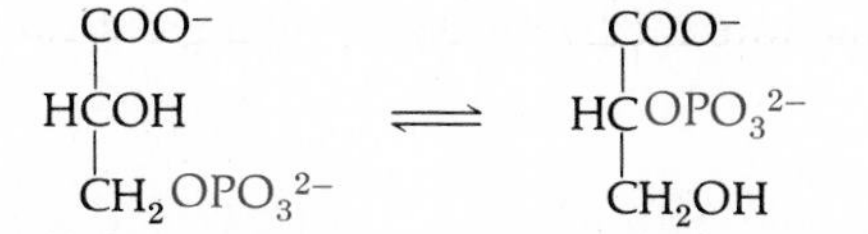

3-Phosphoglycerate $\rightleftharpoons$ 2-Phosphoglycerate

This reaction, catalyzed by *phosphoglyceromutase*, converts the 3-phosphoglycerate formed above to the 2-phosphoglycerate.

$$\underset{\text{3-Phosphoglycerate}}{\overset{\displaystyle COO^-}{\underset{\displaystyle CH_2OPO_3^{2-}}{|}}} \quad \rightleftharpoons \quad \underset{\text{2-Phosphoglycerate}}{\overset{\displaystyle COO^-}{\underset{\displaystyle CH_2OH}{|}}}$$

The mutase (MW = 65,000) is a dimer of identical subunits; the fetal muscle form differs from that of the adult, although both forms persist in adult tissues such as heart and bone.

The mutase requires 2,3-bisphosphoglycerate for its action in a manner similar to that of phosphoglucomutase (Chap. 18). Thus, enzyme and 2,3-bisphosphoglycerate react to give a phosphoenzyme and either the 2- or the 3-phosphoglycerate.

$\quad$ E—OH + 2,3-bisphosphoglycerate $\rightleftharpoons$ E—O—PO_3^{2-} + 2- or 3-phosphoglycerate

At high [3-phosphoglycerate], the 3-phosphoglycerate and phosphoenzyme react to give free enzyme and 2,3-bisphosphoglycerate, which, in turn, yields the 2-phosphoglycerate. The opposite reaction occurs at high [2-phosphoglycerate]. At equilibrium the ratio of the 2- to the 3-phosphate is 50:1. Only small quantities of 2,3-bisphosphoglycerate are required to maintain the reaction and are provided by *2,3-bisphosphoglycerate kinase*, which catalyzes the following reaction.

$\quad$ 3-Phosphoglycerate + ATP $\longrightarrow$ 2,3-bisphosphoglycerate + ADP

2-Phosphoglycerate $\xrightleftharpoons{\text{Mg}^{2+} \text{ or Mn}^{2+}}$ Phosphoenolpyruvate + H₂O

In the next reaction of glycolysis, 2-phosphoglycerate is dehydrated by the action of *enolase* to give phosphoenolpyruvate (PEP), the phosphate ester of the enol tautomer of pyruvate.

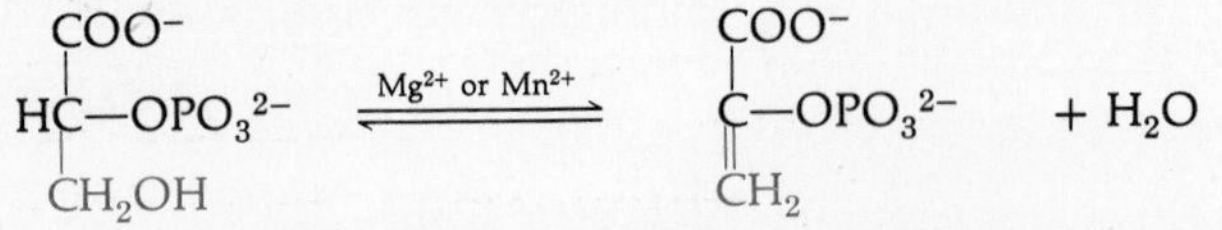

$$
\begin{array}{ccc}
\text{COO}^- & & \text{COO}^- \\
| & \xrightleftharpoons{\text{Mg}^{2+} \text{ or Mn}^{2+}} & | \\
\text{HC}-\text{OPO}_3{}^{2-} & & \text{C}-\text{OPO}_3{}^{2-} \quad + \text{H}_2\text{O} \\
| & & \| \\
\text{CH}_2\text{OH} & & \text{CH}_2 \\
\textbf{2-Phosphoglycerate} & & \textbf{Phosphoenolpyruvate}
\end{array}
$$

At equilibrium (Table 17.1) the ratio of 2-phosphoglycerate to PEP is 2:1; however, PEP, but not 2-phosphoglycerate, contains a high-energy phosphate bond. The $\Delta G°$ for hydrolysis of the phosphate ester of the enol group is about -12 kcal/mol. This is more than sufficient energy to permit synthesis of ATP from PEP in the next step of glycolysis.

Enolase (MW = 88,000) is a dimer with identical subunits. It requires either Mg^{2+} or Mn^{2+} in its active site and binds substrates via the metal. Thus, F^-, a potent inhibitor of glycolysis that causes accumulation of phosphoglycerates, inhibits enolase by forming magnesium fluorophosphate, which is bound at the active site.

competive inhibitor

Phosphoenolpyruvate + ADP $\xrightleftharpoons{\substack{\text{K}^+ \\ \text{Mg}^{2+}}}$ Pyruvate + ATP

This reaction, catalyzed by *pyruvate kinase,* leads to the synthesis of ATP and pyruvate from PEP and ADP.

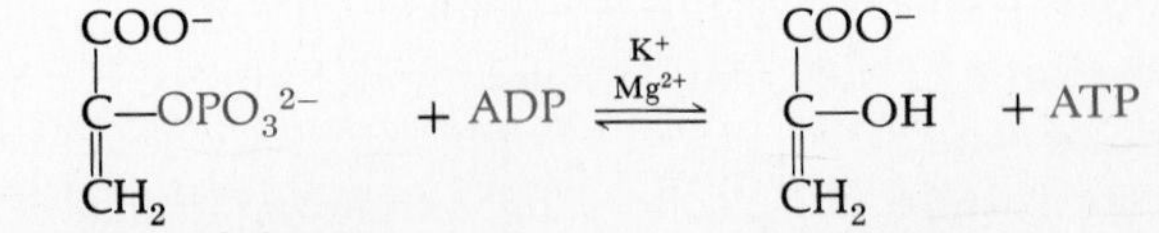

$$
\begin{array}{ccc}
\text{COO}^- & & \text{COO}^- \\
| & & | \\
\text{C}-\text{OPO}_3{}^{2-} \quad + \text{ADP} & \xrightleftharpoons{\substack{\text{K}^+ \\ \text{Mg}^{2+}}} & \text{C}-\text{OH} \quad + \text{ATP} \\
\| & & \| \\
\text{CH}_2 & & \text{CH}_2 \\
\textbf{Phosphoenolpyruvate} & & \textbf{Enolpyruvate}
\end{array}
$$

The reaction is essentially irreversible (Table 17.1) and proceeds with $\Delta G° = -7.5$ kcal/mol. Since 2 mol of PEP are formed per mole of glucose utilized, 2 mol of ATP are also produced per mole of glucose. Kinase activity is dependent on $[\text{K}^+]$, which increases the affinity of PEP for enzyme: Mg^{2+} is required since a Mg^{2+}-ATP complex is the actual substrate.

Pyruvate kinase exists in different forms in different tissues and has a molecular weight equal to 190,000 to 250,000, the exact value depending on the source. All forms of the enzyme, however, contain four identical subunits. The three-dimensional structure of the cat muscle enzyme (MW = 240,000) reveals three domains per subunit, one of which is similar structurally to triose phosphate isomerase. This suggests possible evolutionary relationships among certain glycolytic enzymes.

Three major forms of the kinase are known; muscle and brain contain the

M type, and hepatocytes the L type, whereas most other tissues contain the A type. Each type appears to be under separate genetic control and has quite different catalytic properties, in accord with its differing roles in different tissues.

Pyruvate kinase is an allosteric enzyme and, like phosphofructokinase, its activity is regulated by several means. In general, its activity is high when a net flux of glucose to pyruvate or lactate is required, and low during gluconeogenesis, since wastage of ATP due to substrate cycling must be prevented. Figure 17.6 shows the kinetic behavior of the M and L forms and illustrates, in part, how allosteric effectors regulate activity. The M form normally has hyperbolic kinetics, but ATP, an allosteric effector, inhibits the reaction, producing pronounced sigmoidal kinetics with an increase in K_m for PEP. In contrast, the L form normally has sigmoidal kinetics with substrates; this is further exaggerated by the negative effectors, ATP or alanine. Fructose-1,6-bisphosphate, a positive effector, overcomes the negative effectors, permitting hyperbolic kinetics. Thus, ATP or alanine increases the K_m for PEP, whereas fructose-1,6-bisphosphate decreases the K_m for PEP. Therefore, in liver, the action of the 1,6-bisphosphate is an example of positive feed forward regulation, and the actions of ATP and alanine are examples of feedback regulation.

The L form of pyruvate kinase is also influenced by the diet and by hormones. Kinase levels decrease on fasting, a condition that promotes gluconeogenesis, and increase on a carbohydrate-rich diet, a condition that requires little gluconeogenesis. In addition, insulin increases the [kinase] in liver, whereas glucagon, which stimulates hepatic gluconeogenesis, reduces the activity of the kinase. The molecular basis for glucagon action resides in the reversible phosphorylation-dephosphorylation of the kinase. Thus, glucagon stimulates formation of cAMP and cAMP activates cAMP-dependent protein kinase (Chap. 18), which, in turn, phosphorylates pyruvate kinase. The phosphorylated form is less active than the dephosphorylated form and has a higher K_m for PEP. This type of monocyclic cascade is shown schematically in Fig. 17.7. The phosphorylated form of pyruvate kinase is also more susceptible to inhibition by alanine and ATP. Moreover, its activity is virtually zero in the absence of fructose-1,6-bisphosphate, a positive effector. Dephosphorylation of the kinase is catalyzed by a phosphoprotein phosphatase, whose activity is also regulated by cAMP-dependent protein kinase, as described in Chap. 18. Thus, the net effect

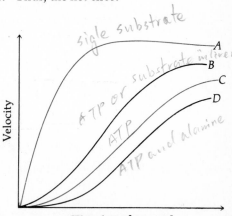

Figure 17.6 Slightly schematic representation of the control of pyruvate kinase. For the muscle enzyme, A represents the enzyme with its normal substrates; B represents the effect of ATP at a single elevated concentration. For the liver enzyme, B represents the enzyme with its normal substrates; C, the effect of elevated ATP; D, the consequence of an increase in both ATP and alanine; while A indicates the effect of fructose bisphosphate on the system containing enzyme and substrates. Varying the concentrations of fructose bisphosphate, ATP, and alanine in the presence of one another alters the curves in the directions shown.

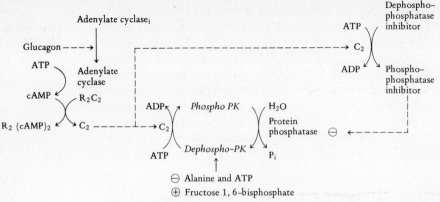

$\ominus$ Alanine and ATP
$\oplus$ Fructose 1, 6–bisphosphate

Figure 17.7 Cascade regulation of pyruvate kinase (PK). The cAMP-dependent protein kinase (C_2), which is activated as the result of glucagon activation of adenylate cyclase, phosphorylates PK as well as the inhibitor of the protein phosphatase that dephosphorylates phospho-PK. The activation of inactive cAMP-dependent protein kinase (R_2C_2) and the regulation of phosphatase activity are described in Chap. 18. $-$ and $+$ refer to negative and positive effectors for the enzymes indicated.

of glucagon is to diminish the flux of glucose to lactate, in accord with its other actions in stimulating gluconeogensis.

Pyruvate + NADH + H$^+$ $\rightleftharpoons$ Lactate + NAD$^+$

The last reaction of the glycolytic pathway results in formation of L-lactate from pyruvate but also results in reoxidation of the NADH produced in the formation of 3-phosphoglycerate earlier in the path.

$$H_3C-\overset{\overset{\displaystyle O}{\|}}{C}-COO^- + NADH + H^+ \rightleftharpoons H_3C-\overset{\overset{\displaystyle OH}{|}}{CH}-COO^- + NAD^+$$
$$\text{L-Lactate}$$

NAD$^+$ is in limited supply in cells, and if the NADH produced were not reoxidized, anaerobic glycolysis would stop when all NAD$^+$ were reduced to NADH. Thus, reduction of pyruvate to lactate catalyzed by *lactate dehydrogenase* is coupled to the oxidation of 3-phosphoglyceraldehyde to 1,3-bisphosphoglycerate as follows:

$$\text{P}_i + \text{3-Phosphoglyceraldehyde} \quad\overset{}{\underset{}{\diagdown}}\quad NAD^+ \quad\overset{}{\underset{}{\diagdown}}\quad \text{Lactate}$$
$$\text{1,3-Bisphosphoglycerate} \quad\overset{}{\underset{}{\diagup}}\quad \underset{+\text{H}^+}{NADH} \quad\overset{}{\underset{}{\diagup}}\quad \text{Pyruvate}$$

The equilibrium position of the lactate dehydrogenase reaction is far to the right; hence, glycolysis results in the accumulation of 2 mol of lactate per mole of glucose. Lactate is a blind alley in metabolism and, once formed, it can only be reconverted to pyruvate. Since lactate is free to diffuse out of cells, however, it can leave muscle and be transported by the circulation to liver, where it can be reconverted to pyruvate, which can then be utilized further in metabolism.

406

Lactate dehydrogenase (MW = 140,000) contains four subunits, which may be of two types that differ in sequence. The M form is characteristic of skeletal muscle and the H form of heart muscle, but hybrids of the two types occur with subunit structures M_4, M_3H, M_2H_2, MH_3, and H_4, the exact amounts of each differing with the tissue. H_4 is strongly inhibited by pyruvate, a property that renders it useful to a tissue like heart, which removes lactate from blood, converts it to pyruvate, and then metabolizes the pyruvate via the citric acid cycle with production of ATP. In contrast, M_4 is not inhibited by pyruvate; hence, it is useful in permitting the large bursts of anaerobic glycolysis required by skeletal muscle on exercise (Chap. 18).

The Efficiency of Glucose Metabolism and the Net Yield of ATP

The net yield of ATP in anaerobic glycolysis is the difference between the ATP produced and consumed in the process. Only four reactions in glycolysis involve ATP, as follows.

$$\text{Glucose} + \text{ATP} \longrightarrow \text{glucose-6-phosphate} + \text{ADP}$$

$$\text{Fructose-6-phosphate} + \text{ATP} \longrightarrow \text{fructose-1,6-bisphosphate} + \text{ADP}$$

$$\text{2 1,3-Bisphosphoglycerate} + \text{2ADP} \longrightarrow \text{2 3-phosphoglycerate} + \text{2ATP}$$

$$\text{2 Phosphoenolpyruvate} + \text{2ADP} \longrightarrow \text{2 pyruvate} + \text{2ATP}$$

Two moles of ATP are consumed in the first two reactions, and 4 moles of ATP are produced by the last two reactions. Thus, there is a net gain of 2 mol of ATP per mol of glucose converted to pyruvate.

The actual $\Delta G°$ for glycolysis under physiological conditions is unknown, but about 48 kcal/mol are produced on oxidation of glucose to pyruvate under standard conditions. Since hydrolysis of 2 mol of ATP yields about 24 kcal under physiological conditions, the efficiency of glycolysis is about 50 percent, i.e., about one-half the potentially available energy may be realized as ATP incident to the formation of 2 mol of pyruvate from 1 mol of glucose.

The net yield of ATP from glucose in a respiring system, however, is far greater than that from glycolysis. Table 17.2 lists the reactions that lead to ATP production from glucose in a respiring system. As noted above, 2 mol of ATP are utilized in the reactions from glucose to fructose-1,6-bisphosphate, whereas 4 mol of ATP are produced by the combined reactions involving triose phosphates and PEP. In contrast to anaerobic glycolysis, however, the 2 mol of

TABLE 17.2

Total ATP Production from Glucose in a Respiring System

Reaction sequence	ATP yield
Glucose → fructose-1,6-bisphosphate	−2
2 Triose phosphate → 2 3-phosphoglycerate	+2
$2NAD^+ \to 2NADH \to 2NAD^+$	+6
2 Phosphoenolpyruvate → 2 pyruvate	+2
2 Pyruvate → 2 acetylCoA $+ 2CO_2$	
$2NAD^+ \to 2NADH \to 2NAD^+$	+6
2 AcetylCoA $\to 4CO_2$	+24
NET: $C_6H_{12}O_6 + 6O_2 \to 6CO_2 + 6H_2O$	+38

NADH produced in the glycolytic pathway on oxidation of glyceraldehyde-3-phosphate yield an additional 4 to 6 mol of ATP by mitochondrial oxidation, the exact amount depending upon the substance that is reduced cytoplasmically and then carried into the mitochondria for further oxidation (Chap. 15). An additional 6 mol of ATP are generated from the NADH produced by the mitochondrial oxidation of pyruvate to acetyl CoA, and mitochondrial oxidation of the resulting acetyl CoA produces 24 mol of ATP (Chap. 15). Thus, complete glucose oxidation can be considered to proceed by the following overall reaction.

$$C_6H_{12}O_6 + 6O_2 + 38ADP + 38P_i \longrightarrow 6CO_2 + 6H_2O + 38ATP$$

It is difficult to calculate the efficiency of this process since it is uncertain what value to assign to the free energy of formation of ATP in the cell. But assuming that the $\Delta G°$ for hydrolysis of ATP is -12 kcal/mol, a total of about 456 kcal of the 686 kcal potentially available may be conserved in the process. The efficiency, therefore, is about 65%.

Alcoholic Fermentation

The reactions of glycolysis leading to pyruvate may also proceed in bacteria and yeast, but in contrast to animals, other means than the lactate dehydrogenase reaction may be used to reoxidize NADH. For example, yeast degrades pyruvate to acetaldehyde and CO_2 in a reaction catalyzed by *pyruvate decarboxylase,* and the resulting acetaldehyde is reduced to ethanol by *alcohol dehydrogenase,* an NAD-dependent enzyme, as follows.

$$CH_3COCOOH \longrightarrow CH_3CHO + CO_2$$
$$CH_3CHO + NADH^+ + H^+ \rightleftharpoons CH_3CH_2OH + NAD^+$$

Thus, ethanol rather than lactate accumulates since acetaldehyde replaces pyruvate as the oxidant for the NADH that arises from oxidation of glyceraldehyde-3-phosphate. $NaHSO_3$ prevents alcoholic fermentation by combining with acetaldehyde, to give a bisulfite addition compound that is not a substrate for alcohol dehydrogenase. NADH may be reoxidized, however, by reduction of dihydroxyacetone phosphate to glycerol by glycerol phosphate dehydrogenase, as discussed above.

$$\text{Dihydroxyacetone phosphate} + NADH + H^+ \rightleftharpoons \text{glycerol-3-phosphate} + NAD^+$$

The glycerol-3-phosphate is hydrolyzed by a phosphatase to glycerol, which accumulates in this type of fermentation.

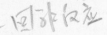

ANAPLEROSIS: PYRUVATE AND OXALOACETATE METABOLISM

Pyruvate occupies a central position in metabolism, since it participates in many other metabolic reactions, as summarized in Fig. 17.8. Its oxidation to acetyl CoA was considered in Chap. 15. Of the other metabolic fates of pyruvate, its special metabolic interrelationships with 4-carbon intermediates of the citric acid cycle are particularly noteworthy. Many intermediates of the cycle are used for the synthesis of other substances; e.g., oxaloacetate and α-ketogluta-

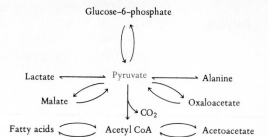

Figure 17.8 Fates of pyruvate in mammals.

rate are precursors of aspartate and glutamate, respectively. The drain of these intermediates would ultimately prevent operation of the cycle were they not replenished. Pyruvate serves this function by a metabolic process called *anaplerosis,* which is defined as any reaction that can restore the concentration of a crucial but depleted intermediate. Moreover, other intermediates of the citric acid cycle, such as oxaloacetate, malate, and citrate, can also be converted to pyruvate or phosphoenolpyruvate and thus provide important sources for gluconeogenesis.

Oxaloacetate Formation from Pyruvate

The major anaplerotic reaction in animal cells, plants, and microorganisms results in synthesis of oxaloacetate from pyruvate, bicarbonate, and ATP and is catalyzed by mitochondrial *pyruvate carboxylase.*

$$CH_3COCOO^- + HCO_3^- + ATP \longrightarrow {}^-OOC-CH_2-COCOO^- + ADP + P_i$$
Pyruvate   **Oxaloacetate**

Animal carboxylases are composed of four identical subunits, which have molecular weights equal to 110,000 to 130,000, depending on the source. Each subunit contains an active site, including tightly bound Mn^{2+} and covalently bound *biotin,* and an allosteric site, which binds acetyl CoA. Biotin is covalently attached to the ϵ-NH$_2$ group of a lysyl residue by an enzyme that catalyzes the following reaction.

$$Enz + ATP + biotin \rightleftharpoons Enz\text{-}biotin + AMP + PP_i$$

The carboxylation of pyruvate occurs in two steps, each of which is catalyzed by a distinct subsite of the active site.

$$Enz\text{-}biotin + HCO_3^- + ATP \rightleftharpoons Enz\text{-}biotin\text{-}CO_2^- + ADP + P_i$$
$$Enz\text{-}biotin\text{-}CO_2^- + pyruvate \rightleftharpoons oxaloacetate + Enz\text{-}biotin$$

The structure of the biotin-CO_2^- complex is

1¹-*N*-Carboxybiotinyl-enzyme

409

The $\Delta G°$ for hydrolysis of carboxybiotin is about -4.7 kcal/mol, more than sufficient for it to serve as a carboxylating agent.

Acetyl CoA is required for carboxylase activity, and in its absence the first of the two steps above cannot occur. Pyruvate, however, also enhances the rate of the first step, and longer-chain acyl CoA derivatives can replace acetyl CoA.

In plants and microorganisms other types of anaplerotic reactions occur. *Phosphoenolpyruvate carboxylase* catalyzes the reaction

$$\text{Phosphoenolpyruvate} + CO_2 + H_2O \longrightarrow \text{oxaloacetate} + P_i$$

Certain microorganisms employ *phosphoenolpyruvate carboxytransphosphorylase.*

$$\text{Phosphoenolpyruvate} + CO_2 + P_i \longrightarrow \text{oxaloacetate} + PP_i$$

Phosphoenolpyruvate Formation from Oxaloacetate

In circumstances where citric acid cycle intermediates are plentiful, they can be used in gluconeogenesis, the process of replenishing glucose by reversal of glycolysis. The pyruvate kinase–catalyzed reaction, however, is essentially irreversible, as discussed above, and must be bypassed; this is accomplished by synthesis of phosphoenolpyruvate, which is then freely available for synthesis of glucose by reversal of glycolysis. The enzyme involved is *pyruvate-carboxykinase,* which catalyzes the reaction *(phosphoenolpyruvate)*

$$\text{Oxaloacetate} + GTP \longrightarrow \text{phosphoenolpyruvate} + GDP + CO_2$$

which proceeds with $\Delta G° = -4$ kcal/mol. The GTP for the reaction is supplied by the mitochondria from the succinate thiokinase reaction (Chap. 15) or by the action of *nucleoside diphosphate kinase*, which catalyzes the reaction

$$ATP + GDP \rightleftharpoons GTP + ADP$$

The carboxykinase (MW $= 75,000$) is primarily in cytosol.

Pyruvate Formation from Malate

Malic enzyme, found in two different forms in cytosol or mitochondria, catalyzes the following reaction.

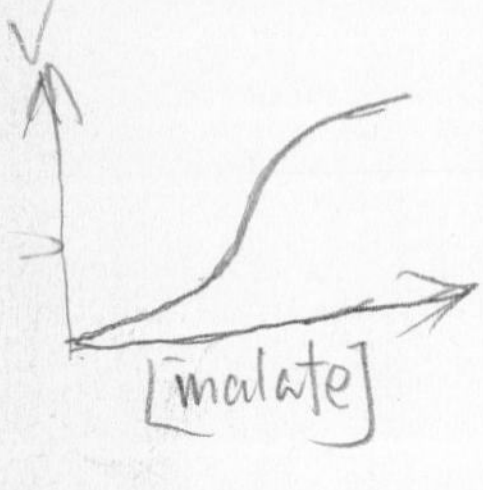

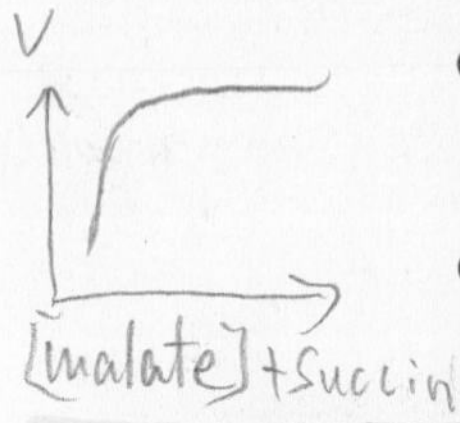

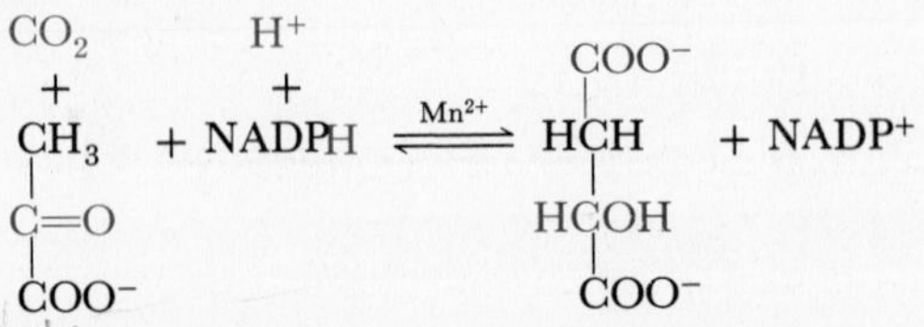

$$
\begin{array}{ccccccc}
CO_2 & & H^+ & & & COO^- & \\
+ & & + & & & | & \\
CH_3 & + NADPH & & \underset{Mn^{2+}}{\rightleftharpoons} & & HCH & + NADP^+ \\
| & & & & & | & \\
C{=}O & & & & & HCOH & \\
| & & & & & | & \\
COO^- & & & & & COO^- &
\end{array}
$$

Its major function is probably formation of NADPH, or the reverse of the reaction as written. Mitochondrial malic enzyme, but not the cytosolic form, is an allosteric enzyme showing sigmoidal kinetics and positive cooperativity as a function of [malate]. Succinate is a positive effector, producing hyperbolic kinetics as a function of [malate] and decreasing the K_m for malate.

ATP citrate lyase of cytosol catalyzes the reaction

$$\text{Citrate} + \text{ATP} + \text{CoA} \longrightarrow \text{acetylCoA} + \text{oxaloacetate} + \text{ADP} + \text{P}_i$$

This reaction, in effect, is the reversal of the citrate synthase reaction of the citric acid cycle in mitochondria. Although the lyase provides oxaloacetate for the cytosol, it also functions to produce acetyl CoA for fatty acid synthesis (Chap. 20).

GLUCONEOGENESIS

Formation of Glucose from Lactate

Gluconeogenesis, the replenishment of carbohydrate from noncarbohydrate precursors, cannot proceed by simple reversal of glycolysis. For the overall process of lactate to glucose, under standard conditions, $\Delta G° = +48$ kcal/mol; thus, for this set of reactions to occur with a net loss of free energy, an energy input greater than 48 kcal/mol is required. Moreover, one enzyme, pyruvate kinase, catalyzes its reverse reaction so slowly that, on both kinetic and thermo-dynamic grounds, some alternative means is needed for formation of phospho-enolpyruvate (PEP), the product of the reversal of the pyruvate kinase reaction. Consideration of the reactions of glycolysis from PEP to glucose reveals that once PEP is formed, the consecutive equilibria of the other reactions would drive PEP back to glucose if the controls were removed from fructose-1,6-bisphosphatase.

The reactions that ensure synthesis of PEP and permit reversal of glycolysis are provided by the consecutive actions of *pyruvate carboxylase* and *pyruvate carboxykinase*, which, as discussed above, catalyze the following reactions, respectively.

$$\text{Pyruvate} + \text{CO}_2 + \text{ATP} \longrightarrow \text{oxaloacetate} + \text{ADP} + \text{P}_i$$

$$\text{Oxaloacetate} + \text{GTP} \longrightarrow \text{phospho enolpyruvate} + \text{GDP} + \text{CO}_2$$

SUM: $\text{Pyruvate} + \text{CO}_2 + \text{ATP} + \text{GTP} \longrightarrow \text{phospho enolpyruvate} + \text{ADP} + \text{GDP} + \text{P}_i$

Since conversion of lactate or pyruvate to glucose occurs aerobically, when O_2 is in good supply, ATP and GTP are readily available from the citric acid cycle and mitochondrial oxidative phosphorylation (Chap. 15). The coupled actions of these two enzymes, however, are complicated in many animal species by the fact that oxaloacetate is made by the carboxylase in mitochondria and is not free to leave, as such, whereas the carboxykinase is in the cytosol. This problem appears to be circumvented by the series of reactions shown in Fig. 17.9, which permit oxaloacetate to leave the mitochondria as malate or aspartate and then be reconverted to oxaloacetate in the cytosol.

If lactate is the gluconeogenic precursor, it is converted in the cytoplasm to pyruvate by the lactate dehydrogenase reaction with production of **NADH**. Pyruvate thus formed enters the mitochondria and gives rise, by the pyruvate carboxylase reaction, to oxaloacetate, which by transamination with glutamate gives α-ketoglutarate and aspartate. The last two compounds then leave the mitochondria and in the cytosol undergo transamination to give glutamate and

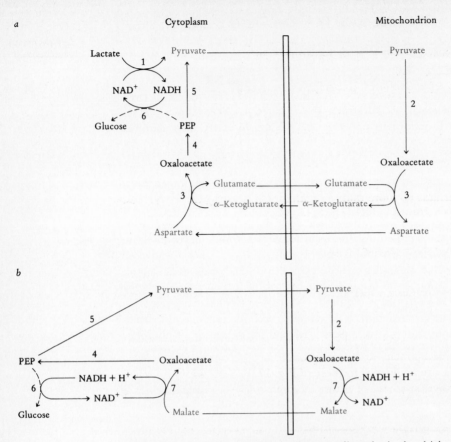

Figure 17.9 The mechanisms for coupling cytosolic and mitochondrial enzymes for gluconeogenesis. The enzymes involved are: 1 lactate dehydrogenase; 2 pyruvate carboxylase; 3 aspartate: glutamate aminotransferase; 4 pyruvate carboxykinase; 5 pyruvate kinase; 6 the enzymes of glycolysis that permit conversion of PEP to glucose, including glyceraldehyde-3-phosphate dehydrogenase, which converts NAD to NADH, reversibly; and 7 malate dehydrogenase. Note that both the aminotransferase and malate dehydrogenase are in cytosol as well as the mitochondria, and are distinct isoenzymic forms.

oxaloacetate. Cytosolically generated glutamate reenters the mitochondria for continuation of the mitochondrial transamination, whereas cytosolic oxaloacetate is converted to PEP by the pyruvate carboxykinase and the PEP is converted to glucose by reversal of the glycolytic reactions. PEP conversion to glucose also results in reoxidation of the NADH produced in the lactate dehydrogenase reaction, a vital part of the entire process, since NAD^+ is in low amounts, and without its reformation from NADH, the entire process would stop.

Also shown in Fig. 17.9 is the mechanism for gluconeogenesis from pyruvate. The pyruvate is taken up by mitochondria and converted to oxaloacetate, which is then reduced to malate with formation of NAD^+. Malate is free to leave the mitochondria and in the cytosol regenerates oxaloacetate with production of the NADH that is required to permit glucose synthesis from PEP, which is formed from oxaloacetate by the pyruvate carboxykinase reaction in the cytosol. This same mechanism under aerobic conditions permits the NADH that is produced cytoplasmically during glycolysis by the glyceraldehyde 3-phosphate dehydrogenase reaction to be utilized in mitochondrial oxidative phosphorylation (Chap. 15).

In contrast to glycolysis, which catalyzes the overall reaction

$$\text{Glucose} + 2\text{ADP} + 2\text{P}_i \longrightarrow 2\,\text{lactate} + 2\text{ATP}$$

the reverse process by the mechanism described above is

$$2\,\text{Lactate} + 6\text{ATP} \longrightarrow \text{glucose} + 6\text{ADP} + 6\text{P}_i$$

Since only 6 mol of ATP are required per mole of glucose formed, resynthesis of glucose requires perhaps no more than 64 kcal/mol, suggesting that resynthesis of glucose is energetically efficient, despite its complexity.

Formation of Acetyl CoA from Glucose

The acetyl CoA formed in mitochondria from pyruvate (Fig. 17.1) is required for fatty acid synthesis, which occurs in the cytosol, as discussed in detail in Chap. 20. It is appropriate to note here, however, that acetyl CoA, like oxaloacetate, is not free to leave the mitochondria; thus the acetyl CoA must be transformed to a transportable substance and then recovered in the cytosol, with return of a carrier component to the mitochondria. The processes involved, shown in Fig. 17.10, are analogous to those of Fig. 17.9. In this case, however, the mechanism involves synthesis of citrate by *citrate synthase*. Citrate, as well as

Figure 17.10 Formation of acetyl CoA in the cytosol from glucose. The enzymes specifically noted are 1 glyceraldehyde-3-phosphate dehydrogenase, 2 pyruvate dehydrogenase, 3 malate dehydrogenase, 4 citrate synthase, and 5 ATP-citrate lyase.

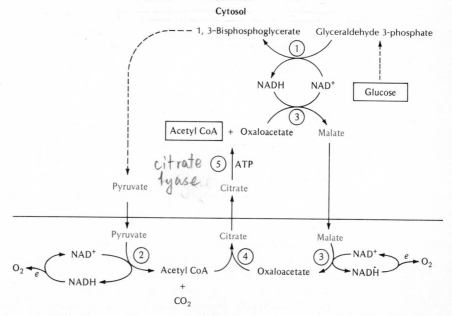

malate, is freely permeable to the mitochondrial membrane but entry and exit of these compounds is a facilitated process using symport mechanisms (Chap. 15) and involves obligatory exchanges, such as malate in–citrate out and malate out–P_i in. In the cytosol, citrate is converted to acetyl CoA and oxaloacetate by citrate lyase, discussed above, and the acetyl CoA is used for cytosolic fatty acid synthesis. The oxaloacetate, however, reenters the mitochondria after conversion to malate. The malate from cytosolic oxaloacetate may also give rise to pyruvate via malic enzyme, which, as discussed above, is in the cytosol, and the pyruvate enters the mitochondria to reinitiate acetyl CoA formation. The pyruvate that enters the mitochondria can then form oxaloacetate via the pyruvate carboxylase reaction. Thus, the total cost to the cell to deliver one molecule of acetyl CoA to the cytosol is two molecules of ATP, one used in the citrate lyase reaction and the other in the pyruvate carboxylase reaction.

REGULATION OF GLYCOLYSIS AND GLUCONEOGENESIS

Gross observation indicates that the processes of glycolysis and gluconeogenesis operate smoothly, in accordance with physiological requirements. A muscle at rest does not release lactate into the circulation, nor does the liver oxidize the lactate influx, despite its high concentration of lactate dehydrogenase. Adults on a high carbohydrate ration store the surplus carbohydrate first as liver and muscle glycogen and then as neutral triacylglycerols made in liver and in adipose tissue. However, during exercise, glycolysis markedly accelerates in muscle; the lactate formed is removed from the circulation by the liver, where it is converted to glycogen. Within hours after the last meal, liver glycogen is utilized as a source of blood glucose, the only fuel acceptable to the brain; the liver ceases to glycolyze, relies on fatty acid oxidation for its own energy, and begins to convert amino acids to glucose. These phenomena indicate that the rates of the enzymic systems that together constitute a pathway are not determined solely by the amounts of enzymes and their substrates. Nor indeed is there evidence for operation of the many futile cycles that might be suggested by examination of Fig. 17.1. Indeed, the overall coupling of the total glycolytic and gluconeogenic pathways

$$\text{Glucose} \longrightarrow 2 \text{ lactate} \longrightarrow \text{glucose}$$

would constitute a giant futile cycle, the net accomplishment of which would be hydrolysis of 4ATP per cycle. It is evident instead that cellular flow-through rates are normally commensurate with need, with relatively little waste.

The controls that ensure that the rates of glycolysis and gluconeogenesis are commensurate with physiological needs are of many kinds; they also vary in an individual with time, from organ to organ, from species to species, and are dependent upon the dietary and hormonal status. The following are the major factors that must be considered to understand mechanisms that control these processes.

1. The maximal, observed rates of glycolysis and gluconeogenesis are much less than expected on the basis of the substrate and enzyme concentrations

in a tissue. Thus, for normal rat liver, glycolysis proceeds at no more than 10 percent the possible rate were the enzymes that are rate limiting, phosphofructokinase and pyruvate kinase, operating maximally and glucose were in good supply. Gluconeogenesis acts at no more than 25 percent the rate expected if pyruvate carboxylase and PEP carboxykinase were operating maximally. Even in muscle, the rate of glycolysis seldom reaches the rate expected on the basis of enzyme and substrate concentrations.

2. **The major controls of glycolysis and gluconeogenesis operate on only a few of the enzymes in the pathway.** Thus, the rate of glycolysis is determined primarily by regulation of the three kinases involved, hexokinase, phosphofructokinase, and pyruvate kinase, each of which is an allosteric enzyme, with specific effectors whose concentrations vary depending on metabolic needs. Similarly, the rates of gluconeogenesis are dependent primarily on allosteric regulation of pyruvate carboxylase and fructose 1,6-bisphosphatase.

3. **The enzymes that are regulated to control the rates of glycolysis do not participate in gluconeogenesis, and vice versa.** The major established specific controls of glycolysis and gluconeogenesis, each of which was considered above, are shown in Fig. 17.11.

Figure 17.11 Some controls operative in glycolysis and gluconeogenesis. The numbers indicate the enzymes that act in each reaction. Enzymes 1 to 13 are the same as those listed in Table 17.1 and Fig. 17.4. The other enzymes are: 14 fructose-2,6-bisphosphokinase; 15 pyruvate dehydrogenase; 16 pyruvate carboxylase; and 17 pyruvate carboxykinase.

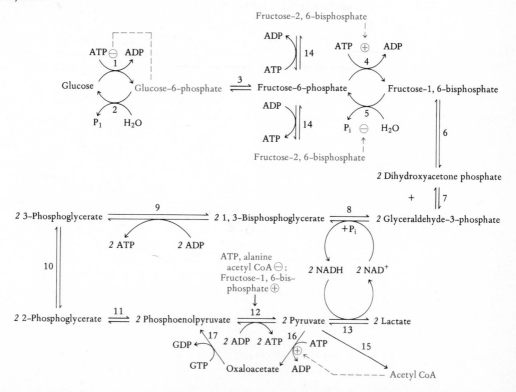

4. **The cell does not appear to attain a steady state, in which the flux of metabolic intermediates through the system remains constant.** This is illustrated by Fig. 17.12, which shows that the concentration of NADH with time in a glycolyzing yeast extract is not constant, but oscillates. This behavior is largely the result of continuing changes in the velocity of the phosphofructokinase reaction and to a lesser extent the pyruvate kinase reaction as substrates, products, and allosteric effectors of these kinases change in concentration. Plots of the concentration of other glycolytic intermediates would also be expected to show similar oscillations.

5. **The major allosteric effectors acting on enzymes of glycolysis or gluconeogenesis are often substrates or products of these pathways.** For example, ATP is a negative effector for phosphofructokinase and pyruvate kinase, whereas fructose-1,6-bisphosphate is a positive effector for pyruvate kinase (Fig. 17.7). Moreover, fructose-2,6-bisphosphate, which is formed under the influence of glucagon when glucose is in good supply, as discussed above (Fig. 17.5), activates phosphofructokinase and inhibits fructose 1,6-bisphosphatase, thus enhancing glycolysis and reducing gluconeogenesis. The regulatory effects of the 2,6-bisphosphate undoubtedly play a major role in controlling the rates of the reciprocal processes of glucose degradation and resynthesis.

6. **The concentrations of the enzymes involved in gluconeogenesis change from fetal to adult life.** During fetal life adequate, continuous supplies of glucose and other nutrients are ensured by placental transfer from the maternal to the fetal circulation, and there is little need for gluconeogenesis or energy storage. At birth, glucose-6-phosphatase, fructose-1,6-bisphosphatase, pyruvate carboxykinase, and various aminotransferases (Chap. 24), each of which is essential for gluconeogenesis, appear in liver.

7. **The dietary and hormonal status of an animal influences the rates of glycolysis and gluconeogenesis.** On fasting, the liver reserves of glycogen are adequate to maintain blood [glucose] for only a few hours; thereafter, gluconeogenesis becomes operative. In contrast, high-carbohydrate diets containing very little fat result in elevation of liver pyruvate kinase, certain enzymes of the phosphogluconate pathway (see below), and the fatty acid

Figure 17.12 Oscillatory behavior of NAD+-NADH in a cell-free extract of yeast. As the supply of glucose neared exhaustion, trehalose, which is fermentable after hydrolysis, was introduced and the concentration of NADH was monitored by following its absorbance. [*From K. Pye and B. Chance, Proc. Natl. Acad. Sci. U.S.A.* **55**:*888 (1966).*]

140mM Trehalose

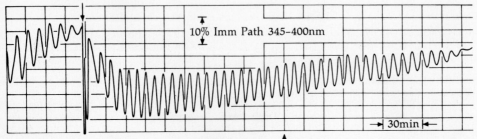

10% Imm Path 345–400nm

→ 30min ←

NAD Reduction ↑

synthesis pathway, enhancing the capacity of liver to convert glucose to fatty acids.

On prolonged fasting, the rate of synthesis of *glucocorticoid hormones* by the adrenal cortex is enhanced, thus stimulating synthesis of pyruvate carboxykinase, glucose-6-phosphatase, and the aminotransferases that are required for gluconeogenesis. The increased levels of the aminotransferases in liver ensure conversion of amino acids to pyruvate, α-ketoglutarate, and oxaloacetate and at the same time provide an increased [alanine], which inhibits pyruvate kinase.

Fasting also causes secretion of epinephrine and glucagon, which stimulate cAMP formation via activation of adenylate cyclase and production of cAMP-dependent protein kinase that phosphorylates pyruvate dehydrogenase (Chap. 15), pyruvate kinase, and fructose-2,6-bisphosphate kinase, each of which has a very low activity or requires positive effectors for activity. Thus, hormones influence the activity of the two enzymes of pyruvate metabolism that catalyze critical reactions in the pathway. Moreover, they alter the activity of fructose-2,6-bisphosphate kinase, which forms an allosteric effector that activates phosphofructokinase, and thus glycolysis, and inhibits fructose-1,6-bisphosphatase and gluconeogenesis. These same hormones also enhance hydrolysis of triacylglycerols (Chap. 20) in adipose tissue, resulting in release of fatty acids into the circulation and thereby ensuring the liver a supply of acetyl CoA for activation of pyruvate carboxylase. Liver glucokinase levels also fall on fasting, minimizing the competition of liver with other tissues for glucose released into the circulation by the liver itself as well as by the kidney and the intestine.

8. **Amino acids are the major sources of carbon in gluconeogenesis.** Degradation of tissue proteins, primarily those of skeletal muscle, provides the amino acids for gluconeogenesis in malnourished animals. The entry points of the carbon chains of the amino acids into gluconeogenesis occur in the citric acid cycle, as described in Chap. 25.

9. **The availability of O_2 results in diminution or abolition of glycolysis.** This phenomenon, called the *Pasteur effect,* and first recognized in microorganisms, results from the fact that effectors of the kinases that act as brakes on glycolysis, e.g., ATP, are in good supply as the result of oxidative phosphorylation and the citric acid cycle.

THE PHOSPHOGLUCONATE OXIDATIVE PATHWAY

This is an alternative pathway for the oxidation of glucose in the cytoplasm that employs distinct enzymes from those of glycolysis. Glucose is oxidized completely to CO_2 and H_2O by the combined reactions of the pathway, but the major functions of the pathway are different: (1) It is the major source of the NADPH that is required in many metabolic reactions, e.g., fatty acid synthesis (Chap. 20), and (2) it leads to the synthesis and disposal of the pentoses that are essential components of many substances, e.g., nucleotides and the nucleic acids (Chaps. 26).

This pathway starts with glucose-6-phosphate and neither requires nor gen-

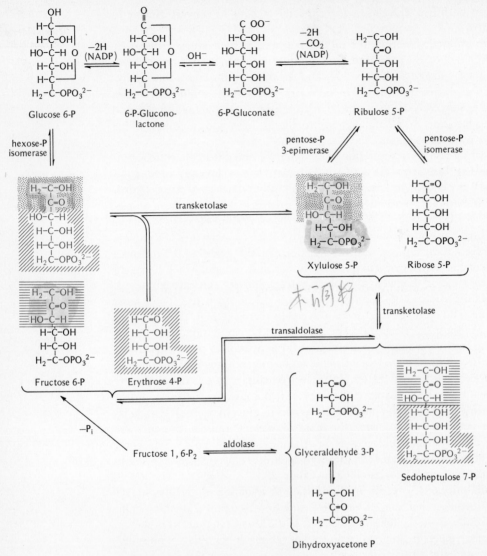

Figure 17.13 Reactions of the phosphogluconate oxidative pathway. [*From B. L. Horecker, p. 69, in A. Kornberg, B. L. Horecker, L. Cornudella, and J. Oró (eds.), Reflections on Biochemistry, Pergamon, New York, 1976.*]

erates ATP. Its operation, summarized in Fig. 17.13, cannot be easily visualized as a consecutive set of transformations to produce six molecules each of CO_2 and H_2O from glucose-6-phosphate, although, in effect, this is what is accomplished. To understand operation of the pathway requires that the reactions involved and the participating enzymes be considered first.

Glucose-6-phosphate → 6-Phosphogluconolactone → 6-Phosphogluconate

The first two reactions of the pathway are (1) oxidation of glucose-6-phosphate by *glucose-6-phosphate dehydrogenase*, which employs $NADP^+$ as co-substrate, and (2) conversion of the 6-phosphogluconolactone produced by the dehydrogenase into 6-phosphogluconate by the action of a specific *lactonase*.

$$
\begin{array}{ccc}
\text{HCOH} & \overset{\text{glucose-6-}}{\underset{\text{phosphate}}{\text{HCOH}}} \; C=O & \text{COO}^- \\
\text{HCOH} & \text{HCOH} & \overset{lactonase}{} \; \text{HCOH} \\
\text{HOCH} \; \underset{\text{dehydrogenase}}{\overset{O}{\rightleftharpoons}} & \text{HOCH} \; O \rightleftharpoons & \text{HOCH} \\
\text{HCOH} & \text{HCOH} & \text{HCOH} \\
\text{HC} & \text{HC} & \text{HCOH} \\
\text{H}_2\text{COPO}_3{}^{2-} & \text{H}_2\text{COPO}_3{}^{2-} & \text{H}_2\text{COPO}_3{}^{2-} \\
+ & + & \\
\text{NADP}^+ & \text{NADPH} + \text{H}^+ & \\
\textbf{Glucose-6-phosphate} & \textbf{6-Phosphoglucono-} & \textbf{6-Phosphogluconate} \\
& \textbf{δ-lactone} &
\end{array}
$$

The equilibrium position of the two reactions is far to the right.

The dehydrogenase (MW = 135,000) is a dimer of identical subunits and exhibits negative cooperativity with substrates, being inhibited by the product NADPH; its reaction is freely reversible. The lactonase reaction equilibrium is far to the right. Although the lactone hydrolyzes spontaneously to gluconate, the rate is enhanced by lactonase.

6-Phosphogluconate $\rightarrow$ D-Ribulose-5-phosphate + CO$_2$

The next step of the pathway is catalyzed by *6-phosphogluconate dehydrogenase*, which leads to formation of CO$_2$ and the first pentose in the pathway, ribulose-5-phosphate.

$$
\begin{array}{ccc}
\text{NADP}^+ & \text{NADPH} + \text{H}^+ & \\
+ & + & \\
\text{COO}^- & \text{COO}^- & \text{CO}_2 \\
& & + \\
\text{HCOH} \;\rightleftharpoons\; & \text{HCOH} \;\overset{\text{Mn}^{2+}}{\rightleftharpoons}\; & \text{CH}_2\text{OH} \\
\text{HOCH} & \text{C}=O & \text{C}=O \\
\text{HCOH} & \text{HCOH} & \text{HCOH} \\
\text{HCOH} & \text{HCOH} & \text{HCOH} \\
\text{H}_2\text{COPO}_3{}^{2-} & \text{H}_2\text{COPO}_3{}^{2-} & \text{H}_2\text{COPO}_3{}^{2-} \\
\textbf{6-Phosphogluconate} & \textbf{3-Keto-6-phospho-} & \textbf{D-Ribulose-} \\
& \textbf{gluconate} & \textbf{5-phosphate}
\end{array}
$$

3-Keto-6-phosphogluconate appears to be an intermediate in the reaction. The rat liver dehydrogenase (MW = 102,000) contains two identical subunits. The dehydrogenase is markedly inhibited at normal intracellular [NADPH] ($K_i^{\text{NADPH}} = 20 \; \mu M$), since its $K_m^{\text{NADP}^+} = 15 \; \mu M$. Its activity is expressed maximally only when NADPH is consumed in other metabolic reactions in the cytosol.

The ribulose-5-phosphate is further metabolized by a series of reactions involving interconversions of 3-, 4-, 5-, 6-, and 7-carbon monosaccharides that are catalyzed by the consecutive actions of several enzymes.

Ribulose-5-phosphate → Ribose-5-phosphate or Xylulose-5-phosphate

Two different enzymes act to metabolize ribulose-5-phosphate.

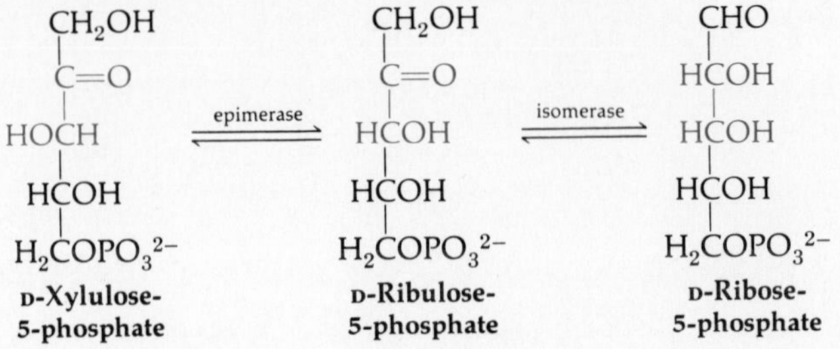

$$
\begin{array}{ccccc}
\text{CH}_2\text{OH} & & \text{CH}_2\text{OH} & & \text{CHO} \\
| & & | & & | \\
\text{C}=\text{O} & & \text{C}=\text{O} & & \text{HCOH} \\
| & \xrightleftharpoons{\text{epimerase}} & | & \xrightleftharpoons{\text{isomerase}} & | \\
\text{HOCH} & & \text{HCOH} & & \text{HCOH} \\
| & & | & & | \\
\text{HCOH} & & \text{HCOH} & & \text{HCOH} \\
| & & | & & | \\
\text{H}_2\text{COPO}_3{}^{2-} & & \text{H}_2\text{COPO}_3{}^{2-} & & \text{H}_2\text{COPO}_3{}^{2-} \\
\text{D-Xylulose-} & & \text{D-Ribulose-} & & \text{D-Ribose-} \\
\text{5-phosphate} & & \text{5-phosphate} & & \text{5-phosphate}
\end{array}
$$

Ribulose-5-phosphate isomerase acts to form ribose-5-phosphate, which is required for synthesis of nucleotides and nucleic acids (Chaps. 26 and 27). The further utilization of ribose-5-phosphate for these functions is a major drain of substrate from the pathway. *Ribulose-5-phosphate epimerase* forms xylulose-5-phosphate and serves with the isomerase to initiate degradation of ribose-5-phosphate arising from either nucleic acid degradation or glucose-6-phosphate, as described above.

Xylulose-5-phosphate + Ribose-5-phosphate → Sedoheptulose-7-phosphate + Glyceraldehyde-3-phosphate

The two products of the epimerase and isomerase reactions are the substrates for the next enzyme, *transketolase,* which catalyzes the following reaction:

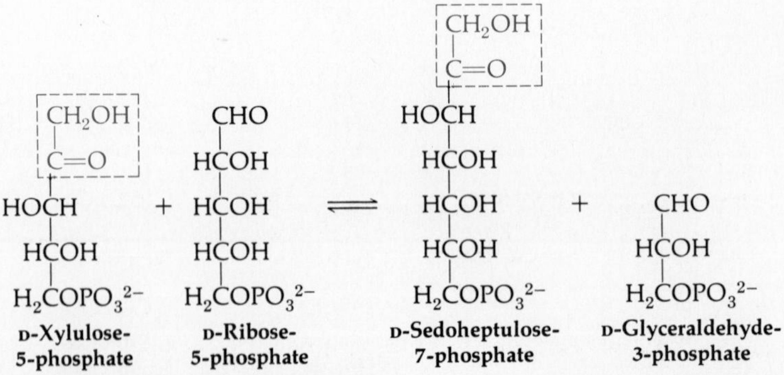

$$
\begin{array}{ccccc}
& & & \text{CH}_2\text{OH} & \\
& & & | & \\
& & & \text{C}=\text{O} & \\
\text{CH}_2\text{OH} & \text{CHO} & & | & \\
| & | & & \text{HOCH} & \\
\text{C}=\text{O} & \text{HCOH} & & | & \\
| & | & & \text{HCOH} & \\
\text{HOCH} & + & \text{HCOH} & \rightleftharpoons & \text{HCOH} & + & \text{CHO} \\
| & | & & | & & | \\
\text{HCOH} & \text{HCOH} & & \text{HCOH} & & \text{HCOH} \\
| & | & & | & & | \\
\text{H}_2\text{COPO}_3{}^{2-} & \text{H}_2\text{COPO}_3{}^{2-} & & \text{H}_2\text{COPO}_3{}^{2-} & & \text{H}_2\text{COPO}_3{}^{2-} \\
\text{D-Xylulose-} & \text{D-Ribose-} & & \text{D-Sedoheptulose-} & & \text{D-Glyceraldehyde-} \\
\text{5-phosphate} & \text{5-phosphate} & & \text{7-phosphate} & & \text{3-phosphate}
\end{array}
$$

The enzyme also catalyzes reactions involving other ketose and aldose substrates, one of which is described below.

Transketolase (MW = 70,000), which is composed of two identical subunits, utilizes thiamine pyrophosphate as coenzyme and requires Mg^{2+}. Its reaction mechanism is similar to that of pyruvate decarboxylase (Chap. 15). In effect, it transfers a "glycoaldehyde" moiety from xylulose-5-phosphate to ribose-5-phosphate to give the 7-carbon sedoheptulose-7-phosphate and glyceraldehyde-3-phosphate.

Sedoheptulose-7-phosphate + Glyceraldehyde-3-phosphate →
Fructose-6-phosphate + Erythrose-4-phosphate

The products of the transketolase reaction are the substrates for *transaldolase,*
which catalyzes the reaction

CH₂OH			CH₂OH	

D-Sedoheptulose-7-phosphate + D-Glyceraldehyde-3-phosphate ⇌ D-Fructose-6-phosphate + D-Erythrose-4-phosphate

This reaction, in effect, returns carbon atoms of pentose to the glycolytic pathway via fructose-6-phosphate.

Xylulose-5-phosphate + Erythrose-4-phosphate →
Fructose-6-phosphate + Glyceraldehyde-3-phosphate

Erythrose-4-phosphate produced from the action of transaldolase is a substrate for another interconversion catalyzed by transketolase.

D-Xylulose-5-phosphate + D-Erythrose-4-phosphate ⇌ D-Fructose-6-phosphate + D-Glyceraldehyde-3-phosphate

Cyclic Nature of the Phosphogluconate Oxidative Pathway

The discussion above has presented all the enzymes and intermediates
unique to this pathway. For its operation, however, four of the enzymes concerned in glycolysis are also required. These are *triose phosphate isomerase* to
catalyze interconversion of glyceraldehyde-3-phosphate and dihydroxyacetone
phosphate; *aldolase* to catalyze formation of fructose-1,6-bisphosphate from the
two triose phosphates; *fructose-1,6-bisphosphatase* to hydrolyze fructose-1,6-bis-
phosphate to fructose-6-phosphate; and *hexose phosphate isomerase* to convert fruc-
tose-6-phosphate to glucose-6-phosphate. These will be recognized as the en-

zymes involved in the normal formation of hexose from triose by reversal of the glycolytic sequence.

With the foregoing reactions in mind, it is possible to reconstruct a system into which hexose continually enters and from which CO_2 emerges as the sole carbon compound. This can be visualized by consideration of the set of balanced equations shown in Table 17.3, which, in sum, describe the conversion of 6 mol of hexose phosphate to 5 mol of hexose phosphate and 6 mol of CO_2. Note that the only reaction in which CO_2 is evolved is the oxidation of 6-phosphogluconic acid.

At first glance this reaction sequence appears complex, but it can more readily be understood from the following considerations. The *oxidizing* mechanism is accomplished entirely by steps 1 and 2, in which $NADP^+$ is reduced and CO_2 is evolved. Oxidation of 6 mol of hexose by steps 1 and 2 results in delivery of 12 pairs of electrons to $NADP^+$, the requisite amount for total oxidation of 1 mol of glucose to 6 mol of CO_2. As a consequence of step 2, 6 mol of ribulose-5-phosphate remain. If this is employed for nucleotide formation, etc., it need only undergo isomerization to ribose-5-phosphate. However, if the carbon of the 6 mol of pentose were rearranged to form 5 mol of hexose, the net

TABLE 17.3 Reactions of the Phosphogluconate Oxidative Pathway

Step	Enzyme	Reaction	Carbon balance
1	Glucose-6-phosphate dehydrogenase	6 Glucose-6-phosphate + $6NADP^+ \rightarrow$ 6 6-phosphogluconate + 6NADPH + $6H^+$	$6(6) \rightarrow 6(6)$
2	Phosphogluconate dehydrogenase	6 6-Phosphogluconate + $6NADP^+ \rightarrow$ 6 ribulose-5-phosphate + 6NADPH + $6H^+$ + $6CO_2$	$6(6) \rightarrow 6(5) + 6(1)$
3	Pentose epimerase	2 Ribulose-5-phosphate $\rightarrow$ 2 xylulose-5-phosphate	$2(5) \rightarrow 2(5)$
	Pentose isomerase	2 Ribulose-5-phosphate $\rightarrow$ 2 ribose-5-phosphate	$2(5) \rightarrow 2(5)$
	Transketolase	2 Xylulose-5-phosphate + 2 ribose-5-phosphate $\rightarrow$ 2 sedoheptulose-7-phosphate + 2 glyceraldehyde-3-phosphate	$2(5) + 2(5) \rightarrow 2(7) + 2(3)$
4	Transaldolase	2 Sedoheptulose-7-phosphate + 2 glyceraldehyde-3-phosphate $\rightarrow$ 2 erythrose-4-phosphate + 2 fructose-6-phosphate	$2(7) + 2(3) \rightarrow 2(4) + 2(6)$
5	Pentose epimerase	2 Ribulose-5-phosphate $\rightarrow$ 2 xylulose-5-phosphate	$2(5) \rightarrow 2(5)$
	Transketolase	2 Xylulose-5-phosphate + 2 erythrose-4-phosphate $\rightarrow$ 2 glyceraldehyde-3-phosphate + 2 fructose-6-phosphate	$2(5) + 2(4) \rightarrow 2(3) + 2(6)$
6	Triose isomerase	Glyceraldehyde-3-phosphate $\rightarrow$ dihydroxyacetone phosphate	$(3) \rightarrow (3)$
7	Aldolase	Dihydroxyacetone phosphate + glyceraldehyde-3-phosphate $\rightarrow$ fructose-1,6-bisphosphate	$(3) + (3) \rightarrow (6)$
8	Phosphatase	Fructose-1,6-bisphosphate $\rightarrow$ fructose-6-phosphate + P_i	$(6) \rightarrow (6)$
9	Hexose isomerase	5 Fructose-6-phosphate $\rightarrow$ 5 glucose-6-phosphate	$5(6) \rightarrow 5(6)$
NET:		6 Glucose-6-phosphate + $12NADP^+ \rightarrow$ 5 glucose-6-phosphate + $6CO_2$ + 12NADPH + $12H^+$ + P_i	$6(6) \rightarrow 5(6) + 6(1)$

achievement would be the complete oxidation of one of the original 6 mol of hexose. This transformation of the carbon of the pentoses to hexose is initiated in steps 3 and 4, with a balance of 2 mol of tetrose and 2 mol of pentose. Step 5 employs the latter to yield an additional 2 mol of hexose and 2 of triose. Steps 5 to 8 convert the two trioses to one hexose. In each series the hexose formed is fructose-6-phosphate. Finally, isomerization of the latter yields 5 mol of glucose-6-phosphate, which can now reenter the sequence at step 1, etc.

The relative amounts of glucose catabolized via glycolysis and the phosphogluconate pathway can be estimated by measuring the rate of CO_2 production from specifically labeled glucose, since the CO_2 arising from the combined actions of glycolysis and the citric acid cycle is derived from C-1 and C-6 of glucose, whereas that from the phosphogluconate pathway arises only from C-1. By this means, skeletal muscle has been found to catabolize glucose solely by glycolysis, whereas liver may consume about 30 percent of its glucose via the phosphogluconate pathway. An even larger fraction is catabolized by the phosphogluconate pathway in mammary gland, testis, adipose tissue, leukocytes, and adrenal cortex. The advantages of the pathway to an organism other than ribose synthesis include: (1) no ATP is required, (2) there is no dependence on 4-carbon dicarboxylic acids from the citric acid cycle, and (3) $NADP^+$ is the exclusive electron acceptor. The last is most significant, since most of the NADPH produced is the source of reducing power for cytosolic synthesis of fatty acids and steroids (Chap. 20). This is in accord with the fact that the phosphogluconate pathway operates actively in tissues that synthesize these substances.

Formation of ribose phosphate is also an important function of the phosphogluconate pathway. Its synthesis can also occur anaerobically, however, by the consecutive actions of certain enzymes of the phosphogluconate pathway and glycolysis as shown in the following reactions.

Fructose-6-phosphate + glyceraldehyde-3-phosphate $\xrightarrow[\text{ketolase}]{\text{trans-}}$

erythrose-4-phosphate + xylulose-5-phosphate (1)

Fructose-6-phosphate + erythrose-4-phosphate $\xrightarrow[\text{aldolase}]{\text{trans-}}$

glyceraldehyde-3-phosphate + sedoheptulose-7-phosphate (2)

Sedoheptulose-7-phosphate + glyceraldehyde-3-phosphate $\xrightarrow[\text{ketolase}]{\text{trans-}}$

ribose-5-phosphate + xylulose-5-phosphate (3)

2 Xylulose-5-phosphate $\longrightarrow$ 2 ribulose-5-phosphate $\longrightarrow$ 2 ribose-5-phosphate (4)

SUM: 2 Fructose-6-phosphate + glyceraldehyde-3-phosphate $\longrightarrow$ 3 ribose-5-phosphate

The relative importance of these reactions as compared to the phosphogluconate oxidative pathway is uncertain.

[Handwritten annotations: transaldolase, $3 + 7 \longrightarrow 4 + 6$; $3 + 7 \longrightarrow 5 + 5$ transketolase; $3 + 6 \longrightarrow 4 + 5$; $6 + 4 \longrightarrow 3 + 7$; $7 + 3 \longrightarrow 5 + 5$; $3x$ Glu-6-Pi + 2NADP$^+$ + H$_2$O $\longrightarrow$ Ribose-5-Pi + CO$_2$ + 2NADPH + 2H$^+$]

SOME SPECIALIZED PATHWAYS OF CARBOHYDRATE METABOLISM IN PLANTS AND MICROORGANISMS

The Methylglyoxal Shunt

Methylglyoxal is produced in microorganisms by *methylglyoxal synthase* as follows:

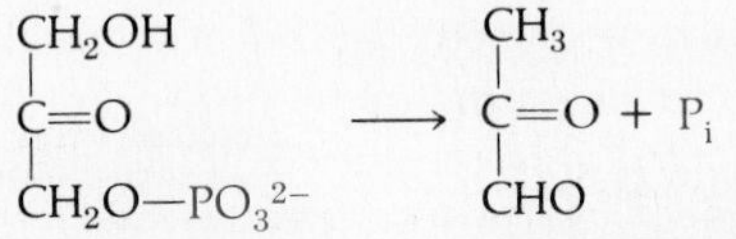

$$
\begin{array}{ccc}
CH_2OH & & CH_3 \\
| & & | \\
C{=}O & \longrightarrow & C{=}O + P_i \\
| & & | \\
CH_2O{-}PO_3{}^{2-} & & CHO \\
\text{Dihydroxyacetone} & & \text{Methylglyoxal} \\
\text{phosphate} & &
\end{array}
$$

and may be converted to D-lactate by *glyoxylase* (Chap. 16). These enzymes could provide a shunt around the low ATP-yielding steps in glycolysis by the following reactions.

$$
\begin{array}{lr}
\text{Glucose} + 2ATP \longrightarrow \text{fructose-1,6-bisphosphate} + 2ADP & (1) \\
\text{Fructose-1,6-bisphosphate} \longrightarrow 2 \text{ dihydroxyacetone phosphate} & (2) \\
2 \text{ Dihydroxyacetone phosphate} \longrightarrow 2 \text{ methylglyoxal} + 2P_i & (3) \\
2 \text{ Methylglyoxal} \longrightarrow 2 \text{ lactate} & (4)
\end{array}
$$

Whereas no ATP is produced by the process, lactate becomes available for further utilization, e.g., in fatty acid synthesis, amino acid synthesis, and utilization in the citric acid cycle.

Glyoxylase is found in many animal tissues, and methylglyoxal is formed from dihydroxyacetone phosphate in liver. The methylglyoxal is converted to D-lactate whereas L-lactate is produced in glycolysis. The significance of these reactions in animals is unclear.

The Glyoxylic Acid Pathway

Plants contain *isocitrate lyase* in subcellular organelles called *glyoxysomes*, and isocitrate lyase catalyzes the reaction

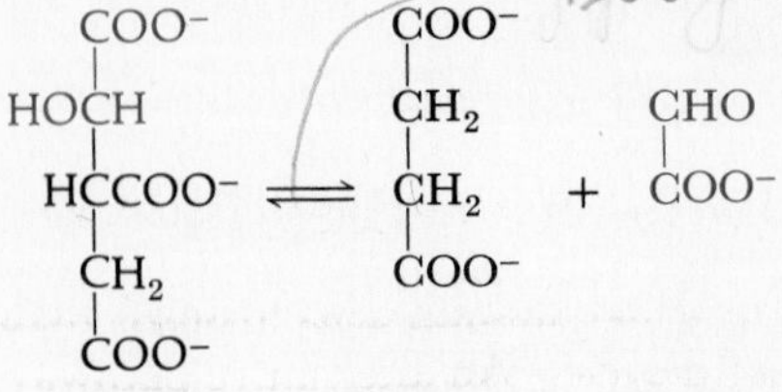

$$
\begin{array}{cccc}
COO^- & COO^- & & \\
| & | & & \\
HOCH & CH_2 & & CHO \\
| & | & & | \\
HCCOO^- \rightleftharpoons & CH_2 & + & COO^- \\
| & | & & \\
CH_2 & COO^- & & \\
| & & & \\
COO^- & & & \\
\textbf{Isocitrate} & \textbf{Succinate} & & \textbf{Glyoxylate}
\end{array}
$$

Glyoxylate so formed can initiate the glyoxylate pathway, shown in Fig. 17.14, by the action of another plant enzyme, *malate synthase,* which converts acetyl CoA and glyoxylate to malate.

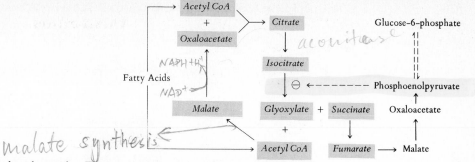

Figure 17.14 The glyoxylate cycle. The colored compounds are participants in the glyoxylate cycle. Oxidation of fatty acids can provide acetyl CoA, which enters the cycle at two points. Each turn of the cycle results in generation of one molecule of succinate. The latter can be oxidized to oxaloacetate by reactions of the citric acid cycle (Chap. 15); the oxaloacetate is decarboxylated and phosphorylated to phosphoenolpyruvate, and the latter is converted to glucose-6-phosphate by reversal of the glycolytic sequence (Fig. 17.3). Thus, four molecules of acetyl CoA are required to make two molecules of succinate, which yield one molecule of hexose plus two molecules of CO_2. — indicates the inhibitory influence of phosphoenolpyruvate on functioning of isocitrate lyase.

$$
\begin{array}{c}
CH_3 \\
| \\
C{=}O \\
| \\
S{-}CoA
\end{array}
\;+\;
\begin{array}{c}
CHO \\
| \\
COO^-
\end{array}
\;\rightleftharpoons\;
\begin{array}{c}
COO^- \\
| \\
CH_2 \\
| \\
HCOH \\
| \\
COO^-
\end{array}
\;+\; CoA{-}SH
$$

Acetyl   **Glyoxylate**   **Malate**
CoA

The net result of this cycle is the production of succinate from acetyl CoA in the overall process

$$2\text{ AcetylCoA} + NAD^+ \longrightarrow \text{succinate} + 2\text{CoA} + NADH + H^+$$

The succinate may be anaplerotic, giving rise to oxaloacetate, or gluconeogenic, giving rise to PEP. Thus, net carbohydrate synthesis from fatty acids via acetyl CoA, which is impossible in animals, is achieved. The process is particularly important in "oil seeds," which are rich in triacylglycerols. These serve not only as a source of energy, but as a source of carbon for synthesis of cell constituents in the developing plant embryo before photosynthesis becomes operative.

Acetyl Phosphate Metabolism in Microorganisms

Most microorganisms, like animal cells, oxidize pyruvate to acetyl CoA, which enters the citric acid cycle. An alternative pathway is the *phosphoroclastic* cleavage in *E. coli*.

$$CH_3COCOO^- + P_i \longrightarrow CH_3COOPO_3^{2-} + HCOO^-$$

Pyruvate   **Acetyl**   **Formate**
                            phosphate

Acetyl phosphate may be converted to acetyl CoA by *phosphotransacetylase*.

$$\text{Acetyl phosphate } + \text{CoA} \rightleftharpoons \text{acetylCoA} + P_i$$

In some microorganisms, the phosphate of acetyl phosphate can be used, in place of ATP, for the phosphorylation of various hexoses.

Propionic Acid Biosynthesis in Microorganisms

In propionibacteria the initial event in propionic acid synthesis is an unusual transfer of CO_2 from methylmalonyl CoA to pyruvate, catalyzed by a biotin-containing enzyme, *methylmalonyl-oxaloacetate transcarboxylase*.

| **Pyruvate** | **Methylmalonyl CoA** | **Oxaloacetate** | **Propionyl CoA** |

The oxaloacetate formed is reduced to succinate by reversal of the usual citric acid cycle reactions, via malate and fumarate. A *transthioesterase* catalyzes transfer of the CoA moiety from propionyl CoA to succinate, forming free propionate and succinyl CoA.

The final step in this process is the regeneration of methylmalonyl CoA from succinyl CoA. This is catalyzed by *methylmalonyl CoA mutase,* which utilizes, as coenzyme, *dimethylbenzimidazole cobamide* (Chap. M22), a derivative of vitamin B_{12}. This reaction sequence can be summarized as follows:

$$\text{Pyruvate} + \text{methylmalonylCoA} \rightleftharpoons \text{oxaloacetate} + \text{propionylCoA} \tag{1}$$

$$\text{Oxaloacetate} + 4(\text{H}) \rightleftharpoons \text{succinate} + H_2O \tag{2}$$

$$\text{Succinate} + \text{propionylCoA} \rightleftharpoons \text{succinylCoA} + \text{propionate} \tag{3}$$

$$\text{SuccinylCoA} \rightleftharpoons \text{methylmalonylCoA} \tag{4}$$

SUM: $\text{Pyruvate} + 4(\text{H}) \longrightarrow \text{propionate} + H_2O$

The metabolism of propionate by animal tissues, essentially by a reversal of this pathway, is discussed in Chap. 20.

Fates of Acetaldehyde

It was indicated earlier that acetaldehyde, formed by decarboxylation of pyruvate, is reduced to ethanol in the course of anaerobic fermentation. In certain aerobic fermentations, as in vinegar production, acetic acid is the terminal product. This may arise by hydrolysis of acetyl CoA or acetyl phosphate or by oxidation of acetaldehyde. Various enzymes are known that catalyze the latter reaction, including glyceraldehyde phosphate dehydrogenase. In mammals, this oxidation is catalyzed by *hepatic aldehyde oxidase*, by *xanthine oxidase*, and by an NAD-linked *aldehyde dehydrogenase*.

Books

Hanson, R. W., and M. A. Mehlman (eds.): *Gluconeogenesis,* Wiley-Interscience, New York, 1976.

Stanbury, J. B., J. B. Wyngaarden, and D. S. Fredrickson (eds.): *The Metabolic Basis of Inherited Disease,* 4th ed., McGraw-Hill, New York, 1978.

Review Articles

Clark, M. G., and H. A. Lardy: Regulation of Intermediary Carbohydrate Metabolism, *MTP Int. Rev. Sci.* (1) **5**:223–266 (1975).

Colowick, S. P.: The Hexokinases, in P. D. Boyer (ed.): *The Enzymes,* 3d ed., vol. IX, part B, Academic, New York, 1976, pp. 1–48.

Engström, L.: The Regulation of Liver Pyruvate Kinase by Phosphorylation-Dephosphorylation, *Curr. Top. Cell. Regul.* **13**:29–52 (1978).

Everse, J., and N. O. Kaplan; Lactate Dehydrogenases: Structure and Function, *Adv. Enzymol.* **37**:61–134 (1973).

Frenkel, R.: Regulation and Physiological Functions of Malic Enzymes, *Curr. Top. Cell. Regul.* **9**:157–181 (1975).

Goldhammer, A. R., and H. H. Paradies: Phosphofructokinase: Structure and Function, *Curr. Top. Cell. Regul.* **15**:109–141 (1979).

Harris, J. I., and M. Waters: *Glyceraldehyde 3-Phosphate Dehydrogenase,* in P. D. Boyer (ed.): *The Enzymes,* 3d ed., vol., XIII, Academic, New York, 1976, pp. 1–49.

Hue, L.: The Role of Futile Cycles in the Regulation of Carbohydrate Metabolism in the Liver, *Adv. Enzymol.* **52**:247–330 (1981).

Ottaway, J. H., and J. Mowbray: The Role of Compartmentation in the Control of Glycolysis, *Curr. Top. Cell. Regul.* **12**:107–208 (1977).

Roseman, S.: The Transport of Carbohydrates by a Bacterial Phosphotransferase System, *J. Gen. Physiol.* **54**:138s–184s (1968).

Utter, M. F., R. E. Barden, and B. L. Taylor: Pyruvate Carboxylase: An Evaluation of the Relationships Between Structure and Mechanisms and Between Structure and Catalytic Activity, *Adv. Enzymol.* **42**:1–72 (1975).

Weinhouse, S.: Regulation of Glucokinase in Liver, *Curr. Top. Cell. Regul.* **11**:1–50 (1976).

Wilson, J. E.: Brain Hexokinase, the Prototype Ambiquitous Enzyme, *Curr. Top. Cell. Regul.* **16**:1–54 (1980).

Wood, H. G., and R. E. Barden: Biotin Enzymes, *Annu. Rev. Biochem.* **46**:385–413 (1977).

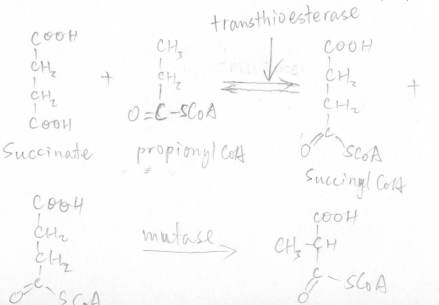

Chapter 18

Carbohydrate Metabolism II

Glycogen metabolism and its control.
Hexose interconversions. Blood glucose
and regulation of glucose metabolism.
Biosynthesis of oligosaccharides.
Biological roles of oligosaccharides.
Microbial cell walls.

This chapter considers the metabolism and functional roles of polysaccharides, ubiquitous components of all organisms, in which they serve as reserve stores of energy, as structural materials, or as participants in a variety of cellular processes. After considering polysaccharides as energy reserves and the metabolism of glycogen in detail, the metabolic interconversions of glucose and other monosaccharides will be described, followed by a summary of oligosaccharide biosynthesis and function in glycoconjugates (Chap. 8).

POLYSACCHARIDES AS ENERGY RESERVES: GLYCOGEN

Microorganisms have few energy reserves unless their growth is limited, whereupon synthesis of storage polysaccharides begins. Plants form sucrose in their leaves for transport to nonphotosynthetic tissues and store starch in many tissues for use in the dark (Chap. 19). Animals receiving their nutrition in pulses (meals), accumulate the polysaccharide glycogen, primarily in liver and muscle. If liver glycogen declines, hepatic and renal gluconeogenesis ensures a continuing supply of glucose to nervous tissue, where it is essential at all times, and to skeletal muscle, where it can be stored as glycogen while at rest and used to initiate anaerobic glycolysis during contraction. Storage of monosaccharides would disturb the osmotic balance of cells, but storage of polysaccharides, which have a small osmotic effect, entails minimal disturbance.

All storage polysaccharides are synthesized or degraded by consecutive addition or removal, respectively, of monosaccharide residues to or from the nonreducing ends of the molecule. Over a given time at fixed enzyme concentrations, more monosaccharides can be added to or released from a population

Any cross-references coded M refer to the companion text, *Principles of Biochemistry: Mammalian Biochemistry*.

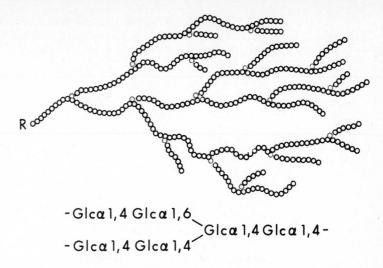

Figure 18.1 Schematic representation of the branched structure of animal glycogen. R indicates the reducing end of the molecule. Each circle represents a glucose residue in α1,4-glycosidic linkage with other glucose residues except at the branch points which are formed by α1,6 glycosidically linked residues, as shown in the structure at the bottom. The highly branched structure above has a molecular weight of about 50,000. Since glycogens have molecular weights in the millions, the highly branched network in the molecule can be readily imagined.

$$-\text{Glc}\alpha 1,4\ \text{Glc}\alpha 1,6$$
$$\text{Glc}\alpha 1,4\ \text{Glc}\alpha 1,4-$$
$$-\text{Glc}\alpha 1,4\ \text{Glc}\alpha 1,4$$

of branched molecules than from unbranched molecules. Therefore, the highly branched, multitiered structure of animal glycogen (Fig. 18.1), which contains an average of 1 α1,6-linked branch for every 12 residues in linear α1,4 linkage, is more advantageous than a linear polysaccharide. Indeed, 8 to 10 percent of all glucose in a glycogen molecule is always available at the nonreducing termini. This is especially useful in muscle, which on contraction, may require a 10^3- to 10^4-fold increased supply of ATP per minute.

Among higher animals, lipids account for more than 90 percent of the energy stores, whereas polysaccharide occurs only in the form of relatively small amounts of glycogen. Glycogen occurs in most tissues, including adipose tissue, heart muscle, kidney, and brain, but skeletal muscle contains about two-thirds of total body glycogen and liver most of the remainder.

Skeletal muscle glycogen occurs as dense spherical granules in close proximity to the sarcoplasmic reticulum at the level of thin filaments (Chap. M8). These granules are tightly bound to many of the enzymes of glycogen metabolism as well as to myosin, actin, and calsequestrin of muscle (Chap. M8), as shown in Fig. 18.2. Granular glycogen is composed of molecules with molecu-

Figure 18.2 Gel electrophoretic analysis in sodium dodecyl sulfate (see Fig. 4.5) of the polypeptide chains of the muscle proteins and the enzymes of glycogen metabolism that are isolated together tightly bound to glycogen granules. Molecular weight markers: myosin heavy chains (MW = 200,000) and protein phosphatase (MW = 35,000). [*From B. Caudwell, J. F. Antonin, and P. Cohen, Eur. J. Biochem.* **86**:511–518 (1978).]

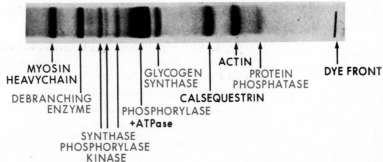

lar weights of many millions, and only a portion can be easily solubilized, perhaps by cleavage of glycosidic bonds; the soluble fraction is polydisperse with MW = 0.2 to 10×10^6.

The metabolism of only one storage polysaccharide, glycogen, will be considered in this chapter, since it plays a central role in the carbohydrate metabolism of higher animals. Special attention is given to skeletal muscle, but it is likely that glycogen metabolism in other tissues is very similar; notable differences, e.g., in liver, will be considered separately.

GLYCOGEN METABOLISM

General Aspects

Several aspects of glycogen metabolism, glycolysis, and gluconeogenesis, shown in Fig. 18.3, reveal important relationships among these pathways. Two enzymes, *glycogen phosphorylase* and *glycogen synthase,* act directly in the degradation and synthesis, respectively, of glycogen, although their activities are regulated by a series of other enzymes and allosteric effectors in a multicyclic cascade (Chap. 10). With the exception of the hydrolysis of glucose-6-phosphate to glucose and P_i, the scheme in Fig. 18.3 applies to all animal cells. Glucose-6-phosphate hydrolysis is unique to liver, kidney, and intestinal mucosa, the only cells that release glucose into the blood. Although the intermediates and enzymes of glycogen metabolism are essentially the same in liver and muscle, glycogen metabolism differs somewhat in the two tissues. Resting muscle accumulates no more than the equivalent of about a 1 percent glycogen solution, even when glucose is plentiful, and must degrade glycogen with extraordinary rapidity when the [ATP] declines during contraction. Liver accumulates the equivalent of a 5 to 8 percent solution of glycogen at normal or elevated blood [glucose] and degrades glycogen for release into the circulation when the [glucose] is below normal. Replenishment of blood glucose by liver can occur rapidly but never as fast as the almost explosive release of glucose-1-phosphate in contracting muscle.

Anaerobic glycolysis generates two molecules of ATP, net, per glucose molecule converted to lactate (Fig. 18.3). Few animal tissues, however, normally generate significant amounts of ATP by anaerobic glycolysis, since the O_2 supply and the steady-state ATP requirements of most mammalian tissues are virtually constant. Only a few tissues, e.g., the retina, have a significant rate of aerobic glycolysis and discharge lactate into the circulation, even at normal blood [glucose] and [O_2]. The prominent exception is skeletal muscle, in which the ATP requirement on repeated contraction far exceeds that possible by utilizing the available supplies of O_2 and oxidizable metabolites; under these conditions, anaerobic glycolysis is of the greatest significance as an ATP source. From Fig. 18.3, it is seen that the glycosidic bond energy of glycogen is preserved during anaerobic glycolysis; only P_i is required to permit conversion of glycogen to glucose-6-phosphate. Thus, the anaerobic yield is three ATP molecules per glucose equivalent from glycogen compared with two ATP molecules when initiated from glucose. The incorporation of one glucose residue into glycogen

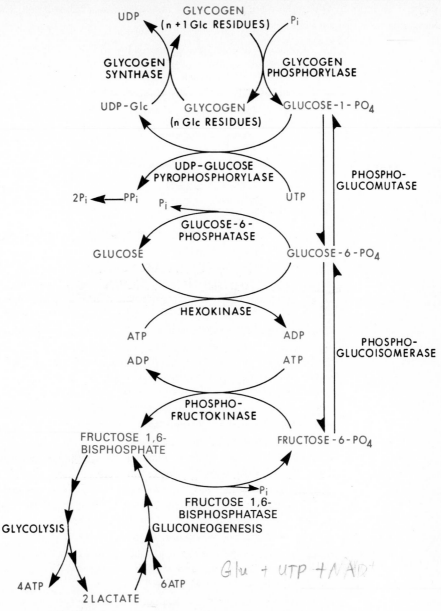

Figure 18.3 Metabolic interrelationships among glycogen metabolism, glycolysis, and gluconeogenesis. Reversible, parallel arrows denote reactions catalyzed by the same enzyme. Curved arrows indicate distinct enzymes for opposing reactions.

commencing from free glucose requires utilization of one molecule each of UTP and ATP. Thus, the net yield from the process glucose → glycogen → lactate is one ATP per glucose equivalent. However, the ATP expenditure for glycogen synthesis occurs in *resting muscle* when O_2 is abundant, permitting storage of this quick energy reserve, whereas the yield of three ATP per glucose equivalent occurs in *contracting muscle* during anaerobic glycolysis.

Glycogen Phosphorolysis: Phosphorylase, Debranching Enzyme, and Phosphoglucomutase

Glycogen degradation proceeds by phosphorolysis and is catalyzed by *glycogen phosphorylase*, leading to glucose-1-phosphate formation as follows.

(Glucose)$_n$

α-D-Glucose 1-phosphate

(Glucose)$_{n-1}$

The equilibrium constant ($K_{eq} = 0.28$ at pH 6.8) for the reaction slightly favors formation of glycogen under physiological conditions, but the [P$_i$]/[Glc-1-P] greatly exceeds the K_{eq} in vivo. Thus, phosphorylase, under physiological conditions, provides a regulated supply of glucose-1-phosphate from glycogen and does not act in glycogen synthesis.

Phosphorylase removes glucose sequentially as glucose-1-phosphate from each branch of the glycogen molecule until four glucose residues remain. Further degradation of the branch requires the action of lysosomal *oligo(α1,4 → α1,4)glucantransferase*, which, as shown in Fig. 18.4, transfers a triglucoside attached to the α1,6-linked glucose to a different chain, extending it by three residues and leaving a single α1,6-linked glucose at the branch point. The

Figure 18.4 Action of phosphorylase and debranching enzyme near a branch point of the glycogen molecule, represented by the circles as in Fig. 18.1. Phosphorylase removes glucose sequentially as glucose-1-phosphate from the nonreducing end of a chain until 4 residues remain in the branch. The glucantransferases and glucosidase activities of debranching enzyme then act in the second and third reactions, respectively.

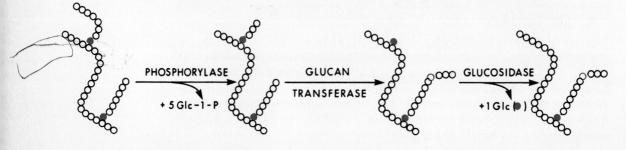

glucantransferase, which appears to be a multifunctional enzyme (Chap. 10) with two distinct active sites on a single polypeptide chain, then acts as an *α1,6 glucosidase* to remove the α1,6-linked glucose at the branch point. This permits the resulting linear α1,4-linked glucose of glycogen to be further degraded by phosphorylase. Thus, the transferase, often called *debranching enzyme,* removes branches from glycogen in two distinct steps. The action of phosphorylase yields glucose-1-phosphate, the major product of glycogen degradation, but about 8 percent of the glucose residues in glycogen are released as free glucose by the debranching enzyme. The debranching enzyme is present in smaller amounts than other enzymes in glycogenolysis and may be rate-limiting.

Glucose-1-phosphate enters the glycolytic pathway by the action of *phosphoglucomutase,* which catalyzes the reaction

$$\text{Glucose-1-phosphate} \xrightleftharpoons{\text{Mg}^{2+}} \text{glucose-6-phosphate}$$

The ratio of the [Glc-6-P] to [Glc-1-P] at equilibrium is 94:6. The reaction proceeds only in the presence of glucose-1,6-bisphosphate, which reacts with the mutase as follows.

$$\text{E}\text{—OH} + \text{Glc-1,6-bisphosphate} \rightleftharpoons \text{E}\text{—O}\text{—PO}_3^{2-} + \text{Glc-1- or 6-phosphate}$$

A single serine hydroxyl group in the enzyme is phosphorylated by the bisphosphate, yielding a mixture of glucose-1- and 6-phosphates. Increasing the [glucose-1-phosphate] in the presence of phosphoenzyme perturbs the equilibrium and gives rise to glucose-6-phosphate and vice versa. Glucose-1,6-bisphosphate is formed by *phosphoglucokinase,* which catalyzes the reaction

$$\text{Glucose-1-phosphate} + \text{ATP} \rightleftharpoons \text{glucose-1,6-bisphosphate} + \text{ADP}$$

Regulation of Glycogenolysis in Skeletal Muscle

The rate of glycogen phosphorolysis is regulated by several different means, but each ultimately results in alteration of the enzymic activity of phosphorylase. In general, phosphorylase is activated when glycogenolysis is required to generate ATP (Fig. 18.2); this occurs within seconds of the onset of muscle contraction. When the demand for ATP stops, phosphorylase activity falls to near zero. The principal feature of the regulation of phosphorylase activity in all tissues is its interconversion into two chemically and functionally distinct forms, designated a and b, in response to hormonal stimuli and, in the case of muscle, to nervous stimulation. Moreover, both forms of phosphorylase are allosteric enzymes, whose activities are controlled by positive and negative effectors that are metabolites of glycogenolysis or glycolysis, e.g., ATP, glucose-6-phosphate, and AMP. Finally, phosphorylases a and b are interconvertible enzymes of a multicyclic cascade (Chap. 10), as shown in Fig. 18.5, that is interlocked with other interconvertible enzymes for glycogen synthesis. To understand the operation of this apparently complex regulatory cascade requires consideration of each of its interconvertible enzymes, the converting enzymes, the allosteric effectors, and the kinds of factors that initiate the cascade. Reference to Chap. 10 for a discussion of the general properties of enzyme cascades may be useful to the reader.

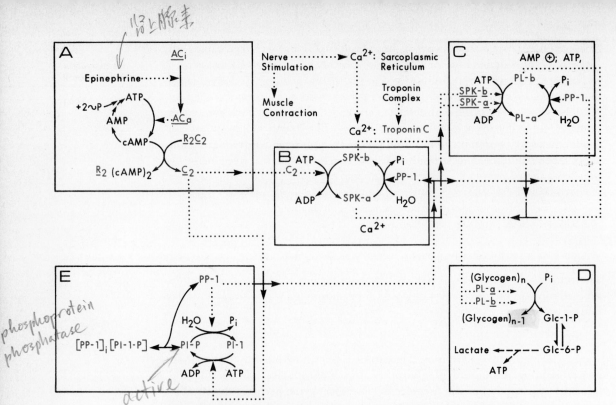

Figure 18.5 The multicyclic cascade that regulates muscle glycogen phosphorolysis. (*a*) Initiation of the cascade by epinephrine converts inactive adenylate cyclase (AC_i) to active enzyme (AC_a), which forms cAMP from ATP. cAMP then activates cAMP-dependent protein kinase, R_2C_2, to give catalytically active C subunits and inactive $R_2(cAMP)_2$ complexes. C subunits act in other cycles of the cascade. (*b*) The synthase-phosphorylase kinase (SPK) cycle. Active C subunits convert low-activity SPK-b to SPK-a, the high-activity form, that acts as a converter enzyme in (*c*). SPK-a also acts in (*c*) on nerve stimulation, which activates SPK-b by Ca^{2+} and troponin C release from the sarcoplasmic reticulum and contractile units. (*c*) The phosphorylase cycle. SPK-a or -b convert the low-activity, AMP-dependent phosphorylase (PL-b) to the high-activity a form (PL-a). (*d*) The events of glycogen phosphorolysis catalyzed by PL-a or PL-b. (*e*) The phosphoprotein phosphatase cycle. Catalytic C subunits from (*a*) convert inactive phosphatase inhibitor (PI) to active phosphatase inhibitor (PI-P), which combines with phosphoprotein phosphatase (PP-1) to inhibit its catalytic action on the cycle in (*b*), where it can convert SPK-a to SPK-b and in the cycle in (*c*), where it converts PL-a to PL-b. In resting muscle, PP-1 is released from inhibition by PI-P and dephosphorylates SPK-a as well as PL-a.

Phosphorylase a and b Phosphorylase is an interconvertible enzyme in one cycle of the glycogenolytic cascade shown in Fig. 18.5. Phosphorylase (MW = 194,800) contains two identical subunits (841 residues) and exists in two forms, a and b, which are interconverted by the converter enzymes *synthase-phosphorylase kinase* (SPK) and *phosphoprotein phosphatase* (PP-1).

$$\text{Phosphorylase b} \underset{\underset{P_i \qquad H_2O}{\xleftarrow{\hspace{1cm}} PP\text{-}1 \xleftarrow{\hspace{1cm}}}}{\overset{\overset{ATP \qquad ADP}{\xrightarrow{\hspace{1cm}} SPK \xrightarrow{\hspace{1cm}}}}{}} \text{Phosphorylase a}$$

SPK catalyzes phosphorylation of a hydroxyl group of serine in each subunit of b to give phosphoserine in a; PP-1 specifically hydrolyzes the phosphoserine in a to give P_i and regenerate b. SPK and PP-1 are participants in yet other cycles in the multicyclic cascade (Fig. 18.5), as discussed below.

Phosphorylases a and b differ somewhat in conformation, largely as the result of the state of phosphorylation. The a and b forms are also allosteric enzymes and respond differently to positive and negative effectors. In addition, each form exists in an enzymically inactive T (tight) state and an active R (relaxed) state (Chap. 10). In general, however, the b form is more responsive to effectors than the a form; e.g., AMP is an obligatory allosteric effector for b but can stimulate the a form to a small extent.

Phosphorylases a and b contain several binding sites for substances that influence their activities, some of which are shown in the three-dimensional structure of a monomer in the a form (T state) in Fig. 18.6. The sites and the ligands they bind as well as the functional effects of binding are listed in Table 18.1. A striking feature of the molecule is its possession of a catalytic face and a control face. The catalytic face includes the active site with pyridoxal phosphate, an essential coenzyme for both the a and b forms, and a glycogen bind-

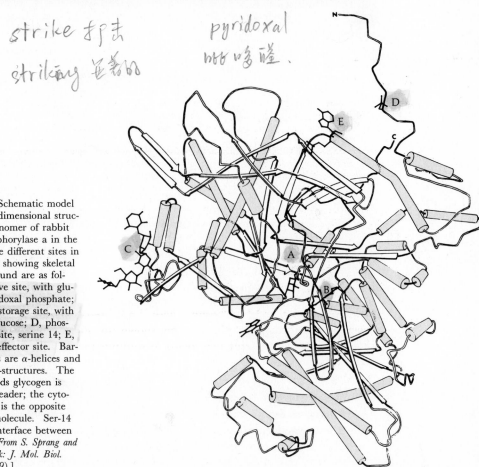

Figure 18.6 Schematic model of the three-dimensional structure of a monomer of rabbit muscle phosphorylase a in the T state. The different sites in the molecule showing skeletal structures bound are as follows: A, active site, with glucose; B, pyridoxal phosphate; C, glycogen storage site, with penta-α1,4-glucose; D, phosphorylation site, serine 14; E, AMP/ATP effector site. Barrel structures are α-helices and arrows are β-structures. The face that binds glycogen is nearest the reader; the cytoplasmic face is the opposite side of the molecule. Ser-14 is near the interface between monomers. [*From S. Sprang and R. J. Fletterick: J. Mol. Biol.* **131**:*523 (1979).*]

TABLE 18.1

Binding of Ligands with Phosphorylase a and b

Ligand	Type of site	Conformer stabilized	Function
Glycogen	Active	—	Binds substrate
Glucose-1-P	Active	R	Binds substrate
P_i	Active	R	Binds substrate
UDP-Glc	Active	R	Inhibits a and b
Pyridoxal phosphate	Active	—	Essential for catalysis
Glucose	Active	T	Inhibits a
AMP	Activator	R	Activates b obligatorily
ATP	Activator	T	Inhibits b
Glucose-6-P	Activator	T	Inhibits b
Phosphoryl	Serine hydroxyl	R	Stabilizes a
Glycogen	Storage	R	Activates b
Purines	Inhibitor	T	Inhibits a

ing (storage) site. The catalytic face is thought to bind to the glycogen granule, as opposed to the control face on the opposite side of the molecule that is oriented to the cytoplasm, where SPK and PP-1 can act and allosteric effectors can bind. The catalytic site is located between two domains of the monomer and contains covalently bound pyridoxal phosphate in aldimine linkage through its carbonyl group to the ε-amino group of Lys-679. The phosphate group of the coenzyme acts as a proton acceptor during catalysis, in accord with complete loss of activity on removal of pyridoxal phosphate. Ser-14 is the phosphorylation site and is located at the surface, close to the interface between dimers, where it is readily accessible to SPK and PP-1 and can influence the monomer-monomer interactions that help determine the different conformations of the a and b forms. The activator site for AMP is far removed from the active site on the control face. Binding of AMP to this site activates the b form by stabilization of its active R state, which has a lower K_m for substrates than the T state. ATP and glucose-6-phosphate also bind the activator site, but are negative effectors since they stabilize the inactive T state. Glucose, which is formed by action of debranching enzyme, inhibits both forms by binding at the active site and stabilizing the inactive T state. UDP-glucose, a substrate for glycogen synthesis, inhibits competitively, but purines such as caffeine inhibit by binding at the active site of the T state.

Resting muscle contains primarily phosphorylase b, whereas muscle after prolonged contraction contains only the a form. Muscle stimulated intermittently contains both forms. Resting muscle requires little ATP; thus little glycogenolysis occurs and the activity of the b form is very low. The [AMP] in resting muscle is sufficient to activate b allosterically, and glycogen and P_i are in good supply, but the [ATP] and [glucose-6-phosphate] counteract the activating effect of AMP. Moreover, UDP-glucose, a competitive inhibitor of both a and b, is also at elevated concentrations. On muscular contraction, however, the [ATP], [glucose-6-phosphate], and [UDP-glucose] fall, permitting AMP to activate b, and glycogenolysis is initiated. If contraction continues, the b form is converted to the a form as the result of the increased activity of SPK and the decreased activity of PP-1 (see below) and the a form continues catalysis largely uninfluenced by AMP.

436

Synthase-Phosphorylase Kinase a and b (SPK) These enzymes are interconvertible in cycle B of the cascade shown in Fig. 18.5. The rabbit muscle enzyme $(MW = 1.3 \times 10^6)$ contains four different subunits firmly bound noncovalently. Its subunit structure is $(\alpha\beta\gamma\delta)_4$; the α, β, γ, and δ chains have $MW = 145,000, 128,000, 45,000,$ and $16,790$, respectively. The δ chain is identical to calmodulin (Chap. 14), the Ca^{2+} binding protein associated with the Ca^{2+}-dependent regulation of several enzymes, and is similar to the muscle Ca^{2+}-binding protein troponin C (Chap. M8), to which it is about 50 percent homologous in amino acid sequence. The α, β, and δ subunits regulate the catalytic activity of SPK, which resides in the γ subunit.

Although SPK is the converter enzyme for phosphorylase b to a, it is itself an interconvertible enzyme, existing in two forms, a and b, through the actions of *cAMP-dependent protein kinase* (C_2) and phosphoprotein phosphatase (PP-1), the same converter enzyme for dephosphorylation of phosphorylase a.

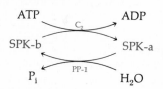

The cAMP-dependent kinase, designated R_2C_2, contains two regulatory (R) subunits and two catalytic subunits (C); cAMP binds the R subunits, releasing catalytically active C subunits, as follows:

$$cAMP + R_2C_2 \rightleftharpoons [cAMP]_2R_2 + C_2$$

The activity of the protein kinase is under hormonal control, as shown in Fig. 18.5a. Thus, in resting muscle, epinephrine (Chaps. M12 and M18) stimulates activation of *adenylate cyclase*, which leads to the production of cyclic AMP (Chap. G12). Cyclic AMP in turn activates cAMP-dependent protein kinase, which catalyzes phosphorylation of SPK-b to give SPK-a. Protein kinase catalyzes phosphorylation of a serine residue in each of the subunits of the α and β chains of SPK-b to give SPK-a. Phosphorylation of the β subunit is sufficient to convert SPK-b to SPK-a, but phosphorylation of α subunits also occurs and aids in regulating dephosphorylation, as discussed below.

The activities of SPK-a and SPK-b are highly dependent on Ca^{2+}, which binds to four sites in each of the four δ chains. SPK-a is about 15 times as active as SPK-b at saturating $[Ca^{2+}]$, reflecting the activating effects of the phosphorylated β chains in SPK-a. Moreover, activation of the a form occurs at much lower $[Ca^{2+}]$ than activation of the b form. However, the b form, when activated by binding Ca^{2+} to its δ chain, is further activated by either free calmodulin and Ca^{2+} or troponin C and Ca^{2+}. This type of activation is caused by binding of calmodulin-Ca^{2+} or troponin C-Ca^{2+} complexes to the α and β subunits of SPK-b; SPK-a is not activated by this means. Since the [troponin C] exceeds [calmodulin] in muscle, activation of SPK-b in vivo is thought to be mediated by troponin C, although SPK-b itself contains δ chains (calmodulin).

The foregoing mechanisms for activation of SPK by Ca^{2+} and calmodulin-Ca^{2+} or troponin C-Ca^{2+} complexes reveals the close coupling of glycogenolysis to muscle contraction. SPK-b is the major species of SPK in resting muscle and

ambient [Ca⁺]

myosin

actin

filament丝

its activity is very low at the low ambient [Ca^{2+}]. Its activity is immediately enhanced on nervous stimulation of muscle contraction, however, by release of Ca^{2+} from the sarcoplasmic reticulum (Chap. M8). Moreover, its activity is even further enhanced by troponin C-Ca^{2+} complexes formed on stimulation of contraction; Ca^{2+}-troponin complexes bind more tightly to troponin I, the inhibitory component of the troponin complex that inhibits interaction of myosin and actin. When actin and myosin are released from inhibition by troponin I, muscle contracts (Chap. M8). The chemical coupling of glycogenolysis and muscular contraction is in accord with the fact that the granules of glycogen, with firmly bound phosphorylase, SPK, and other glycogenolytic enzymes, are found at the level of thin filaments during muscle contraction. Thus, electrical stimulation of muscle contraction instantaneously activates SPK-b without involvement of cyclic-AMP protein kinase. Of course, under physiological conditions, epinephrine also initiates conversion of SPK-b to SPK-a via protein kinase and provides a high-SPK-a activity at the ambient [Ca^{2+}], independent of troponin C-Ca^{2+} activation.

Phosphoprotein Phosphatase The regeneration of the dephosphorylated forms of phosphorylase and SPK is catalyzed by the same enzyme, *phosphoprotein phosphatase* (PP-1). Rabbit skeletal muscle PP-1 (MW = 125,000) dephosphorylates Ser-14 of phosphorylase a; it also dephosphorylates the β subunit of SPK-a about 100 times more rapidly than the α subunit. Other phosphatases also dephosphorylate these enzymes but are less well understood than PP-1, which is likely of major importance physiologically. Since PP-1 regenerates two enzymes in different cycles of the glycogenolytic cascade (Fig. 18.5), its activity must be carefully controlled; this is achieved in several ways. (1) As a nonspecific phosphatase, capable of acting on several cellular phosphoproteins, there is competition among various substrates for PP-1; e.g., SPK-a competitively inhibits the dephosphorylation of phosphorylase a. Several other phosphoproteins formed by cAMP-dependent protein kinase may also act in this fashion. (2) Allosteric effectors of phosphoprotein substrates for PP-1 alter the conformation of the substrates and thereby alter their susceptibility to dephosphorylation by PP-1. Thus, AMP and glucose-1-phosphate inhibit dephosphorylation of phosphorylase *a*, whereas glucose-6-phosphate, glucose, and glycogen stimulate dephosphorylation. The R$_2$ (cAMP)$_2$ complex resulting from activation of adenylate cyclase (Fig. 18.5*a*) also binds phosphorylase a, thereby inhibiting its dephosphorylation by PP-1. (3) Perhaps most important is the inhibition of PP-1 by specific protein inhibitors. Several *phosphoprotein phosphatase inhibitors* (PI) have been found, but that designated PI-1 is physiologically important. PI-1 (MW = 20,000) is present in significant amounts in skeletal muscle and its K_i for PP-1 is $1.6 \times 10^{-9} M$. PI-1 exists in two forms, a dephosphorylated form that does not inhibit PP-1 and a phosphorylated form (PI-P) that is the potent inhibitor. Cyclic AMP-dependent protein kinase phosphorylates PI to form PI-P, as indicated in Fig. 18.5*e*. Thus, stimulation of glycogenolysis by activation of cAMP-dependent protein kinase provides an inhibitor for PP-1, ensuring that the action of PP-1 does not reverse the activating effects of cAMP on glycogenolysis. Indeed, in skeletal muscle the fraction of PI as PI-P increases from 30 to 70 percent on epinephrine administration. The action of PP-1 on

SPK-a is also regulated by phosphorylation of the α and β subunits of SPK. Thus, phosphorylation of the β subunits is solely responsible for converting SPK-b to SPK-a, but phosphorylation of the α subunits regulates the capacity of PP-1 to dephosphorylate the β subunits. Another phosphoprotein phosphatase, PP-2, is required to dephosphorylate α subunits before PP-1 dephosphorylates the β subunits, converting SPK-a to SPK-b.

GLYCOGEN SYNTHESIS

Glycogen Synthase and Branching Enzymes

Two enzymes are required for glycogen synthesis. *Glycogen synthase* transfers glucose from UDP-glucose into $\alpha 1,4$ linkage with glucose at the nonreducing ends of the branches of glycogen (Fig. 18.1).

$$\text{UDP-Glc} + \text{Glc}\alpha 1,4\text{Glc}\alpha 1,4\cdots \xrightarrow{\text{Glycogen synthase}} \text{Glc}\alpha 1,4\text{Glc}\alpha 1,4\text{Glc}\alpha 1,4\cdots + \text{UDP}$$

At equilibrium, glycogen synthesis is favored by a factor of about 250. UDP-glucose is synthesized by *UDP-glucose pyrophosphorylase,* which catalyzes reaction of UTP with glucose-1-phosphate as follows:

Uridine diphosphate glucose (UDP-glucose)

The PP$_i$ is derived from the terminal phosphates of UTP and is degraded by pyrophosphatases to render irreversible the otherwise freely reversible reaction. Glucose-1-phosphate is made available by the action of phosphoglucomutase on glucose-6-phosphate.

Glycogen synthesis requires preformed glycogen, since glycogen synthase employs free glucose or small $\alpha 1,4$-linked glucose oligomers only at very high, unphysiological concentrations. The precursors for initiating glycogen synthesis *de novo* are unknown, but glycogen may be a glycoconjugate, containing small amounts of protein or lipid. If so, its biosynthesis may require initially synthesis of an oligosaccharide attached to protein or lipid, on which, as discussed below, the extensive network of $\alpha 1,4$-linked glucose residues in glycogen can be formed by glycogen synthase.

Synthesis of the branches of glycogen (Fig. 18.1) requires the action of *branching enzyme, amylo-($\alpha 1,4 \to \alpha 1,6$)transglucosylase.* This enzyme cleaves the linear $\alpha 1,4$-linked glucose polymer formed by glycogen synthase to give hexaglucosides or heptaglucosides, which are then transferred to form an $\alpha 1,6$ linkage with a glucose residue 8 to 12 residues down the chain, as follows:

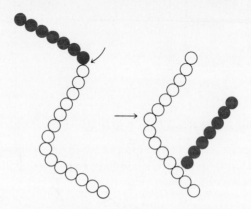

Glycogen synthase then continues to add glucose residues to each branch.

Regulation of Glycogen Synthesis

The principal feature of the regulation of glycogen synthesis is the same as that for regulation of glycogenolysis, i.e., the interconversion of phosphorylated and dephosphorylated forms of interconvertible enzymes by converter enzymes of a multicyclic cascade in response to hormonal stimuli, as shown in Fig. 18.7.

Glycogen synthase is central in the cascade (Fig. 18.7c) and exists in an I form (*i*ndependent of glucose-6-phosphate) and a D form (*d*ependent on glucose-6-phosphate). The interconversion of the two forms is catalyzed by the same enzymes that act in the glycogenolytic cascade (Fig. 18.5), *synthase-phosphorylase kinase* (SPK-a and SPK-b) and *phosphoprotein phosphatase* (PP-1), whose activities are under hormonal control mediated by cAMP.

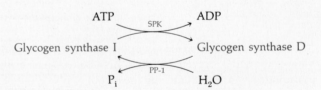

Glycogen synthase (MW $= 350,000$) is a tetramer with identical subunits, and like phosphorylase, it is tightly bound to glycogen granules. Also like phosphorylase, the activity of the synthase is controlled by phosphorylation and allosteric effectors, but the consequences of phosphorylation to activity are just the *reverse* of those with phosphorylase. In resting muscle, the nonphosphorylated, active I form predominates; contracting muscle contains the phosphorylated low-activity D form, whose activity is expressed only at sufficient [glucose-6-phosphate].

Glycogen synthase is phosphorylated by three other kinases in addition to SPK, and different forms of synthase D containing up to four phosphate groups per molecule are found in vivo. The extent of phosphorylation and the serine

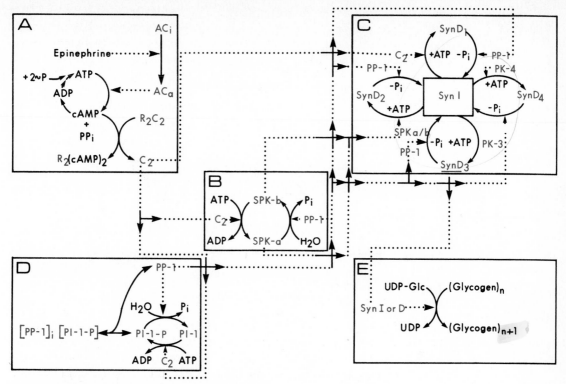

Figure 18.7 The multicyclic cascade for regulation of skeletal muscle glycogen synthesis. (*a*) Initiation of the cascade by epinephrine, as in Fig. 18.5*a*. C subunits of cAMP-dependent protein kinase act on three other cycles of the cascade as indicated by the dotted arrows. (*b*) The synthase-phosphorylase kinase cycle as in Fig. 18.5*b*. (*c*) The glycogen synthase cycle. Glycogen synthase I (Syn I) is converted to four different phosphorylated forms by different phosphokinases to give four different species of Syn D, which unlike Syn I, is highly dependent on the allosteric effector glucose-6-phosphate. Syn D_1 is formed by catalytic subunits C, Syn D_2 by SPK, and Syn D_3 and Syn D_4 by PK-3 and PK-4, two phosphokinases, whose regulatory controls are unknown. Syn I and Syn D, depending on the state, act to catalyze incorporation of additional glucose into glycogen in (*d*). (*e*) The phosphoprotein phosphatase 1 (PP-1) cycle. Catalytic subunits C_2 phosphorylate inactive PP-1 inhibitor, PI, to give active inhibitor PI-P, which inactivates PP-1 by forming the complex [PP-1][PI-P], thereby preventing PP-1 from catalyzing dephosphorylation of SPK-a in (*b*) or each of the four forms of Syn D (*c*). Thus, initiation of the cascade via the path (*a*) → (*b*) → (*c*) ensures phosphorylation of Syn I to Syn D, limiting glycogen synthesis on muscle contraction, and at the same time the path (*a*) → (*b*) → (*d*) prevents reconversion of Syn D to Syn I by inactivation of PP-1. When contraction ceases, inhibition of the cascade by the reactions in (*a*) stop and phosphatases are relieved of their inhibition and act to convert SPK-a to SPK-b and Syn D to Syn I, conditions that permit glycogen synthesis.

hydroxyl groups phosphorylated are different for different kinases. Phosphorylations of synthase by SPK and C_2 are initiated by epinephrine via cAMP (Fig. 18.7), but the factors that initiate and regulate phosphorylation by PK-3 and PK-4 are unknown.

Phosphorylation of glycogen synthase increases the K_m for UDP-glucose in the absence of glucose-6-phosphate but has no effect at saturating [glucose-6-phosphate]. Phosphorylation also decreases the affinity of the enzyme for activator. Thus, as shown in Fig. 18.8*a*, a much higher [glucose-6-phosphate] is required for activation at high, than at low, levels of phosphorylation. At low [glucose-6-phosphate], the highly phosphorylated forms are virtually inactive. The effectiveness of several negative effectors that counteract glucose-6-phosphate, e.g., ATP, ADP, UTP, and P_i, is also influenced by phosphorylation as shown in Fig. 18.8*b*.

Dephosphorylation of synthase D is catalyzed by the same phosphoprotein phosphatase (PP-1) that acts in glycogenolysis. Thus, inhibition of PP-1, initiated by phosphorylation of PI to give the inhibitor PI-P, is dependent on cAMP-dependent protein kinase, thereby preventing dephosphorylation of the D form and essentially stopping glycogen synthesis unless the [glucose-6-phosphate] is sufficiently high.

Coupling of Glycogen Synthesis and Glycogenolysis

Figure 18.9 shows the coupling of the glycogenolytic and glycogen synthesis cascades. Both cascades are triggered by activation of adenylate cyclase by epinephrine; this leads to formation of cAMP, which activates cAMP-dependent protein kinase, R_2C_2. The active catalytic form (C_2) of cAMP-dependent protein kinase then converts, via phosphorylation, SPK-b to SPK-a, as well as glycogen synthase I to D. At this stage of the cascade, glycogen synthase is

Figure 18.8 (*a*) Glycogen synthase activity as a function of the extent of phosphorylation of the synthase and [glucose-6-phosphate]. Activity was measured at a constant [UDP-glucose] = 0.2 m*M*; [glucose-6-phosphate] was varied as indicated in the figure from 0 to 5 m*M*. (*b*) Glycogen synthase activity as a function of the phosphorylation state of the synthase and the concentration of small molecule effectors. The small molecule effectors stimulate or inhibit the synthase activity and can be considered to alter the location of the activity-phosphorylation curves in the sense shown. The dark arrows indicate the shifts in the activity-phosphorylation state curves by small molecule effectors and the hormones insulin, epinephrine, and glucagon. [*From P. J. Roach and J. Larner, J. Biol. Chem. 251:1920 (1976).*]

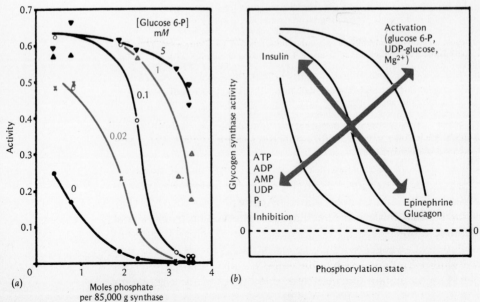

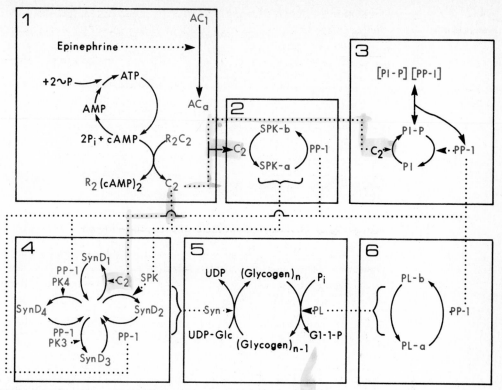

Figure 18.9 Coupling of the glycogen phosphorolysis and glycogen synthesis cascades. The symbols are the same as in Figs. 18.5 and 18.7; the individual cycles operate as described in the text and in the legends to Figs. 18.5 and 18.7. On muscle contraction epinephrine initiates the operation of the cycles as shown in 1; activation of phosphorolysis occurs via the path 1 to 2 to 5 and to inactivation of synthesis by the path 1 to 4 or 1 to 2 to 4. Activation of the cycles also inhibits action of phosphatases, as shown in 3, to prevent reversal of phosphorolysis and inhibition of glycogen synthesis.

converted to the low-activity D form by C_2, ensuring diminished glycogen synthesis, whereas conversion of phosphorylase from the cAMP-dependent b form to the high-activity cAMP-independent a form requires the action of SPK-b or SPK-a in a Ca^{2+}-dependent reaction. Thus, activation of glycogenolysis via phosphorylase a requires one more cycle of the cascade than termination of glycogen synthesis. The sensitivity for regulating a reaction in a multicyclic cascade increases with the number of cycles in the cascade (Chap. 10). Thus, the additional cycle permits finer regulation of glycogenolysis than glycogen synthesis. There is not an equivalent need, however, to regulate synthesis and degradation equally well, since glycogen synthesis in muscle proceeds at a rate no more than 0.3 percent of the highest rate of glycogenolysis.

The activating effects initiated by epinephrine stimulation are prevented from being reversed by phosphoprotein phosphatase (PP-1), since the catalytic C_2 subunits of cAMP-dependent protein kinase phosphorylate PI-1 to PI-P, a specific inhibitor of PP-1. Production of the inhibitor by C_2 early in the cascade prevents reversal of the phosphorylation reactions that produce high-activity SPK-a and phosphorylase *a*, and the low-activity glycogen synthase D.

443

Glycogen is the glycolytic fuel of skeletal muscle and provides the metabolites for ATP synthesis. In contrast, glycogen is not a major energy source for the liver, which consumes mostly fatty acids. Liver stores glucose as glycogen when glucose or other glycogenic substances are abundant and liberates it to the blood for use by other tissues when the blood [glucose] is lowered. Between 2 and 8 percent of the weight of the liver is glycogen in a normally fed animal and the liver [glycogen] remains fairly constant; even so, glycogen is being continuously formed and degraded. Moreover, the maximal possible rates of glycogen synthesis and degradation are almost equal in liver, in contrast to muscle, where the maximal rate of degradation is more than 300-fold that of synthesis. The liver of a fasted animal is virtually depleted of glycogen, but refeeding promptly restores its normal concentration. During the alternation between feeding and short fasting, the blood [glucose] remains within normal limits because of very large changes in the rates of hepatic synthesis and breakdown of glycogen.

Substances that increase the liver glycogen of fasting animals, and thereby the availability of glucose-6-phosphate, are said to be *glycogenic* or *glucogenic,* essentially synonymous terms. Glycogenic substances include not only hexoses but also glycogenic amino acids (Chap. 25), glycerol, intermediates of glycolysis, and sugar alcohols such as sorbitol or inositol. In each instance, the glycogenic substance must be converted to UDP-glucose via glucose-6-phosphate.

Liver glycogen is under endocrine control. Epinephrine and glucagon administered to an animal cause a rapid reduction in liver [glycogen], with resultant *hyperglycemia* (elevated blood [glucose]). This is a protective mechanism in time of stress by ensuring muscle an abundance of glucose without diminishing the amount of glucose required by nervous tissue. Insulin deficiency results in diminution of liver glycogen. The mechanisms of insulin action are complex (Chap. M16), but it is well-established that insulin stimulates glycogen synthesis, and usually to a greater extent in muscle than in liver. Adrenal steroids (Chap. M18) tend to increase liver glycogen by augmenting the supply of glycogenic amino acids. Excessive thyroid hormone, α-adrenergic agents, vasopressin, and angiotensin decrease liver glycogen.

Two glucogenic substances of major importance are lactate and pyruvate, the first diffusable substances produced by muscle on breakdown of glycogen. Thus, the sequence of events that ensues after administration of epinephrine to a fasting animal is (1) a fall in muscle glycogen, (2) an increase in blood lactate and alanine, and (3) a rise in liver glycogen. The alanine is made in muscle from pyruvate and is reconverted to pyruvate in liver (Chap. 25).

Most of the activities associated with muscle glycogen metabolism are also found in liver; thus the metabolic pathways and many of the regulatory mechanisms for the enzymes as described above for skeletal muscle also occur in liver. But there are notable differences between some liver and muscle enzymes and their regulatory controls. Liver phosphorylase a and b are interconvertible via phosphorylase kinase and phosphoprotein phosphatase, but the liver and muscle phosphorylases from the same species differ as shown by poor immunological cross-reactivity. This is consistent with the fact that genetic deficiencies in liver phosphorylase are without effect on the muscle enzyme and vice versa, as discussed below (Table 18.2). AMP is a positive effector of liver phosphorylase b,

TABLE 18.2 Hereditary Disorders of Glycogen Metabolism

Type	Enzymic defect	Glycogen structure	Organ	Eponymic name	Suggested clinical name
Ia	Glucose-6-phosphatase	Normal	Liver, kidney, intestine	von Gierke's disease	Glucose-6-phosphatase-deficient hepatorenal glycogenosis
Ib	Glucose-6-phosphate translocase	Normal	Liver	—	—
II	α-1,4-Glucosidase	Normal	Generalized	Pompe's disease	α-1,4-Glucosidase-deficient generalized glycogenosis
III	Amylo-1,6-glucosidase	Abnormal; outer chains missing or very short	Liver, heart, muscle, leukocytes	Cori's disease	Debrancher-deficient limit dextrinosis
IV	Amylo-(1,4 ⟶ 1,6)-transglycosylase	Abnormal; very long inner and outer unbranched chains	Liver, probably other organs	Andersen's disease	Brancher-deficiency amylopectinosis
V	Muscle glycogen phosphorylase	Normal	Skeletal muscle	McArdle-Schmid-Pearson disease	Myophosphorylase-deficient glycogenosis
VI	Liver glycogen phosphorylase	Normal	Liver, leukocytes	Hers' disease	Hepatophosphorylase-deficient glycogenosis
VII	Muscle phosphofructokinase	Normal	Muscle	—	Muscle phosphofructokinase deficiency
VIII	Liver phosphorylase kinase	Normal	Liver	—	Hepatic phosphorylase kinase deficiency

but glucose may be equally important in regulating the activity of phosphorylase action. Glucose is thought to bind the a form, altering its conformation and thereby enhancing its dephosphorylation by phosphoprotein phosphatase. Such glucose effects could be responsible, at least in part, for the diminished rate of glycogen phosphorolysis in liver at elevated blood [glucose].

Phosphorylation of phosphorylase b may not be catalyzed by a kinase analogous to synthase-phosphorylase kinase (SPK) of muscle, since some evidence suggests that liver phosphorylase b and glycogen synthase D, both of which are phosphorylated by synthase-phosphorylase kinase in muscle, are converted to these forms by separate enzymes in liver. Indeed, a Ca^{2+}-calmodulin-dependent, cyclic AMP-independent, glycogen synthase kinase has been purified from rat liver that rapidly converts synthase I to D but does not act on muscle phosphorylase b. The phosphorylase kinase of liver, however, must also be controlled by [Ca^{2+}], since the α-adrenergic agents vasopressin and angiotensin cause phosphorylase a formation in liver cells without increasing [cAMP] or activating cAMP-dependent protein kinase. Liver cells treated with these agents mobilize intracellular Ca^{2+}, which may be responsible for stimulation of phosphorylase kinase.

Hepatic glycogen synthesis utilizes I and D forms of glycogen synthase under the control, at least in part, of the Ca^{2+}-calmodulin-dependent glycogen synthase kinase and phosphoprotein phosphatases. The D form is active only in the presence of glucose-6-phosphate, a positive effector, and ATP is a negative effector.

Hereditary Disorders of Glycogen Metabolism

Table 18.2 lists the known human hereditary disorders of glycogen metabolism. Each is characterized by the accumulation of glycogen in affected tissues and is called a *glycogen storage disease.* Each disease results from the deficiency of one enzyme of glycogen metabolism, presumably as the result of a mutation in the structural gene for the enzyme involved. The inheritance of each disease is autosomal recessive, except for type VIII, which is sex-linked and fully manifest only in males. Collectively, these diseases illustrate the critical roles of liver glycogen in maintaining blood glucose, and of muscle glycogen as an energy source for intensive muscular work.

Type Ia storage disease is characterized by low glucose-6-phosphatase levels. Liver cannot convert glucose-6-phosphate to glucose and generate sufficient glucose to maintain normal blood levels; thus severe hypoglycemia results. Individuals with type Ib disease are clinically indistinguishable from those with type Ia disease, but they have normal glucose-6-phosphatase levels. These individuals are deficient in *glucose-6-phosphate translocase,* one of the four components in the multifunctional glucose-6-phosphatase system that mediates penetration of glucose-6-phosphate into the lumen of the endoplasmic reticulum (Chap 17). The other components of the system appear to function normally, but the defect is the same as in type Ia disease, since the net effect is the inability to convert glucose-6-phosphate to glucose. Type II disease is caused by a deficiency of a lysosomal enzyme, α*1,4-glucosidase* (maltase), which hydrolyzes the

α1,4-glucosidic bonds in glycogen. Together, the lysosomal enzymes degrade macromolecules, including glycogen, as a normal part of cellular turnover. Thus, type II disease is characterized by the accumulation in many cells, particularly the myocardium, of large vacuoles containing glycogen. Type III disease is a deficiency in debranching enzyme and the glycogen from afflicted individuals has very short outer branches (Fig. 18.1), resulting from phosphorylase action. Glycogen in liver may increase to twice the normal levels, with corresponding hepatic enlargement. In contrast, in type IV disease, which results from a deficiency of the branching enzyme, the liver glycogen has very long outer branches (Fig. 18.1), and early death from liver failure is often observed. Type V disease, caused by deficiency in muscle phosphorylase, has a late onset, with symptoms not usually evident until about 20 years of age. Afflicted individuals have severe muscle cramps on exercise, and the blood [lactate] fails to rise after exercise. Type VI disease is characterized by low liver phosphorylase activity, but clinical symptoms are usually mild. Type VII disease, caused by deficiency of phosphofructokinase of muscle, is clinically identical to type V disease. Type VIII results from a deficiency of liver phosphorylase kinase, but the clinical course of the disease is usually not severe.

BLOOD GLUCOSE AND REGULATION OF GLUCOSE METABOLISM

Glucose is continuously delivered to all tissues by the blood, which normally contains 70 to 90 mg glucose per 100 ml. Slightly higher values occur after meals, but fasting results in little or no decline in blood [glucose] for several days.

The dependence of various tissues on glucose varies widely. Glucose is essential for the central nervous system since it is the major energy source that effectively crosses the blood-brain barrier. Muscle and other tissues can derive much of their energy from other sources, such as acetoacetate (Chap. 20) and branched-chain amino acids (Chap. 25). Heart muscle is relatively inert to fluctuations in blood [glucose], since it effectively removes lactate and fatty acids from the blood. The loss of glucose in the urine is normally very small, regardless of diet.

Carbohydrate absorption is a discontinuous process, and although its contribution of glucose to the blood is large, it is also very variable. The continuing source of blood glucose is hydrolysis in liver, kidney, and intestine of glucose-6-phosphate derived from glycogenolysis and other precursors.

The blood [glucose] is the resultant of the relative rates of glucose production by liver, kidney, and intestine; glucose absorption from the gut; and glucose utilization by all tissues. Such tissues as liver and muscle assimilate glucose more rapidly at high than at low blood [glucose]. Elevated blood [glucose] favors formation of glycogen and triacylglycerols (Chap. 20), especially in liver and adipose tissue.

Insulin administration results in a prompt fall in blood [glucose] by augmenting the transport of glucose from the extra- to the intracellular space. Syn-

hypophyseal

垂体的

thesis of glycogen in muscle and of fatty acids by liver and adipose tissue is enhanced. Conversely, in *diabetes mellitus,* in which insulin is produced in inadequate amounts by the pancreatic B cells or fails to act normally (Chap. M16), blood [glucose] increases, overall glucose utilization is decreased, and glucose is excreted in the urine. Glucose continues to be utilized by the brain and heart, however, at a rate dependent on blood [glucose]. Thus, hyperglycemia serves a useful function in these tissues, but because of lack of insulin, liver and skeletal muscle of a diabetic individual are incapable of utilizing glucose normally, a situation aptly described as "starvation in the midst of plenty."

At least four groups of hormones tend to elevate blood [glucose]. Hypophyseal somatotropin (growth hormone) and adrenocorticosteroids are antagonistic to the action of insulin. Pancreatic glucagon and epinephrine from the adrenal medulla rapidly increase blood [glucose]; certain adrenal corticosteroids also elevate blood [glucose] but their effects are slower and more prolonged. Hyperthyroid individuals exhibit mild diabetes with very low liver [glycogen].

Hyperglycemia may result from overproduction or underutilization of blood glucose. Underutilization may be caused by insulin lack, resistance to insulin action, or excessive secretion of hypophyseal hormones, conditions that prevent normal glucose uptake by cells. Gluconeogenesis is thus enchanced, evident by increases in the amounts of pyruvate carboxylase, phosphoenolpyruvate carboxykinase, and fructose-1,6-bisphosphatase, the enzymes required for glucose synthesis from pyruvate (Chap. 17). Adrenal cortical hormones are responsible for increasing the levels of these enzymes as well as that of glucose-6-phosphatase, which increases the rate of glucose formation from glucose-6-phosphate. By inhibiting protein synthesis and enhancing aminotransferase activities in liver, the adrenal steroids increase the supply of glycogenic amino acids. Liver, which is releasing glucose into the blood, stops synthesizing fatty acids from acetyl CoA (Chap. 20) and, instead, utilizes them as its prime energy source. Fatty acids so utilized are derived from the liver as well as adipose tissue, in which lipolysis is enhanced by both insulin lack and elevated adrenal steroids. In turn, hepatic fatty acid oxidation provides the acetyl CoA required by pyruvate carboxylase, the NADH necessary for reduction of 1,3-diphosphoglycerate to glyceraldehyde-3-phosphate, and the ATP required for the various steps in conversion of pyruvate to glucose-6-phosphate (Chap. 17).

Insulin has effects other than enhancing glucose uptake by cells. It suppresses synthesis of pyruvate carboxylase, phosphoenolpyruvate carboxykinase, and fructose bisphosphatase. In addition, insulin stimulates increased levels of glucokinase and glycogen synthase and enhances the rates of glucose oxidation by the phosphogluconate path and fatty acid synthesis from glucose or pyruvate. The details of the hormonal control of glucose metabolism are presented in Chaps. M16 and M18.

The capacity of an animal to dispose of administered glucose is referred to as *glucose tolerance.* Tolerance is measured by orally administering a fixed amount of glucose (1 g per kg body weight) and measuring the blood [glucose] with time. The blood [glucose] of normal fasting individuals rises to about 140 mg/dl in about 1 hour and then falls to normal levels in about 3 hours. Hyperglycemic subjects show greatly elevated blood [glucose] that increases on glucose administration to even higher levels; over 3 to 4 hours the blood [glucose] does not decrease to the normal range.   Such tests are useful for diagnosis of defects in glucose metabolism.

449

**Carbohydrate
Metabolism II**

HEXOSE INTERCONVERSIONS

Glucose can serve as the precursor for the synthesis of all other monosaccharides in animal and microbial cells.   A number of monosaccharides other than glucose, however, are dietary constituents or are produced endogenously by the normal turnover of cellular glycoconjugates.   These monosaccharides have two general metabolic fates.   Some are glycogenic and are utilized as energy sources after conversion to intermediates of glycolysis or of the phosphogluconate pathway.   Many are also transformed to nucleotide monosaccharides, which are the immediate precursors for oligosaccharide biosynthesis.   For example, fructose, derived from hydrolysis of sucrose, a major dietary carbohydrate, is converted to fructose-1- and 6-phosphates, which serve not only as energy sources but also as specific precursors for the synthesis of nucleotide derivatives of three amino sugars.

Three types of nucleotide sugars are formed from monosaccharides; most are UDP-sugars analogous to UDP-glucose shown above, but GMP and CMP sugars are also formed, e.g., GDP-mannose and CMP-N-acetylneuraminic acid.

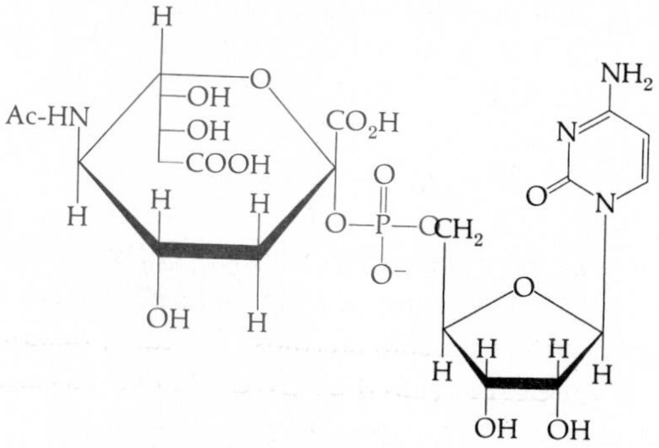

CMP- N-acetylneuraminic acid

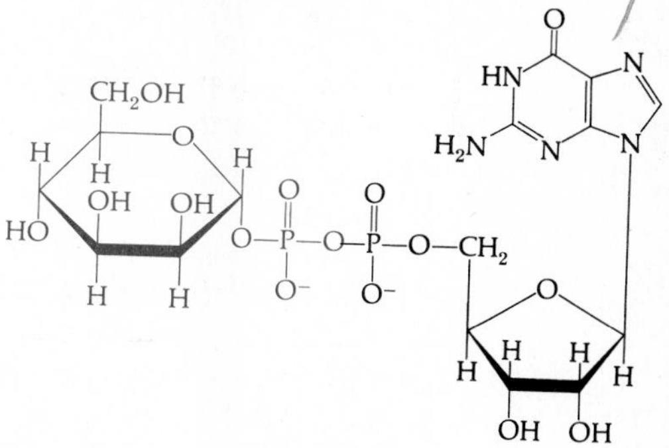

GDP- mannose

Figure 18.10 summarizes the metabolic interconversions of the hexoses and is a useful reference when considering the metabolism of each of the monosaccharides described below.

Galactose

Ingested galactose, derived primarily from milk lactose (Chap. 6), is phosphorylated in many mammalian tissues by *galactokinase*.

$$\text{Galactose} + \text{ATP} \rightleftharpoons \text{galactose-1-phosphate} + \text{ADP}$$

The fetal and infant liver enzymes have lower K_m values for substrates and about 5 times the V_{max} of the adult enzyme.   Galactose-1-phosphate can be returned to the mainstream of glucose metabolism by the consecutive actions of *phosphogalactose uridyltransferase* and *UDP-glucose epimerase*, as follows.

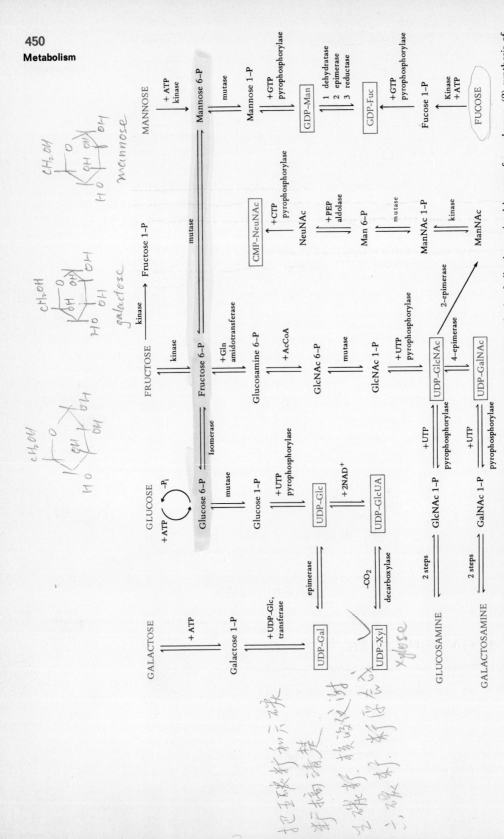

Fig. 18.10 Hexose interconversions, showing some of the pathways for (1) synthesis of all other animal hexoses from glucose, (2) synthesis of glucose-6-phosphate from other hexoses, and (3) synthesis of the nucleotide sugars required for oligosaccharide biosynthesis.

$$\text{Galactose-1-phosphate} + \text{UDP-glucose} \underset{\text{transferase}}{\rightleftharpoons} \text{UDP-galactose} + \text{glucose-1-phosphate}$$

$$\text{UDP-galactose} \underset{\text{epimerase}}{\rightleftharpoons} \text{UDP-glucose}$$

Galactose-1-phosphate is not converted to UDP-galactose by reaction with UTP via a pyrophosphorylase analogous to that for UDP-glucose synthesis. Hereditary lack of the uridyltransferase is the cause of one type of human *galactosemia*, a disorder detectable at infancy and characterized by an inability to utilize the galactose derived from milk lactose (Chap. 6), with resulting growth failure, cataract formation, and mental retardation. If maintained on a galactose-free diet, individuals with this disease may live to adulthood without symptoms of galactosemia and obtain UDP-galactose for oligosaccharide synthesis solely from UDP-glucose.

Another type of galactosemia results from a deficiency of galactokinase, but the clinical symptoms, mainly cataract formation, are milder than those of transferase deficiency. Galactitol is a major factor in producing cataracts in both types of galactosemia and is the reduction product of galactose formed by the action of *polyol:NADP-oxidoreductase*.

$$
\begin{array}{ccc}
\text{CHO} & & \text{CH}_2\text{OH} \\
| & \xrightarrow{\text{NADPH} + \text{H}^+} & | \\
[\text{CHOH}]_4 & & [\text{CHOH}]_4 \\
| & & | \\
\text{CH}_2\text{OH} & & \text{CH}_2\text{OH} \\
\textbf{Galactose} & & \textbf{Galactitol}
\end{array}
$$

Polyol reductase reduces a variety of aldehydes and probably does not normally use galactose, which is a poor substrate. At the high [galactose] in galactosemia, however, it is used, and damaging amounts of galactitol accumulate in the lens.

Fructose

Fructose may be phosphorylated by nonspecific kinases to give fructose-6-phosphate, but most ingested fructose is phosphorylated by *fructokinase*, which catalyzes the reaction

$$\text{Fructose} + \text{ATP} \rightleftharpoons \text{fructose-1-phosphate} + \text{ADP}$$

Fructose-1-phosphate cannot be converted directly to either fructose-6-phosphate or fructose-1,6-bisphosphate but enters the mainstream of glucose metabolism by the action of *fructose-1-phosphate aldolase*.

$$\text{Fructose-1-phosphate} \rightleftharpoons \text{dihydroxyacetone phosphate} + \text{glyceraldehyde}$$

Dihydroxyacetone phosphate is an intermediate of glycolysis but the glyceraldehyde must be oxidized in mitochondria by *glyceraldehyde dehydrogenase* to give glycerate, which is then phosphorylated by glycerate kinase to yield 2-phosphoglycerate, another intermediate of glycolysis (Chap. 17).

Hereditary deficiency of fructokinase causes *essential fructosuria*, a benign human disorder, in which ingested fructose is excreted in the urine and only slowly utilized by tissues. Deficiency of fructose-1-phosphate aldolase, however, called *hereditary fructose intolerance*, causes severe hypoglycemia and vomiting after fructose ingestion. The hypoglycemia is caused by the high cellular [fructose-1-phosphate], which inhibits glycogen phosphorylase.

Amino Sugars: *N*-Acetylglucosamine, *N*-Acetylgalactosamine, and Sialic Acids

Fructose-6-phosphate is the monosaccharide precursor of *N*-actylglucosamine, *N*-acetylgalactosamine, and the sialic acids. The amide nitrogen of glutamine is transferred to fructose-6-phosphate in a reaction catalyzed by *glutamine: fructose-6-phosphate amidotransferase*.

Fructose-6-phosphate + glutamine ⟶ glucosamine-6-phosphate + glutamate

The reaction is essentially irreversible and is the committed step in amino sugar synthesis. The resulting glucosamine-6-phosphate is then acetylated with acetyl CoA to give *N*-acetylglucosamine-6-phosphate, which is converted to *N*-acetylglucosamine-1-phosphate.

Glucosamine-6-phosphate $\xrightarrow{\text{acetyl CoA}}$ *N*-acetylglucosamine-6-phosphate $\underset{\text{mutase}}{\rightleftharpoons}$ *N*-acetylglucosamine-1-phosphate

Glucosamine-6- and 1-phosphates are also formed by phosphorylation of dietary glucosamine by nonspecific kinases. Glucosamine is not glycogenic and does not give rise to hexoses convertible to glycogen. *N*-Acetylglucosamine-1-phosphate is converted to UDP-*N*-acetylglucosamine by a *UDP-N-acetylglucosamine pyrophosphorylase* in a reaction analogous to that for UDP-glucose synthesis,

N-Acetylglucosamine-1-phosphate + UTP ⇌ UDP-*N*-acetylglucosamine + PP$_i$

UDP-*N*-acetylgalactosamine is formed by the action of *UDP-N-acetylglucosamine epimerase*.

UDP-*N*-acetylglucosamine ⇌ UDP-*N*-acetylgalactosamine

Exogenous *N*-acetylgalactosamine can also be converted to the 1-phosphate and hence to UDP-*N*-acetylgalactosamine by a pyrophosphorylase.

UDP-*N*-acetylglucosamine is also a precursor of *N*-acetylneuraminic acid (one type of sialic acid) as follows:

UDP-*N*-acetylglucosamine $\underset{\text{2-epimerase}}{\overset{-\text{UDP}}{\rightleftharpoons}}$ *N*-acetylmannosamine $\underset{\text{kinase}}{\overset{+\text{ATP}}{\rightleftharpoons}}$ *N*-acetylmannosamine-1-phosphate + ADP

The *N*-acetylmannosamine-1-phosphate is converted to the 6-phosphate, which then participates in the equivalent of an aldol condensation with phosphoenolpyruvate catalyzed by *N-acetylneuraminate aldolase*, to give *N*-acetylneuraminic acid 9-phosphate.

Phosphoenol-
pyruvic acid

N-Acetylmannosamine-
6-phosphate

N-Acetylneuraminic acid
9-phosphate

The 9-phosphate is hydrolytically removed to give P_i and N-acetylneuraminic acid, which is a precursor of other sialic acids, such as N-glycolylneuraminate. The latter is formed by the action of a monooxygenase.

$$\text{N-acetylneuraminate} + O_2 + \text{H-donor} \xrightarrow{Fe^{2+}} \text{N-glycolylneuraminate} + H_2O$$

Enzymic acetylation of N-acetylneuraminate with acetyl CoA yields the 7-O and 9-O-acetylated sialic acids.

Synthesis of the CMP-sialic acids, the only nucleotide sugars that are nucleoside monophosphate derivatives, is catalyzed by *sialic acid pyrophosphorylase*.

$$\text{CTP} + \text{Sialic acid} \longrightarrow \text{CMP-sialic acid} + PP_i$$

Little is known about the regulation of nucleotide sugar synthesis in higher animals, but probably all pathways leading to their formation are self-regulating. Although the exact mechansims are unclear, UDP-N-acetylglucosamine appears to regulate its own synthesis in liver by negative feedback control and inhibits the amidotransferase that converts fructose-6-phosphate to glucosamine-6-phosphate. The rates of synthesis of UDP-N-acetylgalactosamine and CMP-sialic acid are also controlled by this means. In cornea, CMP-sialic acid also inhibits the 2-epimerase that converts UDP-N-acetylglucosamine to N-acetylmannosamine.

Mannose and Fucose

Ingested mannose is converted by nonspecific kinases to mannose-6-phosphate, which is then converted to fructose-6-phosphate by *phosphomannose isomerase*, an enzyme similar to phosphoglucose isomerase of glycolysis. Mannose-6-phosphate is also converted to mannose-1-phosphate by *phosphomannose mutase* of liver and other tissues; the 1-phosphate serves as a precursor of GDP-mannose, which is synthesized by the action of *GDP-mannose pyrophosphorylase*.

$$\text{Mannose-1-phosphate} + \text{GTP} \rightleftharpoons \text{GDP-mannose} + PP_i$$

Ingested fucose is converted to fucose-1-phosphate by *fucose kinase* in a reaction utilizing ATP. Fucose is one of the few sugars that is not converted to other monosaccharides, its principal fate being incorporation into oligosaccharides via GDP-fucose, which is formed by *GDP-fucose pyrophosphorylase*.

$$\text{Fucose-1-phosphate} + \text{GTP} \longrightarrow \text{GDP-fucose} + \text{PP}_i$$

GDP-fucose is also synthesized from GDP-mannose in animal cells by three distinct enzyme activities (R = GDP structure).

GDP-ᴅ-mannose → GDP-4-keto-6-deoxy-ᴅ-mannose → GDP-4-keto-6-deoxy-ʟ-galactose → GDP-ʟ-fucose

The enzymes acting are (1) a *4,6-dehydratase*, (2) a *3,5-epimerase*, and (3) a *4-reductase*.

Glucuronic Acid and Xylose

Glucuronic acid arising from ingestion, by the degradation of proteoglycans or by the hydrolysis of UDP-glucuronic acid, participates in a series of carbohydrate interconversions called the *glucuronic acid oxidation pathway*, as shown in Fig. 18.11. Glucuronate is reduced by *glucuronate reductase* to gulonate, which is converted to ascorbic acid by *gulonolactone oxidase* in plants and in many mammals other than primates and guinea pigs (Chap. M22). Gulonate is also oxidized to β-keto-ʟ-gulonate in all animals examined by *gulonate dehydrogenase*. The β-keto-gulonate is decarboxylated by β-keto-*gulonate decarboxylase* to give CO_2 and ʟ-xylulose. ʟ-Xylulose is reduced to xylitol, which is converted to ᴅ-xylulose. ᴅ-*Xylulose kinase* then converts ᴅ-xylulose to ᴅ-xylulose-5-phosphate, which enters the pentose phosphate pathway (Chap. 17). The glucuronate oxidation path is yet another glucose oxidation pathway that bypasses anaerobic glycolysis and the citric acid cycle. The CO_2 is released from carbon 6 of glucose in contrast to the phosphogluconate pathway in which CO_2 arises from carbon 1 of glucose (Chap. 17).

The glucuronic acid pathway is not indispensible in humans, since individuals with the hereditary disease *idiopathic pentosuria*, in which the path is blocked, suffer no ill effects from the defect. These individuals cannot convert ʟ-xylulose to xylitol, and as a consequence, large amounts of ʟ-xylulose are excreted in the urine, especially after glucuronate ingestion. UDP-glucuronate is synthesized by oxidation of UDP-glucose.

$$\text{UDP-glucose} + 2\text{NAD}^+ \longrightarrow \text{UDP-glucuronate} + 2\text{NADH} + 2\text{H}^+$$

One enzyme catalyzes the two consecutive oxidation steps. Free xylose is apparently not utilized by animals nor rapidly absorbed by the gut. The UDP-

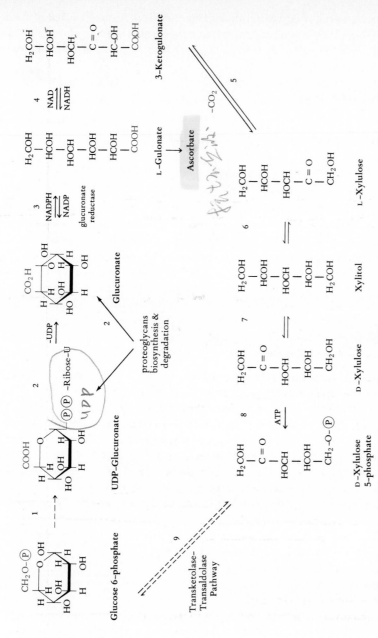

Figure 18.11 The glucuronic acid oxidation pathway, which shows the conversion of glucuronate to glucose-6-phosphate via pentose intermediates. The steps in 1 are given in Fig. 18.10 and those in 9 in Chap. 17. The CO_2 formed at step 7 is derived from carbon 6 of glucose-6-phosphate. Ascorbate is synthesized from gulonate (Chap. M22) in plants and in most animals. The deficiency of the enzyme catalyzing step 6 leads to the human disease idiopathic pentosuria.

xylose for oligosaccharide biosynthesis (Chap. M6) is formed by decarboxylation of UDP-glucuronate in a reaction requiring NAD.

$$\text{UDP-glucuronate} \longrightarrow \text{UDP-xylose} + CO_2$$

UDP-xylose is an allosteric regulator of the oxidation of UDP-glucose to UDP-glucuronate in cornea.

BIOSYNTHESIS OF OLIGOSACCHARIDES

Glycosyltransferases

About 80 different kinds of glycosidic linkages are known in the glycoconjugates of higher animals; each is formed by two of the ten monosaccharides found in glycoconjugates, or by one monosaccharide in glycosidic linkage to a protein or a lipid. Examples of these structures are given in Chaps. 6, 21, and M6 and later in this chapter. Biosynthesis of glycosides occurs in all cells, although synthesis of some is specific to cells that make a unique glycoconjugate, e.g., collagen (Chap. M6). Since the hydrolysis of a glycoside such as maltose, proceeds with $\Delta G° = -4$ kcal/mol, its synthesis requires expenditure of an equivalent amount of energy. This energy is provided by nucleotide sugars, which are the donor substrates for *glycosyltransferases,* the enzyme class that catalyzes synthesis of all glycosides by reactions of the following general form.

Glycosyl$_1$ -phosphonucleotide + glycose$_2$ $\longrightarrow$

Nucleotide sugar   **Acceptor**
donor substrate   **substrate**

glycosyl$_1$-O-glycose$_2$ + nucleoside phosphate

Glycoside

Each of the nucleotide sugars is synthesized as described above; the nucleoside phosphate in the donor substrates is UDP for N-acetylglucosaminyl, N-acetylgalactosaminyl, galactosyl, glucosyl, glucuronyl, and xylosyltransferases; GDP for fucosyl and mannosyltransferases; and CMP for sialyltransferases. Iduronic acid is also found in mammalian glycoconjugates, but it is not incorporated as such from a nucleoside phosphosugar; it is formed by epimerization of glucuronic acid previously incorporated into an oligosaccharide (Chap. M6).

The acceptor substrate specificities of all glycosyltransferases are highly specific. Thus, each distinctive glycosidic bond in an oligosaccharide is formed by one transferase, e.g., the Siaα2-3Gal, Siaα2-6Gal, and Siaα2-6GalNAc linkages are each formed by a specific sialyltransferase. Little is known about the regulation of oligosaccharide biosynthesis, but presumably a given sequence of sugars is synthesized because of the presence of the requisite glycosyltransferases at the site of synthesis.

All glycosyltransferases are bound to membranes of the rough or smooth endoplasmic reticulum or the Golgi apparatus, where oligosaccharide synthesis occurs. Nucleotide sugars are synthesized in the cytoplasm, except for CMP-sialic acid, which is synthesized in the nucleus. Some nucleotide sugars must pass through subcellular membranes to serve as substrates during oligosaccharide synthesis, but little is known of the transport mechanisms.

Glycoconjugates are degraded intracellularly by lysosomal *glycosidases,* which hydrolyze one or at most a few types of linkages. Hereditary deficiencies of some of these enzymes lead to the *mucopolysaccharidoses* (Chap. M6), storage diseases similar to type II glycogen storage disease, considered above.

Biosynthesis of O-Linked Oligosaccharides

The biosynthesis of the ABO blood group active pentasaccharide of submaxillary gland mucin (Chap. M6), shown in Fig. 18.12, illustrates features of the synthesis of an *O*-linked oligosaccharide in glycoproteins. On the average, one out of every three residues in mucin is a serine or threonine residue, each of which is covalently bound to oligosaccharides, synthesized by the consecutive actions of five different glycosyltransferases. Initially, *N*-acetylgalactosamine is glycosidically linked to the hydroxyl group of a serine or threonine residue by an *N*-acetylgalactosaminyltransferase. Subsequently, a *galactosyltransferase,* a *sialyltrans-*

Figure 18.12 The pathway for the biosynthesis of the pig submaxillary mucin *O*-linked oligosaccharide. The path proceeds from structure (1) to structure (5). The alternative structures (7), (8), and (9) are dead-end products since they cannot be converted by the required transferases to give the final product, (5). The solid bar indicates that the reaction cannot proceed, and the dashed bar, that the reaction is very slow.

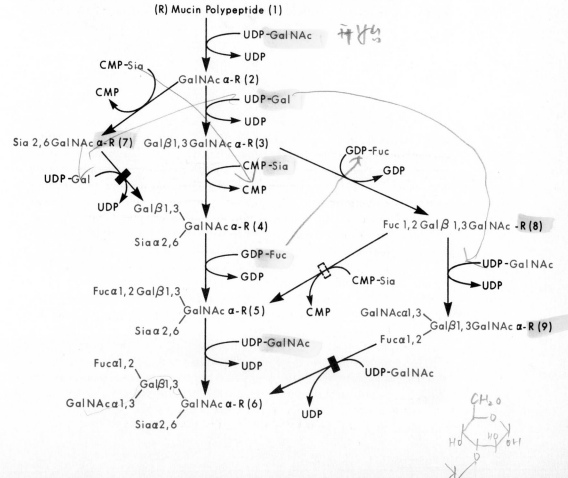

ferase, a *fucosyltransferase*, and another *N-acetylgalactosaminyltransferase* give rise to the pentasaccharide. The final product is formed only if the transferases act sequentially in the order given. Some evidence suggests that the transferases do not always act in this order and give rise to variably small amounts of other oligosaccharides that are not acceptors for any of the other transferases. Thus, after GalNAcα-O-mucin is formed, it can be sialylated by the sialyltransferase to give Siaα-2-6GalNAcα-O-mucin, which cannot be further glycosylated. Indeed, small amounts of this disaccharide are found in mucin. Other tri- and tetrasaccharides, as shown in Fig. 18.12, are also "dead end" products and cannot be used subsequently by the requisite transferase to give the pentasaccharide. Thus, in contrast to protein and nucleic acid synthesis, synthesis of oligosaccharides is often incomplete, giving rise to variably small amounts of structurally different oligosaccharides in the same molecule. This leads, at least in part, to the structural *microheterogeneity* observed in the oligosaccharides of many glycoconjugates.

Biosynthesis of *N*-Linked Oligosaccharides

The biosynthesis of "complex" and "high mannose" type oligosaccharide chains linked to the amide nitrogen of asparagine in glycoproteins (Chap. 6) is more complex than that of O-linked oligosaccharides and occurs in four stages: (1) chain initiation and elongation involving synthesis of a lipid-linked oligosaccharide containing *N*-acetylglucosamine, mannose, and glucose; (2) transfer of the oligosaccharide from the lipid to protein; (3) processing of the oligosaccharide bound to protein by removal of specific mannose and glucose residues; and (4) completion of the chain by addition of residues at the nonreducing ends of the chain. The details of each of the first three stages, shown in Fig. 18.13, appear to be the same in all normal mammalian cells.

Synthesis of Lipid-Linked Oligosaccharides The lipid on which the oligosaccharide chain is initially formed is dolichol phosphate, an isoprenoid phosphate, which in vertebrate tissues contains 18 to 20 isoprene units with two internal trans double bonds (the remainder are cis) and a saturated terminal isoprene unit.

$$H\left[CH_2-\underset{\underset{CH_3}{|}}{C}=CH-CH_2\right]_{18-20}-CH_2-\underset{\underset{CH_3}{|}}{CH}-CH_2-CH_2-O-PO_3^{2-}$$

Dolichol phosphate

Dolichol phosphate, like other isoprenoids and cholesterol (Chap. 21), is derived from hydroxymethylglutaryl CoA (HMGCoA) via farnesylpyrophosphate and involves the action of *trans*- and *cis*-prenyl transferases on the outer mitochondrial membrane. Inhibitors of HMGCoA reductase, such as 25-hydroxycholesterol, inhibit dolichol as well as cholesterol biosynthesis (Chap. 21). Dolichol phosphate is formed from free dolichol by action of a CTP-requiring kinase. Enzymes that dephosphorylate dolichol phosphate and pyrophosphate are found in liver.

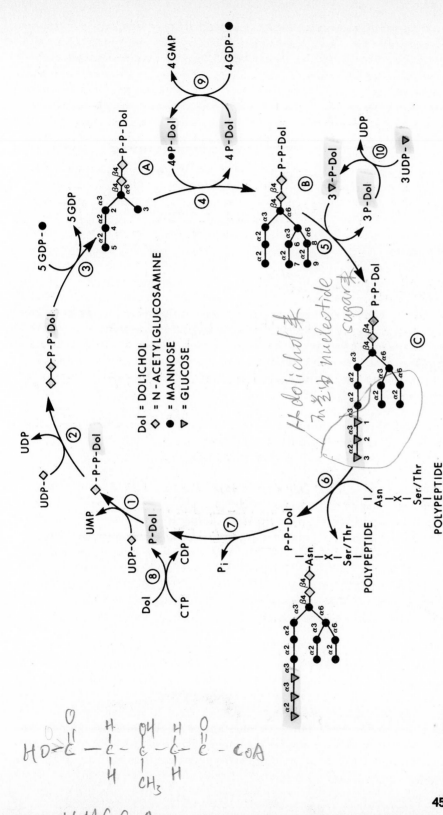

Figure 18.13 Oligosaccharide chain initiation, elongation, and transfer to polypeptide in *N*-linked glycoprotein biosynthesis. Monophosphoryldolichol is derived from dolichol pyrophosphate by hydrolysis (step 7) or from phosphorylation of dolichol by CTP (step 8); it reacts in step 1 to give *N*-acetylglucosamine pyrophosphoryldolichol. A second *N*-acetylglucosamine is added in step 2, and in step 3 five different mannosyltransferases act consecutively to transfer mannose from GDP-mannose to give intermediate A. The latter is elongated at step 4 by action of four different mannosyltransferases that employ mannose phosphoryldolichol, which is produced from GDP-mannose in step 9, to form intermediate B. The latter is elongated at step 5 by addition of three glucose residues derived from glucose monophosphoryldolichol, produced as shown in step 10, to give C, the final oligosaccharide-dolichol intermediate that is transferred to polypeptides containing the recognition sequence Asn-X-Ser/Thr destined to become glycoproteins. The five mannose residues in step 3 are added sequentially in the order 1 to 5 as indicated, and those in step 4 in the order 6 to 9. The three glucose residues in step 5 are added sequentially in the order 1 to 3. Monosaccharides are shown in symbols as indicated. One product of step 6, pyrophosphoryldolichol, is reutilized to start another cycle in step 7.

459

Different glycosyltransferases act sequentially to initiate and elongate the oligosaccharide on membrane-bound dolichol phosphate as shown in Fig. 18.13. The first transferase forms *N*-acetylglucosaminyldiphosphoryldolichol, on which the other monosaccharides are added stepwise by other glycosyltransferases.

UDP—GlcNAc + ⁻O—P—O—Dol ⟶ N-Acetylglucosaminyldiphosphoryldolichol + UMP

N-Acetylglucosaminyldiphosphoryldolichol

tunicamycin
衣霉素,
能抑制鸟苷磷酸相关的
某粒合成(含N-乙酰
葡粒胺的转共移)

The antibiotic *tunicamycin* blocks *N*-linked oligosaccharide synthesis by specifically inhibiting this reaction. The second *N*-acetylglucosamine residue and the next five mannose residues added to *N*-acetylglucosaminylpyrophosphoryldolichol are derived from UDP-*N*-acetylglucosamine and GDP-mannose, respectively. The remaining four mannose residues, however, and the three glucose residues are incorporated by glycosyltransferases that do not employ nucleotide sugars but use either mannosyl or glucosyl*mono*phosphoryldolichol as donor substrate. These monophosphates are synthesized from GDP-mannose or UDP-glucose and dolichol phosphate as shown for the mannose derivative.

derivative
衍生物

GDP—Man + ⁻O—P—O—Dol ⟶ Mannosylphosphoryldolichol + GDP

Mannosylphosphoryldolichol

A mutant lymphoma cell line that lacks the enzyme for synthesis of mannosylphosphoryldolichol cannot add the final four mannose and three glucose residues. All of the dolichol-linked intermediates and the glycosyltransferases involved in their formation appear to be on the cytoplasmic side of the endoplasmic reticulum. It is unclear how the oligosaccharide-dolichol is transferred into the lumen of the endoplasmic reticulum, where the next stage of synthesis occurs.

Transfer of Oligosaccharide to Protein A *protein-oligosaccharyltransferase* of the endoplasmic reticulum transfers the oligosaccharide from oligosaccharide-dolichol to protein (Fig. 18.13). The asparagine residue to which the oligosaccharide is transferred must be in the sequence Asn-X-Thr/Ser, where X can be any other amino acid. However, only about 30 percent of all known Asn-X-Thr/Ser sequences in animal proteins are glycosylated. Present evidence suggests that to serve as an acceptor for the oligosaccharide, the Asn-X-Thr/Ser sequence must be part of a turn or loop at the surface of the protein, most likely a β-turn (Chap. 5). Moreover, transfer of oligosaccharides devoid of the terminal three glucose residues apparently cannot occur rapidly, suggesting that these glucose residues are required for transfer of oligosaccharide from dolichol to protein.

Processing of _N_-Linked Oligosaccharides After incorporation into protein, the oligosaccharide is processed (Fig. 18.14) by removal of glucose and mannose units from nonreducing ends by the actions of _α-glucosidases_ and _α-mannosidases_. Two integral membrane α-glucosidases on the lumenal side of the endoplasmic reticulum are involved; one removes the terminal glucose and the other removes

Figure 18.14 Schematic representation of the processing and completion of the oligosaccharide chain of an _N_-linked oligosaccharide of a membrane-bound glycoprotein. The oligosaccharide, transferred to a nascent polypeptide chain on the lumenal side of the rough endoplasmic reticulum (Fig. 18.13), is processed by removal of glucose (steps 1 and 2) and then mannose residues (steps 3 to 5). At step 6, _N_-acetylglucosamine is added by a specific transferase, whose action is required before final processing in step 7, to remove the final two mannose residues. The fully processed chain after step 8 is then elongated and completed by sequential addition of fucose, galactose, and sialic acid in the Golgi apparatus as shown. The polypeptide chain has begun to fold before step 1. Monosaccharides are represented by symbols as shown.

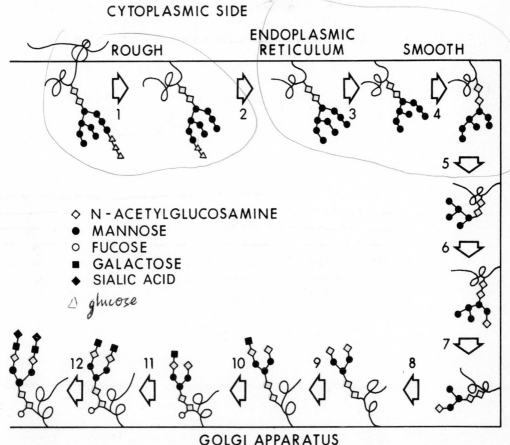

CYTOPLASMIC SIDE

ENDOPLASMIC
ROUGH   RETICULUM   SMOOTH

◇ N - ACETYLGLUCOSAMINE
● MANNOSE
○ FUCOSE
■ GALACTOSE
◆ SIALIC ACID
△ glucose

GOLGI APPARATUS

the next two glucose residues.   Four mannose residues are then removed in an ordered sequence by more than one α-mannosidase bound to membranes of Golgi apparatus.   The order of removal is C, A, D, B in the following reaction.

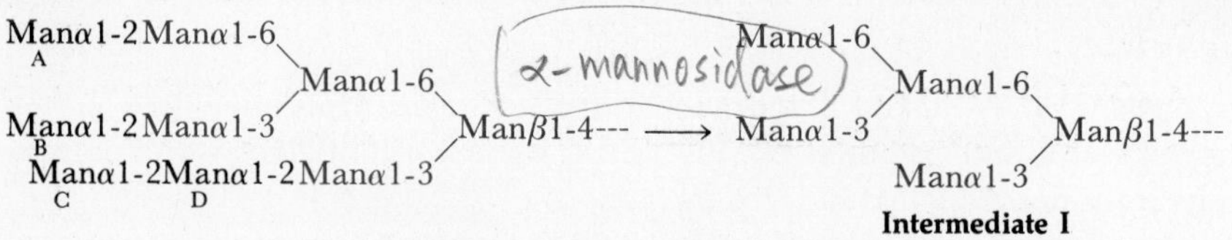

$$
\begin{array}{l}
\text{Man}\alpha1\text{-}2\text{Man}\alpha1\text{-}6 \\
\quad\quad\quad\quad\quad\quad\quad A \\
\quad\quad\quad\quad\quad\quad\quad\quad\quad\quad \text{Man}\alpha1\text{-}6 \\
\text{Man}\alpha1\text{-}2\text{Man}\alpha1\text{-}3 \\
\quad\quad\quad\quad\quad\quad\quad B \quad\quad\quad\quad\quad\quad\quad\quad \text{Man}\beta1\text{-}4\text{---} \\
\text{Man}\alpha1\text{-}2\text{Man}\alpha1\text{-}2\text{Man}\alpha1\text{-}3 \\
\quad\quad\quad\quad\quad\quad\quad C \quad\quad\quad D
\end{array}
$$

α-mannosidase (handwritten annotation)

$$
\xrightarrow{\quad\quad}
\begin{array}{l}
\text{Man}\alpha1\text{-}6 \\
\quad\quad\quad\quad\quad \text{Man}\alpha1\text{-}6 \\
\text{Man}\alpha1\text{-}3 \\
\quad\quad\quad\quad\quad\quad\quad\quad\quad \text{Man}\beta1\text{-}4\text{---} \\
\text{Man}\alpha1\text{-}3
\end{array}
$$

Intermediate I

Further processing of intermediate I cannot occur without the action of an *N-acetylglucosaminyltransferase,* which adds *N*-acetylglucosamine to give the intermediate II, the substrate for another specific *α-mannosidase* that liberates the final two mannose residues, as follows:

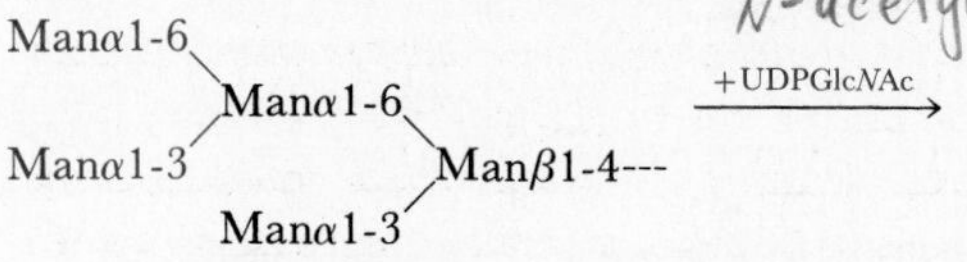

$$
\begin{array}{l}
\text{Man}\alpha1\text{-}6 \\
\quad\quad\quad\quad \text{Man}\alpha1\text{-}6 \\
\text{Man}\alpha1\text{-}3 \\
\quad\quad\quad\quad\quad\quad\quad \text{Man}\beta1\text{-}4\text{---} \\
\text{Man}\alpha1\text{-}3
\end{array}
$$

Intermediate I

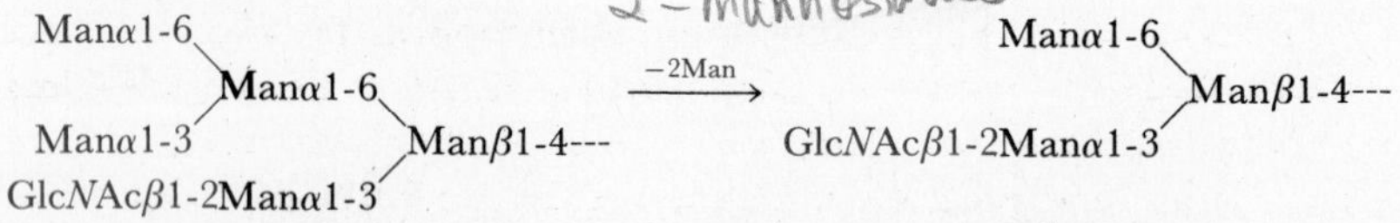

N-acetylglucosaminyltransferase (handwritten annotation)

$$\xrightarrow{\text{+UDPGlcNAc}}$$

$$
\begin{array}{l}
\text{Man}\alpha1\text{-}6 \\
\quad\quad\quad\quad \text{Man}\alpha1\text{-}6 \\
\text{Man}\alpha1\text{-}3 \\
\quad\quad\quad\quad\quad\quad\quad \text{Man}\beta1\text{-}4\text{---} \\
\text{GlcNAc}\beta1\text{-}2\text{Man}\alpha1\text{-}3
\end{array}
$$

Intermediate II

α-mannosidase (handwritten annotation)

$$\xrightarrow{-2\text{Man}}$$

$$
\begin{array}{l}
\text{Man}\alpha1\text{-}6 \\
\quad\quad\quad\quad\quad\quad\quad \text{Man}\beta1\text{-}4\text{---} \\
\text{GlcNAc}\beta1\text{-}2\text{Man}\alpha1\text{-}3
\end{array}
$$

Intermediate III

biantennary 双天线的 分支的 √ (handwritten annotation)

This is an important control in oligosaccharide processing, since cultured mutant animal cells devoid of the *N*-acetylglucosaminyltransferase, but with normal levels of the mannosidase, cannot further process oligosaccharides.   Intermediate III is the final processed oligosaccharide and is then elongated by the addition of other sugars to the terminal residues.

Completion of the Oligosaccharide Chain   Intermediate III, produced by processing, is converted to a completed biantennary oligosaccharide (Fig. 18.14) by the consecutive actions of *N*-acetylglucosaminyl, galactosyl, fucosyl, and sialyl transferases, each of which employs nucleotide sugars as donor substrates.   The exact order of addition of the final residues is unclear, and it is likely that several sequences are possible.   However, certain transferases must act before the others.   Thus, for the sequence, Siaα2-6Galβ1-4GlcNAcβ1-2···, *N*-acetylglucosamine must first be incorporated by a specific *N*-acetylglucosaminyltransferase followed by action of a specific galactosyltransferases and, finally, a sialyltransferase, as indicated in Fig. 18.14.   Fucose is incorporated into some oligosaccharides in α1,3 and/or α1,6 linkage as in the following structure.

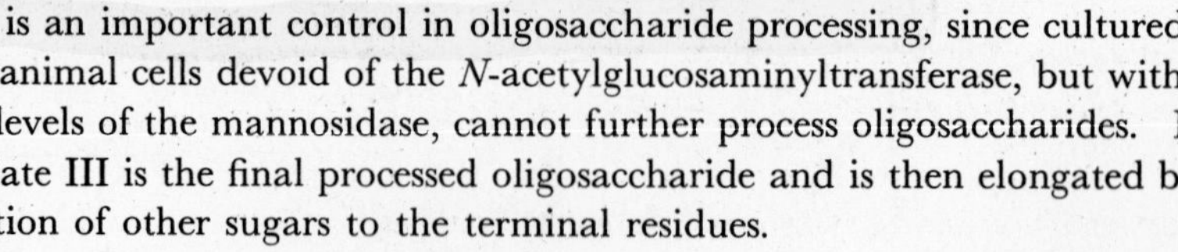

$$
\begin{array}{l}
\text{Sia}\alpha2\text{-}6\text{Gal}\beta1\text{-}4\text{GlcNAc}\beta1\text{-}2\text{Man}\alpha1\text{-}3 \\
\text{Gal}\beta1\text{-}4 \\
\quad\quad\quad\quad\quad\quad\quad\quad\quad\quad\quad\quad\quad\quad\quad\quad \text{Man}\beta1\text{-}4\text{GlcNAc}\beta1\text{-}4 \\
\quad\quad\quad \text{GlcNAc}\beta1\text{-}2\text{Man}\alpha1\text{-}6 \\
\quad\quad\quad\quad\quad\quad\quad\quad\quad\quad\quad\quad\quad\quad\quad\quad\quad\quad\quad\quad \text{GlcNAc-Asn} \\
\text{Fuc}\alpha1\text{-}3 \quad\quad\quad\quad\quad\quad\quad\quad\quad\quad\quad\quad\quad\quad \text{Fuc}\alpha1\text{-}6
\end{array}
$$

The previous action of a transferase may prevent the subsequent action of another, just as in *O*-linked oligosaccharide synthesis; e.g., fucose linked α1,3 in the above structure prevents addition of sialic acid onto the terminal galactose, and fucose cannot be added in $\alpha1,3$ linkage to *N*-acetylglucosamine in the branch containing $\alpha2,6$-linked sialic acid. Thus, the specificities of the glycosyltransferases contribute to the structural microheterogeneity of *N*- as well as *O*-linked oligosaccharides.

Little is known about the biosynthesis of tri- and tetraantennary oligosaccharides, but undoubtedly lipid-linked intermediates are involved.

BIOLOGICAL ROLES OF GLYCOPROTEINS AND GLYCOLIPIDS

Influence of Oligosaccharides on the Properties of Glycoconjugates

The properties of glycoconjugates rich in carbohydrate (>50 percent by weight) may be markedly influenced by their oligosaccharide components. Thus, the proteoglycans of the extracellular matrix (Chap. M6) interact with extracellular and cellular structures mainly through their carbohydrate groups. The mucins of the gastrointestinal and urogenital tracts have a high sialic acid or sulfated polysaccharide content (Chap. M6) that provides them with a net negative charge and permits them to form the viscoelastic gels that serve as biological lubricants. The oligosaccharides of glycolipids (Chap. 21), present in small amounts in the outer layer of the plasma membrane of most cells, serve as hydrophilic head groups of these amphiphiles (Chap. 7). In contrast, the biological activities of most glycoproteins, especially those with a low carbohydrate content, are rarely affected by their oligosaccharides, whether they act as enzymes, hormones, carriers, lectins, etc. The oligosaccharides in these types of glycoproteins, however, appear to serve other vital functions.

Oligosaccharides as Recognition Markers: Lectins

A major and general role of oligosaccharides in glycoproteins and glycolipids is to serve as recognition markers. Because of their multiplicity of structures, the oligosaccharides carry the biological information that aids in regulation of several cellular processes. The oligosaccharides on animal plasma cell membranes, all of which are on the external surface of the cell, account for only 1 to 8 percent of the weight of the membrane, but they play a major role in the interaction of the cell with other cells, and with its environment. Thus, cellular processes such as growth, motility, morphogenesis, endocytosis, and transformation appear to be regulated, at least in part, by the cell surface, and many observations suggest that the integral glycoproteins and glycolipids of the membrane bilayer are involved. Moreover, oligosaccharides can aid in transport and binding of glycoproteins to specific subcellular organelles. To understand the role of oligosaccharides in these processes requires consideration of carbohydrate-binding proteins, called *lectins*, which are widely distributed in all living things that have been examined.

The lectins, first recognized in the seeds of plants, have multiple binding sites and agglutinate animal cells. Many plant lectins have been purified and their structures and binding properties have been well-characterized. For example, the jackbean lectin, *concanavalin A*, binds oligosaccharide ligands containing nonreducing, terminal mannose; *peanut lectin* binds to galactose, and *wheat germ agglutinin* binds to *N*-acetylglucosamine. Some lectins bind two sugars about equally well, but most lectins usually bind one sugar tightly and other sugars more weakly, the strength of binding depending on the structural complimentarity between the binding site and the sugar. The binding sites of lectins often recognize the pentultimate residue as well as the nonreducing terminal residue of a ligand, e.g., peanut lectin binds galactosides, but those in the sequence Galβ1-3GalNAc- are bound more tightly.

The best characterized animal lectins are those of liver, especially that for binding galactosides. The rabbit *galactose lectin* is a glycoprotein (10 percent carbohydrate by weight) with the subunit structure a_4b_2; a and b have MW = 48,000 and 40,000, respectively, and combine to give active species with MW = 270,000 and higher multiples thereof. *N*-Acetylgalactosamine terminated ligands bind even better than those with galactose, but other sugars bind less well; binding of all sugars is absolutely dependent on Ca^{2+}. The galactose lectin is an integral, transmembrane protein and forms receptors (Chap. 14) on the surfaces of hepatocytes. One function of these receptors is to bind and internalize plasma glycoproteins with terminal galactose; the internalized glycoproteins are subsequently degraded by lysosomes. Most plasma glycoproteins (Chap. M1) have sequences such as Siaα2-3 or 2-6Gal or Siaα2-3GalNAc at the termini of their oligosaccharide chains and remain in the circulation for hours to many days, the exact time depending on the protein. After a certain time, however, terminal sialic acid residues are removed, presumably by sialidase action in various organs, exposing galactose or *N*-acetylgalactosamine, and the glycoproteins are cleared from circulation by the liver. Internalization is a threshold process since exposure of up to three or four galactose residues per molecule permits binding, but not internalization; molecules with a greater number of exposed galactose residues readily bind and are internalized. Internalization occurs via coated pits (Chap. 21), and coated vesicles in the cytoplasm deliver the galactose ligands to the lysosomes, which also contain the galactose lectin on the cytoplasmic side of their membranes. Although the lectin participates in receptor-mediated adsorptive endocytosis, several aspects of its behavior are unclear. Most of the lectin in hepatocytes is bound to the lumenal side of membranes of Golgi apparatus and the rough and smooth endoplasmic reticulum, and to the cytoplasmic side of lysosomes, but the function of this pool of lectin is unknown. Moreover, the lectin on the plasma membrane is reutilized after endocytosis, but not by exchanging with the internal, subcellular pool, which seems to be functionally separate from the cell-surface lectin.

Other lectins from liver have also been purified and characterized. One, from avian and amphibian hepatocytes, binds glycoproteins with terminal *N*-acetylglucosamine. It is an integral-membrane glycoprotein on the cytoplasmic surface of the plasma membrane and contains a single polypeptide chain (MW = 26,500) that aggregates to higher oligomers. In its known amino acid sequence (207 residues), residues 24 to 48 are hydrophobic and may be involved in anchoring the protein to the membrane. Binding activity is Ca^{2+}-dependent and highly specific for *N*-acetylglucosamine-terminated ligands. The function of the lectin appears to be similar to the galactose lectin of mammalian liver. Birds and amphibians have plasma proteins with oligosaccharides that terminate primarily in the sequence Galβ1-4GlcNAc-. Thus, removal of galactose by endogenous β-galactosidases exposes N-acetylglucosamine, and the degalactosylated plasma proteins bind cell-surface receptors formed by the lectin and are then internalized for transport to lysosomes, where they are degraded.

Mammalian hepatocytes also contain a Ca^{2+}-dependent fucose binding lectin that is an integral-membrane glycoprotein on the cell surface. Reticuloendothelial cells of liver, but not hepatocytes, contain yet another lectin, which binds N-acetylglucosamine and mannose-terminated glycoproteins equally well. Although this lectin and the fucose lectin form receptors on cell surfaces, their functions are unclear.

Evidence that lectins serve as recognition markers in cellular differentiation and morphogenesis has come from studies with a eukaryotic slime mold (Dictyostelium discoideum) and embryonic chick muscle. The slime mold grows and multiplies as a unicellular amoeba, but on starvation it differentiates to aggregating and adhesive cells to form a multicellular organism. Two different lectins, designated discoidin I and II, both specific for binding galactosides, gradually accumulate as cells commence to differentiate. Only about 2 percent of the lectins are on the cell surface, but binding of galactoside ligands to the cells increases the cell-surface lectin content. Moreover, cell-cell adhesion is blocked by galactose and other inhibitors of the lectin, and mutants of discoidin I with defective binding properties do not differentiate past the aggregating stage, indicating an inability of the cells to adhere to one another. These findings suggest that cell-cell adhesion occurs, at least in part, by the binding of galactosides on one cell to lectins on another.

In developing embryonic chicken muscle, two galactoside-binding lectins also increase more than 10-fold between 8 to 12 days of embryonic age, remaining at high levels for 3 to 4 days, and then decline to undetectable levels in the adult. Lectin I, the lectin in major amount, is intracellular in myoblasts and during differentiation becomes associated with the surface of myotubes before disappearing in adult muscle. The same lectin is found in liver and pancreas in adults. Lectin II is abundant in developing kidney but is still detectable in the adult organ. It is especially abundant in the intestine, where it is formed late in development and increases in adulthood, where it becomes concentrated along with mucin in secretory vesicles of goblet cells. Galactoside-binding lectins similar if not identical to lectin I of chickens are also found in beef heart and lung and the electric organ of eels, but their roles in these tissues are unknown.

Carbohydrates are also recognition markers for adherence of bacteria and viruses to host cells. Thus, Rhizobium trifolii, a nitrogen-fixing symbiotic bacterium for clover, binds to root hairs through reaction of 2-deoxyglucose in its capsular polysaccharide with a 2-deoxyglucose-specific lectin in the hair. After binding, the organism is internalized to form nitrogen-fixing nodules in the root. This symbiosis is specific, since each species of Rhizobium is symbiotic with a particular species of legume (Chap. 22). In contrast, some pathogenic bacteria contain lectins that bind specific oligosaccharides on the surfaces of their host cells; e.g., certain Mycoplasma adhere via sialic acid on host cells prior to infection; different species infect different animals as well as different host organs, suggesting considerable variation in the microbial lectins and the nature of the sialic acid ligands on host cells. Indeed, some viruses also adhere via highly specific sialic acid sequences, e.g., influenza virus and Sendai virus. One strain of human influenza virus adheres only to cells containing sialic acid in the sequence Siaα2-6Galβ1-4GlcNAc, and an equine strain only through Siaα2-3Galβ1-3-GalNAc. Sendai virus binds primarily through Siaα2-8Sia linkages of cell surface gangliosides (Chap. 21). Other bacteria bind via other monosaccharides, e.g., *Escherichia coli* and *Salmonella typhi*, bind to mannose residues on intestinal cells, whereas *Vibrio cholerae* binds to fucose. Bacterial toxins, including cholera (Chaps. M10 and M12) and tetanus, also bind to cell surfaces via sialic acid-rich gangliosides.

A biological role for carbohydrates as recognition markers has been implicated in several other systems. The mating of a haploid yeast (*Hansenula wingeii*) is dependent on mannose-rich cell surface glycoproteins. Fertilization of the brown alga (*Fucus serratus*) depends upon binding of fucose and mannose residues on the egg surface to lectins on the sperm. Cell-surface carbohydrates also act as recognition markers for an α-galactoside-binding lectin in the fertilized egg of the amphibian, *Xenopus laevis*, thereby aiding assembly of extracellular substances on the cell surface.

The protein of the extracellular matrix in mammals, called *fibronectin* (Chap. M6), has a marked influence on cell shape and cell adherence. Fibroblasts in culture do not flatten and adhere to surfaces if extracellular fibronectin is absent. Fibronectin binds the oligosaccharides of the proteoglycans (Chap. M6), particularly those of heparin sulfate as well as cell-surface glycoconjugates and, in this respect, behaves as a lectin. Cells grown on tunicamycin, a specific inhibitor of N-linked oligosaccharide synthesis, reduces spreading and adhesion. These factors influence the migratory properties of cells, particularly in connective tissue (Chap. M6). Other cells, however, such as hepatocytes in culture, also adhere to sugars anchored to an artificial matrix; rat hepatocytes, which contain cell-surface galactose lectin, bind only galactose, whereas chicken hepatocytes, which contain only N-acetylglucosamine lectin, bind N-acetylglucosamine specifically.

Cells in culture also change dramatically when treated with plant lectins. Thus, concanavalin A binds oligosaccharides on lymphocytes or fibroblasts, and is visible as "patches" on the cell surface when suitably labeled; the patches then coalesce at one region of the membrane to form "caps." Cells so treated with lectin then divide, accompanied by dramatic increases in protein and nucleic acid synthesis. Some lectins stimulate mitosis of some cells and not others; e.g., concanavalin A stimulates thymus-derived (T) lymphocytes but not bone marrow (B) lymphocytes (Chap. M4). Moreover, transformed cells bind different lectins, or more of a given lectin than the parental cells. These studies suggest that cell-surface carbohydrates play a role in such processes as mitosis and transformation and, in some manner, can markedly influence a variety of intracellular reactions, but knowledge of the molecular events involved is unclear.

Oligosaccharides in Transport and Secretion of Proteins: Lysosomal Enzymes

Proteins are synthesized on ribosomes in the cytoplasm, in the mitochondrion, or on the cytoplasmic surface of the rough endoplasmic reticulum (Chap.

29). Irrespective of their site of synthesis, all proteins must be delivered to a specific site in the cell or secreted from the cell into the circulation. The molecular mechanisms for "sorting" proteins after synthesis are unclear, but at least one group of enzymes, the lysosomal enzymes, are directed and bound to the lysosomal membrane by recognition markers on their oligosaccharide groups.

Membrane-bound proteins, except for some in mitochondria, are inserted into the lumen of the endoplasmic reticulum (Chap. 29) and glycosylated, as illustrated on Fig. 18.15. Some of the completed glycoproteins remain there, whereas others are delivered to other cell organelles, such as the plasma membrane, the Golgi apparatus, mitochondria, etc., where they may remain as integral membrane-bound components of these structures and serve roles essential to the organelle. Glycosylation of glycoproteins is a continual process, with

Figure 18.15 Diagrammatic representation of the sorting and distribution of newly synthesized glycoproteins. Different types of glycoproteins, indicated by the three symbols ○, ●, and ⊘, are synthesized on ribosomes, enter the lumen of the rough endoplasmic reticulum, and traverse the continuum of membranes to the Golgi apparatus. Sorting of the mixture then occurs, and those molecules destined for the same region are contained in buds that then form coated and smooth vesicles. Vesicles appear to be also formed from smooth endoplasmic reticulum. Vesicles containing membrane-bound proteins destined for the plasma membrane (A) fuse with the inner surface of the plasma membrane, open, and leave the membrane-bound proteins (●) on the cell surface. Other vesicles (B) traverse to the plasma membrane but after opening allow the proteins to be secreted (⊘). Lysosomal enzyme-containing vesicles (C) fuse with lysosomal membranes, permitting the enzymes to anchor to the membrane through the mannose-6-phosphate-specific lectin.

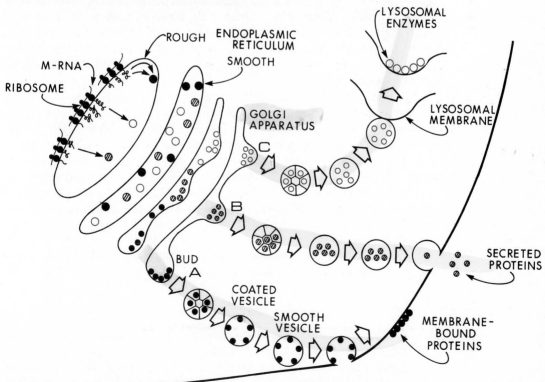

synthesis of lipid-linked oligosaccharides and transfer of oligosaccharide to protein occurring in the rough endoplasmic reticulum and processing and final elongation of the chain occurring in the smooth endoplasmic reticulum and the Golgi apparatus (Figs. 18.13 and 18.14). The soluble glycoproteins to be secreted also traverse the continuum of membranes that appears to connect the rough and smooth endoplasmic reticulum with the Golgi apparatus, where they are also glycosylated as required. The newly synthesized membrane-bound and soluble glycoproteins in the membranous network are then sorted from one another, collecting together on the lumenal side of the membranes. Those to be secreted appear to collect in one region; those destined for other organelles are collected in other regions. The membranes then form buds, which develop into small vesicles containing a specific group of glycoproteins. These vesicles appear to be coated with clathrin (Chap. 14) as in adsorptive endocytosis. The coated vesicles then find their way to a specific organelle and in the process shed their clathrin coat. If destined for secretion, the vesicles bind with and fuse to the inner plasma membrane, where by exocytosis, the proteins leave the cell. Glycosylation was once thought to be essential for proteins destined for secretion, but since secretion of some glycoproteins devoid of their carbohydrate occurs in the presence of tunicamycin, a specific inhibitor of *N*-linked oligosaccharide synthesis, this concept is no longer accepted. Glycosylation is required, however, for lysosomal enzymes to reach and bind to the lysosome.

All lysosomal enzymes, i.e., glycosidases, lipases, proteases, etc., contain *N*-linked oligosaccharides rich in mannose and mannose-6-phosphate. The mannose-6-phosphate serves to bind these enzymes to lysosomal membranes, which contain specific receptors for binding mannose-6-phosphate. The beef liver receptor protein, which has been purified to homogeneity (MW = 80,000), is a typical animal lectin, or carbohydrate-binding protein. The role of mannose-6-phosphate was first recognized in studies with human fibroblasts from individuals with an inherited deficiency in a specific lysosomal hydrolase, e.g., *α-iduronidase,* which causes one type of *mucopolysaccharidosis* (Chap. M6). Thus, if iduronidase from normal individuals is incubated with fibroblasts from iduronidase-deficient individuals, the enzyme is taken up by the cells, bound to lysosomes, and the deficiency is corrected. Removal of the phosphate by phosphatases, or incubation of the cells with mannose-6-phosphate, prevents enzyme uptake. Cell-surface receptors specific for binding mannose-6-phosphate are responsible for the uptake and internalization of the enzyme, but this is not the normal mechanism of delivery to lysosomes, since the enzymes do not leave the cell under physiological conditions.

Mannose-6-phosphate is incorporated into lysosomal enzymes by the action of *N-acetylglucosamine-1-phosphotransferase,* which catalyzes the reaction of UDP-*N*-acetylglucosamine with mannose-rich oligosaccharides as follows:

$$\text{UDP-Glc}N\text{Ac} + \text{Man}\alpha1\text{-2Man}\cdots \longrightarrow \text{Glc}N\text{Ac}\alpha\text{-1-P-6-Man}\alpha1\text{-2Man}\cdots + \text{UMP}$$

Terminal *N*-acetylglucosamine is then removed by *N-acetylglucosaminylphosphodiesterase,* as follows:

GlcNAcα-1-P-6-Manα1-2Man··· → **P-6-Manα1-2Man···**   + GlcNAc

The mannose-6-phosphate content of a lysosomal hydrolase may vary from one to five residues per molecule, and five different mannose residues in lysosomal β-glucuronidase have been shown to be sites of phosphorylation. Moreover, all lysosomal hydrolases have a high mannose content with chains similar to structure 3 in Fig. 18.14 and are devoid of sialic acid and galactose, suggesting that they are not fully processed and completed, as described above. Therefore, the mannose-6-phosphate may serve not only to block the processing steps involving removal of mannose, but also to direct the enzyme to the lysosome, where it is bound.

The human hereditary disease, called *I-cell disease,* a particularly severe, but rare type of mucopolysaccharidosis that results in early death, is caused by a deficiency of the 1-phosphotransferase. Fibroblasts from individuals with I-cell disease have very low levels of all hydrolases, although the culture medium in which they grow is rich in many lysosomal hydrolases. Moreover, the secreted enzymes have a low mannose content and are devoid of mannose-6-phosphate. Unlike normal lysosomal enzymes they contain sialic acid and galactose, suggesting that the oligosaccharides were fully processed and completed, as described above (Fig. 18.14).

Since lysosomal enzymes do not span the vesicular membrane, it is unclear how the mannose-6-phosphate directs a vesicle to lysosomes or how it becomes available for binding to the lysosomal membrane. Perhaps other markers on the cytoplasmic side of a vesicular membrane also aid in directing vesicles to their appointed destinations, even though oligosaccharides may play an important role.

MICROBIAL CELL WALLS

The mechanical properties of a bacterial cell wall must be suited to the needs of the organism. Bacterial cell walls must not only withstand severe osmotic shock, but also grow with the cell, rupture at the moment of cell division, and rapidly reseal thereafter. A complex envelope encases bacteria, and, in *E. coli* and other gram-negative organisms, it is formed by an inner cytoplasmic membrane with a typical bilayer structure as in eukaryotes and a membrane-like, dense outer layer, formed by a number of covalently combined lipids, polysaccharides, and proteins. A thin layer of *peptidoglycan,* which, in effect, is a single bag-shaped macromolecule of peptides covalently linked to polysaccharides in a regular repeating array, forms the cell wall between the inner and outer membranes.

The structures of bacterial cell wall peptidoglycans vary enormously from organism to organism and cannot be considered in detail here; however, the structure and the biosynthesis of one type of bacterial peptidoglycan is presented to illustrate the chemistry and metabolism of this important component of the bacterial envelope and the mechanism of action of some antibiotics that are bacteriocidal.

Peptidoglycan Structure and Biosynthesis

The repeating unit of the peptidoglycan is shown in Fig. 18.16. The carbohydrate moiety is an alternating unit of N-acetylglucosamine and N-acetylmuramic acid (3-O-lactyl-N-acetylglucosamine; abbrev., MurNAc), and the peptide moiety contains both D and L amino acids. Great variation may occur in the composition of the peptide moiety, which serves to cross-link peptidoglycan repeating units, as shown in Fig. 18.17, for *E. coli* and *Staphylococcus aureus*. Pentaglycine cross-links with lysine and alanine residues in *S. aureus* peptidoglycan, whereas an amino group of diaminopimelic acid forms a peptide bond with the terminal carboxyl group of alanine in the peptidoglycan cross-link of *S. aureus*.

Biosynthesis of peptidoglycans proceeds in three stages: synthesis of the peptide moiety of the peptidoglycan, synthesis of the peptidoglycan repeating units, and cross-linking of the repeat units to form the cell-wall polysaccharide. Figure 18.18 shows the reactions in the first stage, which starts with synthesis of N-acetylmuramic acid and then proceeds with the consecutive addition of the amino acids in the peptide moiety. The synthesis of D-alanyl-D-alanine is highly sensitive to the antibiotic *cycloserine*, which is structurally similar to D-alanine.

$$
\begin{array}{cc}
\begin{array}{c}
CO_2H \\
| \\
H-C-NH_2 \\
| \\
CH_3
\end{array}
&
\begin{array}{c}
O=C\!-\!\!-\!\!-NH \\
|\qquad\;| \\
H-C-NH_2\;\; | \\
|\qquad\;| \\
CH_2\!-\!\!-\!\!-O
\end{array}
\\
\text{D-Alanine} & \text{Cycloserine}
\end{array}
$$

In the next stage (Fig. 18.18) the N-acetylmuramylpeptide is transferred to a membrane-bound lipid intermediate, *undecaprenol phosphate*, a C_{55} isoprenol similar to dolichol phosphate.

$$
H\left[CH_2-\underset{\underset{CH_3}{|}}{C}=CH-CH_2\right]_{10}-CH_2-\underset{\underset{CH_3}{|}}{C}=CH-CH_2-O-\underset{\underset{O^-}{\overset{\overset{O}{\|}}{P}}}{}-O^-
$$

Undecaprenol phosphate

The peptidoglycan precursor is linked at C-1 through a pyrophosphate bridge to undecaprenol. In this form it reacts next with UDP-N-acetylglucosamine to

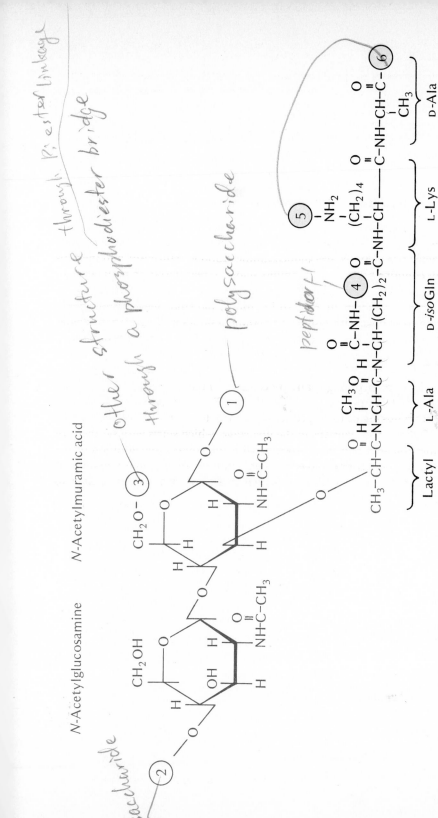

polysaccharide

other structure through R; ester linkage

through a phosphodiester bridge

polysaccharide

peptidoglycan

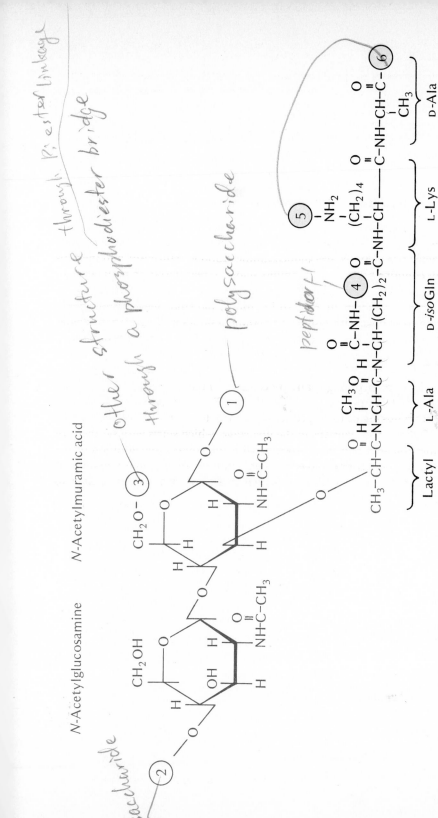

Figure 18.16 The repeating unit of the peptidoglycan of bacterial cell walls. Circled numbers indicate groups involved in joining one repeating unit to another or to other substituents: 1 and 2 are linked to the polysaccharide chains; 3 may be linked to other structures through a phosphodiester bridge; 4 is usually a H atom but sometimes forms a peptide bond with another amino acid; 5 may be bound to 6 in peptide linkage in different repeat units often through a short peptide chain.

Carbohydrate
Metabolism II

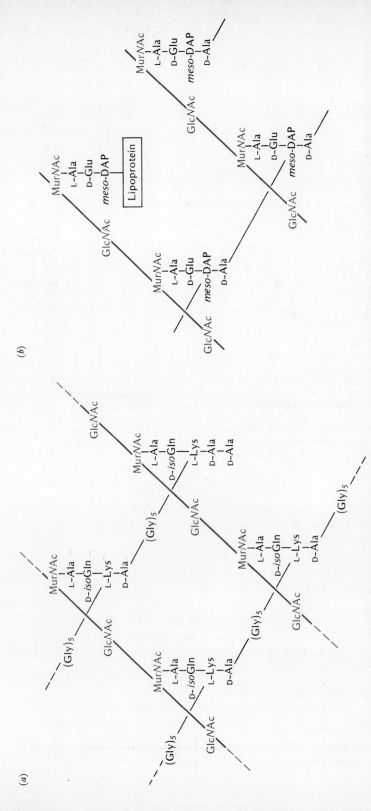

Figure 18.17 The structures of the peptidoglycans of (*a*) *S. aureus* and (*b*) *E. coli*. The structures shown are only parts of the macromolecular three-dimensional network in the polymeric molecule that, in effect, encloses the cell. The disaccharide repeat units are linearly arranged in two dimensions and are cross-linked in the third dimension by peptide bonds formed between the peptide moieties of the peptidoglycan. DAP = diaminopimelic acid; MurNAc = *N*-acetylmuramic acid.

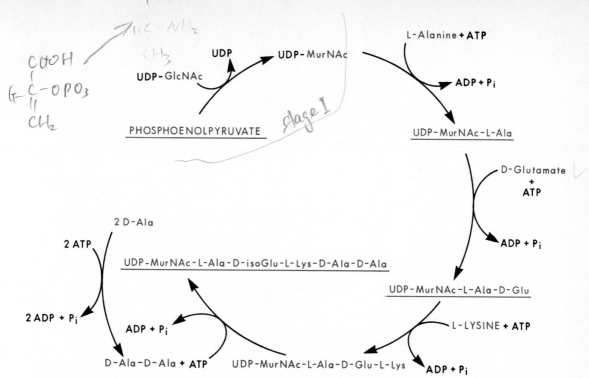

Figure 18.18 The biosynthesis of the *N*-acetylmuramylpeptide of peptidoglycan in *S. aureus*. The process starts with synthesis of UDP-GlcNAc and synthesis of UDP-MurNAc and is followed by the sequential addition of five amino acid residues in ATP-dependent processes. All reactions occur on the bacterial membrane. The inner core contains *O*-pyrophosphorylethanolamine in an unknown linkage to carbohydrate and gives rise to phosphate on hydrolysis. In D-isoGlu the peptide linkage is from the γ-carboxyl group to the α-NH₂ group of lysine.

form the disaccharide pentapeptide, still linked to undecaprenol. In an ATP-dependent reaction, the α-carboxyl group of the D-glutamic acid residue is then amidated to form a D-isoglutamine residue.

To grow the bridging unit, as in *S. aureus* (Fig. 18.19), five glycine residues are then transferred, seriatim, from glycyl tRNA (Chap. 28), first to the α-amino group of the lysine residue and then to the amino group of the preceding glycine.

The final step in peptidoglycan synthesis is a transpeptidation reaction catalyzed by D-*alanyl*-D-*alanine carboxypeptidase,* as shown in Fig. 18.20. In this step the terminal amino group of the pentaglycine chain displaces the COOH-terminal D-alanine and forms a peptide bond between the two chains.

Penicillin Action

The last step in peptidoglycan synthesis is blocked by the penicillins, antibiotics with the general structure

473

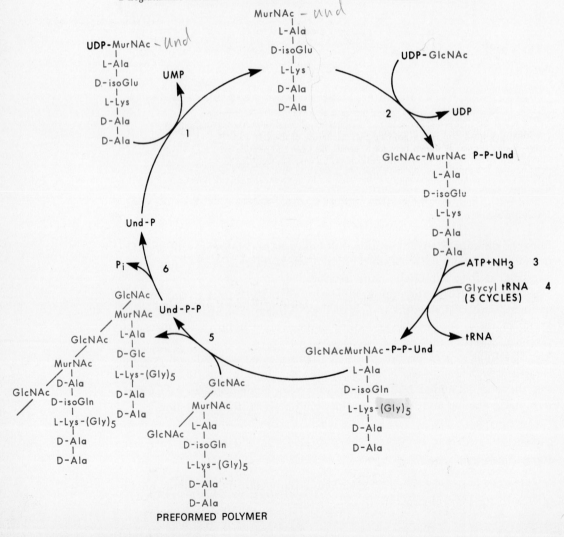

Penicillin

When R is a benzyl group ($C_6H_5CH_2$—), the antibiotic is benzylpenicillin, a derivative widely used therapeutically. Penicillin is thought to inactivate irreversibly the carboxypeptidase by forming a penicilloyl-enzyme intermediate.

Figure 18.19 Biosynthesis of peptidoglycan repeating units of peptidoglycan in *S. aureus;* Und-P = undecaprenol phosphate. At step 3, five cycles of the reaction with glycyl-tRNA form the pentaglycyl group attached to L-lysine, and in another independent reaction the α-COOH group of D-glutamic acid is amidated to give a D-isoglutamine residue.

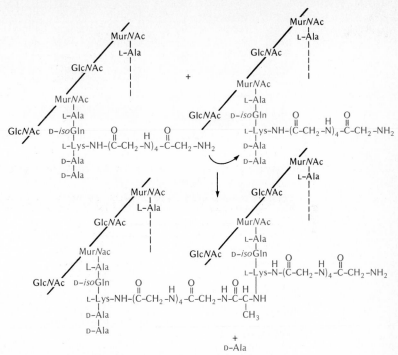

Figure 18.20 The cross-linking of peptidoglycan repeating units in formation of bacterial cell walls. After incorporation of the *N*-acetylglucosaminyl-*N*-acetylmuraminyl peptidyl peptidoglycan into a growing polysaccharide of the cell wall (Fig. 18.19), cross-linking occurs, as shown here for *S. aureus*, between the peptide groups of a peptidoglycan by an enzyme-catalyzed transamidation reaction resulting in displacement of a D-alanine residue from the terminus of an adjacent peptidoglycan repeating unit.

Normally, the transpeptidation proceeds with formation of a D-alanyl-acyl-enzyme intermediate.

$$\text{Enz} + \text{—D-Ala—D-Ala—COOH} \rightleftharpoons \text{D-Ala} + \text{—D-Ala—}\overset{\overset{\displaystyle O}{\|}}{C}\text{—Enz}$$

$$\text{—D-Ala—}\overset{\overset{\displaystyle O}{\|}}{C}\text{—Enz} + H_2N\text{—Gly—} \rightleftharpoons \text{—D-Ala—}\overset{\overset{\displaystyle O}{\|}}{C}\text{—}\overset{\overset{\displaystyle H}{|}}{N}\text{—Gly—} + \text{Enz}$$

Penicillin structurally resembles D-alanyl-D-alanine since it is a cyclic peptide composed of two amino acids, D-valine and L-cysteine; it binds at the active site to form an inactive penicilloyl-enzyme covalent complex similar to the D-alanyl-enzyme complex, thus inhibiting cell-wall synthesis.

Penicillin   **+ Enz** $\rightleftharpoons$ **Penicilloyl-enzyme complex**

475

lethal

detriment

The peptide bond in penicillin that is broken on acylation is much more reactive than that in D-alanyl-D-alanine because of the strain in the four-membered ring. Penicillin is an ideal antibiotic, since there is no known reaction in mammals analogous to that inhibited by penicillin, and very high doses that are lethal to infectious bacteria are without detriment to the host. Regrettably, bacterial mutants resistant to penicillin have emerged. Resistance results from the production of *lactamase*, an enzyme that rapidly hydrolyzes the lactam ring of penicillin, thereby rendering it therapeutically inactive.

Antibiotics other than cycloserine and penicillin inhibit the assembly of the peptidoglycan repeating units into the bacterial cell-wall polymer. *Vancomycin* and *ristocetin* are thought to inhibit the incorporation of undecaprenol phosphoryl sugar intermediates into peptidoglycan. *Bacitracin* is thought to inhibit dephosphorylation of the undecaprenolpyrophosphate at step 6 in Fig. 18.19.

REFERENCES

Books

Cohen, P. (ed.): Recently Discovered Systems of Enzyme Regulation by Reversible Phosphorylation, in *Molecular Aspects of Cellular Regulation*, vol. I, Elsevier, New York, 1981.

Review Articles

Barondes, S. H.: Lectins: Their Multiple Endogenous Cellular Functions, *Annu. Rev. Biochem.* **50:**207–231 (1981).

Beyer, T. A., J. E. Sadler, J. I. Rearick, J. C. Paulson, and R. L. Hill: Glycosyltransferases and Their Use in Assessing Oligosaccharide Structure and Structure-Function Relationships, *Adv. Enzymol.* **43:**23–175 (1981).

Chock, P. B., S. G. Rhee, and E. P. Stadtman: Interconvertible Enzyme Cascades in Cellular Regulation, *Annu. Rev. Biochem.* **49:**813–843 (1980).

Cohen, P.: The Role of cAMP-Dependent Protein Kinase in the Regulation of Glycogen Metabolism in Mammalian Skeletal Muscle, *Cur. Topics Cell. Reg.* **14:**117–195 (1978).

Cohen, P., C. B. Klee, C. Piston, and S. Shivish: Calcium Control of Muscle Phosphorylase Kinase through the Combined Action of Calmodulin and Troponin, *Ann. N.Y. Acad. Sci.* **356:**151–161 (1980).

Fletterick, R. J., and N. B. Madsen: The Structure and Related Functions of Phosphorylase a, *Annu. Rev. Biochem.* **49:**31–61 (1980).

Hers, H. G.: The Control of Glycogen Metabolism in the Liver, *Annu. Rev. Biochem.* **45:**167–189 (1976).

Hubbard, C., and R. J. Ivatt: Synthesis and Processing of Asparagine-Linked Oligosaccharides, *Annu. Rev. Biochem.* **50:**555–583 (1981).

Klee, C. B., T. H. Crouch, and P. G. Richman: Calmodulin, *Annu. Rev. Biochem.* **49:**489–515 (1980).

Krebs, E. G., and J. A. Beavo: Phosphorylation-Dephosphorylation of Enzymes, *Annu. Rev. Biochem.* **48:**923–959 (1979).

Lis, H., and N. Sharon: The Biochemistry of Plant Lectins, *Annu. Rev. Biochem.* **42:**541–574 (1973).

Neufeld, E., and G. Ashwell: Carbohydrate Recognition Systems for Receptor-Mediated Pinocytosis, in W. J. Lennarz (ed.): *Biochemistry of Glycoproteins and Proteoglycans*, Plenum, New York, 1980, pp. 241–266.

Pearse, B. M. F., and M. S. Bretscher: Membrane Recycling by Coated Vesicles, *Annu. Rev. Biochem.* **50:**85–101 (1981).

Schachter, H.: Glycoprotein Biosynthesis, in M. I. Horowitz and W. Pigman (eds.): *Glycoconjugates,* vol. II, Academic, New York, 1978, pp. 88–185.

Schroepfer, G. J.: Sterol Biosynthesis, *Annu. Rev. Biochem.* **50:**585–621 (1981).

Strominger, J. L., K. Izaki, M. Matsuhashi, and D. J. Tipper: Peptidoglycan Transpeptidase and D-Alanine Carboxypeptidase: Penicillin-Sensitive Reactions, *Fed. Proc.* **26:**9–22 (1967).

Chapter 19

Carbohydrate Metabolism III

Photosynthesis.

The prime source of energy in the biosphere is the light absorbed by chlorophyll-containing cells, which is utilized to fix CO_2 into carbohydrate.

$$6CO_2 + 6H_2O \xrightarrow{\text{light}} C_6H_{12}O_6 + 6O_2$$

This process is, in effect, the reverse of the oxidation of glucose. Were it completely efficient, the minimal free energy required would be that which can be derived from glucose oxidation, viz., $+686$ kcal/mol. When a suspension of algae is illuminated in the absence of CO_2 and placed in the dark and $^{14}CO_2$ is then admitted, fixation of $^{14}CO_2$ into carbohydrate proceeds for a brief time. Thus, the process of CO_2 fixation per se is not, strictly speaking, light-dependent. Considering first the chemical events by which CO_2 fixation into carbohydrate is accomplished makes it easier to understand how electromagnetic energy is converted into the forms of chemical energy used for that endergonic process.

The chemical energy directly utilized for converting CO_2 into carbohydrate is a mixture of ATP and reducing power as NADPH:

$$6CO_2 + 12NADPH + 12H^+ + 18ATP \longrightarrow 1 \text{ hexose} + 18ADP + 18P_i + 12NADP^+$$

Thus, the energy of 18 ATP is used to drive to the left the system

$$\text{Glucose} + 12NADP^+ \rightleftharpoons 6CO_2 + 12NADPH + 12H^+$$

for which equilibrium otherwise favors reaction to the right, as written. In the plant cell, the primary photosynthetic apparatus utilizes the energy of absorbed light to drive the synthesis of both NADPH and ATP:

$$H_2O + NADP^+ \xrightarrow{\text{light}} NADPH + H^+ + 1/_2O_2 \qquad (1)$$

$$ADP + P_i \xrightarrow{\text{light}} ATP \qquad (2)$$

Any cross-references coded M refer to the companion text, *Principles of Biochemistry: Mammalian Biochemistry.*

In higher plants, these processes are accomplished in *chloroplasts,* organized cell bodies 3 to 10 μm long and 0.5 to 2.0 μm in diameter, of which there are 50 to 200 in the typical cell. Within a chloroplast are 10 to 100 somewhat cylindrical structures, the *grana* (Fig. 19.1), each of which is a stack of flattened disks

Figure 19.1 (*a*) An electron micrograph of a single granum of the chloroplast from *Zea mays.* [*Courtesy of Dr. L. K. Shumway.*] (*b*) Schematic representation of the ultrastructure of a chloroplast. There are about 50 grana, 0.3 to 0.6 μm in diameter, in an average chloroplast. Each granum is a stack of thylakoids, which are closed sacs. The stroma lamellae interconnect with the membranes of other grana; only one such connection is shown. The thylakoid membrane is about 6 nm thick; the inner thylakoid space, between the roughly parallel walls, is about 15 nm across. [*Adapted from S. Wolfe, Biology of the Cell, Wadsworth, Belmont, Calif., 1972.*]

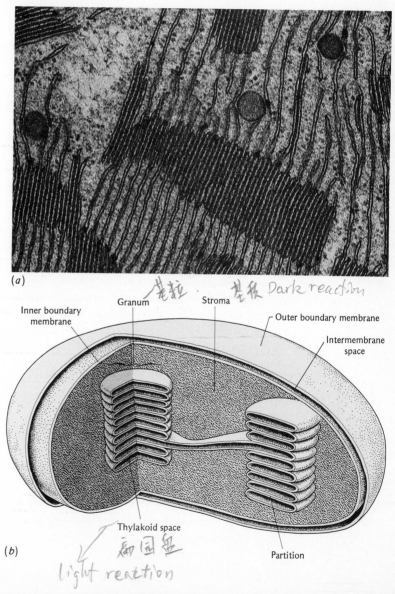

(*a*)

Inner boundary membrane

Granum 光粒 Stroma 基质 *Dark reaction*

Outer boundary membrane

Intermembrane space

Thylakoid space 类圆盘

light reaction

(*b*)

Partition

called *thylakoids* which are connected by *stroma lamellae,* or membranes. The chloroplast envelope is a double membrane; the outermost, like the outer membrane of a mitochondrion, is a lipid bilayer that structurally resembles the endoplasmic reticulum and offers little barrier to passage of ions or molecules of up to about 10,000 daltons. On the other side of the intermembrane space is the true chloroplast membrane, relatively impermeable not only to large molecules but to protons, hydroxyl ions, most charged molecules, and even many small neutral molecules.

Most of the materials transported across the chloroplast membrane occurs by the operation of three *translocases* (see below). Their general properties are similar to those in mitochondria (Chap. 15). (1) The *phosphate translocase* facilitates entry of P_i into the *stroma* space in exchange for movement outward of either dihydroxyacetone phosphate or 3-phosphoglycerate, both of which can also exchange in either direction for each other. (2) The *dicarboxylate translocase* permits each of the following dicarboxylate compounds to exchange across the chloroplast membrane for any of the others: oxaloacetate, malate, succinate, fumarate, aspartate, and glutamate. (3) The *adenine nucleotide translocase* is essentially unidirectional, facilitating ATP entry in exchange for ADP exit but not the reverse.

It appears likely that CO_2 crosses the membrane rather than H_2CO_3 or HCO_3^-. There is about 20 times as much HCO_3^- as H_2CO_3 in the cytosol, and it is thought that the abundant carbonic anhydrase (Chap. M4) ensures an adequate rate of dehydration of HCO_3^- and H_2CO_3 to CO_2 so that the latter can diffuse at a rate sufficient to support photosynthesis.

Inside the chloroplast itself, the energy of absorbed light quanta is utilized by the complex assemblage of proteins in the organized structure of the thylakoid membrane to drive synthesis of ATP and reduction of $NADP^+$; the final step in each of these processes occurs on the stroma side of the thylakoid membrane. Thus, although the thylakoid membranes probably originate by budding from the chloroplast inner membrane, these two membranes have distinctive properties. In the stroma fluid or attached to the stroma side of the thylakoid membrane are all the enzymes required to utilize ATP and NADPH for fixation of CO_2. The end product of this process is dihydroxyacetone phosphate, which leaves the chloroplast by the translocase in exchange for P_i or 3-phosphoglycerate and is converted in the cytosol to hexose phosphate and then to sucrose or starch.

FIXATION OF CO_2

The Dark Reaction

After introduction of $^{14}CO_2$ for brief periods into suspensions of photosynthesizing algae, the first compound to become radioactive is 3-phosphoglyceric acid, labeled predominantly in the carboxyl carbon. The reaction in which it is formed is catalyzed by *ribulose-1,5-bisphosphate carboxylase;* this is the primary dark reaction of the photosynthetic process.

$$
\begin{array}{c}
CH_2OPO_3^{2-} \\
| \\
C=O \\
| \\
HCOH \\
| \\
HCOH \\
| \\
CH_2OPO_3^{2-}
\end{array}
+ \ ^{14}CO_2 + H_2O \longrightarrow
\begin{array}{c}
CH_2OPO_3^{2-} \\
| \\
HCOH \\
| \\
^{14}COOH
\end{array}
+
\begin{array}{c}
COOH \\
| \\
HCOH \\
| \\
CH_2OPO_3^{2-}
\end{array}
$$

Ribulose-
1,5-bisphosphate

2 3-Phosphoglycerate

Presumed enzyme-bound intermediates are

$$
\begin{array}{c}
CH_2O\!-\!PO_3^{2-} \\
| \\
HOC \\
\| \\
HOC \\
| \\
HCOH \\
| \\
CH_2O\!-\!PO_3^{2-}
\end{array}
+ CO_2 \longrightarrow
\begin{array}{c}
CH_2O\!-\!PO_3^{2-} \\
| \\
HOC\!-\!COOH \\
| \\
HOC\!-\!OH \\
| \\
HCOH \\
| \\
CH_2O\!-\!PO_3^{2-}
\end{array}
\longrightarrow
\begin{array}{c}
CH_2O\!-\!PO_3^{2-} \\
| \\
HOC\!-\!COOH \\
| \\
C=O \\
| \\
HCOH \\
| \\
CH_2O\!-\!PO_3^{2-}
\end{array}
$$

I   II   III

spinach
菠菜.

octameric
8聚体

Synthetic keto acid (III) is converted by the enzyme to two molecules of 3-phosphoglycerate more rapidly than V_{max} for the reaction with ribulose-1,5-bisphosphate.

The enzyme from spinach consists of eight α subunits (MW = 55,000) plus eight β subunits (MW $\simeq$ 12,500). The enzyme is visible by electron microscopy as a cubic structure protruding into the stroma from the thylakoid membrane. The octameric α_8 from all plant sources catalyzes the carboxylation reaction; the β subunits are regulatory. In the presence of Mg^{2+} and at high enzyme concentrations, the optimal pH of the intact enzyme shifts from rather alkaline levels, at which the α_8 form is maximally active, to about pH 7.8, where its $K_m^{CO_2} = 12$ to $20 \, \mu M$. Some photosynthetic bacteria like *Rhodospirillum rubrum* contain only α_8 and lack β subunits. In plants the gene that codes for the α subunits is in chloroplast DNA, whereas that for the β subunit is in nuclear DNA (Chap. 29). This enzyme constitutes as much as 25 percent of the total protein of the leaf and is, therefore, the most abundant enzyme in nature.

CARBOHYDRATE SYNTHESIS

At saturating levels of light, corn leaves can fix CO_2 at maximal rates when $[CO_2]$ is maintained at about 4 to 10 μM, or about one-half the $K_m^{CO_2}$ of the carboxylase at its optimal pH, $[Mg^{2+}]$, etc. The activity of the enzyme at its high concentration in the plant, however, can account for the observed rates of CO_2 fixation in most plants.

Ribulose-1,5-bisphosphate is formed in a reaction catalyzed by *phosphoribulokinase:*

$$\text{Ribulose-5-phosphate} + \text{ATP} \xrightarrow{Mg^{2+}} \text{ribulose-1,5-bisphosphate} + \text{ADP}$$

The synthesis of hexose from 3-phosphoglycerate could be completed by the enzymes of glycolysis. However, if all the 3-phosphoglycerate were converted to hexose by reversal of glycolysis, no ribulose bisphosphate would be available to serve as acceptor for CO_2 in subsequent fixation reactions. This situation may be viewed as the converse of that considered in the operation of the phosphogluconate pathway (Chap. 18), viz., to provide a regenerative system by means of which hexose can be accumulated and ribulose-1,5-bisphosphate recovered for the next carboxylation reaction. This is achieved by concerted action of the enzymes of glycolysis and those of the phosphogluconate pathway. These reactions, the *Benson-Calvin pathway*, are summarized in the balanced equations of Table 19.1, which describes the formation of 1 mol of dihydroxyacetone phosphate by fixation of 3 mol of CO_2.

In reactions (2) and (3) the six molecules of phosphoglycerate are reduced to glyceraldehyde-3-phosphate. Three of these are converted to dihydroxyacetone phosphate in reaction (4). One of these represents the net gain of the process. By reactions (5) to (11), the three molecules of glyceraldehyde-3-phosphate left from reaction (3) and the two remaining molecules of dihydroxyacetone phosphate are converted into three molecules of ribulose-5-phosphate, utilizing enzymes similar to those which participate in the phosphogluconate pathway. Phosphorylation by ATP, reaction (12), completes the sequence, yielding ribulose-1,5-bisphosphate to repeat the process.

The summary equation in Table 19.1 reveals how energy is provided in order to fix CO_2 into carbohydrate. Synthesis of the 2 mol of dihydroxyacetone

TABLE 19.1 Triose Accumulation and Pentose Regeneration in Photosynthesis

Step	Enzyme	Reaction
(1)	Carboxylase	3 Ribulose-1,5-bisphosphate + $3CO_2 \rightarrow$ 6 3-phosphoglycerate
(2)	Phosphoglycerate kinase	6 3-Phosphoglycerate + ATP $\rightarrow$ 6 1,3-bisphosphoglycerate
(3)	Glyceraldehyde-3-phosphate dehydrogenase	6 1,3-Bisphosphoglycerate + 6NADPH + $6H^+ \rightarrow$ 6 glyceraldehyde-3-phosphate + $6NADP^+$ + $6P_i$
(4)	Triose isomerase	3 Glyceraldehyde-3-phosphate $\rightarrow$ 3 dihydroxyacetone phosphate
(5)	Aldolase	Glyceraldehyde-3-phosphate + dihydroxyacetone phosphate $\rightarrow$ fructose-1,6-bisphosphate
(6)	Fructose bisphosphatase	Fructose-1,6-bisphosphate $\rightarrow$ fructose-6-phosphate + P_i
(7)	Transketolase	Fructose-6-phosphate + glyceraldehyde-3-phosphate $\rightarrow$ erythrose-4-phosphate + xylulose-5-phosphate
(8)	Aldolase	Erythrose-4-phosphate + dihydroxyacetone phosphate $\rightarrow$ sedoheptulose-1,7-bisphosphate
(9)	Heptulose bisphosphatase	Sedoheptulose-1,7-bisphosphate $\rightarrow$ sedoheptulose-7-phosphate + P_i
(10)	Transketolase	Sedoheptulose-7-phosphate + glyceraldehyde-3-phosphate $\rightarrow$ ribose-5-phosphate + xylulose-5-phosphate
(11)	Pentose epimerase	2 Xylulose-5-phosphate + ribose-5-phosphate $\rightarrow$ 3 ribulose-5-phosphate
(12)	Phosphoribulokinase	3 Ribulose-5-phosphate + 3ATP $\rightarrow$ 3 ribulose-1,5-bisphosphate + 3ADP
	SUM:	$3CO_2$ + 9ATP + 6NADPH + $6H^+ \rightarrow$ dihydroxyacetone phosphate + 9ADP + $6NADP^+$ + $8P_i$

phosphate needed for hexose formation requires that the photochemical system provide 12 mol of NADPH and 18 mol of ATP; fixation of one CO_2 requires two NADPH and three ATP. The validity of this reaction scheme has been shown by reconstruction of the entire process with purified enzymes.

After dihydroxyacetone phosphate has crossed into the cytosol on the P_i translocase (Fig. 19.2), it is converted to usable carbohydrate by the enzymes of glycolysis. Half is reconverted to glyceraldehyde-3-phosphate by isomerase; the mixture converts to fructose-1,6-bisphosphate, and hydrolysis then yields fructose-6-phosphate, which is converted to glucose-6-phosphate and thence to glucose-1-phosphate. The latter may be converted either to starch via ADP-glucose (page 488) or to sucrose by way of UDP-glucose (page 487). Formation of either of the latter requires investment of ATP; for example,

$$\text{ATP} + \text{glucose-1-phosphate} \longrightarrow \text{ADP-glucose} + \text{PP}_i$$

There is need, therefore, for additional ATP in the cytosol, but ATP generated photosynthetically by the thylakoid system (see below) cannot leave the chloroplast. If the rate of ATP formation by aerobic glycolysis and mitochondrial respiration is insufficient, a special means for exporting the equivalent of ATP from the stroma is available, as shown in Fig. 19.3. By this arrangement, NADH as well as ATP becomes available in the cytosol. If the reducing power is in excess, it can be returned to the stroma either by a malate/oxaloacetate cycle or a malate/aspartate cycle analogous to that across mitochondrial membranes (Fig. 15.13).

Figure 19.2 Summary of photosynthesis of sucrose from CO_2. Reduction of $NADP^+$ and synthesis of ATP occur at the thylakoid membrane. CO_2 fixation and net synthesis of dihydroxyacetone phosphate occur in the stroma. Synthesis of hexose, sucrose, and starch occurs in the cytosol; this is discussed later. [*Modified from H. W. Heldt, Horiz. Biochem. Biophys 2:218 (1976).*]

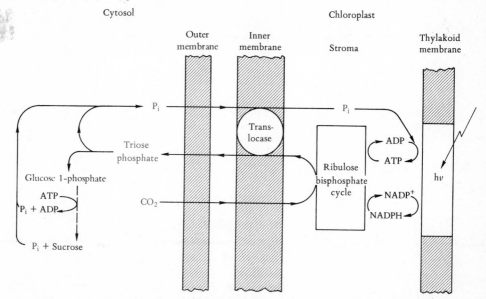

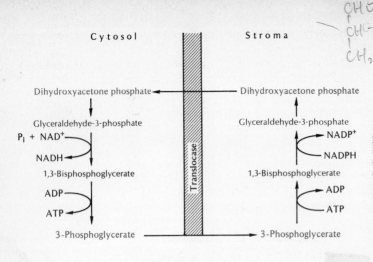

Figure 19.3 The dihydroxyacetone phosphate shuttle. Oxidation of glyceraldehyde-3-phosphate in the cytosol generates NADH and ATP, available for diverse synthetic purposes. The 3-phosphoglycerate returns to the stroma where, using photosynthetically generated ATP and NADPH, it is reconverted to dihydroxyacetone phosphate to repeat the process.

In Bacteria Most bacteria have a Benson-Calvin cycle. However, in some photosynthetic bacteria, the first detectable ^{14}C-labeled compound is alanine, suggesting that the initial event may be formation of pyruvate from CO_2 and a C_2 compound presumed to be acetyl CoA, followed by a transamination reaction (Chap. 22). The concomitant appearance of label in glutamate suggests similar formation of α-ketoglutarate from CO_2 and succinyl CoA. A tentative scheme for the pathway by which CO_2 fixed by these mechanisms can be accumulated as hexose is shown in Fig. 19.4. In photosynthetic bacteria, glyceraldehyde-3-phosphate is linked to NAD rather than NADP as in algal and plant chloroplasts.

Role of C_4 Dicarboxylic Acids in Some Plants Tropical grasses and various cotyledonous organisms tolerate extremes of climate, efficiently fix CO_2 at low ambient concentrations, and have low rates of photorespiration (see below). These plants include some of the latest of the angiosperms to evolve, e.g., cane, sorghum, maize, and most of the hardier weeds. Characteristically, when the leaves of these plants are exposed to $^{14}CO_2$, the first compounds labeled are

Figure 19.4 Tentative scheme of a pathway for accumulation of hexose in some photosynthetic bacteria.

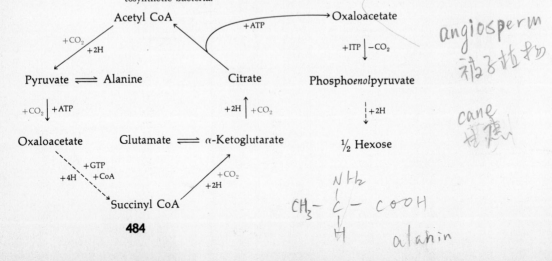

484

oxaloacetate, malate, and aspartate.   Near the air surface of such leaves are the *mesophyll cells,* which engage in photosynthesis but sometimes have sparsely distributed chloroplasts and loosely packed grana.   The vascular bundle is surrounded by bundle sheath cells, which are rich in chloroplasts with densely packed grana.   This arrangement makes possible a transport system (Fig. 19.5) that maintains $[CO_2]$ in the bundle sheath cells of at least $10\ \mu M$, twice that possible when air is in equilibrium with cell fluid at pH 7.5 at 20°C.   Moreover, at the distance of the bundle sheath cells from the surface and with the intervening barriers to diffusion, the concentration of CO_2 in those cells would otherwise be no more than $1\ \mu M$, almost an order of magnitude too low to sustain the ribulose-1,5-bisphosphate reaction.   The C_4 fixation system serves, in effect, as a pump that maintains $[CO_2]$ in the bundle sheath cells at greater than $10\ \mu M$, so that light intensity rather than $[CO_2]$ limits the rate of photosynthesis.

The components of the actual systems vary somewhat among C_4 plants.   In the most frequent arrangement, aspartate is exchanged for alanine, as shown in Fig. 19.6.   Aspartate is converted to oxaloacetate in the cytoplasm in some species and in others in the mitochondria.   In yet others, malate from the mesophyll cells is exchanged for pyruvate.   This is the simplest system, but it involves delivery of reducing power as well as of CO_2.

Photorespiration   A rather surprising aspect of plant metabolism is the marked enhancement of respiration (O_2 consumption) that occurs when most plants are illuminated.   This apparently wasteful phenomenon is due to the fact that *ribulose-1,5-bisphosphate carboxylase* is also an *oxygenase* and can catalyze the reaction

$$\text{Ribulose-1,5-bisphosphate} + O_2 \longrightarrow \text{3-phosphoglycerate} + \text{2-phosphoglycolate}$$

V_{max} for the carboxylation is about 4 times that for the oxygenation when both are measured under ideal conditions.   By increasing [NADH] and reducing [ribulose-1,5-bisphosphate], illumination activates the oxygenation reaction to the same extent as it does the carboxylation reaction.   At 25°C and pH 8.4, $K_m^{CO_2}$ is $12\ \mu M$ and $K_m^{O_2}$ about $300\ \mu M$.   Hence, in air the carboxylation reaction proceeds only 2.5 times as rapidly as the oxygenation.

The phosphoglycolate thus formed is not converted to carbohydrate; it is hydrolyzed to $HO-CH_2-COOH$ (glycolate) and oxidized by a peroxisomal enzyme to $O{=}CH-COOH$ (glyoxylate); some of the latter is transaminated to form glycine (Chap. 22).   Glycine can also serve as a precursor of serine.

Figure 19.5 General metabolic plan of C_4 plants.   In the mesophyll cells, CO_2 is fixed into a C_4 compound that is transported to the bundle sheath cells, where decarboxylation maintains a $[CO_2]$ adequate for efficient operation of ribulose-1,5-bisphosphate carboxylase.

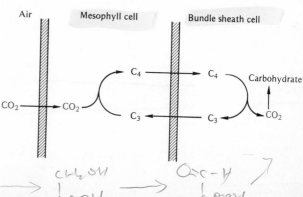

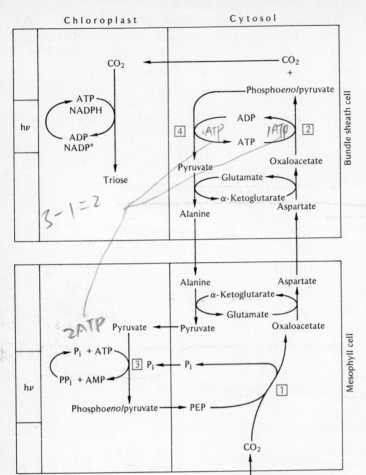
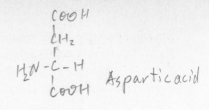

Figure 19.6 One arrangement for metabolic transport in a C_4 plant. The energy source is ATP, synthesized in the chloroplast of the mesophyll cell, which is used to make phospho*enol*pyruvate in the *pyruvate–phosphate dikinase* reaction [reaction (3)]. In the cytosol of that cell, CO_2 is fixed by phospho*enol*pyruvate carboxylase [reaction (1)] into oxaloacetate, which is then transaminated to aspartate; the latter moves to the cytosol of the bundle sheath cell. The amino group is lost by transamination, and the re-formed oxaloacetate is decarboxylated by phospho*enol*pyruvate carboxykinase to CO_2 and phospho*enol*pyruvate [reaction (2)]. The CO_2 is then available to the stroma photosynthetic system for carbohydrate synthesis. Before returning to the mesophyll cell, the phospho*enol*pyruvate donates its phosphate to ADP, catalyzed by pyruvate kinase [reaction (4)], and the resultant pyruvate accepts from glutamate the amino group earlier surrendered by aspartate. Alanine returns to the mesophyll cell, where it is reconverted to pyruvate, which is available to repeat the cycle. The several small cycles require no energy input, the two ATP equivalents used in reaction (3) sufficing to drive the entire process.

Although photorespiration was not a problem for the earliest photosynthesizing cells, since there was so little O_2 in the atmosphere, it is perhaps surprising that this inefficient feature of this enzyme has survived. The slender margin by which the enzyme succeeds is evident in a mutant of *Chlamydomonas reinhardii*, the carboxylase of which is twice as active an oxygenase and one-third as active a carboxylase as the wild type. This strain is incapable of growth by photosynthetic accumulation of carbohydrate, and glucose must be provided in the medium.

Photorespiration constitutes an enormous drain on the worldwide photosynthetic yield of food crops and trees. A great advantage of the C_4 plants, described earlier, is that their carboxylase operates in an environment in which $[CO_2]$ is from 3 to 10 times greater than might otherwise be the case while $[O_2]$ is diminished somewhat, so that carboxylation is favored over oxygenation. This is particularly significant for tropical grasses that must cope with hot environments. The activation energy E_a for carboxylation is only two-thirds that for oxygenation; hence, as temperature rises, oxygenating activity will increase more than carboxylating activity.

486

Sucrose Formation As hexose is accumulated in the cytosol, glucose-1-phosphate is used for formation of UDP-glucose by a pyrophosphorylase (Chap. 18); this is followed by two consecutive reactions.

$$\text{UDP-glucose} + \text{fructose-6-phosphate} \longrightarrow \text{sucrose-6-phosphate} + \text{UDP} \qquad (1)$$

$$\text{Sucrose-6-phosphate} \xrightarrow{\text{H}_2\text{O}} \text{sucrose} + \text{P}_i \qquad (2)$$

$\Delta G°$ for hydrolysis of the $1 \rightarrow 2$ bond between the anomeric carbons of sucrose is unusually high for a glycoside, -6.6 kcal/mol. Since $\Delta G°$ for hydrolysis of UDP-glucose is about -7.5 kcal/mol, reaction (1) is favored in the direction shown, but it is hydrolysis of the 6-phosphate in reaction (2) that ensures net sucrose synthesis and supplies P_i for entry into the chloroplasts. *Sucrose-6-phosphate synthetase* exhibits sigmoidal kinetics with respect to both substrates; hence its action is allosterically regulated.

Little sucrose is retained in the photosynthesizing cells of higher plants. Most of the sucrose is delivered to the nonphotosynthesizing tissues, a process most dramatic in the sugar cane and the sugar beet.

Homopolysaccharide Synthesis

Bacteria Diverse, nonphotosynthesizing bacteria can utilize sucrose or maltose for homopolysaccharide formation. Some bacterial transglycosylase reactions are

$$n \text{ Maltose} \underset{\textit{Escherichia coli}}{\overset{\text{amylomaltase}}{\rightleftharpoons}} \text{amylose} + n \text{ glucose}$$

$$n \text{ Sucrose} \underset{\textit{Neisseria perflava}}{\overset{\text{amylosucrase}}{\rightleftharpoons}} \text{amylose} + n \text{ fructose}$$

$$n \text{ Sucrose} \underset{\textit{Leuconostoc mesenteroides}}{\overset{\text{dextransucrase}}{\rightleftharpoons}} \text{dextran} + n \text{ fructose}$$

$$n \text{ Sucrose} \underset{\textit{Bacillus megaterium}}{\overset{\text{levansucrase}}{\rightleftharpoons}} \text{levan} + n \text{ glucose}$$

Dextran is a linear polymer containing isomaltose, Glcα—6Glc, as the repeating unit. Levan contains fructose residues in 2,6-fructosidic linkage. Extracellular accumulation of these polymers results in dental plaque formation and thus accounts for the high correlation between sucrose ingestion and dental caries.

Plants In higher plants and algae, starch formation proceeds from a nucleoside diphosphate glucose. The formation, from sucrose, of UDP-glucose and ADP-glucose could be accomplished by reversal of the sucrose synthetase reaction, for which $K_{eq} = 0.15$.

$$\text{ADP} + \text{sucrose} \rightleftharpoons \text{ADP-glucose} + \text{fructose}$$

$$\text{UDP} + \text{sucrose} \rightleftharpoons \text{UDP-glucose} + \text{fructose}$$

The UDP-glucose is used for cellulose formation and for diverse transformations already considered in Chap. 18. ADP-glucose is the substrate for starch formation, which is otherwise analogous to glycogen synthesis from UDP-glucose in animals.   It is not clear what fraction of the ADP-glucose is formed by this mechanism.   Some starch-synthesizing cells also contain a *sucrase* that catalyzes hydrolysis of sucrose to glucose and fructose, making both available to those pathways leading to formation of glucose-1-phosphate.   Other cells contain a *sucrose phosphorylase*, which catalyzes the reaction

$$\text{Sucrose} + P_i \rightleftharpoons \alpha\text{-D-glucose-1-phosphate} + \text{fructose}$$

Starch formation in photosynthetic cells begins from glucose-1-phosphate without sucrose as an intermediate.

Regulation of Polysaccharide Formation in Plants and Bacteria

The immediate precursor for synthesis of storage polysaccharides, such as plant starch and glycogen-like polymers in diverse bacteria, is ADP-glucose.   Since in these cells ADP-glucose is not known to participate in other reactions, the committed step leading to polymer formation is

$$\text{ATP} + \text{glucose-1-phosphate} \rightleftharpoons \text{ADP-glucose} + PP_i$$

Polymer synthesis is controlled by regulation of the *pyrophosphorylase* that catalyzes this reaction.   Control is exercised differently in various classes of organisms, as shown in Table 19.2.

Plants exhibit positive feed-forward control of the pyrophosphorylase by 3-phosphoglycerate, the first product of photosynthetic CO_2 fixation. V_{max} is thereby increased ten- to a hundredfold in various species; in barley, as little as 0.007 mM 3-phosphoglycerate suffices for maximal activation.   Other glycolytic intermediates, particularly fructose-6-phosphate, have similar but lesser effects. P_i, in contrast, is a negative effector, with K_i varying among species from 0.02 to 0.2 mM.

Enterobacterial pyrophosphorylases are stimulated by fructose bisphosphate and NADPH, which accumulate when growth is limited by the availability of any nutrient except glucose itself.   *Rhodospirillum rubrum*, a photosynthetic organism that cannot metabolize external glucose but lives anaerobically by utilizing intermediates of the tricarboxylic acid cycle, has a pyrophosphorylase that is stimulated by pyruvate.

TABLE 19.2   Activators and Inhibitors of ADP-Glucose Pyrophosphorylases from Various Sources

Source	Primary activator	Secondary activator	Inhibitor
Leaves of higher plants, green algae	3-Phosphoglycerate	Fructose-6-P, fructose-bis-P, phospho*enol*pyruvate	P_i
E. coli, Aerobacter aerogenes, A. cloacae, Salmonella typhimurium, Citrobacter freundii, E. aurescens	Fructose-bis-P, NADPH, pyridoxal-P	2-Phosphoglycerate, glyceraldehyde-3-phosphate, phospho*enol*pyruvate	AMP
Arthrobacter viscosus		Pyruvate	P_i, AMP, ADP
Agrobacterium tumefaciens	Fructose-6-P	Ribose-5-P	P_i, AMP, ADP
Rhodopseudomonas capsulata		Deoxyribose-5-P	
Rhodospirillum rubrum	Pyruvate		
Serratia marcescens			AMP

In many cases, the enzyme shows a sigmoidal velocity curve with respect to ATP itself. In general, the activators listed in Table 19.2 cause a marked decrease in K_m for both ATP and glucose-1-phosphate. The strong stimulation of the pyrophosphorylases of the *Enterobacteria* by pyridoxal phosphate suggests that as in animal glycogen phosphorylase, this cofactor probably serves in the active site (Chap. 18). The advantage to the cell of the fact that P_i and AMP serve as inhibitors of these pyrophosphorylases will be obvious.

THE PHOTOSYNTHETIC PROCESS

Molecular Absorption of Light Quanta

The fundamental event in photosynthesis is the absorption of light energy by chlorophyll. Chlorophyll a, a magnesium porphyrin with a fifth isocyclic ring in which one pyrrole ring is partially reduced, is the principal chlorophyll of algae and higher plants. Both acidic side chains are esterified, one as a methyl ester and the other as a phytyl ester. The non-ionic magnesium atom is held by two covalent and two coordinate linkages. In chlorophyll b, ring II bears a formyl instead of a methyl group. *Pheophytin* lacks the magnesium atom of chlorophyll.

Chlorophyll a

In order to understand how light energy is used to drive the synthesis of ATP and NADPH, it is necessary to consider some of the fundamental principles of molecular absorption of light.

The energy of a photon (one quantum of electromagnetic radiation) is given by

$$E = \frac{hc}{\lambda}$$

where h is Planck's constant $(6.6 \times 10^{-27}$ erg/s), c is the velocity of light $(3 \times 10^{10}$ cm/s), and λ is the wavelength. By converting ergs to electronvolts and expressing λ in nanometers, we have

$$E = \frac{1235}{\lambda} \quad \text{electronvolts}$$

[handwritten: $= \frac{1235}{675} = 1.84\,eV = 92.4\,k$]

For example, for red light at 675 nm, one quantum corresponds to 1.84 eV; 1 mol of such quanta (1 einstein) corresponds to 42.4 kcal.

A molecule can absorb light quanta only of specific wavelength; this is usually displayed as the *absorption spectrum*. The absorbed energy is then part of the molecule, which is said to be *activated*. The absorbed photon must correspond to the energy required to perturb an electron, raising it to an orbital more remote from the atomic nucleus than in the ground state. Chlorophyll has a continuing system of alternating single and double bonds. In structures of this type, the electrons in the bonds formed by overlap of carbon p orbitals, e.g., six per benzene ring, must be assigned to the entire molecule rather than to specific interatomic bonds, and when light is absorbed, it is an electron in this π-electron system that is in a higher energetic state. When a photon is absorbed, the molecule can exist in a transient excited state, the *singlet*, with a half-life of about 1 ns; it returns to the ground state by loss of energy as heat, by emission of a light quantum (fluorescence), or by ejection of the electron in a more conventional "chemical" electron transfer. The primary photochemical event of photosynthesis occurs from the singlet state.

[handwritten margin note: coulombic $\frac{1}{6}\frac{1}{\kappa}$ / exciton $\frac{1}{6}\frac{1}{\kappa}\frac{x}{8}$]

Activated molecules can also transfer their excess energy with high efficiency to adjacent molecules of the same or other chemical species. This occurs in two ways. If molecules of the same kind are appropriately oriented and at 1 to 2 nm distance, transfer occurs as *exciton* energy; for a given electronic transition, coulombic interactions between the transition dipole moments of the neighboring molecules give rise to a set of discrete exciton energy levels. If the molecules are 5 to 10 nm apart, energy exchange occurs by *induced resonance energy transfer,* a process equivalent to what might occur if an excited molecule were to decay by fluorescence and a second molecule were to absorb the fluoresced quantum although, in fact, no actual light is emitted and absorbed. In this instance, vibration of an electronic dipole of the second molecule is initiated in a manner analogous to the vibrations induced in a tuning fork by the close presence of another tuning fork already vibrating; it is a feasible mode of energy transfer if the fluorescence spectrum of the excited molecule significantly overlaps the absorption spectrum of the second molecule and is complete within 1 to 10 ps.

The Thylakoid Membrane

The use of light energy to drive the synthesis of ATP and the reduction of NADP$^+$ occurs in the *thylakoid* membrane, or *lamella*. As shown in Fig. 19.1, each thylakoid is surrounded by a lipid bilayer membrane and stacks of such thylakoids are the *granum*.

The major classes of lipids of the thylakoid membrane, which occur in proportions of about $10:2:1$, respectively, are: (1) monogalactosylglycerides and digalactosylglycerides; (2) a mixture of phosphoglycerides, including phosphatidylcholine, phosphatidylethanolamine, phosphatidylinositol, and phosphatidylglycerol (Chap. 7); and (3) a sulfolipid, sulfoquinovosyl diglyceride.

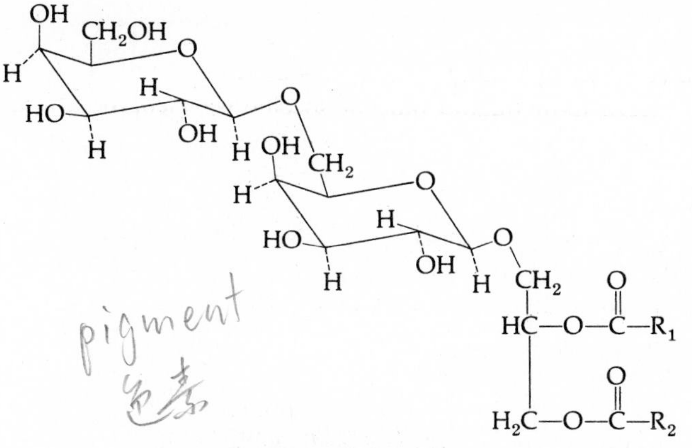

Digalactosyl glyceride

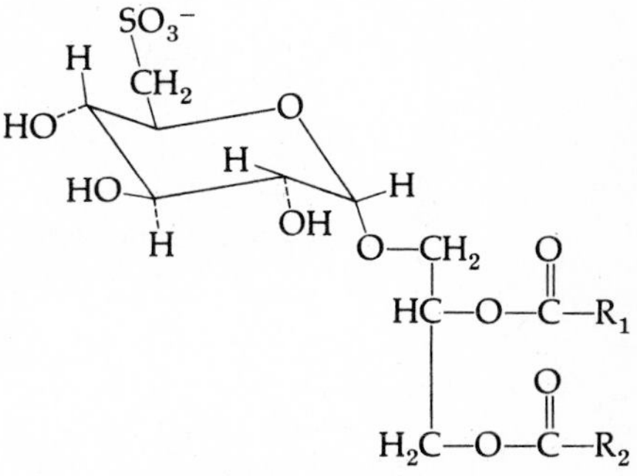

Sulfoquinovosyl diglyceride

Photosynthetic Pigment—Protein Complexes

Most, perhaps all, of the chlorophyll of all photosynthetic species is noncovalently bound to specific proteins. Most of these pigment complexes function as light-harvesting antennae to capture the light and transfer the energy to a photochemical reaction center. Only a small part of the chlorophyll (Chl) of an organism acts in the reaction center as an energy transducer. Two Chl molecules form a "special pair" which accept a captured quantum from the antenna and then transfer an electron to an acceptor, producing a charge separation and hence chemical potential energy (see below). Special pairs are named as P (for pigment) and the maximal absorption wavelength in the red, e.g., P_{700}, P_{680}. Each reaction center with its associated antenna and immediate electron carriers [e.g., quinones, (FeS) centers, cytochromes, etc.] is termed a *photosystem*. Algae and higher plants have two photosystems; photosynthetic bacteria have only one.

Green Plants All plants apparently possess a photosynthetic apparatus which is tripartite, comprising photosystem I, photosystem II, and a light-harvesting array. Each part is composed of more than one kind of pigment-protein complex. Photosystems I and II contain up to half of the photosynthetic pigment molecules; the remainder is in the light-harvesting array. The kinds of pigments in the array vary in different classes of plants, each possessing its own kind of accessory pigments (see below). In green plants, the array is composed of Chl a, Chl b, and carotenoid in a form termed the *light-harvesting Chl a/b-protein*.

Preparations of the Chl a/b-protein from most species contain one or two kinds of major polypeptides (MW $\approx$ 25,000) and a slightly smaller, minor peptide component. Each polypeptide is bound to 6 to 7 Chl molecules (3 or 4 Chl a and 3 Chl b) and a smaller amount of carotenoid.

The compositions of photosystems I (P_{700}) and II (P_{680}) are similar throughout the plant kingdom. Photosystem I has been obtained with a molar ratio of Chl a to P_{700} between 80 to 1 and 20 to 1; a ratio of 40 to 1 being usually obtained. Nevertheless, there is only one Chl-protein band on gel electrophoresis with a molecular weight of 100,000 to 135,000 and two different apoproteins of 60,000 to 70,000. P_{680} complexes have not been well-characterized.

The chlorophyll-protein complexes from higher plants are extremely hydrophobic, are integral components of the membrane, and are extractable only with detergent solutions. Indeed, some of the polypeptides are soluble in certain organic solvents. Like many other membrane-bound proteins, the proportion of hydrophobic amino acids is high.

Bacteria The chlorophyll-protein presently known in greatest detail has been obtained from the green photosynthetic bacterium *Chlorobium limicola*. This complex (MW $\approx$ 150,000) consists of three identical subunits, each containing seven molecules of *bacteriochlorophyll* (BChl). The latter differs from chlorophyll a principally in having an acetyl group (CH_3—CO—) rather than a vinyl group (CH_2=CH—) on ring I of the porphin nucleus; also the double bond at carbon atoms 3 and 4 is reduced, resulting in a shift of the absorption maximum almost 100 nm to the red. The seven BChl-ring systems are all caged inside a hollow cylinder formed by about 15 antiparallel strands of β-pleated-sheets, as shown in Fig. 19.7. The BChl molecules, each about 1.2 nm from two others, can readily participate in exciton energy transfer. The BChl-protein of *Chlorobium* is water-soluble unlike the pigment complexes of higher plants.

Accessory Pigments Photoactivation of chlorophyll a is the photochemical event of photosynthesis. Chlorophyll solutions appear green; viz., chlorophyll a absorbs light in the blue (400 to 450 nm) and in the red (640 to 680 nm) regions but is almost transparent in the green and yellow regions of the spectrum. In intense light, this is of no physiological consequence. In dim light, however, when photosynthesis is limited by the amount of light, it may be expected that the plant would be unable to benefit from a large fraction of the impinging radiation (450 to 600 nm) if chlorophyll a were its only functional pigment. However, this is not the case.

The weak absorbance of chlorophyll a in the green and yellow regions is particularly evident in plants that live submerged in water. Light of longer wavelengths is increasingly absorbed as a function of depth; more than 99 percent of light of wavelengths greater than 600 nm is absorbed in the upper 30 ft. This problem has been solved by the presence of additional pigments. In diatoms, the brown algae, and the dinoflagellates, additional light-gathering pigments are chlorophyll c and a carotenoid, *peridinin*.

Peridinin

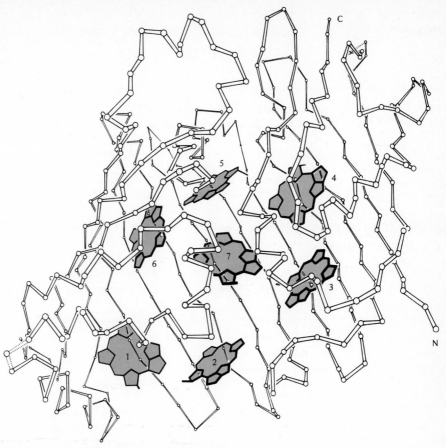

Figure 19.7 Schematic diagram showing the arrangement of the polypeptide backbone and chlorophyll core of one subunit of the bacteriochlorophyll-protein from *Chlorobium limicola*. Presumed α-carbon positions are indicated by circles. The connectivity of the polypeptide chain is in some instances uncertain. For clarity, the magnesium atoms, the chlorophyll-ring substituents, and the phytyl chains, except for the first bond, are omitted. The direction of view is from the threefold axis, which is horizontal, toward the exterior of the molecule. [*Courtesy of Professor B. W. Matthews.*]

Dinoflagellates, e.g., *Gonyaulax polyedra* (the organism responsible for "red tides"), contain a water-soluble protein (MW ≈ 32,000) to which are bound four molecules of peridinin and one of chlorophyll a. The complex exhibits a broad absorption from 400 to 550 nm, thus enabling the organism to absorb wavelengths of light that are poorly absorbed by chlorophyll a.

A different group of pigments, structurally related to mammalian bile pigments (Chap. M3), serves the same function in red and blue-green algae. These pigments are present as chromoproteins in *phycobilisomes,* organized bodies about 32 by 48 nm with a flattened surface attached to the stroma surface of the thylakoid or photosynthetic lamellae. There is one phycobilisome for approximately 2000 chlorophyll molecules. Phycobilisomes contain several kinds of phycobiliproteins; the major ones are *phycoerythrin, phycocyanin,* and *allophycocyanin.*

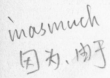

Phycoerythrobilin

Each of these proteins consists of two subunits, α (10,000 to 20,000 daltons) and β (14,000 to 30,000 daltons), the precise size depending on the species; the most common assembly form is $(\alpha\beta)_3$ or $(\alpha\beta)_6$. Each subunit carries three molecules of its pigment. Thus, phycoerythrin $(\alpha_6\beta_6\gamma)$ carries a total of 36 molecules of phycoerythrobilin with two molecules of phycourobilin on its single γ subunit. The pigments are open-chain tetrapyrroles covalently bound to the protein. The relative amounts of the three types of pigment proteins in certain red algae, for example, are 84 percent phycoerythrin, 11 percent phycocyanin, and 5 percent allophycocyanin.

Figure 19.8b illustrates how such systems make possible the use of different regions of the electromagnetic spectrum for photosynthesis in three kinds of photosynthetic organisms.

The Photochemical Event

As already noted (page 491), less than 1 percent of the chlorophyll molecules participate in the *reaction center*. The remainder of the chlorophyll-protein complexes and the diverse accessory pigments constitute a light-harvesting antenna. The energy of photons absorbed anywhere in the complex of 100 to 250 of the various pigment molecules is rapidly transferred to the special chlorophyll at the reaction center with an efficiency of more than 90 percent, thereby permitting the unit to function effectively both in bright and dim light, e.g., at 0.1 percent of the light level at noon of a clear day. This very high efficiency strongly indicates that the entire unit is in a highly ordered array. It follows that the specific chlorophyll receiving this energy must be an *energy sink;* its absorption spectrum must have a principal maximum at a wavelength longer than those of the other components of the system. Inasmuch as longer-wavelength light has a lower energy per quantum, it must be transferred successively by induced resonance energy transfer (page 490) from pigments which absorb at shorter to those which absorb at longer wavelengths (red light). This is indeed the case, as shown for the pigments of a phycobilisome in Table 19.3. At full sunlight, it is the turnover time of a subsequent rate-limiting chemical event

TABLE 19.3 The Energy-Transfer Sequence of a Phycobilisome

	Phycoerythrin $\longrightarrow$	R-Phycocyanin $\longrightarrow$	Allophycocyanin $\longrightarrow$	Chlorophyll a
Absorption maximum, nm	545	553, 618	650	670
Fluorescence maximum, nm	575	636	660	685

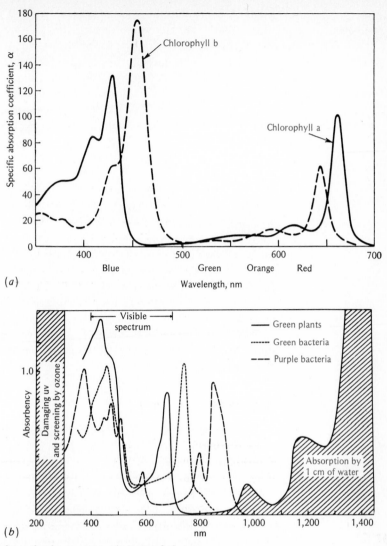

Figure 19.8 Enhancement of light absorption by accessory pigments of photosynthetic cells. (*a*) The absorption spectra of chlorophyll a and b, dissolved in ether. [*Adapted from H. H. Seliger and W. D. McElroy, Light: Physical and Biological Action, Academic Press Inc., New York, 1965, p. 225.*] (*b*) The increased availability of energy for photosynthesis due to light absorption by accessory pigments in regions of the spectrum where chlorophyll is relatively transparent. [*From R. K. Clayton, Molecular Physics in Photosynthesis, Blaisdell, New York, 1966.*]

that limits the photosynthetic rate; this is about 5 ms. Only in dim light is the photosynthetic rate linearly proportional to light intensity.

The specific role of the chlorophyll a at the reaction center is to effect a photochemical separation of oxidizing and reducing power. Such a mechanism can be represented schematically as

$$\frac{A^0}{\underset{B^0}{Chl}} \xrightarrow{h\nu} \frac{A^0}{\underset{B^0}{Chl^*}} \longrightarrow \frac{A^-}{\underset{B^+}{Chl}} \tag{1}$$

where A is a potential electron acceptor and B is a potential electron donor situated on either side of a chlorophyll (Chl) molecule; Chl* is the photoactivated form of Chl, A^- is the reduced form of A, and B^+ is the oxidized form of B. The simplest mechanism to accomplish this is ejection of an electron from the photoactivated chlorophyll, forming a chlorophyll free radical (Chl^+). The electron is accepted by A as the chlorophyll radical regains an electron from B on its other side. It will be recalled that capture of a photon at 675 nm would permit a separation of A^- and B^+ of 1.8 eV if the process were 100 percent efficient.

$$\begin{array}{ccccc} A^0 & & A^0 & A^- & A^- \\ \hline Chl^0 & \xrightarrow{h\nu} & Chl^* \longrightarrow & Chl^+ \longrightarrow & Chl^0 \\ \hline B^0 & & B^0 & B^0 & B^+ \end{array} \qquad (2)$$

A possible wasteful back reaction between A^- and B^+ or Chl^+ is largely circumvented either by rapidly moving the electron from A^- to the next electron acceptor in the sequence or by rapidly reducing Chl^+. In what follows, it is assumed that equation (2) represents a satisfactory model of the operating mechanism. It will be apparent that reaction (2) cannot be repeated until A^- has been reoxidized and B^+ again reduced. In thylakoid membranes and bacterial chromatophores, this is accomplished by an organized electron-transfer system resembling that of the mitochondrial inner membrane, a system of electron carriers so arranged that passage of the stream of electrons makes synthesis of ATP possible.

Bacterial Photosynthesis

It is instructive to consider photosynthesis in bacteria before examining it in algae and higher plants. The process will be easier to understand if one accepts some simple evolutionary concepts. The earliest cells may be assumed to have lived in a relatively rich nutrient broth. Glycolytic generation of ATP and of reducing power became established in such organisms, which dwelt in a reducing atmosphere free of O_2. The earliest organized photochemical units seem to have served for transduction of electromagnetic energy into a form of chemical energy easily convertible to ATP rather than to reducing power. If the medium already provided glucose, there was little advantage to acquisition of a mode for glucose formation. As the supply of external carbohydrate dwindled, however, ability to synthesize carbohydrates became advantageous. The form of CO_2 fixation into carbohydrates known to us is the network of reactions summarized in Table 19.1, which indicates that to accumulate 1 hexose molecule 18 molecules of ATP and 12 of NADH must be provided. The medium already contained sources of reducing power and with the appearance of the earliest photochemical machine for production of ATP or its equivalent, e.g., inorganic pyrophosphate, development of the precursors of photosynthetic bacteria became possible. These organisms thrived because they could utilize two essentially independent processes:

$$H_2A + NAD^+ \longrightarrow A + NADH + H^+ \qquad (1)$$

$$ADP + P_i \xrightarrow{light} ATP \qquad (2)$$

The simplest form of H_2A is H_2 itself; organisms that possess hydrogenase can use hydrogen to reduce NAD^+. Others use inorganic compounds such as S^{2-} or $S_2O_3^{2-}$, as noted earlier. Such organisms as the various *Rhodospirillum* species, the green bacteria, etc., which possess all the enzymes of the CO_2-fixing system described earlier, use chemical reducing power and photosynthetic ATP to accumulate the carbohydrate required for growth and metabolism. It is noteworthy that many of these organisms utilize the same NAD-linked glyceraldehyde-3-phosphate dehydrogenase for glycolysis in the dark as for CO_2 fixation in the light.

The Bacterial Photosynthetic Center The photosynthetic apparatus of these organisms is simpler than that of higher plants. In essence, at the photoreaction center an electron is ejected from a compound with a midpoint potential E_0' somewhat higher than that of mitochondrial cytochrome c, with sufficient energy to reduce a chromophore molecule. Because bacterial chromatophores contain only one photosystem (see below) and have less antenna chlorophyll per reaction center than chloroplasts in higher plants, it has been possible to purify the reaction center itself from several photosynthetic bacteria, e.g., *Rhodospirillum rubrum* and *Rhodopseudomonas*. In most cases the reaction center consists of a molecule of a somewhat nonpolar protein (MW $\approx$ 73,000) built of three nonidentical subunits of molecular weight of approximately 21,000, 24,000, and 28,000. To this protein are bound two molecules of bacteriochlorophyll, one molecule of a bacteriochlorophyll dimer $(BChl)_2$, two molecules of bacteriopheophytin (bacteriochlorophyll lacking Mg^{2+}), one atom of Fe^{2+}, two molecules of ubiquinone (page 501), and one molecule of a carotenoid such as spheroidene.

CH₃O

Spheroidene

The carotenoid of the reaction center is not part of the light-harvesting arrangement and is sometimes not the same carotenoid as that in the bulk carotenoids of the membrane. The reaction-center carotenoid appears to protect activated chlorophyll should the latter be unable to transfer an electron to other components of the center and be in danger of self-destruction by chemical decomposition. In such circumstances, the chlorophyll transfers energy to the carotenoid, which then decays harmlessly to the ground state.

Absorption of a photon by the center initiates the process of *cyclic photophosphorylation*. As described below, activated chlorophyll serves as both the initial electron donor and the ultimate electron acceptor in the cycle. Each photon absorbed causes one electron to go around the circuit. As this occurs, two protons are secreted to the outside. Their return through the coupling factor ATPase generates one ATP.

Within 3 ps of absorption by the reaction center of a photon at 530 nm, an absorption band of bacteriopheophytin, bleaching is observed in the near infrared at 865 nm, the absorption maximum of $(BChl)_2$. This is interpreted to mean energy transfer from activated bacteriopheophytin to $(BChl)_2$, probably with formation of $BChl^+ \cdot I^-$. [I is an intermediate between $(BChl)_2$ and a quinone, and it is one of the two bacteriopheophytin molecules and may also involve a BChl molecule.] In less than 200 ps thereafter, I^- donates an electron to the iron-ubiquinone complex, which, being reduced to $(Fe^{2+} \cdot UQ_2)^-$, is the primary electron acceptor of this system. Within a few microseconds, the oxidized chlorophyll receives an electron from a molecule of cytochrome c, and the $(BChl)_2$ free radical returns to the ground state. As shown in Fig. 19.9, electron transport then proceeds through quinones and a b-type cytochrome to cytochrome c_2, to complete the circuit. One molecule of ATP is formed incident to this electron transport, the mechanism again involving vectorial proton secretion and a Ca^{2+}-sensitive ATPase, as in mitochondria and chloroplasts. This will be developed further (page 502 ff.). Since a light quantum at 890 nm, the absorption maximum of bacteriochlorophyll, is equivalent to 32 kcal/mol, it easily suffices to drive formation of one molecule of ATP.

Electron Transport in Bacterial Photosynthesis Like eukaryotic photosynthetic organisms, photosynthetic bacteria fix CO_2 into carbohydrate, utilizing reduced pyridine nucleotides and ATP to drive the process. ATP is available, as we have seen, by cyclic photophosphorylation, but the organisms do not liberate

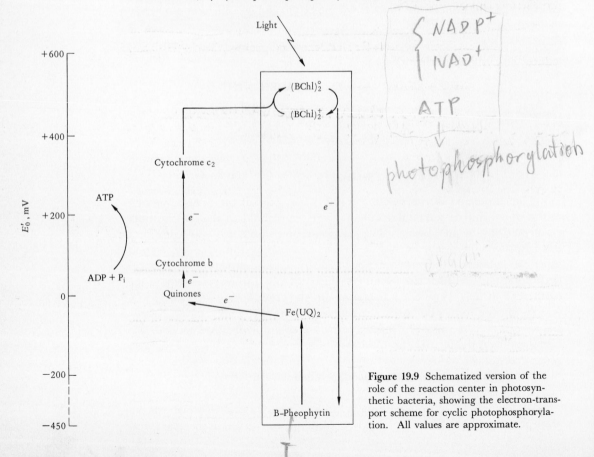

Figure 19.9 Schematized version of the role of the reaction center in photosynthetic bacteria, showing the electron-transport scheme for cyclic photophosphorylation. All values are approximate.

O$_2$ as photosynthesis proceeds; they are unable to use light energy to photolyze water and use its hydrogen as a reducing agent as the eukaryotes do. Hence, an external source of chemical reducing power is necessary. Many reductants that can serve, e.g., succinate, and S^{2-}, do so despite the fact that E_0' for their oxidations is insufficient to reduce NAD$^+$. This is possible because of the equivalence—in the photosynthetic membrane, as in mitochondria—of ΔE, $\Delta \Psi$, and $\sim$P (Chap. 15). Anaerobically in the dark these membranes will also catalyze

$$\text{Succinate} + \text{NAD}^+ + \text{ATP} \longrightarrow \text{fumarate} + \text{NADH} + \text{ADP} + P_i$$

In the light, addition of external ATP is unnecessary; the photoactivated membrane system can catalyze

$$\text{Succinate} + \text{NAD}^+ \xrightarrow{h\nu} \text{fumarate} + \text{NADH} + \text{H}^+$$

Finally, certain photosynthetic microorganisms can utilize H$_2$ gas itself as metabolic reductant. The responsible enzyme, *hydrogenase*, as found in *Clostridium pasteurianum*, effects the reduction of the iron-sulfur protein (ferredoxin) of this species, and, in turn, electrons may be transferred to a flavoprotein that reduces NAD$^+$. A similar system exists in blue-green algae, although, as noted later, they can also utilize H$_2$O photosynthetically as a reductant for NADP$^+$.

The Photochemical Process in Algae and Higher Plants

The adaptation that freed plants from dependence on external sources of chemical reductants is the remarkable photochemical process whereby H$_2$O can reduce the pyridine nucleotides.

$$2\text{H}_2\text{O} + 2\text{NADP}^+ \longrightarrow \text{O}_2 + 2\text{NADPH} + 2\text{H}^+$$

In algae and higher plants, therefore, the photochemical system not only provides ATP, it also drives the reduction of NADP$^+$ by H$_2$O. $\Delta E_0'$ for this reaction is 1.2 V, equivalent to about +55 kcal/mol of NADP$^+$. Since 2 einsteins of light at 675 nm corresponds to about 2×1.85 eV, or 84 kcal, they could support such a process if means were available. The actual mechanism uses four light quanta, about 168 kcal, to accomplish this feat, by operating two electron-transport systems, as shown in Fig. 19.10.

Photosynthetic Electron Transport

The Hill reaction provides an important insight into the nature of *photosynthetic electron transport*. Illumination of chloroplasts permits reduction of artificial acceptors such as quinones, ferricyanide, or methemoglobin with evolution of O$_2$ without requiring CO$_2$ or pyridine nucleotide. Indeed, this was one of the earliest indications that CO$_2$ reduction is not intrinsic to the mechanism of O$_2$ evolution. The overall process is

$$2\text{H}_2\text{O} + 2 \text{ quinone} \xrightarrow{\text{light}} \text{O}_2 + 2 \text{ hydroquinone}$$

It is now evident that the apparatus of algae and the chloroplasts of all higher plants is indeed constructed as shown schematically in Fig. 19.10, that O$_2$ is produced and NADP$^+$ reduced as indicated, that ATP is formed incident to

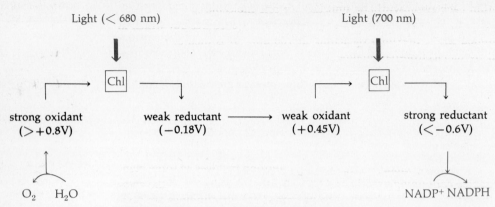

Figure 19.10 General model for the cooperative functioning of two photochemical-reaction centers in plant chloroplasts. Arrows represent the direction of electron flow. Figures in parentheses are E_0' values.

these concerted electron transfers, and that the system has also retained its ancient ability to conduct cyclic photophosphorylation. Details of this plan, as now understood, are shown in Fig. 19.11.

Emerson's observation

Photosynthetic Reaction Centers in Chloroplasts The concept of two distinct reaction centers arose from the observation that light at wavelengths longer than 700 nm is inefficient in driving photosynthesis in chloroplasts, whereas a

Figure 19.11 Tentative scheme for electron transport in a plant photosynthetic unit. Arrows show the direction of electron flow. Broken arrows indicate the electron path during cyclic phosphorylation.

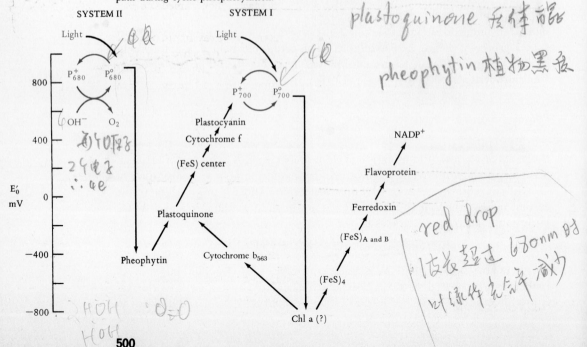

plastoquinone 反体醌

pheophytin 植物黑质

red drop
波长超过 680nm 时
时绿体光合率减少

500

mixture of light at 700 and 680 nm, applied together or alternately, was decidedly more effective than the sum of the two applied alone (*the enhancement effect*). Within each photosynthetic unit there is one molecule of Chl a that is not chemically distinguishable by ordinary extraction procedures but which in its native milieu shows maximal bleaching at 700 nm. It is then evident as a free radical by its electron-paramagnetic-resonance signal. On titration with ferricyanide, this bleachable Chl a molecule exhibits $E_0' = 0.47$ V. This "pigment," designated P_{700}, is the photoreaction center of system I. As in bacteria, P_{700} is almost certainly a dimer of Chl a.

The photocenter of system II is thought to be analogous to P_{700}, i.e., it is a dimer of chlorophyll but it absorbs at 680 nm. Hence it is termed P_{680}. It has many similarities to the bacterial reaction center, e.g., pheophytin as I and plastoquinone-Fe as the primary acceptor. O_2 evolution involves, in some manner, a manganese-containing protein and Cl^-. Neither the identity of the chlorophyll at the center nor the substances in its immediate surroundings are presently known.

Electron-Transport Chains Figure 19.11 summarizes current understanding of the electron-transport chains of the photosynthetic apparatus. All the participating molecules are either firmly lodged in the thylakoid membrane or loosely bound to it. Only the adenine and pyridine nucleotides are dissolved in the stroma fluid, free to combine and dissociate with those enzymes with which they react. For discussion, this system may conveniently be considered as comprising four segments, those leading to and from each of the reaction centers.

System II The aspect of photosynthesis that is least understood is the production of O_2. The usually cited value of E_0' for this 4-electron oxidation of H_2O by activated P_{680} is a composite of the different values for each of the four 1-electron steps. No intermediates have been observed, and it is assumed that four H_2O or OH^- ions bind to the reactive site and depart only as O_2 and H_2O, analogous to the reduction of O_2 by cytochrome oxidase. The immediate oxidant of the water is presently unknown.

System II to P_{700} Electrons from system II enter a pathway reminiscent of that in mitochondria, viz., quinone to an (FeS) center to cytochrome c to a copper protein to the terminal acceptor. It seems that at least one molecule each of plastoquinones A and C is required; these are in equilibrium with a larger pool of quinones, listed in Fig. 19.12.

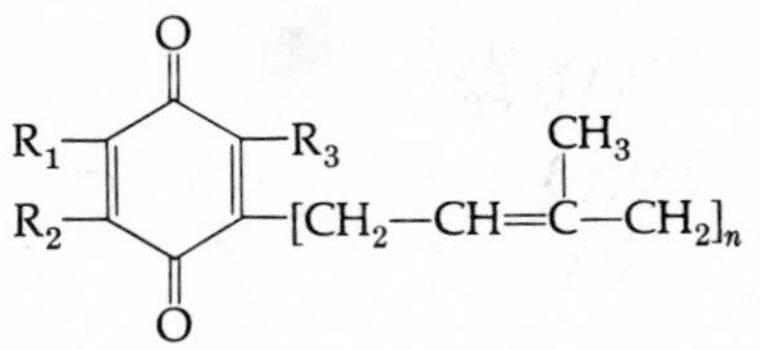

Figure 19.12 Structures of some quinones found in chloroplasts. Ubiquinones and plastoquinones are as shown above; the vitamins K are naphthoquinones (Chap. M23).

	R_1	R_2	R_3	n
Ubiquinones	CH_3O-	CH_3O-	CH_3	6–10
Plastoquinones	CH_3-	CH_3-	H	6–10
Vitamins K	H	H	CH_3	4–9

At least three very different electron carriers are on the path from plasto-quinone to oxidized P_{700}. One is a high-potential form of (FeS) protein, $E'_0 = 290$ mV, that is tightly bound to the membrane and easily reduced by externally added hydroplastoquinones. A second is *cytochrome f* with an absorption maximum at 554.5 nm; this substance is universally present in photosynthetic tissue of green plants. Reduction of cytochrome f is blocked by antimycin a. Cytochrome f is hydrophobic and is an integral protein of the photosynthetic membrane. Finally, chloroplasts of plants (but not bacteria) contain *plastocyanin* (MW $\approx$ 10,400) with one atom of copper. The copper is bonded to the imidazole nitrogen atoms of two histidine residues, a cysteine sulfur, and a methionine sulfur. It is conventional to indicate plastocyanin as the electron carrier between cytochrome f and P_{700}, by analogy with mitochondria and because E'_0 for plastocyanin is 40 mV greater than that for cytochrome f. Electron passage along this chain generates ATP, analogous to ATP formation by complex III of mitochondrial electron transport (Chap. 15).

Reduction of NADP$^+$ Illumination of P_{700}, $E'_0 = +470$, can reduce compounds with a midpoint potential of -700 mV, viz., a span of about 1200 mV. The immediate electron acceptor from P_{700} appears to be a chlorophyll molecule which then passes the electron to membrane-bound (FeS)$_4$ proteins, $E'_0 = -590$ and -550 mV. Electrons pass from them to the (FeS)$_2$ ferredoxins. The rapid passage among the many (FeS) centers presumably prevents charge recombination of A$^-$ and Chl$^+$ (see Fig. 5.10).

In iron-deficient cells of many species, flavodoxins (page 496) appear to replace ferredoxin in this system; flavodoxins are always present instead of ferredoxin in some red algae. Finally, electrons flow via the flavoprotein, *ferredoxin-NADP reductase,* to NADP$^+$, thereby completing the process whereby H$_2$O is used to reduce NADP$^+$ to NADPH.

Photosynthetic Phosphorylation

Cyclic Photophosphorylation in Plants As a minimum, 18 ATP and 12 NADPH are required to achieve hexose synthesis. If the modified system employing a C$_4$ shuttle is operative, 12 additional ATP are required. Moreover, the cell has numerous other requirements for ATP. As shown in Fig. 19.11, system I also affords a mechanism for cyclic photophosphorylation analogous to that in photosynthetic bacteria. It is uncertain whether the electron return begins with soluble ferredoxin, but there is then interposed a distinct cytochrome b$_{562}$, from which electrons flow to the plastoquinone pool and via cytochrome f and plastocyanin back to P_{700}, with ATP formation occurring in association with the process. If essentially all the NADP$^+$ is reduced to NADPH, it feeds back to hinder flow of electrons out of photosystem I, but cyclic electron transport can usefully proceed unhindered if there is a supply of ADP and P$_i$. It is also possible that the modified system I, which serves for cyclic photophosphorylation, is entirely separate and independent of those system I units which are coupled to system II.

Photophosphorylation Electron flow through the photosynthetic electron-transport chain results in generation of ATP. Coupling of the two processes is not as tight as in mitochondria; O_2 evolution and NADP reduction do not have an absolute requirement for the presence of ADP + P_i. However, in the presence of a system for reoxidation of NADPH, uncoupling agents effect a substantial increase in the rate of photosynthetic O_2 evolution and NADP reduction, occasionally by a factor of 3 or 4. Hence, the coupling is highly effective. Presumably, such an arrangement is physiologically advantageous. Whereas mitochondrial oxidation of NADH is futile if ATP is not to be generated, the somewhat looser coupling of the chloroplast permits the apparatus to ensure the cell ample supplies of both ATP and NADPH, somewhat independently of each other.

The mechanism of photophosphorylation closely resembles that of mitochondrial electron transport, and indeed the arguments supporting the proton-motive hypothesis (page 348) are even more compelling for the chloroplast system. Some common attributes include the following:

1. Oligomycin and phlorhizin are inhibitors of both electron transport and phosphorylation.

2. Antimycin a blocks the oxidation of b-type cytochromes.

3. All inhibitors of electron transport prevent phosphorylation.

4. The same lipophilic phenols act as uncouplers.

5. Both processes depend upon the integrity of the membrane.

Electron transport in the photosynthetic membrane engenders a ΔpH of about 3.0 pH units by secretion of protons from the stroma into the minute volume of the thylakoid matrix. It will be recalled that, in the parallel events associated with mitochondrial electron transport, protons move into the mitochondrial intermembrane space and out of the mitochondrion and thus are greatly diluted and buffered. Thus the behavior of thylakoids more nearly resembles that of the inside-out submitochondrial particles prepared by sonication, the secreted protons being retained in a minute volume, so that ΔpH as well as $\Delta\Psi$ is substantial. During thylakoid phosphorylation, ΔpH can account for 150 to 200 mV, while $\Delta\Psi$ may be about 100 mV, ensuring $\Delta\bar{\mu}H^+$ in excess of 250 mV. The role of ΔpH has been demonstrated in the following manner. Chloroplasts in the dark are briefly exposed to a medium at pH 4.5 containing a permeant ion such as succinate, thereby acidifying the thylakoid interior. When the chloroplasts (still in the dark) are transferred to a relatively alkaline solution, pH 8.3, containing ADP and P_i, ΔpH is temporarily 4 pH units or 240 mV. The system then generates 100 ATP molecules per molecule of chlorophyll, a prodigious performance.

The placement of several components within the thylakoid membrane is known: cytochrome f, plastocyanin, and the oxygen generator are on the inside of the thylakoid; ferredoxin-NADP reductase and its subtrates are on the stroma side. The membrane is studded on the stroma side with a coupling factor, CF_1 (MW = 325,000), that markedly resembles that of mitochondria. It also consists of pairs of five nonidentical subunits, α, β, γ, δ, and ϵ.

In the illuminated chloroplasts, protons are secreted across the membrane at one or more segments of the electron-transport system; they return from the matrix to the stroma by way of a confined passage through CF_0 and the reversible ATPase on the outer surface of the thylakoid membrane, at which site $\Delta\bar{\mu}H^+$ drives ATP synthesis.

The actual rate of photosynthetic production of ATP in a living cell, related to the concurrent production of NADPH, remains uncertain. Carbohydrate formation requires a minimal ratio of 3 ATP : 2 NADPH. Other requirements of the cell increase the need for ATP. Chloroplast preparations can be manipulated to yield this 3 : 2 ratio, but it is not known whether this reflects production of more than one ATP molecule as an electron pair traverses systems I and II or whether it reflects a variable rate of cyclic photophosphorylation.

Regulation of Photosynthesis

The CO_2-fixing system operates with a series of automatic self-adjustments, fixing CO_2 efficiently under atmospheric conditions at relatively low light flux and avoiding wasteful reactions in the dark. The control mechanisms are inherent in the properties of several participating enzymes. The high concentration of ribulose-bisphosphate carboxylase and its high turnover number ensure that light rather than $[CO_2]$ is limiting for most of the daylight hours, although the growth of most plants can be stimulated by augmenting $[CO_2]$.

Particularly elegant are those controls which are based on the intrinsic properties of the illuminated thylakoid membrane; it provides reducing power to the stroma while simultaneously increasing its pH as protons are secreted to the thylakoid interior. At the same time, illumination occasions a conformational change in the ATPase that converts it to the active synthetase form. At least three participating enzymes contain sulfhydryl groups that are readily oxidized by O_2 in the dark, resulting in their inactivation, viz., phosphoribulokinase, fructose-1,6-bisphosphatase, and both the NAD^+- and NADP-linked glyceraldehyde-3-phosphate dehydrogenases. Most striking is fructose-1,6-bisphosphatase, which consists of perhaps 15 percent cysteine residues. In vitro, each can be reactivated rapidly by addition of a sulfhydryl compound. The specific reductant has not been identified, but shortly after illumination the reducing power furnished by the electron-transport system is utilized to reactivate this group of enzymes.

In addition, ribulose-1,5-bisphosphate carboxylase activity is governed by its substrate, a negative effector at high concentration while NADPH is a strong positive effector. Thus, at 1.0 mM ribulose-bisphosphate, the enzyme is inert at $[NADPH]/[NADP^+] = 0.5$ yet maximally active at a ratio of 2.0 and substrate concentration of 0.2 mM. Since in illuminated chloroplasts NADP is maintained largely in the reduced state, the enzyme can be fully active. These relations are particularly important because oxygenase and carboxylase activities are similarly affected. In the dark, the enzyme does not waste ribulose-1,5-bisphosphate.

In chloroplasts that utilize the C_4 pathway, yet additional controls are operative (Fig. 19.13). Both pyruvate phosphate dikinase and phospho*enol*pyruvate carboxykinase are inhibited by their products, the former by AMP, phospho*enol*pyruvate, and P_i, the latter by oxaloacetate. Further, the dikinase

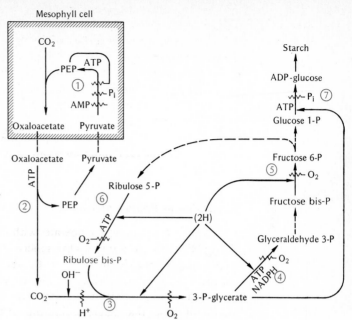

Figure 19.13 Some regulatory controls operative in photosynthesis in a C_4 plant. Shown is the general pattern of the flow of carbon. The enzymes that are control sites are 1: pyruvate phosphate dikinase; 2: phospho*enol*pyruvate carboxykinase; 3: ribulose-bisphosphate carboxylase; 4: glyceraldehyde-3-phosphate dehydrogenase; 5: fructose-bisphosphatase; 6: 5-phosphoribulokinase; and 7: ADP-glucose pyrophosphorylase. Where O_2 is indicated as inhibitor, it signifies aerobic oxidation of enzyme sulfhydryl groups; reactivation is effected by a reductant (2H) provided by the electron-transport system, as are the ATP and NADPH required for the diverse reactions so indicated. PEP = phospho*enol*pyruvate. See text for details.

activity disappears, with a half-life of 15 min, when plants are transferred to the dark and reappears at the same rate in the light. Up to 2000 footcandles, the level of enzymic activity is roughly proportional to light intensity. The inactivated dikinase is restored by an enzyme-catalyzed reduction of a disulfide bond, the reducing power again deriving from the membrane electron-transport system.

The rise in stroma pH upon illumination, due to proton secretion into the thylakoid space, also acts as a light-dark metabolic switch. No CO_2 fixation is evident if the stroma pH falls below 7.4, whereas it is maximal at about pH 8.1. This is due to a significant decrease in $K_m^{CO_2}$ for the carboxylase as pH rises and a marked increase in V_{max} for fructose-1,6-bisphosphatase.

In many photosynthetic bacteria, the regeneration of ribulose-bisphosphate by 5-ribulose phosphate kinase affords a key control; the enzyme manifests sigmoidal kinetics for inhibition by AMP that is offset by both ATP and NADH but not NAD^+. Thus, the response of the thylakoid membrane to illumination acts as an automatic switch that turns on the enzymes required to fix CO_2 into carbohydrate.

Chloroplasts and Mitochondria The multitude of similarities between mitochondria and chloroplasts suggests a common evolutionary ancestry, with the chloroplast necessarily the older. One group of organisms may represent the current progeny of an old transitional form. The purple nonsulfur bacteria are the only organisms which use the same electron-transport chain for both respiration and photosynthesis; this is shown in Fig. 19.14.

The Quantum Yield of Photosynthesis The scheme in Fig. 19.11 requires at least four quanta at each of the two reaction centers to achieve evolution of one

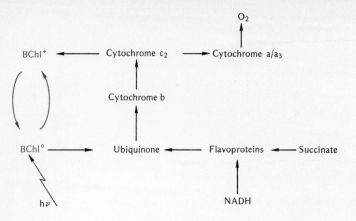

Figure 19.14 Participation of a common electron-transport chain for photosynthesis and respiration in purple bacteria.

O_2, with formation of two NADPH, viz., a total of eight quanta. Assuming the average energy of the absorbed photons to be 40 kcal/einstein, a total of 320 kcal would be used per O_2 formed. Assuming also that $P/2e$ along the transport chain is only 1, then absorption of 6×320 kcal = 1920 kcal results in formation of 12 NADPH, 6 O_2, and 12 ATP. To obtain the additional 6 ATP needed for hexose formation, 12 quanta must be absorbed at P_{700} for cyclic phosphorylation, or another 480 kcal, for a grand total of 2400 kcal. Thus, the overall process has an energy efficiency close to 28 percent since 686 kcal minimally are required for formation of glucose from CO_2 and H_2O.

REFERENCES

Books

Barber, J. (ed.): *Topics in Photosynthesis,* vol. 1–3, Elsevier, Amsterdam, 1976–1979 (a continuing series).

Clayton, R. K.: *Photosynthesis: Physical Mechanisms and Chemical Patterns,* Cambridge University Press, Cambridge, 1980.

Clayton, R. K., and W. R. Sistrom (eds.): *The Photosynthetic Bacteria,* Plenum, New York, 1978.

Encyclopedia of Plant Physiology, New Series, vol. 5, 1978; vol. 6, 1979 (various articles in this continuing series), Springer-Verlag, New York.

Govindjee (ed.): *Bioenergetics of Photosynthesis,* Academic, New York, 1975.

Gregory, R. P. F.: *Biochemistry of Photosynthesis,* 2d ed., Wiley, New York, 1977.

Hatch, M. D., and N. K. Boardman: *Photosynthesis,* vol. 8: *The Biochemistry of Plants,* Academic, New York, 1981.

Olson, J. M., and G. Hind (eds.): *Chlorophyll-Proteins, Reaction Centers, and Photosynthetic Membranes,* Brookhaven Symposia in Biology, no. 28, Brookhaven National Laboratory, Upton, N.Y., 1977.

Review Articles

Anderson, J. M.: The Molecular Organization of the Thylakoid, *Biochim. Biophys. Acta* **416:**191–235 (1975).

Avron, M.: Energy Transduction in Chloroplasts, *Annu. Rev. Biochem.* **46:**143–155 (1977).

Bogorad, L.: Phycobiliproteins and Complementary Chromatic Adaptation, *Annu. Rev. Plant Physiol.* **26:**369–401 (1975).

Crofts, A. R., and P. M. Wood: Photosynthetic Electron-Transport Chains of Plants and Bacteria and Their Roles as Proton Pumps, *Curr. Top. Bioenerg.* **7:**175–244 (1978).

Dutton, P. L., and D. F. Wilson: Redox Potentiometry in Mitochondrial and Photosynthetic Bioenergetics, *Biochim. Biophys. Acta* **346:**165–212 (1974).

Gantt, E.: Phycobilisomes, *Annu. Rev. Plant Physiol.* **32:**327–347 (1981).

Hatch, M. D., and C. R. Slack: Photosynthetic CO_2-Fixation Pathways, *Annu. Rev. Plant Physiol.* **21,** 141–162, 1970.

Heldt, H. W.: Transfer of Substrates across the Chloroplast Envelope, *Horiz. Biochem. Biophys.* **2:**199–229 (1976).

Kelly, G. J., E. Latzko, and M. Gibbs: Regulatory Aspects of Photosynthetic Carbon Metabolism, *Annu. Rev. Plant Physiol.* **27:**181–206 (1976).

Lorimer, G. H.: The Carboxylation and Oxygenation of Ribulose 1,5-bisphosphate: The Primary Events in Photosynthesis and Photorespiration, *Annu. Rev. Plant Physiol.* **32:**349–400 (1981).

Sauer, K.: Photosynthesis—The Light Reactions, *Annu. Rev. Phys. Chem.* **30:**155–178 (1979).

Turner, J. F., and D. H. Turner: The Regulation of Carbohydrate Metabolism, *Annu. Rev. Plant Physiol.* **26:**159–186 (1975).

Zelitch, I.: Pathways of Carbon Fixation in Green Plants, *Annu. Rev. Biochem.* **44:**123–145 (1975).

Chapter 20

Lipid Metabolism I: Triacylglycerols

For optimal growth and maintenance the mammal requires small amounts of the lipid-soluble vitamins (Chap. M23) and certain unsaturated fatty acids. With these exceptions, lipid is apparently not required in the diet; however, lipids are the most concentrated source of energy to the organism, yielding, per gram, over twice as many calories as carbohydrates or proteins and supplying one-third to one-half of the caloric value of the average American diet. Further, consumed foodstuffs in excess of immediate needs are mainly converted to triacylglycerols and stored in the adipose tissue, later to be mobilized and used when needed.

The hydrolytic processes of digestion of the lipids are described elsewhere (Chap. M10). Here we shall be concerned with the absorption, transport, and metabolism of the fatty acids and the triacylglycerols. The metabolism of other major lipids is described in the following chapter.

INTESTINAL ABSORPTION OF LIPIDS

Studies with isotopically labeled triacylglycerols indicate that approximately 40 percent of fed triacylglycerols are hydrolyzed to glycerol and fatty acids, 3 to 10 percent are absorbed as triacylglycerol, and the remainder is partially hydrolyzed, mainly to 2-monoacylglycerols. The water-soluble glycerol leaves the intestine by the portal route.

Fatty acids of chain length less than 10 carbon atoms are absorbed predominantly in nonesterified form by the portal route and consequently are presented directly to the liver. Since only milk is rich in fatty acids of shorter chain length, this is important in infant nutrition.

Any cross-references coded M refer to the companion text, *Principles of Biochemistry: Mammalian Biochemistry*.

Long-chain fatty acids (more than 14 carbon atoms), whether fed as triacylglycerols or as free fatty acids, appear in the lymph almost quantitatively as regenerated triacylglycerols in *chylomicra*. These particles, approximately 1 μm in average diameter, are lipoproteins containing chiefly lipid and a small amount of protein. Thus, during absorption, *resynthesis* of triacylglycerols occurs. The chylomicra enter the blood via the thoracic duct and accessory channels, chiefly at the angle of the left jugular and subclavian veins. In general, lipids that are liquid at body temperature are efficiently digested and absorbed. Lipids containing mainly saturated fatty acids, with melting points significantly above body temperature, are poorly digested and absorbed unless mixed with lower-melting lipids.

Factors Affecting Absorption Not all the factors responsible for partition of fatty acids of different chain length between the lymph and the blood are known. However, differences in water solubility, protein interaction, micelle formation, and enzymic specificity with respect to triacylglycerol resynthesis (see below) affect the different routes of absorption. Of the factors which influence the passage through the intestinal mucosa, the most important are the detergents in the intestinal lumen.

When bile is totally excluded from the intestinal tract as a result of severe liver dysfunction, biliary obstruction, or biliary fistula, lipid absorption is markedly impeded. As a result, the lipid content of *acholic* feces is elevated, with an abundance of insoluble calcium salts of fatty acids. The absence of bile pigment results in characteristic clay-colored stools. The lipid-soluble vitamins (A, D, E, and K) are also poorly absorbed.

The role of bile salts in fatty acid absorption is associated with their formation of mixed micelles with monoacylglycerols, soaps, and other lipid-soluble material. These micelles migrate through the water layer at the mucosal surface.

Although associated with lipid during passage across the mucosal barrier, the bile acids do not enter the lymphatic circulation. Instead they enter the portal blood, from which they are removed by the liver and returned to the duodenum. From this enterohepatic circuit relatively little of the bile salts is lost, little appears in the peripheral blood, and about 500 mg appears per day in human feces.

A second factor that influences the absorption of lipids from the intestinal lumen is the metabolic activity of the intestinal mucosa. Enzymic systems of the mucosa convert free fatty acids and mono- and diacylglycerols to triacylglycerols. These systems are described in Chap M10.

Some selectivity is exhibited by the intestinal mucosa in regard to the absorption of sterols, particularly those of plant origin. Of the principal dietary sterols, only cholesterol (but not its esters) crosses the intestinal wall readily and is absorbed via the chylous route.

PLASMA LIPOPROTEINS AND LIPID TRANSPORT

Inasmuch as most lipids are virtually insoluble in aqueous media, the transport of these substances in the blood plasma is accomplished in association with specific proteins. There is practically no lipid in blood plasma that is not associated with protein. These lipoprotein transport systems serve to supply the tissues with lipids, mainly triacylglycerols, for oxidation or for storage. These proteins are fabricated in the intestinal mucosa (about 20 percent) and liver (about 80 percent).

Normal human plasma in the postabsorptive state contains some 500 mg of total lipid per 100 ml. Approximately 120 mg of this lipid is triacylglycerol and about 220 ± 20 mg is cholesterol, of which some two-thirds is esterified with fatty acids and one-third is free sterol. Phosphoglycerides constitute about 160 mg/dl; the concentration of phosphatidylcholine is higher than that of phosphatidylethanolamine. Free fatty acids are present in small amounts. (See also Table M1.1.)

A number of lipoprotein fractions can be separated by repeatedly centrifuging plasma at high speeds after appropriately increasing the plasma density by addition of salts or ^2H_2O. These lipoproteins are characterized by their flotation constants S_f, which are analogous to the sedimentation constants of

TABLE 20.1 Lipoproteins of Human Plasma

Property	Chylomicra	Very low density (VLDL)	Low density (LDL)	High density (HDL)	Very high density (VHDL)
Density	<0.95	0.95–1.006	1.006–1.063	1.063–1.210	>1.21
S_f	>400	12–400	0–12		
Diameter, Å	300–5000	300–750	200–250	100–150	100
Electrophoretic fraction	———	β	β	α	α
Amount, mg/100 ml plasma	100–250	130–200	210–400	50–130	290–400
Approximate composition, %					
Protein	2	9	21	33	57
Phosphoglyceride	7	18	22	29	21
Cholesterol:					
Free	2	7	8	7	3
Ester	6	15	38	23	14
Triacylglycerol	83	50	10	8	5
Fatty acids	———	1	1	———	———
Apoproteins					
Major*	B, CI, CII, CIII	B, CI, CII, CIII, E	B	AI, AII	AI, AII
Minor	AI, AII	AI, AII, D		CI, CII, CIII, D, E	
Lipid characteristic	Mainly triacylglycerol	Mainly triacylglycerol; phosphatidylcholine and sphingomyelin main P components	High in cholesteryl linoleate	High in phosphatidylcholine and cholesteryl linoleate	

*Protein components that are free of lipid are called apo-B, apo-CI, etc.

ordinary proteins, and by the densities at which they separate. Some of the properties of the lipoprotein fractions are given in Table 20.1. The fractions of lowest density (high S_f) are richest in triacylglycerols and poorest in protein.

The five fractions that are generally recognized are the chylomicra and four fractions characterized by their density as *very low density lipoprotein (VLDL), low-density lipoprotein (LDL), high density lipoprotein (HDL), and very high density lipoprotein (VHDL)*. These fractions each contain more than one kind of apoprotein. The chylomicra are the lipoproteins of largest diameter; the protein portion, synthesized in the intestinal mucosa, involves the same proteins as the VLDL made in the liver.

Electron micrographs of the isolated lipoprotein fractions show that they are spheres of decreasing diameter with increasing density. The lipoproteins consist of a liquid core of hydrophobic lipids—triacylglycerols, cholesterol esters, etc.—whereas the outer portion in contact with the plasma contains such amphiphilic lipids as phosphatidylcholine. The hydrophobic portions of the protein components are largely in the interior and the hydrophilic parts are predominantly on the outside of the particles. Thus, these plasma proteins play the major role in the transport of the water-insoluble lipids. Even in chylomicra, the small amount of protein aids in stabilizing the droplets.

Several polypeptide chains of the apoproteins of the various lipoprotein fractions have been sequenced: apo-CI (57 residues), apo-CIII (79 residues), apo-AI (245 residues), and apo-AII (two identical chains of 77 residues, linked by a disulfide bond). The finding that some of these apoproteins are associated with more than one fraction listed in Table 20.1 indicates that these fractions undergo a changing composition (and hence density and S_f) as lipid is delivered to the tissues. In a 24-h period about 70 to 150 g of exogenous and endogenous triacylglycerols are transported from the intestine and the liver to other tissues. The transport of cholesterol esters is presented in Chap. 21.

Some Defects in Plasma Lipoproteins

Abetalipoproteinemia This rare genetic disorder is characterized by absence of plasma β-lipoproteins of density less than 1.063 and is associated with extensive nerve demyelination. Apo-B is absent from chylomicra, VLDL, and LDL. The levels of plasma triacylglycerols and cholesterol are very low. This provides strong evidence that apo-B is essential for normal absorption, synthesis, and transport of triacylglycerols and cholesterol from intestine and liver. Lipid accumulates in the mucosal cells of the intestinal villi. More than 80 percent of the erythrocytes are *acanthocytes,* erythrocytes having numerous projecting spines (Gk. *akantha,* "a thorn"), indicating that membrane formation is abnormal.

Tangier Disease or Familial High-Density-Lipoprotein Deficiency This rare genetic disorder is characterized by an almost complete absence of plasma HDL as well as by excessive deposition of cholesterol esters in many tissues. There is a marked reduction in plasma cholesterol and phosphoglycerides, but triacylglycerols are normal or elevated. The disease reflects the consequences of the absence of apo-AI, the major protein component of HDL and VHDL.

In brief, triacylglycerols are transported from the intestine in chylomicra and VLDL and delivered mainly to the liver and to the adipose tissue but also to heart, lung, and other organs. At the cell membranes of these tissues, *lipoprotein lipase* hydrolyzes the triacylglycerols; fatty acids and glycerol are absorbed and may be directly metabolized or, in adipose tissues, resynthesized to triacylglycerols for storage. Of all mammalian tissues, only brain does not utilize fatty acids for oxidation.

Lipemia

In the preabsorptive state, the mesenteric and thoracic duct lymph is a clear, watery fluid. Shortly after introduction of a fatty meal into the duodenum, the lymphatic channels, previously seen with difficulty, become distended with a milky fluid rich in triacylglycerols. The fat of the chyle is present in chylomicra. The discharge of this chyle into the venous blood results in a rapid rise in the lipid content of the plasma, occasionally sufficient to result in a milky opalescence. The increase in amount of blood lipid which transiently follows ingestion of fat is called *absorptive lipemia*. The half-time for clearance of human plasma lipid after a meal is normally about 1 h.

Lipoprotein Lipase Intravenous injection of heparin (Chap. M1) markedly accelerates the elimination of turbidity of postabsorptive lipemic plasma in vivo. This is due to activation and release of *lipoprotein lipase* into blood. This enzyme is derived from both liver and adipose tissue and appears in plasma only during lipemia. The activity of the enzyme from adipose tissue is enhanced by apo-CII and by phosphatidylcholine. Moreover, serum albumin, as the acceptor of the fatty acids liberated by the lipase action, plays a major role in the transport of free fatty acids, which are present in a concentration of 8 to 30 mg/dl of plasma (Table M1.1). These fatty acids have a high metabolic turnover rate.

Familial Hyperchylomicronemia In this rare autosomal recessive genetic disorder, there is a deficiency of the lipoprotein lipase derived from adipose tissue that can be elicited by heparin administration. In affected individuals, there is massive chylomicronemia and the levels of triacylglycerols are usually in excess of 2 g/dl of plasma.

Rare cases of total deficiency of apo-CII also manifest severe *hypertriacylglycerolemia*. Other types of hyperlipoproteinemias are classified on the basis of increased concentrations or deficiency of one or more types of plasma lipoproteins. All these rare genetic disorders are associated with high levels of triacylglycerols in plasma.

THE LIPIDS OF THE BODY

In the normal mammal at least 10 to 20 percent of the body weight is lipid, the bulk of which is *triacylglycerol*. This lipid is distributed in all organs, particularly in adipose tissue, in which droplets of triacylglycerols may represent more than 90 percent of the cytoplasm of the cells. Body lipid is a reservoir of potential chemical energy. About 100 times more energy is stored as mobilizable lipid than as mobilizable carbohydrate in the normal human being. This lipid is stored in a relatively water-free state in the tissues in contrast to carbohydrate, which is heavily hydrated. A normal man, weighing 70 kg, is estimated to have 15 kg of triacylglycerols in the tissues.

Much of the lipid of mammals is located subcutaneously, where it serves as an insulator against excessive heat loss to the environment. This function is best exemplified in marine mammals, whose water environment is both colder than body temperature and a far better thermal conductor than air. The subcutaneous lipid depots also insulate against mechanical trauma.

The paucity of depot lipid during the intrauterine life of the mammalian fetus is of interest in relation to the functions of depot lipid during adult life. The fetus derives its nutrition across the placenta from the maternal circulation and does so continuously, in contrast to the adult, who eats intermittently. The fetus, which requires no long-term energy reservoirs, resides in a thermoregulated environment, is well protected by amniotic fluid and maternal tissues against mechanical blows, and thus requires little depot lipid. It is only shortly before term that the fetus acquires its depot lipid.

Characteristics of Depot Lipid The composition of depot lipid varies in different species, but within any species the composition is fairly uniform. More than 99 percent of the lipid of human adipose tissue is triacylglycerol, regardless of anatomical location. In general, depot lipid is richer in saturated fatty acids than liver lipid. The lipid that is deposited subcutaneously is usually as saturated as is compatible with the liquid state. The more nearly saturated a sample of lipid, the higher the energy yield available from oxidation. Thus, mammals deposit under their skins that type of lipid which is richest in chemical potential energy and still liquid at body temperature.

Metabolic Aspects of Body Lipid When an animal is excessively nourished, the quantity of body lipid increases, and, conversely, during fasting the amount of body lipid decreases. These changes involve primarily deposits of triacylglycerols in adipose tissue. It is possible, however, to adjust food intake so that the quantity of body lipid is constant over a long period of time, and indeed in most adult animals there is some regulation of appetite so that the lipid content of the body does not change rapidly.

Depot lipid is continuously being mobilized, new lipid is continuously being deposited, and the constancy of the quantity of depot lipid is the result of a relatively precise adjustment of the rates of these two processes. In the steady state the half-life of depot lipid in the rat is about 8 days. This means that almost 10 percent of the fatty acids in the depot lipid is replaced daily by new fatty acid. In the liver of the rat, the fatty acids have a half-life of about 2 days; in the brain, 10 to 15 days.

OXIDATION OF FATTY ACIDS

Triacylglycerols enter cells probably to only a limited extent. Hydrolysis by lipoprotein lipases mobilized in the plasma or by the hormone-activated

Carnitine
内积汉

lipoprotein lipase of adipose tissue (page 536) results in the formation of fatty acids bound to serum albumin and of glycerol. Glycerol enters the glycolytic pathway via formation of glycerol-3-phosphate by the action of ATP and glycerokinase.

In outline, the oxidation of fatty acids is as follows. Fatty acids enter the cell and must be activated by formation of a CoA derivative. The fatty acyl CoA compounds do not enter the mitochondria, the sole site of fatty acyl CoA oxidation. Carnitine serves as a carrier of the fatty acyl group into the matrix of the mitochondria, where the fatty acyl CoA derivatives are re-formed, releasing carnitine. The fatty acyl CoA derivatives are oxidized by a sequence of reactions in which the fatty acyl chain is shortened by two carbon atoms at a time (β oxidation). All the intermediates in these reactions are CoA derivatives.

Beginning with a fatty acid thioester of coenzyme A, the shortening of the fatty acid chain by two carbon atoms requires four successive reactions as shown below; COSCoA indicates that the acyl derivatives are linked to the thiol group of CoA.

$$RCH_2CH_2CH_2\overset{O}{\overset{\|}{C}}-SCoA + FAD \xrightarrow[\text{dehydrogenase}]{\text{acyl CoA}} RCH_2CH{=}CH\overset{O}{\overset{\|}{C}}-SCoA + FADH_2 \quad (1)$$

$$RCH_2CH{=}CH\overset{O}{\overset{\|}{C}}-SCoA + H_2O \underset{\text{hydratase}}{\overset{\text{enoyl CoA}}{\rightleftharpoons}} RCH_2\overset{OH}{\overset{|}{C}}HCH_2\overset{O}{\overset{\|}{C}}-SCoA \quad (2)$$

$$RCH_2\overset{OH}{\overset{|}{C}}HCH_2\overset{O}{\overset{\|}{C}}-SCoA + NAD^+ \underset{\text{dehydrogenase}}{\overset{\beta\text{-hydroxyacyl CoA}}{\rightleftharpoons}} RCH_2\overset{O}{\overset{\|}{C}}-CH_2\overset{O}{\overset{\|}{C}}-SCoA + NADH + H^+ \quad (3)$$

$$RCH_2\overset{O}{\overset{\|}{C}}-CH_2\overset{O}{\overset{\|}{C}}-SCoA + HSCoA \overset{\text{thiolase}}{\rightleftharpoons} RCH_2\overset{O}{\overset{\|}{C}}-SCoA + CH_3\overset{O}{\overset{\|}{C}}-SCoA \quad (4)$$

Reaction (1) is a dehydrogenation catalyzed by a flavoprotein to yield the 2,3-trans-unsaturated derivative; reaction (2) is the hydration of the double bond to form the 3-hydroxy compound; reaction (3) is the dehydrogenation involving NAD to yield the 3-keto derivative; and reaction (4) is the reaction of the β-ketoacyl CoA with free CoA to yield acetyl CoA and a fatty acyl derivative of CoA that is shorter by two carbon atoms than the original fatty acyl CoA. Successive repetitions of this sequence of four reactions result in the complete degradation of an even-numbered-carbon-atom fatty acid to acetyl CoA. The process degrades a fatty acid containing an odd number of carbon atoms to successive molecules of acetyl CoA and one of propionyl CoA.

Each of the reactions can now be considered further.

Activation Reactions

Metabolism of fatty acids begins with their transformation to the corresponding acyl CoA derivatives; this is accomplished in two ways.

In the first, *acyl CoA synthetases* catalyze formation of the CoA derivatives.

$$RCOOH + HSCoA + ATP \underset{}{\overset{Mg^{2+},\ K^+}{\rightleftharpoons}} RCOSCoA + AMP + PP_i$$

Several such enzymes are known; they are named according to the length of the carbon chain of the compound that reacts most rapidly, e.g., *acetyl CoA synthetase* (acts on C_2 and C_3 fatty acids), *octanoyl CoA synthetase* (C_4 to C_{12} fatty acids), and *dodecanoyl CoA synthetase* (C_{10} to C_{18} fatty acids). The mechanism of such reactions has already been discussed (page 320). These acyl CoA synthetases are found in the endoplasmic reticulum and in the *outer* mitochondrial membrane. An acyl CoA synthetase that utilizes GTP, instead of ATP, is present in the mitochondrial matrix. This enzyme activates long-chain fatty acids, and GDP and P_i are reaction products.

In mitochondria.

long chain F.A + CoA + GTP

↓

→ *long chain acyl CoA + P_i*

+ GDP

A second mechanism for synthesis of acyl CoA derivatives of short-chain fatty acids is the transfer reaction catalyzed by *thiophorases*.

$$\text{Succinyl CoA} + \text{RCOOH} \underset{}{\overset{\text{thiophorase}}{\rightleftharpoons}} \text{succinic acid} + \text{RCOCoA}$$

Activation of fatty acids occurs mainly by synthetase reactions in such animal tissues as liver, heart, and kidney. The thiophorase reaction is most important in extrahepatic tissues in generating acetoacetyl CoA (page 539).

Role of Carnitine in Intracellular Fatty Acyl Transport

Carnitine (L-3-hydroxy-4-trimethylammonium butyrate) serves as a carrier of acyl groups into and out of mitochondria.

$$(CH_3)_3\overset{+}{N}-CH_2-\underset{|}{\overset{OH}{CH}}-CH_2-COO^-$$

Carnitine

Two *acyl transferases* are involved; one catalyzes the acylation of carnitine with short-chain fatty acids and is termed *acetyl CoA carnitine acetyl transferase*. The second enzyme catalyzes acylation of carnitine with long-chain fatty acids and has been named *palmitoyl CoA carnitine palmitoyl transferase*.

$$\text{Acetyl CoA} + \text{carnitine} \rightleftharpoons \text{acetylcarnitine} + \text{CoA}$$
$$\text{Palmitoyl CoA} + \text{carnitine} \rightleftharpoons \text{palmitoylcarnitine} + \text{CoA}$$

The above reactions are reversible; K_{eq} is close to 1, indicating that the *O*-acyl bond of carnitine is a high-energy bond. The enzymes are found both in the cytoplasm and outer membranes of the mitochondria, and in the inner surface of the inner mitochondrial membrane. Thus the carnitine derivatives transport acyl groups into the mitochondria, where the acyl groups are re-transferred to reform the acyl CoA derivatives, which are oxidized in the mitochondria. In effect, the cytoplasmic and mitochondrial pools of CoA are separate.

Enzymes of Fatty Acid Oxidation

Acyl CoA Dehydrogenases These FAD enzymes catalyze the formation of the trans-2,3-unsaturated fatty acyl CoA derivatives (reaction 1 above). Three such enzymes have been isolated and are named for the most rapidly utilized substrates, viz., *butyryl CoA dehydrogenase*, *octanoyl CoA dehydrogenase*, and *hexadeca-noyl CoA dehydrogenase*. The reaction catalyzed is

Saturated fatty acyl CoA + FAD $\longrightarrow$

$$2,3\text{-unsaturated fatty acyl CoA} + FADH_2 \quad (1)$$

As indicated previously (Chap. 16), hydrogen atoms must be transferred from the flavin moiety of these fatty acyl dehydrogenases to an electron-transferring flavoprotein, ETF, which in turn directs the electrons toward cytochrome b.

Enoyl CoA Hydratases (Crotonases) These enzymes catalyze reversibly the hydration of the trans-unsaturated fatty acyl CoA [reaction (2) above].

$$2,3\text{-Unsaturated fatty acyl CoA} + H_2O \rightleftharpoons \text{L-3-hydroxyacyl CoA} \quad (2)$$

One enzyme is most active with crotonyl CoA, $CH_3CH_2{=}CH_2COCoA$ (hence the name *crotonase*); its activity decreases progressively with increasing chain length of the substrate. A second type of enoyl CoA hydratase acts predominantly on medium- and long-chain derivatives.

β-Hydroxyacyl CoA Dehydrogenase Substrates of different chain length are all attacked by a single enzyme.

$$\text{L-3-Hydroxyacyl CoA} + NAD^+ \rightleftharpoons 3\text{-ketoacyl CoA} + NADH + H^+ \quad (3)$$

Thiolases The reaction catalyzed by thiolases [reaction (4)] involves a thiolytic cleavage by CoA with formation of acetyl CoA.

$$C_n\text{ 3-ketoacyl CoA} + CoA \rightleftharpoons C_{n-2}\text{ fatty acyl CoA} + \text{acetyl CoA} \quad (4)$$

The known thiolases exhibit different chain-length specificity. They are thiol enzymes, and an acyl-S-enzyme is an intermediate in the two-step reaction

$$C_n\text{ 3-ketoacyl CoA} + HS\text{-Enz} \rightleftharpoons C_{n-2}\text{ fatty acyl S-Enz} + \text{acetyl CoA}$$
$$C_{n-2}\text{ fatty acyl S-Enz} + CoA \rightleftharpoons C_{n-2}\text{ fatty acyl CoA} + HS\text{-Enz}$$

Although the overall reaction is reversible, the equilibrium position is greatly in the direction of cleavage. The equilibrium constant is 6×10^4 for formation of 2 mol of acetyl CoA from acetoacetyl CoA.

In summary, the shortening of a fatty acyl CoA derivative by two carbon atoms can be represented by the equation

$$RCH_2CH_2CH_2COSCoA + FAD + NAD^+ + HSCoA \longrightarrow$$
$$RCH_2COSCoA + CH_3COSCoA + FADH_2 + NADH + H^+$$

The acetyl CoA generated by fatty acid degradation mixes with acetyl CoA arising from other reactions, including the oxidative decarboxylation of pyruvate (Chap. 15), as well as from the degradative reactions of some amino acids (Chap. 25). The numerous fates of acetyl CoA are described elsewhere.

In addition to acetyl CoA, the products of fatty acid catabolism are the reduced coenzymes NADH and $FADH_2$, which are oxidized by the steps outlined in Chap. 15. Energy for useful work accrues to the organism in two ways as a result of fatty acid degradation. The four hydrogen atoms removed from

Reaction	~P formed, mol
Palmitoyl CoA + $7O_2 \longrightarrow$ 8 acetyl	35
8 Acetyl + $16O_2 \longrightarrow 16H_2O + 16CO_2$ (8 revolutions of tricarboxylic acid cycle, yielding 12 ~ P per revolution)	96
Total	131

the fatty acyl chain in the two dehydrogenase reactions are oxidized via the electron-transport chain. The oxidation of $FADH_2$ and NADH results in formation of ATP, estimated as 5 mol of energy-rich phosphate per mole of O_2 used for production of acetyl CoA by the degradation sequence described above. Thus 1 mol of O_2 is consumed for each mol of acetyl CoA generated. Oxidation of the acetyl CoA produced, via the tricarboxylic acid cycle, yields an additional 12 mol of ATP per mol of acetyl CoA oxidized (page 344). Net energy yield from the oxidation of 1 mol of palmitoyl CoA is calculated in Table 20.2. However, to form palmitoyl CoA from palmitate, 2 equiv of ATP are utilized. Assuming ΔG of -12 kcal/mol of ATP under physiological circumstances, this represents a conservation of about 1550 kcal of chemical energy as ATP (129 mol resulting from the complete oxidation of 1 mol of palmitate). This energy yield is about 60 percent of the 2400 kcal released when 1 mol (256 g) of palmitic acid is oxidized to CO_2 and H_2O in a bomb calorimeter.

The relationship of the cyclic reactions involving fatty acid oxidation to the tricarboxylic acid cycle is shown in Fig. 20.1 for all even-numbered saturated fatty acids.

Figure 20.1 Relationship of fatty acid oxidation to the citric acid cycle.

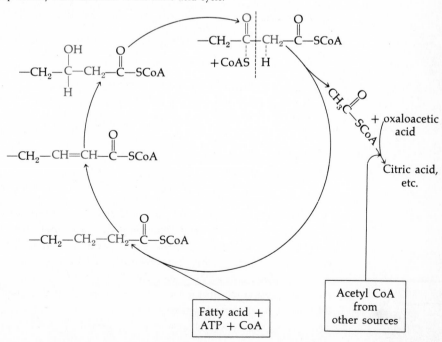

Oxidation of Monoenoic and Polyenoic Acids

The commonly occurring monoenoic acids, oleate (C_{18}) and palmitoleate (C_{16}), are oxidized in the same manner as the saturated fatty acids. However, after removal of three equivalents of acetyl CoA, the remainder is a $\Delta^{3\text{-}cis}$-enoyl CoA derivative, whereas the normal intermediate in β-oxidation is a $\Delta^{2\text{-}trans}$-derivative, as described above. A $\Delta^{3\text{-}cis}:\Delta^{2\text{-}trans}$-*enoyl CoA isomerase* converts the derivatives to normal substrates for the *enoyl CoA hydrase*. Further stepwise removal of acetyl CoA units then proceeds as already described for saturated fatty acyl CoA derivatives.

In the case of polyunsaturated fatty acids, e.g., the C_{18} acid containing $\Delta^{6\text{-}cis}$ and $\Delta^{9\text{-}cis}$ double bonds, two cycles of normal β-oxidation of the CoA derivatives yield the $\Delta^{2\text{-}cis}$, $\Delta^{5\text{-}cis}$ CoA derivative. *Enoyl CoA hydratase* yields the D-3-hydroxy CoA derivative which is not a substrate for the L-3-*hydroxy CoA dehydrogenase*. A *3-hydroxyacyl CoA epimerase* catalyzes conversion of the D-isomer to the L-isomer, as shown.

$$CH_3[CH_2]_7CH{=}CH{-}CH_2{-}\underset{\underset{\displaystyle OH}{|}}{CH}{-}CH_2{-}\underset{\underset{\displaystyle O}{\|}}{C}{-}CoA \overset{epimerase}{\rightleftharpoons}$$

D-3-Hydroxy-$\Delta^{5\text{-}cis}$-enoyl CoA

$$CH_3[CH_2]_7CH{=}CH{-}CH_2{-}\overset{\overset{\displaystyle OH}{|}}{CH}{-}CH_2{-}\underset{\underset{\displaystyle O}{\|}}{C}{-}CoA$$

L-3-Hydroxy-$\Delta^{5\text{-}cis}$-enoyl CoA

The roles of the three additional enzymes which are necessary for oxidation of a dienoic (or polyenoic) acid may be shown in outline below, where A is $\Delta^{3\text{-}cis}:\Delta^{2\text{-}trans}$-*enoyl CoA isomerase; B, $\Delta^{2\text{-}cis}$-enoyl CoA hydrase;* and *C, 3-hydroxyacyl CoA epimerase.* Monoenoic and dienoic acids are oxidized at comparable rates.

$$C_{18:2(9\,cis,\,12\,cis)} \xrightarrow{-3C_2} C_{12:2(3\,cis,\,6\,cis)} \xrightarrow{A} C_{12:2(2\,trans,\,6\,cis)} \xrightarrow{-C_2}$$
$$C_{10:1(4\,cis)} \xrightarrow{-C_2} C_{8:1(2\,cis)} \xrightarrow{B} \text{3-D-hydroxy } C_{8:0} \xrightarrow{C}$$
$$\text{3-L-hydroxy } C_{8:0} \xrightarrow{-C_2} C_{6:0} \xrightarrow{} 3C_2$$

α Oxidation of Fatty Acids

Although β oxidation is the major fate of fatty acids, two other types of oxidation also occur, α and ω oxidation. α Oxidation of long-chain fatty acids to 2-hydroxy acids and then to fatty acids with one carbon atom less than the original substrate have been demonstrated in the microsomes of brain and other tissues. Long-chain 2-hydroxy fatty acids are constituents of brain lipids (Chap. M7), e.g., the C_{24} cerebronic acid (2-hydroxylignoceric acid). These hydroxy fatty acids can be converted to the 2-keto acids, followed by oxidative decarboxylation, resulting in the formation of long-chain fatty acids with an odd number of carbon atoms:

$$RCH_2CH_2CH_2COOH \longrightarrow RCH_2CH_2CHOH{-}COOH \longrightarrow$$
$$RCH_2CH_2CO{-}COOH \longrightarrow RCH_2CH_2COOH + CO_2$$

The initial 2-hydroxylation step is catalyzed by a mitochondrial *monoxygenase* that requires O_2, Mg^{2+}, NADPH, and a heat-stable cofactor. Conversion of the α-hydroxy acid to CO_2 and the next lower unsubstituted acid appears to occur in the endoplasmic reticulum and to require O_2, Fe^{2+}, and ascorbate.

Oxidation of Phytanic Acid Phytanic acid, 3,7,11,15-tetramethylhexadecanoate, an oxidation product of phytol (Chap. 7), is present in animal fat, cow's milk, and foods derived from milk. The phytol presumably originates from plant sources, as it is a substituent of chlorophyll (Chap. 19) and the side chain of vitamin K_2 (Chap. M23).

Large amounts of *phytanic acid* accumulate in the tissues and serum of individuals with *Refsum's disease,* an autosomal recessive genetic disorder affecting the nervous system because of an inability to oxidize this acid. As much as 20 percent of the serum fatty acids and 50 percent of the hepatic fatty acids may be phytanic acid, whereas there is less than 1 μg/ml in normal serum. Diets low in animal fat and milk products appear to relieve some of the symptoms of Refsum's disease.

The presence of the 3-methyl group in phytanic acid blocks β oxidation. In the mitochondria of normal individuals, α-hydroxylation by *phytanate α-hydroxylase* is followed by oxidation by *phytanate α-oxidase* to yield CO_2 and *pristanic acid,* 2,6,10,14-tetramethylpentadecanoic acid, which readily undergoes β oxidation after conversion to its CoA derivative.

Phytanic acid → α-hydroxyphytanic acid

Pristanic acid

The dashed lines show the successive points of cleavage of pristanoyl CoA by β oxidation to yield 3 equiv each of propionyl CoA (see below) and acetyl CoA and, finally, isobutyryl CoA, which is converted to succinyl CoA (Chap. 25). In Refsum's disease, there is a lack of the *phytanate α-hydroxylase.*

ω Oxidation of Fatty Acids

Fatty acids of medium chain length and, to a lesser extent, long-chain fatty acids may initially undergo ω oxidation to ω-hydroxy fatty acids that are subsequently converted to α,ω-dicarboxylic acids. This series of reactions has been observed with liver microsomal enzymes. The initial reaction is catalyzed by a monooxygenase that requires NADPH, O_2, and cytochrome P_{450}. Once formed, the dicarboxylic acid may be shortened from either end of the molecule by the β oxidation sequence described previously.

Propionate Metabolism

Oxidation of a fatty acid with an even number of carbon atoms results in complete degradation to acetyl CoA. Oxidation of an odd-numbered-carbon-atom fatty acid also yields successive molecules of acetyl CoA and 1 equiv of propionyl CoA. Propionate or propionyl CoA is also produced by the oxidation of isoleucine, valine, methionine, and threonine (Chap. 25). Metabolism of propionyl CoA occurs in the mitochondria of liver, cardiac and skeletal muscle, kidney, and other tissues. Formation of propionyl CoA from propionate is catalyzed by acetyl CoA *synthetase* (page 515). The major pathway of propionyl CoA metabolism is summarized in the following three equations; the net result is the formation of succinyl CoA, an intermediate of the tricarboxylic acid cycle.

$$CH_3CH_2COCoA + ATP + CO_2 \underset{\text{carboxylase; Mg}^{2+}}{\overset{\text{propionyl CoA}}{\rightleftharpoons}}$$

$$ADP + P_i + HOOC—CH(CH_3)—COCoA \quad (1)$$
$$\text{D-Methylmalonyl CoA}$$

$$\text{D-MethylmalonylCoA} \underset{\text{racemase}}{\overset{\text{methylmalonyl CoA}}{\rightleftharpoons}} \text{L-methylmalonylCoA} \quad (2)$$

$$\text{L-MethylmalonylCoA} \underset{\text{mutase}}{\overset{\text{methylmalonyl CoA}}{\rightleftharpoons}} \text{succinylCoA} \quad (3)$$

Propionyl CoA carboxylase of human liver has an $\alpha_4\beta_4$ structure; each α subunit contains biotin bound through an amide linkage to ε-amino groups of lysine residues, as in other biotin enzymes (Chap. 17). The carboxylase reaction involves formation of an enzyme-biotin-CO_2 complex similar to that of the acetyl CoA carboxylase (page 523).

Methylmalonyl CoA racemase labilizes the α-hydrogen atom, followed by uptake of a proton from the medium, thus catalyzing interconversion of D- and L-methylmalonyl CoA.

Methylmalonyl CoA mutase utilizes a coenzyme derived from vitamin B_{12} (Chap. M22). When [2-^{14}C]methylmalonyl CoA was converted by the mutase, the label (marked * below) was found in the 3 position of succinyl CoA, indicating an intramolecular transfer of the entire thioester group, —COSCoA, rather than migration of the carboxyl carbon.

$$\begin{array}{ccc}
\text{COOH} & & \text{COOH} \\
| & & | \\
{}^2\overset{*}{\text{C}}\text{HCH}_3 & \rightleftharpoons & {}^3\overset{*}{\text{C}}\text{H}_2 \\
| & & | \\
{}^1\text{COSCoA} & & {}^2\text{CH}_2 \\
& & | \\
& & {}^1\text{COSCoA}
\end{array}$$

Methylmalonyl CoA **Succinyl CoA**

Vitamin B_{12} coenzyme, or 5′-deoxyadenosylcobalamin, has a complex structure with a corrin core which resembles the porphyrin ring system in having four pyrrole-like rings but in which a pair of these rings (A and D) is joined directly instead of through a methene bridge. An atom of divalent cobalt is coordinated to the four nitrogen atoms of the corrin ring system. Two other major components of the coenzyme are nucleotides: one containing the unusual 5,6-dimethylbenzimidazole and the other the 5'-deoxyadenosyl group. The latter is directly linked by a unique carbon-to-cobalt bond. The structure of the coenzyme is shown below; the biosynthesis of the vitamin and its other properties are presented in Chap. M22.

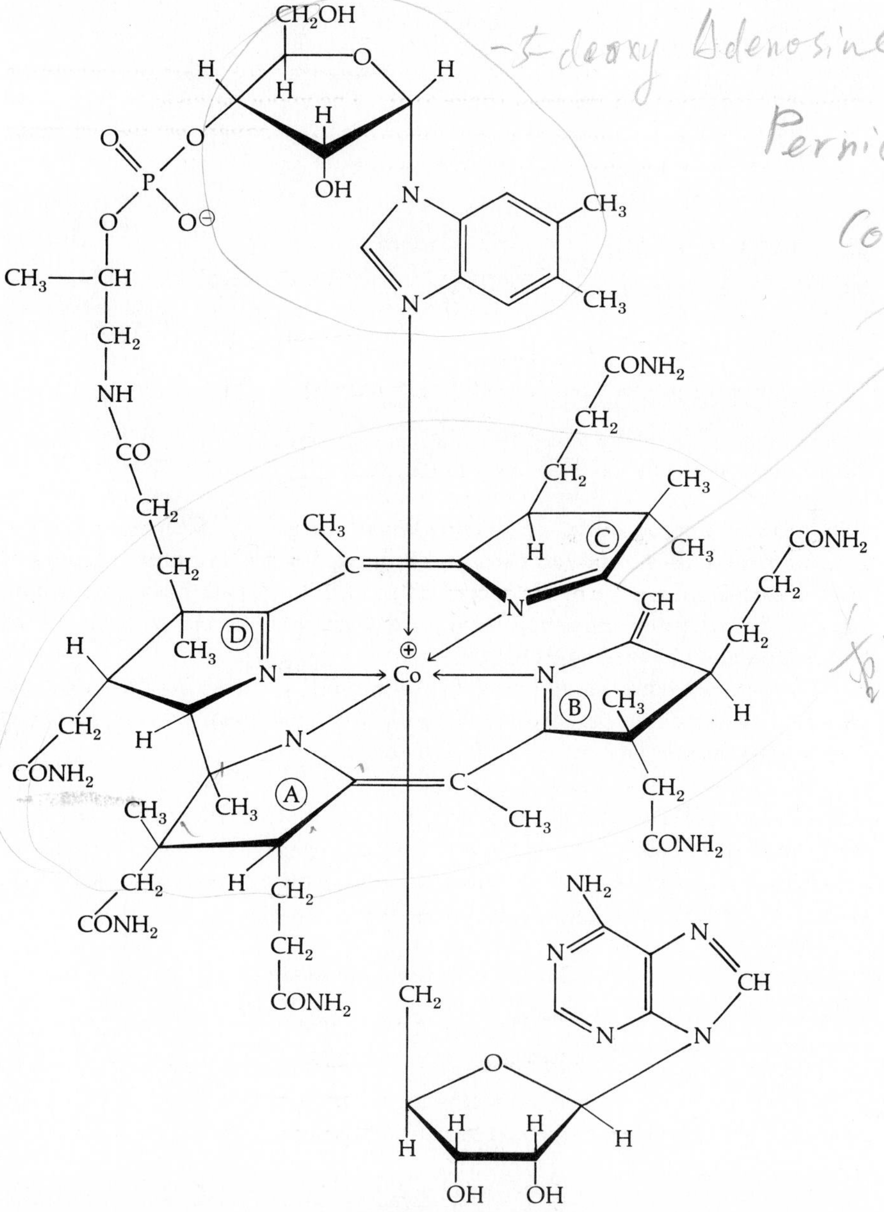

5'-Deoxyadenosylcobalamin

The role of the coenzyme is to remove a hydrogen from one carbon atom by transferring it directly to an adjacent carbon atom, simultaneously effecting the exchange of a second (R) substituent. The H and R are not released into solution.

$$\underset{2}{-\overset{\overset{\displaystyle H}{|}}{C}}\underset{3}{-\overset{\overset{\displaystyle R}{|}}{C}}- \longrightarrow \underset{2}{-\overset{\overset{\displaystyle R}{|}}{C}}\underset{3}{-\overset{\overset{\displaystyle H}{|}}{C}}-$$

At equilibrium, formation of succinyl CoA is favored by a ratio of about 20:1 over methylmalonyl CoA. In patients with vitamin B_{12} deficiency, both propionate and methylmalonate are excreted in the urine in abnormally large amounts.

Two inheritable types of methylmalonic acidemia (and aciduria) are associated in young children with failure to grow and with mental retardation. In one type the mutase protein is absent or defective since addition of the cobamide coenzyme to liver extracts does not restore the activity of the mutase. In the other type, feeding large doses of vitamin B_{12} relieves the acidemia and aciduria, and addition of the cobamide coenzyme to liver extracts restores the activity of the mutase; in these cases there is limited ability to convert the vitamin to the coenzyme (Chap. M22).

Another inheritable disorder of propionate metabolism is due to a defect in propionyl CoA carboxylase, resulting in propionic acidemia (and aciduria). The finding that such individuals, as well as those with methylmalonic acidemia, are capable of oxidizing some propionate to CO_2, even in the absence of propionyl CoA carboxylase, led to elucidation of a second pathway of propionate oxidation. 3-Hydroxypropionate is formed by successive actions of an acyl CoA dehydrogenase followed by enoyl CoA hydratase, as described for oxidation of fatty acids generally (page 516). Conversion of 3-hydroxypropionate to malonic semialdehyde is catalyzed by an NAD-requiring 3-*hydroxypropionate dehydrogenase*. The malonic semialdehyde is converted to acetyl CoA either by direct oxidative decarboxylation [reaction(1)] or by oxidation to malonyl CoA [reaction (2)], followed by decarboxylation [reaction (3)].

$$O{=}CHCH_2COOH + NAD^+ + CoA \longrightarrow acetyl\ CoA + NADH + H^+ + CO_2 \quad (1)$$
Malonic semialdehyde

$$O{=}CHCH_2COOH + acetyl\ CoA + NAD^+ \longrightarrow acetate + malonyl\ CoA + NADH + H^+ \quad (2)$$

$$Malonyl\ CoA \longrightarrow acetyl\ CoA + CO_2 \quad (3)$$

Studies of the above disorders of propionate metabolism reveal yet another pathway, viz., formation of 2-methylcitrate, which is present in the urine as a major product of propionate metabolism whether the propionic acidemia is due to propionyl CoA carboxylase deficiency or secondary to methylmalonic acidemia. Methylcitrate could be formed by the condensation of propionyl CoA and oxaloacetate, analogous to the condensation of acetyl CoA and oxaloacetate to form citrate (Chap. 15).

SYNTHESIS OF FATTY ACIDS

Mammals can synthesize the major portion of the fatty acids required for growth and maintenance. Saturated fatty acids as well as the common singly unsaturated fatty acids are rapidly and abundantly formed from acetyl CoA. Thus any substance capable of yielding acetyl CoA is a potential source of carbon atoms for fatty acid synthesis. The sources of acetyl CoA are the glucose of the diet ingested in excess of immediate energy needs and of the capacity to store polysaccharides, and amino acids that are not required for other functions. Animals can accumulate fat on a fat-free, high-carbohydrate diet.

Biosynthesis of saturated fatty acids occurs in all organisms and in mammals, mainly in adipose tissue, mammary glands, and liver. In contrast to fatty acid oxidation, which is exclusively mitochondrial, the major site of lipogenesis is the cytosol. Further, fatty acid synthesis requires NADPH, bicarbonate, and Mn_2^+, substances unnecessary for oxidation of fatty acids. We shall consider first synthesis in the cytoplasm by the major pathway of lipogenesis and subsequently two other types of synthesis, one concerned with elongation of medium-chain fatty acids in mitochondria and the other with elongation of polyunsaturated fatty acyl CoA derivatives by enzymes located in the membranes of the endoplasmic reticulum.

Palmitate Synthesis

Palmitate synthetase catalyzes the following overall reaction.

$$CH_3-\overset{\overset{\displaystyle O}{\|}}{C}-SCoA + 7^-OOC-CH_2-\overset{\overset{\displaystyle O}{\|}}{C}-SCoA + 14NADPH + 14H^+ \longrightarrow$$
$$\text{Malonyl CoA}$$

$$CH_3-[CH_2]_{14}-COO^- + 7CO_2 + 8CoA + 14NADP^+ + 6H_2O$$
$$\text{Palmitate}$$

The carbon atoms of acetyl CoA become the terminal CH_3-CH_2-group (carbons 15 and 16) of palmitate whereas the remaining carbon atoms (1 through 14) in palmitate are derived from carbon atoms 1 and 2 of malonyl CoA. The seven molecules of CO_2 result from decarboxylation of the terminal COO^- group of malonyl CoA. The final product of the reaction, palmitate, is the source of all other saturated and monounsaturated fatty acids in mammals and of all fatty acids in microorganisms.

Palmitate synthetases differ structurally among different organisms. In *Escherichia coli* and plants, the synthetase is a multienzyme complex that can be dissociated into seven distinct enzymes, each of which catalyzes one step in the overall synthesis of palmitate. In contrast, the synthetase of yeast, avian and mammalian liver, and mammary gland is a multifunctional enzyme containing two subunits (α and β) which, together, catalyze each of the steps required for palmitate synthesis. The complete synthesis of palmitate from acetyl CoA, malonyl CoA, and NADPH is achieved by the same overall set of reactions in *E. coli*, yeast, and liver, but in the mammalian and avian enzymes all of the reactions are catalyzed on the two subunits of this multifunctional enzyme.

a complex

Acetyl CoA Carboxylase Before describing the individual reactions catalyzed by palmitate synthetase, it is necessary to consider first the biosynthesis of malonyl CoA, which provides 14 of the 16 carbon atoms of palmitate. Malonyl CoA is synthesized by the overall reaction

(handwritten top annotation: CH₃—CH with COO and CO—SCoA)

$$CH_3\text{—}COSCoA + HCO_3^- + ATP \longrightarrow \ ^-OOC\text{—}CH_2\text{—}\overset{\overset{\displaystyle O}{\|}}{C}\text{—}SCoA + ADP + P_i$$

catalyzed by the biotin-containing enzyme, *acetyl CoA carboxylase*. The carboxyl-ase also catalyzes formation of methylmalonyl CoA from propionyl CoA, but at about one-fourth the rate of malonyl CoA synthesis from acetyl CoA.

The carboxylase from *E. coli* and plants is a multienzyme complex that may be dissociated into three components: (1) *biotin carboxyl carrier protein* (BCP) (MW = 45,000), containing two identical subunits, each of which has one mole-cule of biotin covalently linked to an ε-amino group of a lysine residue; (2) *biotin carboxylase* (BC) (MW = 98,000), an enzyme with two identical subunits; and (3) *transcarboxylase* (TC) (MW = 130,000), an enzyme with two pairs of subunits of molecular weight 35,000 and 30,000, respectively. Malonyl CoA is synthe-sized in two steps by the action of the two enzymes, each of which employs the biotin carrier protein as one substrate. The carboxylase catalyzes carboxylation of BCP as follows:

(handwritten margin note: the only one step that needs ATP-)

$$BCP + HCO_3^- + ATP \xrightarrow{\ BC\ } BCP\text{—}COO^- + ADP + P_i \qquad (1)$$

to yield *carboxybiotin carboxyl carrier protein* (BCP—COO⁻), in which the carboxy-biotin groups have the structures shown earlier (Chap. 17). The *transcarboxylase* catalyzes the synthesis of malonyl CoA from acetyl CoA and BCP—COO⁻ as follows:

$$BCP\text{—}COO^- + CH_3CO\text{—}SCoA \xrightarrow{\ TC\ } \ ^-OOC\text{—}CH_2\text{—}CO\text{—}SCoA + BCP \qquad (2)$$

The free energy of cleavage of carboxybiotin protein, $\Delta G° = -4.7$ kcal/mol at pH 7.0, is sufficient to allow the compound to act as a carboxylating agent in reaction (2) as well as in other reactions with suitable acceptors. The exergonic nature of the cleavage also explains the requirement for ATP for formation of the carboxybiotin protein.

In contrast to the acetyl CoA carboxylase of *E. coli,* the carboxylases from animals and yeast are multifunctional enzymes in which the BCP, the carboxyl-ase, and the transcarboxylase activities are present in a single biotin-containing polypeptide chain (MW $\approx$ 220,000). At least two such chains are present in the active carboxylases.

Control of acetyl CoA carboxylase activity The activity of the acetyl CoA carboxylase of liver and adipose tissue depends on the presence of citrate and isocitrate. In its inactive form the enzyme consists of protomers of 400,000 to 500,000 daltons, the protomer comprising four subunits. Activation by citrate or isocitrate results in polymerization of the enzyme to a filamentous form of 5 to 10 million daltons. Formation of carboxybiotin results in depolymerization to protomers unless citrate or isocitrate is present.

(handwritten margin note: polymerization 聚合 Ku(5/7/13))

Inasmuch as malonyl CoA has no other metabolic fate than its role in palmitate formation, control of the carboxylase controls fatty acid synthesis without interfering with other pathways of metabolism.

Thus, citrate and isocitrate, participants in the tricarboxylic acid cycle, serve as regulatory factors by activating acetyl CoA carboxylase, which cata-lyzes the rate-limiting step in fatty acid synthesis in mammals. It is noteworthy that the enzyme modifiers control V_{max} rather than K_m. Note that excess citrate in mitochondria passes freely into the cytoplasm. Other regulatory factors in lipid metabolism are considered later.

Conversion of Malonyl CoA to Palmitate   The conversion of malonyl CoA to palmitate by *palmitate synthetase* proceeds in seven discrete steps, each of which is essentially identical in the *E. coli* and mammalian synthetases.   Since each reaction in the *E. coli* synthetase has been studied with a single purified enzyme, it is convenient to consider first palmitate synthesis as revealed by examination of the *E. coli* multienzyme complex.   Initially, the acetyl group of one molecule of acetyl CoA is transferred to a serine hydroxyl group of a transacylase and then to the single sulfhydryl group of a 4′-phosphopantetheine group that is covalently bound to the hydroxyl group of serine in the *acyl carrier protein* (ACP) (MW = 8847).

All acyl intermediates are bound as thiol esters in the same fashion as to CoA, which also contains 4′-phosphopantetheine as its acyl-binding site.

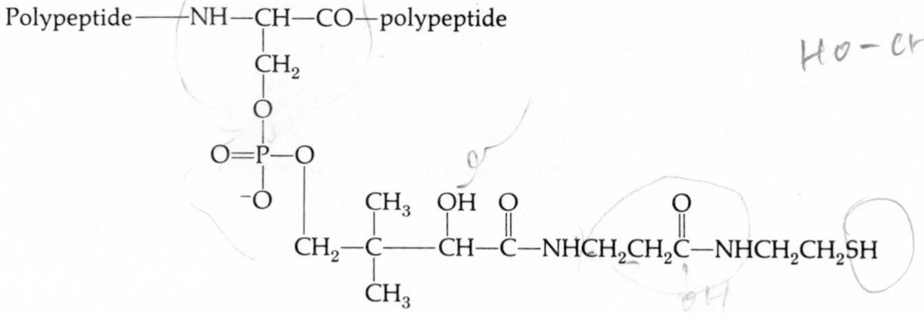

Polypeptide——NH—CH—CO—polypeptide

$$HO-CH_2-\overset{\overset{NH_2}{|}}{\underset{\underset{H}{|}}{C}}-COOH$$

serine

Acyl carrier protein

Each of the subsequent steps in the synthesis of palmitate results in the stepwise elongation of the carbon chain of the acetyl group of acetyl-ACP, as shown below.   ACP is written as HS—ACP to indicate its active acylation site.

Reaction (1).   *Acetyl transacylase*   serine hydroxyl group

(a) $CH_3CO—S—CoA + OH—Ser—Enz \rightleftharpoons CH_3CO—O—Ser—Enz + CoASH$

In this step, the acetyl group is transferred to a specific serine hydroxyl of the enzyme.

(b) $CH_3CO—O—Ser—Enz + HS—ACP \rightleftharpoons CH_3CO—S—ACP—Enz—Ser—OH$

The acetyl group is transferred to the sulfhydryl group of the ACP part of the enzyme.

(c) $CH_3CO—S—ACP + HS—Enz \text{ (cond)} \rightleftharpoons CH_3CO—S—Enz \text{ (cond)} + HS—ACP$

Acetyl is now delivered to a sulfhydryl group of the condensing activity (cond) [see reaction (3)].

Reaction (2).   *Malonyl transacylase*

$HOOCCH_2CO—S—CoA + HS—ACP—Enz \rightleftharpoons$
$$HOOCCH_2CO—S—ACP—Enz + CoASH$$

Reaction (3). Condensation, *β-ketoacyl-ACP synthetase*

$$HOOCCH_2CO—S—ACP—Enz + CH_3CO—S—Enz \text{ (cond)} \rightleftharpoons$$
$$CO_2 + CH_3COCH_2CO—S—ACP—Enz + HS—Enz \text{ (cond)}$$

Reaction (4). *β-Ketoacyl-ACP reductase*

$$CH_3COCH_2CO—S—ACP—Enz + NADPH + H^+ \rightleftharpoons$$
$$CH_3CHOHCH_2CO—S—ACP—Enz + NADP^+$$

Reaction (5). *β-Hydroxyacyl-ACP dehydratase*

$$CH_3CHOHCH_2CO—S—ACP—Enz \rightleftharpoons H_2O + CH_3CH{=}CHCO—S—ACP—Enz$$

Reaction (6). *2,3-trans-Enoylacyl reductase*

$$CH_3CH{=}CHCO—S—ACP—Enz + NADPH + H^+ \rightleftharpoons$$
$$CH_3CH_2CH_2CO—S—ACP—Enz + NADP^+$$

Reaction (7). *Transacylase*

The butyryl group formed by reaction (6) is transferred to the sulfhydryl group of the synthetase, as in reaction (1c), to repeat the cycle. This process is continued until synthesis of the palmitoyl (C_{16}) group is completed.

$$CH_3CH_2CH_2CO—S—ACP—Enz + HO—Ser—Enz \text{ (cond)} \rightleftharpoons$$
$$CH_3CH_2CH_2CO—O—Ser—Enz \text{ (cond)} + HS—ACP—Enz$$

When the process is completed by formation of the palmitoyl group, a *deacylase* of animal tissues liberates palmitate.

$$CH_3[CH_2]_{14}CO—O—Ser—Enz + H_2O \longrightarrow CH_3[CH_2]_{14}COO^- + HO—Ser—Enz$$
$$\textbf{Palmitate}$$

In most microorganisms, the palmitoyl moiety is transferred to CoA.

$$CH_3[CH_2]_{14}CO—O—Ser—Enz + CoA \rightleftharpoons$$
$$CH_3[CH_2]_{14}CO—S—CoA + HO—Ser—Enz$$
$$\textbf{Palmitoyl CoA}$$

Seven cycles are required in the multifunctional mammalian synthetase to incorporate each of the seven 2-carbon units derived from malonyl CoA. The enzyme accepts only acetyl CoA and malonyl CoA; fatty acyl derivatives of intermediate chain length are not acted upon by this system. Moreover, the process terminates only with the formation of palmitate. Palmitoyl CoA inhibits the palmitate synthetase by dissociating it into the two subunits which are inactive, and NADPH activates the enzyme by promoting association of the subunits. This multifunctional enzyme molecule provides a highly efficient, integrated mechanism for the sequential synthetic process. It also ensures the formation of palmitate inasmuch as only acetyl and malonyl can enter and the deacylase is specific for the release of palmitate.

Our knowledge of this multifunctional enzyme, *palmitate synthetase,* of eukaryotic cells is presently more extensive for the yeast enzyme, which has an $\alpha_6\beta_6$ structure with subunits of approximately 180,000 and 185,000, respectively. Genetic and enzymic studies have shown that the two subunits are distinct multifunctional polypeptides. The α subunit comprises the ACP, the β-ketoacyl synthetase, and the β-ketoacyl reductase; the β subunit contains the acetyl transacylase, malonyl transacylase (identical in yeast with palmitoyl transacylase), the dehydratase, and the enoylacyl reductase. Thus each enzymic activity is present in equimolar amount.

Induction of Formation of the Enzymes of Lipogenesis In adipocytes and liver, synthesis of the enzymes of lipogenesis is induced by a high-carbohydrate diet, which evokes a high level of plasma insulin and a low level of glucagon. In tissue cultures of differentiating rat preadipocytes, addition of insulin and adrenal steroids accelerates protein synthesis to produce as much as a 50-fold coordinate increase of *ATP citrate lyase, acetyl CoA carboxylase, palmitate synthetase,* and *malic enzyme.* The first three enzymes are essential for fatty acid synthesis and the reaction of the malic enzyme is one of the two major sources of the NADPH required for lipogenesis. For storage, much of the synthesized palmitate (C_{16}) is first converted to stearate (C_{18}) and then to oleate (C_{18}; one double bond). Synthesis of the enzymes responsible for these latter conversions is also induced by insulin, adrenal steroids, and triiodothyronine.

Under conditions of prolonged carbohydrate deprivation or starvation, the amounts of all of these enzymes of lipogenesis decrease. Thus, aside from immediate hormonal effects on the relative balance of utilization and storage of carbohydrates and triacylglycerols, there are longer-term effects on the amounts of the enzymes catalyzing lipogenesis.

Some of the differences between the mitochondrial oxidation system and the cytoplasmic synthetase system are noted in Table 20.3.

Mitochondrial Elongation of Fatty Acids

Mitochondria contain an enzyme complex that catalyzes *elongation* of preformed fatty acids by successive addition of acetyl CoA units. Fixation of CO_2 is not required, and the products are essentially C_{18}, C_{20}, C_{22}, and C_{24} fatty acids. Indeed, there are three distinct condensing enzymes which catalyze conversion of C_{16} to C_{18}, C_{18} to C_{20}, and C_{20} to C_{22} and C_{24}.

The postulated sequence of chain elongation by four steps is

$$CH_3COSCoA + RCH_2COSCoA \xrightleftharpoons{\text{thiolase}} RCH_2COCH_2COSCoA + CoA \qquad (1)$$

$$RCH_2COCH_2COSCoA + NADH + H^+ \xrightleftharpoons[\text{dehydrogenase}]{\beta\text{-hydroxyacyl}} RCH_2CHOHCH_2COSCoA + NAD^+ \qquad (2)$$

$$RCH_2CHOHCH_2COSCoA \xrightleftharpoons[\text{dehydrase}]{\beta\text{-hydroxyacyl}} RCH_2CH{=}CHCOSCoA + H_2O \qquad (3)$$

$$RCH_2CH{=}CHCOSCoA + NADPH + H^+ \xrightleftharpoons[\text{reductase}]{\text{enoyl}} RCH_2CH_2CH_2COSCoA + NADP^+ \qquad (4)$$

TABLE 20.3 Comparison of Compounds Involved in Fatty Acid Metabolism

Step or component	Degradation	Synthesis
SH component	CoA	Acyl carrier protein
Intermediate SH derivative	Acetyl CoA	Malonyl ACP + acetyl ACP
Keto $\longleftrightarrow$ hydroxy	NAD, L-β-hydroxybutyryl CoA	NADPH, D-β-hydroxybutyryl-ACP
Crotonyl $\longleftrightarrow$ butyryl	FAD, electron-transport system	NADPH, fatty acyl–ACP

The first three steps are equivalent to the last three reactions for fatty acid oxidation (page 515 ff.) although they are probably not catalyzed by the same enzymes. However, reaction (4) is catalyzed by an NADPH-requiring enoyl reductase instead of the specific FAD-requiring acyl CoA dehydrogenases, which catalyze an irreversible desaturation of fatty acids (page 515). The synthetic complex is active with unsaturated fatty acids also; this is considered below (page 531).

Microsomal Elongation of Fatty Acids

The microsomal mechanism for elongation of fatty acids utilizes both saturated and unsaturated fatty acyl CoA derivatives, via the malonyl CoA pathway rather than with acetyl CoA. The fatty acid is first converted to acyl CoA and then reacts with malonyl CoA, followed by reduction with a NADPH-requiring enzyme. The intermediates are similar to those in the palmitate synthetase except that they are not bound to an ACP. Elongation occurs most rapidly with C_{10} to C_{16} fatty acyl derivatives and with C_{18} unsaturated compounds (also see Chap. 16).

Sources of Reduced Nucleotides

For lipogenesis to function, there must be available sources of NADPH for the cytoplasmic and mitochondrial malonyl CoA synthetic pathway and of NADH and NADPH for the mitochondrial and microsomal chain-elongation pathways. In the cytoplasm, several sources of reduced nucleotides are available.

1. Significant sources of cytoplasmic NADPH are the dehydrogenations of the phosphogluconate oxidative pathway. This links fatty acid synthesis to the oxidative pathway of glucose metabolism, particularly since mitochondria do not readily oxidize external NADPH.

2. Cytoplasmic NADH arises in the triose phosphate dehydrogenase reaction (Chap. 17).

3. A third source is from operation of the citric acid cycle. Citrate formed in mitochondria and transferred to the extramitochondrial compartment is cleaved by *ATP citrate lyase* (page 411). The oxaloacetate formed reacts with NADH to yield malate, which in the reaction catalyzed by malic enzyme (page 410) yields pyruvate, NADPH, and CO_2. This sequence would again couple the NADH arising from glycolysis to the synthesis of cytoplasmic NADPH. This is considered in greater detail later (page 535).

All three possible modes of providing reduced nucleotides for cytoplasmic lipogenesis reemphasize the linking of fatty acid synthesis to carbohydrate oxidation. Also, oxidation of acetyl CoA via the citric acid cycle [reaction(3)] depends on a source of oxaloacetate. This may arise by carboxylation of pyruvate derived from glycolysis or from alanine by transamination (Chap. 23). For fatty acid elongation in the mitochondria, reduced pyridine nucleotides

TABLE 20.4

System	Cell fraction	Substrates	Cofactors	Product	Some Aspects of Pathways of Fatty Acid Synthesis
De novo formation	Cytoplasm	Acetyl CoA + malonyl CoA	NADPH, ACP	$C_{16:0}$	
Elongation	Mitochondria	Long-chain acyl CoA (sat. or unsat.) + acetyl CoA	NADPH, NADH	$C_{n+2:y}$ acyl CoA ($y = 0$–6)	
	Microsomes	Long-chain acyl CoA ($C_{10:0}$–$C_{16:0}$) + malonyl CoA	NADPH	$C_{n+2:0}$ acyl CoA	
		Unsat. acyl CoA ($C_{18:3} >$ $C_{18:2} >$ $C_{18:1}$) + malonyl CoA	NADPH	Unsat. C_{n+2} acyl CoA	
Condensation	Soluble complex	C_n acyl CoA ($n = 8, 10, 16, 24$)	NADH, ATP, Mn^{2+}	C_{2n} acyl CoA	

Source: Adapted from J. A. Olson, *Annu. Rev. Biochem.* **35**:559 (1966).

are provided by functioning of the citric acid cycle, with transhydrogenation from NADH as the immediate source of NADPH.

Table 20.4 indicates some of the salient features of pathways of fatty acid synthesis.

FATTY ACID INTERCONVERSIONS

Although the content of the diet varies considerably in the degree of unsaturation and chain length of the fatty acids, the average composition of tissue fatty acids tends to remain constant.

Shortening and Lengthening of the Carbon Skeleton

When isotopically labeled palmitate (C_{16}) is fed to rats, isotope is recovered not only in the palmitate of the body lipid, but also in stearate (C_{18}) and myristate (C_{14}). Similarly, feeding labeled stearate leads to labeling of palmitate. Hence the carbon skeleton of ingested saturated fatty acids may be altered by the gain or loss of two carbon atoms at a time, as expected from the reactions of degradation and elongation previously described.

Formation of Monoenoic Acids

The Δ^9-monounsaturated fatty acids are synthesized by a microsomal-bound desaturase complex consisting of cytochrome b_5, *NADH-cytochrome b_5 reductase* (Chap. 16), and a desaturase. *Stearoyl CoA desaturase* catalyzes formation of oleoyl CoA. Palmitoleoyl CoA is formed from palmitoyl CoA. Molecular O_2 is obligatory for this reaction, as it is also for all other mammalian desaturase reactions (see below).

The function of the Δ^9-desaturase system is clearly to provide a mixture of triacylglycerols with low melting points. Triacylglycerols containing only saturated fatty acids are solids at body temperature, e.g., for tristearin, the melting point is 73°C, whereas for oleodistearin the melting point is 43.5°, and for dioleomonostearin the melting point is 23.5°. Those triacylglycerols with high melting points are poorly incorporated into transport lipoproteins and are poorly hydrolyzed by lipases.

In the adipose tissue of mammals the most abundant fatty acid is oleate (Table 7.3). Thus some of the palmitate is converted to stearate by the microsomal elongation system, and the latter is converted to oleate by the specific *stearoyl CoA desaturase*. As shown in Fig. 20.2, other longer-chain, monoenoic acids are formed by addition of C_2 units, but these are present in only small amounts.

As already noted (page 527), there is hormonal induction of the synthesis of the *stearoyl CoA desaturase*. In the rat this enzyme has a half-life of only 3 to 4 h; thus induction of its synthesis by insulin and other hormones exerts an important control on lipogenesis. When corn oil, which is rich in linoleate, is fed to rats, the synthesis of the desaturase is repressed. A high level of dietary carbohydrate or saturated fatty acids induces formation of the desaturase. These

Figure 20.2 Biosynthesis of some fatty acids in mammals. The larger numbers indicate the position of the double bonds; the subscripts indicate the number of carbon atoms, with the number of double bonds to the right of the colon. All of the double bonds in these acids are *cis*.

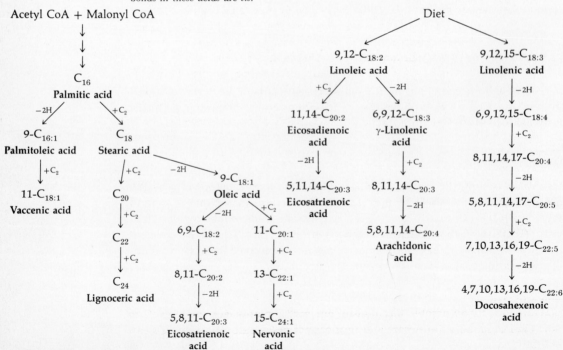

findings are in accord with many observations that synthesis and storage of triacylglycerols with low melting points are favored.

531

Lipid Metabolism I:
Triacylglycerols

Formation and Transformations of Polyenoic Acids

Animal tissues contain a variety of polyunsaturated fatty acids. Of these, one series can be fabricated by the animal *de novo*. These are the fatty acids of which all the double bonds lie between the seventh carbon from the terminal methyl group and the carboxyl group. As seen from Fig. 20.2 such fatty acids can be made by desaturation and chain elongation, starting with oleic acid. However, polyunsaturated fatty acids in which one or more double bonds are situated within the terminal seven carbon atoms cannot be made *de novo*. Such polyunsaturated acids are essential in the diet.

There are four series of polyunsaturated acids in the mammal; two are derived from dietary linoleic and linolenic acids, and two are synthesized from the monounsaturated acids oleic and palmitoleic acids, these in turn being formed from the corresponding saturated acids (see above). The four series can be recognized by the distance between the terminal methyl group (ω) and the nearest double bond.

Linoleic family: $CH_3-[CH_2]_4-CH=CH-$ ($\omega - 6$)

Linolenic family: $CH_3-CH_2-CH=CH-$ ($\omega - 3$)

Palmitoleic family: $CH_3-[CH_2]_5-CH=CH-$ ($\omega - 7$)

Oleic family: $CH_3-[CH_2]_7-CH=CH-$ ($\omega - 9$)

All other polyunsaturated acids can be made from these four precursors by reaction sequences in which the chain is alternately elongated and desaturated. Elongation can occur by operation of the microsomal complex that utilizes malonyl CoA. Desaturation is accomplished also in the microsomal fraction by a mechanism similar to, but not identical with, that concerned with formation of the monounsaturated acids. Conjugated double bonds are not formed in animal tissues. The formation of arachidonic acid is presented in more detail in Fig. 20.3; Fig. 20.2 summarizes the operation of the various pathways that account for the diverse polyunsaturated fatty acids of animal tissues.

Desaturation and elongation reactions occur more extensively in liver than in extrahepatic tissues. Fasting and diabetes are marked by an inhibition of desaturation pathways.

Although oxidation of polyunsaturated fatty acids occurs, these acids are not an important source of energy. Oxidation occurs in mitochondria as previously described.

Arachidonate and homo-γ-linolenate are the prime sources for formation of the prostaglandins (Chap. M13).

In the aerobic system of some microorganisms, O_2 and NADPH are required, as well as ferredoxin, a specific NADPH reductase (a flavin enzyme), and a desaturase specific for stearoyl ACP. In plants, desaturation of the ACP derivatives is by an aerobic mechanism, but higher plants lack the long-chain acyl CoA–ACP acyl transferase and thus cannot desaturate palmitoyl or stearoyl CoA.

$$CH_3-[CH_2]_4-CH=CH-CH_2-CH=CH-[CH_2]_7-COOH$$

Linoleic acid

$$\downarrow -2H$$

$$CH_3-[CH_2]_4-CH=CH-CH_2-CH=CH-CH_2-CH=CH-[CH_2]_4-COOH$$

γ-Linolenic acid

$$\downarrow +C_2$$

$$CH_3-[CH_2]_4-[CH=CH-CH_2]_3-[CH_2]_5-COOH$$

Homo-γ-linolenic acid

$$\downarrow -2H$$

$$CH_3-[CH_2]_4-[CH=CH-CH_2]_4-[CH_2]_2-COOH$$

Arachidonic acid

Figure 20.3 Formation of arachidonic acid from dietary linoleic acid.

An anaerobic pathway in microorganisms consists of formation of a medium-chain β-hydroxyacyl-ACP derivative, followed by dehydration, then by elongation reactions catalyzed by the fatty acid synthetase complex, of which the dehydrase, termed *β-hydroxydecanoyl thioester dehydrase,* is a component. The dehydrase from *E. coli* catalyzes either *trans-α,β* or *cis-β, γ* dehydration, with the former predominating. The enzyme exhibits a high degree of chain-length specificity, being most active with the C_{10} β-hydroxy thioester. The formation of palmitoleic acid from octanoyl-ACP in *E. coli* proceeds as follows:

$$CH_3[CH_2]_6CO-SACP \longrightarrow\!\!\!\longrightarrow CH_3[CH_2]_6COCH_2CO-SACP \xrightarrow{\text{NADPH}}$$

Octanoyl-ACP   **β-Ketodecanoyl-ACP**

$$CH_3[CH_2]_6CHOHCH_2CO-SACP \longrightarrow CH_3[CH_2]_5CH=CHCH_2CO-SACP \longrightarrow\!\!\!\longrightarrow$$

β-Hydroxydecanoyl-ACP   **β,γ-Decenoyl-ACP**

$$CH_3[CH_2]_5C=CH[CH_2]_7CO-SACP \longrightarrow CH_3[CH_2]_5CH=CH[CH_2]_7COOH$$

Palmitoleoyl-ACP   **Palmitoleic acid**

Several different acyl-ACP dehydrases are present in *E. coli,* with specificity being determined by chain length.

SYNTHESIS OF TRIACYLGLYCEROLS

In the tissues and body fluids of the animal organism, fatty acids are present largely as esters. Synthesis of triacylglycerols occurs primarily in liver and in adipose tissue from the CoA derivatives of fatty acids via phosphatidic acid. Free fatty acids are converted to acyl CoA derivatives by one of several chain-length-dependent *fatty acid CoA ligases* in microsomes.

$$\text{Fatty acid} + \text{CoA} + \text{ATP} \longrightarrow \text{fatty acyl CoA} + \text{AMP} + PP_i$$

1-Acylglycerol-3-phosphate is then formed from acyl CoA and L-glycerol-3-phosphate to give 1-acylglycerol-3-phosphate in a reaction catalyzed by *sn-glycerol-3-phosphate acyltransferase.* 1-Acylglycerol-3-phosphate reacts with a second acyl

532

CoA in a reaction catalyzed by *lysophosphatidic acid acyltransferase* to give phosphatidic acid.

$$
\begin{array}{c}
\mathrm{H_2COH} \\
\mathrm{HOCH} \\
\mathrm{H_2COPO_3^{2-}}
\end{array}
\quad
\xrightarrow[\quad\quad]{\;R'COCoA\;\;\searrow\;\;CoA\;}
\quad
\begin{array}{c}
\mathrm{H_2COOCR'} \\
\mathrm{HOCH} \\
\mathrm{H_2COPO_3^{2-}}
\end{array}
\quad
\xrightarrow[\quad\quad]{\;R''COCoA\;\;\searrow\;\;CoA\;}
\quad
\begin{array}{c}
\mathrm{H_2COOCR'} \\
\mathrm{R''COOCH} \\
\mathrm{H_2COPO_3^{2-}}
\end{array}
$$

L-Glycerol-3-phosphate · 1-Acylglycerol-3-phosphate (lysophosphatidic acid) · L-Phosphatidic acid

Phosphatidic acid is a precursor of both the triacylglycerols and some phosphoglycerides (Chap. 21). Glycerol-3-phosphate is derived from free glycerol, which is phosphorylated by *glycerokinase* and ATP, or by reduction of dihydroxyacetone phosphate.

Hydrolysis of phosphatidic acid by a *phosphatidic acid phosphatase* yields a 1,2-diacylglycerol, which in turn reacts with another mole of acyl CoA in a reaction catalyzed by *diacylglycerol* acyltransferase to form a neutral triacylglycerol.

$$
\text{L-Phosphatidic acid}
\xrightarrow[\quad\quad]{\;H_2O\;\;\searrow\;\;P_i\;}
\text{L-1,2,diacylglycerol}
\xrightarrow[\quad\quad]{\;acylCoA\;\;\searrow\;\;CoA\;}
\text{triacylglycerol}
$$

In addition to the above pathway for formation of phosphatidic acid, there is acylation of dihydroxyacetone phosphate by *dihydroxyacetone phosphate acyltransferase*.

$$
\begin{array}{c}
\mathrm{H_2COH} \\
\mathrm{C{=}O} \\
\mathrm{H_2COPO_3^{2-}}
\end{array}
\quad
\xrightarrow[\quad\quad]{\;RCOCoA\;\;\searrow\;\;CoA\;}
\quad
\begin{array}{c}
\mathrm{H_2COOCR} \\
\mathrm{C{=}O} \\
\mathrm{H_2COPO_3^{2-}}
\end{array}
\quad
\xrightarrow[\quad\quad]{\;NADPH + H^+\;\;\searrow\;\;NADP^+\;}
\quad
\begin{array}{c}
\mathrm{H_2COOCR} \\
\mathrm{HOCH} \\
\mathrm{H_2COPO_3^{2-}}
\end{array}
$$

Dihydroxyacetone phosphate · Acyldihydroxyacetone phosphate · Lysophosphatidic acid

The *acyl transferase* is found in both peroxisomes and microsomes and is specific for saturated fatty acids. Reduction to lysophosphatidic acid is accomplished by a microsomal *oxidoreductase*.

Note that the direct diacylation of glycerol-3-phosphate leads to random incorporation of saturated and unsaturated fatty acids whereas the stepwise acylation via dihydroxyacetone phosphate yields the asymmetric pattern shown by many natural sources of triacylglycerols and phosphoglycerides; e.g., the phosphatidylcholine of plasma, liver, kidney, and brain is predominantly of the asymmetric pattern. However, the phosphatidylcholine of lung surfactant contains palmitate in both positions.

Intestinal mucosa synthesizes triacylglycerols from free fatty acids and mono- and diacylglycerols. The path of synthesis of triacylglycerols from free fatty acids and of 1,2-diacylglycerols is undoubtedly the same as that given above. However, the incorporation of monoacylglycerols into the triacylglycerols that appear in the chyle is unique to intestinal mucosa. A microsomal system from intestine catalyzes the reaction

$$\text{Monoacylglycerol} + \text{fatty acyl CoA} \longrightarrow \text{diacylglycerol} + \text{CoA}$$

Liver microsomes fail to catalyze this reaction under the same conditions.

In *Clostridium butyricum* and *E. coli*, lysophosphatidic and phosphatidic acids can be synthesized by acylation of glycerol-3-phosphate by acyl-ACP derivatives. Indeed, the latter are better utilized for these syntheses than the corresponding acyl CoA derivatives.

REGULATION OF LIPID METABOLISM

The prime source of carbon for fatty acid synthesis is carbohydrate. When the latter provides a surfeit of calories, fatty acid synthesis is augmented. If the caloric excess is as fatty acids, these are deposited in tissues primarily as triacylglycerols. A continuing channeling of dietary fatty acids into formation of complex lipids also occurs, but this is minor in comparison to either synthesis of triacylglycerols, interconversions of fatty acids, or fatty acid oxidation. Although both synthesis and oxidation of fatty acids occur simultaneously, these processes do not proceed at equal rates; one or the other may predominate depending on the nutritional state of the animal.

A number of factors normally regulates the rate of fatty acid synthesis, the conversion of excess carbohydrate to fatty acids, and the storage of lipid. The entry of glucose into most cells depends upon action of insulin (Chap. 17), which thus controls the availability of excess carbohydrate for glycogen synthesis. Since the capacity to store glycogen is limited, the consequent increased formation and accumulation of citrate and isocitrate activates acetyl CoA carboxylase (page 527), the rate-limiting enzyme in fatty acid synthesis. In addition, citrate lyase synthesis is augmented by diets high in carbohydrate, as well as by thyroxine, thereby increasing the rate of formation from citrate of acetyl CoA, the initial substrate for fatty acid synthesis. Acetyl CoA also stimulates pyruvate carboxylation (anaplerosis, Chap. 17) and inhibits pyruvate decarboxylation, thus accelerating ultimate provision of NADPH required for fatty acid synthesis. Phosphoenolpyruvate carboxylase activity is also enhanced in fasting, diabetes, and by cortisol administration.

A key enzyme in regulating the overall rates of fatty acid degradation and synthesis is *isocitrate dehydrogenase* (Chap. 15) of the citric acid cycle. The activity of this mitochondrial enzyme is subject to allosteric regulation by ATP and ADP. When the ATP level is high, the dehydrogenase is inhibited and citrate accumulates and leaves the mitochondria, thus making citrate available in the cytoplasm for degradation by *ATP citrate lyase* to acetyl CoA and hence for lipogenesis. When the ATP level is low and ADP accumulates in the mitochondria, the activity of the isocitrate dehydrogenase is greatly enhanced since ADP is a positive effector of the enzyme. Thus the activity of the entire cycle is enhanced, and little citrate leaves the mitochondrion. Thus, the regulated activity of the isocitrate dehydrogenase is one of the principal switches that determines whether fatty acids are to be oxidized or synthesized.

In contrast to the ability to convert carbohydrate to fatty acids, mammals cannot effect a *net* conversion of fatty acids into carbohydrate. This is largely because formation of acetyl CoA and CO_2 from pyruvate is irreversible (Chap. 15).

As already noted, lipogenesis occurs primarily in the cytoplasm from acetyl CoA, whereas formation of acetyl CoA from pyruvate takes place in mitochondria. Translocation of acetyl CoA occurs predominantly by formation of citrate; acetyl CoA is regenerated in the cytoplasm by the *ATP citrate lyase* reaction. Acetyl groups can also be transferred via the carnitine shuttle (page 515).

The conversion of glucose to palmitate can be represented by the overall summation for supplying the necessary acetyl CoA.

$$4 \text{ Glucose} + 14\text{NAD}^+ + 14\text{ADP} + 14\text{P}_i + \text{O}_2 + 8\text{CoA} \longrightarrow$$
$$8 \text{ acetyl CoA} + 14\text{NADH} + 14\text{H}^+ + 8\text{CO}_2 + 14\text{ATP} + 2\text{H}_2\text{O}$$

The net synthesis of palmitic acid from acetyl CoA can be summarized as

$$8 \text{ Acetyl CoA} + 7\text{ATP} + 14\text{NADPH} + 14\text{H}^+ \longrightarrow$$
$$\text{palmitic acid} + 7\text{ADP} + 7\text{P}_i + 8\text{CoA} + 14\text{NADP}^+ + 6\text{H}_2\text{O}$$

Although the amount of carbon balances, the oxidation of glucose yields NADH whereas synthesis of palmitic acid requires NADPH, which must be supplied primarily by the oxidation of glucose-6-phosphate to yield ribulose-5-phosphate + 2NADPH (Chap. 17). In liver and adipose tissue, when lipogenesis predominates, the phosphogluconate cycle is very active. The ribulose-5-phosphate is recovered by operation of the transaldolase and transketolase reactions.

$$(a) \quad \text{Pyruvate}_m + \text{ATP} + \text{CO}_2 \longrightarrow \text{oxaloacetate}_m + \text{ADP} + \text{P}_i \qquad (1)$$

$$\text{Oxaloacetate}_m + \text{acetyl CoA}_m \longrightarrow \text{citrate}_m + \text{CoA} \qquad (2)$$

$$\text{Citrate}_m \longrightarrow \text{citrate}_c \qquad (3)$$

$$(b) \quad \text{Citrate}_c + \text{ATP} + \text{CoA} \longrightarrow \text{oxaloacetate}_c + \text{acetyl CoA}_c + \text{ADP} + \text{P}_i \qquad (4)$$

$$\text{Oxaloacetate}_c + \text{NADH}_c + \text{H}^+ \longrightarrow \text{malate}_c + \text{NAD}^+_c \qquad (5)$$

$$\text{Malate}_c + \text{NADP}^+_c \longrightarrow \text{pyruvate}_c + \text{NADPH}_c + \text{H}^+ + \text{CO}_2 \qquad (6)$$

NET: $\text{Acetyl CoA}_m + 2\text{ATP} + \text{NADH}_c + \text{NADP}^+_c \longrightarrow$
$$\text{acetyl CoA}_c + 2\text{ADP} + 2\text{P}_i + \text{NAD}^+_c + \text{NADPH}_c$$

Some of the required NADPH is undoubtedly derived by the combined operation of the preceding two sets of reactions, (a) and (b). The subscripts *m* and *c* designate location of the metabolite in the mitochondrial and cytoplasmic compartments, respectively.

If, indeed, the NADPH were to arise in this manner, then the sum of the above reactions indicates utilization of two ATP to achieve transhydrogenation from one NADH to one NADPH. The ATP required for the process is probably overestimated. In fact, data indicate a general requirement for one ATP for transhydrogenation from NADH to NADPH.

$$\text{NADH} + \text{NADP}^+ + \text{ATP} \longrightarrow \text{NAD}^+ + \text{NADPH} + \text{ADP} + \text{P}_i$$

The process is dependent on the transport of pyruvate, citrate, and malate across the mitochondrial membrane.

Additional factors modulate reactions (a) and (b). Acetyl CoA stimulates pyruvate carboxylation [reaction (1)] and inhibits pyruvate decarboxylation, thus accelerating ultimate provision of NADPH by the above reactions and promoting fatty acid synthesis.

The energy equivalents for conversion of carbohydrate to fat can be calculated by assuming that $1NADH \approx 3ATP$; $NADPH \approx 1NADH + 1ATP \approx 4ATP$; $1FADH_2 \approx 2ATP$; 1 acetyl $CoA \approx 12ATP$. From these values, the synthesis of tripalmitin requires 500 equiv of ATP whereas its subsequent complete oxidation yields 409 equiv of ATP, an overall efficiency of about 80 percent. Moreover, this does not take into account any further inefficiency in the translocation reactions of (a) and (b). Nevertheless, although the animal pays a price for its ability to store energy as fat, the importance of conserving excess calories for future needs is obvious.

MOBILIZATION OF DEPOT AND LIVER LIPID

The triacylglycerols stored in adipose tissue constitute the principal reservoir of substrate for oxidative metabolism. Mobilization, distribution, and oxidation of these fatty acids can occur rapidly enough to support as much as 50 percent of the oxidative metabolism of the body. This occurs in a variety of circumstances, e.g., starvation, exposure to cold, exercise, reproduction, and growth. The common stimulus for fatty acid mobilization in these diverse conditions is provided through increased secretion of lipid-mobilizing hormones. The major site of degradation of fatty acids to acetoacetate is the liver (page 538). Thus, the transport of lipid from the depots to the liver, and vice versa, is of considerable metabolic significance.

Mobilization of Depot Lipid

Even in the animal in caloric balance, a considerable fraction of the depot lipid is mobilized daily, enters the bloodstream, and is delivered to the various organs. Rapid mobilization to the liver is particularly striking; here the lipid may be stored temporarily or degraded by the reactions discussed previously. Lipid leaving adipose cells is largely transported as free fatty acids bound to serum albumin (page 514). The release of fatty acids is accompanied by the appearance in plasma of glycerol as well, although in amounts significantly less than the equivalent of fatty acids. The fatty acids released from adipose cells derive from stored triacylglycerols; hydrolysis is catalyzed by a distinct hormone-sensitive *lipoprotein lipase*, which differs from the plasma membrane enzyme of the adipose cells that serves for the assimilation of triacylglycerols (page 512). The activity of the intracellular enzyme(s) is controlled by cyclic AMP via the action of various hormones (see below).

Hormonal Influences on Lipid Mobilization A number of hormones induce lipid mobilization from the depots. In general, the sequence of events is (1) conversion of triacylglycerols to free fatty acids within the adipose tissue; (2) release of these fatty acids into the blood, a process requiring the presence of serum albumin; (3) transport of fatty acids by the circulation to diverse organs and tissues, with utilization for oxidation or for synthesis of triacylglycerols and phosphoglycerides; (4) discharge of triacylglycerols and phosphoglycerides from the liver, as lipoproteins, with resultant lipemia; and (5) discharge from the liver of ketone bodies (see below) formed from fatty acids.

The hormones active in lipid mobilization from adipose tissue include epinephrine and norepinephrine, adrenal steroids, glucagon, and the hypophyseal hormones—vasopressin, thyrotropin, adrenocorticotropin, luteotropin, somatotropin, and the β- and γ-lipotropins. Presumably, all of these agents function by activation of fat cell *adenylate cyclase*. These hormones stimulate lipolysis and inhibit the stimulation by insulin of triacylglycerol synthesis in adipose tissue.

The release of fatty acids from human adipose tissue is strikingly increased when adrenal medullary tumors (*pheochromocytoma*) result in excessive epinephrine secretion; plasma concentrations of unesterified fatty acids many times normal may be evident. The action of adrenocorticotropin on adipose tissue is direct and independent of the trophic influence of this hormone on the adrenal cortex (Chap. M18). Other hormones listed above all exert this type of direct action on adipose tissue. The wide diversity of humoral agents stimulating lipid release from adipose tissue provides an explanation for the hyperlipemia and liver lipid deposition present in various experimental and clinical conditions (see below).

Hepatic Poisons and Lipid Mobilization Chlorinated aliphatic and aromatic hydrocarbons are potent hepatotoxic agents that produce lipid accumulation in the liver. Poisoning by carbon tetrachloride, because of its volatility and wide industrial use, is often encountered in clinical medicine.

Fatty liver is frequently seen in chronic infectious diseases, such as tuberculosis, in metabolic disturbances, including starvation, and in early alcoholic liver cirrhosis (page 544). The most striking cases were observed in patients with severe, untreated diabetes in the preinsulin era. As indicated below, the liver of the diabetic person is less than normally competent to synthesize fatty acids but is normally able to degrade fatty acids. The fatty liver is attributed to excessive migration of depot lipid to the liver, and the lipemia observed in diabetes is taken to represent in part lipid in transit from the depots. The augmenting effect of hormones on lipid mobilization from depots suggests that most clinical cases of fatty liver arise from excessively rapid mobilization of depot lipid. On the other hand, liver lipid accumulation in circumstances of liver poisoning may also reflect a diminished capacity of injured hepatic cells to degrade fatty acids.

Mobilization of Liver Lipid

With the discovery of insulin, it became possible to maintain totally pancreatectomized dogs in reasonably good health. However, even when adequate insulin was supplied to control the diabetes, the animals developed severe fatty livers. These fatty livers could be corrected or prevented by addition of phosphatidylcholine or choline to the diet. Also, fatty livers are produced in rats by administration of a diet poor in choline and low in protein; this fatty liver can be cured or prevented by administration of choline.

The fatty liver observed in these experimental animals arises by a different mechanism from that previously considered. Choline is a constituent of phosphatidylcholine, and the capacity of the liver to generate phosphatidylcholine is dependent on a supply of choline or a supply of methyl groups from S-adenosylmethionine to permit choline synthesis (Chap. 21). The term *lipotropic substance* has been assigned to compounds capable of preventing or correcting the fatty liver of choline deficiency.

These findings have suggested that although fatty acids are delivered to the liver in a variety of forms, e.g., unesterified or esterified to glycerol or to cholesterol, and are, in addition, synthesized in the liver from carbohydrate or amino acid precursors, they leave the liver not only in these forms, as lipoproteins, but also as choline-containing phosphoglycerides of lipoproteins. Indeed, the liver is the major site of synthesis of the plasma phosphoglycerides. According to this view, when the availability of choline or methyl groups for its synthesis is restricted, the rate of phosphatidylcholine synthesis decreases and consequently the rate at which fatty acids are discharged from the liver falls below normal. If other processes continue at a normal rate, an accumulation of lipid in the liver results.

Some of the effects of methyl-group deficiency undoubtedly relate to depressed levels of carnitine, which requires three methyl groups for its synthesis (Chap. 24). In carnitine depletion of experimental animals, there is a decreased rate of long-chain fatty acid oxidation and accumulation of triacylglycerols in the tissues; however, oxidation of short-chain fatty acids is unimpaired.

THE KETONE BODIES AND KETOSIS

In liver mitochondria fatty acids are oxidized to acetyl CoA. A portion of the acetyl CoA is oxidized via the tricarboxylic acid cycle to CO_2 and a portion is converted via 3-hydroxy-3-methylglutaryl CoA to acetoacetate by the following reactions. The sequence is initiated by head-to-tail condensation of two molecules of acetyl CoA catalyzed by *acetyl CoA-acetyl transferase* in the mitochondria.

$$2 \text{ Acetyl CoA} \rightleftharpoons \text{acetoacetyl CoA} + \text{CoA}$$

The major fate of acetoacetyl CoA in liver mitochondria is conversion to 3-hydroxy-3-methylglutaryl CoA. In the cytosol, the latter is an intermediate in the formation of cholesterol (Chap. 21).

$$\text{Acetoacetyl CoA} + \text{acetyl CoA} + H_2O \longrightarrow \underset{\underset{\text{CH}_3}{|}}{\text{HOOCCH}_2\text{CCH}_2\text{COCoA}} + \text{CoA}$$

3-Hydroxy-3-methylglutaryl CoA

Cleavage of 3-hydroxy-3-methylglutaryl CoA by *hydroxymethylglutaryl CoA lyase* yields acetoacetate.

$$\text{3-Hydroxy-3-methylglutaryl CoA} \longrightarrow \text{acetoacetate} + \text{acetyl CoA}$$

The reduction of acetoacetate is effected by a specific inner-membrane-bound mitochondrial *β-hydroxybutyric acid dehydrogenase* and yields D-β-hydroxybutyrate. In contrast, reduction of acetoacetyl CoA catalyzed by β-hydroxyacyl CoA dehydrogenase (page 516) yields L-β-hydroxybutyryl CoA.

$$\text{CH}_3\text{COCH}_2\text{COO}^- + \text{NADH} + \text{H}^+ \underset{\text{dehydrogenase}}{\rightleftharpoons} \text{CH}_3\text{CHOHCH}_2\text{COO}^- + \text{NAD}^+$$

Acetoacetate   D-β-Hydroxybutyrate

The β-hydroxybutyrate dehydrogenase of bovine heart mitochondria can be isolated as an inactive protein; activity is restored by addition of phosphatidylcholine.

Although this reaction is reversible, wide variations in the amounts and in the ratio of the two acids in the circulation are encountered. Acetoacetate, β-hydroxybutyrate, and acetone (which arises from acetoacetate by nonenzymic decarboxylation) are traditionally but inaccurately termed *ketone bodies.*

The release of acetoacetate and β-hydroxybutyrate by liver into the blood and their transport to extrahepatic tissues are continuing, normal processes. The total ketone body concentration in blood, expressed as β-hydroxybutyrate, is normally below 3 mg/dl, and the average total daily excretion in the urine is approximately 20 mg. This is because of efficient mechanisms for removal of β-hydroxybutyrate and acetoacetate from blood by peripheral tissues, especially by skeletal and cardiac muscle, which derive a sizable fraction of their total energy requirements from this nutrient. Normally, adult brain depends entirely on glucose for its energy; however, under conditions of starvation, as well as in the fetus and neonate, the brain derives a considerable part of its energy from these acids.

In order to be utilized in the peripheral tissues, β-hydroxybutyrate is oxidized to acetoacetate and the latter is reconverted in the mitochondria to its CoA derivative by a specific *thiophorase* which is absent in the liver.

Acetoacetate + succinyl CoA ⇌ acetoacetyl CoA + succinate

Acetoacetyl CoA is cleaved by *thiolase* (page 516), yielding two molecules of acetyl CoA, which then enters the citric acid cycle.

The importance of this pathway is indicated by the fact that the normal human liver is potentially capable of making the equivalent of half its weight as acetoacetate each day!

The roles of acetyl CoA and 3-hydroxy-3-methylglutaryl CoA (HMGCoA) in lipid metabolism are depicted in Fig. 20.4. It should be recalled that some of these processes occur in the mitochondria (or on its membranes) whereas others, fatty acid synthesis and cholesterol formation (Chap. 21), occur in the endoplasmic reticulum of the cytoplasm.

In circumstances of limited utilization of carbohydrate and/or substantial mobilization of fatty acids to the liver, there is a markedly diminished rate of operation of two of the three pathways for metabolizing acetyl CoA, viz., the citric acid cycle and fatty acid synthesis. The result is a channeling of acetyl CoA into 3-hydroxy-3-methylglutaryl CoA. This results in increased formation of acetoacetate, β-hydroxybutyrate, and acetone and elevates their concentration in the blood above normal, resulting in *ketonemia.* If the blood level exceeds

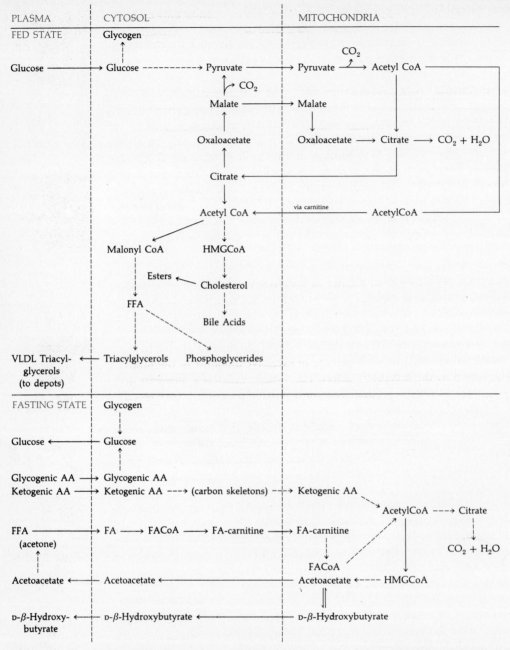

Figure 20.4 A summary of the major metabolic pathways occurring in the liver mitochondria and cytosol during the fed state and the fasting state, as well as the predominant flow of certain metabolites to and from the plasma and the liver. Although many of these processes take place during both of these situations, it is the *dominant* situation that is emphasized in this figure. In the fed state, as well as in the fasting state (as shown), some amino acids, which are present in excess of other requirements, contribute carbon skeletons to form glucose (glycogenic amino acids) and others to form acetyl CoA (ketogenic amino acids); these processes are described in Chap. 25. These conversions of amino acids to glucose, to acetyl CoA, or both are greatly increased in the fasting state. The conversion of acetyl CoA to cholesterol and the formation of bile acids from the latter are described in Chap. 21. Not shown in this summary is the generation of NADPH, from the conversion of malate to pyruvate and from glucose oxidation via the pentose cycle; NADPH is essential for fatty acid synthesis and for cholesterol formation. Formation and utilization of ATP are also omitted. FFA = free fatty acids; FA = fatty acids; HMG CoA = 3-hydroxy-3-methylglutaryl CoA; AA = amino acids.

the renal threshold and appreciable amounts of ketone bodies appear in the urine, *ketonuria* results. Whenever a marked degree of ketonemia and ketonuria exists, the odor of acetone is likely to be detected in the exhaled air. This triad of ketonemia, ketonuria, and acetone odor of the breath is termed *ketosis*.

Causes of Ketosis Any circumstance associated with diminished availability of carbohydrate will accentuate utilization of fatty acids. This occurs within a few hours after meals or after any excessive utilization of carbohydrate. In *starvation*, glycogen stores are depleted and survival depends largely on energy derived from depot lipid. Mobilization of this lipid is reflected in lipemia. Oxidation of fatty acids in the liver proceeds more rapidly than usual, with augmented production of ketone bodies. In addition there is a deficit of oxalo-acetate and thus a decreased formation of citrate. The low level of oxaloacetate is accentuated because it is being utilized for glucose formation from amino acids in the liver (Chap. 25). This further impairs operation of the citric acid cycle.

Ketosis incident to starvation is most frequently encountered clinically in gastrointestinal disturbances in infancy or pregnancy. Other circumstances in which excessive lipid and diminished carbohydrate are being metabolized may also lead to ketosis, e.g., renal glucosuria (Chap. M5) and abrupt replacement of a normal diet by one low in carbohydrate and very rich in lipid.

An important cause of ketosis is *diabetes mellitus*. In the diabetic, in contrast to the above situations, glucose is present in excessive amounts in the fluids of the body; however, the metabolic defect, viz., insulin deficiency, prevents utilization of glucose at a normal rate. Normally, insulin depresses cAMP levels in adipose tissue, thus reducing the activity of hormone-sensitive *triacylglycerol lipase*. When insulin levels are low, as in diabetes, the lipase is more active; hence, there is increased formation and release of fatty acids. From the point of view of the effect upon lipid metabolism, diabetes and starvation resemble each other. Despite the hyperglycemia in diabetes, glucose is not being catabolized at a normal rate in muscle or in liver. Excessive mobilization of depot lipid leads to lipemia and fatty liver. Generation of acetoacetate occurs at an excessive rate in the liver while resynthesis of fatty acids is inhibited. If acetoacetate is formed at a rate in excess of the capacity of the extrahepatic tissues to utilize it, ketosis will develop. In diabetic individuals with severe ketosis, urinary excretion of ketone bodies may be as high as 5000 mg/day and the blood concentration may reach 90 mg/dl, in contrast to a normal value <3 mg.

Consequences of Ketosis The complications encountered when more ketone bodies are formed than can be utilized relate primarily to the mode of excretion of acetoacetate and β-hydroxybutyrate. These two substances are excreted in large part as the sodium salts. Depletion of the body fluids of Na⁺ and other cations leads to acidosis (Chap. M5), often termed *ketoacidosis*. Coincident with excretion of these salts (as well as glucose, in the diabetic patient), large quantities of fluid are lost in the urine. A tendency to nausea and emesis may cause additional fluid loss, while depression of the central nervous system, leading ultimately to profound coma with areflexia, interferes with normal drinking of water. All these factors complicate the acidosis by producing severe dehydration. Ketoacidosis is of frequent occurrence in the untreated juvenile diabetic because of failure to secrete sufficient insulin. In late-onset diabetes after age 40, severe ketoacidosis is less frequent (Chap. M16).

DIETARY STATUS AND HEPATIC METABOLISM OF FATTY ACIDS

From the preceding descriptions it is evident that there is normally exquisite control of oxidation of fatty acids via the citric acid cycle and of the formation of acetoacetate to be delivered to the peripheral tissues. These mitochondrial processes are linked to events in the cytosol, where fatty acids are synthesized from acetyl CoA. Cholesterol is also made in the cytosol from acetyl CoA (Chap. 21). It is useful to review the changes that occur in the fed state, after digestion of a meal containing carbohydrate and triacylglycerols, and in the fasting state. Some of these changes are summarized in Table 20.5 (see also Fig. 20.4).

From Table 20.5 and earlier discussion, it is evident that the balance between insulin and glucagon plays a major role in the shifts between glucose and fatty acid metabolism. High insulin levels repress the hormone-sensitive lipase of adipose cells and stimulate glucose utilization (Chap. 18) and lipogenesis,

TABLE 20.5

Effect of Dietary Status on Deposition, Mobilization, and Utilization of Fatty Acids—Relationships to Carbohydrate Metabolism

Fed state	Fasting state
1. Blood glucose level rises to 150 mg/dl.	1. Blood glucose level falls to 80 mg/dl; in prolonged starvation to 50–60 mg/dl.
2. Pancreatic β cells release insulin to a plasma concentration of 100 μunits/ml.	2. Insulin release is slowed and plasma concentration drops to below 10 μunits/ml.
3. Release of glucagon from pancreatic α cells drops to a plasma concentration of 80 pg/ml.	3. Glucagon increases to a plasma concentration of 120–150 pg/ml.
4. Muscle shifts to glucose as major source of energy.	4. Muscle shifts to fatty acids and some amino acids as major fuel sources.
5. Liver converts excess glucose to triacylglycerols; these are exported via VLDL to adipocytes.	5. Lipogenesis ceases; oxidation of fatty acids and production of acetoacetate increases.
6. Lipoprotein lipase of adipocytes hydrolyzes triacylglycerols of VLDL to fatty acids which are incorporated into triacylglycerols for storage.	6. Hormone-sensitive lipase is stimulated to convert triacylglycerols to fatty acids; the latter are transported by albumin to liver and peripheral tissues. In liver, an increasing proportion of fatty acids is converted to acetoacetate.
7. Insulin stimulates protein synthesis, including enzymes of lipogenesis.	7. As supplies of glycogen and glucose are exhausted, there is proteolysis of liver and muscle proteins; glycogenic amino acids (Chap. 25) are converted by liver to plasma glucose.

and, in conjunction with steroids from the adrenal cortex, stimulate synthesis of the enzymes of lipogenesis (page 527).

When the supply of glucose diminishes, the insulin level falls, the glucagon level increases, and lipogenesis from carbohydrate essentially ceases. Glucose has to be spared for brain metabolism since, as already noted, under normal conditions this organ requires glucose. Many amino acids are glucogenic; that is, the carbon skeleton can be converted to glucose (Chap. 25) and this augments the glucose supply. The increased glucagon stimulates release of fatty acids to plasma. The activity of the tricarboxylic acid cycle in the mitochondria decreases, since oxaloacetate, which is being diverted to glucose formation in the cytosol, falls to a low level. This results in two crucial changes: (1) citrate, which exerts an important control on lipogenesis by its stimulation of acetyl CoA carboxylase (page 524), is exported to the cytoplasm in gradually diminished amounts; this effectively shuts off the committed step in lipogenesis. (2) The acetyl CoA, which is formed in the liver mitochondria by degradation of fatty acids, is now increasingly diverted to form 3-hydroxy-3 methylglutaryl CoA and its product, acetoacetate.

When glucose again becomes ample, after refeeding, the process is reversed. The citric acid cycle again becomes operative, because of ample oxaloacetate; the high level of ATP inhibits isocitrate dehydrogenase and citrate is exported to the cytoplasm; and the acetyl CoA carboxylase is activated, increasing formation of malonyl CoA. It has been indicated that malonyl CoA exerts an inhibitory effect on the *carnitine acyltransferase*, which is responsible for conversion of fatty acyl CoA derivatives to fatty acyl carnitine for transfer to the mitochondria. If this is so, it would help to explain the sparing effect of carbohydrate on fatty acid oxidation and the diversion of acetyl CoA to form triacylglycerols. In any case, with ample glucose, lipogenesis is increased and fatty acid oxidation is diminished.

In summary, there are continuous changes in liver metabolism reflecting the nutritional state of the animal, shifting from carbohydrate as the predominant source of energy in the fed state to fatty acid oxidation when the supply of carbohydrate falls. The major regulatory effectors in the liver are the hormones insulin and glucagon (Table 20.5). As discussed later (Chap. 25), the metabolism of amino acids also plays a prominent part in these cycles, since not only does the oxidation of certain amino acids supply considerable energy to the extrahepatic tissues, but many others can be used by the liver to maintain the level of blood glucose. It should be noted that the heart does not alter its metabolism significantly. Both in the fed and fasting states, it is largely a consumer of fatty acids, ketone bodies, and certain amino acids (Chap. 25) for its energy.

METABOLISM OF ETHANOL AND ALCOHOLISM: RELATIONSHIP TO LIPID METABOLISM

Consumption of ethanol in modest amounts, aside from its psychic effects, simply contributes additional calories to the diet by virtue of its oxidation to acetate and conversion to acetyl CoA, which enters the usual pathways for this substance. In larger quantities, ethanol consumption creates a number of metabolic problems with ultimate tissue damage, the effects varying considerably depending on individual tolerance, diet, and other factors.

Alcohol, being freely water-soluble, is rapidly absorbed—approximately 20 percent from the stomach and 80 percent from the intestines. A single dose is detectable in the blood in about 5 min, with a maximum reached in $\frac{1}{2}$ to 2 h. Absorption is retarded, however, by simultaneous or earlier consumption of fatty foods. There is no tissue storage of ethanol, and it reaches all organs of the body. Large consumption over a prolonged period produces inflammation of stomach, intestines, and pancreas with impairment of digestion and absorption, as well as secondary effects on the central nervous sytem, frequently associated with nutritional deficiencies produced by low consumption of protein and low intake of vitamins. The Wernicke-Korsakoff syndrome of polyneuropathy is relieved by administration of thiamine, but folate and niacin deficiency may also occur.

Alcohol dehydrogenase of liver is the primary enzyme involved in ethanol metabolism.

$$\text{Ethanol} + \text{NAD}^+ \rightleftharpoons \text{acetaldehyde} + \text{NADH} + \text{H}^+$$

As already noted (Chap 16) acetaldehyde is oxidized to acetate, mainly in liver, by three different enzymes: the iron-molybdoflavoprotein *aldehyde oxidase,* the iron-molybdoflavoprotein *xanthine oxidase,* and the NAD-linked *aldehyde oxidase.* The acetyl CoA that is formed is then utilized in usual fashion. The production of both acetyl CoA and NADH and consequent formation of ATP spare fatty acids that tend to accumulate. The high levels of NADH may result in reduction of pyruvate to lactate, which produces a mild lactic acidosis and hence a lower than normal blood glucose level because of a lower rate of gluconeogenesis from pyruvate (Chap 17). Because of diminished availability of glucose, there is excessive production of ketone bodies in some alcoholics; this mimics the ketosis of diabetics with many of the consequences already described above.

Associated with chronic alcoholism is excessive mobilization of lipid and ultimate development of a fatty liver. The production of fatty hepatosis appears to be a combination of several factors: sparing action of ethanol oxidation on utilization of liver triacylglycerols, excessive mobilization of triacylglycerols from adipose tissue to the liver caused in part by action of ethanol in triggering release of lipolytic hormones, and failure to synthesize sufficient lipoprotein for transport of triacylglycerols because of alterations of amino acid availability. The fatty hepatosis reveals parenchymal cells heavily infiltrated with triacylglycerols. In approximately 10 percent of patients with severe alcoholism this leads to *cirrhosis,* excessive production of fibrous tissue that destroys the normal architecture of the liver.

REFERENCES

See list following Chap. 21.

Chapter 21

Lipid Metabolism II

Phosphoglycerides. Sphingolipids.
Cholesterol metabolism and its controls.

THE PHOSPHOGLYCERIDES

The most polar lipids, the phosphoglycerides, are amphiphiles (Chap. 7) with ionic hydrophilic head groups and hydrophobic tails. They are major constituents of membrane systems (Chap. 13), as well as important components of bile, serum lipoproteins, and lung surfactant. Studies of the turnover of various phosphoglycerides in diverse sites in the animal have shown that the half-lives of these lipids range from less than 1 day for liver phosphatidylcholine to more than 200 days for brain phosphatidylethanolamine. Thus, some phosphoglycerides serve mainly as structural components with slow turnover, whereas others are metabolically active and participate in a variety of functions.

The pathways of phosphoglyceride biosynthesis in eukaryotic cells are summarized in Fig. 21.1. Most of the individual reactions are catalyzed by enzymes that are bound to membranes of the endoplasmic reticulum or to the mitochondria. Synthesis of the phosphoglycerides occurs asymmetrically on the cytoplasmic side of the endoplasmic reticulum, indicating that after synthesis there is considerable transmembrane movement within the membranes of the cell.

Phosphatidates

The phosphatidates are the major class of phosphoglycerides. These derivatives of glycerol-3-phosphate usually contain a nitrogenous base—serine, ethanolamine, or choline—which can be synthesized by the organism. The most abundant phosphatidate, phosphatidylcholine, can be synthesized *de novo* from phosphatidylethanolamine and from dietary choline.

Any cross-references coded M refer to the companion text, *Principles of Biochemistry: Mammalian Biochemistry.*

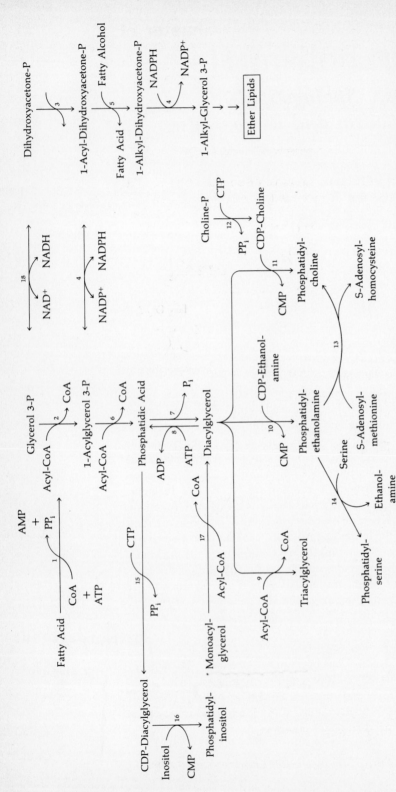

Figure 21.1 Some of the major pathways and enzymes of phosphoglyceride synthesis. 1. Fatty acid CoA ligase (AMP); 2. *sn*-glycerol-3-P acyltransferase; 3. dihydroxyacetone-P acyltransferase; 4. acyl(alkyl)dihydroxyacetone-P oxidoreductase; 5. alkyldihydroxyacetone-P synthase; 6. lysophosphatidic acid acyltransferase; 7. phosphatidic acid phosphatase; 8. diacylglycerol kinase; 9. diacylglycerol acyltransferase; 10. diacylglycerol ethanolaminephosphotransferase; 11. diacylglycerol cholinephosphotransferase; 12. choline-P cytidyltransferase; 13. phosphatidylethanolamine *N*-methyltransferase; 14. phosphatidylethanolamine serinetransferase; 15. phosphatidic acid cytidyltransferase (CDP-diacylglycerol synthase); 16. phosphatidylinositol synthase; 17. monoacylglycerol acyltransferase; 18. glycerol-3-P dehydrogenase. P = phosphate; ether lipids = plasmalogens. [*From R. M. Bell, L. M. Ballas, and R. A. Coleman, J. Lipid Res. 22:391 (1981).*]

Phosphatidic Acid A key substance in some types of phosphatidate synthesis is *phosphatidic acid*. Its formation from glycerol-3-phosphate or from dihydroxyacetone phosphate and 2 mol of fatty acyl CoA has been described (page 533). In addition, there are other routes for formation of phosphatidic acid from triacylglycerols. Overall, these pathways can be outlined as follows.

The 1,2-diacylglycerol is converted to phosphatidic acid by a *diacylglycerol kinase*.

$$\text{L-1,2-Diacylglycerol} + \text{ATP} \longrightarrow \text{L-phosphatidic acid} + \text{ADP}$$

Enzymes of brain tissue and intestinal mucosa convert the monoacylglycerol to L-lysophosphatidic acid, which is then transformed to phosphatidic acid.

$$\alpha\text{-Monoacylglycerol} + \text{ATP} \longrightarrow \text{ADP} + \quad
\begin{array}{l}
\text{H}_2\text{COOR}' \\
\text{HOCH} \\
\text{H}_2\text{COPO}_3{}^{2-}
\end{array}
\xrightarrow{+\text{R}''\text{COCoA}}$$

L-Lysophosphatidic acid

$$\begin{array}{l}
\text{H}_2\text{COOCR}' \\
\text{R}''\text{COOCH} \\
\text{H}_2\text{COPO}_3{}^{2-}
\end{array}$$

L-Phosphatidic acid

Formation of Phosphatidylethanolamine Initially, ethanolamine is phosphorylated by an *ethanolamine kinase* to form phosphoethanolamine.

$$\text{Ethanolamine} + \text{ATP} \xrightarrow{\text{Mg}^{2+}} {}^-\text{HO}_3\text{—P—OCH}_2\text{CH}_2\text{NH}_3{}^+ + \text{ADP}$$

Phosphoethanolamine

A *cytidyl transferase* then catalyzes reaction of phosphoethanolamine with CTP to yield cytidine diphosphoethanolamine.

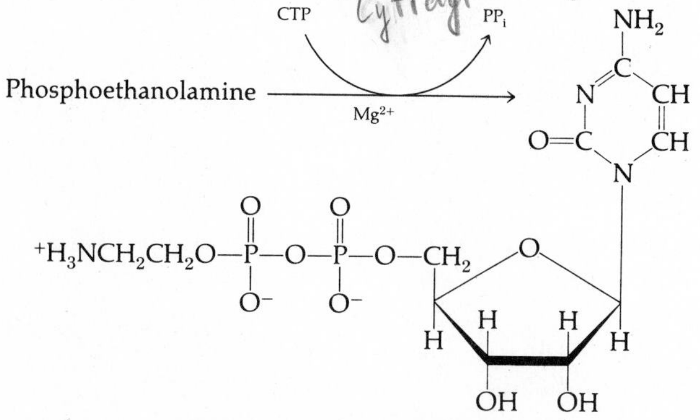

Cytidine diphosphoethanolamine

CDP-ethanolamine is then converted to the phosphatidate by a *phosphotransferase*-catalyzed reaction.

CDP-ethanolamine + L-1,2-diacylglycerol $\xrightarrow{Mg^{2+}}$

$$\begin{array}{l} H_2COOCR \\ R'COOCH \\ \qquad\qquad O \\ \qquad\qquad \| \\ H_2CO-P-OCH_2CH_2NH_3^+ \\ \qquad\qquad | \\ \qquad\qquad O^- \end{array}$$

\+ CMP

Phosphatidylethanolamine

prevalent
盛行

The 1,2-diacylglycerol may arise from triacylglycerol or by the action of a phosphatase on phosphatidic acid. The latter route is the more prevalent since mammalian phosphatidates, including phosphatidylcholine (see below), usually contain a saturated fatty acid at R and an unsaturated fatty acid at R'. As already noted (page 533), this would be expected for phosphatidic acid or for the diacylglycerol synthesized via dihydroxyacetone phosphate.

Formation of Phosphatidylcholine This occurs in stages by successive transfer of three methyl groups from *S*-adenosylmethionine (page 612) to phosphatidylethanolamine.

Phosphatidylethanolamine + 3 *S*-adenosylmethionine $\longrightarrow$ phosphatidylcholine + 3 *S*-adenosylhomocysteine

3 SAM

3 SAH

These methylations, catalyzed by *N-methyl transferases,* occur primarily in the liver and only with phosphatidylethanolamine as substrate and represent the mechanism for endogenous formation of choline. The microsomal enzyme(s) form monomethylethanolamine-, dimethylethanolamine-, and choline-containing phosphatidates.

The *de novo* pathway is estimated to provide only about 20 percent of the phosphatidylcholine made in the liver. Thus a dietary source of choline is essential (Chaps. 20 and M22). Phosphatidylcholine formation from free choline involves utilization of cytidine diphosphocholine as the key intermediate; its structure is similar to that of CDP-ethanolamine, and it is synthesized by analogous reactions.

Phosphocholine formation is catalyzed by *choline kinase*.

$$Choline + ATP \xrightarrow{Mg^{2+}} \begin{array}{c} O \\ \| \\ {}^-O-P-OCH_2CH_2\overset{+}{N}(CH_3)_3 \\ | \\ O^- \end{array} + ADP$$

Phosphocholine

Formation of CDP-choline from CTP and phosphocholine is analogous to the formation of CDP-ethanolamine from CTP and phosphoethanolamine. The *CTP-phosphocholine:cytidyl transferase* is thought to be the regulated step in formation of phosphatidylcholine, since its activity is stimulated by diphosphatidylglycerol (cardiolipin).

Cytidine triphosphate + phosphocholine $\xrightarrow{Mg^{2+}}$ cytidine diphosphocholine + PP$_i$

The PP$_i$ is derived from CTP.

A *choline phosphotransferase* catalyzes reaction of cytidine diphosphocholine (CDP-choline) with 1,2-diacylglycerol to yield phosphatidylcholine and cytidine monophosphate.

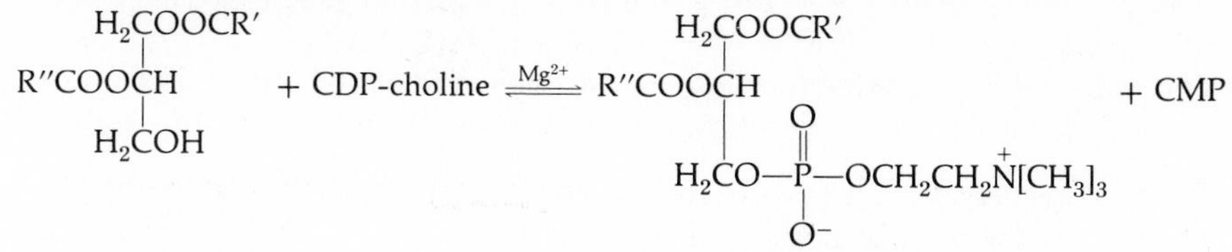

$$
\begin{array}{c}
\text{H}_2\text{COOCR}' \\
\text{R}''\text{COOCH} \qquad + \text{ CDP-choline} \xrightarrow{\text{Mg}^{2+}} \\
\text{H}_2\text{COH}
\end{array}
\qquad
\begin{array}{c}
\text{H}_2\text{COOCR}' \\
\text{R}''\text{COOCH} \qquad\qquad\qquad\quad + \text{ CMP} \\
\text{H}_2\text{CO}-\overset{\overset{\text{O}}{\|}}{\underset{\underset{\text{O}^-}{}}{\text{P}}}-\text{OCH}_2\text{CH}_2\overset{+}{\text{N}}[\text{CH}_3]_3
\end{array}
$$

L-1,2-Diacylglycerol   **Phosphatidylcholine**

The phosphatidylcholine of plasma lipoproteins contains predominantly linoleate in the 2 position. This is the acyl group transferred to form cholesterol esters (page 561).

Formation of Phosphatidylserine   In mammals phosphatidylserine is formed by an exchange reaction, catalyzed by *phosphatidylethanolamine:serine transferase.*

$$\text{Phosphatidylethanolamine} + \text{L-serine} \xrightarrow{\text{Ca}^{2+}} \text{phosphatidylserine} + \text{ethanolamine}$$

Phosphatidylserine is decarboxylated by *phosphatidylserine decarboxylase,* an enzyme containing pyridoxal phosphate.

$$\text{Phosphatidylserine} \longrightarrow \text{phosphatidylethanolamine} + \text{CO}_2$$

The net result of the exchange reaction and the decarboxylation is the conversion of serine to ethanolamine. This provides additional ethanolamine for the *de novo* synthesis of phosphatidylcholine, as described above. In *Escherichia coli,* *phosphatidylserine decarboxylase,* a membrane-bound enzyme, contains an NH$_2$-terminal pyruvoyl residue, which is essential for the action of the enzyme.

In *E. coli* and other bacteria, phosphatidylserine is formed via cytidine diphosphodiacylglycerol (CDP-diacylglycerol), which in turn is derived from L-phosphatidic acid by *CDP-diacylglycerol synthetase.*

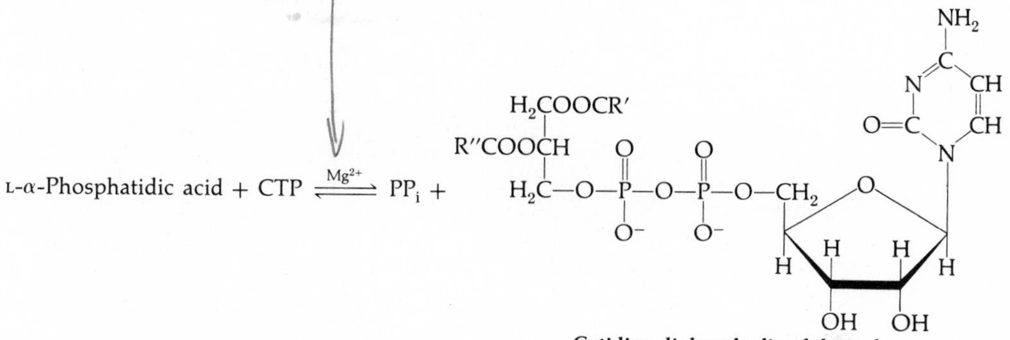

$$\text{L-}\alpha\text{-Phosphatidic acid} + \text{CTP} \xrightarrow{\text{Mg}^{2+}} \text{PP}_i +$$

Cytidine diphosphodiacylglycerol

The next step results in formation of phosphatidylserine.

$$\text{CDP-diacylglycerol} + \text{L-serine} \longrightarrow \text{phosphatidylserine} + \text{CMP}$$

In these bacteria, phosphatidylserine is decarboxylated to yield phosphatidylethanolamine, as above.

CDP-diacylglycerol is also formed in mammalian tissues and provides the major route for synthesis of phosphatidylinositides and phosphatidylglycerols.

Formation of Phosphatidylinositides and Phosphatidylglycerols These nitrogen-free derivatives of glycerol phosphate are synthesized by transferases.

$$\text{CDP-diacylglycerol} + \text{myoinositol} \xrightarrow{\text{Mn}^{2+}} \text{phosphatidylinositol} + \text{CMP}$$

The di- and triphosphoinositides (Chap. 7) are formed by successive phosphorylations of the monoinositide first in the 4 and then in the 5 position by two specific kinases in the presence of ATP and either Mg^{2+} or Mn^{2+}. The triphosphoinositide is particularly abundant in the myelin membranes of nervous tissue.

The phosphatidylglycerols (Chap. 7) are synthesized by a similar reaction.

$$\text{CDP-diacylglycerol} + \text{glycerol-3-phosphate} \longrightarrow$$
$$\text{3-phosphatidyl-1'-glycerophosphate} + \text{CMP} \quad (1)$$

$$\text{3-Phosphatidyl-1'-glycerophosphate} \xrightarrow{\text{phosphatase}} \begin{array}{ll} H_2\text{COOCR} & H_2\text{COH} \\ R'\text{COOCH} & \text{HCOH} + P_i \end{array} \quad (2)$$

$$H_2C-O-\overset{\displaystyle O}{\underset{\displaystyle O^-}{P}}-O-CH_2$$

3-Phosphatidyl-1'-glycerol

Diphosphatidylglycerol (cardiolipin) (Chap. 7) is formed from two molecules of phosphatidylglycerol. CDP-diglyceride is required as a cofactor. Cardiolipin represents 10 percent or more of the lipids in the mitochondrial membrane.

$$\text{2 Phosphatidylglycerol} \longrightarrow \text{cardiolipin} + \text{glycerol}$$

Formation of Phosphatidates by Transfer and Exchange In mammals CDP-diacylglycerol can be formed from phosphatidic acid as described above for *E. coli*. The liponucleotide can also be formed from the CDP-monoacylglycerol by a specific *arachidonoyl CoA: CDP-monoacylglycerol acyltransferase* in liver microsomes. The result is specific incorporation of arachidonic acid at the 2 position to form the CDP-diacylglycerol. Indeed, human platelets contain primarily the 1-stearyl-2-arachidonoylphosphatidylinositol. The arachidonic acid is liberated from this phosphatidate during blood clotting (Chap. M1) and is used for formation of thromboxane and of prostaglandins (Chap. M13).

There appear to be other transferases that are specific for transfer or exchange of certain fatty acids from their CoA derivatives to phosphatidates or CDP-monoacylglycerol precursors. For example, in lung some of the alveolar cells produce a substance known as *surfactant* which contributes to the stability of the alveoli and thus aids in preventing lung collapse (*atelectasis*). The predomi-

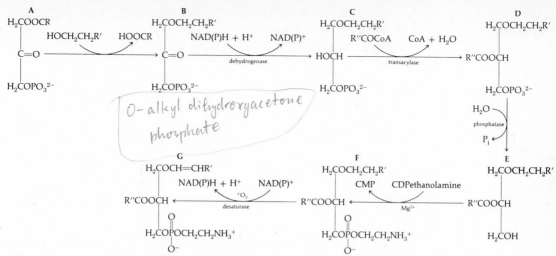

Figure 21.2 Formation of a plasmalogenic monoacylglycerol. A, monoacyl dihydroxy-acetone phosphate; B, 1-alkyl dihydroxyacetone phosphate; C, O-alkylglyceryl phosphate; D, 1-alkyl-2-acylglyceryl phosphate; E, 1-alkyl-2-acylglycerol; F, 1-alkyl-2-acylphosphatidylethanolamine; G, ethanolamine plasmalogen, or 1-alkenyl-2-acylphosphatidalethanolamine.

nant lipid in surfactant is dipalmitoylphosphatidylcholine which is made by a transferase specific for palmitoyl CoA and lysophosphatidylcholine.

Although many exchange reactions of one acyl moiety for another in phosphatidates have been observed, the specificity and number of such enzymes are still largely unknown. This also applies to the plasmalogens described below, which may have specific acyl groups in the 2 position.

Plasmalogens (Ether Lipids)

The acyl group of monoacyl dihydroxyacetone phosphate is exchanged for a fatty alcohol (formed by reduction of fatty acid) with retention of the α-hydrogen atoms and the oxygen of the alcohol. The nature of this unusual replacement reaction, in which a new bond of different character is formed, is unknown. The O-alkyl dihydroxyacetone phosphate is then reduced to O-alkylglyceryl phosphate with NADPH or NADH as hydrogen donor. This is followed by acylation via a fatty acyl CoA. The 1-alkyl-2-acylglyceryl phosphate is dephosphorylated and converted to an alkyl analog of the phosphoglycerides containing choline, ethanolamine, or serine by the reactions previously described. The 1-alkyl-2-acylglyceryl phosphoethanolamine is converted to the plasmalogen by a specific desaturase in the presence of O_2 and NADH or NADPH. The overall process is summarized in Fig. 21.2.

Action of Phospholipases

The phospholipases have been useful in the study of the structure of phosphoglycerides as well as indicating routes of their degradation. The susceptible bonds in phosphatidylcholine are indicated by capital letters.

551

$$\begin{array}{c} \overset{1}{H_2}CO \overset{A}{-} OCR \\ R'CO \overset{B}{-} O\overset{2}{CH} \\ \overset{}{H_2}\overset{3}{CO} \overset{C}{-} \overset{O}{\underset{O^-}{\overset{||}{P}}} \overset{D}{-} OCH_2CH_2\overset{+}{N}[CH_3]_3 \end{array}$$

hemolytic
溶血的

The enzymes liberating an acyl group from a phosphoglyceride are phospholipases A. For hydrolysis at bond A, the enzyme is phospholipase A_1; for that at bond B, phospholipase A_2.

Phospholipase A_2 yields lysophosphatidylcholine, a strong detergent and a potent hemolytic agent. Phospholipase A_2 also cleaves bond B in phosphatidylethanolamine and in the choline- and ethanolamine-containing plasmalogens. The action of phospholipase A_2 provides either arachidonic or homo-γ-linolenic acids, the initial substrates for prostaglandin synthesis (Chap. M13). Some snake venoms contain similar Ca^{2+}-requiring phospholipases.

Lysophospholipase of pancreas and other tissues catalyzes hydrolysis of the single ester bond (A) in lysophosphatidylcholine or lysophosphatidylethanolamine.

$$\text{Lysophosphatidylcholine} + H_2O \xrightarrow{Ca^{2+}} \text{glycerylphosphocholine} + \text{fatty acid}$$

Phospholipase B of animal tissues hydrolyzes both bonds A and B. *Phospholipase C*, which cleaves bond C, is found in the α toxin of *Clostridium welchii* and other strains of clostridia and bacilli. These enzymes require Zn^{2+}, Ca^{2+}, or other divalent cations. A phospholipase C has also been found in liver lysosomes (see below).

$$\text{Phosphoglyceride} + H_2O \longrightarrow$$
$$\text{1,2-diacylglycerol} + \text{phosphorylated nitrogenous base}$$

Phospholipase D in plant tissues catalyzes transphosphatidylation reactions as well as hydrolytic cleavage of the terminal diester bond (at *D*) of glycerophosphatidates containing choline, ethanolamine, serine, or glycerol, with the formation of phosphatidic acid.

$$\text{Phosphatidylcholine} + H_2O \longrightarrow \text{phosphatidic acid} + \text{choline} \qquad (1)$$
$$\text{Phosphatidylcholine} + ROH \rightleftharpoons \text{phosphatidyl—OR} + \text{choline} \qquad (2)$$

Both reactions are activated by Ca^{2+}. Reaction (2), a transphosphatidylation, could provide a mechanism for synthesis and turnover of phosphoglycerides.

Transphosphatidylations, reaction (2), in which free serine, ethanolamine, or choline exchange with the bases of phosphoglycerides are also catalyzed by brain microsomal fractions. Present evidence suggests that such reactions may be the major pathway for phosphatidylserine synthesis.

Turnover of Phosphoglycerides In liver lysosomes there appear to be two routes of degradation of phosphatidates by enzymes optimally active near pH 4.5.[1] One route involves the successive action of phospholipases A_1 and A_2 to liberate glycerylphosphocholine, followed by formation of glycerophosphate and choline. The other route involves hydrolysis by phospholipase C with maximal activity towards phosphatidylinositol but also active towards the phosphatidates containing choline, glycerol, serine, and ethanolamine in order of decreasing rate. The released diacylglycerol (and the monoacylglycerol) can be reutilized as already described (Chap. 20).

SPHINGOLIPIDS

Formation of Sphingosines The long-chain, aliphatic bases, e.g., sphinganine (dihydrosphingosine) and sphingenine (sphingosine), are synthesized by many animal tissues.

(i) 24ATP

Synthase

$$CH_3[CH_2]_{14}-\overset{O}{\underset{}{C}}-SCoA + \underset{NH_2}{\underset{|}{C}}HCH_2OH \xrightarrow[\text{phosphate-requiring}]{\text{pyridoxal}} CH_3[CH_2]_{14}-\overset{O}{\underset{}{C}}-\underset{NH_2}{\underset{|}{C}}H-CH_2OH + CO_2 + CoA \quad (1)$$

$$\text{Palmitoyl CoA} \qquad \text{L-Serine} \qquad \qquad \text{3-Dehydrosphinganine}$$

$$\text{3-Dehydrosphinganine} \xrightarrow[\text{reductase}]{\text{NADPH + H}^+ \quad \text{NADP}^+} CH_3-[CH_2]_{14}-\underset{NH_2}{\underset{|}{C}}H-\overset{OH}{\underset{}{C}}H-CH_2OH + NADP^+ \quad (2)$$

(stereospecific)

$$\text{D-Sphinganine}$$

Note that the carboxyl group of serine is lost as CO_2 in reaction (1) and that carbon atoms 1 and 2 and the amino group of the product are derived from the α- and β-carbon atoms of serine. The condensing reaction (1) functions with CoA derivatives of C_{14} to C_{18} acids and the reduction reaction with C_{14} to C_{18} 3-dehydrosphinganines; these yield the known C_{16} and C_{18} sphinganines.

Sphingenine can be formed by reactions (1) and (2) from the CoA derivative of hexadecenoic acid and, presumably, by dehydrogenation of sphinganine to yield this trans compound. Alternatively, desaturation of the fatty acyl CoA takes place before condensation with serine and may be the major route; the reaction is catalyzed by a *flavin dehydrogenase:*

$$\text{D-Sphinganine} \xrightarrow[]{\text{FAD} \quad \text{FADH}_2} CH_3[CH_2]_{12}-CH=CH-\underset{NH_2}{\underset{|}{C}}H-\overset{OH}{\underset{}{C}}H-CH_2OH \quad (3)$$

$$\text{D-Sphingenine}$$

D-Sphingosine (4-sphingenine)

Formation of Ceramides The ceramides, fatty acid amides of sphingenine, are formed by a *ceramide: N-acyl transferase* from fatty acyl CoA derivatives.

$$\text{D-Sphingenine} \xrightarrow[]{\text{RCOCoA} \quad \text{CoA}} CH_3[CH_2]_{12}-CH=CH-\underset{HN-\overset{O}{\underset{}{C}}-R}{\underset{|}{C}}H-\overset{OH}{\underset{}{C}}H-CH_2OH$$

$$\text{Ceramide}$$

Formation of Sphingomyelins There are several routes for the synthesis of sphingomyelins. One accounts for most of the synthesis and is catalyzed by a *CDP-choline: ceramide-choline phosphotransferase*, found in liver mitochondria and in homogenates of brain and spleen.

$$\text{Ceramide} \xrightarrow[\text{CDP-choline}]{\quad\quad\text{CMP}\quad} CH_3[CH_2]_{12}-CH=CH-\underset{\underset{\underset{O}{\parallel}}{\underset{HN-C-R}{\mid}}}{CH}-\underset{\underset{OH}{\mid}}{CH}-CH_2O-\underset{\underset{O^-}{\overset{\overset{O}{\parallel}}{\mid}}}{P}-OCH_2CH_2\overset{+}{N}(CH_3)_3$$

Sphingomyelin

Another route is as follows:

Phosphatidylcholine + ceramide $\rightleftharpoons$ sphingomyelin + diglyceride

A minor route of sphingomyelin formation involves a two-step sequence.

CDP-choline + sphingenine $\longrightarrow$ sphingenylphosphorylcholine + CMP   (1)

Sphingenylphosphorylcholine + fatty acylCoA $\longrightarrow$ sphingomyelin + CoA   (2)

GLYCOSPHINGOLIPIDS

The cerebrosides, ceramide oligosaccharides, cerebroside sulfatides, and gangliosides (Chap. 7) are present in small amounts in the membranes of a wide variety of tissues. Nervous tissues are particularly rich in gangliosides. The biological functions of the glycosphingolipids are not completely understood, but certain viruses, such as influenza, adhere to cell surfaces through viral receptors that bind specific sialic acid groups in gangliosides. Cholera and diphtheria toxins also bind to host cells through sialic acid groups in gangliosides. The specificity of the ABO and the Lewis blood group antigens (Chap. M3) on cells such as erythrocytes are determined by the carbohydrate groups of ceramide oligosaccharides.

Glycolipids are synthesized by the sequential actions of a series of glycosyltransferases as shown in Fig. 21.3. The simplest glycolipids, gluco- or galactocerebrosides (ceramide monosaccharides), are synthesized by the transfer of a monosaccharide from the appropriate nucleotide sugar to the C-1 hydroxyl group of a ceramide, the lipid component of all glycosphingolipids. Further addition of monosaccharides from nucleotide sugars produces the more complex glycolipids with two, three, four, or more sugars in glycosidic linkage. The synthetic pathways are similar to those for the synthesis of the O-linked oligosaccharides in glycoproteins (Chap. 18). Indeed, some glycosyltransferases appear to use glycolipids and glycoproteins equally well as substrates, e.g., certain *sialyltransferases*, although some glycosyltransferases are specific for glycolipid substrates. Synthesis of the sulfatides also requires the transfer of sulfate from 3'-phosphoadenosine-5'-phosphosulfate to specific hydroxyl groups on the carbohydrate.

The main feature that distinguishes gangliosides from other types of glycolipids is the presence of sialic acid, which is often N-acetylneuraminic acid

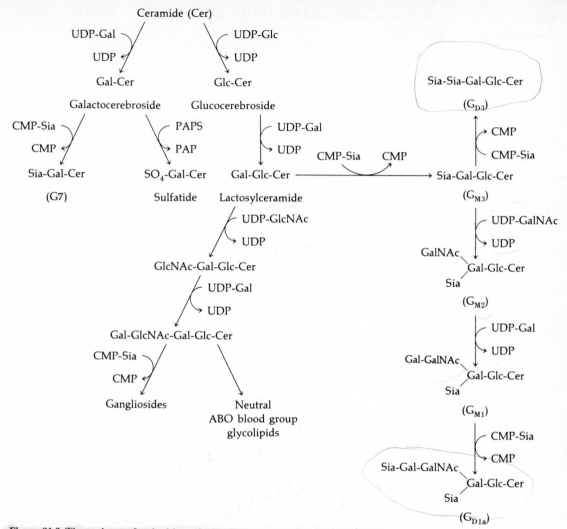

Figure 21.3 The pathways for the biosynthesis of some glycosphingolipids. Glycosyltransferases act sequentially to add monosaccharides stepwise to ceramide or mono-, di-, etc., ceramide saccharides. The disialyl linkage in GD₃ is Siaα2-8Sia. Similar pathways lead to several other glycosphingolipids, whose structures are not shown. PAPS = 3′-phosphoadenosine-5′-phosphosulfate.

(Chap. 18). Gangliosides are commonly abbreviated by the letter G, with a subscript M, D, or T to designate that they contain one, two, or three sialic acid residues, respectively, and a number or letter to distinguish different members of a group from one another. The major human brain gangliosides are essentially all derivatives of G_{Dla}, with sugar residues in the following linkages.

$$\text{Sia}\alpha2\text{-3Gal}\beta1\text{-3GalNAc}\beta1\text{-4}$$
$$\text{Gal}\beta1\text{-4}\beta O\text{-Cer}$$
$$\text{Sia}\alpha2\text{-3}$$

G_{Dla} → most common ganglioside in brain tissue of vertibrate

NANA

555

Other gangliosides in the brain and many other tissues contain N-acetylglucosamine rather than N-acetylgalactosamine, however, and are derivatives of the ceramide oligosaccharide Galβ1-4GlcNAcβ1-3Galβ1-4Glcβ0-Cer. Neutral ceramide oligosaccharides with blood group antigenic activities are derivatives of the ceramide oligosaccharide GalNAcβ1-3Galβ1-3Galβ1-4Glcβ0-Cer.

Glycolipid Catabolism

The glycosphingolipids, sulfatidates, and sphingomyelins are being continuously synthesized, as described in the preceding section, and are simultaneously also undergoing degradation by hydrolytic enzymes present in the lysosomes. Since concentrations remain essentially constant, normally there is a dynamic steady state. Absence or partial deficiency of any one of the specific hydrolases leads to an accumulation, particularly in the nervous system, of one or more of these compounds. This causes a neurological deterioration that usually occurs after the first months of life and leads to early death, with the exception of the adult form of Gaucher's disease and Fabry's disease.

Figure 21.4 shows the pathways of hydrolysis leading to ceramide. The site of action of the specific missing hydrolase and the genetically determined disorder that ensues are indicated. Most of the enzymes involved are *glycosidases*,

Figure 21.4 Catabolism of typical glycosphingolipids by human lysosomal glycosidases, sphingolipase, and sulfatase. Diseases associated with a hydrolase deficiency are indicated, and the block in catabolism in each disease is indicated by a solid bar. PC is phosphocholine.

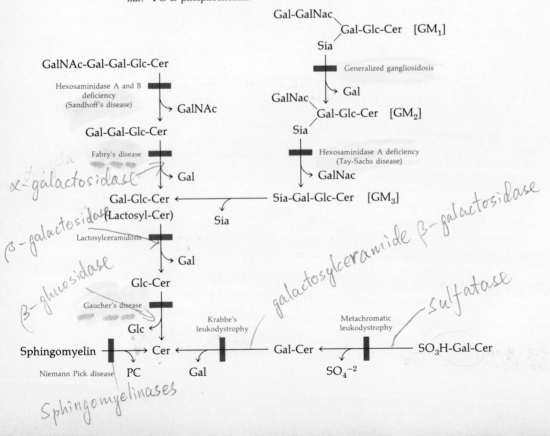

lysosomal enzymes with high specificity for the types of sugars that form the susceptible glycosidic bond and the configuration of the bond.

Sphingomyelinases hydrolyze sphingomyelin to ceramide and phosphocholine. The enzymes are normally found in the spleen, liver, and kidney, as well as brain and other tissues. In the adult form of *Niemann-Pick disease,* sphingomyelin accumulates mainly in spleen and liver, whereas in the infantile form, there is mental retardation and early death. The two forms of the disease may be due to lack of one of two different isoenzymes.

A specific *sulfatase,* absent in *metachromatic leukodystrophy,* normally hydrolyzes sulfatidates to inorganic sulfate and galactosylceramide; the latter is hydrolyzed by *galactosylceramide β-galactosidase,* which is lacking in *Krabbe's leukodystrophy.* These leukodystrophies are characterized by accumulation of the sulfatidates in the former disorder and the galactosylceramide in the latter. Abnormally high concentrations of both occur in the white matter of the brain.

Lack of *hexosaminidase A* (*β-N-acetylgalactosaminidase*) leads to *Tay-Sachs disease,* in which there are excessive amounts of the *Tay-Sachs ganglioside,* GalNAc-Gal-Glc-Cer, in the brain and spleen. In a rare form of the Tay-Sachs disorder, *Sandhoff's disease,* both *hexosaminidases A and B* activities are lacking (see Fig. 21.4). Both activities are normally present in the same enzyme, which consists of two subunits. In Tay-Sachs disorder the α subunit is lacking, whereas in Sandhoff's disease the β subunit is absent.

In the adult form of *Gaucher's disease,* resulting from absence of β-glucosidase, glucosylceramide accumulates in reticuloendothelial cells, mainly in liver, spleen, and bone marrow. In the infantile form, the storage is primarily in the brain.

A specific *α-galactosidase* is absent in *Fabry's disease.* Excessive amounts of ceramide trihexoside (Fig. 21.4) are present, primarily in the kidney but also in the intestine and lymph. There is progressive functional impairment of the kidney and other organs.

A specific *β-galactosidase,* is absent in *lactosylceramidosis,* leading to accumulation of lactosylceramide in brain, liver, bone marrow, and other tissues.

In the rare *Farber's disease,* the hydrolase that liberates sphingenine and a fatty acid from ceramide is lacking.

amniotic

Essentially all the above disorders can be detected by assay for the responsible enzyme in samples of cells obtained from the amniotic fluid (*amniocentesis*). Thus, diagnostic tests can be performed on fetuses suspected of possessing the defective gene responsible for the disorder.

In *I-cell disease,* there is a deficiency of a large number of lysosomal enzymes: glycosidases, proteinases, and others. Studies on normal fibroblasts in tissue culture and on those from individuals with this defect have led to some understanding of the formation and processing of these enzymes (page 469).

Metabolism of Sphinganine The formation of sphinganine and its role in the formation of the glycosphingolipids have been described above. Relatively little is known regarding the degradation of these compounds aside from the hydrolytic reactions yielding ceramide (Fig. 21.4). Sphinganine is metabolized in brain, liver, kidney, and other tissues. The initial step is a phosphorylation followed by a lyase reaction.

$$\text{Sphinganine} + \text{ATP} \xrightarrow[\text{kinase}]{\text{Mg}^{2+}} \text{sphinganine-1-phosphate} + \text{ADP}$$

$$\text{Sphinganine-1-phosphate} \xrightarrow[\text{phosphate}]{\text{pyridoxal}} \text{palmitaldehyde} + \text{ethanolamine phosphate}$$

CHOLESTEROL METABOLISM AND ITS CONTROL

Since cholesterol is present in all animal and some plant tissues, all animals ingest this sterol. Cholesterol is a universal and essential constituent of animal cell membranes (Chap. 13) and can be made by essentially all mammalian cells although the liver is responsible for more than 80 percent of its synthesis. Most of the dietary and biosynthetic cholesterol in the mammal is ultimately converted to the bile acids and excreted in the stool.

Biosynthesis of Cholesterol

All carbon atoms of endogenous cholesterol are derived from the acetyl group of acetyl CoA. Some of the enzymes involved are associated with the endoplasmic reticulum, others are water-soluble proteins of the cytoplasm. The initial reaction involves formation of 3-hydroxy-3-methylglutaryl CoA which is derived from 3 mol of acetyl CoA (page 538). Note that this synthesis is distinct from the similar process in mitochondria leading to acetoacetate formation.

3-Hydroxy-3-methylglutaryl CoA is the immediate precursor of *mevalonic acid,* formed in a reaction catalyzed by a sulfhydryl-containing, NADPH-requiring enzyme, *3-hydroxy-3-methylglutaryl CoA reductase.* The reaction involves reduction of the esterified carboxyl group of hydroxymethylglutaryl CoA in the enzyme-bound substrate in two steps (Fig. 21.5); this is the committed step of cholesterol synthesis.

Sterol biosynthesis occurs by condensation of six 5-carbon units, each derived from mevalonate. The sequence of reactions between mevalonate and squalene (the precursor of sterols) is shown in Fig. 21.5. Mevalonate is phosphorylated by ATP in two successive reactions catalyzed by two distinct enzymes, with sequential formation of 5-phospho- and 5-pyrophosphomevalonate. The latter is converted to isopentenyl pyrophosphate with ATP as cofactor by elimination of the carboxyl and 3-hydroxy groups. The ATP is thought to act as a nucleophile, attacking the 3-hydroxy group, with displacement of ADP and formation of P_i. Isopentenyl pyrophosphate is rapidly isomerized to dimethylallyl pyrophosphate by addition of a proton to the terminal double-bond methylene carbon, followed by elimination of a proton from C-2. These two isomers are key substances in polyisoprenoid synthesis and are the

Figure 21.5 Sequence of reactions between 3-hydroxy-3-methylglutaryl CoA and squalene. P = phosphate; PP = pyrophosphate; IPP = isopentenyl pyrophosphate; DPP = dimethylallyl pyrophosphate. Enzymes catalyzing the above reactions have been designated as follows: 1, *3-hydroxy-3-methylglutaryl CoA reductase;* 2, *mevalonate kinase;* 3, *phosphomevalonate kinase;* 4, *pyrophosphomevalonate decarboxylase;* 5, *isopentenyl pyrophosphate isomerase;* 6, *geranyl pyrophosphate synthase;* 7, *farnesyl pyrophosphate synthase;* 8, *presqualene synthase;* 9, *squalene synthase.*

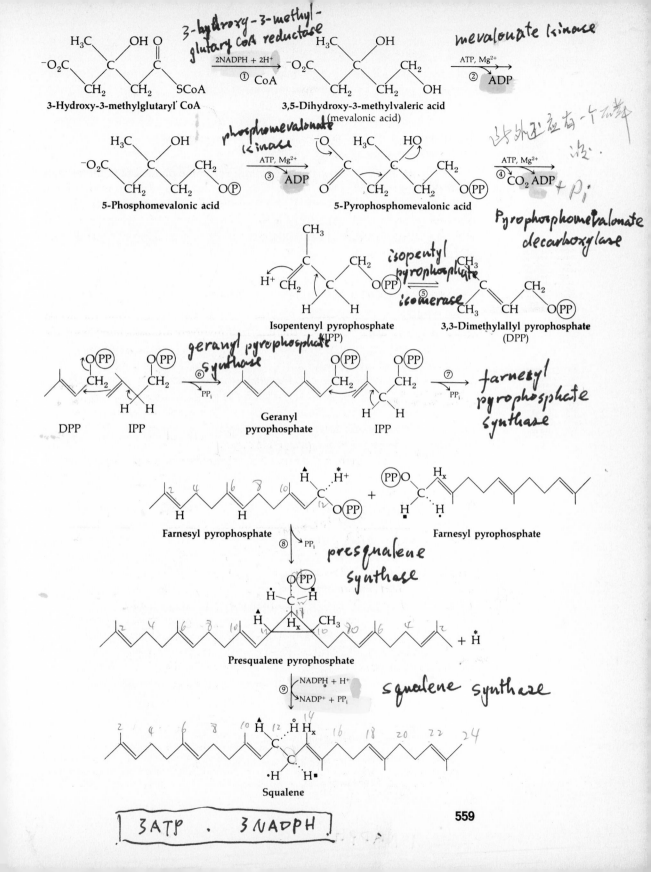

3-hydroxy-3-methyl-glutaryl CoA reductase

mevalonate kinase

phosphomevalonate kinase

3-Hydroxy-3-methylglutaryl CoA

$2NADPH + 2H^+$
① CoA

3,5-Dihydroxy-3-methylvaleric acid
(mevalonic acid)

ATP, Mg^{2+}
② ADP

5-Phosphomevalonic acid

ATP, Mg^{2+}
③ ADP

5-Pyrophosphomevalonic acid

ATP, Mg^{2+}
④ CO_2 ADP $+ P_i$

此外还应有一个磷酸基.

Pyrophosphomevalonate decarboxylase

Isopentenyl pyrophosphate
(IPP)

isopentyl pyrophosphate isomerase
⑤

3,3-Dimethylallyl pyrophosphate
(DPP)

geranyl pyrophosphate synthase
⑥

DPP   IPP

Geranyl pyrophosphate   IPP

⑦ farnesyl pyrophosphate synthase

Farnesyl pyrophosphate

Farnesyl pyrophosphate

⑧ PP_i

presqualene synthase

Presqualene pyrophosphate

⑨ $NADPH + H^+$
$NADP^+ + PP_i$

squalene synthase

Squalene

3 ATP . 3 NADPH

559

basis for subsequent reactions involving carbon-to-carbon bond biosynthetic reactions in sterol formation. The two compounds condense to form geranyl pyrophosphate. Repetition of this reaction between the latter and another mole of isopentenyl pyrophosphate leads to formation of farnesyl pyrophosphate.

In the absence of any cofactor, except Mg^{2+}, two molecules of farnesyl pyrophosphate are condensed by microsomal enzymes to presqualene pyrophosphate, accompanied by loss of one H atom from C-1 of one of the two molecules of farnesyl pyrophosphate. Presqualene pyrophosphate, which contains three asymmetric centers at the three carbon atoms of the cyclopropane ring, is rearranged and reduced with NADPH as coenzyme to the symmetrical C_{30} terpene, squalene. The H atom, lost in the previous reaction, is now replaced by direct transfer from NADPH. Note that the formation of squalene is completely anaerobic.

Squalene is transformed into the tetracyclic steroidal configuration by the pathway depicted in Fig. 21.6. The enzyme *squalene epoxidase* catalyzes conversion of squalene to the 2,3 oxide, the oxygen being derived from "activated molecular oxygen" of unknown identity. The oxide then undergoes an anaerobic cyclization, catalyzed by *squalene oxide cyclase*, to lanosterol. This enzymic

Figure 21.6 The conversion of squalene to lanosterol. 1 = *squalene epoxidase;* the enzyme contains two components, an NADPH-cytochrome c reductase and a nonheme iron-protein. 2 = *squalene oxidocyclase,* which acts in a fully concerted process. There are no known cofactors.

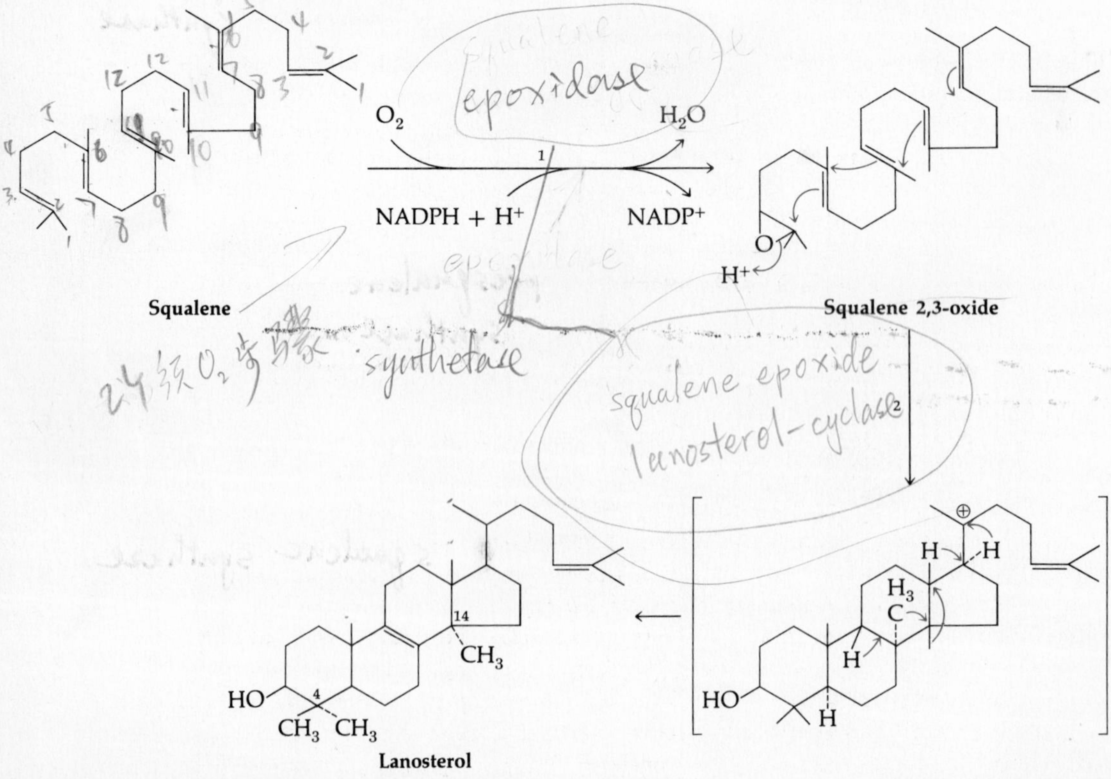

system in liver cyclizes squalene oxide to lanosterol exclusively.   Other cyclizing systems occur in other species, including plants, to account for the formation of other tetracyclic triterpenes and sterols.   The cyclization of squalene 2,3-oxide is postulated to be initiated by attack of a proton on the oxide ring and is followed by concerted electron shifts leading to ring closures (Fig. 21.6) and formation of a transient carbonium ion at C-20.   Lanosterol can be derived from squalene oxide by a series of concerted hydride and methyl shifts and elimination of the proton from C-9.

The conversion of lanosterol to cholesterol (Fig. 21.7) requires the removal of three methyl groups, the saturation of the double bond in the side chain, and the migration of the double bond from position 8,9 in lanosterol to position 5,6. Some of these reactions can occur in several sequences; however, the oxidation of the methyl groups occurs in a fixed order.   The predominant pathway is shown in Fig. 21.7.

Cholesterol of Body Fluids and Tissues

The cholesterol in lymph and plasma is mainly in lipoproteins.   Approximately two-thirds of the cholesterol in plasma is esterified with a fatty acid. Formation of cholesteryl esters is due primarily to a plasma enzyme, *phosphatidylcholine:cholesterol acyl transferase*, which catalyzes the transfer of a fatty acyl group from the 2 position of phosphatidylcholine to the 3-hydroxyl group of cholesterol.   The enzyme is active only in the presence of apo-Al (Table 20.1).

Phosphatidylcholine + cholesterol $\longrightarrow$ lysophosphatidylcholine + cholesterol ester

This enzyme, which contains 24 percent carbohydrate, is responsible for most of the cholesterol esters found in plasma, since individuals lacking the plasma acyl transferase (a rare inborn error) possess almost no plasma cholesterol esters. Feeding labeled cholesterol to these patients results in the appearance of a small amount of labeled cholesterol esters in the plasma, indicating that some esterification occurs during intestinal absorption.   The diminished concentration of cholesterol esters in liver disease may be due to a deficiency of phosphatidylcholine production or to lack of the plasma transferase, since the enzyme is made in the liver.   Plasma cholesterol esters contain mainly linoleate, in accordance with the transfer from phosphatidylcholine of an unsaturated fatty acyl group at carbon 2.   Phosphatidylcholine is regenerated from lysophosphatidylcholine by reacylation with linoleoyl CoA.

linoleic
$C_{18}H_{32}O_2 \Delta 9,12.$

The liver serves both as the chief source and the chief agent for disposal of plasma cholesterol, a portion of that removed from the blood appearing in the bile.   Though sparingly soluble in water, cholesterol readily dissolves in the aqueous bile in the micelles of bile salts and phosphoglycerides.   In the gallbladder, both water and bile salts are reabsorbed by the cholecystic mucosa, and if this process continues excessively, cholesterol crystals separate from the bile. Either biliary stasis or inflammatory disease of the gallbladder can lead to this situation.   Concretions made up chiefly of cholesterol crystals are found in the *calculi* of the biliary tract, the disease being termed *cholelithiasis* (Chap. M10).

Cholesterol enters the intestinal tract by excretion across the intestinal mucosa as well as via the bile.   In the lumen of the gut a portion is reduced

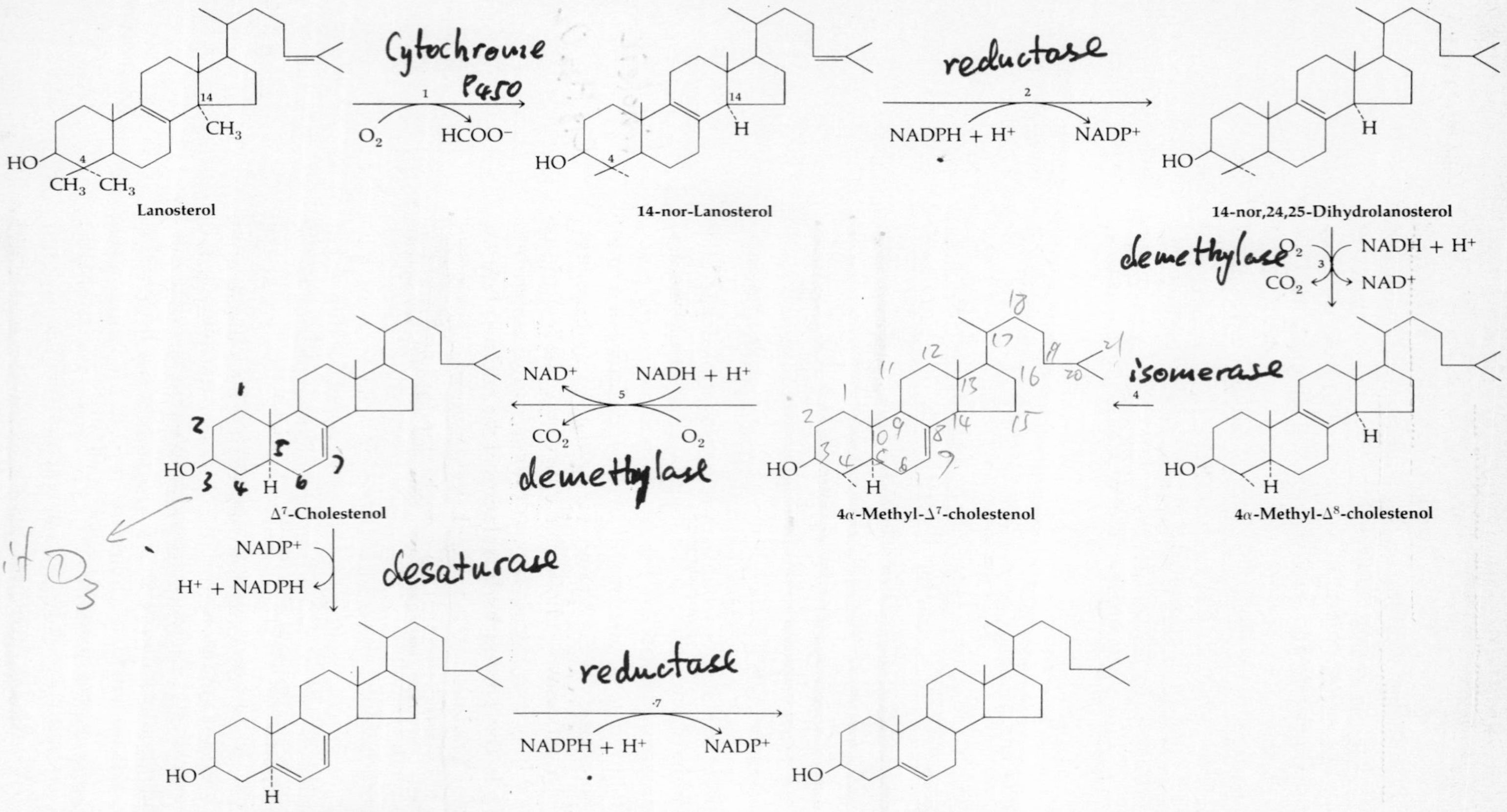

Figure 21.7 The major pathway for conversion of lanosterol to cholesterol. The enzymes involved are as follows: 1, *cytochrome P$_{450}$*; 2, *reductase*; 3, *demethylase*; 4, *isomerase*; 5, *demethylase* (may be the same as 3); 6, a desaturase involving cytochrome b$_5$ (appears to be similar to *stearoyl CoA desaturase* of liver); 7, *reductase*. Note that the 4-α-methyl group is removed first and that between steps 3 and 5 the remaining 4-methyl group epimerizes to α. The demethylase steps, 3 and 5, are complex and the components of the enzyme system are unknown, although the methyl group is apparently first oxidized to carboxyl and then lost as CO_2.

562

microbially to coprostanol and cholestanol via the steps indicated below and thereby is excluded from reabsorption. These two stanols, together with choles- terol, constitute the bulk of the fecal sterols. Certain of these transformations, e.g., from cholestenone to cholestanol, also occur in the liver.

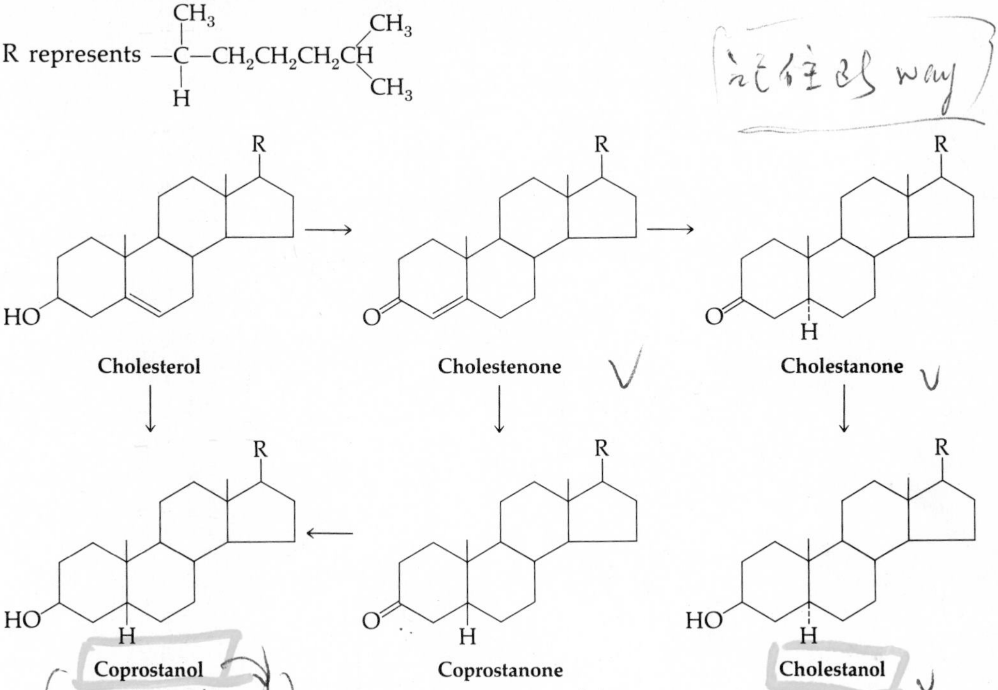

R represents
$$-\underset{\underset{H}{\mid}}{\overset{\overset{CH_3}{\mid}}{C}}-CH_2CH_2CH_2\underset{\overset{\mid}{CH_3}}{\overset{CH_3}{CH}}$$

(את לייזל way)

Cholesterol → Cholestenone → Cholestanone

Coprostanol ← Coprostanone   Cholestanol

(5β-cholestano l)

Formation of Bile Acids

Approximately 80 percent of the cholesterol is transformed by liver tissue into various bile acids. Quantitatively, cholic and chenodeoxycholic acids are the main products derived from cholesterol; the pathways for their biosynthesis are depicted in Fig. 21.8. In this process, hydroxylation of the nucleus of choles- terol is completed before degradation of the side chain is finished. The micro- somal hydroxylating systems are dependent on cytochrome P_{450} and *NADPH- cytochrome P_{450} reductase* (Chap. 16).

Glycocholic acid (choloylglycine) is formed by the following two reactions:

$$\text{Cholic acid} + \text{ATP} + \text{CoA} \longrightarrow \text{Choloyl CoA} + \text{AMP} + \text{PP}_i \qquad (1)$$

$$\text{Choloyl CoA} + \text{glycine} \longrightarrow \text{choloylglycine} + \text{CoA} \qquad (2)$$

Taurocholic acid is formed by the same enzyme from choloyl CoA and taurine $(H_2NCH_2-CH_2SO_3H)$. Analogous derivatives are formed with deoxycholate and chenodeoxycholate. The CoA derivatives [reaction (1)] are made by a microsomal enzyme. The nonspecific *choloyl CoA: N-acyl transferase* [reaction (2)] is a soluble enzyme.

The bile salts, excreted in high concentration (0.5 to 1.5 percent) in the bile, are reabsorbed in the small intestinal mucosa, enter the portal blood, and thus return to the liver. About 500 mg of total bile acid is lost in the feces daily.

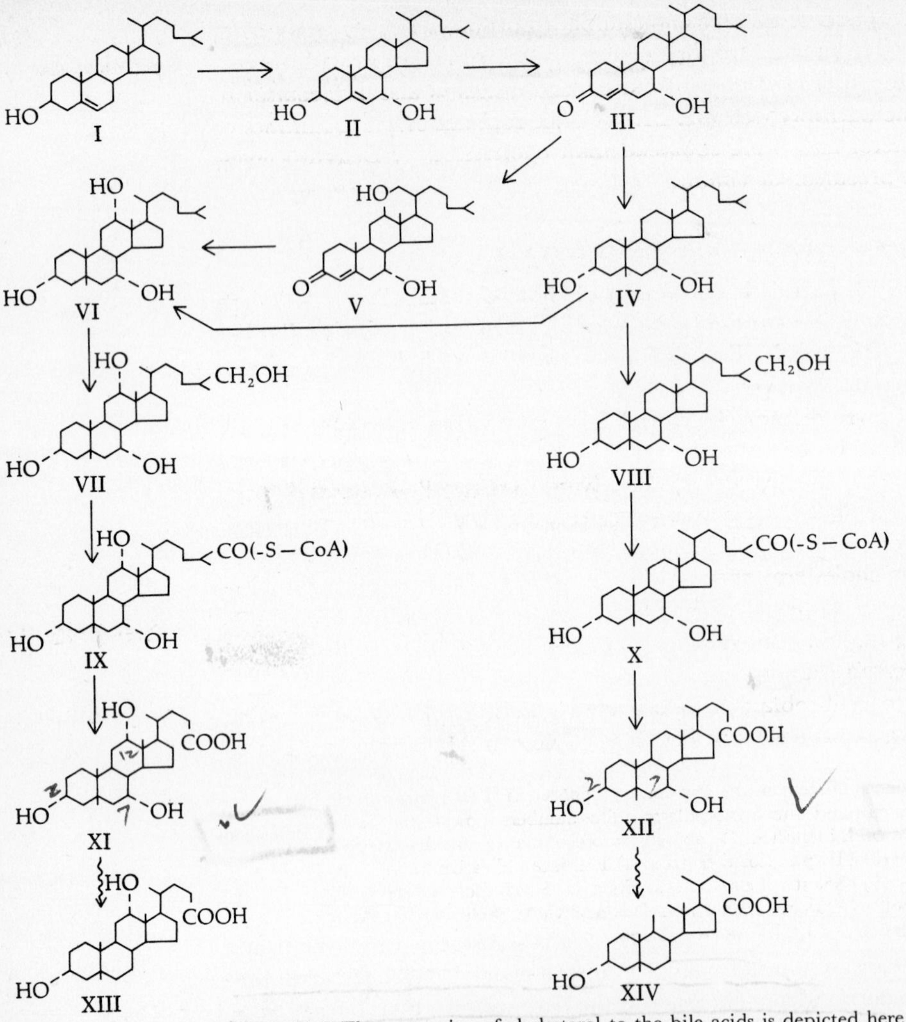

Figure 21.8 The conversion of cholesterol to the bile acids is depicted here in schematic form. I = cholesterol; II = 7α-hydroxycholesterol (5-cholestene-3β,7α-diol); III = 7α-hydroxy-4-cholesten-3-one; IV = 5β-cholestane-3α,7α-diol; V = 7α,12α-dihydroxy-4-cholesten-3-one; VI = 5β-cholestane-3α,7α,12α-tetrol; VII = 5β-cholestane-3α,7α,12α-26-tetrol; VIII = 5β-cholestane-3α,7α,26-triol; IX = 3α,7α,12α-trihydroxy-5β-cholestanoic acid; X = 3α,7α-dihydroxy-5β-cholestanoic acid; XI = cholic acid; XII = chenodeoxycholic acid; XIII = deoxycholic acid; XIV = lithocholic acid. → = reactions catalyzed by liver enzymes; ⤳ = reactions catalyzed by microbial enzymes. [*From H. Danielsson and J. Sjövall, Annu. Rev. Biochem. 44:234 (1975).*]

The larger portion, which is returned to the liver, is involved in the *enterohepatic circulation*. In addition to chenodeoxycholic and cholic acid, as the bile salts, deoxycholic and lithocholic acids, which are produced by microbial enzymes in the gut, are added to the enterohepatic circulation. Further discussion of the bile acids is presented in Chap. M10.

Other Products of Cholesterol Metabolism

Cholesterol is the source of the steroid hormones formed in the gonads and the adrenal cortex. These pathways of biosynthesis are presented in Part M3.

564

During absorption of dietary cholesterol, a portion is dehydrogenated to 7-dehydrocholesterol by the intestinal mucosa. This transformation is also effected by skin and other tissues. 7-Dehydrocholesterol is also a precursor of cholesterol in one pathway (see Fig. 21.7). Skin contains stored quantities of 7-dehydrocholesterol and more squalene than cholesterol. 7-Dehydrocholesterol serves as a precursor of vitamin D (Chap. M23).

Control of Cholesterol Metabolism and Its Disorders

Many factors play a role in controlling cholesterol metabolism. Although the liver and the intestine are the major sites of cholesterol synthesis, all other tissues can make this sterol. Except for nervous tissue, cholesterol in tissues exhibits a continuous turnover.

The liver receives dietary cholesterol from chylomicra remnants and excretes it into bile. Liver also secretes cholesterol and triacylglycerols into plasma; these are transported by VLDL. Note that apo-B is essential (page 511). The triacylglycerols are removed by adipose tissue, leaving cholesterol mainly as its esters with the VLDL now converted to LDL. When cells are actively taking up cholesterol, they synthesize specific LDL receptors which are clustered in specific depressions on the cell surfaces as so-called coated pits (Chap. 14). The highest concentrations of these receptors are in the adrenal cortex and the corpus luteum.

Cultured human fibroblasts have provided a useful model for investigating the sequence of events, as shown in Fig. 21.9. These are (1) binding of LDL to

Figure 21.9 The sequence of steps in the low-density lipoprotein (LDL) pathway of cholesterol transfer in cultured human fibroblasts. The numbers indicate the sites at which mutations have been identified: (1) abetalipoproteinemia; (2) familial hypercholesterolemia (FH); (3) FH, receptor defective; (4) FH, internalization defect; (5) Wolman's disease; (6) cholesterol ester storage disease. HMG CoA reductase = 3-hydroxy-3-methylglutaryl CoA reductase; ACAT = acyl CoA: cholesterol acytransferase. [*From M. S. Brown and J. L. Goldstein, Harvey Lect. 73:163 (1979).*]

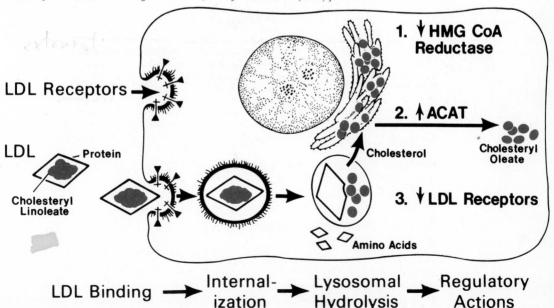

the protein receptors; (2) internalization of receptor-bound LDL through endocytosis of the pits to form vesicles which fuse with the lysosomes; (3) degradation by lysosomes resulting in hydrolysis both of the cholesteryl linoleate by the acid lysosomal lipase and of the apoproteins of the LDL (mainly apo-B) by the proteolytic enzymes of the lysosomes; (4) return of the liberated amino acids to the pool of free amino acids and entry of the free cholesterol into the cell membranes; and (5) repression by the accumulated cholesterol of both the biosynthesis and the activity of the microsomal *3-hydroxy-3-methylglutaryl-CoA* (*HMG CoA*) *reductase*. There is also repression of the synthesis of LDL receptors. At the same time, there is activation of the intracellular *acyl CoA: cholesterol acyltransferase,* which synthesizes cholesteryl oleate for storage of most of the excess cholesterol. Only free cholesterol is incorporated into membranes. Excess free cholesterol leaves the cell, is bound to the plasma HDL, and returned to the liver. It appears that HDL plays a major role in acting as a scavenger to prevent excess accumulation and deposition of cholesterol in blood vessels.

The importance of plasma LDL and its role in cholesterol metabolism is indicated in the genetic disorder *abetalipoproteinemia* (Chap. 20), in which apo-B and plasma LDL are lacking. Fresh skin slices of these patients manifest rates of cholesterol synthesis as much as five times higher than in normal tissues, yet there is a very low plasma level of cholesterol and this is unavailable to body cells.

Familial Hypercholesterolemia This genetically dominant disorder manifests in both heterozygotes and homozygotes a high level of plasma cholesterol and of LDL. Fibroblasts and other cells from such individuals are deficient in receptors for plasma LDL. The allelic genetic defects are of three types: (1) there is no binding of LDL; (2) there is much reduced binding of LDL; and (3) there is normal binding but no internalization of LDL. In consequence, entry of cholesterol esters into the cells is impaired, and the intracellular feedback by cholesterol on the HMG CoA reductase is lacking. This leads to excessive rates of cholesterol synthesis in liver and other tissues and results in hypercholesterolemia. The heterozygote for the condition has about half the normal number of LDL receptors per cell and manifests an LDL level in the plasma about 2.5 times normal. The homozygote lacks receptors for LDL and has about 6 times the amount of LDL in plasma.

Although in these conditions, the major route of cholesterol absorption is defective, there is some removal of plasma cholesterol by a "scavenger" pathway. This is performed by nonspecific pinocytosis by reticuloendothelial cells. It is estimated that about one-third of plasma cholesterol is removed by this secondary pathway.

About 2 individuals per 1000 of the population of the United States are heterozygous for the familial hypercholesterolemia and are prone to atherosclerosis (arterial deposition of cholesterol plaques) between the ages of 35 and 55. The incidence of the homozygous state is about one per million; its bearers usually die before 20 years of age. In addition to coronary artery occlusion, both types may manifest *xanthomatosis,* the presence of multiple *xanthomas* of the skin, tendon sheaths, and bone. Xanthomas are deposits of LDL-derived cholesterol esters usually present in macrophages ("foam cells").

Cholesterol Ester Storage Disease and Wolman's Disease   These disorders involve a deficiency of the lysosomal lipase that normally hydrolyzes the cholesterol esters carried by LDL; as a result, these esters accumulate within lysosomes throughout the body.   There is also accumulation of triacylglycerols, particularly in the very severe but rare Wolman's disease, a condition usually fatal in the early months of life.

Diet and Cholesterol Metabolism   Both fasting and cholesterol feeding markedly reduce the *HMG CoA reductase* activity in liver.   A feedback mechanism controls cholesterol synthesis, since both the synthesis and the activity of this enzyme are inhibited by cholesterol.   Enzymic activity becomes elevated on refeeding after a period of fasting.   An increase in dietary carbohydrate or triacylglycerol augments cholesterol synthesis from acetyl CoA.

The feeding of cholesterol, which limits mevalonate formation, also inhibits conversion of administered mevalonate to cholesterol.   It has been suggested that the second feedback site of dietary cholesterol occurs at a step just prior to cyclization of squalene to lanosterol, inasmuch as cells continue to produce from farnesyl pyrophosphate such compounds as dolichol (Chap. 18) and ubiquinone (Chap. 15).

Alterations in the level of blood cholesterol have been noted in response to changes in the degree of saturation of dietary fatty acids (Chap. M21).   The more saturated the fatty acids of the diet, the higher the serum cholesterol concentration.   The basis for this effect is unknown.

The liver plays a primary role in the degradation of cholesterol.   The rate of cholesterol conversion to the bile acids will influence the level of cholesterol excretion by the liver into the bile and hence the quantity of cholesterol absorbed from the intestine.   Thus this cholesterol, as well as that of dietary origin, can influence the rate of cholesterol synthesis.

Hormonal Control of Cholesterol Synthesis   Conversion of HMG CoA to mevalonate, the major committed step of cholesterol synthesis, is modulated by controls which are in turn regulated by several hormones.   The regulation is of two types—quick response by altering the activity of the reductase and longer-term response by controlling the amount of the reductase.   Both processes are under the influence of a variety of hormones, mainly glucagon and insulin, but also thyroid hormones (Chap. M14) and glucocorticoids (Chap. M18).

The regulation of the reductase is controlled by a cascade system, probably initially by cyclic AMP, which involves more than one step, but ultimately on a microsomal *HMG CoA reductase kinase* that phosphorylates the reductase in the presence of ATP and Mg^{2+}.   The kinase also exists in inactive and active forms and is regulated by *HMG CoA reductase kinase*.   These interrelationships are shown diagrammatically in Fig. 21.10.

In cultured hepatocytes insulin increases dephosphorylation of both the reductase kinase and the reductase by a protein phosphatase, thus increasing cholesterol formation stemming from HMG CoA, whereas glucagon increases phosphorylation, thus activating the kinase and decreasing the activity of the reductase.   These two hormones also inversely affect the mechanisms that control the total amount of the reductase in the liver.   Thus insulin has the overall

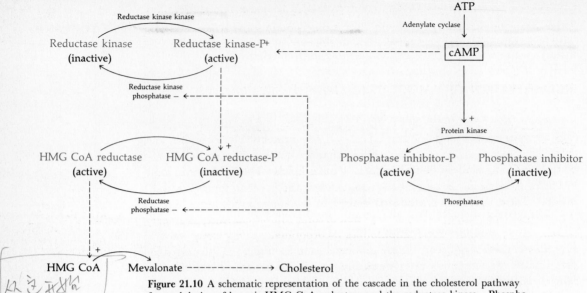

Figure 21.10 A schematic representation of the cascade in the cholesterol pathway for modulation of hepatic HMG CoA reductase and the reductase kinase. Phosphorylated forms of the enzymes are designated P. The enzymically active form of HMG CoA reductase is dephosphorylated, whereas the active forms of the reductase kinase and the phosphatase inhibitor protein are phosphorylated. cAMP is presumed to modulate the cascade by increasing the activity of the cAMP-dependent reductase kinase and to inhibit reductase kinase and HMG CoA reductase phosphatases by increasing the activity of the phosphatase inhibitor protein. [*Adapted from Z. H. Beg, J. A. Stonig, and H. B. Brewer, Jr., Proc. Natl. Acad. Sci. U.S.* **76**:*4375 (1979).*]

effect of increasing both lipogenesis and cholesterol synthesis and glucagon exerts the opposite influence in both cases.

A further feedback control is exerted by mevalonate. When it is fed to animals or added to hepatocytes or fibroblasts, it suppresses HMG CoA reductase activity.

It should be reemphasized that HMG CoA formation leading to synthesis of the ketone bodies is a mitochondrial process, whereas HMG CoA which is used for cholesterol synthesis occurs in the endoplasmic reticulum from acetyl CoA generated by the *ATP citrate lyase* reaction. There is evidence, however, that HMG CoA can be formed from acetoacetate in the cytoplasm, particularly in the neonatal rat, by the following reactions.

$$\text{Acetoacetate} + \text{CoA} + \text{ATP} \xrightarrow{\text{acetoacetyl CoA synthetase}} \text{acetoacetyl CoA} + \text{AMP} + \text{PP}_i \quad (1)$$

$$\text{Acetoacetyl CoA} + \text{CoA} \xrightarrow{\text{thiolase}} 2 \text{ acetyl CoA} \quad (2)$$

$$\text{Acetoacetyl CoA} + \text{acetyl CoA} \xrightarrow{+\text{H}_2\text{O}} \text{HMG CoA} + \text{CoA} \quad (3)$$

The presence in the cytoplasm of the enzymes for catalysis of these three reactions would explain the mild hypercholesterolemia and hypertriacylglycerolemia associated with human pregnancy, since these temporary hyperlipidemic responses coincide with a moderate ketosis. A similar situation occurs in diabetics, who are prone to atherosclerosis. The excessive mobilization of ketone bodies from the mitochondria would permit diversion of some acetoacetate for formation of cholesterol and fatty acids in the cytosol.

INCORPORATION OF AMPHIPATHIC LIPIDS IN MEMBRANES

As already described (Chap. 13), the membranes of cells and their organelles are a complex mosaic of amphipathic lipids, proteins, glycoproteins, and other substances. Many of the enzymes that are present in membranes are bound to specific kinds of phosphatidates or other lipids. The lipids are in some instances essential for the activity of these enzymes and in others for their stabilization. These lipids are synthesized mainly in the endoplasmic reticulum (or the derived microsomes), with participation of some soluble enzymes, as already noted. How, then, are these amphiphiles incorporated appropriately into the various membranes of cells? These problems seem even more perplexing since amphiphiles are frequently asymmetrically distributed between the membrane bilayers (Chap. 13).

Membranes are not made *de novo*. They grow at appropriate times by expansion of existing membranes (Chap. 13). Thus the external plasma membrane expands prior to cell division; mitochondrial membranes grow before these bodies divide, etc. Present evidence indicates two processes that can effect an expansion of existing membranes by incorporation of new amphiphiles, such as phosphatidates, cholesterol, gangliosides, etc. The first process occurs by the pinching-off of small vesicles from the membranes of the smooth endoplasmic reticulum or the Golgi apparatus and migration to a preexisting membrane, followed by binding and fusion to that membrane. The incorporated lipids then flow laterally to achieve appropriate distribution in the membrane. The second process involves protein-mediated transfer of lipids. Proteins have been isolated from liver and brain that can bind simultaneously to small vesicles and to cellular membranes and then mediate specific transfer or exchange of phosphatidates, sphingomyelins, cholesterol, etc. These exchange proteins may be responsible for both specific net transfers and exchanges, as well as for aiding the asymmetric distribution of the lipids in outer and inner membranes.

Studies with isotopically labeled membrane lipids in unilamellar vesicles and with intact erythrocytes show exchange of the erythrocyte components with phosphatidylcholine, other phosphatidates, and sphingomyelin (more slowly) in the presence of an exchange protein isolated from bovine liver. As expected, there is more rapid exchange of lipids with the membrane outer bilayer than with the inner. Assessment of the number, specificity, and mechanism of exchange of such proteins should increase understanding of intracellular transfer of amphiphiles from the sites of synthesis to specific membranes. Moreover, such proteins may play a role in transferring excessive amounts of certain lipids from membranes for delivery to lysosomes for hydrolysis. The existence of the sphingolipodystrophies and similar disorders shows that excessive amounts of essential membrane components must be destroyed, especially so in the cells of the nervous system, which do not divide after birth.

REFERENCES

Books

Ansell, G. B., J. N. Hawthorne, and R. M. C. Dawson (eds.): *Form and Function of Phospholipids,* 2d ed., Elsevier, New York, 1973.

Mead, J. F., and A. J. Fulco: *The Unsaturated and Polyunsaturated Fatty Acids in Health and Disease,* Thomas, Springfield, Ill., 1976.

Paoletti, R., and D. Kritchevsky (eds.): *Advances in Lipid Research,* Academic, New York (a continuing series).

Rommel, K., and R. Bohmer (eds.): *Lipid Absorption: Biochemical and Clinical Aspects,* University Park Press, Baltimore, 1976.

Stanbury, J. B., J. B. Wyngaarden, and D. S. Fredrickson (eds.): *The Metabolic Basis of Inherited Disease,* 4th ed., McGraw-Hill, New York, 1978 (See Part 4, *Disorders Characterized by Evidence of Abnormal Lipid Metabolism,* and Chap. 21, *Disorders of Propionate, Methylmalonate, and Cobalamin Metabolism.*)

Taylor, W. (ed.): *The Hepatobiliary System,* Plenum, New York, 1976.

Review Articles

Beg, Z. H., and H. B. Brewer, Jr.: Regulation of Liver 3-Hydroxy-3-methylglutaryl CoA Reductase, *Curr. Top. Cell. Regul.* **20:**139–184 (1981).

Bell, R. M., L. M. Ballas, and R. A. Coleman: Lipid Topogenesis, *J. Lipid Research* **22:**391–403 (1981).

Bell, R. M., and R. A. Coleman: Enzymes of Glycerolipid Synthesis in Eukaryotes, *Annu. Rev. Biochem.* **49:**459–488 (1980).

Björntorp, P., and J. Östman: Human Adipose Tissue: Dynamics and Regulation, *Adv. Metabol. Disord.* **5:**277–327 (1971).

Bloch, K.: The Biological Synthesis of Cholesterol, *Science* **150:**19–23 (1965).

Bloch, K., and D. Vance: Control Mechanisms in the Synthesis of Saturated Fatty Acids, *Annu. Rev. Biochem.* **46:**263–298 (1977).

Brady, R. O.: Sphingolipidoses, *Annu. Rev. Biochem.* **47:**687–713 (1978).

Brown, M. S., and J. L. Goldstein: Familial Hypercholesterolemia: Model for Genetic Receptor Disease, *Harvey Lect.* **73:**163–201 (1979).

Danielsson, H., and J. Sjövall: Bile Acid Metabolism, *Annu. Rev. Biochem.* **44:**233–253 (1975).

Dawson, G.: Glycolipid Biosynthesis, pp. 255–284, and Glycolipid Catabolism, pp. 285–336, in M. L. Horowitz and W. Pigman (eds.): *The Glycoconjugates,* vol. II, Academic, New York, 1978.

Fishman, P. H., and R. O. Brady: Biosynthesis and Function of Gangliosides, *Science* **194:**906–915 (1976).

Jackson, R. L., J. D. Morrisett, and A. M. Gotto, Jr.: Lipoprotein Structure and Metabolism, *Physiol. Rev.* **56:**259–316 (1976).

Krebs, H. A.: The Regulation of Release of Ketone Bodies by the Liver, *Adv. Enzyme Reg.* **4:**339–354 (1966).

Masoro, E. J.: Lipids and Lipid Metabolism, *Annu. Rev. Physiol.* **39:**301–321 (1977).

McGarry, J. D., and D. W. Foster: Regulation of Hepatic Fatty Acid Oxidation and Ketone Body Production, *Annu. Rev. Biochem.* **49:**395–420 (1980).

Numa, S., and S. Yamashita: Regulation of Lipogenesis in Animal Tissues, *Curr. Top. Cell Regul.* **8:**197–246 (1975).

Van den Bosch, H.: Phosphoglyceride Metabolism, *Annu. Rev. Biochem.* **43:**243–277 (1974).

Volpe, J. J., and P. R. Vagelos: Mechanisms and Regulation of Biosynthesis of Saturated Fatty Acids, *Physiol. Rev.* **56:**339–417 (1976).

Wood, H. G., and R. E. Barden: Biotin Enzymes, *Annu. Rev. Biochem.* **46:**385–413 (1977).

Chapter 22

Amino Acid Metabolism I: Plants and Microorganisms

Fixation of nitrogen and ammonia. Fixation of sulfur. Synthesis of amino acids.

Collectively, plants and microorganisms fix 10^{12} tons of CO_2 per year into the mixture of organic compounds characteristic of the living world. The overall process is

$$CO_2 + SO_4{}^{2-} + N_2(\text{or } NO_3{}^-) + P_i \xrightarrow{\text{light}}$$
$$O_2 + \text{proteins, nucleic acids, polysaccharides, lipids, etc.}$$

The metabolism and life cycles of plants and microorganisms operate largely to ensure growth and reproduction; they do not exhibit the relatively long dynamic steady states characteristic of the mature animal. The primary, energy-utilizing process is the photosynthetic, reductive fixation of CO_2 into carbohydrate (Chap. 19); synthesis of all other organic compounds is initiated with metabolic intermediates derived from that process. The nitrogen and sulfur required for amino acid synthesis also occur in the environment in oxidized states, $SO_4{}^{2-}$, N_2, and $NO_3{}^-$, and must be reduced before they can be incorporated into amino acids. From the amino acids thus formed are derived all other nitrogen-containing compounds. This chapter presents the process by which nitrogen and sulfur are reduced to metabolically available forms and are combined with intermediates of carbohydrate metabolism to synthesize the 20 amino acids characteristic of all living forms.

FIXATION OF NITROGEN

The reduction of N_2 to ammonia, commonly termed *nitrogen fixation*, is accomplished only by microorganisms. Plants cannot reduce N_2 but live in sym-

Any cross-references coded M refer to the companion text, *Principles of Biochemistry: Mammalian Biochemistry.*

biosis with the bacteria of their root nodules, usually of the genus *Rhizobium,* and thereby enrich the nitrogen content of the soil. Indeed, the symbiotic and photosynthetic N_2-fixing systems are largely responsible for the total of approximately 10^8 tons of N_2 fixed annually. Free-living nonphotosynthetic bacteria, e.g., *Azotobacter, Klebsiella,* and *Clostridium,* contribute relatively little to the total, although studies with such organisms have provided much of the information on the biochemistry of N_2 fixation.

It is estimated that 10 to 15 percent of the nitrogen fixed in freshwater is done by blue-green algae (cyanobacteria) using N_2 that is reduced to NH_3 by the same type of nitrogenase system (see below) as is found in most nodule bacteria. All the N_2 reduction occurring in the oceans is performed by blue-green algae; the amount is uncertain since it occurs only in shallow coastal waters, where other nutrients are available.

The basic features of biological N_2 fixation (Fig. 22.1) are: (1) *nitrogenase,* the cardinal enzyme complex in the process; (2) a strong reductant such as reduced ferredoxin (page 364) or flavodoxin (Fig. 5.10); (3) ATP; (4) a system for regulation of the rate of ammonia production and one for assimilation, since biosynthesis of the nitrogenase complex ceases when ammonia accumulates; and (5) protection of the N_2 fixation system from molecular oxygen, which inactivates nitrogenase and competes for reductant (in aerobic bacteria).

Reduction of Nitrogen

The stoichiometry of the overall process of nitrogen fixation is

$$N_2 + 6e^- + 12ATP + 12H_2O \longrightarrow 2NH_4^+ + 12ADP + 12P_i + 4H^+$$

"Nitrogenase" can also catalyze the reduction of compounds containing triple bonds, e.g., N_2O, HN_3, HCN, other nitriles, and some non-nitrogen-containing compounds. Indeed, reduction of acetylene is the most convenient assay for nitrogenase.

The origin and nature of the electron donors vary among the different groups of nitrogen-fixing organisms. Aerobic bacteria, e.g., *Azotobacter,* develop reducing power, as well as ATP, for nitrogen fixation from their own carbohydrate metabolism via NADP-linked substrates. Photosynthetic bacteria and blue-green algae are capable of photochemical production of strong reductants.

The nitrogenase complex comprises two components; neither is active in the absence of the other. The sizes vary with the microbial source: component I (nitrogenase itself; MW $\approx$ 210,000 to 240,000) with four identical peptide chains each with an Mo-Fe coenzyme of unknown structure and an (Fe_4S_4) group; component II (nitrogenase reductase; MW $\approx$ 55,000 to 60,000) with two

Figure 22.1 Overall representation of nitrogen fixation.

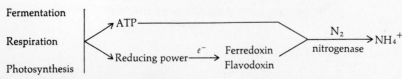

II : I = 2 : 1

identical chains, each with an (Fe_4S_4) group. The complex has two units of component II for each component I.

Reducing power is furnished by NADPH via a flavoprotein, which reduces either ferredoxin (page 364) or flavodoxin (Fig. 5.10). This, in turn, reduces component II, which then reduces the Mo-Fe protein. Four ATP are hydrolyzed for each pair of electrons transferred. The mechanism by which ATP is used is unknown. Mg^{2+} is required, but Mn^{2+} or Fe^{2+} are also functional.

Both components of the nitrogenase are irreversibly poisoned by O_2. The apparent function of the leghemoglobins, or legume hemoglobins, supplied to the nodules of *Rhizobium* in the legumes, is to bind O_2 and maintain pO_2 below 0.01 mmHg. In the blue-green algae the walls of the heterocysts separate the nitrogenase from the O_2 produced by photosynthesis.

*legume
豆科植物*

In the reduction of N_2 no free intermediates are formed, but it appears that the process is stepwise, presumably with the intermediates $HN{=}NH$ and $H_2N{-}NH_2$ (hydrazine). This is supported by the observation that in the reduction of acetylene, $HC{\equiv}CH$, some ethylene, $H_2C{=}CH_2$, is observed.

The complexity of nitrogen fixation is indicated by genetic studies on mutants of *Klebsiella pneumoniae*. Some 14 genes, clustered together, are required for this organism to grow on N_2; absence of the protein product of any one of these genes results in growth failure. Four genes are involved in formation of the Mo-Fe coenzyme of component I; one codes for the protein of component II, but two additional gene products are needed to render component II active; other genes code for proteins involved in electron-transport, and still others are involved in regulation of the process.

The species of *Rhizobium* is specific for a given plant species; e.g., alfalfa, soybean, clover, etc. Attachment of the bacteria to the host plant is by bacterial surface polysaccharides, which bind to the host plant *lectins*. (Chap. 18). Infection threads are produced and result in formation of a nodule. Within the nodule, the plant cells are filled with the bacteria, which become attached to membranes. The legume supplies much of the energy (ATP, NADPH) for N_2 fixation, as well as the leghemoglobin which protects the nitrogenase from the deleterious effects of O_2.

Control of nitrogenase activity is exerted in two ways. One is a coarse control, in which enzyme synthesis is repressed by excess of ammonia; the other is a fine control, in which the activity of the nitrogenase is regulated by ADP. Should the amount of ADP increase to about twice that of ATP, further utilization of ATP by nitrogenase is completely inhibited.

Reduction of Nitrate and Nitrite

nitrate 硝酸盐 nitrite 亚硝酸盐

Rainwater contains nitrate generated by electric discharges in the atmosphere. A variety of microorganisms and most higher plants can reduce nitrate to NH_3 in two enzymic steps. The first enzyme, *nitrate reductase,* catalyzes the two-electron reduction of nitrate to nitrite. In bacterial and plant cells, the electron donor is NADH; in fungi, NADPH. The nitrate reductases of fungi and *Chlorella* are flavoproteins.

Nitrate reductase (MW = 800,000) is a tetramer of dimers consisting of one subunit of molecular weight 150,000 and one of molecular weight 50,000. The enzyme contains one Mo atom in a coenzyme, one FAD, and several Fe_4S_4 groups per basic dimer. The electron path appears to be

$$NADPH \xrightarrow{e} FAD \xrightarrow{e} Fe_4S_4 \xrightarrow{e} Mo \xrightarrow{e} NO_3^- \longrightarrow NO_2^-$$

to catalyze

$$NADPH + H^+ + NO_3^- \longrightarrow NADP^+ + NO_2^- + H_2O$$

Molybdenum undergoes cyclic changes from Mo^{5+} to Mo^{4+} during reduction of NO_3^-.

Nitrite reductase catalyzes the 6-electron reduction of NO_2^- to NH_3 with NADPH as reductant. The *Neurospora* enzyme contains FAD and *siroheme* (see below) as prosthetic groups. An Fe_2S_2 center is present in spinach nitrite reductase. Presumably the electron pathway is similar to that in sulfite reductase (page 579). The net reaction is

$$3NADPH + NO_2^- + 4H^+ \longrightarrow NH_3 + 3NADP^+ + 2H_2O$$

The siroheme iron is the binding site for NO_2^-; no intermediates dissociate until NH_3 appears.

The reductant for nitrite reductase is ferredoxin in plants and algae; in green plants, the ferredoxin can be reduced photosynthetically or by NADPH in the presence of a flavoprotein enzyme, *ferredoxin*-NADPH *oxidoreductase*. Thus, the iron-sulfur protein ferredoxin can serve as the electron donor to the iron-sulfur grouping in nitrite reductase; the electron is then transferred to nitrite via siroheme. Some fungal nitrite reductases contain a bound flavin that facilitates reduction of the siroheme by a reduced pyridine nucleotide.

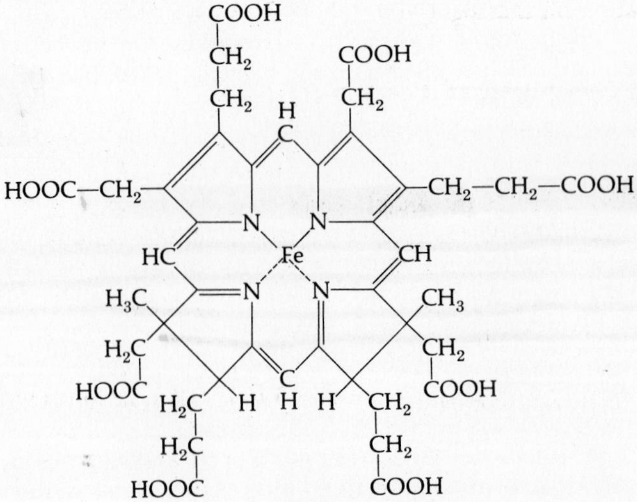

Siroheme

Siroheme is an iron tetrahydroporphyrin of the isobacteriochlorin type, viz., two adjoining pyrrole rings are fully reduced, and each pyrrole bears one propionic and one acetic acid side chain; hence it is a derivative of uroporphyrin (Chap. 24).

FIXATION OF AMMONIA

575

Amino Acid
Metabolism I:
Plants and
Microorganisms

Ammonia, of whatever origin, can be combined into organic linkage by three major reactions that occur in all organisms. These reactions result in formation of glutamate, glutamine, and carbamoyl phosphate. Utilization of the nitrogen of carbamoyl phosphate is limited to two pathways, one contributing a single nitrogen atom in the synthesis of pyrimidines (Chap. 26) and the other donating one nitrogen atom to the synthesis of arginine (page 616). *Essentially all other nitrogen atoms of amino acids or of other organic compounds are derived directly or indirectly from glutamate or the amide group of glutamine.* Although NH_3 can be utilized instead of glutamine in some enzymic reactions, glutamine is preferred in most instances.

In bacteria such as *Escherichia coli* and *Bacillus megaterium* and in plants, there is a *glutamate dehydrogenase* that specifically uses NADPH:

$$\alpha\text{-Ketoglutarate} + NH_4^+ + NADPH \rightleftharpoons \text{glutamate} + NADP^+ + H_2O \quad (1)$$

The glutamate is used by *glutamine synthetase* to fix a second molecule of NH_3.

$$\text{Glutamate} + NH_3 + ATP \longrightarrow \text{glutamine} + ADP + P_i \quad (2)$$

Glutamine can then be utilized by *glutamate synthase* to form more glutamate.

$$\alpha\text{-Ketoglutarate} + \text{glutamine} + NADPH + H^+ \longrightarrow 2\text{ glutamate} + NADP^+ \quad (3)$$

Most of the glutamate synthesized by *E. coli* is undoubtedly made by reaction (3); however, some glutamate must be first synthesized by reaction (1) to provide an initial source of glutamate to make glutamine by reaction (2). Reactions (2) and (3) can then proceed in cyclic fashion, with a net production of glutamate from α-ketoglutarate.

Glutamate Dehydrogenase, Reaction (1)

The enzyme from *E. coli* is a hexamer of six identical subunits (MW = 50,000). Mutants lacking an active dehydrogenase must be supplied with a small amount of glutamate for growth.

There is a large diversity of glutamate dehydrogenases in various organisms, and the controls influencing their activities are of different kinds. The enzymes of mammals and other vertebrates are described in Chap. 23. *Neurospora crassa* can be grown in a medium containing glucose, a trace of biotin, and inorganic salts, including $(NH_4)_2SO_4$. In this medium, the organism possesses a glutamate dehydrogenase that is specific for NADPH, as in reaction (1), and presumably is responsible for synthesis of the glutamate necessary for growth. Glutamate synthase has not been detected. The NADPH-specific dehydrogenase is similar to that of *E. coli*. It contains six identical subunits of 452 amino acid residues each. The known sequence shows some homology to that of the vertebrate glutamate dehydrogenases.

When glutamate is added to a culture of *N. crassa* and the amount of glucose decreased, synthesis of the NADPH enzyme is repressed and the enzyme disappears; it is replaced by a glutamate dehydrogenase that is specific for NAD. This enzyme presumably functions by oxidizing part of the glutamate supplied in the medium, to provide α-ketoglutarate to the tricarboxylic acid cycle and NADH for oxidative phosphorylation to make ATP. The distinctive NAD^+-linked enzyme (MW = 480,000) has only four identical subunits. Thus, in this organism, control of nitrogen incorporation and of energy requirements is regulated by coordinate biosynthesis or repression of two distinctive glutamate dehydrogenases. Some higher plants also have NAD^+- and $NADP^+$-specific dehydrogenases but their roles and controls are unknown.

Glutamate Synthase, Reaction (3)

This enzyme as obtained from *E. coli* (MW = 800,000) has two types of subunits: one contains nonheme iron, FAD, and FMN; the other subunit binds NADPH. Mutants lacking an active synthase require large amounts of glutamate for growth. Hence the reaction catalyzed by this enzyme is responsible for most of the synthesis of glutamate in this organism. Inasmuch as higher plants contain only relatively small amounts of an NADPH-specific glutamate dehydrogenase and a large amount of glutamate synthase, a similar situation prevails in these species.

Glutamine Synthetase, Reaction (2)

Regulation of the activity of the glutamine synthetase of *E. coli* is complex; a schematic representation of this regulation is given in Fig. 22.2. The enzyme (MW = 600,000) consists of 12 identical subunits and is inactivated by the transfer from ATP of a 5'-adenylyl group to form a phosphodiester linkage with the phenolic hydroxyl group of a specific tyrosine residue of each subunit. This reaction, termed *adenylylation*, is catalyzed by *ATP:glutamine synthetase adenylyltransferase* (MW ≈ 130,000). Reactivation is accomplished by a phosphorolytic cleavage of the adenylyltyrosyl bond to form ADP and the active enzyme; this process is catalyzed by the same adenylyltransferase in the presence of a different form of a regulatory protein, designated P_{II} (MW = 50,000).

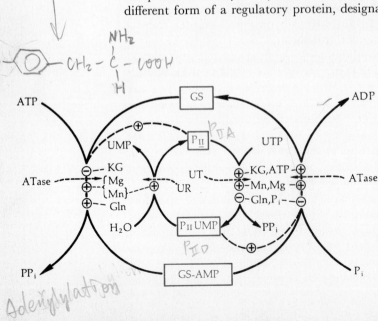

Figure 22.2 Schematic representation of the interrelationship between the covalent modifications of the P_{II} regulatory protein and glutamine synthetase (GS) and the control of these modifications by various metabolites.
ATase = ATP:glutamine synthetase adenylyltransferase; UT = uridylyltransferase; UR = uridyl-removing enzyme; KG = α-ketoglutarate; Gln = glutamine; + = stimulation; − = inhibition. P_{II} is P_{IIA} in text and P_{II} UMP is P_{IID} in text, viz., the forms of P_{II} stimulating adenylylation and deadenylylation, respectively. [*From E. R. Stadman and A. Ginsburg, p. 755, in P. D. Boyer (ed.), The Enzymes, vol. X, 3d ed., Academic, New York, 1974.*]

Adenylyltransferase is therefore a bifunctional catalyst responsible for both adenylylation and deadenylylation of glutamine synthetase. Which reaction is catalyzed is determined by the regulatory protein P_{II}. The latter is interconvertible into two forms, an unmodified form that stimulates adenylylation and therefore has been designated P_{IIA} and a modified form, P_{IID}, that stimulates only the deadenylylation. In the conversion of P_{IIA} to P_{IID}, a uridylyl group from UTP is transferred to the P_{II} protein by a *specific uridylyltransferase* (MW = 160,000) which requires ATP, α-ketoglutarate, and either Mg^{2+} or Mn^{2+}.

Thus, the two functions of the same enzyme, viz., *ATP:glutamine synthetase adenylyltransferase (ATase)*, involving a common substrate, *glutamine synthetase · AMP*, are regulated by UTP, α-ketoglutarate, and P_{II}, thereby preventing a futile cycle which would result in a significant loss in free energy:

$$\text{ATP} + \text{glutamine synthetase} \xrightarrow{\text{ATase, } P_{IIA}, \text{ Me}^{2+}} \text{glutamine synthetase} \cdot \text{AMP} + \text{PP}_i \longrightarrow 2\text{P}_i$$

$$\text{Glutamine synthetase} \cdot \text{AMP} + \text{P}_i \xrightarrow{\text{ATase, } P_{IID}, \text{ Me}^{2+}} \text{glutamine synthetase} + \text{ADP}$$

SUM: $\text{ATP} \longrightarrow \text{ADP} + \text{P}_i$

Synthesis of glutamine synthetase and its activity are negatively affected by the presence of ultimate products whose biosynthesis utilizes the amide group of glutamine. These numerous compounds include tryptophan, histidine, CTP, AMP, GMP, glucosamine-6-phosphate, NAD, asparagine, glutamate, and carbamoyl phosphate. In addition, the enzyme is inhibited by glycine, alanine, and serine.

This complex control mechanism operative in *E. coli* glutamine synthetase is not universal among bacteria. Thus, the purified enzyme from *Bacillus subtilis* is not controlled by adenylylation, nor is it a substrate of *E. coli* glutamine synthetase adenylyltransferase. Glutamine synthetases of mammalian tissues are discussed in Chap. 23.

Glutamine occupies a central position in nitrogen metabolism. It is not only a building block in the synthesis of proteins but also serves as a source of nitrogen in the biosynthesis of the compounds noted above and participates in the glutamate synthase reaction discussed earlier. Coupling of the glutamate synthase reaction with glutamine synthesis and various transamination reactions (page 582) provides a major pathway for the ATP-dependent, essentially irreversible synthesis of virtually all amino acids.

$$\text{Glutamine} + \alpha\text{-ketoglutarate} + \text{NADPH} + \text{H}^+ \xrightarrow{\text{glutamate synthase}} 2 \text{ glutamate} + \text{NADP}^+ \quad (1)$$

$$\text{Glutamate} + \text{ATP} + \text{NH}_3 \xrightarrow{\text{glutamine synthetase}} \text{glutamine} + \text{ADP} + \text{P}_i \quad (2)$$

$$\text{Glutamate} + \text{RCOCOOH} \xrightarrow{\text{aminotransferase}} \text{RCHNH}_2\text{COOH} + \alpha\text{-ketoglutarate} \quad (3)$$

SUM: $\text{ATP} + \text{NH}_4^+ + \text{RCOCOOH} + \text{NADPH} \longrightarrow$
$$\text{RCHNH}_2\text{COOH} + \text{ADP} + \text{P}_i + \text{NADP}^+$$

Other roles of glutamine are discussed elsewhere.

Carbamoyl phosphate formation, catalyzed by *carbamoyl phosphate synthetase*, is the initial step for the synthesis of both arginine and pyrimidines in distinct pathways. Yeast and *Neurospora* contain two different enzymes which catalyze the following reactions:

$$NH_3 + CO_2 + 2ATP \longrightarrow H_2N{-}\overset{\overset{\displaystyle O}{\|}}{C}OPO_3^{2-} + 2ADP + P_i \qquad (1)$$

$$Glutamine + CO_2 + 2ATP \longrightarrow glutamate + H_2N{-}\overset{\overset{\displaystyle O}{\|}}{C}OPO_3^{2-} + 2ADP + P_i \quad (2)$$

Carbamoyl phosphate

The first reaction is thought to proceed via formation of enzyme-bound carboxyl phosphate, $HO{-}CO{-}O{-}PO_3^{2-}$ which reacts with NH_3 to form enzyme-bound *carbamoyl*, $-CO{-}NH_2$; the latter, in turn, reacts with ATP to form carbamoyl phosphate, which dissociates from the enzyme.

In contrast, the enzymes from both *E. coli* and *S. typhimurium*, which are composed of two nonidentical subunits, will catalyze both reactions. The heavy subunit (MW = 110,000) of the *S. typhimurium* enzyme (MW = 150,000) catalyzes the synthesis of carbamoyl phosphate from NH_3 but not glutamine. Addition of the light subunit (MW = 45,000) restores full ability to utilize glutamine as well. The enzyme is activated by ornithine and inhibited by UMP; these effects appear adequate to control the supply of carbamoyl phosphate according to the need for both arginine and pyrimidines. These effectors act on the affinity of the enzyme for ATP.

Carbamoyl phosphate synthetase activity with NH_3 as substrate is about one-half that with one-tenth the concentration of glutamine. This suggests the basis for the unique role of glutamine in reactions in which synthesis proceeds by transamidation rather than amide synthesis from NH_3. At cellular pH, around 99 percent of "ammonia" exists as NH_4^+. Use of glutamine therefore permits use of uncharged $-NH_2$, avoiding the toxic high concentrations of NH_4^+ that might otherwise be required. Higher plants and animals have evolved to permit utilization of glutamine for many such processes.

Carbamoyl phosphate synthetases of animals are discussed in Chap. 23.

FIXATION OF SULFUR

Formation of cysteine by plants and microorganisms is dependent upon a source of H_2S. The environment, however, generally provides SO_4^{2-}, occasionally $S_2O_3^{2-}$ (thiosulfate) and, in some instances, elemental sulfur. Some microorganisms can reduce SO_4^{2-}, $S_2O_3^{2-}$, and elemental S, utilizing these in a manner similar to NO_3^-, as terminal acceptors of an electron-transport chain, in lieu of O_2. In such instances, electrons are delivered from reduced pyridine nucleotides via flavoproteins and cytochromes, permitting ATP formation. Higher plants, however, use SO_4^{2-} for amino acid synthesis.

The initial step in SO_4^{2-} utilization is the formation of a compound that also serves as a general agent for esterification of sulfate with alcoholic and phenolic compounds and polysaccharides in most living forms, including mammals, viz., *3'-phosphoadenosine-5'-phosphosulfate*.

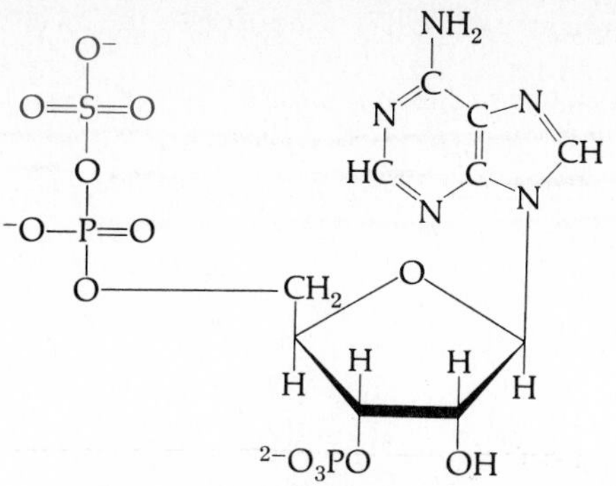

3′-Phosphoadenosine-5′-phosphosulfate

adenosine
腺苷

This compound is formed in a two-step process. The first is catalyzed by *ATP-sulfurylase*. The enzyme is specific for ATP or dATP; Mg^{2+} is optimal for bacteria and Mn^{2+} for mammals.

$$SO_4^{2-} + ATP \xrightarrow{Mg^{2+}} \text{adenosine-5′-phosphosulfate (adenylyl sulfate)} + PP_i$$

The mechanism of the above reaction appears to be of the sequential type in which there is a nucleophilic displacement by sulfate of the β-phosphorus atom of the enzyme-bound ATP-Mg^{2+} complex with formation of adenylyl sulfate and PP_i. Although the equilibrium lies to the left, it is assumed that the hydrolysis of pyrophosphate drives formation of adenylyl sulfate.

In the second step adenylyl sulfate is phosphorylated at the 3′ position by *adenylyl kinase*.

$$\text{Adenylyl sulfate} + ATP \xrightarrow{Mg^{2+}} \text{3′-phosphoadenosine-5′-phosphosulfate} + ADP + H^+$$

Sulfate Reduction

Sulfite is formed from 3′-phosphoadenosine-5′-phosphosulfate (PAPS) in bacteria by the thioredoxin enzyme system (Chap. 26).

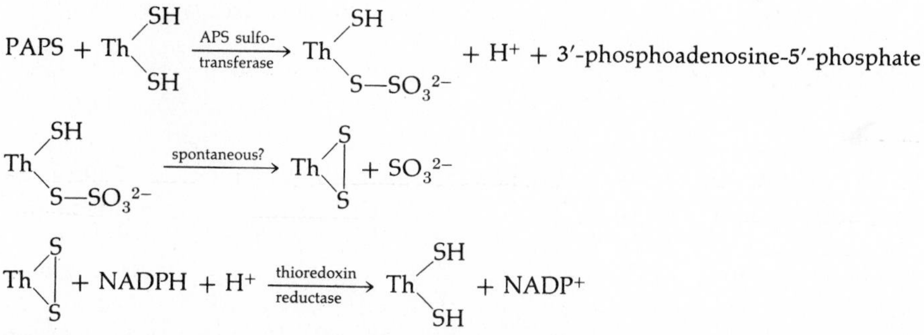

In the above

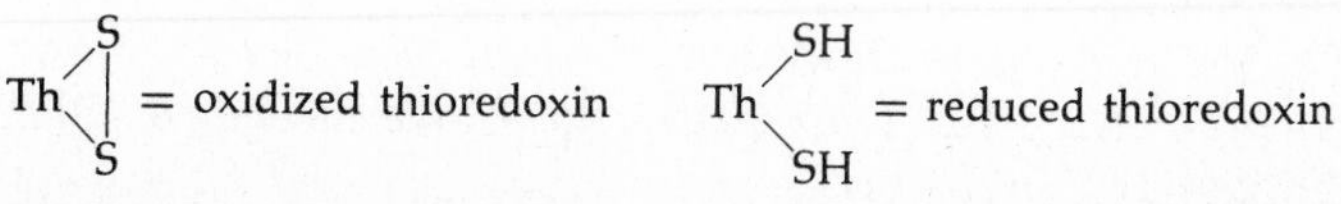

Adenylyl sulfate, rather than PAPS, is reduced to the level of SO_3^{2-} by a similar mechanism in plants and algae.

Reduction of sulfite to sulfide is catalyzed by sulfite reductase, an enzyme whose synthesis is repressed by cysteine and cystine. This 6-electron reduction by *sulfite reductase* is among the most complex known. It catalyzes the overall reaction

$$3NADPH + 3H^+ + SO_3^{2-} \longrightarrow 3NADP^+ + S^{2-} + 3H_2O$$

As isolated from *E. coli* the enzyme (MW = 670,000) has a subunit structure $\alpha_8\beta_4$. Each α dimer bears one FAD and one FMN, while each β chain contains one Fe_4S_4 group and one *siroheme* (page 574).

Electron flow occurs along the linear sequence

$$NADPH \xrightarrow{e} FAD \xrightarrow{e} FMN \longrightarrow Fe_4S_4 \xrightarrow{e} heme \xrightarrow{e} SO_3^{2-}$$

The NADPH donates an electron pair to FAD, from which it is rapidly transferred to the FMN. If only one NADPH is added to the enzyme per mole of heme, the stable form of the chain is $FAD \rightarrow FMNH \cdot \rightarrow Fe_4S_4 \rightarrow heme \rightarrow Fe^{2+}$. In the presence of excess NADPH, all components are fully reduced. The heme is the binding site for SO_3^{2-}. No intermediary reduced form of the sulfur appears until 3 equiv of NADPH have been oxidized and S^{2-} is released. The process is slow with a turnover rate of about six per second.

AMINO ACID SYNTHESIS

General Considerations

From the intermediates of carbohydrate metabolism and the three forms in which ammonia is fixed (see above), plant cells can synthesize all 18 of the other amino acids. This is accomplished by processes that vary from a single reaction to a lengthy reaction sequence. For a few amino acid syntheses, simple equilibrium conditions ensure adequate synthetic rates. Thus, the transamination to form aspartate has an equilibrium constant close to 1.

$$\text{Glutamate + oxaloacetate} \rightleftharpoons \alpha\text{-ketoglutarate + aspartate}$$

This reaction proceeds to the right as long as there is a supply of oxaloacetate and the α-ketoglutarate is reconverted to glutamate. Several other amino acid syntheses have this character; others, however are more complex.

A continuing supply of all the amino acids is ensured by the fact that most pathways are essentially irreversible; i.e., they proceed with a substantial loss of free energy. In general, this is accomplished by reactions in which ATP is utilized and in effect hydrolyzed to ADP + P_i. Even more effective are those instances in which, overall, $ATP \rightarrow AMP + PP_i$, since the pyrophosphate is irreversibly hydrolyzed ($PP_i \rightarrow 2P_i$). In other instances, synthesis is ensured by a reductive reaction, usually employing a pyridine nucleotide, in which equilibrium strongly favors such reduction. Information comes largely from study of the enzymes of a few bacterial species, particularly *E. coli*, because of their relatively easy availability. As noted in Chap. 16, these cells generally utilize

NADH as a reductant. It was in later evolution that specificity for NADPH in synthetic pathways arose. Hence, when the reductant is NADH for a bacterial enzyme, it is likely that the corresponding plant enzyme utilizes NADPH for the same function.

Many microorganisms are entirely self-sufficient in that they can synthesize all the amino acids. In contrast, the *Lactobacilli,* which flourish in milk, the proteins of which provide all the amino acids, are practically incapable of synthesizing amino acids *de novo* and remain entirely dependent upon their environment to provide amino acids.

Presumably, all plant cells can synthesize their full complement of amino acids. The pathways to be described in the following pages were established chiefly for bacteria, yeast, and molds. Pathways in higher plants are generally similar, but relatively little is known of these enzymes.

Control Mechanisms

Individual reaction steps of amino acid syntheses are frequently regulated by one of the control mechanisms noted in Chap. 10. In many instances (Table 22.1), the amino acid produced is a negative allosteric effector of the enzyme

TABLE 22.1

Some Examples of End-Product Inhibition in the Biosynthesis of Amino Acids

Amino acid	Reaction inhibited	Organism or tissue
Arginine	Glutamic acid → N-acetylglutamic acid	E. coli
	N-Acetylglutamic acid → N-acetylglutamic acid 5-phosphate	Micrococcus glutamicus
	N-Acetylglutamic-γ-semialdehyde → N-acetylornithine	E. coli
	Ornithine → citrulline	E. coli
Cysteine	Homocysteine → cystathionine	Rat liver
Histidine	ATP + 5-phosphoribosylpyrophosphate → phosphoribosyl ATP	S. typhimurium
Isoleucine	Threonine → α-ketobutyric acid	E. coli
Leucine	α-Ketoisovaleric acid + acetylCoA → α-isopropylmalic acid	Neurospora
Lysine	Aspartic acid → β-aspartyl phosphate	E. coli
	Aspartic semialdehyde → 2,3-dihydropicolinic acid	E. coli
Methionine	O-Acetylhomoserine + cysteine → cystathionine + acetic acid	Neurospora, yeast
	Homoserine → O-succinylhomoserine	S. typhimurium
	Aspartic acid → β-aspartyl phosphate	E. coli
Proline	Glutamic acid → glutamic acid semialdehyde	E. coli
Serine	Phosphoserine → serine *phosphatase*	Rat liver
Threonine	Aspartic acid → β-aspartyl phosphate	E. coli
	Aspartic-β-semialdehyde → homoserine	Rhodopseudomonas spheroides
	Homoserine → 4-phosphohomoserine	E. coli
Tryptophan	5-Phosphoshikimic acid → anthranilic acid	E. coli
Valine	Pyruvic acid → α-acetolactic acid	Aerobacter aerogenes

responsible for catalysis of the "committed step" in the pathway leading to its own formation. In one instance, there is a committed step that precedes a branch point leading to two different amino acids, viz., lysine and threonine. This is regulated in differing ways in various organisms. Thus, formation of β-aspartyl phosphate (page 586) is catalyzed by three different enzymes in *E. coli*: one sensitive to lysine, a second to threonine, and the third to methionine. In other organisms, a single enzyme is present that is weakly inhibited by each amino acid alone but strongly inhibited when both are present (concerted feedback inhibition).

notice

Concerted Feed back inhibition

Significantly slower regulation is achieved by repression, by an amino acid, of the synthesis of the enzyme that catalyzes the committed step in its own synthetic pathway. Indeed, the presence of the product may result in repression of formation of the enzymes catalyzing all the reactions at the committed step and beyond (Chap. 28).

Examples of these types of control mechanisms will be found in the material that follows. Their operation can be appreciated from the dramatic differences in the behavior of bacteria growing in media of varying composition. An organism that grows in a medium containing inorganic salts and carbohydrate does so by synthesizing all its amino acids. If it is placed in a medium containing all 20 amino acids, they are accepted from the medium and used for protein synthesis and the cell makes little or no amino acid for itself.

Transamination

The major mechanism for moving the amino group of glutamate to other carbon chains is called *transamination;* the reaction shown above (page 580) for aspartate synthesis is a prototype. The enzymes are *aminotransferases* and are specified by the two amino acids involved, e.g., *glutamate-aspartate aminotransferase*. In almost all instances, glutamate is one of the reacting partners. The coenzyme is always pyridoxal phosphate, which alternates with pyridoxamine phosphate in each reaction cycle. Only two amino acids, threonine and lysine, do not derive their amino groups by transamination reactions.

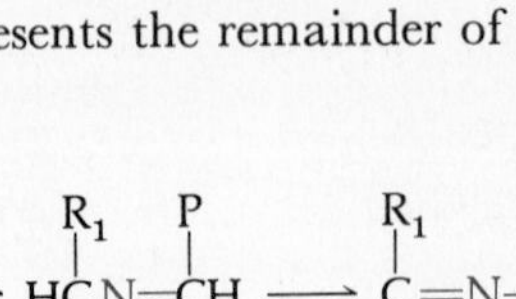

Vit B6

Pyridoxal phosphate 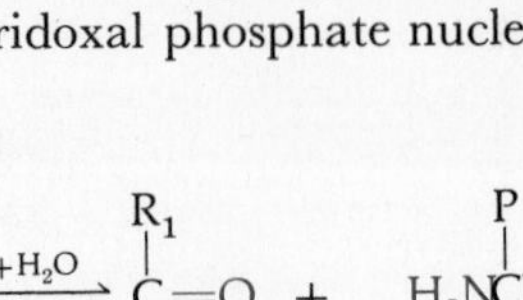   **Pyridoxamine phosphate**

In a general way, the following events occur in a single cycle of enzymic activity. P represents the remainder of the pyridoxal phosphate nucleus bound to enzyme.

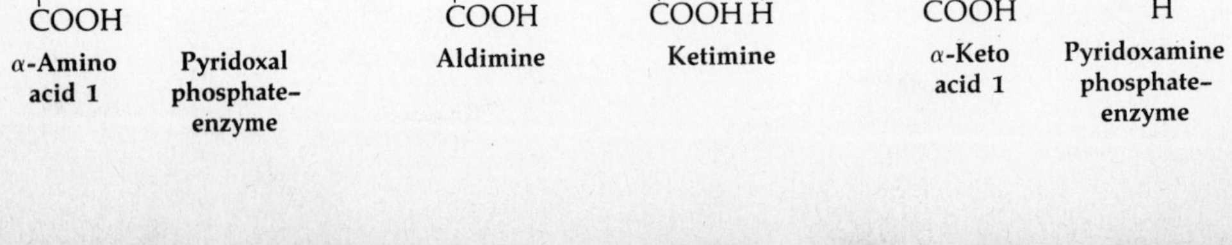

$$
\begin{array}{ccccccc}
\overset{R_1}{\underset{\underset{COOH}{|}}{HC}NH_2} + O=\overset{P}{\underset{}{CH}} & \underset{+H_2O}{\overset{-H_2O}{\rightleftharpoons}} & \overset{R_1}{\underset{\underset{COOH}{|}}{HC}}N=\overset{P}{CH} & \rightleftharpoons & \overset{R_1}{\underset{\underset{COOH\,H}{|}}{C}}=N-\overset{P}{CH} & \underset{-H_2O}{\overset{+H_2O}{\rightleftharpoons}} & \overset{R_1}{\underset{\underset{COOH}{|}}{C}}=O + H_2N\overset{P}{\underset{H}{CH}}
\end{array} \quad (1)
$$

α-Amino acid 1   Pyridoxal phosphate–enzyme   Aldimine   Ketimine   α-Keto acid 1   Pyridoxamine phosphate–enzyme

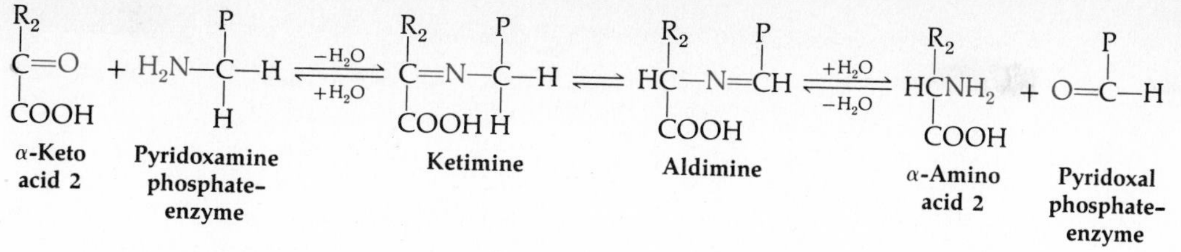

$$\underset{\substack{\alpha\text{-Keto}\\\text{acid 2}}}{\overset{R_2}{\underset{\text{COOH}}{\text{C}=\text{O}}}} + \underset{\substack{\text{Pyridoxamine}\\\text{phosphate-}\\\text{enzyme}}}{\overset{P}{\underset{\text{H}}{\text{H}_2\text{N}-\text{C}-\text{H}}}} \underset{+H_2O}{\overset{-H_2O}{\rightleftharpoons}} \underset{\text{Ketimine}}{\overset{R_2\quad P}{\underset{\text{COOH H}}{\text{C}=\text{N}-\text{C}-\text{H}}}} \rightleftharpoons \underset{\text{Aldimine}}{\overset{R_2\quad P}{\underset{\text{COOH}}{\text{HC}-\text{N}=\text{CH}}}} \underset{-H_2O}{\overset{+H_2O}{\rightleftharpoons}} \underset{\substack{\alpha\text{-Amino}\\\text{acid 2}}}{\overset{R_2}{\underset{\text{COOH}}{\text{HCNH}_2}}} + \underset{\substack{\text{Pyridoxal}\\\text{phosphate-}\\\text{enzyme}}}{\overset{P}{\text{O}=\text{C}-\text{H}}} \quad (2)$$

Multispecific *aminotransferases* may be present in microorganisms. Two aminotransferases have been purified from *E. coli*; one is primarily an *aspartate aminotransferase,* although it also utilizes aromatic amino acids. The other is an *aromatic aminotransferase.* The two enzymes differ in the V_{max} and K_m values with their common substrates and pyridoxal phosphate, in heat stability, and in pH optima with the amino acid substrates. The enzymes are similar in composition; each consists of two subunits.

Knowledge of aminotransferases from plant tissue is relatively limited. As in microorganisms, the two most active transaminations occur between α-ketoglutarate and aspartate or alanine.

Alanine   Alanine is made by transamination with pyruvate, which, like α-ketoglutarate, arises in the main pathway of carbohydrate metabolism.

$$\text{Glutamate} + \text{pyruvate} \rightleftharpoons \alpha\text{-ketoglutarate} + \text{alanine}$$

Serine   Serine formation (Fig. 22.3) begins with oxidation of 3-phosphoglycerate by NAD^+ to yield 3-phosphohydroxypyruvate. The latter is transaminated with glutamate, forming 3-phosphoserine, which is then hydrolyzed by serine phosphatase. In bacterial systems, serine acts as a negative effector on both *3-phosphoglycerate dehydrogenase* and *the phosphatase.*

$$\text{HO-CH}_2\text{-}\overset{NH_2}{\underset{}{\text{CH}}}\text{-COOH}$$

Glycine   Glycine is made by removal of the β-carbon of serine:

$$\underset{\text{Serine}}{\text{HOCH}_2\text{CHNH}_2\text{COOH}} \rightleftharpoons \underset{\text{Glycine}}{\text{NH}_2\text{CH}_2\text{COOH}} + \text{"C}_1\text{"}$$

$$N^5 N^{10}\text{-methylene THF}$$

This process yields the glycine needed for protein synthesis and other metabolic processes; it is also the source of active C_1 compounds at the oxidation levels of CH_3OH, $HCHO$, or $HCOOH$. Their transformations make possible the addition to diverse compounds of the groups $-CH_3$, $-CH_2OH$, and $-CHO$, respectively. The metabolism of C_1 compounds is considered in Chap. 23.

Figure 22.3 Pathway of serine biosynthesis.

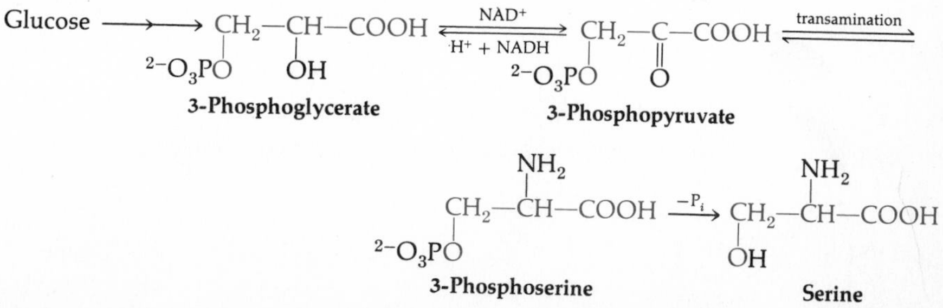

$$\text{Glucose} \longrightarrow \underset{\text{3-Phosphoglycerate}}{\overset{2-}{\text{O}_3\text{PO}}\ \underset{\text{OH}}{\text{CH}_2\text{-CH-COOH}}} \underset{H^+ + NADH}{\overset{NAD^+}{\rightleftharpoons}} \underset{\text{3-Phosphopyruvate}}{\overset{2-}{\text{O}_3\text{PO}}\ \underset{\text{O}}{\text{CH}_2\text{-C-COOH}}} \overset{\text{transamination}}{\rightleftharpoons}$$

$$\underset{\text{3-Phosphoserine}}{\overset{2-}{\text{O}_3\text{PO}}\ \underset{}{\text{CH}_2\text{-}\overset{NH_2}{\text{CH}}\text{-COOH}}} \overset{-P_i}{\longrightarrow} \underset{\text{Serine}}{\underset{\text{OH}}{\text{CH}_2\text{-}\overset{NH_2}{\text{CH}}\text{-COOH}}}$$

Asparagine Asparagine synthesis in bacteria is catalyzed by *asparagine syn-thetase.*

$$\underset{\text{Aspartic acid}}{\text{HOOCCH}_2\overset{\overset{\displaystyle NH_2}{|}}{\text{CH}}\text{COOH}} + \text{ATP} + \text{NH}_3 \xrightarrow{\text{Mg}^{2+}} \underset{\text{Asparagine}}{\text{H}_2\text{N}-\overset{\overset{\displaystyle O}{\|}}{\text{C}}\text{CH}_2\overset{\overset{\displaystyle NH_2}{|}}{\text{CH}}\text{COOH}} + \text{AMP} + \text{PP}_i$$

A similar pathway presumably operates in most plants which utilize aspartate and ammonia. In some plants, however, there is an additional pathway in which β-cyanoalanine is formed from cyanide and cysteine

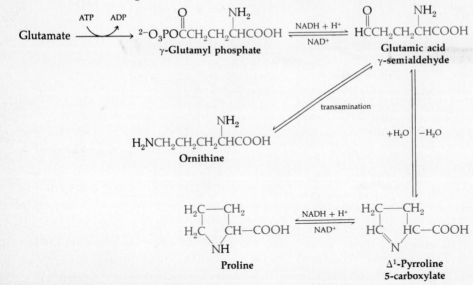

Many of these plants, e.g., the common vetch and *Lathyrus,* also form cyanide-containing glycosides. The relative importance of this pathway for synthesis of asparagine is unclear. In may prove to be a means for detoxifying excess cyanide in these plants. Certain plants synthesize large amounts of asparagine, e.g., lupine seeds may accumulate more than 20 percent of the dry weight of the seedlings as asparagine under certain conditions of growth.

Proline and Ornithine In most microorganisms the 5-carbon chain of glutamic acid serves for formation of proline and ornithine (Fig. 22.4). The sequence begins by reduction of the γ-carboxyl group of glutamic acid. To achieve this, a sequence analogous to the reversible interconversion of glyceraldehyde-3-phosphate and 3-phosphoglyceric acid in the glycolytic sequence is utilized, resulting

Figure 22.4 Interrelationships in the synthesis of ornithine and proline.

in glutamic-γ-semialdehyde. When ring closure is effected, Δ^1-pyrroline 5-carboxylic acid is formed, which, on reduction by NADH, results in formation of proline. Thus, overall direction is given by the net hydrolysis of ATP and two reductive steps.

In ornithine formation in *E. coli*, glutamic acid is *N*-acetylated to form α-*N*-acetylglutamic acid before the reaction with ATP (Fig. 22.5). This ensures that there can be no spontaneous ring closure when the α-*N*-acetylglutamyl phosphate is reduced to the corresponding γ-aldehyde. After transamination, the acetyl group is removed by hydrolysis (Fig. 22.5*a*). Thus, to the energy of reduction and ATP hydrolysis there is added the net hydrolysis of acetyl CoA to drive the process. Higher plants, however, are more economical. Although possessing a *glutamate acetylase*, they also have an *ornithine-glutamate transacetylase*, which catalyzes transfer of the acetyl group from α-*N*-acetylornithine directly to a new molecule of glutamate [Fig. 22.5*b*, reaction (4)], completing the synthesis of one molecule of ornithine and initiating synthesis of the next one. This pathway permits net reduction of energy cost to the cell since, once initiated, there is no further requirement for acetyl CoA.

Ornithine is not incorporated into proteins; it is merely an intermediate in the synthesis of arginine. Hence in bacteria it is arginine that represses formation of the enzymes of the pathway beginning with acetylation of glutamate, and it is arginine also that operates as a direct feedback negative effector on those enzymes. However, it is proline rather than arginine that similarly inhibits reduction of glutamate itself to the corresponding semialdehyde.

A direct conversion of ornithine to proline and NH_3 demonstrated in *Clostridium* is catalyzed by *ornithine cyclase*. Enzyme-bound ornithine is converted to 2-oxo-5-aminopentanoic acid by oxidation with enzyme-bound NAD^+ and release of NH_3. The bound oxoaminopentanoic acid undergoes ring closure to form Δ^1-pyrroline 2-carboxylic acid, and the pyrroline ring is then reduced to proline using the bound NADH as the reductant. The proline is then released

Figure 22.5 Pathways for synthesis of ornithine in (*a*) microorganisms and (*b*) higher plants. 1 = acetylase, 2 = an aminotransferase, 3 = acetylornithinase, and 4 = ornithine-glutamate transacetylase.

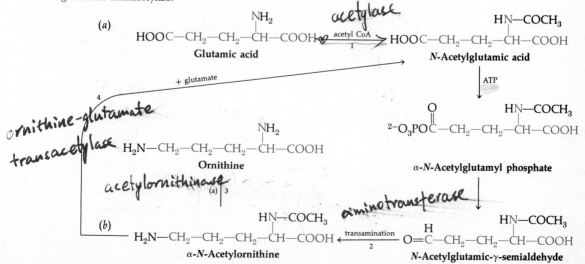

from the enzyme. Note that pyrroline 2-carboxylic acid is the intermediate, rather than the 5-derivative as in Fig. 22.4.

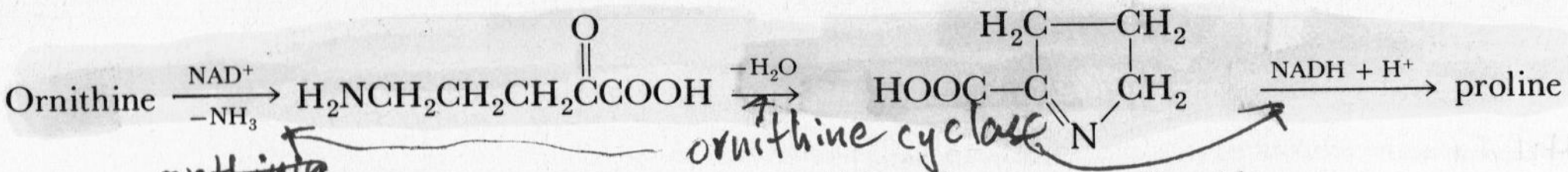

[handwritten: ornithine]

$$\text{Ornithine} \xrightarrow[-NH_3]{NAD^+} H_2NCH_2CH_2CH_2\overset{\overset{O}{\|}}{C}COOH \xrightarrow{-H_2O} \text{[}\Delta^1\text{-Pyrroline 2-carboxylic acid]} \xrightarrow{NADH + H^+} \text{proline}$$

2-Oxo-5-aminopentanoate Δ^1-Pyrroline 2-carboxylic acid

[handwritten: ornithine cyclase]

Cysteine Several enzymes catalyze cysteine formation in microorganisms. In some, cysteine synthase, with pyridoxal phosphate as coenzyme, catalyzes the reaction.

[handwritten: HS–CH₂–CH–COOH with NH above]

$$\underset{\text{L-Serine}}{\overset{OH \quad NH_2}{CH_2-CH-COOH}} + H_2S \rightleftharpoons \underset{\text{L-Cysteine}}{\overset{SH \quad NH_2}{CH_2-CH-COOH}} + H_2O$$

The equilibrium of this reaction, however, lies to the left. Other microorganisms and plants utilize H_2S for cysteine formation by the sequence

$$\text{Serine} + \text{acetylCoA} \rightleftharpoons O\text{-acetylserine} + \text{CoA} \tag{1}$$
$$O\text{-Acetylserine} + H_2S \rightleftharpoons \text{cysteine} + \text{acetate} + H_2O \tag{2}$$

Reaction (1) is catalyzed by *serine transacetylase;* reaction (2) by *O-acetylserine sulfhydrylase.* The two enzymes are present in a complex consisting of one molecule of the transacetylase and two of the sulfhydrylase. The nature of reaction (2) suggests that more than these two enzymes may be present in the complex. Since the overall process involves net hydrolysis of one molecule of acetyl CoA per molecule of cysteine formed, the sequence is favorable for cysteine formation.

In mammals, cysteine is formed from methionine (page 604).

Syntheses from Aspartic Acid In bacteria and plants, three amino acids, methionine, threonine, and isoleucine, derive all or part of their carbon atoms from aspartate. In bacteria, but not in fungi, yeasts, or, probably, plants, lysine is also synthesized from aspartate. The first step in all these syntheses is formation of β-aspartyl phosphate by ATP in a reaction catalyzed by *aspartate kinase.* In *E. coli* there are three distinct aspartate kinases with very different properties. *Aspartate kinase I* is inhibited by threonine in a feedback mechanism; *aspartate kinase II* is repressed by methionine; and *aspartate kinase III* is allosterically inhibited by lysine.

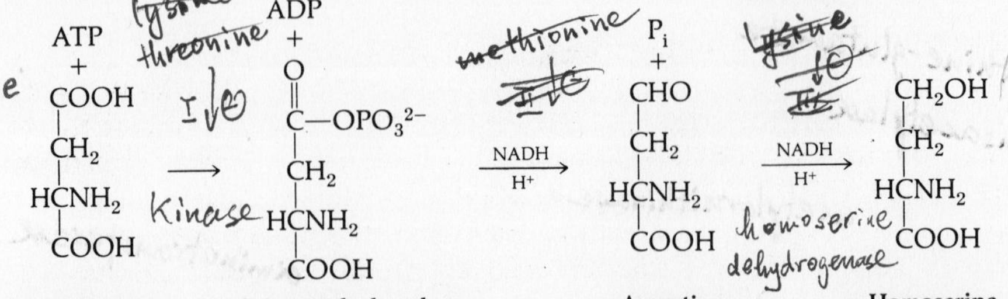

[handwritten: HOOC–CH₂–CH–COOH with NH above]

[handwritten left margin: I ⊖ — threonin ; II ⊖ — methionine ; III ⊖ — lysine]

$$\underset{\text{Aspartic acid}}{\overset{COOH}{\underset{COOH}{\overset{|}{\underset{|}{\overset{CH_2}{\underset{HCNH_2}{|}}}}}}} \xrightarrow[\text{Kinase}]{\overset{ATP}{\underset{}{+}} \quad \overset{ADP}{\underset{}{+}}} \underset{\beta\text{-Aspartyl phosphate}}{\overset{O}{\underset{COOH}{\overset{\|}{\underset{|}{\overset{C-OPO_3^{2-}}{\underset{HCNH_2}{\overset{|}{\underset{|}{CH_2}}}}}}}}} \xrightarrow[H^+]{NADH} \underset{\substack{\text{Aspartic} \\ \beta\text{-semialdehyde}}}{\overset{CHO}{\underset{COOH}{\overset{|}{\underset{|}{\overset{CH_2}{\underset{HCNH_2}{|}}}}}}} \xrightarrow[H^+]{NADH} \underset{\text{Homoserine}}{\overset{CH_2OH}{\underset{COOH}{\overset{|}{\underset{|}{\overset{CH_2}{\underset{HCNH_2}{|}}}}}}}$$

[handwritten: lysine, threonine, I↓⊖; methionine, II↓ Pᵢ; leucine, lysine, III; homoserine dehydrogenase]

Formation of aspartic β-semialdehyde resembles that of glutamic semialdehyde. Conversion of aspartic β-semialdehyde to homoserine is catalyzed by *homoserine dehydrogenase*. Homoserine is the branch point at which the pathways of methionine, threonine, and isoleucine diverge.

Methionine Synthesis of methionine begins with conversion of homoserine to O-succinylhomoserine by acyl transfer from succinyl CoA and catalyzed by *homoserine succinylase*. The enzyme is subject to feedback inhibition by methionine; this provides the specific control step in methionine synthesis. Reaction of O-succinyl-homoserine with cysteine is catalyzed by *cystathionine γ-synthase* to form cystathionine, a mixed thioether, which is converted by *cystathionine β-lyase* to homocysteine, pyruvate, and NH_3. In *Neurospora*, *N*-acetylhomoserine may function in the synthetic pathway in lieu of O-succinylhomoserine; the former results from a reaction between homoserine and acetyl CoA, catalyzed by *homoserine transacetylase*.

O-Succinyl-homoserine + Cysteine $\xrightarrow[\text{—succinate}]{\text{Cystathionine } \gamma\text{-synthase}}$ Cystathionine $\xrightarrow[\text{—pyruvate} \atop \text{—NH}_3]{\text{Cystathionine } \beta\text{-lyase}}$ Homocysteine

In *E. coli, methionine synthase* (*homocysteine methyltransferase*) catalyzes transfer of a methyl group to homocysteine to complete the synthesis of methionine. The prosthetic group of the enzyme is reduced methylcobalamin (Chap. M22). N^5-Methyltetrahydrofolate (page 609) provides the methyl group.

Threonine Synthesis of this amino acid begins with the phosphorylation of homoserine by ATP, catalyzed by *homoserine kinase*. *Threonine synthase*, which requires pyridoxal phosphate, catalyzes conversion of O-phosphohomoserine to threonine. The reaction results in conversion of a compound with a γ-hydroxyl group into one with a β-hydroxyl group.

$^{2-}O_3P-O-CH_2-CH_2-CH-COOH$ (O-Phosphohomoserine) $\xrightarrow[\text{phosphate}]{\text{pyridoxal}}$ $H_3C-CH-CH-COOH + P_i$ (Threonine)

Isoleucine Threonine is the source of four of the six carbon atoms of isoleucine (Fig. 22.6). The initial step is catalyzed by *threonine deaminase*, which utilizes pyridoxal phosphate as coenzyme, and yields α-ketobutyrate. α-Ketobutyrate condenses with "acetaldehyde," supplied as α-hydroxyethylthiamine pyrophosphate (page 321), to yield α-aceto-α-hydroxybutyrate, which is reduced by a pyridine nucleotide in a mutase reaction to form α,β-dihydroxy-β-methyl-valerate. A pinacol rearrangement (step 4), rare in biological reactions, followed by transamination from glutamate completes the pathway. Control of this pathway is effected at the level of threonine deaminase since the remaining

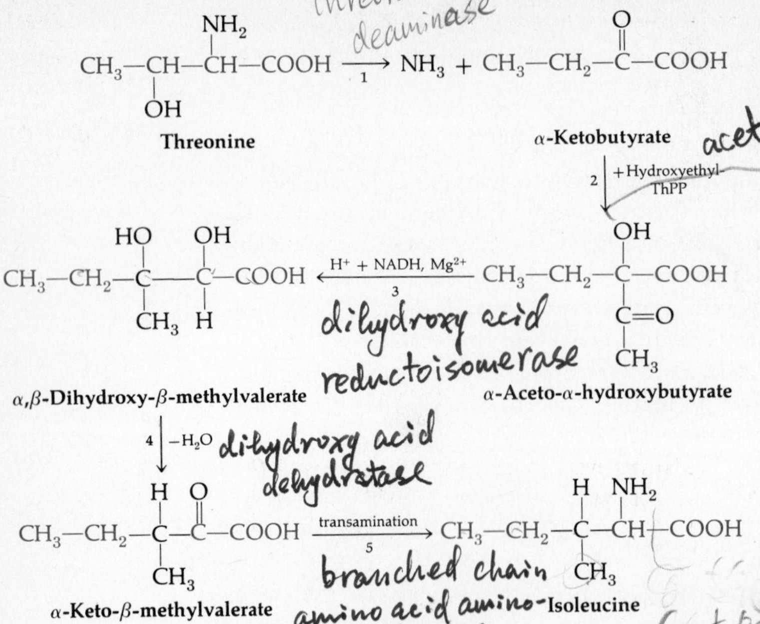

$$CH_3-CH(NH_2)-CH(OH)-COOH \xrightarrow{1} NH_3 + CH_3-CH_2-C(=O)-COOH$$

Threonine *(Threonine deaminase)*   **α-Ketobutyrate**

(acetolactate synthase or CH₃–CH–ThPP)

$$2 \downarrow \text{+Hydroxyethyl ThPP}$$

α-Aceto-α-hydroxybutyrate:
CH_3-CH_2-C(OH)(-C(=O)CH_3)-COOH

$$\xleftarrow{3}\; H^+ + NADH, Mg^{2+}$$

α,β-Dihydroxy-β-methylvalerate
CH_3-CH_2-C(OH)(CH_3)-CH(OH)-COOH
dihydroxy acid reductoisomerase

$$4 \downarrow -H_2O \quad \text{dihydroxy acid dehydratase}$$

α-Keto-β-methylvalerate
CH_3-CH_2-CH(CH_3)-C(=O)-COOH
$$\xrightarrow[5]{\text{transamination}}$$
CH_3-CH_2-CH(CH_3)-CH(NH_2)-COOH **Isoleucine**
branched chain amino acid aminotransferase

Figure 22.6 Pathway of isoleucine synthesis. Enzymes catalyzing reactions are indicated by numbers adjacent to arrows: 1 = threonine dehydratase; 2 = acetolactate synthase; 3 = dihydroxy acid reductoisomerase; 4 = dihydroxy acid dehydratase; 5 = branched-chain amino acid aminotransferase. ThPP = thiamine pyrophosphate.

steps are catalyzed by enzymes that also function in valine synthesis. The deaminase is both inhibited and repressed by isoleucine.

Valine Synthesis of valine is similar to that of isoleucine. Indeed, four of the enzymes are common to both pathways and are probably present in the cell as a multienzyme complex bound to the particulate portion of the cell. α-Hydroxyethylthiamine pyrophosphate condenses with pyruvate to form α-acetolactic acid (Fig. 22.7). *Dihydroxy acid reductoisomerase*, requiring Mg^{2+} and utilizing NADH, catalyzes transformation to α, β-dihydroxyvalerate. This enzyme complex consists of *acetolactate mutase* and a *reductase*; the former catalyzes another pinacol rearrangement (step 3) of α-acetolactate to α-keto-β-hydroxy-

Figure 22.7 Pathway of valine synthesis. Enzymes catalyzing reactions are indicated by numbers adjacent to arrows: 1 = acetolactate synthase; 2 = dihydroxy acid reductoisomerase; 3 = dihydroxy acid dehydratase; 4 = branched-chain amino acid aminotransferase. ThPP = thiamine pyrophosphate. Enzymes 1 to 4 are the same enzymes as 2 to 5, respectively, in Fig. 22.6.

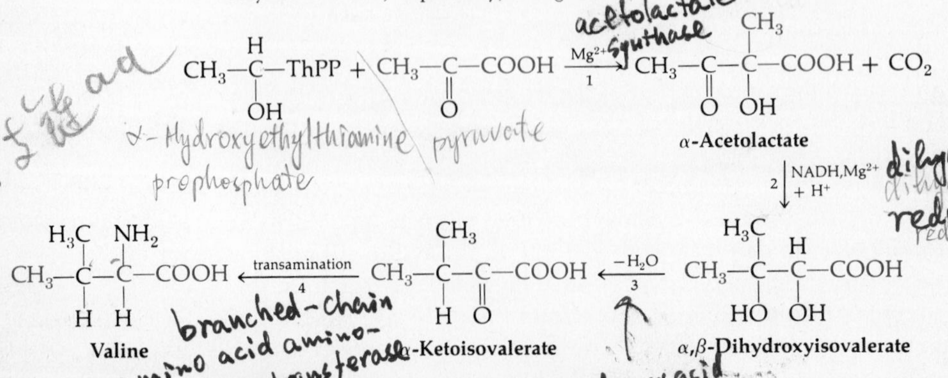

$$CH_3-C(H)(OH)-ThPP + CH_3-C(=O)-COOH \xrightarrow[1]{Mg^{2+}} CH_3-C(=O)-C(CH_3)(OH)-COOH + CO_2$$

α-Hydroxyethylthiamine pyrophosphate   *pyruvate*   **α-Acetolactate**

$$2 \downarrow \begin{array}{c} NADH, Mg^{2+} \\ + H^+ \end{array} \quad \text{dihydroxy acid reductoisomerase}$$

Valine
CH_3-C(CH_3)(H)(NH_2)-COOH
$$\xleftarrow{\text{transamination}}{}_{4}$$
α-Ketoisovalerate CH_3-C(CH_3)(=O...)
$$\xleftarrow[3]{-H_2O}$$
α,β-Dihydroxyisovalerate
CH_3-C(CH_3)(OH)-C(OH)-COOH
branched-chain amino acid aminotransferase   *dihydroxy acid dehydratase*

588

isovaleric acid, which is converted to α,β-dihydroxyisovaleric acid by the reductase. A dehydration reaction then yields α-ketoisovalerate, which transaminates with glutamate to form valine.

Leucine Synthesis of leucine (Fig. 22.8) begins with α-ketoisovalerate, which is also the immediate precursor of valine. Condensation occurs with acetyl CoA, in a manner reminiscent of the formation of citrate. Subsequent steps, leading to formation of α-ketoisocaproate, are analogous to the formation of α-ketoglutarate from citrate. Transamination completes the reaction sequence. In *Neurospora*, the first enzyme in this sequence, α-*isopropylmalate synthase*, is repressed by leucine; the synthesis of subsequent enzymes of this pathway is induced by the product of the first reaction, and the synthesis of the inducer is controlled by feedback inhibition.

Lysine Lysine synthesis occurs by two different pathways. One, termed the *diaminopimelic acid pathway,* is the route of lysine synthesis in bacteria, certain lower fungi, algae, and higher plants. The other, termed the *aminoadipic acid pathway,* occurs in other classes of lower fungi, in higher fungi, and in *Euglena*.

In the diaminopimelate pathway (Fig. 22.9), the carbon chain of lysine is synthesized from aspartic β-semialdehyde (page 586) and pyruvate in an aldol condensation catalyzed by a *condensing* enzyme, *dihydropicolinate synthase,* with loss of two molecules of water to form 2,3-dihydrodipicolinate. The latter is reduced to Δ^1-piperideine-2,6-dicarboxylate by a NADPH-requiring Δ^1-*piperideine-2,6-dicarboxylate dehydrogenase;* on hydrolysis, α-amino-ϵ-ketopimelate is formed. In *E. coli*, succinylation with succinyl CoA followed by transamination and desuccinylation yields L,L-α,ϵ-diaminopimelate. The latter, by the action of an *epimerase,* is converted to *meso*-diaminopimelate, which is decarboxylated to ly-

Figure 22.8 Pathway of leucine synthesis. Enzymes catalyzing reactions are indicated by numbers adjacent to arrows: 1 = α-isopropylmalate synthase; 2 = α-isopropylmalate isomerase; 3 = isopropylmalate dehydrogenase; 4 = decarboxylase; 5 = branched-chain amino acid aminotransferase.

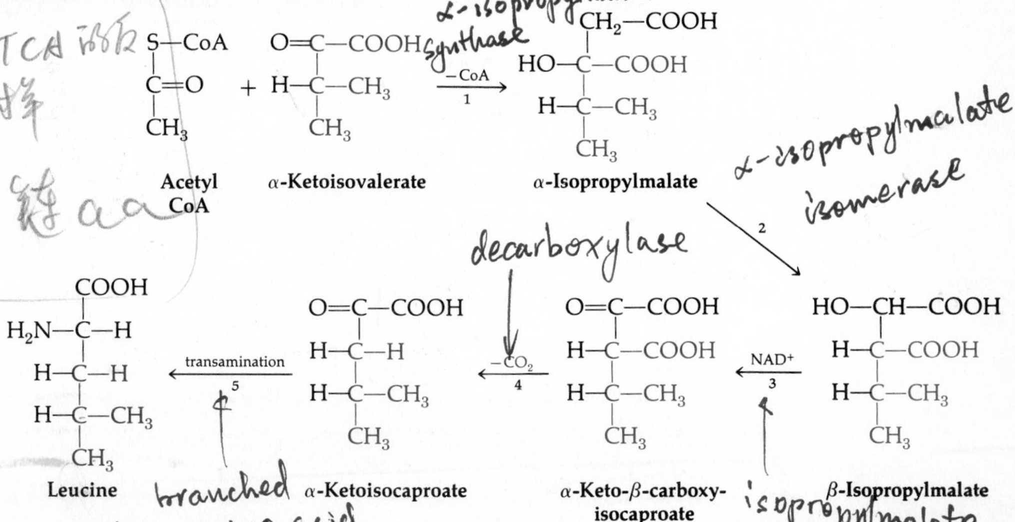

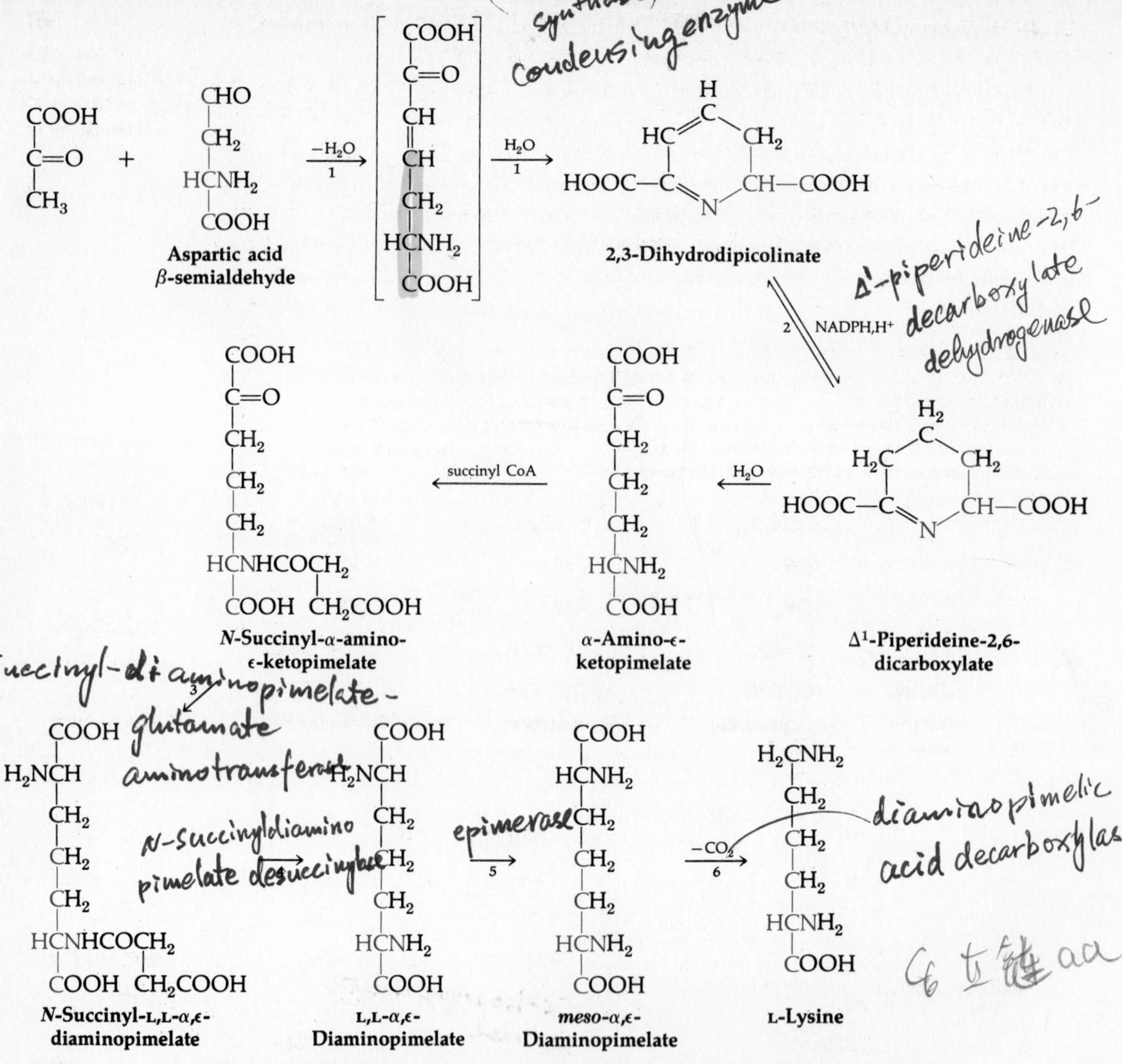

Aspartic acid
β-semialdehyde

2,3-Dihydrodipicolinate

N-Succinyl-α-amino-
ε-ketopimelate

α-Amino-ε-
ketopimelate

Δ^1-Piperideine-2,6-
dicarboxylate

succinyl CoA

N-Succinyl-L,L-α,ε-
diaminopimelate

L,L-α,ε-
Diaminopimelate

meso-α,ε-
Diaminopimelate

L-Lysine

Figure 22.9 Diaminopimelic acid pathway of lysine synthesis. Numbers at arrows indicate enzymes that have been delineated or suggested as catalysts in the pathway. 1 = condensing enzyme (dihydropicolinate synthase); 2 = $\triangle^1$-piperideine-2,6-dicarboxylate dehydrogenase; 3 = N-succinyldiaminopimelate-glutamate aminotransferase; 4 = N-succinyldiaminopimelate desuccinylase; 5 = epimerase; 6 = diaminopimelic acid decarboxylase.

sine by a pyridoxal phosphate–requiring *dicarboxylase* specific for the meso isomer. In *B. megaterium*, acetyl is used as the blocking group in lieu of succinyl.

At least two control points are present in the above sequence in *E. coli* (Fig. 22.9). One is the allosteric regulation by lysine of aspartate kinase III, which catalyzes the first step leading to aspartic β-semialdehyde (page 586). A peculiar property of aspartate kinase III is that it is also sensitive to leucine, as well as to a synergistic inhibition by both leucine and lysine. The second control point is at the condensation of the semialdehyde with pyruvate, the "branching point" toward lysine synthesis, which is inhibited by lysine. This control thus affects the first reaction in the sequence that leads only to lysine.

The aminoadipate pathway of lysine synthesis (Fig. 22.10) utilizes acetyl CoA and α-ketoglutarate, which condense to form homocitrate. The latter, in reactions analogous to those of the citric acid cycle (Chap. 15), yields *cis*-homoaconitate, homoisocitrate, oxaloglutarate, and α-ketoadipate. Transamination yields α-aminoadipate, which, on reduction by an NADPH-requiring enzyme, forms α-aminoadipic δ-semialdehyde. Condensation of the latter with glutamate produces ϵ-*N*-(L-glutaryl-2)-L-lysine, termed *saccharopine* because it is

Figure 22.10 Aminoadipic acid pathway of lysine biosynthesis. Numbers at arrows indicate enzymes that have been indicated or delineated as catalysts in the pathway. 1 = homocitrate synthase; 2 = *cis*-homoaconitase; 3 = homoaconitate hydratase; 4 = homoisocitrate dehydrogenase; 5 = α-aminoadipate–glutamate aminotransferase; 6 = α-aminoadipate semialdehyde dehydrogenase; 7 = aminoadipate semialdehyde-glutamate reductase; 8 = saccharopine dehydrogenase.

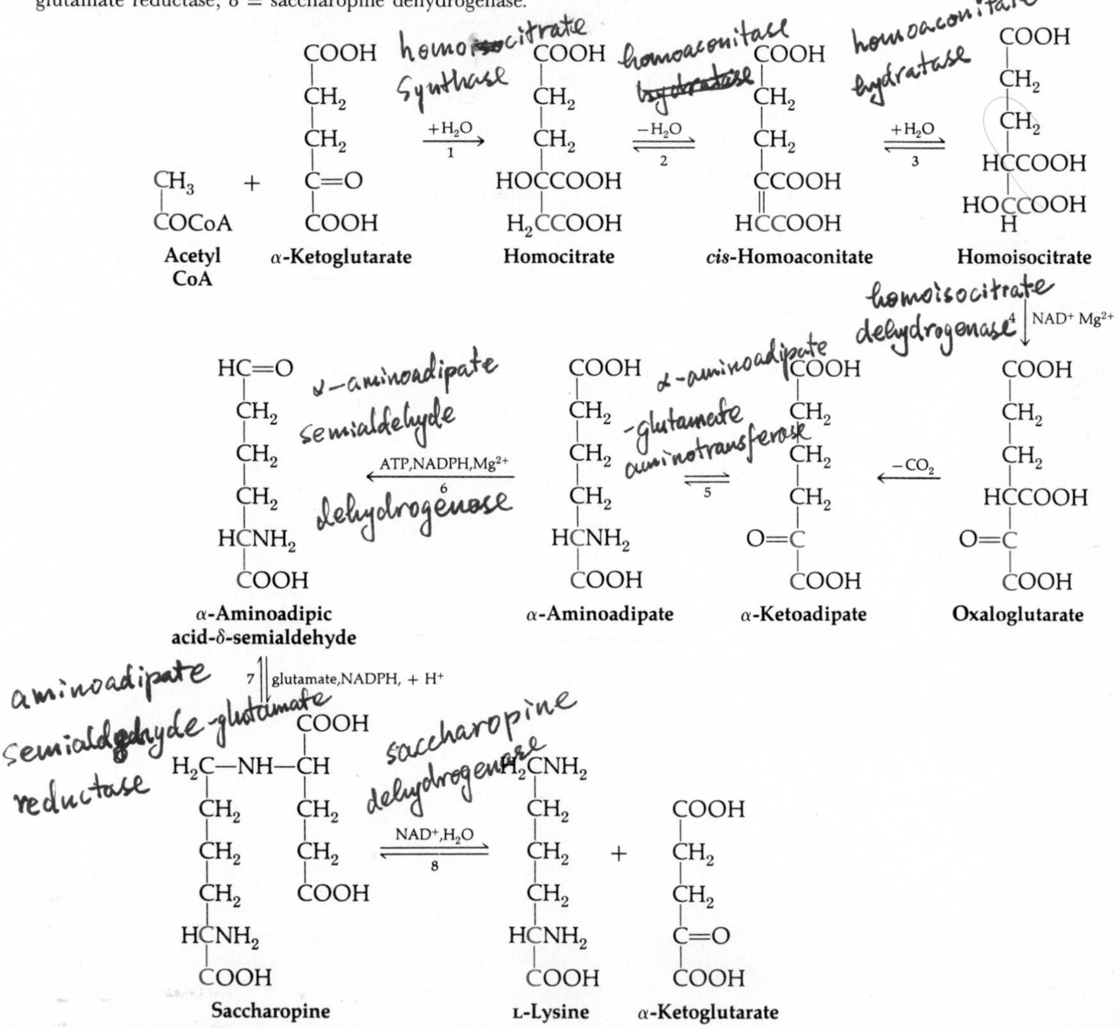

an intermediate of lysine synthesis in *Saccharomyces*. Reductive and hydrolytic reactions convert saccharopine to lysine and α-ketoglutarate.

Phenylalanine and Tyrosine The aromatic ring of these amino acids is derived from the four carbon atoms of erythrose-4-phosphate; the other two carbon atoms, as well as those of the side chains, are provided by phosphopyruvate. Studies with labeled precursors and identification of the compounds that accumulate in the media of mutants blocked at various steps of the synthetic pathway have elucidated the pathways depicted in Fig. 22.11. As indicated in the figure, chorismic (Gk., "to branch") acid is a branch point in the synthesis of the aromatic amino acids; conversion to anthranilate leads to tryptophan synthesis (see below), whereas transformation to prephenate provides a precursor of phenylalanine and tyrosine. The latter arises in mammals by hydroxylation of phenylalanine (Chap. 23); this reaction also occurs in certain microorganisms.

Two enzymes catalyzing formation of the first product of the synthetic pathway in Fig. 22.11, 3-deoxy-D-*arabino*-heptulosonic acid 7-phosphate, have been separated from extracts of *E. coli*. One is inhibited by phenylalanine and by tryptophan, the other by tyrosine, which also represses synthesis of the tyrosine-sensitive enzyme. In *B. subtilis, 3-deoxy-D-arabino-heptulosonic acid 7-phosphate synthase* exhibits properties of an allosteric system that is inhibited by prephenate and by chorismate, providing another feedback control mechanism. Thus, this first step in the multibranched pathways for synthesis of the aromatic amino acids is subject to multiple controls. In addition, the first two reactions specific for phenylalanine synthesis, viz., those catalyzed by *chorismate mutase* and *prephenate dehydratase,* respectively (Fig. 22.11), are subject to feedback inhibition by phenylalanine in *Salmonella typhimurium.* Reactions (2) to (6) in Fig. 22.11 are catalyzed by a multienzyme complex (MW = 200,000) that has been isolated from several species of fungi. The five component enzymes are covalently linked. Reaction (9), catalyzed by *anthranilate synthase* in *B. subtilis,* also catalyzes the following reaction.

$$NH_4^+ + \text{chorismic acid} \xrightarrow{\text{Mg}^{2+}} \text{anthranilic acid} + H_2O$$

Tryptophan The pathway of tryptophan synthesis in higher plants is unknown. In *E. coli*, tryptophan synthesis (Fig. 22.12) begins with anthranilate, which is formed in the sequence of reactions leading to the other aromatic amino acids (Fig. 22.11). Two of the enzymes involved, *indoleglycerol phosphate synthase* and *tryptophan synthase,* have been obtained in crystalline form. The former, MW = 44,000, is a single polypeptide chain. *Tryptophan synthase*

Figure 22.11 Pathways of synthesis of phenylalanine, tyrosine, and anthranilic acid, a precursor in tryptophan synthesis (page 594). Enzymes catalyzing reactions are indicated by numbers adjacent to arrows, which also serve as reaction numbers: 1 = 3-deoxy-2-keto-D-*arabino*-heptulosonic acid 7-phosphate synthase; 2 = dehydroquinate synthase; 3 = 5-dehydroquinate dehydratase; 4 = shikimate dehydrogenase; 5 = shikimate kinase; 6 = 3-enoylpyruvylshikimate-5-phosphate synthase; 7 = chorismate synthase; 8 = chorismate mutase; 9 = anthranilate synthase; 10 = prephenate dehydratase; 11 = prephenate dehydrogenase; 12 = phenylalanine aminotransferase; 13 = tyrosine aminotransferase.

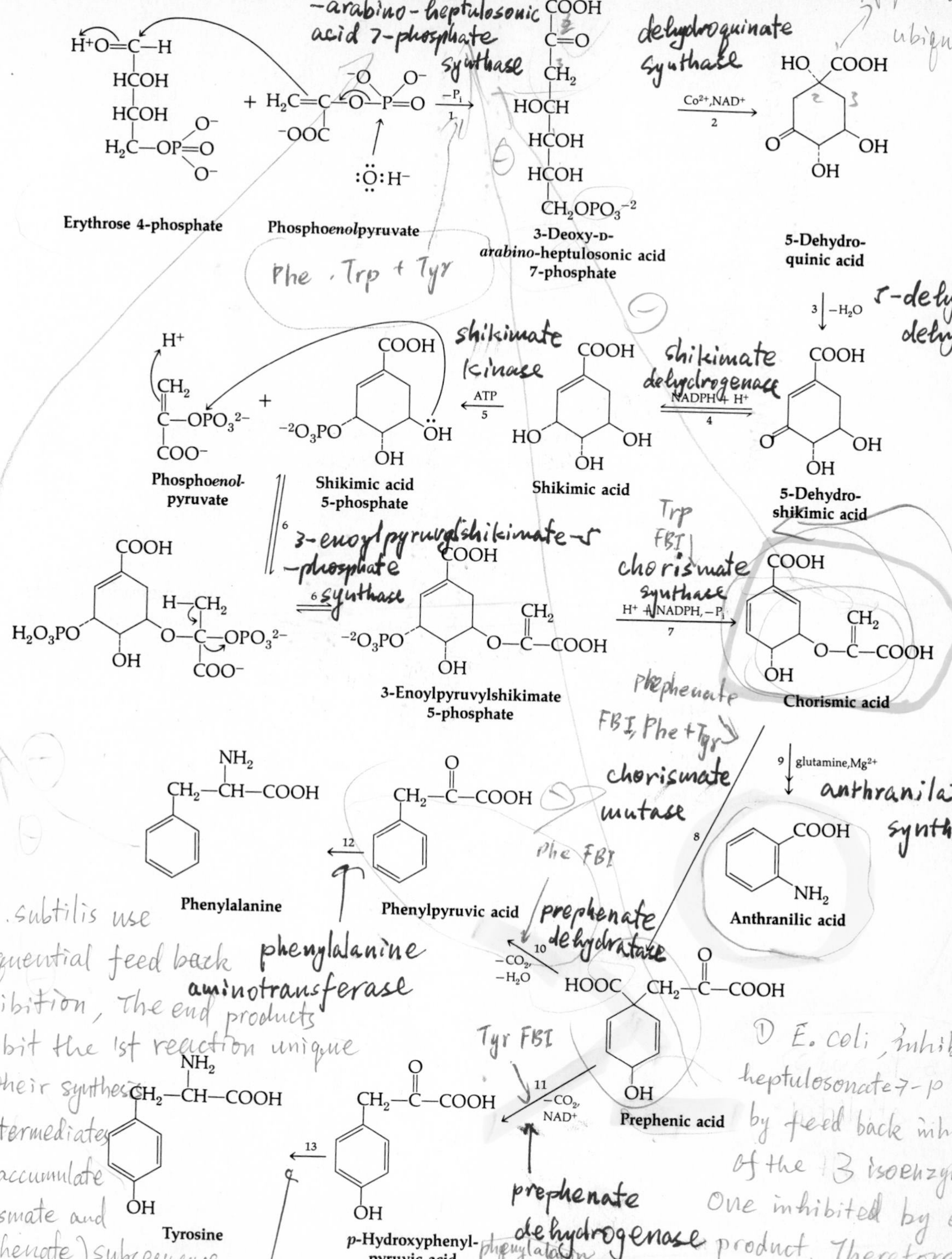

3-deoxy-2-keto-D-arabino-heptulosonic acid 7-phosphate synthase

Erythrose 4-phosphate + **Phosphoenolpyruvate** $\xrightarrow[1]{-P_i}$ **3-Deoxy-D-arabino-heptulosonic acid 7-phosphate**

Phe, Trp + Tyr

dehydroquinate synthase $\xrightarrow[2]{Co^{2+}, NAD^+}$ **5-Dehydroquinic acid**

plastoquinone ubiquinone

$\xrightarrow[3]{-H_2O}$ 5-dehydroquinate dehydratase

5-Dehydroshikimic acid

shikimate dehydrogenase $\xleftarrow[4]{NADPH + H^+}$

Shikimic acid

shikimate kinase $\xleftarrow[5]{ATP}$ **Shikimic acid 5-phosphate**

Phosphoenolpyruvate +

3-enoylpyruvylshikimate-5-phosphate synthase 6

3-Enoylpyruvylshikimate 5-phosphate

Trp FBI
chorismate synthase $\xrightarrow[7]{H^+ + NADPH, -P_i}$

Chorismic acid

prephenate FBI, Phe + Tyr
chorismate mutase

9 | glutamine, Mg^{2+} anthranilate synthase

Anthranilic acid

Phe FBI

Phenylalanine $\xleftarrow[12]{}$ **Phenylpyruvic acid** $\xleftarrow[10]{-CO_2 \\ -H_2O}$ prephenate dehydratase

phenylalanine aminotransferase

Tyr FBI

Tyrosine $\xleftarrow[13]{}$ **p-Hydroxyphenylpyruvic acid** $\xleftarrow[11]{-CO_2 \\ NAD^+}$ prephenate dehydrogenase

Prephenic acid

tyrosine aminotransferase

(2) B. subtilis use sequential feed back inhibition, The end products inhibit the 1st reaction unique to their synthesis, the intermediates that accumulate chorismate and prephenate) subsequence inhibit the heptulosonate-P synthase (aldolase)

Eryth-4-P + PEP ⟶ heptulosonate-7-Pi ⟶ chorismate

(1) E. coli, inhibits the heptulosonate 7-P synthetase by feed back inhibition of the 3 isoenzymes One inhibited by each product, Therefore the presence of all three (Tyr, Phe, Trp) are required to completely inhibit this step

phenylalanine ⟶ tyrosine prephenate

593

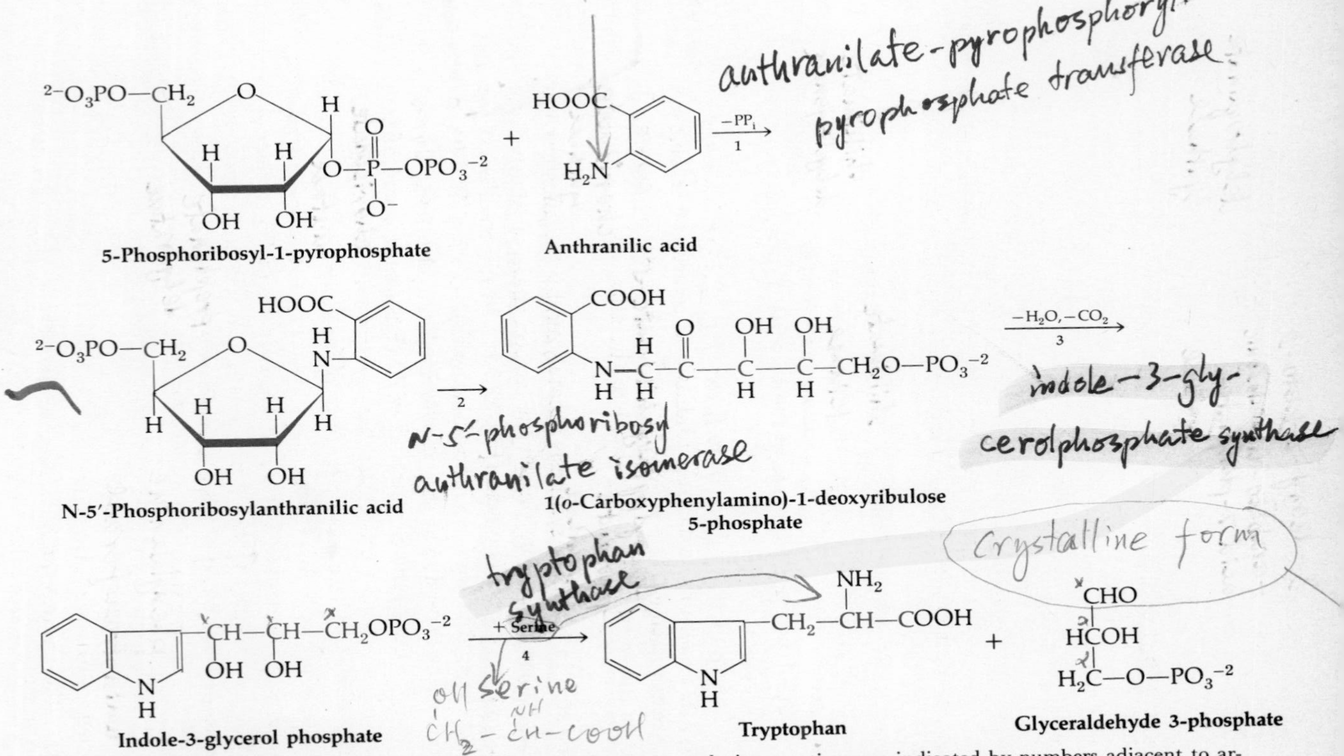

5-Phosphoribosyl-1-pyrophosphate

+

Anthranilic acid

$^{2-}O_3PO-CH_2$

HOOC

H_2N

from Gln. via transamination

anthranilate-pyrophosphoryl-
pyrophosphate transferase

$\xrightarrow[1]{-PP_i}$

N-5'-Phosphoribosylanthranilic acid

N-5'-phosphoribosyl
anthranilate isomerase

$\xrightarrow{2}$

1(o-Carboxyphenylamino)-1-deoxyribulose
5-phosphate

$\xrightarrow[3]{-H_2O, -CO_2}$

indole-3-gly-
cerolphosphate synthase

Indole-3-glycerol phosphate

tryptophan
synthase

+ Serine
$\xrightarrow{4}$

or Serine
$CH_2-CH-COOH$

Crystalline form

CHO
HCOH
$H_2C-O-PO_3^{-2}$

Glyceraldehyde 3-phosphate

+

NH_2
$CH_2-CH-COOH$

Tryptophan

Figure 22.12 Pathway of tryptophan synthesis of *E. coli.* Enzymes catalyzing reactions are indicated by numbers adjacent to arrows. 1 = anthranilate-pyrophosphorylpyrophosphate transferase; 2 = N-5'-phosphoribosylanthranilate isomerase; 3 = indole-3-glycerolphosphate synthase; 4 = tryptophan synthase.

(MW = 148,000), a pyridoxal phosphate–dependent enzyme, catalyzes the final reactions in the sequence leading to the synthesis of tryptophan. The reactions that can be catalyzed by this enzyme include

$$\text{Indole-3-glycerol phosphate} \rightleftharpoons \text{indole} + \text{D-glyceraldehyde 3-phosphate} \quad (1)$$

$$\text{Indole} + \text{L-serine} \xrightarrow{\text{pyridoxal phosphate}} \text{L-tryptophan} + H_2O \quad (2)$$

$$\text{Indole-3-glycerol phosphate} + \text{L-serine} \rightleftharpoons$$
$$\text{L-tryptophan} + \text{D-glyceraldehyde 3-phosphate} \quad (3)$$

It is probable that formation of tryptophan from indoleglycerol phosphate reaction (3) occurs via reactions (1) and (2) with indole as an enzyme-bound intermediate.

Studies of genetic alterations of the structure and biosynthetic controls of *tryptophan synthase* have provided important insights of these mechanisms (Chap. 28). The enzyme from *E. coli* has an $\alpha_2\beta_2$ structure. Reaction (1) is catalyzed by α alone but only at a fraction of the rate of the associated complex; reaction (2) is catalyzed by β_2 alone.

The activity of anthranilate synthase from *E. coli* (Fig. 22.11) is inhibited by tryptophan; in yeast, the amino acid both inhibits and represses the enzyme. As mentioned previously (page 592), tryptophan also represses the enzyme in *E. coli* that catalyzes the initial step in synthesis of the aromatic amino acids.

Histidine The first step in histidine synthesis is catalyzed by the allosteric enzyme *ATP phosphoribosyltransferase*, resulting in the carbon chain of ribose being linked to a nitrogen atom of adenylic acid (Fig. 22.13). In the reaction of the C-1 carbon atom of phosphoribosyl pyrophosphate with N-1 of ATP with expulsion of PP_i, there occurs inversion of the α configuration to the newly formed (β) bond. In the next reaction, the purine ring of AMP is opened, and after a nitrogen atom has been provided from glutamine, the structure breaks apart to yield imidazole glycerol phosphate, in which the imidazole ring of histidine is fully formed and attached to a 3-carbon chain; 5-aminoimidazole-4-carboxamide ribonucleotide, an intermediate in purine synthesis (Chap. 26), is released.

In the synthesis of imidazole glycerol phosphate, the side chain and connecting two carbons of the ring derive from the five carbons of the ribose of 5-phosphoribosyl-1-pyrophosphate. The —N=C— adjoining stems from the pyrimidine portion of the fused purine nucleus. Since the carbon atom of this fragment originates, during purine synthesis, from the formyl of N^{10}-formyltetrahydrofolate (page 610), it derives from the β-carbon of serine or other sources of C_1 units. The final N atom is provided by the amide nitrogen of glutamine. Thus, the histidine-synthesizing system utilizes a portion of an existing purine nucleus but leaves behind a fragment (aminoimidazole carboxamide ribonucleotide), which is reconverted to purines (Chap. 26). In the last step, a primary hydroxyl group of histidinol is oxidized by 2 equiv of NAD^+ to the corresponding carboxyl group by consecutive oxidations on the surface of a single enzyme.

The above complex pathway for histidine synthesis was established by studies of *E. coli* and *Salmonella*. The sequence has a number of metabolic controls, and each of the nine steps has been identified with one of the nine genes of

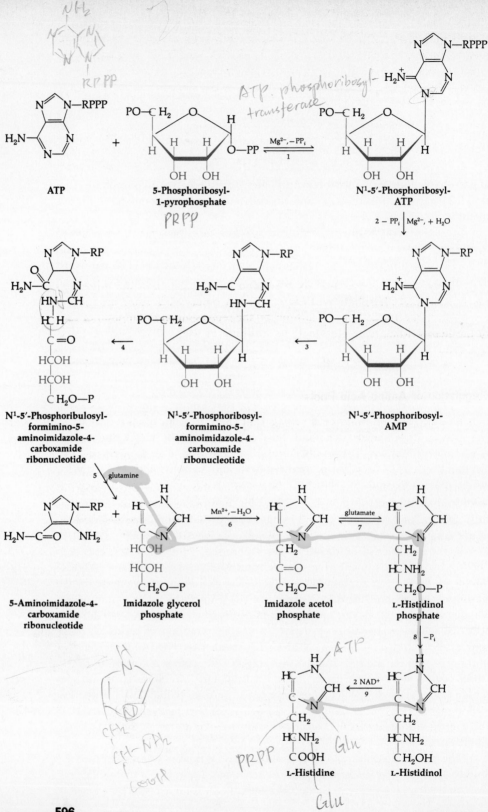

ATP

5-Phosphoribosyl-
1-pyrophosphate

N^1-5'-Phosphoribosyl-
ATP

N^1-5'-Phosphoribulosyl-
formimino-5-
aminoimidazole-4-
carboxamide
ribonucleotide

N^1-5'-Phosphoribosyl-
formimino-5-
aminoimidazole-4-
carboxamide
ribonucleotide

N^1-5'-Phosphoribosyl-
AMP

5-Aminoimidazole-4-
carboxamide
ribonucleotide

Imidazole glycerol
phosphate

Imidazole acetol
phosphate

L-Histidinol
phosphate

L-Histidine

L-Histidinol

the histidine operon. The first reaction is inhibited specifically by histidine, which regulates the activity of the *ATP phosphoribosyltransferase*. In addition, the entire enzyme system is subject to *coordinate repression;* i.e., the presence of excess histidine in the culture medium represses the synthesis of all nine enzymes catalyzing steps in histidine synthesis.

Arginine Synthesis of arginine commences with ornithine, which has been described (page 584). There remains to indicate only the genesis of the guanidino group. This has been studied more closely in animal (Chap. 23) than in plant systems; the sequence is identical with that shown below.

$$\text{Ornithine} + \text{Carbamoyl phosphate} \xrightarrow[\text{}]{P_i} \text{citrulline} \xrightarrow[\text{aspartate}]{\text{ATP} \quad \text{AMP}+\text{PP}_i} \text{argininosuccinate} \xrightarrow[\text{}]{\text{fumarate}} \text{arginine}$$

Regulatory control is exercised by repression and inhibition of the mitochondrial carbamoyl phosphate synthetase, which is balanced in some measure by the positive effector action of ornithine, by the repression and inhibition of each of the steps leading from glutamate to ornithine, as well as inhibition of citrulline synthesis.

Regulation of Amino Acid Pools

Examination of living cells reveals that there is at all times a very small pool of free amino acids. The basis for this in animal cells will be discussed in the following chapter. In microbial and plant cells, amino acids rarely undergo degradation but are used for protein synthesis and for construction of a myriad of other nitrogenous substances. However, none is present in excessive amounts. Even in a microorganism that has been derepressed with respect to some key enzyme, feedback inhibitions prevent waste of much starting material or energy. Only the derepressed enzyme is then present in excess. Remarkably, the evolutionary process has worked so well that there is always available to these cells, when they have available energy and starting materials, a balanced supply of each of the 20 common amino acids. It is the concentration of each amino acid which determines the rate of its own synthesis and which, in turn, by this mechanism regulates the rate of its removal from the amino acid pool for other synthetic purposes.

Figure 22.13 Pathway of histidine synthesis. RPPP = ribose triphosphate, RP = ribose-5-phosphate, PP_i = pyrophosphate, P = orthophosphate. Enzymes catalyzing reactions are indicated by numbers adjacent to arrows: 1 = ATP phosphoribosyl transferase; 2 = pyrophosphohydrolase; 3 = phosphoribosyl-AMP cyclohydrolase; 4 = phosphoribosyl formimino-5-aminoimidazole carboxamide ribonucleotide isomerase; 5 = glutamine amidotransferase; 6 = imidazole glycerol phosphate dehydratase; 7 = L-histidinol phosphate-glutamate aminotransferase; 8 = histidinol phosphate phosphatase; 9 = histidinol dehydrogenase.

REFERENCES

Books

Bender, D. A.: *Amino Acid Metabolism,* Wiley, New York, 1975.

Greenberg, D. M. (ed.): Metabolic Pathways, vol. III: *Amino Acids and Tetrapyrroles,* 3d ed., 1969; vol. VII: *Metabolism of Sulfur Compounds,* 1975, Academic, New York.

Meister, A.: *Biochemistry of the Amino Acids,* 2d ed., vols. I and II, Academic, New York, 1965.

Review Articles

Roberts, G. P., and W. J. Brill: Genetics and Regulation of Nitrogen Fixation, *Annu. Rev. Microbiol.* **35:**207–235 (1981).

Stadtman, E. R., and A. Ginsburg: The Glutamine Synthetase of *Escherichia coli:* Structure and Control, pp. 755–807, in P. D. Boyer (ed.): *The Enzymes,* 3d ed., vol. 10, Academic, New York, 1974.

Umbarger, H. E.: Amino Acid Biosynthesis and Its Regulation, *Annu. Rev. Biochem.* **47:**533–606 (1978).

Winter, H. C., and R. H. Burris: Nitrogenase, *Annu. Rev. Biochem.* **45:**409–426 (1976).

Chapter 23

Amino Acid Metabolism II: Mammals

Essential and nonessential amino acids.
Metabolism of one-carbon compounds.
Metabolism of ammonia and the urea cycle.
Overall aspects of amino acid metabolism.

In animals, as in plants and microorganisms, the principal fate of amino acids is incorporation into proteins. In addition, synthesis of a variety of nitrogen-containing compounds utilizes various amino acids as precursors. The ability to perform these syntheses depends upon the continuing availability of a pool of amino acids, balanced with respect to the requirements for each of the 20 individual amino acids. No cell contains a storage form of amino acids in any way analogous to glycogen or triacylglycerols, nor can cells synthesize a protein molecule if one of its normal constituent amino acids is lacking (Chaps. 28 and 29).

In plants and microorganisms, as seen in the previous chapter, this balanced mixture is provided by their own synthetic mechanisms, self-regulated by feedback controls such that the supply of each of the 20 amino acids equals the demand. However, animals, lacking photosynthetic ability, depend upon ingestion of plants, microorganisms, or other animals for the carbohydrate and lipid whose oxidation and other modes of metabolism furnish them with the energy and precursors they require for synthesis of diverse metabolic and structural materials. The quantity of food ingested daily relates largely to that caloric supply. Animal cells are unable to synthesize, *de novo*, half of the 20 amino acids, nor can they synthesize that group of compounds metabolically needed in small quantities, termed *vitamins* (Chaps. M22 and M23). Given an abundance of other amino acids, the α-keto acid corresponding to the nutritionally essential amino acids can be substituted for these amino acids in the diet. It is the carbon skeleton of each that cannot be synthesized by animals.

It is not, however, lack of photosynthetic ability that occasions this absolute demand for certain amino acids and vitamins. Hypothetically, given an adequate supply of carbohydrate, were animals able to degrade all ingested amino

Any cross-references coded M refer to the companion text, *Principles of Biochemistry: Mammalian Biochemistry.*

acids to NH_3, retention of the amino acid-synthesizing capabilities of their primitive ancestors should not have been incompatible with animal life. In fulfilling the requirement for calories as carbohydrate or fat, animals ingest plants, microorganisms, or other animals that necessarily contain all the amino acids and vitamins as well. This fact permitted loss of the ability to synthesize various amino acids and vitamins with very little penalty, as animals evolved to inhabit a remarkable diversity of environments. That this was not entirely without penalty, however, is attested by the catalog of deficiency diseases, including *kwashiorkor* (Chap. M21) and *pellagra* (Chap. M22), when human beings have ingested one of various unbalanced diets for prolonged periods of time.

There are also metabolic consequences. Even in the very young growing animal, the food supply may well provide more amino acids per 1000 kcal than required for the net syntheses in which it engages. Some animals experience a long life period during which no *net* synthesis is occurring; e.g., an adult human being, maintaining constant weight, may ingest from 20 to 150 g of mixed amino acids per day in the course of consuming the requisite 2000 kcal. This range is made possible by mechanisms that permit utilization of some of the ingested materials and disposal of the remainder. In experimental situations and in relatively rare ordinary experience, the mixture of amino acids in the diet may be limited and inappropriate for the various ongoing syntheses in which the animal engages. This requires that, to the extent possible, this mixture be rebalanced, utilizing at least the nitrogen of the ingested amino acids to synthesize those amino acids for which animals have retained this capability. Finally, there may be relatively brief periods of no food supply to which the animal must make some biologically acceptable adjustment.

It is the purpose of this chapter to examine these general aspects of amino acid metabolism in mammals.

TRANSPORT OF AMINO ACIDS

The action of the proteolytic enzymes in the stomach and small intestine results in the hydrolysis of most of the dietary protein to amino acids (Chap. M10). Absorption of amino acids occurs chiefly in the small intestine and is an energy-requiring process resembling in several respects the active transport of glucose (Chap. 17). There are carrier-mediated ATP- and Na^+-dependent transport systems for each of four groups of free amino acids (neutral, basic, acidic, and proline). In addition there is in some tissues (intestine, brain, and kidney) a system, termed the *γ-glutamyl cycle*, that utilizes glutathione, γ-glutamylcysteinylglycine (Chap. 24), for the transport of certain amino acids. The operation of the γ-glutamyl cycle, which is shown in Fig. 23.1, proceeds in the following manner.

The glutamyl residue of intracellular glutathione is transferred to the amino acid to be transported in a reaction catalyzed by membrane-bound γ-glutamyl transferase. The products are the γ-glutamylamino acid now present in the cytosol and cysteinylglycine. The latter is hydrolyzed by the action of a specific peptidase. In the next step the γ-glutamyl amino acid is cleaved by *γ-glutamylcyclotransferase* to the free amino acid and 5-oxoproline (pyrrolidone

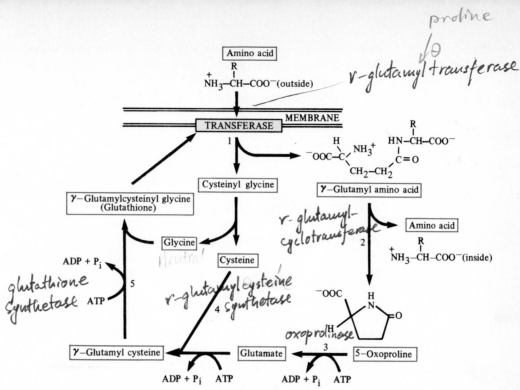

proline

γ-glutamyl transferase

γ-glutamyl-cyclotransferase

γ-glutamyl cysteine synthetase

glutathione synthetase

oxoprolinase

Figure 23.1 The γ-glutamyl cycle for the transport of certain amino acids. Reaction 1 is catalyzed by *γ-glutamyl transferase*, reaction 2 by *γ-glutamylcyclotransferase*, reaction 3 by *5-oxoprolinase*, reaction 4 by *γ-glutamylcysteine synthetase*, and reaction 5 by *glutathione synthetase*.

carboxylic acid). These three sequential reactions result in the transport of one molecule of amino acid into the cell with the utilization of the energy of hydrolysis of the peptide bonds of glutathione. If the process is to be continuous, glutathione must be regenerated. This is accomplished in a sequence of three reactions. In the first, 5-oxoproline is hydrolyzed to L-glutamate by *5-oxoprolinase*. The L-glutamate reacts with L-cysteine in a reaction catalyzed by *γ-glutamylcysteine synthetase* (Chap. 24), forming γ-glutamylcysteine. Finally, glutathione is regenerated by the action of *glutathione synthetase*. The glutathione can now participate in another round of the cycle. Note that three molecules of ATP are required for the resynthesis of glutathione and hence for the transport of one molecule of amino acid into the cell. The γ-glutamyl transferase is most active with glutamine, cysteine, and other neutral amino acids but less active toward aspartate and certain of the branched-chain and aromatic amino acids. It is inactive with proline.

AMINO ACIDS NUTRITIONALLY ESSENTIAL FOR HUMAN BEINGS

In the course of evolution, animals lost the ability to synthesize the carbon chains of certain of the α-keto acids corresponding to amino acids normally present in most proteins. Amino acids that cannot be synthesized at a rate adequate to meet metabolic requirements have been termed *essential amino acids* (Table 23.1). This classification, originally based on growth behavior of the albino rat, is corroborated by studies of nitrogen balance (Chap. M21) in human infants and young adults.

The terms *essential* and *nonessential* relate, in retrospect, to a nutritional concept determined under unnatural, laboratory conditions. The list is, in reality, a classification of amino acids according to whether their carbon skeletons can or cannot be synthesized *de novo* by human beings and other animals. Actually, one must add tyrosine and cysteine to the list of essential amino acids since tyrosine is formed in one step directly from phenylalanine (page 606) and cysteine derives its sulfur uniquely from dietary methionine (page 604). Hence, the apparent quantitative phenylalanine requirement is truly a requirement for phenylalanine plus tyrosine, and the quantitative daily requirement for methionine is for methionine plus cysteine. The status of arginine is rather ambiguous since, as will be seen later, the entire arginine molecule can be synthesized by human beings and other animals.

The list of nonessential amino acids proves to be rather limited; viz., alanine and aspartic and glutamic acids, which can be made via transamination (pages 604 ff.) from α-keto acids arising in the citric acid cycle; proline, which is made from glutamic acid (page 607); serine, made from an intermediate in glycolysis (page 607); glycine synthesized from serine (page 607); and the amino acid amides glutamine and asparagine, made, respectively, from glutamate (page 614) and aspartate (page 614). Thus, these are the amino acids whose syntheses are simplest and whose immediate precursors are metabolically always present in all species. At the same time, these amino acids are abundant in proteins and their absence in a diet of natural foodstuffs is impossible. Conversely, when deliberately excluded from an experimental diet, the "requirement" for the sum of all "essential" amino acids increases somewhat to provide some nitrogen for the synthesis of this "nonessential" group. The nutritional consequences of a diet of natural foodstuffs that is relatively low in lysine, tryptophan, or methionine are considered in Chap. M21.

Hepatic Amino Acid Metabolism

The liver and, to a lesser extent, the kidney are the principal sites of amino acid metabolism in mammals. The detailed processes by which amino acids are transformed to diverse nitrogenous compounds other than α-amino acids are presented in Chap. 24. We will be concerned here with (1) the synthesis of the nonessential amino acids and (2) the disposal of surplus amounts of both the carbon chains of ingested amino acids and of nitrogen.

TABLE 23.1

Classification of the Amino Acids with Respect to Their Growth Effect in the White Rat	Essential	Nonessential
	Arginine,* histidine, isoleucine, leucine, lysine, methionine, phenylalanine, threonine, tryptophan, valine	Alanine, asparagine, aspartic acid, cystine, glutamic acid, glutamine, glycine, proline, serine, tyrosine

*Arginine can be synthesized by the rat but not rapidly enough to meet the demands of *normal* growth.

Transamination The process of transamination, described in detail in the preceding chapter, is catalyzed by pyridoxal phosphate-dependent enzymes termed *aminotransferases* and takes the following general form:

$$\underset{\text{L-Amino acid}}{\text{R}\overset{\text{H}}{\underset{\text{COO}^-}{-\text{C}-\text{NH}_3^+}}} + \underset{\alpha\text{-Ketoglutarate}}{\overset{\text{H}_2\text{C}-\text{COO}^-}{\underset{\text{COO}^-}{\text{CH}_2}}} \rightleftharpoons \underset{\alpha\text{-Keto acid}}{\text{R}\overset{\text{O}}{\underset{\text{COO}^-}{-\text{C}}}} + \underset{\text{L-Glutamate}}{\overset{\text{H}_2\text{C}-\text{COO}^-}{\underset{\text{COO}^-}{\text{CH}_2}}}$$

The mechanism by which pyridoxal phosphate participates in aminotransferase reactions has already been presented (Chap. 22), as have the nomenclature and specificity of these enzymes. In mammals, all amino acids, except lysine and threonine, can participate in transmination. In addition to glutamate, the usual donor of amino groups in transamination reactions, asparagine and glutamine also participate. In these instances, the products are, respectively, α-ketosuccinamic acid and α-ketoglutaramic acid, the ω-amides of the corresponding α-keto acids. These are then converted to NH_3 and the keto acid by specific ω-*amidases*. Thus,

$$\text{L-Glutamine} + \alpha\text{-keto acid} \rightleftharpoons \alpha\text{-ketoglutaramic acid} + \text{L-}\alpha\text{-amino acid} \quad (1)$$

$$\alpha\text{-Ketoglutaramic acid} + \text{H}_2\text{O} \xrightarrow{\omega\text{-amidase}} \alpha\text{-ketoglutaric acid} + \text{NH}_3 \quad (2)$$

The aminotransferases for aspartate, alanine, and glutamine are present in both cytosol and mitochondria as isoenzymic forms. These isoenzymes exhibit different catalytic and physical properties and substrate specificity. The aspartate aminotransferases of pig heart mitochondria and cytosol are each of molecular weight of approximately 90,000 and contain two identical subunits. The two enzymes differ in amino acid sequence but are similar in conformation. The aminotransferases of cytosol are generally more active than those of mitochondria. Amino group transfer occurring in the cytosol results in formation of glutamate. The latter enters the mitochondria, where it donates its amino group to oxaloacetate by the action of mitochondrial aminotransferase to yield aspartate, an amino-group donor in the formation of urea (page 617).

Glutamate can also be deaminated directly to yield ammonia and α-ketoglutarate by the action of *glutamate dehydrogenase* (page 613). Since glutamate is the only amino acid for which a specific and highly active dehydrogenase exists, combination of the activity of the transaminases with that of glutamate dehydrogenase has the same effect as though there were individual dehydrogenases for each of the other amino acids. Thus, removal of ammonia from amino acids is regulated mainly by control of glutamate dehydrogenase (page 613).

Inasmuch as the equilibrium constant for all transamination reactions is close to 1, their rates are determined solely by the concentrations of the substrates and the amount of enzyme present. The concentration and activity of specific aminotransferases in liver can be influenced markedly by some hormones. Thus, the rate of synthesis of *tyrosine aminotransferase* in liver can be increased by the administration of adrenal cortical steroids (Chap. M18), insulin (Chap. M16), glucagon (Chap. M16), or cyclic AMP.

METABOLIC ORIGINS OF THE NUTRITIONALLY NONESSENTIAL AMINO ACIDS

The nutritionally nonessential amino acids are all made by rather simple metabolic reaction sequences from readily available materials. In normal circumstances, animals ingesting proteins as sources of the essential amino acids have little demand for directed synthesis of the nonessential ones.

Glutamic Acid With α-ketoglutarate always present, the abundant reversible aminotransferases described above suffice to ensure a supply of glutamate. This amino acid also arises in the further metabolism of histidine, proline, 5-oxoproline, ornithine, and glutamine (Chap. 25). However, these compounds are in turn synthesized from glutamate (Chap. 22). There are additional sources of glutamate, viz., from the reductive amination of α-ketoglutarate and NH_3, catalyzed by *glutamate dehydrogenase,* and from the reaction catalyzed by *glutamate synthase* (Chap. 22). This process, the formation of glutamate, is a central event in the nitrogen metabolism of plants and microorganisms and also occurs in animals. However, the NH_3 originates predominantly from the reverse process or from the hydrolysis of glutamine, and the latter is formed from glutamate itself and the NH_3 derived from its oxidation. Hence, although these reactions do indeed occur, they should not be regarded as a mammalian equivalent of glutamate synthesis in plants; the closest equivalent is the transamination process.

Aspartic Acid and Alanine These two amino acids derive directly from transamination of oxaloacetate and pyruvate, respectively, with glutamate.

Cysteine The sulfur atom of cysteine is obtained uniquely from the essential amino acid methionine. If sufficient methionine is fed, there is no dietary requirement for additional cysteine. The reactions involved in cysteine biosynthesis, all occurring in the liver, are the following:

1. Demethylation of methionine to homocysteine (homolog of cysteine).

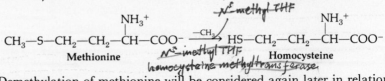

Demethylation of methionine will be considered again later in relation to transmethylation (page 611).

2. Homocysteine condenses with serine in a reaction catalyzed by *cystathionine β-synthase,* a pyridoxal phosphate-requiring enzyme; the product is cystathionine.

$$\text{}^-\text{OOC}-\underset{\underset{\text{NH}_3{}^+}{|}}{\text{CH}}-\text{CH}_2-\text{CH}_2-\text{S}\,\text{H} + \text{HO}-\text{CH}_2-\underset{\underset{}{|}}{\overset{\overset{\text{NH}_3{}^+}{|}}{\text{CH}}}-\text{COO}^- \longrightarrow$$

Cystathionine
β-synthase

Homocysteine   Serine

$$\text{}^-\text{OOC}-\underset{\underset{\text{NH}_3{}^+}{|}}{\text{CH}}-\text{CH}_2-\text{CH}_2-\text{S}-\text{CH}_2-\overset{\overset{\text{NH}_3{}^+}{|}}{\text{CH}}-\text{COO}^- + \text{H}_2\text{O}$$

Cystathionine

3. Cleavage of cystathionine is catalyzed by *cystathionine γ-lyase,* which utilizes pyridoxal phosphate.

cystathionine γ-lyase

$$\text{Cystathionine} + \text{H}_2\text{O} \longrightarrow \text{}^-\text{OOC}-\underset{\underset{\text{NH}_3{}^+}{|}}{\text{CH}}-\text{CH}_2-\text{SH} + \text{CH}_3-\text{CH}_2-\underset{\underset{\text{O}}{\|}}{\text{C}}-\text{COO}^- + \text{NH}_3$$

Cysteine   **α-Ketobutyric acid**

The net effect of the above three reactions is transfer of the sulfhydryl group of homocysteine to the carbon chain of serine.

Cysteine is an allosteric inhibitor of cystathionine γ-lyase and also suppresses the synthesis of cystathionine β-synthase. The latter effect may explain the sparing action of cysteine (or cystine) on the dietary requirement for methionine for the growth of rats on synthetic rations. If sufficient cysteine (or cystine) is provided, the methionine requirement is less than half that required in the absence of cysteine. Suppression by the latter of hepatic synthesis of cystathionine β-synthase would decrease channeling of homocysteine into formation of cystathionine.

Cystathionine is the key intermediate in the above reactions; its only function in mammals is to serve as an intermediate in the transfer of sulfur from methionine to cysteine. Cystathionine γ-lyase deficiency is a genetic defect in which there is a reduced affinity of the enzyme for pyridoxal phosphate; it leads to persistent excretion of large amounts of cystathionine in the urine. A genetic defect in cystathionine synthase produces *homocystinuria,* i.e., the excessive excretion of homocystine, and is associated with severe mental retardation.

Although the metabolic equivalence of cysteine and cystine is established, no enzymic system catalyzing their interconversion is known in mammalian tissue. As described below, oxidized glutathione can function in a nonenzymic oxidation of cysteine to cystine.

Little or no free cystine is present in cells. The cystine of proteins is formed by oxidation of cysteine residues after their incorporation into polypeptide chains (Chap. 28). However, in the presence of O_2 and cations such as Fe^{2+} or Cu^{2+}, cystine may be formed from cysteine nonenzymically. If this occurs, ready reversal can be catalyzed by *glutathione reductase.* Glutathione (GSH, below) reacts nonenzymically with any disulfide, e.g., cystine, to form a mixed disulfide.

$$\text{GSH} + \text{R}'\text{SSR}'' \rightleftharpoons \text{GSSR}'' + \text{R}'\text{SH}$$

A second molecule of glutathione, reacting with the mixed disulfide, yields oxidized glutathione (GSSG).

$$GSH + GSSR'' \longrightarrow GSSG + R''SH$$

Glutathione reductase is a flavoprotein that catalyzes the reaction

$$GSSG + NADPH + H^+ \rightleftharpoons 2GSH + NADP^+$$

In this manner, any cystine (R'SSR'' in the above reactions) which may form is reduced back to cysteine for use by the cell.

Tyrosine Tyrosine biosynthesis in mammals occurs by hydroxylation of phenylalanine, an essential amino acid. Much of the dietary requirement for phenylalanine is, in fact, due to the need for tyrosine. If the latter is fed, the dietary requirement for phenylalanine is reduced substantially. In this sense, tyrosine bears the same relationship to phenylalanine as cysteine does to methionine. In normal metabolism, the only known fate of phenylalanine, other than utilization for protein synthesis, is its conversion to tyrosine.

The *phenylalanine hydroxylase* (*phenylalanine-4-monooxygenase*) system in mammalian liver is a mixed function oxygenase (Chap. 16) that utilizes NADPH and the electron donor *tetrahydrobiopterin,* a reduced pteridine derivative, in the two-step reaction

L-Phenylalanine + tetrahydrobiopterin + $O_2 \xrightarrow{\text{phenylalanine hydroxylase}}$ L-tyrosine + dihydrobiopterin + H_2O　(1)

Dihydrobiopterin + NADPH + $H^+ \longrightarrow$ tetrahydrobiopterin + $NADP^+$　(2)

Reaction (1) is catalyzed by *phenylalanine hydroxylase* and reaction (2) by *dihydropteridine reductase*. The latter permits reutilization of the coenzyme, which functions to transfer reducing equivalents from NADPH, the ultimate electron donor, to the electron acceptor, one of the oxygen atoms of O_2.

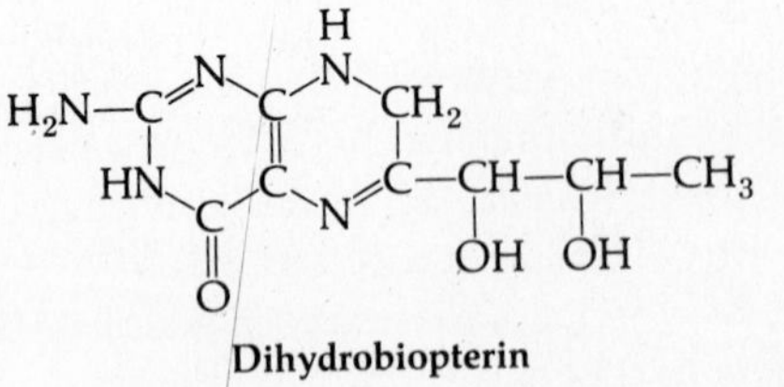

Dihydrobiopterin 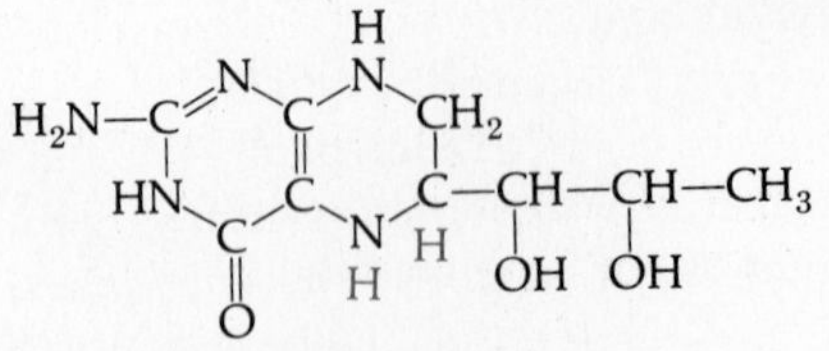　　　　　　　　　　Tetrahydrobiopterin

The activity of rat phenylalanine hydroxylase is modulated by a mechanism involving phosphorylation (activation) and dephosphorylation (inactivation) of the enzyme catalyzed, respectively, by cAMP-dependent protein kinase and a phosphoprotein phosphatase.

Hereditary lack of phenylalanine hydroxylase results in *phenylketonuria*. The recessive gene is carried by about 1 in every 80 individuals of European origin. In the absence of this enzyme, minor pathways of phenylalanine metabolism, little used in normal individuals, become prominent. Transamination from phenylalanine yields phenylpyruvate, of which as much as 1 to 2 g per day may be excreted. Severe mental retardation is evident early in children with phenylketonuria; restriction of their dietary intake of phenylalanine reduces the blood level of phenylalanine, abolishes excretion of phenylpyruvate, and prevents, in considerable degree, the mental retardation. The accumulation of phenylpyruvate leads also to formation and urinary excretion of phenyllactate, *o*-hydroxyphenyllactate, and phenylacetate, the last as phenylacetylglutamine.

Proline and Ornithine Glutamic acid serves as the precursor for synthesis of *proline* and *ornithine* (Chap. 22). The latter is not present in proteins but is the precursor of arginine (page 617). The 5-carbon chain of glutamic acid is converted to γ-glutamyl phosphate, which is reduced to glutamic semialdehyde. If the latter is transaminated, ornithine results; if, however, ring closure occurs, Δ¹-pyrroline-5-carboxylic acid is formed. Reduction of the latter by NADPH gives proline.

Serine The carbon chain of serine derives from 3-phosphoglycerate formed during glycolysis (Chap. 22). The latter can be oxidized to 3-phosphohydroxypyruvate in a NAD^+-requiring reaction catalyzed by *3-phosphoglycerate dehydrogenase*. Transamination from glutamate yields 3-phosphoserine; hydrolysis by *serine phosphatase* yields serine. 3-Phosphoglycerate dehydrogenase and serine phosphatase are inhibited by serine; this provides a means of regulating serine formation. Serine may also be formed from glycine (see below).

The above enzymes of serine biosynthesis, viz., the dehydrogenase, transaminase, and phosphatase, are under hormonal influence. Testosterone (Chap. M17) administration significantly increases the activities of these enzymes in liver, kidney, and the prostate. The transaminase of liver is elevated following administration of adrenal cortical steroids (Chap. M18). The diverse metabolic fates of serine (Chap. 25) and its role as a primary source of 1-carbon units (see below) add significance to the mechanisms that regulate its synthesis.

Glycine The major source of glycine is from serine in a reaction catalyzed by a pyridoxal phosphate-requiring *serine transhydroxymethylase,* resulting in the transfer of the β-carbon atom of serine to tetrahydrofolate (see below).

$$\text{L-Serine} + \text{tetrahydrofolate} \underset{}{\overset{Mn^{2+}}{\rightleftharpoons}} \text{glycine} + N^5, N^{10}\text{-methylenetetrahydrofolate}$$

Since this reaction is reversible, it also provides an additional pathway for the biosynthesis of serine.

Although serine is the *primary* source of 1-carbon units (see below), an additional source is the α-carbon atom of glycine in the reaction catalyzed by *glycine synthase* (Chap. 25). The significance of the contribution of the latter to the 1-carbon supply is not established.

METABOLISM OF ONE-CARBON COMPOUNDS

In the conversion of serine to glycine, tetrahydrofolic acid (THF) serves as the acceptor of the β-carbon of serine (see above). THF is the metabolically significant form of the vitamin folic acid (Chap. M22), which functions as a carrier of C-1 groups (see Fig. 23.2).

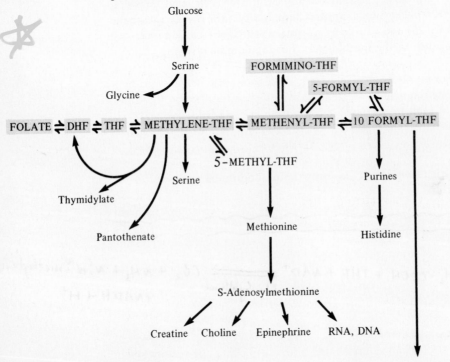

Tetrahydrofolic acid

2-Amino-4-hydroxy 6-methylpterin   *p*-Aminobenzoic acid   Glutamic acid

Pteroic acid

Pteryolglutamic acid (tetrahydrofolic acid)

Tetrahydrofolate is formed from dihydrofolate in a reaction catalyzed by the NADP-linked enzyme *dihydrofolate reductase*.

Figure 23.2 Tetrahydrofolate derivatives as Cl carriers in biosynthetic reactions. DHF = dihydrofolate; THF = tetrahydrofolate. [*Courtesy of Professor J. Rabinowitz.*]

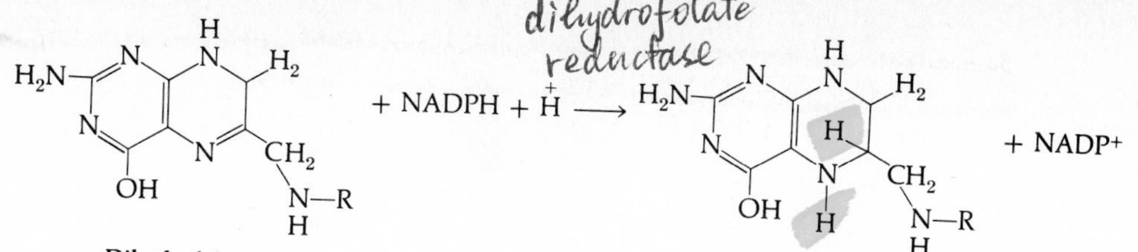

dihydrofolate reductase

+ NADPH + H⁺ ⟶   + NADP+

Dihydrofolate

Tetrahydrofolate

The activity of this enzyme is central in providing adequate quantities of tetrahydrofolate. The strong and specific competitive inhibition of this enzyme by structural analogs of its substrate, dihydrofolate, has provided a group of folic acid antagonists which have been useful clinically in restraining the rapid proliferation of specific types of malignant cells (Chap. 26).

As noted above, tetrahydrofolate can accept a 1-carbon unit from serine and thereby be converted to N^5,N^{10}-methylene THF. The N^5,N^{10}-methylene THF can be oxidized to N^5,N^{10}-methenyl THF by the NADP-linked *N^5,N^{10}-methylene THF dehydrogenase*.

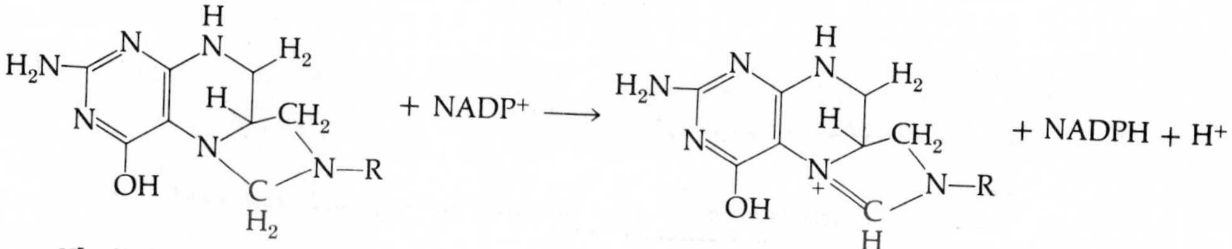

+ NADP+ ⟶   + NADPH + H⁺

N^5,N^{10}-Methylene THF

N^5,N^{10}-Methenyl THF

Hydrolysis of N^5,N^{10}-methenyl THF by the enzyme *N^5,N^{10}-methenyl THF cyclohydrolase* yields two other one-carbon folate derivatives, N^{10}-formyl THF, the folate derivative that is utilized for formyl transfer in many synthetic pathways, and N^5-formyl THF. The oxidation state of the carbon is the same in all three derivatives.

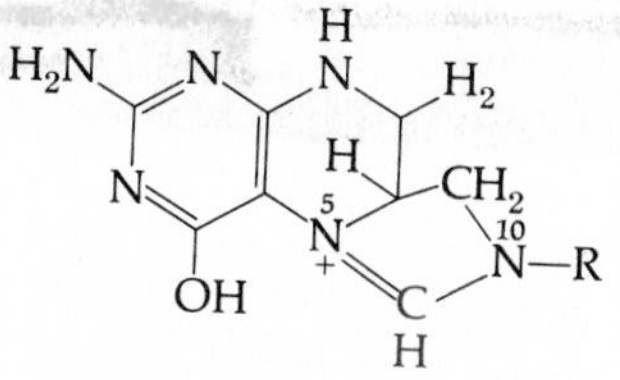

N^5,N^{10}-Methenyl THF

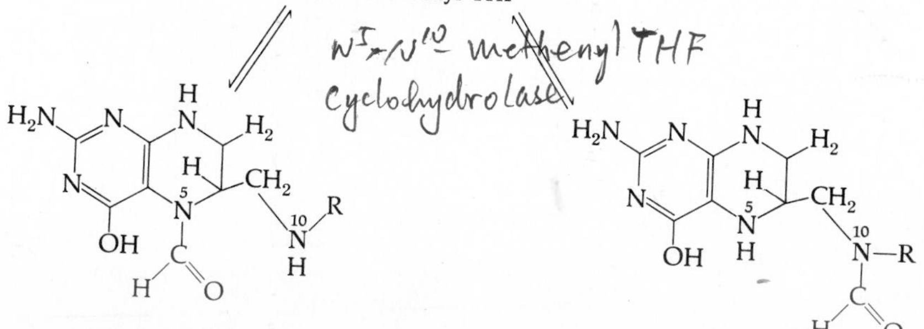

N^5-N^{10} methenyl THF cyclohydrolase

N^5-Formyl THF

N^{10}-Formyl THF

609

N^{10}-formyl THF can also be formed directly from formic acid in a reaction requiring ATP and catalyzed by N^{10}-formyl THF synthetase.

ATP   ADP + P_i

THF   + HCOO⁻   ⟶ N^{10}-Formyl THF

N^5,N^{10}-methylene THF can be reduced to the 5-methyl derivative, the most abundant form of folate in mammalian liver. Reduction is effected by the flavoprotein N^5,N^{10}-methylene THF reductase.

N^5,N^{10}-Methylene THF   + NADH + H⁺ ⇌ N^5-Methyl THF   + NAD⁺

In some reactions, a formyl group is transferred directly from a metabolite to the N^5 position of tetrahydrofolic acid. In these instances, a second reaction requiring ATP must occur to form the N^5,N^{10}-methenyl THF, which then is converted to the N^{10} derivative or may be reduced to the N^5,N^{10}-methylene compound. In a limited number of cases, THF may accept a formimidoyl group, —CH=NH, at the N^5 position in a reaction catalyzed by a *formimino transferase*. This is converted by a *cyclodeaminase* to the N^5,N^{10}-methenyl compound, which is utilized in the usual manner. Thus, tetrahydrofolate serves as a carrier for a 1-carbon unit at three levels of oxidation, viz., —CH$_3$, —CH$_2$OH(—CH$_2$—), and —CHO(=CH—), corresponding to methanol, formaldehyde, and formic acid, respectively, as well as the formimidoyl group, —CH=NH.

Three of the above enzymes, viz., N^{10}-formyl THF synthetase, N^5,N^{10}-methylene THF dehydrogenase, and N^5,N^{10}-methenyl THF cyclohydrolase, can be obtained as separate enzymes from bacterial sources. In contrast, these three enzymic activities in mammals (sheep liver) reside in a single multifunctional protein (MW = 213,000) composed of two apparently identical subunits. A similar multifunctional enzyme may occur in yeast, since mutation produces simultaneous loss or dramatic reduction of all three of the above enzymic activities.

A second complex has been isolated from pig liver in which a single type of subunit appears responsible for two other tetrahydrofolate enzymic activities, viz., the formimino transferase reaction and the cyclodeaminase reaction.

As noted previously (page 607), the primary source of 1-carbon units is serine in the reaction catalyzed by serine transhydroxymethylase (page 607). This enzyme is present in approximately equal amounts in the form of isoenzymes in the mitochondria and cytosol of rat liver. Although folate and tetra-hydrofolate are present both in mitochondria and in cytosol, apparently there is no folate transport across the mitochondrial inner membrane. In contrast, gly-cine and serine appear to be readily transported across this membrane. Since each of the two serine transhydroxymethyl transferases catalyzes the transfer of a 1-carbon unit, it has been proposed that they are involved in the transport of these units across the inner mitochondrial membrane, thus providing a *1-carbon shuttle*. This would be analogous to other shuttle phenomena across the inner mitochondrial membrane, e.g., the role of cytoplasmic and mitochondrial isoen-zymes of malate dehydrogenase in the transfer of reducing equivalents (Chaps. 12 and 15).

The various roles of the 1-carbon derivatives of tetrahydrofolate are illus-trated in Fig. 23.2. N^{10}-Formyl THF is the C-1 donor in purine (Chap. 26) and histidine (Chap. 22) biosynthesis. It is also a substrate in the synthesis of the formylmethionyl tRNA, required for initiation of polypeptide synthesis in pro-karyotes (Chap. 28). N^5,N^{10}-Methylene THF is required for the synthesis of thymidylate (Chap. 26), pantothenate (Chap. M22), and serine (page 607). The role of N^5-methyl THF in the *de novo* synthesis of the methyl groups of methio-nine is described in Chap. 22.

Genesis of Methyl Groups

Several enzymic systems have been described for the biosynthesis of methi-onine (Chap. 22). In mammals, methionine synthesis is catalyzed by a cobalamine-containing enzyme, N^5-*methyl THF:homocysteine methyltransferase*. The role of cobalamine and the mechanism of the transmethylation are discussed in Chap. M22.

$$HS-CH_2-CH_2-\overset{NH_3^+}{\underset{H}{C}}-COO^- \; + \; N^5\text{-Methyl THF}$$

Homocysteine N^5-Methyl THF

$$CH_3-S-CH_2-CH_2-\overset{NH_3^+}{\underset{H}{C}}-COO^- \; + \; THF$$

Methionine   THF

Transmethylation The transfer of the methyl group of methionine to appropri-ate acceptors is of general metabolic significance. The active form of methio-nine that functions in methylation is *S*-adenosylmethionine, a sulfonium form of methionine, with a free energy of hydrolysis comparable to that of the pyrophos-phate linkage of ATP. Synthesis of *S*-adenosylmethionine is catalyzed by *methi-onine adenosyl transferase*.

$$\text{L-Methionine} + \text{ATP} \longrightarrow$$

(structure of S-Adenosylmethionine, with COO⁻, HC—NH₃⁺, CH₂, CH₂, H₃C—S⁺—CH₂, adenine and ribose portions)

$$+ \text{PP}_i + \text{P}_i$$

S-Adenosylmethionine

The reaction is unusual in that methionine activation is coupled to cleavage of ATP at C-5′ with the formation of inorganic tripolyphosphate. The latter, which remains bound to the enzyme, is then cleaved to P_i and PP_i, the P_i being derived from the γ-phosphate of ATP.

In the presence of the appropriate specific *methyltransferases (methylferases)*, the methyl group of *S*-adenosylmethionine may be transferred, for example, to guanidinoacetic acid to form creatine (Chap. 24), to phosphatidylethanolamine to form phosphatidylcholine (Chap. 21), to norepinephrine to form epinephrine (Chap. M18), or to nicotinamide to form N^1-methylnicotinamide (Chap. M22). In each case the other product formed from *S*-adenosylmethionine is *S*-adenosyl-L-homocysteine, which is cleaved to adenosine and homocysteine.

In animals, resynthesis of methionine derives from the fact that choline, liberated from phosphatidylcholine, may be oxidized in two stages to betaine; the latter then transmethylates to homocysteine with formation of methionine.

$$(CH_3)_3\overset{+}{N}CH_2-CH_2OH \xrightarrow{\text{FAD} \quad \text{FADH}_2} (CH_3)_3\overset{+}{N}CH_2-CHO \xrightarrow{\text{NAD}^+ \quad \text{NADH} + \text{H}^+} (CH_3)_3\overset{+}{N}CH_2-COO^-$$

Choline   Betaine aldehyde   Betaine

$$(CH_3)_3\overset{+}{N}-CH_2-COO^- + HS-CH_2-CH_2-\overset{\overset{\displaystyle NH_3^+}{|}}{C}H-COO^- \longrightarrow$$

Betaine   Homocysteine

$$(CH_3)_2N-CH_2-COO^- + H_3C-S-CH_2-CH_2-\overset{\overset{\displaystyle NH_3^+}{|}}{C}H-COO^-$$

Dimethylglycine   Methionine

If homocysteine (as homocystine) is fed to rats on a methionine-deficient diet, the methylation of homocysteine provides sufficient methionine for normal growth. Although homocystine is not present in the normal diet, and generally homocysteine formed by transmethylation is not remethylated but is used for cysteine synthesis (page 605), it can arise from *S*-adenosylhomocysteine, a product of methionine metabolism (Chap. 25). Accordingly, in the animal economy, dietary methionine is the major source of methyl groups that, in turn, derive primarily from serine via N^5,N^{10}-methylenetetrahydrofolate (page 607).

In the human being, amino acids are deaminated at the rate of approximately 70 g per day. This occurs principally through the aminotransferases coupled to the action of *glutamate dehydrogenase*. The latter, which is present in mitochondria, catalyzes the reaction

$$\text{L-Glutamate} + NAD^+ \text{ (or } NADP^+) + H_2O \rightleftharpoons$$
$$\alpha\text{-ketoglutarate} + NADH \text{ (or } NADPH) + NH_4^+$$

The glutamate dehydrogenase of liver (MW = 336,000) is composed of six identical subunits of known amino acid sequence. It can utilize either pyridine nucleotide; it is an allosteric enzyme, stimulated by ADP and GDP and inhibited by ATP, GTP, and pyridoxal 5'-phosphate. Each subunit contains separate binding sites for substrate, coenzyme, and the purine nucleotide effectors that modify the activity of the enzyme.

Normally the blood level of NH_4^+ is low, ranging from 0.1 to 0.5 mg/dl, and, indeed, higher levels are toxic, particularly in nervous tissue. Therefore NH_4^+ must be very rapidly removed either by reutilization or excretion. Reutilization of ammonia is accomplished in two ways: formation of glutamate from α-ketoglutarate and amidation of glutamate to form glutamine. Excretion of ammonia also occurs in two ways. Aquatic lower vertebrates (fish and some amphibians) excrete ammonia directly; terrestrial vertebrates convert ammonia to urea (mammals) or uric acid (reptiles and birds).

Oxidative Deamination of Amino Acids Although the glutamate dehydrogenase reaction provides the major route for the deamination of amino acids, an additional minor route is by means of the L-*amino acid oxidase* present in liver and kidney. The general reaction is

$$R-CHNH_2-COO^- + FMN + H_2O \rightleftharpoons R-CO-COO^- + NH_3 + FMNH_2$$
$$FMNH_2 + O_2 \longrightarrow FMN + H_2O_2 \quad \xrightarrow{catalase} \quad H_2O + \tfrac{1}{2}O_2$$

The peroxide formed is decomposed by *catalase* (Chap. 16), present in peroxisomes. L-Amino acid oxidases from snake venom and certain microorganisms utilize FAD as the prosthetic group. L-Amino acid oxidase catalyzes oxidation of all naturally occurring L-amino acids except serine, threonine, and the dicarboxylic amino acids.

The role of the highly potent D-*amino acid oxidase* of liver and kidney is enigmatic. This cytoplasmic enzyme (Chap. 16) utilizes FAD as the prosthetic group and is present in the peroxisomes of kidney, brain, and liver. It catalyzes oxidation of the unnatural D antipode of a large number of amino acids; however, D-amino acids are unknown in mammalian metabolism. This enzyme does provide a means of oxidizing the D-amino acids of bacterial cell walls should any be absorbed from the intestine. D-Amino acid oxidase may play a role in oxidizing glycine since this amino acid is a good substrate.

$$^+H_3N-CH_2-COO^- + O_2 + H_2O \longrightarrow O=\overset{H}{\underset{}{C}}-COO^- + NH_4^+ + H_2O_2$$
Glycine   Glyoxylic acid

Monoamine and diamine oxidases catalyze the aerobic oxidation of a wide variety of important physiological amines, e.g., epinephrine, norepinephrine, and dopamine (Chap. M18), to the corresponding aldehydes and NH_3. Although the amount of each individual amine is small, the total of such activity may contribute significantly to the pool of ammonia.

Nonoxidative Deamination of Amino Acids A group of pyridoxal phosphate-dependent *dehydratases* catalyzes removal of the amino groups of serine, cysteine, homoserine, threonine, and, perhaps, homocysteine (Chap. 25); in each instance NH_4^+ and the corresponding α-keto acids are formed. Rather unusually, amino group removal from histidine by this mechanism results in formation of the corresponding acrylic acid (Chap. 25). Other examples of nonoxidative removal of amino groups of amino acids will be found in Chap. 25.

Formation of Glutamate, Glutamine, and Asparagine Formation of glutamate by reversal of the glutamate dehydrogenase reaction is theoretically possible, inasmuch as the NADPH is almost entirely in the reduced form at all times. However, the enzyme is present in mitochondria and, as already noted (page 613), its activity is strongly regulated by nucleotide effectors. This suggests that the primary role of the enzyme is to supply α-ketoglutarate to the tricarboxylic acid cycle and NADH for oxidative phosphorylation when the ADP level is high and the ATP level is low. The α-ketoglutarate can also function in transamination to remove amino groups from various amino acids when these are present in excess in order to supply energy or to furnish their carbon skeletons for gluconeogenesis (see Chap. 25 for further discussion). Under these circumstances, the glutamate dehydrogenase reaction becomes the major source of NH_4^+.

The principal fate of NH_4^+ is the synthesis of *glutamine*. This reaction is catalyzed in liver and brain by *glutamine synthetase* (MW = 350,000).

$$\text{Glutamic acid} + NH_3 + ATP \xrightarrow{\text{Mg}^{2+} \text{ or Mn}^{2+}} \text{glutamine} + ADP + P_i$$

The reaction proceeds in two steps. An initial reaction involves glutamate and ATP with formation of an activated enzyme-bound intermediate, ADP-γ-glutamyl phosphate. In the second step, this intermediate reacts with NH_4^+ to form glutamine, ADP, and P_i. Glutamine synthetase of mammalian tissues is an octamer of eight identical subunits and appears to be associated with the endoplasmic reticulum. The enzyme from animal cells differs significantly from that of microbial sources and apparently is not subject to the complex allosteric controls regulating the *Escherichia coli* enzyme (Chap. 22).

Glutamine is constantly leaving the liver and is delivered to all other organs, including the brain. Hydrolysis is effected by a widely distributed *glutaminase*.

$$\text{Glutamine} + H_2O \xrightarrow{\text{glutaminase}} \text{glutamic acid} + NH_3$$

Hence, in addition to the availability of glutamine for protein synthesis, its use as the equivalent of unprotonated NH_3 for delivery as a nontoxic form of NH_3 makes glutamine a temporary store of NH_3.

Asparagine can be formed in mammalian tissues from glutamine in a reaction catalyzed by the glutamine-dependent asparagine synthetase.

$$^-OOC-CH_2-\underset{\underset{NH_3^+}{|}}{CH}-COO^- + H_2N-\underset{\underset{\parallel}{O}}{C}-CH_2-CH_2-\underset{\underset{NH_3^+}{|}}{CH}-COO^- + ATP + H_2O \xrightarrow{Mg^{2+}}$$

Aspartic acid   **Glutamine**

$$H_2N-\underset{\underset{\parallel}{O}}{C}-CH_2-\underset{\underset{NH_3^+}{|}}{CH}-COO^- + {}^-OOC-CH_2-CH_2-\underset{\underset{NH_3^+}{|}}{CH}-COO^- + AMP + PP_i$$

Asparagine   **Glutamic acid**

Glutamine is poorly replaced by NH_3 in the above reaction. The enzyme is a *glutamine amidotransferase,* and its properties resemble those of glutamine-dependent *carbamoyl phosphate synthetase* (page 616) and *NAD synthetase* (Chap. 26).

UREA SYNTHESIS: THE UREA CYCLE

In the human being as much as 90 percent of urinary nitrogen is in the form of urea. The remainder is in the form of NH_4^+, creatinine, and uric acid (Table 23.2). Urea is synthesized from arginine almost exclusively in the liver, the only organ known to contain *arginase* (see Fig. 23.3). Cleavage of arginine

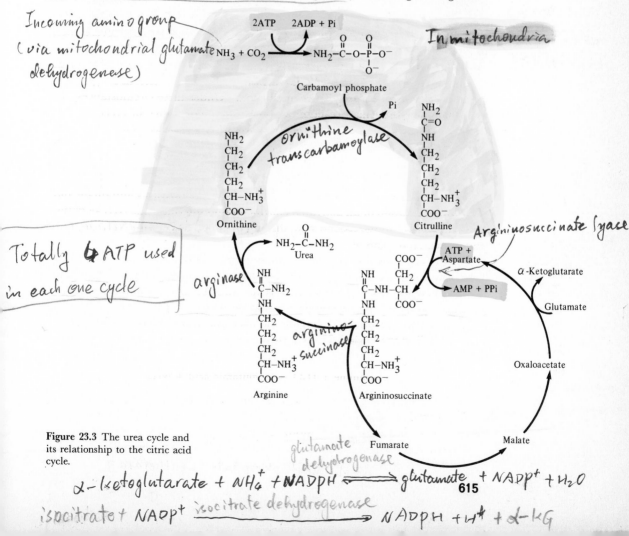

Incoming amino group (via mitochondrial glutamate dehydrogenase)

In mitochondria

Totally 4 ATP used in each one cycle

Figure 23.3 The urea cycle and its relationship to the citric acid cycle.

$$\alpha\text{-ketoglutarate} + NH_4^+ + NADPH \rightleftharpoons glutamate + NADP^+ + H_2O$$

glutamate dehydrogenase

isocitrate + NADP^+ \xrightarrow{isocitrate\ dehydrogenase} NADPH + H^+ + \alpha\text{-}KG

615

TABLE 23.2

Nitrogen Excretion in Humans	Product	g/24 h
	Urea	30
	NH_4^+	0.7
	Creatinine	1.0–1.8
	Uric acid	0.5–1.0

by arginase yields urea and ornithine. The urea is transported via the blood to the kidneys and excreted in the urine. Ornithine enters the mitochondria, where it is converted to citrulline by addition of a carbamoyl group derived from carbamoyl phosphate in a reaction catalyzed by *ornithine transcarbamoylase*. The citrulline is condensed with aspartate to form argininosuccinate, which is then cleaved in the cytosol by *argininosuccinate lyase* to form arginine and fumarate. A pathway from fumarate to aspartate shown in Fig. 23.3 links the urea cycle to the tricarboxylic acid cycle.

The overall reaction accomplished by the urea cycle is

$$NH_4^+ + HCO_3^- + 3ATP + 2H_2O + aspartate \longrightarrow urea + 2ADP + AMP + fumarate + 2P_i + PP_i$$

Note that incorporation of one equivalent of NH_4^+ into urea occurs *via* synthesis of carbamoyl phosphate. Fixation of the second equivalent is a consequence of incorporation of NH_4^+ into aspartate and requires the reductive amination by glutamate dehydrogenase of α-ketoglutarate to form glutamate, which then transaminates to oxaloacetate. The NADPH generated by the mitochondrial NADPH-specific isocitric dehydrogenase (Chap. 15) provides the requisite reducing equivalents for glutamate synthesis.

Synthesis of Carbamoyl Phosphate Two different enzymes catalyze carbamoyl phosphate synthesis; they have been designated *carbamoyl phosphate synthetase I* and *carbamoyl phosphate synthetase II*. The former is located in liver mitochondria and apparently is not present in other tissues. This enzyme utilizes ammonia exclusively as the nitrogen donor and requires *N*-acetyl-L-glutamic acid as a positive allosteric effector.

carbamoyl phosphate synthetase I

$$NH_4^+ + HCO_3^- + 2ATP \xrightarrow{Mg^{2+}} H_2N-CO-OPO_3^{2-} + 2ADP + P_i$$

The major function of this enzyme is to provide carbamoyl phosphate for the synthesis of arginine and hence of urea. Studies of the mechanism of the above reaction suggest that the first mole of ATP utilized activates bicarbonate by converting it to an enzyme-bound carboxyphosphate. The latter complex reacts with NH_4^+ to form an enzyme-bound carbamate, with liberation of P_i. Reaction with a second mole of ATP yields carbamoyl phosphate, ADP, and the free enzyme.

Carbamoyl phosphate synthetase II requires glutamine as the nitrogen donor, is independent of *N*-acetylglutamate, and catalyzes the reaction

$$L\text{-Glutamine} + HCO_3^- + 2ATP + H_2O \xrightarrow[K^+]{Mg^{2+}} H_2N-CO-OPO_3^{2-} + 2ADP + P_i + L\text{-glutamate}$$

carbamoyl phosphate synthetase II

This enzyme, which is inhibited by UTP, is present in the cytoplasm of cells of rapidly growing tissues, including tumors, and provides carbamoyl phosphate for pyrimidine biosynthesis (Chap. 26).

Synthesis of Arginine The first step in arginine synthesis utilizes carbamoyl phosphate and ornithine and is catalyzed by *ornithine transcarbamoylase* with formation of citrulline.

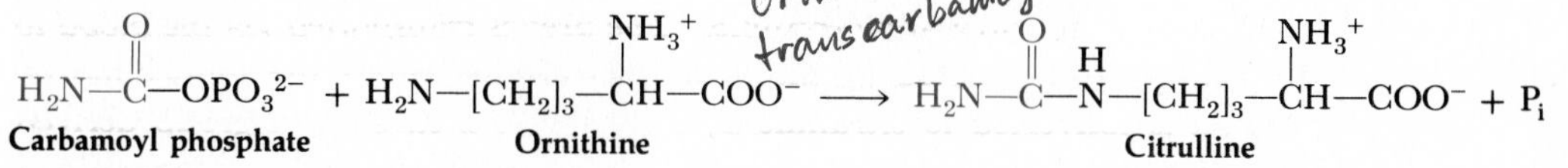

$$\underset{\text{Carbamoyl phosphate}}{\overset{\overset{\displaystyle O}{\|}}{H_2N-C-OPO_3^{2-}}} + \underset{\text{Ornithine}}{\overset{\overset{\displaystyle NH_3^+}{|}}{H_2N-[CH_2]_3-CH-COO^-}} \xrightarrow{\text{ornithine transcarbamoylase}} \underset{\text{Citrulline}}{\overset{\overset{\displaystyle O}{\|}\quad\overset{\displaystyle H}{|}\quad\quad\overset{\displaystyle NH_3^+}{|}}{H_2N-C-N-[CH_2]_3-CH-COO^-}} + P_i$$

The transcarbamoylase is tightly associated with carbamoyl phosphate synthetase I (above) as part of a single multifunctional protein (Chap. 26). These tight associations mitigate the lability of carbamoyl phosphate, which decomposes rapidly.

The next reaction, catalyzed by *argininosuccinate synthetase*, requires ATP and Mg^{2+} and involves a condensation of citrulline and aspartic acid.

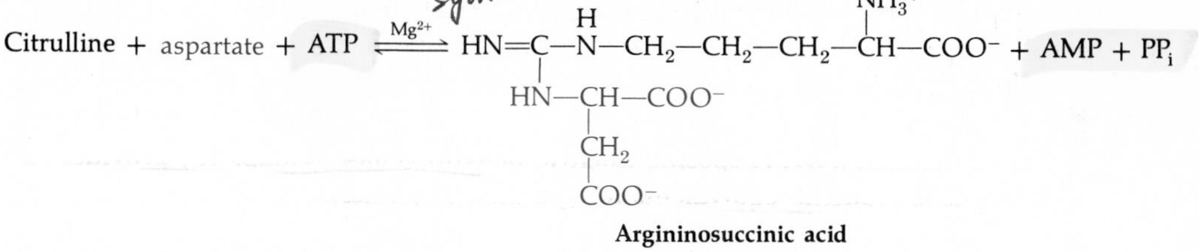

$$\text{Citrulline} + \text{aspartate} + \text{ATP} \underset{}{\overset{Mg^{2+}}{\rightleftharpoons}} \underset{\text{Argininosuccinic acid}}{\begin{array}{l} \overset{\displaystyle H}{\underset{\displaystyle |}{}} \qquad\qquad\qquad \overset{\displaystyle NH_3^+}{|} \\ HN=C-N-CH_2-CH_2-CH_2-CH-COO^- \\ \ \ |\\ HN-CH-COO^- \\ \qquad\ \ | \\ \qquad\ \ CH_2 \\ \qquad\ \ | \\ \qquad\ \ COO^- \end{array}} + \text{AMP} + PP_i$$

Argininosuccinate is cleaved by *argininosuccinase* to yield arginine and fumarate.

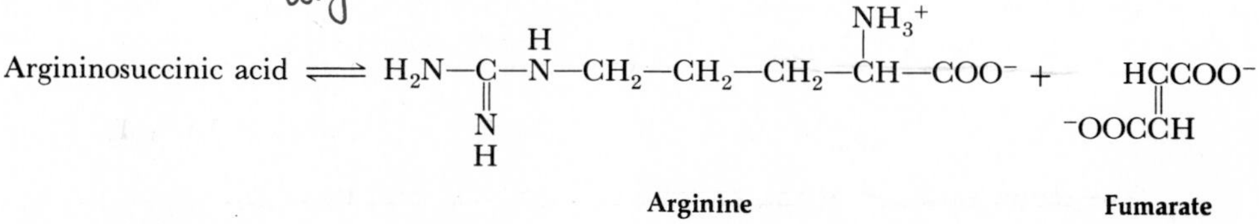

$$\text{Argininosuccinic acid} \rightleftharpoons \underset{\text{Arginine}}{\begin{array}{l} \overset{\displaystyle H}{|} \qquad\qquad\qquad \overset{\displaystyle NH_3^+}{|} \\ H_2N-C-N-CH_2-CH_2-CH_2-CH-COO^- \\ \quad\ \| \\ \quad\ N \\ \quad\ | \\ \quad\ H \end{array}} + \underset{\text{Fumarate}}{\begin{array}{l} HCCOO^- \\ \ \| \\ {}^-OOCCH \end{array}}$$

This reaction is neither hydrolytic nor oxidative and is readily reversible. The enzyme is present in kidney, liver, and brain. Exchange studies with citrulline labeled in the ureido group with ^{18}O showed a transfer of isotope to AMP, suggesting than an enzyme-bound adenylylcitrulline intermediate is involved in the synthetase reaction.

Hydrolysis of Arginine Arginine is hydrolyzed by *arginase* (MW = 120,000). As noted above, arginase is present uniquely in the liver. It is composed of four subunits, each bound to one atom of Mn^{2+}.

$$\underset{\text{Arginine}}{\overset{\displaystyle HN}{\underset{\displaystyle H}{\overset{\displaystyle \parallel}{H_2N-C-N}}-CH_2-CH_2-CH_2-\overset{\displaystyle NH_3^+}{\underset{\displaystyle |}{CH}}-COO^-}} \xrightarrow{+H_2O}$$

$$\underset{\text{Ornithine}}{H_2N-CH_2-CH_2-CH_2-\overset{\displaystyle NH_3^+}{\underset{\displaystyle |}{CH}}-COO^-} + \underset{\text{Urea}}{H_2N-\overset{\displaystyle O}{\overset{\displaystyle \parallel}{C}}-NH_2}$$

Defects in Urea Synthesis Inborn errors of each of the five enzymes of urea metabolism have been described. All result in *hyperammonemia* and *mental retardation*. The most severe cases involve deficiencies in carbamoyl phosphate synthetase and ornithine transcarbamoylase. In the latter, an additional finding is *orotic aciduria,* due to the diversion of carbamoyl phosphate to the pyrimidine biosynthetic pathway (Chap. 26). The others are *citrullinemia,* resulting from a defect in argininosuccinate synthetase, and *argininosuccinic aciduria,* which is a consequence of a deficient argininosuccinate lyase. These defects are characterized by high blood level and renal excretion of citrulline and argininosuccinate, respectively. *Hyperargininemia* is the result of a defect in arginase which results in elevated levels of arginine in blood and urine. Since urea excretion persists in patients with all these disorders, the enzymic defects may only be partial. Presumably a complete lack of any of these enzymes would be lethal.

Control of the Urea Cycle

The urea cycle is controlled by the positive effector *N*-acetylglutamate. In the absence of *N*-acetylglutamate, carbamoyl phosphate synthetase, which catalyzes the first and the rate-limiting step in the pathway, is inactive. After the ingestion of large amounts of amino acids (as after a protein meal), carbamoyl phosphate synthetase activity in liver increases in parallel with the increase in the level of *N*-acetylglutamate, which is derived from glutamate.

$$\text{Glutamate} + \text{acetyl CoA} \longrightarrow N\text{-acetylglutamate} + \text{CoA}$$

N-acetylglutamate is a precursor of ornithine and hence of arginine and its synthesis is inhibited by arginine (Chap. 22).

Under conditions of rapid growth and protein synthesis, arginine is rapidly consumed and its steady-state concentration is relatively low. However, the concentration of *N*-acetylglutamate is still sufficiently high to stimulate synthesis of carbamoyl phosphate. Since the concentration of ornithine is also low under these conditions, the carbamoyl phosphate enters the pyrimidine biosynthetic pathway (Chap. 26) required for nucleic acid synthesis rather than the urea cycle.

In summary, when glutamate, aspartate, and their respective amides are in excess, glutamate dehydrogenase functions in the direction of deamination (page 613). When these amino acids are in short supply and other amino acids are in excess, e.g., glycine, serine, cysteine, etc., glutamate dehydrogenase functions in the direction of amination (page 614). The synthesized glutamate can then be used to form aspartate by transamination for use in urea synthesis, to form acetylglutamate for stimulation of carbamoyl phosphate synthesis (see above) and to form glutamine. It is by these means that there is maintained a proper balance of all the amino acids, the continuous and essential operation of the tricarboxylic acid cycle, and the disposal of excess ammonia to form urea by the liver. The bloodstream serves as the channel by which the amino acids are distributed among the various organs, each of which has different demands which vary depending on its metabolic needs (see below and Chap. 25).

OVERALL ASPECTS OF AMINO ACID METABOLISM

The metabolism of amino acids by bacteria growing logarithmically in a rich nutrient broth is in marked contrast to that of mammals. Bacteria accept from the medium those amino acids needed for maximal protein synthesis. Little degradation of protein occurs, and the removal of amino acids from the medium is balanced by the sum of amino acids incorporated into protein plus those utilized for synthesis of the diverse nitrogeneous compounds found in such organisms. Little or no nitrogen, except for occasional hydrolytic enzymes, is returned to the medium, and amino acids not needed do not enter the cell.

No comparable situation occurs during mammalian life. At all stages there is a continuing entry and loss of nitrogen compounds. Amino acids and lesser amounts of other nitrogenous compounds enter the body and are processed, and the metabolic products are excreted. During infancy and childhood and convalescence from a debilitating disease, the intake of nitrogen exceeds the output (*positive nitrogen balance*); the opposite situation may prevail in senescence or, relatively briefly, during starvation or certain wasting diseases (*negative nitrogen balance*). However, even in infancy or senescence the daily departure from *nitrogen equilibrium* is usually only a small fraction of the total amount of nitrogen metabolized.

Whether derived from the diet or from endogenous sources, amino acids are channeled into the following pathways: (1) incorporation into protein; (2) incorporation into small peptides; (3) utilization of the nitrogen for synthesis of "nonessential" amino acids; (4) utilization for synthesis of nitrogenous compounds that are not amino acids; (5) removal of the α-amino group, with subsequent formation of urea and oxidation of the corresponding α-keto acid (Chap. 25). The last process constitutes the mammalian "solution" to the variability of protein intake. Although the product is different, the excretion of uric acid by birds and of NH_3 by sharks serves the same end—disposal of the nitrogen ingested in excess of requirements.

There is, however, a fixed requirement for a minimal daily intake of each of the essential amino acids. Young animals exhibit maximal growth and adults remain in nitrogen balance when fed a diet consisting only of ample amounts of the essential amino acids, vitamins, and a source of calories. The omission of any one essential amino acid, or of all of them, results promptly in a negative nitrogen balance of about 4 to 5 g of nitrogen per day. This response appears to relate to the behavior of those enzyme systems that catalyze the synthesis of assorted nitrogenous compounds such as purines, pyrimidines, nicotinic acid, creatine, etc. These activities are presumably each rate-controlled, but at their own committed steps; hence they withdraw nitrogen from the amino acid pool regardless of the amino acids derived from the diet. Should a single essential amino acid be missing from the diet (an experimental artifact), then, since a certain amount of protein (serum albumin, digestive enzymes, hormones) undergoes continuing hydrolysis, that amino acid will be withdrawn from the pool thus generated and utilized for a synthetic pathway other than protein, e.g., tryptophan to nicotinic acid (Chap. 24). The remaining amino acids will suffer their normal fate. Hence, negative nitrogen balance is of the same order whether only one or all of the essential amino acids is missing.

Maximal negative nitrogen balance is elicited by starvation. This is occasioned not by the need of nitrogenous compounds but by the fact that as body proteins are hydrolyzed, amino acids with carbon chains that can be oxidized with generation of ATP, i.e., those which can be degraded to yield citric acid cycle intermediates, are then thus utilized. The surplus nitrogen generated is then removed by urea synthesis. The maximal extent of this process is of the order of 30 g of nitrogen per day.

It is apparent from the increment in total body protein during growth and convalescence that dietary amino acids are utilized for net protein synthesis. However, the total synthesis of protein under these circumstances substantially exceeds that net increment. Even in the adult in nitrogen equilibrium, there is continual synthesis of digestive enzymes that may be lost in the feces, protein hormones, which are made in endocrine glands and degraded elsewhere in the body, formation and degradation of plasma proteins, and replacement of the entire protein complement of those cell populations of relatively short existence, e.g., erythrocytes, leukocytes, and cells of the gastrointestinal mucosa.

Proteins that are normally found extracellularly also turn over at varying rates (Chap. M1). In general, soluble proteins that leave their sites of synthesis are replaced by new molecules relatively more rapidly than proteins which remain as intracellular components. Thus, in human beings in nitrogen balance, the half-life of serum proteins is approximately 10 days. In contrast, however, the principal extracellular protein of connective tissue, collagen, exhibits almost no significant incorporation of labeled amino acids in adult animals, and labeled amino acids, incorporated when the animal is young and growing, do not disappear from this protein.

In human beings, the rate of protein synthesis is 0.6 to 1.0 g of nitrogen per kilogram per day. Thus, a 70-kg adult man synthesizes and degrades about 400 g of protein in a 24-h period. This can be compared with the fact that an average daily American diet provides about 100 g of amino acids and, at any instant, the total amount of free amino acids in the body fluids is of the order of 30 g. Thus, approximately 300 g of total body protein is recycled per day. The factors controlling protein synthesis in a general way are presented in Chap. 28. It is evident that there is no general control of protein synthesis but independent controls for the synthesis of each protein, so that synthesis occurs at a rate that will satisfy physiological requirements. Hence, for each protein, synthesis exactly balances its degradation. However, the factors that regulate the *rates* of protein synthesis and degradation are not fully understood. Apparently it is not the supply of amino acids that is primarily rate-limiting; net protein synthesis cannot be increased by augmenting the amino acid supply when the latter is adequate. The influence of various endocrine factors on protein synthesis will be discussed in Part M3. In contrast to the energy- and information-requiring apparatus for protein synthesis, intracellular protein degradation probably occurs primarily by simple hydrolysis catalyzed by proteolytic enzymes present in the lysosomes (Chap. 13) and by other cytoplasmic enzymes.

REFERENCES

Review Articles

Benkovic, S. J.: On the Mechanism of Action of Folate- and Biopterin-Requiring Enzymes, *Annu. Rev. Biochem.* **49:**227–251 (1980).

Blakley, R. L.: The Biochemistry of Folic Acid and Related Pteridines, in A. Neuberger and E. L. Tatum (eds.): *Frontiers of Biology,* vol. 13, Wiley, New York, 1969.

Wellner, D., and A. Meister: A Survey of Inborn Errors of Amino Acid Metabolism and Transport in Man, *Annu. Rev. Biochem.* **50:**911–968 (1981).

Meister, A., and S. S. Tate: Glutathione and Related γ-Glutamyl Compounds: Biosynthesis and Utilization, *Annu. Rev. Biochem.* **45:**559–604 (1976).

Ratner, S.: Enzymes of Arginine and Urea Synthesis, *Adv. Enzymol.* **39:**1–90 (1973).

Chapter 24

Amino Acid Metabolism III

Synthesis of amides and oligopeptides. Detoxification. Transamidination. Transmethylation. Porphyrin synthesis. Amino acid decarboxylation. Polyamines.

The principal fate of amino acids is their incorporation into proteins (Chaps. 28 and 29). Amino acids are also required for synthesis of compounds that are utilized for other roles in metabolism. Models of such functions have already been encountered in the synthesis of amino acids themselves, e.g., glutamic acid as a contributor of an amino group by transamination, glutamine as an ammonia donor, methionine as a source of methyl groups, and serine as a source of biologically active —CH_2OH and —CHO groups. The utilization of specific amino acids in the synthesis of purines and pyrimidines is described in Chap. 26. The present chapter provides other examples of the contributions of amino acids to the synthesis of diverse nitrogenous compounds.

SYNTHESIS OF OLIGOPEPTIDES

All living cells synthesize a variety of amino acid amides (Chap. 23) and oligopeptides. In most cases the energy required is directly derived from ATP, but by several different mechanisms.

Glutathione (γ-Glutamylcysteinylglycine)

This tripeptide is widely distributed in all living things at intracellular concentrations ranging from 0.1 to 10 mM. Its various functions include (1) destruction of endogenous peroxides and free radicals; (2) maintenance of SH groups in proteins; (3) a coenzyme for enzymes, e.g., glyoxylase; (4) detoxification of foreign compounds; (5) participant in disulfide interchange reactions; and (6) transport of amino acids across membranes (Chap. 23).

Any cross-references coded M refer to the companion text, *Principles of Biochemistry: Mammalian Biochemistry.*

Synthesis of glutathione requires two enzymes, both of which have been highly purified from several animal tissues. *γ-Glutamylcysteine synthetase* catalyzes the first step in the synthesis

$$\text{Glutamate} + \text{cysteine} + \text{ATP} \xrightleftharpoons{\text{Mg}^{2+}} \gamma\text{-glutamylcysteine} + \text{ADP} + \text{P}_i$$

and *glutathione synthetase* the second,

$$\gamma\text{-Glutamylcysteine} + \text{glycine} + \text{ATP} \xrightleftharpoons{\text{K}^+} \gamma\text{-glutamylcysteinylglycine} + \text{ADP} + \text{P}_i$$

The activity of each synthetase is stimulated by Mg^{2+} or K^+, as indicated. Both enzymes appear to form an acyl phosphate-enzyme intermediate analogous to that of glutamine synthetase. Glutathione inhibits γ-glutamylcysteine synthetase competitively, suggesting that glutathione inhibits its own synthesis by nonallosteric feedback inhibition. Each enzyme is not absolutely specific for its amino acid substrates; thus they catalyze synthesis of glutathione analogs, e.g., the ophthalmic acids in the eye (Chap. M9).

Methionine sulfoximine is a powerful inhibitor of both enzymes as well as glutamine synthetase and is thought to be a transition state analog (Chap. 11) of the γ-glutamyl phosphate-enzyme intermediate.

Sulfoximine

Methionine sulfoximine was first recognized as the active convulsant in flour treated with the bleaching agent NCl_3, which converted protein-bound methionine to the sulfoximine.

Carnosine and Anserine

Carnosine (β-alanylhistidine) and anserine (β-alanyl-N^1-methylhistidine) of brain and muscle (Chaps. M7 and M8) are synthesized by *carnosine synthetase* by a mechanism analogous to those for formation of acyl CoA (Chap. 20) and aminoacyl-tRNA (Chap. 28) derivatives.

$$\beta\text{-Alanine} + \text{ATP} + \text{Enz} \longrightarrow \text{Enz-}\beta\text{-alanyl -AMP} + \text{PP}_i$$
$$\text{Enz-}\beta\text{-alanyl -AMP} + \text{histidine} \longrightarrow \text{carnosine} + \text{AMP} + \text{Enz}$$

N^1-Methylhistidine replaces histidine in anserine synthesis. Unlike glutamine and glutathione synthesis, synthesis of these dipeptides is rendered essentially irreversible by the rapid hydrolysis of PP_i by pyrophosphatases.

Gramicidin S

Many oligopeptides containing up to 20 residues are synthesized by fungi and bacteria by mechanisms different from the usual ribosome-dependent processes of protein synthesis (Chap. 28). Synthesis of the antibiotic *gramicidin S* from *Bacillus brevis* illustrates one mechanism for synthesis of such oligopeptides. Gramicidin S is a cyclic peptide containing four different L-amino acids and D-phenylalanine.

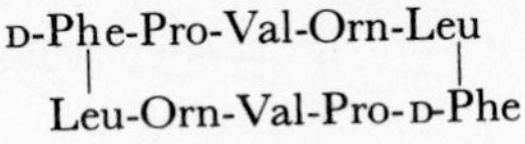

$$\text{D-Phe-Pro-Val-Orn-Leu}$$
$$\text{Leu-Orn-Val-Pro-D-Phe}$$

Gramicidin S

Two enzymes in the multienzyme complex *gramicidin synthetase* are required. Enzyme I reacts with each of the four L-amino acids and ATP to give an amino acyl thiol ester derived from a thiol group of the enzyme as follows:

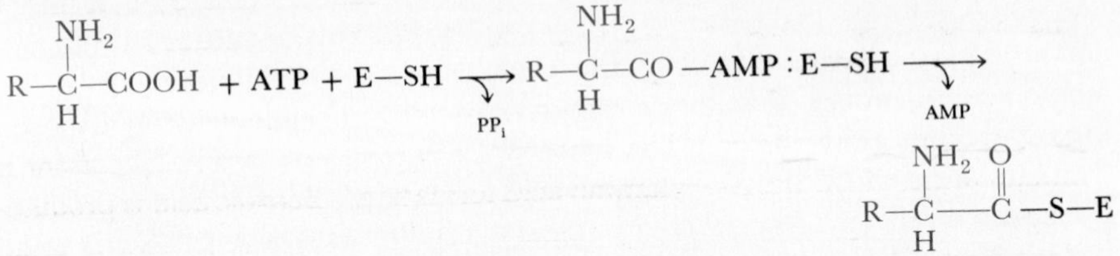

Enzyme II forms an analogous thiol ester but only with L- or D-phenylalanine. Initiation of synthesis commences by reaction of prolyl enzyme I reacting with phenylalanyl enzyme II as follows:

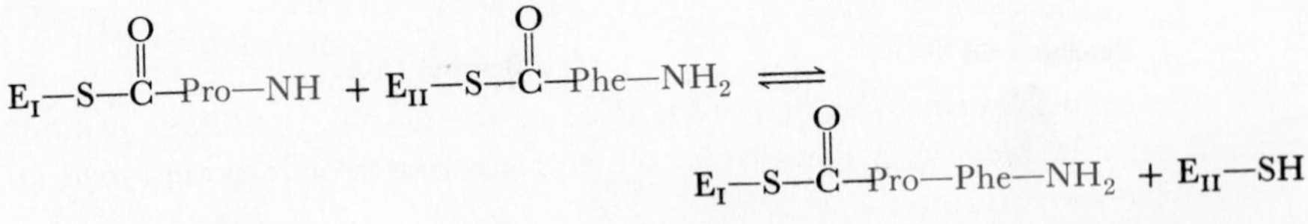

Elongation of the peptide occurs by subsequent reaction of the above product of initiation with E_I charged with the other requisite amino acids as follows:

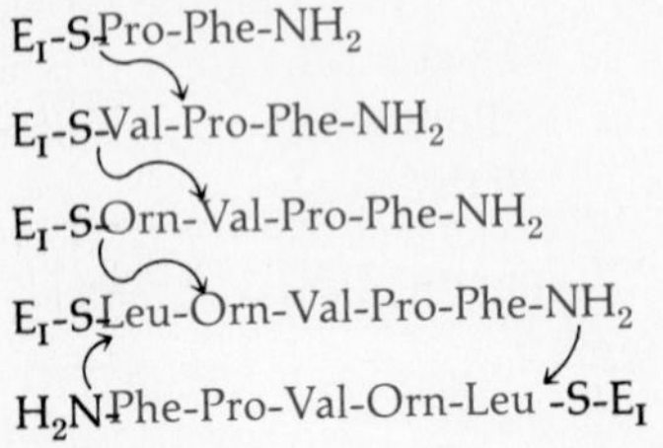

$$E_I\text{-S-Pro-Phe-NH}_2$$
$$E_I\text{-S-Val-Pro-Phe-NH}_2$$
$$E_I\text{-S-Orn-Val-Pro-Phe-NH}_2$$
$$E_I\text{-S-Leu-Orn-Val-Pro-Phe-NH}_2$$
$$H_2N\text{-Phe-Pro-Val-Orn-Leu-S-}E_I$$

"Head to tail" interaction of two pentapeptidyl thiol ester E_I derivatives then results in cyclization of the peptide with formation of gramicidin S. Omission of any one amino acid terminates the progress of elongation. Clearly, gramicidin synthetase contains the template for correct ordering of the peptide sequences. Analogous processes in higher organisms are unknown and synthesis of such oligopeptides as oxytocin and vasopressin (Chap. M20) does not proceed by this means.

Many compounds are encountered by living cells that cannot be transformed into substances readily excreted in the urine or the bile. Such substances include catabolites of various amino acids or other nitrogen-containing organic compounds. Other substances are so-called xenobiotics, foreign organic compounds not normally produced in metabolism, but introduced by ingestion, inhalation, or touch. Many of these compounds are potentially toxic to living things and must be "detoxified" by reactions that render them less reactive, or make them more water-soluble and thus more amenable to excretion. A major means for detoxification of such substances is by cytochrome P_{450}-catalyzed monoxygenation (Chap. 16). Other means include formation of *N*-acylamino acids and mercapturic acids.

A wide variety of aromatic carboxylic acids are detoxified in the liver by reaction with amino acids. The synthesis of *hippuric acid*, normally present in human urine, illustrates *N*-acylamino acid formation. Benzoic acid is produced as a minor end product of phenylalanine metabolism and is present in many plant foods. It reacts to form benzoyl CoA in a reaction catalyzed by liver *acyl CoA : transferase*.

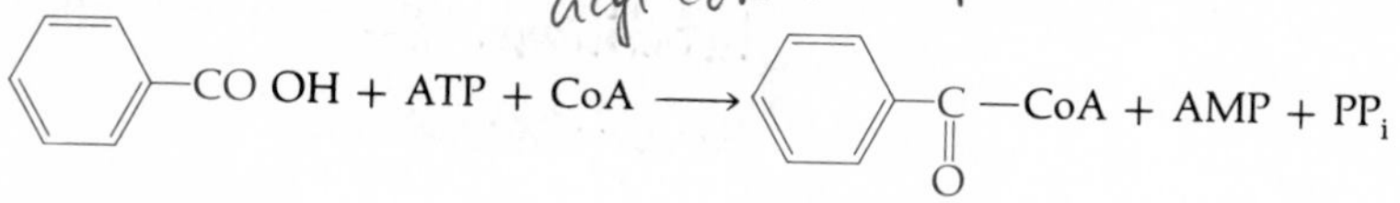

Benzoic acid + ATP + CoA $\longrightarrow$ Benzoyl CoA + AMP + PP_i

Benzoic acid — CO OH

Benzoyl CoA — C—CoA

Hippuric acid is formed in a second reaction catalyzed by liver mitochondrial *acyl CoA : glycine transferase*.

BenzoylCoA + glycine $\longrightarrow$ C—N—CH_2COOH + CoA

Hippuric acid

Injected benzoic acid markedly enhances hippuric acid synthesis. Phenylacetic acid arising in metabolism yields phenaceturic acid in a similar manner.

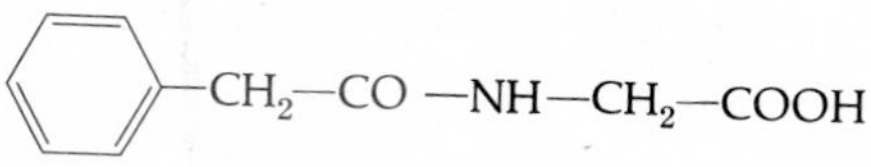

—CH_2—CO —NH—CH_2—COOH

Phenaceturic acid

Some aromatic acids may also be conjugated with glutamine by *acyl CoA : glutamine N-transferase*. Thus phenylacetic acid, also derived normally from phenylalanine metabolism, forms phenylacetyl CoA, which then is conjugated with glutamine to give phenylacetylglutamine.

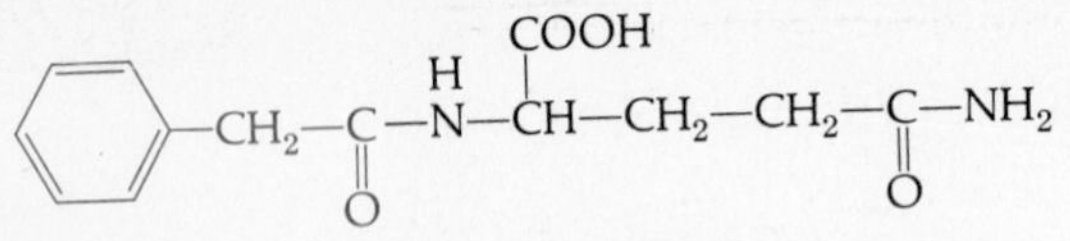

Phenylacetylglutamine

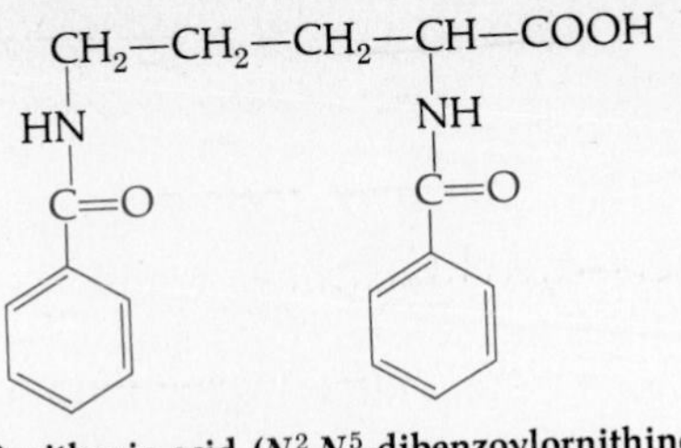

Ornithuric acid (N^2,N^5-dibenzoylornithine)

In some birds, the CoA derivatives of both benzoic and phenylacetic acids are conjugated with the α and δ amino groups of ornithine to form compounds such as ornithuric acid (shown above).

Another means for detoxification involves formation of *mercapturic acids*, the generic name for a vast array of substances that are formed from xenobiotics, including many carcinogens. Mercapturic acid synthesis commences by conjugation of xenobiotics with glutathione in reactions catalyzed by *glutathione-S-transferases*, a group of enzymes that were first recognized to bind many substances and were thus called *ligandins*. An enormous number of aliphatic, aromatic, and heterocyclic compounds bearing an electrophilic carbon atom are substrates for the transferases. Reaction with 1-chloro-2,4-dinitrobenzene serves as an example.

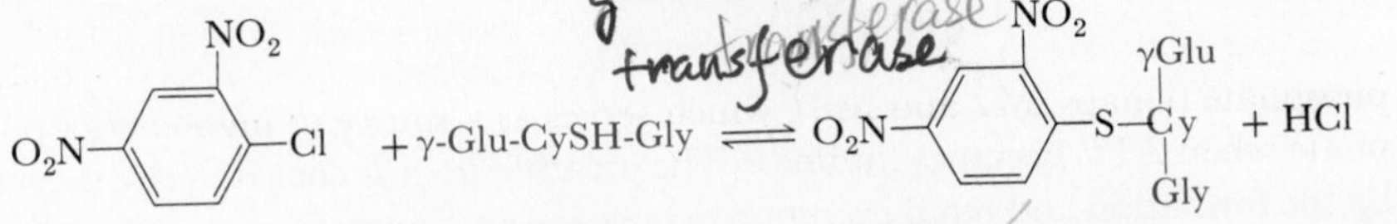

2,4-Dinitrochlorobenzene   **Glutathione**   **Glutathione adduct**

The γ-glutamyl and glycyl residues are then removed by γ-glutamyltranspeptidase and an unidentified peptidase, respectively, to give the S-cysteinyl thiol ester, which on acetylation by reaction with acetyl CoA yields the mercapturic acid.

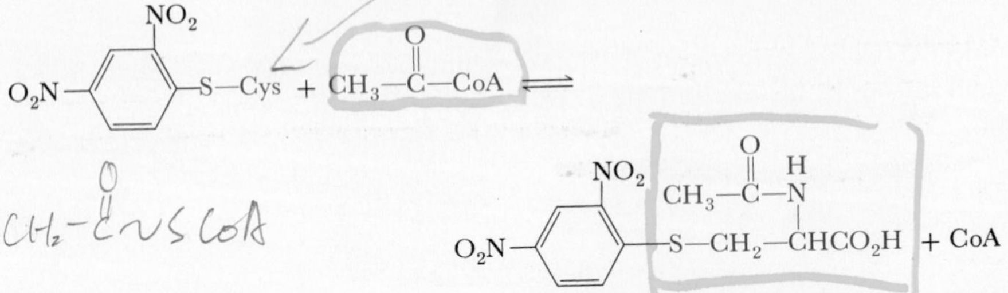

2,4-Dinitrophenyl mercapturic acid

Many electrophiles that are substrates for the transferase are epoxide carcinogens generated by the cytochrome P_{450}-linked, mixed-function oxidases (Chap. 16).

626

Glutathione-*S*-transferases are widely distributed and represent about 5 percent of the protein in the cytosol of many organs, including liver, intestine, and kidney, those organs most likely to come into direct contact with xenobiotics. Moreover, these enzymes are found in elasmobranchs, teleosts, crustaceans, planaria, and insects, as well as mammals. Six different electrophoretic forms of the transferase have been identified in human liver, but all appear to be basic proteins with two subunits of equal size (MW = 25,000). In addition to their catalytic activity, they also bind strongly to many metabolites that are poorly soluble in water, e.g., bilirubin derived from heme catabolism, which is produced at the rate of about 250 mg/day (Chap. M3) and when not excreted causes jaundice. It is believed that the transferases serve to store such compounds as bilirubin intracellularly and permit them to remain in solution before conjugation as glucuronides (Chap. M3) and excretion in the bile and urine. Many other hydrophobic substances including sterols, hydrocarbons, and drugs are also bound by the transferases. Other compounds that are highly reactive, e.g., alkylating agents, react directly with the transferases and form stable covalent linkages to them. This provides yet another mechanism for detoxification of xenobiotics by the transferases. Thus, the transferases are multifunctional in detoxification.

TRANSAMIDINATION—CREATINE BIOSYNTHESIS

Mammalian muscle and brain contain substantial quantities of creatine phosphate (Chaps. M7 and M8), which serves as a source of high-energy phosphate when ATP becomes limited. The creatine pool is continuously depleted by the formation and renal excretion of creatine and creatinine, an anhydride of creatine.

Creatine   Creatinine

Humans excrete about 2 g of creatinine plus creatine per day; the latter must be replaced in the diet or by biosynthesis. Two enzymes are involved in creatine biosynthesis. Mitochondrial *arginine:glycine amidinotransferase* catalyzes the following transamidination reaction.

Arginine   Glycine

Ornithine   Guanidinoacetic acid

The second reaction is catalyzed by cytosolic *S-adenosylmethionine : guanidinoacetate-N-methyltransferase* as follows:

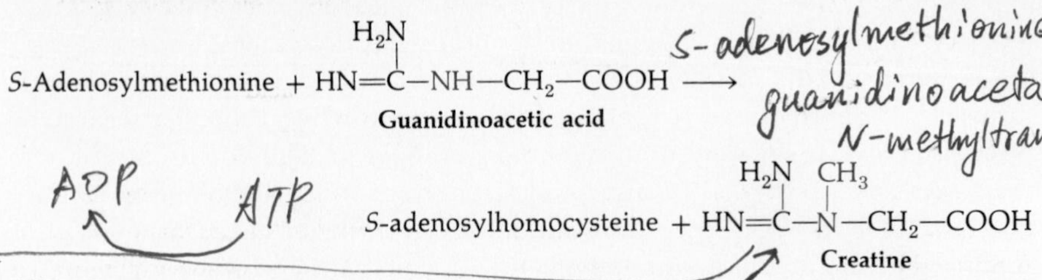

$$\text{S-Adenosylmethionine} + \underset{\text{Guanidinoacetic acid}}{HN=\overset{\overset{\displaystyle H_2N}{|}}{C}-NH-CH_2-COOH} \longrightarrow$$

$$\text{S-adenosylhomocysteine} + \underset{\text{Creatine}}{HN=\overset{\overset{\displaystyle H_2N \quad CH_3}{|}}{C}-N-CH_2-COOH}$$

[handwritten annotations: S-adenosylmethionine: guanidinoacetate-N-methyltransferase; ADP; ATP]

[handwritten left margin structure with label "phosphocreatine" and arrow to "creatinine"]

The amidinotransferase forms an intermediate on reaction with arginine that contains "active urea," in which the amidine group to be transferred to glycine is transiently attached to a thiol group of the enzyme. In humans, both enzymes are found in the liver and the pancreas. In general, tissues that synthesize creatine deliver it to the blood, which transports it to various organs, e.g., muscle and brain, where it is converted to creatine phosphate. This separation of functions permits separate regulation of synthesis and utilization.

Creatine synthesis is subject to negative feedback regulation by repression in mammals. Ingestion of creatine by rats markedly suppresses the level of renal transamidinase activity. That this is a physiological effect is evidenced by the fact that endogenously synthesized creatine is as effective a metabolite repressor as exogenous creatine. The importance of this control mechanism is evident if the controlled step is bypassed by injecting guanidinoacetic acid. Excessive creatine synthesis then results in a fatty liver and impaired growth, both of which can be prevented by the administration of additional methionine.

TRANSMETHYLATION   *[handwritten: Ser ⟶ THF ⟶ methionine ⟶ SAM]*

Transfer of methyl groups is catalyzed by *methylferases* or *methyltransferases*. In most instances, the methylating agent is *S*-adenosylmethionine, derived from methionine and ATP, as described previously (Chap. 23). As indicated in Table 24.1, methyl groups can be transferred to sulfur, nitrogen, carbon, or oxygen. Although *S*-adenosylmethionine serves as a general methyl-group donor to numerous acceptors, formation of the methyl group of methionine is limited to a few reactions. The major source of the methyl group of methionine is N^5, N^{10}-methylenetetrahydrofolic acid (Chap. 23), which in turn derives from a limited number of metabolites, notably serine (Chap. 23), that can donate the 1-carbon functional group to tetrahydrofolate.

Protein Methylation

Methyl groups from *S*-adenosylmethionine are transferred to the side chains of three different amino acid residues in a variety of proteins. Three different classes of *protein methylases* are widely distributed in animal tissues. *Methylase I* catalyzes formation of methylarginine, which is found in histones, myosin, a nuclear protein, and myelin. There are three forms of methylarginine, each

TABLE 24.1

Substrate	Product	Methylating agent	Some Examples of Transmethylation
Homocysteine	Methionine	N^5-Methyltetrahydrofolic acid Methyl-B_{12} Betaine (from choline)	
N-Acetyl-5-hydroxytryptamine	Melatonin	S-Adenosylmethionine	
Carnosine	Anserine	S-Adenosylmethionine	
β-Sitosterol*	Ergosterol	S-Adenosylmethionine	
†	Vitamin B_{12}	S-Adenosylmethionine	
Epinephrine	Metanephrine	S-Adenosylmethionine	
Guanidinoacetic acid	Creatine	S-Adenosylmethionine	
Histamine	N-Methylhistamine	S-Adenosylmethionine	
Nicotinic acid*	Trigonelline	S-Adenosylmethionine	
Nicotinamide	N^1-Methylnicotinamide	S-Adenosylmethionine	
Norepinephrine	Epinephrine	S-Adenosylmethionine	
Phosphatidylethanolamine	Phosphatidylcholine	S-Adenosylmethionine	

*Only in plants.

† The substrate that is methylated has not been identified; experimental data were obtained with microorganisms grown on a medium containing labeled methylating agent, followed by degradation of product (vitamin B_{12}) for location of labeled methyl groups (Chap. M22).

containing methylated guanidinium groups with the following structures, where R = [CH$_2$]$_3$—CH(NH$_2$)—COOH.

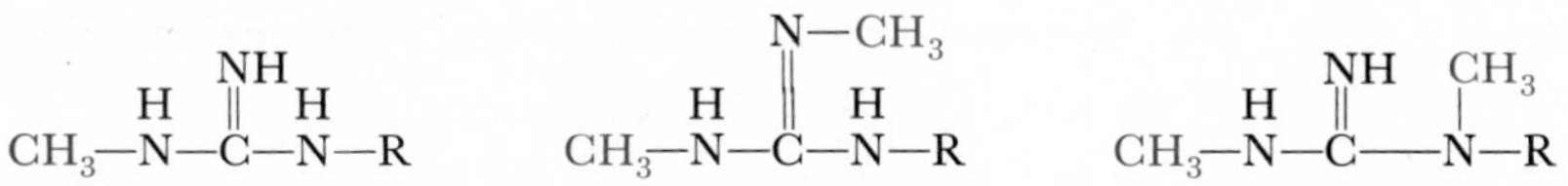

NH
H ‖ H
CH$_3$—N—C—N—R
N^G-monomethyl

N—CH$_3$
H ‖ H
CH$_3$—N—C—N—R
N^G,N'^G-dimethyl

NH CH$_3$
H ‖ |
CH$_3$—N—C—N—R
$N^G,N^δ$-dimethyl

Methylase II catalyzes formation of methyl esters of free carboxyl groups in proteins, whereas *methylase III* catalyzes formation of ε-N-mono-, di-, and trimethyllysines. Methyl esters are not known to be widely distributed in eukaryotes, although they occur in proteins involved in chemotaxis in *Salmonella*. Methyllysines are widely distributed, occurring in such proteins as myosin, actin, histones, ribosomal proteins, opsin, and yeast and plant cytochrome c. Specific methylases may be required for synthesis of the mono-, di-, and trimethyllysines as well as the different methylarginine derivatives. Each methylase acts on protein substrates and is inactive on free amino acids. A methylase III that is specific for methylation of the single ε-N-trimethyllysine residue in yeast cytochrome c has been isolated from *Saccharomyces cerevisiae*. A fourth transferase activity that forms N^3-methylhistidine in myosin has also been identified in crude muscle preparations.

The biological significance of methylation of proteins is unclear. An FAD-dependent ε-*alkyllysinase* (ε-alkyl-L-lysine oxygen oxidoreductase) from kidney acts on methylated histones giving rise to formaldehyde and demethylated histones. It has been suggested that methylation and demethylation of proteins may be associated with regulation of enzyme activities, as documented for protein phosphorylation, since methylating and demethylating activities change in a reciprocal manner during growth and in tumor cells. Methylation of nucleic acids is considered in Chap. 27.

Formation of Carnitine

Most mammals obtain the carnitine that is required for transfer of acetyl and fatty acyl groups from cytosol to mitochondria and vice versa (Chap. 20) from the diet, but they also synthesize it endogenously. The carbon and nitrogen atoms of carnitine are derived from ϵ-N-trimethyllysine of methylated proteins, synthesized as discussed above. Free lysine can give rise to trimethyllysine in *Neurospora* but not in rats or humans. Trimethyllysine administered to mammals is largely excreted in the urine (>97 percent) and is poorly taken up by various organs. Protein-bound trimethyllysine is freed by hydrolysis of protein and, in the tissue where it is formed, is converted to carnitine as shown in Fig. 24.1. Trimethyllysine is first hydroxylated by a *mitochondrial dioxygenase*, and the β-hydroxyl derivative so formed is then cleaved to give γ-butyrobetaine aldehyde. The aldehyde is oxidized to γ-butyrobetaine and the latter hydroxylated by a cytosolic dioxygenase to form carnitine. Ascorbate is the most effective reducing agent for the dioxygenases, whereas α-ketoglutarate and Fe^{2+} are highly specific for these enzymes. Pyridoxal phosphate is a cofactor for the aldolase. Each of the enzymes involved has been identified in several human tissues, including liver, kidney, muscle, heart, and brain.

Fate of Methyl Groups

Transmethylation from S-adenosylmethionine, a sulfonium compound, to most acceptors proceeds with a favorable change in free energy. Most compounds that serve as methyl donors in living systems also appear to be com-

Figure 24.1 The biosynthesis of carnitine in humans. The enzymes involved are (1) trimethyllysine β-hydroxylase, (2) β-hydroxyltrimethyllysine aldolase, (3) γ-trimethylaminobutyraldehyde dehydrogenase, and (4) γ-butyrobetaine hydroxylase.

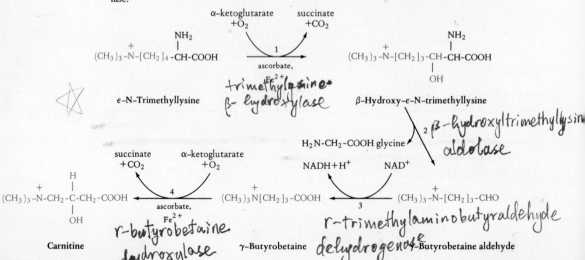

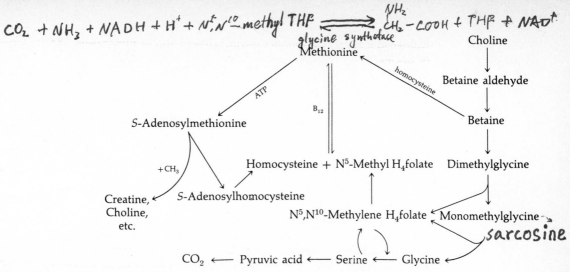

$$CO_2 + NH_3 + NADH + H^+ + N^5,N^{10}\text{-methyl THF} \rightleftharpoons \underset{\text{glycine synthetase}}{\overset{NH_2}{\underset{}{CH_2-COOH}}} + THF + NAD^+$$

Figure 24.2 Aspects of methyl-group metabolism. H_4 folate = tetrahydrofolic acid.

pounds of this type. Thus, the enzymic methylation of homocysteine by betaine, with methyl-group migration from a quaternary nitrogen, requires no additional energy source. In contrast, dimethylglycine and sarcosine cannot serve as methylating agents; their methyl groups are removed by oxidation and transferred to tetrahydrofolic acid (Chap. 23).

As noted earlier, choline can be oxidized to betaine (Chap. 23), and one methyl group can then be transferred to homocysteine when homocysteine is available. Otherwise, betaine may lose all its methyl groups, sequentially, by oxidation. Each is oxidized to the level of HCHO, transferred to tetrahydrofolic acid, and thus returned to the C_1 pool in the sequence choline → betaine aldehyde → betaine → dimethylglycine → monomethylglycine (sarcosine) → glycine.

When methionine or sarcosine labeled with ^{14}C in their methyl groups is administered, the isotope is found in the expected positions in the various compounds already cited. However, a significant fraction of administered isotope also appears as $^{14}CO_2$, and this is the ultimate fate of this carbon, except for that lost in the urine in small amounts as creatinine, N^1-methylnicotinamide, etc. The most likely pathway by which C_1 compounds can be oxidized to CO_2 is by recombination of N^5,N^{10}-methylene-tetrahydrofolic acid with glycine to form serine (page 607), which is converted to pyruvate by *serine dehydratase* (page 650). Oxidation of the pyruvate via the citric acid cycle would result in formation of CO_2 from the carbon that had once been in the C_1 pool. Some of the relationships considered above are shown in Fig. 24.2; reference should also be made to Fig. 23.2.

THE BIOSYNTHESIS OF TETRAPYRROLES: HEME SYNTHESIS

The four major classes of tetrapyrroles in living things, the hemes (Chap. M4), chlorophylls (Chap. 19), bilins (Chap. 19), and corrins (Chap. M22) share

631

a common biosynthetic pathway (Fig. 24.3), each being derived from δ-amino-levulinic acid. Many steps in the synthesis of chlorophylls, corrins, and bilins remain to be established; thus only heme synthesis, which is better understood, is considered here.

The enzymes for heme biosynthesis have been found in all animal tissues examined except the mature enucleated mammalian erythrocyte, which, lacking mitochondria, is devoid of some enzymes in the pathway. Seven steps are now recognized for *de novo* synthesis of heme, as shown in Fig. 24.4. The first as well as the last three steps in the path occur in mitochondria, whereas the other steps are catalyzed by cytosolic enzymes.

Synthesis of Protoporphyrin

The immediate precursor of the porphyrin ring is δ-aminolevulinic acid (ALA), which is formed from glycine and succinyl CoA in the reaction catalyzed by mitochondrial *δ-aminolevulinic acid synthetase*. The rat liver enzyme (MW = 110,000) is composed of two identical subunits and requires pyridoxal phosphate and Mg^{2+} as cofactors. As shown in the following reactions, where E—CHO is ALA synthetase and with CHO representing the aldehyde group of pyridoxal, glycine reacts first with the aldehyde group of the enzyme-bound pyridoxal phosphate, and the resulting adduct reacts with succinyl CoA, to give enzyme-bound α-amino-β-ketoadipic acid, which then decarboxylates to give ALA.

$$E—CHO + H_2N—CH_2—CO_2H \xrightarrow[Mg^{2+}]{-H_2O} E—CH{=}N—CH_2—CO_2H$$

Glycine

$$E—CH{=}N—CH_2—CO_2H \xrightarrow{-H^+} E—\overset{\ominus}{C}{=}N—CH_2—CO_2H$$

$$E—\overset{\ominus}{C}{=}N—CH_2—CO_2H + HO_2C—CH_2—CH_2—\overset{O}{\overset{\|}{C}}—CoA \longrightarrow$$

Succinyl CoA

$$E—\overset{H}{C}{=}N—\overset{CO_2H}{\underset{\|}{\underset{O}{C}}}{-}[CH_2]_2—CO_2H + CoA$$

$$\overset{H}{E}—C{=}N—CH—\overset{CO_2H}{\underset{\|}{\underset{O}{C}}}{-}[CH_2]_2—CO_2H \xrightarrow[+H_2O]{-CO_2}$$

$$E—CHO + H_2N—CH_2—\overset{O}{\overset{\|}{C}}—CH_2—CH_2—CO_2H$$

δ-Aminolevulinic acid

As discussed below (Regulation of Heme Synthesis), ALA synthetase is a key enzyme in regulation of heme biosynthesis. Synthesis of ALA is the only step in the path that utilizes energy; therefore it requires a functional tricarboxylic acid cycle to provide succinyl CoA. All other steps are thermodynamically favored and are essentially irreversible.

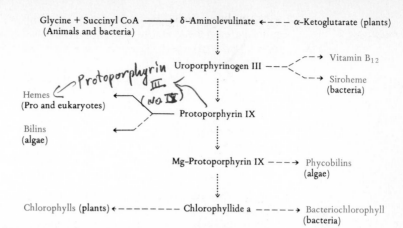

Glycine + Succinyl CoA ——→ δ–Aminolevulinate ←--- α–Ketoglutarate (plants)
(Animals and bacteria)

Protoporphyrin III (No IX) [handwritten]

Uroporphyrinogen III ---→ Vitamin B$_{12}$

Siroheme
(bacteria)

Hemes
(Pro and eukaryotes)

Protoporphyrin IX

Bilins
(algae)

Mg–Protoporphyrin IX ---→ Phycobilins
(algae)

Chlorophylls (plants) ←-------- Chlorophyllide a ----→ Bacteriochlorophyll
(bacteria)

Figure 24.3 Overall tetrapyrrole biosynthetic pathways. The multibranched pathways lead to synthesis of the hemes, corrins (vitamin B$_{12}$), bilins, and chlorophylls, as indicated. Dashed lines represent incompletely understood paths; dotted lines represent paths with several intermediates.

Figure 24.4 Pathway of heme biosynthesis. δ-Aminolevulinic acid (ALA) is synthesized in the mitochondria by the ALA synthetase, the first and the rate-limiting enzyme in the pathway. ALA leaves the mitochondria to be converted to porphobilinogen (PBG); four molecules of PBG then condense to form a porphyrin ring. The next three steps involve reactions that alter the substituents on the pyrrole rings to give protoporphyrin IX, which binds Fe^{2+} in the last, ferrocheletase-catalyzed reaction, to give heme. A = —CH$_2$—CO$_2$H; P = —CH$_2$—CH$_2$—CO$_2$H; V = —CH$_2$=CH$_2$; M = —CH$_3$.

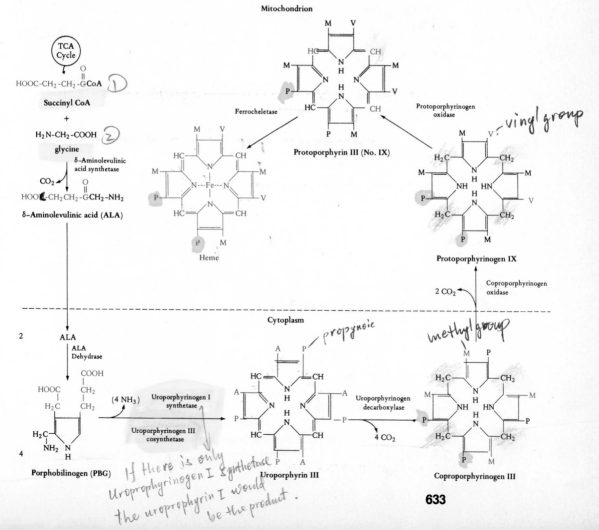

633

ALA can also be synthesized in rat mitochondria by a transamination reaction catalyzed by L-*alanine*: γ,δ-*dioxovalerate aminotransferase,* as follows:

$$CH_3-\underset{\underset{NH_2}{|}}{CH}-CO_2H \; + \; HC-C-[CH_2]_2-CO_2^- \longrightarrow$$

Alanine γ,δ-**Dioxovalerate**

$$H_2N-CH_2-\overset{O}{\overset{||}{C}}-[CH_2]_2-CO_2^- \; + \; CH_3-\overset{O}{\overset{||}{C}}-CO_2^-$$

δ-**Aminolevulinate**   **Pyruvate**

The enzyme requires pyridoxal phosphate and has a high turnover number. Since a source of the γ,δ-dioxovalerate is unknown, the significance of the reaction for heme synthesis is unclear. Plants do not have ALA synthetase and appear to synthesize ALA by a transamination reaction, but it is not known whether γ,δ-dioxovalerate is involved (Fig. 24.3).

Synthesis of Porphobilinogen

The second step in heme synthesis (Fig. 24.3), formation of porphobilinogen from ALA, is catalyzed by cytosolic δ-*aminolevulinic acid dehydrase* (MW = 280,000). After leaving the mitochondrion, two molecules of ALA condense to form one molecule of porphobilinogen with the elimination of two water molecules.

The labeled (•) carbon atoms originate from the α-carbon atoms of glycine.

Synthesis of the Porphyrin Ring and Heme

Porphobilinogen is converted in the cytosol to uroporphyrinogen III and coproporphyrinogen III (Fig. 24.4). Four molecules of porphobilinogen combine to form uroporphyrinogen III in a reaction catalyzed by two enzymes, *uroporphyrinogen I synthetase* and *uroporphyrinogen III cosynthetase*. The I synthetase when acting alone in vitro catalyzes formation of urophorphyrinogen I, a compound of no known physiological function, but in the presence of the III-cosynthetase, synthesis of the intermediate, uroporphyrinogen III, occurs

Uroporphyrinogen III

Uroporphyrinogen I

The two porphyrinogens differ only in the positions of the substituents on pyrrole ring D. Other porphyrinogen isomers are also synthesized by in vitro systems, but they too are of unknown biological significance. It is unlikely that such isomers are produced in significant amounts in vivo. How the synthetase and cosynthetase act to ensure synthesis of the correct intermediate is unknown.

Uroporphyrinogen III is converted to coproporphyrinogen III in a decarboxylation reaction catalyzed by cytosolic *uroporphyrinogen decarboxylase*. In this reaction the four acetate groups on each of the pyrrole rings are decarboxylated, giving rise to the four ring methyl groups. Coproporphyrinogen III then enters the mitochondria, where it is acted upon by *coproporphyrinogen oxidase* to form protoporphyrinogen IX. In this step (Fig. 24.4) propionic acid groups on both the B and the C rings are converted to vinyl groups, with the production of CO_2. The oxidase has an absolute requirement for dioxygen, but has no known cofactors. Protoporphyrinogen IX is converted by mitochondrial *protoporphyrinogen oxidase* to protoporphyrin IX with formation of the four methene bridges that link the four pyrrole rings. The nature of these oxidases has not been elucidated.

The final step (Fig. 24.4) that leads to formation of heme is catalyzed by *ferrochelatase*, an enzyme on the inner mitochondrial membrane that acts in the presence of reducing agents such as ascorbate or glutathione to insert Fe^{2+} into protoporphyrin IX. Although Fe^{2+} can combine slowly with protoporphyrin IX nonenzymically to give heme, the process is catalyzed in vivo.

Present evidence suggests that the pathway for heme synthesis is the same whether the heme is destined for combination with hemoglobin, myoglobin, catalase, a cytochrome, or any one of the many other heme proteins.

Regulation of Heme Synthesis

The rate of heme synthesis is controlled primarily by ALA synthetase. Two mechanisms serve to regulate the rate of heme synthesis and each involves a process that influences the concentration of mitochondrial ALA synthetase (Fig. 24.5). The biological half-life of the synthetase is very short: 60 to 70 min in rat liver. Like many mitochondrial proteins, the synthetase is encoded by nuclear genes, synthesized on cytoplasmic ribosomes, and then translocated from the

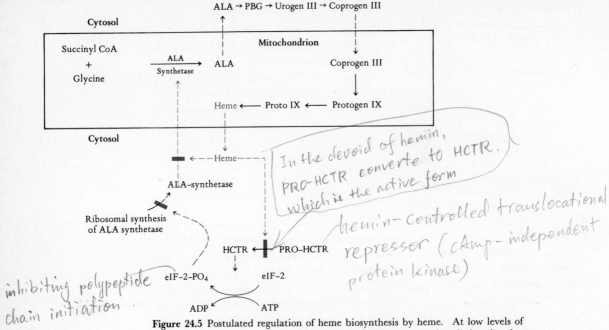

In the devoid of hemin, PRO-HCTR converte to HCTR, which is the active form

hemin-controlled translocational repressor (cAmp-independent protein kinase)

inhibiting polypeptide chain initiation

Figure 24.5 Postulated regulation of heme biosynthesis by heme. At low levels of heme a cAMP-independent protein kinase (HCTR) is activated and phosphorylates eIF-2, an initiation factor of protein synthesis (Chap. 29). Phosphorylated eIF-2 inhibits ribosomal synthesis of ALA synthetase. Heme also blocks translocation of ALA synthetase into the mitochondrion. Thus, depending on the variation in concentration of the heme pool, which is estimated to be 10^{-7} to $10^{-6}\,M$, ALA synthetase concentrations in mitochondria are controlled to regulate heme synthesis. Decreased heme concentrations repress heme synthesis, whereas elevated heme levels decrease mitochondrial ALA synthetase concentration by inhibiting its translocation. Abbreviations are PBG, porphobilinogen; urogen III, uroporphyrinogen; coprogen III, coproporphyrinogen III; protogen IX, protoporphyrinogen IX; proto IX, protoporphyrin IX.

hemin 可抑制紅血球

cytoplasm to the mitochondria (Chap. 29). Heme appears to act at two different steps in these processes. First, at low concentrations of heme the rate of ribosomal synthesis of the synthetase is repressed, and, second, at high concentrations of heme the rate of translocation of ALA synthetase into mitochondria is inhibited. In reticulocytes, protein synthesis stops in the absence of hemin. The lack of hemin is thought to allow the activation of *hemin-controlled translocational repressor,* which is a cyclic AMP-independent protein kinase. A molecular mechanism for activation of the repressor is unknown, but after activation, the repressor phosphorylates eIF-2, the initiation factor that mediates binding of met-tRNA$_f$ to 40S ribosomal subunits. Phosphorylated eIF-2 inhibits polypeptide chain initiation, perhaps by inhibiting binding of met-tRNA$_f$ to 40S subunits (Chap. 29). Hemin also acts in an unknown fashion to block translocation of ALA synthetase into mitochondria. Drugs such as allylisopropylacetamide induce synthesis of hepatic ALA synthetase in rats, and when such drugs are injected into rats along with hemin, the cytosolic level of ALA synthetase increases markedly, whereas the mitochondrial level decreases.

Many compounds stimulate heme synthesis by inducing ALA synthetase.

636

Lipid-soluble compounds and steroids act by stimulating the synthesis of the apoprotein of cytochrome P_{450}, which is involved in hydroxylation of these compounds. The demand for heme is thereby increased and ALA synthetase levels increase. The mitochondrial pool of heme is 10^{-7} to 10^{-8} M, and even a small diminution of the pool decreases the rate of heme synthesis by repressing synthesis of ALA synthetase. Many other factors may modify the induction of ALA synthetase. Glucose blocks the induction by unknown mechanisms. Iron administration reduces the heme pool by enhancing heme degradation and thereby induces the synthetase.

AMINO ACID DECARBOXYLATION

Amino acid decarboxylases are widely distributed in nature; each of these enzymes, with the exception of that catalyzing decarboxylation of histidine requires pyridoxal phosphate as coenzyme and catalyzes removal of a carboxyl group as CO_2. The reaction mechanism for α decarboxylation involves Schiff base formation between pyridoxal phosphate and the amino acid, followed by decarboxylation, as a proton from the medium replaces the carboxyl carbon in its attachment to the α-carbon atom. Equilibrium lies far to the right, as written.

$$R-CH_2-CHNH_2-COOH \longrightarrow R-CH_2-CH_2NH_2 + CO_2$$

Some metabolically significant α-amino acid decarboxylases are presented below.

γ-Aminobutyric Acid

γ-Aminobutyric acid, which serves as a neurotransmitter in brain (Chap. M7), is the product of the L-*glutamate α-decarboxylase* activity of this organ.

$$\underset{\textbf{Glutamic acid}}{HOOC-CH_2-CH_2-CHNH_2-COOH} \xrightarrow{\text{L-glutamate } \alpha\text{-decarboxylase}} \underset{\textbf{γ-Aminobutyric acid}}{HOOC-CH_2-CH_2-CH_2NH_2} + CO_2$$

In its subsequent metabolism, γ-aminobutyrate transaminates to α-ketoglutarate with formation of succinic semialdehyde. The latter is oxidized to succinate, which enters the tricarboxylic acid cycle.

β-Alanine

The only β-amino acid of physiological significance is β-*alanine*. The latter is present in many tissues and in plasma as the free amino acid and as a component of carnosine and of anserine and in coenzyme A as part of the pantothenic acid moiety.

In animals, although some β-alanine arises from pyrimidine degradation (Chap. 26), an alternative, perhaps more important, pathway is the following:

$$CH_3CH_2COSCoA \xrightleftharpoons{-2H} CH_2=CHCOSCoA \xrightleftharpoons{+H_2O}$$

Propionyl CoA   **Acrylyl CoA**

$$HOCH_2CH_2COSCoA \xrightleftharpoons{-CoA} HOCH_2CH_2COOH \xrightleftharpoons{-2H}$$

β-Hydroxypropionyl CoA   **β-Hydroxypropionic
acid**

$$OHCCH_2COOH \xrightleftharpoons[\text{transamination}] CH_2CH_2COOH$$

Malonic semialdehyde

$$\overset{NH_2}{\underset{|}{}}$$
CH_2CH_2COOH

β-Alanine

Decarboxylation of Aromatic Amino Acids

General *aromatic L-amino acid decarboxylase* in mammalian tissues catalyzes decarboxylation of histidine, tyrosine, tryptophan, phenylalanine, 3,4-dihydroxyphenylalanine, and 5-hydroxytryptophan. In addition, there are specific decarboxylases for some of the individual aromatic amino acids.

Histamine The general aromatic amino acid decarboxylase has low activity toward histidine. A specific enzyme present in mast cells, a major site of histamine formation, is the chief catalyst for histidine decarboxylation. Many other tissues, e.g., lung, liver, muscle, and gastric mucosa, have a high histamine content owing to synthesis of this amine *in situ*. Histamine is a powerful vasodilator and in excessive concentrations may cause vascular collapse. The base is liberated in persons in traumatic shock and in localized areas of inflammation. Histamine stimulates secretion of both pepsin and acid by the stomach (Chap. M10). *Diamine oxidase*, a widely distributed flavoprotein, converts histamine to the corresponding aldehyde and NH_3. Some undegraded histamine is excreted in the urine as the N-acetyl and as the N^1-methyl derivatives; the latter is the major metabolite of histamine in humans.

Histidine decarboxylase of a *Lactobacillus* contains covalently bound pyruvate, which is essential for its activity. The enzyme is synthesized as an inactive proenzyme with a single polypeptide chain and is converted to active enzyme by NH_4^+ or K^+ above pH 7. Activation results in cleavage of a polypeptide chain with release of NH_3 from a seryl residue and formation of two subunits, one with a terminal pyruvoyl residue, as follows:

H$_2$N-Ser-Glu···Thr-Thr-Ala-Ser $\quad$ H$_2$O $\quad$ NH$_3$ $\quad$ H$_2$N-Ser-Glu···Thr-Thr-Ala-Ser-COOH

HOOC-Tyr············Phe-Ser $\xrightarrow[\substack{K^+, \text{ pH } 7.6 \\ 37°}]{}$ HOOC-Tyr············Phe-pyruvoyl

Proenzyme $\qquad\qquad\qquad\qquad\qquad\qquad$ **Active enzyme**

Similar mechanisms may account for formation of covalently bound pyruvate in other bacterial and mammalian enzymes, but this remains to be established.

Hydroxytyramine and Related Compounds A specific hydroxylase, termed *tyrosinase*, catalyzes hydroxylation of tyrosine in the liver and in melanin-forming cells, to yield 3,4-dihydroxyphenylalanine (dopa). The latter is then decarboxylated by *aromatic amino acid decarboxylase* to dihydroxyphenylethylamine (*o*-hydroxy-tyramine or dopamine), an intermediate in melanin formation (Fig. 24.6) (also see Chap 25). The decarboxylase is present in kidney and adrenal tissue as well as in sympathetic ganglia and nerves. The activity of the general aromatic L-amino acid decarboxylase in liver, which also catalyzes conversion of dopa to dopamine, is subject to end-product feedback inhibition by dopa, whereas the activity of the decarboxylating enzyme in the other tissues mentioned above is not altered by dopa administration. This may be of some significance since it has been observed that the efficacy of L-dopa, now widely used in the treatment of Parkinson's disease, increases with continued administration. By a reduction of decarboxylase activity in the liver without any effect on the enzyme in brain, more L-dopa would be available to the brain for decarboxylation to dopamine, an inhibitory neurotransmitter and believed to be the effective therapeutic agent in Parkinson's disease (Chap. M7).

In human beings and rats, administration of hydroxytyramine leads to excretion in the urine of homoprotocatechuic acid (3,4-dihydroxyphenylacetic acid) and 3,4-dihydroxyphenylethanol, as well as their methylated derivatives, homovanillic acid and 3-methoxy-4-hydroxyphenylethanol, respectively. The structures of these compounds are shown in Fig. 24.6. Dopa also serves as precursor in norepinephrine and epinephrine biosynthesis (Chap. M18).

Tryptophan Derivatives The aromatic amino acid decarboxylase of liver has weak activity toward tryptophan. The resulting product, tryptamine, has no specific role but when oxidized by *monoamine oxidase* yields an aldehyde that is readily oxidized by *aldehyde oxidase* to indoleacetic acid. This compound is pres-

Figure 24.6 Formation and fate of hydroxytyramine and related compounds.

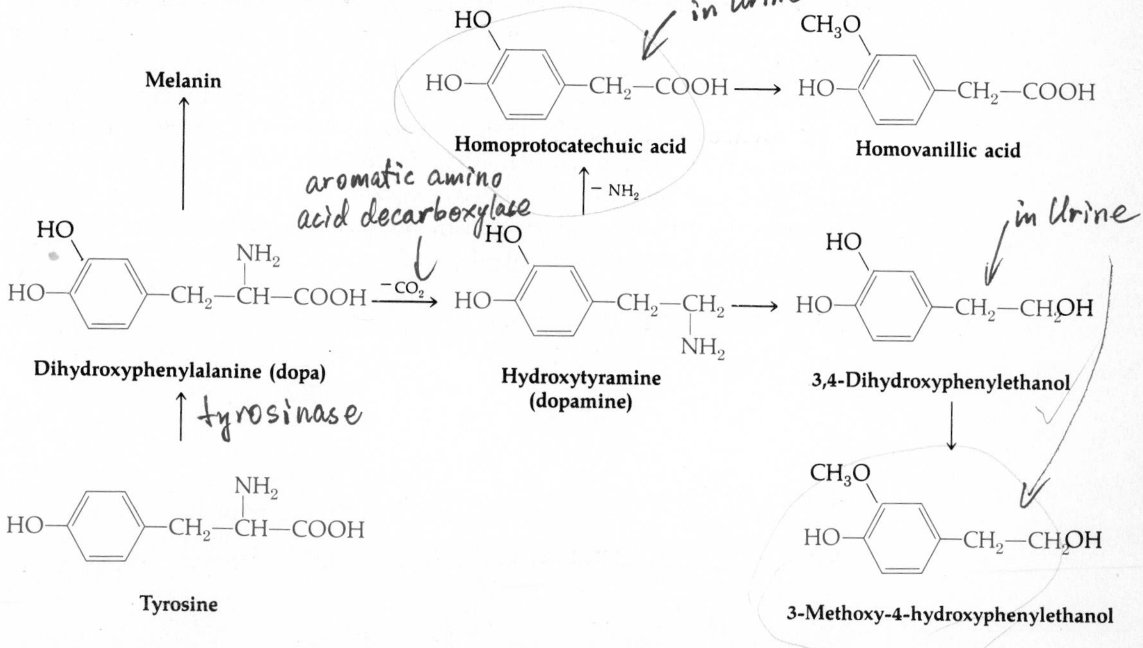

serotonine

ent in small quantities in normal urine. However, a major route of indoleacetic acid formation, both in plants, where it serves as an *auxin* (plant growth hormone), and in animals, is by oxidative decarboxylation of indolepyruvic acid, formed from tryptophan by transamination.

Tryptophan also is hydroxylated by a *tryptophan 5-monooxygenase,* present in brain, to 5-hydroxytryptophan. The enzyme requires, as a reductant cofactor, tetrahydropteridine, which is generated during the reaction in a manner analogous to that of the phenylalanine-hydroxylating enzyme system in liver (Chap. 23). Decarboxylation of 5-hydroxytryptophan by the enzyme *5-hydroxytryptophan decarboxylase,* also present in brain as well as in kidney, yields 5-hydroxytryptamine, or *serotonin.*

The role of serotonin as a neurohumoral agent in human beings is considered in Chap. M7. This amine is also present in intestinal tissue, blood platelets, and mast cells and is a potent vasoconstrictor. It is also a constituent of many venoms, e.g., wasp and toad venom. *N*-Methylated derivatives of serotonin, e.g., bufotenine, are rather widely distributed among amphibia and cause central nervous system damage in mammals.

Serotonin and its major metabolic products, 5-hydroxyindoleacetic acid and 5-methoxyindoleacetic acid, are present in urine. The former can arise by a NAD-dependent oxidation of 5-hydroxytryptophal, which arises in turn from serotonin by action of *monoamine oxidase.* Inhibition of serotonin metabolism by administration of monoamine oxidase inhibitors leads to increased formation of the *N*-acetyl and *N*-methyl derivatives of serotonin.

5-Hydroxytryptophol also is a metabolite of serotonin as a result of the action of monoamine oxidase and is present in human urine as such, as the 5-methoxy derivative, and as the conjugates 5-hydroxytryptophol-*O*-sulfate and 5-hydroxytryptophol-*O*-glucosiduronide. After oral administration of labeled serotonin, more than 80 percent of the labeled products in the urine was 5-hydroxyindoleacetic acid; the total 5-hydroxytryptophol excretion represented approximately 2 percent. 5-Methoxytryptophol is present in the pineal gland and has been shown to inhibit sexual development in the female rat.

Approximately 7 mg of 5-hydroxyindoleacetic acid is excreted in normal urine per day. It has been estimated that 3 percent of the dietary tryptophan is metabolized via this pathway. In patients with *malignant carcinoid,* as much as 400 mg of 5-hydroxyindoleacetic acid are excreted daily; its excretion also rises in a disorder of tryptophan metabolism, *Hartnup disease* (Chaps. 25, M5, and M10). Liver microsomes also catalyze hydroxylation of indole-containing compounds to the corresponding 6-hydroxy derivatives; the possible metabolic significance of this is not known. Certain of these relationships are shown in Fig. 24.7.

Numerous other instances of amino acid decarboxylation have been described and will be cited elsewhere in this book. However, not all decarboxylations occur while the amino acids are in the free state. Decarboxylation of serine occurs only when serine is in ester linkage in phosphatidylserine (Chap. 21), with formation of phosphatidylethanolamine. Somewhat analogous is the formation of coenzyme A, in which pantoic acid (Chap. M22) forms the pantoyl amide of cysteine, which is then phosphorylated and decarboxylated to form 4′-phosphopantetheine (Chap. 26).

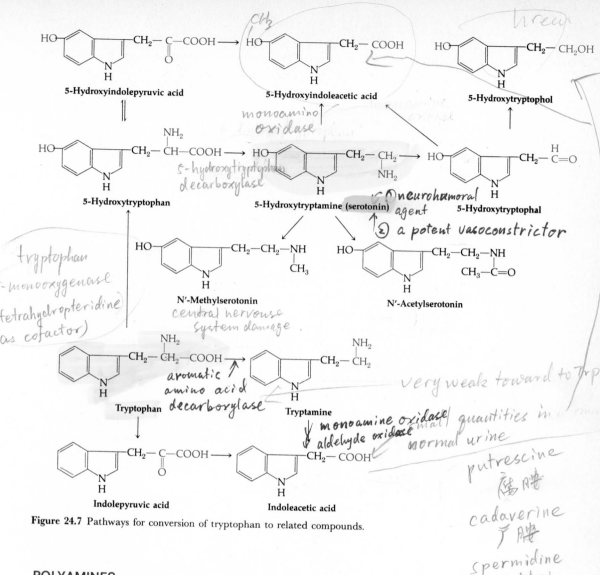

Figure 24.7 Pathways for conversion of tryptophan to related compounds.

Handwritten annotations on figure:
- CH₃
- urea
- monoamino oxidase
- tryptophan 5-monooxygenase (tetrahydropteridine as cofactor)
- 5-hydroxytryptophan decarboxylase
- ① neurohumoral agent
- ② a potent vasoconstrictor
- central nervouse system damage.
- aromatic amino acid decarboxylase
- monoamine oxidase / aldehyde oxidase
- very weak toward to Trp. small quantities in normal / normal urine
- putrescine 腐胺
- cadaverine 尸胺
- spermidine 亚精胺

POLYAMINES

Putrescine, cadaverine, spermidine, and spermine, whose structures are shown in Fig. 24.8, are ubiquitous polyamines in biological materials. Putrescine and cadaverine have long been recognized as products of the bacterial decomposition of living matter and are well known for their unpleasant odors. Their wide distribution in living things, however, has only been recognized recently.

The relative amounts of polyamines in different materials are listed in Table 24.2. Prokaryotes generally have higher amounts of putrescine than spermidine and lack spermine. Eukaryotes contain little putrescine but have both spermidine and spermine. Rapidly proliferating tissues, such as tumors, have especially high concentrations of polyamines. Cells and secretions of the male reproductive system are also rich in spermine and spermidine. Polyamines

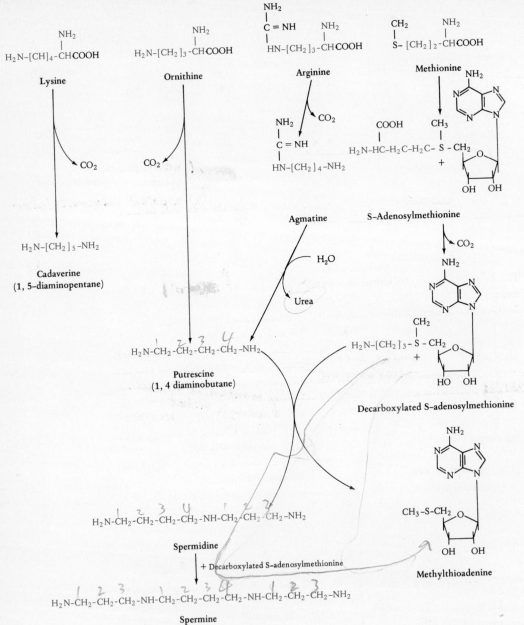

Figure 24.8 Synthesis of polyamines.

are often found associated with nucleic acids; thus they are packaged with DNA in *Escherichia coli* bacteriophage T4 (Chap. 30) and account for about 40 percent of the total cations in the phage. The high affinity of polyamines for nucleic acids and their ability to stabilize DNA against denaturation and shearing are due to electrostatic interactions between the cationic amines and the anionic nucleic acids. Polyamines also interact with several macromolecules

TABLE 24.2

	Putrescine	Spermidine	Spermine	Concentrations* of Polyamines in Various Biological Materials
E. coli	13.1	4.7	0	
Brain (human)	0.015	0.23	0.10	
Seminal plasma (human)	0.23	0.11	3.04	
Prostate (rat)		7.73	4.77	
Liver (rat)	0.01	0.61	0.82	
Hepatoma (rat)	0.29	0.92	0.69	

*μmol/g wet weight.

and organelles; thus they stabilize spheroplasts from bacteria as well as mitochondria against lysis by hypotonic media.

Biosynthesis of Polyamines

The biosynthesis of polyamines is shown in Fig. 24.8. Bacterial putrescine may arise by decarboxylation of either ornithine or arginine, although ornithine decarboxylation is usually the major path. The action of *ornithine decarboxylase* gives rise directly to putrescine whereas that of *arginine decarboxylase* yields agmatine, which is then converted to putrescine and urea by *agmatine ureohydrolase*. Both decarboxylases require pyridoxal phosphate as cofactors and are subject to feedback inhibition and repression by putrescine and spermidine. In animal cells, putrescine is formed solely by ornithine decarboxylase, whose properties differ from the bacterial enzyme, although it also requires pyridoxal phosphate as a cofactor. Cadaverine is formed in bacteria by the action of *lysine decarboxylase* in a reaction analogous to ornithine decarboxylation.

Methionine and putrescine are biosynthetic precursors of spermidine and spermine. S-Adenosylmethionine formed from methionine and ATP (Chap. 23) is acted upon by *S-adenosylmethionine decarboxylase* to give decarboxylated S-adenosylmethionine. The decarboxylases from bacterial and animal cells are devoid of pyridoxal phosphate but both enzymes contain a covalently linked pyruvoyl residue; the E. coli enzyme contains one pyruvoyl residue per molecule (MW = 113,000) whereas the rat liver enzyme contains one pyruvoyl residue in each of its two subunits (MW = 42,000). The pyruvoyl groups are required for enzymic activity. The bacterial and animal S-adenosylmethionine decarboxylases also differ in other properties; e.g., the bacterial enzyme requires Mg^{2+}, whereas the eukaryotic enzyme requires putrescine for activity. Spermidine is synthesized from decarboxylated S-adenosylmethionine and putrescine in an unusual reaction catalyzed by *aminopropyl transferase*, an enzyme with no known cofactors. Rather than transfer the methyl group as in transmethylation reactions utilizing S-adenosylmethionine (Chap. 23), the aminopropyl group on the sulfonium sulfur of decarboxylated S-adenosylmethionine is transferred by the aminopropyl transferase. Bacteria are devoid of spermine and accordingly contain only one aminopropyl transferase. Animals, however, contain two aminopropyl transferases, one catalyzing spermidine synthesis and another, the synthesis of spermine from spermidine and decarboxylated S-adenosylmethionine (Fig. 24.8).

Metabolism of Polyamines

Putrescine and spermidine are found in small amounts (0.1 to 0.6 nmol/ml) in human plasma and cerebrospinal fluid; spermine and cadaverine are present in even smaller amounts. Urinary excretion of total polyamines is also low, amounting to only 5 to 10 mg per 24 h. Most of the urinary polyamines exist as monoacetyl derivatives, whose site of synthesis is unknown.

Amine oxidase activities have been found in various bacteria and in plant and animal tissues, but their role in turnover of the polyamines is unclear. A copper-containing *diamine oxidase* oxidizes spermine and spermidine with the production of H_2O_2 and NH_3 as follows:

$$H_2N[CH_2]_3NH[CH_2]_4NH[CH_2]_3NH_2 + 2H_2O \xrightarrow{+2O_2}$$
$$OHC[CH_2]_2NH[CH_2]_4NH[CH_2]_2CHO + 2NH_3 + 2H_2O_2$$

$$H_2N[CH_2]_3NH[CH_2]_4NH_2 + H_2O \xrightarrow{+O_2}$$
$$OHC[CH_2]_2NH[CH_2]_4NH_2 + NH_3 + H_2O_2$$

A liver peroxisomal *polyamine oxidase* of rat catalyzes the interconversions of polyamines; spermine is converted to spermidine and aminopropionaldehyde whereas spermidine is converted to putrescine and aminopropionaldehyde. The significance of these catabolic enzymes is unclear.

Biological Functions

The polyamines have been implicated in a wide variety of physiological processes. A possible role of polyamines in growth is suggested by a variety of observations. Bacterial mutants are known that require polyamines for growth at a normal rate. In addition, ornithine decarboxylase levels are normally low in quiescent animal cells but increase dramatically in rapidly proliferating cells under a variety of conditions that promote increased rates of protein synthesis, e.g., in embryonic and malignant growth, in regenerating liver, virus infection, or administration of growth hormone, cortisol, glucagon, prolactin, or dibutyryl cyclic AMP to cell cultures. In these cases, ornithine decarboxylase is rapidly synthesized and degraded; in regenerating rat liver its half-life is 10 min, a much shorter time than that known for any other animal enzyme. The polyamines generated by the high levels of ornithine decarboxylase repress synthesis of the decarboxylase; thus it would appear that the rapid turnover of the enzyme is a mechanism to ensure appropriate polyamine concentrations under conditions of cell proliferation. Since the rise in the decarboxylase and polyamine levels often precedes increases in DNA and RNA synthesis and polyamines are often found in physical association with DNA, it has been suggested that polyamines directly control RNA and DNA synthesis, but experimental confirmation of such a relationship has not been obtained. In addition to a possible role in cell growth, polyamines have been implicated in a wide variety of other cellular processes, such as platelet aggregation, stimulation of lipolysis, activation of phosphorylase b, and modification of protein kinase activities, to name only a few affected processes; but whether such effects are of importance in vivo remains to be established.

Polyamines participate in clot formation of rodent seminal vesicle secretions. Prostate glands and seminal fluids are rich in polyamines (Table 24.2), which are covalently cross-linked into five different proteins of rat seminal plasma. Cross-linking of the proteins is catalyzed by a transglutaminase in seminal plasma that is similar in its catalytic action to factor XIII of blood coagulation (Chap. M1) and another type of transglutaminase that is widely distributed in many animal tissues. These enzymes catalyze formation of cross-links between glutamine and lysine residues in proteins as follows:

$$\text{—Gln—}\overset{\overset{\displaystyle O}{\|}}{C}\text{—NH}_2 + \text{H}_2\text{N—Lys—} \rightleftharpoons \text{—Glu—}\overset{\overset{\displaystyle O}{\|}}{C}\text{—}\overset{\overset{\displaystyle H}{}}{N}\text{—Lys—} + \text{NH}_3$$

The γ-carboxamide group of glutamine and the ε—NH$_2$ group of lysine thus react to form an ε-(γ-glutamyl)lysine cross-link. The prostate enzyme, however, catalyzes cross-linking via polyamines.

$$\text{—Gln—}\overset{\overset{\displaystyle O}{\|}}{C}\text{—NH}_2 + \text{H}_2\text{N—R—NH}_2 + \text{H}_2\text{N—}\overset{\overset{\displaystyle O}{\|}}{C}\text{—Gln—} \rightleftharpoons$$

$$\text{—Gln—}\overset{\overset{\displaystyle O}{\|}}{C}\text{—}\overset{\overset{\displaystyle H}{}}{N}\text{—R—}\overset{\overset{\displaystyle H}{}}{N}\text{—}\overset{\overset{\displaystyle O}{\|}}{C}\text{—Gln—} + 2\text{NH}_3$$

Clots so formed participate in formation of the vaginal plugs in rodents. Proteins in human lymphocytes treated with mitogens also contain monosubstituted polyamines linked through γ-glutamyl residues, but the biological significance of such groups is unknown.

REFERENCES

Books

Arias, I. M., and W. B. Jakoby (eds.): *Glutathione: Metabolism and Functions*, Raven, New York, 1976.

Review Articles

Folk, J. E.: Transglutaminases, *Annu. Rev. Biochem.* **49**:517–532 (1980).

Granick, S., and S. I. Beale: Hemes, Chlorophylls, and Related Compounds: Biosynthesis and Metabolic Regulation, *Adv. Enzymol.* **46**:33–203 (1978).

Jakoby, W. B.: The Glutathione Transferases: A Group of Multifunctional Detoxification Proteins, *Adv. Enzymol.* **46**:383–414 (1978).

Jänne, J., H. Pösö, and A. Raina: Polyamines in Rapid Growth and Cancer, *Biochim. Biophys. Acta* **473**:241–293 (1977).

Paik, W. K., and S. Kim: Protein Methylation: Chemical, Enzymological and Biological Significance, *Adv. Enzymol.* **42**:227–286 (1975).

Walker, J. B.: Creatine: Biosynthesis, Regulation and Function, *Adv. Enzymol.* **50**:177–242 (1979).

Chapter 25

Amino Acid Metabolism IV: Fates in Mammals

phlorhizin
花佳甘.

When mammals are presented with amounts of amino acids larger than needed for synthesis of proteins and other nitrogenous compounds, most of the surplus cannot be stored. Instead, each of the amino acids is degraded by pathways that transform most of the carbon chain either into pyruvate or directly to acetyl CoA and thus into pathways common to the metabolism of carbohydrates and fatty acids. The metabolism of the excess amino acids contributes significantly to satisfying the energy requirements of the tissues not only immediately after a meal but also in the postabsorptive state. Further, there are important differences in the ways by which different organs take up or supply various amino acids to the blood.

Degradation of amino acids to supply energy is not a significant aspect of the life of the photosynthetic plant, which makes amino acids at a rate commensurate with growth requirements, as does a culture of bacteria in a medium that supports growth. However, many microorganisms can be grown in media in which one amino acid or a few constitute the major source of energy and of carbon and nitrogen, and from which can be fabricated all the materials essential for life. These bacteria possess catabolic pathways for amino acids that are analogous to those of vertebrates.

GLYCOGENIC AND KETOGENIC AMINO ACIDS

The administration of a single amino acid to diabetic animals or to animals treated with phlorhizin will increase urinary excretion of either glucose or ketone bodies. Fasted animals can be utilized to ascertain whether an administered amino acid produces an increment in liver glycogen.

Any cross-references coded M refer to the companion text, *Principles of Biochemistry: Mammalian Biochemistry.*

Transformations that lead to pyruvate, oxaloacetate, or α-ketoglutarate make possible net glucose formation, whereas pathways that lead to acetyl CoA or to acetoacetate are termed *ketogenic*. The pathways for glycogenesis (Chap. 17) and for ketogenesis (Chap. 20) have been described previously. On this basis, amino acids can, in general, be classified with respect to whether they are glycogenic and or ketogenic (Fig. 25.1).

The initial step in the metabolism of most amino acids is removal of the α-amino group by transamination or oxidation. Hence, it is largely the fate of the corresponding α-keto acids that is described in the following discussions.

AMINO ACID METABOLISM IN HUMAN TISSUES

The levels of free amino acids in plasma (Chap. M1; Table M1.1) show relatively little variation. These steady-state concentrations represent a balance between utilization by various tissues and release from endogeneous protein and amino acid stores. Inasmuch as the liver is the site of urea formation and muscle contains more than half of the free amino acid stores, these two organs play major roles in determining the turnover of amino acids.

Figure 25.1 The steps at which the carbon skeletons of the amino acids can enter the tricarboxylic acid cycle. Those which enter as acetyl CoA or acetoacetyl CoA are ketogenic (in red); those which enter elsewhere are potentially glycogenic. Note that some carbon skeletons are fragmented to yield more than one product, e.g., isoleucine and tyrosine. Aspartate can form oxaloacetate by transamination, but after formation and cleavage of argininosuccinate (page 617), its carbon skeleton enters the cycle as fumarate.

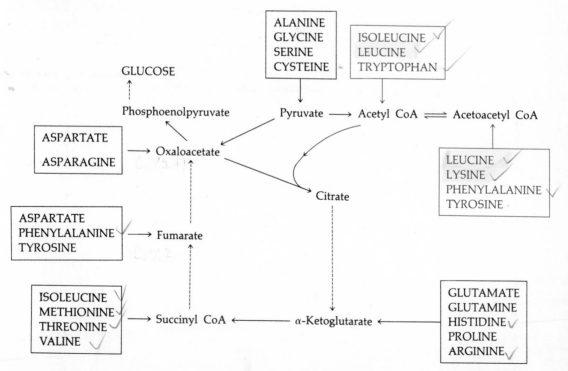

Splanchnic
内脏

inosine
次黄苷

*NH₃ from branched
amino acid. leucine
valine., isoleucine*

In the postabsorptive state, 10 to 12 h after a meal, there is a considerable release of amino acids from skeletal and cardiac muscle, alanine and glutamine representing more than half of the free amino acids released. At the same time there is an uptake of these and other amino acids by liver as well as other splanchnic tissues, i.e., upper gastrointestinal tract, spleen, and pancreas. Glutamine and asparagine are predominantly taken up and utilized by the gut, as are the two corresponding dicarboxylic acids. In contrast, alanine is released from the gut. Some of the relationships are shown in Fig. 25.2.

As already noted, the liver is the primary site of gluconeogenesis from amino acids and other precursors (Chap. 17). The alanine derived from cardiac and skeletal muscle, gut, kidney, and other tissues is converted to pyruvate by transamination and then converted to glucose. In the extrahepatic tissues, some of the pyruvate derived from glycolysis is converted to alanine and delivered to the blood. Hence there is an "alanine-glucose cycle." The amino groups required to form alanine by transamination are presumably derived largely from the branched-chain amino acids, leucine, valine, and isoleucine, whose derived α-keto acids are largely oxidized in skeletal and cardiac muscle (page 658). A portion of the amino groups may also be obtained from the deamination of adenylate to form inosinate during active muscular contraction (Chap. M8); some of the ammonia is incorporated by the action of glutamate dehydrogenase into glutamate, which can then be converted to glutamine or transaminated to form alanine from pyruvate.

The great differences in utilization or disposal of amino acids by various mammalian organs and tissues have only recently come under detailed investigation. A few of the additional findings can be briefly summarized. Extrahepatic tissues, such as muscle and brain, are the major sites of catabolism of the branched-chain amino acids. The uptake by brain of branched-chain amino acids, particularly valine, is greater than that of all other amino acids. Further, rat brain can oxidize these amino acids at almost four times the rate of muscle or liver on a unit weight basis. However, because of the greater mass of cardiac and skeletal muscles, most of the branched-chain amino acids are metabolized by these tissues. Kidney shows a net uptake of glutamine, proline, and glycine

Figure 25.2 Interorgan transfer of amino acids in the normal postabsorptive state. The important role of alanine in transfer from muscle and gut to the liver is shown. As indicated in the text, the branched-chain amino acids are mainly metabolized in muscle; other amino acids are mainly metabolized in the liver. [*Adapted from P. Felig, Annu. Rev. Biochem.* 44:993 (1975).]

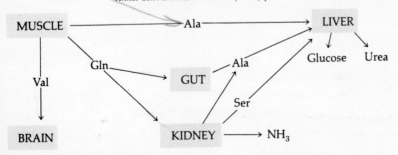

and a significant output of alanine and serine. Indeed, it appears that the kidney is a major supplier of serine to liver and other tissues.

As already emphasized and as will be further described (Part M3), hormones have a major influence on amino acid metabolism. Insulin stimulates protein synthesis and amino acid transport into cells, thus lowering amino acid concentrations in the blood (Chap. M16). The adrenal steroids (glucocorticoids) (Chap. M18) decrease protein synthesis in muscle and augment hepatic uptake of amino acids and hepatic gluconeogenesis. These steroids also stimulate the synthesis of various hepatic enzymes involved in amino acid catabolism, e.g., tyrosine-glutamate aminotransferase (page 661), tryptophan dioxygenase (page 668), and others.

METABOLIC FATES OF INDIVIDUAL AMINO ACIDS

Alanine, Glutamate, and Aspartate *glycogenic*

Removal of the amino group of each of these three amino acids yields α-keto acids that are glycogenic. Glutamate may also be transaminated to one of many keto acids or be oxidized by *glutamate dehydrogenase* (page 613); aspartate and alanine are transaminated to form oxaloacetate and pyruvate, respectively.

Ornithine, Proline, and Hydroxyproline

Ornithine and proline, formed originally from glutamate (page 607), are reconvertible to glutamic acid. *Proline oxidase* of mitochondria effects oxidation to Δ^1-pyrroline 5-carboxylic acid. [This enzyme is distinct from the NADPH-linked reductase responsible for proline formation (page 607).] Spontaneous hydrolysis yields glutamic semialdehyde, which can also be formed from ornithine by transamination. Oxidation to glutamate completes these sequences.

Hydroxyproline is oxidized by a microsomal, cytochrome b_5–linked oxidase in liver and by L-amino acid oxidase in kidney to Δ^1-pyrroline-3-hydroxy-5-carboxylic acid, which, by reactions analogous to those described for proline metabolism, leads to *erythro*-L-γ-hydroxyglutamate. Transamination forms α-keto-γ-hydroxyglutarate, which is cleaved in an aldolase type of reaction to glyoxylic acid and pyruvic acid (Fig. 25.3, pathway *A*.) The fate of glyoxylic acid will be described below (page 651). In brain, hydroxyglutamate is decarboxylated to 2-hydroxy-4-aminobutyrate, which, in turn, yields malate (pathway *B*).

Glutamine

Glutamine is converted to glutamate and NH_3 by glutaminase (page 614). Two widely distributed glutaminase isoenzymes have been described, one of which is P_i-dependent. Renal glutaminase makes possible the conservation of cations by the organism and the maintenance of the normal pH of body fluids (Chap. M5).

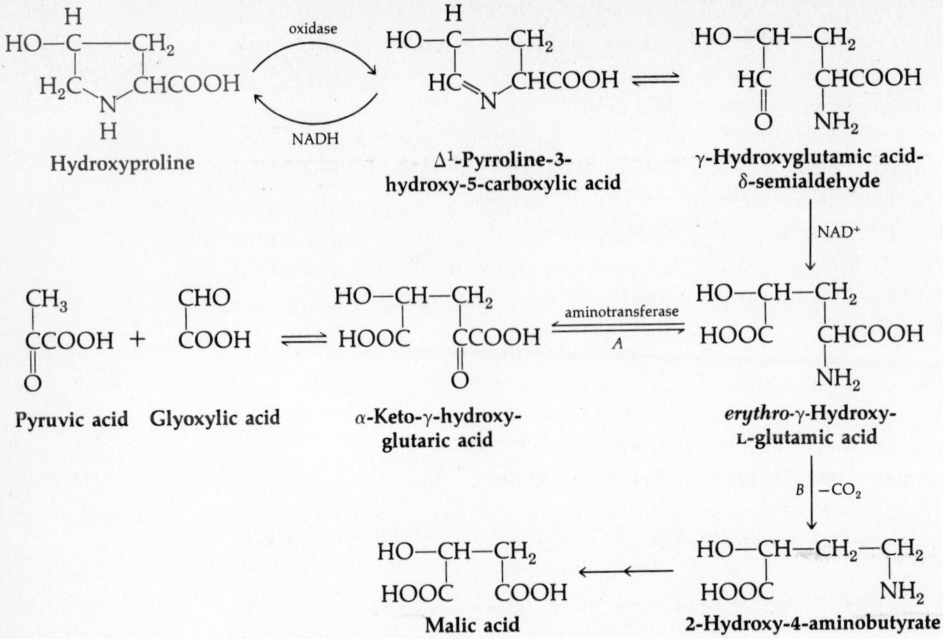

$$\text{Hydroxyproline} \xrightleftharpoons[\text{NADH}]{\text{oxidase}} \Delta^1\text{-Pyrroline-3-hydroxy-5-carboxylic acid} \rightleftharpoons \gamma\text{-Hydroxyglutamic acid-}\delta\text{-semialdehyde}$$

Hydroxyproline — HO—C(H)—CH₂ / H₂C, N(H), CHCOOH

Δ¹-Pyrroline-3-hydroxy-5-carboxylic acid — HO—C(H)—CH₂ / HC=N, CHCOOH

γ-Hydroxyglutamic acid-δ-semialdehyde — HO—CH—CH₂ / HC(=O), CHCOOH, NH₂

$$\text{Pyruvic acid} + \text{Glyoxylic acid} \rightleftharpoons \alpha\text{-Keto-}\gamma\text{-hydroxyglutaric acid} \xrightleftharpoons[A]{\text{aminotransferase}} erythro\text{-}\gamma\text{-Hydroxy-L-glutamic acid}$$

Pyruvic acid — CH₃—C(COOH)=O

Glyoxylic acid — CHO—COOH

α-Keto-γ-hydroxyglutaric acid — HO—CH—CH₂ / HOOC, C(COOH)=O

erythro-γ-Hydroxy-L-glutamic acid — HO—CH—CH₂ / HOOC, CHCOOH, NH₂

(via NAD⁺)

B | —CO₂

2-Hydroxy-4-aminobutyrate — HO—CH—CH₂—CH₂ / HOOC, NH₂

Malic acid — HO—CH—CH₂ / HOOC, COOH

Figure 25.3 Metabolic fates of hydroxyproline in mammals.

Glutamine contributes to the synthesis of a variety of substances, e.g., purines and pyrimidines (Chap. 26), hexosamines (Chap. 17), and specific amino acids (Chap. 23), by transfer of its amide group.

Asparagine

$$NH_2\!-\!\overset{O}{\overset{\|}{C}}\!-\!CH_2\!-\!\overset{NH_2}{\overset{|}{C}H}\!-\!COOH \xrightarrow{\ asparaginase\ } {}^-OOC\!-\!CH_2\!-\!\overset{NH_2}{\overset{|}{C}H}\!-\!COO^-$$

Present knowledge indicates that asparagine metabolism is restricted to (1) utilization for protein synthesis, (2) conversion to aspartate and NH₃ by asparaginase, and (3) transamination to give α-ketosuccinamic acid (page 603).

Serine

$$H_2N\!-\!\overset{O}{\overset{\|}{C}}\!-\!CH_2\!-\!\overset{NH_2}{\overset{|}{C}}\!-\!COO^- \xrightarrow[\alpha KG \quad Glu]{transamination} H_2N\!-\!\overset{O}{\overset{\|}{C}}\!-\!CH_2\!-\!\overset{O}{\overset{\|}{C}}\!-\!COO^- \xrightarrow{\omega\text{-}amidase} HOOC\!-\!CH_2\!-\!\overset{O}{\overset{\|}{C}}\!-\!C$$

The conversion of serine to glycine and N^5,N^{10}-methylenetetrahydrofolate has been described (Chap. 23). The major fate of serine is an α,β dehydration reaction catalyzed by *serine-threonine dehydratase*, so named because both amino acids are substrates for the mammalian enzyme.

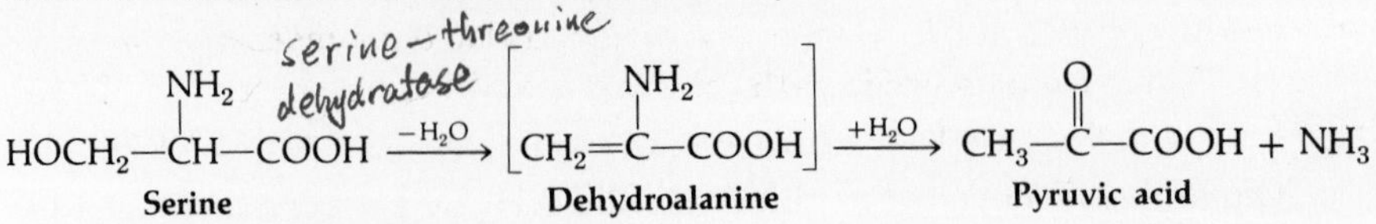

$$HOCH_2\!-\!\overset{NH_2}{\overset{|}{C}H}\!-\!COOH \xrightarrow[-H_2O]{\text{serine-threonine dehydratase}} \left[CH_2\!=\!\overset{NH_2}{\overset{|}{C}}\!-\!COOH \right] \xrightarrow{+H_2O} CH_3\!-\!\overset{O}{\overset{\|}{C}}\!-\!COOH + NH_3$$

Serine   **Dehydroalanine**   **Pyruvic acid**

The dehydratase of sheep liver consists of two subunits (MW = 41,000 and 47,000). The smaller one contains a residue of α-ketobutyrate in an amide bond at the NH₂-terminal end of the peptide chain. This residue is presumably derived from a threonyl residue. The keto group forms a Schiff base with the amino acid substrate. Thus, this enzyme resembles several other enzymes of amino acid metabolism in having a nonpyridoxal phosphate-carbonyl cofactor. In contrast, the serine dehydratase of *Escherichia coli* utilizes pyridoxal phosphate as cofactor.

Some degradation of serine may also occur by transamination with formation of β-hydroxypyruvic acid; the latter may give rise to glucose and glycogen. The contributions of serine to cysteine formation (Chap. 23) and to phosphoglyceride synthesis (Chap. 21) have been described.

Glycine $CO_2 + NH_3 + N^5,N^{10}$-methylene THF $\underset{\text{Gly. Synthase}}{\xrightarrow{+ NADH^+ + H^+}} CH_2$-COOH $+ THF + NAD^+$

Under ordinary circumstances, it is likely that glycine is mainly utilized for incorporation into proteins and for formation of a variety of compounds, e.g., glutathione, creatine and porphyrins (Chap. 24), purine nucleotides (Chap. 26), and various conjugates, such as hippurate and glycocholate (Chap. 21). There are, however, a number of pathways for disposal of glycine. It can be converted to serine by reversal of the *serine transhydroxymethylase* reaction (Chap. 23). Reversal of the *glycine synthase* reaction leads to formation of CO_2, NH_3, and N^5, N^{10}-tetrahydrofolate.

D-*Amino acid oxidase* (Chap. 16) catalyzes the following reaction:

$$H_2NCH_2COOH + O_2 + H_2O \longrightarrow O{=}CHCOOH + NH_3 + H_2O_2$$

Glycine   **Glyoxylic acid**

Oxidation of glyoxylate yields oxalate, of interest because of the genetic disorder *primary oxaluria,* in which there is increasing deposition of calcium oxalate in the kidneys and other tissues (*oxalosis*). Progressive renal failure with early death because of uremia is common in this disorder.

Primary oxaluria occurs because of a reduction in the activity of the cytoplasmic enzyme α-*ketoglutarate: glyoxylate carboligase,* which catalyzes the following reaction:

$$O{=}CHCOO^- + {}^-OOC{-}CH_2{-}CH_2{-}\overset{\overset{\displaystyle O}{\|}}{C}{-}COO^- \xrightarrow[CO_2]{Mg^{2+}} {}^-OOC{-}CH_2{-}CH_2{-}\overset{\overset{\displaystyle O}{\|}}{C}{-}\overset{\overset{\displaystyle OH}{|}}{C}H{-}COO^-$$

Glyoxylate   **α-Ketoglutarate**   **α-Hydroxy-β-ketoadipate**

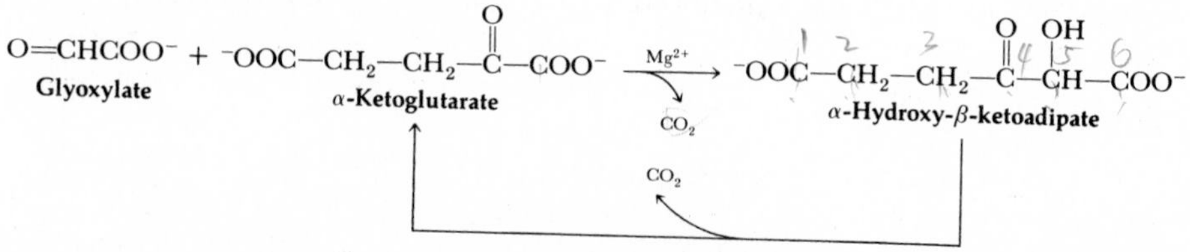

This reaction is dependent on thiamine pyrophosphate. It is likely that the α-hydroxy-β-ketoadipate is oxidatively decarboxylated to regenerate α-ketoglutarate, thus forming a cyclic pathway for disposal of glyoxylate. The latter can also arise from degradation of hydroxyproline (Fig. 25.3) or by transamination of glycine, presumably a minor pathway.

Cysteine

The ultimate metabolic fate of cysteine is formation of inorganic sulfate and pyruvate. This involves initial removal of the amino group by transamination, initial oxidation of the organically bound sulfur, or initial concurrent removal of both the sulfur and the amino group, respectively. These pathways are considered in turn and are shown in Fig. 25.4.

Transamination Transamination of cysteine in liver yields β-mercaptopyruvate, which undergoes desulfuration to form pyruvate. This sulfur may appear as H_2S because of the presence of reducing agents, e.g., glutathione (RSH, below).

$$\underset{\text{Cysteine}}{HSCH_2-\underset{\overset{|}{NH_2}}{CH}-COOH} \xrightleftharpoons{\text{transamination}} \underset{\beta\text{-Mercaptopyruvate}}{HSCH_2-\overset{\overset{O}{\|}}{C}-COO^-} \longrightarrow$$

$$\underset{\text{Pyruvate}}{CH_3-\overset{\overset{O}{\|}}{C}-COO^-} + [S] \xrightarrow{2RSH} H_2S + RSSR$$

Oxidative Pathway The major pathway of cysteine catabolism in animals involves its oxidation to cysteine sulfinic acid. The reaction is catalyzed by an Fe^{2+}-containing dioxygenase. Both O atoms of the sulfinic acid are derived from O_2.

$$\underset{\text{Cysteine}}{HS-CH_2-\underset{\overset{|}{NH_2}}{CH}-COOH} \xrightarrow{+O_2} \underset{\text{Cysteine sulfinate}}{^-O_2S-CH_2-\underset{\overset{|}{NH_2}}{CH}-COOH} \xrightarrow{\text{transamination}}$$

$$\underset{\beta\text{-Sulfinylpyruvate}}{^-O_2S-CH_2-\overset{\overset{}{\underset{\underset{O}{\|}}{C}}}{C}-COO^-} \longrightarrow \underset{\text{Pyruvate}}{CH_3-\overset{\overset{}{\underset{\underset{O}{\|}}{C}}}{C}-COO^-} + SO_3^{2-} \longrightarrow SO_4^{2-}$$

Aspartate transaminase catalyzes transamination of cysteine sulfinate with α-ketoglutarate to yield pyruvate and sulfite. A direct desulfination of cysteine sulfinate has also been described, catalyzed by a liver preparation with formation of alanine and sulfite. The SO_3^{2-} is then oxidized to SO_4^{2-}, as described below (page 653).

Sulfate, the ultimate form in which ingested sulfur is excreted, is also utilized for synthesis of 3'-phosphoadenosine-5'-phosphosulfate, which participates in synthesis of diverse sulfate esters.

Cysteine sulfinate is also utilized for the synthesis of several sulfur-containing compounds, including taurine and isethionic acid (Fig. 25.4), a normal metabolite of taurine. Isethionic acid and taurine are present in significant quantities in nerve tissue, but their role is unknown. Taurine participates in the formation of the bile acid taurocholic acid.

Cysteamine is present in CoA; it is liberated upon hydrolysis of this coenzyme. Its fate is shown in Fig. 25.4.

Removal of Sulfur Desulfuration of the β-mercaptopyruvate, arising from transamination of cysteine, is catalyzed by cytoplasmic *β-mercaptopyruvate trans-*

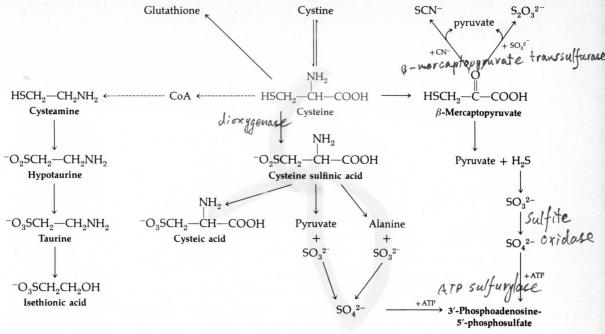

Figure 25.4 Major metabolic pathways of cysteine and its sulfur in mammals.

sulfurase, present in most tissues, particularly in liver and kidney. This enzyme catalyzes transfer of the sulfur atom of β-mercaptopyruvate to nucleophiles such as SO_3^{2-}, CN^-, and RSH to form, respectively, $S_2O_3^{2-}$, SCN^-, and $R\!-\!S\!-\!SH$. In the presence of excess thiol, the $RS\!-\!SH$ is reduced to form H_2S and disulfide.

A second sulfur transferase, *thiosulfate sulfur transferase (rhodanese),* is especially abundant in the mitochondria of liver and kidney. The enzyme catalyzes the transfer of divalent sulfur from a variety of donors, e.g., thiosulfate and organic persulfides, to acceptors such as sulfite, with formation of thiosulfate, and to cyanide, with formation of thiocyanate. The physiological functioning of this enzyme in cyanide detoxification, with formation of thiocyanate, has been demonstrated. The divalent sulfur atoms of the donor molecules are attached to a sulfhydryl group of the enzyme before transfer to the acceptor.

Cysteine can be cleaved directly to pyruvate, NH_3, and H_2S in some bacteria by *cysteine desulfhydrase;* this enzyme has not been found in animal tissues.

$$Cysteine + H_2O \longrightarrow pyruvate + H_2S + NH_3$$

Sulfide can be oxidized to SO_3^{2-} and SO_4^{2-} in liver and kidney; the enzyme systems are in the mitochondria. The mode of conversion of sulfide to sulfite in animal tissues has not been elucidated. Oxidation of sulfite to sulfate is catalyzed by mitochondrial *sulfite oxidase* (page 379).

The importance of sulfite oxidation in normal human metabolism is suggested by discovery of children lacking hepatic and renal sulfite oxidase. The urine contains large amounts of thiosulfate and sulfite but virtually no sulfate. Severe neurological abnormalities are evident at birth with progressive

653

deterioration and death at an early age. No alterations in levels of several sulfate esters can be detected in the tissues and urine, suggesting that accumulation of sulfite and/or its by-products, rather than a sulfate deficiency, is responsible for the pathological effects.

Inorganic and ester SO_4^{2-} are major products of sulfur metabolism, with the former constituting approximately 80 percent of the 24-h urinary excretion of total sulfur. Sulfate esters of numerous metabolites, e.g., steroids, aromatic hydroxy compounds, oligosaccharides, and a small quantity of organic sulfur, e.g., cystine and taurine (see below), constitute the remainder.

Other pathways of cysteine metabolism are also operative in the liver but are of lesser significance. The pyridoxal phosphate-requiring *cystathionine γ-synthase*, that also catalyzes conversion of cystathionine to homoserine plus cysteine in mammalian and microbial metabolism has been described (page 605). In the presence of cystine, this enzyme catalyzes the following reaction series.

$$\underset{\text{Cystine}}{\text{HOOC}\!-\!\underset{\underset{NH_2}{|}}{CH}\!-\!CH_2S\!-\!SCH_2\!-\!\underset{\overset{|}{NH_2}}{CH}\!-\!COOH} \xrightarrow{\text{pyruvate + NH}_3} \underset{\text{Thiocysteine}}{\text{HOOC}\!-\!\underset{\underset{NH_2}{|}}{CH}\!-\!CH_2S\!-\!SH} \quad (1)$$

Cystathionine γ-synthase

$$\underset{\text{Thiocysteine}}{\text{HOOC}\!-\!\underset{\underset{NH_2}{|}}{CH}\!-\!CH_2S\!-\!SH} + \underset{\text{Cysteine}}{\text{HSCH}_2\!-\!\underset{\overset{|}{NH_2}}{CH}\!-\!COOH} \longrightarrow \text{cystine} + H_2S \quad (2)$$

SUM: Cysteine $\longrightarrow$ pyruvate $+$ NH$_3$ $+$ H$_2$S

Three hereditary disorders of cystine metabolism have been described, viz., cystinuria, cystathionuria, and homocystinuria. *Cystinuria* is characterized by an abnormally high urinary excretion of cystine (Chap. M5), as well as of several diamino acids, and is due to a defect in the renal transport system for these amino acids, with consequent failure of renal tubular reabsorption. The intestinal transport of these amino acids is also defective in such individuals (Chaps. M5 and M10).

Cystathioninuria results from an imbalance between the cystathionine synthase (page 604) and cystathionase (page 605) reactions. Cystathioninurias can be divided into two categories, those related to augmented synthesis of cystathionine and those based on its decreased degradation. The rate of cystathionine synthesis probably depends upon the availability of the substrate, homocysteine, as well as on the tissue concentration of cystathionine synthase. In persons deficient in N^5-methyltetrahydrofolate-homocysteine methyltransferase (page 611), the resulting excess homocysteine is channeled into cystathionine accumulation. The second type of cystathioninuria is associated with a reduction of cystathionase activity, which also occurs in vitamin B$_6$ deficiency, although both cystathionine synthase and cystathionase utilize pyridoxal phosphate as a cofactor. The genetically defective apoenzyme may exhibit diminished capacity to bind the coenzyme, as well as inability to catalyze the reaction.

Homocystinuria results from defects in homocysteine metabolism based upon any of several genetic disorders that can impair the N^5-methyltetrahydrofolate-homocysteine methyltransferase reaction (page 611). Those include (1) a defect in the synthesis of the apoenzyme; (2) deficient synthesis of the substrate, methyltetrahydrofolate; (3) inadequate concentrations of the active form of the coenzyme, methyl cobalamin (Chap. M22); and (4) a deficiency in N^5,N^{10}-methylenetetrahydrofolate reductase (page 610). Since the last enzyme is necessary for the synthesis of methyltetrahydrofolate, the result would be a secondary deficiency of the substrate rather than the enzyme or coenzyme for the methyl transferase reaction.

Methionine

Some of the principal reactions of methionine metabolism are outlined in Fig. 25.5. The major metabolic roles of methionine are (1) utilization for protein synthesis; (2) conversion to S-adenosylmethionine, the prime methyl-group donor (page 612); and (3) conversion via the transsulfuration pathway to cysta-

Figure 25.5 Some reactions of methionine metabolism.

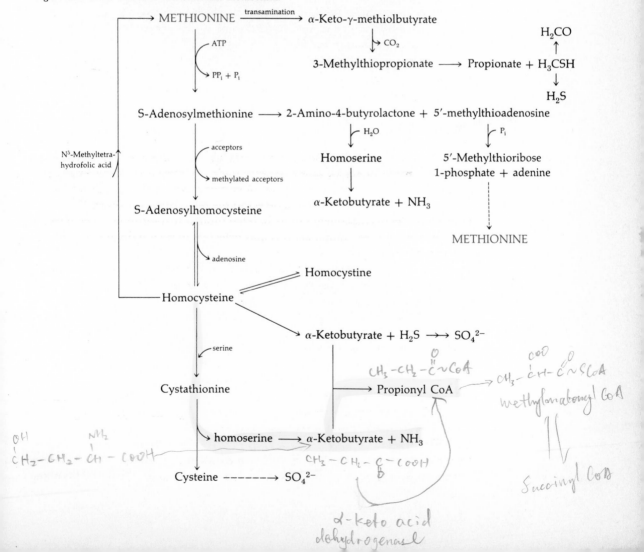

thionine, cysteine, and other derivatives. The last two of these are related; methionine is converted sequentially to *S*-adenosylmethionine, *S*-adenosylhomocysteine, and homocysteine. The last can be converted irreversibly to cystathionine, which is cleaved to α-ketobutyrate and cysteine (page 605). Oxidative decarboxylation of α-ketobutyrate results in formation of propionyl CoA, the subsequent metabolism of which has been considered previously (page 520).

Methionine can be degraded directly in liver and by bacteria to 3-methylthiopropionate via α-keto-γ-methiolbutyrate by transamination. The methylthiopropionate presumably yields acrylate or propionate; formaldehyde and H_2S are the immediate products from methylmercaptan (Fig. 25.5).

When excess methionine is fed to animals, growth is retarded. This has been ascribed to a limited ability to utilize all the methionine via the usual pathways. The result is an increased production of the highly toxic H_2S and H_3CSH. The *hypermethioninemia* observed in premature infants, in patients with liver disease, and in other situations is usually accompanied by various toxic manifestations which may be largely due to slow disposal of H_2S and H_3CSH.

An additional catabolic pathway for *S*-adenosylmethionine yields 2-amino-4-butyrolactone (homoserine lactone), and 5′-methylthioadenosine, which also arises in spermidine synthesis, in which *S*-adenosylmethionine participates (page 642). The methylthioadenosine is then cleaved to 5′-methylthioribose-1-phosphate and adenine.

acrylate

Ade

S-Adenosylmethionine $\longrightarrow$ $H_3C-S-CH_2$ ... **5′-Methylthioadenosine** (OH OH)

$+$ $CH_2-CH_2-\overset{NH_2}{CH}-C{=}O$ (O)

2-Amino-4-butyrolactone

$\downarrow$ $+H_2O$

$HOCH_2-CH_2-\overset{NH_2}{CH}-COOH$

Homoserine

Adenine $\longleftarrow$ P_i

$H_3C-S-CH_2$ ... OPO_3^{2-} (OH OH) $\dashrightarrow$ Methionine

5′-Methylthioribose-1-phosphate

The catalytic complex probably consists of two enzymes, *S-adenosylmethionine alkyl transferase* and *methylthioadenosine phosphorylase*. Hydrolysis of the aminobutyrolactone yields homoserine, which is converted to α-ketobutyrate and NH_3 in the cystathionine-γ-lyase-catalyzed reaction (page 605).

The 5′-methylthioribose-1-phosphate is converted in liver and other cells to methionine by unknown reactions in which the methylthio group and carbon atoms 2, 3, 4, and 5 of the ribose moiety are retained. Since methylthioadenosine is a byproduct of spermidine formation (Chap. 24), the pathway serves to conserve the methylthio group by reformation of methionine. The process would be of particular significance in rapidly dividing cells when relatively large amounts of polyamines are made.

Ethylene Formation The plant hormone ethylene, which initiates fruit ripening and also influences growth and development, is derived from methionine via S-adenosylmethionine. The latter is split *anaerobically*, presumably by a step involving pyridoxal phosphate (PLP).

$$S\text{-Adenosylmethionine} \xrightarrow{\text{PLP}}$$

$$\underset{\textbf{1-Aminocyclopropane-1-carboxylic acid}}{\begin{array}{c} H_2C \\ | \quad\backslash \\ \quad\;\; C \\ | \quad/ \quad \backslash \\ H_2C \quad\;\; COO^- \end{array}\;\; NH_3^+} \qquad + \text{ 5'-methylthioadenosine}$$

The cyclopropane derivative is fragmented by an *aerobic* process that may involve hydrogen peroxide.

$$\begin{array}{c} H_2C \\ | \quad\backslash \\ \quad\;\; C \\ | \quad/ \quad \backslash \\ H_2C \quad\;\; COO^- \end{array}\;\; NH_3^+ \xrightarrow[\text{or } H_2O_2]{O_2 + 2H} H_2C = CH_2 + HCOO^- + CO_2 + NH_4^+$$
$$\qquad\qquad\qquad\qquad\qquad\quad \textbf{Ethylene} \qquad \textbf{Formate}$$

Studies with labeled methionine show that C_1 yields CO_2, C_2 formate, and C_3 and C_4 ethylene. The process appears to be a major fate of methionine in fruit-bearing plants. Injection of either methionine or the aminocyclopropane carboxylic acid increases ethylene production and hastens fruit ripening, as does exposure to ethylene itself.

Threonine

Three pathways are known for threonine degradation in mammals, viz., (1) conversion to α-ketobutyric acid by *serine-threonine dehydratase* (page 650); (2) cleavage to glycine and acetaldehyde by *threonine aldolase;* and (3) dehydrogenation and decarboxylation to yield aminoacetone. The rate of the first reaction in liver increases following either threonine feeding or injection of an adrenal cortical steroid. The acetaldehyde formed in the second reaction is a source of acetyl CoA via oxidation to acetate. The third reaction occurs in liver; 2-amino-3-ketobutyrate is a presumed intermediate, which probably decarboxylates spontaneously. Aminoacetone is converted to 2-ketopropanol either by *monoamine oxidase* or by transamination.

$$\underset{\textbf{Aminoacetone}}{CH_3-\overset{\overset{\displaystyle O}{\|}}{C}-CH_2NH_2} + O_2 \xrightarrow{\text{monoamine oxidase}} \underset{\textbf{2-Ketopropanol}}{CH_3-\overset{\overset{\displaystyle O}{\|}}{C}-CH_2OH} + NH_3 + H_2O$$

The product is converted to pyruvate by a dehydrogenase that utilizes NADP or NAD, or to methylglyoxal and then to D-lactate by *glyoxalase*. The aminoacetone pathway is an important route for the catabolism of threonine in vertebrates.

Threonine metabolism leading to propionyl CoA, via α-ketobutyrate (the first reaction above), could make possible the potential appearance of three of the four carbon atoms of threonine as glucose, whereas the threonine aldolase reaction should provide two carbons to glucose, via glycine, and two carbons to ketone bodies, via acetaldehyde. However, ketone-body formation in vivo from threonine has not been reported.

Valine, Leucine, and Isoleucine

The catabolic fates of the branched-chain, aliphatic amino acids, valine, leucine, and isoleucine, exhibit several features in common. These amino acids are metabolized not only in liver but also in extrahepatic tissues, e.g., heart, kidney, brain, intestine, and skeletal muscle and provide important sources of energy for these tissues. Initial transamination with α-ketoglutarate as acceptor yields the corresponding α-keto acids. The occurrence of *hypervalinemia* in one child indicates the absence of a specific aminotransferase which was deficient in the leukocytes whereas transamination of leucine and isoleucine was normal. A single aminotransferase acts on leucine and isoleucine.

Each of the three branched-chain α-keto acids is decarboxylated oxidatively by a mitochondrial *branched-chain α-keto acid dehydrogenase* to the acyl CoA derivatives with one less carbon atom. CoA, NAD, and thiamine pyrophosphate are essential.

$$RCOCOOH + CoA\text{-}SH + NAD^+ \longrightarrow RCOS\text{-}CoA + CO_2 + NADH + H^+$$

All three branched-chain α-keto acids are oxidized by the same enzyme complex which also acts on pyruvate and α-ketobutyrate (a product of both threonine and methionine metabolism), but not on α-ketoglutarate. The branched-chain dehydrogenase strongly resembles the specific pyruvate and α-ketoglutarate dehydrogenases (Chap. 15) in containing the FAD-enzyme *dihydrolipoyl dehydrogenase* and *dihydrolipoyl transacylase*. The mechanism of action of all these dehydrogenases is similar. In contrast to pyruvate dehydrogenase, the branched-chain α-keto acid dehydrogenase complex is not regulated by phosphorylation-dephosphorylation; however, it is inhibited by the end products of the reaction, i.e., NADH and any of the branched-chain acyl CoA derivatives.

In the autosomal recessive hereditary disorder *maple syrup urine disease,* all three branched-chain α-keto acids are present in the urine, producing the characteristic odor. There are also elevated levels of the three branched-chain amino acids in the blood and urine of the affected children who may survive for a few years but show severe mental retardation and at autopsy extensive failure of myelinization. The defect is in the branched-chain α-keto acid dehydrogenase complex.

Each of the branched-chain CoA derivatives formed from the α-keto acids is dehydrogenated by mitochondrial FAD enzymes that appear to be distinct from the similar enzymes that catalyze dehydrogenation of the short straight-chain CoA derivatives that function in fatty acid metabolism (Chap. 20).

Valine

Oxidation of valine (Fig. 25.6) leads to methylmalonyl CoA, which is converted to succinyl CoA (page 520) and completely oxidized, or provides three of

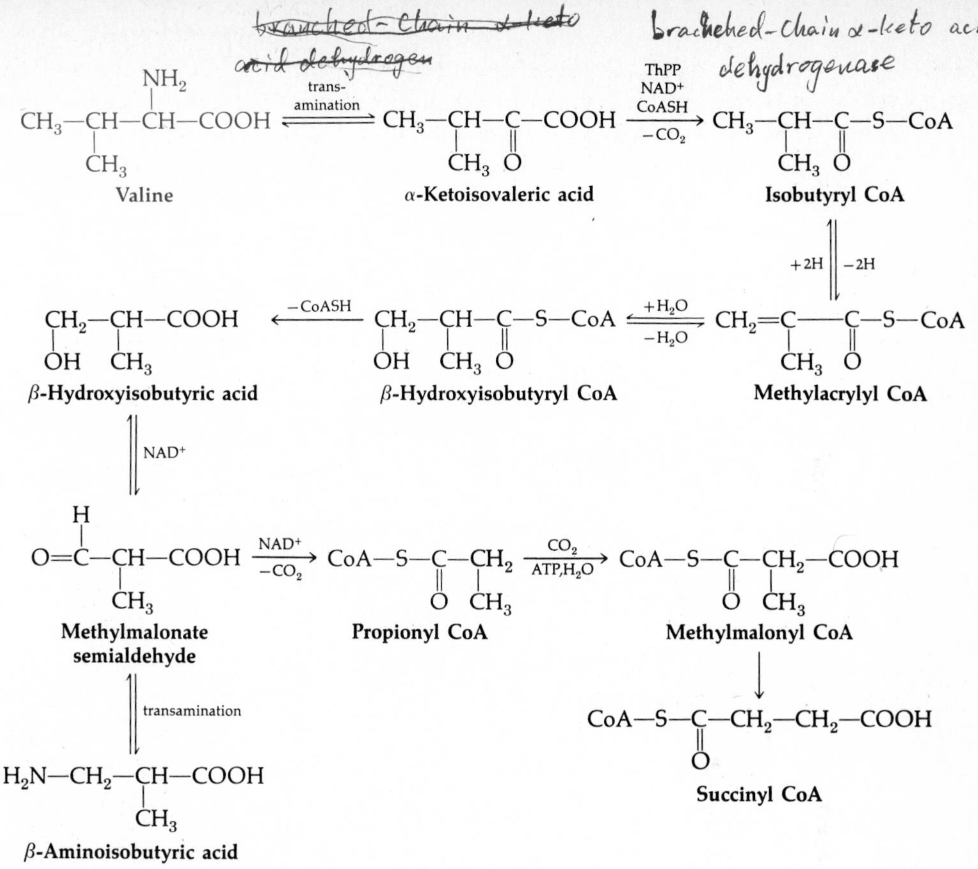

Figure 25.6 Metabolic fate of valine.

five carbon atoms for gluconeogenesis. Formation of methylmalonate semial-dehyde (Fig. 25.6) followed by transamination of the latter would yield β-aminoisobutyric acid, which is also a product of pyrimidine metabolism (Chap. 26). β-Aminoisobutyric acid is present in the urine of some individuals in amounts between 200 to 300 mg/day, either because of an inherited trait or as a consequence of disease.

Leucine

Oxidation of leucine (Fig. 25.7) terminates with formation of one molecule of acetoacetic acid and one molecule of acetyl CoA, in accord with the finding that, in the diabetic animal, 1 mol of leucine yields 1.5 mol of acetoacetic acid. The conversion of β-methylcrotonyl CoA to β-methylglutaconyl CoA is a bio-tin-requiring reaction with a mechanism similar to that of other biotin-CO_2 reactions (page 409). Noteworthy also is the production of 3-hydroxy-3-meth-ylglutaryl CoA, the intermediate in formation of acetoacetyl CoA.

A metabolic defect, *isovaleric acidemia*, is specific for leucine metabolism and is characterized by high levels of isovaleric acid in the plasma. The condition can be fatal and, if untreated by restriction of leucine intake, leads to mental retardation and episodes of convulsion. During periods of remission, the isovaleric acid produced from leucine is excreted mainly as the glycine conju-gate. The metabolic lesion is the result of reduced activity of the specific

659

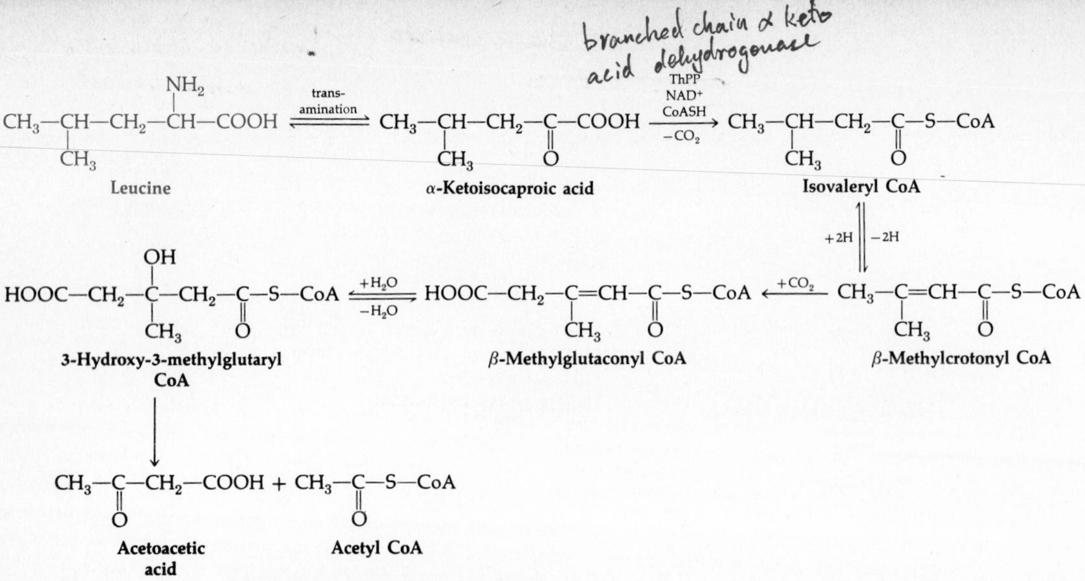

$$\text{CH}_3\text{—CH—CH}_2\text{—CH—COOH} \underset{\text{amination}}{\overset{\text{trans-}}{\rightleftharpoons}} \text{CH}_3\text{—CH—CH}_2\text{—C—COOH} \xrightarrow[-CO_2]{\text{CoASH}} \text{CH}_3\text{—CH—CH}_2\text{—C—S—CoA}$$

NH₂ above the CH, CH₃ below; ThPP NAD⁺ above arrow

Leucine   α-Ketoisocaproic acid   Isovaleryl CoA

+2H ‖ −2H

$$\text{HOOC—CH}_2\text{—C—CH}_2\text{—C—S—CoA} \underset{-H_2O}{\overset{+H_2O}{\rightleftharpoons}} \text{HOOC—CH}_2\text{—C=CH—C—S—CoA} \xleftarrow{+CO_2} \text{CH}_3\text{—C=CH—C—S—CoA}$$

3-Hydroxy-3-methylglutaryl CoA   β-Methylglutaconyl CoA   β-Methylcrotonyl CoA

$$\text{CH}_3\text{—C—CH}_2\text{—COOH} + \text{CH}_3\text{—C—S—CoA}$$

Acetoacetic acid   Acetyl CoA

Figure 25.7 Metabolic fate of leucine.

isovaleryl-CoA dehydrogenase catalyzing the conversion of isovaleryl CoA to β-methyl-crotonyl CoA (Fig. 25.7).

β-Methylcrotonyl acidemia is characterized by excretion not only of this acid but also of its glycine conjugate. Presumably, the enzyme, β-*methylcrotonyl CoA carboxylase*, is defective since administration of large amounts of biotin produces a remission of the disorder.

Interconversion of β-Leucine and Leucine In addition to the major pathways of valine and leucine metabolism shown in Fig. 25.6 and 25.7, another pathway is present in liver. Isobutyryl CoA, derived from valine and phytanic acid (Chap. 20), is converted to leucine, presumably via β-ketoisocaproic acid to β-leucine in the following postulated steps.

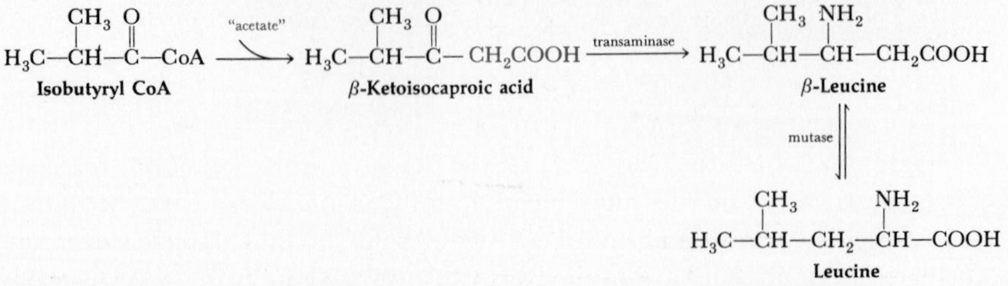

$$\text{H}_3\text{C—CH—C—CoA} \xrightarrow{\text{"acetate"}} \text{H}_3\text{C—CH—C—CH}_2\text{COOH} \xrightarrow{\text{transaminase}} \text{H}_3\text{C—CH—CH—CH}_2\text{COOH}$$

Isobutyryl CoA   β-Ketoisocaproic acid   β-Leucine

mutase

$$\text{H}_3\text{C—CH—CH}_2\text{—CH—COOH}$$

Leucine

Interconversion of β-leucine and leucine is catalyzed by a cobamide-dependent enzyme, *leucine 2,3-aminomutase*, in the same fashion as the conversion of methyl-malonyl CoA to succinyl CoA (Chap. 20). Patients with pernicious anemia, deficient in the B_{12} coenzyme, show an elevated level of β-leucine and a lower level of leucine in plasma.

Inasmuch as leucine is essential in the diet, the amount of leucine that is furnished by the above pathway is inadequate for physiological needs.

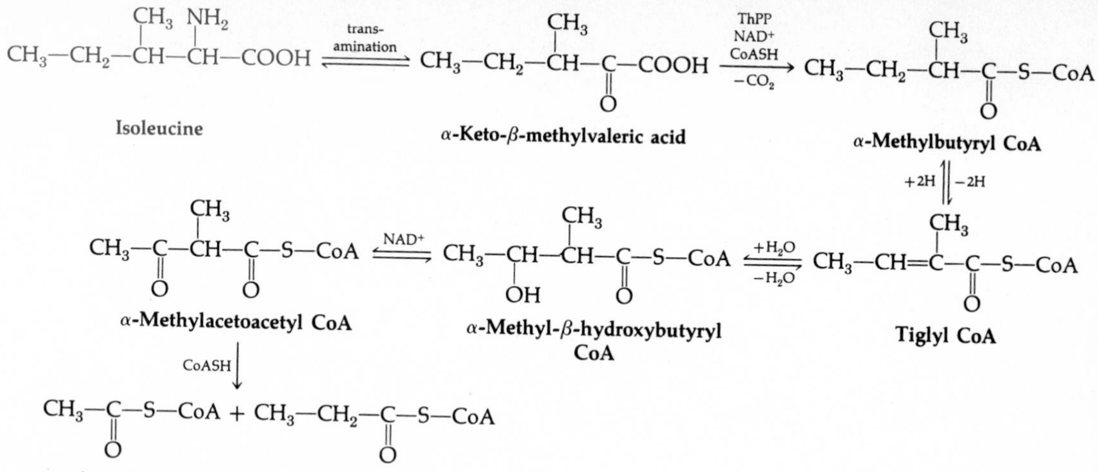

CH$_3$—CH$_2$—$\overset{\overset{\displaystyle CH_3}{|}}{CH}$—$\overset{\overset{\displaystyle NH_2}{|}}{CH}$—COOH $\underset{\text{amination}}{\overset{\text{trans-}}{\rightleftharpoons}}$ CH$_3$—CH$_2$—$\overset{\overset{\displaystyle CH_3}{|}}{CH}$—$\underset{\overset{\displaystyle \|}{O}}{C}$—COOH $\xrightarrow[-CO_2]{\overset{\text{ThPP}}{\underset{\text{CoASH}}{\overset{\text{NAD}^+}{}}}}$ CH$_3$—CH$_2$—$\overset{\overset{\displaystyle CH_3}{|}}{CH}$—$\underset{\overset{\displaystyle \|}{O}}{C}$—S—CoA

Isoleucine   α-Keto-β-methylvaleric acid   α-Methylbutyryl CoA

+2H ‖ −2H

CH$_3$—$\underset{\overset{\displaystyle \|}{O}}{C}$—$\overset{\overset{\displaystyle CH_3}{|}}{CH}$—$\underset{\overset{\displaystyle \|}{O}}{C}$—S—CoA $\overset{\text{NAD}^+}{\rightleftharpoons}$ CH$_3$—$\overset{\overset{\displaystyle CH_3}{|}}{CH}$—$\underset{\overset{\displaystyle |}{OH}}{CH}$—$\underset{\overset{\displaystyle \|}{O}}{C}$—S—CoA $\underset{-H_2O}{\overset{+H_2O}{\rightleftharpoons}}$ CH$_3$—CH=$\overset{\overset{\displaystyle CH_3}{|}}{C}$—$\underset{\overset{\displaystyle \|}{O}}{C}$—S—CoA

α-Methylacetoacetyl CoA   α-Methyl-β-hydroxybutyryl CoA   Tiglyl CoA

CoASH ↓

CH$_3$—$\underset{\overset{\displaystyle \|}{O}}{C}$—S—CoA + CH$_3$—CH$_2$—$\underset{\overset{\displaystyle \|}{O}}{C}$—S—CoA

Acetyl CoA   Propionyl CoA

Figure 25.8 Metabolic fate of isoleucine.

Isoleucine

Metabolism of isoleucine (Fig. 25.8) is concluded with formation of one molecule each of acetyl CoA and propionyl CoA, in accord with evidence that this amino acid is both weakly glucogenic and weakly ketogenic.

Phenylalanine

In normal individuals, almost all the phenylalanine not utilized for protein synthesis is converted to tyrosine by the action of phenylalanine hydroxylase in the liver (page 606). Only in phenylketonuria is metabolism diverted to formation of phenylpyruvate, which accumulates or is reduced to phenyllactate; a small fraction of the latter is oxidized to phenylacetate, which is excreted as phenylacetylglutamine. Pathways of phenylalanine and tyrosine metabolism are depicted in Fig. 25.9.

In plants, *phenylalanine amino-lyase* catalyzes the following reaction.

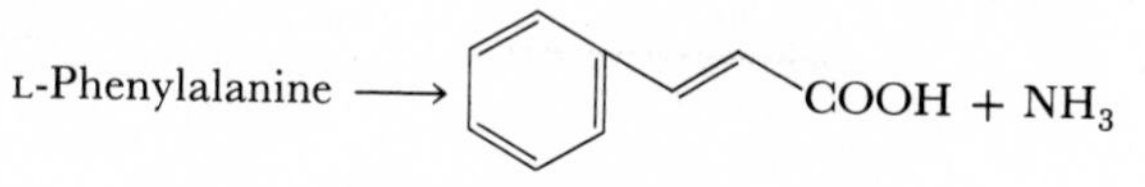

L-Phenylalanine $\longrightarrow$ ⟨benzene ring⟩—CH=CH—COOH + NH$_3$

trans-Cinnamic acid

The lyase does not utilize pyridoxal phosphate but presumably operates in the same fashion as the similar *histidine ammonia-lyase* (page 666). Cinnamic acid is a major source of the aromatic ring for formation of many compounds including lignins, tannins, many alkaloids, plant pigments, and others.

Tyrosine

The initial event in hepatic tyrosine metabolism is transamination, catalyzed by tyrosine-glutamate aminotransferase, an inducible enzyme whose synthesis increases significantly in the liver after administration of tyrosine or of

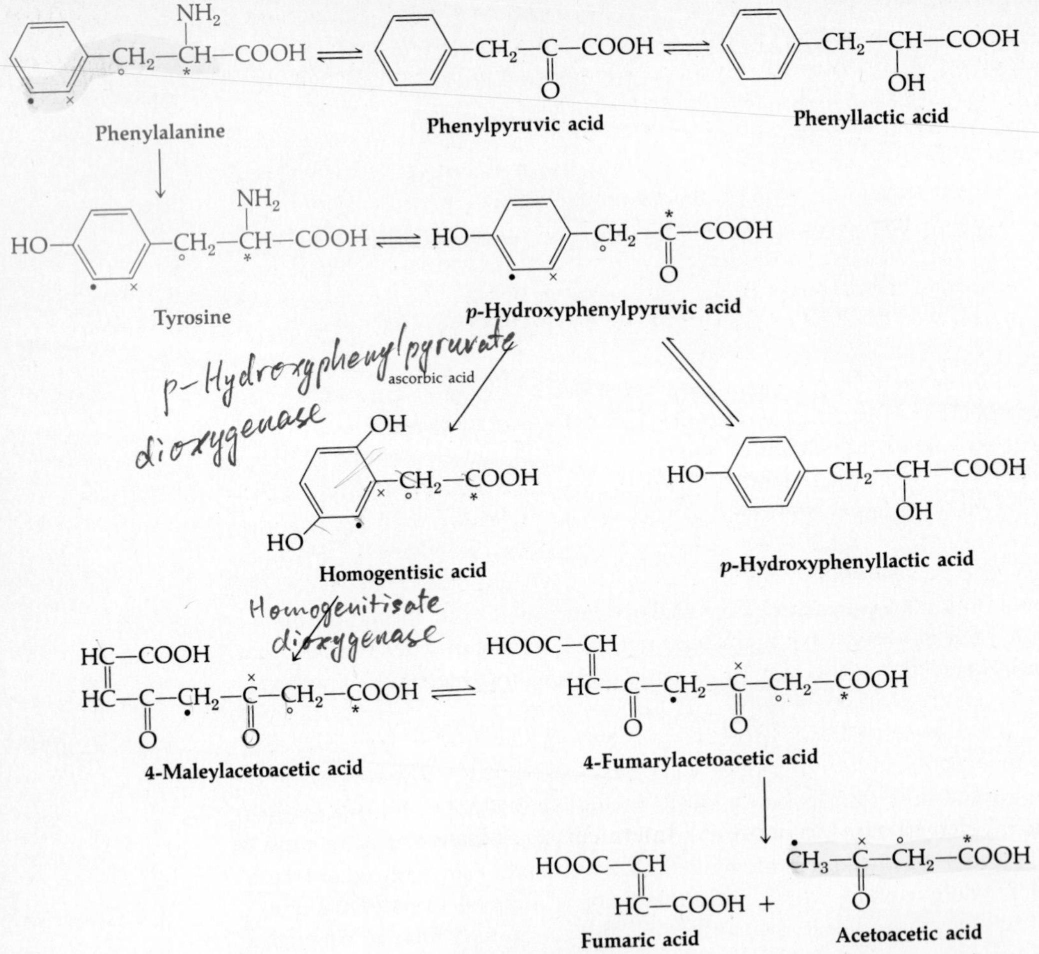

CH₂—CH—COOH (NH₂) ⇌ —CH₂—C—COOH (O) ⇌ —CH₂—CH—COOH (OH)

Phenylalanine   **Phenylpyruvic acid**   **Phenyllactic acid**

HO—⟩—CH₂—CH—COOH (NH₂) ⇌ HO—⟩—CH₂—C—COOH (O)

Tyrosine   **p-Hydroxyphenylpyruvic acid**

p-Hydroxyphenylpyruvate dioxygenase

ascorbic acid

OH —CH₂—COOH / HO

HO—⟩—CH₂—CH—COOH (OH)

Homogentisic acid   **p-Hydroxyphenyllactic acid**

Homogenitisate dioxygenase

HC—COOH / HC—C—CH₂—C—CH₂—COOH ⇌ HOOC—CH / HC—C—CH₂—C—CH₂—COOH

4-Maleylacetoacetic acid   **4-Fumarylacetoacetic acid**

HOOC—CH / HC—COOH +   CH₃—C—CH₂—COOH (O)

Fumaric acid   **Acetoacetic acid**

Figure 25.9 Principal pathways of phenylalanine and tyrosine metabolism. Isotopic tracer studies of the conversion of phenylalanine to acetoacetate have shown that (1) the α-carbon atom (*) of phenylalanine becomes the carboxyl carbon of acetoacetate, (2) carbon atom 2 of the aromatic ring (×) is a precursor of the carbonyl carbon of acetoacetate, (3) carbon atom 1 or 3 (•) of the ring is a precursor of the terminal carbon atom of acetoacetate, and (4) the β-carbon atom of phenylalanine (°) becomes the α-carbon atom of acetoacetate. Thus, a shift of the side chain occurs during formation of homogentisic acid.

adrenal steroids (Chap. M18), insulin, or glucagon (Chap. M16). The level of activity of this enzyme is decreased following injection of hypophyseal somatotropin (Chap. M20). *p-Hydroxyphenylpyruvate dioxygenase,* a copper-containing protein, catalyzes oxidation of *p*-hydroxyphenylpyruvate to homogentisic acid. Comparison of the structures of substrate and product indicates the complexity of the reaction: hydroxylation of the ring, oxidation, decarboxylation, and migration of the side chain. The activity of this enzyme is low in fetal liver and increases slowly following birth. This may account in part for the urinary excretion of relatively large amounts of the hydroxyphenyllactate metabolites, formed by the action of *aromatic α-keto acid reductase,* in premature infants or in

662

normal infants who are fed tyrosine. Ascorbic acid is essential for the normal activity of p-hydroxyphenylpyruvate dioxygenase, since ascorbic acid–deficient guinea pigs and infants excrete substantial amounts of p-hydroxyphenylpyruvic acid in response to an administered dose of tyrosine.

Oxidation of homogentisic acid (Fig. 25.9) is catalyzed by *homogentisate dioxygenase,* which requires Fe^{2+} and a sulfhydryl compound, e.g., reduced glutathione, for its action. Scission of the aromatic ring between the side chain and the adjacent hydroxyl groups occurs, with one of the O_2 atoms becoming the 3-carbonyl oxygen of the product, maleylacetoacetate, and the other the carbonyl oxygen of the carboxyl group. As indicated by the isotopic tracer studies (Fig. 25.9), there is a shift of the side chain during the formation of homogentisate.

Failure to oxidize homogentisate is seen in *alkaptonuria,* a hereditary disorder. The livers of alkaptonuric individuals lack homogentisate oxidase but contain all the earlier and later enzymes of the tyrosine catabolic pathway. Excretion of homogentisic acid is apparent since the urine, when made slightly alkaline, rapidly darkens on exposure to air due to oxidation of homogentisic acid to a quinone, which polymerizes to a melanin-like material. With the passage of time abnormal pigmentation of cartilage and other connective tissue (*ochronosis*) may become apparent in alkaptonurics.

Only one case of *tyrosinosis* has been reported. The metabolic block appears to have been due to lack of *p-hydroxyphenylpyruvate dioxygenase,* with consequent excessive excretion of p-hydroxyphenylpyruvic acid.

p-Hydroxyphenyllactic acid, arising from tyrosine metabolism (Fig. 25.9), is metabolized via dehydration and subsequent β oxidation to form p-hydroxybenzoic acid. The latter is a precursor of ubiquinone (page 337), via an intermediate, 5-demethoxyubiquinone-9. Intermediate products are as yet unidentified.

p-Hydroxyphenyllactic acid   p-Hydroxybenzoic acid   5-Demethoxyubiquinone-9

$$R = -(CH_2-CH=\overset{\overset{\textstyle CH_3}{\textstyle |}}{C}-CH_2)_9H$$

Ubiquinone-9

Conversion of tyrosine to hydroxytyramine was described previously (page 638). In that sequence, the initial formation of dihydroxyphenylalanine is of prime significance for catecholamine synthesis in brain (Chap. M7) and in the adrenal medulla (Chap. M18).

Tyrosine as a Precursor of Melanin In the basal layer of the epidermis are melanoblasts, in which the dark pigment melanin is synthesized from tyrosine by the copper-containing *tyrosinase*. This enzyme catalyzes oxidation of tyrosine to dihydroxyphenylalanine (dopa) and to dopa quinone (Fig. 25.10). Tyrosinase can also catalyze the oxidation of dopa. *Hallachrome*, so named because it was isolated from a polychaete worm, *Halla parthenolapa*, is formed nonenzymically by dismutation of dopa quinone and the leuko compound.

$$\text{Dopa quinone} + \text{leuko compound} \longrightarrow \text{dopa} + \text{hallachrome}$$

The oxidation of indole 5,6-quinone to melanin is probably spontaneous. Melanin (Gk. *melas,* "black") is a group of polymers of random structure. The color of skin depends upon the distribution of melanoblasts, the melanin concentration, and perhaps its state of oxidation, since melanin can be reduced with ascorbic acid or hydrosulfite from a black to a tan form.

Melanin is also normally found in the retina, the ciliary body, the choroid, the substantia nigra of brain, and the adrenal medulla. Melanoblasts occasionally give rise to highly malignant melanomas, which may or may not be pigmented. When stained with dopa, nonpigmented melanotic tumors rapidly darken and are therefore called *amelanotic melanomas*. Darkening of the human skin is initiated by ultraviolet irradiation of tyrosine, leading to formation of dopa. In genetic *albinism,* either melanin-forming cells or tyrosinase, or both, may be entirely absent.

Lysine

Lysine does not participate in transamination. Lysine degradation in liver and other tissues occurs in mitochondria primarily by a pathway with interme-

Figure 25.10 Melanin formation from tyrosine.

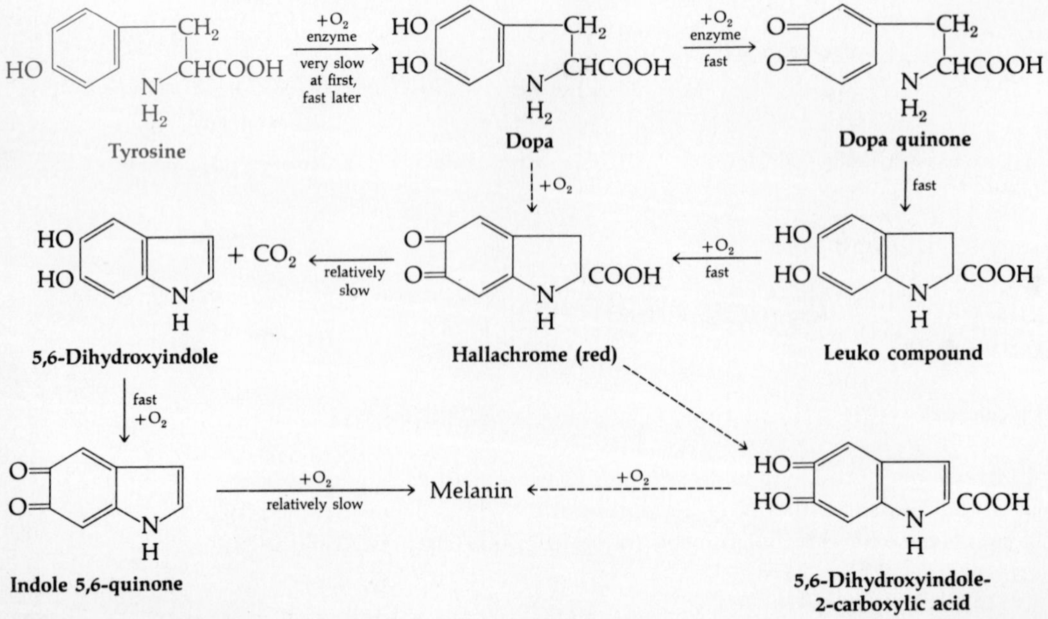

diates identical to those of lysine synthesis in fungi (Chap. 22). Saccharopine offers a direct route to α-aminoadipic-ε-semialdehyde (Fig. 25.11). This semialdehyde also arises in a second pathway of lysine degradation initiated by oxidative deamination, catalyzed by L-amino acid oxidase of liver. The resultant keto acid cyclizes spontaneously to yield Δ^1-piperideine 2-carboxylic acid; reduction by an enzyme that utilizes either NADH or NADPH leads to pipecolic acid. The latter is oxidized to Δ^1-piperideine 6-carboxylic acid, which, in the presence of ATP and Mg^{2+}, can be oxidized by liver mitochondria to α-aminoadipic acid via the semialdehyde. Liver contains an enzyme catalyzing a pyridine nucleotide-dependent oxidation of α-aminoadipic acid-ε-semialdehyde to α-aminoadipic acid. Metabolism of the latter proceeds by transamination to

Figure 25.11 Pathways for the metabolic degradation of lysine.

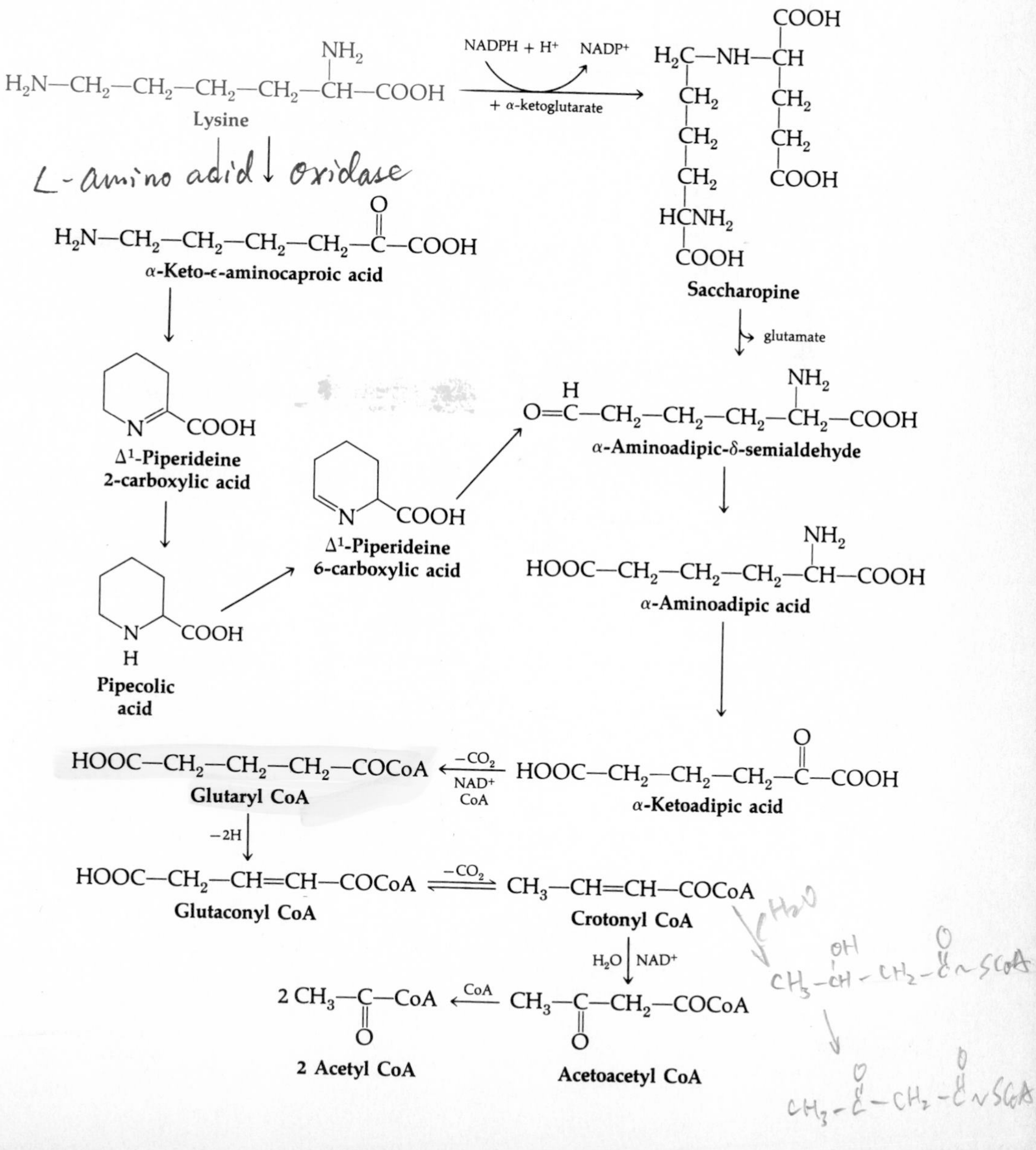

α-ketoadipic acid, which, by oxidative decarboxylation, yields glutaryl CoA; the latter is metabolized as indicated in Fig. 25.11.

A third pathway of lysine degradation in mammalian tissue, as well as in yeast, proceeds through acylated intermediates and thus avoids cyclization of the keto derivative of lysine into the pipecolic acid pathway. An enzyme in liver catalyzes formation of ε-*N*-acetyllysine. The acyl residue is retained through steps leading to α-hydroxy-ε-acetamidocaproic acid (Fig. 25.12), which on deacylation leads to glutarate. Alternatively, deacylation occurring at an earlier step would permit metabolism via the pipecolate pathway (Fig. 25.11).

A rare genetic disorder, in which there is mental and physical retardation and other defects, is associated with a low rate of saccharopine formation, *hyperlysinemia* and *hyperlysinuria;* this suggests that the saccharopine pathway is the predominant one in human tissues.

Arginine

Arginine catabolism in mammals to ornithine and urea has been described (Chap. 23), as has the metabolic fate of ornithine.

Histidine

Although histidine can transaminate to form the corresponding imidazole pyruvate, the major pathway of histidine degradation is initiated by *histidine ammonia-lyase,* which catalyzes the α,β removal of a molecule of NH_3 with formation of urocanate (Fig. 25.13).

Histidine ammonia-lyase from rat liver is a sulfhydryl enzyme and has a dehydroalanine residue ($-NH-\overset{\overset{\displaystyle CH_2}{\|}}{C}-CO-$) at the active site. Histidine ammonia-lyase is lacking in individuals with the hereditary disorder *histidinemia,* resulting in elevated blood and urine levels of histidine.

Urocanic acid hydratase catalyzes transformation of urocanate to imidazolone 3-propionate. This reaction involves the elements of water and an internal oxidation and reduction. The enzyme contains tightly bound NAD^+. *Imidazo-*

Figure 25.12 Acylated intermediates in lysine degradation.

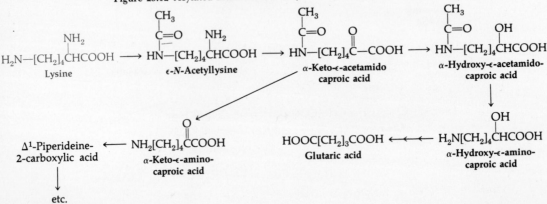

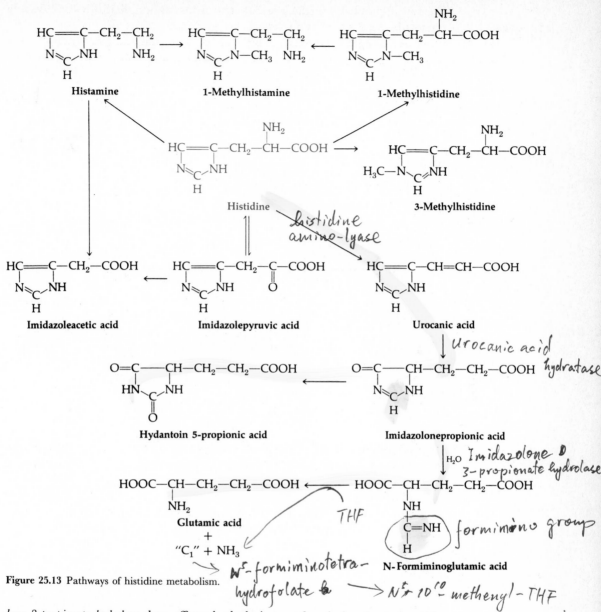

Figure 25.13 Pathways of histidine metabolism.

lone 3-propionate hydrolase then effects hydrolysis to α-formiminoglutamic acid. The formimino group can be transferred, by a specific transferase, to the N^5 position of tetrahydrofolate. The N^5-formiminotetrahydrofolate hydrolyzes, cyclizes to the N^5,N^{10}-methenyl derivative, and can yield the N^{10}-formyl- or the N^5, N^{10}-methylenetetrahydrofolate. Thus the C_2 of the histidine imidazole ring returns to the C_1 pool. Animals or human beings deficient in folic acid or in vitamin B_{12} excrete excessive amounts of formiminoglutamic acid, suggesting that a major pathway for metabolism of the formimino group is conversion to a methyl group.

A small amount of imidazolone 3-propionate may also be oxidized to hydantoin 5-propionate, which is excreted in the urine. Pathways of histidine metabolism are depicted in Fig. 25.13.

Tryptophan

Tryptophan is degraded in mammals primarily by two pathways: One involves oxidation of tryptophan to 5-hydroxytryptophan followed by decarboxylation to 5-hydroxytryptamine (serotonin). This pathway was described previously, as well as that leading to indoleacetic acid (page 640). The other pathway involves oxidation of tryptophan to kynurenine, which is converted to a series of intermediates and by-products (Fig. 25.14), all but one of which (glutaryl CoA) appear in the urine in varying amounts, and the sum of which accounts approximately for the total metabolism of tryptophan.

Oxidation of tryptophan to *N*-formylkynurenine is catalyzed by *tryptophan 2,3-dioxygenase*. Each mole of enzyme contains 2 mol of ferroheme. The dioxygenase is widely distributed in nature and has been highly purified from liver cytosol and from a *Pseudomonas*.

Tryptophan dioxygenase is not present in the livers of normal rats up to the tenth postnatal day but can be induced prematurely by adrenal cortical steroids (Chap. M18). Once the enzyme is present, administration of either the steroids or tryptophan, elevates the level of enzymic activity. The elevation of enzyme levels by steroids occurs by stimulating synthesis of the specific messenger RNA (mRNA) that codes (Chap. 29) for synthesis of the apoenzyme of the dioxygenase. In contrast, the elevated dioxygenase levels induced by tryptophan administration are due to a decreased rate of apoenzyme degradation while synthesis proceeds normally.

In *Hartnup disease,* a hereditary disorder associated with mental retardation, only a small fraction of ingested tryptophan is oxidized because of a defect in intestinal transport. Hydroxylation of kynurenine is catalyzed by *kynurenine 3-oxygenase,* which requires NADPH and O_2. 3-Hydroxykynurenine is found in mammalian urine as the glucosiduronate, the *O*-sulfate, and the *N*-α-acetyl derivative. A pyridoxal phosphate–requiring enzyme, *kynureninase,* can catalyze cleavage of both kynurenine and 3-hydroxykynurenine to anthranilic and 3-hydroxyanthranilic acids, respectively. Alanine is the other reaction product in both instances. 3-Hydroxykynurenine is split approximately twice as rapidly as kynurenine. The latter also transaminates; the corresponding keto acid undergoes ring closure to yield kynurenic acid. This compound is dehydroxylated to quinaldic acid. Quinaldic acid excretion accounts for approximately 30 percent of ingested kynurenic acid in human subjects.

3-Hydroxyanthranilic acid oxidation is catalyzed by mitochondrial *3-hydroxyanthranilate oxygenase,* present in liver and kidney. The enzyme requires Fe^{2+} and a sulfhydryl compound; O_2 is incorporated into the product, α-amino-β-carboxymuconic-δ-semialdehyde. This compound represents a branch point in tryptophan degradation; it may be converted to quinolinic and nicotinic acids or follow a pathway to glutaric acid and thus to acetyl CoA. The semialdehyde may also be decarboxylated, followed by cyclization with formation of picolinic acid. These transformations are shown in Fig. 25.14.

Quinolinic acid reacts with 5-phosphoribosyl-1-pyrophosphate, with ac-

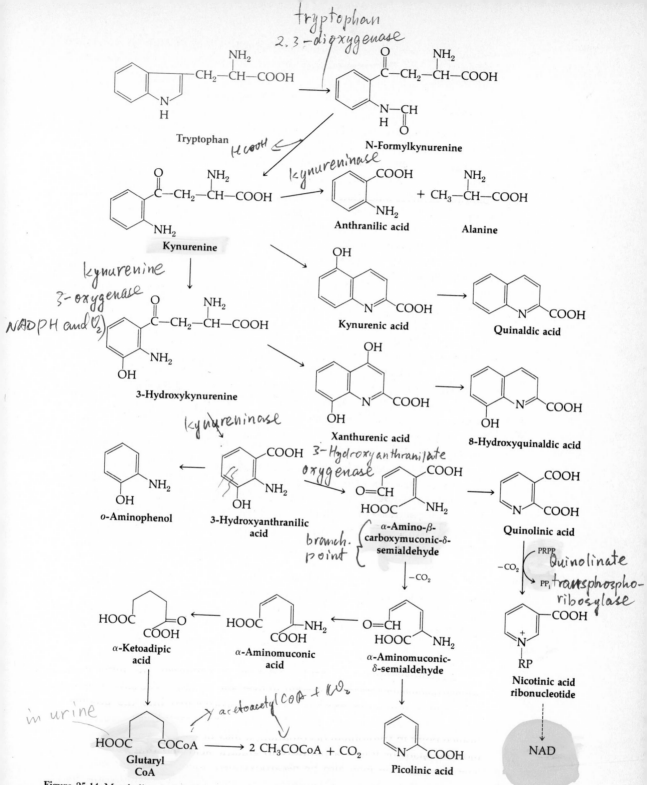

Figure 25.14 Metabolic relationships between tryptophan and its metabolites. PRPP = 5-phosphoribosyl-1-pyrophosphate; RP = ribose-5-phosphate.

estrogen
雌激素

companying loss of CO_2, to form nicotinic acid ribonucleotide in a reaction catalyzed by *quinolinate transphosphoribosylase*. Nicotinic acid ribonucleotide is directly utilized for NAD synthesis (Chap. 26).

Both rats and human beings deficient in pyridoxine excrete abnormally large amounts of kynurenine and xanthurenic acid. The latter is also excreted in excessive quantity in pregnancy; this may be related to elevated blood levels of estrogenic and progestational hormones, since estrogen has been reported to inhibit the conversion in vitro of kynurenine to kynurenic acid. Xanthurenic acid may lose a hydroxyl group with formation of 8-hydroxyquinaldic acid. Indoleacetic acid is present in the urine of mammals as indoleaceturic acid, formed by conjugation with glycine. A possible precursor is tryptamine, produced in the large intestine by bacterial decarboxylation of tryptophan. The microorganisms of the large intestine can further degrade indoleacetic acid to yield indole (R=H), indoxyl (R=OH), skatole (R=CH$_3$), and skatoxyl (R=CH$_2$OH), as shown in the structure below.

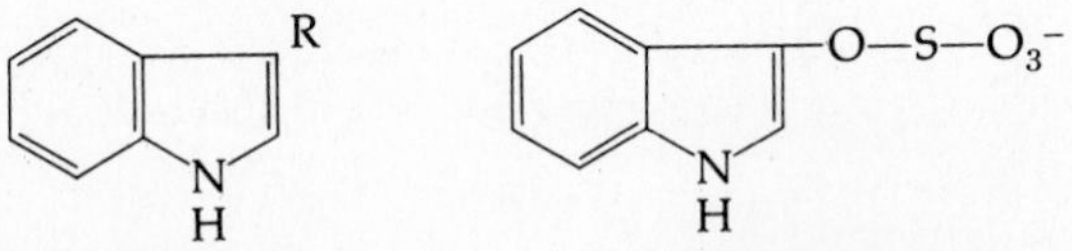

Indole (R = H)   Indoxylsulfate

Skatole and indole contribute to the unpleasant odor of feces. Small amounts of indoxyl and skatoxyl enter the circulation from the gut, are conjugated either with sulfate or glucuronic acid in the liver, and are excreted in the urine as ester sulfates or as glucosiduronates. Indican is the potassium salt of indoxylsulfate.

Tryptophanase of *E. coli* is composed of four identical subunits and catalyzes reversibly the cleavage of tryptophan, and several α,β elimination and β replacement reactions. Pyridoxal phosphate is the coenzyme.

$$\text{Tryptophan} \rightleftharpoons \text{indole} + \text{pyruvate} + NH_3$$
$$\text{Serine} \longrightarrow \text{pyruvate} + NH_3$$
$$\text{Cysteine} + H_2O \longrightarrow \text{pyruvate} + NH_3 + H_2S$$
$$\text{Serine} + \text{indole} \longrightarrow \text{tryptophan} + H_2O$$
$$\text{Cysteine} + \text{indole} \longrightarrow \text{tryptophan} + H_2S$$

REFERENCES

Books

Bender, D. A.: *Amino Acid Metabolism*, Wiley, New York, 1975.

Greenberg, D. M. (ed.): *Metabolic Pathways*, 3d ed., vol. VII, *Metabolism of Sulfur Compounds*, Academic, New York, 1975.

Munro, H. N., and J. B. Allison (eds.): *Mammalian Protein Metabolism*, vols. I and II, Academic, New York, 1964.

Stanbury, J. B., J. B. Wyngaarden, and D. S. Fredrickson (eds.): *The Metabolic Basis of Inherited Disease*, 4th ed., McGraw-Hill, New York, 1978.

Walser, M., and J. R. Williamson (eds.): *Metabolism and Clinical Implications of Branched-chain Amino and Ketoacids,* Elsevier/North-Holland, New York, 1981.

Review Articles

Adams, E., and L. Frank: Metabolism of Proline and Hydroxyprolines, *Annu. Rev. Biochem.* **49:**1006–1061 (1980).

Braunstein, A. E.: Amino Group Transfer, pp. 379–481, in P. D. Boyer (ed.): The Enzymes, 3d ed., vol. 9, Academic, New York, 1973.

Felig, P.: Amino Acid Metabolism in Man, *Annu. Rev. Biochem.* **44:**993–995 (1975).

Knox, W. E.: Sir Archibald Garrod's "Inborn Errors of Metabolism," I: Cystinuria; II: Alkaptonuria, *Am. J. Hum. Genet.* **10:**3–32, 95–124 (1958).

Nyhan, W. L.: Inheritable Abnormalities of Amino Acid Metabolism, in A. Neuberger and L. L. M. Van Deenum (eds.): *Comprehensive Biochemistry,* vol. 19A, pp. 305–468, Elsevier, Amsterdam, 1981.

Snell, E. E.; Pyruvate-containing Enzymes, *Trends Biochem. Sci.* **2:**131–135 (1977).

Snell, K.: Muscle Alanine Synthesis and Hepatic Gluconeogenesis, *Trends Biochem. Sci.* **8:**205–213 (1979).

Truffa-Bachi, P., and G. N. Cohen: Amino Acid Metabolism, *Annu. Rev. Biochem.* **42:**113–134 (1973).

Wellner, D., and A. Meister: A Survey of Inborn Errors of Amino Acid Metabolism and Transport in Man, *Annu. Rev. Biochem.* **50:**911–968 (1981).

Chapter 26

Metabolism of Purine and Pyrimidine Nucleotides

The chemistry of the purines and pyrimidines was presented in Chap. 8 and the diverse roles of various nucleotides have been discussed in preceding chapters. This chapter will be concerned with the origin and metabolic fate of each of the nucleotides. The biosynthesis of the nucleic acids and their participation in protein synthesis are considered in the following chapters.

The purine and pyrimidine nucleotides can be synthesized *de novo* from nonucleotide precursors (amino acids, CO_2, and NH_3), or they can be synthesized directly from preformed purine and pyrimidine bases by what have been termed *salvage pathways*. The word *salvage* should not be taken to indicate that these pathways are of lesser significance. Indeed, in certain types of cells they are the major routes of purine and pyrimidine nucleotide biosynthesis.

SYNTHESIS OF PURINE AND PYRIMIDINE NUCLEOTIDES

De novo Pathway of Purine Nucleotide Synthesis

The general precursors of the purine ring were first established by administering isotopically labeled nutrients to pigeons, which excrete their waste nitrogen in the form of uric acid. Chemical degradation of the excreted uric acid yielded the following general picture of the origin of the purine nucleus.

Carbon atoms 2 and 8 derive from formate or the 1-carbon unit arising from various compounds, e.g., serine and glycine (Chap. 23). Carbon atom 6 originates from CO_2. Glycine contributes carbon atoms at positions 4 and 5 and the nitrogen at 7; the nitrogen atom at position 1 derives from aspartic acid; and glutamine amide nitrogen contributes the nitrogen at positions 3 and 9.

Any cross-references coded M refer to the companion text, *Principles of Biochemistry: Mammalian Biochemistry*.

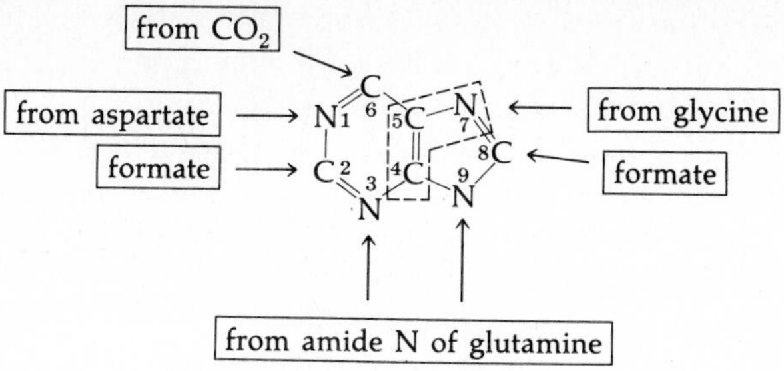

The general route of *de novo* pathways for purine biosynthesis is essentially the same in many species (mammals, birds, yeast, and bacteria) that have been studied. In essence, this process consists of a series of successive reactions by which the purine ring is assembled on carbon 1 of ribose-5-phosphate, leading directly to the purine ribonucleotides. Neither free purines nor nucleosides appear as intermediates in this sequence.

Synthesis of Inosine-5′-Phosphate 5-Phosphoribosyl-1-pyrophosphate (PRPP) is the initial reactant in purine nucleotide biosynthesis. This compound, which is also essential for the synthesis of pyrimidine nucleotides, is formed from ribose-5-phosphate and ATP in an unusual reaction that involves transfer of pyrophosphate rather than phosphate. The synthesis of PRPP by *PRPP synthetase* proceeds according to the following reaction:

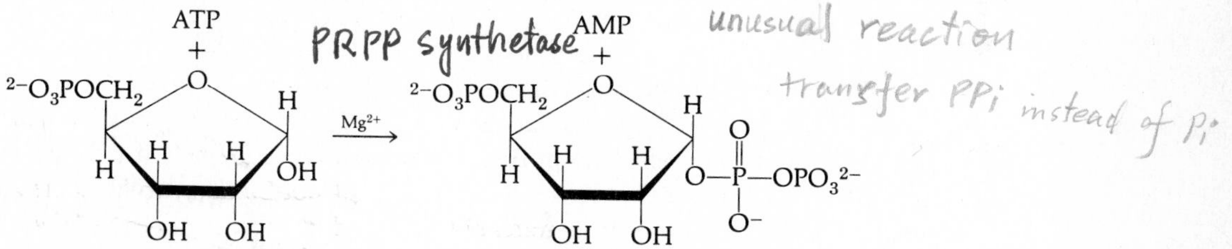

Ribose-5-phosphate **α-5-Phosphoribosyl-1-pyrophosphate**

In the first step of the pathway, the amino sugar *5-phosphoribosyl-1-amine* is formed by condensation of PRPP and glutamine, followed by removal of glutamate from the product. This reaction, catalyzed by the enzyme *amidophosphoribosyl transferase,* is the "committed" step in purine nucleotide biosynthesis and is subject to feedback inhibition by purine nucleotides.

5-Phosphoribosyl-1-pyrophosphate + glutamine $\overset{Mg^{2+}}{\rightleftharpoons}$

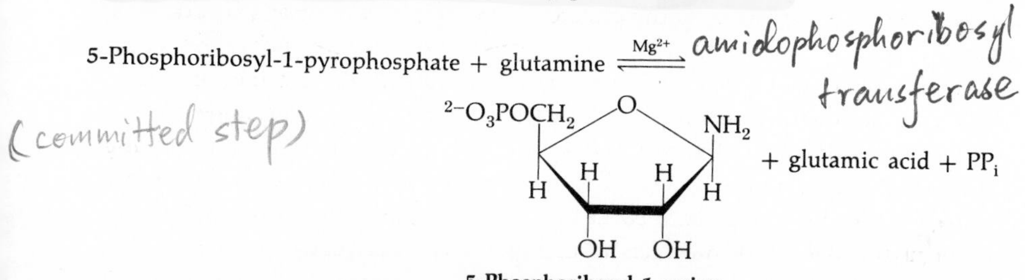

5-Phosphoribosyl-1-amine

The pyrophosphate bond in PRPP is in α linkage whereas the configuration at C-1 in 5-phosphoribosyl-1-amine is β. Thus, the displacement of pyrophosphate by the amide group of glutamine is accompanied by an inversion of spatial configuration at C-1 and the β configuration of the purine nucleotides is established.

The individual steps in the assembly of a purine nucleotide beginning with 5-phosphoribosyl amine are shown in Fig. 26.1. Of these reactions, several are subject to the action of specific inhibitors. The glutamine analogs *azaserine* and *6-diazo-5-oxo-L-2-aminohexanoic acid* inhibit the synthesis of 5-phosphoribosyl amine and the formation of formylglycinamidine ribonucleotide, both of which involve transfer of the amide N of glutamine to another carbon chain. In the latter case, reaction of azaserine with the *formylglycinamide ribonucleotide amidotransferase* results in acylation of an essential thiol group of a cysteine residue to form an inactive thioether derivative.

diazo group
$= N^{\ominus} = N^{\oplus}$ 重氮基

$$\overset{-}{N}=\overset{+}{N}=CH-\overset{\overset{O}{\|}}{C}-O-CH_2-\overset{\overset{NH_2}{|}}{CH}-COOH$$
Azaserine (*O*-diazoacetyl-L-serine)

$$\overset{-}{N}=\overset{+}{N}=CH-\overset{\overset{O}{\|}}{C}-CH_2-CH_2-\overset{\overset{NH_2}{|}}{CH}-COOH$$
6-Diazo-5-oxo-L-2-aminohexanoic acid

Figure 26.1 Pathway of purine nucleotide biosynthesis: synthesis of inosine-5'-phosphate.

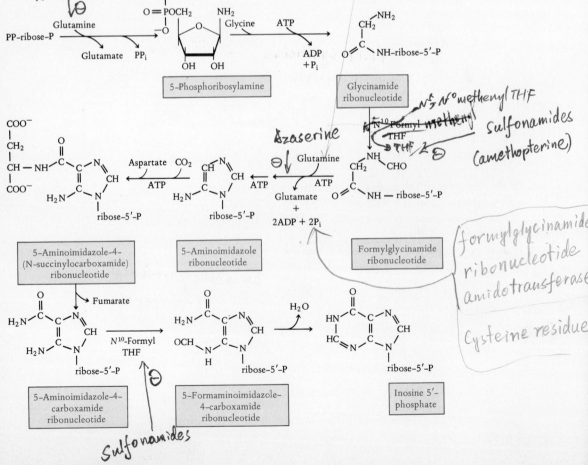

Azaserine

Azaserine

N^5, N^{10} *methenyl THF*
N^{10}-Formyl *methenyl* THF
THF
sulfonamides
(amethopterine)

formylglycinamide
ribonucleotide
amidotransferase

Cysteine residue

Sulfonamides

The *sulfonamides*, as well as *amethopterin* (*methotrexate*) and *aminopterin*, inhibit different early steps in the formation of formylglycinamide ribonucleotide and 5-formaminoimidazole-4-carboxamide ribonucleotide. Both of these formylation reactions require a folic acid coenzyme, whose synthesis is prevented by these antibiotics (Chap. 23).

Synthesis of Adenosine-5′-Phosphate and Guanosine-5′-Phosphate The synthesis of adenosine-5′-phosphate and guanosine-5′-phosphate from inosinate is shown in Fig. 26.2. Synthesis of adenylate occurs through the initial formation of adenylosuccinate, with participation of aspartate and GTP, the latter yielding GDP and P_i. Nonhydrolytic cleavage of adenylosuccinate then produces adenosine-5′-phosphate and fumarate.

This reaction is analogous to the cleavage of 5-aminoimidazole-4-*N*-succinocarboxamide ribonucleotide and is probably catalyzed by the same enzyme, *adenylosuccinase*.

[handwritten: inosinate 次黄苷酸]

Figure 26.2 Pathway of purine nucleotide biosynthesis: synthesis of adenosine-5′-phosphate and guanosine-5′-phosphate.

[handwritten annotations in figure: (irreversible); Azaserine; 6-Diazo-5-oxo-L-2-aminohexanoic acid]

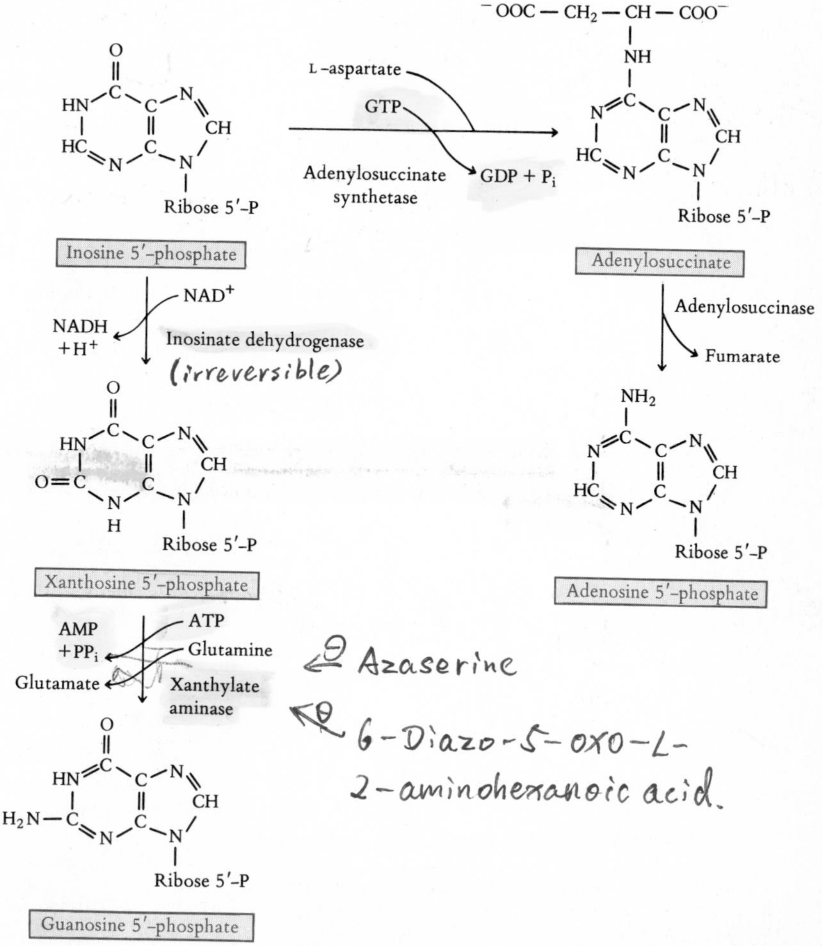

The synthesis of guanosine-5′-phosphate from inosinate proceeds first by oxidation to xanthosine-5′-phosphate; this irreversible reaction is catalyzed by *inosinate dehydrogenase.* The next step, catalyzed by *guanylate synthetase,* requires glutamine and ATP. Like other reactions requiring glutamine, the formation of guanosine-5′-phosphate is irreversible and is inhibited by azaserine and 6-diazo-5-oxo-L-2-aminohexanoic acid.

Salvage Pathways of Purine Nucleotide Synthesis

As described above, adenosine-5′-phosphate and guanosine-5′-phosphate are formed via inosinate. However, purine nucleotides can also be formed from free purines and from purine nucleosides, derived from the breakdown of nucleic acids or nucleotides.

Free purines can react directly with PRPP to yield nucleoside 5′-monophosphates. The reversible reactions shown below are catalyzed by two distinct enzymes, an *adenine phosphoribosyltransferase* and a *hypoxanthine-guanine phosphoribosyltransferase:*

adenine phosphoribosyl-transferase

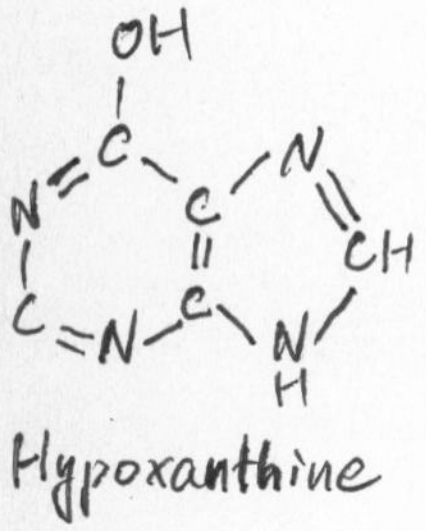

Adenine + PRPP $\rightleftharpoons$ adenosine-5′-phosphate + PP_i

Guanine + PRPP $\rightleftharpoons$ guanosine-5′-phosphate + PP_i

Hypoxanthine + PRPP $\rightleftharpoons$ inosine-5′-phosphate + PP_i

hypoxanthine-guanine phosphoribosyltransferase

Other salvage pathways involve conversion of free purines to nucleosides and nucleosides to nucleotides.

Known reactions for purine nucleoside formation, catalyzed by *purine nucleoside phosphorylase,* are:

Hypoxanthine + ribose-1-phosphate $\rightleftharpoons$ inosine + P_i

Hypoxanthine + deoxyribose-1-phosphate $\rightleftharpoons$ deoxyinosine + P_i

Guanine + ribose-1-phosphate $\rightleftharpoons$ guanosine + P_i

Guanine + deoxyribose-1-phosphate $\rightleftharpoons$ deoxyguanosine + P_i

Adenosine and deoxyadenosine are not substrates for purine nucleoside phosphorylase; however, they can react after deamination to inosine or deoxyinosine.

Conversion of the adenosine to adenosine-5′-phosphate is catalyzed by *adenosine kinase:*

adenosine kinase

Adenosine + ATP $\longrightarrow$ adenosine-5′-phosphate + ADP

Deoxyadenosine can also be phosphorylated. In contrast, neither inosine nor guanosine is a substrate for this enzyme.

In summary, the formation of purine nucleotides in mammalian tissues can occur by two pathways. The first is the *de novo* route that involves synthesis from simpler acyclic precursors. The second is the salvage route, in which purines or nucleosides, originating from the extracellular or intracellular degradative processes, are reutilized by conversion to nucleotides; this pathway is of particular significance in the regulation of purine nucleotide metabolism (see below).

Hypoxanthine

Conversion of the purine nucleoside monophosphates to diphosphates is catalyzed by specific *kinases*.

Adenosine-5'-phosphate + ATP $\xrightarrow{\text{kinase}}$ adenosine-5'-diphosphate + ADP

Guanosine-5'-phosphate + ATP $\longrightarrow$ guanosine-5'-diphosphate + ADP

These reactions are driven by regeneration of ATP and by the further phosphorylation of the diphosphate by the nonspecific *nucleoside diphosphate kinase*.

Adenosine-5'-diphosphate + ATP $\rightleftharpoons$ adenosine-5'-triphosphate + ADP

Guanosine-5'-diphosphate + ATP $\rightleftharpoons$ guanosine-5'-triphosphate + ADP

Regulation of Purine Nucleotide Synthesis

As already noted (page 673), the committed step in purine nucleotide biosynthesis is the formation of 5-phosphoribosylamine. The amidophosphoribosyl transferase, which catalyzes this reaction, appears to be controlled through an allosteric mechanism that depends on interaction of PRPP, the substrate, as a positive effector, and the end products of the pathway, purine nucleotides, as negative effectors. Increases in PRPP concentration within the physiological range lead to activation of the enzyme, while an increase in the concentration of purine nucleotides leads to inhibition. The enzyme exhibits synergistic inhibition with mixtures of adenine and guanine nucleotides, which suggests separate modifier sites for the two nucleotides.

Inosinate is the precursor for both adenosine-5'-phosphate and guanosine-5'-phosphate (page 675). The latter inhibits formation of xanthosine-5'-phosphate, and adenosine-5'-phosphate inhibits synthesis of adenylosuccinic acid. In effect, these feedback mechanisms serve to prevent further formation of adenosine-5'-phosphate or guanosine-5'-phosphate when either is present in excessive amounts, whether they arise by *de novo* synthesis or from a salvage pathway. Some of the regulatory steps in purine nucleotide synthesis are shown in Fig. 26.3.

De novo Pathway of Pyrimidine Nucleotide Synthesis

Unlike the purine nucleotides, the synthesis *de novo* of pyrimidine nucleotides proceeds by synthesis of the pyrimidine nucleus followed by attachment to ribose phosphate (Fig. 26.4). The key intermediate is *orotic acid.*

The *de novo* pathway leading to the formation of orotic acid and pyrimidine nucleotides begins with the synthesis of carbamoyl phosphate. As previously noted (Chap. 23), there are two distinct *carbamoyl phosphate synthetases* in mammalian tissues. The one associated with the formation of citrulline in the arginine pathway and present predominantly in liver mitochondria (Chap. 23) utilizes NH_3; the second enzyme, present in the cytoplasm of all animal tissues capable of forming pyrimidine nucleotides, utilizes glutamine and is the site of regulation of pyrimidine nucleotide synthesis.

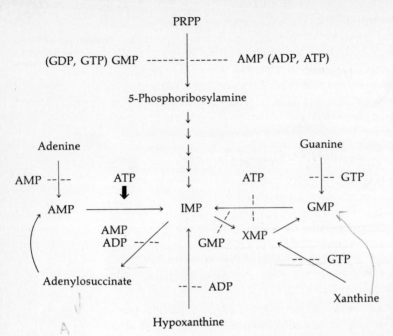

PRPP

(GDP, GTP) GMP -------|------- AMP (ADP, ATP)

5-Phosphoribosylamine

Adenine   Guanine

AMP --|--   ATP   ATP   --|-- GTP

  AMP   ⬇   ⟶   IMP ⟷   GMP

        AMP   XMP
        ADP --/--   GMP

                                    --|-- GTP

Adenylosuccinate   --|-- ADP   Xanthine

Hypoxanthine

Figure 26.3 Interrelationships and control mechanisms in purine nucleotide biosynthesis. ↓ indicates activation; --- indicates inhibition.

As might be expected, the carbamoyl phosphate synthetase reaction is inhibited by the glutamine analogs azaserine and 6-diazo-5-oxo-L-aminohexanoic acid (page 674).

Aspartate transcarbamoylase (ATCase) catalyzes the second step in the *de novo* pathway of pyrimidine nucleotide synthesis, the synthesis of carbamoyl aspartic acid.

The equilibrium for formation of carbamoyl aspartic acid is strongly in

Figure 26.4 Pathway of orotate biosynthesis.

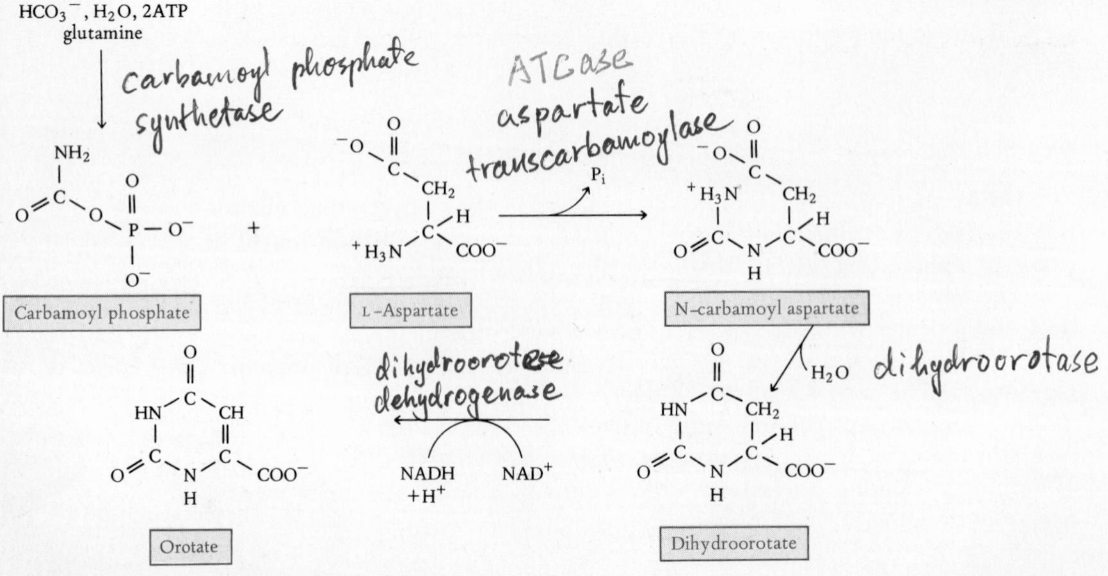

HCO_3^-, H_2O, 2ATP
glutamine

carbamoyl phosphate synthetase

ATCase
aspartate transcarbamoylase

Carbamoyl phosphate   L-Aspartate   N-carbamoyl aspartate

dihydroorotate dehydrogenase H_2O *dihydroorotase*

Orotate   NADH +H⁺   NAD⁺   Dihydroorotate

678

a kind of negative cooperativity

favor of synthesis. This is the committed step in pyrimidine biosynthesis in *Escherichia coli* and is subject to end-product feedback inhibition by cytidine triphosphate, an ultimate product of pyrimidine biosynthesis; the mechanism of this regulation has been discussed (Chap. 11). In contrast, the ATCase activity of various animal tissues is unaffected by cytidine triphosphate and other pyrimidine nucleoside phosphates and appears not to be a regulatory site.

Ring closure, catalyzed by *dihydroorotase*, yields L-dihydroorotic acid. The dihydroorotic acid is then oxidized to orotic acid by *dihydroorotate dehydrogenase*.

The second stage of pyrimidine nucleotide biosynthesis involves the condensation of orotic acid with PRPP to yield orotidine-5′-phosphate, with the elimination of pyrophosphate (Fig. 26.5). The reaction, catalyzed by *orotate phosphoribosyl transferase,* is analogous to the initial step in purine nucleotide synthesis and produces the β configuration of the pyrimidine nucleotides. The orotidine-5′-phosphate is then decarboxylated by *orotidine-5′-phosphate decarboxylase* to yield uridine-5′-phosphate. Antimetabolites of orotate and uridine inhibit these reactions. *5-Azaorotate* is a competitive inhibitor of orotate in the condensation with PRPP, and *6-azauridylate* blocks the decarboxylation of orotidylate.

The six enzymes required for the *de novo* synthesis of UMP in animals are present in three proteins. The first is a multifunctional enzyme (MW = 210,000) which contains three of the enzymic activities: *carbamoyl phosphate synthetase, aspartate transcarbamoylase,* and *dihydroorotase.* The second protein contains only *dihydroorotate dehydrogenase,* and the third (MW = 55,000) contains both *orotate phosphoribosyl transferase* and *orotidylate decarboxylase.* The latter has been found to be defective in patients with *orotic aciduria* (page 681). In contrast to animals, the six enzymes required for the *de novo* biosynthesis of UMP in prokaryotes are separate and distinct proteins. Carbamoyl phosphate synthetase and aspartate transcarbamoylase, the first two enzymes of the sequence, are each composed of two different polypeptides, but none of the six proteins is part of a multienzyme complex.

Figure 26.5 Synthesis of uridine-5′-phosphate from orotate.

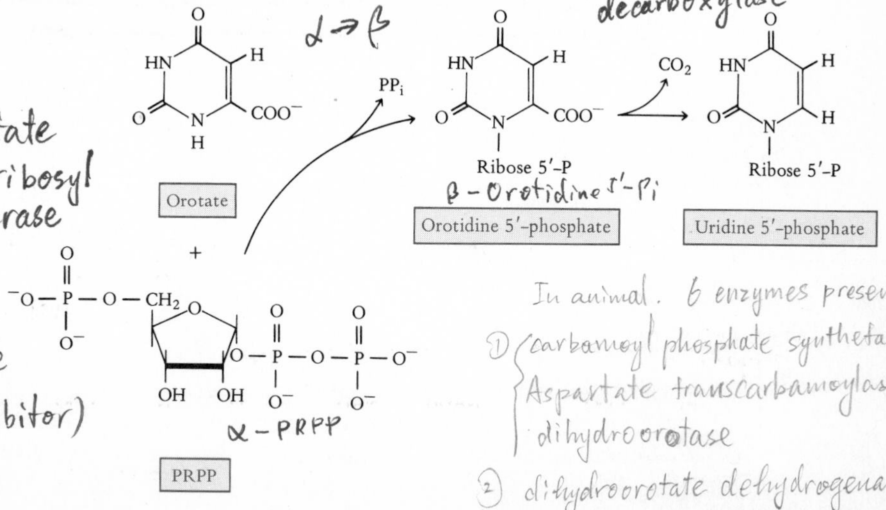

Handwritten annotations:

In E. coli, CTP inhibits ATCase
In animal. ATCase is unaffected by CTP

6-azauridylate
orotidine-5′-phosphate decarboxylase

orotate phosphoribosyl transferase
5-Azaorotate (competitive inhibitor)

α→β   PPi   CO2

Orotate

Ribose 5′-P
β-Orotidine 5′-Pi
Orotidine 5′-phosphate

Ribose 5′-P
Uridine 5′-phosphate

α-PRPP
PRPP

In animal. 6 enzymes present in 3 protein
① carbamoyl phosphate synthetase
  Aspartate transcarbamoylase
  dihydroorotase
② dihydroorotate dehydrogenase
③ orotate phosphoribosyl transferase
  orotidylate decarboxylase

Uracil
피리딘 염기합성

Like purine nucleotides, pyrimidine nucleotides can also be formed from free pyrimidines and pyrimidine nucleosides by salvage pathways. Uracil can be converted to UMP by direct reaction with PRPP catalyzed by *uracil phosphoribosyl transferase,* found only in microbial cells. Cytosine is not a substrate for this enzyme.

uracil phosphoribosyl transferase

$$\text{Uracil} + \text{PRPP} \rightleftharpoons \text{uridine-5'-phosphate} + \text{PP}_i$$

UMP can also be generated from uridine by the *uridine-cytidine kinase* reaction

uridine-cytidine kinase

$$\text{Uridine} + \text{ATP} \longrightarrow \text{uridine-5'-phosphate} + \text{ADP}$$

Whereas uridine and cytidine are substrates, orotidine is not. Finally, uridine can be synthesized from uracil in a reaction catalyzed by *uridine phosphorylase,* which in the reverse direction leads to its phosphorolysis.

uridine phosphorylase

$$\text{Uracil} + \text{ribose-1-P} \rightleftharpoons \text{uridine} + \text{P}_i$$

Cytidine is not known to be cleaved phosphorolytically; however, upon deamination by *cytidine deaminase* to form uridine, it becomes a substrate for the phosphorylase. A similar but distinct enzyme, *thymidine phosphorylase,* will be considered below.

Pyrimidine Nucleoside Monophosphate Conversion to Triphosphate

Uridine-5'-phosphate is converted to uridine-5'-diphosphate by the specific *uridylate kinase* and then to the corresponding triphosphate by *nucleoside diphosphate kinase.*

$$\text{Uridine-5'-phosphate} + \text{ATP} \longrightarrow \text{uridine-5'-diphosphate} + \text{ADP}$$

$$\text{Uridine-5'-diphosphate} + \text{ATP} \longrightarrow \text{uridine-5'-triphosphate} + \text{ADP}$$

In the only known pathway for formation of a cytidine nucleotide, uridine triphosphate is aminated to yield the corresponding cytidine triphosphate. Ribose-PPP in structures below represents ribose-5'-triphosphate.

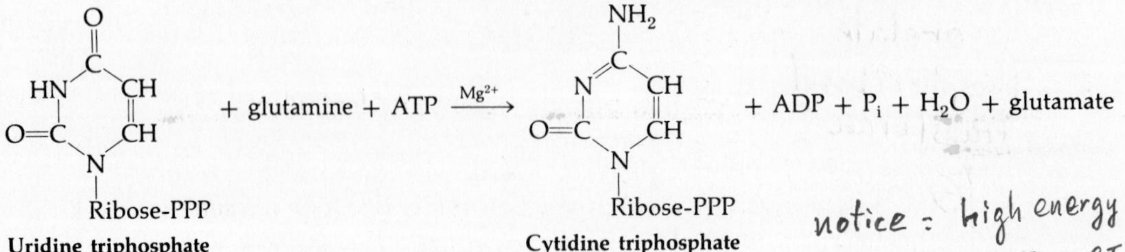

Uridine triphosphate

UTP

Cytidine triphosphate

CTP

notice = high energy level
UTP → CTP

Cytidine triphosphate can also be generated from cytidine-5'-monophosphate by consecutive transfers of phosphate from ATP with the intermediate formation of cytidine-5'-diphosphate in reactions catalyzed by the specific *cytidylate kinase* and *nucleoside diphosphate kinase.*

A human hereditary disorder of pyrimidine metabolism known as *orotic aciduria* is characterized by accumulation and urinary excretion of orotic acid. Children with this disorder do not grow normally and suffer from a *megaloblastic anemia.* Both the phosphoribosyl transferase and the decarboxylase leading to formation of UMP are much reduced in activity, a finding which led to the discovery that both of these activities are on the same polypeptide chain. Administration of uridine or cytidine restores normal growth, abolishes the anemia, and reduces the excretion of orotic acid.

As noted above, in the *de novo* pathway for formation of purine nucleotides, all intermediates are derivatives of ribose-5-phosphate, whereas in pyrimidine nucleotide biosynthesis the pyrimidine ring is formed before coupling with ribose phosphate. Although purines contain a pyrimidine ring fused to an imidazole, the precursors of the two ring systems are different. Note also that the initial nucleotides formed in both cases, inosinic acid and orotidylic acid, are not major constituents of nucleic acids.

FORMATION OF DEOXYRIBONUCLEOTIDES

Purine and pyrimidine deoxyribonucleotides are generated by reduction of the ribose moieties of the corresponding ribonucleotides. The direct reduction of ribonucleotides to deoxyribonucleotides was first inferred from studies in which [^{14}C]cytidine labeled in both the base and sugar was administered to rats. The finding that the ratio of the specific radioactivities of the cytosine and deoxyribose in the isolated DNA was the same as in the injected cytidine suggested that the transformation of ribose to deoxyribose takes place without scission of the pyrimidine-ribose bond.

The mechanism whereby this transformation occurs was first elucidated in studies with extracts of *E. coli.* A single enzyme, *ribonucleoside diphosphate reductase,* catalyzes the reduction of all four ribonucleoside diphosphates, ADP, GDP, CDP, and UDP, to the corresponding deoxy derivatives (dADP, dGDP, dCDP, and dUDP, respectively).

Ribonucleoside diphosphate + Enz − (SH)$_2$ $\longrightarrow$
deoxyribonucleoside diphosphate + Enz − (S-S)

In this reaction, the hydroxyl group at the 2′ position of the ribose moiety of a ribonucleoside diphosphate is replaced by a hydrogen *with retention of configuration at the 3′ carbon atom.*

The immediate source of reducing power is a pair of sulfhydryl groups on the enzyme, which after oxidation are restored by a second enzyme. Two different coenzymes may be involved. One, *thioredoxin,* is a heat-stable protein (MW = 12,000) with two sulfhydryl groups that are oxidized to form a disulfide bridge. The thioredoxin is restored to the reduced form by NADPH and *thioredoxin reductase,* a flavoprotein that contains 2 mol of flavin adenine dinucleotide per mole of enzyme (MW = 68,000). The second hydrogen donor is reduced glutathione acting in the presence of the heat-stable protein *glutaredoxin.* The oxidized glutathione is restored to the active reduced state by the NADPH-linked *glutathione reductase* (Chap. 24).

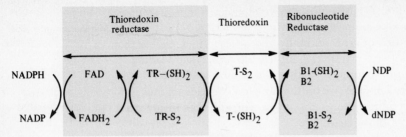

Figure 26.6 The thioredoxin system as a reductant in the ribonucleoside diphosphate reductase system. [From *L. Thelander and P. Reichard, Annu. Rev. Biochem.* **48**:133, (1979). *Reproduced with permission,* © 1979 by Annual Reviews Inc.]

The electron transport chain from NADPH leading to the reduction of ribonucleoside diphosphates is shown in Fig. 26.6. A similar scheme applies to glutaredoxin. Note that while thioredoxin is reduced directly by thioredoxin reductase, glutaredoxin is reduced *via* glutathione and glutathione reductase.

The ribonucleoside diphosphate reductase of *E. coli* consists of two subunits, *B1* and *B2*. B1 (MW = 160,000) is made up of two identical polypeptides, each containing a substrate binding site and two kinds of effector binding sites. One confers substrate specificity, and the other regulates overall catalytic efficiency. B1 also contains the thiol groups essential for activity. Protein B2 (MW = 78,000), also made up of two identical polypeptides, contains two non-heme iron atoms and an organic free radical delocalized over the aromatic ring of a tyrosine residue. Hydroxyurea, an inhibitor of ribonucleoside diphosphate reductase, does so by inactivating the free radical.

The catalytic site of the enzyme is formed by the interaction of B1 and B2; B1 contributes the thiol groups and binds the substrate, and B2 is the source of the free radical (Fig. 26.7).

The activity of ribonucleoside diphosphate reductase is allosterically regulated by nucleoside triphosphates, some acting as stimulators and others as inhibitors. Thus, the reduction of both CDP and UDP is strongly stimulated by ATP and the reductions of ADP and GDP are stimulated by dGTP and

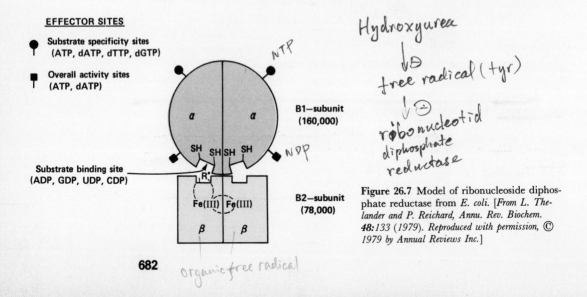

EFFECTOR SITES

● Substrate specificity sites
  (ATP, dATP, dTTP, dGTP)

■ Overall activity sites
  (ATP, dATP)

Substrate binding site
(ADP, GDP, UDP, CDP)

B1—subunit
(160,000)

B2—subunit
(78,000)

Figure 26.7 Model of ribonucleoside diphosphate reductase from *E. coli.* [From L. Thelander and P. Reichard, Annu. Rev. Biochem. **48**:133 (1979). *Reproduced with permission,* © 1979 by Annual Reviews Inc.]

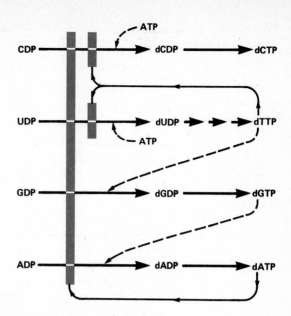

obligatory 필수적인 3가지 조건/작용

Figure 26.8 Allosteric regulation of deoxyribonucleotide synthesis by ribonucleoside diphosphate reductase of *E. coli.* ⟶ indicates positive effects: ▌ indicates negative effects.

dTTP. dATP inhibits the reduction of all four ribonucleoside diphosphates (Fig. 26.8).

The total system of effectors and inhibitors permits a balanced supply of the reduced diphosphates and hence of the triphosphates, the immediate substrates for the synthesis of DNA (Chap. 27).

The ribonucleoside diphosphate reductase system isolated from various animal cells has properties similar to that from *E. coli*. However, a different ribonucleoside diphosphate reductase has been found in various species of *Lactobacillus*. The enzyme, in this case, consists of only a single polypeptide (MW = 76,000). Its major characteristics are as follows: (1) The preferred substrates are the ribonucleoside triphosphates, ATP, GTP, CTP, and UTP. (2) Dihydrolipoate can serve as a hydrogen donor for the reduction, although the actual reductant has not been identified. (3) *5,6-Dimethylbenzimidazole cobamide coenzyme* (vitamin B_{12} coenzyme, Chap. M22) is an obligatory component. (4) Mg^{2+} and ATP (and dATP) stimulate the reduction of CTP and inhibit the reduction of UTP and GTP. (5) Various deoxyribonucleoside triphosphates act as effectors, stimulatory and inhibitory, in a pattern that differs in detail from that of the *E. coli* enzyme but is similar in principle.

Formation of Thymine Nucleotides

Deoxyuridine-5'-phosphate (dUMP), the immediate precursor of dTMP, is formed by hydrolysis of deoxyuridine-5'-triphosphate (dUTP) in a reaction catalyzed by *deoxyuridine triphosphate diphosphohydrolase* (dUTPase).

dUTP diphosphohydrolase

Deoxyuridine-5'-triphosphate + H_2O ⟶ deoxyuridine-5'-phosphate + PP_i

This reaction also serves to prevent incorporation of uracil into DNA by eliminating dUTP as a substrate for DNA polymerase (Chap. 27).

Thymine

deoxyribose 5-Pi

dUmp

683

Deamination of dCMP catalyzed by *deoxycytidylic acid aminohydrolase* also generates dUMP.

$$\text{Deoxycytidine-5'-phosphate} + H_2O \longrightarrow \text{deoxyuridine-5'-phosphate} + NH_3$$

This enzyme, present in liver, also catalyzes the deamination of the methyl- and hydroxymethyldeoxycytidylate (see below) to form thymidylate and hydroxymethyluridylate, respectively. dCTP is an allosteric activator, and dTTP is an inhibitor of the enzyme.

Methylation of dUMP, catalyzed by *thymidylate synthetase* is the *de novo* route of thymidylate synthesis.

$$\text{Deoxyuridine-5'-phosphate} + N^5,N^{10}\text{-methylenetetrahydrofolate} \longrightarrow$$
$$\text{thymidine-5'-phosphate} + \text{dihydrofolate}$$

$$\text{Dihydrofolate} + NADPH + H^+ \longrightarrow \text{tetrahydrofolate} + NADP^+$$

SUM: $\text{Deoxyuridine-5'-phosphate} + NADPH + H^+ \longrightarrow \text{thymidine-5'-phosphate} + NADP^+$

Tetrahydrofolate serves both as a carbon carrier and as a direct hydrogen donor in this complex reaction. The dihydrofolate formed is then reduced to tetrahydrofolate in an NADP-linked reaction catalyzed by *dihydrofolate reductase* (Chap. 23).

An analogous reaction occurs in the formation of 5-hydroxymethyl-deoxycytidine-5'-phosphate from dCMP in cells of *E. coli* infected with bacteriophages T2, T4, or T6 (Chap. 30), although no reduction is involved.

$$N^5,N^{10}\text{-Methylenetetrahydrofolate} +$$

Deoxycytidine-5'-phosphate

$$\text{tetrahydrofolate} +$$

5-Hydroxymethyldeoxycytidine-5'-phosphate

Fluorodeoxyuridylate is an inhibitor of the thymidylate synthetase, and the folate antagonists aminopterin and methotrexate are potent inhibitors of dihydrofolate reductase (Chap. 23).

dTMP can also be synthesized from thymine by a salvage pathway involving *thymidine phosphorylase* and *thymidine kinase*.

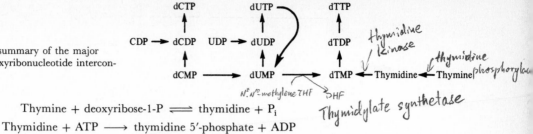

Figure 26.9 A summary of the major pyrimidine deoxyribonucleotide interconversions.

$$\text{Thymine} + \text{deoxyribose-1-P} \rightleftharpoons \text{thymidine} + P_i$$

$$\text{Thymidine} + \text{ATP} \longrightarrow \text{thymidine 5'-phosphate} + \text{ADP}$$

The level of *thymidine kinase* increases markedly during infection with certain viruses or under conditions of increased growth rate, e.g., during liver regeneration. This enzyme is allosterically regulated, being inhibited by dTTP and stimulated by a variety of deoxyribonucleoside diphosphates.

Formation of the deoxyribonucleotide di- and triphosphates from the monophosphates occurs by means of specific deoxyribonucleotide kinases and the nonspecific nucleoside diphosphate kinase, as described for the ribonucleotides (page 677). A summary of the major pyrimidine deoxyribonucleotide interconversions is given in Fig. 26.9.

Regulation of Nucleotide Synthesis

From the foregoing it is evident that separate regulatory mechanisms are operative in controlling formation of purine nucleotides (page 672), pyrimidine nucleotides (page 677), and the deoxyribonucleotides (page 681). The ribonucleoside triphosphates are utilized in the formation of all types of RNA (Chap. 28), and the deoxyribonucleoside triphosphates are the immediate precursors for synthesis of DNA (Chap. 27). The observation that the deoxyribonucleotide precursors exist in very low concentrations in cells suggests that deoxynucleotide synthesis may be a rate-limiting step in DNA synthesis. Thus, balanced cellular growth and multiplication require the synthesis of large amounts of the various nucleoside triphosphates, which must be provided in appropriate proportions. The immediate source of energy for formation of all these compounds is ATP, and regulation of its formation depends on the energy needs of the cell (Chap. 15).

SYNTHESIS OF NUCLEOTIDE COENZYMES

All the nucleotides found in RNA play other important metabolic roles. Thus, reactions have been encountered that involve adenylate, uridylate, cytidylate, and thymidylate or their respective 5'-di- and triphosphates. The biosynthesis and roles of these compounds have been considered in previous pages. Consideration has also been given to nucleotides that contain moieties not found in nucleic acids, e.g., flavin, nicotinamide, and pantothenic acid.

Flavin Nucleotides *Riboflavin*, or *7,8-dimethyl-10-(1'-D-ribityl)-isoalloxazine* (Chap. M22), is an essential dietary constituent for mammals. As already discussed, it functions in a mono- or dinucleotide form as the prosthetic group of a number of enzymes. *Flavin mononucleotide,* riboflavin-5'-phosphate, is formed from riboflavin and ATP in a reaction catalyzed by a *flavokinase.*

$$\text{Riboflavin} + \text{ATP} \longrightarrow \text{flavin mononucleotide (FMN)} + \text{ADP}$$

Flavin adenine dinucleotide is formed from the mononucleotide by a reversible reaction catalyzed by *flavin nucleotide pyrophosphorylase.*

$$\text{Flavin mononucleotide} + \text{ATP} \rightleftharpoons \text{flavin adenine dinucleotide (FAD)} + \text{PP}_i$$

Pyridine Nucleotides *Nicotinamide adenine dinucleotide (NAD)* (Chap. 15) contains *nicotinamide,* an important dietary constituent for mammals (Chap. M22). The *de novo* pathway for formation of nicotinic acid (niacin) from tryptophan (Chap. 25) yields nicotinic acid mononucleotide directly.

In mammalian cells, nicotinic acid reacts with PRPP to form nicotinic acid mononucleotide, which then condenses with ATP to form desamido-NAD. The latter is converted to NAD by reaction with glutamine and ATP.

$$\text{Nicotinic acid} + \text{PRPP} \rightleftharpoons \text{nicotinic acid mononucleotide} + \text{PP}_i \qquad (1)$$

$$\text{Nicotinic acid mononucleotide} + \text{ATP} \rightleftharpoons \text{desamido-NAD} + \text{PP}_i \qquad (2)$$

$$\text{Desamido-NAD} + \text{glutamine} + \text{ATP} \rightleftharpoons$$
$$\text{NAD} + \text{glutamic acid} + \text{PP}_i + \text{AMP} \qquad (3)$$

The *NAD synthetase* catalyzing reaction (3) is strongly inhibited by azaserine.

Erythrocytes and, presumably, other tissues can synthesize nicotinamide mononucleotide from nicotinamide by the reaction

$$\text{Nicotinamide} + \text{PRPP} \rightleftharpoons \text{nicotinamide mononucleotide (NMN)} + \text{PP}_i \qquad (4)$$

Reaction (2) is catalyzed by NAD *pyrophosphorylase,* which can also catalyze the reaction

$$\text{NMN} + \text{ATP} \rightleftharpoons \text{NAD} + \text{PP}_i \qquad (5)$$

No enzyme capable of catalyzing direct synthesis of nicotinamide from nicotinic acid has been found in plant or animal cells. However, nicotinamide can arise from nicotinic acid by consecutive operation of reactions (1) to (3), followed by the action of *nicotinamide adenine dinucleotide glycohydrolase* (NADase), which catalyzes the hydrolysis of NAD at the *N*-glycosyl linkage between ribose and nicotinamide in the following manner.

$$\text{NAD} + \text{Enz} \rightleftharpoons \text{nicotinamide} + \text{Enz-adenosine-5'-pyrophospho-5-ribose}$$

$$\text{Enz-adenosine-5'-pyrophospho-5-ribose} + \text{H}_2\text{O} \longrightarrow$$
$$\text{Enz} + \text{adenosine-5'-pyrophospho-5-ribose}$$

SUM: $\text{NAD} + \text{H}_2\text{O} \longrightarrow \text{nicotinamide} + \text{adenosine-5'-pyrophospho-5-ribose}$

Adenosine-diphosphate-ribose is hydrolyzed to adenosine-5'-phosphate and ribose-5-phosphate, which then follow the usual metabolic routes of these compounds.

The enzyme *nucleotide pyrophosphatase* can catalyze the hydrolysis of the pyrophosphate linkage in NAD, NADP, FAD, ATP, and thiamine pyrophosphate. Its physiological role is unknown.

$$NAD + ATP \longrightarrow NADP + ADP$$

Coenzyme A The complete structure of *coenzyme A* is given in Chap. 15. The *pantothenic acid* (pantoyl-β-alanine) portion of the molecule is required in the mammalian diet; its synthesis in microorganisms is described in Chap M22.

Pantothenic acid also occurs in nature in combination with β-mercaptoethylamine (cysteamine) as *pantetheine.*

pantetheine
泛酰巯基乙胺

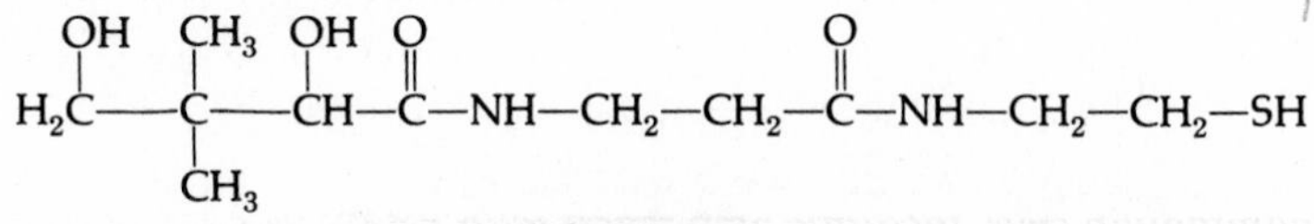

$$H_2C-\underset{\underset{CH_3}{|}}{\overset{\overset{CH_3}{|}}{C}}-\overset{\overset{OH}{|}}{CH}-\overset{\overset{O}{||}}{C}-NH-CH_2-CH_2-\overset{\overset{O}{||}}{C}-NH-CH_2-CH_2-SH$$

Pantetheine

Pantetheine is an intermediate in the pathway of CoA formation in mammalian liver and some microorganisms, as shown in the following reactions:

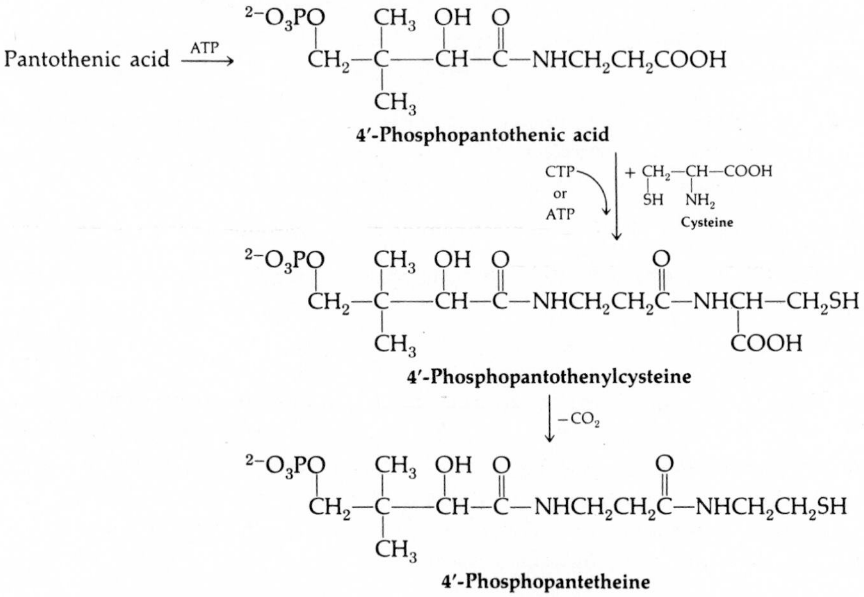

Pantothenic acid $\xrightarrow{\text{ATP}}$ $^{2-}O_3PO-CH_2-\underset{\underset{CH_3}{|}}{\overset{\overset{CH_3}{|}}{C}}-\overset{\overset{OH}{|}}{CH}-\overset{\overset{O}{||}}{C}-NHCH_2CH_2COOH$

4'-Phosphopantothenic acid

CTP or ATP $\Big\downarrow$ $+ \underset{\underset{SH}{|}}{CH_2}-\underset{\underset{NH_2}{|}}{CH}-COOH$ Cysteine

$^{2-}O_3PO-CH_2-\underset{\underset{CH_3}{|}}{\overset{\overset{CH_3}{|}}{C}}-\overset{\overset{OH}{|}}{CH}-\overset{\overset{O}{||}}{C}-NHCH_2CH_2\overset{\overset{O}{||}}{C}-NH\underset{\underset{COOH}{|}}{CH}-CH_2SH$

4'-Phosphopantothenylcysteine

$\Big\downarrow -CO_2$

$^{2-}O_3PO-CH_2-\underset{\underset{CH_3}{|}}{\overset{\overset{CH_3}{|}}{C}}-\overset{\overset{OH}{|}}{CH}-\overset{\overset{O}{||}}{C}-NHCH_2CH_2\overset{\overset{O}{||}}{C}-NHCH_2CH_2SH$

4'-Phosphopantetheine

4'-Phosphopantetheine + ATP $\rightleftharpoons$ dephosphoCoA + PP$_i$

The 3'-phosphate of the adenosine moiety of CoA is lacking in dephosphoCoA; the latter is converted to CoA by a specific *dephosphoCoA kinase.*

$$DephosphoCoA + ATP \longrightarrow CoA + ADP$$

Inhibitors of Nucleotide Synthesis

Inasmuch as rapidly dividing cells, e.g., tumor cells and bacteria, have high requirements for nucleotides for the biosynthesis of nucleic acids whereas adult tissues grow slowly, growth of tumors and bacteria can be inhibited by blocking nucleotide synthesis. As already noted (page 675) sulfonamides block formation of folic acid in organisms dependent on this process. Since folic acid is required for formylation in two steps of purine nucleotide synthesis and for thymidylate synthesis, formation of nucleic acids and various nucleotide coenzymes is prevented.

The folic acid antagonists aminopterin and amethopterin block the reduction of dihydrofolate to tetrahydrofolate by the specific reductase (page 675). Again, this prevents formylation at two steps in purine nucleotide synthesis and in the formation of thymidylate.

Since transfer of amide groups from glutamine is required in purine nucleotide synthesis, azaserine (page 674) and similar compounds block synthesis of purine nucleotides and hence formation of nucleic acids.

Some halogenated pyrimidines, e.g., *5-fluorodeoxyuridine,* block DNA synthesis by inhibiting thymidylate synthetase (page 684). Other halogenated pyrimidines and purines act as mutagens (Chap. 27). 5-Azaorotate and 6-azauridylate inhibit the *de novo* synthesis of uridylate (page 679).

DEGRADATION OF PURINE AND PYRIMIDINE NUCLEOTIDES

Enzymic Digestion of Nucleic Acids

In addition to the provision of nucleotides by the biosynthetic pathways discussed in the preceding pages, these compounds are present as constituents of the nucleic acids found in the diet. Unlike the proteins, the digestion of which begins in the stomach, the nucleic acids are unaffected by gastric enzymes, and their digestion occurs mainly in the duodenum. The pancreas forms *nucleases,* which are secreted in the pancreatic juice. Pancreatic *ribonuclease* hydrolyzes only RNA, liberating pyrimidine mononucleotides and oligonucleotides terminating in a pyrimidine nucleoside 3'-phosphate. Pancreatic *deoxyribonuclease* specifically hydrolyzes DNA to oligonucleotides (Chap. 8). The intestinal mucosa forms *diesterases,* which then hydrolyze the oligonucleotides to mononucleotides.

Degradation of Mononucleotides Mononucleotides are hydrolyzed by intestinal *phosphatases* or *nucleotidases,* yielding nucleosides and P_i. Little is known concerning the individuality or specificity of such enzymes, although it is probable that many separate enzymes exist. Specific *5' nucleotidases* cleave nucleoside 5'-phosphates but do not attack the isomeric nucleoside 3'- or 2'-phosphates.

Adenosine-5'-phosphate is deaminated by the irreversible action of *adenylate deaminase,* which is present in large amounts in muscle (Chap. M8) and other tissues.

$$\text{Adenosine-5'-phosphate} + H_2O \longrightarrow \text{inosine-5'-phosphate} + NH_4^+$$

Inosinate found in muscle reflects the high activity of the abundant adenylate deaminase rather than the rate of *de novo* purine synthesis.

Deamination of adenosine and deoxyadenosine to form inosine and deoxyinosine is catalyzed by *adenosine deaminase.* Absence of this enzyme results in a severe immunodeficiency disease possibly because of the specific inhibition of

DNA replication in lymphocytes (Chap. M2). Accumulation of deoxyadenosine as a consequence of the loss of adenosine deaminase leads to a greatly increased level of dATP by the successive action of deoxyadenosine kinase, adenylate kinase, and nucleoside diphosphate kinase. The high level of dATP in turn allosterically inhibits production of the other deoxynucleotides by ribonucleoside diphosphate reductase (page 683). The tissue-specific effect may be explained by the finding that lymphoid tissues are particularly active in the phosphorylation of deoxyadenosine. A similar explanation can also be applied to the immunodeficiency which accompanies an inherited deficiency in purine nucleoside phosphorylase (page 676).

Extracts of various tissues, e.g., spleen, liver, kidney, bone marrow, cleave the N-glycosyl linkage of nucleosides. The metabolism of these compounds probably occurs mainly in these tissues. The so-called nucleosidases have not been extensively investigated, and knowledge of these enzymes is fragmentary; however, a specific *pyrimidine nucleosidase* has been identified which catalyzes the reaction

$$\text{Uridine} + H_2O \longrightarrow \text{uracil} + \text{ribose}$$

Phosphorolysis of the N-glycosyl linkage of nucleosides also occurs in reactions catalyzed by the specific nucleoside phosphorylases (page 676).

Degradation of Purines

In mammals, most of the nitrogen in the rings of administered adenine, guanine, xanthine, or hypoxanthine appears in the urine in the form of uric acid or allantoin. The purine ring is therefore not completely degraded; only small amounts of urea or ammonia are derived from this source.

Adenine deaminase and *guanine deaminase* are specific deaminases which act hydrolytically to yield hypoxanthine and xanthine.

$$\text{Adenine} + H_2O \longrightarrow \text{hypoxanthine} + NH_3$$

$$\text{Guanine} + H_2O \longrightarrow \text{xanthine} + NH_3$$

Guanine deaminase is present in liver, kidney, spleen, etc. Adenine deaminase is present in microorganisms and invertebrates but not in mammalian tissue.

Xanthine oxidase (Chap. 16), a molybdenum-containing flavoprotein, catalyzes the oxidation of both hypoxanthine and xanthine.

$$\text{Hypoxanthine} + O_2 + H_2O \longrightarrow \text{xanthine} + H_2O_2$$

$$\text{Xanthine} + O_2 + H_2O \longrightarrow \text{uric acid} + H_2O_2$$

Uric acid formation occurs in the liver and intestinal mucosa. In most mammals uric acid is oxidized by a liver enzyme, *urate oxidase*, a copper-protein (Chap. 16).

$$\text{Uric acid} + 2H_2O + O_2 \longrightarrow \text{allantoin} + CO_2 + H_2O_2$$

The stages of purine degradation from adenine and guanine to uric acid and allantoin are summarized in Fig. 26.10.

Like birds, human beings and other primates lack urate oxidase; hence, uric acid is the main end product of purine metabolism in these species. Other

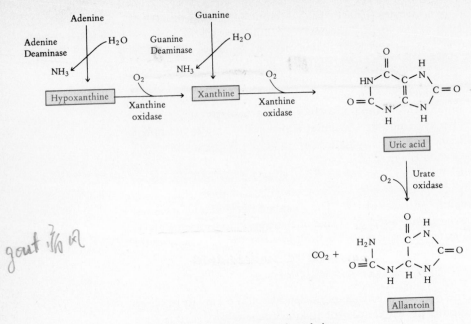

Figure 26.10 Pathway of purine degradation.

mammals, which have urate oxidase, excrete allantoin as the final product of purine metabolism. The Dalmatian hound, like other dogs, possesses urate oxidase in the liver yet excretes uric acid as the end product of purine metabolism. The failure to oxidize urate in these dogs is thought to be a consequence of a defect in urate transport across cell membranes in the liver and kidney. The pig, which is deficient in guanine deaminase, excretes guanine as well as allantoin. Indeed, guanine gout has been reported in this species because of the deposition of guanine crystals in the joints and is analogous to human gout, in which monosodium urate may accumulate in cartilage (see below). Guanine is the terminal product of purine metabolism in spiders.

A summary of the interconversions of purine derivatives (Fig. 26.11) shows that reactions may occur at the level of nucleotides, nucleosides, or free purines. Not all these reactions necessarily occur in every tissue or in every species, but there is evidence for each pathway shown.

Uric Acid Production in Human Beings

Uric acid production and excretion proceed at a relatively constant rate in human beings when the diet is free from purines. This uric acid is derived from endogenous purine metabolism and, like many body constituents, reflects a steady state in which the rates of purine synthesis and purine catabolism, measured by the excretion of uric acid, are approximately equal. Some foods, such as milk, cheese, and eggs, are low in purines, whereas foods rich in nucleic acids, e.g., liver and thymus, are high in purines.

Gout The concentration of uric acid in normal plasma is about 2 to 7.5 mg/dl, with an average for adults of approximately 4.1 mg/dl for females and about

690

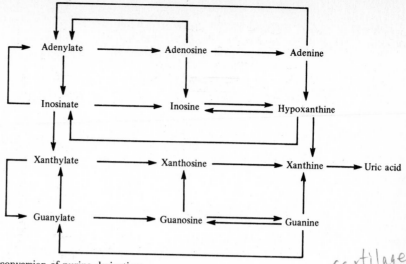

Figure 26.11 Summary of interconversion of purine derivatives.

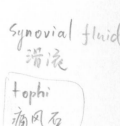

5.0 mg/dl for males. The solubility of monosodium urate in body fluids is 6.4 mg/dl. When the serum urate level exceeds this value, *hyperuricemia* (elevated blood uric acid) is said to exist. Hyperuricemia is often associated with a characteristic form of acute arthritis termed *gout* in which crystals of monosodium urate can be demonstrated in leukocytes in synovial fluid. In chronic gout, large amounts of monosodium urate are deposited as tophi in cartilage, soft tissues, tendons, and synovial membranes. When sodium urate deposits occur in renal interstitial tissue, they produce renal damage. Uric acid crystals may also form in the collecting tubules, renal pelvis, or ureters, leading to urinary obstruction by calculi (stones).

Only about 5 percent of cases of gout occur in females. Although the genetics is complex, there appears to be a familial incidence. Hyperuricemia is frequently observed in the asymptomatic male relatives of gouty individuals. Most patients with gout or hyperuricemia have the *primary* form of the disease, i.e., the condition is not associated with some other acquired disorder. *Secondary* hyperuricemia may develop in the course of another disease or as a result of treatment with certain drugs. About 10 percent of patients with primary gout overproduce uric acid for unknown reasons. In the other 90 percent the defect appears to be an impaired renal clearance of urate.

Individuals suffering from the hyperuricemia of gout and other conditions have been treated with *allopurinol,* an analog of hypoxanthine.

Allopurinol   **Alloxanthine**

Since the compound is a specific inhibitor of xanthine oxidase, there is a gradual decrease in the levels of blood and urinary uric acid. In addition, there is a decrease in total purine nucleotide production in patients treated with this drug. The inhibition of purine nucleotide biosynthesis is probably a consequence of increased reutilization of hypoxanthine for nucleotide synthesis and the resultant decrease in PRPP, inasmuch as both the increase in purine nucleotides and decrease in PRPP lead to a decrease in the activity of amidophosphoribosyl transferase (page 673).

Lesch-Nyhan Syndrome In this X chromosome-linked disorder, associated with excessive uric acid production, *hypoxanthine-guanine phosphoribosyltransferase* (page 676) is less than 1 percent of normal. The disorder is characterized by mental deficiency, aggressive behavior, self-mutilation, excessive purine production, renal stones, renal failure, and a three- to sixfold overexcretion of uric acid. The enzyme defect has been shown in erythrocytes, leukocytes, skin fibroblasts, kidney, brain, and liver. Since normal children excrete daily about 10 mg of uric acid per kilogram of body weight and the hyperuricemic ones an average of about 47 mg/kg, the difference of 37 mg/kg per 24 h affords an estimate of the amount of purines normally salvaged by the enzyme.

Inasmuch as formation of phosphoribosylamine is strongly inhibited by purine nucleotides (page 673), the deficiency of the hypoxanthine-guanine phosphoribosyl transferase leads to a lack of salvage and hence to excessive production of purines, principally in the liver. Under normal circumstances, therefore, the salvage pathway is clearly of great importance in maintaining a balance between the *de novo* and salvage pathways. Allopurinol can be used to control the gouty symptoms of patients with the Lesch-Nyhan syndrome.

Further Degradation of Purines

In animals other than mammals, purine metabolism may proceed through further degradative reactions as a result of the action of the enzymes *allantoinase* and *allantoicase*. The urea formed is hydrolyzed to NH_3 and CO_2 in some species, because of the presence of intestinal microorganisms that contain *urease*.

A summary of the biological variation of the end products of purine metabolism is given in Table 26.1. Birds and some reptiles, which do not synthesize urea, direct almost all excess amino acid nitrogen into the formation of glycine,

TABLE 26.1

Product excreted	Animal group	Final Excretory Products of Purine Metabolism
Uric acid	Primates, Dalmatian dog, birds, some reptiles (snakes and lizards)	
Allantoin	Mammals other than primates, some reptiles (turtles), gastropod mollusks	
Allantoic acid	Some teleost fishes	
Urea	Most fishes, amphibia, fresh-water lamellibranch mollusks	
Ammonia	Some marine invertebrates, crustaceans, etc.	

ureotelic
排尿素的生物

aspartic acid, and glutamine. Total purine formation greatly exceeds actual requirements for purine nucleotides, and uric acid is the major end product of all nitrogen metabolism in these species. Species that excrete nitrogen mainly as uric acid are called *uricotelic,* in contrast to *ureotelic* animals, which excrete nitrogen primarily in urea.

Arginase (Chap. 23) is present in the livers of vertebrates that have a ureotelic metabolism but not in the livers of animals that have a uricotelic metabolism. Thus, the end products of purine and amino acid metabolism depend on the survival of a small group of enzymes. Lack of arginase diverts amino acid nitrogen into purines in uricoteles. It may be recalled that nitrogen from glycine, glutamine, and aspartic acid is directly utilized in formation of purine nucleotides. By the reactions of transamination and glutamine formation (Chap. 22), additional amino nitrogen and NH_3 can be furnished for purine formation and hence for disposal in the form of uric acid.

Because the degradation of purines is much less complete in higher animals, it is apparent that certain enzymes have been lost during animal evolution, e.g., urate oxidase, allantoinase, allantoicase, and urease.

Degradation of Pyrimidines

Deamination of cytidine to form uridine by *cytidine deaminase* has been demonstrated in a number of animal tissues and in bacteria. Cytosine and methylcytosine are deaminated to yield, respectively, uracil and thymine. The presently known pathways for the degradation of uracil and thymine are shown in Fig. 26.12.

The catabolism of uracil and thymine is initiated by *reduction* reactions to give the dihydro compounds dihydrouracil and dihydrothymine, which are then hydrolyzed by *hydropyrimidine hydrase* to the β-ureido compounds. Further hydrolysis yields the β-amino acids. The utilization and fate of β-alanine have been discussed (Chap. 25). β-Aminoisobutyrate may yield methylmalonate; the fate of the CoA derivative of this compound has also been discussed (Chap. 20). β-Aminoisobutyrate is found in the urine of some individuals in amounts up to 200 to 300 mg per day, either because of an inherited trait or as a consequence of disease. It is excreted in increased amounts after administration of diets rich in DNA. Increased levels may also occur in the urine of patients with tumors.

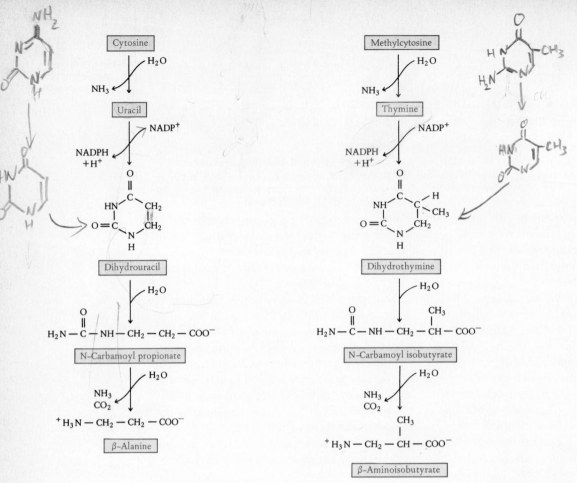

Figure 26.12 Pathways of pyrimidine degradation.

REFERENCES

Books

Davidson, J. N.: *The Biochemistry of the Nucleic Acids,* 8th ed., R. P. L. Adams, R. H. Burdon, A. M. Campbell, and R. S. Smellie (eds.): Academic, New York, 1977.

Henderson, J. F.: Regulation of Purine Biosynthesis, *Am. Chem. Soc. Monog.* **170:** (1972).

Henderson, J. F., and A. R. P. Paterson: *Nucleotide Metabolism,* Academic, New York, 1973.

Hoffee, P. A., and M. E. Jones: Purine and Pyrimidine Nucleotide Metabolism, in *Methods in Enzymology,* vol. LI, Academic, New York, 1978.

Kornberg, A.: *DNA Replication,* Freeman, San Francisco, 1980.

Seegmiller, J. E.: Diseases of Purine and Pyrimidine Metabolism in P. K. Bondy (ed.): *Duncan's Diseases of Metabolism,* 7th ed., vol. 1, Saunders, Philadelphia, 1974, pp. 655–774.

Stanbury, J. B., J. B. Wyngaarden, and D. S. Fredrickson (eds.): *The Metabolic Basis of Inherited Disease,* 4th ed., McGraw-Hill, New York, 1978. (See Part VI, "Disorders of Purine and Pyrimidine Metabolism.")

Benkovic, S. J.: Mechanism of Folate-Enzymes, *Annu. Rev. Biochem.* **49:**227–251 (1980).

Cozzarelli, N. R.: The Mechanism of Action of Inhibitors of DNA Synthesis, *Annu. Rev. Biochem.* **46:**641–668 (1977).

Hitchings, G. H.: Indications for Control Mechanisms in Purine and Pyrimidine Biosynthesis as Revealed by Studies with Inhibitors, *Adv. Enzyme Regul.* **9:**121–129 (1974).

Jones, M. E.: Pyrimidine Nucleotide Biosynthesis in Animals: Genes, Enzymes, and Regulation of UMP Biosynthesis, *Annu. Rev. Biochem.* **49:**253–279 (1980).

Murray, A. W., D. C. Elliott, and M. R. Atkinson: Nucleotide Biosynthesis from Preformed Purines in Mammalian Cells: Regulatory Mechanisms and Biological Significance, *Prog. Nucleic Acid Res. Mol. Biol.* **10:**87–119 (1970).

Thelander, L., and P. Reichard: Reduction of Ribonucleotides, *Annu. Rev. Biochem.* **48:**133–158 (1979).

Part

4

Molecular Genetics

Chapter 27

The Gene and Its Replication

Nature of the gene.   DNA synthesis and repair.   Recombination.   Modification and restriction of DNA.   Molecular cloning.

Adequate understanding of the molecular basis of inheritance requires answers to the following questions: (1) What is the nature of the gene?   (2) How are genes replicated so that at cell division each daughter cell acquires the same complement of genetic material as the parent cell?   (3) How are the genes expressed; i.e., what are the mechanisms by which the genes determine the chemical, metabolic, and morphological characteristics of the individual cell or organism?

THE GENE AS DNA

A variety of findings demonstrate unambiguously that the gene is DNA.

1. The DNA content of a cell is an absolute constant for each species, consonant with the role of DNA as the carrier of genetic information.

2. Mutation of genes can be produced by ultraviolet light.   The action spectrum producing mutations closely corresponds to the absorption spectrum of nucleic acid.   Similarly, other physical and chemical agents known to be mutagenic can alter nucleic acids.

3. Virulent strains of pneumococci have capsular polysaccharides.   Avirulent noncapsulated cells may be transformed into encapsulated organisms by the introduction of DNA isolated from a capsular-forming organism.   This effect is called *transformation*.   Thus minute amounts of DNA from the capsular type III pneumococcus induce formation of type III polysaccharide in a type II, acapsular, strain.   In a similar manner, DNA from diverse bacterial strains can transform variant members of

Any cross-references coded M refer to the companion text, *Principles of Biochemistry: Mammalian Biochemistry*.

the same species with respect to other characteristics. The DNA acts as a genetic determinant in all cases since the transformed strain continues to reproduce thereafter as the induced type in the absence of further added transforming factor.

4. Most bacterial viruses (bacteriophages) and many animal viruses contain only DNA and protein. In certain bacteriophages (Chap. 30) transfer of genetic information to progeny virus has been shown to be due entirely to the DNA. The protein serves only to enable the DNA to enter the host cell and is not further utilized.

All the above information has established that the genetic material is DNA. Therefore, DNA must possess the fundamental properties of the gene. It must direct the synthesis of an exact replica of itself, and it must possess the ability to direct the formation of enzymes and other proteins. The details of these processes will be considered in this and the following three chapters.

In brief, replication of DNA is a *semiconservative process* (see below) that begins at a fixed point and proceeds by the simultaneous copying of both strands of the duplex. The monomeric unit utilized in this process is a *deoxyribonucleoside-5'-triphosphate*. Polymerization of nucleotides occurs by attack of the 3'-hydroxyl group of the growing DNA strand on the nucleotidyl phosphorus of a deoxyribonucleoside-5'-triphosphate with formation of a 3',5'-phosphodiester bond and the elimination of PP_i. The strand opposite to the growing strand serves as a template to direct the sequence in which nucleotides are polymerized, adenine pairing with thymine and guanine with cytosine.

Expression of the genetic material is achieved by the synthesis from ribonucleoside-5'-triphosphates of an RNA transcript that is complementary to one of the two strands of the DNA duplex. In this process, three kinds of RNA are formed, viz., *transfer RNA (tRNA)*, *ribosomal RNA (rRNA)*, and *messenger RNA (mRNA)* (Chap. 8). The last is translated into protein, with the aid of tRNA on particles containing rRNA, termed *ribosomes*.

SEMICONSERVATIVE REPLICATION OF DNA

As already noted (Chap. 8), DNA molecules are double helical structures in which the two strands are bound by hydrogen bonds between specific amino and keto groups: adenine (A) to thymine (T), and guanine (G) to cytosine (C). On the basis of this structure, it was suggested that replication might proceed by separation of the two strands and that new chains of nucleotides complementary to each of the strands would be formed. Thus, each single strand would serve as a template to form a new DNA strand, which is not identical but complementary to the template. When the process is complete, two new double-stranded molecules of DNA would have been formed, one strand serving as template and the other strand being newly synthesized.

The earliest support for this hypothesis was an experiment in which cells of *Escherichia coli* were cultured in a medium that labeled their DNA completely with ^{15}N. The bacteria were then washed carefully and permitted to undergo one synchronous cell division in a medium in which all the nitrogen was in the form of ^{14}N. During this cell division, the total DNA of the culture doubled.

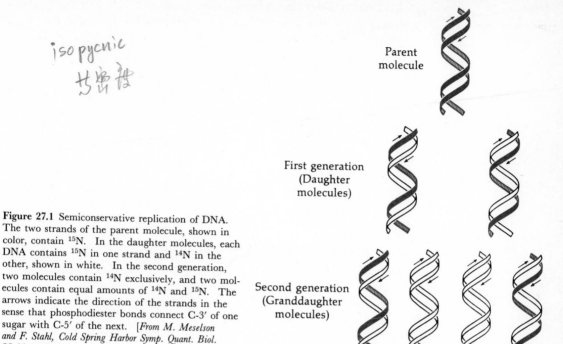

isopycnic
苏密度

Parent molecule

First generation (Daughter molecules)

Second generation (Granddaughter molecules)

Figure 27.1 Semiconservative replication of DNA. The two strands of the parent molecule, shown in color, contain ^{15}N. In the daughter molecules, each DNA contains ^{15}N in one strand and ^{14}N in the other, shown in white. In the second generation, two molecules contain ^{14}N exclusively, and two molecules contain equal amounts of ^{14}N and ^{15}N. The arrows indicate the direction of the strands in the sense that phosphodiester bonds connect C-3′ of one sugar with C-5′ of the next. [*From M. Meselson and F. Stahl, Cold Spring Harbor Symp. Quant. Biol.* **23**:20 (1953).]

By isopycnic centrifugation in a concentrated solution of CsCl (Chap. 8) it was found that the DNA of the culture did not consist of two types, that is, DNA with ^{15}N and DNA with ^{14}N, but that all the DNA behaved as though it were a hybrid containing equal parts of both ^{14}N and ^{15}N. This is the result to be expected from the mechanism of replication described above. During replication, each double-stranded ^{15}N-labeled DNA separated to yield two single strands. Each of the latter served as a template on which ^{14}N-labeled DNA was synthesized, yielding a hybrid double-stranded molecule with one strand containing ^{15}N and the other containing ^{14}N. This process, termed *semiconservative replication,* is shown schematically in Fig. 27.1.

Topology of DNA Replication

Radioautographic and electron microscopic observations of replicating DNA molecules have demonstrated that replication occurs simultaneously on both strands, starting at a single fixed origin and proceeding in one or in both directions. In the case of the circular DNA molecule that constitutes the chromosome of *E. coli* (Table 8.4), replication is bidirectional. The average rate at which the two growing points, or *replication forks,* move during replication of this chromosome is approximately 45 kilobases (kb) per minute per fork at 37°C. Since there are 10 base pairs per turn of the helix, the rate of unwinding of the parental duplex at each fork must be approximately 4500 turns per minute (Fig. 27.2).

Inasmuch as unwinding at one fork requires that the parental duplex rotate in a direction counter to that necessary for unwinding at the opposite fork, a

chromosome
染色体

unwinding
解链

701

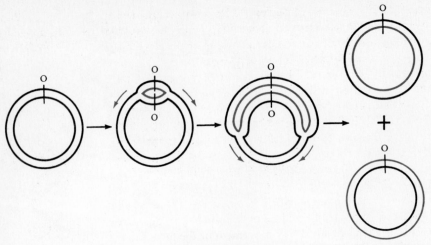

Figure 27.2 Bidirectional replication of a circular DNA molecule starting at a fixed origin (O). The colored line indicates newly synthesized strands, and the arrows indicate the direction of movement of the two replication forks. For simplicity, the helical structure of the duplex DNA is not shown.

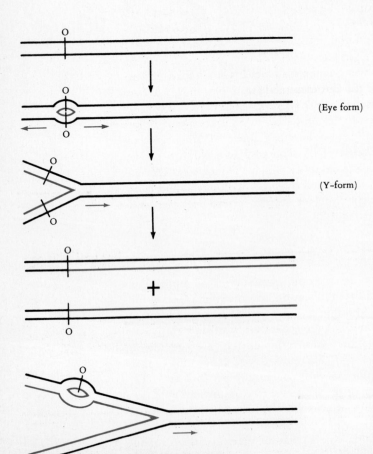

(Eye form)

(Y–form)

Figure 27.3 Bidirectional replication of linear DNA from bacteriophage T7. Replication begins at an origin (O) located 6.5 kb from one end and generates both "eye" and "Y" forms. The bottom diagram illustrates activation of an origin in one of the daughter duplexes before completion of replication of the parental DNA molecule.

swivel (or swivels) must exist in the parental duplex. Such a swivel may result from the action of enzymes termed *topoisomerases* (page 708), which are capable of removing superhelical turns from supercoiled circular DNA molecules (Chap. 8) by the cleavage and resynthesis of phosphodiester bonds.

Replication of a linear DNA molecule (that from bacteriophage T7, Chap. 30) also proceeds bidirectionally from a single origin, which is located at 6.5 kb from one end. Because of the asymmetry of the origin relative to the ends of the molecule, an "eye" form, which is first produced by the two replication forks, opens into a "Y" form when the one fork reaches the end closest to the origin, leaving a single fork to complete the replication (Fig. 27.3).

The origin in one or both daughter duplexes may be activated before replication of the parental T7 DNA molecule is complete, so that more than two forks may act simultaneously in a single replication intermediate. This allows a more rapid increase in the population of replicas of the single parental DNA molecule without any increase in the rate of fork movement or the formation of origins at other positions in the T7 DNA molecule. A similar advanced activation of an origin has been observed under certain conditions in the circular *E. coli* chromosome.

The rate of fork movement during replication of eukaryotic chromosomes is more than an order of magnitude less than that in *E. coli,* yet their very long DNA molecules (see Table 8.4) can be replicated in even less time than is required to replicate an *E. coli* chromosome. This rapid rate of replication is achieved by a serial array of origins, each of which creates two forks that move in opposite directions. The topological problems of replication, e.g., requirement for a swivel to unwind the parental duplex, are similar to those for the bidirectional replication of the *E. coli* chromosome (Fig. 27.4).

In different animals, and even in nuclei from different cells of the same

Figure 27.4 Replication of a very long eukaryotic DNA molecule. A serial array of origins is shown, each of which generates two replication forks that move in opposite directions.

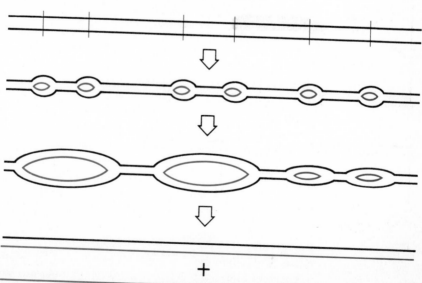

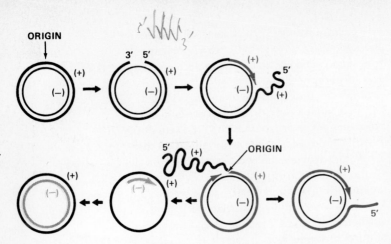

Figure 27.5 Unidirectional replication of a closed circular duplex DNA molecule by a mechanism in which a "rolling circle" results from covalent extension of a parental strand in a circular duplex chromosome. [*From A. Kornberg, DNA Replication, Freeman, San Francisco, 1980, p. 397.*]

animal, the spacing between adjacent origins varies considerably. For example, in the cleavage nuclei of fertilized eggs of *Drosophila melanogaster* engaged in rapid cell division the average spacing is only 7 to 8 kb. Hence, two forks proceeding bidirectionally from each origin at a rate of only 2 kb/min per fork could complete replication of even the longest chromosomal DNA molecule in *Drosophila* in a few minutes. In contrast, the average spacing between origins in *Drosophila* cells grown in culture is 40 kb. In these cells the rate of DNA replication is approximately 200-fold lower than that in cleavage nuclei.

An interesting example of unidirectional replication involving only a single replication fork is the *rolling circle*. This mechanism has been observed in certain viruses (bacteriophages φX174 and lambda) (see Chap. 30) and has been identified as the means by which the gene that specifies ribosomal RNA (Chap. 8) is amplified in the oocytes of the South African clawed toad, *Xenopus laevis*. In this process, cleavage of a single phosphodiester bond in a closed circular duplex DNA molecule provides a primer (see below) for chain extension at the 3'-hydroxyl terminus. Replication around the circle thus causes displacement of the 5'-phosphoryl end. This displaced strand can then be replicated in the 5'-to-3' direction. In essence, the rolling circle allows a tandem array of repeated sequences to be formed from a single sequence (Fig. 27.5).

ENZYMES OF DNA REPLICATION

DNA Polymerases

Investigation of the mechanism of DNA synthesis in extracts of *E. coli* led to the identification and subsequent isolation of an enzyme, *DNA polymerase* (now designated *DNA polymerase I*), that catalyzes the polymerization of nucleotides at the specific direction of a DNA template. This enzyme has served as the prototype for all known template-directed nucleic acid polymerases (Chaps. 28, 29, and 30).

Prokaryotic DNA Polymerases DNA polymerase I of *E. coli* consists of a single polypeptide chain (MW = 109,000) containing one atom of Zn^{2+} per molecule of enzyme. The synthesis of a single phosphodiester bond catalyzed by DNA

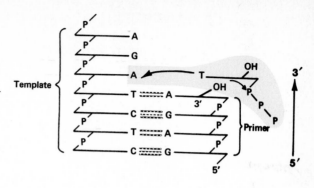

Figure 27.6 Model of DNA template-primer at which DNA polymerase I can catalyze the polymerization of nucleotides in the 5′-to-3′ direction. Synthesis of a phosphodiester bond between the primer and the incoming dTTP (in color) is shown. [*From A. Kornberg, Science* **163**:*1410 (1969). Copyright* © *1969 by The American Association for the Advancement of Science.*]

polymerase I proceeds according to the reaction

$$(\text{dNMP})_n + \text{dNTP} \underset{}{\overset{\text{Mg}^{2+}}{\rightleftharpoons}} (\text{dNMP})_{n+1} + \text{PP}_i$$

where $(\text{dNMP})_n$ refers to a polymer composed of n deoxyribonucleotide residues and dNTP refers to a deoxyribonucleoside triphosphate. Polymerization proceeds exclusively in the 5′-to-3′ direction by the addition of mononucleotide units from the deoxyribonucleoside-5′-triphosphate to the 3′-hydroxyl terminus of the $(\text{dNMP})_n$ *primer*. The reaction consists of the nucleophilic attack of the 3′-hydroxyl group of the primer terminus on the α-phosphorus of the deoxyribonucleoside triphosphate, with the elimination of PP_i. In addition to the primer, a *template* is absolutely required and directs the enzyme in its selection of the specific deoxyribonucleoside triphosphate according to the conventional base-pairing rules (Fig. 27.6). The product is therefore a faithful replica of the template. In the reverse direction, the enzyme can catalyze the phosphorolysis of DNA, although the rapid hydrolysis of PP_i by *inorganic pyrophosphatase* would tend to minimize the physiological significance of the reverse reaction.

In addition to synthesis and phosphorolysis, DNA polymerase I can also catalyze the hydrolysis of phosphodiester bonds. Under appropriate conditions, the enzyme exhibits two different hydrolytic activities. One specifically hydrolyzes single-stranded DNA from the 3′ end of the chain in the 3′-to-5′ direction (3′ → 5′ exonuclease), and a second hydrolyzes double-stranded DNA from the 5′ end in the 5′-to-3′ direction (5′ → 3′ exonuclease). DNA polymerase I can be cleaved by the action of proteases (subtilisin or trypsin) to yield a "large" fragment (MW = 76,000), which contains both the polymerase and 3′ → 5′ exonuclease activities and a "small" fragment (MW = 36,000), containing only the 5′ → 3′ exonuclease activity. The small fragment corresponds to the amino-terminal portion of the molecule. Thus, DNA polymerase I is a multifunctional enzyme containing at least two distinct enzymic activities in a single polypeptide chain.

The polymerase and 5′ → 3′ exonuclease activities of DNA polymerase I can act coordinately to catalyze the simultaneous incorporation of nucleotides at the 3′-hydroxyl end and the release of nucleotides from the 5′ end of a single strand break (a nick) in duplex DNA. The net effect of this concerted 5′ → 3′ polymerase and 5′ → 3′ exonuclease action is to propagate the nick along the DNA molecule (Fig. 27.7). This nick displacement, or *nick translation,* is the mechanism whereby such photoproducts as thymine dimers are excised during

705

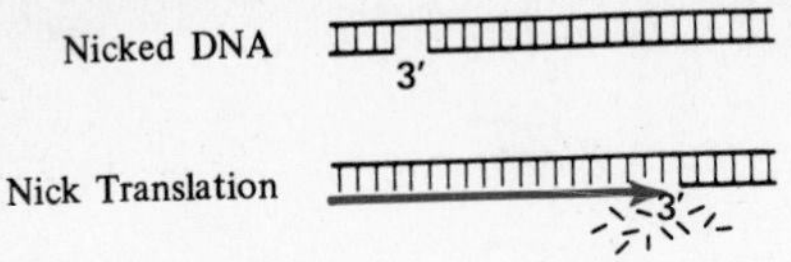

Figure 27.7 Action of DNA polymerase I at a single strand scission (a "nick") in duplex DNA leading to translation of the nick.

repair of ultraviolet-induced damage to DNA (page 715). Removal of an initiating RNA segment from newly synthesized DNA chains also occurs in this manner (page 713).

As noted above, the $3' \rightarrow 5'$ exonuclease activity of DNA polymerase I specifically hydrolyzes single-stranded DNA. It will, in addition, catalyze the removal of unpaired nucleotides at the 3' termini of duplex DNA to generate a fully base-paired primer template, which is no longer a substrate for exonuclease action. Thus, the $3' \rightarrow 5'$ exonuclease acting coordinately with polymerase may remove mismatched (and hence unpaired) nucleotides inserted by polymerase action and in this way perform a proofreading function during DNA replication. Indeed, certain mutants of bacteriophage T4 with a defect in the gene that codes for the phage-induced DNA polymerase (Chap. 30) show a greatly increased frequency of mutations (Chap. 28) throughout the T4 genome and, in one instance, this *mutator* effect is a consequence of a defective $3' \rightarrow 5'$ exonuclease associated with the mutant DNA polymerase. In another case, the mutator phenotype is the result of a reduced specificity in base selection by the polymerase, rather than a defective $3' \rightarrow 5'$ exonuclease.

In addition to DNA polymerase I, which is present in greatest abundance, *E. coli* contains DNA polymerases II and III. *DNA polymerase II* (MW = 120,000) closely resembles DNA polymerase I in the reaction it catalyzes (see page 705) but does not have an associated $5' \rightarrow 3'$ exonuclease activity. *DNA polymerase III* (MW = 180,000) is composed of three subunits (α, ϵ, θ). However, it can be isolated in an even more complex holoenzyme form (page 711) containing three additional subunits (Table 27.1). Like DNA polymerase I, DNA polymerase III possesses both $5' \rightarrow 3'$ and $3' \rightarrow 5'$ exonuclease activities.

Mutants of *E. coli* with an abnormally thermolabile DNA polymerase III are unable to grow or replicate their DNA at 42°C although they are able to do so at 30°C; hence this enzyme is required for DNA replication in vivo (page 711). Certain mutants with a defect in DNA polymerase I also show an impairment in DNA replication in vivo (page 711). As yet, no specific function has been ascribed to DNA polymerase II.

Eukaryotic DNA Polymerases Like *E. coli* and other prokaryotes, eukaryotic cells contain a multiplicity of DNA polymerases; however, unlike those of the former, the eukaryotic polymerases are devoid of $3' \rightarrow 5'$ or $5' \rightarrow 3'$ exonuclease activity. Their polymerization mechanism appears to be the same as that for the prokaryotic polymerases. The preponderant enzyme (80 to 90 percent of

TABLE 27.1

Subunit Structure of DNA Polymerase III Holoenzyme

Subunit	α	ϵ	θ	β	γ	δ
Gene	*dna*E (*pol*C)	Unknown	Unknown	*dna*N	*dna*X	*dna*Z
Molecular weight	140,000	25,000	10,000	40,000	52,000	25,000

706

the total) is termed *DNA polymerase* α (MW ≈ 300,000) and contains four or five subunits; it is the eukaryotic DNA polymerase most clearly associated with chromosomal replication. A second DNA polymerase, β, consists of only a single polypeptide chain (MW ≈ 45,000) and is associated principally with DNA repair. Both of these enzymes are localized in the nucleus. The third polymerase, *DNA polymerase* γ (MW ≈ 140,000) is found in mitochondria and is presumably responsible for the replication of mitochondrial DNA (Chap. 16). However, a small amount of the enzyme is found in the nucleus and may be utilized for the replication of certain viruses (for example, adenovirus, Chap. 30).

Retroviruses (*Rous sarcoma, avian myeloblastosis, Rauscher leukemia,* see Chap. 30) contain within the virion a unique type of RNA-dependent DNA polymerase, termed *reverse transcriptase.* This enzyme can catalyze the synthesis of a DNA copy of the viral RNA, which can then be integrated into the host genome. In the *provirus* state, the viral genome can be expressed, leading either to virus proliferation or to tumor formation.

The reverse transcriptase isolated from the *avian myeloblastosis virus* has been the most thoroughly studied of this class of enzymes. It contains two atoms of Zn^{2+} per molecule of enzyme (MW = 160,000) with subunits of 65,000 (α) and 95,000 (β) daltons. Reverse transcriptase activity is associated only with the α subunit; the function of the β subunit is unknown.

The viral RNA-dependent DNA polymerases differ most markedly from the cellular α and β polymerases in their capacity to utilize as a primer template the RNA extracted from the virion. This RNA is a large 70S complex composed of four 35S units and 12 to 20 4S RNA segments. One of these is a tryptophanyl tRNA or prolyl tRNA (Chap. 28) that serves as a primer in the reaction. The reverse transcriptases also have an intrinsic ribonuclease activity, termed *RNase H,* that specifically degrades the RNA component of the RNA-DNA hybrid synthesized by action of the polymerase.

DNA Ligases

DNA-joining enzymes, or *ligases,* catalyze the synthesis of phosphodiester bonds between preformed polynucleotides. They are required for the repair (page 715) and recombination of DNA (page 719) and for discontinuous DNA replication (page 710). The enzyme from *E. coli,* consisting of a single polypeptide chain (MW = 75,000), has been characterized most extensively.

The synthesis of phosphodiester bonds catalyzed by DNA ligase proceeds by a mechanism which differs from that of DNA polymerase action. Synthesis occurs between directly adjacent 5′-phosphoryl and 3′-hydroxyl groups in duplex DNA, coupled to the cleavage of the pyrophosphate bond of NAD^+, in a sequence of three steps (Fig. 27.8): (1) transfer of the adenylyl group of NAD^+ to the ε-amino group of a lysine residue in the enzyme to form a covalent enzyme-adenylate (phosphoamide bond) intermediate with elimination of nicotinamide mononucleotide; (2) activation of the 5′-phosphoryl terminus of the DNA by transfer of the adenylyl group from the enzyme to form the pyrophosphoryl derivative, DNA-adenylate; and (3) phosphodiester bond synthesis by attack of the 3′-hydroxyl terminus of the DNA on the activated 5′-phosphate with the release of AMP.

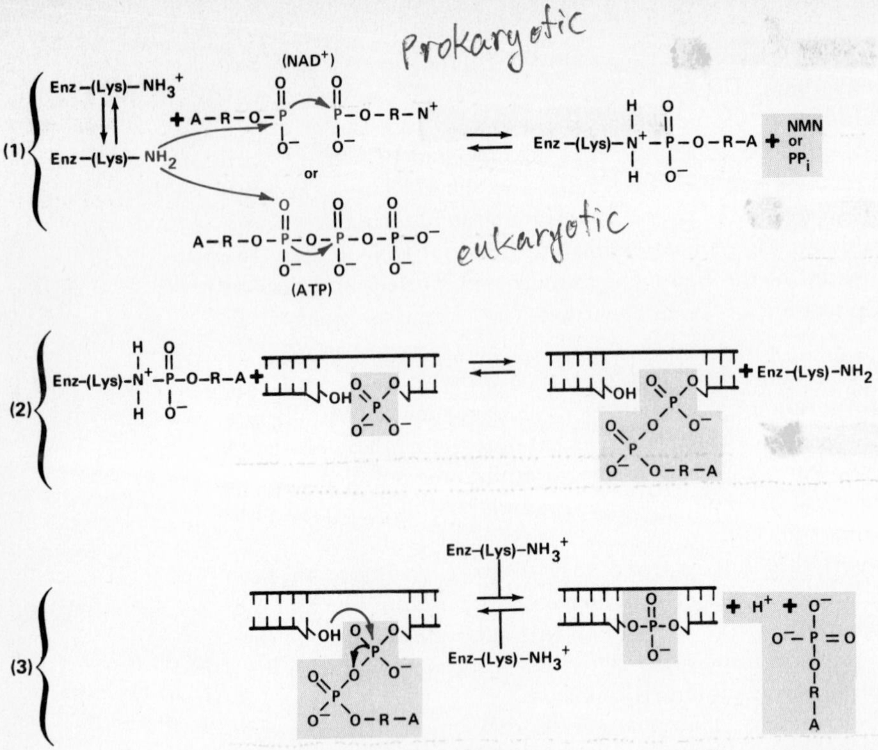

Figure 27.8 Mechanism of DNA ligase reaction. NMN = nicotinamide mononucleotide. [*From I. R. Lehman, Science 186:792 (1974). Copyright © 1974 by The American Association for the Advancement of Science.*]

In contrast to bacteria which possess only a single DNA ligase, two different DNA ligases are present in eukaryotes. Both are localized in the nucleus and appear to catalyze phosphodiester bond synthesis by the same three-step mechanisms utilized by the prokaryotic enzymes; however, ATP rather than NAD+ serves as cofactor.

An analogous enzyme, RNA ligase (MW = 41,000), is induced upon infection of *E. coli* with bacteriophage T4 (Chap. 30). RNA ligase catalyzes the intra- as well as intermolecular joining of both RNA and DNA chains. The mechanism of phosphodiester bond synthesis is the same as that for DNA ligase with the exception that a complementary strand is not required; the abutment of ends is carried out at acceptor and donor binding sites on the enzyme. Like the T4-induced DNA ligase, ATP serves as the cofactor. Although RNA ligase might be expected to function in the repair of RNA chains in a manner analogous to the repair of DNA chains by DNA ligase, there is as yet no evidence that this is indeed the case.

DNA Topoisomerases

DNA topoisomerases are enzymes that catalyze the concerted breakage and rejoining of phosphodiester bonds within DNA chains. In doing so, they catalyze the conversion of one DNA topological isomer (*topoisomer*) to another, and hence the name topoisomerase (Chap. 8). One class of topoisomerase (type I)

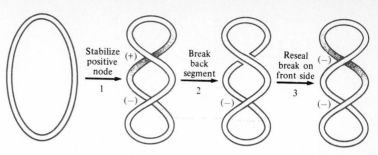

Figure 27.9 A model for the super-coiling of closed circular duplex DNA by DNA gyrase. The enzyme maintains and acts at the upper node; the (+) and (−) symbols refer to the nodes. [*From N. R. Cozzarelli, Science* **207**:*953 (1980). Copyright* © *by The American Association for the Advancement of Science.*]

Stabilize positive node
1

Break back segment
2

Reseal break on front side
3

typified by the ω protein of *E. coli* (MW = 110,000) catalyzes the relaxation of negatively supercoiled DNA in a reaction that involves transient breaks on one strand at a time with the formation of a covalent enzyme-DNA intermediate. No added cofactor is required. A second class of topoisomerase (type II) catalyzes the breakage and rejoining of DNA strands coupled to the hydrolysis of ATP to ADP and P_i with the transient introduction of double-strand breaks.

The bacterial *DNA gyrases* (*E. coli* and *Micrococcus luteus*), which catalyze the introduction of negative supercoils into DNA, are the best understood of the type II topoisomerases. The DNA gyrase of *E. coli* is a tetramer with subunits of 105,000 (α) and 95,000 (β); the protomeric structure is $\alpha_2\beta_2$ (MW = 400,000). Although the mechanism by which gyrase introduces negative supercoils has not yet been established with certainty, a plausible model of its action is shown in Fig. 27.9. Gyrase first binds to DNA in such a way that the bound segments cross to form a right-handed node. This action stabilizes a positive supercoil and induces a negative supercoil. Gyrase then introduces a double-strand break in the DNA at the back of the right-handed node and passes the front segment through the break, inverting the handedness and thus the sign of the node. Gyrase could then release one of the two crossing segments of the negative node, the negative supercoils would distribute along the DNA, and the cycle could begin again.

Essentially all duplex DNA exists naturally in a negatively supertwisted form and it is this topoisomeric form of DNA that undergoes semiconservative replication. Thus, DNA gyrase is indispensable for this process. In fact, specific inhibitors of DNA gyrase, *nalidixic acid, coumermycin,* and *novobiocin,* are potent inhibitors of DNA replication in *E. coli.*

DNA Binding and Unwinding Proteins

Proteins variously termed *single-stranded DNA binding proteins* or *helix destabilizing proteins* bind tightly and cooperatively to single-stranded DNA, but weakly to duplex DNA. Inasmuch as transient melting occurs in regions of duplex DNA at physiological temperatures, particularly those rich in adenine and thymine (Chap. 8), binding proteins can associate with such transiently single-stranded regions within the duplex and displace the position of equilibrium between duplex and single-stranded forms toward the latter. As a result, localized separation of the two strands may occur at a temperature considerably below the melting temperature (Fig. 27.10). In this manner the single-stranded DNA binding proteins may facilitate unwinding of the double helix in advance

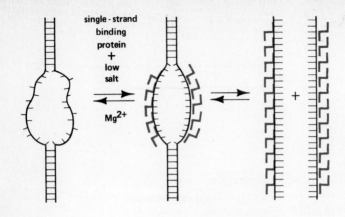

single - strand
binding
protein
+
low
salt

Mg²⁺

Figure 27.10 Action of a single-stranded DNA binding protein in displacing the equilibrium between the duplex and single-stranded forms of DNA. [Adapted from A. Kornberg, DNA Replication, Freeman, San Francisco, 1980, p. 280.]

of the replication fork (see, for example, Fig. 27.2). Indeed, mutants of *E. coli* with a defect in the gene that codes for single-stranded DNA binding protein are correspondingly defective in DNA replication.

The *DNA helicases* catalyze the unwinding of duplex DNA coupled to the hydrolysis of ATP to ADP and P_i with the consumption of 1 or 2 equiv of ATP per base pair hydrolyzed. The mechanism whereby the energy of ATP hydrolysis is coupled to the catalytic separation of DNA strands is unknown; however, it may be analogous to other reactions in which ATP hydrolysis facilitates translocation of a protein or other macromolecule on a polymeric surface, for example, the sliding of myosin on actin (Chap. M8) and the transfer of aminoacyl tRNA to ribosome-bound mRNA (Chap. 28). The DNA helicases probably function in concert with single-stranded DNA binding proteins to promote unwinding of the DNA duplex at a replication fork. In fact, a defect in one of the helicases of *E. coli*, known as the *rep protein*, leads to a reduced rate of movement of replication forks in vivo.

Discontinuous DNA Replication

The polymerization of nucleotides by all known DNA polymerases is uniquely in the 5'-to-3' direction. Moreover, there is an absolute requirement for a 3'-hydroxyl-terminated primer onto which nucleotides can be polymerized. To explain the semiconservative replication of a duplex DNA molecule, it is therefore necessary to understand (1) how the two strands of the DNA duplex, which are of opposite chemical polarity (5' to 3' and 3' to 5'), are synthesized simultaneously and (2) how the synthesis of DNA chains is initiated, i.e., how the primer is formed.

The first question has been answered by the finding that the most recently synthesized (nascent) DNA in *E. coli* is in the form of short segments 100 to 1000 nucleotide residues in length termed *Okazaki fragments*. Similar segments, 100 to 400 residues long, are found in animal cells. These nascent DNA chains are then incorporated into the main body of the replicating DNA molecule by the action of DNA ligase. Inasmuch as the 5'-to-3' synthesis of Okazaki fragments can proceed in the overall 3'-to-5' direction (Fig. 27.11), the discontinuous mode of replication can explain the simultaneous replication in the 5'-to-3' direction

710

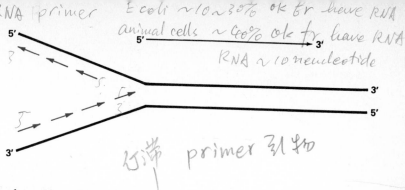

Handwritten margin notes (top): RNA primer · E.coli ~10~30% ok for have RNA · animal cells ~60% ok for have RNA · RNA ~10 nucleotide · 切断 primer 引物

Figure 27.11 Discontinuous replication of DNA. Note that since replication must occur in the 5'-to-3' direction on both strands of the duplex, polymerization cannot occur simultaneously on both strands. Thus, a single-stranded region must exist transiently in the region of the replication fork.

of both strands of the DNA duplex. Note that only the "lagging" strand, e.g., that synthesized in the overall 3'-to-5' direction, need be replicated discontinuously; the leading strand can be replicated without interruption.

Insight into the mechanism of DNA chain initiation has come from in vitro studies of viral DNA replication, described below.

Multienzyme Complexes of DNA Replication

The complexity of duplex DNA replication in *E. coli* is indicated by the large number of genetic loci that are required for this process. Each of these loci is associated with one or more *conditional* mutations that, unlike the wild-type allele, permit DNA replication and hence growth at 30°C but prohibit it at 42°C. Thus far, 13 such *dna* genes have been identified: *dna*A, *dna*B, *dna*C, *dna*E, *dna*G, *dna*I, *dna*J, *dna*K, *dna*L, *dna*N, *dna*P, *dna*X, and *dna*Z. The products of genes *dna*B, C, E, G, L, N, X, and Z are required for chain elongation; the products of genes *dna*A, J, K, P, and possibly C are believed to act in the initiation of chromosomal replication. Another gene, *dna*T, may function in the termination of a round of replication. The *dna*B product forms part of the priming system (the *primosome*) needed for the initiation of DNA chains, and the *dna*E, *dna*N, *dna*X, and *dna*Z genes code for the α, β, δ, and γ subunits, respectively, of the DNA polymerase III holoenzyme (Table 27.1). In addition to the products of the *dna* genes, DNA ligase (*lig*), DNA polymerase I (*pol*A), single-stranded DNA binding protein (*SSB*), and DNA gyrase (the β and α subunits specified by the *cou* and *nal*A genes, respectively) are also indispensable for normal DNA replication.

Insight into the function of each of these gene products in the discontinuous replication of the *E. coli* chromosome has come from studies of the replication of the small bacterial viruses M13 and φX174. These and related bacteriophages contain a single molecule of circular single-stranded DNA with 5500 to 6500 nucleotide residues which during the initial phase of their replication are converted to a closed circular duplex molecule (RFI) (Chap. 30). In particular, such studies have revealed the complex manner in which DNA chains are started *de novo*. In each instance initiation of a DNA chain is accomplished by the synthesis of a short RNA primer. In the case of M13, synthesis of the primer is catalyzed by RNA polymerase (Chap. 28) and occurs at a specific site on the M13 DNA; in the case of φX174 a multienzyme primosome is involved (Table 27.2).

Handwritten margin notes (right): allele 等位基因 (一种基因可有几种变异形式) · indispensable 必不可少 的

TABLE 27.2

Protein Requirements for the In Vitro Conversion of M13 and φX174 Single-Stranded DNAs to the Duplex RFI Form

Stage	M13	φX174
Prepriming	Single-stranded DNA binding protein	Single-stranded DNA binding protein Protein i n n' n'' dnaB protein dnaC protein
Priming	RNA polymerase	dnaG protein (primase) dnaB protein
Elongation	DNA polymerase III holoenzyme	DNA polymerase III holoenzyme
Excision of primer	DNA polymerase I	DNA polymerase I
Ligation	DNA ligase	DNA ligase

[handwritten: primosome binding site; allosteric e... of nucle...; triphos...; ligation 连接; hairpin 发夹]

Replication of M13 DNA

Conversion of the circular single-stranded DNA of phage M13 to the duplex form (RFI) occurs as shown in Fig. 27.12. The single-stranded circle is first coated with binding protein except for a small duplex hairpin region. The latter is transcribed by RNA polymerase (Chap. 28) to produce a short RNA primer. Indeed, M13 replication in vivo is specifically blocked by the RNA polymerase inhibitor rifampicin (Chap. 28). DNA polymerase III holoenzyme

Figure 27.12 Scheme for conversion of phage M13 circular single strands to closed circular duplex DNA by the action of *E. coli* replication proteins. A duplex hairpin region is indicated at the site at which RNA synthesis is initiated by analogy to promotors of transcription in duplex DNA (Chap. 28). See text for details. [*From R. Schekman, A. Weiner, and A. Kornberg, Science, 186:987 (1974). Copyright © 1974 by The American Association for the Advancement of Science.*]

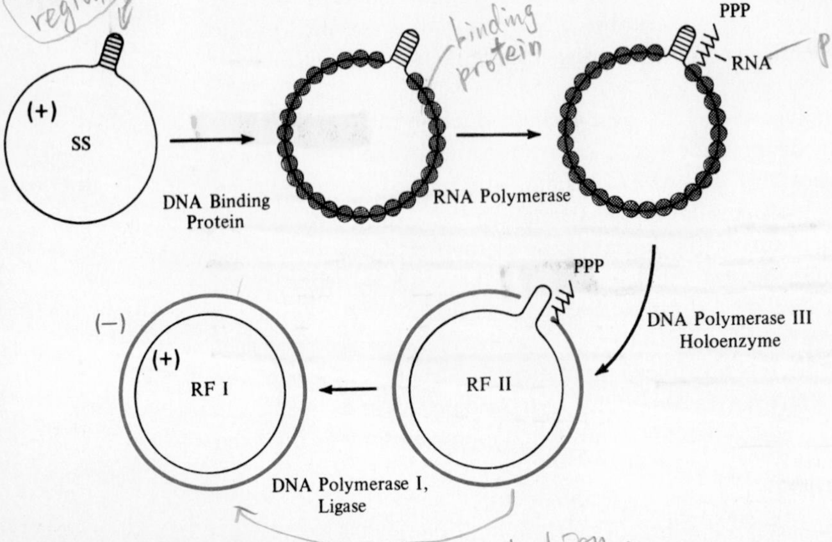

[handwritten annotations: SS — single-stranded; hairpin region; binding protein; PPP — RNA primer; nick translation]

extend, 仲. 拉开

(page 706) then extends the primer to form the complementary strand. The primer is excised and the resulting gap filled through the process of nick translation catalyzed by DNA polymerase I (page 704). Finally, the circular duplex DNA is covalently closed by DNA ligase.

Replication of φX174 DNA

nick 切口

In contrast to the replication of M13 DNA, which is initiated by RNA polymerase and requires only DNA polymerase III holoenzyme, single-stranded DNA binding protein, DNA polymerase I, and DNA ligase, conversion of single-stranded circular φX174 DNA does not involve RNA polymerase but instead requires a group of eight proteins in addition to DNA polymerase III holoenzyme and single-stranded DNA binding protein. These proteins may be associated in vivo as a replication complex, or *replisome*. Inasmuch as several of these proteins, in particular, the products of the *dna*B, C, and G genes, are essential for *E. coli* DNA replication in vivo, conversion of single-stranded φX174 DNA to the duplex form as observed in vitro may reflect features of the replication of the circular chromosome of *E. coli*, in particular, the manner in which the Okazaki fragments (page 710) are initiated and elongated.

Figure 27.13 illustrates the stages in conversion of the circular single-

Figure 27.13 Scheme for conversion of φX174 circular single strands to closed circular duplex DNA. See text for details. [*From K. Arai, R. Low, J. Kobori, J. Shlomai, and A. Kornberg, J. Biol. Chem. 256:5273 (1981).*]

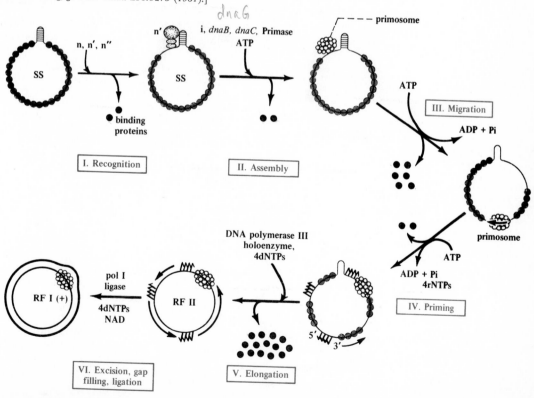

assembly
集合 装配

stranded φX174 DNA (SS) to the supercoiled duplex replicative form (RFI). Stage I consists of the recognition by protein n' of a specific nucleotide sequence at the origin of φX174 DNA coated with single-stranded DNA binding protein, which serves as the site for assembly of the primosome. Proteins n and n″ are also required at this point. Assembly of the primosome at the n' recognition site is completed upon binding of the *dna*B, *dna*C, *dna*G (primase), and *i* proteins (stage II). The assembled primosome then migrates processively (e.g., without dissociation) in a direction opposite to DNA elongation, coupled to the hydrolysis of ATP catalyzed by protein n' (stage III); single-stranded DNA binding protein dissociates from the DNA during migration of the primosome. The hexameric *dna*B protein then responds to the allosteric effects of a nucleoside triphosphate to induce changes in the structure of the single-stranded φX174 DNA at preferred sequences, thereby enabling the primase to synthesize short primers for the initiation of DNA synthesis (stage IV). The primers are extended by the action of DNA polymerase III holoenzyme (stage V), and the covalently closed circular duplex is generated by the action of DNA polymerase I and DNA ligase, as in the case of M13 replication (page 712) (stage VI). A comparison of the enzymic requirements for conversion of M13 and φX174 single-stranded DNAs to their respective duplex RFI forms is given in Table 27.2.

Replication of the *E. coli* Chromosome

A model for the replication fork of the *E. coli* chromosome (Fig. 27.2) that is derived from the foregoing analysis of φX174 DNA replication is shown in Fig. 27.14. The leading strand is synthesized by DNA polymerase III holoenzyme

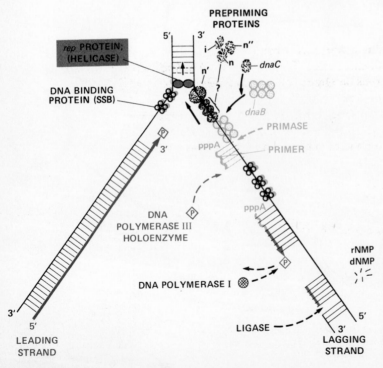

Figure 27.14 Model for the action of the primosome at the replication fork of the *E. coli* chromosome. The primosome assembled at or near the DNA replication origin migrates on the lagging strand without dissociation, coupled to the movement of the replication fork. [*From K. Arai, R. L. Low, and A. Kornberg, Proc. Natl. Acad. Sci. USA* **78**:*707 (1981)*.]

by a relatively continuous mechanism; synthesis of the lagging strand is discontinuous. The primosome, once assembled at or near the origin of replication, migrates processively coupled to movement of the replication fork. Multiple, short RNA primers synthesized by primase at numerous preferred regions are elongated into Okazaki fragments by DNA polymerase III holoenzyme. Unwinding of the duplex at the replication fork is promoted by single-stranded DNA binding protein and a helicase (page 715); movement of the primosome may also facilitate unwinding of the duplex. The coupled excision of RNA primers and filling in of the resultant gap by DNA polymerase I and the joining of the nascent DNA chains by DNA ligase complete the process.

Although it is now reasonably clear how DNA chains are initiated, no information is presently available concerning the mechanism of generation of a replication fork at a unique site on the chromosome and hence the onset of a cycle of DNA replication.

REPAIR OF DNA

Damage to the genome after any chemical and physical insult can be repaired by a variety of mechanisms. The best-understood is that responsible for the repair of damage to DNA resulting from ultraviolet irradiation. Exposure of DNA to ultraviolet irradiation produces thymine dimers in which adjacent thymine residues are fused by formation of a cyclobutane ring (Fig. 27.15). Less frequently, dimers are formed between adjacent thymine and cytosine residues. The pyrimidine dimer causes a distortion in the helix beyond which replication cannot proceed and hence must be removed.

Excision Repair

In *E. coli*, removal of thymine dimers is accomplished primarily by an excision process that proceeds in three steps (Fig. 27.16): (1) The first step is cleavage of the DNA helix at or near the thymine dimer by a specific endonuclease. This *uv endonuclease* is composed of three subunits (MWs = 115,000, 84,000, and 27,000), the products of the *uvrA, uvrB,* and *uvrC,* genes, respectively. (2) The second involves simultaneous excision of an oligonucleotide containing the thymine dimer and reincorporation of nucleotides at the gap thus created by

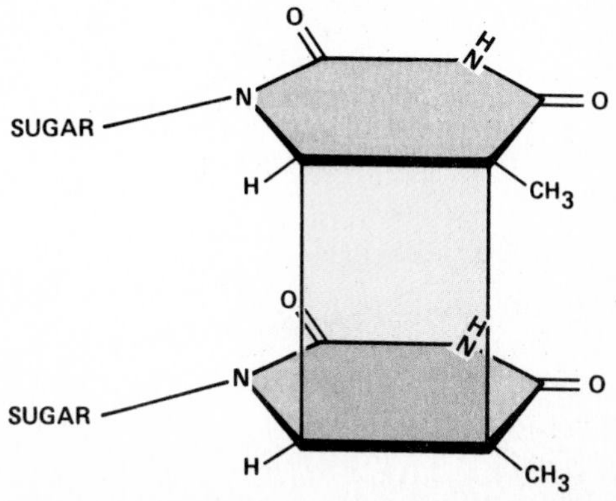

Figure 27.15 Structure of a thymine dimer. Adjacent thymidylate residues in a DNA chain are fused through a cyclobutane ring by bonds between carbon atoms 5 and 6 in each thymine.

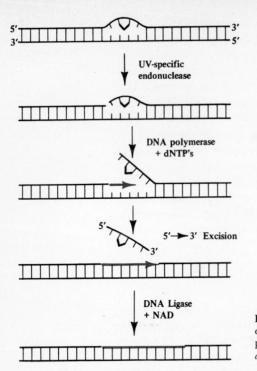

① UV· endonuclease

② 5'→3' exonuclease and DNA polymerase I

③ DNA ligase

Figure 27.16 Excision-repair of DNA containing a thymine dimer. Note the process of nick translation catalyzed by DNA polymerase I. [*From R. B. Kelly, M. R. Atkinson, J. A. Huberman, and A. Kornberg, Nature* **224**, *491 (1969).*]

the 5' → 3' exonuclease and polymerase activities of DNA polymerase I. This is the process of nick translation. ③ The third step is formation of a phospho-diester bond between the newly synthesized segment and the remainder of the helix, catalyzed by DNA ligase. An alternative mode of excision repair observed in *Micrococcus luteus* involves cleavage of the *N*-glycosylic bond linking one

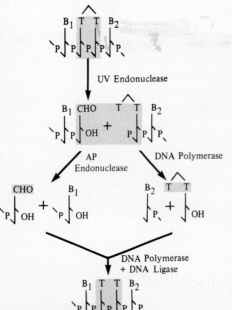

① uv-specific endonuclease cleavage at or near the Thymidi dimer

② excision of thymine dimer and nick transplant

③ DNA ligase connectes the new strand and remainder helix.

Figure 27.17 Excision-repair of thymine dimer-containing DNA by creation of an apyrimidinic site. Cleavage by a UV endonuclease of a glycosylic bond linking the thymine dimer to the DNA results in the generation of an apyrimidinic site with an aldehydic group on the deoxyribose. An AP endonuclease releases the deoxyribose phosphate, and DNA polymerase I excises the thymine dimer. The resulting gap is filled in by DNA polymerase I and the nick is sealed by the action of DNA ligase. For simplicity, only the strand containing thymine dimer is shown.

of the thymine dimers to the deoxyribose backbone of the DNA, followed by hydrolysis of the phosphodiester bond at the site of glycosylic bond cleavage (See Fig. 27.17).

Thus far, excision enzymes have not been observed in eukaryotes. However, several rare hereditary diseases in human beings are linked to defects in excision-repair. These include *xeroderma pigmentosum, ataxia-telangiectasia, Fanconi's anemia,* and *Bloom's syndrome.* Patients with xeroderma pigmentosum are abnormally sensitive to sunlight and show a very high incidence of skin cancer. When cells cultured from the skin of these patients are exposed to ultraviolet irradiation, the rate of excision of thymine dimers is markedly reduced compared with cells derived from normal individuals. A detailed analysis of *xeroderma pigmentosum* patients has shown them to fall within seven separate complementation groups, suggesting that at least seven different genes are required for the excision of thymine dimers. Thus, it would appear that human skin cells, like *E. coli,* require a multienzyme complex for the initial steps in excision-repair.

Bypass Repair

A second mode of repair of ultraviolet-induced lesions involves bypass of the thymine dimer either by recombination (page 719) or by an "error-prone" transdimer synthesis.

Recombinational Repair Recombinational repair, also termed *postreplication repair,* is of particular importance at a replication fork where excision of a thymine dimer in a strand that has unwound from the duplex and is being replicated would lead to the nonrepairable scission of the single strand (see Fig. 27.18). Thus, in ultraviolet-irradiated bacteria in which dimer excision lags behind replication, a gap is generated in the template strand opposite each thymine dimer when strand elongation blocked by the dimer resumes beyond it, presumably at the next site of initiation of a nascent DNA chain. Recombinational transfer of a DNA segment from the parental to the progeny strand can fill in the gap, and the discontinuity in the parental strand can be removed by repair replication. Finally, the bypassed thymine dimer can be removed by excision-repair. Postreplication repair has thus far been convincingly demonstrated only in bacteria.

Error-Prone, or Mutagenic, Repair Error-prone, or mutagenic, repair is a complex process induced by ultraviolet irradiation as well as other agents that damage DNA or interrupt its replication. Although its mechanism is not understood, it is believed to involve the indiscriminate insertion of nucleotides opposite the thymine dimer, thus bypassing the lesion but leading in turn to a high frequency of mutations at the site of the lesion (Fig. 27.17).

Photoreactivation

Photoreactivation represents the simplest mode of repair of DNA containing thymine dimers. The photoreactivating enzyme binds to the region of the DNA containing the dimer to form an enzyme-DNA chromophore that absorbs

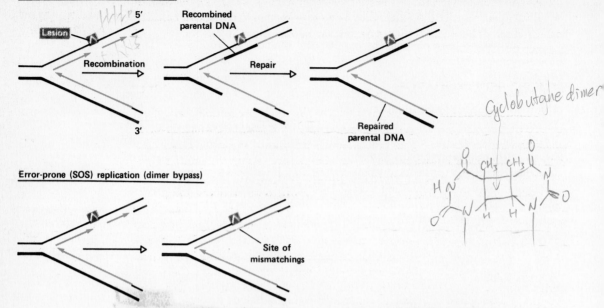

Postreplication recombination repair

5′

Lesion

Recombination →

Recombined
parental DNA

Repair →

3′

Repaired
parental DNA

Error-prone (SOS) replication (dimer bypass)

Site of
mismatchings

Cyclobutane dimer

Figure 27.18 Scheme for the bypass repair of DNA by recombination or by error-prone repair of the sequence containing the lesion. [*From A. Kornberg, DNA Replication, Freeman, San Francisco, 1980, p. 617.*]

visible light (300 to 600 nm) and then catalyzes cleavage of the cyclobutyl pyrimidine dimer. As a result, the monomeric thymine residues are restored and the enzyme dissociates from the DNA. Photoreactivating enzymes have been demonstrated in cells as diverse as *E. coli* and human lymphocytes.

Base Excision

Removal of a damaged base from DNA can be achieved by cleavage of the *N*-glycosylic bond linking the base to the deoxyribose backbone of the DNA, thereby creating an *apurinic* or *apyrimidinic* site. For example, uracil residues may be incorporated into DNA in place of thymine, or they may be derived from cytosine by spontaneous deamination. The latter results in a change from a G-C to an A-T base pair. To prevent this potentially mutagenic event the specific *uracil N-glycosylase* (MW = 35,000) of *E. coli* removes the uracil, leaving an apyrimidinic site in the DNA. Other glycosylases catalyze the release of hypoxanthine and 3-methyladenine to generate apurinic sites. Such lesions may also result from spontaneous cleavage of the purine glycosylic bond. The apurinic and apyrimidinic (AP) sites are repaired by a class of enzymes termed *AP endonucleases,* which cleave the phosphodiester bond at the AP site. The latter can then be removed and the resulting gap repaired by the action of DNA polymerase I and DNA ligase, as in the case of excision-repair of thymine dimers. As noted previously (page 716), excision-repair of thymine dimer-containing DNA in *M. luteus* also proceeds by cleavage of a glycosylic bond linking the thymine dimer to the DNA (Fig. 27.17).

718

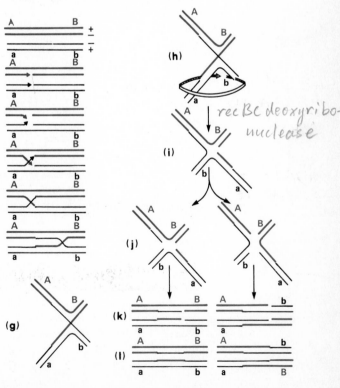

Another mode of repair of apurinic lesions in DNA is by direct insertion of the missing purine. Such *insertases* have been identified in *E. coli* and in cultured human cells. The mechanism of formation of the *N*-glycosylic bond is unknown.

GENETIC RECOMBINATION

Genetic recombination is the process whereby new linkage relationships are established between genes. It is of two types: *general recombination,* which involves exchanges between homologous chromosomes essentially anywhere along their length, and *site-specific recombination,* in which exchanges occur at a limited number of specific sites. Because general recombination permits the generation of individuals with new genotypes, it is a major driving force in evolution (Chap. 31); it also contributes to the generation of immunologic diversity (Chap. M2). Site-specific recombination is responsible for the integration of viral into host genomes (Chap. 30). A widely accepted model for general recombination is shown in Fig. 27.19. According to this model, two homologous double helices are aligned (a) and adjacent phosphodiester bonds are cleaved (b). The free ends thus created leave the complementary strands to which they had been hydrogen bonded and become associated with the complementary strands in the homologous double helix (c and d). The result of this reciprocal strand transfer and subsequent ligation is to establish a physical connection between the two DNA molecules (e). Continuing strand transfer by the two polynucleotide chains involved in the crossover can occur simultaneously with the rotation of

Figure 27.19 A model for general genetic recombination. Details of the model are given in the text. [*From H. Potter and D. Dressler, Cold Spring Harbor Symp. Quant. Biol.* **43**:*969 (1979).*]

the four double helical arms around their cylindrical axes, allowing the point of linkage between the two molecules to move (f and g). This process, which has been termed *branch migration*, can lead to the development of regions of heterozygosity during recombination. Maturation of the resultant structure (h to l) along two related pathways involving isomerization and endonucleolytic cleavages then yields two different pairs of recombinant chromosomes.

Although the details have not yet been clarified, two enzymes have been identified that participate in this reaction sequence. One, the product of the *rec*B and *rec*C genes of *E. coli*, termed the *rec*BC *deoxyribonuclease*, is composed of two subunits (MWs = 140,000 and 128,000) and has both endo- and exonuclease activities. It requires ATP for phosphodiester bond hydrolysis, and it may be involved in the terminal processing steps of the sequence (Fig. 27.19, steps h to l). Mutations in the *rec*B or *rec*C genes cause a 10- to 100-fold decrease in recombination proficiency. The second enzyme, specified by the *rec*A gene and termed the *rec*A *protein* (MW = 38,000), catalyzes the exchange of strands between a single-stranded segment of DNA and a homologous duplex; this is coupled to the hydrolysis of ATP to ADP and P$_i$ (Fig. 27.20). Such a reaction is analogous to the generation of the crossover structure (structure e of Fig. 27.19) and the subsequent branch migration that leads to extensive regions of heterozy-

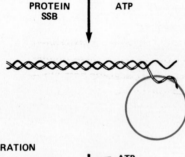

STRAND ASSIMILATION

recA
PROTEIN   ATP
SSB

BRANCH MIGRATION

recA
PROTEIN   ATP
SSB
                ADP + Pi

Figure 27.20 Role of the *rec*A protein of *E. coli* in promoting the exchange of homologous DNA strands. In the first step, termed strand assimilation, which requires ATP but not its hydrolysis, one end of the duplex DNA forms a limited number of base pairs with the circular single-stranded DNA. In the second step, branch migration, coupled to ATP hydrolysis, results in formation of a duplex circle composed of the circular single strand paired with one of the two strands of the linear duplex and displacement of the other strand. SSB: single-stranded DNA binding protein. [*From M. Cox and I. R. Lehman, Proc. Natl. Acad. Sci. USA* **78**:3433 (1981).]

gosity. Indeed, mutations in the recA gene eliminate general recombination in *E. coli*. Thus far, enzymes comparable to the recA protein and the recBC deoxyribonuclease have not been observed in eukaryotes.

The site-specific integration of the prophage form of bacteriophage λ (Chap. 30) into the *E. coli* chromosome does not make use of the recA or recBC enzymes. Instead, it is accomplished by an as yet undefined mechanism that involves a type I topoisomerase (page 708) coded for by the viral genome, acting together with as many as four bacterial gene products.

MODIFICATION AND RESTRICTION OF DNA

As described in Chap. 8, DNA may contain methylated and otherwise modified bases. One of the first of these to be recognized was 5-hydroxymethylcytosine, which replaces cytosine in the DNA of bacteriophages T2, T4, and T6 (Chap. 30). The hydroxymethylcytosine is further modified by glucosylation in patterns characteristic for each of these DNAs (Table 27.3).

The replacement of cytosine with glucosylated hydroxymethylcytosine can be accounted for by the induction of a series of enzymes upon infection of *E. coli* with phages T2, T4, and T6. These enzymes include (1) dCMP hydroxymethylase, which converts dCMP to 5-hydroxymethyl dCMP (Chap. 26); this can in turn be converted to the deoxyribonucleoside triphosphate; (2) a dCTP pyrophosphatase that degrades dCTP (and dCDP) to dCMP, rendering dCTP unavailable for DNA replication while regenerating dCMP as a substrate for dCMP hydroxymethylase; and (3) a family of glucosyl transferases that catalyze the transfer of glucosyl units from UDP-glucose to hydroxymethylcytosine (HMC) in DNA.

$$\text{UDP-glucose} + \text{HMC-DNA} \rightleftharpoons \alpha\text{-glucosyl-HMC-DNA} + \text{UDP}$$
$$\text{UDP-glucose} + \text{HMC-DNA} \rightleftharpoons \beta\text{-glucosyl-HMC-DNA} + \text{UDP}$$
$$\text{UDP-glucose} + \alpha\text{-glucosyl-DNA} \rightleftharpoons \text{gentiobiosyl-HMC-DNA} + \text{UDP}$$

Modification of T2, T4, and T6 phages by substitution of glucosylated hydroxymethylcytosine for cytosine serves to protect the DNA of these phages from cleavage by endonucleases that rapidly and completely degrade host DNA to nucleotides; these then serve for the synthesis of viral DNA.

A second and more general modification of DNA entails methylation of the N-6 amino group of adenine (and less frequently the C-5 group of cytosine) within the DNA. This occurs by transfer of a methyl group from *S*-adenosyl-

	Percent of total hydroxymethylcytosine			TABLE 27.3
	T2	T4	T6	Glucosylation of Hydroxymethyl-cytosine Residues of the DNAs of Bacteriophages T2, T4, and T6
Unglucosylated	25	0	25	
α-Glucosyl	70	70	3	
β-Glucosyl	0	30	0	
α-Gentiobiosyl (α-glucosyl-β-glucosyl)	5	0	72	

721

methionine to adenine or cytosine residues within a specific DNA sequence, catalyzed by the appropriate *transmethylase* (Chap. 23). Methylation of the DNA in this manner renders it immune to the corresponding *restriction endonuclease* (see below) that produces double-strand scissions in DNAs lacking the necessary modification. The restriction-modification system is therefore capable of degrading a foreign DNA that may have gained access to the cell while at the same time protecting its own DNA, thus providing it with a remarkable immunity to invasion of the cell by, for example, viral nucleic acids.

Restriction endonucleases have been identified in a wide variety of bacterial species, and several have been purified to homogeneity. They fall into two classes: *type I,* which requires Mg^{2+}, ATP, and S-adenosylmethionine to promote phosphodiester bond cleavage, and *type II,* which requires only Mg^{2+}. The type I enzymes utilize S-adenosylmethionine to methylate a specific binding site; however, cleavage of the DNA occurs at phosphodiester linkages remote from the binding site. The products are heterogeneous in size and the nucleotide sequences at their termini have not been determined. Aside from introducing a conformational change, the role of ATP in the action of the type I enzyme is unclear.

The nucleotide sequences at the sites of cleavage and of modification of more than 200 of the type II restriction endonucleases have been determined; several of these are listed in Table 27.4. An important generalization derived from knowledge of these sequences is that the sites of cleavage are symmetrical. They display 180° (twofold) rotational symmetry about an axis perpendicular to the DNA helix axis. In some cases, the sites of cleavage are staggered; the products therefore have overlapping 3'-hydroxyl and 5'-phosphoryl termini. Inasmuch as one of these enzymes, the *EcoRI endonuclease,* is known to be composed of two subunits (MW = 28,000), the twofold symmetry of the cleavage site may reflect a symmetrical orientation of these subunits.

Because of their capacity to cleave unmodified DNAs in a highly specific way, the restriction nucleases have become valuable reagents in the determination of nucleotide sequences in DNA (Chap. 8). They have also been employed in the genetic mapping of eukaryotic and viral chromosomes (Chaps. 29 and 30). The use of restriction nucleases in the construction and cloning of hybrid DNA molecules is described on page 724.

In addition to the highly specific restriction-modification system described above, there are transmethylases that catalyze the methylation by S-adenosylmethionine of adenine and cytosine residues in DNA, forming 6-methylaminopurine and 5-methylcytosine to an extent much greater than that of the restriction-modification system; their function is unknown, but it has been proposed that these modifications may serve to regulate transcription in eukaryotes (Chap. 29).

PLASMIDS

Bacteria often possess small supercoiled DNA molecules termed *plasmids* that are separate from the genome and range in size from 2 to 200 kb. These molecules, which are usually present in about 20 copies per cell, are capable of

TABLE 27.4

Recognition Sequences of Some Restriction Endonucleases†

Enzyme	Sequence at restriction site
EcoRI	5′ GA*AT TC CT TA*AG 5′
HindII	5′ GT Py Pu A*C CA*Pu Py T G 5′
BglII	5′ A*GATC T T CTAGA* 5′
HaeII	5′ Pu G CGC*Py Py C*GCG Pu 5′
HaeIII	5′ GG C*C CC*G G 5′

†The base modified by the specific methylase is indicated by the asterisk. A* refers to N^6-methyladenine and C* is 5-methylcytosine. The nomenclatural system for the restriction enzymes is that of H. Smith and D. Nathans, *J. Mol. Biol.* **81:**419 (1973). Thus *Eco*RI is the enzyme from *E. coli* bearing the plasmid with drug resistance transfer factor I; *Hind*II is from *Haemophilus influenzae* Rd; *Bgl*II is from *Bacillus globiggi*, and *Hae*II and *Hae*III are from *Haemophilus aegyptius*. Arrows show points of cleavage.

autonomous replication. Since they commonly contain genes that lead to the inactivation of a variety of antibiotics, they confer antibiotic resistance and hence a selective advantage on the bacteria that harbor them. They may also bear genes for restriction and modification enzymes, for example, the *Eco*RI endonuclease and transmethylase (page 722).

One class of plasmids, the *transposons,* are capable of being integrated into the host genome. Integration, which can occur at many sites in the genome, depends upon terminally repeated sequences within the transposon that are complementary to and can therefore hybridize with *insertion sequences* that are widely dispersed throughout the host genome. Transposons and insertion sequences can be inserted, excised, and reinserted at different sites by an unknown mechanism that does not require the *rec*A gene (page 720) and thus does not involve homologous recombination.

In the course of their transposition, transposons and insertion sequences are believed to generate rearrangements within the host genome. Since insertion of a transposable element within a gene may inactivate the gene, such insertions can generate spontaneous mutations. Transposable elements have also been identified in eukaryotes (Chap. 29).

homogeneity

丁氢 — 丧亿

propagation

繁殖

DNA segments from bacterial chromosomes can be inserted into bacteriophage DNAs during site-specific recombination in vivo (page 721). In this way, homogeneous populations of DNA for the study of bacterial genes can be obtained. Homogeneity results from the fact that all molecules in the population are descendants, via replication, of a single DNA molecule; i.e., the DNA has been cloned. DNA segments from any source may also be inserted in vitro into bacteriophage λ DNA or into the DNA of bacterial *plasmids*. The resulting hybrid DNA molecules can then be propagated and cloned in vivo. This methodology permits the isolation of individual DNA segments (genes) from any source in large amounts and in homogeneous form.

Several methods have been developed for the construction of such hybrid DNA molecules. One of these is illustrated in Fig. 27.21, in which the autonomously replicating DNA molecule (the vector) is a plasmid that confers resistance to the antibiotic *tetracycline* and hence provides a phenotype permitting selection of cells harboring the plasmid and the DNA segment to be cloned. The circular DNA of the plasmid is first cleaved at a single site by the *Eco*RI endonuclease (Table 27.4) to generate termini that are both identical and self-complementary.

$$
\begin{array}{cccc}
& \downarrow & & \\
-\text{GAATTC}- & & -\overset{3'}{\text{G}} & \overset{5'}{\text{AATTC}}- \\
& \longrightarrow & & + \\
-\text{CTTAAG}- & & -\underset{5'}{\text{CTTAA}} & \underset{3'}{\text{G}}- \\
& \uparrow & &
\end{array}
$$

The DNA to be inserted into the plasmid is also cleaved to generate a set of the similarly terminated segments. These segments are then annealed to the plas-

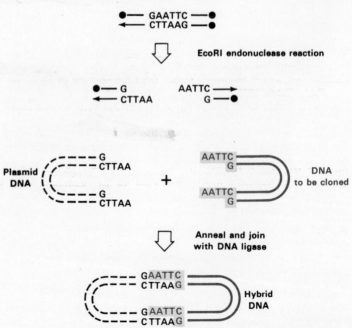

Figure 27.21 Construction in vitro of a hybrid DNA molecule. The plasmid DNA containing a single restriction site and the DNA to be cloned are subjected to the action of *Eco*RI endonuclease. The cleaved DNAs are annealed and ligated to yield the hybrid DNA molecule. [*After P. Wensink, D. J. Finnegan, J. E. Donelson, and D. S. Hogness, Cell 3:315 (1974).*]

mid DNA through their self-complementary termini, and a covalent linkage is established by the action of DNA ligase. An obvious requirement of this method is that successful insertion not inactivate the required functions of the vector, i.e., the capacity for autonomous replication and expression of the selected phenotype, in this instance, tetracycline resistance.

Other methods involve the creation of cohesive termini by the addition, by means of the enzyme *terminal deoxynucleotidyl transferase*, of deoxyadenylate and deoxythymidylate sequences to the 3′-hydroxyl termini of the plasmid DNA and the DNA to be cloned. After annealing, the two DNAs are joined via their complementary deoxyadenylate-deoxythymidylate "tails." Any gaps at the juncture of the two molecules are filled in with DNA polymerase I and covalently sealed with DNA ligase. Cohesive termini may also be generated at the ends of DNA molecules by the enzymic ligation of chemically synthesized overlapping ends that make up a restriction endonuclease recognition sequence.

Expression of eukaryotic genes occurs in *E. coli*. For example, the gene from yeast that specifies one of the enzymes of histidine biosynthesis, *imidazole glycerol phosphate dehydratase* (Chap. 22), has been cloned in a histidine-requiring mutant of *E. coli* and found to permit growth of the mutant in a medium lacking histidine. Other eukaryotic genes that have been cloned and expressed in *E. coli* are those specifying the hormones somatostatin (Chap. M10), insulin (Chap. 16), and growth hormone (Chap. M20). Cloning and expression in *E. coli* of the gene for human somatostatin are described in the following chapter.

REFERENCES

Books

Boyer, P. D. (ed.): *The Enzymes, Nucleic Acids, Part A,* 3d ed., vol. XIV, Academic, New York, 1982.

Cohn, W. (ed.): *Progress in Nucleic Acid Research and Molecular Biology,* vols. 1–, Academic, New York, 1963–. A continuing series.

DNA Replication and Recombination, Cold Spring Harbor Symp. Quant. Biol., vol. 43, 1979.

Goulian, M., and P. Hanawalt (eds.): *DNA Synthesis and its Regulation,* Benjamin, Menlo Park, Calif., 1975.

Hanawalt, P. C., E. C. Friedberg, and C. F. Fox (eds.): *DNA Repair Mechanisms,* Academic, New York, 1978.

Kornberg, A.: *DNA Replication,* Freeman, San Francisco, 1980.

Molineux, I., and M. Kohiyama: *DNA Synthesis: Present and Future,* NATO Advanced Study Institutes Series, Series A, Plenum, New York, 1978.

Stahl, F. W.: *Genetic Recombination,* Freeman, San Francisco, 1979.

Watson, J. B.: *Molecular Biology of the Gene,* 3d, ed., Benjamin, New York, 1976.

Review Articles

Champoux, J. J.: Proteins that Affect DNA Conformation, *Annu. Rev. Biochem.* **47:**449–479 (1978).

Cozzarelli, N. R.: The Mechanism of Action of Inhibitors of DNA Synthesis, *Annu. Rev. Biochem.* **46:**641–648 (1977).

Cozzarelli, N. R.: DNA Gyrase and the Supercoiling of DNA, *Science* **207:**953–960 (1980).

DePamphilis, M. L., and P. M. Wassarman: Replication of Eukaryotic Chromosomes: A Close-up of the Replication Fork, *Annu. Rev. Biochem.* **49:**627–666 (1980).

Friedberg, E. C., K. H. Cook, J. Duncan, and K. Mortelmans: DNA Repair Enzymes in Mammalian Cells, in K. C. Smith (ed): *Photochemical and Photobiological Reviews,* vol. 2, Plenum, New York, 1977, pp. 263–322.

Hanawalt, P. C., P. K. Cooper, A. K. Ganesan, and C. A. Smith: DNA Repair in Bacteria and Mammalian Cells, *Annu. Rev. Biochem.* **48:**783–836 (1979).

Kornberg, A.: The Enzymatic Replication of DNA, *CRC Crit. Rev. Biochem.* **7:**23–49 (1979).

Lehman, I. R.: DNA Ligase: Structure, Mechanism and Function, *Science,* **186:**790–797 (1974).

Lehman, I. R., and D. Uyemura: DNA Polymerase I: Essential Replication Enzyme, *Science* **193:**963–969 (1976).

Ogawa, T., and T. Okazaki: Discontinuous DNA Replication, *Annu. Rev. Biochem.* **49:**421–457 (1980).

Radding, C. M.: Genetic Recombination: Strand Transfer and Mismatch Repair, *Annu. Rev. Biochem.* **47:**847–880 (1978).

Sheinin, R., and J. Humbert: Some Aspects of Eukaryotic DNA Replication, *Annu. Rev. Biochem.* **47:**277–316 (1978).

Söderhäll, S., and T. Lindhal: DNA Ligases of Eukaryotes, *FEBS Lett.* **67:**1–8 (1976).

Tomizawa, J. I., and G. Selzer: Initiation of DNA Synthesis in *Escherichia coli, Annu. Rev. Biochem.* **48:**999–1034 (1979).

Wang, J. C., R. I. Gumport, K. Janaherean, K. Kirkegaard, L. Kleven, M. L. Kotewiz, and Y. C. Tse: DNA Topoisomerases, in *Mechanistic Studies of DNA Replication and Genetic Recombination,* Academic, New York, 1980.

Chapter 28

Gene Expression and Its Control: Prokaryotes

THE GENE AND PROTEIN SYNTHESIS

This chapter will consider the mechanisms involved in protein synthesis. Since proteins are synthesized from free amino acids, the major questions can be stated as follows: (1) How is energy made available for formation of peptide bonds? (2) How are the sequences of nucleotides in DNA translated into the specific sequences of amino acids in proteins, i.e., how is the gene expressed? (3) What are the controlling mechanisms that determine when the genetic information is to be utilized?

Before considering protein synthesis in detail, the main features will be summarized briefly. The process can be divided into three stages:

1. Synthesis by the enzyme RNA polymerase of an RNA that is a complementary copy of one of the two strands of the DNA template; this is the process of *transcription*. Three kinds of RNA are synthesized: messenger RNA (mRNA), ribosomal RNA (rRNA), and transfer RNA (tRNA).

2. *Processing* of the primary RNA transcript to yield functional mRNA, rRNA, and tRNA.

3. *Translation* of mRNA on ribosomes, cytoplasmic particles containing rRNA and proteins, with the aid of tRNA, to yield a polypeptide chain.

Two fundamental features of the translation process should be noted: (1) Synthesis of a polypeptide chain begins at the amino terminus, and chain elongation proceeds by addition of one amino acid residue at a time. The time

Any cross-references coded M refer to the companion text, *Principles of Biochemistry: Mammalian Biochemistry.*

sequence for adding residues to the growing polypeptide chain is thus colinear with the sequence of amino acid residues in these chains, starting with the amino-terminal residue. (2) The mRNA is "read" in successive groups of three adjacent nucleotides (*codons*), starting from a fixed point, the triplet codon specifying which of the 20 amino acids should next be added to the growing polypeptide chain.

In discussing the details of protein synthesis, we shall begin with a consideration of the translation mechanism, proceed to the derivation of the genetic code, and then consider the mechanism of transcription and its control. Here, we shall be concerned principally with prokaryotes. Those features of gene expression that are unique to eukaryotes will be treated in the following chapter.

COMPONENTS OF THE TRANSLATION MECHANISM

Transfer RNA (tRNA)

Each tRNA molecule consists of a single chain of approximately 80 nucleotides, many of which are modified (see Table 8.6). The nucleotide sequences of more than 100 kinds of tRNA from plants, animals, and bacteria are known; on the basis of this information, every tRNA can be folded into a two-dimensional cloverleaf containing a high frequency of paired bases (Fig. 28.1). Nucleotides

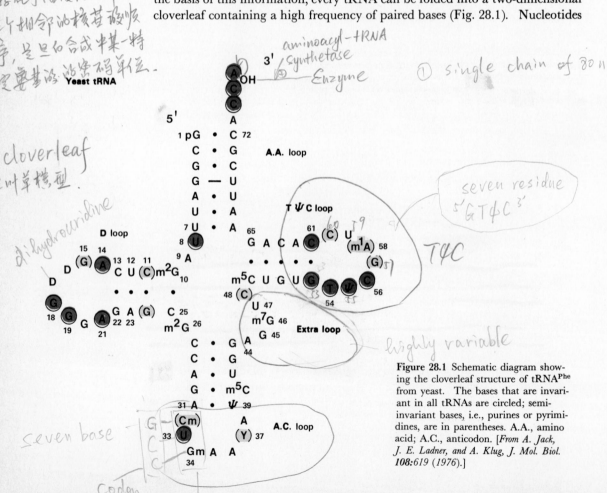

Figure 28.1 Schematic diagram showing the cloverleaf structure of tRNA^{Phe} from yeast. The bases that are invariant in all tRNAs are circled; semi-invariant bases, i.e., purines or pyrimidines, are in parentheses. A.A., amino acid; A.C., anticodon. [*From A. Jack, J. E. Ladner, and A. Klug, J. Mol. Biol.* **108**:619 (1976).]

that are not hydrogen-bonded form five characteristic structures in the clover-leaf: (1) A tail of four residues at the 3′ end of the molecule, of which the three terminal nucleotides are always CCA; the carboxyl group of the appropriate amino acid is enzymically linked to the 3′-hydroxyl group of the terminal adenylate residue by aminoacyl-tRNA synthetases (see below); (2) the TψC loop, consisting of seven residues, which invariably includes the sequence $5'GT\psi C3'$; (3) a highly variable extra loop which usually consists of 4 or 5 nucleotides but may have as many as 21; (4) the *anticodon* loop, which consists of seven bases including the sequence opposite to the codon sequence; e.g., the codon $5'GCC3'$ of the mRNA pairs with the anticodon $3'CIG5'$; and (5) the D loop, consisting of 8 to 12 nucleotides and containing dihydrouridine at a relatively high frequency. In addition to providing the site for attachment of an amino acid and the anticodon function, these non-hydrogen-bonded structures are believed to play a role in binding the tRNA to the ribosome (the TψC loop) and to the specific aminoacyl-tRNA synthetases (the D loop).

Since crystals can be formed from a mixture of different tRNAs, the tertiary structures of all tRNA molecules are probably very similar. The three-dimensional structure of yeast phenylalanine tRNA, shown in Fig. 28.2, retains the double-helical stem regions that appear in the two-dimensional cloverleaf arrangement. However, additional hydrogen bonds fold the cloverleaf into an L-shaped structure. The anticodon is located at one end of the L, about 70 Å

Figure 28.2 Schematic diagram of the three-dimensional structure of tRNA^Phe from yeast showing the chain folding and interaction between bases. The ribose phosphate backbone is shown as a continuous line. The long straight line indicates base pairs in the double-helical stem; shorter lines represent unpaired bases and the dotted lines represent base pairs outside the helices. [*From A. Jack, J. E. Ladner, and A. Klug, J. Mol. Biol.* **108**:*619 (1976).*]

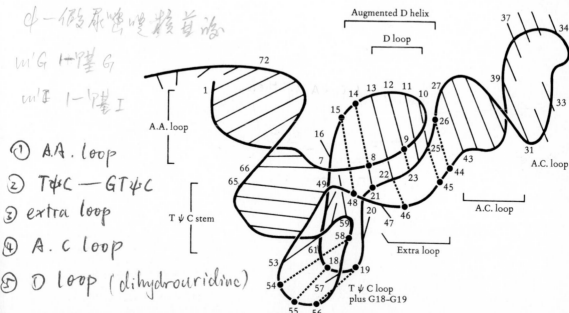

cognate 同族炎
自族训

from the amino acid–acceptor CCA group at the opposite end, while the D and TψC loops interact to form the corner of the L. In many respects the complex folding of the tRNA chain is reminiscent of the folding of a polypeptide chain and is consistent with the multiple highly specific interactions of tRNAs both with proteins and with other nucleic acids, in particular those present in the ribosome.

Amino Acid Activation and Attachment to tRNA The overall reaction for this process is shown in Fig. 28.3. Part of the energy required for protein synthesis is provided by ATP in the esterification of each amino acid to its cognate tRNA. The first step consists of the formation of an enzyme-bound aminoacyl-adenylate complex [reaction (1)], in which the amino acid carboxyl forms an anhydride with the phosphate of AMP. This is followed by transfer of the aminoacyl moiety to a specific tRNA [reaction (2)]; the carboxyl group of the amino acid is linked by an ester bond to the 3'-hydroxyl group of the ribose of the terminal adenylate residue of tRNA. Both steps are catalyzed by a single enzyme, an *aminoacyl-tRNA synthetase*, which is specific for the amino acid as well as the tRNA acceptor and requires Mg^{2+}. By this process each amino acid is activated and linked to a specific set of cognate tRNAs.

3 group of amino-
acyl-tRNA synthetase

① single-chain enzymes
ile. val. leu.

② identical subunits
met (4) ser.(2)

③ dissimilar subunits
Gly.

Many synthetases have been obtained in purified form. Some are single-chain enzymes, e.g., those for isoleucine, valine, and leucine, with molecular weights around 100,000. Some consist of identical catalytically active subunits, e.g., for methionine, with four subunits (MW $\approx$ 45,000 each), and for serine, with two subunits (MW $\approx$ 45,000 each). A third group comprises synthetases with dissimilar subunits, e.g., for glycine with two each of molecular weight 33,000 and two each of molecular weight 80,000, and for tryptophan, which has an $\alpha_2\beta_2$ structure but a molecular weight of 27,000 for all subunits.

Figure 28.3 Aminoacylation of tRNA. R' refers to the specific amino acid to be linked to its cognate tRNA ($tRNA^{R_1}$) in the reaction catalyzed by the specific aminoacyl tRNA synthetase (Enz^{R_1}).

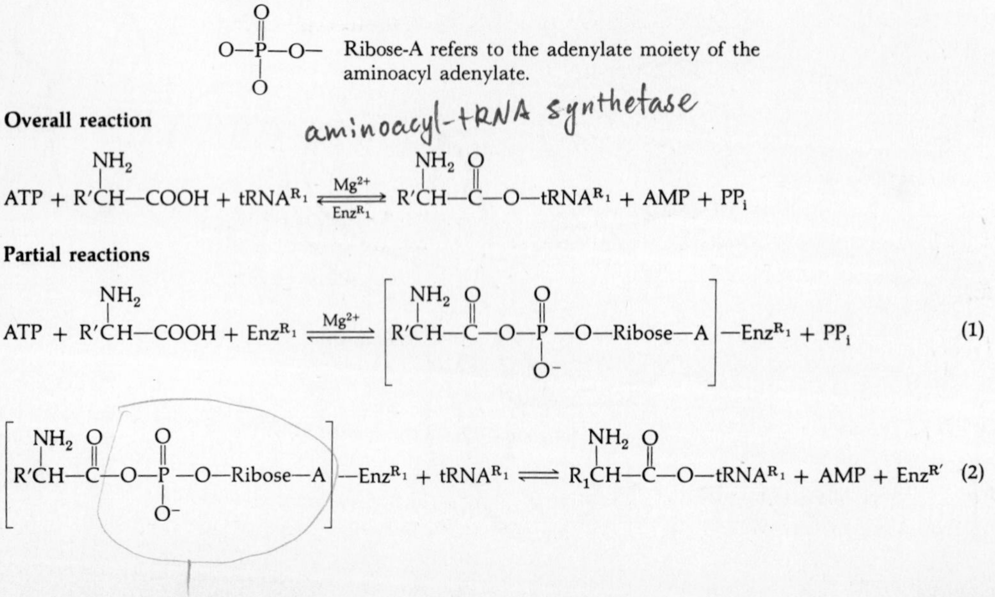

$$O-\overset{\overset{\displaystyle O}{\|}}{\underset{\underset{\displaystyle O}{|}}{P}}-O-$$

Ribose-A refers to the adenylate moiety of the aminoacyl adenylate.

aminoacyl-tRNA synthetase

Overall reaction

$$ATP + R'\overset{\overset{\displaystyle NH_2}{|}}{CH}-COOH + tRNA^{R_1} \underset{Enz^{R_1}}{\overset{Mg^{2+}}{\rightleftharpoons}} R'\overset{\overset{\displaystyle NH_2}{|}}{\underset{}{CH}}-\overset{\overset{\displaystyle O}{\|}}{C}-O-tRNA^{R_1} + AMP + PP_i$$

Partial reactions

$$ATP + R'\overset{\overset{\displaystyle NH_2}{|}}{CH}-COOH + Enz^{R_1} \overset{Mg^{2+}}{\rightleftharpoons} \left[R'\overset{\overset{\displaystyle NH_2}{|}}{CH}-\overset{\overset{\displaystyle O}{\|}}{C}-O-\overset{\overset{\displaystyle O}{\|}}{\underset{\underset{\displaystyle O^-}{|}}{P}}-O-Ribose-A \right]-Enz^{R_1} + PP_i \qquad (1)$$

$$\left[R'\overset{\overset{\displaystyle NH_2}{|}}{CH}-\overset{\overset{\displaystyle O}{\|}}{C}-O-\overset{\overset{\displaystyle O}{\|}}{\underset{\underset{\displaystyle O^-}{|}}{P}}-O-Ribose-A \right]-Enz^{R_1} + tRNA^{R_1} \rightleftharpoons R_1\overset{\overset{\displaystyle NH_2}{|}}{CH}-\overset{\overset{\displaystyle O}{\|}}{C}-O-tRNA^{R_1} + AMP + Enz^{R'} \qquad (2)$$

AMP

The accuracy of translation is determined by the specificity of the amino-acyl-tRNA synthetase reaction. The specific unique sequence of each of the many proteins made in the same cell is, in fact, an indication of the low incidence of error in this reaction. The operation of this precise mechanism is illustrated by isoleucyl-tRNA synthetase; it must discriminate between isoleucine and valine, which differ by only a single methylene group. Some specificity is achieved at the level of substrate binding ($K_m^{\text{isoleucine}} = 5\ \mu M$, $K_m^{\text{valine}} = 0.4\ \text{m}M$). However, whereas the enzyme-isoleucyl-adenylate complex reacts with its cognate tRNA to form isoleucyl-tRNA, the enzyme-valyl-adenylate is rapidly hydrolyzed in the presence of the isoleucine-specific tRNA to yield free valine and AMP.

Although only a single aminoacyl-tRNA synthetase exists for each of the 20 amino acids commonly found in proteins, there may be several tRNAs for one amino acid. Each synthetase, however, can recognize all the tRNAs for a single amino acid. The 20 amino acids commonly found in proteins are the only ones with specific tRNAs. The other amino acid residues of many proteins result from various modifications of the polypeptide chain.

Ribosomes

The template mRNA and the many aminoacylated tRNAs meet on the ribosome, where assembly of a polypeptide chain takes place. A ribosome consists of a large and a small subunit, each of which contains RNA and protein. In eukaryotes, most of the ribosomes are attached to the membranes of the endoplasmic reticulum, where they form the *rough endoplasmic reticulum.*

In prokaryotes, the functional ribosome has a sedimentation coefficient of 70 S and a mass of 3×10^6 daltons. The 70-S particle is composed of a small (30 S) and a large (50 S) subunit. The 30-S subunit (MW = 0.9×10^6) consists of a 16-S rRNA molecule of 1520 nucleotide residues and 21 different proteins ranging in molecular weight from 10,700 to 65,000. The 50-S subunit (MW = 2×10^6) contains two RNA species, a 23-S RNA of 3100 nucleotides and a 5-S RNA of 120 nucleotides, and 34 proteins with molecular weights ranging from 9000 to 28,500.

The ribosome in eukaryotes (80 S) has a mass of 4.5×10^6 daltons. Like the prokaryotic ribosome, it is composed of a small (40 S) and a large (60 S) subunit. The 40-S subunit contains an 18-S rRNA molecule of 2000 nucleotide residues and approximately 30 proteins. The 60-S subunit (MW = 3×10^6) contains a 28-S rRNA of 4000 nucleotides, 5.8-S RNA, and a 5-S rRNA composed of 121 nucleotides; approximately 50 different proteins are present in this particle. A comparison of prokaryotic and eukaryotic ribosomes is given in Table 28.1.

Ribosomal RNAs contain modified bases; however, modification is not as extensive as in tRNA. Methylated bases and pseudouridine are found in the 16-S and the 23-S rRNAs of *Escherichia coli*; methylated 2′ oxygens of the ribose of the 28-S ribosomal RNAs are found in eukaryotes. The 5-S rRNAs contain only the four usual nucleotides.

The 30-S ribosomal subunits of *E. coli* have been reconstituted from the 21 separate proteins and the 16-S rRNA. Similar reconstitution of the 50-S sub-

TABLE 28.1

Comparison of Prokaryotic and Eukaryotic Ribosomes

	Prokaryotes	Eukaryotes
	70-S Ribosome (MW = 3×10^6)	80-S Ribosome (MW = 4.5×10^6)
	30-S Subunit (MW = 0.9×10^6)	40-S Subunit (MW = 1.5×10^6)
	16-S RNA (MW = 0.6×10^6)	18-S RNA
	1520 Nucleotides	~2000 Nucleotides
	21 Proteins (S1 to S21)	~30 Proteins
	50-S Subunit (MW = 2.0×10^6)	60-S Subunit (MW = 3.0×10^6)
	23-S RNA (MW = 1.2×10^6)	28-S RNA (~ 4000 nucleotides)
	~3100 Nucleotides	5.8-S RNA
	5-S RNA	5-S RNA
	120 Nucleotides	121 Nucleotides
	34 Proteins (L1 to L34)	~50 Proteins

unit has also been accomplished. In both cases, the order of addition of the individual ribosomal subunits during reconstitution is of critical importance, indicating that the steps in assembly of a ribosomal particle must occur in a specific sequence.

Various techniques have been used to study the structure of the ribosomal subunits, e.g., fluorescence spectroscopy, low-angle scattering of neutrons, the cross-linking of proteins with bifunctional reagents, and immune electron microscopy. Of these, immune electron microscopy has been most informative. Antibodies against individual ribosomal proteins are used to locate the protein on the surface of the ribosome, and the points of contact between the protein and its antibody are identified by electron microscopy. As indicated in Fig. 28.4, the ribosomal proteins have very specific locations.

The topography of binding between proteins and ribosomal RNAs is also known in part, particularly for the 30-S subunit. Because of the complex interactions between the RNA and proteins, and between the proteins, it is more profitable to consider the functions of topographical domains in the ribosomes instead of attempting to assign unit functions to the individual proteins. Significant information regarding the function of certain domains will be discussed in the context of the description of the translation process.

Messenger RNA

Studies of the kinetics of β-galactosidase synthesis in *E. coli* led to the discovery that control of protein synthesis is determined by the rate of synthesis of a short-lived messenger RNA. In contrast to the constant presence of tRNA and rRNA, mRNAs do not accumulate. Hence, mRNA represents only a small percentage of the total RNA of the cell, most of which is rRNA and tRNA (see below).

The size of a prokaryotic mRNA depends upon the size of the polypeptide chain to be synthesized. Thus, for a triplet code (see below) a polypeptide of 100 amino acid residues (MW = 11,000) requires 300 nucleotides in the coding sequence, whereas a polypeptide of molecular weight 110,000 requires 3000 nucleotides in the coding sequence. In addition, mRNAs may contain non-

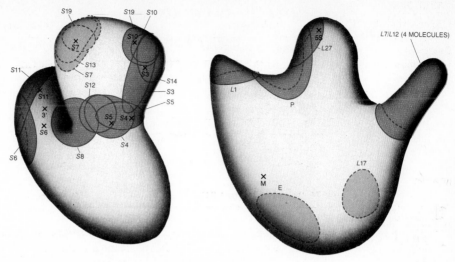

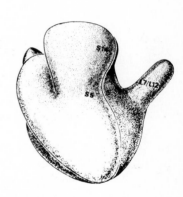

Figure 28.4 Map of proteins in the *E. coli* ribosome shows their location in the small (30-S) subunit (*left*) and the large (50-S) subunit (*right*). The orientation of both subunits in the monomeric ribosome is shown below. Ribosomal proteins are indicated by a prefix indicating a small (S) or a large (L) subunit protein and a number denoting each protein, e.g., S3 is a small subunit protein and L27 is a large subunit protein. The patches were mapped by immune electron microscopy; they are the sites on the surface of the ribosome where antibodies bind to the protein that was used as the antigen. Each protein makes one such appearance at the surface of a subunit, except for protein S19, which evidently makes two appearances, and protein S4, which is under S5 and S12. The crosses were mapped by neutron diffraction. Five additional sites are shown. They are M, the site at which the ribosome can be anchored to intracellular membrane; 5S, the location of 5S RNA, a part of the large subunit; 3′, an end of the 16S ribosomal RNA; P, the site at which successive amino acids are linked to make a polypeptide chain, and E, the site where the newly synthesized polypeptide chain emerges from the ribosome. [*Reprinted with permission of the author from J. A. Lake, The Ribosome, Scientific American:* **245**:*84–97, August, 1981.*]

coding, or *leader*, sequences at their 5′ termini. Some mRNAs code for more than one polypeptide chain and are referred to as *polygenic, or multicistronic, mRNAs*. Such mRNAs contain noncoding intergenic regions as well as leader sequences. The function of the noncoding leader and intergenic sequences is considered below.

There are sequences in DNA that are complementary to the three types of RNA, viz., tRNA, rRNA, and mRNA. Thus, all three types of RNA are synthesized on DNA templates. In essence, the role of DNA as the genetic material is to provide templates for self-replication and for the synthesis of all the types of RNA required for gene expression. At any instant, mRNA comprises only 3 to 4 percent of the total cellular RNA of *E. coli* with rRNA and tRNA representing 80 to 85 percent and 10 to 15 percent of the total, respectively. Nevertheless, only about 0.2 percent of the nucleotides of the *E. coli* genome is used for synthesis of approximately 60 tRNAs and approximately 0.5 percent for synthesis of rRNA. Thus, more than 99 percent of the DNA serves as a template for mRNA synthesis.

Having described the three components required for protein synthesis, viz., a mRNA with a leader sequence, tRNAs, and ribosomes, we now consider how this process occurs. The translation of mRNA into a polypeptide sequence may be envisaged as consisting of three consecutive phases: (1) *initiation* of the peptide chain, a highly specific process, particularly since many mRNA molecules are multicistronic; (2) the repetitive addition of single amino acid residues from aminoacyl-tRNAs in the *elongation* of the chain; and (3) the specific *termination* of the peptide chain and its release from the ribosome.

Initiation and Polypeptide Synthesis

There are two species of methionine-specific tRNA; one designated as $tRNA_m^{Met}$ acts as an adapter for methionine residues required within the polypeptide chain; the second, $tRNA_f^{Met}$, has a unique role in the initiation of polypeptide chain synthesis. Both are acted on by the same methionyl-tRNA synthetase to yield the charged form of $tRNA^{Met}$, that is, met-$tRNA^{Met}$. In prokaryotes, the amino group of met-$tRNA_f^{Met}$ can be formylated (met-$tRNA_m^{Met}$ cannot) by N^{10}-formyltetrahydrofolic acid in the presence of the specific *trans-formylase* to produce N-formylmet-$tRNA_f^{Met}$ (fmet-$tRNA_f^{Met}$). The latter functions in polypeptide chain initiation, as indicated below (Fig. 28.5). Although

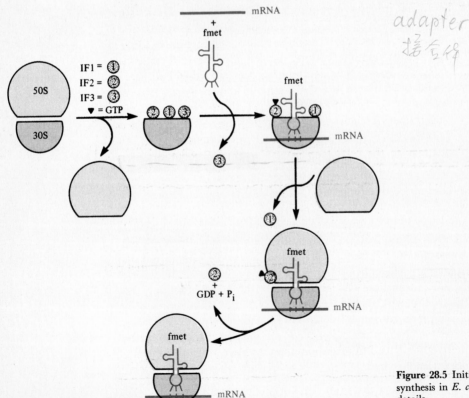

Figure 28.5 Initiation of peptide synthesis in *E. coli*. See text for details.

all polypeptides in prokaryotes are initiated with formylmethionine at the amino terminus, the formyl group is subsequently removed by a *deformylase;* however, in some instances the methionine is removed by the action of an aminopeptidase.

In addition to fmet-tRNA$_f^{Met}$, mRNA, and 50-S and 30-S ribosomal subunits, the initiation process in prokaryotes requires three proteins not normally present in the ribosome. These are called *initiation factors* (IF) and designated IF1 (MW = 9000), IF2 (which consists of two subunits, IF2a, MW = 115,000, and IF2b, MW = 90,000), and IF3 (MW = 21,000).

In eukaryotes, the initiator-tRNA for cytoplasmic polypeptide synthesis is also specific for methionine; however, the met-tRNA^{Met} is not formylated. It reacts with the initiation factors termed eIF1 (MW = 15,000), eIF2 (MW = 150,000), and eIF3 (MW ≈ 700,000), the 40-S ribosomal subunit, and the mRNA. Additional factors eIF4A, 4B, 4C, 4D, 5, and CO-eIF-1 have been identified which facilitate the initiation of eukaryotic translation. In contrast, polypeptide synthesis in mitochondria and chloroplasts is the same as in prokaryotes; i.e., initiation occurs via an fmet-tRNA$_f^{Met}$ and a 30-S ribosomal subunit that subsequently reacts with a 50-S subunit.

The process begins with the attachment of fmet-tRNA$_f^{Met}$ to an IF2-GTP complex, which then binds to the 30-S subunit to which IF3 and the mRNA are attached (Fig. 28.5). Binding to the 30-S subunit is facilitated in an as yet undetermined way by IF1. IF3 serves a dual function: not only does it facilitate the binding of mRNA to the 30-S subunit, but it also functions as an antiassociation factor that inhibits premature association between the 30-S and 50-S subunits to form the 70-S unit. IF3 binds to the 30-S subunit near the 3′ end of the 16-S rRNA in this subunit and is thought to facilitate the binding of mRNA by mediating an interaction between the 3′ end of the 16-S rRNA and the 5′ end of the mRNA. In fact, homologies have been identified between the sequence at the 3′ end of 16-S rRNA and the sequence in the leader region at the 5′ end of several mRNAs. These homologies permit base pairing between the two RNAs, and, indeed, a bimolecular complex of this kind has been isolated from *E. coli* ribosomes (Fig. 28.6). Since there are sequences in 5-S RNA that are complementary to the TψC loop in tRNA, the 5-S RNA may serve some function in tRNA binding.

The final 30-S complex, which also contains IF1, combines with the 50-S subunit after release of IF3 to form the 70-S ribosome-initiator complex, with the release of IF2-GDP, P$_i$, and IF1. This initiator complex contains two tRNA binding sites termed the P (peptidyl) site and the A (aminoacyl) site. The fmet-tRNA$_f^{Met}$ lies in the P site, with its anticodon (3′ ← UAC-5′) bound to the codon for methionine (5′-AUG → 3′). Thus, the reading frame is defined by specific interactions of the ribosome-fmet-tRNA$_f^{Met}$ with mRNA.

Polypeptide Chain Elongation

In the first step, the aminoacyl-tRNA of the amino acid that is specified by the next codon in the mRNA (moving in the 5′ → 3′ direction in the mRNA) is bound in the A site of the 70-S ribosome. This reaction requires GTP and the two elongation factors EF-Tu (MW = 47,000) and EF-Ts (MW = 36,000) and

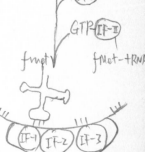

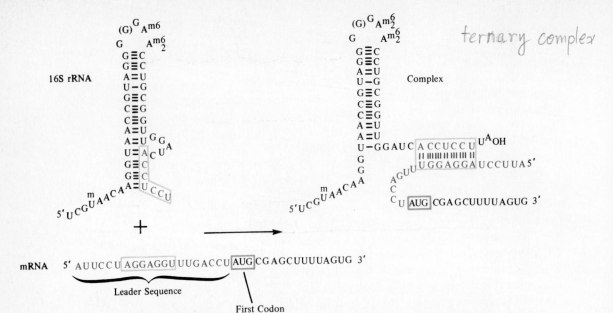

ternary complex

Figure 28.6 Formation of hydrogen-bonded complex between 3′ end of 16-S RNA and 5′ end of mRNA. [*From J. A. Steitz and K. Jakes, Proc. Natl. Acad. Sci. USA 72:4734 (1975).*]

proceeds as shown in Fig. 28.7. A complex of EF-Tu and EF-Ts (EF-Tu · EF-Ts) reacts with GTP to form EF-Tu · GTP, with the release of EF-Ts. EF-Tu · GTP reacts with the aminoacyl-tRNA to form the ternary complex, aminoacyl-tRNA · EF-Tu · GTP. The latter is then transferred to the A site of the ribosome having fmet-tRNA$_f^{Met}$ bound to its P site. GTP is hydrolyzed and EF-Tu · GDP is released. EF-Tu · GDP reacts with EF-Ts to regenerate the EF-Tu · EF-Ts complex with the displacement of GDP. The result of this sequence of reactions is a 70-S ribosome with aminoacyl-tRNA positioned in the A site and fmet-tRNA$_f^{Met}$ bound at the P site. A similar sequence occurs in eukaryotes with the comparable elongation factors designated as EF-1α (MW = 56,000) and EF-1β (MW = 30,000).

Peptidyl transferase (a component of the 50-S ribosomal subunit) catalyzes transfer of the formylmethionine moiety (or the peptidyl chain in later stages) to the amino group of the newly bound aminoacyl-tRNA occupying the A site to form a dipeptidyl-tRNA. Thus, the polypeptide chain grows by adding one residue at a time to the carboxyl end of the chain. At this stage an uncharged tRNA remains in the P site and the nascent chain, attached to the tRNA of the last amino acid added, is in the A site.

In a reaction requiring another protein factor, EF-G (EF-2 in eukaryotes) and GTP, the ribosome undergoes translocation by one codon (three bases). In this process the dipeptidyl-tRNA is shifted from the A site to the P site, the uncharged tRNA$_f^{Met}$ is released, and GTP is hydrolyzed to GDP and P$_i$. Hydrolysis of GTP enables EF-G to be released from the ribosome so that it can act catalytically. The cycle is now repeated, achieving addition of the next amino acid, etc.

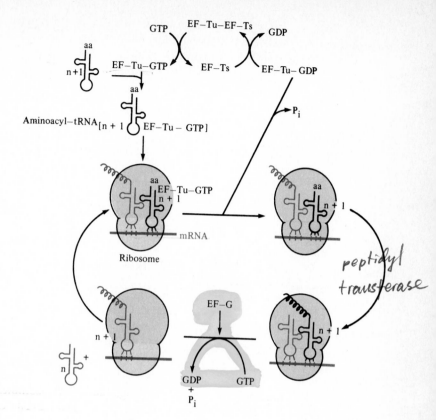

Figure 28.7 Polypeptide chain elongation in *E. coli*. See text for details. [*From Y. Kaziro, p. 88, in A. Kornberg, B. L. Horecker, L. Cornudella, and J. Oró (eds.): Reflections in Biochemistry, Pergamon, New York, 1976.*]

Polypeptide Chain Termination

Termination of polypeptide chain elongation occurs where specific termination signals appear in the mRNA, i.e., at one or more of the following three trinucleotide sequences or *triplets:* UAA, UAG, and UGA. These are referred to as *chain-termination* or *nonsense codons* (see below), since whenever they occur, they cause termination of chain elongation and release of the polypeptide chain. This results from hydrolysis of the peptidyl-tRNA linkage in the P site. The polypeptide is released, the tRNA leaves the P site, and the ribosome then dissociates into its 30S and 50S subunits. In prokaryotes, termination is mediated by three proteins, RF1, RF2, and RF3. RF1 (MW = 44,000) promotes termination in response to UAA or UAG, RF2 (MW = 47,000) to UAA or UGA. RF3 stimulates the action of RF1 and RF2; it also has an intrinsic GTPase activity. A single protein termination factor RF (MW = 56,500) is operative in eukaryotic protein synthesis.

In the overall process, a total of 3 equiv of ATP is expended in the synthesis of a single peptide bond. Thus, one ATP is required for the activation of an amino acid to form the aminoacyl tRNA, and two GTPs are hydrolyzed to GDP and P_i during the elongation phase. Since initiation requires an additional GTP, 4 equiv of ATP are expended in the synthesis of the first polypeptide bond.

In almost every case, a given mRNA is translated simultaneously by large numbers of ribosomes, yielding structures called *polyribosomes,* or *polysomes.* In

737

bacteria, translation begins at the 5' end of the mRNA, while its own synthesis continues at the 3' end. In the case of multicistronic mRNA, ribosomes and completed polypeptide chains are released at each termination signal and synthesis of the next peptide chain requires reinitiation at the beginning of each message.

Inhibitors of Translation

As in the investigation of other complex metabolic processes, specific inhibitors have played a major role in separating the steps in the biosynthesis of nucleic acids and proteins. Some of these inhibitors are synthetic compounds; others were first isolated from the culture filtrates of various organisms as antibiotics in efforts to control infectious diseases or to inhibit the growth of malignant tissue. Substances that inhibit formation of purine and pyrimidine nucleotides indirectly block the formation of nucleic acids; they are described in Chap. 26. Inhibitors of mRNA synthesis that act by blocking DNA-dependent RNA polymerases will be discussed below. The inhibitors to be considered in this section are antibiotics that affect the translation of mRNA. Because they react specifically with prokaryotic ribosomes, several of these agents have found widespread use in the treatment of infectious diseases.

Puromycin, which has a structure analogous to an aminoacyl tRNA (see below), inhibits protein synthesis by mimicking the aminoacyl adenosine portion of aminoacyl-tRNA and competing with the aminoacyl-tRNA as an acceptor in the transpeptidation reaction. Consequently, the growing peptide (or the formylmethionine) is transferred to the amino group of puromycin, and the newly formed peptidyl-puromycin dissociates from the ribosomes. Thus, normal chain growth is aborted, and incomplete peptide chains bearing carboxyl-terminal puromycin are released from the ribosomes.

Streptomycin inhibits protein synthesis by binding to the 30-S subunit; the 30-S-streptomycin complex forms a less efficient or more unstable initiation complex that dissociates, thus aborting the translation process. Streptomycin binding to the 30-S subunit also modifies the efficiency and fidelity of matching aminoacyl-tRNAs to their respective codons in the A site. Resistance to, or even dependence on, streptomycin results from a mutational change in a single polypeptide of the 30-S ribosomal subunit.

Puromycin   Aminoacyl-tRNA

Tetracyclines inhibit polypeptide chain elongation by blocking entry of aminoacyl-tRNA to the A site; the nascent polypeptide chain remains in the P site and can react normally with puromycin.

Chloramphenicol blocks the peptidyl synthetase activity of the 50-S subunit of 70-S ribosomes.

Cycloheximide is thought to act in the same manner as chloramphenicol on the 60-S subunit of eukaryotic 80-S ribosomes.

Erythromycin binds to the 50-S ribosomal subunit and appears to block the translocation step, thereby "freezing" the peptidyl-tRNA in the A site. Mutations causing resistance to erythromycin are known to affect one of the proteins of the 50-S ribosomal subunit of *E. coli*.

Fusidic acid is a steroid that affects the translocation step in eukaryotic ribosomes after formation of the peptide bond, possibly by preventing cleavage of GTP in the EF-2-mediated cleavage-translocation reaction.

Diphtheria toxin specifically inhibits eukaryotic but not prokaryotic protein synthesis by inactivating elongation factor EF-2. This process is described in Chap. 29.

THE GENETIC CODE

Inasmuch as there are only four bases in DNA and 20 different amino acids in proteins, it is obvious that more than one base is required to specify an amino acid. A code consisting of pairs, or doublets, of bases could specify only 16 amino acids (4^2); hence it was predicted that the *codon*, i.e., the nucleotide sequence that defines an amino acid, is a triplet.

The earliest experimental proof for this idea resulted from an examination of the effects of the dye *proflavine* on the rII locus of bacteriophage T4 (Chap. 30).

$$H_2N \quad\quad N \quad\quad NH_2$$

Proflavine

Because it can intercalate between the bases of the DNA duplex, proflavine can cause the deletion or insertion of a nucleotide in a gene during its replication. Such a genetic alteration leads to formation of nonfunctional proteins because the reading of the codons is shifted from the point of nucleotide deletion (or insertion). Hence, the amino acid sequence is completely altered from the point of nucleotide deletion (or insertion) to the COOH end of the peptide chain, as shown for the deletion in the arbitrary example below.

Normal reading:	Phe-	Lys-	Pro-	Phe-	Gln-	Lys-	Gly-	Asn
	UUU	AAA	CCC	ŲUC	CAG	AAG	GGU	AAU
After deletion:	Phe-	Lys-	Pro-		Ser	-Arg	-Arg	-Val

Deletion of three adjacent nucleotides, however, leads to the formation of a protein in which only a single amino acid is deleted, but the reading frame would remain undisturbed for the synthesis of the remainder of the protein. For three nonadjacent deletions (or insertions) of nucleotides, the amino acid sequence would be altered only between the first and last of these changes; the remainder of the amino acid sequence would be unchanged. It was, in fact, observed that in the genome of bacteriophage T4, incorporation by recombination in the rII gene of three nucleotide deletions (or insertions) gave a functional protein, whereas one or two nucleotide deletions (or insertions) did not. This result indicated that the code must be read in groups of three nucleotides; i.e., the codon is a triplet.

Deciphering the Genetic Code

The elucidation of the genetic code, i.e., the determination of the trinucleotide sequence that defines each amino acid, was made possible by the observation that the synthetic polynucleotide poly(U) (Chap. 8) could serve as an mRNA for the synthesis of polyphenylalanine in extracts of *E. coli* capable of carrying out protein synthesis. Of 18 amino acids tested, only polyphenylalanine synthesis occurred. Thus, assuming a triplet code, UUU is the codon for phenylalanine.

Subsequently, the genetic code was deciphered by three types of experiments:

1. The use of polyribonucleotides synthesized with the enzyme polynucleotide phosphorylase. This enzyme catalyzes the reaction

$$n\text{NDP} \rightleftharpoons (\text{NMP})_n + n\text{P}_i$$

where NDP is any ribonucleoside diphosphate and $(\text{NMP})_n$ is a polymer of n ribonucleoside monophosphate residues. As noted above, poly(U) leads to the synthesis of polyphenylalanine. Poly(C) and poly(A) yield polyproline and polylysine, respectively. Such homopolymers as well as random copolymers synthesized from mixtures of ribonucleoside diphosphates were used to ascertain the base composition of the codons, corresponding to a given amino acid; however, this did not permit determination of the sequence of bases within a codon.

2. The binding of trinucleotides of known sequence and their cognate aminoacyl-tRNAs to ribosomes. Poly(U) can combine with ribosomes to form a complex that binds only phenylalanyl-tRNA. Similarly, a single trinucleotide of known sequence bound to ribosomes will induce the binding of a specific aminoacyl-tRNA. Thus, $5'\text{UUG}^{3'}$ promotes binding of leu-tRNA^{Leu}, etc. The most effective trinucleotides are those possessing at one terminus free 3'- and 2'-hydroxyl groups and a phosphate group at the 5' end. While trinucleotide binding studies produced unambiguous results in many cases, other assignments of codons were dubious because of the low level of binding and the requirement for nonphysiological conditions, particularly the use of high Mg^{2+} concentrations.

3. Unambiguous codon assignments with the aid of copolymers of defined nucleotide sequence. Consider an RNA copolymer in which the dinucleotide XY is repeated. This will form a linear array of the two alternating triplets **XYX** and *YXY,* that is, . . . **X**-*Y*-**X**-*Y*-**X**-*Y*-**X**-*Y*-**X**, etc., and should therefore code for a polypeptide containing an alternating sequence of two amino acids. For example, when X and Y are U and G, a polypeptide with an alternating sequence of cysteine and valine is synthesized. Similar experiments were performed with polynucleotides consisting of repeating trinucleotides or repeating tetranucleotides. The results of many such experiments together with the trinucleotide binding studies have established all the triplet codons that correspond to each of the 20 amino acids.

Experiments with defined copolymers also allowed a definition of the initiation codon and led to a better understanding of the proper conditions necessary for accurate translation in cell-free systems. For example, the copolymer . . . G-A-U-G-A-U-G-A-U-G-A-U . . . will code for polymethionine (AUG AUG AUG . . .) and polyaspartate (GAU GAU GAU . . .) at high Mg^{2+} concentrations (0.05 M). The third reading frame (UGA UGA UGA . . .) does not yield a polypeptide because UGA is a termination codon. However, at lower Mg^{2+} concentrations (5 mM), little or no polypeptide synthesis is observed unless fmet-tRNA$_f^{Met}$ is present in addition to the met-tRNA^{Met}. In this case, the only polypeptide synthesized in appreciable yield is fMet-Met-Met-. . . . Similarly, . . . G-U-G-U-G-U-G-U. . . , which at high Mg^{2+} concentrations yields the polypeptide composed of alternating cysteine and valine residues cited above, yields virtually no polypeptide at low Mg^{2+} unless fmet-tRNA$_f^{Met}$ is present. In this case the polypeptide synthesized is fMet-Cys-Val-Cys-Val-, etc. This indicates that GUG can also code for the initiating methionine, although it codes for valine when it is an internal codon. This conclusion has since been confirmed by determination of the nucleotide sequences of many natural mRNAs.

The complete codon dictionary is given in Table 28.2. Note that of the 64

TABLE 28.2

**Dictionary of
Amino Acid
Codons***

First position	Second position				Third position
	U	C	A	G	
U	Phe	Ser	Tyr	Cys	U
	Phe	Ser	Tyr	Cys	C
	Leu	Ser	(CT)	(CT)	A
	Leu	Ser	(CT)	Trp	G
C	Leu	Pro	His	Arg	U
	Leu	Pro	His	Arg	C
	Leu	Pro	Gln	Arg	A
	Leu	Pro	Gln	Arg	G
A	Ile	Thr	Asn	Ser	U
	Ile	Thr	Asn	Ser	C
	Ile	Thr	Lys	Arg	A
	Met (CI)	Thr	Lys	Arg	G
G	Val	Ala	Asp	Gly	U
	Val	Ala	Asp	Gly	C
	Val	Ala	Glu	Gly	A
	Val (CI)	Ala	Glu	Gly	G

*The first position refers to the initial nucleotide of the mRNA triplet bearing a 5'-OH or 5'-phosphate; the third nucleotide of the triplet bears a 3'-phosphate connecting to the next triplet. (CI) refers to a chain-initiating codon for the NH$_2$ end of a peptide chain. (CT) refers to a chain-terminating codon for the COOH end of a peptide chain.

TABLE 28.3

Multiple tRNAs for Glycine	tRNA for glycine	Anticodon in the tRNA	Glycine codons recognized by the tRNA
	tRNA$_{GGG}^{GlyI}$	$3'CCC5'$	$5'GGG3'$
	tRNA$_{GGA/G}^{GlyII}$	$3'CCU5'$	$5'GGA3'$ $5'GGG3'$
	tRNA$_{GGU/C}^{GlyIII}$	$3'CCA5'$	$5'GGU3'$ $5'GGC3'$

possible triplets, all but three are translatable either as an initiator amino acid or as internal amino acids. UAA, UAG, and UGA are not translatable into amino acids; instead they signal chain termination. Several triplets code for the same amino acid, a situation termed *degeneracy* in cryptography, i.e., a multiplicity of valid solutions to a code. Although the code is degenerate, it is not *ambiguous;* no codon, with the exception of GUG, codes for more than one amino acid. As indicated above, GUG is an initiation signal for fmet at the beginning of an mRNA but codes for valine in the interior of the sequence.

In many cases, the degeneracy is accommodated by multiple tRNAs for a single amino acid, each with a different anticodon. In other instances, a given tRNA can translate more than one codon. Both cases are illustrated in Table 28.3. Thus, tRNA^{GlyIII} can translate GGU and GGC but cannot translate the other two glycine codons, whereas tRNA^{GlyII} exhibits the opposite specificity. On the other hand, tRNA$_{GGG}^{GlyI}$ can translate only the GGG codon. Given the existence of the other two glycine tRNAs, tRNA^{GlyI} would appear unnecessary for polypeptide synthesis.

When a single anticodon can recognize more than one internal codon, the difference between the codons that are recognized is invariably in the 3'-terminal position of the codon. This characteristic feature can be accounted for by the so-called *wobble hypothesis.* In essence, this hypothesis allows the third base in the codon to form unconventional base pairs that approximate the configuration of the standard A-U and G-C pairs. Table 28.4 indicates the base pairing permitted by the wobble hypothesis. In general, the predictions of the hypothesis have proved correct as the anticodon sequences for more tRNAs have become known. Note, however, that one exception to these pairing rules is found in Table 28.3, viz., the anticodon in tRNA^{GlyIII} $3'CCA5'$ is recognized by the glycine codon $5'GGC3'$.

TABLE 28.4

Base Pairs Permitted by the Wobble Hypothesis	Base in third position of anticodon	Base in third position of codon
	G	U or C
	C	G
	A	U
	U	A or G
	I	A, U, or C

742

With the exception noted below, the code is probably universal. This conclusion comes indirectly from analyzing changes in amino acid sequence resulting from mutations in such diverse proteins as the α chain of tryptophan synthetase of *E. coli* (page 751) and of human hemoglobin (Chap. 29). It is also directly supported by the determination of base sequences in many natural mRNAs of various species.

Information derived from studies of homologous proteins, which show that large portions of the amino acid sequences have been retained even in distantly related species, also supports the universality of the genetic code. For example, the eukaryotic cytochromes c of various species of animals, plants, and fungi all have approximately one-third of the residues of the sequences in common (Chap. 31). This indicates a conservatism in which the same protein has retained essential features of its amino acid sequence for a long period of evolution (Chap. 31). It is difficult to visualize this degree of retention of structure if the code had changed continuously during evolution.

If the code were not essentially invariant, a change in the tRNA for cysteine to tryptophan, for example, would lead at each site along each peptide chain to the incorporation of tryptophan instead of cysteine. Enzymes that require sulfhydryl groups for activity would be inert, disulfide bridges could not be formed, and the introduction of the bulky side chains of tryptophan would make folding the polypeptide chain into its normal conformation difficult. Such a mutation leading to the wrong anticodon would obviously be lethal in haploid cells and certainly deleterious if not lethal in diploid cells. Misreading of the code by suppressor gene products (see below) must be regarded as a relatively rare event that has survival value only when it leads to some new advantage for the cell.

An exception to the universality of the genetic code is to be found in the genome of mitochondria. DNA sequence analysis of human mitochondria has shown that there are at least four differences from the universal genetic code: AUA, normally the codon for isoleucine, codes for methionine; UGA, normally one of the three termination signals, codes for tryptophan. AGA and AGG, normally codons for arginine, are termination signals; and AUA or AUU are sometimes used as initiation codons in place of AUG. The basis for these changes is unknown, however; one consequence of this divergence in the genetic code is the genetic isolation of the mitochondrion. A mRNA containing any of these codons will obviously specify different amino acids depending on which side of the mitochondrial membrane it is translated.

There is no clear information on the origin of the genetic code. No chemical logic links the anticodon of tRNA or the codon of mRNA with the amino acid that is thus coded. Nor is there any obvious reason why some amino acids are represented by only one or two codons and others by as many as six codons. Nevertheless, once this code was established, it must have become effectively immutable. Its essential universality is the strongest evidence that all living organisms derive from a single ancestral form.

The mechanisms of DNA replication and RNA formation and of translation into protein are essentially the same in all organisms. Evolutionary changes with the development of enormous diversity of organisms have taken place not by major alterations of these biosynthetic processes but by the formation of new genes leading to synthesis of new enzymes or other proteins and hence to new structures and functions. These problems will be considered below.

Mutations

Alterations of the base sequence in either of the two strands of DNA will lead to permanently inherited changes, e.g., mutations, since subsequent DNA replication will perpetuate the changes. Any alteration of the DNA of the genome will in turn be reflected in the mRNA and hence in the synthesized protein or in the rRNA and various types of tRNA.

Point mutations, which reflect a change in a single base or base pair, result in a single base change in the mRNA and thus the replacement of one amino acid by another in the sequence of a peptide chain.

Point mutations, which are by far the most common genetic change, are produced by errors during DNA replication or repair (Chap. 27). They may also result from exposure of DNA in vivo or in vitro to mutagenic agents such as nitrous acid, which deaminates the bases, changing C to U and A to I (read as G). Similarly, hydroxylamine (NH_2OH) deaminates C to U. Some analogs of purines or pyrimidines or the respective nucleosides or nucleotides that become incorporated into DNA also generate point mutations since they can lead to errors in base pairing. Examples include uracil, substituted at the 5 position with I, Br, or Cl, 8-azaguanine, and 6-thioguanine. In all cases, changes observed in the amino acid sequence of proteins isolated from organisms treated with these agents are consonant with the predicted codon changes (Table 28.2).

Certain acridine dyes, e.g., proflavine, which can intercalate between bases in the strands of DNA, produce nucleotide deletions (or insertions) rather than point mutations (page 739). Deletions of amino acids are known in homologous proteins of various species, and even in human hemoglobin (Chap. 29). Although the mechanism of such "natural" deletions is unknown, presumably they are the result of breaks and losses of nucleotides in a DNA chain. Such deletions in the genome can be visualized directly by heteroduplex mapping (see Fig. 29.4).

Suppression of Mutations

A mutation at a given site in the genome can be reversed by a second mutation. Such reversions to the normal, or wild, phenotype may occur by simply changing the altered codon back to its original sequence. In some cases, however, the second mutation occurs at a different site; such second-site reversions are known as *suppressor mutations*, one class of which results from mutations in the genes that code for tRNA. Let us consider the following three types of mutations in a coding sequence: (1) A wild-type codon is changed to a terminator or nonsense codon, causing premature termination of translation, i.e., a *nonsense mutation*. (2) A wild-type codon is changed to the codon for another amino acid, resulting in an amino acid substitution in the polypeptide, a *missense mutation*. (3) A base is inserted into the coding sequence, resulting in a shift in reading frame during translation, *a frame-shift mutation.* Each of these types of mutation can be suppressed by suppressor genes that result from a change in the base sequence of a tRNA. Examples of each are given below.

Nonsense Suppression A change in sequence that results in a change in tyrosine codons from $^5{}'UAC^{3'}$ and $^5{}'UAU^{3'}$ to $^5{}'UAG^{3'}$ are nonsense mutations causing premature chain termination. A mutation in the tRNA^{Tyr} gene that results in a change in the anticodon from $^3{}'AUG^{5'}$ to $^3{}'AUC^{5'}$ will permit the terminator codon $^5{}'UAG^{3'}$ to be translated as tyrosine and hence suppresses the nonsense mutation.

Missense Suppression Introduction of A at the 5' position of the glycine codon $^5{}'GGG^{3'}$ gives the arginine codon $^5{}'AGG^{3'}$. This missense mutation can be suppressed by a mutation of the tRNA^{GlyI} gene, which results in a change of the anticodon from $^3{}'CCC^{5'}$ to $^3{}'UCC^{5'}$, thereby permitting the mutant tRNA^{GlyI} to recognize the arginine condon. As indicated in Table 28.3, the glycine codon $^5{}'GGG^{3'}$ can be translated by another tRNA for glycine (tRNA^{GlyII}), and hence the mutation in the tRNA^{GlyI} gene will not prevent translation of $^5{}'GGG^{3'}$ at other loci.

Frame-Shift Suppression Frame-shift mutations that result from the insertion of an extra base pair can also be suppressed by mutations in tRNA genes. Thus, mutation in a tRNA^{Gly} gene yields an extra base in the anticodon loop of the tRNA so that $^3{}'CCC^{5'}$ is converted to $^3{}'CCCC^{5'}$. Addition of the extra base allows the tRNA to read four rather than three bases in the mRNA and is believed to result in "pulling" the mRNA across the ribosomal surface in a jump of four rather than three nucleotides. This will correct for the shift in the reading frame that resulted from the insertion of an extra base pair.

Thus far, suppressors of deletion mutations have not been observed.

TRANSCRIPTION

Transcription of the genome is mediated by *DNA-dependent RNA polymerases.* Addition to prokaryotic cells of a specific inhibitor of RNA polymerase, e.g., actinomycin D (page 746), which blocks the chain elongation phase of RNA synthesis by intercalating into the DNA duplex at G-C base pairs, blocks the synthesis of mRNA, rRNA, and tRNA. **RNA polymerase is therefore responsible for all cellular RNA synthesis.**

RNA polymerase catalyzes the initiation, elongation, and termination of polyribonucleotide chains, requiring all four ribonucleoside triphosphates as substrates. The reaction is absolutely dependent upon the presence of a divalent metal ion (Mg^{2+} or Mn^{2+}) and requires DNA or a polydeoxyribonucleotide as a template. The RNA polymerase reaction may be written simply as

$$n\text{NTP} \xrightarrow{\text{DNA}} (\text{NMP})_n + n\text{PP}_i$$

Unlike DNA polymerase (Chap. 27), the DNA template is conserved to be

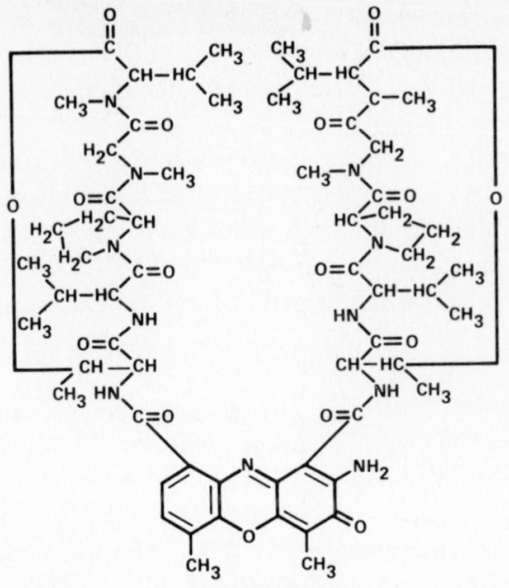

Actinomycin D

reused repeatedly and serves a purely catalytic role; it is not a primer to be extended and incorporated into the product.

In eukaryotic cell, there are RNA polymerase I, II, III [handwritten marginalia]

In contrast to bacteria, which contain a single kind of RNA polymerase, eukaryotic cells each contain several distinct polymerases that are localized in different subcellular fractions and are responsible for synthesis of the different types of RNA. These will be discussed more fully in Chap. 29.

Structure of Bacterial RNA Polymerases The RNA polymerase of *E. coli* has been the most extensively studied of these enzymes. It is a high-molecular-weight protein (480,000) consisting of four major subunits, designated β', β, α, and σ (MWs = 160,000, 150,000, 36,500, and 86,000, respectively) and present in the ratio 1:1:2:1. RNA polymerase containing only α, β, and β' subunits is referred to as the *core polymerase*. Active enzyme can be reconstituted from the separated subunits. Reconstitution occurs in the ordered sequence

$$2\alpha \longrightarrow \alpha_2 \xrightarrow{\ \beta\ } \alpha_2\beta \xrightarrow{\ \beta'\ } \underset{\text{Core polymerase}}{\alpha_2\beta\beta'} \xrightarrow{\ \sigma\ } \alpha_2\beta\beta'\sigma$$

binary [handwritten marginalia]
α₂ββ'. [handwritten marginalia]

Steps in the DNA-Directed Synthesis of RNA

The DNA-directed synthesis of RNA by RNA polymerase can be divided into four stages that constitute the *transcription cycle* (Fig. 28.8). The steps are (1) binding to template, (2) chain initiation, (3) chain elongation, and (4) chain termination and enzyme release.

Binding to Template The first step in the transcription cycle involves interaction of the *RNA polymerase holoenzyme* (core enzyme containing the σ subunit) with the DNA template to form a binary complex that can then bind the ribo-

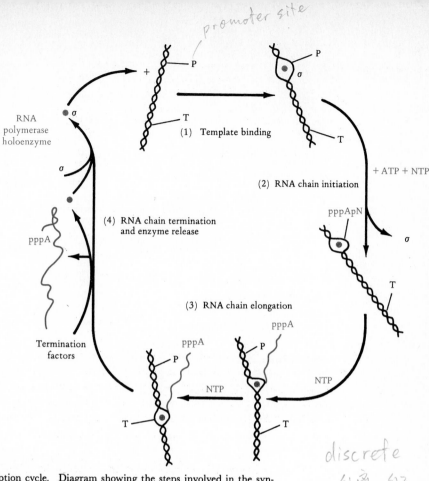

promoter site

Figure 28.8 The transcription cycle. Diagram showing the steps involved in the synthesis of an RNA transcript by a prokaryotic RNA polymerase. P and T refer to promoter and terminator sequences, and NTP refers to ribonucleoside triphosphates. [*From M. J. Chamberlin, p. 17, in R. Losick and M. Chamberlin (eds.): RNA Polymerase, Cold Spring Harbor Laboratory, Cold Spring Harbor, N.Y., 1976.*]

discrete
生盈. 生盈

nucleoside triphosphates and initiate synthesis of an RNA chain. Although RNA polymerase shows a significant affinity for all regions of a DNA molecule, it binds with an extremely high affinity ($K_{ass} = 2 \times 10^{11} M^{-1}$) at discrete sequences on only one of the two strands of the DNA template termed *promoter sites*. It is at these sites that transcription of a given gene begins. The σ subunit reduces the nonspecific binding of RNA polymerase to DNA by three orders of magnitude and in this way may facilitate the rapid and efficient location of promoter sites by the enzyme. Binding of RNA polymerase to promoter sites results in a very limited melting (strand separation) of the DNA helix at those sites to generate what is termed an *open promoter complex*. Thus, it would appear that the enzyme must open a limited number of base pairs at the starting site to gain access to the base-pairing information on the strand to be transcribed.

Tight binding of RNA polymerase to DNA renders approximately 40 base pairs resistant to hydrolysis by pancreatic DNase (Chap. 8). This property has permitted isolation of the promoters for a number of DNA templates, and the nucleotide sequences of several of these have been determined. As shown in Fig.

The σ subunit may facilitate the rapid and efficient location of promotor sites by the enzyme.

duces

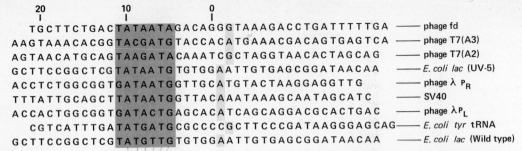

```
        20              10               0
        |               |                |
       TGCTTCTGAC TATAATA GACAGGGTAAAGACCTGATTTTTGA ——— phage fd
  AAGTAAACACGG TACGATG TACCACATGAAACGACAGTGAGTCA ——— phage T7(A3)
  AGTAACATGCAG TAAGATA CAAATCGCTAGGTAACACTAGCAG ——— phage T7(A2)
  GCTTCCGGCTCG TATAATG TGTGGAATTGTGAGCGGATAACAA ——— E. coli lac (UV-5)
  ACCTCTGGCGGT GATAATG GTTGCATGTACTAAGGAGGTTG ——— phage λ P_R
  TTTATTGCAGCT TATAATG GTTACAAATAAAGCAATAGCATC ——— SV40
  ACCACTGGCGGT GATACTG AGCACATCAGCAGGACGCACTGAC ——— phage λP_L
  CGTCATTTGA TATGATG CGCCCCGCTTCCCGATAAGGGAGCAG——— E. coli tyr tRNA
  GCTTCCGGCTCG TATGTTG TGTGGAATTGTGAGCGGATAACAA ——— E. coli lac (Wild type)
```

Figure 28.9 Promoter fragments protected by RNA polymerase against nuclease digestion. [*From W. Gilbert, p. 194, in R. Losick and M. Chamberlin (eds.): RNA Polymerase, Cold Spring Harbor Laboratory, Cold Spring Harbor, N.Y., 1976.*]

28.9, a common heptanucleotide sequence exists five or six nucleotides to the left of the start of transcription. It orients the RNA polymerase holoenzyme to begin transcription from left to right and is in the middle of the region protected by the holoenzyme against DNase digestion. The high AT content of the sequence may be an important factor in facilitating the melting that will expose the template strand for transcription.

Repressor proteins, which suppress transcription at specific promoter sites (see below), block formation of open promoter complexes. In the cases studied most extensively, repressors bind tightly to specific sequences at or near the promoter region termed *operator sites* (see below) and repress initiation of transcription by directly blocking access of RNA polymerase to the promoter.

Chain Initiation and Elongation Initiation of RNA chains by RNA polymerase occurs *de novo* without the aid of a primer. Either ATP or GTP can serve specifically as the initiating nucleotide to yield a dinucleoside tetraphosphate by the following reaction:

$$\text{ppp}^A_G + \text{pppX} \xrightarrow{\text{DNA}} \text{ppp}^A_G\text{pX} + \text{PP}_i$$

This dinucleotide is then elongated by the sequential addition of ribonucleoside monophosphates to the 3′-hydroxyl terminus in the chain-elongation phase of the reaction:

$$\text{ppp}^A_G\text{pX} + n\text{pppY} \xrightarrow{\text{DNA}} \text{ppp}^A_G\text{pX(pY)}_n = n\text{PP}_i$$

Rifampicin, which binds to the β subunit of RNA polymerase, completely inhibits the initiation reaction.

Rifamycin B
(Rifampicin)

Streptolydigin, which also binds to the β subunit, blocks chain elongation. It would thus appear that the β subunit is involved in both initiation and elongation in RNA synthesis. The action of actinomycin D in preventing chain growth has been described (page 745).

Once chain elongation has begun, the σ subunit dissociates from the holoenzyme and can interact rapidly with other core RNA polymerase molecules, if they are available. Chain elongation probably proceeds with the melting of base pairs only in that region of the DNA duplex being transcribed. The interaction between DNA template and the RNA transcript is a transient one, and the fully base-paired DNA helical structure is re-formed when the RNA transcript has been completed. Transcription, therefore, unlike DNA replication, is a fully conservative process.

Chain Termination and Enzyme Release Termination of the synthesis of RNA chains by RNA polymerase can occur at specific termination sequences in vitro, termination being accompanied by release of the enzyme. In other instances, termination of certain transcripts in vitro appears not to occur at the sites observed in vivo but continues into adjacent regions. In some, but not all, instances in which this has been examined, termination at the appropriate sequence in vitro can be induced by addition of a protein factor termed *rho*. This factor, which is not normally associated with the RNA polymerase holoenzyme, is a hexameric protein (MW of subunits = 50,000). It has a weak affinity for DNA but binds tightly to RNA and in its presence exhibits ATPase activity. The role of rho in the control of transcription termination is discussed below.

PROCESSING OF RNA TRANSCRIPTS

Most RNA transcripts are processed by posttranscriptional modification. This may be of five kinds: (1) endonucleolytic cleavage of a long primary transcript, which may be multicistronic; (2) modification of nucleosides within the transcript; this occurs primarily in tRNA and to a lesser extent in rRNA (Chap. 8); (3) addition of nucleotides to the 3' terminus of the transcript; (4) addition of nucleotides to the 5' terminus of the transcript; (5) excision of "intervening" sequences from the interior of the transcript, e.g., splicing. Of these, only endonucleolytic cleavage and nucleoside modification are found in prokaryotes. All five types of processing reactions occur in eukaryotes (Chap. 29).

Ribosomal RNA

In *E. coli* the genes for rRNA are arranged in a long transcriptional unit with the sequences specifying the 16-S, 23-S, and sometimes the 5-S RNA in tandem. The primary transcript is a 30-S RNA (MW = 2.1×10^6), which is hydrolyzed into chains of 0.6×10^6 and 1.2×10^6 daltons together with a number of undefined fragments. The former are further processed to yield the 16-S and 23-S rRNA molecules present in the 30-S and 50-S prokaryotic ribosomes. The 5-S RNA may also be generated from the 30-S precursor by endonucleolytic cleavage, possible at the 3' terminus. The nucleotide sequences corresponding to the 16-S RNA appear to be at the 5' end of the precursor molecule, and cleavage of the precursor begins before completion of the 30-S primary transcript. The initial cleavages of the precursor RNA are catalyzed by *RNase III*, initially identified as an RNase that specifically hydrolyzes double-stranded RNA. The sites of cleavage within the precursor may therefore be hydrogen-bonded duplex regions within the single-stranded precursor, although specific nucleotide sequences may be involved as well. In mutants of *E. coli* defective in RNase III, ribosomal RNA accumulates as the 30-S precursor. The ribosomal enzymes *RNase M16* and *RNase M23* are responsible for the formation of the mature 16-S and 23-S rRNAs from their immediate precursors. Another enzyme, *RNase M5,* is responsible for the formation of 5-S rRNA.

Transfer RNA

In prokaryotes, all tRNAs that have been examined are initially transcribed in the form of larger precursors. Inasmuch as the tRNA genes often occur in clusters, these transcripts may contain more than one kind of tRNA. The sequence of the precursor tRNA for tRNA[Tyr] is given in Fig. 28.10.

The precursor for two tRNAs (tRNA[Pro] and tRNA[Ser]) specified by genes of bacteriophage T4 (Chap. 30) is an example of a precursor containing two different tRNA species. The eight tRNA genes specified by this virus are transcribed

Figure 28.10 Nucleotide sequence of the precursor to tRNA[Tyr] from *E. coli* showing the residues released in the conversion to mature tRNA[Tyr]. The nucleotides enclosed in the boxes at the 5' and 3' termini are those removed during the cleavage steps that yield the mature tRNA. [*From S. Altman, Cell 4:21 (1975).*]

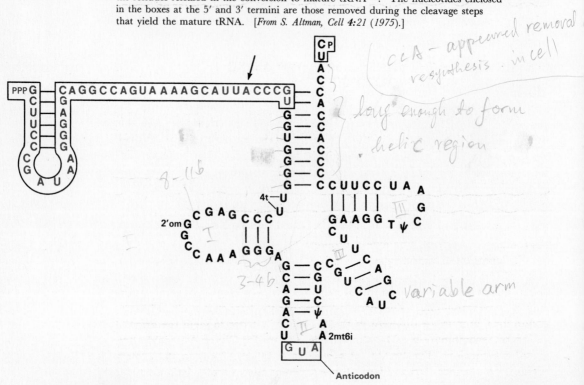

as one long polynucleotide; it is subsequently cleaved into dimeric precursors, which are then further processed to yield the mature tRNAs. Although the *E. coli* tRNA genes code for the CCA sequence at the 3′ end of the mature tRNA, this does not appear to be the case for the T4 phage genes. The CCA sequence in this instance is added posttranscriptionally by the action of a specific tRNA-CMP-AMP *pyrophosphorylase* supplied by the host.

Processing of tRNA precursors requires several nucleases, of which one, *RNase P*, is known to cleave the precursors to create the 5′ end of the mature tRNA. This enzyme, though highly specific with regard to its site of action on a given precursor, appears to be involved in producing the correct 5′ termini in all *E. coli* tRNAs. It contains a tightly bound RNA of approximately 300 nucleotides which may be involved in substrate recognition.

Other processing ribonucleases act at intergenic sequences and at the 3′ termini. Mutants of *E. coli* defective specifically in the enzyme that catalyzes transfer of an isopentenyl residue to the appropriate adenylate residue in tRNA^{Tyr} (Fig. 8.16) accumulate a tRNA^{Tyr} precursor, suggesting that modification is required for proper processing of tRNA precursors.

Messenger RNA

Translation of mRNAs in *E. coli* begins at their 5′ termini before the complete transcript has been synthesized; hence they are not processed prior to translation. However, multicistronic bacteriophage RNA transcripts have been detected that are subsequently cleaved to generate several small mRNAs. The effective nuclease appears to be RNase III, the enzyme responsible for processing of rRNA precursors (page 750).

COLINEARITY OF THE GENETIC MAP AND AMINO ACID SEQUENCE IN PROKARYOTES

Implicit in the foregoing presentation is the concept that the sequence of nucleotides in DNA is colinear with the sequence of amino acid residues in proteins. This fundamental thesis can be illustrated by examining gene-protein relationships in the tryptophan synthetase α chain of *E. coli* (page 595).

Many mutants of *E. coli* are known in which the activity of the α protein of tryptophan synthetase is altered. This protein and many of its inactive variants have been isolated in pure form, and the amino acid substitutions determined. Such mutant strains can be detected because of the development of tryptophan dependence for growth (inactive α protein).

In a peptide obtained from the α protein, a Gly I is replaced by Glu or Arg (Fig. 28.11), the enzyme containing either of these residues being totally inactive. Recombination between the two mutant types yields the wild-type active enzyme containing Gly. This demonstrates that the base that is altered in the codon for Gly must be different in the two cases in order to yield the original codon by recombination. Mutations in the codons for Glu or Arg yield additional substitutions (Fig. 28.11). A Gly II at a different locus in the protein yields, by mutations, Asp and Cys. Since different amino acid substitutions are

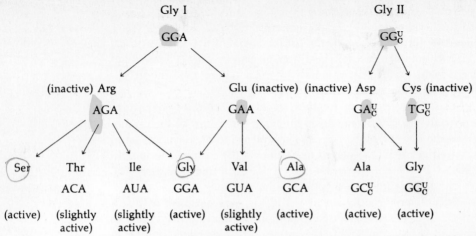

Figure 28.11 The amino acid substitutions observed at two sites, Gly I and Gly II, that are critical for synthesis of active tryptophan synthetase protein. The associated nucleotide triplets for the mRNA were deduced from known codon relationships, all of which differ by single nucleotides. The codons for Gly I and Gly II differ since they yield different substitutions. The third base in the codon for Gly II can be either U or C. The terms *active* and *inactive* refer to the enzyme formed in these strains. [*From the studies of C. Yanofsky and coworkers.*]

obtained by point mutations at I and II, it is evident that the codons at I and II differ. The deduced codons are indicated for each substituted amino acid (Fig. 28.11).

When the position of Gly I is occupied by small neutral residues, Gly, Ala, or Ser, the enzyme is active. With Thr, Ile, or Val in this position, there is slight activity, but activity is lost when an ionic residue, Arg or Glu, is present.

Extensive genetic mapping studies of mutants of the gene for tryptophan synthetase α protein in conjunction with studies of the sequence of the protein have demonstrated the strict colinearity of the amino acid sequence and genetic

Figure 28.12 Genetic map of the A gene of tryptophan synthetase and the corresponding amino acid changes in the protein, illustrating colinearity. The positions of these changes in the amino acid sequence are also indicated. The terms on the top line (B51, etc.) represent the experimental designation of the mutant in which the specific change was detected. [*From C. Yanofsky, G. R. Drapeau, J. R. Guest, and B. C. Carlton, Proc. Natl. Acad. Sci. USA 57:296 (1967).*]

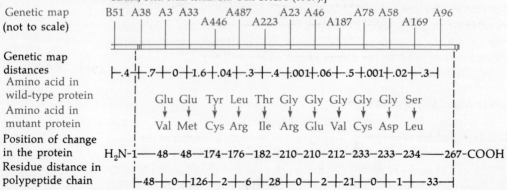

752

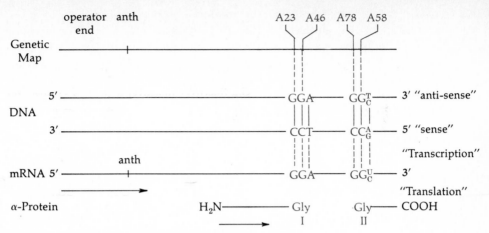

Figure 28.13 A diagrammatic representation of the colinear relationships of the tryptophan operon, the DNA and RNA nucleotide sequences, and the amino acid sequence of the α protein. Mapping of the mutants for Gly I and Gly II and the various recombinants has yielded the mapping relationships with regard to the amino acid sequence of the protein. The operator region is at the anthranilate synthetase (anth) end, and transcription for mRNA starts at this end, as does protein synthesis. [*From J. R. Guest and C. Yanofsky, Nature* **210**:799 (1966).]

map. Nucleotide sequence analysis of the gene specifying the α protein, as expected, has confirmed the colinearity of amino acid sequence and nucleotide sequence. The relationships of the mutational sites, amino acid sequence, and substitutions in the tryptophan synthetase α protein are shown in Fig. 28.12. In contrast to prokaryotes, amino acid sequence is not colinear with nucleotide sequence in eukaryotes. Instead, intervening sequences are present in the genome that are transcribed and subsequently excised during processing of the RNA transcript. This phenomenon is described in Chap. 29.

From the colinearity of the DNA and protein sequences, the known sequences of ribonucleotides that constitute the code, and from the direction of synthesis of mRNA (page 735 ff.) and of the protein (page 734 ff.), the interrelationships shown in Fig. 28.13 have been deduced. To recapitulate briefly, mRNA is synthesized, beginning with the 3′-ended strand of DNA serving as template and in antiparallel sequence. In effect, transcription results in an mRNA nucleotide sequence that is identical with that of the 5′-ended DNA strand except for the presence of U instead of T in the latter. The protein is synthesized from the NH_2 terminus by transfer of one amino acid residue at a time from the respective aminoacyl-tRNA bearing an anticodon that is complementary and antiparallel to the triplet codon of the mRNA.

CONTROL OF PROTEIN SYNTHESIS

Regulation of gene expression may occur at any one or several of the many steps in the pathway from gene to polypeptide product. Of these, modulation of the frequency of transcription is the best-understood mode of controlling the rate of polypeptide synthesis. The frequency of transcription of a given coding

753

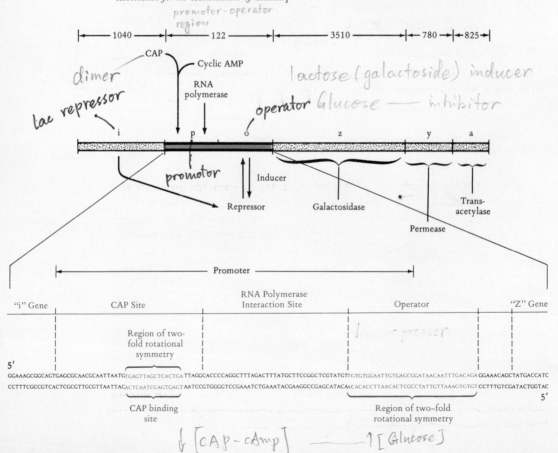

sequence in most instances depends upon the frequency of initiation. This is clearly illustrated in the case of the genes controlling the enzymes responsible for lactose metabolism in *E. coli*. These genes are situated next to each other in the chromosome and form a coordinately regulated unit known as an *operon*. One of the control elements, the *operator*, is situated at the 5′ end of the operon and is the site at which the repressor (page 755) is bound. The second regulatory element, the *promoter*, is directly adjacent to the operator and, as noted previously (page 755), is the RNA polymerase binding site.

[handwritten margin notes: situate 好位于 好处于; operon 操纵子]

The Lactose Operon

The lactose operon of *E. coli* is a single transcriptional unit of 5277 nucleotides that contains at the 5′ end a promoter-operator region of 122 nucleotides, followed by sequences (*structural genes*) coding for *β-galactosidase, galactoside permease*, and *thiogalactoside transacetylase* (Fig. 28.14). *β*-Galactosidase catalyzes

Figure 28.14 The *lac* operon of *E. coli*. The promoter (p) and operator (o) regions have been greatly expanded. The numbers at the top of the figure refer to the number of base pairs in each of the regions of the operon. The *i* gene specifies the *lac* repressor, the *z* gene, *β*-galactosidase, the *y* gene, galactoside permease, and the *a* gene, thiogalactoside transacetylase. [*Modified from R. C. Dickson, J. Abelson, W. M. Barnes, and W. S. Reznikoff, Science* **187**:27 (*1975*). *Copyright* © *1975 by the American Association for the Advancement of Science.*]

Figure 28.15 Possible arrangement of the CAP factor and the RNA polymerase subunits on the *lac* promoter. [*From W. Gilbert, p. 199, in R. Losick and M. Chamberlin (eds.): RNA Polymerase, Cold Spring Harbor Laboratory, Cold Spring Harbor, N.Y., 1976.*]

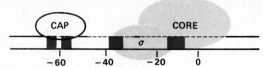

the hydrolysis of β-galactosides, including lactose; the permease controls the rate of entry of β-galactosides into the *E. coli* cell. The function of the transacetylase is unknown.

In the presence of lactose, or other galactosides termed *inducers,* the rate of synthesis of the polypeptide coded for by the lactose operon can be increased more than a 1000-fold. This does not occur if glucose is present in addition to the inducer. Glucose thus inhibits the induced synthesis of the enzymes of lactose catabolism. The lactose operon therefore functions to facilitate the use of lactose as a secondary carbon and energy source, after exhaustion of glucose from the medium.

As noted above, the rate of synthesis of the polypeptides coded by the lactose operon is controlled by regulation of the frequency of the initiation of transcription. This occurs as follows. Initiation of transcription requires both the RNA polymerase holoenzyme and a second protein termed *CAP* (catabolite activator *p*rotein), which must have cAMP bound to it. CAP is a dimer of identical subunits of molecular weight 22,500. Binding of the CAP-cAMP complex to the CAP site in the promoter of the operon is a prerequisite for the formation of an open complex (page 754) between RNA polymerase and the promoter (Figs. 28.14 and 28.15). The level of CAP-cAMP complex is determined by the cAMP concentration, which decreases with increasing glucose concentration. If the concentration of CAP-cAMP decreases beyond a certain level, the RNA polymerase cannot form the open complex required for initiation of transcription. The presence of glucose therefore prevents transcription of the lactose operon. Even at the high cAMP concentrations that exist in the absence of glucose (concentrations sufficient to ensure that the CAP site is occupied by the CAP-cAMP complex), formation of the open complex between RNA polymerase and the promoter is prevented if another protein, the *lac* repressor, is bound to the operator site. This site is located between the promoter and the β-galactosidase gene, although it appears to exhibit some overlap with the promoter (Figs. 28.14 and 28.15). The *lac* repressor consists of four identical 37,000-dalton subunits. Binding the tetramer to the operator site prevents binding of the RNA polymerase. Each subunit of the tetramer also contains a site at which the inducers, i.e., the β-galactosides, can bind. When these sites are occupied by an inducer, the repressor can no longer bind to the operator site. Consequently, inducers free the operator site, and the RNA polymerase can then initiate transcription.

The sequence of base pairs in the operator exhibits a twofold rotational symmetry in a segment of 35 base pairs of 120 Å, shown in Fig. 28.14. The *lac* repressor tetramer appears to be dumbbell-shaped and exhibits molecular dimensions of 45 by 60 by 120 to 140 Å. This shape can be accounted for by three twofold axes of symmetry, so that the long axis of the repressor can be aligned with the helical axis of the DNA, resulting in a twofold symmetry axis in the tetramer that matches that of the operator (Fig. 28.16).

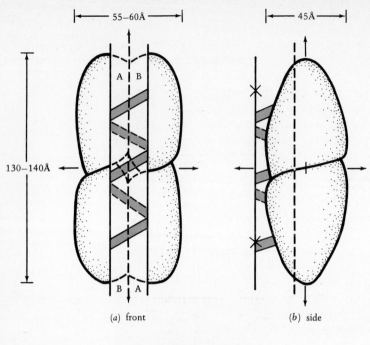

(a) front   (b) side

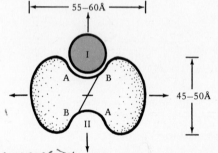

(c) top

(handwritten) ① constitutive expression the mutation in the operator or in the structural gene for the repressor

(handwritten marginal) ②

① Mutations in the operator or in the structural gene for the repressor can prevent binding of the repressor to the operator, thereby permitting unregulated expression of the operon, i.e., *constitutive expression.* ② Mutations in the repressor can also prevent binding of the inducer to the repressor and thereby make repression irreversible; i.e., expression of the operon becomes noninducible. ③ Mutations that prevent cAMP synthesis or alter CAP prevent transcription of the lactose and other operons, thereby dissociating the expression of these operons from control by the metabolic state of the cell. ④ It is thus apparent that modulation of initiation of transcription of the lactose operon can be affected by the synergistic action of two small molecules (cAMP and a β-galactoside) that cause changes of state in the promoter via their effects on two proteins, viz., CAP and *lac* repressor.

E. coli can regulate the rate of synthesis of the enzymes required for tryptophan biosynthesis in response to changes in the intracellular concentration of tryptophan. When grown in the presence of excess tryptophan, the enzyme levels are only about 5 percent of those when the bacterium is grown in a medium lacking tryptophan, i.e., conditions in which tryptophan must be synthesized. Thus, in this and similar instances, *E. coli* responds to its nutritional environment by either turning on or turning off synthesis of the appropriate biosynthetic enzymes. For tryptophan this is achieved by transcriptional control of the tryptophan (*trp*) operon.

Like the *lac* operon, the *trp* operon of *E. coli* is a single transcriptional unit specifying the structure of the five polypeptides that catalyze the reactions required for tryptophan biosynthesis beginning with chorismate (Fig. 28.17). The intact *trp* mRNA is composed of approximately 7000 nucleotides and contains a *leader* sequence of 162 nucleotides preceding the sequences that code for the polypeptide products of the operon. A promoter-operator regulatory region precedes and partially overlaps the site of initiation of transcription and determines whether RNA polymerase transcribes the operon. Transcription is initiated at the promoter, which is at the operator end of the operon. The five polypeptide products are synthesized sequentially, coordinately, and in equimolar amounts, by translation of the polycistronic *trp* mRNA as it is being synthesized. However, unlike the *lac* operon, the *trp* repressor protein cannot bind to the operator unless tryptophan, the corepressor, is present. Thus, in the absence of tryptophan, *trp* repressor cannot bind to the operator and the cell is said to be fully *derepressed*; i.e., transcription, hence translation, of the *trp* operon

Figure 28.17 The *trp* operon of *E. coli*, showing the estimated lengths of the nucleotide sequences and the polypeptide products of the structural genes. p_1 and o refer to the principal promoter and operator; a is the attenuator, and p_2 is a low-efficiency internal promoter. [*From K. Bertrand, L. Korn, F. Lee, T. Platt, C. L. Squires, C. Squires, and C. Yanofsky, Science* **189**:22 (1975). *Copyright © 1975 by the American Association for the Advancement of Science.*]

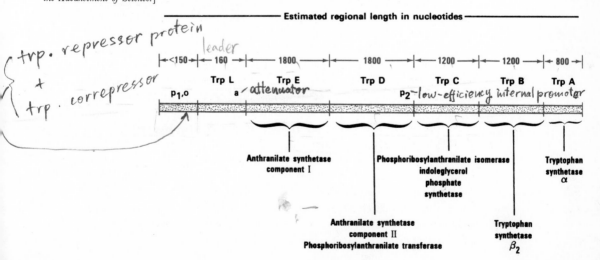

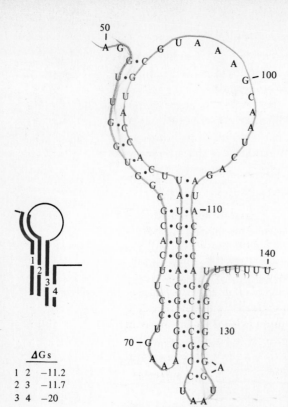

ΔGs

1 2	−11.2
2 3	−11.7
3 4	−20

Figure 28.18 Proposed secondary structures in *E. coli* terminated *trp* leader RNA. The calculated free energies (ΔGs) of formation of the stem and loop structures are given in kilocalories per mole. [*From D. Oxender, G. Zurawski, and C. Yanofsky, Proc. Natl. Acad. Sci. USA 76:5524 (1979).*]

occurs at a maximal rate. In the presence of tryptophan, the cell is in a *repressed* state. Repressor is bound to the operator, and as a consequence the synthesis of tryptophan biosynthetic enzymes is coordinately *repressed* and their level declines.

In addition to the promoter-operator region, the *trp* operon contains a second regulatory site, the *attenuator,* a sequence in the leader region, beyond the site at which transcription is initiated, and preceding the first structural gene *trpE*. Transcription, initiated at the promoter, is either terminated at the attenuator to produce a 140-residue leader transcript or allowed to proceed into the five structural genes of the operon. The availability of tryptophan-charged tRNA^{Trp} is the signal that triggers transcription termination at the attenuator by controlling translation of the transcript of the leader region. The charged tRNA^{Trp} is presumably required for translation of the segment of the leader transcript that codes for a 14-residue peptide containing adjacent tryptophan residues. The 3′ half of the terminated *trp* leader transcript exhibits extensive secondary structure (Fig. 28.18), and the ability to form this secondary structure is essential for normal regulation of transcription termination at the attenuator. A model for attenuation of the *E. coli trp* operon based on these features is shown in Fig. 28.19. Under conditions of excess tryptophan the ribosome (shaded circle) translating the newly transcribed leader RNA will synthesize the complete leader peptide. During this synthesis, the ribosome will mask regions 1 and 2 of the RNA (Fig. 28.19) and prevent the formation of stem and loop 1.2 or 2.3.

mask
遮蔽

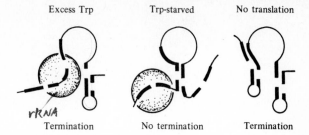

Figure 28.19 Model for attenuation of the *E. coli trp* operon. See text for details. [*From D. Oxender, G. Zurawski, and C. Yanofsky, Proc. Natl. Acad. Sci. USA* **76**:5524 (1979).]

Stem and loop 3.4 will be free to form and signal the RNA polymerase molecule transcribing the leader region to terminate transcription. Under conditions of tryptophan starvation, charged tRNA^Trp will be limiting and the ribosome will stall at the adjacent tryptophan codons in the leader peptide coding region. Because only region 1 is masked, stem and loop 2.3 will be free to form as regions 2 and 3 are synthesized. Formation of stem and loop 2.3 will exclude the formation of stem and loop 3.4, which is required as the signal for transcriptional termination. Consequently, RNA polymerase will continue transcription into the structural genes. Under conditions in which the leader peptide is not translated, for example, upon starvation for amino acids occurring early in the leader peptide, stem and loop 1.2 will be free to form as regions 1 and 2 are synthesized. Formation of stem and loop 1.2 will prevent the formation of stem and loop 2.3 and thereby permit the formation of stem and loop 3.4. This will signal transcription termination.

In addition to control of expression of the *trp* operon by the regulation of mRNA synthesis, flow of substrate through the biosynthetic pathway is also regulated. Tryptophan binds to and thereby causes the feedback inhibition of the enzyme complex that catalyzes the first reaction of the pathway. Control of tryptophan synthesis by feedback inhibition is obviously more rapid than control by regulation of transcription and hence specific enzyme synthesis.

SYNTHESIS AND CLONING OF GENES

Synthesis of the Gene for Tyrosine Suppressor tRNA

The development of methods for the chemical synthesis of short (8 to 15 residues) oligodeoxyribonucleotides and their enzymic ligation has permitted the total synthesis of the gene that specifies the tyrosine suppressor tRNA of *E. coli* (page 745).

The initial product of the gene is known to contain 126 nucleotides; however, this precursor polynucleotide undergoes enzymic cleavage to generate the mature tRNA, consisting of only 86 nucleotides (see Fig. 28.10 for the closely related tRNA^Tyr). The remainder of the gene is composed of a 56-nucleotide-long promoter at one end and a 25-nucleotide terminator sequence at the other. Thus, the functional tyrosine tRNA gene, whose total synthesis has been achieved, is a duplex DNA molecule 207 nucleotides in length (Fig. 28.20). Its entire nucleotide sequence was established by first determining the ribonucleotide sequence of the tRNA precursor and then determining the

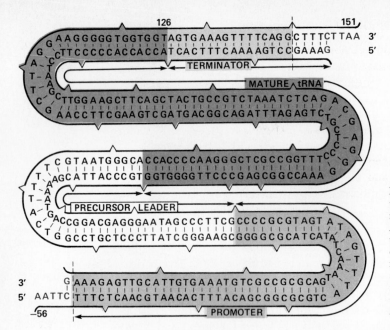

Figure 28.20 Nucleotide sequence of the DNA corresponding to the gene for the tyrosine suppressor tRNA of *E. coli.* The promoter sequence extends from residue −56 through residue 1, the tRNA precursor extends from residue 1 through residue 126, and the terminator sequence extends from residue 127 through residue 151. [*Courtesy of Professor H. G. Khorana.*]

deoxyribonucleotide sequence of the regulatory elements (promoter and terminator regions).

The strategy employed in the synthesis of the portion of the gene specifying the tRNA precursor involved three basic steps:

1. The synthesis by chemical means of short deoxyribonucleotide segments corresponding to both strands of the DNA duplex. Synthesis of these oligonucleotides was accomplished both by stepwise polymerization of mononucleotides and by the condensation of short (di-, tri-, and tetra-nucleotide) oligonucleotides to generate oligonucleotide chains of the desired length.

2. With the use of DNA ligase, joining of several of the segments at a time to form a total of four DNA duplexes. In each case, the segments were chosen so that there were complementary single-stranded termini that would permit them to be annealed in the proper order and then enzymically ligated.

3. Joining the duplexes, again by means of single-strand overlaps at their termini, to form the DNA duplex corresponding to the tyrosine tRNA precursor. Completion of the gene was accomplished by synthesis of the promoter and terminator sequences and their fusion at the appropriate ends to the 126-nucleotide long "structural" portion of the molecule.

The synthetic tyrosine suppressor tRNA gene was fully functional after insertion into bacteriophage λ bearing a *nonsense* mutation (page 745) as judged by the restoration of the capacity of the mutant bacteriophage to proliferate in the appropriate host cell.

somatostatin 生长激素释放抑制因子

The gene for the human peptide hormone somatostatin (Chap. M16) has been synthesized chemically, inserted into an appropriate plasmid vector, and cloned in *E. coli*. The somatostatin gene is expressed and high yields of the hormone have been obtained in this manner. The procedure is diagrammed in Fig. 28.21.

With the amino acid sequence as a guide, codons were chosen that corresponded to each of the 14 amino acid residues of the hormone. The polynucleotide that was synthesized contained in addition to the nucleotide sequence of the somatostatin gene a promoter and the initiation and termination signals that are functional in *E. coli*. In addition, a single-stranded cohesive sequence was added to each end of the synthetic duplex to facilitate insertion into the vector.

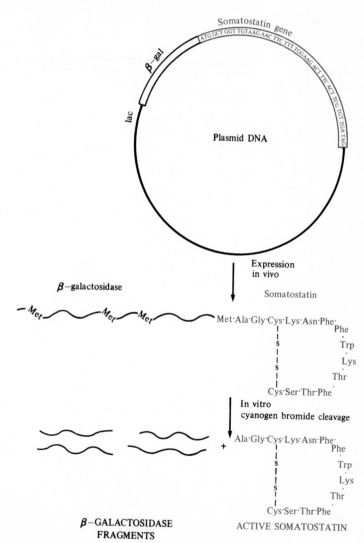

Figure 28.21 Cloning and expression of the gene for human somatostatin. See text for details. [*Adapted from K. Itakura, T. Hirose, R. Crea, A. D. Riggs, H. L. Heyneker, F. Boliver, and H. W. Boyer, Science* **198**:*1056 (1977). Copyright* © *1977 by the American Association for the Advancement of Science.*]

The plasmid vector was constructed by inserting a restriction fragment containing the *lac* promoter, operator, and most (98 percent) of the gene for β-galactosidase (page 754). The synthetic somatostatin gene was inserted at the end of the β-galactosidase gene, and the codon for methionine (ATG) was placed between the two. Upon transformation of *E. coli* with this plasmid there was transcription from the plasmid *lac* promoter. This was followed by translation of the mRNA to yield a chimeric polypeptide consisting of β-galactosidase and somatostatin. The polypeptide was treated with cyanogen bromide, which cleaves the polypeptide at the carboxyl-terminal side of all methionine residues (Chap. 5). Inasmuch as somatostatin lacks methionine, it is stable under the conditions necessary to release it from the chimeric protein.

Similar methods have been used for the synthesis of the larger polypeptide hormones, viz., human growth hormone (Chap. M20) and insulin (Chap. M16). In these instances, because of their relatively large size, chemical synthesis of the nucleotide sequence corresponding to the gene was impractical. However, a "c-DNA" copy of the gene was generated by isolation of the specific mRNA and synthesis of a DNA copy by reverse transcriptase (Chap. 27). The bacterial control elements, i.e., promoter, operator, and ribosomal binding sites, were then added to permit proper transcription and translation in *E. coli*.

REFERENCES

See list following Chap. 29.

Chapter 29

Gene Expression and Its Control: Eukaryotes

haploid

repetitive

Many of the features of gene expression described in Chap. 28 pertain to both prokaryotes and eukaryotes. There are, however, aspects of gene expression that are unique to eukaryotes. In large part, these result from profound differences between prokaryotes and eukaryotes in cellular and genomic organization. (1) Eukaryotes contain subcellular organelles encompassed by membranes. These organelles include the nucleus, mitochondria, chloroplasts, each with its own genetic complement: the endoplasmic reticulum, the Golgi apparatus, and lysosomes (Chap. 13). In particular, the existence of a nuclear membrane imposes a separation of transcription and translation and introduces the problem of transmembrane transport, i.e., transport of RNA transcripts from the nucleus to the cytoplasm and of proteins from the cytoplasm to the nucleus. (2) Although there is an enormous variation, the genome size, i.e., the amount of DNA in the haploid set of chromosomes, of eukaryotes is invariably greater than that of prokaryotes. Moreover, the DNA of eukaryotes, as a consequence of its specific association with histones (Chap. 8), is physically organized into chromatin. (3) Eukaryotic genomes contain repetitive sequences as well as single-copy sequences (Chap. 8). The repetitive sequences are of two types: short, highly repetitive sequences and moderately repetitive sequences. Prokaryotes contain only single-copy sequences. (4) Eukaryotic, unlike prokaryotic, genes are not continuously colinear with mRNA, but are instead interrupted by *intervening* sequences that are transcribed but not translated. (5) Eukaryotic mRNAs show a considerably greater stability than prokaryotic mRNAs. This enhanced stability offers a greater opportunity for the regulation of gene expression at the level of translation.

Any cross-references coded M refer to the companion text, *Principles of Biochemistry: Mammalian Biochemistry.*

Unlike prokaryotes, eukaryotes contain more than a single chromosome. However, several lines of evidence indicate that each chromosome contains but a single DNA molecule. Autoradiographic analysis of plant and mammalian chromosomes labeled with tritiated thymidine and then allowed to divide through several generations shows that the DNA labeled initially is maintained as a single "unit." Viscoelastic-retardation measurements (Chap. 8) indicate that the largest DNA molecules present in *Drosophila melanogaster* are of a size commensurate with the mass of DNA in the largest chromosome. Sedimentation-velocity analysis and electron microscopy of yeast chromosomes have also detected intact DNA molecules of the same size as the chromosomes of this organism. Thus, a 1-cm-long DNA molecule must be packaged in a 1-nm-long chromosome. This very tight packing is mediated partly through association of the DNA with histones to form chromatin (Chap. 8). Other proteins, in addition to histones, are found in association with DNA in chromosomes. These nonhistone chromosomal proteins are exceedingly diverse in properties and no doubt serve important structural, enzymic, and regulatory roles in the chromatin complex.

Nucleosomal Structure of Chromatin

Purified chromatin contains equivalent weights of histones and DNA with an equivalent number of basic amino acid residues and DNA phosphate groups. Thus, the DNA-histone interaction is probably stabilized by charge neutralization.

Five types of histones, H1, H2A, H2B, H3, and H4, are found in chromatin (Chap. 8). They are present in equimolar amounts, with the exception of histone H1, of which there is half the amount of the others. When extracted from chromatin by gentle procedures, the histones are found in specific associations; H3 is complexed with H4 as a tetramer, $(H3)_2(H4)_2$, and H2A and H2B are associated in an oligomer of the type $(H2B-H2A)_n$. Upon chemical cross-linking of histones to their nearest histone neighbors in chromatin, the four histones, H3, H4, H2A, and H2B, form an octamer consisting of the $(H3)_2(H4)_2$ tetramer and two molecules each of H2A and H2B. Moreover, the octamers can further associate to form a chain of repeating octamers (Fig. 29.1).

X-ray analyses of nuclei and of isolated chromatin have indicated that histones and DNA are arranged within a structure that is repeated at intervals of 110 Å. If chromatin is hydrolyzed with micrococcal nuclease, a series of discrete fragments is generated that are multiples of a DNA repeating unit of about 200 base pairs. Moreover, when viewed in the electron microscope

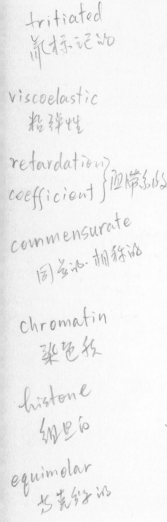

(a) OCTAMER

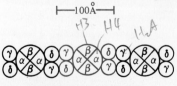

(b) CHAIN OF OCTAMERS

Figure 29.1 Possible arrangements of histones (*a*) within the octamer and (*b*) in chains of octamers. α, β, γ, and δ represent histones H4, H3, H2A, and H2B, respectively.

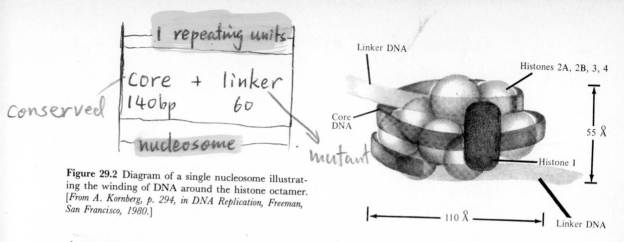

Handwritten annotations: "1 repeating units", "Core + linker", "140bp   60", "nucleosome", "Conserved", "mutant"

Figure labels: Linker DNA, Histones 2A, 2B, 3, 4, Core DNA, 55 Å, Histone I, 110 Å, Linker DNA

Figure 29.2 Diagram of a single nucleosome illustrating the winding of DNA around the histone octamer. [*From A. Kornberg, p. 294, in DNA Replication, Freeman, San Francisco, 1980.*]

chromatin fibers appear to consist of a chain of "beads" on a string. The isolated beads, termed *nucleosomes*, have a diameter of approximately 110 Å and a molecular weight of 256,000 and consist of a histone octamer of MW 109,000 and 200 base pairs of DNA.

Exhaustive digestion of chromatin with micrococcal nuclease yields a *core* particle containing a length of protected DNA of approximately 140 base pairs. The remaining 60 base pair segment of DNA which is relatively accessible to nuclease digestion serves as a *linker* between cores and is attached to histone H1.

To summarize, the nucleosome is a spherical structure approximately 110 Å in diameter and consisting of a histone octamer about which is wound some 200 base pairs of DNA. The latter is composed of a 140 base pair core attached to a 60 base pair linker (Fig. 29.2).

The nucleosome structure is universal; the chromatin of all eukaryotic organisms that have been examined shows repeat patterns upon nuclease digestion. The length of the repeating units varies in different organisms (170 to 200 base pairs), but this variation appears largely in the internucleosome linkers, whereas the 140 base pair core appears to be constant and hence conserved. Inasmuch as nucleosomes are present in condensed mitotic metaphase chromosomes as well as interphase chromatin, a higher order of packing of nucleosomes must exist to account for the differential folding of chromosomes during mitosis.

Modification of Histones

Despite the extraordinary conservation in evolution of amino acid sequences among the histones (Chap. 8), they may undergo extensive modification. These modifications occur primarily in the basic NH_2-terminal sequences of the core histones and they reduce the net positive charge, weakening the interaction of histone with DNA. Lysine residues in the NH_2-terminal region are acetylated on the ϵ-NH_2 groups immediately after the synthesis of histones H2A, 2B, 3, and 4, and the acetyl groups are removed as the histones bind to DNA. Preexisting histones are also acetylated and deacetylated during DNA replication.

Modification of histone H1 occurs primarily by phosphorylation of four serine residues in the COOH-terminal basic region of the protein and appears to

Handwritten margin notes: "bead 有孔小珠", "mitotic 有丝分裂的", "metaphase 中期", "①weakening the interaction of histone with DNA", "②the acetyl groups are removed as the histones bind to DNA"

765

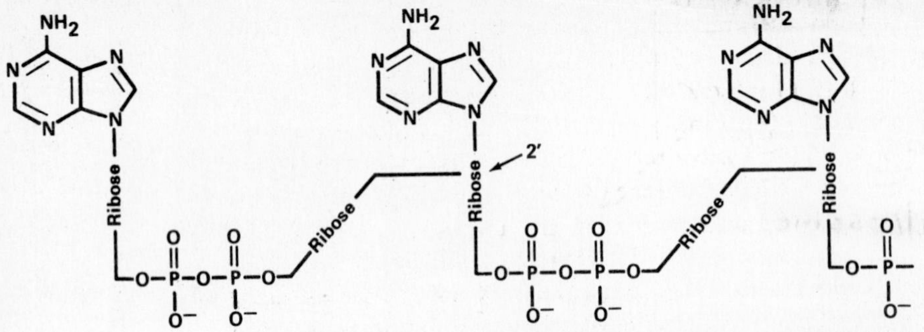

Figure 29.3 Structure of poly(ADP-ribose). [*Modified from Y. Nishizuka, K. Ueda, K. Yoshihara, H. Yamamura, M. Takeda, and O. Hayaishi, Cold Spring Harbor Symp. Quant. Biol. 34:781 (1969).*]

be synchronized with the cell cycle. Two serine residues are phosphorylated during the period of active DNA replication (Chap. 27) and two additional sites are modified just prior to mitosis. Immediately after mitosis H1 is dephosphorylated. It thus appears that acetylation of lysine residues in H2A, H2B, H3, and H4 and phosphorylation of two serine residues in H1 may be essential for DNA replication. Phosphorylation of the last two serine residues in H1 may be required for chromosome condensation during mitosis. One additional serine residue in H1 is phosphorylated in response to hormonally induced increases in cyclic AMP levels. Increased acetylation of the arginine-rich histones has also been noted as a consequence of hormone stimulation. These modifications may result in changes in the structure of chromatin which permit its activation for transcription (see below).

In addition to acetylation and phosphorylation, histones may undergo polyadenosine diphosphate ribosylation. The *poly(ADP-ribose) synthetase*, which exists in close association with chromatin and requires DNA for activity, catalyzes the transfer of the ADP-ribose moiety of NAD to histones H1 and H2B as well as certain nonhistone chromosomal proteins to form polymers of chain lengths up to 50 units.

$$nNAD + \text{histone} \rightleftharpoons n \text{ nicotinamide} + (ADP\text{-}ribose)_n\text{--histone}$$

The polymers are composed of repeating ADP-ribose units in which the internucleotide linkages are formed between C-1' of one ribose and C-2' of the adjacent one (Fig. 29.3). They are bound in ester linkage through the ribose moiety of the ADP-ribosyl unit to the γ-carboxyl group of one of the glutamate residues of histone H1 and to the γ-carboxyl groups of three glutamate residues and the carboxyl group of the COOH-terminal lysine of histone H2B.

In addition to the modifications described above, some histones may be methylated. Among the methylated residues are ϵ-N-mono-, di-, and trimethyllysine, ω-N-methylarginine, and 3-methylhistidine.

SEQUENCE ARRANGEMENTS IN EUKARYOTIC DNA

In prokaryotes, essentially all the nucleotide sequences in the chromosome occur only once per haploid genome; i.e., they are nonrepetitive. In eukaryotes, on the other hand, a given nucleotide sequence may be repeated many times. The fraction of the genome that is repetitive may vary from one organism to another; however, this fraction is generally appreciable and often accounts for more than half the genome. This fundamental difference between prokaryotic and eukaryotic genomes is most easily observed by examination of the kinetics of reassociation of denatured DNAs from the two sources (see Fig. 8.11). When prokaryotic DNA is fragmented into small segments less than 0.5 kb long and denatured and the resulting single strands permitted to reassociate to form double-stranded molecules, the kinetics of reassociation obey a single second-order rate equation, indicating that essentially all the sequences in the prokaryotic genomes occur as a single copy.

An example of the precise nature of the reassociation is provided by examining the reaction between a strand of wild-type bacteriophage λ DNA (Table 8.4) and the complementary strand obtained from a mutant of λ in which a segment of the DNA has been deleted. The heteroduplex formed between these strands will have a single-stranded loop containing the DNA in the wild-type λ strand that is not present in the deletion mutant. Such heteroduplex regions can easily be recognized in the electron microscope (Fig. 29.4). The extent of the deletion and its position can be determined by measuring the relevant contour lengths.

Figure 29.4 Visualization of heteroduplex structures formed between one strand from phage λ⁺ (wild-type) DNA and the complementary strand from phage λ bearing a deletion. (*a*) Electron micrograph. (*b*) Tracing of the heteroduplex molecule showing the deletion loops. (*c*) Schematic drawing of the heteroduplex molecule. [*Courtesy of Dr. Ronald Davis.*]

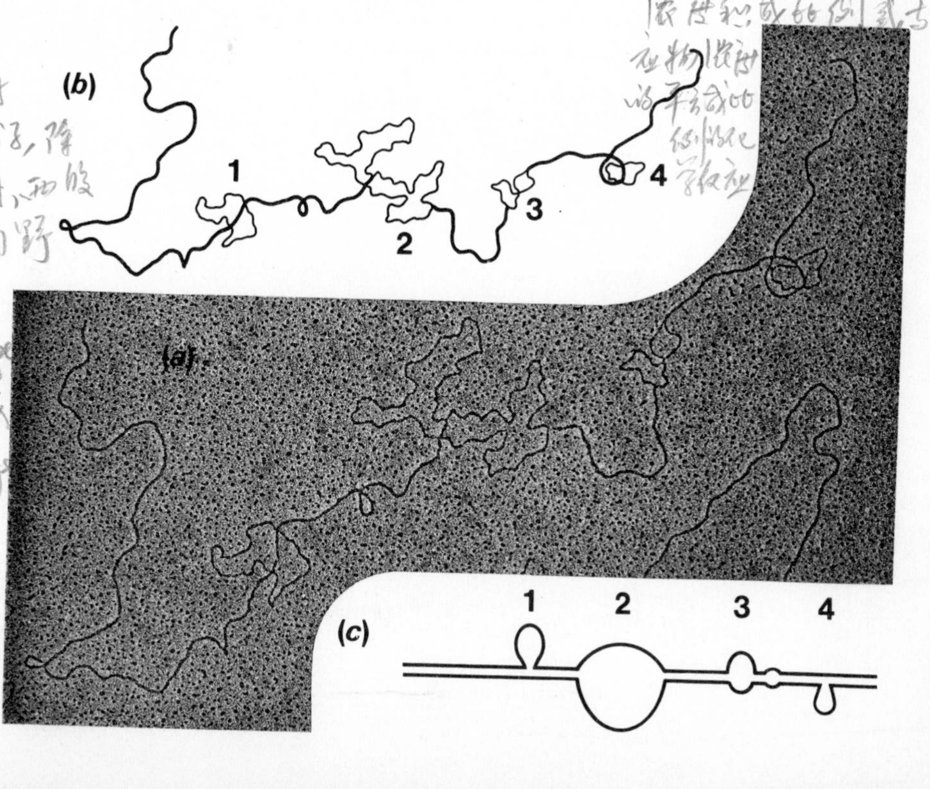

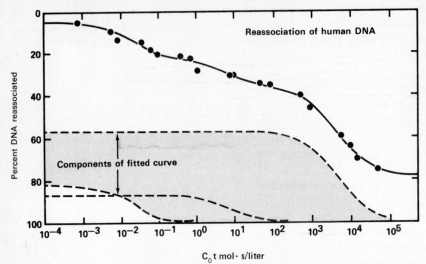

Figure 29.5 Reassociation kinetics of human DNA. The curves show the three kinds of sequences: highly repetitive, moderately repetitive, and nonrepetitive. [*From F. E. Arrighi and G. F. Saunders, p. 113, in R. F. Pfeiffer (ed.), Modern Aspects of Cytogenetics: Heterochromatin in Man, Shattauer Verlag, Stuttgart, 1972.*]

The reassociation kinetics for eukaryotic DNAs are more complex than those for prokaryotes because they contain a mixture of repetitive and nonrepetitive sequences. Often the reassociation curves are triphasic, indicating the existence of three classes of sequences: *highly repetitive, moderately repetitive,* and *nonrepetitive* (Fig. 29.5).

Highly Repetitive Sequences

The highly repetitive sequences seen in reassociation plots can be physically separated from the bulk of the chromosomal DNA by isopycnic centrifugation

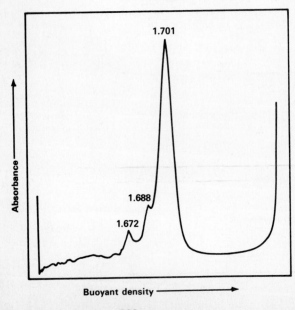

Figure 29.6 Separation of satellite DNAs from main-band DNA of *Drosophila melanogaster* by CsCl density-gradient centrifugation. The peak at a density of 1.701 g/cm^3 is main-band DNA, and the peaks at densities of 1.672 and 1.688 g/cm^3 represent the two satellite DNAs. [*From A. J. Peacock, D. Brutlag, E. Goldring, R. Appels, C. W. Hinton, and D. L. Lindsley, Cold Spring Harbor Symp. Quant. Biol. 38:405 (1973).*]

in CsCl (Chap. 8). This is most readily accomplished by shearing the DNA into double-stranded segments of sizes from 20 to 60 kb before centrifugation. Under these conditions, the DNA distributes into a *main band* and a set of smaller bands termed *satellite bands* (Fig. 29.6).

In human DNA, four satellite DNAs have been observed that constitute 6 percent of the chromosomal DNA, or a total of 174,000 kb per genome. The predominant satellite DNAs generally contain very simple sequences that are repeated manyfold. For instance, in the example shown in Fig. 29.6, the sequence

$$-ACAAACT-$$
$$-TGTTTGA-$$

in the satellite DNA with a buoyant density of 1.672 g/cm³ is repeated 12 × 10⁶ times. The satellite DNAs are generally found in the centromeric region of chromosomes and may act in pairing and segregation.

Moderately Repetitive Sequences

Most of the moderately repetitive DNA sequences are interspersed with nonrepetitive single-copy sequences. Thus, in the chromosomes of *Xenopus laevis*, a short repetitive sequence of about 0.3 kb is interspersed with nonrepetitive sequences of approximately 1 kb. The function of the interspersed moderately repetitive sequences is not known, but it is suspected that they play some regulatory role in gene expression. Some moderately repetitive sequences are however arranged in tandem arrays, as described below.

Genes for the 18-S and 28-S Ribosomal RNAs The repeating unit in the array of genes that code for eukaryotic ribosomal RNAs consists of a region that is transcribed to yield a single RNA transcript which is subsequently processed (see below) plus a spacer region that is not transcribed (Fig. 29.7). The 28-S ribosomal RNA gene may also contain an intervening sequence (see below). In human DNA, the length of the transcribed region is 13 kb, that of the nontranscribed spacer is approximately 30 kb, and the repeating unit is 43 kb. In

Figure 29.7 Organization of repeating units of ribosomal RNA. Transcription units are indicated by the colored regions; nontranscribed spacers are white. [*From B. Lewin, p. 884, in Gene Expression 2, 2d ed., Wiley, New York, 1980.*]

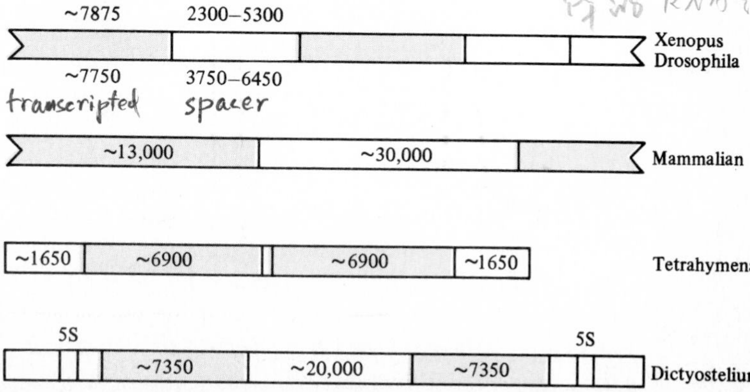

coalesce 揉合,熔合

tandem
串联到 彼列

discrepancy
差异,不一致

all cases that have been examined there are several hundred such repeating units per eukaryotic genome; they are located in regions of one or more specific chromosomes. The DNA of these chromosomes is usually looped off from the main chromosomal fiber mass in the form of highly extended threads, which coalesce with specific proteins to give rise to the *nucleolus*.

Genes for 5-S Ribosomal RNA The 5-S ribosomal RNA is transcribed from a gene that differs from that specifying the 18-S and 28-S ribosomal RNAs. The number of copies ranges from 165 in *D. melanogaster* to 24,000 in *X. laevis*. Although they are located at many sites on different chromosomes, each site contains many copies in tandem, with a nontranscribed spacer, as in the 18-S and 28-S ribosomal RNA genes.

Genes for Transfer RNA Each of the transfer RNA genes appears to be arranged in tandem, again with a spacer. In lower eukaryotes, e.g., yeast, intervening sequences are present within the gene. There are approximately 1000 tRNA gene copies per eukaryotic genome.

Histone Genes The five histone genes (H1, H2A, H2B, H3, and H4) are repeated several hundred times as a tandem cluster within the eukaryotic genome and are separated from each other by spacers of variable lengths.

Nonrepetitive Sequences

In eukaryotes, the number of nonrepetitive, single-copy sequences coding for proteins represents only a small fraction of the total single-copy sequences as determined by reassociation kinetics. Although the basis for this discrepancy has not been entirely resolved, it is apparent that in eukaryotes, structural genes occupy considerably more DNA than in prokaryotes. This is because the coding sequences of eukaryotes are often interrupted by noncoding intervening sequences. Furthermore, the regulation of transcription in eukaryotes is much more complex than in prokaryotes and as a consequence requires more DNA. Thus, DNA sequences at some distance from the transcription initiation site play a critical role in determining the frequency of initiation of transcription.

Transposable Sequences

The genome of eukaryotes, like that of the prokaryotes, is not static. In both yeast (*Saccharomyces cerevisiae*) and in *Drosophila*, repetitive sequences have been observed that undergo frequent transposition and are not fixed in their location in the genome. In *Drosophila*, three such families of dispersed genes have been identified. All bear a direct repetition of a short sequence at each end of the unit which may be involved in the recombinational event required for transposition of the unit (Chap. 27). One of these dispersed gene families, termed *copia*, is approximately 5 kb in length with a direct repeat of about 300 base pairs. The number of copies of *copia* varies from 60 to 180 depending upon the particular *Drosophila* strain.

As described above, DNA in eukaryotes is tightly packed into the nucleosomes of chromatin. Because of the packing, promoters may be sterically unavailable to RNA polymerase. Furthermore, even if the promoters were available, the highly folded state of the DNA within the nucleosome might not allow sufficient degrees of freedom for the RNA polymerase to form the "open" complex required for the initiation of transcription (Chap. 28). Although it is not known how chromatin is "activated" for transcription, there are indications that a change in chromatin structure precedes the onset of transcription.

Eukaryotic RNA Polymerases

RNA polymerases have been isolated from calf thymus, human tissue culture cells, rat liver, and yeast. In each case, three distinct enzymes are present in association with the nucleus. All three are of high molecular weight (500,000 to 700,000) and contain an extremely complex array of subunits.

RNA polymerase I of calf thymus consists of six subunits (MWs = 197,000, 126,000, 51,000, 44,000, 25,000, and 16,500) in the ratio 1:1:1:1:2:2. RNA polymerase II of calf thymus is composed of five subunits (MWs = 214,000, 140,000, 34,000, 25,000, and 16,000) RNA polymerase III of *X. laevis* contains nine polypeptides (MW = 155,000, 137,000, 92,000, 68,000, 52,000, 42,000, 33,000, 29,000, and 19,000). Because these enzymes have not been reconstituted from the separated subunits, it is not known whether all are required for RNA polymerase activity. It is clear, however, that the three enzymes with the complement of subunits described above are not sufficient for the accurate initiation and termination of transcription. Other factors are required (see below).

Although the three RNA polymerases are restricted to the nucleus, their locations within the nucleus differ, as do their functions. Thus, RNA polymerase I is found in association with the nucleolus and catalyzes synthesis of rRNA. RNA polymerase II is present in the nucleoplasm and is responsible for the synthesis of mRNA. RNA polymerase III is also present in the nucleoplasm and catalyzes the synthesis of 5-S RNA and tRNAs. The three enzymes differ in their response to α-*amanitin*, the toxic principle of the mushroom *Amanita phalloides*. α-Amanitin specifically inhibits RNA polymerase II, and, in fact, cells resistant to α-amanitin contain a II-type enzyme that is unaffected by the poison. RNA polymerase I is completely resistant to α-amanitin and RNA polymerase III is inhibited, but at a 1000 times greater concentration of α-amanitin than that required to inactivate RNA polymerase II.

Mitochondria possess an RNA polymerase that is clearly distinct from the three nuclear enzymes and is presumably responsible for the synthesis of mitochondrial RNA. The polymerase from mitochondria of *Neurospora crassa* is noteworthy in consisting of a single polypeptide chain (MW = 64,000). Like the bacterial RNA polymerases, it is inhibited by rifampicin (Chap. 28) but is insensitive to α-amanitin, perhaps reflecting a presumed prokaryotic origin of mitochondria (Chap. 31).

Amanitin

RNA polymerase I
association with the nucleolus and catalyzes synthesis of rRNA

RNA polymerase II
in the nucleoplasm. responsible for the synthesis of mRNA

RNA polymerase III
In the nucleoplasm for synthesis of 5-SRNA and tRNA

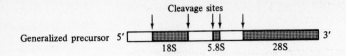

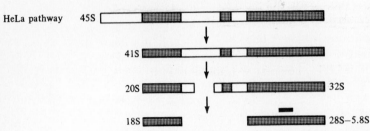

Generalized precursor 5′   18S   5.8S   28S   3′

HeLa pathway   45S

41S

20S   32S

18S   28S–5.8S

Figure 29.8 Processing pathway in human HeLa cells for ribosomal RNA. The precursor RNA molecule has a transcribed spacer at the 5′ end, followed by the 18-S rRNA sequence; a 5.8-S sequence lies within the internal transcribed spacer and the 28-S sequence lies at the 3′ end of the molecule. The time at which the 5.8-S segment becomes associated with the 28-S rRNA is not known. [*From B. Lewin, p. 871, in Gene Expression 2, 2d ed., Wiley, New York, 1980.*]

Synthesis of 18-S and 28-S Ribosomal RNAs

As noted above (page 769), the genes for the 18-S and 28-S rRNAs occur in tandemly repeating units in which each unit consists of a transcribed segment and an adjacent nontranscribed spacer. The transcription of the rRNA genes by RNA polymerase I in the nucleolus yields a large precursor RNA which is then processed by as yet unidentified RNases to yield the 18-S and 28-S products as indicated in Fig. 29.8.

Synthesis of 5-S Ribosomal RNA

Unlike the primary RNA polymerase I transcript of the 18-S and 28-S rRNA genes, larger precursors of 5-S rRNA have not been detected. The tandemly repeated 5-S RNA genes from *X. laevis* have been isolated by molecular cloning (Chap. 27) and can be faithfully transcribed in vitro by RNA polymerase III in the presence of a factor, *TFIIIA*, isolated from *X. laevis* ovaries. It has been observed that sequences within the 5-S rRNA gene are required for and control the site of initiation of transcription. These sequences occupy a region of about 30 base pairs, beginning about 50 base pairs from the 5′ end and terminating at approximately 80 base pairs from that end.

Factor TFIIIA binds specifically to the intragenic control region and is believed to direct the RNA polymerase III to the transcription initiation site.

Structure of mRNA

mRNAs found in the cytoplasm of eukaryotic cells are monocistronic and are relatively short (mean length of approximately 2000 nucleotides). In contrast, the nuclear RNA containing the primary transcript and referred to as *heterogeneous nuclear RNA* is considerably larger in size than cytoplasmic mRNA, is present in much greater abundance, and turns over very rapidly. It also exists in association with proteins in the form of *ribonucleoprotein particles*. A possible function of these particles is the transport of mRNA from the nucleus to the cytoplasm.

Eukaryotic mRNAs consist of three parts: (1) a 5′ cap which precedes the transcribed sequences, (2) the transcribed sequence, and (3) the poly(A) tail.

The 5′ Cap The 5′ terminus of all eukaryotic mRNAs transcribed from nuclear structural genes exhibit a "cap" of the type 7-methyl-G(5′)ppp(5′)XpYp, where X, the first of the transcribed nucleotides in the mRNA (Base$_1$ in Fig. 29.9) is often methylated at 2′-O-, as is the second transcribed nucleotide, Y, though less frequently. Capping occurs in the nucleus shortly after the initiation of transcription. Figure 29.9 shows the structure of the 5′ cap. In contrast to mRNA, the 5′ terminal nucleotide of the 5-S RNA retains the triphosphate group of the nucleoside triphosphate substrate.

The capped nucleotide (i.e., that represented by X) is not only the first transcribed nucleotide in the mRNA, but is very likely the initial nucleotide in the primary transcript; that is, the transcription initiation site is the base pair in the gene that corresponds to X. The 5′ caps facilitate the initiation of translation and may serve to stabilize the mRNA. In this regard, it should be recalled that eukaryotic mRNAs are generally more stable than their prokaryotic counterparts, exhibiting half-lives measured in hours or days instead of minutes.

Capping is the first of the processing reactions which eukaryotic transcripts are known to undergo. The reactions involved in synthesis of the 5′ caps are shown in Fig. 29.10.

Figure 29.9 Structure of the 5′ cap of eukaryotic mRNA. The nucleotide bearing Base$_1$ = X in the text, that bearing Base$_2$ = Y.

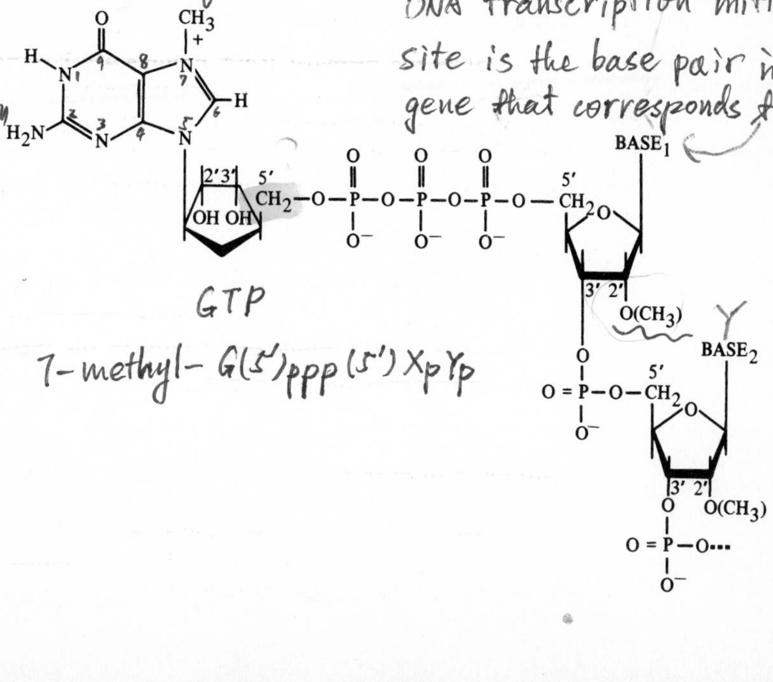

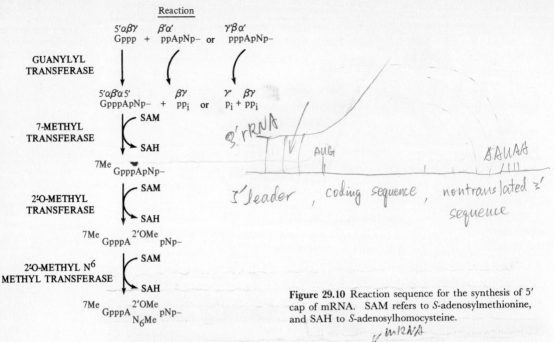

Reaction

$$
\begin{array}{ccccc}
5'\alpha\beta\gamma & \beta'\alpha' & & \gamma'\beta'\alpha' \\
\text{Gppp} + & \text{ppApNp--} & \textbf{or} & \text{pppApNp--}
\end{array}
$$

GUANYLYL TRANSFERASE

$$
\begin{array}{ccccc}
5'\alpha\beta'\alpha 5' & & \beta\gamma & & \gamma' \quad \beta\gamma \\
\text{GpppApNp--} + & \text{pp}_i & \textbf{or} & \text{p}_i + \text{pp}_i
\end{array}
$$

7-METHYL TRANSFERASE

SAM → SAH

$$\text{7Me}\,\text{GpppApNp--}$$

2'-O-METHYL TRANSFERASE

SAM → SAH

$$\text{7Me}\,\text{GpppA}\,^{2'\text{OMe}}\text{pNp--}$$

2'-O-METHYL N^6 METHYL TRANSFERASE

SAM → SAH

$$\text{7Me}\,\text{GpppA}\,^{2'\text{OMe}}_{\text{N}_6\text{Me}}\,\text{pNp--}$$

Figure 29.10 Reaction sequence for the synthesis of 5' cap of mRNA. SAM refers to S-adenosylmethionine, and SAH to S-adenosylhomocysteine.

Transcribed Sequences These consist of a 5' leader, the coding sequence, and a nontranslated 3' sequence. The number of nucleotides between the 5' cap and the AUG initiator codon (the 5' leader) is highly variable in eukaryotic mRNAs (37 to 38 nucleotides in the α-globin mRNAs of rabbits and human beings; 52 nucleotides in their β globin). As in prokaryotes, there is some complementarity between the sequences in the 5' leader of many eukaryotic mRNAs and the conserved sequence at the 3' end of the eukaryotic 18-S rRNA, indicating that this complementarity may be used for ribosome attachment as it is in bacteria (Fig. 28.6). The actual coding sequence is contiguous as in bacterial mRNAs.

The coding sequence is followed by a 3' nontranslated region of variable length (108 nucleotides in the α globin mRNA of human beings; 631 nucleotides in the chicken ovalbumin mRNA). The sequence AAUAA is generally found approximately 20 to 25 residues from the 3' end of the transcribed sequences in the mRNA and may act as a signal for processing, i.e., polyadenylation, of the 3' end of the primary transcript.

The Poly(A) Tail Most eukaryotic mRNAs are terminated by a poly(A) "tail," 40 to 200 residues in length. These are added in the nucleus after transcription has been terminated, possibly by the action of *poly(A) polymerase,* which has been identified in several eukaryotic tissues. The function of the poly(A) tail is unknown; however, it may act to stabilize the mRNA. It should be noted, however, that some mRNAs, e.g., those for histones, are not polyadenylated at their 3' termini.

Intervening Sequences in Eukaryotic mRNA

A remarkable feature of many eukaryotic structural genes is that their sequences are *not* colinear with the sequence in the corresponding mRNA but

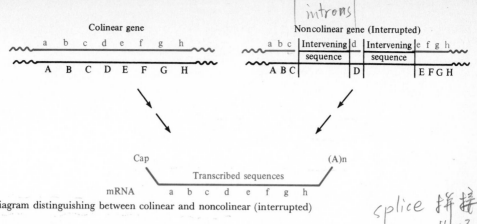

Figure 29.11 Diagram distinguishing between colinear and noncolinear (interrupted) genes.

splice 拼接 後導

contain intervening sequences that are absent from the mRNA from which their protein product is synthesized. Figure 29.11 illustrates the difference between colinear and noncolinear (interrupted) genes. Such genes are called *interrupted* genes because they contain blocks of intervening sequences, termed *introns*, interspersed among blocks of sequences found in the mRNA, the *exons*. Examples of interrupted structural genes are shown in Fig. 29.12. Both intron and exon sequences are colinearly represented in the primary transcript of these genes, the intron sequence being "spliced out" during processing of the primary transcript. As indicated in Fig. 29.13, the sequences at the exon-intron and intron-exon boundaries are to some extent conserved, indicating that they may be involved in the splicing reaction to be considered below.

Although the function of the introns in gene expression is unknown, it is possible that they may divide the structural gene into the coding sequences that form the structural domains in the protein product of the gene. This has led to the hypothesis that interrupted genes are a manifestation of an evolutionary

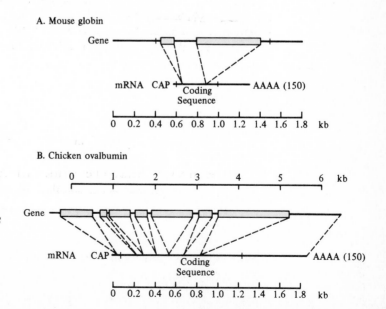

Figure 29.12 Comparison of maps of the genes and mRNAs for (*a*) mouse β globin and (*b*) chicken ovalbumin. The intervening sequences are shown as the colored bars. [*Adapted from J. Abelson, Annu. Rev. Biochem., 48:1035 (1979). Reproduced with permission,* © *1979 by Annual Reviews, Inc.*]

775

	Coding	Intervening	Coding	
chicken	... TCAAAAG|GTAGGC	....	TGCTCTAG|CAACTC ...	ovalbumin 1
	... AAATAAG|GTGAGC	....	AATTACAG|GTTGTTC ...	2
	... AGCTCAG|GTACAG	....	GTATTCAG|TGTGGCA ...	3
	... CCTGCCA|GTAAGT	....	CTTTACAG|GAATAGT...	4
	... ACAAATG|GTAAGT	....	TCTTAAAG|GAATTAT ...	5
	... GACTGAG|GTATAT	....	TGCTCTAG|CAAGAA ...	6
	... TGAGCAG|GTATAT	....	CCTTCCAG|CTTGAGA...	7
mouse & rabbit	... TGGGCAG|GTGAGC	....	CTTTTTAG|GCTGCTG ...	β^{maj} globin 1 / β globin 1
mouse	... TGGGCAG|GTGAGC	....	TCCCACAG|CTCCTG ...	β^{maj} globin 2
rabbit	... CTTCAGG|GTGAGT	....	TCCTACAG|CTCCTG ...	β globin 2
mouse	... CTTCAAG|GTATGC	....		α globin 2
rat	... CAAGCAG|GTACTC	....	TCTTCCAG|GTCATTG ...	Insulin I & II – 1
	... CCACAAG|GTAAGC	....	CCTGGCAG|TGCACA ...	Insulin II – 2
B. mori	... TCTGCAG|GTGAGT	....	TGTTTCAG|TATGTCG ...	Silk fibroin
canonical sequence	AGGT$^{G}_{A}$		TTXCAG	

Figure 29.13 Nucleotide sequences at junctions between coding regions and intervening sequences. [*Adapted from B. Lewin, p. 816, in Gene Expression 2, 2d ed., Wiley, New York, 1980.*]

process in which the exons representing different domains are recombined to form new genes that, in turn, yield new forms of proteins (Chap. 31).

Examples of eukaryotic genes that do not contain intervening sequences are the histone genes of sea urchins and *Drosophila* (page 764) and the mammalian interferon gene (Chap. 30). In these instances, the mRNA and its structural gene are colinear. In fact, most of the structural genes that have been analyzed in yeast and in *Drosophila* are lacking in interrupted sequences. Moreover, even for a gene coding for the same protein in a different species, the picture is inconsistent; e.g., the cytochrome c gene in yeast lacks introns whereas this gene in the rat has a single intron.

The β-Globin Gene The β-globin gene of the mouse contains two intervening sequences. The smaller of the two is between codons 31 and 32 in the nucleotide sequence of the gene and is 116 base pairs long. The larger intervening sequence is between codons 104 and 105 and is 646 base pairs in length. In addition there are 52 base pairs in the 5′ untranslated leader portion and 110 base pairs in the 3′ untranslated region (Fig. 29.12). In the human and rabbit β-globin genes, intervening sequences are found in precisely the same positions as in the mouse, i.e., between codons 31 and 32, and codons 104 and 105. In contrast, the large intron in the human β-globin gene is composed of 900 base pairs and is therefore substantially larger than that in the mouse. It also shows little sequence homology with the intron of the mouse β-globin gene. Thus, location, but not size or sequence of the introns, is rigidly conserved in the β-globin gene.

The Ovalbumin Gene The chicken ovalbumin gene is considerably more complex than the β-globin gene and contains seven intervening sequences (Fig.

29.12). The mRNA for ovalbumin contains a 76–base pair 5′ leader sequence, a coding sequence of 1149 base pairs, and a 3′ untranslated sequence preceding the poly(A) tail. One intervening sequence is located in the 5′ leader region of the mRNA at a position 19 base pairs before the AUG initiation codon. Six intervening sequences are found in the coding region and none is in the 3′ untranslated region. In total, the intervening sequences increase the genomic equivalent of the 1859 base pair ovalbumin in RNA by more than a factor of 3.

The Splicing Reaction Genes that contain introns yield primary transcripts containing the intron sequences and these are excised to yield the mature mRNAs. The process is believed to proceed in two steps: the endonucleolytic cleavage of the precursor RNA at the boundaries of the intron, followed by ligation of the two coding regions (exons) that were previously separated by the introns. These two steps have been resolved in the splicing reactions in vitro of tRNA precursors of yeast containing intervening sequences. However, they remain hypothetical for the mRNA precursors. Genes with multiple introns, e.g., the ovalbumin gene, might be expected to yield multiple intermediates in the process of conversion of the primary transcript to mRNA. Such intermediates have in fact been observed; however, the kinetics of their appearance and disappearance are inadequate to define specific pathways for the conversion.

A clue to the mechanism of splicing of mRNA precursors has come from the analysis of a class of discrete, stable small RNA molecules 90 to 220 nucleotides in length that are found in the nuclei of cells from a wide variety of eukaryotes. The most abundant of these small RNAs exist as a closely related set of RNA-protein complexes termed *small ribonucleoproteins* ($S_{20,w} = 10$ S), which contain seven prominent nuclear proteins with MWs between 12,000 and 32,000. One of these small RNAs, U1, contains sequences that exhibit considerable complementarity to the splice junctions of the precursor RNAs for mRNAs (Fig. 29.13). Thus, this RNA might serve to align the two junctions that are to be spliced. Consistent with this possibility is the high degree of conservation of both the sequences of the small RNA and its associated proteins, its great abundance in metabolically active cells, and its cosedimentation with the larger ribonucleoproteins that contain the large heterogeneous nuclear RNA.

Synthesis of mRNA

RNA polymerase II is responsible for the transcription of both colinear and interrupted structural genes in eukaryotes. Thus, the addition of α-amanitin at an appropriate concentration completely blocks the synthesis of mRNA in vivo. As in the case of the synthesis of 5-S RNA by RNA polymerase III, factors in addition to the purified enzyme are required for the accurate initiation and termination of transcription in vitro. This has been found for both the histone genes of *Drosophila*, which do not contain intervening sequences, and for certain of the "late" genes of adenovirus, type 2 (Chap. 30), which do.

Although a promoter, i.e., the site of the initiation of transcription by RNA polymerase II, has not been defined in eukaryotic genes (see Fig. 28.9 for sequences of prokaryotic promoters), a conserved sequence, 5′-TATAAATA-3′, is located 24 ± 2 base pairs from the 5′-end of virtually all structural genes, both colinear and interrupted, that have been examined. This sequence is thought to be at least part of the promoter for RNA polymerase transcription. Thus, transcription in vitro of a cloned histone gene from which this AT-rich sequence has been deleted results in the appearance of aberrant sites of initiation as well as a decreasing frequency of initiation.

Other sequences have been identified that are necessary for the faithful initiation of transcription in vivo and may form part of a complex eukaryotic promoter.

REGULATION OF GENE EXPRESSION

Some enzymes are present in cells in large amounts and others in minute quantities, although each is represented equally in the genome. Even more dramatically, erythrocytes contain hemoglobin, and fibroblasts contain procollagen, yet all cells of an organism contain the same genetic information. Moreover, there are a number of instances of sequential synthesis of proteins, an example of which is hemoglobin (Chap. M4). Fetal hemoglobin (HbF = $\alpha_2\gamma_2$) is present in the fetus with little or no HbA ($\alpha_2\beta_2$), indicating that β chains are not fabricated at a significant rate. In the newborn, formation of γ chains is repressed and production of β chains is accelerated. It is of interest, therefore, to ascertain what genetic or other factors control gene expression.

Figure 29.14 Steps in gene expression at which regulation can occur.

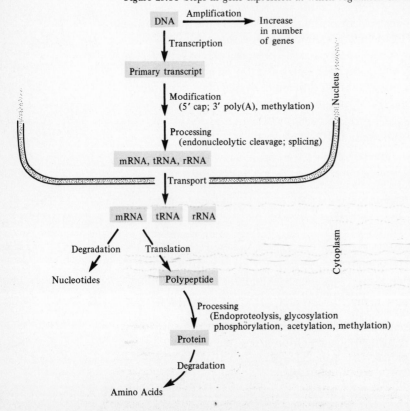

As is apparent from the foregoing discussion, gene expression may in principle be regulated at any of several levels. These include (1) gene amplification or rearrangement, (2) modulation of the frequency of initiation of transcription, (3) processing and transport of the primary transcript to form a functional mRNA, (4) modulation of frequency of translation, and (5) posttranslational processing and transport of the polypeptide product of translation (Fig. 29.14). We have already considered several aspects of the regulation of gene expression at the level of transcription and processing of RNA transcripts. We shall now consider examples of the regulation of gene expression at the other levels cited.

Gene Amplification or Rearrangement

The chorion membrane of *Drosophila* eggs is synthesized by ovarian follicle cells during a brief period of approximately 5 h. During this period, large quantities of the different chorion proteins are synthesized at different times, the synthesis of each protein resulting from the appearance of large quantities of its mRNA at the appropriate time. Just prior to the appearance of the mRNA, the number of chorion genes is amplified ten- to thirtyfold. In this instance, then, increased gene expression is a consequence of differential DNA replication leading in turn to an increase in the number of gene copies.

Regulation of gene expression by genomic rearrangements is exemplified by the rearrangements of the coding sequences for the variable region of immunoglobulin genes and their joining to the coding portion of the constant region. The details of this process are presented in Chap. M2.

Steroid Hormone Induction of Gene Expression

Many steroid hormones appear to act by a common mechanism in which the steroid passively diffuses through the plasma membrane and associates with a specific soluble receptor protein (Chap. M12). The various steroids, e.g., estrogens, androgens, progesterone, and the adrenal cortical steroids, recognize specific receptors which bind the appropriate steroids with very high affinities (K_d of 10^{-10} to $10^{-9} M$). This binding either stabilizes or induces an "activated" state of the steroid receptor, which acquires an increased affinity for binding sites in the chromatin of the nucleus so that the steroid-receptor complexes pass from the cytoplasm into the nucleus, where they bind the chromatin. In this manner the amount of mRNA from specific genes is either increased or decreased. The process is shown diagrammatically in Fig. 29.15.

Studies of the binding of steroid-receptor complexes to the nucleus demonstrate that it is a nonsaturable process and indicate that the number of binding sites in the nucleus far exceeds the total number of receptor molecules per cell (approximately 10^4). Moreover, these binding sites exhibit a relatively low affinity for the steroid-receptor complex with a K_d of approximately $10^{-4} M$.

Inasmuch as the primary effects of steroid hormones on gene expression are highly specific, e.g., a limited set of genes is specifically induced (or repressed), a comparably limited set of high-affinity chromatin binding sites is expected. This discrepancy is explained by the existence of two classes of binding sites: one class consisting of a large number of nonspecific low-affinity sites and the other class, of a small number of specific, high-affinity sites that regulate transcription

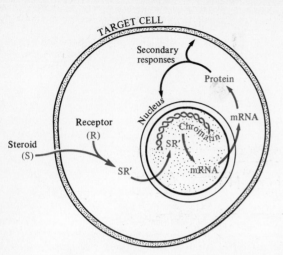

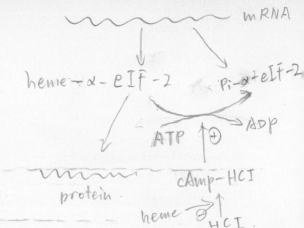

Figure 29.15 Model for the regulation of gene expression by steroid hormones. See text for details.

of a small set of genes. The nonspecific sites, because of their large numbers, mask the high-affinity sites.

The effect of steroids on the expression of several genes is to increase or decrease the amount of mRNA derived from the gene. Clearly, this effect could result either from a primary action by the steroid-receptor on expression of the gene at the level of transcription, processing, or transport of the mRNA, or it could represent a secondary action by the products of genes that exhibit a primary response. It appears that the induction of ovalbumin and conalbumin synthesis in chick oviducts by estrogens is a secondary response in that the induction of their mRNAs is abolished by inhibitors of protein synthesis (cycloheximide, puromycin, etc.). In contrast, the induction by glucocorticoids of the mRNAs specified by certain viruses, in particular the mouse mammary tumor virus, is insensitive to inhibitors of protein synthesis and may, therefore, be an example of a primary hormonal response.

Translational Control of Gene Expression

The regulation of gene expression at the level of translation is exemplified by control of the synthesis of the globin subunits of hemoglobin (Chap. M4). The translation of globin mRNAs in reticulocytes depends upon the presence of heme, the prosthetic group of hemoglobin. In the absence of hemin, an inhibitor of the initiation of translation is activated. This hemin-controlled inhibitor (HCI) is a cAMP-independent protein kinase that inhibits initiation by phosphorylation and hence inactivates the 38,000-dalton α subunit of the initiation protein eIF-2 (Chap. 27). In the presence of an "eIF-2-stimulating protein," eIF-2 forms a ternary complex with GTP and initiator methionyl tRNA, which binds to the 40-S ribosomal subunit to form the 40-S initiator complex. Thus, hemin binds to the regulatory subunit of the kinase and as a consequence eIF-2α is not phosphorylated and consequently not inactivated.

There are other mechanisms of translational control in eukaryotes that appear to act by phosphorylation of eIF-2α. Thus, double-stranded RNA and oxidized glutathione appear to activate HCI-like inhibitors of the initiation of

780

translation that are also associated with a cAMP-independent protein kinase that phosphorylates and thereby inactivates the α subunit of eIF-2α.

Treatment of viral cells with *interferon* induces an enzyme that in the presence of double-stranded RNA catalyzes the synthesis of a potent translational inhibitor, the novel trinucleotide, pppA2'p5'A2'p5'A (containing 2'5' rather than 3'5' phosphodiester linkages), which activates an endonuclease that specifically degrades mRNA. The mode of action of interferon is considered in Chap. 30.

Diphtheria toxin is also an inhibitor of translation. It does so by specifically blocking the elongation phase of protein synthesis. After the intact toxin enters the cell, it is hydrolyzed by a trypsin-like enzyme to yield a fragment that blocks the translocation step of polypeptide chain elongation, leaving the growing peptidyl chain in the A site. The active fragment of the toxin is an enzyme that catalyzes transfer of the ADP-ribose portion of NAD to the elongation factor, EF-2, in a reaction analogous to ADP-ribosylation of histones (page 766), thereby blocking the action of the EF-2. Transfer is to a novel amino acid, *diphthamide*, found in all EF-2s.

Diphthamide, 2-[3-carboxy-
amido-3-(trimethylammonio)propyl]histidine.

The ADP-ribosylation reaction is reversible, and the ADP-ribose can be removed from EF-2 by the addition of high concentrations of nicotinamide.

$$NAD + EF\text{-}2 \xrightleftharpoons[\text{toxin fragment}]{\text{diphtheria}} \text{adenosine diphosphoribosyl-EF-2} + \text{nicotinamide}$$

Posttranslational Transport of Proteins

Although protein synthesis occurs in the cytoplasm, many proteins are localized to the various subcellular membranes in a highly selective manner. Proteins indigenous to one membrane are rarely, if ever, found in another (Chap. 13). Furthermore, many cells are capable of transmembrane secretion of proteins. ①The synthesis of membrane and secreted proteins proceeds by the same mechanism as all other cellular proteins. They must, however, be specifically inserted into the appropriate membrane. ②There are at least two ways in which this can occur. One involves the spontaneous insertion into the lipid bilayer as a consequence of a hydrophobic surface or domain on the protein. An example of such spontaneous insertion is cytochrome b_5 and the cytochrome b_5 reductase that associates with the microsomal electron transport system (Chap. 16).

A second mechanism involves transmembrane insertion during protein synthesis—the process of *vectorial discharge*. In these instances, the NH$_2$-terminal portion of the protein, which is synthesized first, is found on the lumenal side of the endoplasmic reticulum (ER) or the extracellular side of the plasma membrane since these sides are topologically equivalent. Correspondingly, the COOH-terminal portion, synthesized last, resides on the cytoplasmic side of the membrane. Similarly, soluble proteins destined for secretion from the cells are also transferred across the ER membrane but, unlike membrane proteins, are completely discharged into the lumen of the ER.

Proteins that are inserted into membranes during their synthesis are necessarily synthesized by ribosomes bound to the membrane by the growing polypeptide chain. In fact, those ribosomes are also held in place by the specific binding to proteins located in the ER membrane. The ER is highly specialized for the vectorial discharge process and is studded with these membrane-bound ribosomes. Indeed, these ribosomes are densely clustered into specific regions on the ER surface termed the *rough ER*. Portions of ER lacking bound ribosomes are called the *smooth ER* (Chap. 13).

The membrane-bound ribosomes are not a special class of ribosomes but instead represent undifferentiated ribosomes directed to the ER membrane at each round of protein synthesis. This important point has been largely resolved by the *signal hypothesis* and evidence that supports it (Fig. 29.16).

The signal hypothesis holds that synthesis of secreted and vectorially discharged membrane proteins begins on ribosomes free in the cytosol. A sequence of 15 to 30 residues, the *signal sequence*, is present at the NH$_2$ terminus of all proteins destined for vectorial discharge. When this sequence emerges from the ribosome, it is bound by a specific receptor found only in the rough ER. This binding event results in the attachment of the free ribosome bearing the signal sequence to the rough ER, forming a membrane-bound ribosome and translation complex. In addition, the binding results in the formation of a transient pore through which the nascent peptide chain passes as polypeptide synthesis proceeds. The net result is the vectorial transfer of the protein across the rough ER membrane. The signal sequence is removed by a specific protease, the *signal peptidase*, before synthesis of the membrane or secreted protein is completed.

Many features of the signal hypothesis have been confirmed. All vectorially discharged proteins have signal sequences at the NH$_2$ termini. These se-

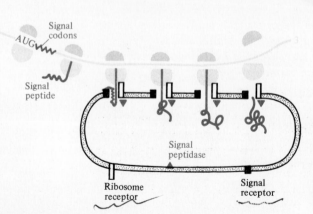

Signal codons

AUG

Signal peptide

Signal peptidase

Ribosome receptor

Signal receptor

Figure 29.16 Diagram illustrating the signal hypothesis. The various stages of the translation of a mRNA for a secreted protein on a membrane-bound polysome are indicated. [*From G. Blobel, P. Walter, C. N. Chang, B. M. Goldman, A. H. Erickson, and V. R. Lingappa, in C. R. Hopkins and C. V. Duncan (eds.), Secretory Mechanisms, Cambridge, London, 1979, p. 9.*]

quences are exceptionally hydrophobic and are removed during protein synthesis by a signal peptidase found only in the rough ER. Such sequences are absent from soluble cytosolic proteins. A specific receptor for the signal sequence, present in the ER membrane, has also been demonstrated. Finally, mutant proteins of *Escherichia coli* with an altered signal sequence are released into the cytosol instead of the membrane. Correspondingly, when the signal sequence is fused genetically to a soluble enzyme, the soluble enzyme is vectorially discharged.

Posttranslational Modification of Proteins

The genetic code is responsible for incorporation of the 20 amino acids generally present in proteins. As already noted, many of these residues are subsequently modified by phosphorylation, methylation, hydroxylation, oxidation, glycosylation, etc. These reactions occur during or after formation of the polypeptide chains. Prosthetic groups may also be covalently attached to specific residues.

Many polypeptide hormones possess a terminal α-carboxamide group, e.g., vasopressin, oxytocin, ACTH, and others (Chap. M10; Part M3). It has been demonstrated for several of these hormones that the precursor is synthesized with a carboxyl terminal sequence containing a neutral amino acid (amidated in the hormone) followed by an additional glycine residue. A pituitary enzyme has been shown to effect an oxidative reaction at the N—C bond of the glycine residue; this is followed by spontaneous hydrolysis to form the carboxamide and liberate glyoxylic acid, as shown below.

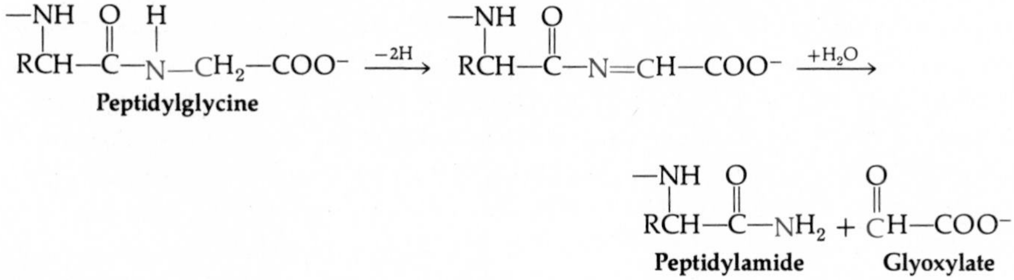

$$\underset{\text{Peptidylglycine}}{\overset{\displaystyle -\text{NH} \quad \text{O} \quad \text{H}}{\text{RCH}-\text{C}-\text{N}-\text{CH}_2-\text{COO}^-}} \xrightarrow{-2\text{H}} \overset{\displaystyle -\text{NH} \quad \text{O}}{\text{RCH}-\text{C}-\text{N}=\text{CH}-\text{COO}^-} \xrightarrow{+\text{H}_2\text{O}}$$

$$\underset{\text{Peptidylamide}}{\overset{\displaystyle -\text{NH} \quad \text{O}}{\text{RCH}-\text{C}-\text{NH}_2}} + \underset{\text{Glyoxylate}}{\overset{\displaystyle \text{O}}{\text{CH}-\text{COO}^-}}$$

Genomes of Subcellular Organelles

Mitochondria and chloroplasts both possess their own genetic complements which are in the form of closed circular duplex DNA molecules. Mitochondrial DNAs have molecular weights ranging from 1 to 5×10^7, depending upon the organism; the DNA of chloroplasts is larger (8 to 12×10^7).

Because of the limited coding capacity of the mitochondrial genome, most of the mitochondrial proteins are encoded by nuclear DNA; however, several are specified by mitochondrial DNA and are synthesized by a transcription-translation system specific to the mitochondrion. In the case of yeast these include three of the seven subunits of *cytochrome oxidase*, two of the subunits of the *oligomycin-sensitive ATPase*, as well as the *cytochrome b apoprotein* (Chap. 16). A similar separation of genes is observed for chloroplast proteins. Thus, the large subunit of *ribulose bisphosphate carboxylase* (Chap. 19) is coded for by the chloroplast DNA and synthesized within that organelle; the small subunit is encoded in the nuclear DNA and synthesized in the cytoplasm.

The arrangement of genetic information within the mitochondrial genome may differ markedly from that of nuclear genomes. Thus, the nucleotide sequence of the 16,569–base pair human mitochondrial genome shows an extraordinarily high degree of packing of genetic information with only a few and, in some instances, no noncoding nucleotides between genes. For example, the gene for subunit II of cytochrome oxidase is immediately contiguous with a tRNA^{Asp} gene. In contrast, in yeast mitochondria, the genes are widely spaced and in some cases possess intervening sequences. The departure of the mitochondrial genetic code from the universal genetic code employed by nuclear genomes has been noted previously (Chap. 28).

MOLECULAR GENETICS OF HEMOGLOBIN

More is presently known concerning the molecular genetics of hemoglobin than any other human protein. It can serve, therefore, to illustrate the molecular basis of genetic disorders in human beings (see also Chap. 31).

The genes coding for the α and β globin chains in human beings are situated on separate chromosomes. The α chain gene, which is duplicated in most human populations, is located on chromosome 16 and the β (as well as the γ and δ) chain gene is on chromosome 11. The nucleotide sequences of the human α, β, and γ mRNAs are known and, as described previously, (page 776) contain 5' caps, 3' poly(A) tails and untranslated sequences at the 5' and 3' ends. The β, δ, and γ genes contain intervening sequences; none has been identified in the human α gene, although such sequences are present in the mouse α gene. The nucleotide sequence of human β globin mRNA is shown in Fig. 29.17.

Genetic disorders of hemoglobin structure and synthesis fall into three categories: (1) Structural hemoglobin variants leading to abnormal forms of hemoglobin. These result from missense mutations, e.g., single-base substitutions leading to amino acid substitutions, deletions, or insertions of one or more bases, changes in reading frame, and premature chain terminations. The physiological consequences of the abnormal forms of hemoglobin resulting from the mutations are discussed in Chap. M4. (2) The *thalassemias,* which are characterized by a reduced rate of synthesis or absence of one of the globin chains, resulting in an imbalanced globin chain synthesis. (3) Hereditary persistence of fetal hemoglobin, in which the normal switch from fetal to adult hemoglobin synthesis does not occur.

Missense Mutations

Most abnormal hemoglobins are the result of missense mutations in the α or β globin structural genes. More than 100 different single amino acid substitutions have thus far been identified in the β chain alone. Some of the abnormal hemoglobins for which these substitutions have been established are described in Chap. M4.

(5') m⁷GpppA-C-A-U-U-U-G-C-U-U-U-C-U-G-A-C-A-C-A-A-C-U-G-U-G-U-U-C-A-C-U-A-G-C-A-A-C-C-U-C-A-A-A-C-A-G-A-C-A-C-C-
   1   10   20   30   40   50

Met	Val¹	His	Leu	Thr	Pro	Glu	Glu	Lys	Ser	Ala	Val	Thr	Ala	Leu	Trp	Gly
A-U-G	G-U-G	C-A-C	C-U-G	A-C-U	C-C-U	G-A-G	G-A-G	A-A-G	U-C-N	G-C-N	G-U-U	A-C-U	G-C-N	N-U-N	U-G-G	G-G-C

         60   70   80   90   100

Lys	Val	Asn	Val²⁰	Asp	Glu	Val	Gly	Gly	Glu	Ala	Leu	Gly	Arg³⁰	Leu	Leu	Val
A-A-G	G-U-G	A-A-C	G-U-G	G-A-U	G-A-A	G-U-U	G-G-U	G-G-U	G-A-G	G-C-C	C-U-G	G-G-C	A-G-G	C-U-G	C-U-G	G-U-G

        110   120   130   140   150

Val	Tyr	Pro	Trp	Thr	Gln	Arg⁴⁰	Phe	Phe	Glu	Ser	Phe	Gly	Asp	Leu	Ser	Thr
G-U-C	U-A-C	C-C-U	U-G-G	A-C-C	C-A-G	A-G-G	U-U-C	U-U-U	G-A-G	U-C-C	U-U-U	G-G-G	G-A-U	C-U-G	U-C-C	A-C-U

        160   170   180   190   200

Pro	Asp	Ala	Val	Met	Gly	Asn	Pro	Lys	Val⁵⁰	Lys	Ala	His	Gly	Lys	Lys	Val
C-C-U	G-A-U	G-C-A	G-U-U	A-U-G	G-G-C	A-A-C	C-C-U	A-A-G	G-U-G	A-A-G	G-C-U	C-A-U	G-G-C	A-A-G	A-A-A	G-U-G

        210   220   230   240   250

Leu	Gly	Ala⁷⁰	Phe	Ser	Asp	Gly	Leu	Ala	His	Leu	Asp	Asn⁸⁰	Leu	Lys	Gly	Thr
C-U-C	G-G-U	G-C-C	U-U-U	A-G-U	G-A-U	G-G-C	C-U-G	G-C-U	C-A-C	C-U-G	G-A-C	A-A-C	C-U-C	A-A-G	G-G-C	A-C-C

        260   270   280   290   300

Phe	Ala	Thr	Leu	Ser	Glu⁹⁰	Leu	His	Cys	Asp	Lys	Leu	His	Val	Asp	Pro¹⁰⁰	Glu
U-U-U	G-C-C	A-C-A	C-U-G	A-G-U	G-A-G	C-U-G	C-A-C	U-G-U	G-A-C	A-A-G	C-U-G	C-A-C	G-U-G	G-A-U	C-C-U	G-A-G

        310   320   330   340   350

Asn	Phe	Arg	Leu	Leu	Gly	Asn	Val	Leu¹¹⁰	Val	Cys	Val	Leu	Ala	His	His	Phe
A-A-C	U-U-C	A-G-G	C-U-C	C-U-G	G-G-C	A-A-C	G-U-G	C-U-G	G-U-C	U-G-U	G-U-G	C-U-G	G-C-C	C-A-U	C-A-C	U-U-U

        360   370   380   390   400

Gly	Lys¹²⁰	Glu	Phe	Thr	Pro	Pro	Val	Gln	Ala	Ala	Tyr¹³⁰	Gln	Lys	Val	Val	Ala
G-G-C	A-A-A	G-A-A	U-U-C	A-C-C	C-C-A	C-C-A	G-U-G	C-A-G	G-C-U	G-C-C	U-A-U	C-A-G	A-A-A	G-U-G	G-U-G	G-C-U

        410   420   430   440   450

Gly	Val	Ala	Asn	Ala¹⁴⁰	Leu	Ala	His	Lys	Tyr	His¹⁴⁶	Term
G-G-U	G-U-G	G-C-U	A-A-U	G-C-C	C-U-G	G-C-C	C-A-C	A-A-G	U-A-U	C-A-C	U-A-A	G-C-U-C-G-C-U-U-U-C-U-U-G-C-U-

        460   470   480   490   500

C-U-C-C-A-A-U-U-U-C-U-A-U-U-A-A-A-C-C-U-U-C-C-U-U-U-G-U-U-C-C-U-A-A-G-U-C-C-A-A-C-U-A-C-U-A-A-A-C-U-
510   520   530   540   550   560

G-G-G-G-A-U-A-U-U-A-U-G-A-A-G-G-G-C-C-U-U-G-A-G-C-A-U-C-U-G-G-A-U-U-C-U-G-C-C-U-A-A-U-A-A-A-A-A-A-C-
570   580   590   600   610

A-U-U-U-A-U-U-U-U-C-A-U-U-G-C-poly(A)
620

Figure 29.17 mRNA and amino acid sequence of human β globin. The numbering above the sequence pertains to the amino acid sequence of the β chain of globin; the numbers below the sequence refer to the complete processed mRNA.

Chain Termination Mutants

Hemoglobin McKees Rocks is an example of a chain termination mutation in the β globin gene leading to a truncated β chain. Codon 145 (see Fig. 29.17) is changed from UAU, which specifies tyrosine, to UAA which is a terminator codon, resulting in a β chain lacking the COOH-terminal two residues. These relationships are illustrated in Table 29.1.

Examples of elongated α globin chains that result from a failure of polypeptide chain termination are hemoglobins Constant Spring, Icaria, and Koya Dora. Hb$_{Con Spr}$, which is carried by approximately 1 person in 20 in Thailand, contains 31 residues beyond the carboxyl-terminal arginine of the normal α chain. The first 141 residues are identical with the normal α chain. Normal

TABLE 29.1

Hemoglobin McKees Rocks, a Truncated β Chain

	Codon number				
	143	144	145	146	147
β_A amino acid	His	Lys	Tyr	His	Term
β_A codon	CAC	AAG	UAU	CAC	UAA
$\beta_{\text{McKees Rocks}}$ codon	CAC	AAG	UA^A_G		
$\beta_{\text{McKees Rocks}}$ amino acid	His	Lys	Term		

α-chain mRNA contains the chain-terminating codon UAA at codon 142, but glutamine is located at residue 142 in $Hb_{\text{Con Spr}}$. Thus, $Hb_{\text{Con Spr}}$ can be plausibly accounted for by a C for U substitution in the UAA codon giving rise to glutamine (CAA) at residue 142, resulting, in turn, in the translation of a segment of mRNA not normally translated.

Hb_{Icaria} and $Hb_{\text{Koya Dora}}$ are similar to $Hb_{\text{Con Spr}}$ except for amino acid 142, which is lysine in Hb_{Icaria} and serine in $Hb_{\text{Koya Dora}}$. Codons for these amino acids like that for glutamine differ from UAA by a single base. Table 29.2 demonstrates these relationships.

Hb_{Wayne} has an elongated α chain containing a sequence of eight amino acids, residues 139 to 146, that is unique and has not been found in normal α, $\alpha_{\text{Con Spr}}$, α_{Icaria}, or $\alpha_{\text{Koya Dora}}$. The amino acid sequence and length of Hb_{Wayne} are explained by deletion of a U in codon 138 and the consequent shift in reading frame. Codon 147 in the shifted frame is UAG, which is a chain terminator (Table 29.3).

Thus the carboxyl-terminal elongations of α and β globins can result from a failure of normal chain termination produced by either a base substitution in the normal UAA termination codon or by a deletion before the termination codon, resulting in a shift in the reading frame. Translation then continues until the next terminator codon is reached.

Thalassemias

Normally the rates of synthesis of the α and β chains of hemoglobin must be virtually identical. The *thalassemias* are characterized by a reduced rate of synthesis of one or more of the globin chains. As a consequence of the imbalance in globin chain production, there is precipitation of the excess chains,

TABLE 29.2 Hemoglobins Constant Springs and Koya Dora, Elongated α Chains

	Residue number													
	1..	137	138	139	140	141	142	143	144	145	146	147	148...	172
α_A	Val	Thr	Ser	Lys	Tyr	Arg								
$\alpha_{\text{Con Spr}}$	Val	.	.	.	.	Arg	Gln	Ala	Gly	Ala	Ser	Val	Ala...	Glu
α_{Icaria}	Val	.	.	.	.	Arg	Lys	.	.	.	.	.	....	.
$\alpha_{\text{Koya Dora}}$	Val	.	.	.	.	Arg	Ser	.	.	.	.	.	....	.
α_{Wayne}	Val	.	.	Asn	Thr	Val	Lys	Leu	Glu	Pro	Arg			
Codon		GUG	UCC	AAA	UAC	CGU	UAA							

TABLE 29.3 Hemoglobin Wayne, an Elongated β Chain

	Codon number											
	137	138	139	140	141	142	143	144	145	146	147	148
α_A amino acid	Thr	Ser	Lys	Tyr	Arg	Term						
α_A codon	ACC	UC(C)	AAA	UAC	CGU	UAA	GCU	GGA	GCC	UCG	GUA	GCA
α_{Wayne} codon	ACC	UCA	AAU	ACC	GUU	AAG	CUG	GAG	CCU	CGG	UAG	
α_{Wayne} amino acid	Thr	Ser	Asn	Thr	Val	Lys	Leu	Glu	Pro	Arg	Term	

Note: (C)-deleted in α Wayne.

lowered hemoglobin levels, and reduced red cell survival. In α *thalassemia* there is a deficiency in α chains, and hence β chains precipitate; in β *thalassemia* the reverse situation is found. In α^0 or β^0 *thalassemias* none of the corresponding chains is produced, while in α^+ and β^+ thalassemia the α and β chains are synthesized in reduced amounts.

α^0 Thalassemia results from a deletion of the α-globin gene. Homozygotes with α^0 thalassemia exhibit a syndrome known as *hydrops fetalis*. These individuals do not synthesize α-globin chains and usually die before birth or soon thereafter. mRNA for α-globin is undetectable in hydrops cells as judged by the inability of RNA extracted from these cells to direct the synthesis of α-globin chains in vitro in the presence of ribosomes and translation factors, although the same RNA can direct synthesis of β-globin chains.

Complementary DNA (cDNA) to α-globin mRNA can be synthesized by the viral reverse transcriptase (Chap. 27) using globin mRNA from normal human cells as template. α-Globin cDNA can then serve as a probe in hybridization tests (page 759) for α-globin gene sequences in cellular DNA. No hybridization can be detected between α-globin cDNA and DNA from hydrops cells, although β-globin cDNA can form hybrids in normal amounts with hydrops DNA. Thus, the α-globin gene is largely or entirely deleted in individuals with α^0 thalassemia. There are two adjacent genes coding for α globin, and both are deleted in homozygotes with α^0 thalassemia. A milder form of thalassemia results when only one of the two α genes is deleted.

In contrast to the α^0 thalassemias the β^0 thalassemias are heterogeneous. They may be of three kinds: (1) the β-globin gene is deleted in a manner analogous to the α^0 thalassemia; (2) the β-globin gene remains intact, but no β-globin mRNA is detectable; and (3) the β-globin gene remains intact and β-globin mRNA is synthesized but is not translated. This may be a consequence of a defect in the mRNA, which does not permit it to be translated, or alternatively, which greatly reduces its stability. In one such instance the β-thalassemia defect is associated with a nucleotide change at a splice junction, thereby preventing normal processing of the primary β-globin transcript.

Hereditary Persistence of Fetal Hemoglobin

This defect is a consequence of a deletion of both the β- and δ-globin genes with the consequent persistence of HbF ($\alpha_2\gamma_2$). The continued synthesis of the γ chain of globin is dependent upon the extent of the β and δ deletion. It also results from a fusion of the γ- to the β-globin genes following genetic recombination. The resulting *hemoglobin Kenya* consists of an NH_2-terminal portion derived from the γ gene and the COOH-terminal portion from the gene for β globin. Synthesis of hemoglobin Kenya again persists and the switch to the adult forms of hemoglobins is either diminished or totally suppressed.

THE CELL CYCLE

The foregoing descriptions of the known mechanisms of DNA replication, transcription, and translation do not imply that all these processes occur concurrently or that all types of cells have a similar life history with respect to these processes. Indeed, studies of different types of differentiated cells of eukaryotes reveal great variations in their life cycles.

Mouse hepatoma cells, grown in culture, can be induced to undergo synchronous divisions. Such cells manifest at least four distinct stages: (1) a period of *pre-DNA synthesis* of 8 to 10 h, during which there is synthesis of RNA and protein; (2) a period of *active DNA synthesis,* also of 8- to 10-hours' duration, during which there is also concurrent formation of RNA and protein; (3) a phase of *post-DNA synthesis* of 3 to 4 h, in which RNA and protein synthesis continues; and (4) *mitosis,* lasting only about 1 h, in which time the chromosomes contract, line up along the spindle fibers, and separate. During mitosis there is little or no detectable synthesis of DNA, RNA, or protein, and polyribosomes cannot be observed. After cell division, this cycle, occupying approximately 18 to 24 h, is then repeated.

An example of an extended period during which there is no DNA synthesis or mitosis is in the development of the large oocytes of Amphibia before the first meiotic division. These cells remain at a stage of development that may last for months when the chromosomes are present as extended fibers of DNA. At many sites along each thread of DNA there are long, extended RNA chains being synthesized by RNA polymerase. These represent predominantly mRNA chains that are being made in preparation for their functional role after fertilization has occurred.

A few additional examples of special types of life cycles of differentiated cells may be mentioned. Mammalian nerve cells of the adult animal do not undergo cell division and do not synthesize DNA; yet such cells continue actively to synthesize RNA and protein and survive for the lifetime of the individual (Chap. M7). Similarly, mature lymphoid cells do not divide, despite the presence of a nucleus, but they continue to synthesize specific antibody protein and function in cell-mediated immunological responses (Chap. M2). In contrast, enucleated mammalian erythrocytes lack DNA and the ability to synthesize proteins. Even with limited metabolic capability, these cells survive for weeks or months, depending on the species (Chap. M3).

REFERENCES

Books

Bunn, H. F., B. G. Forget, and H. M. Ranney: *Human Hemoglobins,* Saunders, Philadelphia, 1977.

Chromatin, *Cold Spring Harbor Symp. Quant. Biol.*, vol. 42, 1977.

Cohn, W. E.: *Progress in Nucleic Acid Research and Molecular Biology,* vols. 1–26, Academic, New York, 1963– (a continuing series).

Davidson, E. H.: *Gene Activity in Early Development,* 2d ed., Academic, New York, 1976.

Hood, L., J. Wilson, and W. B. Wood: *Molecular Biology of Eukaryotic Cells,* Benjamin, New York, 1975.

Lewin, B.: *Gene Expression 2. Eucaryotic Chromosomes,* 2d ed., Wiley, New York, 1980.

Losik, R., and M. Chamberlin (eds.): *RNA Polymerase,* Cold Spring Harbor Laboratory, Cold Spring Harbor, N.Y., 1976.

Nomura, M., A. Tissières, and P. Lengyel (eds.): *Ribosomes,* Cold Spring Harbor Laboratory, Cold Spring Harbor, N.Y., 1974.

Transcription of Genetic Material, *Cold Spring Harbor Symp. Quant. Biol.,* vol. 35, 1970.

Watson, J. B.: *Molecular Biology of the Gene,* 3d ed., Benjamin, New York, 1976.

Weissbach, H., and S. Pestka (eds.): *Molecular Mechanisms of Protein Biosynthesis,* Academic, New York, 1977.

Review Articles

Abelson, J.: RNA Processing and the Intervening Sequence Problem, *Annu. Rev. Biochem.* **48:**1035–1069 (1979).

Altman, S.: Biosynthesis of Transfer RNA in *Escherichia coli, Cell* **4:**21–29 (1975).

Berg, P.: Dissections and Reconstructions of Genes and Chromosomes, *Science* **213:**296–303 (1981).

Bertrand, K., L. Korn, F. Lee, T. Platt, C. L. Squires, C. Squires, and C. Yanofsky: New Features of the Regulation of the Tryptophan Operon, *Science* **189:**22-25 (1975).

Breathnach, R., and P. Chambon: Organization and Expression of Eukaryotic Split Genes Coding for Proteins, *Annu. Rev. Biochem.* **50:**349–383 (1981).

Brimacombe, R., G. Stoeffler, and H. G. Wittmann: Ribosome Structure, *Annu. Rev. Biochem.* **47:**217–249 (1978).

Davidson, E. H., and R. J. Britten: Regulation of Gene Expression: Possible Role of Repetitive Sequences, *Science* **204:**1052–1059 (1979).

Davis, B. D., and P.-C. Tai: The Mechanism of Protein Secretion Across Membranes, *Nature (London)* **283:**433–438 (1980).

Dickson, R. C., J. Abelson, W. M. Barnes, and W. S. Reznikoff: Gene Regulation: The Lac Control Region, *Science* **187:**27–35 (1975).

Ellis, R. J.: Chloroplast Proteins. Synthesis, Transport and Assembly, *Annu. Rev. Plant Physiol.* **32:**111–137 (1981).

Gorski, J., and F. Gannon: Current Models of Steroid Action: A Critique, *Annu. Rev. Physiol.* **38:**425–450 (1976).

Grunberg-Manago, M., R. H. Buckingham, B. S. Cooperman, and J. W. B. Hershey: Structure and Function of the Translational Machinery, *Symp. Soc. Gen. Microbiol.* **28:**27–110 (1978).

Hayaishi, O., and K. Ueda: Poly(ADP-Ribose) and ADP-Ribosylation of Proteins, *Annu. Rev. Biochem.* **46:**95–116 (1977).

Kornberg, R. D.: Structure of Chromatin, *Annu. Rev. Biochem.* **46:**931–954 (1977).

Kreil, G.: Transfer of Proteins Across Membranes, *Annu. Rev. Biochem.* **50:**317–348 (1981).

Long, E. O., and I. B. Dawid: Repeated Genes in Eukaryotes, *Annu. Rev. Biochem.* **49:**727–764 (1980).

McGhee, J. D., and G. Felsenfeld: Nucleosome Structure, *Annu. Rev. Biochem.* **49:**1115–1156 (1980).

Ochoa, S., and C. DeHaro: Regulation of Protein Synthesis in Eukaryotes, *Annu. Rev. Biochem.* **48:**549–580 (1979).

Revel, M., and Y. Groner: Post-Transcriptional and Translational Controls of Gene Expression in Eukaryotes, *Annu. Rev. Biochem.* **47:**1079–1126 (1979).

Rich, A., and U. L. Raj Bhandary: Transfer RNA: Molecular Structure, Sequence and Properties, *Annu. Rev. Biochem.* **45:**805–860 (1976).

Schimmel, P. R., and D. Söll: Aminoacyl-tRNA synthetases: General Features and Recognition of Transfer RNAs, *Annu. Rev. Biochem.* **48:**601–648 (1979).

Shatkin, A. J.: Capping of Eukaryotic mRNAs, *Cell* **9:**645–653 (1976).

Various authors, pp. 1–374, in P. D. Boyer (ed.): *The Enzymes,* 3d ed., vol. X, Academic, New York, 1974; and vols. XIV and XV, 1982.

Weatherall, D. J., and J. B. Clegg: Recent Developments in the Molecular Genetics of Human Hemoglobin, *Cell* **16:**467–469 (1979).

Wool, I. G.: The Structure and Function of Eukaryotic Ribosomes, *Annu. Rev. Biochem.* **48:**719–754 (1979).

Yamamoto, K. R., and B. M. Alberts: Steroid Receptors: Elements for Modulation of Eukaryotic Transcription, *Annu. Rev. Biochem.* **45:**721–746 (1976).

Yanofsky, C.: Attenuation in the Control of Expression of Bacterial Operons, *Nature* **289:**751 (1981).

Chapter 30

Viruses

Viruses were first identified as causative agents of infectious diseases in plants and animals at the end of the nineteenth century. Because of their agricultural and medical importance, they were studied initially by botanists and pathologists. Viruses became of interest to biochemists when it became apparent that because of their relatively simple genomes, they could serve as model systems for the analysis of nucleic acid replication and gene expression (Chaps. 27, 28, and 29). Because the simplest viruses are composed only of nucleic acid and protein, they have also served as models for the study of the assembly of subcellular organelles.

Despite their great diversity, viruses do display unique features that clearly set them apart from other microorganisms: (1) They possess only a single form of nucleic acid, either RNA or DNA. In fact, some are unique in utilizing RNA rather than DNA as their genetic material. (2) They lack ribosomes or other subcellular organelles and are devoid of systems for the generation of energy. Viruses must therefore rely upon the machinery of the cell they infect for the supply of precursor molecules and enzymes for replication and propagation.

In the free state, viruses consist of a nucleic acid component, surrounded by a protein coat. Some viruses are further enveloped by a membrane layer consisting of glycoproteins and lipid. Upon gaining access to the interior of the host cell, the viral nucleic acid may replicate autonomously or may be integrated into the host genome and replicate in concert with the host.

VIRUS MULTIPLICATION CYCLE

Although many viruses with widely different structures are known, their multiplication cycles are very similar, consisting of the following sequence of

Any cross-references coded M refer to the companion text, *Principles of Biochemistry: Mammalian Biochemistry*.

events: (1) The virus particle, referred to as a *virion*, attaches to the surface of the host cell; the cell membrane is then penetrated either by the intact virus, viz., the nucleic acid in its protein coat (many animal and plant viruses), or by the viral nucleic acid alone (bacterial viruses). (2) Viral proteins are synthesized at the direction of the infecting viral nucleic acid and the viral nucleic acid is replicated. The mode of synthesis of the mRNA and the specific pathway of replication depend upon the nature and the quantity of nucleic acid in the virion. (3) New virions are assembled from newly synthesized protein and nucleic acid. (4) Progeny virions are released from the cell either by lysis of the cell or by passage through the cell membrane without cell lysis.

Viruses invariably attach to specific macromolecular groups (*receptors*) located on the surface of their host cells. The ability of a virus to infect a particular cell type is, in fact, often expressed at this stage. Thus, the *T-even bacteriophages* (page 793) attach to specific portions of the lipopolysaccharide of the cell wall of *Escherichia coli,* and *polio virus* multiplies exclusively in selected human and simian cells because only these cells have a specific lipoprotein receptor in the cell membrane.

Both DNA and RNA, in either their single- or double-stranded forms, can serve as the genetic material for the different viruses (Table 30.1). Depending on the nature of the viral nucleic acid, viral mRNA is synthesized according to one of the several routes shown in Fig. 30.1; the plus sign refers to the mRNA sequence and the minus sign to its complement.

Bacteriophages T4, T7, and λ; adenovirus; simian virus 40 (SV40); vaccinia; herpes virus; and hepatitis virus all contain double-stranded DNA, and (+)mRNA is synthesized using sections of both DNA strands as template (Chap. 28). In the case of bacteriophage ϕX174, which contains a single-stranded circular DNA molecule, a complementary copy of the single strand is synthesized, and the resulting duplex is used to generate (+)mRNA. In bacteriophages Qβ and R17 and poliovirus, which contain single-stranded RNA, a complementary copy of the RNA [(−)RNA] is synthesized, which then serves as

TABLE 30.1

Nucleic Acid Composition of Some Viruses	Nucleic acid in virion	Virus	Genome size, kilobases (kb)
	RNA, single-stranded	Bacteriophage Qβ	3
		Polio	7.5
		Rous sarcoma	10
		Rabies	16
		Vesicular stomatitis	16
	RNA, double-stranded	Reovirus	23
	DNA, single-stranded	Minute virus of mice	5
		Bacteriophage ϕX174	5.4
	DNA, double-stranded	Simian virus 40 (SV40)	5.2
		Papilloma (wart)	8
		Adenovirus	36
		Bacteriophage λ	46
		Bacteriophage T4	120
		Herpes	156
		Pox viruses	240

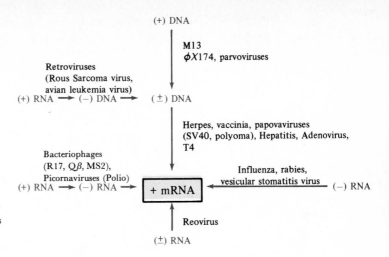

Figure 30.1 Patterns of mRNA synthesis following infection with various viruses. [*Adapted from D. Baltimore, Bacteriol. Rev.* **35**:235 (1971).]

a template for the synthesis of (+)mRNA The RNA-containing tumor virus that produces the *Rous sarcoma* generates a single-stranded DNA copy of the viral RNA (Chap. 28), which in turn generates a double-stranded DNA. The latter then serves as a template for the formation of (+)mRNA. *Reovirus,* whose genome consists of double-stranded RNA, generates a (+)mRNA copy. Finally, the RNA-containing viruses, *influenza, mumps, measles, rabies,* and *vesicular stomatitis virus,* generate a (+)mRNA at the direction of the complementary (−)RNA present within the virion.

A general rule that has emerged from the examination of a large number of viruses is that whenever the route of mRNA synthesis requires a polymerase that is not present in the host cell, that enzyme is carried in the virion itself and enters the host cell together with the viral nucleic acid.

Having described the general features of viral multiplication and mRNA synthesis, we shall now consider the replication of several of the major classes of viruses cited above.

DOUBLE-STRANDED DNA VIRUSES

Bacteriophage T4

Bacteriophage T4 is one of the "T phages" (T1 to T7) that infect *E. coli.* Because of its short replication cycle and the ease of cultivation of its host, T4 has been one of the most thoroughly studied of all viruses. Like other bacteriophages, T4 is composed of only a protein coat and DNA. The virus (Fig. 30.2) has a relatively complex morphology, consisting of an icosahedral head, into which the DNA is tightly packed, and a tail composed of a cylindrical sheath and a hollow inner core; the tail is attached to the head by a short collar. At the base of the tail is an end plate, to which are fastened spikes, or pins, and a group of six thin fibers. Penetration of the viral nucleic acid into the host bacterium occurs in the following sequence. The phage attaches to the bacterium via the tail fibers, and the tips of the tail pins are fixed to the cell wall. The complex series of steps that ensues results in the contraction of the sheath, coupled to the

793

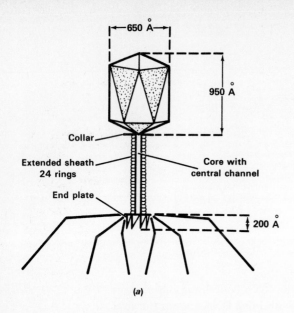

(a)

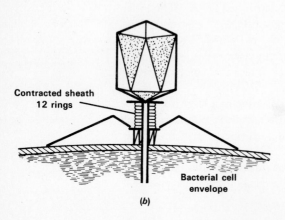

(b)

Figure 30.2 Diagrammatic representation of bacterio-phage T4 showing (a) its component parts and (b) the mechanism of penetration through the bacterial cell wall. [*From L. D. Simon and T. F. Anderson, J. Virol. 32:279 (1967).*]

hydrolysis of ATP to ADP and P_i, and penetration of the cell wall by the core. The DNA is then injected from the head of the virus through the core into the cell.

The DNA of phage T4 is a linear duplex (MW = 120×10^6) containing 1.66×10^5 nucleotide pairs. It is unusual in that cytosine is completely replaced by glucosylated hydroxymethylcytosine (Table 27.3); the latter serves to protect the viral DNA from deoxyribonucleases induced upon infection. The enzymic mechanisms that permit the substitution of glucosylated hydroxymethylcytosine for cytosine have been described (Chap. 27). Another unusual feature of T4 is that in a population of T4 DNA molecules, the sequences are circularly permuted with respect to one another. That is, different molecules begin and end at different points in the sequence. This property accounts for the circularity of the genetic linkage map for this phage, whose DNA is linear (Fig. 30.3). In addition to its circular permutation, each T4 DNA molecule is terminally redundant; i.e., the sequence at one end is repeated at the other.

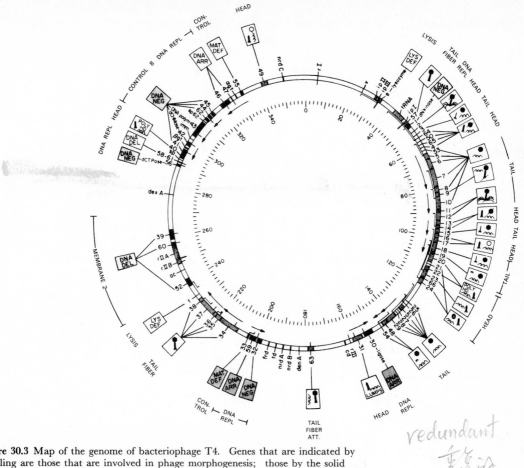

Figure 30.3 Map of the genome of bacteriophage T4. Genes that are indicated by stippling are those that are involved in phage morphogenesis; those by the solid black areas represent genes involved in nucleic acid metabolism. [*From W. B. Wood, in R. C. King (ed), Handbook of Genetics, Plenum, New York, 1973, p. 328.*]

Thus, each molecule contains somewhat more than one genome equivalent of DNA. The extent of the redundancy is about 2 percent of the T4 genome.

A genetic map of T4 is given in Fig. 30.3. Thus far, 135 genes have been identified, which account for approximately 90 percent of the coding capacity of the T4 genome. These genes can be divided into two major classes. One specifies the enzymes required for the degradation of host DNA and the rapid synthesis of T4 DNA, including *deoxyribonucleases, deoxycytidine triphosphatase, deoxynucleotide kinase, deoxycytidylate hydroxymethylase, glucosyl transferases, DNA polymerase, DNA ligase, DNA topoisomerases,* and *single-stranded DNA binding protein.* The second class specifies the proteins required for assembly of the phage particle. It is noteworthy that the various genes, with only a few exceptions, are clustered according to function (Fig. 30.3). In contrast to the genes involved in lactose and tryptophan metabolism in *E. coli* (Chap. 28), T4 gene clusters do not form an operon and their expression is not coordinated controlled by, for example, formation of a polycistronic mRNA.

Multiplication of a T4 phage particle is complete within 20 min at 30°C.

After adsorption of the phage and penetration of the viral DNA, host macromolecular synthesis ceases abruptly and, at approximately 5 min, transcription of certain of the phage genes begins. The transcripts are complementary to only one of the two strands of the infecting T4 DNA molecule and specify a set of *early* enzymes required for replication of the T4 genome, e.g., DNA polymerase and single-stranded DNA binding protein. Shortly after viral DNA synthesis begins, at 7 min postinfection, a new class of transcripts is generated. These postreplicative, or *late*, mRNAs are complementary to the strand opposite that from which the early mRNAs are synthesized and code primarily for the structural proteins of the phage. Synthesis of early mRNAs diminishes in rate or ceases with the onset of late transcription. However, a third class of mRNAs, termed *quasi-late*, which are synthesized at a low rate during the early interval, are generated at a greatly elevated rate during the late period. These quasi-late mRNAs specify several of the enzymes needed for DNA replication, e.g., *deoxyribonucleotide kinase*, as well as components required for assembly of the phage particle.

The mechanisms that regulate the switch from transcription of early to late and quasi-late mRNAs appear to involve modifications of the host RNA polymerase. Thus, ADP-ribosylation of a specific arginine residue of the α-subunit of the RNA polymerase results in loss of the ability of the enzyme to recognize host promoters and hence host-specific transcription ceases. Interaction of the RNA polymerase with the products of genes 33 and 55 produces a shut-off of early enzyme synthesis and the onset of transcription of late genes. The latter requires some synthesis of T4 DNA, which appears to be a consequence of the fact that only DNA with single-strand breaks can serve as a suitable template for late gene transcription. Discontinuous T4 DNA replication presumably provides the requisite single-strand breaks.

The synthesis in vitro of T4 DNA by a set of T4 early enzymes has been accomplished and has clarified several aspects of the process in vivo. Four enzymes, the products of genes 41, 44, 45, and 62, in addition to the T4-specific DNA polymerase, the product of gene 43, and the T4 single-stranded DNA binding protein, the product of gene 32, are required. The gene-44 and -62 proteins form a tight complex that displays DNA-dependent ATPase activity. In the presence of the gene-45 protein the gene 44-62 complex increases the rate of polymerization by the DNA polymerase to rates (800 nucleotides/s) that approach those that occur in vivo (Fig. 30.4). The gene-41 protein appears to be a T4-specific *primase* (Chap. 27) that catalyzes the initiation of DNA chains on a single-stranded DNA template coated with the gene-32 protein, in a manner analogous to the RNA priming of "Okazaki fragments" on the "lagging side" of the replication fork that occurs during replication of the *E. coli* chromo-

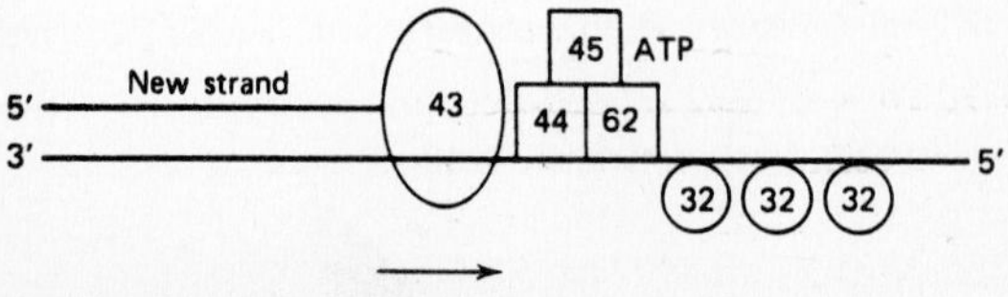

Figure 30.4 Postulated arrangement of T4 replication proteins at the replication fork. The product of gene 43 is the DNA polymerase; the gene 44-62 complex is a DNA-dependent ATPase, the gene 41 product is the RNA primase, and the gene 32 product is a single-stranded DNA binding protein. [*Adapted from B. Alberts, C. F. Morris, D. Mace, N. Sinha, M. Bittner, and L. Moran, in M. Goulian and P. Hanawalt (eds.), DNA Synthesis and its Regulation, Benjamin, Menlo Park, Calif., p. 250 1975.*]

some (see Fig. 27.15). Extensive replication on single-stranded DNA templates generates typical rolling circle forms (Fig. 27.5) in which discontinuous DNA synthesis occurs on long unbranched double-stranded tails.

With the development of a pool of new T4 DNA molecules, assembly of the mature T4 virion begins. The latter contains 36 different structural proteins whose correct assembly requires the participation of 57 genes. Because it can occur with relatively high efficiency in vitro, T4 assembly has served as a model for the morphogenesis of complex supramolecular structures, e.g., subcellular organelles. The assembly process, deduced from the electron microscopy of intermediates and in vitro assembly studies, occurs in a defined sequence of steps determined by interactions of the gene products, viz., the proteins. The pathway consists of three parts leading independently to the tail, head, and tail fibers, which are then combined as shown in Fig. 30.5. Of the 57 genes involved in the process, 7 appear to specify nonstructural accessory proteins that in some manner promote or direct assembly. One of these, the product of gene 22, catalyzes the proteolytic cleavage of a 55,000-dalton precursor of the major capsid protein (MW = 45,000) in the virion.

Once the mature virus is assembled, lysis of the host cell follows with the liberation of several hundred phage particles, a process promoted in part by a phage-induced *endolysin*.

Simian Virus 40

Simian virus 40 (SV40) is one of the *papovaviruses,* other members of which are *polyoma* and *papilloma*. SV40 contains only protein (88 percent) and DNA (12 percent). The DNA of the virion consists of a single closed circular duplex of 5243 base pairs (MW = 3.6×10^6); its complete nucleotide sequence is known. Like other circular DNA molecules, it exists in a supercoiled configuration (Chap. 8). The DNA of SV40 is surrounded by a coat composed of three proteins, VP1, VP2, and VP3; each encoded by the virus. In addition, the viral DNA is associated with cellular histones, specifically, H2A, H2B, H3, and H4 (Chap. 8).

Infection by SV40 of monkey cells in culture (a *permissive* host) results in a lytic cycle leading to lysis of the cell with the liberation of viral progeny. Infection of a *nonpermissive* host, e.g., mouse cells, is abortive; viral progeny are not produced, however, the cells may undergo *oncogenic transformation*.

The course of events during productive infection of monkey kidney cells by SV40 is shown in Fig. 30.6. The first stage, which lasts approximately 8 h, consists of the adsorption of the virus, its entrance into the cell via pinocytotic vesicles, penetration of the nucleus, and removal of the capsid proteins (uncoating). Expression of the SV40 genome then occurs in a regulated temporal sequence. Transcription in the counterclockwise direction of approximately one-half of one of the two strands (E strand) produces the early mRNAs (Fig. 30.7). These mRNAs encode two virus-specific proteins, the large T and small t antigens. The large T antigen (MW = 90 to 100×10^3), which is confined to the nucleus, is required for viral DNA replication and cellular transformation; the small t antigen (MW = 15 to 20×10^3), only found in the cytoplasm, enhances the efficiency of transformation. The *large T* and *small t* mRNAs have 5′

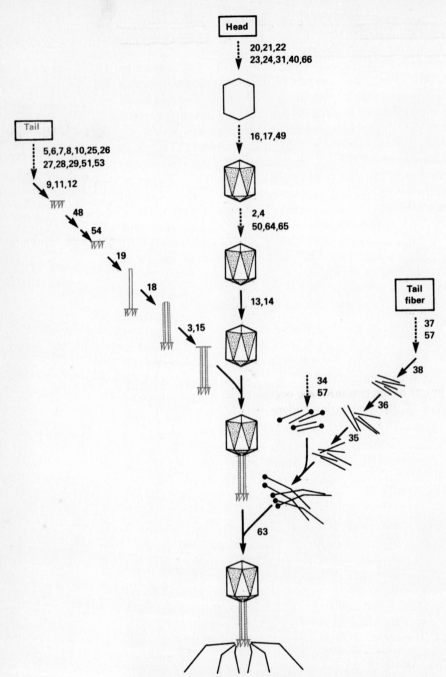

Figure 30.5 Assembly of bacteriophage T4. The numbers refer to the genes required at each step in the process. [*From W. B. Wood and R. J. Bishop, in C. F. Fox and W. S. Robinson (eds.), Virus Research, Academic, New York, 1973, p. 303.*]

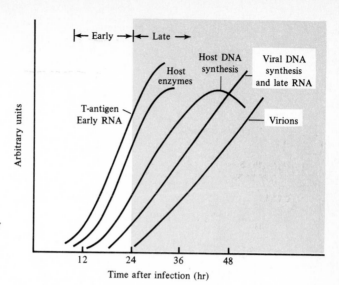

Figure 30.6 Time-course of macromolecular synthesis during productive infection of monkey kidney cells by SV40. [*From S. E. Luria, J. E. Darnell, Jr., D. Baltimore, and A. Campbell, in General Virology, 3d ed., Wiley, New York, 1978, p. 354.*]

termini originating from nucleotide sequences near the origin of DNA replication (*ori*).

Both mRNAs span most of the early region (Fig. 30.7); however, the large T mRNA lacks a nucleotide sequence present in the DNA [between coordinates 0.59 and 0.54, Fig. 30.7; coordinate 0/1.0 is taken as the single site of cleavage by the restriction endonuclease *EcoRI* (Chap. 27)]. Thus, small t appears to be translated from the longer of the two mRNAs, beginning at or near the 5' end and terminating at one of the termination codons known from the DNA sequence to fall between coordinates 0.55 and 0.53. On the other hand, large T is translated from the shorter mRNA, beginning at the same initiator codon and terminating at or near the nucleotide sequence corresponding to coordinate 0.18. Small t, translated from the continuous nucleotide sequence between 0.65 and 0.55, should have a molecular weight of 19,000 to 20,000. Translation of large T mRNA, from which the sequence between coordinates 0.59 and 0.54 has been removed from the primary transcript in a splicing reaction (Chap. 29) but beginning at the same initiator codon and ending at coordinate 0.18, should yield a polypeptide of molecular weight 80,000 to 90,000. Thus, modification of sequence by splicing permits the same segment of DNA to generate alternative mRNAs that can code for different polypeptides. These relationships are shown in Fig. 30.8.

As noted above, infection by SV40 of a nonpermissive host is abortive. The virus initiates the infection, and all the early events described for the lytic cycle occur, up to and including synthesis of large T and small t antigens and induction of host DNA replication. However, viral DNA replication does not occur, and progeny virus are not formed. Under these conditions, most cells survive the infection, and of these survivors a small proportion (less than 10 percent) are *transformed* to malignancy. Such transformed cells manifest heritable alterations in their properties, i.e., growth in a less oriented manner and loss of susceptibility to contact inhibition of growth and, when injected into a suitable host animal, give rise to a tumor. The transformed cells contain all or part

799

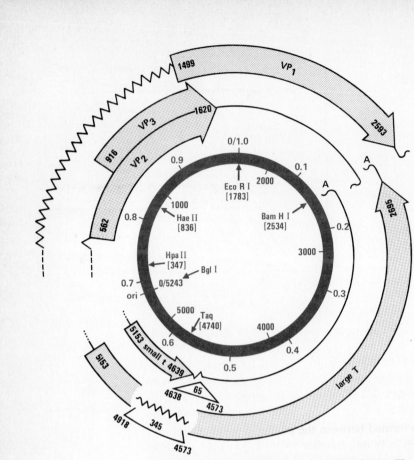

Figure 30.7 Physical and genetic map of SV40 DNA. The inner circle shows the closed circular DNA molecule; indicated within the circle are the nucleotide-pair map coordinates starting and ending at 0/5243. Also shown by small arrows within the circle are the sites at which five restriction endonucleases cleave SV40 DNA once. Arranged around the outside of the circle are the map coordinates, expressed in fractional lengths, beginning at the reference point 0/1.0 (the *Eco*RI endonuclease cleavage site) and proceeding clockwise around the circle. The coding regions for the early and late proteins are shown as stippled arrows extending from the nucleotide pair of the first codon to the nucleotide pair that specifies termination of the protein coding sequence. Each of the coding regions is embedded in a mRNA, the span of which is indicated by dotted or dashed 5′ ends and wavy poly(A) 3′ ends. The jagged, or saw-toothed, portions of each mRNA indicate the portions of the transcript that are spliced in forming the mature mRNAs. [*From P. Berg, Science 213:296–303 (1981).*]

of the viral DNA integrated into the cellular chromosomal DNA and produce mRNAs and proteins coded by the early genes.

SINGLE-STRANDED DNA VIRUSES

Bacteriophage φX174

The small polyhedral bacteriophage φX174 of *E. coli* contains a single-stranded, circular DNA molecule composed of 5370 nucleotides. It contains nine genes (A through H), most of which code for the structural proteins of the

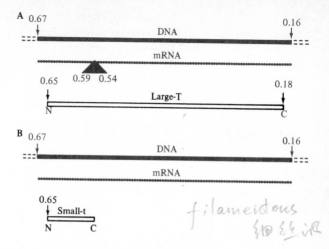

Figure 30.8 Model for the organization and expression of the nucleotide sequences coding for large-T and small-t antigens. [*From L. V. Crawford, C. N. Cole, A. E. Smith, E. Paucha, P. Tegtmeyer, K. Rundell, and P. Berg, Proc. Natl. Acad. Sci. USA 75:117 (1978).*]

filameidous
生物系统讲义

virus. The DNA of the small filamentous bacteriophage M13 (or the related fd) has a similar size and structure. Replication of φX174 DNA occurs by the synthesis of a strand complementary to the parental DNA using host replication proteins (Chap. 27) to generate a closed circular duplex DNA molecule, termed RFI (see Fig. 27.13).

The steps involved in the generation of progeny single strands from φX174 RFI as established in vitro are depicted in Fig. 30.9. The product of the viral gene A (*gene A protein*) cleaves the viral (+) strand of the superhelical RFI at a site which corresponds to the origin of replication as determined in vivo. At the same time a covalent linkage is formed between the gene A protein and DNA. Action of the *rep protein*, a DNA helicase, together with single-stranded DNA

Figure 30.9 DNA replicative steps in φX174 multiplication. [*From A. Kornberg, CRC Crit. Rev. Biochem. 7:23 (1979). Reproduced with permission, © The Chemical Rubber Co., CRC Press, Inc.*]

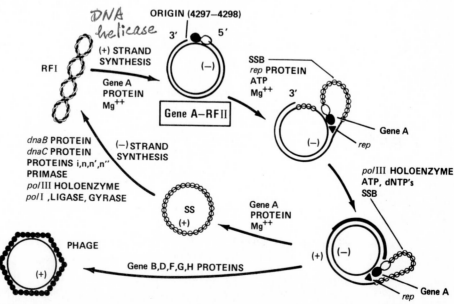

binding protein (Chap. 27), permits unwinding of the duplex. In the presence of DNA polymerase III holoenzyme (Chap. 27), ATP, and the deoxynucleoside triphosphates, a replication fork is established that generates single-stranded circular viral progeny molecules. The latter are encapsulated to generate infective progeny viruses that are released from the cell when an endolysin, specified by one of the viral genes (gene E), produces cell lysis.

Knowledge of the amino acid sequences of the major φX174 coded proteins and the complete nucleotide sequence has enabled the boundaries of each of the genes to be precisely located in the sequence of the DNA. Although the length of φX174 DNA is 5370 nucleotides, approximately 6100 nucleotides are required to code for the nine proteins specified by the virus. The apparent discrepancy has been resolved by the finding that gene D, which specifies a protein whose function is unknown, lies within the sequence of gene E, which codes for the endolysin; i.e., *the two genes are translated from the same nucleotide sequence but in different reading frames* (Chap. 28). Despite the fact that the two genes are translated in two reading frames from a common DNA sequence, they are completely independent as judged by standard genetic criteria. For example, chain-terminating nonsense mutants in gene E produce a completely functional full-sized gene D protein, and normal gene E function is present in gene D nonsense mutants. Similarly, gene B is coded entirely within the sequence of gene A but with different reading frames. It is likely that the size of the φX174 genome is limited by packaging within the virion or other unknown constraint. Therefore, the use of the same nucleotide sequence read in different frames to specify two different proteins provides a highly economical means of circumventing such a size limitation.

RNA VIRUSES

Viruses with (+)RNA Genomes

The RNA bacteriophages of *E. coli* (Qβ, R17, MS2) and the *picornaviruses*, which include the viruses responsible for *poliomyelitis*, the *common cold, hoof-and-mouth disease of cattle*, and *myocarditis*, contain a single-stranded (+)RNA. The bacteriophage RNAs contain 3 kb, while the *picornaviruses* range from 7 to 8 kb. In each case, the virions are spherical particles about 25 nm in diameter with icosohedral symmetry.

RNA Phages Because of their relative simplicity, multiplication of the RNA bacteriophages is understood in greatest detail. After entry of the bacteriophage RNA into the cell, the RNA functions as an mRNA to direct the synthesis of both the major coat, or *capsid*, protein and one of the subunits of the phage-induced RNA *replicase*, the enzyme responsible for the synthesis of phage RNA.

In the case of bacteriophage Qβ, the *replicase* is composed of four subunit polypeptides: subunit α (MW = 70,000) is the 30-S ribosomal protein, S1 (Chap. 28); β (MW = 65,000) is the virus-encoded enzyme; and γ (MW = 47,000) and δ (MW = 35,000) are the initiation factors EF-Tu and EF-Ts, respectively (Chap. 28). The α subunit is required for binding of the

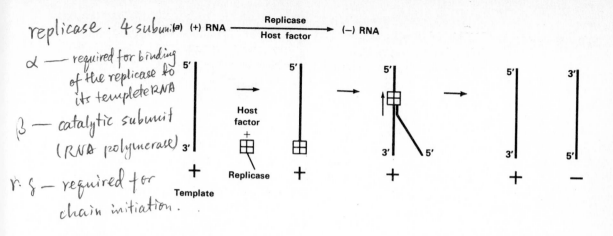

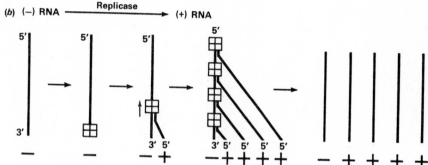

Figure 30.10 Schematic representation of bacteriophage RNA replication. See text for details.

replicase to its template RNA; β is the catalytic subunit, e.g., the RNA polymerase; subunits γ and δ are required for RNA chain initiation.

The process of RNA replication proceeds in two stages (Fig. 30.10). In the first step, the viral (+) strand serves as a template for the synthesis of a complementary (−) strand by the replicase acting in conjunction with a host factor. Although the conventional base pairing rules determine the sequence of the (−) strand, a double-stranded replicative intermediate analogous to that found in ϕX174 DNA replication is not formed. The product is a free single (−) strand with pppG at the 5′ terminus. In the second stage of the reaction, the newly synthesized (−) strand is used by the replicase (without host factor) as template for the synthesis of progeny (+) strands. These contain the sequence . . . pCpCpC at the 3′ terminus; however, a terminal adenylate residue is added by a host-specified terminal addition enzyme just prior to encapsidation of the RNA.

In addition to the RNA replicase and capsid protein, the bacteriophage RNA specifies a third so-called maturation protein that is required for proper assembly of the virion. In contrast to the capsid protein, which is present in 180 copies per virion, there is only a single copy of the maturation protein.

Control of phage protein synthesis appears to be at the level of translation (Chap. 28). Thus, there are three ribosome binding sites on each (+)RNA molecule, and each serves as the starting site for translation of each of the three

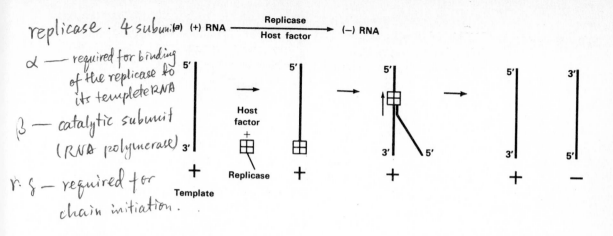

803

proteins coded for by the RNA, e.g., the replicase, capsid protein, and maturation protein. Different amounts of the three proteins are synthesized, and synthesis of each appears to be independently regulated by the accessibility to ribosomes of the initiator AUG sequence. In addition, the RNA replicase, by binding at the initiator site for the capsid protein, can function as a repressor of capsid protein synthesis. In a similar way, the capsid protein can repress synthesis of the RNA replicase.

Poliovirus The poliovirus RNA (MW = 2.6×10^6) is composed of approximately 7500 nucleotides. A poly(A) tail analogous to that found in eukaryotic mRNAs is attached to the 3′ terminus. However, in contrast to eukaryotic mRNAs the 5′ terminus does not bear a 5′ cap. Instead, the 5′-terminal nucleotide sequence pUpUp(A_4)pCpApG . . . is linked through a phosphotyrosine residue to a virus-specific protein. The protein is also found on the viral RNA when it forms part of a replicative intermediate (see below) and is believed to represent an initiating structure, or primer, for viral RNA synthesis.

Poliovirus RNA initiates infection in a manner similar to bacteriophage RNA. The (+)RNA functions as messenger, using the host translational apparatus to direct synthesis of capsid proteins and a polio-specific RNA replicase. The replicase generates many copies of (−)RNA, which then serve as a template for the synthesis of (+)RNA. The latter is then encapsulated by capsid proteins to yield progeny virus (Fig. 30.11). The poliovirus specifies six proteins: the RNA replicase, four different capsid proteins, and a protein (X) of unknown function. Unlike the RNA phages, polio mRNA is translated as one continuous polypeptide chain of 2300 amino acid residues. This polypeptide is progressively cleaved by host (and possibly viral) chymotrypsin-like proteases into six proteins, as illustrated in Fig. 30.12. It is believed that cleavage of VP0 to VP2

Figure 30.11 Schematic representation of poliovirus RNA replication. Note the similarity to bacteriophage RNA replication.

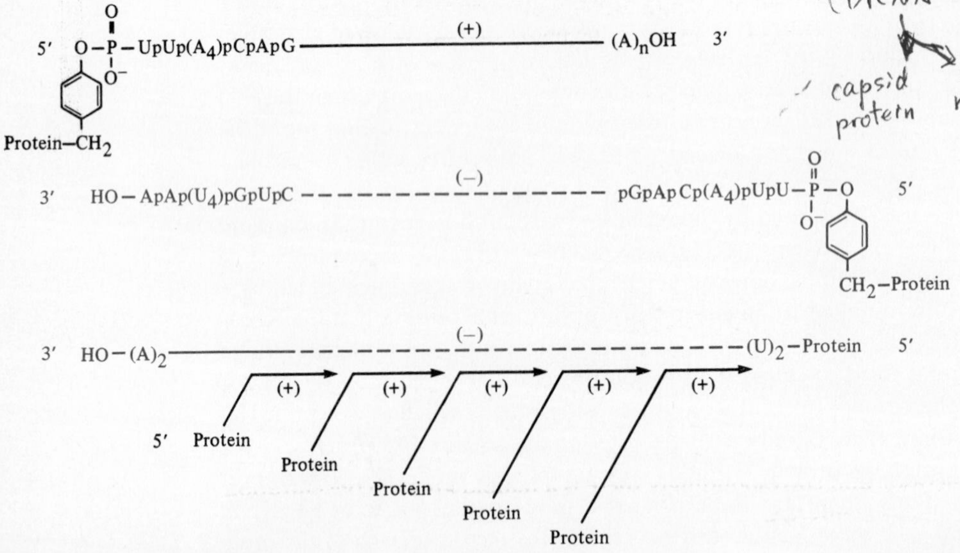

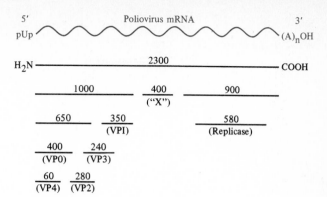

Figure 30.12 Processing by proteolytic cleavage of continuous 2300-residue polypeptide synthesized during poliovirus infection.

and VP4 occurs during assembly of the virions from RNA and the procapsids (VP0, VP1, and VP3).

Also in contrast to the bacteriophage RNA, poliovirus contains only a single ribosomal RNA binding site per (+)RNA molecule. As a consequence, the six proteins formed by proteolysis are synthesized in equimolar amounts. The occurrence of only a single initiator sequence for protein synthesis at or near the 5′ terminus of each mRNA constitutes a fundamental difference between prokaryotic and eukaryotic cells. Thus far, each eukaryotic mRNA sequence has been found to be translated into only one polypeptide chain (Chap. 29).

Viruses with (−)RNA Genomes

Two major classes of viruses, the *rhabdoviruses,* e.g., *vesicular stomatitis virus* (VSV) and *rabies,* and the *paramyxoviruses,* e.g., *Newcastle disease virus* of chickens, *parainfluenza, mumps, measles,* and *distemper* viruses, contain (−)RNA within the virion. These viruses are more complex than those considered above in that they are enveloped by a membrane composed of lipid and glycoprotein.

Of this group, the vesicular stomatitis virus has been examined in greatest detail. This virus is composed of a single RNA molecule, 11.5 kb in length, and five different proteins, termed N (MW = 50,000), present in 700 copies; L (MW = 190,000), 50 copies; NS (MW = 45,000), a small number of copies; G (MW = 65,000), 500 copies; and M (MW = 29,000), 1600 copies. G and M form the viral envelope.

Upon penetration of the membrane of the infected cell, the virus, free of its envelope (the *nucleocapsid*), is released into the cytoplasm. The viral RNA polymerase (proteins L and possibly NS) transcribes the (−) viral RNA. A poly(A) tail and methylguanylate cap are then added to the 3′ and 5′ termini of the mRNA synthesized. A separate molecule of mRNA is made for the synthesis of each protein. The (+)mRNAs then bind to host ribosomes, and viral proteins are synthesized by using the translation machinery of the host cell. The quantity of each mRNA synthesized corresponds approximately to the number of molecules of each protein in the virion. Thus, there are 10 times as many molecules of N (the major capsid protein) per virion than of L protein (RNA polymerase), and correspondingly there is a tenfold greater concentration of N mRNA than L mRNA.

805

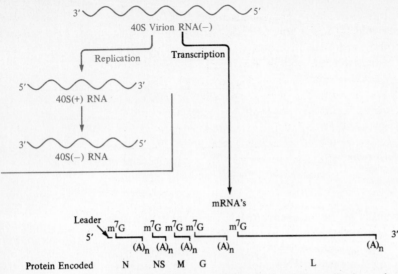

Figure 30.13 Separate pathways of viral RNA replication and mRNA synthesis of vesicular stomatitis virus. [*From S. E. Luria, J. E. Darnell, Jr., D. Baltimore, and A. Campbell, in General Virology, 3d ed., Wiley, New York, 1978, p. 331.*]

Replication of the viral RNA begins after synthesis of new viral proteins has been initiated. First, a (+)RNA strand is synthesized that is the same length as the (−)RNA strand in the virion. The (+)RNA strands then serve as templates for synthesis of new viral RNA strands that are encapsulated by the N protein to generate the nucleocapsid. The separate pathways of replication and transcription are indicated in Fig. 30.13. Assembly of the infectious viral particle is completed when the nucleocapsid is enveloped by components of the cell membrane which provide the G and M proteins. The virus is finally released from the infected cell by budding off from the cell membrane.

Viruses with Single-Stranded Segmented (−)RNA Genomes

One of the best known of these viruses is *influenza*, which is a spherical particle about 110 nm in diameter enveloped in a membrane composed of lipids derived from the host cell and virus-specific proteins. Inside the membrane are eight (−)RNA molecules (MW = 3.4 to 9.8×10^5), each in association with an RNA polymerase.

Upon infection, the (−)RNA molecules are transcribed to yield corresponding lengths of (+)mRNA, which then acquire poly(A) tails at their 3′ termini and methylguanylate caps at their 5′ termini. In contrast, the (−)RNA molecules remain unmodified. The (+)mRNAs are translated on host ribosomes to generate virus-specific proteins and also serve as templates for synthesis of the complementary (−)RNA segments. The mechanism by which this occurs is not understood; nor is it known how the eight discrete (−)RNA segments constituting the complete influenza virus genome are assembled to produce complete infectious virions.

A *hemagglutinin* and a *neuraminidase* are present within the viral envelope and constitute the major viral antigens. Both are glycoproteins, and antibodies towards them protect against infection by influenza virus. The antigenic specificity of influenza virus is highly variable. Mutation in the RNA leads to amino acid replacements in the hemagglutinin and in the neuraminidase, causing changes in their antigenic specificity, a process called *antigenic drift*. Since these two proteins are coded for by separate genes, their antigenic structures vary independently of each other. It is believed, in fact, that the major influenza pandemics that sweep the world at 10- to 12-year intervals result from large changes in antigenic specificity corresponding to changes in either the hemagglutinin or neuraminidase or both. Such changes may be a consequence of recombination between human and animal strains of the virus. Sequence analyses of the neuraminidases from the *A2* and *Hong Kong* strains show them to differ by only a few amino acids. They are also antigenically similar. In contrast, the hemagglutinins from the two strains show marked differences. Indeed, one is closely related to the hemagglutinin of an influenza strain isolated from horses and ducks. Thus, the Hong Kong strain of influenza may have arisen by recombination between the human A2 strain and an animal strain.

Viruses with Double-Stranded Segmented (+)RNA Genomes

Reovirus constitutes a family of spherical virions 60 to 80 nm in diameter containing 10 different molecules of double-stranded (+)RNA with a combined total of 23 kb. The 10 molecules fall into three groups: three molecules containing 4500 base pairs each, three with 2300 base pairs each, and four with 1200 base pairs each. The 10 RNA molecules are believed to be held together in a specific arrangement.

In addition to RNA, the reovirus virion contains eight different proteins. Three have chain lengths of 1500 residues, two have 750 residues, and three have 400 residues. Two of the 400-residue proteins and one of the 750-residue proteins are found in the outer shell of the virion; the remaining proteins are in the core with the RNA. Inasmuch as three bases specify one amino acid residue and each RNA segment is transcribed into a single mRNA molecule, the polypeptide chain lengths correspond to the lengths of the three groups of RNA segments. The close correspondence between genome segment length and polypeptide chain length provides yet another example of the general rule for eukaryotic messengers stated previously (page 772), viz., that each messenger specifies a single protein.

When a reovirus virion infects its host cell, the proteins of the outer shell are removed and the core enters the cytoplasm. Removal of the shell activates the RNA polymerase present in the core of the virion. The polymerase then transcribes all 10 genome segments; however, it does so asymmetrically, i.e., only one of the two RNA strands, and conservatively. There appears to be only a single transcription start signal and one termination signal, located at the very ends of the segment so that the mRNA synthesized is exactly the same length as the genome segment. The mRNAs are released from the virion core, but the genome segments remain inside. The mRNAs are unusual in that they do not contain poly(A) segments at their 3′ termini; however, like the (+)RNA strands in the genome, they contain capped 5′ termini. Like the RNA polymerase, the enzymes responsible for synthesis of the cap are contained within the virion.

Although all the genome segments are transcribed throughout the cycle of infection, they are not transcribed with equal efficiency. The different mRNAs are also translated with different efficiencies. In this manner, the viral proteins can be synthesized in approximately the ratio of their occurrence in the virion.

To summarize, the reovirus virion contains 10 double-stranded RNA segments; each segment produces one mRNA; each messenger synthesizes one type of protein. Replication of the double-stranded RNA segments occurs through the same (+)mRNAs that serve to direct synthesis of viral proteins. Thus, asymmetric, conservative transcription of a virion RNA segment leads to the synthesis of a single-stranded (+)mRNA; this is replicated to generate a new double-stranded RNA segment, which is finally incorporated into the progeny virion. Because the assembly of a set of mRNA segments is random, segments of different parentage can be included within the same nascent core. As a result, cells infected with two reovirus mutants altered in different genome segments produce a high frequency of recombinant genomes.

Retroviruses

The retroviruses are RNA-containing viruses that during their replication cycle integrate into the host genome as a DNA copy of the RNA. Many different retroviruses are known. Although some are oncogenic and hence are referred to as *RNA tumor viruses,* many of the retroviruses do not produce malignancy or other disease. The retroviruses about which most is known are the *C type* viruses, which include the *murine* and *avian leukemia* and *sarcoma viruses.*

The virion, which is approximately spherical (100 to 150 nm in diameter) consists of a core surrounded by a membranous envelope composed of glycoprotein and lipid. It contains from 7 to 10 different types of protein ranging in molecular weight from 11 to 100×10^3, including the RNA-dependent DNA polymerase [reverse transcriptase (Chap. 27)]. The genome of the retrovirus is diploid and consists of two identical 35-S (+)RNA molecules joined at or near their 5′ termini presumably through hydrogen bonds, although this has not been firmly established. The two RNA molecules contain 5′ caps and a 3′ poly(A) segment and are associated in the form of a ribonucleoprotein with a highly basic protein and two phosphoproteins. In the case of the avian sarcoma virus, each of the RNA molecules has a molecule of tRNA^{Trp} firmly bound at a site 101 nucleotides from the 5′ end, which serves as a primer for the synthesis of the DNA strand in the RNA-DNA hybrid product of the reverse transcriptase (see below). In the case of the murine virus, tRNA^{Pro} serves this function.

Four genes have been identified in the avian sarcoma virus genome: *gag,* which encodes the proteins of the viral core; these are generated by processing a larger precursor polypeptide; *pol,* which encodes the reverse transcriptase; *env,* which encodes the glycoproteins of the viral envelope; and *src,* which specifies a protein required for oncogenic transformation. The arrangement of these genes within the retrovirus genome is shown in Fig. 30.14.

5′   m^{7}GpppG$_m$ —— *gag*   *pol*   *env*   *src* —— (pA)$_{200}$pA-OH   3′

Figure 30.14 Arrangement of genes in genome of retroviruses.

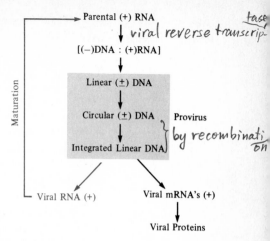

concomitant
机体论

viral reverse transcrip-tase

Provirus } by recombinati-on

bud
芽

Figure 30.15 Life cycle of retroviruses.

Unlike the papovaviruses, e.g., SV40, the RNA tumor viruses do not usually kill the cells in which they replicate, and replication and transformation are often concomitant events. The multiplication cycle of the retroviruses consists of two distinct phases (Fig. 30.15). Phase I consists of the synthesis of the DNA copy of the RNA genome (*the provirus*) and its integration into the host cell DNA. This occurs in the following sequence. (1) The infecting virion binds to a specific cell receptor and penetrates the nucleus, where it is uncoated. (2) A DNA strand complementary to the infecting viral RNA is synthesized by the viral reverse transcriptase and the viral RNA strand is removed, probably by the RNAse H that is part of the reverse transcriptase (Chap. 27). (3) A DNA strand with the same base sequence as the viral RNA is synthesized. (4) The linear duplex DNA generated in these two steps is circularized and integrated into the host cell DNA by recombination. In general, more than one copy of the viral genome is integrated into the host DNA at nonspecific sites. Phase II involves the expression of proviral DNA and the maturation of the new virions: (1) The integrated viral DNA is transcribed by the host RNA polymerase II, and the primary transcript is processed to form viral RNA and three mRNAs, a *gag-pol* mRNA, which is indistinguishable from the intact viral RNA, an *env* mRNA, and a *src* mRNA. Inasmuch as all three species of mRNA contain a common 100-nucleotide sequence at the 5′ end of the chain, they are very likely initiated at the same site, then spliced to form the mature RNAs, which acquire 5′ caps and 3′ poly(A) tails. (2) The mRNAs are translated into proteins which then undergo processing. As noted above, the *gag* proteins are formed by proteolysis of a larger polyprotein. (3) The maturing virions migrate to the cell surface, from which they bud off, acquiring an outer envelope in the process. The infected cell is believed to assume a transformed phenotype after it has divided and transcription has begun.

As noted above, the *src* gene is essential for neoplastic transformation of the host by avian sarcoma virus. The product of the *src* gene is a *protein kinase* (MW = 60,000) that catalyzes the transfer of the γ phosphate of ATP to tyrosine residues of a variety of proteins, including the protein kinase itself. Indeed, the *src* gene product is isolated as a phosphoprotein.

The *src* gene of the *Rous sarcoma virus* is homologous to a cellular gene, denoted *sarc,* present in normal chickens, whose product is a phosphoprotein (MW = 60,000) that possesses protein kinase activity similar to the viral enzyme in that it catalyzes the phosphorylation of tyrosine residues. The similarity between the two proteins has led to the proposal that the Rous sarcoma virus may have arisen as a result of recombination between retrovirus and cellular *sarc* gene sequences.

The level of the viral *src* gene product in transformed cells is approximately 50-fold greater than that of its cellular counterpart in normal cells. Thus, transformation may be a consequence of the increased enzyme level leading in turn to an increased level of phosphorylation of particular target proteins. Alternatively, were the viral enzyme to be altered in its substrate specificity, proteins not normally phosphorylated might undergo modification, again resulting in transformation.

The phosphorylation of cellular proteins has been linked to the induction of cell growth. Thus, the mitogenic action of epidermal growth factor (EGF) (Chap. M20) on epidermal and epithelial cells may be related to an increase in protein kinase activity upon binding of EGF to its receptor in the plasma membrane, which produces an increase in phosphotyrosine levels in a number of membrane proteins, including the receptor itself.

THE INTERFERONS

The interferons are a class of small glycoproteins ranging in molecular weight from 26,000 (mouse and human being) to 38,000 (chicken). They are of two principal types, leukocyte interferon, which consists of a family of approximately 10 different but homologous proteins, and fibroblast interferon, of which there are two representatives.

Upon exposure to interferon, cells develop an antiviral state in which the replication of a wide variety of viruses is inhibited. Interferon can also inhibit cell growth. The production and release of interferon are, in turn, induced by double-stranded RNA, an intermediate in the replication of the RNA viruses.

The antiviral action of interferon proceeds by two mechanisms, both of which result in the inhibition of viral mRNA translation. These are outlined in Fig. 30.16. The first mechanism involves inhibition of the initiation of translation and is analogous to that described earlier for the regulation of hemoglobin synthesis by hemin (Chap. 29). Exposure of cells to interferon results in the induction of a protein kinase, which in the presence of double-stranded RNA catalyzes the phosphorylation and corresponding inactivation of the initiation factor eIF-2. The second mechanism involves the premature degradation of mRNA. Interferon-treated cells contain a double-stranded RNA-dependent enzyme, termed *2-5A synthetase,* that catalyzes the synthesis of the trinucleotide pppA2′p5′A2′p5′A (2-5A), and higher oligomers in which the internucleotide linkage is 2′-5′ rather than 3′-5′. The 2-5A is an activator of an endogenous ribonuclease that degrades the viral mRNA, thereby inhibiting protein synthesis. 2-5A is itself rapidly degraded by an interferon-induced *2-5A exonuclease.* Inasmuch as the activity of the 2-5A activated ribonuclease depends upon the

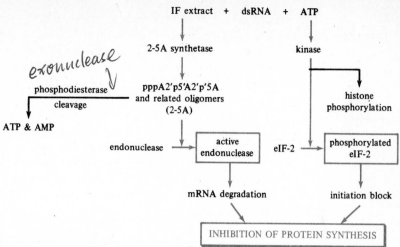

IF extract + dsRNA + ATP

exonuclease

2-5A synthetase   kinase

phosphodiesterase
cleavage

pppA2'p5'A2'p'5A
and related oligomers
(2-5A)

histone
phosphorylation

ATP & AMP

endonuclease → active
endonuclease   eIF-2 → phosphorylated
eIF-2

mRNA degradation   initiation block

INHIBITION OF PROTEIN SYNTHESIS

Figure 30.16 The two pathways involved in the inhibition of protein synthesis in extracts of interferon-treated cells by double-stranded RNA. IF = interferon. [*From B. R. G. Williams and I. M. Kerr, Trends in Biochem. Sci. 5:138 (1980).*]

continued presence of 2-5A, the exonuclease may in some manner regulate the 2-5A-induced inhibition of protein synthesis.

REFERENCES

Books

Davis, B. D., R. Dulbecco, H. N. Eisen, H. S. Ginsberg, and W. B. Wood: *Microbiology,* 2d ed., sec. V, Harper & Row, New York, 1973.

Fenner, F., B. R. McAuslan, C. A. Mims, J. Sambrook, and D. O. White: *The Biology of Animal Viruses,* vols. I and II, Academic, New York, 1974.

Fox, C. F., and W. S. Robinson (eds.): *Virus Research,* Academic, New York, 1973.

Fraenkel-Conrat, H., and R. R. Wagner (eds.): *Comprehensive Virology,* vols. 1–5, Plenum, New York, 1974.

Hershey, A. D. (ed.): *The Bacteriophage Lambda,* Cold Spring Harbor Laboratories, Cold Spring Harbor, N.Y., 1971.

Kornberg, A.: *DNA Replication,* Freeman, San Francisco, 1980.

Lauffer, M. A., F. B. Bank, K. Maramorosch, and K. M. Smith (eds.): *Advances in Virus Research,* vols. 1–26, Academic, New York, 1953– (a continuing series).

Luria, S. E., J. E. Darnell, Jr., D. Baltimore, and A. Campbell: *General Virology,* 3d ed., Wiley, New York, 1978.

Mathews, C. K.: *Bacteriophage Biochemistry,* Van Nostrand, New York, 1971.

Tooze, J., and R. D. Hynes: *The Molecular Biology of Tumour Viruses,* 2d ed., Cold Spring Harbor Laboratories, Cold Spring Harbor, N.Y., 1980.

Tumor Viruses, Cold Spring Harbor Symp. Quant. Biol., vol. 34, 1974.

Viral Oncogenes, Cold Spring Harbor Symp. Quant. Biol., vol. 44, 1980.

Zinder, N. (ed.): *RNA Phages,* Cold Spring Harbor Laboratories, Cold Spring Harbor, N.Y., 1975.

Review Articles

Baglioni, C.: Interferon-Induced Enzymatic Activities and Their Role in the Antiviral State, *Cell* **17**:255 (1979).

811

Baltimore, D.: Expression of Animal Virus Genomes, *Bacteriol. Rev.* **35:**235–241 (1971).

Bishop, J. M.: Retroviruses, *Annu. Rev. Biochem.* **47:**35 (1978).

Casjens, S., and J. King: Virus Assembly, *Annu. Rev. Biochem.* **44:**555–611 (1974).

Dulbecco, R.: Mechanisms of Virus-induced Oncogenesis, in F. O. Schmitt and F. G. Worden (eds.): *The Neurosciences,* Third Study Program, MIT, Cambridge, Mass., 1974, pp. 1057–1065.

Palese, P.: The Genes of Influenza Virus, *Cell* **10:**1 (1977).

Shatkin, A. J.: Animal RNA Viruses: Genome Structure and Function, *Annu. Rev. Biochem.* **43:**643–666 (1974).

Silverstein, S. C., J. K. Christman, and G. Acs: The Reovirus Replicative Cycle, *Annu. Rev. Biochem.* **45:**375–408 (1976).

Williams, B. R. G., and I. M. Kerr: The 2-5A System in Interferon and Control Cells, *Trends in Biochem. Sci.* **5:**138 (1980).

Wood, W. B., and H. R. Revel: The Genome of Bacteriophage T4, *Bacteriol. Rev.* **40:**847–868 (1976).

Chapter 31

Molecular Evolution

Genetic variation of protein structure.
Evolution of proteins. Hereditary
disorders of metabolism.

The amino acid sequence of a protein responsible for a particular biological function is essentially identical in all individual members of the species. Thus, hemoglobin, cytochrome c, insulin, fibrinogen, serum albumin, etc., have sequences characteristic of the species in which they occur. The constancy of the amino acid sequences of individual proteins within a species results from faithful DNA replication, and transcription and translation of DNA sequences into the amino acid sequences of polypeptide chains (Chaps. 28 and 29). However, the differences in the amino acid sequences of a given protein between different species reflect biological evolution, since the genome of a species is the net result of the accumulated mutations in that species during its evolution. Hence, comparison of the sequences of a group of proteins with the same or related functions from different species provides insight into the molecular events that have occurred during evolution. Such information can be obtained by direct sequencing of homologous proteins or by sequencing the DNA of the responsible gene when the protein for which it codes can be identified.

It is an overstatement, however, that the amino acid sequence of a particular protein is always identical in each individual of a species; some percentage, very small or large, of individuals within a species may exhibit small but significant differences in the sequence of a particular protein. This polymorphism in amino acid sequence is a reflection of the continual mutation of the genome of the species and is as essential in the biological evolution of living things as the apparent constancy of the genome of the species. Thus, comparison of the structural variants of a particular protein within a single species reflects the kinds of mutations that continually arise during evolution. If a structural variant provides selective, reproductive advantage to the species, the mutant gene may become, in time, the norm for the genome. Most mutations of structural

Any cross-references coded M refer to the companion text, *Principles of Biochemistry: Mammalian Biochemistry.*

genes provide no selective advantage to the species and do not penetrate the population. Indeed, some mutations lead to an aberrant function of a protein and are the basis of hereditary disorders.

In this chapter we shall be concerned with several aspects of genetically determined changes in protein structure. First, consideration will be given to the information available about the structure-function relationships of a particular protein simply by comparison of the amino acid sequences of that protein from a number of species or from the variants of the proteins within a species. Indeed, nature has performed an elegant series of experiments for the biochemist by providing an extensive set of sequence alterations in proteins that afford considerable insight into the structural basis for their biological functions.

Second, we shall consider aspects of biological evolution that can be deduced from comparison of the amino acid sequences of proteins. From such comparisons and present knowledge of molecular genetics, it is possible to describe the kinds of changes in the genome (DNA) that during evolution have led to altered protein structure and hence to modified functions. Moreover, comparison of sequences provides insight into some fundamental concepts of darwinian evolution, including natural selection, population variation, and the survival value of genetically determined protein variants. Finally, the genetically determined differences in the structures and functions of various proteins, as reflected in abnormal metabolism of human beings, will be considered.

AMINO ACID SUBSTITUTION AND PROTEIN FUNCTION

Genetic variation in the structure of a protein can provide valuable information concerning the role of amino acid residues in the function of a specific enzyme or other protein. A remarkable example is provided by the more than 300 human hemoglobin variants since many have aberrant functional properties. As described in Chap. M4, correlation of the structures and functions of normal and abnormal hemoglobins have given extensive insight into the structural basis of hemoglobin function.

The interpretation of these differences in the effects of amino acid substitutions rests on present knowledge of protein structure and function. The amino- or carboxyl-terminal ends of certain proteins may be entirely dispensable for function. Chemical treatment that alters specific amino acid side chains may produce no demonstrable effect on the function of certain proteins; e.g., conversion of many or all lysine residues to homoarginine residues by reaction with O-methylisourea does not alter the enzymic activity of cytochrome c, ribonuclease, lysozyme, papain, etc. Similarly, other residues of proteins can be acylated, oxidized, or treated in other ways without significant effect on function. In contrast, certain residues in enzymes are absolutely critical for their functions, as in the demonstrations of active sites, elucidated with various types of inhibitors.

As a *minimum*, we may expect that those residues which are in active sites, which are essential for binding substrates, cofactors, or prosthetic groups or which determine the folding and essential conformation of the peptide chains cannot be altered by genetic or chemical modification without profound effects on function. The substitutions for Gly I or Gly II that result in inactive variants of tryptophan synthetase (Chap. 28) are presumably due to conformational effects on the protein since a variety of small nonpolar residues can substitute for Gly but ionic residues cannot. Examination of the conformation of the chains of hemoglobin (Chap. M4) suggests that the regions at which greatest variation in amino acid sequence occurs without the disturbance of function are in some sections of the chains not immediately involved in critical interactions required for binding O_2. This is in keeping with the fact that the hemoglobins of vertebrates possess similar functional properties, and despite substantial differences between them in amino acid sequence and other properties, all these hemoglobins have a very similar conformation that is required for O_2 binding. Indeed, it is now evident that substantial alterations in sequence have occurred but that essential conformation is retained as shown by x-ray analysis, even in hemoglobins of nitrogen-fixing plants, e.g., soybeans and lupines, and those of higher vertebrates.

The most extensive studies presently available are on the sequences of the cytochromes c of more than 70 different species of animals, plants, protists, and fungi. In cytochromes with 103 or 104 residues, about 20 residues have thus far been found to be constant (Fig. 31.1). At all other residue positions in the protein, amino acid replacements of various kinds have been found. In general terms, many types of "conservative" substitutions can occur without major change of functional properties of the protein or enzyme. As examples, we may cite the following: substitution of aliphatic hydrophobic residues for one another (Ile, Val, Leu, Met) and substitution of polar residues for one another (Arg for Lys, Glu for Asp, Gln for Asn). Even ionic residues of opposite charge substitute for one another (Asp or Glu for Lys), presumably because such residues are mainly on the surface of the protein. Substitutions that are "radical," representing different kinds of side chains, can occur without deleterious effect only when the locus is not critical for function or conformation. Comparison of the three-dimensional structure of the cytochromes c from tuna and horse demonstrate no essential change in conformation in keeping with expectation from their functional similarity (see below).

In the instance of the extracellular proteolytic enzyme subtilisin, produced by certain bacilli, comparison of the sequences from two different species shows 83 differences in amino acid sequence of the 274 residues present in common. (There is a one-residue deletion in one type.) Its three-dimensional structure reveals that only two of the differences are in the interior of the protein; the remainder are on the surface. Some substitutions on the surface of the enzyme are conservative, others radical, but no major substitutions are present near the active or binding sites of the enzymes, reflecting the fact that both subtilisins are almost identical in kinetic properties and specificity.

EVOLUTION AND PROTEIN STRUCTURE

Comparison of the proteins of various organisms shows that during the course of evolution (1) proteins that fulfill the same function at all phylogenetic levels, e.g., the enzymes of glycolysis, cytochrome c, etc., whose synthesis is directed by *homologous* genes, have been extensively modified; (2) some proteins

Figure 31.1 — composite amino acid sequence (three panels)

```
                                      N
        D               D             T
        N                             S
        B    D          NDN           S                   DN
  *TS   TB NS   STT   D          DQ           E   NB   D
  SSE   STASE  DESS  ED          TG    N      G   BE   T        T
  EEP   ESVEQ  NPEA  GNT         EA    S      A   EQ   P       N E
  PAPG  QELPP  BAZV  ATV         VMQ   E   Q  TV  QGNEG        BTQ
  PAKAF ZPYAA *SIAK  RRI         ILKAG Z      SC  ZAGGA        SSA
  GKKXKR AGFKK GZKKR GKLF        KFRCA   ACHAL GI GKAAV        QGVG
                                                    ARKLK      HKIAP
                                                     δδδ

  -11               1              10               20              30
```

```
                                           D
                                    D  N   D
                           D        N  B   N
 N                         N        T  T   DB  D
 S                D        B        E  S   NS  S
 Q                E        S        Q  E   BED M
 G  A             Q        QS       G  Q   EQN L      D   T
 N  Y  V   V TT   TP  E    GA       NQV Z  QPT Y      E   E
 S  F  I   IN QS  EA   T P NAAM     TAI G  ZGSMF      I   Z
 A  H  L   YS HQ  SAVS YS TA GBIKK  GVL A  PAVLH      VY  LN
 V  LWGF   FGRKA  GQVKG FAYSK AMKRR AIKWK  YKHFR      KFLKB
      δδ            δ
            40               50               60               70
```

```
                                           D
                                           N
                          D                B
                          N                T T        D
                    DT    N          TT    E S        T N
                    TS    T          ES    Z E        S S
                    SE    E          ZE    V Q        ETES
              T     EQD   Q          VQ    L Z        ASQE
           S  P     SBZZB G          LZ    H I        VAZQ
  S  YV    A  G I   PSPGE ADVI   T   HI    V A Z Q    LCAA
  K  FM    G  V A L KAAAQ VNIV   S   YMKH  L C A A    KKKE
  PXKHI    K TKM I FVGF XKKKZ RKHLL  AFLRK KKKKE
                            δ               δδδ
            80               90               100
```

Figure 31.1 A composite amino acid sequence for all presently known mitochondrial cytochromes c from animals, plants, fungi, and protists including more than 70 different species. A—Ala, B—Asp or Asn, C—cysteine, D—Asp, E—Glu, F—Phe, G—Gly, H—His, I—Ile, K—Lys, L—Leu, M—Met, N—Asn, P—Pro, Q—Gln, R—Arg, S—Ser, T—Thr, V—Val, W—Trp, X—trimethylLys, Y—Tyr, Z—Glu or Gln, δ—deleted (a gap to maximize homology). The asterisk * at position —8 indicates acetylation of the NH_2-terminal residue in all higher plants; * at residue 1 indicates acetylglycine in all chordate cytochromes c. In most species, the heme is attached to the cysteines at residues 14 and 17. The ligands to the heme iron are from His-18 and Met-80. A single residue of trimethyllysine is present at position 72 in the fungi and two such residues in the higher plants at positions 72 and 86. The most conservative regions are in the sequences near the heme attachment and around the ligands to the heme iron. Residues in color are constant in all cytochromes that have been studied. The protist cytochromes show the greatest variations in sequence from those of the other species. [From E. Margoliash, S. Ferguson-Miller, C. H. Kang, and D. L. Brautigan, Fed. Proc. 35:2125 (1976).]

have disappeared; and (3) many new proteins have appeared. Since, in the immediate sense, the structure and distribution of proteins constitute the phenotype of a species, these changes represent the operation of evolution itself.

The sequences of only a relatively small number of proteins are known at various phylogenetic levels. It is clear, however, that the same functional enzymes from phylogenetically remote species retain similar conformation but may have very different sequences. Most significant, however, is the fact that for any functional enzyme, no amino acid substitution has occurred in the active site. Presumably, any substitution that alters its steric fit and substrate-binding capacity or that results in loss of a residue participating in the catalytic process (serine, histidine, sulfhydryl group, etc.) must destroy enzymic activity. Loss of activity might also result from substitutions that alter the conformation of the protein in a manner affecting the structure at the active site. Evidently, it is substitutions of this type or chain-terminating mutations (Chap. 28) that, at some distant time, resulted in loss of specific enzymes, thus making animals dependent on their food supply for the nutritionally essential amino acids, fatty acids, and vitamins.

It is unknown whether vertebrates continue to synthesize similar functionless genetic derivatives of formerly active proteins that once made possible the synthesis of substances essential in the human diet, e.g. tryptophan and thiamine. Not all substitutions at a given site are deleterious with respect to function. Indeed, it is clear that there is considerable tolerance for replacement, deletion, or insertion of amino acids in many regions of various proteins (see Fig. 31.1). In addition to the information from sequence studies of proteins, examples of such tolerance are apparent from (1) the specific immunochemical behavior of similar proteins from different species, (2) the fact that many enzymes can tolerate chemical modification of various residues, (3) the proteolytic removal of a segment of a polypeptide chain without inactivation, and (4) the gross differences in composition of many homologous proteins from various species.

The protein for which most information is available is cytochrome c, as described previously. If cytochrome c is to be useful physiologically, it must not be altered by mutation in a manner preventing its "fit" to both cytochrome oxidase and cytochrome c_1, changing its potential, or interfering with its capacity for attachment of the heme. Since animals, plants, and fungi diverged from some common ancestral form, amino acid replacements have accumulated in the cytochromes c of these species without major effect on their function in mitochondrial electron transport.

The numbers of amino acid replacements in cytochrome c of some species are compared in Table 31.1 for the 104 homologous residues. First, it is evident that the cytochromes of closely related vertebrates differ in none or only a few residues, e.g., human beings and other primates, various ungulates. Second, the greater the apparent taxonomic difference, the more the cytochromes are likely to differ. This is apparent among the vertebrates and in comparing the vertebrates with the plant (wheat), the insect, or an ascomycete (yeast, *Neurospora*). However, yeast and *Neurospora* cytochromes differ almost as much from each other as they do from those of higher organisms. This indicates either that the evolutionary pathway for these two organisms diverged a long time ago or that

TABLE 31.1   Variations in Structure of Cytochrome c

Species	Number of variant amino acid residues compared with human cytochrome*	Cytochromes compared	Number of variant amino acid residues
Chimpanzee	0	Horse and dog	6
Rhesus monkey	1	Horse and pig	3
Kangaroo	10	Pig, cow, and sheep	0
Dog	11	Pig and whale	2
Horse	12	Whale, camel, and guanaco	0
Chicken	13	Horse and kangaroo	7
Rattlesnake	14	Rabbit and pig	4
Tuna fish	21	Rabbit and dog	5
Dogfish	23	Dogfish and tuna	14
Moth (*Samia cynthia*)	31	Moth and tuna	33
Wheat	35	Moth and yeast	48
Neurospora	43	*Neurospora* and wheat	46
Yeast	44	Yeast and *Neurospora*	39

*Most vertebrate cytochromes contain 104 residues. Comparisons are based on these residues and do not include extra residues at the NH_2 terminus in nonvertebrate cytochromes.

the greater reproductive rate of such microorganisms can lead to accumulated differences in sequence in relatively short periods.

All the cytochromes c mentioned (Table 31.1) react with yeast and mammalian cytochrome oxidases. It is evident, therefore, that the principal aspects of the structure and function of cytochrome c have not significantly altered since the evolutionary divergence of animals, plants, and fungi occurred about 1500 million years ago. Since the function of cytochrome c involves its role in mitochondrial terminal oxidation and phosphorylation, presumably a similar age must be assumed for mitochondria. The age of the earth is presently estimated at approximately 4500 million years; hence one of the critical stages in evolution, the development of efficient aerobic metabolism, almost certainly occurred early in evolution. Through studies of other "old" proteins, it is likely that further insight into the evolution and development of the major cellular functions can be obtained.

Although cytochrome c and other proteins have undergone many changes in sequence during the course of evolution, this is not necessarily true for all proteins. Histone H4 (Chap. 8) of calf thymus and of pea seedlings shows only two conservative amino acid replacements in the 102 residues of the sequence. It is clear that selection pressures have tended to conserve the sequence of this protein. This implies not only that essentially each residue in the sequence of histone H4 is critical for its function and that almost all mutations have been deleterious, but also that the specific functions of the protein evolved early and specifically in the evolution of eukaryotic organisms. This interpretation is supported by studies of the DNA sequences in the multiple genes for histone H4 in the sea urchin. These show many changes in those bases in the codons that do not alter the specified residue. Thus, "synonymous changes" in the codons which make a protein of identical sequence are tolerated but not most of the mutations that result in incorporation of a different residue.

The above leads to the view that not all proteins have evolved or are evolving at the same rate. "Old proteins," such as histone H4, must have undergone an evolution leading to a nearly definitive sequence at some early time and have tended to show little change since then. In contrast, proteins that have developed more recently in evolution, particularly in higher organisms, manifest much greater changes in structure in response to changes in selection pressures of the internal and external environments. Hemoglobins, discussed below, are excellent examples of the latter phenomenon.

Changes in the Genome

Many kinds of changes in the DNA are reflected in changes of protein structure. Although many of these alterations have been experimentally produced in different organisms by exposure to x-rays, ultraviolet illumination, or chemical mutagens, the study of protein structure shows that similar changes have occurred during evolution. The nature of the mutagens acting during evolution is open to conjecture, but the genetic changes presented below are those which can lead to protein modifications and hence to altered function.

Point Mutations The substitution of a single amino acid residue as the result of a change in a single base pair in the DNA is the most common change. The protein for which the largest number of point mutations is presently known is human hemoglobin (Chap. M4).

Deletions and Additions Deletion and addition of one or several amino acid residues both within peptide chains and at the ends of chains have been found in human hemoglobin (Chap. M4) and in many homologous proteins from different species. For example, the cytochromes c of vertebrates all have 103 or 104 residues, whereas those of plants and vertebrates have many additional residues at the amino-terminal ends and deletions of single residues within the chain (Fig. 31.1). Deletions or additions within a polypeptide chain must involve the loss or gain of three or some multiple of three nucleotides (Chap. 28) to produce a functional protein. Deletion or addition of residues at the ends of polypeptide chains reflects mutations that have resulted in a different locus for a starting codon or for a termination codon.

Gene Duplication Higher organisms synthesize many more proteins than simpler organisms. This is paralleled by an increase in the total DNA of the genome. The primary process by which this occurred is *complete duplication* of a single gene, an entire chromosome, or the entire chromosome complement (polyploidy, see below). After such duplication, if the progeny are viable, the diploid organism has four rather than two genes for a given protein; in subsequent generations, each pair can follow an independent path of variation. As long as one pair of genes continues to direct synthesis of the original functional protein, the other may undergo relatively drastic mutational changes and, indeed, ultimately become responsible for synthesis of a protein that serves the same function with different specificity or enhanced activity or even serves a different function. If the altered protein provides greater "survival value," it will be retained and the organism will flourish and perhaps come to occupy a somewhat different ecological niche. Gene duplication itself may enhance viability (and hence selective advantage) by providing an increased rate of mRNA formation and thus of protein synthesis, particularly when a protein with multiple, identical subunits must be synthesized.

Changes in protein specificity or function are associated with "radical" substitutions of amino acid residues; it is only in this way that substantial modification of properties of the original protein can be achieved. Yet most radical substitutions produce defective proteins leading to genetic disorders (see below) of lower survival value. In other words, the mechanisms of genetic change that lead to enhanced survival in one situation may lead to diminished survival in another. Conservative substitutions rarely lead to alteration of function.

Evidence for complete gene duplication is striking indeed, and many examples have been recognized. One type is represented by multiple subunit proteins in which the two kinds of subunits are nonidentical but homologous, as in the vertebrate hemoglobins. A second kind of gene duplication is manifested by homologous proteins consisting of single peptide chains such as the markedly different proteases of the serine-histidine-aspartic acid type (see below) and the several examples of homologous polypeptide hormones with strikingly diverse activity. *Isoenzymes* of homologous sequence and conformation with different intracellular localizations provide important instances of gene duplication, e.g., cytosolic and mitochondrial *malate dehydrogenases* and *glutamate-aspartate aminotransferases*.

When the sequences of the α and β chains of HbA, the γ chain of HbF, and the δ chain of HbA_2 are compared, as in Table 31.2, the β, γ, and δ chains are seen to have fewer sequence differences compared to one another than when compared to α chains. Comparison of the α and β chains of human HbA with those of other mammals reveals fewer differences between α chains than between β chains. Moreover, since the α chains are common to HbA, HbF, and HbA_2, it has been suggested that the α chain is the oldest of these and might be homologous with the chain of lamprey oxyhemoglobin, which is monomeric (like myoglobin) in the oxygenated form. This is in contrast to the hemoglobins of other vertebrates, which are tetrameric; for example, $HbA = \alpha_2\beta_2$, $HbF = \alpha_2\gamma_2$, $HbA_2 = \alpha_2\delta_2$. Hence, from the observed number of differences, it has been further suggested that succeeding duplication of genes for the α, γ, and β chains gave rise during evolution to the four chains as follows.

$$\alpha \text{ chain} \begin{cases} \alpha \text{ chain} \\ \gamma \text{ chain} \begin{cases} \gamma \text{ chain} \\ \beta \text{ chain} \begin{cases} \beta \text{ chain} \\ \delta \text{ chain} \end{cases} \end{cases} \end{cases}$$

Thus a complete duplication of an α-chain gene gave rise to two genes, which evolved independently to give, in time, α and γ genes. Similarly, the γ genes later duplicated to give β genes, and the β genes, the δ genes.

TABLE 31.2

Chains compared	Number of variant amino acid residues
β and δ	6
β and γ	36
α and β	78
α and γ	83

Supporting evidence has come from a study of the hemoglobins of the primates. Living representatives of this order, according to the paleontological record and their comparative anatomy, appeared in the following sequence: tree shrews, lemurs, lorises, new world monkeys, old world monkeys, apes, human beings. The α chains of hemoglobins of these genera show relatively few amino acid substitutions. The β chains differ increasingly from those of human beings as one proceeds to more primitive forms; most striking is the fact that the substitutions occur largely at those positions in which human β chains differ from γ chains. Indeed, the hemoglobins of adult shrews and lemurs resemble human fetal hemoglobin, HbF, more closely than human HbA.

Variation in the primary structures of vertebrate hemoglobins is extensive, and this must largely reflect the vast range of adaptation of these species to aerobic life, from the depths of the seas to the highest altitudes, from sedentary to highly mobile species in water, on land, and in the air, and even from free-living pelagic embryos of fresh and salt water to the sheltered embryos of viviparous species and the shell-encased embryos of many reptiles and all birds. These evolutionary adaptations are reflected in the affinity of the particular hemoglobin for oxygen, its responsiveness to the CO_2 content of the blood, and other characteristics (Chap. M4).

Gene duplication leading to distinct proteins that are functional at different stages of development is shown by adult and fetal hemoglobins (Chap. M3) and by human α-fetoprotein and serum albumin (Chap. M1).

Another example of gene duplications leading to the evolution of new functions is provided by the monomeric structures of trypsin, chymotrypsin A, chymotrypsin B, and elastase. The sequences and conformation of these enzymes show considerable homology in structure, trypsin and chymotrypsin A being identical in approximately 40 percent of their sequences. This is an example of a similar function, protein hydrolysis, being retained but with considerable change in specificity with respect to the types of bonds hydrolyzed. In the binding site for the residue adjacent to the scissile bond, trypsin possesses an aspartic acid residue and chymotrypsin a serine residue. This single change reflects their difference in binding specificity. It is evident that for vertebrates dependent on a source of amino acids in the diet, enhanced survival value would result from the appearance of multiple proteinases capable of achieving maximal digestion of dietary protein. Further, thrombin and other proteolytic enzymes involved in blood clotting (Chap. M1) are also homologous with the pancreatic proteinases, reflecting other gene duplications that permit utilization of the same kinds of enzymes for different functions. Similarly, the *cocoonase* of some insects resembles trypsin in its properties, but its function is to digest the fibroin of the

TABLE 31.3 Homology of Sequences around the Reactive Serine and Histidine Residues of Pancreatic Proteinases and Some Related Enzymes

Enzyme	Reactive serine*	Reactive histidine*
Bovine trypsin	Cys-Gln-Gly-Asp-**Ser**-Gly-Gly-Pro-Val	Ser-Ala-Ala-**His**-Cys-Tyr
Bovine chymotrypsin A	Cys-Met-Gly-Asp-**Ser**-Gly-Gly-Pro-Leu	Thr-Ala-Ala-**His**-Cys-Gly
Bovine chymotrypsin B	Cys-Met-Gly-Asp-**Ser**-Gly-Gly-Pro-Leu	Thr-Ala-Ala-**His**-Cys-Gly
Porcine elastase	Cys-Gln-Gly-Asp-**Ser**-Gly-Gly-Pro-Leu	Thr-Ala-Ala-**His**-Cys-Val
Bovine thrombin	Cys-Glu-Gly-Asp-**Ser**-Gly-Gly-Pro-Phe	Thr-Ala-Ala-**His**-Cys-Leu
Sorangium proteinase	Gly-Arg-Gly-Asp-**Ser**-Gly-Gly-Ser-Trp	Thr-Ala-Gly-**His**-Cys-Gly
Streptomyces griseus proteinase	Cys-Gln-Gly-Asp-**Ser**-Gly-Gly-Pro-Val	Thr-Ala-Ala-**His**-Cys-Val

*The reactive residues in the active sites of these enzymes are in boldface. The two sets of sequences are remote from one another in the linear sequences of these enzymes.

cocoon, permitting the emergence of the matured individual. Table 31.3 shows parts of the sequences of several of these enzymes around the reactive histidine and serine residues that participate in active sites of these enzymes.

The polypeptide hormones of the neurohypophysis (Chap. M20) provide another example of the evolution of structure and function. The structures of these hormones are given in Table 31.4. Evolution of these hormones clearly represents single base changes in the codons that have occurred sequentially, yielding single amino acid substitutions. The vasotocin of the frog possesses weak activity both as a vasopressin and as an oxytocin, in contrast to the potent hormones of the mammal and other higher vertebrates. The change from vasotocin to arginine-vasopressin involves a substitution at position 3 of Ile by Phe. Similarly, from vasotocin to oxytocin, the change in position 8 is from Arg to Leu. For the evolution from a single hormone to the presence of two or more hormones in the same species, we must assume a gene duplication, as well as point mutations, in order to have independent production of two or more hor-

TABLE 31.4 Amino Acid Sequences of Some Neurohypophyseal Hormones

Hormone	Residues			Species
	1 2 3 4 5 6 7 8 9			
Ancestral form*	Cys-Tyr- X-X -Asn-Cys-Pro- X -Gly NH$_2$			
Glumitocin		Ile-Ser	Gln	Cartilaginous fishes (rays)
Aspargtocin		Ile-Gln	Val	Cartilaginous fishes (sharks)
Valitocin		Ile-Asn	Leu	Cartilaginous fishes (sharks)
Isotocin		Ile-Ser	Ile	Bony fishes
Mesotocin		Ile-Gln	Ile	Amphibians, reptiles, birds
Vasotocin		Ile-Gln	Arg	Bony fishes, amphibians, reptiles, birds
Oxytocin		Ile-Gln	Leu	Mammals
Vasopressin		Phe-Gln	Arg	Mammals except pig and marsupials
Vasopressin		Phe-Gln	Lys	Pig and marsupials
Phenypressin		Phe-Phe-Gln	Arg	Marsupials

*A disulfide bond links residues 1 and 6 in each hormone. The residues that differ from the presumed ancestral form are indicated in each case. It is supposed that vasotocin is the ancestral form of the vasopressins. [*Adapted from R. Acher, Angew. Chem.* **91**:905 (1979).]

822

mones. Similar interrelationships are found for adrenocorticotropic hormone (ACTH), melanocyte-stimulating hormone (MSH), β-lipotropin (Chap. M20), and the two-chain hormones, viz., thyroid-stimulating hormone, luteinizing hormone, follicle-stimulating hormone, and chorionic gonadotropin, all of which have identical α chains but different β chains (Chap. M20).

The structures of the hormones secretin and glucagon are sufficiently similar to indicate a common genetic origin. This is a particularly striking example both of change of function and of alteration of control of protein synthesis. Glucagon, synthesized in the pancreas, augments blood glucose concentration by increasing glycogenolysis in the liver. Secretin, made in the intestinal mucosa, stimulates the flow of pancreatic juice (Chap. M10). Presumably, gene duplication has led to the evolution of two different hormones, each under separate genetic control and each produced by specific cells.

Another instance of close structural relationship is shown by egg-white *lysozyme* and one of the two components of *lactose synthase* (Chap. 10), the *α-lactalbumin* of bovine milk. Of the 129 residues in lysozyme and the 123 residues in lactalbumin, at least 40 are identical in the corresponding positions and 27 represent conservative replacements; the four disulfide bridges are in corresponding loci in the two proteins. On the basis of this close homology, it is evident that the two proteins were derived by gene duplication, one duplicate continuing to produce lysozyme, the other eventually giving rise by mutations to α-lactalbumin. Clearly, this is an instance of origination of new function, one enzyme catalyzing hydrolysis of a $\beta(1,4)$-glycosidic bond, the other being involved in synthesis of such a bond. Further, α-lactalbumin is a component of an enzyme that originated late in evolution since lactose occurs only in the milk of mammals.

In addition to complete gene duplication, there is also ample evidence for *fusion duplication,* a mutational event in which duplicate structural genes in continuous sequence fuse end to end. After this type of duplication, the gene determines the structure of a protein, in which the sequence from the NH_2- to the COOH-terminal residue is repeated entirely or in part. That is, a gene resulting from fusion duplication determines the sequence of a protein in which approximately the first half of the sequence of the polypeptide chain is identical to approximately the last half. An example of fusion duplication is provided by the α^1 and α^2 chains of the human haptoglobins (Chap. M1). This type of mutation permits elongation of a polypeptide chain (also see Chap. M2 for evolution of immunoglobulin structures).

Gene Association and Gene Fusion Lactose synthase also serves as an example of another phenomenon, the association of two protein components of entirely different structure and origin to form an enzyme of new specificity. Numerous examples of enzymes with multiple subunits have been described in Part 3. Some of these, e.g., lactate or glutamate dehydrogenase, have identical subunits, whereas others have subunits of entirely different structure, e.g., the aspartate transcarbamoylase of *Escherichia coli* with catalytic and regulatory subunits, and bacterial RNA polymerases, cytochrome oxidase, and mitochondrial ATP synthase, each with different kinds of subunits varying in number from four to six. How these associations developed is unknown.

For some enzyme subunits, however, it is clear that when two subunits catalyzing consecutive steps in a reaction sequence became associated, there was an enchancement of efficiency. This is shown by the α and β subunits of the *tryptophan synthetase* of *E. coli* (Chap. 28). Indeed, in *Neurospora*, the functions of the two subunits of the *E. coli* enzyme are present at two sites in a single, larger polypeptide chain. Thus, gene fusion must have occurred, ensuring that the consecutive reactions took place on a single protein surface and that both activities were present in equal amounts. Similarly, gene fusion must have occurred in the case of the *palmitate synthetase* (Chap. 20) of eukaryotes in contrast to the separate enzymes of *E. coli* and other prokaryotes. Gene fusion is also apparent for the first three enzymes of pyrimidine nucleotide synthesis in mammals (Chap. 26). Multifunctional enzymes with several catalytic activities for successive reactions exhibit the phenomenon of "substrate channeling" and prevent the wasteful and possibly deleterious accumulation of metabolic intermediates.

Translocation Genetic studies on many organisms have shown that a portion of one chromosome can be translocated to another. Similar translocations took place during the course of evolution in the case of the duplicated genes for the human hemoglobin subunits (Chap. M4). The two α genes are on one chromosome, whereas the genes for the β and γ chains are in tandem on another chromosome.

Some translocations could be advantageous in producing close association of formerly independent genes; the converse could also be true. Where gene fusion has taken place (see above), the genes producing proteins that exhibit close functional relationship had to be adjacent to one another in the chromosome. On the other hand, when gene duplication occurred and each pair of genes evolved to perform a novel function, the former close association might have been undesirable. Translocation makes possible a separation of genetic and regulatory controls.

Noncoding DNA The discovery that eukaryotic DNA possesses large segments of DNA that do not code for any protein and, indeed, that even genes coding for proteins contain large noncoding regions, *introns* (Chap. 29), provides new dimensions for recognizing evolutionary possibilities. First, noncoding DNA provides a possible reservoir for evolution of new genes or for incorporation of noncoding DNA into preexisting genes to extend or modify sequences by translocation. Second, the presence of introns has suggested a means whereby new large segments of DNA could be incorporated or deleted from the interior of genes, thereby adding or deleting many amino acid residues simultaneously. This would help to explain the observations that many homologous proteins differ in the presence or absence of long segments of amino acid sequences, e.g., trypsin, chymotrypsin, and thrombin (Chap. M1). The above-mentioned possibilities indicate a way whereby evolution of proteins could proceed in substantial "jumps" involving many residues simultaneously rather than only by the relatively slow process of incorporating point mutations, which can substitute only one amino acid residue at a time.

Polyploidy In plants particularly, it is evident that chromosome doubling occurred frequently. Indeed, it is possible to induce artifically fusion of nuclei from two distinct species with a doubling of each of the chromosomes from the two parental types to form a hybrid species. The doubling is essential to allow conjugation of the chromosomes before the reduction division of meiosis (haploid gamete cells) to permit each daughter cell to have a complete complement of chromosomes. The first such case was the crossing of cabbage and radish. Presently, some strains of the cross between wheat (*Triticum*) and rye (*Secale*) have assumed agricultural importance. The new species, *Triticale*, has greater resistance to drought and to cold in climates marginal for wheat, and the flour is superior to that from rye. Although this form of evolution has probably been of some importance in plants, there is no evidence that it has been significant in animal evolution.

Symbiosis and Interspecific Gene Transfer Many examples are known of two distinct species that live in close association with obvious metabolic advantage to both, e.g., the root-nodule bacteria of leguminous plants. It appears that some of the organelles of eukaryotic cells arose from the symbiotic incorporation of free-living unicellular organisms—a blue-green alga being the ancestor of chloroplasts of higher plants and an aerobic bacterium that of mitochondria.

The chloroplasts of eukaryotic red algae contain *phycobiliproteins* (Chap. 19), which are homologous in sequence and are immunologically related to those of prokaryotic blue-green algae. Even the conformation of these proteins has been retained since the α subunits of phycocyanin from one species can be combined with the β subunits of the other to produce hybrid molecules that are indistinguishable from the parent types.

Mammalian cells contain both cytoplasmic and mitochondrial *superoxide dismutases,* enzymes which are absolutely essential for aerobic life and which are absent only from obligate anaerobic organisms (Chap. 15). The mitochondrial enzyme is homologous in sequence and contains Mn like those of some bacteria. Other aerobic bacteria contain an Fe-dismutase homologous in sequence to the Mn-enzyme, indicating a common ancestry of both types of enzymes. The cytoplasmic erythrocyte enzyme is distinct in sequence and contains Cu and Zn.

Similarly, some aerobic bacteria contain a type of cytochrome homologous in sequence and conformation to eukaryotic mitochondrial cytochrome *c*. The concept that these organelles originated from other organisms is supported by the fact that mitochondria and chloroplasts are capable of division and multiplication, and possess a unique complement of DNA (Chap. 29).

The view that chloroplasts and mitochondria originated by symbiotic incorporation of other organisms, followed by loss of some functions in the "new" species, is attractive since organisms with a more versatile metabolic repertoire would have a greater opportunity to survive in a changing environment. The DNA of the yeast mitochondrion codes for about a dozen polypeptide chains, each of which has an unusually high content of hydrophobic residues. Each appears to be a component of an enzyme or electron carrier that is integral to the mitochondrial inner membrane; thus, only three of the polypeptides of cytochrome oxidase are synthesized in the mitochondrion, others being specified in the nucleus and synthesized on cytoplasmic ribosomes. Presumably, the limited mitochondrial genome is the remnant of that of the ancient bacterial ancestor.

The ability of many bacteriophages to perform *transduction,* i.e., to introduce segments of DNA from one organism to another and, indeed, for viruses to become incorporated into the host genome (Chap. 30) suggests additional routes for exchange and enlargement of the genome of existing species.

Loss of Genome As already noted above, proteins (enzymes) have been lost during the course of evolution. It has been demonstrated that a strain of *E. coli* that has lost the ability to make an amino acid, e.g., tryptophan, can grow faster in a tryptophan-containing culture than can a wild type that retains the ability to make tryptophan. Inasmuch as considerable energy in the form of ATP is needed to make the DNA as well as the resulting mRNA and protein, the loss of a gene, when the necessary metabolite is readily available from the environment, confers a selective advantage. It is assumed that, for this reason, higher organisms that have lost the ability to synthesize certain amino acids, vitamins, etc., have gained some enhanced survival value.

Changes in Regulation All the aforementioned changes can contribute to variation in the sequences of proteins and by gene duplication to enlargement of the genome. It should not be forgotten, however, that alterations in the structure or complement of regulatory genes can also have profound effects. Regulatory genes are concerned with the relative and absolute amounts of various proteins and enzymes and indeed of entire metabolic pathways. Inasmuch as in higher organisms the numbers of cells of various types and the size of parts of organs and of the organism itself are all genetically controlled, regulatory genes have profound effects on the phenotypic character of the organism. It will be sufficient to mention the extreme differences in size of various mammals—mice and whales—and the great differences in the size of the brain in relation to body size that distinguish the various species. Yet there are relatively few differences in cell types among mammals; the differences are mainly quantitative. Unfortunately, we have little precise knowledge as yet concerning the genes regulating the differentiation of the unique features of different species of mammals.

EXPRESSION OF EVOLUTIONARY FACTORS

Darwinian evolution is expressed in the form of survival value, i.e., the ability of the offspring of a particular species to compete for a particular ecological niche. Evolution can also be expressed as "descent with modification." Undoubtedly, most modifications in the genome are deleterious; i.e., they have diminished survival value and will ultimately disappear. Some modifications can be neutral. Neutral mutations are most likely to spread into a population when they are associated with changes producing enhanced survival value.

Descent with modification is readily assessed by the *divergence* in amino acid sequence of a homologous protein such as cytochrome c. From the type of data given in Table 31.1 or, better, from the calculated number of mutations in the codons for the amino acid changes, it has been possible to construct a phylogenetic tree for the evolution of this protein, a tree which is consistent with that derived from other evidence.

As already noted (page 818), the rate of divergence varies with different homologous proteins. Even for the evolution of a single protein, cytochrome c, the rate of incorporation of mutations in the amino acid sequence has varied at different times and in different phyla. In any event it is evident that the evolution of proteins has involved not only point mutations but all the types of changes in the genome already discussed.

Biologists have long recognized that *convergent* factors also operate in evolution. This concept can be expressed in terms of analogous structures and mechanisms, as well as in those that also develop some homology. The wings of insects and birds operate by distinct mechanisms and possess entirely different structures, but they both serve the same function. There are many types of enzymes catalyzing reactions that utilize molecular oxygen but perform these reactions by distinct mechanisms, and the enzymes are not structurally related. Thus it cannot be assumed, without additional evidence, that proteins performing the same or a similar function are homologous in structure. Similarly, there are many types of enzymes that catalyze the hydrolysis of peptide bonds. Each of these families of enzymes has distinct mechanisms and structures, although there are many enzymes of homologous structure and mechanism within one family, e.g., the group including the pancreatic proteinases, blood-clotting enzymes, etc.

A few examples are presently known of distinct types of enzymes that presumably developed by convergent evolution. The subtilisins possess a serine-histidine-aspartic acid charge-relay system similar to that of pancreatic proteases, yet the two groups of enzymes have entirely distinct amino acid sequences and, indeed, differ in their three-dimensional structures. The similar mechanism of catalysis must have arisen by convergent evolution. The mitochondrial and cytoplasmic types of superoxide dismutases also illustrate the phenomenon of convergence (page 825).

POPULATION VARIATION AND SURVIVAL VALUE

In the progeny of diploid organisms only identical twins possess the same genome. Within a species there is considerable heterozygosity, i.e., gene variants within each pair of allelomorphic genes. Further, there is some redundance, i.e., gene duplicates, for many genes. Electrophoretic patterns have indicated the presence of isoenzymes in many tissues. Because of sexual mating there is a continuous redistribution of genes in the population of a species. Natural selection operates on the entire population of a species and operates on the entire genome depending on the phenotypic adaptation to a particular set of environmental conditions.

For the human population, a single example will be cited to indicate that genetic changes which in one set of circumstances might be deleterious in another possess selective advantage. Before the development of the insecticide DDT, malaria was the leading cause of death in the world. The malarial parasite, spread by mosquitoes, spends part of its life cycle in the erythrocyte. Certain genetic variants of human hemoglobin (Chap. M4) result in a microcytic anemia, viz., erythrocytes smaller than normal size, effectively impeding the growth of the malarial organism. Thus, individuals *heterozygous* for HbS and HbC in different parts of Africa, HbD_{Punjab} in India, HbE in other parts of Asia, and thalassemias of various types in Asia and around the shores of the Mediterranean all possess enhanced survival value in malarial regions, whereas individuals *homozygous* for normal hemoglobin are more susceptible to malaria. In some cases, individuals homozygous for the modified hemoglobins have a disability that lowers survival value. Of the several hundred known genetic variants of human hemoglobins, only those which conferred an advantage in malarial regions spread throughout a large population.

HEREDITARY DISORDERS OF METABOLISM

For many of the genes in each human cell, there is considerable genetic variability, or polymorphism, which is reflected in height, pigmentation, blood pressure, etc. Some of these differences are innocuous, but others reflect great variations in ability to meet the challenges of different environments or to perform normal metabolic functions. In certain diseases, the genetic difference is so critical that it is expressed in a way that may be disabling in any environment. These are genetic disorders or inborn errors of metabolism of which more than 1500 have been identified. The biochemical bases for many of these remain to be established. Since genes serve only to direct or regulate the synthesis of proteins, it is evident that all genetically transmitted disorders represent aberrations in protein structure.

Many metabolic defects of human beings are due to absence or reduced activity of single specific enzymes. Table 31.5 lists some hereditary disorders in which the lacking or modified enzyme or protein has been identified. Most of these defects are apparent phenotypically only in the homozygous state. In the heterozygote, with one normal and one mutant gene, there is usually sufficient enzyme to meet physiological needs. Clearly, the older concept of *dominance* can be explained on the basis that one normal gene of each pair can support sufficient protein synthesis to serve the needs of the organism. In heterozygotes with genes for the production of one normal and one hemoglobin with an aberrant β chain, for example, HbC, HbS, HbM (Chap. M4), there are present approximately equal amounts of each variety. Similarly, the heterozygotic parents of galactosemic infants have approximately 50 percent of the normal amount of galactose-1-phosphate uridylyltransferase.

Most of the inherited enzymic abnormalities that have been described represent cases in which a complete metabolic block of some type has occurred. The consequences of such a block may be reflected in different ways: (1) there may be a complete absence of the final product of the pathway, e.g., the lack of melanin in albinism; (2) there may be an accumulation of an intermediate metabolite for which there is no significant alternative pathway, e.g., certain glycogen storage disorders, in which glycogen is made but cannot be utilized, or alkaptonuria, in which homogentistic acid accumulates; (3) there may be an accumulation of products of an alternative, and ordinarily minor, pathway of metabolism, as in phenylketonuria, leading to excretion of large amounts of compounds ordinarily produced in trace amounts; (4) a degradative enzyme may be lacking, leading to excessive accumulation of normal components that continue to be fabricated, as in the various glycosphingolipodystrophies; or

TABLE 31.5

Disorder*	Affected enzyme or protein	Some Hereditary Disorders in Human Beings in Which the Specific Lacking or Modified Enzyme or Protein Has Been Identified
Acanthocytosis	β-Lipoproteins (low-density)	
Acatalasia	Catalase	
Acid phosphatase deficiency	Lysosomal acid phosphatase	
Adrenal hyperplasias	See Chap. M18.	
Afibrinogenemia	Fibrinogen	
Agammaglobulinemia	γ-Globulin	
Albinism	Tyrosinase	
Aldosterone synthesis, defect in	18-Hydroxylase	
Alkaptonuria	Homogentisate oxidase	
Analbuminemia	Serum albumin	
Apnea, drug-induced	Pseudocholinesterase	
Argininemia	Arginase	
Argininosuccinic aciduria	Argininosuccinase	
Aspartylglycosaminuria	Specific hydrolase	
Atransferrinemia	Transferrin	
Carnosinemia	Carnosinase	
Cholesterol ester deficiency (Norum's disease)	Phosphatidylcholine-cholesterol acyltransferase	
Citrullinemia	Argininosuccinate synthetase	
Crigler-Najjar syndrome	Glucuronyl transferase	
Cystathioninuria	Cystathionase	
Diabetes (infantile)	Insulin	
Disaccharide intolerance	Sucrase, maltase, or lactase	
Dwarfism (ateliotic)	Somatotropin (growth hormone)	
Fabry's disease	Trihexosylceramide galactosylhydrolase	
Favism, primaquine sensitivity, nonspherocytic hemolytic anemia syndrome	Glucose-6-phosphate dehydrogenase	
Formiminotransferase deficiency	Formiminotransferase	
Fructose intolerance	Fructose-1-phosphate aldolase	
Fructosuria	Hepatic fructokinase	
Fucosidosis	Fucosidase	
Galactokinase deficiency	Galactokinase	
Galactosemia	Galactose-1-phosphate uridylyl transferase	
Gangliosidosis, generalized	β-Galactosidase	
Gargoylism (Hurler's syndrome)	α-L-Iduronidase	
Gaucher's disease	Glucocerebrosidase	
Glycogen storage diseases (nine types)	See Chap. 18.	
Goiter (familial)	Iodotyrosine dehalogenase	
Gout, primary (one form)	Hypoxanthine-guanine phosphoribosyltransferase	
Hemoglobinopathies	Hemoglobins (see Chap. M4)	
Hemolytic anemias	See Chap. M3.	
Hemophilias	See Chap. M1.	
Histidinemia	Histidase	
Homocystinuria	Cystathione synthetase	
β-Hydroxyisovaleric aciduria (and β-methylcrotonylglycinuria)	β-Methylcrotonyl CoA carboxylase	

829

TABLE 31.5

Some Hereditary Disorders in Human Beings in Which the Specific Lacking or Modified Enzyme or Protein Has Been Identified (*Continued*)

Disorder*	Affected enzyme or protein
Hydroxyprolinemia	Hydroxyproline oxidase
Hyperammonemia I	Ornithine transcarbamoylase
Hyperammonemia II	Carbamoylphosphate synthetase
Hyperglycinemia, ketotic form	Propionate carboxylase
Hyperlysinemia	Lysine-ketoglutarate reductase
Hyperoxaluria:	
I glycolic aciduria	α-Ketoglutarate-glyoxylate carboligase
II glyceric aciduria	D-Glycerate dehydrogenase
Hyperprolinemia I	Proline oxidase deficiency
Hyperprolinemia II	Δ^1-Pyrroline-5-carboxylate dehydrogenase
Hypophosphatasia	Alkaline phosphatase
Isovaleric acidemia	Isovalerate CoA dehydrogenase
Krabbe's leukodystrophy	Galactosylceramide β-galactosyl-hydrolase
Lactase deficiency, adult intestinal	Lactase
Lactose intolerance of infancy	Lactase
Lactosyl ceramidosis	Lactosylceramide galactosylhydrolase
Leigh's necrotizing encephalomyelopathy	Pyruvate carboxylase
Lesch-Nyhan syndrome	Hypoxanthine-guanine phosphoribosyltransferase
Lipase deficiency, congenital	Lipase (pancreatic)
Lysine intolerance	L-Lysine: NAD-oxidoreductase
Mannosidosis	α-Mannosidase
Maple syrup urine disease	Keto acid decarboxylase
Metachromatic leukodystrophy	Arylsulfatase A (sulfatidate sulfatase)
Methemoglobinemia	NADH-methemoglobin reductase
Methylmalonic aciduria (two types)	Methylmalonyl CoA carboxymutase
Myeloperoxidase deficiency with disseminated candidiasis	Myeloperoxidase
Neonatal jaundice	Glutathione peroxidase
Orotic aciduria	Orotidylate pyrophosphorylase and orotidylate decarboxylase
Pentosuria	L-Xylulose dehydrogenase
Phenylketonuria	Phenylalanine hydroxylase
Porphyrias	See Chap. M3.
Pulmonary emphysema	α_1-Antitrypsin
Pyridoxine-dependent infantile convulsions	Glutamate decarboxylase
Pyridoxine-responsive anemia	δ-Aminolevulinate synthase
Pyroglutamic aciduria	Pyroglutamate hydrolase
Refsum's disease	Phytanate α-oxidase
Sandhoff's disease	N-Acetyl-β-hexosaminidase A and B
Sanfilippo's, syndrome A	Heparan sulfate sulfatase
Sanfilippo's, syndrome B	N-Acetyl-α-D-glucosaminidase
Sarcosinemia	Sarcosine dehydrogenase
Sucrose intolerance	Sucrase, isomaltase
Sulfite oxidase deficiency	Sulfite oxidase
Tangier disease	Lipoprotein (high-density) deficiency

TABLE 31.5

Disorder*	Affected enzyme or protein	Some Hereditary Disorders in Human Beings in Which the Specific Lacking or Modified Enzyme or Protein Has Been Identified (*Continued*)
Tay-Sachs disease	Hexosaminidase A	
Testicular feminization	Δ^4-5α-Reductase	
Thyroid hormonogenesis (defect in) hereditary familial goiter	Iodotyrosine dehalogenase	
Trypsinogen deficiency disease	Trypsinogen	
Tyrosinemia I	p-Hydroxyphenylpyruvate oxidase	
Tyrosinemia II	Tyrosine aminotransferase	
Valinemia	Valine aminotransferase	
Wilson's disease	Ceruloplasmin	
Wolman's disease	Cholesterol ester hydrolase	
Xanthinuria	Xanthine oxidase	
Xanthurenic aciduria	Kynurenase	

*Page references for discussion of most of these disorders may be found by consulting the index in this or the companion text, *Mammalian Biochemistry*.

(5) the presence of an abnormally large amount of an enzyme resulting in the routing of an excessive quantity of substrate into one pathway at the expense of another. This is the case for a hereditary hemolytic anemia in which there is 50 to 70 times as much *adenosine deaminase* as in normal erythrocytes, thus leading to a reduction in the levels of various adenosine derivatives, including ATP, NADP, etc.

In addition to the genetic abnormalities that have been ascribed to the lack or modification of a particular enzyme or protein, there are many hereditary disorders in which the affected protein has not yet been identified; some of these are listed in Table 31.6. As biochemical knowledge increases, the protein aberrations involved in such disorders will undoubtedly be identified. Furthermore, as quantitative differences in metabolism become further defined, the list will continue to grow. As in the case of some abnormal hemoglobins, where the physiological consequences of the modified protein may be negligible or profound, similar findings can be expected for other altered proteins and enzymes.

TABLE 31.6

Disorder	Biochemical manifestation	Some Disorders in Which the Affected Protein Has Not Been Identified
Congenital steatorrhea	Failure to digest and/or absorb lipid	
Cystic fibrosis	Exocrine gland secretions	
Cystinuria	Excretion of cystine, lysine, arginine, and ornithine	
Cystinosis	Inability to utilize amino acids, notably cystine; aberration of amino acid transport into cells	
DNA repair deficiencies	See Chap. 27.	
Fanconi syndrome	Increased excretion of amino acids and other metabolites	
Hypomagnesemia	Failure of intestinal absorption of Mg^{2+}	
Lactate acidosis (congenital)	Decreased gluconeogenesis; probably several forms	

In addition to the abnormalities listed in Table 31.5, there are defects that are due to the presence of an abnormal number of chromosomes. In *Down's syndrome*, autosome number 21 is present as a triploid rather than in the normal diploid condition. Individuals are also known in whom the Y chromosome is lacking, as in *Turner's syndrome*, in which there is female appearance associated with deafness, mental retardation, cardiac deformities, and failure to reach sexual maturity. In *Klinefelter's syndrome*, the XXY individuals appear to be male but do not undergo sexual maturity and are frequently mentally retarded. Since the X and Y chromosomes carry genes concerned with sexual differentiation, it is not surprising that these individuals present syndromes associated with abnormal sexual development.

Concluding Remarks

In concluding this discussion of human hereditary disorders, some aspects of studies of human genetics warrant comment. In the early part of this century, after Mendel's laws were rediscovered, many genetic studies were made of human traits. It soon became obvious that not all the physical, psychological, and metabolic characteristics ascribed to genetic differences are in fact inherited. The difficulties arise because human families are small and it is impossible to obtain accurate data for more than a few generations. Populations are highly mobile, which frequently makes it difficult or impossible to study all members of a family. Common social or environmental factors, which are largely uncontrolled, may simulate a genetic picture.

These cautionary remarks are necessary because in the study of human genetics many meaningless "pedigrees" have been compiled. In this connection, it may be noted that rickets, a disorder due to deficiency of vitamin D or to inadequate exposure of the skin to ultraviolet radiation (Chap. M23), was long ascribed to a genetic factor. Similarly, goiter, once so prevalent in parts of Switzerland and in the "goiter belt" of the United States, was believed to be inherited. It is now recognized that an environmental factor, deficiency of iodine, was one etiological factor operative in producing the enlarged thyroid gland. Nevertheless, some forms of goiter are inherited (Chap. M14). Clearly, it will be necessary to determine the exact contribution of genetic, nutritional, and other environmental factors involved in each metabolic abnormality. Further, efforts must be devoted to learning how to correct the manifestations of genetic disorders. The successes achieved in galactosemia by withholding galactose from the diet of infants, in phenylketonuria by limiting intake of phenylalanine, and in Refsum's disease by withholding phytanic acid, in diabetes by administering insulin, or in gout by administering an inhibitor of xanthine oxidase indicate approaches to these problems. Another approach to management of such disorders, as yet unsuccessful, has been by injection of the "missing enzyme" which may be taken up by cells containing excessive quantities of a stored metabolite, as in the glycosphingolipodystrophies.

In turn, such success brings the social and moral problems raised by societal practice which knowingly tends to increase the frequency of deleterious genes in the general population. Even transmittal of normal genes into the somatic cells of such individuals by transduction or by transfer of normal somatic cells would not mitigate the latter problem.

REFERENCES

Books

Bryson, V., and H. J. Vogel (eds.): *Evolving Genes and Proteins,* Academic, New York, 1965.

Garrod, A. E.: *Inborn Errors of Metabolism,* reprinted with supplement by H. Harris, Oxford, Fairlawn, N.J., 1963.

Harris, H.: *The Principles of Human Biochemical Genetics,* 2d ed., American Elsevier, New York, 1975.

McKusick, V. A.: *Mendelian Inheritance in Man: Catalogs of Autosomal Dominant, Autosomal Recessive and X-linked Phenotypes,* 4th ed., Johns Hopkins, Baltimore, 1975.

Stanbury, J. B., J. B. Wyngaarden, D. S. Fredrickson, J. L. Goldstein, and M. S. Brown (eds.): *The Metabolic Basis of Inherited Disease,* 5th ed., McGraw-Hill, New York, 1982.

Review Articles

Dixon, G. H.: Mechanisms of Protein Evolution, pp. 147–204, in P. N. Campbell and G. D. Greville (eds.): *Essays in Biochemistry,* vol. 2, Academic, New York, 1966.

Doolittle, R. F.: Protein Evolution, pp. 2–118, in H. Neurath and R. L. Hill (eds.): *The Proteins,* 3d ed., vol. IV, Academic, New York, 1979.

FASEB Conference, Genetics and Biological Evolution: *Fed. Proc.,* **35**(10):2077–2204 (1976).

Raivio, K. O., and J. E. Seegmiller: Genetic Diseases of Metabolism, *Annu. Rev. Biochem.,* **41:**543–576 (1972).

Smith, E. L.: Evolution of Enzymes, pp. 267–339, in P. D. Boyer (ed.): *The Enzymes,* 3d ed., vol. 1, Academic, New York, 1970.

Index

HbCO	carbon monoxide hemoglobin
HbO_2	oxyhemoglobin
HDL	high density lipoprotein
His	histidyl
HMC	5-hydroxymethylcytosine
Hyl	hydroxylysyl
Hyp	hydroxyprolyl
I	inosine residue (hypoxanthine ribonucleoside)
ICSH	interstitial cell-stimulating hormone (same as LH)
IdUA	iduronic acid residue
Ile	isoleucyl
IMP	inosine monophosphate (5'-phosphate of ribosyl inosine)
K_a	association constant
kb	kilobases (DNA or RNA)
kcal	kilocalories
K_d	dissociation constant
K_i	inhibition constant
K_m	Michaelis constant
Leu	leucyl
LDL	low density lipoprotein
LH	luteinizing hormone (same as ICSH)
Lys	lysyl
Man	mannose residue
Met	methionyl
MetHb	methemoglobin
min	minute
mRNA	messenger RNA
MSH	melanocyte-stimulating hormone
MW	molecular weight
NAD^+ (NADH)	nicotinamide adenine dinucleotide (reduced form)
$NADP^+$ (NADPH)	nicotinamide adenine dinucleotide phosphate (reduced form)
NeuAc	N-acetylneuraminic acid
NeuGlyc	N-glycolylneuraminic acid
nm	nanometers
NMN	nicotinamide mononucleotide
pI	isoelectric point
P_i	inorganic phosphate

PABA	p-aminobenzoic acid
PEP	phosphoenolpyruvate
pGlu	pyroglutamyl
Phe	phenylalanyl
PP_i	inorganic pyrophosphate
Pro	prolyl
PRPP	5-phosphoribosyl 1-pyrophosphate
RNA	ribonucleic acid
RNase	ribonuclease
rpm	revolutions per minute
RQ	respiratory quotient
rRNA	ribosomal RNA
s	second
S	Svedberg constant
Ser	seryl
Sia	sialic acid residue
T or Thy	thymine
ThPP	thiamine pyrophosphate
Thr	threonyl
tRNA	transfer RNA
Trp	tryptophanyl
TSH	thyroid-stimulating hormone
Tyr	tyrosyl
U or Ura	uracil
UDP	uridine diphosphate
UDP-glucose	uridine diphosphoglucose
UMP	uridine monophosphate (5'-phosphate of ribosyl uridine)
UTP	uridine triphosphate
V	volts
V_{max}	maximum velocity
Val	valyl
VHDL	very high density lipoprotein
VLDL	very low density lipoprotein
XMP	xanthine monophosphate (5'-phosphate of ribosyl xanthine)
Xyl	xylose residue

Other specialized symbols are indicated in specific tables, e.g., symbols for nucleosides found in transfer RNAs are given in Table G8.6, page G153; single-letter abbreviations for amino acid residues in Table G.3.1, page G32.ff.